Renewable Energy Systems
Modeling, Optimization and Control

Advances in Nonlinear Dynamics and Chaos (ANDC)

Renewable Energy Systems
Modeling, Optimization and Control

Edited by

Ahmad Taher Azar

Faculty of Computers and Artificial Intelligence, Benha University, Benha, Egypt;
College of Computer and Information Sciences, Prince Sultan University,
Riyadh, Saudi Arabia

Nashwa Ahmad Kamal

Faculty of Engineering, Cairo University, Giza, Egypt;
International Group of Control Systems (IGCS), Cairo, Egypt

Series Editors

Ahmad Taher Azar

Sundarapandian Vaidyanathan

ACADEMIC PRESS
An imprint of Elsevier

ELSEVIER

Academic Press is an imprint of Elsevier
125 London Wall, London EC2Y 5AS, United Kingdom
525 B Street, Suite 1650, San Diego, CA 92101, United States
50 Hampshire Street, 5th Floor, Cambridge, MA 02139, United States
The Boulevard, Langford Lane, Kidlington, Oxford OX5 1GB, United Kingdom

Notices
Knowledge and best practice in this field are constantly changing. As new research and experience broaden our understanding, changes in research methods, professional practices, or medical treatment may become necessary.

Practitioners and researchers must always rely on their own experience and knowledge in evaluating and using any information, methods, compounds, or experiments described herein. In using such information or methods they should be mindful of their own safety and the safety of others, including parties for whom they have a professional responsibility.

To the fullest extent of the law, neither the Publisher nor the authors, contributors, or editors, assume any liability for any injury and/or damage to persons or property as a matter of products liability, negligence or otherwise, or from any use or operation of any methods, products, instructions, or ideas contained in the material herein.

British Library Cataloguing-in-Publication Data
A catalogue record for this book is available from the British Library

Library of Congress Cataloging-in-Publication Data
A catalog record for this book is available from the Library of Congress

ISBN: 978-0-12-820004-9

For Information on all Academic Press publications
visit our website at https://www.elsevier.com/books-and-journals

Publisher: Mara Conner
Acquisitions Editor: Sonnini R. Yura
Editorial Project Manager: Charlotte Rowley
Production Project Manager: Sojan P. Pazhayattil
Cover Designer: Mark Rogers

Typeset by MPS Limited, Chennai, India

Contents

30. Optimal design and techno-socio-economic analysis of hybrid renewable system for gird-connected system

Aashish Kumar Bohre, Yashwant Sawle and Parimal Acharjee

31. Stand-alone hybrid system of solar photovoltaics/wind energy resources: an eco-friendly sustainable approach

Faizan Arif Khan, Nitai Pal and Syed Hasan Saeed

List of contributors

Parimal Acharjee Department of Electrical Engineering, National Institute of Technology (NIT), Durgapur, India

Kammogne Soup Tewa Alain Laboratory of Condensed Matter, Electronics and Signal Processing (LAMACETS), Department of Physic, Faculty of Sciences, University of Dschang, Dschang, Cameroon

Najib M. Alfakih State Key Laboratory of Power Transmission Equipment & System Security and New Technology, School of Electrical Engineering, Chongqing University, Chongqing, P.R. China

Mudassar Ali Department of Telecommunication and Information Engineering, University of Engineering and Technology, Taxila, Taxila, Pakistan

Mohamed M. Aly Electrical Engineering Department, Faculty of Engineering, Aswan University, Aswan, Egypt

Karima Amara Electrical Engineering Advanced Technology Laboratory (LATAGE), Mouloud Mammeri University, Tizi Ouzou, Algeria

P. Anandhraj Research Scholar, Anna University

E. Artigao-Andicoberry Renewable Energy Research Institute and DIEEAC-ETSII-AB, Universidad de Castilla-La Mancha, Albacete, Spain

P. Arvind Department of Electrical and Electronics Engineering, Birla Institute of Technology, Mesra, Ranchi, India

Ahmad Taher Azar Faculty of Computers and Artificial Intelligence, Benha University, Benha, Egypt; College of Computer and Information Sciences, Prince Sultan University, Riyadh, Saudi Arabia

Meryeme Azaroual Engineering for Smart and Sustainable Systems Research Center, Mohammadia School of Engineers, Mohammed V University, Rabat, Morocco

Lhoussaine Bahatti IESI Laboratory, ENSET Mohammedia, Hassan II University of Casablanca, Casablanca, Morocco

Naglaa K. Bahgaat Electrical Communication Department, Faculty of Engineering, Canadian International College (CIC), Giza, Egypt

Rachid Bannari Laboratory Systems Engineering, Ensa, Ibn Tofail University Kenitra, Kenitra, Morocco

Agnimitra Biswas Department of Mechanical Engineering, National Institute of Technology Silchar, Silchar, India

Aashish Kumar Bohre Department of Electrical Engineering, National Institute of Technology (NIT), Durgapur, India

David Borge-Diez Department of Electrical and Systems Engineering and Automation, University of León, Campus de Vegazana, S / N, León, Spain

Sourav Chakraborty Department of Electrical and Electronics Engineering, Birla Institute of Technology, Mesra, Ranchi, India

Chakib Chatri Engineering for Smart and Sustainable Systems Research Center, Mohammadia School of Engineers, Mohammed V University in Rabat, Rabat, Morocco

Mohamed Cherkaoui Mohammadia School of Engineers, Mohammed V University in Rabat, Rabat, Morocco

Elmostafa Chetouani Exploitation and Processing of Renewable Energy Team, Laboratory of Electronics, Instrumentation and Energy, Department of Physics, Faculty of Sciences, University of Chouaib Doukkali, El Jadida, Morocco

Antonio Colmenar-Santos Department of Electric, Electronic and Control Engineering, UNED, Juan Del Rosal, 12 – Ciudad Universitaria, Madrid, Spain

M.L. Corradini School of Science and Technology, Mathematics Division, University of Camerino, Italy

Jesús De-León-Morales Department of Electrical Engineering, Faculty of Mechanical and Electrical Engineering, Autonomous University of Nuevo León, San Nicolás de los Garza, Mexico

Hakim Denoun Electrical Engineering Advanced Technology Laboratory (LATAGE), Mouloud Mammeri University, Tizi Ouzou, Algeria

M. Edwin Department of Mechanical Engineering, University College of Engineering, Nagercoil, Anna University Constituent College, Nagercoil, India

Abderrahim El Fadili FST Mohammedia, Hassan II University of Casablanca, Casablanca, Morocco

Ismail El Kafazi Laboratory SMARTILAB, Moroccan School of Engineering Sciences, EMSI Rabat, Rabat, Morocco

Abdelmounime El Magri IESI Laboratory, ENSET Mohammedia, Hassan II University of Casablanca, Casablanca, Morocco

Kamal Elyaalaoui Mohammadia School of Engineers, Mohammed V University in Rabat, Rabat, Morocco

Youssef Errami Laboratory of Electronics, Instrumentation and Energy, Team of Exploitation and Processing of Renewable Energy, Department of Physics, Faculty of Science, University of Chouaib Doukkali, El Jadida, Morocco

Arezki Fekik Akli Mohand Oulhadj University, Bouira, Algeria; Electrical Engineering Advanced Technology Laboratory (LATAGE), Mouloud Mammeri University, Tizi Ouzou, Algeria

Marco A. Flores Instituto de Investigacion en Energia IIE, Universidad Nacional Autónoma de Honduras (UNAH), Tegucigalpa, Honduras

Aldo José Flores-Guerrero Department of Electrical Engineering, Faculty of Mechanical and Electrical Engineering, Autonomous University of Nuevo León, San Nicolás de los Garza, Mexico

Diriba Kajela Geleta Department of Mathematics, Punjabi University, Patiala, India; Department of Mathematics, Madda Walabu University, Bale Robe, Ethiopia

Sunitha George Department of Instrumentation and Control Engineering, Netaji Subhas University of Technology, Dwarka, India

Fouad Giri LAC Laboratory, University of Caen Normandie, Caen, France

Anubhav Goel Department of Polymer and Process Engineering, Indian Institute of Technology Roorkee, Roorkee, India

E. Gómez-Lázaro Renewable Energy Research Institute and DIEEAC-ETSII-AB, Universidad de Castilla-La Mancha, Albacete, Spain

Tulasichandra Sekhar Gorripotu Department of Electrical & Electronics Engineering, Sri Sivani College of Engineering, Srikakulam, India

Rajat Gupta Department of Mechanical Engineering, National Institute of Technology Silchar, Silchar, India

Susana V. Gutiérrez-Martínez Department of Electrical Engineering, Faculty of Mechanical and Electrical Engineering, Autonomous University of Nuevo León, San Nicolás de los Garza, Mexico

Mohamed Lamine Hamida Electrical Engineering Advanced Technology Laboratory (LATAGE), Mouloud Mammeri University, Tizi Ouzou, Algeria

I. Hammou Ou Ali Engineering for Smart and Sustainable Systems Research Center, Mohammadia School of Engineers, Mohammed V University, Rabat, Morocco

A. Honrubia-Escribano Renewable Energy Research Institute and DIEEAC-ETSII-AB, Universidad de Castilla-La Mancha, Albacete, Spain

Amjad J. Humaidi Department of Control and Systems Engineering, University of Technology, Baghdad, Iraq

Ibraheem Kasim Ibraheem Department of Electrical Engineering, College of Engineering, University of Baghdad, Baghdad, Iraq

Rajaa Naji E.L. Idrissi Engineering for Smart and Sustainable Systems Research Center, Mohammadia School of Engineers, Mohammed V University in Rabat, Rabat, Morocco

G. Ippoliti Department of Information Engineering, Marche Polytechnic University, Ancona, Italy

Jagadish Department of Mechanical Engineering, National Institute of Technology Raipur, Raipur, India

D.B. Jani Gujarat Technological University—GTU, Government Engineering College, Dahod, India

Francisco Jurado Department of Electrical Engineering, University of Jaén, Jaén, Spain

Nashwa Ahmad Kamal Faculty of Engineering, Cairo University, Giza, Egypt; International Group of Control Systems (IGCS), Riyadh, Saudi Arabia

Salah Kamel Electrical Engineering Department, Faculty of Engineering, Aswan University, Aswan, Egypt; Department of Electrical Engineering, Faculty of Engineering, Aswan University, Aswan, Egypt

Faizan Arif Khan Department of Electrical Engineering, Indian Institute of Technology (Indian School of Mines), Dhanbad, India

Cheshta Jain Khare Shri G.S. Institute of Technology and Science, Indore, India

Vikas Khare School of Technology Management & Engineering (STME), NMIMS University, Indore, India

Mohammed Kissaoui IESI Laboratory, ENSET Mohammedia, Hassan II University of Casablanca, Casablanca, Morocco

Deepak Kumar Department of Electrical and Electronics Engineering, Birla Institute of Technology, Mesra, Ranchi, India

Vineet Kumar Division of Instrumentation and Control Engineering, Netaji Subhas University of Technology, Dwarka, India

Rachid Lajouad IESI Laboratory, ENSET Mohammedia, Hassan II University of Casablanca, Casablanca, Morocco

Ana-Rosa Linares-Mena Department of Electric, Electronic and Control Engineering, UNED, Juan Del Rosal, 12 – Ciudad Universitaria, Madrid, Spain

Mohamed Maaroufi Engineering for Smart and Sustainable Systems Research Center, Mohammadia School of Engineers, Mohammed V University in Rabat, Rabat, Morocco

Tahir Nadeem Malik Department of Electrical Engineering, University of Engineering and Technology, Taxila, Taxila, Pakistan

Gaurav Manik Department of Polymer and Process Engineering, Indian Institute of Technology Roorkee, Roorkee, India

Mukhdeep Singh Manshahia Department of Mathematics, Punjabi University, Patiala, India

S. Martín-Martínez Renewable Energy Research Institute and DIEEAC-ETSII-AB, Universidad de Castilla-La Mancha, Albacete, Spain

Ahmed S. Menesy Department of Electrical Engineering, Faculty of Engineering, Minia University, Minya, Egypt; State Key Laboratory of Power Transmission Equipment & System Security and New Technology, School of Electrical Engineering, Chongqing University, Chongqing, P.R. China

Shikha Mittal Department of Mathematics, Jesus and Mary College, University of Delhi, New Delhi, India

Amal Amin Mohamed Electrical Engineering Department, Faculty of Engineering, Aswan University, Aswan, Egypt

K. Mohana Sundaram EEE Department, KPR Institute of Engineerimg and Technology, Coimbatore

Enrique-Luis Molina-Ibáñez Department of Electric, Electronic and Control Engineering, UNED, Juan Del Rosal, 12 – Ciudad Universitaria, Madrid, Spain

Muhammad Naeem Department of Electrical Engineering, COMSATS University Islamabad, Wah Campus, Wah Cantt, Pakistan

M. Saranya Nair School of Electronics Engineering, Vellore Institute of Technology, Chennai Campus, Chennai, India

Abdellatif Obbadi Laboratory of Electronics, Instrumentation and Energy, Team of Exploitation and Processing of Renewable Energy, Department of Physics, Faculty of Science, University of Chouaib Doukkali, El Jadida, Morocco

G. Orlando Department of Information Engineering, Marche Polytechnic University, Ancona, Italy

M. Ouassaid Engineering for Smart and Sustainable Systems Research Center, Mohammadia School of Engineers, Mohammed V University, Rabat, Morocco

Mohammed Ouassaid Engineering for Smart and Sustainable Systems Research Center, Mohammadia School of Engineers, Mohammed V University in Rabat, Rabat, Morocco

Nitai Pal Department of Electrical Engineering, Indian Institute of Technology (Indian School of Mines), Dhanbad, India

P. Pandiyan Associate Professor, EEE Department, KPR Institute of Engineering and Technology, Coimbatore

Ramana Pilla Department of Electrical & Electronics Engineering, GMR Institute of Technology, Rajam, Srikakulam, India

Farhan Qamar Department of Telecommunication and Information Engineering, University of Engineering and Technology, Taxila, Taxila, Pakistan

Nouman Qamar Department of Electrical Engineering, University of Engineering and Technology, Taxila, Taxila, Pakistan

K.P.S. Rana Division of Instrumentation and Control Engineering, Netaji Subhas University of Technology, Dwarka, India

Kengne Romanic Laboratory of Condensed Matter, Electronics and Signal Processing (LAMACETS), Department of Physic, Faculty of Sciences, University of Dschang, Dschang, Cameroon

Enrique Rosales-Asensio Department of Electrical Engineering, University of Las Palmas de Gran Canaria, Campus de Tafira S / N, Las Palmas de Gran Canaria, Spain

Francisco Ruiz Instituto de Investigacion en Energia IIE, Universidad Nacional Autónoma de Honduras (UNAH), Tegucigalpa, Honduras

Syed Hasan Saeed Department of Electronics & Communication Engineering, Integral University, Lucknow, India

Smail Sahnoun Laboratory of Electronics, Instrumentation and Energy, Team of Exploitation and Processing of Renewable Energy, Department of Physics, Faculty of Science, University of Chouaib Doukkali, El Jadida, Morocco

Yashwant Sawle School of Electrical Engineering, Vellore Institute of Technology (VIT), Vellore, India

Nitish Sehgal Department of Instrumentation and Control Engineering, Netaji Subhas University of Technology, Dwarka, India

S. Joseph Sekhar Department of Engineering, Shinas College of Technology, University of Technology and Applied Sciences, Shinas, Sultanate of Oman

Ali Selim Electrical Engineering Department, Faculty of Engineering, Aswan University, Aswan, Egypt; Department of Electrical Engineering, University of Jaén, Jaén, Spain

Fernando E. Serrano Instituto de Investigacion en Energia IIE, Universidad Nacional Autónoma de Honduras (UNAH), Tegucigalpa, Honduras; Universidad Tecnolgica Centroamericana (UNITEC), Zona Jacaleapa, Tegucigalpa, Honduras

Hamdy M. Sultan Department of Electrical Engineering, Faculty of Engineering, Minia University, Minya, Egypt

Mahrous A. Taher Department of Electrical Engineering, Faculty of Engineering, Aswan University, Aswan, Egypt

Ahmad Taher Azar Faculty of Computers and Artificial Intelligence, Benha University, Benha, Egypt; College of Computer and Information Sciences, Prince Sultan University, Riyadh, Kingdom of Saudi Arabia

P.R. Thakura Department of Electrical and Electronics Engineering, Birla Institute of Technology, Mesra, Ranchi, India

Sundarapandian Vaidyanathan Research and Development Centre, Vel Tech University, Chennai, India

H.K. Verma Shri G.S. Institute of Technology and Science, Indore, India

R. Villena-Ruiz Renewable Energy Research Institute and DIEEAC-ETSII-AB, Universidad de Castilla-La Mancha, Albacete, Spain

Boaz Wadawa Laboratory of Electronics, Instrumentation and Energy, Team of Exploitation and Processing of Renewable Energy, Department of Physics, Faculty of Science, University of Chouaib Doukkali, El Jadida, Morocco

Aziz Watil IESI Laboratory, ENSET Mohammedia, Hassan II University of Casablanca, Casablanca, Morocco

Nacera Yassa Akli Mohand Oulhadj University, Bouira, Algeria

Preface

Renewable energy systems play a decisive role on the path toward meeting the energy demands of a growing global population and the economic needs of countries at all stages of development if these goals are to be achieved in a climate-friendly environment. Indeed, given contemporary climate change challenges, renewable energy has a dual role to play in both mitigation and adaptation efforts. The search for sustainable energy, therefore, continues to dominate the 21st-century research, business, and policy. In addition, renewable energy systems exhibit changing dynamics, nonlinearities, and uncertainties and challenges that require advanced control and optimization strategies to solve effectively. The use of more efficient control and optimization strategies would not only enhance the performance of these systems but would also reduce the cost per kilowatt-hour produced. This book aims to cross-pollinate the recent advances in the study of renewable energy control systems by bringing together diverse scientific breakthroughs on the modeling, control, and optimization of renewable energy systems by leading energy systems engineering researchers in this field.

About the book

This book "*Renewable Energy Systems: Modeling, Optimization, and Control,*" consisting of 31 contributed chapters, is written by subject experts who have specialization in various topics addressed in this book. The special chapters have been brought out in this book after a rigorous review process in the broad areas of modeling, optimization, and control. Special importance was given to chapters offering practical solutions and novel methods for the recent research problems in the mathematical modeling and applications of renewable energy systems.

Objectives of the book

This book has a special focus on renewable energy control systems. This book will have chapters with a current focus on research and novel solutions to so many problems in that field. This book will cover most of the modeling, control theorems, and optimization techniques, altogether which will solve many scientific issues for researchers in the field of renewable energy and can be considered a good reference for young researchers. Many multidisciplinary applications have been discussed in this book with their new fundamentals, modeling, control, and experimental results. Simply, this book will achieve the gap between different interdisciplinary applications starting from mathematical concepts, modeling, optimization, and up to recent control techniques.

Organization of the book

This well-structured book consists of 31 full chapters.

Book features

- Deals with the recent research problems in the area of renewable energy systems
- Presents various techniques of modeling, optimization, and advanced control techniques for renewable energy
- Contains a good literature survey with a long list of references
- Well written with a good exposition of the research problem, methodology, block diagrams, and mathematical techniques
- Lucidly illustrates with numerical examples and simulations
- Discusses details of engineering applications and future research areas

Audience

This book is primarily meant for researchers from academia and industry, who are working on renewable energy research areas, electrical engineering, control engineering, electronic engineering, mechanical engineering, and computer science. The book can also be used at the graduate or advanced undergraduate level as a textbook or major reference for courses such as power systems, control systems, electrical devices, scientific modeling, and computational science.

Acknowledgments

As the editors, we hope that the chapters in this well-structured book will stimulate further research in renewable energy systems and utilize them in real-world applications.

We hope sincerely that this book, covering many different topics, will be very useful for all readers.

We would like to thank all the reviewers for their diligence in reviewing the chapters.

Special thanks goes to Elsevier, especially the book Editorial team.

Ahmad Taher Azar[1,2] **and Nashwa Ahmad Kamal**[3,4]
[1]*Faculty of Computers and Artificial Intelligence, Benha University, Benha, Egypt,*
[2]*College of Computer and Information Sciences, Prince Sultan University, Riyadh, Saudi Arabia,*
[3]*Faculty of Engineering, Cairo University, Giza, Egypt,*
[4]*International Group of Control Systems (IGCS), Cairo, Egypt*

Chapter 1

Efficiency maximization of wind turbines using data-driven Model-Free Adaptive Control

M.L. Corradini[1], G. Ippoliti[2] and G. Orlando[2]

[1]*School of Science and Technology, Mathematics Division, University of Camerino, Italy*, [2]*Department of Information Engineering, Marche Polytechnic University, Ancona, Italy*

1.1 Introduction

Modeling and control of wind turbines (WTs) is a challenging task (Johnson, Pao, Balas, & Fingersh, 2006), both because their aerodynamics are highly complex and nonlinear and because rotors are forced by stochastic and turbulent wind inflow fields. In general, reducing the cost of wind energy is the key factor driving a successful growth of the wind energy sector, and manufacturers increasingly turn to more refined control systems (Bossanyi et al., 2012; Ozbek & Rixen, 2013) requiring detailed models of the turbine to capture the complete dynamic behavior of the system (Simani & Castaldi, 2013; van der Veen, van Wingerden, Fleming, Scholbrock, & Verhaegen, 2013) to improve efficiency. System identification techniques may allow improvements in controller design through the analysis and understanding of the system dynamics (van der Veen et al., 2013): several contributions have appeared on the identification of linear, time-invariant models of WTs (Hansen, Thomsen, Fuglsang, & Knudsen, 2006; Iribas & Landau, 2009; Iribas-Latour & Landau, 2013; van Baars & Bongers, 1994), but it is apparent that nonlinearity must be considered in control design (Bossanyi, 2000) since WTs are nonlinear systems operating under large wind speed variations. Identification techniques for nonlinear systems and for linear parameter-varying systems are an active research area, though these methods still present significant challenges in terms of reliability and computational complexity, thus making their application under closed-loop conditions troublesome (van Wingerden & Verhaegen, 2009).

In the framework of the discussed identification-based control approaches, a dynamic linearization technique using pseudo-partial derivatives (PPDs) has been recently proposed and coupled with a data-driven control method based on Model-Free Adaptive Control (MFAC). The MFAC algorithm has been recently proposed for a class of general discrete-time nonlinear systems. First discussed in Hou (1994), extended in Hou and Jin (2011), and finally thoroughly formalized in Hou and Jin (2014), MFAC makes use of an equivalent dynamic linearization model obtained adopting a dynamic linearization technique based on PPDs. In addition to a number of interesting features, discussed in Hou and Jin (2011) and ranging from low cost and easy applicability, successful implementation in practical applications, and absence of training phases, it is worth to notice that BIBO stability and closed-loop convergence of the tracking error have been theoretically proved under mild assumptions (Hou & Jin, 2011). In this framework, the proposal has been very recently presented (Liu & Yang, 2019) to couple the so-called prescribed performance control (PPC) (Bechlioulis & Rovinthakis, 2008, 2009) with MFAC, to embed a constraint on the tracking error within the MFAC mechanism. A complement to this interesting approach has been proposed in the considered algorithm. In particular, still retaining the main setup presented in Liu and Yang (2019), the forms of the sliding surface and of the control law have been modified to provide a rigorous proof ensuring the boundedness of the control law. Furthermore, a rigorous stability of the closed-loop system is provided, leading to the definition of suitable constraints on the gain of the sliding-mode-based control term.

To support the theoretical development with significant test data, the proposed approach has been applied to the problem of efficiency maximization of a 5-MW WT operating in region 2 (medium-speed region) using the high-fidelity widely recognized

Renewable Energy Systems. DOI: https://doi.org/10.1016/B978-0-12-820004-9.00007-3

simulation tool FAST (NREL-NWTC, n.d.) (data provided by FAST are comparable to experimental data in studies about WTs). A comparative analysis has been performed, to quantify the improvement in terms of control accuracy of the proposed algorithm with respect to both the approaches described in Hou and Jin (2011) and Liu and Yang (2019).

The chapter is organized as follows. Section 1.2 presents some preliminary issues about the wind control problem in region 2 and the considered control model. The main technical results for the proposed control solution have been reported in Section 1.3. A study using FAST addressing the control of a 5-MW WT operating in the medium-speed region is reported in Section 1.4. The chapter ends with a few comments reported in Section 1.5.

1.2 Problem statement

1.2.1 The problem of optimal power extraction for wind turbines

As widely known, variable-speed WTs operate differently depending on wind speed (Johnson et al., 2006). It is customary to consider three different regions of operation, and the so-called region 2 (or medium wind speed region associated to wind speeds of medium entity) is an operational mode where generator torque control is used with the objective of maximizing wind energy capture. The aerodynamic power extracted by the WT from the wind is given by Bianchi, Battista, and Mantz (2007):

$$P_a = \frac{1}{2}\rho_{\text{air}}\pi R^2 C_p(\lambda, \beta)V^3 \tag{1.1}$$

where R is the WT rotor radius, ρ_{air} is the air density, and V is the wind speed. The torque extracted by the turbine from the wind is given by:

$$T_a = \frac{P_a}{\omega_r} = \frac{\rho_{\text{air}}\pi R^3 C_p(\lambda, \beta)V^2}{2\lambda}. \tag{1.2}$$

The power coefficient $C_p(\lambda, \beta)$ expressed the turbine efficiency in converting the wind energy into mechanical energy (Bianchi et al., 2007), is a nonlinear function (Monroy & Alvarez-Icaza, 2006; Siegfried, 1998), and depends on blade aerodynamic design and WT operating conditions. It depends nonlinearly on the blade pitch angle β (equal to zero since pitch control is not active in region 2) and the so-called tip speed ratio λ, defined as follows (Qiao, Qu, & Harley, 2009):

$$\lambda = \frac{\omega_r R}{V} \tag{1.3}$$

being ω_r the turbine angular shaft speed.

This chapter addresses generator torque control of variable-speed WTs aimed at maximizing the energy capture when operating in region 2 ($\beta = 0$ degrees). The control objective is to achieve and maintain the maximum power coefficient $C_{p\text{max}}$ regardless of wind speed. In other words, the turbine has to be driven at the unique WT shaft rotational speed ensuring the maximum power coefficient $C_{p\text{max}}$ (Bianchi et al., 2007), in view of the definition of λ in Eq. (1.3). According to Eq. (1.1), indeed, when C_p is controlled at the maximum value, the maximum mechanical power is extracted from the wind energy. For the NREL 5-MW WT (Jonkman, Butterfield, Musial, & Scott, 2009), the maximum power coefficient $C_{p\text{max}} = 0.482$ is achieved for a tip speed ratio value of $\lambda_{\text{opt}} = 7.55$. Thus the optimal WT angular shaft speed is given by:

$$\omega_{\text{ro}} = \frac{\lambda_{\text{opt}}V}{r}. \tag{1.4}$$

The mechanical equation governing the turbine can be given as follows (Johnson et al., 2006):

$$J\dot{\omega}_r(t) = T_a(t) - T_e(t)N_g \tag{1.5}$$

This equation models the drivetrain dynamics, where $\omega_r(t)$ is the rotor speed, $T_a(t)$ is the aerodynamic torque and $T_e(t)$ is the electrical generator torque. Moreover, N_g is the high-speed shaft (HSS)-to-low-speed shaft gearbox ratio, and J is the moments of inertia about the rotation axis.

1.2.2 Data-driven Model-Free Adaptive Control

Consider the following discrete-time SISO nonlinear system (Hou & Jin, 2011):

$$y(k + 1) = f(y(k)\dots y(k - n_y), u(k), u(k - 1), \dots, u(k - n_u)) \tag{1.6}$$

where $u(k)$ and $y(k)$ are the system input and output at time k, n_y and n_u are unknown orders, and $f(\ldots)$ is an unknown nonlinear function. The partial form dynamic linearization (PFDL) of the plant (Eq. 1.6) is based on the following assumptions.

Assumption 2.1: *The partial derivatives of $f(\ldots)$ with respect to the control input $u(k)$, $u(k-1)\ldots, u(k-d_u)$ are continuous, d_u being a positive discrete constant known as control input length constant of linearization for the discrete-time nonlinear system.*

Define: $\quad \Delta y(k+1) = y(k+1) - y(k), \quad \Delta u(k-i) = u(k-i) - u(k-i-1), \quad i = 0, 1, \ldots, d_u - 1, \quad \Delta U(k) = [\Delta u(k),$ $\Delta u(k-1), \ldots, \Delta u(k-d_u+1)]$, with $u(k) = 0$ for $k \leq 0$.

Assumption 2.2: *The plant $f(\ldots)$ is generalized Lipschitz, that is, $|\Delta y(k+1)| \leq b||\Delta U(k)||$, $b \in \mathbb{R}_+$, and $||\Delta U(k)|| \neq 0$ $\forall k > 0$*

The PPD-based model of the plant (Eq. 1.6) relies on the following theorem (Hou & Jin, 2011).

Theorem 2.1: *For the nonlinear system [Eq. (1.6)] satisfying Assumptions 2.1 and 2.2, there exists a parameter vector $\Phi(k)$, called the PPD vector, such that the plant can be transformed into the following equivalent PFDL description*

$$\Delta y(k+1) = \Phi(k)^T \Delta U(k) \tag{1.7}$$

where $\Phi(k) = [\phi_1(k), \phi_2(k), \ldots, \phi_{du}(k)]$ and $||\Phi(k)|| \leq b$, where b is a positive constant. Following Hou and Jin (2011), the estimate $\hat{\Phi}(k)$ of the unknown PPD vector $\Phi(k)$ can be derived using the modified projection algorithm starting from the following cost function:

$$J(\hat{\Phi}(k)) = |y(k) - y(k-1) - \hat{\Phi}(k)^T \Delta U(k-1)|^2 + \mu||\hat{\Phi}(k) - \hat{\Phi}(k-1)||^2; \quad \mu > 0 \tag{1.8}$$

obtaining:

$$\hat{\Phi}(k) = \hat{\Phi}(k-1) + \frac{\eta \Delta U(k-1)(\Delta y(k) - \hat{\Phi}(k-1)^T \Delta U(k-1))}{\mu + ||\Delta U(k-1)||^2}; \tag{1.9}$$

$$\hat{\Phi}(k) = \hat{\Phi}(1), \quad \text{if } ||\hat{\Phi}(k)|| \leq \epsilon \text{ or } sign(\hat{\phi}_1(k)) \neq sign(\hat{\phi}_1(1)) \tag{1.10}$$

with $\eta \in (0, 2)$ and ε is a positive design constant.

Remark 2.1: *Due to Eq. (1.10), it can be assumed, without loss of generality, that $\hat{\phi}_1(k) > 0$.*

The control problem addressed in this chapter is the tracking of a constant reference output variable $y*$. As usual, the quantity $e(k) = y* - y(k)$ is the tracking error, whose dynamics can be easily derived from Eq. (1.7):

$$e(k+1) = e(k) + \phi_1(k)\Delta u(k) + \psi(k-1) \tag{1.11}$$

with

$$\psi(k-1) \triangleq \sum_{i=2}^{d_u} \phi_i(k)\Delta u(k-i+1) \tag{1.12}$$

1.3   Control design

Before proceeding to the design of the control input, the following assumption (Hou & Jin, 2011) is required to prove the stability and convergence of the overall control scheme.

Assumption 3.1: *The first element of the PPD vector satisfies* $\phi_1(k) > \delta$ $\forall k$, *where* δ *is a small positive constant.*

Following Liu and Yang (2019), a PPC requirement is considered. In particular, a positive, decreasing, discrete-time sequence $\rho(k)$ is defined as follows:

$$\rho(k+1) = (1 - \theta_1)\rho(k) + \theta_1 \rho_\infty, \quad \theta_1 \in (0, 1) \tag{1.13}$$

with $\rho(0) > \rho_\infty > 0$. It is required that

$$-\rho(k) < e(k) < \rho(k) \tag{1.14}$$

where $e(k) = y* - y(k)$ is the tracking error. According to Liu and Yang (2019), the following transformed error is introduced:

$$\tau(k) = \frac{1}{2}\ln\left(\frac{\rho(k) + e(k)}{\rho(k) - e(k)}\right) \tag{1.15}$$

thus transforming the initial problem containing the constraint on the tracking error into an unconstrained problem. Using Eq. (1.15), the following sliding surface is defined:

$$s(k+1) = s(k) + \tau(k+1) \tag{1.16}$$

differently from Liu−Yang.

Remark 3.1: *In this chapter, a sliding mode based control law will be provided solving the PPC problem. The reason for proposing the sliding surface [Eq. (1.16)] is that a rigorous stability analysis will be provided guaranteeing both the achievement of the tracking performances and the boundedness of the closed-loop variables. Such stability analysis will be performed, differently from (Liu and Yang, 2019) studying the behavior of the "true" plant [Eq. (1.17)] (not the estimated one* $\Delta y(k+1) = \hat{\boldsymbol{\Phi}}(k)^T \Delta \mathbf{U}(k)$*) fed by the proposed sliding mode control law.*

Furthermore, it should be recalled that, since $\boldsymbol{\Phi}(k)$ and $\hat{\boldsymbol{\Phi}}(k)$ are bounded in view of Theorem 2.1, there exists $\lambda_{\min} > 0$ such that for $\lambda > \lambda_{\min}$ it holds (Hou & Jin, 2011)

$$\left|\frac{\hat{\phi}_1(k)}{\lambda + \hat{\phi}_1(k)^2}\right| < M_1 < \frac{0.5}{b}; \quad M_2 \leq \left|\frac{\phi_1(k)\hat{\phi}_1(k)}{\lambda + \hat{\phi}_1(k)^2}\right| \leq 0.5 \tag{1.17}$$

where M_1 and M_2 are positive constants, and there exists $\rho_1 \in (0, 1)$ such that

$$\left|1 - \rho_1\frac{\phi_1(k)\hat{\phi}_1(k)}{\lambda + \hat{\phi}_1(k)^2}\right| < 1 \tag{1.18}$$

Consider the following control input

$$\Delta u(k) = u_M(k) + u_S(k) \tag{1.19}$$

where

$$u_S(k) = \frac{\rho_1|\hat{\phi}_1(k)|}{\lambda + \hat{\phi}_1(k)^2}\Gamma_s(k)sign(s(k)) \tag{1.20}$$

with $\Gamma_s(k) > 0$ to be defined in the following, and

$$u_M(k) = \frac{\rho_1\hat{\phi}_1(k)}{\lambda + \hat{\phi}_1(k)^2}\left(e(k) - \sum_{i=2}^{du}\rho_i\hat{\phi}_i(k)\Delta u(k - i + 1) + \frac{\rho(k+1)}{1 + \xi(k)}\right) \tag{1.21}$$

where $\xi(k) = \exp(-\alpha\tau(k))$, $\alpha \in (0, 1)$ and with $\rho_i \in (0, 1)$ properly chosen positive constants. From the definition of the tracking error, one has:

$$e(k+1) = e(k) - \sum_{i=1}^{du}\phi_i(k)\Delta u(k - i + 1) \tag{1.22}$$

Defining $\Delta s(k+1) = s(k+1) - s(k)$, it follows:

$$\Delta s(k+1) = \tau(k+1) = \frac{1}{2}\ln\left(\frac{\rho(k+1)+e(k+1)}{\rho(k+1)-e(k+1)}\right) \tag{1.23}$$

and after some manipulations one gets:

$$\Delta s(k+1) = \frac{1}{2}\ln\left(\frac{\alpha(k)+M_a(k)-\phi_1(k)u_S(k)}{\beta(k)-M_a(k)+\phi_1(k)u_S(k)}\right) \tag{1.24}$$

where

$$\alpha(k) \triangleq \rho(k+1)\left(1 - \frac{\rho_1\phi_1(k)\hat{\phi}_1(k)}{\lambda+\hat{\phi}_1(k)^2}\frac{1}{1+\xi(k)}\right) \tag{1.25}$$

$$\beta(k) \triangleq \rho(k+1)\left(1 + \frac{\rho_1\phi_1(k)\hat{\phi}_1(k)}{\lambda+\hat{\phi}_1(k)^2}\frac{1}{1+\xi(k)}\right) \tag{1.26}$$

$$M_a(k) \triangleq e(k)\left(1 - \frac{\rho_1\phi_1(k)\hat{\phi}_1(k)}{\lambda+\hat{\phi}_1(k)^2}\right) + \\ + \sum_{i=2}^{du}\left(\frac{\rho_i\phi_1(k)\hat{\phi}_1(k)}{\lambda+\hat{\phi}_1(k)^2}\hat{\phi}_i(k) - \phi_i(k)\right)\Delta u(k-i+1) \tag{1.27}$$

Lemma 3.1: *Consider the plant [Eq. (1.7)] satisfying Assumption 3.1, controlled by the control input [Eq. (1.19)]. Then, there exists $\overline{M}_a > 0$ such that $|M_a(k)| \leq \overline{M}_a\ \forall k$.*

The proof follows the lines of Hou and Jin (2011) and is omitted for brevity.
Defining:

$$\gamma(k) \triangleq \frac{\rho_1\phi_1(k)\hat{\phi}_1(k)}{\lambda+\hat{\phi}_1(k)^2} \tag{1.28}$$

in view of Remark 2.1, Assumption 3.1 and Eq. (1.17), it holds:

$$0 < \gamma(k) < 0.5 \tag{1.29}$$

and introducing the following definitions:

$$\nu_b(k) \triangleq \frac{1+\xi(k)}{1+\xi(k)-\gamma(k)} \tag{1.30}$$

$$\nu_c(k) \triangleq \frac{1+\xi(k)}{1+\xi(k)+\gamma(k)} \tag{1.31}$$

$$\delta(k) \triangleq \frac{1+\xi(k)-\gamma(k)}{1+\xi(k)+\gamma(k)} \tag{1.32}$$

due to Eq. (1.29), $\nu_b(k)$, $\nu_c(k)$, and $\delta(k)$ are bounded as follows:

$$1 \leq \nu_b(k) \leq 2 \tag{1.33}$$

$$\frac{2}{3} < \nu_c(k) \leq 1 \tag{1.34}$$

$$1/3 < \delta(k) < 1 \tag{1.35}$$

Moreover, comparing Eqs. (1.25), (1.26), and (1.30), one has:

$$\alpha(k) = \frac{\rho(k+1)}{\nu_b(k)} \tag{1.36}$$

$$\beta(k) = \frac{\rho(k+1)}{\nu_c(k)} \tag{1.37}$$

As a consequence, $\Delta s(k+1)$ can be written as:

$$\Delta s(k+1) = \frac{1}{2} \ln\left(\frac{\rho(k+1) + (M_a(k) - \phi_1(k)u_S(k))\nu_b(k)}{\rho(k+1) - (M_a(k) - \phi_1(k)u_S(k))\nu_c(k)}\right) \delta(k) \tag{1.38}$$

Theorem 3.1: *Consider the plant [Eq. (1.7)] satisfying Assumption 3.1, controlled by the control input [Eq. (1.19)]. Under the condition:*

$$\left(\overline{M}_a < min\left\{ \frac{\rho_\infty}{\left(6 + \frac{1}{M_2}\right)}, \frac{\left(2 - \frac{2}{3M_2}\right)\rho_\infty}{\left(2 + \frac{1}{M_2}\right)} \right\} \atop \frac{1}{3} < M_2 \le \frac{1}{2} \right) \tag{1.39}$$

for $\lambda > \lambda_{min}$ in Eq. (1.7) with suitable values of $\lambda_{min} > 0$, the gain $\Gamma_s(k)$ can be properly designed such that the sliding variable $s(k)$ is bounded. As a consequence, $\tau(k)$ is bounded and the condition [Eq. (1.14)] is satisfied.

Proof: The proof consists of two steps. First, it will be proved that $s(k)\Delta s(k+1) < 0$, and next, it will be shown that a region exists bounding the sliding variable.

Step 1, $s(k) > 0$.
Define:

$$\eta(k) \triangleq M_a(k) - \frac{\rho_1\phi_1(k)|\hat{\phi}_1(k)|}{\lambda + \hat{\phi}_1(k)^2}\Gamma_s(k)$$

The condition $\Delta s(k+1) < 0$ requires that

$$\begin{pmatrix} \eta(k) < 0 \\ \rho(k+1) + \eta(k)\nu_b(k) > 0 \end{pmatrix} \tag{1.40}$$

Moreover, due to Eq. (1.40), it follows:

$$\rho(k+1) + \eta(k)\nu_b(k) < \rho(k+1) - \eta(k)\nu_c(k) \tag{1.41}$$

Taking the worst case, and defining $\theta(k) \triangleq \frac{\rho_1\phi_1(k)|\hat{\phi}_1(k)|}{\lambda + \hat{\phi}_1(k)^2}$, conditions (1.40) correspond to:

$$\overline{M}_a < \theta(k)\Gamma_s(k) < \frac{\rho(k+1)}{\nu_b(k)} - \overline{M}_a \tag{1.42}$$

and the following strongest condition will be considered instead:

$$\overline{M}_a < \theta(k)\Gamma_s(k) < \frac{\rho(k+1)}{\nu_b(k)} - 3\overline{M}_a \tag{1.43}$$

Recalling that, according to Eq. (1.17), $M_2 \le \frac{\theta(k)}{\rho_1} \le 0.5$, condition (1.43) is fulfilled if:

$$\begin{cases} \Gamma_s(k) > \dfrac{\overline{M}_a}{\rho_1 M_2} \\ \Gamma_s(k) < \dfrac{\rho(k+1) - 6\overline{M}_a}{\rho_1} \end{cases} \tag{1.44}$$

which requires:

$$\left(6 + \frac{1}{M_2}\right)\overline{M}_a < \rho(k+1) \tag{1.45}$$

or, as a stronger condition:

$$\left(6 + \frac{1}{M_2}\right)\overline{M}_a < \rho_\infty \tag{1.46}$$

Step 1, $s(k) < 0$.
Define:

$$\overline{\eta}(k) \triangleq M_a(k) + \theta(k)\Gamma_s(k) \tag{1.47}$$

The condition $\Delta s(k+1) > 0$ requires that:

$$\begin{cases} \overline{\eta}(k) > 0 \\ \rho(k+1) - \overline{\eta}(k)\nu_c(k) > 0 \\ \rho(k+1) + \overline{\eta}(k)\nu_b(k) > 3(\rho(k+1) - \overline{\eta}(k)\nu_c(k)) \end{cases} \tag{1.48}$$

Considering the worst case and definition (1.47), inequalities (1.48) correspond to:

$$\frac{2}{3}\rho(k+1) < \theta(k)\Gamma_s(k) + M_a(k) < \rho(k+1) \tag{1.49}$$

The following strongest condition will be considered instead:

$$\frac{2}{3}\rho(k+1) < \theta(k)\Gamma_s(k) + M_a(k) < a_1\rho(k+1) \tag{1.50}$$

with $\frac{2}{3} < a_1 < 1$.

Taking into account Lemma 3.1 and recalling that, according to Eq. (1.17), $M_2 \le \frac{\theta(k)}{\rho_1} \le 0.5$, from Eq. (1.50), the following conditions on $\Gamma_s(k)$ can be derived:

$$\begin{cases} \Gamma_s(k) < \dfrac{2[a_1\rho(k+1) - \overline{M}_a]}{\rho_1} \\ \Gamma_s(k) > \dfrac{\frac{2}{3}\rho(k+1) + \overline{M}_a}{M_2\rho_1} \end{cases} \tag{1.51}$$

which requires:

$$\left(2 + \frac{1}{M_2}\right)\overline{M}_a < \left(2a_1 - \frac{2}{3M_2}\right)\rho(k+1) \tag{1.52}$$

or, as a stronger condition:

$$\left(2 + \frac{1}{M_2}\right)\overline{M}_a < \left(2a_1 - \frac{2}{3M_2}\right)\rho_\infty \tag{1.53}$$

To have a feasible solution interval for M_2, recalling Eq. (1.17), and considering also Eq. (1.46), one has the final inequalities system on M_2:

$$\begin{cases} \left(6 + \dfrac{1}{M_2}\right)\overline{M}_a < \rho_\infty \\[2mm] \left(2 + \dfrac{1}{M_2}\right)\overline{M}_a < \left(2a_1 - \dfrac{2}{3M_2}\right)\rho_\infty \\[2mm] \dfrac{1}{3} < M_2 \leq \dfrac{1}{2} \end{cases} \tag{1.54}$$

that is,

$$\begin{cases} \overline{M}_a < \min\left\{ \dfrac{\rho_\infty}{\left(6 + \dfrac{1}{M_2}\right)}, \dfrac{\left(2a_1 - \dfrac{2}{3M_2}\right)\rho_\infty}{\left(2 + \dfrac{1}{M_2}\right)} \right\} \\[4mm] \dfrac{1}{3} < M_2 \leq \dfrac{1}{2} \end{cases} \tag{1.55}$$

which corresponds to Eq. (1.39).

Step 2. Consider expression (1.38). Due to conditions derived in *Step 1* when $s(k) > 0$, it holds $\Delta s(k+1) < 0$, that is,

$$\begin{aligned} \Delta s(k+1) &= \frac{1}{2}\ln\left[\frac{N_1(k)}{D_1(k)}\right] = \\ &\frac{1}{2}\ln\left\{\frac{[\rho(k+1) + M_a(k)\nu_b(k) - \theta(k)\Gamma_s(k)\nu_b(k)]\delta(k)}{\rho(k+1) - M_a(k)\nu_c(k) + \theta(k)\Gamma_s(k)\nu_c(k)}\right\} < 0 \\ &\Rightarrow \frac{N_1(k)}{D_1(k)} < 1 \end{aligned} \tag{1.56}$$

Moreover, using Eq. (1.43) and taking the worst case, one has:

$$\begin{aligned} N_1(k) &\geq \\ &\left(\rho(k+1) - 2\overline{M}_a - \left(\frac{\rho(k+1)}{\nu_b(k)} - 3\overline{M}_a\right)\nu_b(k)\right)\delta(k) \\ &\geq \frac{\overline{M}_a}{3} \triangleq N_1^{\min} \end{aligned} \tag{1.57}$$

$$\begin{aligned} D_1(k) &\leq \\ &\left(\rho(k+1) + \overline{M}_a + \left(\frac{\rho(k+1)}{\nu_b(k)} - 3\overline{M}_a\right)\nu_c(k)\right) \\ &\leq 2\rho(k+1) - 2\overline{M}_a \leq 2\rho(k+1) \leq 2\rho(0) \triangleq D_1^{\max} \end{aligned} \tag{1.58}$$

Taking into account Eqs. (1.56)−(1.58), it follows:

$$\frac{\overline{M}_a}{6\rho(0)} = \frac{N_1^{\min}}{D_1^{\max}} < \frac{N_1(k)}{D_1(k)} < 1 \tag{1.59}$$

Consider again expression (1.38). Due to conditions derived in *Step 1* when $s(k) < 0$, it holds $\Delta s(k + 1) > 0$, that is,

$$\Delta s(k + 1) = \frac{1}{2} \ln \left(\frac{N_2(k)}{D_2(k)} \right) =$$

$$\frac{1}{2} \ln \left\{ \frac{[\rho(k + 1) + M_a(k)\nu_b(k) + \theta(k)\Gamma_s(k)\nu_b(k)]\delta(k)}{\rho(k + 1) - M_a(k)\nu_c(k) - \theta(k)\Gamma_s(k)\nu_c(k)} \right\} > 0 \tag{1.60}$$

$$\Rightarrow \frac{N_2(k)}{D_2(k)} > 1$$

Using Eq. (1.50) and taking the worst case, one has:

$$N_2(k) \leq \{\rho(k + 1) + 2\overline{M}_a + 2[\rho(k + 1) - \overline{M}_a]\}$$
$$\leq 3\rho(k + 1) \leq 3\rho(0) \triangleq N_2^{\max} \tag{1.61}$$

$$D_2(k) \geq \rho(k + 1) - a_1\rho(k + 1) \geq (1 - a_1)\rho(k + 1)$$
$$\geq (1 - a_1)\rho_\infty \triangleq D_2^{\min} \tag{1.62}$$

Taking into account Eqs. (1.60)–(1.62), it follows:

$$1 < \frac{N_2(k)}{D_2(k)} < \frac{N_2^{\max}}{D_2^{\min}} = \frac{3\rho(0)}{(1 - a_1)\rho_\infty} \tag{1.63}$$

Collecting the results obtained in both cases $s(k) > 0$ and $s(k) < 0$ one gets:

$$\overline{\alpha} \triangleq \frac{1}{2} \ln \left(\frac{\overline{M}_a}{6\rho(0)} \right) < \Delta s(k + 1)$$

$$< \frac{1}{2} \ln \left(\frac{3\rho(0)}{(1 - a_1)\rho_\infty} \right) \triangleq \overline{\beta} \tag{1.64}$$

where $\overline{\alpha} < 0$ and $\overline{\beta} > 0$. Define

$$\Lambda = \{\overline{\alpha} \leq s(k) \leq \overline{\beta}\} \tag{1.65}$$

It is straightforward to see that, once this region is entered by the sliding variable $s(k)$ at the time instant k, then $s(k + 1) \in \Lambda(k + 1)$. In fact, if $s(k) > 0$ and $s(k) \in \Lambda$, it means that. Since $\overline{\alpha} < \Delta s(k + 1) < 0$, one has $\overline{\alpha} < s(k + 1) < s(k) \leq \overline{\beta}$. Analogously, if $s(k) < 0$ and $s(k) \in \Lambda$, it means that $\overline{\alpha} \leq s(k) < 0$. Since $\overline{\beta} \geq \Delta s(k + 1) > 0$, one has $\overline{\beta} > s(k + 1) > s(k) \geq \overline{\alpha}$.

1.4 Simulation study using FAST

The application of a rotor speed control approach would require the availability of a reference speed ensuring the tracking of the maximum delivered power point, but this is hindered by the strongly nonlinear nature of Eq. (1.2). The approach taken here is to consider a PFDL description of the overall nonlinear dynamics connecting the aerodynamic torque $y(k) = T_a(k)$ (output variable) and the electrical torque $u(k) = T_e(k)$ (input variable) under the effect of the (not manipulable) external input $V(k)$ (wind speed), which basically affects the aerodynamic torque $T_a(k)$ through the nonlinear power coefficient C_p. In such a framework, the WT rotor speed $\omega_r(k)$ acts as an internal (state) variable governing the system dynamics. The aerodynamic torque $T_a(k)$ can be estimated by the torque transmitted through the HSS, which can be measured from a strain gage mounted on the HSS as reported by NREL in NREL-NWTC (2015).

With the described setting, a nonlinear model of the form [Eq. (1.6)] has been considered and a PPD-based model has been derived. The addressed control problem can be now stated as an output tracking problem with respect to the optimal torque in the region 2 control law, built based on instantaneous rotor speed measurements as proposed in Jonkman et al. (2009).

The proposed controller has been tested by intensive simulations using the NREL FAST code. The FAST simulator is a high fidelity aeroelastic simulator of two- and three-bladed horizontal-axis WTs (Jonkman & Buhl, 2005; NREL-NWTC, n.d.) widely adopted for WT design and certification (Buhl & Manjock, 2006; Manjock, 2005). Simulations have been performed using the NREL 5-MW WT, whose parameters have been derived from (Jonkman et al., 2009). The FAST wind data (NREL-NWTC, n.d.) used for validation tests are shown in Fig. 1.1 (mean value 8 m s^{-1}). Initial conditions have been set as $\omega_r(0) \approx 0.9$ rad s^{-1}.

The following parameters have been used for simulation tests: $d_u = 2$, $\rho_1 = 1, \rho_2 = 1$, $\rho_\infty = 10^5$, $\rho(0) = 1 \times 10^7$, $\theta_1 = 0.8$, $\mu = 10^4$, $\lambda = 10^4$, $\eta = 1.5$, $\hat{\Phi}(0) = [50 \; 50]^T$.

Some of the performed tests have been reported in Figs. 1.2−1.4. Fig. 1.2 shows the angular rotor speed, Fig. 1.3 shows the optimal generator-torque (Jonkman et al., 2009) tracking error, and Fig. 1.4 shows the tip speed ratio. A comparison of the proposed control approach with respect to the performances obtained with the control algorithms described in Hou and Jin (2011) and Liu and Yang (2019) has been reported.

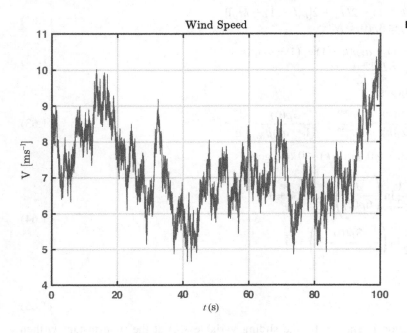

FIGURE 1.1 Wind inflow.

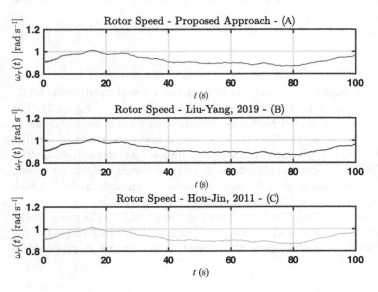

FIGURE 1.2 Rotor speed. (A) Proposed approach; (B) approach proposed in Liu and Yang (2019); and (C) approach proposed in Hou and Jin (2011).

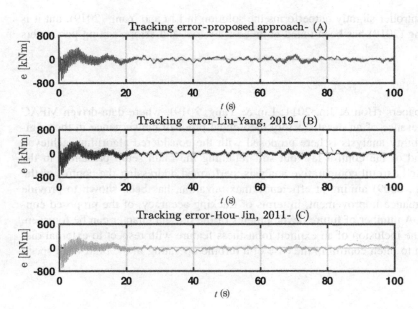

FIGURE 1.3 Tracking error. (A) Proposed approach; (B) approach proposed in Liu and Yang (2019); and (C) approach proposed in Hou and Jin (2011).

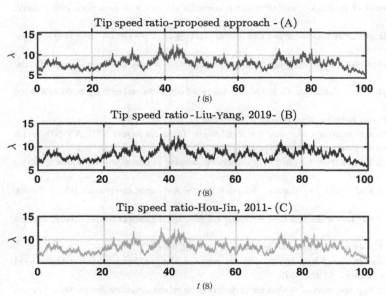

FIGURE 1.4 Tip speed ratio. (A) Proposed approach; (B) approach proposed in Liu and Yang (2019); and (C) approach proposed in Hou and Jin (2011).

TABLE 1.1 Performance comparison: IAE (from time instant $t = 10$ s).

Proposed approach	Liu−Yang (Liu & Yang, 2019)	Hou−Jin (Hou & Jin, 2011)
8.2344e + 08	8.7353e + 08	9.2970e + 08

To analyze the performance of the proposed controller in terms of fulfillment of the prescribed performance requirement, the IAE criterion (i.e., the integral of the absolute value of the tracking error) has been computed for the considered controller starting from the time instant $t = 10$ s to eliminate the transient influence. The results are reported in Table 1.1. It can be noticed that an improvement has been obtained with respect to the control algorithm described in Hou and Jin (2011) of approximately 8% and a comparable performance has been achieved with respect to Liu and Yang (2019).

The reported results show that the proposed controller slightly outperforms the solution in Liu and Yang (2019), but it is worth noting that the algorithm in Liu and Yang (2019) has been tuned at its best, by trial and errors, setting parameters without any methodological support.

1.5 Conclusions

The present study is inspired by very recent papers (Hou & Jin, 2011; Liu & Yang, 2019), where data-driven MFAC controllers have been presented. Due to the presence of an unsatisfactory theoretical proof of convergence in the original paper (Liu & Yang, 2019), a rigorous stability analysis is here proposed with the considered algorithm, achieved modifying the forms of the sliding surface and of the control law but still retaining the main setup presented in the source paper (Liu & Yang, 2019). The presented careful comparative analysis, performed addressing the control of the three-blade NREL 5-MW WT (Jonkman et al., 2009) aiming at efficiency maximization, has been shown to provide very satisfactory results, showing some performance improvement, in terms of tracking accuracy, of the proposed control law with respect to the available literature. A number of future developments of the present chapter can be foreseen, namely the extension to multioutput systems, the inclusion of an explicit robustness feature with respect to external disturbances affecting the plant, and the extension to pitch control in the case of a turbine operating in the high wind speed region.

References

Bechlioulis, C., & Rovinthakis, G. (2008). Robust adaptive control of feedback linearizable mimo nonlinear systems with prescribed performance. *IEEE Transactions on Automatic Control, 53*, 2090−2099.
Bechlioulis, C., & Rovinthakis, G. (2009). Adaptive control with guaranteed transient and steady-state tracking error bounds for strict feedback systems. *Automatica, 45*, 532−538.
Bianchi, F. D., Battista, H. N. D., & Mantz, R. J. (2007). *Wind turbine control systems: Principles, modelling and gain scheduling design*. Berlin: Springer-Verlag.
Bossanyi, E., Savini, B., Iribas, M., Hau, M., Fischer, B., Schlipf, D., ... Carcangiu, C. E. (2012). Advanced controller research for multi-MW wind turbines in the UPWIND project. *Wind Energy, 15*, 119−145.
Bossanyi, E. A. (2000). The design of closed loop controllers for wind turbines. *Wind Energy, 3*, 149−163.
Buhl, M. L. & Manjock, A. (2006). *A comparison of wind turbine aeroelastic codes used for certification*. Technical Report NREL/CP-500-39113. National Renewable Energy Laboratory, Golden, CO. <http://www.nrel.gov/docs/fy06osti/39113.pdf>.
Hansen, M. H., Thomsen, K., Fuglsang, P., & Knudsen, T. (2006). Two methods for estimating aeroelastic damping of operational wind turbine modes from experiments. *Wind Energy, 9*, 179−191.
Hou, Z. (1994). *The parameter identification, adaptive control and model free learning adaptive control for nonlinear systems* (Ph.D. thesis). Northeastern University Shenyang, China.
Hou, Z., & Jin, S. (2011). A novel data-driven control approach for a class of discrete-time nonlinear systems. *IEEE Transactions on Control Systems Technology, 19*, 1549−1558.
Hou, Z., & Jin, S. (Eds.), (2014). *Model-Free Adaptive Control: Theory and applications*. Boca Raton, FL: CRC Press.
Iribas, M., & Landau, I.-D. (2009). Identification of wind turbines in closed-loop operation in the presence of three-dimensional turbulence wind speed: torque demand to measured generator speed loop. *Wind Energy, 12*, 660−675.
Iribas-Latour, M., & Landau, I.-D. (2013). Identification in closed-loop operation of models for collective pitch robust controller design. *Wind Energy, 16*, 383−399.
Johnson, K., Pao, L., Balas, M., & Fingersh, L. (2006). Control of variable-speed wind turbines: standard and adaptive techniques for maximizing energy capture. *IEEE Control Systems Magazine, 26*, 70−81.
Jonkman, J., Butterfield, S., Musial, W., & Scott, G. (2009). *Definition of a 5-MW reference wind turbine for offshore system development*. Technical Report NREL/TP-500-38060. Colorado.
Jonkman, J. M. & Buhl, M. L. (2005). *FAST user's guide*. Technical Report NREL/EL-500−38230. National Renewable Energy Laboratory, Golden, CO. <http://wind.nrel.gov/public/bjonkman/TestPage/FAST.pdf>.
Liu, D., & Yang, G.-H. (2019). Data-driven adaptive sliding mode control of nonlinear discrete-time systems with prescribed performance. *IEEE Transactions on Systems, Man, and Cybernetics: Systems, 49*, 2598−2604.
Manjock, A. (2005). *Design codes FAST and ADAMS® for load calculations of onshore wind turbines*. Technical Report 72042. Germanischer Lloyd WindEnergie GmbH, Hamburg, Germany. GL Wind Order No. 23354/66. <http://wind.nrel.gov/designcodes/papers/GL_Report.pdf>.
Monroy, A., & Alvarez-Icaza, L. (2006). Real-time identification of wind turbine rotor power coefficient. In: *Proceedings of the forty-fifth IEEE conference on decision and control*. (pp. 3690−3695).
NREL-NWTC. (n.d.). NWTC information portal (FAST). <https://nwtc.nrel.gov/FAST>. Accessed 28.04.16.
NREL-NWTC. (2015). Resistant moment of the rotor and of the electric generator. <https://wind.nrel.gov/forum/wind/viewtopic.php?f = 4t = 525>. Accessed 19.03.20.

Ozbek, M., & Rixen, D. J. (2013). Operational modal analysis of a 2.5 MW wind turbine using optical measurement techniques and strain gauges. *Wind Energy, 16*, 367–381.

Qiao, W., Qu, L., & Harley, R. (2009). Control of IPM synchronous generator for maximum wind power generation considering magnetic saturation. *IEEE Transactions on Industry Applications, 45*, 1095–1105.

Siegfried, H. (1998). *Grid integration of wind energy conversion systems*. John Wiley & Sons Ltd.

Simani, S., & Castaldi, P. (2013). Data-driven and adaptive control applications to a wind turbine benchmark model. *Control Engineering Practice, 21*, 1678–1693.

van Baars, G. E., & Bongers, P. M. M. (1994). Closed loop system identification of an industrial wind turbine system: experiment design and first validation results. In: *Proceedings of the thirty-third IEEE conference on decision and control.* (Vol. 1, pp. 625–630).

van der Veen, G., van Wingerden, J., Fleming, P., Scholbrock, A., & Verhaegen, M. (2013). Global data-driven modeling of wind turbines in the presence of turbulence. *Control Engineering Practice, 21*, 441–454.

van Wingerden, J.-W., & Verhaegen, M. (2009). Subspace identification of bilinear and LPV systems for open- and closed-loop data. *Automatica, 45*, 372–381.

Chapter 2

Advanced control design based on sliding modes technique for power extraction maximization in variable speed wind turbine*

Aldo José Flores-Guerrero, Jesús De-León-Morales and Susana V. Gutiérrez-Martínez

Department of Electrical Engineering, Faculty of Mechanical and Electrical Engineering , Autonomous University of Nuevo León, San Nicolás de los Garza, Mexico

2.1 Introduction

Through the years the society and industry have experienced and exploited different sources of energy; however, electrical energy has become one of the most important, due to its socio-economical benefits. Nevertheless, the real demand for electrical energy is increasing due to technological development and increased productivity in industry and energy consumption in homes. Thus large-scale conversion systems are required. Therefore the costs of power generation increase, which leads to the implementation of different power electric plants for satisfying the electric power demand, such as thermal, nuclear, geothermal, and combined cycle.

Among the different power plants, thermal plants have been widely used over the world in the last decades, thanks to the low generation cost, high conversion efficiency, easy installation, and less construction space required.

Regardless, through the years this process has presented several environmental risks because of coal using. The burning of coal and other fossil fuels release pollutants to the atmosphere, such as carbon dioxide and sulfur dioxide, which are harmful and are the cause of greenhouse effect, which leads to global warming, spread of the diseases, and the melting of ice caps.

These environmental risks and the possible energy crisis have led to improve and implement new technologies for electric power generation based on clean energies solutions, such as solar and wind energy, which are the most popular. Among these energy sources, wind power has experienced an important growth in the last decades, due to its socio-economical profits and environmental friendly energy production because no pollution is involved in this process.

According to the information published by the World Wide Energy Association in 2020, the overall capacity of all wind turbines (WTs) installed in the end of 2019 reached 650.8 GW, which can cover more than 6% of the global electricity demand. This growth implies new challenges in engineering and science.

On the other hand, thanks to the advances in power electronics and aerodynamic engineering, the WTs become more attractive for power generation; thus advanced control strategies for power generation are required. These controls should be capable to reduce the cost of wind energy production by increasing turbine efficiency and the life time of components and structures.

---

*Fully documented templates are available in the elsarticle package on CTAN.

Renewable Energy Systems. DOI: https://doi.org/10.1016/B978-0-12-820004-9.00004-8

2.1.1 A description of wind turbines

The circulation of the air masses in the atmosphere (well known as wind) is a consequence of the temperature and pressure gradients, due to the interaction of the energy that comes from the sun with the air masses, generating differences in density between them.

Thanks to the application of aerodynamic principles and mechanical links together with electrical devices, it is possible to obtain electrical energy from the kinetic energy available in the wind. This energy conversion process is carried out by WTs.

The WTs are complex electromechanical devices comprised of many components with different functions, interacting with a changing environment. The main components of a WT, shown in Fig. 2.1, are (1) the rotor, which includes the blades, hub, and aerodynamic surfaces; (2) the drive train, which is composed by the gearbox (if any), low and high speed shafts referred to the rotor and generator, respectively, the mechanical brakes and couplings connecting them; (3) the main frame, which provides support for the mounting and proper alignment of the drive train components, along with the nacelle that protects them from the environmental conditions (Manwell, McGowan, & Rogers, 2010); and (4) the tower, which supports the nacelle at an appropriate height to reduce the influence of turbulence and maximize the wind energy. On the top of the tower, the yaw system keeps the rotor shaft aligned with the wind.

The conversion process is summarized as follows: the blades connected to the rotor by the hub are moved by the wind causing them to spin, producing a rotation in the low speed shaft. This rotation is transmitted to the high speed shaft through the gearbox, which serves to increase the rotational speed.

The high-speed shaft drives the rotational torque to the generator where the conversion process of mechanical energy into electrical energy is carried out. This conversion requires a fixed input speed or power electronic devices to adapt the output energy to the grid requirements (Márquez, Pérez, Marugán, & Papaelias, 2016).

There exists a theoretical limit based on aerodynamic principles that states the amount of wind energy that can be converted into mechanical energy. This limit receives the name of Betz limit, which is less than 59.3% according to the blade element momentum theory. This limit plays a decisive role in the control designing of WTs to produce electrical power.

2.1.2 Wind turbines structures and operation conditions

WTs can be classified according to: (1) output power, (2) electric machine, (3) speed operation, and (4) turbine orientation, as shown in Fig. 2.2.

Regarding output power, the WTs are classified into: (1) small, (2) medium, and (3) large capacity.

The output power depends on the diameter of the turbine and the blade length, which influence the rotor swept area and thus the amount of captured power.

FIGURE 2.1 Wind turbine components.

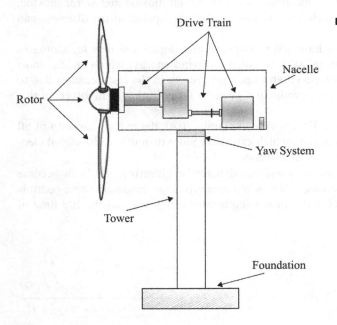

Drive Train

Nacelle

Rotor

Yaw System

Tower

Foundation

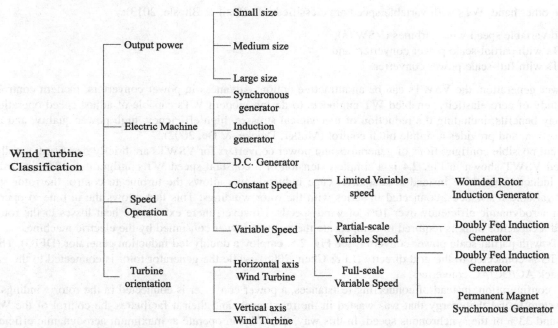

```
                                      ┌─── Small size
                                      │
                 ┌─── Output power ───┼─── Medium size
                 │                    │
                 │                    └─── Large size
                 │                    ┌─── Synchronous
                 │                    │    generator
                 ├─── Electric Machine┼─── Induction
                 │                    │    generator
  Wind Turbine   │                    └─── D.C. Generator
  Classification │
                 │                    ┌─── Constant Speed
                 │                    │
                 ├─── Speed           │                    ┌─── Limited Variable ─── Wounded Rotor
                 │    Operation       │                    │    speed                Induction Generator
                 │                    └─── Variable Speed ──┤
                 │                                          ├─── Partial-scale  ─── Doubly Fed Induction
                 │                                          │    Variable Speed      Generator
                 │                    ┌─── Horizontal axis  │
                 └─── Turbine         │    Wind Turbine     └─── Full-scale      ─── Doubly Fed Induction
                      orientation ────┤                         Variable Speed       Generator
                                      └─── Vertical axis
                                           Wind Turbine                          ─── Permanent Magnet
                                                                                     Synchronous Generator
```

FIGURE 2.2   Wind turbine classification.

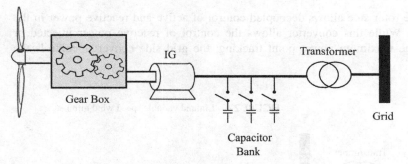

Gear Box   IG   Transformer

Capacitor
Bank

Grid

FIGURE 2.3   Constant speed wind turbine configuration.

On the other hand, to provide electrical power to small stores, farms, residences, and communities, small size WTs (rated below 100 kW) have been developed. Furthermore, medium size WTs with power capacity (less than 1 MW) are commonly used in the industry applications. Finally, large size WTs are used in large utility grids (Manwell et al., 2010).

The WTs can be constructed using different electrical machines. The most widely used in the industry are generators. The most common types are: (1) induction generators, (2) synchronous generators, and (3) DC generators.

The DC generators are usually used in small WTs. However, nearly all large-scale WTs use either induction or synchronous generators. For WTs in grid-connected applications, induction generators are the most used, due to its smooth connection to the electrical grid and the reduced costs (Susperregui, Jugo, Lizarraga, & Tapia, 2014).

Another classification of the WTs is, in terms of speed operation, such as (1) constant speed and (2) variable speed WTs. The choice of whether the rotor speed is constant or variable may have some impact on the overall control design.

An example of a constant speed WT is based on the induction generator (squirrel cage) connected to the grid as seen in Fig. 2.3. This WT operates with very little variations in rotor speed since it is directly connected to the grid, which is operating at a fixed frequency. However, it is necessary an external reactive power support to compensate power consumed by the induction machine. Furthermore, for regulating power at high wind speeds, stall control and blade pitch control are required. Despite constant speed configuration is very simple and robust, energy capture from the wind is suboptimal and reactive power compensation is required.

On the other hand, WTs with variable speed are classified into (Patil & Bhosle, 2013):

- limited variable speed wind turbines (VSWTs);
- VSWTs with partial-scale power converter; and
- VSWTs with full-scale power converter.

In power generation, the VSWTs can be an attractive option. Advances in power converters, modern control theory and the study of aero-elasticity, enabled WT engineers to design modern WTs capable of adjust speed operation. This offers many benefits, including the reduction of mechanical stresses, high efficiency, high power quality, and acoustic noise reduction, and provides a simple pitch control (Muller, Deicke, & De, 2002).

Different possible configurations of generators and power converters for VSWTs are briefly described as follows.

Limited VSWT shown in Fig. 2.4 is a simple extension of a constant speed WT configuration. The inclusion of a wounded induction machine instead of a squirrel cage induction one allows the turbine to control the rotor speed by means of variable resistances, connected in series with the rotor windings. This fact allows the turbine to operate at a maximum aerodynamic efficiency over 10% of wind speeds. However, there exist some heat losses in the rotor resistances and a capacitor bank is required to compensate the reactive power consumed by the electric machine.

VSWTs with partial-scale power converter, see Fig. 2.5, employ a doubly fed induction generator (DFIG). The stator of the DFIG is connected to the grid directly (Li & Chen, 2008), while the generator rotor is connected to the grid by a back-to-back AC/DC/AC converter.

In this configuration, instead of connecting resistances, a power converter is connected in the rotor windings, allowing the system to use the energy that was wasted in the resistances, and then it facilitates the control of the WT rotor speed around 33% of the synchronous speed. In this way, the WT can operate at maximum aerodynamic efficiency for different wind speeds. The main advantage of this configuration is the reduced converter costs due to the total of power passing through the converter, which is about 25% of the machine's full output power (Muller et al., 2002; Singh et al., 2014).

Furthermore, the converter connected in the rotor side allows decoupled control of active and reactive power in the stator, which is suitable for control purposes. While this converter allows the control of reactive power injected or absorbed by the stator windings as well as the maximum power point tracking, the grid side converter controls the power factor (Zhang, Cai, & Guo, 2009).

FIGURE 2.4 Limited variable speed wind turbine.

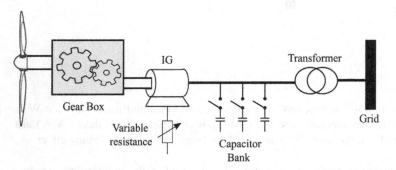

FIGURE 2.5 Variable speed wind turbines with partial-scale power converter.

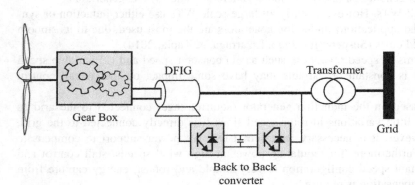

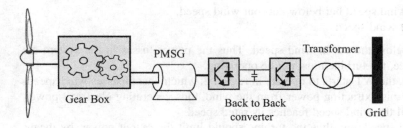

FIGURE 2.6 Variable speed wind turbines with full-scale power converter.

PMSG

Transformer

Gear Box

Back to Back converter

Grid

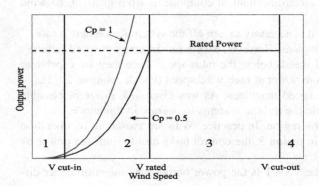

FIGURE 2.7 Wind turbine operation regions.

The DFIG is one of the mostly used electric machines for high-power variable speed wind energy conversion systems, thanks to its advantages, such as flexible operation, decoupled active and reactive power, reduced converter costs as well as power factor control, and improved system efficiency. These advantages have led to an important research and development of control strategies for VSWT based on DFIG (Beltran, Benbouzid, & Ahmed-Ali, 2010; Chen, Guerrero, & Blaabjerg, 2009; Liu, Han, & Wang, 2016; Susperregui et al., 2014).

The VSWTs with full-scale power converter configuration is illustrated in Fig. 2.6. A full-scale power converter is used to transmit the machine's full output power to the grid. In contrast with the VSWTs with partial-scale power converter configuration, the power converter should be larger. However, this increases the costs of the power converter, and its efficiency would determinate the total system efficiency. These WTs commonly employ a permanent magnet synchronous generator, which may allow to avoid the use of gearbox. However, DFIG can be used in this configuration as well. Thanks to the power converter, a good isolation from the grid and decoupled active and reactive power can be achieved. This configuration is extensively used in offshore applications (Singh et al., 2014).

The main difference between the operation speeds is how the aerodynamic efficiency is controlled. The constant speed WT operates at a fixed WT rotor speed, in spite of wind speed variations. Thus the tip speed ratio (TSR) would be changing according to the wind speed. Furthermore, the aerodynamic efficiency depends directly on the TSR and the blade pitch angle, and it will take a maximum value for a given wind speed. On the other hand, VSWTs are capable to vary the WT rotor speed proportionally to the wind speed, keeping a constant TSR, which leads to the aerodynamic efficiency maximization.

Finally, according to the orientation in the rotor axis, the WTs can be classified into: (1) horizontal-axis wind turbines (HAWTs) and (2) vertical-axis wind turbines. Although a WT can be built in either a vertical-axis or horizontal-axis configuration, for power production, the HAWTs are more used in the industry since they dominate the utility-scale WT market. Furthermore, aerodynamic and practical advantages can be found in HAWTs, such as high power output, high efficiency, high reliability, and high operational speed. On the other hand, due to the size of these WTs, transporting, maintenance, and installation costs are high.

2.1.2.1 Operation regions
Depending on the wind speed profile, there are different operation regions in WTs according to the output power, represented in Fig. 2.7. These regions are:

- Region 1: wind velocity is below the cut-in wind speed.
- Region 2: wind velocity is below the rated wind speed.

- Region 3: wind velocity is greater than rated wind speed but below cut- out wind speed.
- Region 4: wind velocity is greater than cut-out wind speed.

In region 1, the wind speed is low, that is, below the cut-in wind speed. Thus the power losses in the system are greater than the available power in the wind; hence, the turbine must not be operating.

When the wind speed reaches the cut-in speed, the WT starts to operate in region 2, which encompasses wind speeds between cut-in and rated speed. Hence, the turbine is extracting power from the wind, thus generating electrical power. However, it is not generating the rated power until the wind speed reaches its rated speed.

In region 3, the wind speed is above the rated speed, and thus the turbine should limit the output power, by means of blade pitch angle control and regulating torque to protect the electromechanical components from gusts in the wind that may cause abrupt torques.

Finally, in region 4, the wind speed crosses the cut-out speed and it is necessary to turn off the system for keeping it safe.

In general, variable and constant WTs have the same operation regions. However, except for the constant speed turbine's design operating point, VSWTs tend to be more efficient for wind speeds below the rated speed since they are capable to operate near by the maximum power efficiency and thus convert more power at each wind speed (Pao & Johnson, 2011).

For this reason, modern large commercial WTs are variable speed machines. As was discussed, different possible configurations of power converters and electrical generators enable the turbine systems to operate in this mode.

VSWTs show different control tasks according to the operation region. In practice, WTs are switching its operation mode between regions 2 and 3. Generally, for high wind speeds in region 3, the control tasks aim to limit the power by means of regulating the torque and controlling blade pitch angle.

One of the main and most challenging control objectives of the VSWT is the power output maximization under different wind speeds in region 2.

As a consequence of the instantaneous changing of the wind, it is desirable to determine the optimal WT rotor speed that ensures the power extraction maximization. Several maximum power point tracking (MPPT) algorithms, such as TSR control, optimal torque (OT) control, and power signal feedback control, among others (Abdullah, Yatim, Tan, & Saidur, 2012), combined with appropriate control techniques, have been proposed over the last decades to achieve the power extraction maximization.

2.1.3 Problem statement

In this section, the problem of power extraction maximization of a horizontal-axis VSWT operating in region 2 through the MPPT will be addressed.

Solutions to solve this problem have been published in the control engineering literature. Among these solutions, there are those based on the standard proportional—integral derivative (Wang, Tse, & Gao, 2011) widely used in the industry, linear quadratic Gaussian (Munteanu, Cutululis, Bratcu, & Ceangă, 2005; Poitiers, Bouaouiche, & Machmoum, 2009) or based on more modern linear control techniques, such as H∞ (Ma, Chen, Liu, & Allgöwer, 2014; Nayeh, Moradi, & Vossoughi, 2020), or the most popular nonlinear control technique feedback linearization (Boukhezzar & Siguerdidjane, 2009; Zhang et al., 2009), neural networks (Kong, Liu, & Lee, 2014), or those based on robust nonlinear control techniques as the backstepping control design (Yahdou, Boudjema, Taleb, & Djilali, 2018).

However, these control strategies when are implemented in VSWT systems have not achieved better performances due to the high nonlinear behavior, the presence of parametric uncertainties and unmodeled dynamics, and many assumptions that must be considered for obtaining simplified mathematical models for control purpose design.

Several works have been published for this purpose. Despite successful development of robust adaptive control, sliding mode control (SMC) technique has proved to be the best choice when uncertainties and unmodeled dynamics are involved in the control design (Levant, 2003).

One of the most attractive methodologies of control design based on sliding mode techniques is the super twisting control, which has been applied recently to VSWT systems. This success is due to its finite-time convergence to the sliding surface and robustness to compensate bounded matched uncertainties (insensitivity) (Beltran, Benbouzid, & Ahmed-Ali, 2012; Liu et al., 2016).

Several SMC strategies have been implemented in WTs. For instance, in Beltran et al. (2010), a high gain observer combined with a second-order sliding mode (SOSM) controller has been designed for achieving the MPPT in a DFIG-VSWT operating in region 2.

Furthermore, a super twisting algorithm (STA) has been designed in Valenciaga and Evangelista (2010) and Valenciaga (2010) for control active and reactive power of a VSWT based on brushless doubly fed reluctance machine.

In addition, an adaptive super twisting control has been applied for control active and reactive power of a DFIG as well (see Dash & Patnaik, 2014).

Moreover, an OT control strategy was developed combined with an STA controller for achieving the maximum power extraction in a VSWT based on DFIG in Beltran et al. (2012).

On the other hand, a super twisting control with variable gain has been introduced in Evangelista, Valenciaga, and Puleston (2013) where the performance of the closed-loop system, in terms of mechanical loads and power tracking, has been achieved.

A control algorithm based on the standard SMC for achieving the torque and reactive power control in a VSWT based on DFIG, and under grid voltages variations, has been developed in Susperregui, Martinez, Tapia, and Vechiu (2013).

Despite the SMC strategies proposed in previous works, the control objectives were limited to the electrical part, without considering the effects that may occur in the mechanical process.

However, to achieve the objective of obtaining the maximum power extraction by means of maximizing the aerodynamic efficiency, the generator torque was selected to track the optimal aerodynamic torque, neglecting the dynamics of the drive train. Then, to achieve this objective, it was assumed that the aerodynamic torque and the electromagnetic torque were equal, which is not a realistic assumption particularly for gusts in the wind since the change of wind speed will not immediately be transferred to the generator side (Liu et al., 2016).

2.1.4 The main contribution

In this work, to deal with the power extraction maximization, a dynamic VSWT model based on DFIG is established, which includes the aerodynamic and electrical parts of the system, as well as external uncertainties and parametric disturbances.

Furthermore, an STA for VSWT system based on DFIG is proposed, to make the WT rotor speed tracks the desired speed (the speed that maximizes the power coefficient C_p) in spite of system uncertainties. Moreover, using sliding mode techniques, a control law is designed to make the rotor current track the reference value for a VSWT, in spite of the external perturbations and uncertainties, and reaching high efficiency.

Besides, from a theoretical point of view, by using quadratic Lyapunov function, sufficient conditions are obtained to ensure the finite-time convergence of the control strategy.

Furthermore, the proposed control strategy is validated on a three-bladed 1.5-MW VSWT using Fatigue, Aerodynamic, Structures and Turbulence (FAST) (Jonkman et al., 2005) and TurbSim (Jonkman & Buhl, 2006) simulator from National Renewable Energy Laboratory (NREL) interfaced with MATLAB−Simulink to emulate a realistic WT with all freedom degrees enabled.

2.1.5 Chapter structure

This chapter is organized as follows. In Section 2.2, a mathematical model describing the behavior of a VSWT is introduced. Furthermore, from the mathematical model obtained of a VSWT, an STA is designed for VSWT system based on DFIG in Section 2.3, to make the WT rotor speed track the desired speed. In Section 2.4, simulation results are given for illustrating the performance of the proposed control algorithm when it is applied to a VSWT. Finally, conclusions and future directions are given.

2.2 Modeling variable speed wind turbine

WT modeling must contain many degrees of freedom (DOFs) to cover with the most important dynamic effects. The WT system is composed into three subsystems: (1) the aerodynamic subsystem, (2) mechanical subsystem, and (3) electrical subsystem (Golnary and Moradi, 2019) which is illustrated in Fig. 2.8. tem.

2.2.1 Aerodynamic subsystem of wind turbine

One of the most important aerodynamic properties of a WT is the aerodynamic efficiency, which is related with the blade geometry (Golnary & Moradi, 2019). It is represented with the power coefficient C_p, as the ratio of the instantaneous turbine power P to the instantaneous power available in the wind P_{Wind}, and is given by

$$C_p = \frac{P}{P_{Wind}}$$

(2.1)

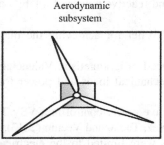
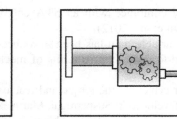
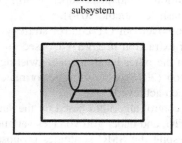

FIGURE 2.8 Wind turbine subsystems.

where P is given by

$$P = 1/2\pi\varphi R^2 C_p v^3 \tag{2.2}$$

and P_{Wind} is

$$P_{Wind} = 1/2\pi\varphi R^2 v^3 \tag{2.3}$$

with φ as the air density, R is the rotor diameter and v is the wind speed.

Thus the captured aerodynamic power that the WT extracts from the wind is a function of two main factors: the wind speed $v(t)$ and the power coefficient of WT C_p.

Furthermore, the power coefficient C_p is a nonlinear function that depends on the blade pitch angle β and TSR λ, that is, $C_p = C_p(\lambda, \beta)$.

On the other hand, the TSR λ of a WT is defined as the ratio between the tip blade speed ω_{rtur} and wind speed v, which is expressed as

$$\lambda = \frac{\omega_{rtur} R}{v} \tag{2.4}$$

The maximum aerodynamic efficiency, called Betz Limit, $Cp_{max} = 16/27$ is the maximum theoretically possible rotor power coefficient. However, in practice, several effects, such as finite blades number and associated tip losses, lead to a decrease in the maximum achievable power coefficient (Manwell et al., 2010).

For control design purposes, the objective is to produce the maximal aero-dynamcal power $P = 1/2\pi\varphi R^2 C_p(\lambda, \beta)v^3$, by means to maximize the power coefficient C_p at λ_{opt} and the blade pitch angle β_{opt} under conditions of region 2. For this end, the maximum value of the power coefficient C_p can be obtained by plotting C_p in terms of TSR λ, and blade pitch angle β, by using an aerodynamic code or by means of approximated nonlinear functions of (λ, β). For instance by'

$$C_p(\lambda, \beta) = c_1 \left(\frac{c_2}{\Gamma(\lambda, \beta)} - c_3\beta - c_4 \right) exp^{-c_5/\Gamma} + c_6\lambda \tag{2.5}$$

where

$$\frac{1}{\Gamma(\lambda, \beta)} = \left(\frac{1}{\lambda + 0.08\beta} \right) - \left(\frac{0.035}{\beta^3 + 1} \right) \tag{2.6}$$

and the coefficient $c_1, \ldots, c_6$ depends on the WT design characteristics.

Furthermore, the following coefficient values are adopted from Liu et al. (2016). $c_1 = 0.5176$, $c_2 = 116$, $c_3 = 0.4$, $c_4 = 5$, $c_5 = 21$, and $c_6 = 0.0068$.

Fig. 2.9 shows the plot of power coefficient C_p versus the TSR λ, for different values of the blade pitch angle β. A maximal value of C_p can be observed that corresponds to $\lambda = 8$ and $\beta = 0$.

Another method for obtaining the power coefficient, is using FAST. For simulating power coefficient, suitable configurations, that is, with all freedom degrees disabled in FAST Elastodyn and a wind step profile input from 5 to 15 m s^{-1} were performed. Fig. 2.10 shows a record of the power coefficient for different (λ, β) values.

These results are in compliance with the Betz limit in WT systems, stating that the maximum power coefficient should be less than 0.59.

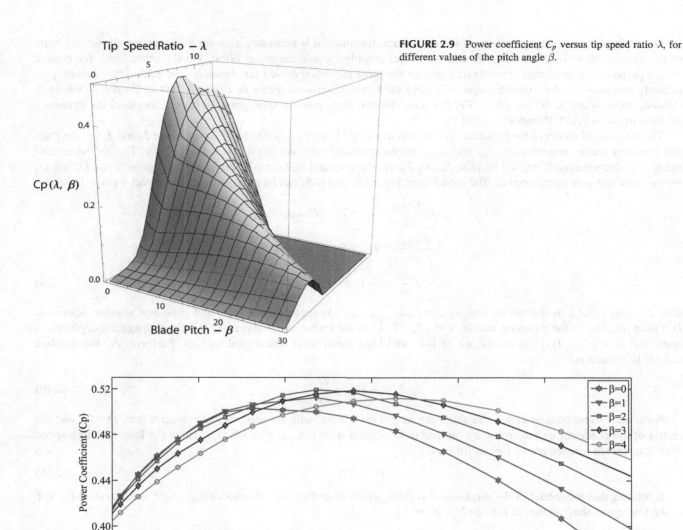

FIGURE 2.9 Power coefficient C_p versus tip speed ratio λ, for different values of the pitch angle β.

FIGURE 2.10 Power coefficient C_p versus tip speed ratio λ, for different values of the pitch angle β.

Remark 2.1: For control design purposes, the optimal values of λ and β will be considered according to Malcolm and Hansen (2006), which states that for WindPACT 1.5-MW WT, the $C_{p_{max}} = 0.5$ occurs at $\lambda = 7$ and $\beta = 2.6$.

It is worth mentioning that the overall aerodynamics calculations will be determined by using AeroDyn module from FAST. AeroDyn is a time-domain WT aerodynamics module that has been coupled into the FAST version 8 multiphysics engineering tool to enable aero-elastic simulation of HAWTs (Jonkman, Hayman, Jonkman, Damiani, & Murray, 2015).

2.2.2 Mechanical subsystem of wind turbine

The purpose of this subsystem is transmitting the mechanical energy obtained by the wind to the electrical subsystem. A complete drive train of a WT consists of all the rotating components: rotor, main shaft, couplings, gearbox, breaks, and generator.

The drive train system of a WT can be represented by means of a set of interconnected rotating masses. Several mass models have been proposed, for instance, we can find models of a six masses, five masses, three masses, two masses, or single lumped mass consisting solely of the rotor and the rigid shaft (Jing, 2012; Muyeen et al., 2007).

Generally, for an accurate analysis of the drive train dynamics, it is necessary a model that contains all rotating parts of the system. However, this will increase the model complexity and computer effort. On the other hand, for control design purposes, reduced mass models that include the most important drive train dynamics are taken into account; particularly two mass models' configuration is widely used in the literature (Petru & Thiringer, 2002; Quinonez-Varela & Cruden, 2008; Wang & Weiss, 2007). Furthermore, the two mass models' configuration is used to model the dynamics of drive trains in FAST (Singh et al., 2014).

The mechanical model of the two-mass WT shown in Fig. 2.11 can be described as follows:where J_r and J_g are the rotor and generator inertia, respectively; ω_{rtur} and ω_{rgen} are the rotor and generator angular velocities; T_a and T_{em} are the aerodynamic and electromagnetic applied torques; T_{ls} and T_{hs} are the low- and high-speed shaft transmitted torques; and D_r and D_g are the rotor and generator damping. The model from Eqs. (2.7) and (2.8) can be represented in a compact form as follows:

$$J_r \frac{d\omega_{rtur}}{dt} = T_a - T_{ls} - D_r \omega_{rtur} \tag{2.7}$$

$$J_g \frac{d\omega_{rgen}}{dt} = T_{hs} - T_{em} - D_g \omega_{rgen} \tag{2.8}$$

$$J \frac{d\omega}{dt} = -D\omega + T - T_s \tag{2.9}$$

with $J = diag\left[J_r, J_g\right]$ is the inertia matrix, $\omega = \begin{bmatrix} \omega_{rtur} & \omega_{rgen} \end{bmatrix}^T$ is vector of the rotor and generator angular velocities, $D = diag\left[D_r, D_g\right]$ is the damping matrix, $T = \begin{bmatrix} T_a & T_g \end{bmatrix}^T$ is the vector of aerodynamic and electromagnetic applied torques, and $T_s = \begin{bmatrix} T_{ls} & T_{hs} \end{bmatrix}^T$ is the vector of low- and high-speed shaft transmitted torques. Furthermore, the gearbox ratio N is defined as:

$$N = \frac{\omega_{rgen}}{\omega_{rtur}} = \frac{T_{ls}}{T_{hs}} \tag{2.10}$$

Now, after some computations, and assuming a rigid drive train, with no gearbox inefficiencies (i.e., $D = 0$), and the inertia of the generator and the rotor are lumped in low-speed shaft (i.e., $J_t = J_r + N^2 J_g$). Then, the low- and high-speed shaft transmitted torques are related as follows:

$$T_{ls} = N T_{hs} \tag{2.11}$$

It follows that the model of the mechanical system, which describes the behavior of the single mass model of a WT in the low speed shaft shown in Fig. 2.12, is given by

$$J_t \frac{d\omega_{rtur}}{dt} = T_a - N T_{em} \tag{2.12}$$

For control design, a single mass rigid drive train model will be considered. However, during FAST simulation, a two-mass flexible drive train model will be set up.

2.2.3 Electrical subsystem of wind turbine

The electrical subsystem is composed for the generator and power control devices. In this work, DFIG model is considered due to its advantages, such as adjustable speed operation and reactive power control (El Azzaoui, Mahmoudi, Bossoufi, & El Ghamrasni, 2016; Muller et al., 2002).

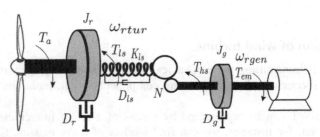

FIGURE 2.11 Two-mass drive train scheme.

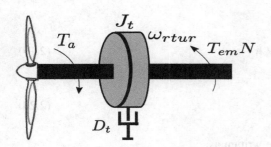

FIGURE 2.12 One-mass drive train scheme.

For control purposes, the generator mathematical model can be written, in synchronously rotating frame $d - q$, as follows:

$$\frac{d\Phi_s}{dt} = - R_s I_s + \omega_s \Im \Phi_s + V_s \tag{2.13}$$

$$\frac{d\Phi_r}{dt} = - R_r I_r + \omega_r \Im \Phi_r + V_r \tag{2.14}$$

where $\Phi_r = \begin{bmatrix} \Phi_{rd}, & \Phi_{rq} \end{bmatrix}^T$ and $\Phi_s = \begin{bmatrix} \Phi_{sd}, & \Phi_{sq} \end{bmatrix}^T$ are the stator and rotor vector fluxes, in the $d-q$ axes, respectively; $I_s = \begin{bmatrix} I_{sd}, & I_{sq} \end{bmatrix}^T$ and $I_r = \begin{bmatrix} I_{rd}, & I_{rq} \end{bmatrix}^T$ are the stator and rotor currents, in the $d-q$ axes; $V_s = \begin{bmatrix} V_{sd}, & V_{sq} \end{bmatrix}^T$ and $V_r = \begin{bmatrix} V_{rd}, & V_{rq} \end{bmatrix}^T$ are the stator and rotor voltage vectors, respectively; ω_s and ω_r are the stator and rotor angular velocities of the generator, linked by the following relation; $\omega_s = \omega_{rgen} + \omega_r$, where ω_{rgen} is the mechanical pulsation of DFIG given in Eq. (2.8), and $\Im$ is the skew matrix given by $\Im = \begin{bmatrix} 0 & 1 \\ -1 & 0 \end{bmatrix}$.

Furthermore, $\omega_r = \omega_s - \omega_{rgen}$, where $\omega_{rgen} = pN\omega_{rtur}$, $\omega_s = 2\pi f_s$, and p is the pole numbers.

Then, expressing the stator and rotor vector flux Φ_s and Φ_r, in terms of the stator and rotor vector currents I_s and I_r, in the $d-q$ axes, is given by

$$\Phi_s = L_s I_s + M I_r \tag{2.15}$$

$$\Phi_r = L_r I_r + M I_s \tag{2.16}$$

where L_s, L_r, and M are the stator, rotor, and mutual inductances, respectively.

Furthermore, the electromagnetic torque T_{em} can be written as follows:

$$T_{em} = pM I_r^T \Im I_s \tag{2.17}$$

Now, by applying the stator field orientation control, to decouple the control of the active and reactive powers (P_s, Q_s), respectively; it follows that the components of the stator flux $\Phi_s = \begin{bmatrix} \Phi_{sd}, & \Phi_{sq} \end{bmatrix}^T$ are given by:

$$\Phi_{sd} = ||\Phi_s||, \quad \Phi_{sq} = 0 \tag{2.18}$$

where $||\Phi_s||$ is the norm of the vector Φ_s.

Assumption 2.1: Consider that voltage and frequency remain approximately constant, and the stator resistance is neglecting, that is, $R_s = 0$.

Remark 2.2: The assumption of the stator resistance is neglecting is a realistic assumption for medium power machines.

Then, $\Phi_{sd} = ||\Phi_s||$ equal to a constant, $\Phi_{sq} = 0$, and form Assumption 2.1, it follows that

$$\omega_s \Im \Phi_s + V_s = 0$$

the components of the stator voltage $V_s = \begin{bmatrix} V_{sd}, & V_{sq} \end{bmatrix}^T$ are given by

$$V_{sd} = 0, \quad V_{sq} = \omega_s ||\Phi_s|| = ||V_s|| \tag{2.19}$$

From Eqs. (2.15) and (2.18), we can write:

$$\Phi_s = ||\Phi_s||\Psi = L_s I_s + M I_r \tag{2.20}$$

where $\Psi = \begin{bmatrix} 1 & 0 \end{bmatrix}^T$,

The equations connecting the stator currents with the rotor currents are written as:

$$I_s = -\frac{M}{L_s} I_r + \frac{||\Phi_s||}{L_s} \Psi \tag{2.21}$$

By replacing flux and stator currents in Eq. (2.16) by Eq. (2.21), we have:

$$\Phi_r = \sigma L_r I_r + ||\Phi_s|| \frac{M}{L_s} \Psi \tag{2.22}$$

where $\sigma = 1 - \frac{M^2}{L_r L_s}$. Then, replacing the expressions of Eq. (2.22) in Eq. (2.14), the vector rotor current I_r is given by:

$$\frac{dI_r}{dt} = \frac{1}{\sigma Lr} \left(V_r - R_r I_r + \omega_r \sigma L_r \Im I_r - \omega_r \frac{M}{L_s} ||\Phi_s|| \Upsilon \right) \tag{2.23}$$

where $\Upsilon = \begin{bmatrix} 0 & 1 \end{bmatrix}^T$. Then, electromagnetic torque can be rewritten as follows:

$$T_{em} = -p \frac{M}{L_s} ||\Phi_s|| I_{rq} \tag{2.24}$$

It follows that the active and reactive power (P_s, Q_s) are presented as:

$$P_s = -\frac{||V_s||M}{L_s} I_{rq}$$

$$Q_s = -\frac{||V_s||M}{L_s} I_{rd} + \frac{||V_s||^2}{\omega_s L_s} \tag{2.25}$$

According to Eqs. (2.24) and (2.25), we conclude that the stator active power and electromagnetic torque depend on the quadrature rotor current and the reactive power depends on the direct rotor current.

The mathematical model describing the behavior of a VSWT system model is given by

$$\frac{dx}{dt} = f(x) + g(x)u \tag{2.26}$$

where $x = \begin{bmatrix} \omega_r, I_{rd}, I_{rq} \end{bmatrix}^T$ is the state vector, the input $u = \begin{bmatrix} V_{rd}, V_{rq} \end{bmatrix}^T$, and $y = \begin{bmatrix} \omega_r, I_{rd} \end{bmatrix}^T$ is the measurable output system with

$$f(x) = \begin{pmatrix} \frac{1}{J_t} \left(T_a + pN \frac{M}{L_s} ||\Phi_s|| I_{rq} \right) \\ -\frac{R_r}{\sigma L_r} I_{rd} + \omega_r I_{rq} \\ -\frac{R_r}{\sigma L_r} I_{rq} - \omega_r \left(I_{rd} + \frac{1}{\sigma L_r} \frac{M}{L_s} ||\Phi_s|| \right) \end{pmatrix}$$

and

$$g(x) = \begin{pmatrix} 0 & 0 \\ \frac{1}{\sigma L_r} & 0 \\ 0 & \frac{1}{\sigma L_r} \end{pmatrix}$$

The above system can be represented in two subsystems as follows:

$$\begin{pmatrix} \dfrac{d\omega_{rtur}}{dt} \\[2mm] \dfrac{dI_{rq}}{dt} \end{pmatrix} = \begin{pmatrix} \dfrac{1}{J_t}\left(T_a + pN\dfrac{M}{L_s}\|\Phi_s\|I_{rq}\right) \\[2mm] -\dfrac{R_r}{\sigma L_r}I_{rq} - \omega_r\left(I_{rd} + \dfrac{1}{\sigma L_r}\dfrac{M}{L_s}\|\Phi_s\|\right) + \dfrac{V_{rq}}{\sigma L_r} \end{pmatrix} \tag{2.27}$$

and

$$\frac{dI_{rd}}{dt} = -\frac{R_r}{\sigma L_r}I_{rd} + \omega_r I_{rq} + \frac{V_{rd}}{\sigma L_r} \tag{2.28}$$

It is clear that the resulting model of the VSWT system is an interconnected nonlinear system, with strong coupled dynamics, which are used to design the control laws to achieve the control objectives based on the sliding mode techniques.

2.2.4   Control objectives for variable speed wind turbine

This work will focus on the power extraction maximization in a horizontal axis VSWT operating in region 2.

The main control objective is to design a controller (generator rotor voltages) for achieving the MPPT ($P = Cp_{max}P_{wind}$), despite wind speed variations.

More precisely, the power coefficient $C_p(\lambda, \beta)$, which depends on the TSR λ and blade pitch angle β, must reach its maximum value Cp_{max} in the presence of wind speed variations or under turbulence to maximize the power extraction $P_{max} = Cp_{max}P_{wind}$, for $Cp_{max} = C_p(\lambda_{opt}) = C_p(\lambda_{opt}, 2.6)$.

The above control objective can be achieved, when the WT is operating under wind conditions associated with region 2, then the blade pitch angle β is fixed to equal to 2.6. Then, it follows that $C_p(\lambda) = C_p(\lambda, 2.6)$.

Furthermore, the TSR λ must be maintained at an optimal point to maximize the power coefficient C_p, that is, $Cp_{max} = C_p(\lambda_{opt}) = C_p(\lambda_{opt}, 2.6)$. Then, it is necessary to control the WT rotor speed $\omega_{rtur} = \omega_{rtur_{opt}}$ taking into account the wind speed variation v, to maintain a constant TSR $\lambda = \lambda_{opt}$. Thus the optimal WT rotor speed is given by:

$$\omega_{rtur_{opt}} = \frac{\lambda_{opt}v}{R} \tag{2.29}$$

Then, the optimal WT rotor speed $\omega_{rtur_{opt}}$ is expressed in terms of ω_r as follows:

$$\omega_r = \omega_s - pN\frac{\lambda_{opt}v}{R} \tag{2.30}$$

It follows that controlling ω_r implies that $\omega_{rtur_{opt}}$ is controlled.

On the other hand, the rotor voltages can also control the stator side reactive power P expressed as:

$$Q_s = \frac{\|V_s\|}{L_s}\left(\frac{\|V_s\|}{\omega_s} - MI_{rd}\right) \tag{2.31}$$

The reactive power Q_s reference value can be specified according to the grid requirements.

The overall control objectives of a VSWT based on an asynchronous DFIG are:

1. The power extraction maximization, by controlling the turbine rotor speed (ω_{rtur}) by means of the q-axis component of generator rotor voltage V_{rq}, to track the optimal rotor speed $\omega_{rtur_{opt}}$.
2. The stator reactive power Q_s regulation to follow the grid specifications, by controlling I_{rd} through the d-axis component of generator rotor voltage V_{rq}, to keep the power factor constant, constant terminal voltages, and dynamics compensation (Sun, Mi, Yu, Wu, & Tian, 2009).

2.3   Sliding mode control design

VSWT model, given in Eq. (2.26), is a highly nonlinear system with strong couplings and uncertainties in the aerodynamic and electrical subsystems. To achieve the control objectives previously defined in Section 2.4, it is necessary to design a control strategy that deals with the system features.

Control under uncertainty conditions is one of the biggest challenges of the modern control theory. As was mentioned in Section 2.1, in spite of successful development of robust control strategies, such as backstepping and adaptive control (Levant, 2003), sliding mode approach is the most suitable choice to work with nonparametric uncertainties and unmodeled dynamics.

This fact is due to the SMC advantages, such as insensitivity property to the bounded perturbations/uncertainties, high accuracy, and finite time convergence. However, the standard SMC, referred to the first-order SMC, is limited to some conditions, such that the sliding surface design is restricted to have relative degree one with respect to the control and the insensitivity property is only in matched bounded perturbations/uncertainties. Furthermore, it shows significant drawback of chattering phenomenon due to the high frequency control switching. This issue could be dangerous for VSWT control applications.

Moreover, control methods have been proposed to deal with these features, such as high gain control and the sliding-sector method (Levant, 2003). However, the accuracy and robustness of the sliding mode were partially lost.

On the other hand, higher-order sliding modes act on the higher-order derivatives of the sliding surfaces instead of influencing to the first derivative as it happens in the standard sliding modes, which can help to reduce the chattering phenomenon and provide higher accuracy. Particularly, the STA well known as SOSM algorithm introduced in Levant (1993) has been exhaustively used to solve control problems; furthermore, observation (Davila & Fridman, 2005) and robust exact differentiation (Levant, 1998; Pisano & Usai, 2007) can be achieved with this technique (Moreno & Osorio, 2012).

The advantages of STA over the other SMC strategies, such as finite time convergence, robustness (the class of perturbations for which finite time stability can be established is much larger since mismatched perturbation are allowed), and chattering attenuation, make this control suitable for achieving control objectives in VSWT systems (Liu et al., 2016; Wang & Liu, 2015). Then, for achieving MPPT in a VSWT and reactive power regulation, a control strategy based on STA will be designed, for driving the WT rotor speed to track the desired speed, and forcing the rotor current to track a desired reference of a horizontal VSWT, in spite of system uncertainties, external perturbations, and uncertainties. Besides, by using a quadratic Lyapunov function, sufficient conditions will be obtained to ensure the finite time convergence of the closed-loop system under the proposed control strategy.

2.3.1 Super twisting algorithm

Consider the following class of nonlinear uncertain systems in the form:

$$\begin{aligned} \dot{x}_1 &= f_1(x) + \rho_1 \\ \dot{x}_2 &= f_2(x) + g_2(x)u + \rho_2 \\ y &= Cx \end{aligned} \tag{2.32}$$

where x_1 and x_2 are the states of the system, u and y represent the input and output of the system respectively, $f_1(x)$, $f_2(x)$ and $g_2(x)$ are smooth nonlinear functions, ρ_1 is the mismatched uncertainties, and ρ_2 represents the matched uncertainty. They include parameter variations, unmodeled dynamics, and external disturbances, and $C = \begin{bmatrix} 1 & 0 \end{bmatrix}$.

The following assumptions are introduced.

Assumption 2.2: The functions $f_i(x)$, $i = 1, 2$ and $g_2(x)$ are Lipschitz with respect to x.

Now, for STA control design, consider the following sliding surface is

$$s = c_1 e_1 + e_2 \tag{2.33}$$

where $e_1 = x_1 - x_1^*$ and $e_2 = \dot{x}_1 - \dot{x}_1^*$, with x_1^* as a desired reference for x_1, $c_1 > 0$. Taking the time derivative of the sliding variable s, it follows that

$$\dot{s} = c_1 f_1(x) + f_2(x) + g_2(x)u + \rho - c_1 \dot{x}_1^* - \ddot{x}_1^* \tag{2.34}$$

where $\rho = c_1 \rho_1 + \rho_2$.

Assumption 2.3: The uncertainty ρ and its derivative $\dot{\rho} = \gamma$ are bounded, that is, there exist known positive constants C_1 and C_2 such that

$$|\rho| \leq C_1, \quad |\gamma| \leq C_2.$$

Assumption 2.4: All the states are available for control design.

Choosing the control input as

$$u = \frac{1}{g_2(x)}\left\{ -c_1 f_1(x) - f_2(x) + c_1 \ddot{x}_1^* + \ddot{x}_1* + v \right\} \tag{2.35}$$

where v is the new control given by

$$v = -2\mathcal{L}|s|^{\frac{1}{2}}sign(s) - \frac{\mathcal{L}^2}{2}\int_0^t sign(s)d\tau \tag{2.36}$$

where $\mathcal{L} > 0$ is the controller gain and $g(x) \neq 0$. Then, the sliding surface dynamic is given by:

$$\dot{s} = v + \rho \tag{2.37}$$

or equivalently

$$\dot{s} = -2\mathcal{L}|s|^{\frac{1}{2}}sign(s) - \frac{\mathcal{L}^2}{2}\int_0^t sign(s)d\tau + \rho \tag{2.38}$$

Now, choosing a following change of variable

$$\begin{aligned} z_1 &= s \\ z_2 &= -\frac{\mathcal{L}^2}{2}\int_0^t sign(s)d\tau + \rho \end{aligned} \tag{2.39}$$

Then, the closed-loop system Eq. (2.39), in new coordinates, is given by:

$$\begin{aligned} \dot{z}_1 &= -2\mathcal{L}|z_1|^{\frac{1}{2}}sign(z_1) + z_2 \\ \dot{z}_2 &= -\frac{\mathcal{L}^2}{2}sign(z_1) + \gamma \end{aligned} \tag{2.40}$$

where $\gamma = \dot{\rho}$.

Now the following result about the finite time convergence of the system in closed loop with the super twisting controller Eqs. (2.35) and (2.36) can be established as follows:

Proposition 2.1: Consider system in Eq. (2.32) and Assumptions 2.2, 2.3, and 2.4 hold. Then, the system Eq. (2.32) in closed loop under the action of the super twisting controller Eqs. (2.35) and (2.36) is such that the tracking error converges in finite time to zero, for $\mathcal{L} > 0$ large enough.

Proof.: See the Appendix.

Remark 2.3: Notice that the super twisting control design, the gains are parametrized in terms of only one gain $\mathcal{L}$, allowing its easy tuning as well as its implementation.

2.3.2 Variable speed wind turbine controller design

A controller for a VSWT system will be design based on the super twisting methodology, to achieve control objectives. Consider the system Eq. (2.27) and let be the tracking error between the actual rotor speed and the desired reference $\omega_{rtur_{opt}}$, is given by $e_1 = \omega_{rtur} - \omega_{rtur_{opt}}$, where $\omega_{rtur_{opt}}$ is obtained from

$$\omega_{rtur_{opt}} = \frac{\lambda_{opt}v}{R} \tag{2.41}$$

Then, the sliding surface is defined as follows:

$$\begin{aligned} s_\omega &= c_1 e_1 + \frac{de_1}{dt} \\ &= c_1\left(\omega_{rtur} - \omega_{rtur_{opt}}\right) - \left(\dot{\omega}_{rtur} - \dot{\omega}_{rtur_{opt}}\right) \end{aligned} \tag{2.42}$$

where $c_1 > 0$. Taking the first-order derivatives of the sliding variable s_ω, and using model Eq. (2.27) and Eq. (2.30) in terms of ω_{rtur}, it follows that

$$\dot{s}_\omega = c_1 \dot{e}_1 + \ddot{e}_1 \tag{2.43}$$

$$\dot{s}_\omega = c_1(\dot{\omega}_{rtur} - \dot{\omega}_{rtur_{opt}}) + \frac{\dot{T}_a}{J_t} + pN\frac{M}{L_s}\frac{\|\Phi_s\|}{J_t}$$
$$\times \left(-\frac{R_r}{\sigma L_r}I_{rq} + \left(pN\omega_{rtur} - \omega_s\right)\left(I_{rd} + \frac{1}{\sigma L_r}\frac{M}{L_s}\|\Phi_s\|\right) + \frac{V_{rq}}{\sigma L_r}\right) \tag{2.44}$$
$$+ \ddot{\omega}_{rtur_{opt}}$$

which is of the form Eq. (2.34). Applying a control V_{rq} of the form Eq. (2.37), it follows that

$$\dot{s}_\omega = \rho_\omega + U_{rq}$$

where ρ_ω contains the parameter perturbation and the input U_{rq} is designed using the STA:

$$\begin{cases} U_{rq} = -2\mathcal{L}_\omega|s_\omega|^{1/2}sign(s_\omega) + w_1 \\ \dot{w}_1 = \frac{\mathcal{L}_\omega^2}{2}sign(s_\omega) \end{cases} \tag{2.45}$$

with $\mathcal{L}_\omega > 0$. The desired d-axis component of rotor current I_{rd}^* is obtained by setting the reactive power equal to zero. Therefore from Eq. (2.31), lead to the rotor reference current

$$I_{rd}^* = \frac{\|V_s\|}{\omega_s M} \tag{2.46}$$

Then, the tracking error between the actual and the desired d-axis component of rotor current I_{rd}^* is given as $e_2 = I_{rd} - I_{rd}^*$, then the sliding surface for the current loop I_{rd} is given by

$$s_{I_{rd}} = I_{rd} - I_{rd}^* \tag{2.47}$$

it follows that

$$\dot{s}_{I_{rd}} = \left(-\frac{R_r}{\sigma L_r}I_{rd} - (pN\omega_{rtur} - \omega_s)I_{rq} + \frac{V_{rd}}{\sigma L_r}\right) - \dot{I}_{rd}^* \tag{2.48}$$

Applying a control V_{rd} such that the following

$$\dot{s}_{I_{rd}} = \rho_{I_{rd}} + U_{rd}$$

is obtained, where $\rho_{I_{rd}}$ contains the parameter perturbation and the input U_{rd} is designed using the STA:

$$\begin{cases} U_{rd} = -2\mathcal{L}_{I_{rd}}|s_{I_{rd}}|^{1/2}sign\left(s_{I_{rd}}\right) + w_2 \\ \dot{w}_2 = -\frac{\mathcal{L}_{I_{rd}}^2}{2}sign\left(s_{I_{rd}}\right) \end{cases} \tag{2.49}$$

with $\mathcal{L}_{I_{rd}} > 0$.

2.4 Simulation results

The performance of the closed-loop system under the action of the proposed controller is verified via simulations, using FAST code developed by NREL interfaced with the popular MATLAB/Simulink platform.

As shown in Fig. 2.13, the main tool boxes of this platform are: FAST simulator, TurbSim, S-Function interface, and MATLAB/Simulink.

FAST simulator has been chosen for the validation of control strategies thanks to detailed structural and dynamics models of WTs. FAST code is a comprehensive aero-elastic simulator capable of predicting both the extreme and fatigue loads of two- and three-bladed HAWT (Jonkman et al., 2005).

TurbSim is a stochastic, full-field, turbulent-wind simulator (Jonkman & Buhl, 2006).

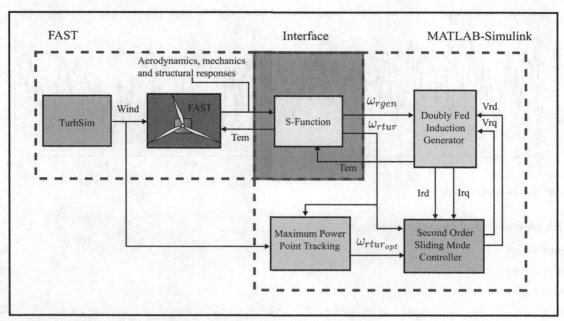

FIGURE 2.13 FAST-interface simulator, wind turbine, and control scheme.

Furthermore, S-function represents an interface between FAST and MATLAB/Simulink, which has been developed to use the FAST equations of motion.

Active controls for determining many aspects of the turbine operation, such as blade pitch angle, variable speed torque, high speed shaft brake, and nacelle yaw control, may be implemented during simulation analysis. Hence, the electrical model, represented by DFIG, grid, and control strategy based on sliding mode techniques, is simulated in the Simulink platform, while the FAST simulator incorporates the complete nonlinear aerodynamic WT motion equations.

During simulation, WT aerodynamics and structural responses to wind-inflow conditions are determined online.

2.4.1   Test conditions

Validation tests to verify the effectiveness of the proposed controller were performed using WindPACT 1.5-MW VSWT based on DFIG, with all DOFs enabled in the FAST model, under 10-min turbulent wind speed profile generated by TurbSim (Jonkman & Buhl, 2006), based on Kaimal turbulence spectra model, as shown in Fig. 2.14.

The overall meteorological boundary conditions from the wind turbulence are shown in Table 2.1.

Regarding the operation regions, the cut-in wind speed of a 1.5-MW WT is at 3 m s^{-1} and the rated wind speed is close to 11 m s^{-1}; therefore the wind speed will vary around 7.2 m s^{-1} to maintain the turbine operating in region 2. The turbine parameters contained in the FAST primary input files and the generator parameters used for simulations tasks are given in Tables 2.2 and 2.3, respectively. Simulations were performed by using Euler integration algorithm with a fixed step of 0.001 s.

The controller gains $\mathcal{L}_\omega$ and $\mathcal{L}_{Ird}$ were deliberately chosen according to conditions of Eq. (2.57) and through analysis and computer simulation, to guarantee its finite time stability. For simulation study, the controller gains are chosen as: $\mathcal{L}_\omega = 42$ and $\mathcal{L}_{Ird} = 150$ and the parameters of the sliding variable as $c_1 = 100$.

2.4.2   Discussion of the simulation results

As clearly shown in Fig. 2.15, a good tracking performance of the WT rotor rotational speed ω_{rtur} and fast time response are achieved. Furthermore, it shows that the proposed control is robust in spite of the disturbances caused by the wind speed turbulence and unmodeled dynamics.

The tracking error between the optimal WT rotor speed and the actual WT rotor speed e_1 is plotted in Fig. 2.16. It can be seen that the tracking error holds close to zero.

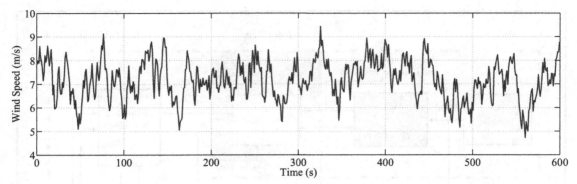

FIGURE 2.14   Turbulent wind profile.

TABLE 2.1 Meteorological boundary conditions.

Turbulence model	IECKAI
Turbulence intensity	15%
IEC turbulence type	NTM
Wind profile type	IEC
Ref Ht	84.288
Mean wind speed	7.2

TABLE 2.2 Wind turbine parameters (Golnary & Moradi, 2019).

Parameters	Value
Blades number	3
Rotor radius	35 m
Hub height	84.3 m
Rated power	1.5 MW
Total inertia (J_t)	2962443.5 kg m^2
Gearbox ratio (N)	87.965
Low speed shaft torsional spring	5.6×10^9 Nm rad^{-1}
Low speed shaft torsional damping	1×10^7 Nm rad^{-1}

TABLE 2.3 DFIG parameters (Golnary & Moradi, 2019).

Parameters	Value
Synchronous speed (ω_s)	314.1593 rad s^{-1}
Stator resistance (R_s)	0.005 Ω
Rotor resistance (R_r)	0.0089 Ω
Stator inductance (L_s)	0.407 mH
Rotor inductance (L_r)	0.299 mH
Mutual inductance (M)	0.016 mH
Pole numbers (p)	2

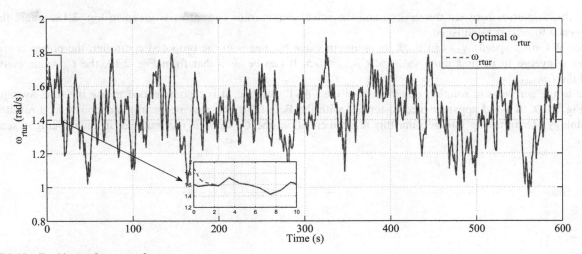

FIGURE 2.15 Tracking performance of ω_{rtur}.

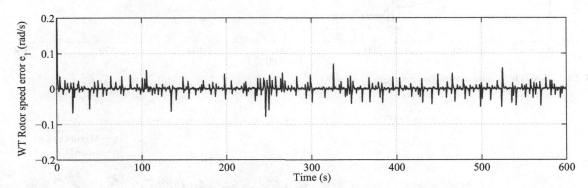

FIGURE 2.16 Wind turbine rotor speed tracking error e_1.

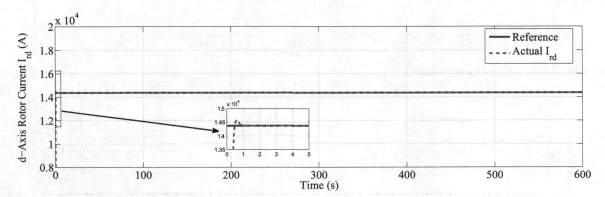

FIGURE 2.17 Regulation performance of current I_{rd}.

Moreover, Fig. 2.17 shows the good regulation performance of the d-axis rotor current I_{rd}, which is selected such that the reactive power is suppressed. These results prove that under the action of the proposed control algorithms, the d-axis rotor current I_{rd}, and the WT rotor rotational speed ω_{rtur} track their respective references values with smooth profiles and chattering free behavior. Notice that a fast response is achieved.

The tracking error between the desired and the actual *d*-axis rotor current e_2 is given in Fig. 2.18, where the fast convergence to zero is established.

Since WT rotor speed ω_{rtur} can track its optimum value by means of the proposed controller, the power coefficient C_p must converge to its maximum value at $Cp_{max} = 0.5$. It can be seen that from Fig. 2.19, the C_p value converges around the Cp_{max}.

Due to the aero-elastic solution obtained from the FAST software, it is recommended to use the time-averaged C_p value Fig. 2.20, to avoid apparent inconsistencies with the Betz limit that may occur instantaneously as a result of the methodology. Then, taking into account this recommendation, the C_p value is recalculated, in time-averaged sense, as follows:

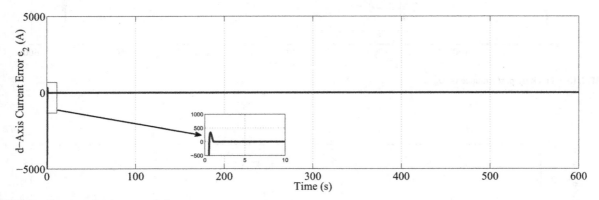

FIGURE 2.18 *d*-Axis rotor current regulation error e_2.

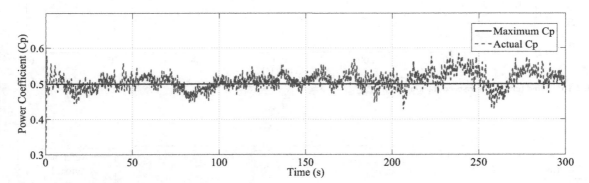

FIGURE 2.19 Power coefficient C_p performance.

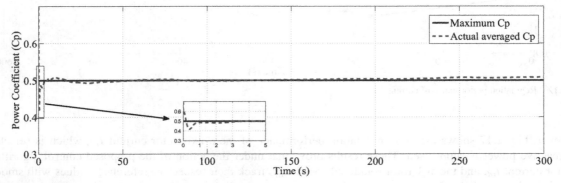

FIGURE 2.20 Time averaged power coefficient C_p performance.

It can be observed that C_p value is now reached and stays close to the Cp_{max}, which yields to the MPPT. According to Eq. (2.2), the maximum power extraction occurs when C_p is hold in Cp_{max} for a given wind speed corresponding in region 2.

Fig. 2.21 shows the tracking of the maximum aerodynamic power P considering the time average of the C_p value, which shows good performances of power extraction maximization.

Finally, the proposed output feedback V_{rd} and V_{rq} is plotted in Figs. 2.22 and 2.23, respectively; as can be seen a low control energy is needed for achieving the control objectives, which produces lower costs to the overall system but also, this leads to protect and keep safe the electrical system operation.

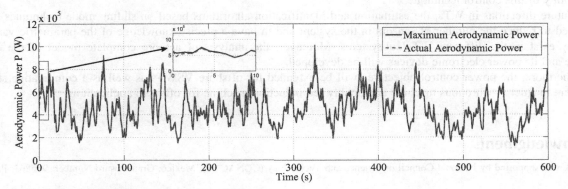

FIGURE 2.21   Maximum aerodynamic power P tracking.

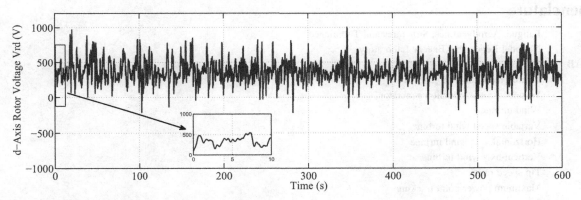

FIGURE 2.22   Controller output d-axis rotor voltage V_{rd}.

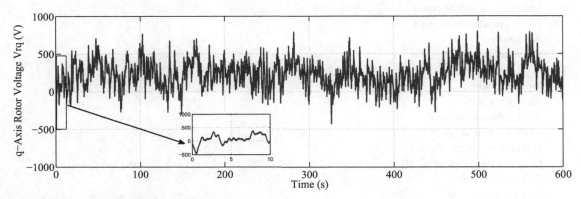

FIGURE 2.23   Controller output q-axis rotor voltage V_{rq}.

2.5 Conclusion and future directions

Control for VSWT system, for maximizing power extraction and regulating the stator reactive power to follow the grid requirements, has been designed using SMC techniques, which is an attractive technique in the area of wind energy technology. Furthermore, an analysis of stability has been presented using a Lyapunov approach, where sufficient conditions have been obtained to ensure the convergences in finite time to zero of the trajectories. Using these sufficient conditions, the controller gains are chosen easily than the ones obtained from the traditional methods.

The efficiency and robustness of the proposed SMC algorithm has been tested under realistic and dynamic scenarios using NREL FAST code enabling almost all the freedom degrees available in the software. Finally, it has been shown that the control objectives to maximize the power extraction have been achieved successfully showing the potential applicability of this control technique.

As future directions in WTs, the estimation and identification algorithms based on sliding mode techniques will be developed to reduce the number of sensors in the system and to have a precise knowledge of the parametric variations that may exist in the different WTs subsystems. Moreover, the analysis of a more complete model of the electric machine and its power electronic devices will be developed.

Furthermore, the power control objectives will be extended for offshore VSWTs as well as a comparative study of permanent magnet synchronous machine and doubly fed induction machine for offshore applications.

Acknowledgments

This work was supported by National Council on Science and Technology (CONACYT), Mexico, Grant/Award Number: 966863; PAICYT-UANL.

Nomenclature

FAST	Fatigue, Aerodynamics, Structures and Turbulence
NREL	National Renewable Energy Laboratory
MATLAB	Matrix Laboratory
DFIG	Doubly fed induction generator
WRIG	Wounded-rotor induction generator
WT	Wind turbine
VSWT	Variable speed wind turbine
HAWT	Horizontal-axis wind turbine
VAWT	Vertical-axis wind turbine
TSR	Tip speed ratio
MPPT	Maximum power point tracking
STA	Super twisting algorithm
SMC	Sliding mode control
C_p	Power coefficient
P	Instantaneous turbine power
P_{Wind}	Power available in the wind
φ	Air density
R	Rotor radius
v	Wind velocity
λ	Tip speed ratio
β	Blade pitch angle
ω_{rtur}	Wind turbine rotor speed
ω_{rgen}	Generator angular speed
ω_s	Synchronous speed
ω_r	Generator rotor angular speed
J_t	Turbine total inertia
T_a	Aerodynamic torque

T_{em}	Electromagnetic torque
T_{ls}	Low speed shaft torque
T_{hs}	High speed shaft torque
D_r, D_g	Rotor and generator external damping
K_{ls}	Low speed shaft torsional spring
N	Gearbox ratio
s, r	Stator and rotor index
d, q	Synchronous reference frame index
Φ	Flux
I	Current
V	Voltage
L	Inductance
M	Mutual inductance
p	Pole number
σ	Leakage coefficient
P, Q	Active and reactive power

References

Abdullah, M. A., Yatim, A., Tan, C. W., & Saidur, R. (2012). A review of maximum power point tracking algorithms for wind energy systems. *Renewable and Sustainable Energy Reviews, 16*, 3220–3227.

Beltran, B., Benbouzid, M. H., & Ahmed-Ali, T. (2010). A combined high gain observer and high-order sliding mode controller for a dfig-based wind turbine. In *2010 IEEE international energy conference* (pp. 322–327). IEEE.

Beltran, B., Benbouzid, M. E. H., & Ahmed-Ali, T. (2012). Second-order sliding mode control of a doubly fed induction generator driven wind turbine. *IEEE Transactions on Energy Conversion, 27*, 261–269.

Boukhezzar, B., & Siguerdidjane, H. (2009). Nonlinear control with wind estimation of a dfig variable speed wind turbine for power capture optimization. *Energy Conversion and Management, 50*, 885–892.

Chen, Z., Guerrero, J. M., & Blaabjerg, F. (2009). A review of the state of the art of power electronics for wind turbines. *IEEE Transactions on Power Electronics, 24*, 1859–1875.

Dash, P., & Patnaik, R. (2014). Adaptive second order sliding mode control of doubly fed induction generator in wind energy conversion system. *Journal of Renewable and Sustainable Energy, 6*, 053143.

Davila, J., Fridman, L., & Levant, A. (2005). Second-order sliding-mode observer for mechanical systems. *IEEE Transactions on Automatic Control, 50*, 1785–1789.

El Azzaoui, M., Mahmoudi, H., Bossoufi, B., & El Ghamrasni, M. (2016). Comparative study of the sliding mode and backstepping control in power control of a doubly fed induction generator. In *2016 international symposium on fundamentals of electrical engineering (ISFEE)* (pp. 1–5). IEEE,.

Evangelista, C., Valenciaga, F., & Puleston, P. (2013). Active and reactive power control for wind turbine based on a mimo 2-sliding mode algorithm with variable gains. *IEEE Transactions on Energy Conversion, 28*, 682–689.

Golnary, F., & Moradi, H. (2019). Dynamic modelling and design of various robust sliding mode controls for the wind turbine with estimation of wind speed. *Applied Mathematical Modelling, 65*, 566–585.

Jing, X. (2012). *Modeling and control of a doubly-fed induction generator for wind turbine-generator systems.*

Jonkman, B. J., & Buhl Jr, M. L. (2006). *TurbSim user's guide.* Technical Report, National Renewable Energy Lab. (NREL), Golden, CO.

Jonkman, J., Hayman, G., Jonkman, B., Damiani, R., & Murray, R. (2015). *Aerodyn v15 user's guide and theory manual.* Golden, CO: NREL.

Jonkman, J. M., Buhl, M. L., Jr, et al. (2005). *Fast user's guide.* Golden, CO: National Renewable Energy Laboratory.

Kobayashi, S., & Furuta, K. (2007). Frequency characteristics of levant's differentiator and adaptive sliding mode differentiator. *International Journal of Systems Science, 38*, 825–832.

Kong, X., Liu, X., & Lee, K. Y. (2014). Data-driven modelling of a doubly fed induction generator wind turbine system based on neural networks. *IET Renewable Power Generation, 8*, 849–857.

Levant, A. (1993). Sliding order and sliding accuracy in sliding mode control. *International journal of control, 58*, 1247–1263.

Levant, A. (1998). Robust exact differentiation via sliding mode technique. *Automatica 34*, 379–384.

Levant, A. (2003). Higher-order sliding modes, differentiation and output-feedback control. *International Journal of Control, 76*, 924–941.

Li, H., & Chen, Z. (2008). Overview of different wind generator systems and their comparisons. *IET Renewable Power Generation, 2*, 123–138.

Liu, X., Han, Y., & Wang, C. (2016). Second-order sliding mode control for power optimisation of dfig-based variable speed wind turbine. *IET Renewable Power Generation, 11*, 408–418.

Ma, M., Chen, H., Liu, X., & Allgöwer, F. (2014). Moving horizon h_∞ control of variable speed wind turbines with actuator saturation. *IET Renewable Power Generation, 8*, 498–508.

Malcolm, D., & Hansen, A. (2006). *WindPACT turbine rotor design study: June 2000–June 2002 (revised)*. Technical Report. National Renewable Energy Lab. (NREL), Golden, CO.

Manwell, J. F., McGowan, J. G., & Rogers, A. L. (2010). *Wind energy explained: Theory, design and application*. John Wiley & Sons.

Márquez, F. P. G., Pérez, J. M. P., Marugán, A. P., & Papaelias, M. (2016). Identification of critical components of wind turbines using fta over the time. *Renewable Energy, 87*, 869–883.

Moreno, J. A., & Osorio, M. (2012). Strict lyapunov functions for the super-twisting algorithm. *IEEE Transactions on Automatic Control, 57*, 1035–1040.

Muller, S., Deicke, M., & De Doncker, R. W. (2002). Doubly fed induction generator systems for wind turbines. *IEEE Industry Applications Magazine, 8*, 26–33.

Munteanu, I., Cutululis, N. A., Bratcu, A. I., & Ceangă, E. (2005). Optimization of variable speed wind power systems based on a LQG approach. *Control Engineering Practice, 13*, 903–912.

Muyeen, S., Ali, M. H., Takahashi, R., Murata, T., Tamura, J., Tomaki, Y., ... Sasano, E. (2007). Comparative study on transient stability analysis of wind turbine generator system using different drive train models. *IET Renewable Power Generation, 1*, 131–141.

Nayeh, R. F., Moradi, H., & Vossoughi, G. (2020). Multivariable robust control of a horizontal wind turbine under various operating modes and uncertainties: A comparison on sliding mode and $h\infty$ control. *International Journal of Electrical Power Energy Systems, 115*, 105474.

Pao, L. Y., & Johnson, K. E. (2011). Control of wind turbines. *IEEE Control systems magazine, 31*, 44–62.

Patil, N., & Bhosle, Y. (2013). A review on wind turbine generator topologies. In *2013 international conference on power, energy and control (ICPEC)* (pp. 625–629). IEEE.

Petru, T., & Thiringer, T. (2002). Modeling of wind turbines for power system studies. *IEEE transactions on Power Systems, 17*, 1132–1139.

Pisano, A., & Usai, E. (2007). Globally convergent real-time differentiation via second order sliding modes. *International Journal of Systems Science, 38*, 833–844.

Poitiers, F., Bouaouiche, T., & Machmoum, M. (2009). Advanced control of a doubly-fed induction generator for wind energy conversion. *Electric Power Systems Research, 79*, 1085–1096.

Quinonez-Varela, G., & Cruden, A. (2008). Modelling and validation of a squirrel cage induction generator wind turbine during connection to the local grid. *IET Generation, Transmission & Distribution, 2*, 301–309.

Singh, M., Muljadi, E., Jonkman, J., Gevorgian, V., Girsang, I., & Dhupia, J. (2014). *Simulation for wind turbine generators—With FAST and MATLAB—Simulink modules*. Technical Report. National Renewable Energy Lab. (NREL), Golden, CO.

Sun, L., Mi, Z., Yu, Y., Wu, T., & Tian, H. (2009). Active power and reactive power regulation capacity study of dfig wind turbine. In *2009 international conference on sustainable power generation and supply* (pp. 1–6). IEEE.

Susperregui, A., Martinez, M. I., Tapia, G., & Vechiu, I. (2013). Second-order sliding-mode controller design and tuning for grid synchronisation and power control of a wind turbine-driven doubly fed induction generator. *IET Renewable Power Generation, 7*, 540–551.

Susperregui, A., Jugo, J., Lizarraga, I., & Tapia, G. (2014). Automated control of doubly fed induction generator integrating sensorless parameter estimation and grid synchronisation. *IET Renewable Power Generation, 8*, 76–89.

Valenciaga, F. (2010). Second order sliding power control for a variable speed-constant frequency energy conversion system. *Energy Conversion and Management, 51*, 3000–3008.

Valenciaga, F., & Evangelista, C. (2010). 2-sliding active and reactive power control of a wind energy conversion system. *IET Control Theory & Applications, 4*, 2479–2490.

Wang, C., & Liu, X. (2015). Sliding mode control for maximum wind energy capture of dfig-based wind turbine. In *The 27th Chinese control and decision conference (2015 CCDC)* (pp. 4485–4489). IEEE.

Wang, C., & Weiss, G. (2007). Integral input-to-state stability of the drive-train of a wind turbine. In *2007 46th IEEE conference on decision and control* (pp. 6100–6105). IEEE.

Wang, J., Tse, N., & Gao, Z. (2011). Synthesis on pi-based pitch controller of large wind turbines generator. *Energy conversion and management, 52*, 1288–1294.

Yahdou, A., Boudjema, Z., Taleb, R., & Djilali, A. B. (2018). Backstepping sliding mode control of a dual rotor wind turbine system. In *2018 international conference on electrical sciences and technologies in Maghreb (CISTEM)* (pp. 1–5). IEEE.

Zhang, L., Cai, X., & Guo, J. (2009). Simplified input-output linearizing and decoupling control of wind turbine driven doubly-fed induction generators. In *2009 IEEE 6th international power electronics and motion control conference* (pp. 632–637). IEEE.

Appendix

Proof of Proposition 2.1: Consider the following change of coordinates

$$\xi_1 = |z_1|^{\frac{1}{2}} sign(z_1)$$

$$\xi_2 = z_2 \tag{2.50}$$

Then, the dynamical system Eq. (2.40), written in terms of the new coordinates, is given by:

$$\dot{\xi}_1 = \frac{1}{2|z_1|^{\frac{1}{2}}} \left\{ (A - KC)\xi + \Gamma \right\} \tag{2.51}$$

where $\xi = \begin{bmatrix} \xi_1 & \xi_2 \end{bmatrix}^T$ is the state vector and $K = S_x^{-1}C^T$ is the gain of the controller, where S_x is positive definite matrix solution of the algebraic Lyapunov equation

$$\mathcal{L}S_x + A^T S_x + S_x A - C^T C = 0 \tag{2.52}$$

Now analyzing the convergence of the tracking error to zero, consider the following candidate Lyapunov function:

$$V(\xi) = \xi^T S_x \xi \tag{2.53}$$

which is quadratic and positive definite, in the new coordinates. Notice that $V(\xi)$ is continuous everywhere and continuous differentiable everywhere except in the set $\zeta = \left\{ (\xi_1, \xi_2) \in R^2 | \xi_1 = 0 \right\}$. Since the trajectories of the system cannot stay in the set ζ before reaching the origin, the time derivative of V can be calculated in the usual way everywhere except when the trajectories intersect in the set ζ. If the trajectories reaches the origin after time T, then it will stay there. Taking the time derivative of $V(\xi)$ along the trajectories of Eq. (2.51), it follows that:

$$\dot{V}(\xi) \leq \frac{1}{2|z_1|^{\frac{1}{2}}} \left[-\mathcal{L}\xi^T S_x \xi + 2\xi^T S_x \Gamma \right] \tag{2.54}$$

From Assumption 2.3, that is, $||\Gamma|| \leq \vartheta ||\xi||$, for $\vartheta > 0$, and taking the norm of the nonlinear term $2\xi^T S_x \Gamma$, we have

$$\dot{V}(\xi) \leq -\frac{\mathcal{L}}{2|z_1|^{\frac{1}{2}}} V(\xi) + \frac{1}{|z_1|^{\frac{1}{2}}} ||S_x|| \vartheta ||\xi||^2 \tag{2.55}$$

Using the fact that there exist positive constants $\lambda_{min}(S_x)$ and $\lambda_{max}(S_x)$ such that:

$$\lambda_{min}(S_x) ||\xi||^2 \leq V(\xi) \leq \lambda_{max}(S_x) ||\xi||^2$$

and taking into account that the inequality $|z_1|^{\frac{1}{2}} \leq ||\xi|| \leq \left\{ \frac{V(\xi)}{\lambda_{max}(S_x)} \right\}^{\frac{1}{2}}$ is satisfied, it follows that

$$\dot{V}(\xi) \leq -\frac{(\mathcal{L} - \mu)}{2|z_1|^{\frac{1}{2}}} V(\xi) \tag{2.56}$$

where $\mu = \frac{2||S_x|| \vartheta}{\lambda_{min}(S_x)}$. Then, it follows that

$$\dot{V}(\xi) \leq -\delta V(\xi)^{\frac{1}{2}} \tag{2.57}$$

where $\delta = \frac{\mathcal{L} - \mu}{2\lambda_{min}^{-\frac{1}{2}}(S_x)}$. Choosing $\mathcal{L}$ sufficiently large such that the inequality $\mathcal{L} > \mu$ holds, then $\dot{V}(\xi)$ is definite negative. This shows that $V(\xi)$ is a strong Lyapunov function and the trajectories converge in finite time. To estimate the convergence time, notice that the solution of the differential equation $\dot{v} = -\delta v^{\frac{1}{2}}$, $v(0) = v_0$; is given by $v(t) = \left\{ v_0^{\frac{1}{2}} - \frac{1}{2}\delta t \right\}^2$. From the comparison principle (Kobayashi & Furuta, 2007), that $V(\xi) < v(t)$ when $V(\xi(0)) < v_0$, then ξ converges to zero in finite time and reaches that value at time given by

$$T = \frac{2V(\xi(0))^{\frac{1}{2}}}{\delta}$$

Thus the state $\xi = \begin{bmatrix} \xi_1 & \xi_2 \end{bmatrix}^T$ converges to zero in finite time. This implies that also the states z_1 and z_2 will converge to zero in finite time. As a result, s and $\dot{s} = -2\mathcal{L}|z_1|^{\frac{1}{2}}sign(z_1) + z_2$ tend to zero.

Chapter 3

Generic modeling and control of wind turbines following IEC 61400-27-1

R. Villena-Ruiz, A. Honrubia-Escribano, E. Artigao-Andicoberry, S. Martín-Martínez and E. Gómez-Lázaro

Renewable Energy Research Institute and DIEEAC-ETSII-AB, Universidad de Castilla-La Mancha, Albacete, Spain

3.1   Introduction

Wind energy is the most utilized of all non-dispatchable renewable energy sources (RES) for electricity generation today. Its installed capacity increased 93 GW globally in 2020, a 53% year-on-year increase despite the impacts of the COVID-19 crisis (GWEC, 2021). China and the United States have led this trend, accounting for more than 73% of new capacity, while Europe installed 14.7 GW of new wind power capacity in 2020 (WindEurope, 2020). A necessary requirement for secure system operation is ensuring adequate performance of non-dispatchable RES, through system observability and controllability (Badrzadeh et al., 2020). As an example, the Spanish Transmission System Operator (TSO) has an observability of 100% of wind and 84% of photovoltaic generation. In addition, 99% of wind and 47% of photovoltaic generation are directly controlled by the TSO (Van Putten, Sewdien, & Almeida de Graaf, 2020).

Under this rapid growth scenario for wind energy, the development of dynamic wind turbine (WT) and wind power plant simulation models for the appropriate integration of this type of non-dispatchable energy generation becomes mandatory. The main applications of these models include connection studies, operational planning, and real-time operations. In this regard, two international working groups, the International Electrotechnical Commission (IEC) and the Western Electricity Coordinating Council (WECC), have developed generic dynamic WT models. These entities addressed the unification and standardization of the behavior of the main WT topologies available in the market, through Standard IEC 61400-27-1 (IEC 61400-27-1, 2020) and "WECC Second Generation of Wind Turbine Models" (WECC, 2014), respectively. Despite both entities having the same goal, IEC prioritized the veracity of the simulation models to achieve very similar responses to field measurements, whereas WECC focused on simplified WT models by reducing the number of parameters, thus also reducing the computational time.

There are four types of generic WT models to cover the most common WT technologies (Sørensen, Andresen, Fortmann, & Pourbeik, 2013). The Type 1 WT is characterized by the use of asynchronous generators directly connected to the grid (typically squirrel cage induction generator). Standard IEC 61400-27-1 defines two Type 1 WT models according to the main technological differences from the point of view of power system stability studies: Type 1A with fixed pitch angle and Type 1B with pitch control for fault-ride through. The Type 2 WT is similar to the Type 1 WT but equipped with a variable rotor resistance (VRR), therefore using a wound rotor asynchronous generator (WRAG). The Type 3 WT consists of a doubly fed induction generator (DFIG), where the stator is directly connected to the grid and the rotor is connected to the grid through a power converter. Type 4 is a full-conversion WT generator that uses a back-to-back AC-to-DC and DC-to-AC voltage source converter to allow the physical generator to operate over a wide range of speeds. The actual generator can be a synchronous machine with a wound rotor, a permanent magnet machine, or an induction machine (Walling, Gursoy, & English, 2012).

To widen the scope of application of generic WT simulation models, the present chapter first addresses a review of the previous works focused on the modeling of IEC generic WT models. Second, we model, simulate, and validate the WT model with the largest market share and which is one of the most technologically advanced, the Type 3 WT (or DFIG) defined by Standard IEC 61400-27-1. The specific DFIG control approaches regarding active power, reactive power, and crowbar control are also addressed. The Type 3 WT model will be submitted to voltage dips of different duration and magnitude under different loading conditions, and the responses will be validated against field

Renewable Energy Systems. DOI: https://doi.org/10.1016/B978-0-12-820004-9.00006-1

measurements of an operating WT. Moreover, the current chapter will showcase the results obtained when deploying both a multidisciplinary software tool, MATLAB/Simulink (ML), and one of the leading electric power system analysis software tools today, DIgSILENT PowerFactory (PF).

Further to the introduction, the chapter is structured as follows: Section 3.2 summarizes the main contributions on the modeling of dynamic WT models found in the scientific literature and highlights the need to conduct the validation of IEC models using specialized power system simulation software tools. Section 3.3 addresses the generic IEC Type 3 WT, from both a technical and modeling point of view, in addition to describing the validation methodology developed by Standard IEC 61400-27-2 and followed to evaluate the WT model's behavior. Section 3.4 analyzes the results obtained from the different validation test cases performed. Finally, Section 3.5 summarizes the main conclusions drawn from the analyses conducted.

3.2 Literature review

A look at the scientific literature shows that there are several contributions concerning the validation of private and highly specific WT models developed by different manufacturers. In Timbus et al. (2011), a WT simulation model belonging to a specific vendor is studied. The simulated results of a specific DFIG WT model are compared to the measured responses of an actual 2-MW DFIG WT under voltage dips in Seman, Niiranen, Virtanen, and Matsinen (2008). In Akhmatov et al. (2010), the validation of a Siemens WT simulation model under unbalanced faults is performed. In Trilla et al. (2011), a complete DFIG WT simulation model using field measurements of an actual DFIG WT is validated. Finally, in Chang, Hu, Tang, and Song (2018), the validation of a detailed WT simulation model using an analytical method, rather than field measurements, is conducted.

In light of the above, it can be seen that few studies simulating and validating IEC generic WT models have been published. This is partly because Standard IEC 61400-27-1 was published in 2015 and its second edition, including updated information, was released in 2020. Furthermore, most of the IEC-based works published used general or multi-disciplinary software tools, mainly ML, to perform the validation tasks. Lorenzo-Bonache, Honrubia-Escribano, Jiménez-Buendía, Molina-García, and Gómez-Lázaro (2017) studied the transient response of the IEC Type 3 WT model under several voltage dips implemented in ML, while the operation of the active and reactive power control models of the generic Type 3 WT model was also analyzed using this software tool in Lorenzo-Bonache, Villena-Ruiz, Honrubia-Escribano, and Gómez-Lázaro (2017). In addition to the Type 4 WT model developed by the IEC, Lorenzo-Bonache, Honrubia-Escribano, Jiménez-Buendía, and Gómez-Lázaro (2018) also validated the Type 4 WT model developed by the WECC (2014), using ML in both cases. This software tool is also used in Honrubia-Escribano, Jiménez-Buendía, Gómez-Lázaro, and Fortmann (2018) to conduct the validation of an IEC-based Type 3 WT simulation model, as well as in Villena-Ruiz et al. (2019b) and Jiménez-Buendía, Villena-Ruiz, Honrubia-Escribano, Molina-García, and Gómez-Lázaro (2019) to submit the IEC and the WECC Type 3 WT models, respectively, to the technical requirements established by a national grid code.

There is therefore a clear lack of studies addressing the modeling and validation process of IEC generic WT models using specialized software tools. Indeed, it is extremely important to conduct this type of study employing specialized power system simulation software tools since these are the types of tools used by TSOs and distribution system operators. Moreover, these specialized tools allow for larger and more complex actual power systems to be modeled.

In the present chapter, PF is the specialized tool employed. PF is regarded as one of the most advanced software tools in the field of electrical engineering. Moreover, it has positioned itself as one of the world's leaders in wind power systems. Nonetheless, only two publications addressing the analysis of IEC generic WT models using PF are found in the scientific literature, (Göksu et al., 2016; Göksu, Altin, Fortmann, & Sørensen, 2016). However, Göksu et al. (2016) focused on the responses of a generic Type 3 WT model with power factor controller at plant level during changes in the set point values rather than during voltage dips. Meanwhile, Göksu et al. (2016) used PF to perform a compatibility analysis between the generic IEC and WECC WT simulation models.

3.3 Modeling, simulation and validation of the Type 3 WT model defined by Standard IEC 61400-27-1

The present section defines the main electrical and mechanical components of a Type 3 WT and the two types of this model distinguished in the IEC 61400-27-1 guidelines (IEC 61400-27-1, 2020). Moreover, it describes the different dynamic and control models that form part of the IEC Type 3 WT and how these are related. Finally, it addresses the validation methodology defined by Standard IEC 61400-27-2 (IEC 61400-27-2, 2020) and shows the characteristics of the validation test cases performed.

3.3.1 IEC Type 3 WT model

When conducting transient stability analyses of generic IEC WT simulation models, a series of modeling assumptions, listed in IEC 61400-27-1 (2020), must be taken into account. Among these assumptions are that wind speed is assumed to be constant during the simulation and also that the IEC WT models are intended for fundamental frequency-positive sequence responses and are to be used for balanced short circuits or changes in the reference values. In this sense, since voltage dips are one of the most critical situations faced by WTs, this chapter focuses on the assessment of the responses of the IEC generic Type 3 WT model under this type of grid disturbance. Moreover, the Type 3 WT model was chosen for study as it is one of the most technically advanced WTs (Fortmann, 2014) and is currently the most widely installed worldwide.

As mentioned in Section 3.1, the Type 3 WT is equipped with a DFIG, and its main electrical and mechanical components are shown in Fig. 3.1. The low-speed side of the WT, which is the WT rotor, is connected to the high-speed side, that is, to the asynchronous generator, through a gearbox. The back-to-back power converter consists of the generator side converter, the line side converter, and the direct current link and is usually rated from 30% to 40% of the nominal power of the WT. The power converter also contains the direct current capacitor. Moreover, depending on the strategy used to tackle voltage disturbances (Honrubia-Escribano, Gómez-Lázaro, Fortmann, Sørensen, & Martín-Martínez, 2018), Type 3 WTs may include a chopper and/or a crowbar protection system. In this sense, Standard IEC 61400-27-1 distinguishes two types of generic Type 3 WT simulation models:

1. Type 3A WT, the generator model of which does not include the crowbar protection system.
2. Type 3B WT, the generator system of which includes the crowbar protection system. This protection system short circuits the rotor terminals for a certain period of time to protect the generator side converter against high induced currents during voltage dips. During this time, the DFIG behaves as a squirrel cage induction generator, that is, as a Type 1 WT (Salles, Hameyer, Cardoso, Grilo, & Rahmann, 2010; Buendía & Gordo, 2012).

In the present chapter, the WT modeled, simulated, and validated is the generic IEC Type 3B WT—hereinafter referred to as Type 3 WT—since the field measurements used to perform the validation of the simulation model correspond to an actual DFIG WT equipped with a crowbar protection system.

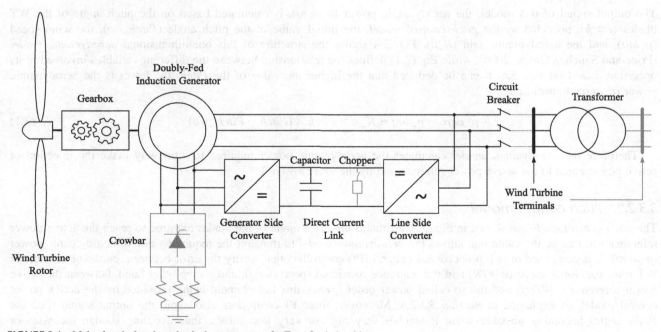

FIGURE 3.1 Main electrical and mechanical components of a Type 3 wind turbine.

3.3.2 Modeling of the generic Type 3 WT model

A total of six main submodels make up the modular structure of the generic Type 3 WT defined by IEC 61400-27-1, as shown in Fig. 3.2. Four of these are dynamic models, while the remaining two are control models: *electrical system model, generator system model, mechanical model, aerodynamic model, pitch control model*, and *generator control model*. The *generator control model* is, in turn, composed of four control submodels: active power control model (*P control model*), reactive power control model (*Q control model*), *current limitation model*, and reactive power limitation (*Q limitation model*). The structure of all these models, as defined by Standard IEC 61400-27-1, is detailed in IEC 61400-27-1 (2020).

As explained in Section 3.2, most of the IEC-based scientific contributions focus on the analyses of the responses of generic WT models when simulated in multidisciplinary software tools. To widen the scope of these studies and enrich the scientific literature on this topic, the present chapter analyzes the simulated responses of a generic Type 3 WT model when implemented not only in a multidisciplinary software tool, ML, but also in a specialized software tool in the field of power system analysis, PF. To perform the validation of the WT model, the fundamental frequency positive sequence data series of the field measurements are compared with the simulated results obtained.

A comprehensive parameter analysis of the IEC generic Type 3 WT model is performed in Lorenzo-Bonache, Honrubia-Escribano, et al. (2017). Moreover, this work, together with other contributions, such as Lorenzo-Bonache, Villena-Ruiz, et al. (2017) and Lorenzo-Bonache, Honrubia-Escribano, Fortmann, and Gómez-Lázaro (2019), shows the different dynamic and control models that form part of the IEC Type 3 WT when simulated in ML. Therefore intended to complement these works and provide a different modeling approach to the generic WT model, this chapter describes the structures of some of the submodels that form part of this WT when modeled in PF.

The modular structure—or composite frame—of the generic Type 3 WT model once modeled in PF, which is equivalent to that shown in Fig. 3.2, is shown in Fig. 3.3. Some additional measurement and electrical devices, as well as dynamic models, needed to be used to conduct the simulations in this case, and their functionality is described in Villena-Ruiz, Honrubia-Escribano, Fortmann, and Gómez-Lázaro (2020). Indeed, all these additional devices, together with the AC current source through which the active and reactive currents are injected into the grid, form the so-called *electrical system model* (Fig. 3.2).

The operation of the other control and dynamic models that form part of the IEC Type 3 WT is described in the following subsections, based on the structure shown in Fig. 3.3.

3.3.2.1 Aerodynamic model

The output signal of this model, the aerodynamic power (*p_aero*), is calculated based on the pitch angle of the WT blades (*Pitch*), provided by the *pitch control model*, the initial value of the pitch angle (*Pitch_wo*), the wind speed (*p_init*), and the aerodynamic gain (*K_a*). Fig. 3.4 shows the structure of this one-dimensional *aerodynamic model* (Price and Sánchez-Gasca, 2006), while Eq. (3.1) defines the relationship between the different variables involved in its operation, based on Fig. 3.4. It can be deduced that the higher the value of the *Pitch*, the lower is the aerodynamic power (*p_aero*) transmitted.

$$p\_aero = p\_init - K\_a \times Pitch \times (Pitch - Pitch\_wo) \tag{3.1}$$

Therefore the *aerodynamic model* calculates the aerodynamic power required to ultimately make the injection of active power equal to the active power reference set by the user (*pWTref*).

3.3.2.2 Pitch control model

The *pitch control model*, as shown in Fig. 3.5, estimates the *Pitch* angle of the blades required to reach the active power reference set, that is, the value that allows the *aerodynamic model* to transmit the required value of aerodynamic power (*p_aero*). It is composed of two proportional integral (PI) controllers that rectify the error between, on the one hand, the WT rotor rotational speed (*wWTR*) and the reference rotational speed (*wref*), and on the other hand, between the active power reference (*pWTref*) and the so-called power order (*pord*), this last element being provided by the active power control model, as explained in Section 3.3.2.4. Moreover, these PI controllers also avoid the output signal (i.e., the *Pitch* angle) becoming unstable when it reaches very high or very low values, thus avoiding sudden increases or decreases in the rotor speed.

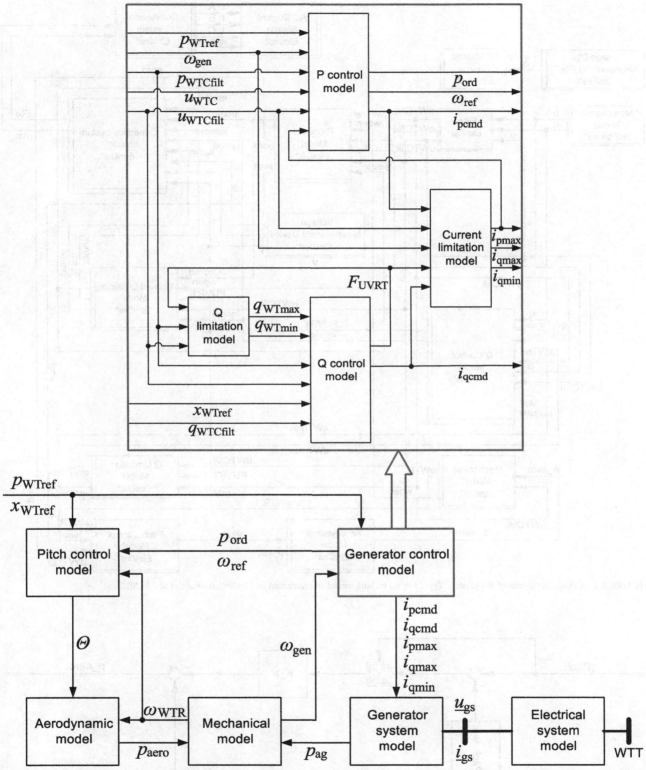

FIGURE 3.2 Modular structure of the generic International Electrotechnical Commission Type 3 wind turbine. *Adapted from IEC 61400-27-1 (2020). Wind turbines—Part 27-1: Electrical simulation models for wind power generation—Wind turbines. International Electrotechnical Commission.*

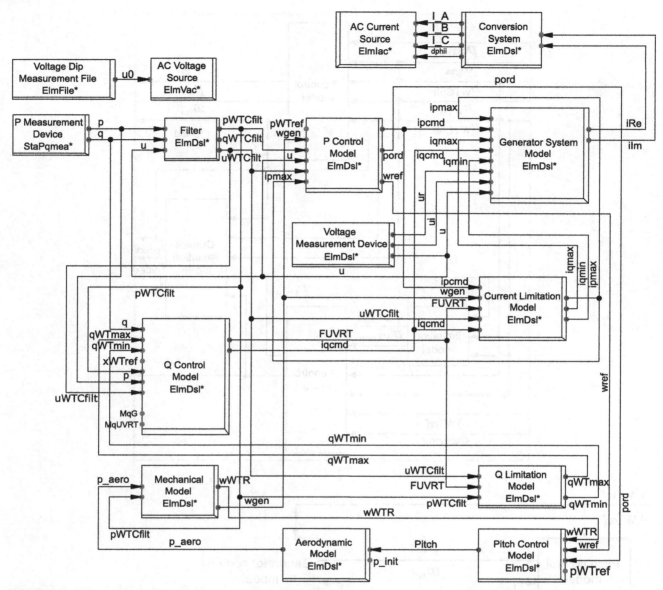

FIGURE 3.3 Composite frame of the generic Type 3 wind turbine model implemented in PF (Villena-Ruiz et al., 2020).

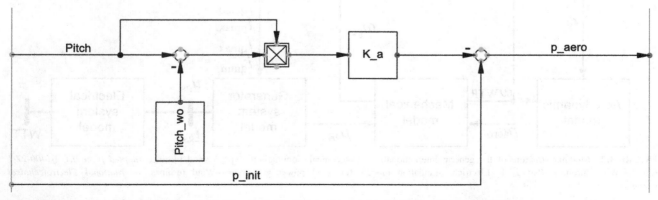

FIGURE 3.4 Aerodynamic model of the generic Type 3 wind turbine implemented in PF (Villena-Ruiz et al., 2020).

(A)

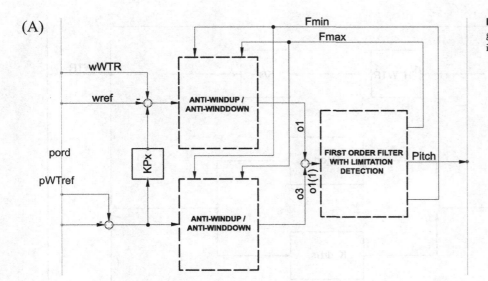

FIGURE 3.5 Pitch control model of the generic Type 3 wind turbine implemented in PF (Villena-Ruiz et al., 2020).

(B)

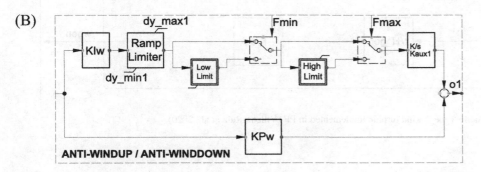

(C)

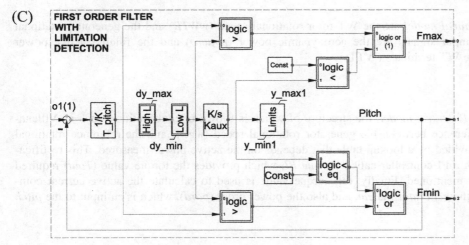

3.3.2.3 Mechanical model

Standard IEC 61400-27-1 uses a two-mass mechanical model (IEC 61400-27-1 2020; Honrubia-Escribano et al., 2012). Its structure is shown in Fig. 3.6, implemented in PF. This model represents the low-speed and high-speed sides of the WT, that is, the WT rotor and the induction generator, modeled through their corresponding inertia constants, H_WTR and H_gen. The drive train, that is, the interaction between both sides, is modeled as a spring and a damper, represented by their stiffness (K_drt) and damping (C_drt) coefficients.

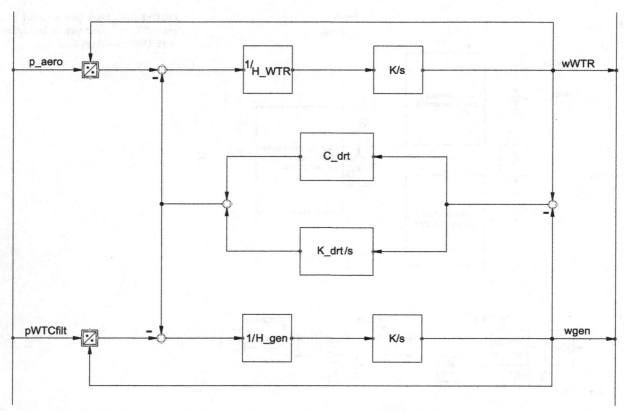

FIGURE 3.6 Mechanical model of the generic Type 3 wind turbine implemented in PF (Villena-Ruiz et al., 2020).

The output signals of the *mechanical model* are the WT rotor rotational speed (*wWTR*) and the generator rotational speed (*wgen*), while the required input signals are the aerodynamic power (*p_aero*) and the filtered active power (*pWTCfilt*), which is measured at the WT terminals (WTTs).

3.3.2.4   P control model

The active power control model or *P control model*, the structure of which is shown in contributions, such as Villena-Ruiz et al. (2020), rectifies the difference between the generator rotational speed (*wgen*) and the reference rotational speed (*wref*). The latter signal is provided by a lookup table that depends on the active power measured. This rectification of the error is achieved through a PI controller called *Torque PI*, which provides the torque value (*Tout*) required to correct the difference previously mentioned. Finally, this torque value is used to calculate the active current command (*ipcmd*), which is an input to the *generator system*, and also the power order (*pord*), which is an input to the *pitch control model*.

3.3.2.5   Q control model

The reactive power reference (*xWTref*) is an input signal to the reactive power control model and is set by the user, representing the desired reactive power at the WTTs (IEC 61400-27-1, 2020; Lorenzo-Bonache et al., 2017). Therefore, this reactive power control model—or *Q control model*—is able to control the injection of reactive power at the WTTs through the calculation of the reactive current command (*iqcmd*). Moreover, it injects different values of reactive power depending on the operation stage of the WT: normal, fault, or post-fault operation. This is achieved by calculation of the undervoltage ride through flag signal (*FUVRT*), which is precisely the signal that indicates the operation stage. Its structure is described in Villena-Ruiz et al. (2020).

3.3.2.6 Q limitation model

The values of the dynamic signals *qWTmax* and *qWTmin*, which represent the maximum and minimum reactive power allowed at the WTTs, respectively, are provided by the reactive power limitation, or *Q limitation model*, as shown in Fig. 3.7. These signals are inputs to the reactive power control model and depend on the voltage measured, the operation stage of the WT, and the active power measured.

3.3.2.7 Current limitation model

The maximum value of active current, as well as the maximum and minimum values of reactive current allowed at the WTTs (*ipmax*, *iqmax*, and *iqmin*), is provided by the *current limitation model*. All of these are dynamic values that depend on the active and reactive current commands, the voltage, the generator rotational speed, and the operation stage of the WT (normal, fault or post-fault operation). The structure of the *current limitation model* is described in IEC 61400-27-1 (2020) and Villena-Ruiz et al. (2020).

3.3.2.8 Generator system

As explained in Section 3.3.1, the difference between Type 3A and Type 3B WTs lies in the protection system implemented. Type 3B WT, the *generator system* of which includes the crowbar protection system, is the chosen topology modeled in this chapter (see Fig. 3.8). At the simulation level, the crowbar protection system multiplies the active and reactive power commands (*ipcmd* and *iqcmd*) by zero, when the voltage variation is above a specific threshold (see Fig. 3.8A).

The *generator system* is also composed of the so-called *Reference Frame Rotation* (see Fig. 3.8B), which is a submodel that coordinates the phase of the current with the phase of the measured voltage, providing the final values of active current (*iRe*) and reactive current (*iIm*) that will be injected into the grid through the AC current source employed (see Fig. 3.3).

3.3.3 Simulation and validation of the generic Type 3 WT model

Standard IEC 61400-27-1 defines two scenarios where the IEC validation methodology developed can be applied to the generic WT models (IEC 61400-27-1, 2020): first, under changes in the active and reactive power reference values, and second, under voltage dips. To carry out the validation tasks and assess the accuracy of the behavior of the generic WT

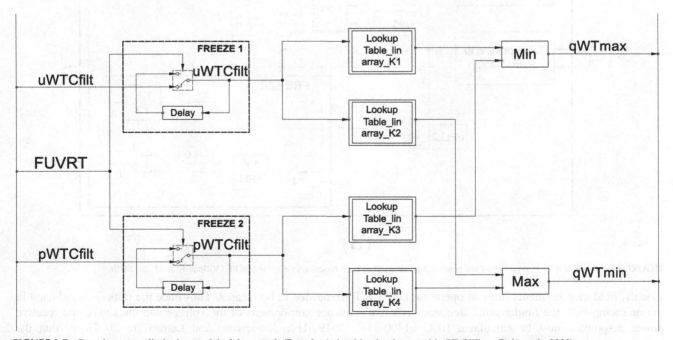

FIGURE 3.7 Reactive power limitation model of the generic Type 3 wind turbine implemented in PF (Villena-Ruiz et al., 2020).

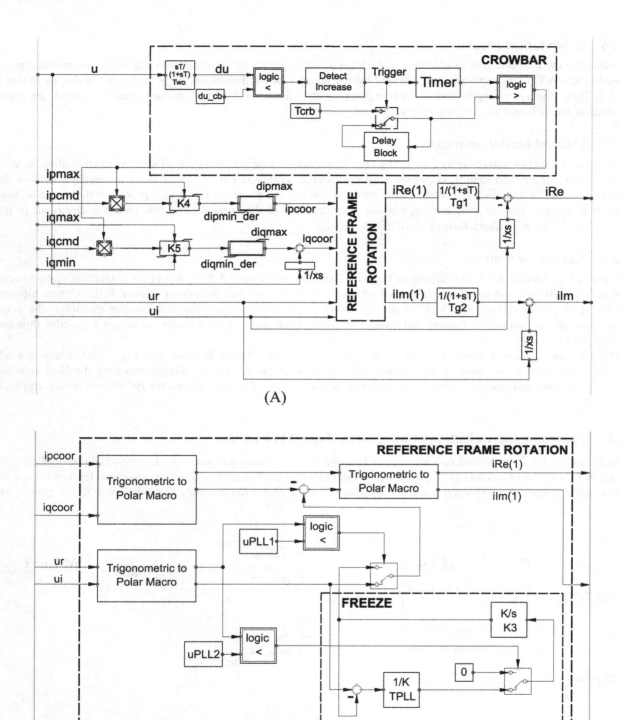

FIGURE 3.8 Generator system model of the generic Type 3 wind turbine model implemented in PF (Villena-Ruiz et al., 2020).

models, field measurements from an operating DFIG WT are needed in both cases. Thus once the tests are conducted in the operating WT, the fundamental frequency positive sequence components of the voltage and the active and reactive power responses must be calculated (IEC 61400-21-1, 2019; Uski-Joutsenvuo and Lemström, 2007), so that the responses of the operating WT and the simulated responses of the IEC WT model can be compared.

However, as stated in Section 3.3.1, this chapter studies and validates the generic Type 3 WT under voltage dips since these are one of the most critical situations faced by WTs (Honrubia-Escribano, Jiménez-Buendía, et al., 2018). Therefore the same voltage dip as that conducted and measured in the operating DFIG WT must be reproduced in the generic WT simulation model. In this sense, two different approaches can be followed to perform the validation of the generic Type 3 WT model (Honrubia-Escribano et al., 2019):

- *Full-system approach*: In this case, the complete physical test bench used to conduct the field tests must be implemented and simulated along with the WT simulation model, so that the dip and duration of the voltage dip can be defined by setting the proper values to the resistance and reactance parameters. Therefore slight differences between the measured and the simulated voltage dip profiles may exist.
- *Play-back approach*: In this case, a voltage-dependent source must be implemented along with the WT simulation model, so that the voltage dip signal measured during the field tests can be set as input to this voltage source. This allows the measured and simulated voltage profiles to be exactly the same.

In the present chapter, the two validation approaches are used to perform the analyses. When modeling the IEC Type 3 WT in PF, the play-back validation approach is followed, while the full-system approach is followed when modeling and simulating the IEC WT model in ML.

Once the voltage dip has been applied to the generic Type 3 WT model following one of the two validation approaches, its simulation responses must be measured. The field data series and the simulated data series must then be compared to obtain the time series of the errors for both the active and the reactive power. Finally, the validation errors or validation performance indicators defined below must be estimated. For this purpose, Standard IEC 61400-27-2 defines three different time windows within a certain voltage dip profile (IEC 61400-27-2, 2020), as shown in Fig. 3.9.

The three time windows are as follows: (1) pre-fault window (W_{pre}), which starts 1000 ms before the fault initiation (t_{begin}); (2) fault window (W_{fault}), which lasts from the onset (t_{fault}) to the clearance of the fault (t_{clear}); and (3) post-fault window (W_{post}), which lasts 5000 ms after the voltage dip clearance, which marks the end of the comparison window (t_{end}).

Furthermore, two quasisteady (QS) state subwindows, also shown in Fig. 3.9, are defined, so that the transient periods that appear in the simulated responses of the generic WT model can be discarded in the evaluation of the accuracy of its responses (Bech, 2014; Meuser & Brennecke, 2015; Sørensen, Andresen, Bech, Fortmann, & Pourbeik, 2012; Zhao et al., 2015). Thus a period of 140 ms is not considered at the start of the fault window, so that the fault QS state subwindow ($W_{faultQS}$) lasts from that point ($t_{faultQS}$) to the clearance of the voltage dip (t_{clear}). A period of 500 ms is also discarded at the start of the post-fault window, which defines the post-fault QS state subwindow (W_{postQS}). This second subwindow lasts from that point ($t_{clearQS}$) to the end of the voltage dip profile (t_{end}).

Therefore as mentioned above, the error time series (x_{error}) for the active and the reactive power must be obtained by comparing the simulated responses of the WT model (x_{sim}) with the field measurements of the operating DFIG WT (x_{mea}), as shown in Eq. (3.2).

$$x_{error}(n) = x_{sim}(n) - x_{mea}(n) \tag{3.2}$$

The validation errors must now be calculated according to Eqs. (3.3)–(3.5). In Eqs. (3.2)–(3.5), n indicates the indices of the vectors, while N is the total number of samples used. These performance indicators will be calculated during

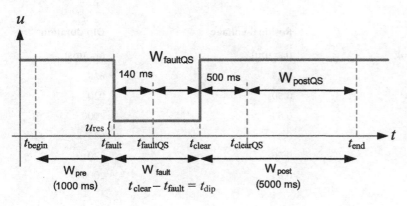

FIGURE 3.9 Voltage dip windows for International Electrotechnical Commission wind turbine model validation (Villena-Ruiz, Lorenzo-Bonache, Honrubia-Escribano, Jiménez-Buendía, & Gómez-Lázaro, 2019a).

the fault and post-fault windows since a correct initialization of the generic model results in very low values of the pre-fault errors and allows them to be neglected.

$$x_{\text{ME}} = \frac{\sum_{n=1}^{N} x_{\text{error}}(n)}{N} \tag{3.3}$$

$$x_{\text{MAE}} = \frac{\sum_{n=1}^{N} |x_{\text{error}}(n)|}{N} \tag{3.4}$$

$$x_{\text{MXE}} = \max(|x_{\text{error}}(n)|) \tag{3.5}$$

The mean error [ME, see Eq. (3.3)], is the main value of the error and is related to the steady-state performance of the model. The mean absolute error [MAE, see Eq. (3.4)] is the mean value of the absolute error and is also related to the steady-state performance of the model, albeit based on the mean deviation. Finally, the maximum absolute error [MXE, see Eq. (3.5)] provides information on the transient performance of the generic WT model.

Table 3.1 shows the time windows over which the validation performance indicators are calculated, according to Fig. 3.9, depending on whether the QS state subwindows are discarded or not.

In the present chapter, the IEC generic Type 3 WT was subjected to three different voltage dip tests under two different loading conditions, full load and partial load. Thus a total of six test cases were considered for the validation of the WT model by applying the methodology defined by Standard IEC 61400-27-2, described in the previous paragraphs. Table 3.2 summarizes the characteristics of all these tests.

Fig. 3.10 shows the voltage dip profiles at full and partial load conditions. Although the dip and duration of the three different faults are the same for both loading conditions (see Table 3.2), there are slight differences between them after being conducted, and hence all of them are shown. In Fig. 3.10A and B, the solid lines represent FL_1 and PL_1 test cases, the dashed lines represent FL_2 and PL_2 test cases, and FL_3 and PL_3 test cases are represented by the dashed-dotted lines.

As indicated at the beginning of the present section, when the generic WT model is modeled in PF and the playback validation approach is followed, the field data of the voltage dip profiles exactly match the voltage dips

TABLE 3.1 Time periods considered for the calculation of the validation errors (IEC 61400-27-2, 2020).

Validation error	Fault window	Post-fault window
Mean error	$[t_{\text{fault}}\ t_{\text{clear}}]$	$[t_{\text{clear}}\ t_{\text{end}}]$
Mean absolute error	$[t_{\text{faultQS}}\ t_{\text{clear}}]$	$[t_{\text{clear}}\ t_{\text{end}}]$
Maximum absolute error	$[t_{\text{faultQS}}\ t_{\text{clear}}]$	$[t_{\text{clearQS}}\ t_{\text{end}}]$

TABLE 3.2 Validation test cases performed.

Loading conditions	Load	Residual voltage	Dip duration
	p_init (pu)	u_{res} (pu)	t_{dip} (ms)
Full load, FL_1	1.00	0.25	625
Full load, FL_2	1.00	0.50	920
Full load, FL_3	1.00	0.85	2000
Partial load, PL_1	0.27	0.25	625
Partial load, PL_2	0.21	0.50	920
Partial load, PL_3	0.26	0.85	2000

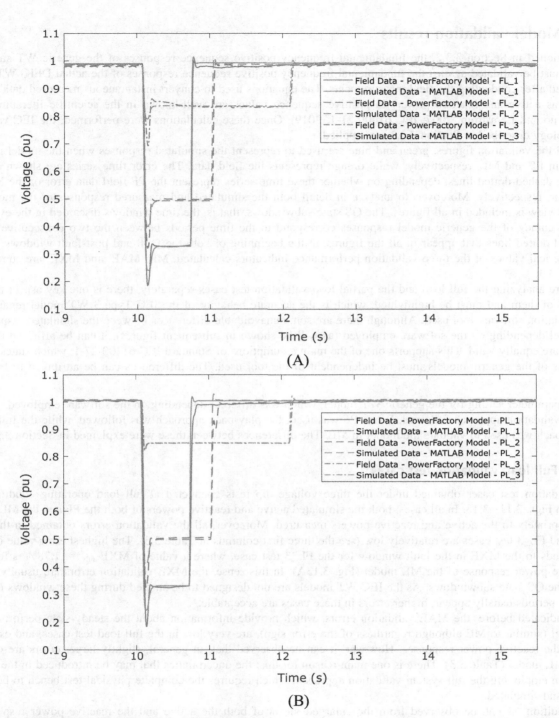

FIGURE 3.10 Voltage dip tests for full load and partial load test cases.

reproduced in the generic WT model. This is observed in Fig. 3.10, where the green lines represent both the field data and the voltage dip profile reproduced in the WT model implemented in PF. Fig. 3.10 also shows the slight differences between the green and the blue lines, that is, between the measured and the simulated voltage data series. These differences are because the blue lines represent the voltage dip profiles simulated in the generic WT when it is modeled in ML, that is, when the full-system validation approach is followed.

3.4 Model validation results

As mentioned in Section 3.3.2, the fundamental frequency positive sequence responses of the generic WT simulation model must be validated against the fundamental frequency positive sequence responses of the actual DFIG WT model, calculated after conducting the field measurements. The equations used to convert instantaneous measured data (voltage as well as active and reactive power) to positive sequence values are well-known in the scientific literature (Uski-Joutsenvuo and Lemström, 2007; IEC 61400-21-1, 2019). Once these calculations were performed, the IEC validation methodology described in Section 3.3.3 was applied.

In all the validation figures, green and blue are used to represent the simulated responses when the model is implemented in PF and ML, respectively, while orange represents the field data. The error time series are shown as green and blue dashed-dotted lines, depending on whether these time series represent the PF-field data error or the ML-field data error, respectively. Moreover, to analyze in detail both the simulated and measured responses of the models, an enlarged view is included in all figures. The QS state subwindows, that is, the time windows discarded in the evaluation of the accuracy of the generic model responses correspond to the time periods between the two consecutive vertical solid and dotted lines that appear in all the figures, at the beginning of both the fault and post-fault windows. Finally, the numerical values of the three validation performance indicators calculated, ME, MAE, and MXE, are summarized in Table 3.3.

Before analyzing the full load and the partial load validation test cases separately, there is one key aspect that concerns all of them and must be highlighted, which is the accurate behavior of the IEC Type 3 WT model regardless of the simulation software tool used. Although there are some unavoidable differences between the simulated responses of the model depending on the software employed (as will be shown in subsequent figures), it can be affirmed that both of them are equally valid. This supports one of the main assumptions of Standard IEC 61400-27-1, which states that the responses of the generic models must be independent of the tool used. The differences can be attributed to two main aspects:

1. The parameter setting for the generic WT model, which was different depending on the software employed.
2. The validation approach followed. When PF was used, the play-back approach was followed, while the full-system approach was used to validate the model in ML. The differences between these were explained in Section 3.3.3.

3.4.1 Full load validation test cases

The validation test cases obtained under the three voltage dip tests conducted at full load operating conditions are shown in Figs. 3.11−3.13. In all cases, both the simulated active and reactive powers of both the PF and the ML models fit appropriately to the active and reactive powers measured. Moreover, all the validation errors obtained in the FL_1, FL_2, and FL_3 test cases are relatively low (see the three first columns in Table 3.3). The highest error value obtained corresponds to the MXE in the fault window for the FL_3 test case, where a value of $MXE_{fault} = 11.78\%$ is found for the active power response of the ML model (Fig. 3.13A). In this sense, the MXE validation errors are usually located around the QS state subwindows. As the IEC WT models are not designed to be studied during these windows in which transient periods usually appear, higher errors in these cases are acceptable.

As indicated before, the MAE validation errors, which provide information about the steady-state performance of the model (similar to ME although regardless of the error sign), are very low in the full load test cases, and especially low for the reactive power responses. However, it can be observed that, in general, slightly larger errors are obtained for the ML model (Table 3.3). There is one main reason for this: the uncertainties that may be introduced in the simulation when employing the full-system validation approach, which requires the complete physical test bench to be implemented and simulated.

In addition, as can be observed from the enlarged views of both the active and the reactive power responses in Figs. 3.11−3.13, there are significant differences between the simulated and the field data, especially when the fault occurs. These differences are mainly due to the difficulties that arise when developing a generic crowbar protection model that can be adapted to the manufacturer's crowbar models. As mentioned in Section 3.3.2.8, at simulation level, the crowbar system multiplies by zero the active and reactive current commands when the voltage variation is above a specific threshold, and consequently, there is a sudden reduction in the active and reactive power signals when the fault occurs and when it is later cleared.

In addition, it should be noted that, as can be observed in Figs. 3.11A, 3.12A, and 3.13A after the fault clearance, an accurate correlation in both the amplitude and the phase shift between the measured and the simulated active power responses of both the PF and the ML models exists. The oscillations are caused by the two-mass mechanical model

TABLE 3.3 Validation errors of test cases performed.

	FL_1		FL_2		FL_3		PL_1		PL_2		PL_3	
	PF	ML	PF	ML	PF	ML	PF	ML	PF	ML	PF	ML
Active power												
ME_{fault}(%)	−1.11	−2.12	0.48	0.29	−1.19	−3.18	−0.51	0.67	2.14	3.80	0.00	0.63
MAE_{fault}(%)	1.38	0.50	2.50	2.54	2.57	3.60	0.77	1.35	1.44	3.37	0.36	0.78
MXE_{fault}(%)	6.98	1.30	6.98	10.76	10.27	11.78	1.62	2.44	4.95	11.76	0.82	1.59
ME_{post}(%)	−1.79	−1.81	1.38	0.50	−0.49	−1.50	1.53	0.04	2.08	1.64	−2.79	−2.39
MAE_{post}(%)	2.24	4.47	2.35	5.10	2.03	2.42	2.20	1.82	3.02	2.96	2.86	2.45
MXE_{post}(%)	4.14	8.97	6.00	11.45	5.10	6.41	1.75	3.32	4.39	6.29	4.52	4.02
Reactive power												
ME_{fault}(%)	−2.18	0.23	0.54	−1.26	0.72	−2.19	−0.06	1.17	−0.48	−1.92	−0.66	−1.38
MAE_{fault}(%)	0.84	1.66	0.69	1.54	0.92	2.58	0.65	1.84	0.84	2.98	0.65	1.44
MXE_{fault}(%)	9.48	6.03	2.31	5.66	4.74	4.28	4.60	2.88	4.94	7.93	1.89	1.88
ME_{post}(%)	−0.52	−1.63	0.52	−0.88	−0.35	−0.71	1.31	−2.04	1.51	−1.90	0.07	−1.65
MAE_{post}(%)	1.27	1.73	1.44	1.24	1.14	0.80	2.07	2.52	1.69	2.82	0.75	1.79
MXE_{post}(%)	3.94	3.50	4.71	2.63	2.04	2.77	1.22	3.84	1.99	4.31	2.33	5.00

ME, mean error; *MAE*, mean absolute error; *ML*, MATLAB/Simulink; *MXE*, maximum absolute error; *PF*, DIgSILENT PowerFactory.

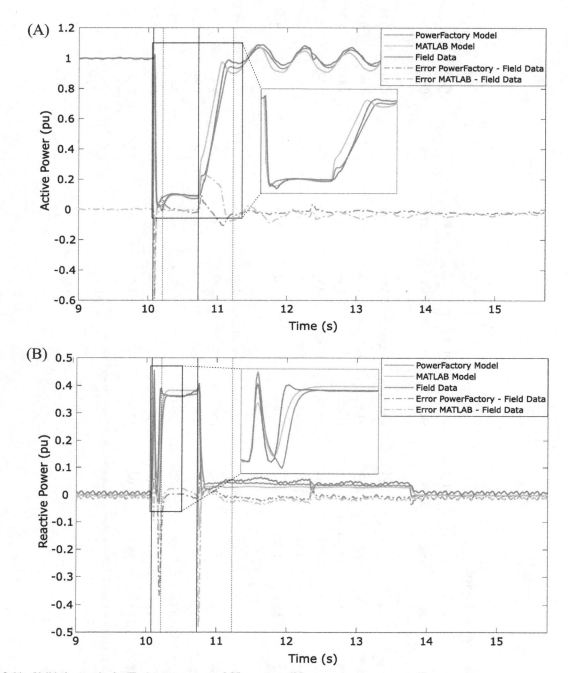

FIGURE 3.11 Validation results for FL_1 test case: $u_{res} = 0.25$ pu, $t_{dip} = 625$ ms.

operation, which attempts to stabilize the rotational speeds of the high-speed and low-speed sides after the voltage dip. Therefore this accurate correlation is due to the optimal fitting of the mechanical model parameters. However, there are still some minor differences between the measured and simulated data during these oscillations. These are mainly due to the greater complexity of the drive train models included in the actual WTs, which usually consider more than two masses.

Finally, it is worth noting that the fault and post-fault injection of reactive power into the grid conducted by the actual DFIG WT is accurately simulated by both the PF and ML models in all cases, as can be seen in Figs. 3.11B, 3.12B, and 3.13B.

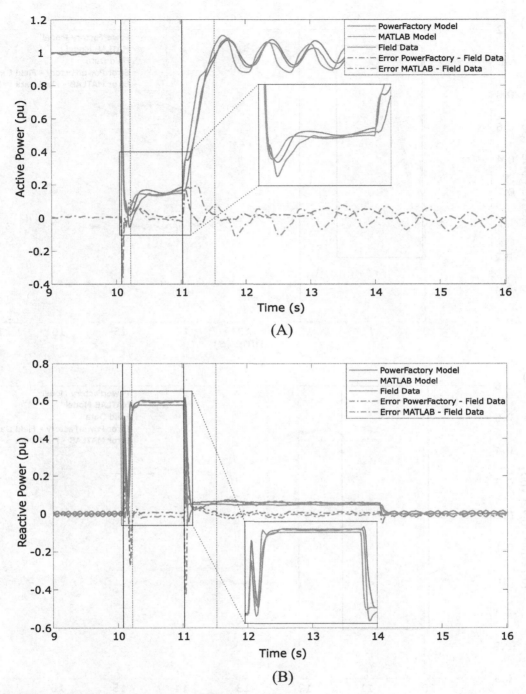

(A)

(B)

FIGURE 3.12 Validation results for FL_2 test case: $u_{res} = 0.50$ pu, $t_{dip} = 920$ ms.

3.4.2 Partial load validation test cases

The validation test cases obtained under the three voltage dip tests conducted at partial load operating conditions are shown in Figs. 3.14−3.16. As in the case of the WT operating at full load conditions, the simulated active and reactive powers fit properly to the measured active and reactive powers.

The validation performance indicators obtained for the PL_1, PL_2, and PL_3 tests are also very low (see the last three columns in Table 3.3) for both the PF and ML models. Moreover, as in the previous cases, the highest error value

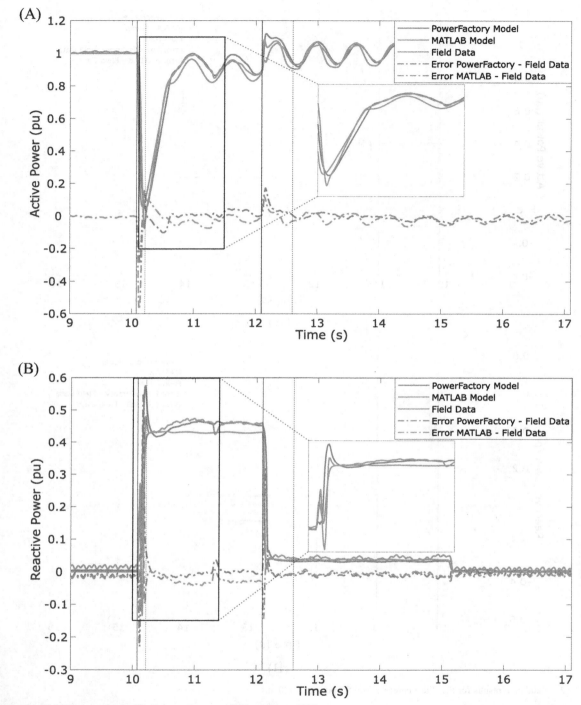

FIGURE 3.13 Validation results for FL_3 test case: $u_{res} = 0.85$ pu, $t_{dip} = 1000$ ms.

obtained corresponds to the MXE in the fault window for the PL_2 test case, where a value of $MXE_{fault} = 11.76\%$ is found for the active power response of the ML model (Fig. 3.15A).

The MAEs obtained, very low in all the partial load validation cases, indicate that the generic Type 3 WT model offers an accurate steady-state response when implemented in both the multidisciplinary and the specialized simulation software tools (see Table 3.3).

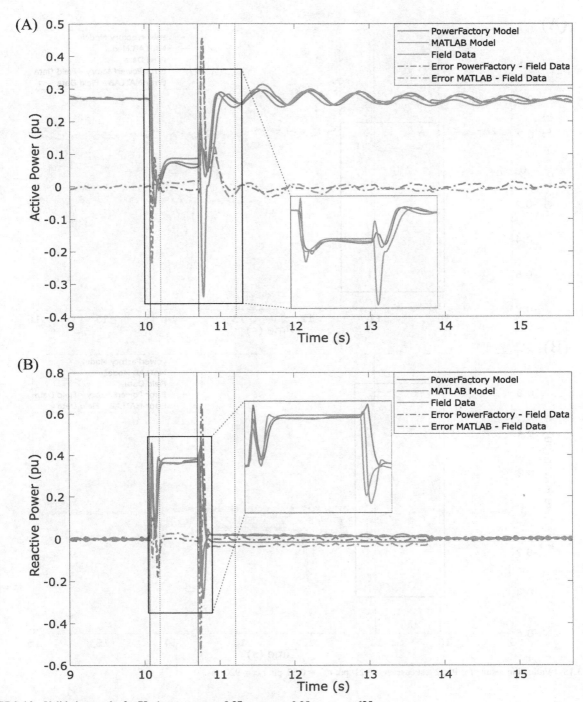

FIGURE 3.14 Validation results for PL_1 test case: $p = 0.27$ pu, $u_{res} = 0.25$ pu, $t_{dip} = 625$ ms.

Nevertheless, when comparing with the previous test cases at full load, larger differences between the field data and the simulated responses can be found in this case, especially in the active power responses once the fault is cleared. The subsynchronous operation of the induction machine is mainly responsible for this. When the WT is operating at partial load condition and a voltage dip occurs, the crowbar protection system is activated to protect the rotor side converter against high induced currents. The crowbar activation causes the rotor terminals to be short-circuited and makes the induction machine of the actual WT suddenly consume active power during a short period of time. This effect can be clearly seen in Figs. 3.14A and 3.15A, and especially in their enlarged views, where a negative peak of active power

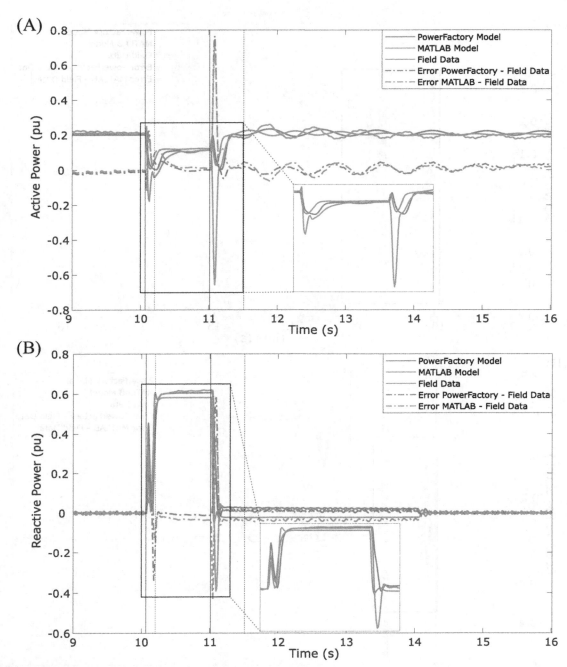

FIGURE 3.15 Validation results for PL_2 test case: $p = 0.21$ pu, $u_{res} = 0.50$ pu, $t_{dip} = 920$ ms.

is observed for the field data (orange line), once the fault is cleared. Therefore during this period of time, the induction generator behaves as a motor. However, it is clear that the generic Type 3 WT model is unable to represent this active power consumption behavior (see the green and blue lines in Figs. 3.14A and 3.15A). The inability of the IEC model to represent the subsynchronous behavior of the induction machine is due to the generator system being implemented as a simplified model with significant differences regarding the highly complex operation of the actual generator.

Despite this, and given that generic models were designed to obtain sufficiently accurate responses when conducting RMS transient stability analyses using highly simplified dynamic and control models, it can be concluded that the Type 3 WT model offers a satisfactory response.

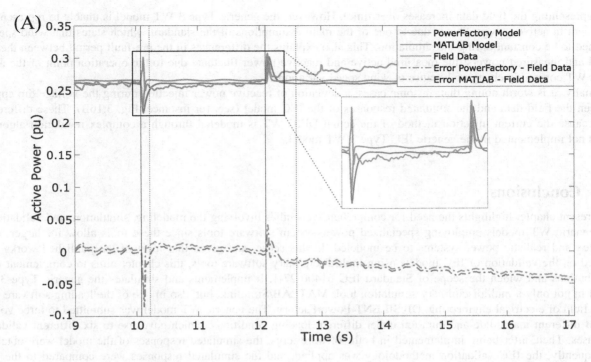

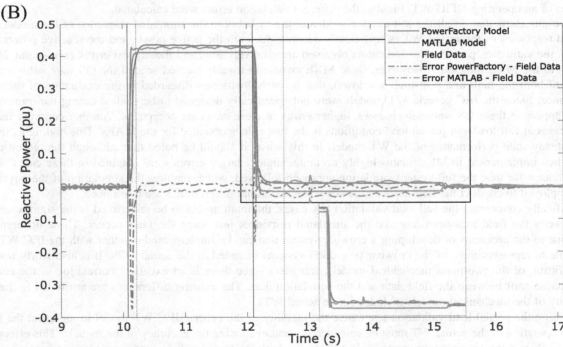

FIGURE 3.16 Validation results for PL_3 test case: $p = 0.26$ pu, $u_{res} = 0.85$ pu, $t_{dip} = 2000$ ms.

It should also be noted that, in the last partial load validation test case, PL_3, as shown in Fig. 3.16, the voltage dip is less deep and there is no crowbar activation. In this case, the subsynchronous operation of the induction machine is not observed, and there is hence no consumption of active power (see Fig. 3.16A). Very low values of validation errors were therefore obtained in this case during the fault window for active power.

As can also be observed in Fig. 3.16A, the rise in the value of the validation performance indicators during the post-fault period is mainly due to the wind speed increment in the field measurements (see in Fig. 3.16A how the orange line representing the field data increases over time). However, the generic Type 3 WT model is unable to represent this increment in active power. This is due to one of the main assumptions of the standard, which states that wind speed is assumed to be constant during the simulation. This also explains the differences in the pre-fault period between the simulated and measured powers: the measured active and reactive power fluctuate due to the operation point of the actual DFIG WT constantly changing because of wind speed variations.

Finally, it is worth noting that, in some cases, a deviation of reactive power injection during the voltage dip appears between the field data and the simulated responses of the IEC model (see, for instance, Fig. 3.16B). These differences are because the current injection method of the actual DFIG WT is modeled through a complex regulation algorithm that is not implemented in the generic IEC Type 3 WT model.

3.5 Conclusions

The present chapter highlights the need for comprehensive studies involving the modeling, simulation, and validation of IEC generic WT models employing specialized power system software tools since these tools allow for larger, more complex and realistic power systems to be modeled. In this sense, and given that most of the published works have focused on the validation of IEC models using multidisciplinary software tools, this chapter aims to complement these contributions and widen the scope of Standard IEC 61400-27-1. It implements and simulates the generic Type 3 WT model in not only a multidisciplinary simulation tool, MATLAB/Simulink, but also in one of the leading software tools in the field of electrical engineering, DIgSILENT PowerFactory. The generic WT model was submitted to three voltage dips of different magnitude and duration under different loading conditions, which gave rise to six different validation test cases. Then, after being implemented in both software tools, the simulated responses of the model were obtained. Subsequently, the IEC validation methodology was applied, and the simulated responses were compared to the field responses of an operating DFIG WT. Finally, the different validation errors were calculated.

The results show that, in all the test cases considered, the fit between the simulated responses of the model and the measured responses of the actual WT is sufficiently accurate for both the active power and the reactive power. In line with this, the validation performance indicators obtained are relatively low, and the highest errors correspond, in nearly all cases, to the MXE indicators. However, these MXE errors are usually located around the QS state subwindows (at the start of both the fault and post-fault windows), that is, the subwindows discarded in the evaluation of the models' performance. Since the IEC generic WT models were not specifically designed to be studied during the transient periods that appear in these QS state subwindows, higher errors in these cases are acceptable. Another common feature of the test cases at full load and partial load conditions is the low values obtained for the MAEs. This highlights the satisfactory steady-state performance of the WT model. In this sense, it should be noted that, although the response of the model when implemented in ML remains highly accurate, slightly larger errors were obtained in these cases. There is one key reason for this: the full-system validation approach followed, which requires the simulation of the physical test bench employed to conduct the field tests and hence some additional uncertainties are introduced.

Specifically concerning the full load validation test cases, the main aspect to be underlined is the significant difference between the field measurements and the simulated responses just when the fault occurs. These differences are mainly due to the problems of developing a crowbar system that can be implemented together with the IEC WT model and is able to represent most of the crowbar protection systems included in the actual WTs. It is also worth noting the optimal fitting of the two-mass mechanical model parameters since there is an excellent correlation in the amplitude and the phase shift between the field data and the simulation data. The existing differences are attributed to the higher complexity of the mechanical models included in the actual WTs.

Regarding the partial load validation test cases, the inability of the generic IEC WT model to represent the subsynchronous operation of the actual WT must be considered when evaluating the accuracy of the models. This effect is particularly visible in the active power graphics just when the fault is cleared, where a power consumption peak appears. This negative peak is caused by the activation of the crowbar protection system, which makes the induction machine of the actual WT operate in a subsynchronous mode for a short period, behaving as a motor. The IEC model is unable to emulate such behavior because of the simplifications introduced in the generator system modeled.

Finally, it must be noted that the assumption of the IEC models regarding the wind speed, that is considered to be constant during the simulation, is noticeable before the onset of the voltage dip, during the pre-fault window. During these periods, the operation points of the actual WT constantly fluctuate and these oscillations cannot be emulated by the simplified WT model. Moreover, the differences in the reactive power injection between the simulation and field data can be attributed to the greater complexity of the injection algorithm used by the actual DFIG WT.

Considering that the aim of the IEC-developed WT models is to use simplified simulation models to reproduce the response of actual WTs with reasonable precision, in addition to being models that can be implemented in any simulation software tool, it can be concluded that the generic Type 3 WT model offers a highly satisfactory response when modeled in both ML and PF, thus fulfilling its aim.

References

Akhmatov, V., Andresen, B., Nielsen, J. N., Jensen, K. H., Goldenbaum, N. M., Thisted, J., & Frydensbjerg, M. (2010). Unbalanced short-circuit faults: Siemens wind power full scale converter interfaced wind turbine model and certified fault-ride-through validation. In: *European wind energy conference & exhibition* (p. 9).

Badrzadeh, B., Emin, Z., Hillberg, E., Jacobson, D., Kocewiak, L., Lietz, G., ... Val, M. (2020). Escudero, The need or enhanced power system modelling techniques and simulation tools. *Cigre Science & Engineering, 17*, 30–46.

Bech, J. (2014). Siemens experience with validation of different types of wind turbine models. In: *Proceedings of the IEEE power and energy society general meeting, Washington, DC, USA* (pp. 27–31).

Buendía, F.J., & Gordo, B.B. (2012). Generic simplified simulation model for DFIG with active crowbar. In: Proc. 11th wind integration workshop (p. 6).

Fortmann, J. (2014). *Modeling of wind turbines with doubly fed generator system* (Ph.D. thesis). Department for Electrical Power Systems, University of Duisburg-Essen.

Chang, Y., Hu, J., Tang, W., & Song, G. (2018). Fault current analysis of type-3 WTs considering sequential switching of internal control and protection circuits in multi time scales during LVRT. *IEEE Transactions on Power Systems, 33*(6), 6894–6903.

Göksu, Ö., Altin, M., Fortmann, J., & Sørensen, P. E. (2016). Field validation of IEC 61400-27-1 wind generation Type 3 model with plant power factor controller. *IEEE Transactions on Energy Conversion, 31*(3), 1170–1178.

Göksu, Ö., Sørensen, P. E., Morales, A., Weigel, S., Fortmann, J., & Pourbeik, P. (2016). Compatibility of IEC 61400-27-1 and WECC 2nd generation wind turbine models. In: *15th international workshop on large-scale integration of wind power into power systems as well as on transmission networks for offshore wind power plants.*

Honrubia-Escribano, A., Gómez-Lázaro, E., Vigueras-Rodríguez, A., Molina-García, A., Fuentes, J., & Muljadi, E. (2012). Assessment of DFIG simplified model parameters using field test data. In: *IEEE power electronics and machines in wind applications* (pp. 1–7).

GWEC. (2021). *Global Wind Report 2021.* Global Wind Energy Council.

Honrubia-Escribano, A., Gómez-Lázaro, E., Fortmann, J., Sørensen, P., & Martín-Martínez, S. (2018). Generic dynamic wind turbine models for power system stability analysis: A comprehensive review. *Renewable and Sustainable Energy Reviews, 81*, 1939–1952.

Honrubia-Escribano, A., Jiménez-Buendía, F., Gómez-Lázaro, E., & Fortmann, J. (2018). Field validation of a Standard Type 3 wind turbine model for power system stability, according to the requirements imposed by IEC 61400-27-1. *IEEE Transactions on Energy Conversion, 33*(1), 137–145.

Honrubia-Escribano, A., Jiménez-Buendía, F., Sosa-Avendaño, J. L., Gartmann, P., Frahm, S., Fortmann, J., Sørensen, P. E., & Gómez-Lázaro, E. (2019). Fault-ride trough validation of IEC 61400-27-1 type 3 and type 4 models of different wind turbine manufacturers. *Energies, 12*(16), 3039.

IEC 61400-21-1. (2019). *Wind energy generation systems - Part 21-1: Measurement and assessment of electrical characteristics -.* International Electrotechnical Commission.

IEC 61400-27-1. (2020). *Ed. 2. Wind turbines—Part 27-1: Electrical simulation models for wind power generation—Wind turbines.* International Electrotechnical Commission.

IEC 61400-27-2. (2020). *Wind energy generation systems - Part 27-2: Electrical simulation models - Model validation.* International Electrotechnical Commission.

Jiménez-Buendía, F., Villena-Ruiz, R., Honrubia-Escribano, A., Molina-García, Á., & Gómez-Lázaro, E. (2019). Submission of a WECC DFIG wind turbine model to Spanish operation procedure 12.3. *Energies, 12*(19), 3749.

Lorenzo-Bonache, A., Honrubia-Escribano, A., Jiménez-Buendía, F., Molina-García, Á., & Gómez-Lázaro, E. (2017). Generic Type 3 wind turbine model based on IEC 61400-27-1: Parameter analysis and transient response under voltage dips. *Energies, 10*(9), 1441.

Lorenzo-Bonache, A., Honrubia-Escribano, A., Jiménez-Buendía, F., & Gómez-Lázaro, E. (2018). Field validation of generic type 4 wind turbine models based on IEC and WECC guidelines. *IEEE Transactions on Energy Conversion, 34*(2), 933–941.

Lorenzo-Bonache, A., Honrubia-Escribano, A., Fortmann, J., & Gómez-Lázaro, E. (2019). Generic Type 3 WT models: Comparison between IEC and WECC approaches. *IET Renewable Power Generation, 13*(7), 1168–1178.

Lorenzo-Bonache A., Villena-Ruiz R., Honrubia-Escribano A., & Gómez-Lázaro E. (2017). Operation of active and reactive control systems of a generic Type 3 WT model. In: *11th IEEE international conference on compatibility, power electronics and power engineering (CPE-POWERENG)* (pp. 606–610).

Meuser, M., & Brennecke, M. (2015). Analysis and comparison of national and international validation methods to assess the quality of DG simulation models. In: *International ETG congress 2015; Die Energiewende-Blueprints for the new energy age, VDE* (pp. 1–7).

Price, W., & Sánchez-Gasca, J. (2006). Simplified wind turbine generator aerodynamic models for transient stability studies. In: *2006 IEEE PES power systems conference and exposition* (pp. 986–992).

Salles, M. B., Hameyer, K., Cardoso, J. R., Grilo, A., & Rahmann, C. (2010). Crowbar system in doubly fed induction wind generators. *Energies, 3*(4), 738–753.

Seman, S., Niiranen, J., Virtanen, R., & Matsinen, J. P. (2008). Low voltage ride-through analysis of 2 MW DFIG wind turbine—Grid code compliance validations. In: *IEEE Power and energy society general meeting* (pp. 1–6). doi:10.1109/PES0.2008.4596687.

Sørensen, P., Andresen, B., Fortmann, J., & Pourbeik, P. (2013). Modular structure of wind turbine models in IEC 61400-27-1. In: *IEEE power & energy society general meeting* (pp. 1–5).

Sørensen, P. E., Andresen, B., Bech, J., Fortmann, J., & Pourbeik, P. (2012). Progress in IEC 61400-27: Electrical simulation models for wind power generation. In: *11th international workshop on large-scale integration of wind power into power systems as well as on transmission networks for offshore wind power plants.* IEEE.

Timbus, A., Korba, P., Vilhunen, A., Pepe, G., Seman, S., & Niiranen, J. (2011). Simplified model of wind turbines with doubly-fed induction generator. In: *10th international workshop on large-scale integration of wind power into power systems as well as on transmission networks for offshore wind power farms* (p. 6).

Trilla, L., Gomis-Bellmunt, O., Junyent-Ferre, A., Mata, M., Sánchez Navarro, J., & Sudria-Andreu, A. (2011). Modeling and validation of DFIG 3-MW wind turbine using field test data of balanced and unbalanced voltage sags. *IEEE Transactions on Sustainable Energy, 2*(4), 509–519. Available from https://doi.org/10.1109/TSTE.2011.2155685.

Uski-Joutsenvuo, S., & Lemström, B. (2007). Dynamic wind turbine and farm models for power system studies. VTT Tech. Res. Centre Finland, Espoo, Finland, Tech. Rep. VTT.

Van Putten, J., Sewdien, V., & Almeida de Graaf, S. (2020). Requirements and capabilities of power electronic interfaced devices for enhancing system operation. Tech. Rep.

Villena-Ruiz, R., Honrubia-Escribano, A., Fortmann, J., & Gómez-Lázaro, E. (2020). Field validation of a standard Type 3 wind turbine model implemented in DIgSILENT-PowerFactory following IEC 61400-27-1 guidelines. *International Journal of Electrical Power & Energy Systems, 116*, 105553.

Villena-Ruiz, R., Lorenzo-Bonache, A., Honrubia-Escribano, A., Jiménez-Buendía, F., & Gómez-Lázaro, E. (2019a). Implementation of IEC 61400-27-1 Type 3 model: Performance analysis under different modeling approaches. *Energies, 12*(14), 2690.

Villena-Ruiz, R., Jiménez-Buendía, F., Honrubia-Escribano, A., Molina-García, Á., & Gómez-Lázaro, E. (2019b). Compliance of a generic Type 3 WT model with the Spanish grid code. *Energies, 12*(9), 1631.

Walling, R., Gursoy, E., & English, B. (2012). Current contributions from Type 3 and Type 4 wind turbine generators during faults. In: *IEEE PES T&D* (pp. 1–6).

WECC REMTF. (2014). WECC second generation of wind turbine models guidelines. Tech. Rep. WECC.

WindEurope. (2020). *Wind energy in Europe 2020 Statistics and the outlook for 2021-2025.* Brussels, Belgium: Wind Europe.

Zhao, H., Wu, Q., Margaris, I., Bech, J., Sørensen, P. E., & Andresen, B. (2015). Implementation and validation of IEC generic type 1A wind turbine generator model. *International Transactions on Electrical Energy Systems, 25*(9), 1804–1813.

Chapter 4

Development of a nonlinear backstepping approach of grid-connected permanent magnet synchronous generator wind farm structure

Youssef Errami, Abdellatif Obbadi and Smail Sahnoun
Laboratory of Electronics, Instrumentation and Energy, Team of Exploitation and Processing of Renewable Energy, Department of Physics, Faculty of Science, University of Chouaib Doukkali, El Jadida, Morocco

4.1 Introduction

In recent years with the augment in population and industrialization, there has been increasing attention in renewable energy sources, which are unlimited in nature and eco-friendliness. Also, wind energy is envisaged to play a considerable position in the upcoming energy industry to become one of the most main alternatives and to get the position of conventional fossil resources (Arslan, Julián Aristizábal, & Arshad, 2019; Basak, Bhuvaneswari, & Rahul, 2020). On the other hand, in different wind power generation devices, variable velocity wind generation structure (VV-WGS) is preferable than constant speed structures because of their low torque peak and high-energy efficiency (Chen, David, Dionysios, & Aliprantis, 2013; Li et al., 2012; Melo & Chien, 2014). Also, VV-WGS can achieve dependability at all wind velocities and the maximum effectiveness, enhanced electrical grid trouble elimination characteristics, and the minimization of the flicker trouble. Among all the power generators employed in producing wind energy, the use of permanent magnet synchronous generator (PMSG) in the global VV-WGS is skyrocketing recently due to its better performance in power density, low systematic cost, efficiency and reliability, variable velocity mode, and full power control capacity (Bakhtiari & Nazarzadeh, 2020; Errami, Benchagra, Hillal, Ouassaid, & Maaroufi, 2012). At the same time, PMSG with elevated pole numbers has been employed to remove the gearbox saving construction, operation, and maintenance costs (Errami, Obbadi, Sahnoun, Ouassaid, & Maaroufi, 2020; Gul, Gao, & Lenwari, 2020). To command the VV-WGS with synchronous generator, power electronic converter devices are frequently employed as the connection between the VV-WGS and the electrical grid (Liu et al., 2019). The typical topology of the PMSG with WGS is depicted in Fig. 4.1. This interface consists of a side converter of generator, an inverter, and a dc link. The capacitor decoupling, furnishes the chance of independent command for every power converter. The rectifier is utilized to command the torque, the velocity, or the power for generator. Then, in normal employment, the converters are utilized to transmit the power of the generator in the type of fluctuating pulse and unstable voltage to the unchangeable pulse as well as unchanging voltage electrical grid (Alizadeh & Yazdani, 2013; Blaabjerg & Ma, 2013; Zhang, Zhao, Qiao, & Qu, 2014). Besides, the inverter is controlled to maintain an invariable dc-link voltage and to regulate the reactive and active power to follow the network requirements. To guarantee steady and protected operation of existing utility grid, the power quality is a major concern for the VV-WGS and has received world-wide attention. In addition, the power generation from the wind is reliant with the turbine velocity. Then, for each velocity of the wind, if this device is ready to rotate at an optimal value, then the wind system can operate in maximum power point tracking (MPPT) (Elkhatib, Aitouche, Ghorbani, & Bayart, 2014; He, Li, & Harley, 2013; Kim, Van, Lee, Song, & Kim, 2013). So, the current progress has been purposeful principally on command methods for the highest production of power. Diverse control schemes for VV-WGS have been presented in the literature. Vector control approach (VCA) is the most popular method

Renewable Energy Systems. DOI: https://doi.org/10.1016/B978-0-12-820004-9.00008-5

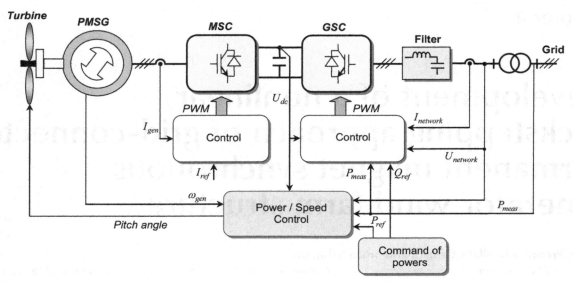

FIGURE 4.1 Scheme of wind generation structure with permanent magnet synchronous generator.

used in the PMSG-based WECS. Shariatpanah, Fadaeinedjad, and Rashidinejad (2013), Li, Haskew, Swatloski, and Gathings (2012), and Errami, Ouassaid, and Maaroufi (2012a) have proposed an MPPT for WECS with PMSG. The MPPT strategy allows process of the turbine at its greatest power coefficient through varying wind velocity conditions to extract the maximum possible wind power.

Chinchilla, Arnaltes, and Burgos (2006) have presented a grid-connected WECS with PMSG and two controlled converters. These converters are controlled by VCA along with space vector modulation method. Nguyen and Lee (2013) and Calle, Busquets-Monge, Kouro, and Wu (2013) have presented low-voltage ride through method for WECS with PMSG through the network fault and if the utility network completely distributed. Then, VCA consists of properly adjusted proportional integral (PI) controllers. It decomposes the currents into reactive and active components and regulates them individually with PI regulators. Then, the characteristics are significantly relied on the variations of the PI parameters and this technique necessitates precise information of system parameters. Also, the major disadvantages of VC are the complexity in the tuning of the PI regulator parameters and its sensitivity to uncertainty of the model. For that reason, direct control methods (DCM) have been proposed as alternative approaches. Rajaei, Mohamadian, and Yazdian Varjani (2013) have proposed DTC of synchronous generator using a rectifier of Vienna and the consequence influences of this configuration voltage on flux and torque. Errami, Obbadi, Ouassaid, and Maaroufi (2018) and Zhang et al. (2014) have discussed MPPT technique with DTC. Zhang, Zhao, Qiao, and Qu (2015) have presented a DTC method for WECS-based PMSG. Freire and Cardoso (2014) have discussed a fault-tolerant generator drive utilizing direct control approaches to provide a WECS structure-based PMSG and ensuring fault-tolerant capabilities. So, DCM are characterized by fast dynamic response, small reliance on parameter, and no coordinate change. Besides, these methods accomplish torque control with reduced fluctuation. Then, their performances can deteriorate because of the enlarged undulations in powers or torque/flux at changeable switching frequency and established state. Also, their performance declines if the speed operation is very low.

This work introduces a nonlinear backstepping command (NBC) for a network-connected WGS with PMSG to ensure maximum power capture. This method possesses a feedback design that allows controlling both the speed and the current by a unique control means. Indeed, the basic idea of the NBC is the stabilization of a virtual control situation and it is founded upon a breakdown process of the complex nonlinear control design into recursive, smaller, and simpler subsystems. So, the droop control rule is realized progressive. Besides, this approach of nonlinear control is generally attractive due to its capacity to deal with uncertainties, systems with nonlinearity, perturbations, and errors of modeling (Foo & Rahman, 2009; Tong & Li, 2012). Sizhan, Jinjun, Linyuan, and Yangque (2013) proposed fuzzy tracking regulator-based NBC for an output feedback. Aounallah, Essounbouli, Hamzaoui, and Bouchafaa (2018) presented a fractional-order control for a WECS-based DFIG to ensure maximum power extraction. Wang, Wai, and Liao (2013) have presented an NBC for the inverter using in high-voltage direct current (HVDC). Consequently, a unity power factor (UPF) and a stable HVDC bus for electrical network utilization were accomplished. Karthikeyan, Kummara, Nagamani, and Saravana Ilango (2011) and Khan, N., Hinchey, and Rahman (2013) have presented NBC to

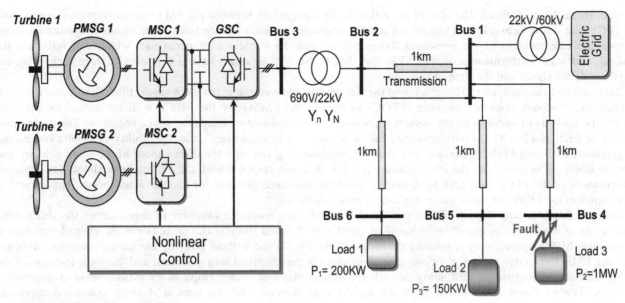

FIGURE 4.2 Configuration of the wind farm.

maximize the power extracted of a WECS. The constitution of presented device, in this chapter, is depicted in Fig. 4.2. The capacitor is employed to disassociate the command for both converters. The rectifiers are used to ensure the maximum tracking of power (MTP) algorithm via commanding the rotational speeds of the turbine devices. The inverter is used to command the voltage of dc link to be invariable with wind speed modifications and to synchronize the power produced by the WGS with the electrical network. Furthermore, it must possess the aptitude to adjust the power that the WECS interchanges with the power grid and to reach the UPF of dispositive. In the sections that follow, the chapter first presents the modeling of the wind turbine (WT) and the PMSGs in Section 4.2. Then, Section 4.3 introduces the NBC for a WECS. Simulation results are presented in Section 4.4. Finally, the chapter concludes with the outline of the major points.

4.2   Related work

The power quality from wind system has received world-wide and increasing penetration wind power will count toward a considerable portion of future electrical power systems. In addition, among the significant research themes in the WECS sector, there is the study of the recent and various command methods that have been applied to wind power system-based PMSG to attain good transient, steady-state performances and that can preserve maximum tracking in power regardless of the consequences of the parameter modifications of device, incertitude in the electric and mechanical models, and changes of wind velocity.

Errouissi, Al-Durra, and Debouza (2018) deal with a PI regulator to improve the transitory performance for the regulation current of WECS-based PMSG. Specifically, this approach reposes of perturbation observer (PO) with command and linearization of feedback method to guarantee nominal transitory performance recuperation with model incertitude. Indeed, this technique uses two supplementary components. The first part represents an antiwindup compensator, whereas the second component employs the modified data of the step to annul the result of the swift step modify in the required power on the transitory reaction. So, this change in the proposed PI regulator allows assuring nothing steady-state error with no surrendering the nominal transitory performance precise with the regulator of situation feedback in defiance the existence of exterior perturbation and model incertitude. Besides, in command saturation, the adopted PI current regulator has a considerable aptitude to deal with the windup circumstance, engendering an excellent transitory performance.

Errami, Obbadi, Sahnoun, et al. (2020) presented sliding mode approach (SMA) useful to WGS-based PMSG. The control method is exploited to maintain the turbine speeds at a select value achieved from the MTP technique, to adjust dc-link voltage and to reach UPF. Moreover, pitch control method is utilized to avert overcharging in the event of elevated wind velocity and to anticipate WT destruction. Stability theory of Lyapunov has been employed to demonstrate the stability of the controllers. The idea of the article is the comparison between the SMA and conventional vector control (VC) with utility network fault conditions and the probable existence of incertitude. The results of simulation demonstrate the effectualness of the presented techniques against the variations of parameter, which can influence the dynamic and static performances of the WFS. But, SMA also VC approach have a few disadvantages regarding the overshoot of the signal and the time of response.

Bakhtiari and Nazarzadeh (2020) examined the design procedure of the extended Kalman filter (EKF) and the feedforward and feedback optimal controller (FFOC) to control the PMSG for the MPPT with numerous disturbances. Indeed, the analysis of stability on the system is presented and employed to design the speed regulator. The design procedures of EKF and FFOC are uncomplicated methods and easy to implement. A comparability of results concerning the proposed EKF and FFOC technique and usual PI regulator is given with the disturbances of the wind velocity and faults in electrical network. So, the PI regulator has a low dynamic response with the elevated overshoot. The dynamic performances of the FFOC and EKF have been tracked the reference velocity for diverse wind speeds, but a tracking error shows in the LQG controller under the steady-state conditions.

Roberto, Solsona, and Busada (2016) presented a Luenberger nonlinear observer to approximate the mechanical variables as of the electrical quantities in variable speed WECS-based PMSG. The objective of the control approach is to guarantee MPPT operation by regulating the torque of the PMSG and without utilizing mechanical detectors to determine the PMSG velocity. Indeed, the controller approximates the electrical torque, speed, and the rotor location of the PMSG. Besides, approximates are finding through a nonlinear observer, which employs the measurement of the electric variables. The evaluations are provided to the regulator to construct the mechanical detector command approach. Simulations were provided to depict the qualities of the presented technique. It was verified that the performance of the system is analogous to that acquired using regulators and necessitates a mechanical detector. But, this model has a small disadvantage to attempt the fast modification of the mechanical torque. Also, it is characterized by electrical torque undulations, which are produced by the current source rectifier and the elevated switching frequency.

Fernando, Kenne, and Lagarrigue (2016) proposed an adaptive control scheme for MPPT of stand-alone PMSG wind systems. The strategy used is the adjustment of a neural network identifier (NNI) to evaluate the mechanical torque of the WT and the wind velocity is determined utilizing the optimal mechanical torque. Besides, a backstepping regulator (BR) is obtained to control the rotor velocity to the finest velocity value. A real-time active nonlinear learning rule of the weight vector is employed for the network of neural. The regulator stability is obtained via analysis of Lyapunov. The composition of the NNI and the BR offers improved speed control of WT. The simulation results with parameter uncertainties are presented and discussed to confirm the soundness of the employed method.

Matraji, Al-Durra, and Errouissi (2018) proposed adaptive second-order SMA for WECS-based PMSG. A second-order SMA is founded on a modified adaptive super twisting approach and a feedback linearization control. The aim of this technique is to oblige the generator to transfer the required power to the electrical network using the control of the winding current and to attain an excellent transitory performance during the functioning range of the structure. So, fast transient response is achieved and the chattering problem is attenuated. Besides, this method does not necessitate the information of actual parameter limits of device.

Yang et al. (2019) proposed an adaptive fractional order PID (AFOPID) control for WECS-based PMSG to assure MPPT. The combinatory result of PMSG nonlinearities, uncertainties of parameter, and stochastic wind velocity modification is aggregated into a disturbance. Moreover, the estimate of the perturbation is utilized as a supplementary command signal, which is completely compensated by a fractional order PID regulator to attain overall robust command consistence, an improved tracking performance, high reliability, and simple structure compared to that of PID regulator. Also, this technique does not necessitate an accurate generator model and a completely decoupled control of d-axis current and mechanical rotation could be accomplished. Then, AFOPID approach can successfully handle diverse modeling uncertainties and complex nonlinearities.

Yang, Zhang, Zhang, Wang, and Li (2018) presented an adaptive integral sliding mode direct power control approach for voltage source converter. Indeed, the strategy consists of the direct power controller-based integral sliding mode regulator and PI to assure power regulation objective. Besides, an ac-line parameter adaptive method by error cancellation is employed to enhance the robustiousness and the transitory performance of the structure. So, the parameters of the regulator are adapted conforming to the real functioning situations so that the needed performance can be achieved. The simulation results show adequate performance of the presented approach.

Chen et al. (2019) examined a nonlinear adaptive pitch angle regulator based on observers for WT with PMSG to restrict the retrieved wind power around rated power. Indeed, the proposed technique used a designed PO to estimate and to compensate nonlinearities of system, unknown parameter incertitude, and unknown troubles of the WECS-based PMSG. So, two second-order POs and one third-order status and PO (SPS) are used. Besides, the proposed technique does not necessitate complete status measurements or the precise model of system. The objective of the work is the comparability between the NAPAC and the feedback-linearizing regulator, VC through PI controllers, and PI regulator with gain scheduling. The comparisons of simulation results, with diverse situation, prove the effectiveness of the presented NAPAC technique.

Errami, Obbadi, and Sahnoun (2020) presented a technique of combining adaptive sliding mode observer (ASMO) using active perturbation rejection command (APRC) or the WECS-based PMSG to guarantee the maximum power point. Indeed, SMA, which is insensible to the change of parameter of system and exterior perturbation, is employed to examine the speed and position of generator. So, ASMO can efficiently examine the information of rotor generator location and feeble the natural chattering of usual SM. Besides, APRC can improve the antiperturbations aptitude once the device is with linear status feedback. Results of simulation demonstrate that technique has strong robustness and more correct speed tracking.

4.3   Mathematical model of wind turbine generator

4.3.1   The wind turbine system

The total power generated by the WT can be calculated by Errouissi et al. (2018):

$$P_t = \frac{1}{2}\rho A C_p(\lambda, \beta)v^3 \tag{4.1}$$

ρ is the density of the air; A is the surface swept by the propeller; C_p is the power coefficient; β is the pitch angle, and v is the wind velocity (in m s^{-1}). The specific wind speed (SWS) λ is (Errami , Ouassaid, & Maaroufi, 2015):

$$\lambda = \frac{\omega_t R}{v} \tag{4.2}$$

R is the blade length and ω_t is the turbine rotation speed. The mechanical torque T_t is:

$$T_t = \frac{1}{2}\rho A C_p(\lambda, \beta)v^3 \frac{1}{\omega_t} \tag{4.3}$$

$C_p(\lambda, \beta)$ is defined as (Errami, Obbadi, Sahnoun, et al., 2020):

$$C_p(\lambda, \beta) = \frac{1}{2}\left(\frac{116}{\lambda_m} - 0.4\beta - 5\right)e^{-\left(\frac{21}{\lambda_m}\right)} \quad \frac{1}{\lambda_m} = \frac{1}{\lambda + 0.08\beta} - \frac{0.035}{\beta^3 + 1} \tag{4.4}$$

The highest value of C_P, that is, $C_{p-\max} = 0.41$, is attained if $\beta = 0$ and $\lambda_{opt} = 8.1$. Fig. 4.3 depicts C_P versus λ characteristics of the turbine, achieved for diverse values of β. Also, for a particular blade angle and wind velocity, the maximum power can be generated at a specified shaft velocity of a turbine. Therefore to realize the power efficacy maximization, the turbine TSR should be sustained at λ_{opt} despite modifications of wind speed (Errami, Hillal, Benchagra, Ouassaid, & Maaroufi, 2012). Accordingly, the MPPT algorithm can be achieved by modifying the speed

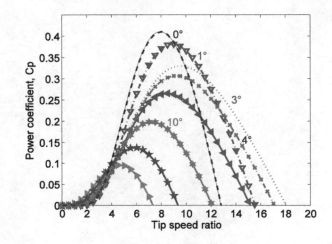

FIGURE 4.3 C_p characteristics as a function of λ for different values of β.

of the generator to maintain a best C_P at which the highest power is generated from accessible wind power (Bakhtiari & Nazarzadeh, 2020). Fig. 4.4 explains the curves of the highest power extracted and the extracted power of the turbine at diverse speeds (Errami, Obbadi, Sahnoun, et al., 2020). This figure illustrates that the MPPT technique can be described as a function of PMSG velocity.

4.3.2 PMSG modeling

The stator voltage equation for generator can be formulated by (Errami, Obbadi, Sahnoun, et al., 2020):

$$\begin{bmatrix} v_{md} \\ v_{mq} \end{bmatrix} = \begin{bmatrix} R_g + p.L_d & -\omega_e L_q \\ \omega_e L_d & R_g + pL_d \end{bmatrix} \begin{bmatrix} i_d \\ i_q \end{bmatrix} + \begin{bmatrix} 0 \\ \omega_e \psi_f \end{bmatrix} \tag{4.5}$$

where v_{mq}, v_{md} are stator voltage; i_q, i_d are current of the stator; L_q, L_d are inductances of the PMSG; R_g is the resistance of stator; and ψ_f is the permanent magnetic flux. The electric rotary velocity of the generator ω_e is written as:

$$\omega_e = p_n \omega_t \tag{4.6}$$

where p_n is the pole pair number of the PMSG and ω_t is the mechanical speed. If the generator is supposed to have identical inductances in q-axis and d-axis ($L_q = L_d = L_s$), the electromagnetic torque of PMSG is (Errouissi et al., 2018):

$$T_{ge} = \frac{3}{2} p_n \left(\psi_f i_q \right) \tag{4.7}$$

Also, Eq. (4.5) can be expressed as:

$$v_{md} = -\omega_e L_s i_q + \left(R_g i_d + L_s \frac{di_d}{dt} \right) \tag{4.8}$$

$$v_{mq} = \omega_e L_s i_d + \omega_e \psi_f + \left(R_g i_q + L_s \frac{di_q}{dt} \right) \tag{4.9}$$

The mechanical equation of the system is:

$$J \frac{d\omega_t}{dt} = T_{ge} - T_t - F_v \omega_t \tag{4.10}$$

where J is the inertia moment; F_v is the coefficient of the viscous friction; and T_t denotes the mechanical torque of the turbine.

FIGURE 4.4 Characteristic of power at diverse wind velocity.

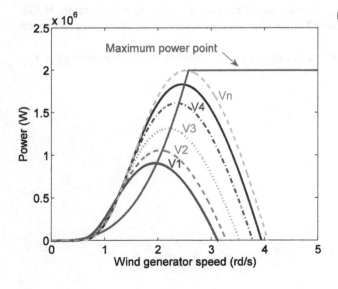

4.4   Control schemes of wind farm

4.4.1   MPPT technique

The reference speed of the generator resultant to the maximum power generated at a certain wind velocity is obtained by the MTP technique. Indeed, as the velocity of the wind is inferior than the rated velocity, the turbine is commanded for MPPT. Also, to produce maximum power, the turbine has to be worked to deliver the maximum mechanical power, P_{t_max}, by controlling the speed of turbine in diverse wind velocity below rated power of the WECS. In addition, if the speed of turbine can for all time be controlled to make turbine system operate with λ_{opt}, disregarding the wind velocity, then the WT works at the peak of its C_P curve and the coefficient of power attains its highest value. As a result, the MTP regulator configuration is utilized to produce the reference velocity adjustment of generator side converter, which can be calculated as follows:

$$\omega_{t-opt} = \frac{v\lambda_{opt}}{R} \tag{4.11}$$

Consequently, the P_{t_max} characteristic is provided as function of ω_{t-opt}, the velocity attributed to the PMSG side:

$$P_{t\_max} = K\omega_{t-opt}^3 \tag{4.12}$$

where $K = \frac{1}{2}\rho A C_{Pmax}\left(\frac{R}{\lambda_{opt}}\right)^3$. K is based on the aerodynamics of the blade and parameters of WT. The MTP algorithm system calculates this reference velocity ω_{t-opt} and an optimal value of SWS λ_{opt} can be preserved. Consequently, the highest power can be extracted. According to the wind speed, the MTP approach controls the electrical power produced, conveying the turbine working points onto the "maximum power point," similar to that in Fig. 4.4. As a result, we can obtain the highest power P_{t_max} by changing the velocity of the generator during wind speed deviations and underrated power of the WGS. ω_{t-opt} is the suggestion speed of the generator, which is generated by an MTP algorithm and is useful to the velocity control system of each generator side converter.

4.4.2   Nonlinear control of WFS

4.4.2.1   Generator side converters control

The command scheme is based on nonlinear backstepping theory. This technique utilizes the implicit command variable to construct the original system, which is easy and with high order. It generates a corresponding error changeable, which can be stabilized by vigilantly selecting correct inputs of control. So, the last control variables can be resulting using suitable functions of Lyapunov. On the other hand, according to Eqs. (4.7) and (4.8), the speed of each PMSG can be commanded by adjusting i_q. As a result, to create a backstepping speed controller, the tracking error of the speed error is choosen as (Roberto et al., 2016):

$$\delta_\omega = \omega_{t-opt} - \omega_t \tag{4.13}$$

Taking the derivative of δ_ω, we have:

$$\frac{d\delta_\omega}{dt} = \frac{d\omega_{t\_opt}}{dt} - \frac{d\omega_t}{dt} \tag{4.14}$$

According to Eqs. (4.7), (4.10), and (4.14), the derivative of the speed error δ_ω can be represented:

$$\frac{d\delta_\omega}{dt} = \frac{d\omega_{t\_opt}}{dt} - \frac{1}{J}\left[\frac{3}{2}p_n\left(\psi_f i_q\right) - T_t - F_v\omega_t\right] \tag{4.15}$$

To found the function stabilization, a first function of Lyapunov candidate is constructed as (Roberto et al., 2016; Slotine and Li, 1991):

$$B_\omega = \frac{1}{2}\delta_\omega^2 \tag{4.16}$$

The Lyapunov function derivative is given as:

$$\frac{dB_\omega}{dt} = \delta_\omega\frac{d\delta_\omega}{dt} = \delta_\omega\frac{d\omega_{t\_opt}}{dt} - \frac{\delta_\omega}{J}\left[\frac{3}{2}p_n\left(\psi_f i_q\right) - T_t - F_v\omega_t\right] \tag{4.17}$$

Besides, we can rewrite Eq. (4.17) as:

$$\frac{dB_\omega}{dt} = -k_\omega \delta_\omega^2 + \frac{\delta_\omega}{J}\left(Jk_\omega\delta_\omega + T_t - \frac{3}{2}p_n\psi_f i_q + F_v\omega_t + J\frac{d\omega_{t\_opt}}{dt}\right) \tag{4.18}$$

where k_ω is the loop reaction invariable. Also, the satisfactory condition for the overall stability of the system is (Kortabarria et al., 2014):

$$\frac{dB_\omega}{dt} < 0 \tag{4.19}$$

Furthermore, the q-axis current and d-axis current components are identified as the virtual control variables. i_{d-ref} is preset to zero to diminish the copper failure. So, the generator speed convergence to ω_{t_opt} is insured if one defines the equations:

$$i_{d-ref} = 0 \tag{4.20}$$

$$i_{q-ref} = \frac{2}{3p_n\psi_f}(Jk_\omega\delta_\omega + J\frac{d\omega_{t\_opt}}{dt} + F_v\omega_t + T_t) \tag{4.21}$$

to adjust i_d and i_q to their suggestion values, we choose the errors of status:

$$\delta_d = i_{d-ref} - i_d \tag{4.22}$$

$$\delta_q = i_{q-ref} - i_q \tag{4.23}$$

where i_{q-ref}, i_{d-ref} are the desired values of q-axis and d-axis current components, respectively. So:

$$\frac{d\delta_d}{dt} = \frac{di_{d-ref}}{dt} - \frac{di_d}{dt} = -\frac{1}{L_s}(v_{md} + L_s\omega_e i_q - R_g i_d) \tag{4.24}$$

$$\frac{d\delta_q}{dt} = \frac{di_{q-ref}}{dt} - \frac{di_q}{dt} = \frac{di_{q-ref}}{dt} - \frac{1}{L_s}(v_{mq} - \omega_e\psi_f - R_g i_q - L_s\omega_e i_d) \tag{4.25}$$

Using Eqs. (4.22)–(4.24), it can be obtained:

$$\delta_d\frac{d\delta_d}{dt} = \delta_d\left[-\frac{1}{L_s}(v_{md} + L_s\omega_e i_q - R_g i_d)\right]$$
$$= -\theta_d\delta_d^2 + \frac{\delta_d}{L_s}\left[-v_{md} - L_s\omega_e i_q + R_g i_d + L_s\theta_d\delta_d\right] \tag{4.26}$$

$$\delta_q\frac{d\delta_q}{dt} = \delta_q\left[\frac{di_{q-ref}}{dt} - \frac{di_q}{dt}\right]$$
$$= -\theta_q\delta_q^2 + \frac{\delta_q}{L_s}\left[L_s\frac{di_{q-r}}{dt} - v_{mq} + \omega_e\psi_f + R_g i_q + L_s\omega_e i_d + \theta_q L_s\delta_q\right] \tag{4.27}$$

where θ_d and θ_q are the closed-loop constants. To found the function of stabilization, we choose a second function candidate of Lyapunov as:

$$H = B_\omega + \frac{1}{2}\delta_q^2 + \frac{1}{2}\delta_d^2$$
$$= \frac{1}{2}\delta_\omega^2 + \frac{1}{2}\delta_q^2 + \frac{1}{2}\delta_d^2 \tag{4.28}$$

Also, the time derivate of H is:

$$\frac{dH}{dt} = \delta_\omega\frac{d\delta_\omega}{dt} + \delta_d\frac{d\delta_d}{dt} + \delta_q\frac{d\delta_q}{dt} \tag{4.29}$$

With Eqs. (4.18), (4.26), and (4.27), Eq. (4.29) can be rewritten as:

$$\frac{d\mathrm{H}}{dt} = -k_\omega \delta_\omega^2 + \frac{\delta_\omega}{J}\left(T_t - \frac{3}{2}p_n\psi_f i_q + F_v\omega_t + Jk_\omega\delta_\omega + J\frac{d\omega_{t\_opt}}{dt}\right)$$

$$-\theta_d\delta_d^2 + \frac{\delta_d}{L_s}\left[-v_{gd} - L_s\omega_e i_q + R_g i_d + L_s\theta_d\delta_d\right] \tag{4.30}$$

$$-\theta_q\delta_q^2 + \frac{\delta_q}{L_s}\left[L_s\frac{di_{q-ref}}{dt} - v_{mq} + \omega_e\psi_f + R_g i_q + L_s\omega_e i_d + \theta_q L_s\delta_q\right]$$

Consequently, to ensure that $\frac{d\mathrm{H}}{dt}$ is consistently negative and to secure the overall asymptotic stability, NBC rules are selected as follows:

$$v_{q-ref} = \omega_e\psi_f + R_g i_q + L_s\omega_e i_d + L_s\frac{di_{q-ref}}{dt} + L_s\theta_q\delta_q \tag{4.31}$$

$$v_{d-ref} = L_s\theta_d\delta_d - L_s\omega_e i_q + R_g i_d \tag{4.32}$$

According to Eqs. (4.20), (4.21), (4.31), and (4.32), we can rewrite Eq. (4.30) as:

$$\frac{d\mathrm{H}}{dt} = -k_\omega\delta_\omega^2 - \theta_d\delta_d^2 - \theta_q\delta_q^2 \le 0 \tag{4.33}$$

Because $\frac{d\mathrm{H}}{dt}$ is a negative specified function, this may imply that δ_ω, δ_q, and δ_d go to zero symptotically. At last, PWM is employed to generate the command signal to realize the NBC for a single PMSG side rectifier. Fig. 4.5 depicts the control structure for each PMSG side converter.

4.4.2.2 Pitch angle control

The turbine has two functioning zones conforming to the velocity of the wind. In the first zone where the wind velocity is inferior than the threshold of the wind velocity, the velocity for each turbine is regulated at the best value and the maximum energy is produced from the turbine. The turbine supervisor keeps the C_P at its maximum. The pitch angle regulator is not active, and β is fixed at zero degree. In the second zone, where the velocity of the wind surpasses its reference value, the pitch angle is activated for restricting the aerodynamic power extracted with the turbine and C_p diminishes to restrict the turbine velocity. Thus the turbine functions at lower efficacy less than $C_{p-\max}$. Consequently, the power of the PMSG is restricted at the rated value. Also, the rotating speed of the PMSG can be maintained at its

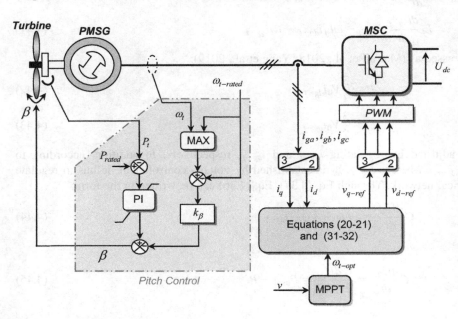

FIGURE 4.5 Schematic representation of backstepping approach for individual machine side converter.

reference value. The pitch is used to avoid the damage of turbines from too much wind speed and β will augment until the turbine system is at the rated velocity. The scheme of pitch angle system is depicted in Fig. 4.5 where P_t is the power, which is extracted from every turbine system.

4.4.2.3 Control of inverter

The dynamic voltage equations of the inverter side can be represented as follows (Fernando et al., 2016; Errami, Hillal, et al., 2012):

$$\frac{di_{dg-f}}{dt} = \frac{1}{L_f}(e_d - v_d + \omega L_f i_{qg-f} - R_f i_{dg-f}) \tag{4.34}$$

$$\frac{di_{qg-f}}{dt} = \frac{1}{L_f}(e_q - v_q - \omega L_f i_{dg-f} - R_f i_{qg-f}) \tag{4.35}$$

where L_f is the inductance of the filter; R_f is the resistance of the filter; e_d, e_q are inverter d-axis and q-axis voltage components; v_d, v_q are grid voltage components; i_{dg-f}, i_{qg-f} are currents of electrical network; and ω is the power grid angular pulse. Reactive and active powers can be achieved as:

$$P = \frac{3}{2}(v_d i_{dg-f} + v_q i_{qg-f}) \tag{4.36}$$

$$Q = \frac{3}{2}(v_d i_{qg-f} - v_q i_{dg-f}) \tag{4.37}$$

Thus the equation of dc-link system can be found as:

$$\frac{dU_{dc}}{dt} = \frac{3}{2C}\left(\frac{v_d}{U_{dc}}i_{dg-f} + \frac{v_q}{U_{dc}}i_{qg-f}\right) - \frac{i_0}{C} \tag{4.38}$$

where U_{dc} is the voltage of dc link; i_0 is the electrical network side transmission line current; and C is the dc-link capacitor. If we consider that the d-axis of the dq reference frame is directed alongside the electrical network voltage, so:

$$v_d = V_0 \text{ and } v_q = 0 \tag{4.39}$$

According to Eqs. (4.36), (4.37), and (4.39):

$$L_f \frac{di_{dg-f}}{dt} = e_d - V_0 + \omega L_f i_{qg-f} - R_f i_{dg-f} \tag{4.40}$$

$$L_f \frac{di_{qg-f}}{dt} = e_q - \omega L_f i_{dg-f} - R_f i_{qg-f} \tag{4.41}$$

In addition, the powers can be expressed as (Matraji et al., 2018; Yang et al., 2019):

$$P = \frac{3}{2}V_0 i_{dg-f} \tag{4.42}$$

$$Q = \frac{3}{2}V_0 i_{qg-f} \tag{4.43}$$

Accordingly, the powers can be adjusted by regulating i_{dg-f} and i_{qg-f}, respectively. In addition, according to Eq. (4.43), $i_{qg-f-ref}$ is created by $Q_{grid-ref}$, whereas $i_{dg-f-ref}$ is established by voltage controller of dc bus to regulate the active power supplied to the electrical network. Through Eq. (4.37), Eq. (4.36) can be written in the form:

$$C\frac{dU_{dc}}{dt} = -i_0 + \frac{3}{2}\frac{V_0}{U_{dc}}i_{dg-f} \tag{4.44}$$

So:

$$U_{dc}C\frac{dU_{dc}}{dt} = \frac{C}{2}\frac{dU_{dc}^2}{dt} = P_{total} - P \tag{4.45}$$

where P_{total} is the total power supplied by the generators. Accordingly, the active power reference is supplied by voltage regulator of dc bus. On the other hand, with Eq. (4.42), Eq. (4.45) can be written as follows:

$$\frac{dU_{\text{dc}}^2}{dt} = \frac{2}{C}\left(P_{\text{total}} - \frac{3}{2}V_0 i_{dg-f}\right) \tag{4.46}$$

The state variable tracking error is:

$$\delta_{\text{bus}} = U_{\text{dc}-ref}^2 - U_{\text{dc}}^2 \tag{4.47}$$

On the other hand:

$$\frac{d\delta_{\text{bus}}}{dt} = \frac{dU_{\text{dc}-ref}^2}{dt} - \frac{2}{C}\left(P_{\text{total}} - \frac{3}{2}V_0 i_{dg-f-ref}\right) \tag{4.48}$$

The major control purpose is to adjust the voltage of dc bus. As a result, i_{dg-f} is employed as virtual accommodate variable so as to adjust δ_{bus}. The function of Lyapunov can be chosen as:

$$H_{\text{bus}} = \frac{1}{2}\delta_{\text{bus}}^2 \tag{4.49}$$

With Eq. (4.48), the derivative of the time for H_{bus} is:

$$
\begin{aligned}
\frac{dH_{\text{bus}}}{dt} &= \delta_{\text{bus}}\frac{d\delta_{\text{bus}}}{dt} \\
&= \delta_{\text{bus}}\left[\frac{dU_{\text{dc}-ref}^2}{dt} - \frac{2}{C}\left(P_{\text{total}} - \frac{3}{2}V_0 i_{dg-f-ref}\right)\right] \\
&= -k_{\text{bus}}\delta_{\text{bus}}^2 + \delta_{\text{bus}}\left[\frac{dU_{\text{dc}-ref}^2}{dt} - \frac{2}{C}\left(P_E - \frac{3}{2}V_0 i_{dg-f-ref}\right) + k_{\text{bus}}\delta_{\text{bus}}\right]
\end{aligned}
\tag{4.50}
$$

where k_{bus} is the feedback invariable of the closed loop. As a result, the convergence of U_{dc} to the $U_{\text{dc}-ref}$ is achieved if one delineates the next equation:

$$i_{dg-f-ref} = \frac{1}{V_0}\left[\frac{2}{3}P_{\text{total}} - \frac{C}{3}\left(\frac{dU_{\text{dc}-ref}^2}{dt} + k_{\text{bus}}\delta_{\text{bus}}\right)\right] \tag{4.51}$$

According to Eq. (4.51), we can rewrite Eq. (4.50) as:

$$\frac{dH_{\text{bus}}}{dt} = -k_{\text{bus}}\delta_{\text{bus}}^2 \leq 0 \tag{4.52}$$

So, the system stability is ensured. To control the powers with quadrature and direct electrical grid current components, the error is defined using i_{qg-f} and i_{dg-f}

$$\delta_{d-f} = i_{dg-f-ref} - i_{dg-f} \tag{4.53}$$

$$\delta_{q-f} = i_{qg-f-ref} - i_{qg-f} \tag{4.54}$$

The reference $i_{qg-f-ref}$ is obtained from the required factor of power. Consequently, using Eqs. (4.51) and (4.53), Eq. (4.48) can be rewritten as follows:

$$\frac{d\delta_{\text{bus}}}{dt} = -k_{\text{bus}}\delta_{\text{bus}} + \frac{3}{C}V_0\delta_{d-f} \tag{4.55}$$

Besides, according to Eqs. (4.40) and (4.41), it can be obtained:

$$\frac{d\delta_{d-f}}{dt} = \frac{di_{dg-f-ref}}{dt} - \frac{di_{dg-f}}{dt} = \frac{di_{dg-f-ref}}{dt} - \frac{1}{L_f}(e_d - V_0 - R_f i_{dg-f} + L_f \omega i_{qg-f}) \tag{4.56}$$

$$\frac{d\delta_{q-f}}{dt} = \frac{di_{qg-f-ref}}{dt} - \frac{di_{qg-f}}{dt} = \frac{di_{qg-f-ref}}{dt} - \frac{1}{L_f}(e_q - L_f \omega i_{dg-f} - R_f i_{qg-f}) \tag{4.57}$$

Combining Eqs. (4.53) and (4.54) with Eqs. (4.56) and (4.57), we obtain:

$$
\begin{aligned}
\delta_{d-f}\frac{d\delta_{d-f}}{dt} &= \delta_{d-f}\left[\frac{di_{dg-f-ref}}{dt} - \frac{1}{L_f}(e_d - V_0 - R_f i_{dg-f} + L_f \omega i_{qg-f})\right] \\
&= -k_{d-f}\delta_{d-f}^2 + \frac{\hat{I}\mu_{d-f}}{L_f}\left[L_f \frac{di_{dg-f-ref}}{dt} - e_d + V_0 + R_f i_{dg-f} - L_f \omega i_{qg-f} + k_{d-f}L_f\delta_{d-f}\right]
\end{aligned}
\tag{4.58}
$$

$$
\begin{aligned}
\delta_{q-f}\frac{d\delta_{q-f}}{dt} &= \delta_{q-f}\left[-\frac{di_{q-f}}{dt}\right] \\
&= -k_{q-f}\delta_{q-f}^2 + \frac{\delta_{q-f}}{L_f}\left[-\delta_q + R_f i_{q-f} + L_f \omega i_{d-f} + k_{q-f}L_f\delta_{q-f}\right]
\end{aligned}
\tag{4.59}
$$

where k_{d-f} and k_{q-f} are the positive configuration invariables of the closed loop. To establish the stabilizing function, considering the new Lyapunov function with uses variables δ_{d-f} and δ_{q-f} as:

$$H_f = \frac{1}{2}\delta_{bus}^2 + \frac{1}{2}\delta_{d-f}^2 + \frac{1}{2}\delta_{q-f}^2 \tag{4.60}$$

With the derivative of the function of Lyapunov (Eq. 4.60):

$$\frac{dH_f}{dt} = \delta_{bus}\frac{d\delta_{bus}}{dt} + \delta_{d-f}\frac{d\delta_{d-f}}{dt} + \delta_{q-f}\frac{d\delta_{q-f}}{dt} \tag{4.61}$$

Also, with Eqs. (4.55), (4.58), and (4.59), we can rewrite Eq. (4.61) as:

$$
\begin{aligned}
\frac{dH_f}{dt} &= -k_{bus}\delta_{bus}^2 - k_{d-f}\delta_{d-f}^2 - k_{q-f}\delta_{q-f}^2 \\
&\quad + \frac{\delta_{d-f}}{L_f}\left[\frac{3L_f}{C}V_0\delta_{bus} + L_f\frac{di_{dg-f-ref}}{dt} - \delta_d + R_f i_{dg-f} - L_f \omega i_{qg-f} + V_0 + k_{d-f}L_f\delta_{d-f}\right] \\
&\quad + \frac{\delta_{q-f}}{L_f}\left[-\delta_q + R_f i_{qg-f} + L_f \omega i_{dg-f} + k_{q-f}L_f\delta_{q-f}\right]
\end{aligned}
\tag{4.62}
$$

To guarantee the stability of system, a negative value for $\frac{dH_f}{dt}$ must be chosen. Consequently, according to Eq. (4.62), NBC rules are selected as follows:

$$v_{dg-f-r} = \frac{3L_f}{C}V_0\delta_{bus} + L_f\frac{di_{dg-f-ref}}{dt} + R_f i_{dg-f} - L_f \omega i_{qg-f} + V_0 + L_f k_{d-f}\delta_{d-f} \tag{4.63}$$

$$v_{qg-f-r} = R_f i_{qg-f} + L_f \omega i_{dg-f} + L_f k_{q-f}\delta_{q-f} \tag{4.64}$$

According to Eqs. (4.63) and (4.64), Eq. (4.62) can be rewritten as follows:

$$\frac{dH_f}{dt} = -k_{bus}\delta_{bus}^2 - k_{d-f}\delta_{d-f}^2 - k_{q-f}\delta_{q-f}^2 \leq 0 \tag{4.65}$$

Because the derivative of H_f is negative, it can involve that $\delta_{bus}, \delta_{q-f}$ and δ_{d-f} go to zero symptotically. Also, the dc-bus structure voltage control tracking is attained. At last, PWM is employed to generate the adjustment signal to implement the NBC for the inverter. Fig. 4.6 depicts the diagram of the control technique for grid-side converter.

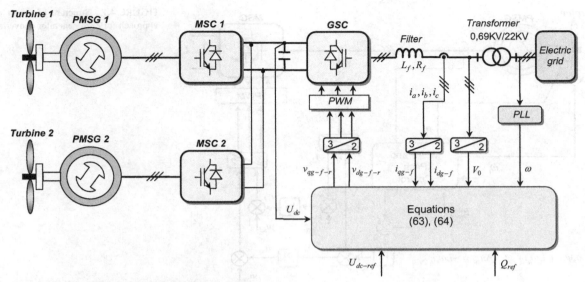

FIGURE 4.6 Structure of the nonlinear control in WFS with backstepping.

4.4.3 Vector control technique of WFS

4.4.3.1 Regulator of PMSG side

We can rewrite Eq. (4.5) as:

$$v_{md} = \left(R_g i_d + L_d \frac{di_d}{dt}\right) - \omega_e L_q i_q \tag{4.66}$$

$$v_{mq} = \left(R_g i_q + L_q \frac{di_q}{dt}\right) + \omega_e L_d i_d + \omega_e \psi_f \tag{4.67}$$

The term in the bracket of Eqs. (4.66) and (4.67) is utilized as the equation of state involving the current and voltage in the loop of d- and q-axes, but the else terms are used as components of compensation. Consequently, it is achievable to decouple the individual loops of regulation. The vector control technique (VCT) for the every rectifier is illustrated in Fig. 4.7 by employing dual loop, which is closed for regulation. In the interior quicker loop, the current regulators are utilized to control currents of d-axis and q-axis of stator to follow the reference, but a velocity regulator is utilized in the exterior loop to control the velocity of the PMSG to track the directive value ω_{t-opt}.

Constitution of current regulators

Eqs. (4.66) and (4.67) are employed to find the function of transfer for the $d-q$-axis. Then:

$$\frac{I_d(p)}{V_{md}(p)} = \frac{1}{R_g + pL_s} \tag{4.68}$$

$$\frac{I_q(p)}{V_{mq}(p)} = \frac{1}{R_g + pL_s} \tag{4.69}$$

Consequently, the structure of the d and q interior current loops of the generator side regulator are illustrated in Fig. 4.8 (Errami, Obbadi, & Sahnoun, 2020). If we consider the loop of d-axis and using a PI corrector, then $C_{\mathrm{PMSG}}(p)$ is:

$$C_{\mathrm{PMSG}}(p) = k_p + \frac{k_i}{p} \tag{4.70}$$

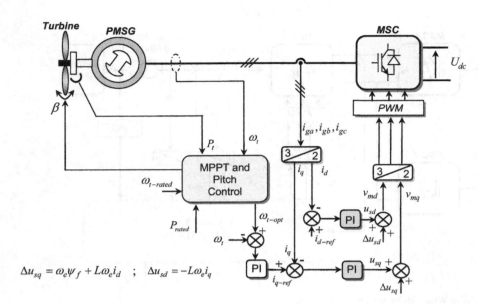

FIGURE 4.7 Structure of vector control approach for each generator converter.

$$\Delta u_{sq} = \omega_e \psi_f + L\omega_e i_d \quad ; \quad \Delta u_{sd} = -L\omega_e i_q$$

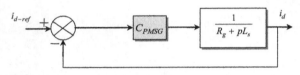

(A : Structure of the d interior current loops of the generator side regulator)

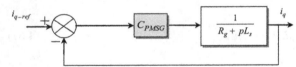

(B: Structure of the q interior current loops of the generator side regulator)

FIGURE 4.8 Current inner loop of every generator side regulator in q-axis and d-axis.

$$\text{FTOL}(p)_d = \left(k_p + \frac{k_i}{p}\right)\left(\frac{1}{R_g + pL_s}\right) \tag{4.71}$$

where k_p is the proportional coefficient and k_i is the integral coefficient. In addition, the function of transfer for open-loop (FTOL) is:

To attain the best poursuing performance of the interior loop for the current, the zero point of the PI corrector needs to compensate the pole point of $\text{FTOL}(p)_d$. Thus

$$\frac{k_p}{k_i} = \tau_d = \frac{L_s}{R_g} \tag{4.72}$$

Combining Eq. (4.71) with Eq. (4.72), the FTOL is:

$$\text{FTOL}(p)_d = \frac{k_i}{pR_g} \tag{4.73}$$

Finally, the function of transfer for close loop (FTCL) is:

$$\text{FTCL}(p)_d = \frac{1}{1 + \frac{R_g}{k_i}p} \tag{4.74}$$

k_p and k_f can be designed by:

$$k_i = \frac{R_g}{\tau_{bf}} = \frac{30R_g^2}{L_s} \tag{4.75}$$

$$k_p = k_i\tau_d$$

Velocity regulation

Eq. (4.10) can be written as:

$$\omega_t(p) = \frac{1}{Jp + F_v}(T_{ge}(p) - T_t(p)) \tag{4.76}$$

The structure of the velocity external loop utilized to adjust the velocity of every turbine and to achieve the MPPT is depicted in Fig. 4.9. So, the FTOL of velocity is:

$$\text{FTOL}(p)_{\omega_t} = \left(k_{p\omega} + \frac{k_{i\omega}}{p}\right)\left(\frac{1}{F_v + Jp}\right) \tag{4.77}$$

To obtain the best poursuing performance of the velocity loop, the nil point of the PI controller requires the compensation of the pole point for the $\text{FTOL}(p)_{\omega_t}$ (Yan, Lin, Feng, & Zhu, 2014). Therefore the FTCL of velocity is:

$$\text{FTCL}(p)_{\omega_t} = \frac{\omega_t}{\omega_{t-opt}} = \frac{pk_{p\omega} + k_{i\omega}}{Jp^2 + (k_{p\omega} + F_v)p + k_{i\omega}} \tag{4.78}$$

While the typical equation of the second-order system is:

$$\frac{1}{\omega_n^2}p^2 + \left(\frac{2\xi}{\omega_n}\right)p + 1 \tag{4.79}$$

where ω_n is the corner pulse and ξ is the damping coefficient of second-order system.

By comparing Eq. (4.79) and the denominator of Eq. (4.78), the velocity regulator gains can be calculated by:

$$\frac{J}{k_{i\omega}} = \frac{1}{\omega_n^2} \tag{4.80}$$

$$\frac{F_v + k_{p\omega}}{k_{i\omega}} = \frac{2\xi}{\omega_n} \tag{4.81}$$

If ω_n and ξ are chosen, then:

$$k_{i\omega} = J\omega_n^2 \tag{4.82}$$

$$k_{p\omega} = \frac{2\xi k_{i\omega}}{\omega_n} - F_v \tag{4.83}$$

4.4.3.2 Control technique for the inverter

The structure of the inverter on the basis of the VCT is illustrated in Fig. 4.10. The control technique for inverter is developed based on the decoupled control approach. It consists of two control loops. Loop of q-axis is employed for the regulation of the reactive power while d-axis loop is utilized to adjust the dc-link voltage. The voltage of dc link is

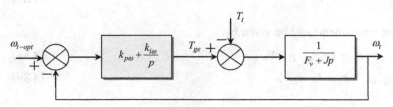

FIGURE 4.9 Structure of velocity external loop for every PMSG side regulator.

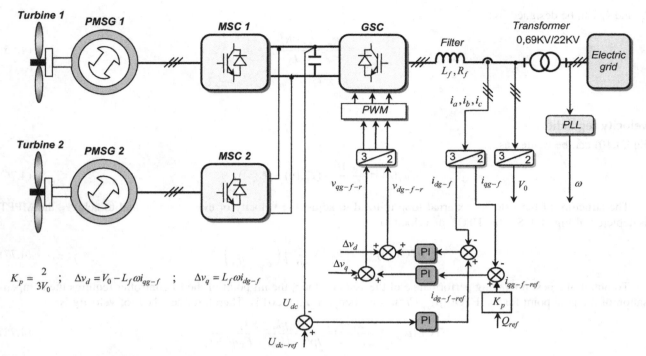

FIGURE 4.10 Schematic representation of vector control technique for inverter.

measured and evaluated against the voltage of the reference ($U_{dc-ref} = 1600\ V$) at which the voltage of dc link has to be preserved constant. Indeed, the inside loops of control adjust d-axis current and q-axis current. The reactive power is regulated to a value of zero to achieve UPF.

Reactive and active power regulation

Based on Eqs. (4.42) and (4.43), reactive and active power regulation can be realized by regulating quadrature and direct components of current electrical grid (Errami, Ouassaid, et al., 2012). Moreover, $i_{qg-f-ref}$ is created with the reactive power Q_{ref} conforming to Eq. (4.43) and it fixed to zero to accomplish the control of UPF, whereas $i_{dg-f-ref}$ is determined with controller of voltage of dc bus to adjust the active power supplied to the power grid. According to Eqs. (4.40) and (4.41)

$$L_f \frac{di_{dg-f}}{dt} + R_f i_{dg-f} = e_d - V_0 + \omega L_f i_{qg-f} \tag{4.84}$$

$$L_f \frac{di_{qg-f}}{dt} + R_f i_{qg-f} = e_q - \omega L_f i_{dg-f} \tag{4.85}$$

To compensate the cross-coupling consequence owing to the filter of the output in the rotary synchronously reference frame, the dissociating voltages are added to the outputs of current controllers. Consequently, we utilize: e'_d and e'_q as:

$$L_f \frac{di_{dg-f}}{dt} + R_f i_{dg-f} = e'_d \tag{4.86}$$

$$L_f \frac{di_{qg-f}}{dt} + R_f i_{qg-f} = e'_q \tag{4.87}$$

After that, the inverter q-axis and d-axis voltage components can be given by:

$$e_{d-inv} = e'_d - L_f \omega i_{qg-f} + V_0 \tag{4.88}$$

$$e_{q-inv} = e'_q + L_f \omega i_{dg-f} \tag{4.89}$$

Also, e'_d and e'_q are determined by:

$$e'_d = K_P(i_{dg-f-ref} - i_{dg-f}) + K_I \text{â} \langle (i_{dg-f-ref} - i_{dg-f}) dt \tag{4.90}$$

$$e'_q = K_P(i_{qg-f-ref} - i_{qg-f}) + K_I \text{â} \langle (i_{qg-f-ref} - i_{qg-f}) dt \tag{4.91}$$

Fig. 4.10 illustrates the control schematic representation of inverter using the previous approach. Indeed, the scheme of the d and q interior current loops of the network side regulator is illustrated in Fig. 4.11. In consideration of the d-axis loop, the regulator $C_{\text{grid}}(p)$ habitually utilizes the PI category. Owing to choosing the PI approach, $C_{\text{grid}}(p)$ can be chosen as:

$$C_{\text{grid}}(p) = k_{pf} + \frac{k_{if}}{p} \tag{4.92}$$

where k_{pf} is the proportional coefficient and k_{if} is the integral coefficient. The FTOL exclusive of disturbance is provided by:

$$\text{FTOL}(p)_{d-\text{grid}} = \left(k_{pf} + \frac{k_{if}}{p} \right) \left(\frac{1}{R_f + pL_f} \right) \tag{4.93}$$

To achieve the best pursuing performance of the interior loop for the current, the zero point of the PI regulator needs to compensate the pole point of the $\text{FTOL}(p)_{d-\text{grid}}$. Therefore we utilize:
So, the $\text{FTOL}(p)_{d-\text{grid}}$ is:

$$\text{FTOL}(p)_{d-\text{grid}} = \frac{k_{if}}{pR_f} \tag{4.95}$$

Moreover, the $\text{FTCL}(p)_{d-f}$ is:

$$\text{FTCL}(p)_{d-f} = \frac{1}{\frac{R_f}{k_{if}}p + 1} = \frac{1}{\tau'_{bf}p + 1} \tag{4.96}$$

The time constant τ'_{bf} characterizes the first-order responses. Also, it can be adapted by adjusting k_{if} and k_{pf}. Therefore PI parameters of the interior loop can be finding as:

$$k_{if} = \frac{R_f}{\tau_{df}} = \frac{R_f^2}{L_f} \tag{4.97}$$

$$k_{pf} = k_{if}\tau_{df}$$

dc-Link control

Conforming to the abovementioned study, the dc-link voltage's scheme is shown in Fig. 4.12. The $\text{FTCL}(p)_{dc}$ is:

$$\text{FTCL}(p)_{dc} = \frac{1}{k_{ic}} \frac{k_{ic} + k_{pc}p}{1 + \frac{k_{pc}}{k_{ic}}p + \frac{2}{3}\frac{U_{dc-ref}}{V_0}\frac{C}{k_{ic}}p^2} \tag{4.98}$$

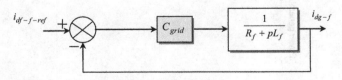

(A : d interior current loops of the grid -side regulator)

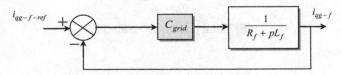

(B : q interior current loops of the grid -side regulator)

FIGURE 4.11   Configuration of the inner loops for the current

$$\frac{k_{pf}}{k_{if}} = \tau_{df} = \frac{L_f}{R_f} \tag{4.94}$$

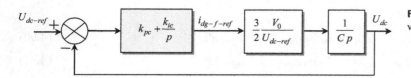

FIGURE 4.12 Schematic representation of the dc-link voltage external loop regulator.

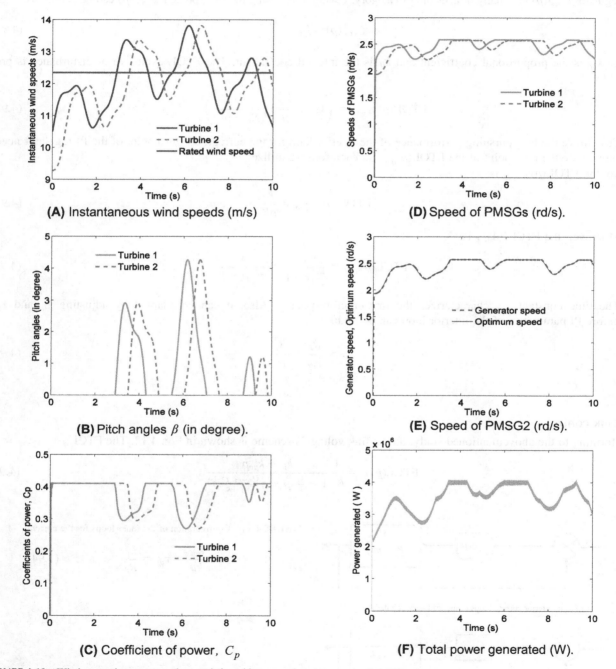

(A) Instantaneous wind speeds (m/s)

(B) Pitch angles β (in degree).

(C) Coefficient of power, C_p

(D) Speed of PMSGs (rd/s).

(E) Speed of PMSG2 (rd/s).

(F) Total power generated (W).

FIGURE 4.13 Wind generation structure characteristics with proposed strategy.

PI parameters k_{ic} and k_{pc} can be attained using comparison with standard second-order system (Errami, Obbadi, & Sahnoun, 2020). Then:

$$k_{ic} = \frac{2C}{3} \frac{U_{dc-ref}}{V_0} \omega_0'^2 \tag{4.99}$$

$$k_{pc} = \frac{4C}{3} \frac{U_{dc-ref}}{V_0} \xi' \omega_0' \tag{4.100}$$

where ω_2' is the corner pulse and ξ' is the damping coefficient of the second-order system. The scheme of the current and regulators of the dc-link voltage is depicted in Fig. 4.10.

4.5 Simulation result analysis

Performance evaluation under normal operation

To verify the effectiveness of the presented NBC, simulations were completed using MATLAB/Simulink and the constitution is illustrated in Figs. 4.2, 4.5, and 4.6. The parameters of the WFS are given in the Appendix. First, we present the control performance under changing conditions of the wind. Second, we present the behavior of the WFS during the voltage dip. Moreover, in the simulation, to control the generator side converters, i_{d-ref} is fixed at zero. To control the grid side inverter, Q_{ref} is fixed at zero. In addition, the reference of the dc-link voltage is $U_{dc-ref} = 1600$ V. The frequency of electrical grid is 50 Hz. The rated velocity of the wind is $v_n = 12.4$ m s^{-1}. Fig. 4.13 depicts the wind velocity for the two WTs, the pitch angle for every turbine, the coefficients of power, the power generated by the system, and the rotational velocity for every turbine. As can be seen, if the wind velocity is below their nominal values v_n, the rotor velocities of the generators change conforming to the velocity of the wind variation directly so as to ensure MPPT. Also, the coefficients of power are sustained at their maximal C_{p-max}, by NBC of PMSG-side converters, in spite of wind velocity fluctuations. The pitch angles are inactive and the angles β are remained constant at zero degree. The evolution of the rotating speed of turbine 2 and the reference ω_{t-opt} is illustrated in Fig. 4.13E. The NBC can make the WT rotor velocity follow the reference of speed in varying velocity of wind. So, the value of ω_t is adjusted to attain an optimal value for MPPT operation. However, if the speed of the wind surpasses the rated value, the pitch angles are activated for restricting the aerodynamic power captured by the turbines and they limit the rotational speeds at their maximum value. Also, the turbine works at lower effectiveness because Cp is lower than C_{p-max}. As a result, the WFS power is restricted at the rated value (Fig. 4.13F). The total extracted power is depicted in Fig. 4.13F. So, the extracted power varies conforming to the velocities of turbines and the pitch angles. Fig. 4.14 presents the voltage of the dc link. U_{dc} is very close to $U_{dc-ref} = 1600$ V. The overlarged view related to the three-phase voltage and current of electrical network is given in Fig. 4.15. So, the synchronization of the output line current and voltage is confirmed by the waveforms of figure. UPF of WGS is attained relatively and it depends not only on the deviation of the speed of the wind but only on the reactive power reference (Q_{ref}).

Performance of the control systems with electrical network faults

To assess the control capacity and performance of the presented NBC method for WGS, the configuration has been simulated with short-circuit fault. Fig. 4.2 illustrates the simulation configuration of the system with the electrical grid. A fault is likely to observe at 0.6 s and endure for 200 ms. Figs. 4.16, 4.17, 4.18 evaluate the responses of system for

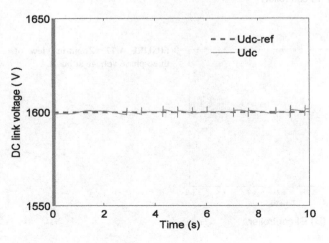

FIGURE 4.14 Voltage of dc link (V).

the presented NBC and classic VC, once the fault incident is considered to occur at bus 4 of load 3. Figs. 4.16 and 4.17 illustrate the diagrams and a closer examination of the three-phase voltage at bus 4 during the default through and once the fault. From these results, it is extremely apparent to see that the three-phase voltage variations were more rigorously distorted with the PI controllers. In opposition, by NBC, the oscillations' amplitude is reduced and the voltage of the electrical network can be recuperated quickly after the fault disappearance. So, the presented NBC allows reduced overshoot and faster response. Fig. 4.18 depicts the distinction in performance of the frequency between the two techniques of control. It is very clear to see that a highest frequency fluctuation is diminished from 2 to <0.21 Hz by NBC. In addition, the excursion of the frequency is detained earlier in the NBC method. Therefore the variations in frequency are extensively reduced for the dip of voltage, and the system with NBC is little sensitive to faults in electric grid.

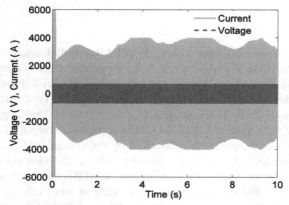

FIGURE 4.15 Three-phase voltage and current of electrical network.

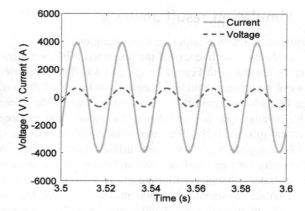

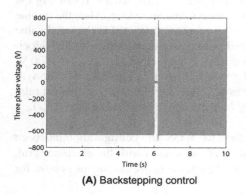

(A) Backstepping control

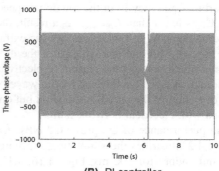

(B) PI controller

FIGURE 4.16 Three phase voltage at bus 4.

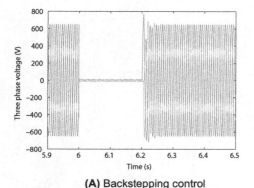

(A) Backstepping control

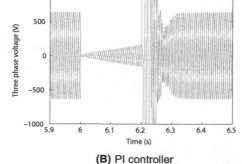

(B) PI controller

FIGURE 4.17 Zoom-in view of three-phase voltage at bus 4.

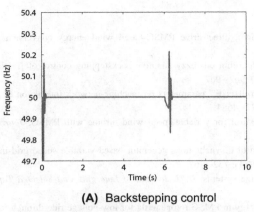

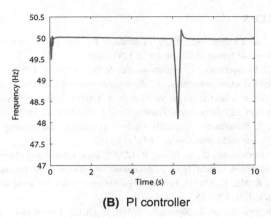

(A) Backstepping control

(B) PI controller

FIGURE 4.18 Frequency response of the WFS under network faults.

4.6   Conclusions

In this chapter, the nonlinear control strategies for the WFS-based PMSG are introduced. These methods are used to adjust the power exchange between the WFS and the electric network with MPPT approach. The presented approach is based on the NBC method. The complete derivation for the command techniques has been given and the existence conditions of the backstepping approach are established by utilizing the theory of Lyapunov stability. The efficiency of the proposed method has been assessed under both grid-normal and -fault conditions. The performance of the designed strategies is highlighted by comparison with the PI controllers. The presented nonlinear command permits minor overshoot and it is little sensitive to grid faults.

Appendix

Tables A1 and A2.

TABLE A1 Parameters of the power synchronous generators.

Parameter	Value
P_r rated power	2 MW
ω_{t-r} rated mechanical speed	2.57 rad s$^{a^{1}}$
R_s stator resistance	0.008 Î©
L_s stator d-axis inductance	0.0003 H
ψ_f permanent magnet flux	3.86 Wb
p_n pole pairs	60

TABLE A2 Parameters of the turbine.

Parameter	Value
ρ the air density	1.08 kg m$^{a^{3}}$
A area swept by blades	4775.94 m^2
v_n base wind speed	12.4 m s$^{a^{1}}$

References

Alizadeh, O., & Yazdani, A. (2013). A strategy for real power control in a direct-drive PMSG-based wind energy conversion system. *IEEE Transactions on Power Delivery, 28*(3), 1297−1305.

Aounallah, T., Essounbouli, N., Hamzaoui, A., & Bouchafaa, F. (2018). Algorithm on fuzzy adaptive backstepping control of fractional order for doubly-fed induction generators. *IET Renewable Power Generation, 12*(8), 962−967.

Arslan, H., JuliÃ¡n AristizÃ¡bal, A., & Arshad, A. (2019). Effect of PI controllersâ€™ parameters on machine-network interaction of gridconnected PMSG system. *IET Generation, Transmission & Distribution, 13*(20), 4677−4684.

Bakhtiari, F., & Nazarzadeh, J. (2020). Optimal estimation and tracking control for variable-speed wind turbine with PMSG. *Journal of Modern Power Systems and Clean Energy, 8*(1), 159−167.

Basak, R., Bhuvaneswari, G., & Rahul, R. (2020). Low voltage ride-through of a synchronous generator based variable speed grid-interfaced wind energy conversion system. *IEEE Transactions on Industry Applications, 56*(1), 752−762.

Blaabjerg, F., & Ma, K. (2013). Future on power electronics for wind turbine systems. *IEEE Journal of Emerging and Selected Topics in Power Electronics, 1*(3), 139−152.

Calle, S., Busquets-Monge, S., Kouro, S., & Wu, B. (2013). Use of stored energy in PMSG rotor inertia for low-voltage ride-through in back-to-back NPC converter-based wind power systems. *IEEE Transactions on Industrial Electronics, 60*(5), 1787−1796.

Chen, H., David, N., Dionysios, C., & Aliprantis. (2013). Analysis of permanent-magnet synchronous generator with vienna rectifier for wind energy conversion system. *IEEE Transactions on Sustainable Energy, 4*(1), 154−163.

Chen, J., Yang, B., Duan, W., Shu, H., An, N., Chen, L., & Yu, T. (2019). Adaptive pitch control of variable-pitch PMSG based wind turbine. *Applied Science., 9*(2019), 1−20.

Chinchilla, M., Arnaltes, S., & Burgos, J. C. (2006). Control of permanent-magnet generators applied to variable-speed wind-energy systems connected to the grid. *IEEE Transactions on Energy Conversion, 21*(1), 130−135.

Elkhatib, K., Aitouche, A., Ghorbani, R., & Bayart, M. (2014). Fuzzy scheduler fault-tolerant control for wind energy conversion systems. *IEEE Transactions on Control Systems Technology, 22*(1), 119−131.

Errami, Y., Ouassaid, M., & Maaroufi, M. (2012). Control of grid connected PMSG based variable speed wind energy conversion system. *International Review on Modelling and Simulations, 5*(2), 655−664, ISSN:1974−9821.

Errami, Y., Benchagra, M., Hillal, M., Ouassaid, M., & Maaroufi, M. (2012). MPPT strategy and direct torque control of PMSG used for variable speed wind energy conversion system. *International Review on Modelling and Simulations, 5*(2), 887−898, ISSN: 1974-9821.

Errami, Y., Hillal, M., Benchagra, M., Ouassaid, M., & Maaroufi, M. (2012). Nonlinear control of MPPT and grid connected for variable speed wind energy conversion system based on the PMSG. *Journal of Theoretical and Applied Information Technology, 39*(2), 204−217.

Errami, Y., Ouassaid, M., & Maaroufi, M. (2015). Modelling and optimal power control for permanent magnet synchronous generator wind turbine system connected to utility grid with fault conditions. *World Journal of Modelling and Simulation, 11*(2), 123−135, In this issue.

Errami, Y., Obbadi, A., Ouassaid, M., & Maaroufi, M. (2018). Direct torque control strategy applied to the grid connected wind farm based on the PMSG and controlled with variable structure approach. *International Journal of Power and Energy Conversion, 9*(1), 58−88.

Errami, Y., Obbadi, A., Sahnoun, S., Ouassaid, M., & Maaroufi, M. (2020). Variable structure power control under different operating conditions of PM synchronous generator wind farm connected to electrical network. *International Journal of Power and Energy Conversion, 11*(1), 22−59.

Errami, Y., Obbadi, A., & Sahnoun, S. (2020). Control of PMSG wind electrical system in network context and during the MPP tracking process. *International Journal of Systems, Control and Communications, 11*(2), 200−225.

Errouissi, R., Al-Durra, A., & Debouza, M. (2018). A novel design of pi current controller for PMSG-based wind turbine considering transient performance specifications and control saturation. *IEEE Transactions on Industrial Electronics, 65*(11), 8624−8634.

Fernando, J. L., Kenne, G., & Lagarrigue, L. F. (2016). A novel online training neural network-based algorithm for wind speed estimation and adaptive control of PMSG wind turbine system for maximum power extraction. *Renewable Energy, 86*(2016), 38−48.

Foo, G., & Rahman, M. F. (2009). Direct torque and flux control of an IPM synchronous motor drive using a backstepping approach. *IET Electric Power Applications, 3*(5), 413−421.

Freire, N. M. A., & Cardoso, A. J. M. (2014). A fault-tolerant direct controlled PMSG drive for wind energy conversion systems. *IEEE Transactions on Industrial Electronics, 61*(2), 821−834.

Gul, W., Gao, Q., & Lenwari, W. (2020). Optimal design of a 5 MW double-stator single-rotor PMSG for offshore direct drive wind turbines. *IEEE Transactions on Industry Applications, 56*(1), 216−225.

He, L., Li, Y., & Harley, R. (2013). Adaptive multi-mode power control of a direct-drive PM wind generation system in a microgrid. *IEEE Journal of Emerging and Selected Topics in Power Electronics, 1*(4), 217−225.

Karthikeyan, A., Kummara, S., Nagamani, C., & Saravana Ilango, G. (2011). Power control of grid connected doubly fed induction generator using adaptive backstepping approach. In *IEEE international conference on environment and electrical engineering (EEEIC)* (pp. 1−4).

Khan, Rabbi, N., S.F., Hinchey, M.J., & Rahman, M.A. (2013). Adaptive backstepping based maximum power point tracking control for a variable speed marine current energy conversion system. In *IEEE Canadian conference of electrical and computer engineering (CCECE)* (pp. 1−5).

Kim, K., Van, T., Lee, D., Song, S.-H., & Kim, E. ,-H. (2013). Maximum output power tracking control in variable-speed wind turbine systems considering rotor inertial power. *IEEE Transactions on Industrial Electronics, 60*(8), 3207−3217.

Kortabarria, I., Andreu, J., MartÃnez de AlegrÃa, I., JimÃ©nez, J., GÃ¡rate, J., & Robles, E. (2014). A novel adaptative maximum power point tracking algorithm for small wind turbines. *Renewable Energy, 63*(2014), 785−796.

Li, S., Haskew, T., Swatloski, R., & Gathings, W. (2012). Optimal and direct-current vector control of direct-driven PMSG wind turbines. *IEEE Transactions on Power Electronics*, *27*(5), 2325–2337.

Li, S., Timothy, A., Haskew., Richard, P., Swatloski., & Gathings, W. (2012). Optimal and direct-current vector control of direct-driven PMSG wind turbines. *IEEE Transactions on Power Electronics*, *27*(5), 2325–2337.

Liu, G., Hu, J., Tian, G., Xu, L., & Wang, S. (2019). Study on high voltage ride through control strategy of PMSG-based wind turbine generation system with SCESU. *Journal of Engineering*, *2019*(17), 4257–4260.

Matraji, I., Al-Durra, A., & Errouissi, R. (2018). Design and experimental validation of enhanced adaptive second-order SMC for PMSG-based wind energy conversion system. *International Journal of Electrical Power & Energy Systems*, *103*(2018), 21–30.

Melo, D., & Chien, L. (2014). Synergistic control between hydrogen storage-system and offshore wind farm for grid operation. *IEEE Transactions on Sustainable Energy*, *5*(1), 18–27.

Nguyen, T., & Lee, D. (2013). Advanced fault ride-through technique for PMSG wind turbine systems using line-side converter as STATCOM. *IEEE Transactions on Industrial Electronics*, *60*(7), 2842–2850.

Rajaei, A., Mohamadian, M., & Yazdian Varjani, A. (2013). Vienna-rectifier-based direct torque control of PMSG for wind energy application. *IEEE Transactions on Industrial Electronics*, *60*(7), 2919–2929.

Roberto, F., Solsona, J., & Busada, C. (2016). Nonlinear observer-based control for PMSG wind turbine. *Energy*, *113*, 248–257.

Shariatpanah, H., Fadaeinedjad, R., & Rashidinejad, M. (2013). A new model for PMSG-based wind turbine with yaw control. *IEEE Transactions on Energy Conversion*, *28*(4), 929–937.

Sizhan, Z., Jinjun, L., Linyuan, Z., & Yangque, Z. (2013) Improved DC-link voltage control of PMSG WECS based on feedback linearization under grid faults. In *IEEE applied power electronics conference and exposition (APEC)* (pp. 2895–2899).

Slotine, J. E., & Li, W. (1991). *Applied nonlinear control*. Englewood Cliffs, NJ: Prentice Hall.

Tong, S., & Li, Y. (2012). Adaptive fuzzy output feedback tracking backstepping control of strict-feedback nonlinear systems with unknown dead zones. *IEEE Transactions on Fuzzy Systems*, *20*(1), 168–180.

Wang, G., Wai, R., & Liao, Y. (2013). Design of backstepping power control for grid-side converter of voltage source converter-based high-voltage dc wind power generation system. *IET Renewable Power Generation*, *7*(2), 118–133.

Yang, B., Yu, T., Shu, H., Han, Y., Cao, P., & Jiang, L. (2019). Adaptive fractional order PID control of PMSG based wind energy conversion system for MPPT using linear observers. *International Transactions on Electrical Energy Systems*, *29*(1), 1–18.

Yang, W., Zhang, A., Zhang, H., Wang, J., & Li, G. (2018). Adaptive integral sliding mode direct power control for VSC MVDC system converter stations. *International Transactions on Electrical Energy System*, *28*(4), 1–15.

Yan, J., Lin, H., Feng, Y., & Zhu, Z. Q. (2014). Control of a grid-connected direct-drive wind energy conversion system. *Renewable Energy*, *66* (2014), 371–380.

Zhang, Z., Zhao, Y., Qiao, W., & Qu, L. (2014). A space-vector modulated sensorless direct-torque control for direct-drive PMSG wind turbines. *IEEE Transactions on Industry Applications*, *50*(4), 2331–2341.

Zhang, Z., Zhao, Y., Qiao, W., & Qu, L. (2015). A discrete-time direct-torque control for direct-drive PMSG-based wind energy conversion systems. *IEEE Transactions on Industry Applications*, *51*(4), 3504–3514.

Further reading

Soliman., Mahmoud, A., Hasanien., Hany, M., Azazi., Haitham, Z., ... Mahmoud, S. A. (2019). An adaptive fuzzy logic control strategy for performance enhancement of a grid-connected PMSG-based wind turbine. *IEEE Transactions on Industrial Informatics*, *15*(6), 3163–3173.

Xiong, L., Wang, J., Mi, X., & Waseem Khan, M. (2018). Fractional order sliding mode based direct power control of grid-connected DFIG. *IEEE Transactions on Power Systems*, *33*(3), 3087–3096.

Wang, S., Li, S., Gu, R., Ma, L., & Li, M. (2019). Adaptive sliding mode based active disturbance rejection control method for a direct-driven wind power conversion system. *The Journal of Engineering*, *2019*(22), 8365–8369.

Chapter 5

Model predictive control-based energy management strategy for grid-connected residential photovoltaic–wind–battery system

Meryeme Azaroual, Mohammed Ouassaid and Mohamed Maaroufi

Engineering for Smart and Sustainable Systems Research Center, Mohammadia School of Engineers, Mohammed V University, Rabat, Morocco

5.1 Introduction

5.1.1 Motivations

In the last decade the interest of the countries worldwide turned to the use of renewable energy sources (RES) to generate electricity. This is thanks to the advantages of sustainable energy sources, mainly solar and wind energies, such as low cost and environmental protection. In contrast to the conventional sources, RES can reduce the world's reliance on nonrenewable fuels by using hybrid power systems (Askarzadeh, 2016; Luna, Diaz, Nelson, & Graells, 2016; Ma, Wu, & Hao, 2018; Negi & Mathew, 2014; Pachori & Suhane, 2014; Paul, Hossain, Ghosh, Mandal, & Kamalasadan, 2018; Tang, Li, & Lai, 2018; Zhang, Fu, & Zhu, 2018). Consequently, they can be considered the most convenient source that provides electricity to the power grid system. Because of the discontinuous nature of renewable energy resources, they cannot generate uninterrupted power to satisfy the load demand; therefore the integration of the battery storage system is required to match appropriately the shortage. Many strategies and advanced methods of optimization and control are applied to cover the random nature of RES, satisfy the energy demand, and improve energy efficiency. Therefore the incorporation of RES in the grid-connected system or isolated system has gained an increased interest (Aktas, Erhan, & Özdemir, 2018; Hossain, Pota, Squartini, & Abdou, 2019; Mbungu, Naidoo, & Bansal, 2016; Momoh, 2012) by using many approaches to schedule optimally the energy flow of the hybrid system. Besides, many hybrid systems are implemented and operated on batteries in different applications (industrial, residential, and commercial). However, the current challenge is to improve the optimal energy management by using the appropriate method. In this vein, numerous research efforts have been done on designing energy controllers for grid-connected systems to satisfy the load demand and reduce the electricity cost. The most commonly used approaches to solve the optimization problems are the deterministic methods such as linear programming (LP), quadratic programming, dynamic programming, and the stochastic methods as the genetic algorithm (GA), particle swarm optimization (PSO). Nevertheless, those methods cannot handle the unpredicted time delays and control using only the determined time horizon. Therefore the model predictive control (MPC) approach is the appropriate technique to tackle different problems that the previous strategies cannot handle such as the unexpected deviation of the renewable output power and/or the load demand, weather conditions, or any other external disturbances. Generally, the MPC aims to minimize the penalty function under different constraints with an optimization of the comfort level, energy, or cost.

Renewable Energy Systems. DOI: https://doi.org/10.1016/B978-0-12-820004-9.00014-0

5.1.2 Contributions

Most of the studies on grid-connected systems have designed different energy management strategies on maximizing different energy sources such as photovoltaic (PV), wind, and battery, but there is a gap in reducing energy cost and the interaction between the power generation, the load demand, and the grid. Likewise, the profit from managing the excess energy sold to the main grid has been discussed. The optimal control considered in these studies is an open-loop strategy due to the absence of feedback that makes the system susceptible to disturbances in RES and load. Therefore the major addition of this study is to apply an MPC strategy and to compare it with an open-loop control to validate the developed method. Meanwhile, other objectives applied to the model are to encourage the employment of RES and to benefit from the feed-in tariffs (FITs). As such the battery degradation cost can be higher, the MPC should allow the battery to buy energy from the grid during low price hours. Therefore the system can prevent simultaneously the decreasing lifetime of the battery due to the sudden charging/discharging decisions caused by the successive fluctuations in renewable power generation and demand. The closed-loop control will be able to adjust the RE and the load when disturbances appear, which makes the proposed approach useful for different real-time applications.

Accordingly, this chapter proposes a novel optimal energy management strategy (EMS) to assess the potential of energy and cost-savings under a residential attractive FIT. The optimization aims to minimize the energy consumption from the utility grid under time-of-use (TOU) tariff and to profit from the selling price during peak price period. A typical Moroccan household contains PV, wind, and battery systems, and the main grid is considered as a case study. With this in mind, the contributions of this work can be summarized as follows:

- The developed strategy allows for residential customers to reduce the energy purchased from the national grid during high price periods using the TOU tariff while coordinating between the production and consumption of their household demand.
- The model permits to benefit from the excess energy generations of PV, wind turbine (WT), and battery storage system to be sold to the grid. Therefore an attractive PV and wind FITs are applied to earn more revenue.
- An MPC strategy is performed as a closed-loop control to predict the future conduct of the system and to adjust PV, WT, and load disturbances. A comparison with open-loop control using LP algorithm is performed to demonstrate the robustness of the proposed method in terms of cost-savings. Besides, the proposed strategy can penalize the usage of the battery storage system by analyzing the state of charge (SOC) of the battery bank.

Then, the simulation results reveal that the developed energy management method schedules the energy flows of the power sources optimally and achieves the maximum profit.

5.1.3 Organization of the chapter

The remainder of this chapter is organized as follows: Section 5.2 presents a review of the related works. In Section 5.3 the configuration of the hybrid power system and the mathematical modeling of each component are described. The detailed structure of the proposed EMS and the model of the open-loop strategy based on LP are presented. The developed MPC approach and the mathematical formulation of the strategy are stated in Section 5.4. Finally, simulation, discussion results, and suggestions for future work are presented in the last section.

5.2 Related works

In Hossain, Pota, Squartini, Abdou, et al. (2019), a modified adaptive accelerated PSO (MAAPSO) is applied as a control strategy for grid-connected PV−wind-distributed generation. Then, a comparison between the proposed method for control and the proportional integral derivatives controller is performed to demonstrate that MAAPSO has better performance under different scenarios. The study is addressed to make the system able to handle some issues concerning the power grid in terms of flexibility. In Bendary and Ismail (2019), the authors propose a controlled energy managing system for the charging process of the battery system to be well managed using the GA approach. The studied system is an off-grid hybrid system with a PV generator, WT, fuel cell, and batteries to supply the load demand and achieve the charging balance of the battery bank. The simulation results in MATLAB software prove the efficiency of the proposed technique. Maleki, Rosen, and Pourfayaz (2017) design a grid-connected PV, wind generator, fuel cell, and solar-thermal collector system to supply a household. A modified heuristic approach using the PSO is developed to get the optimal operation of the hybrid system optimization. A comparison with another optimization method (GA) is performed in terms of efficiency and simulation time. As an outcome, the simulation results show that using the designed hybrid system, the daily cost is reduced in comparison with other cases using heat and power systems. Furthermore, the results show that the proposed PSO approach has the best results than the GA. The objective of the work in Chakir et al. (2020) is to manage the energy flows by choosing a suitable battery to prolong the battery bank life span. Indeed, the results of simulation using MATLAB/Simulink show that the load demand is satisfied in different cases considering the variation of the weather and also the behavior of the residents. Abdelshafy, Jurasz, Hassan, and Mohamed (2020) present novel energy management for a grid-connected double storage system (DSS) composed of PV, WT, battery bank, and pumped-storage. The excess of PV and wind output power is absorbed by the batteries to cover the load's requirements. Thus the developed strategy succeeds to minimize the electricity cost and CO_2 emissions and the objective function is solved using the nondominated sorting GA. Furthermore, a comparative study of the studied model with the DSS and a PV/wind/pumped-storage hydropower system is done to demonstrate the robustness of the proposed methodology. The electricity cost is reduced by 22.2%; therefore the energy imported from the main grid is minimized.

In Seal, Boulet, and Dehkordi (2019), a new EMS-based MPC is performed for comfort in the residential sector. The system consists of a PV system, battery storage, and a heat pump. The designed model is responsible for controlling the energy flow between the PV system, battery storage, and the main grid using the TOU program to specify the energy cost of the system while achieving the predefined comfort. Then, a comparison with a rule-based strategy of management is established to prove the effectiveness of the MPC approach and its capacity to achieve the desired objective. In Yan et al. (2019), an EMS is implemented using the MPC for a microgrid containing a PV system, a battery bank, water tank, and other electrical equipment. The study aims to reduce the energy cost and have better self-consumed using load shifting. The results of the penalty function prove the validity of the proposed method. In Lupangu, Justo, and Bansal (2020), an optimal control applying the MPC method is designed for a hybrid PV−battery storage system. The objective function of the optimization problem aims to maximize the benefits from different scenarios of electricity markets (variable electricity tariff). The proposed strategy can predict the SOC of the battery and electricity price while compensating the reactive power over the control scheme. Then, a simulation is realized in MATLAB/Simulink under different cases to validate the efficiency of the proposed schedule management. Besides, it has been demonstrated that the designed control ensures the best performances. In Yilmaz, Sezgin, and Gol (2020), an optimal microgrid central controller is designed to control the real-time operation of a microgrid with real data of a PV system and battery bank. The study is based on an MPC that provides the optimal control under the operational and economic constraints. The aim is to reduce the energy cost of the microgrid using the mixed-integer LP (MILP) algorithm to solve the optimization problem while taking into account the aging effect of the battery storage. In Marín et al. (2019), the authors present a microgrid based on the MPC for an optimal management system. The strategy is used to minimize the energy cost imported from the grid and maximize the self-consumption of the RE resources, while defining the power reference followed by the lower level real-time controller. Due to the intermittent nature of the PV system, the required demand power is compensated by the battery power or grid power. Moreover, the results of the simulation show that when comparing with a basic EMS, the proposed method achieves the minimum energy costs and reduces the peak load. An EMS of a grid-tied system composed of PV panels, batteries (Li-ion), fuel cell, hydrogen tank, and electrolyzer based on the MPC approach is designed in Aguilera Gonzalez et al. (2019). The methodology is able to predict the future output behavior of the battery system while satisfying the load demand by the hybrid system. Furthermore, the penalty function minimizes the electricity consumed from the utility grid. In Acevedo-Arenas et al. (2019), an MPC strategy using the evolutionary algorithms is implemented for the optimal schedule of renewable energy production and demand response. The results reveal that the proposed algorithm can minimize the energy cost of the demand and maximize renewable energy for different seasonal data.

In Bonthu, Aguilera, Pham, Phung, and Ha (2019), MPC control is investigated for a grid-tied PV−battery storage system to optimize the energy cost of a residential commercial building. This study presents an optimal strategy using an MILP algorithm to lower the electricity bill by taking into consideration the dynamic efficiency of power converters. The results depict that the developed strategy gives benefits concerning cost-saving with minimum loss in power conversion. The system designed in Trifkovic, Sheikhzadeh, and Nigim (2013) is an isolated hybrid PV−wind−electrolyzer−fuel hybrid system. The implemented strategy is the MPC using different control systems that can guarantee the balance between renewable energy and storage system. The component of the model is designed in MATLAB/Simulink environment with two stages of the control. Simulation results are discussed to illustrate the efficacy of the proposed power management strategy. In Lagorse, Simões, Marcelo, and Miraoui (2009), a supervisory control system using MPC was proposed for the optimal operation and management of small-scale PV−wind hybrid system. The control focused on controlling the energy flow of the hybrid system based on multiagent system techniques. Another attempt of using MPC control is studied in Khan, Pasupuleti, Al-Fattah, and Tahmasebi (2019). The hybrid system consists of a solar PV and battery storage system. The purpose of the study is to realize an optimal generation from the sources while providing controls for the battery charging for a smooth PV output. The obtained results prove the robustness of the system to handle the variation in RES and load demand. In Garcia-Torres and Bordons (2015), the authors develop an optimal control based on MPC for a microgrid integrating RES and energy storage system. Because of the problem of the time response and degradation of the battery bank system, the usage of hydrogen in the storage system is suggested as a solution for the mentioned issues. As a result, the proposed MPC strategy, which is performed under different constraints, can minimize the degradation of the storage system and maximize the benefits of the saving cost of the microgrid. In the same vein, Dongol, Feldmann, and Schmidt (2018) studied an MPC model for a residential grid-connected PV system to have an economic benefit from the battery storage system. The purpose is to reduce the electricity grid imports as well as to shave the peaks of the household demand. A mathematical model of the MPC approach is presented with an analysis of the simulation results to show the ability of the MPC to follow the variation of the weather forecast and operate the system to handle this deviation by managing the battery storage.

However, among the aforementioned works, the energy management strategies are implemented without as an open-loop control. Thus they cannot efficiently handle the control when load demand and source disturbances occur, which can deteriorate the performance of the hybrid system. To tackle this issue in grid-connected hybrid system, a closed-loop control technique-based MPC method is adopted in this work. Indeed, the proposed strategy ensures better performance in terms of robustness and convergence. Furthermore, the proposed approach gives feedback to the optimization model to predict the future response of the system and the solution is updated appropriately.

5.3 The architecture of original grid-tied PV−WT−battery and optimal control strategy

5.3.1 Subsystems

The system considered in this chapter is composed of a PV system, a WT system, a battery storage, and the grid. Fig. 5.1 presents the structure of the grid-connected residential hybrid system. P_{PV}, P_{WT}, P_B, P_G, and P_{Load} are the output power of the PV, wind generators, the battery bank (charge and discharge), the grid, and the load, respectively.

5.3.2 PV generator

Solar panels convert sunlight into electricity based on the PV effect. To produce the maximum energy from solar panels, they function using the maximum power point tracking method. The hourly power output P_{PV} generated from the PV array of a given area is formulated as follows (Hossain, Pota, Squartini, Zaman, & Muttaqi, 2019):

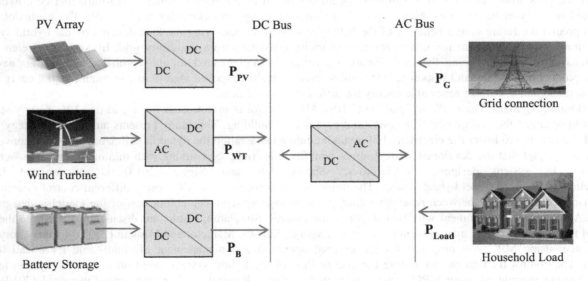

FIGURE 5.1 Residential grid-tied power system.

$$P_{\text{PV}} = \eta_{\text{PV}} \times \text{SI} \times A_{\text{PV}}(1 - 0.005(t_0 - 25)) \tag{5.1}$$

where η_{PV}, SI, and A_{PV} are the efficiency of solar generation, solar irradiation (kWh m^{-2}), and PV array area (m^2), respectively.

5.3.3 Wind generator

The hourly power output of a WT depends on the wind speed. Indeed, WT power is a function of the cubic of velocity (Hossain et al., 2019). A conversion of the average hourly wind speed corresponding to the values at the hub height is required. To convert to the hub height, the applied equation is usually formulated as follows:

$$\nu = \nu_0 \times \left(\frac{h}{h_0}\right)^{\beta} \tag{5.2}$$

where ν denotes the wind speed (m s^{-1}); h and h_0 are the height and the reference height (m), respectively; ν_0 is the wind speed at h_0; and β is the power law exponent ranging.

The electric wind power produced depends on the wind speed and can be presented as follows:

$$P_{\text{EW}} = \begin{cases} 0 \\ P_r \times \dfrac{v^3 - v_c^3}{v_r^3 - v_c^3} \\ P_r \end{cases} \begin{matrix} \text{if } "v_f < v" \text{ or } "v \le v_c" \\ \text{if } "v_c \le v \le v_r" \\ \text{if } "v_r \le v \le v_f" \end{matrix} \tag{5.3}$$

Then, the equation of the total output power of WT can be expressed as follows:

$$P_{\text{WT}} = P_{\text{EW}} \times \eta_{\text{WT}} \tag{5.4}$$

where P_r and v_r refer to the rated power and rated wind speed, respectively; v, v_c, and v_f are the wind speed, the cut-in wind speed, and the cut-off wind speed, respectively; and η_{WT} is the efficiency of the wind generator.

5.3.4 Battery storage system

In this work the SOC of the battery should change dynamically, so that the battery can be charged by PV, WT, and grid or discharged to supply the household demand. Eq. (5.5) represents the dynamic SOC of the energy storage system based on the initial SOC$_{(t)}$ (Khawaja, Giaouris, Patsios, & Dahidah, 2017).

$$\text{SOC}_{(t+1)} = \text{SOC}_{(t)} + \Delta t \times n_{\text{ch}} \times \left(P_{\text{ch}(t)}\right) - \Delta t \times n_{\text{dis}} \times \left(P_{\text{dis}(t)}\right) \tag{5.5}$$

where $P_{\text{ch}}(t)$ and $P_{\text{dis}}(t)$ are the power flow of charging and discharging, respectively; n_{ch} and n_{dis} denote the charging and discharging, respectively; and Δt is the sampling time.

$$n_{\text{ch}} = \frac{\eta_{\text{ch}}}{E_{\text{nom}}} \tag{5.6}$$

$$n_{\text{dis}} = \frac{1}{\eta_{\text{dis}} \times E_{\text{nom}}} \tag{5.7}$$

where E_{nom} is the nominal energy of the battery and η_{ch} and η_{dis} are the battery charging and discharging efficiency, respectively.

5.3.5 Utility grid and electricity tariff

In this work the main grid is modeled to be able to concurrently providing and receiving power from the battery, PV, and wind generators. The electricity tariff program employed in this study is the TOU tariff in which electricity price varies during peak and off-peak periods depending on the customer usage (industry, residential, etc.) and the electricity market regulation. In Morocco, the National Office of Electricity and Drinking Water (ONEE) is considered as the main power supply utility company and applies a simple tariff (ρ_f) for the residential customer at each specified consumption range.

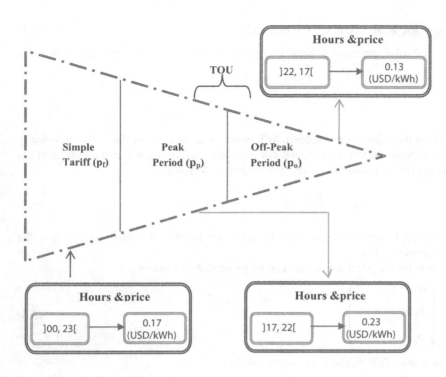

As a part of the measures implemented at the national level, aiming at strengthening energy efficiency, the ONEE has set up a new TOU tariff intended for low-voltage customers for domestic use, an industrial and agricultural driving force whose average monthly consumption exceeds 500 kWh.

Fig. 5.2 shows the prices of the different time intervals $\rho(t)$. Each interval refers to the daily electricity tariff of the imported power from the grid. There are two types of $\rho(t)$, the simple tariff (ρ_f) and the TOU tariff that contains the off-peak and peak periods as follows:

$$\rho(t) = \left\{ \begin{array}{c} \rho_f \\ \text{or} \\ \rho_p; \rho_o \end{array} \right\} \tag{5.8}$$

The currency of Morocco is the Moroccan Dirham (MAD), and it is converted to the US Dollar (USD) with 1 USD = 9.329 MAD.

5.4 Energy management strategy and the model of the open-loop control

Fig. 5.3 describes the energy management for the studied grid-tied household with a PV system, WT system, and battery storage system. The arrows show the hourly power flow direction in the hybrid system. $P_{\text{PV-B}}$, $P_{\text{PV-L}}$, and $P_{\text{PV-G}}$ represent the power flow from the PV system to charge the battery, feeding the load demand, and exporting the surplus power to the national grid under FIT, respectively. For the control variable $P_{\text{WT-B}}$, $P_{\text{WT-L}}$, and $P_{\text{WT-G}}$ are the power flow from the WT to charge the battery bank, satisfying energy demand, and exporting the excess power to the main grid under the corresponding FIT, respectively. While $P_{\text{G-L}}$ is the power flow imported from the utility grid to supply the load, the power flows $P_{\text{B-L}}$ and $P_{\text{B-G}}$ represent the discharging of the battery to meet the load demand and the surplus power sold to the grid, respectively. In addition, the inverters and chargers like the converter (DC/DC) and the bi-directional inverter (DC/AC) are required for the hybrid system to convert the generated powers.

5.4.1 Energy management strategy

The flowchart in Fig. 5.4 shows the process flow for the optimal control strategy. The PV and WT productions are compared at any instant with the load demand $P_L(t)$. If the RE production is less than load demand, the battery is discharged to meet the demand by $P_{\text{B-L}}(t)$. Furthermore, if the PV and WT productions are still insufficient to satisfy the demand

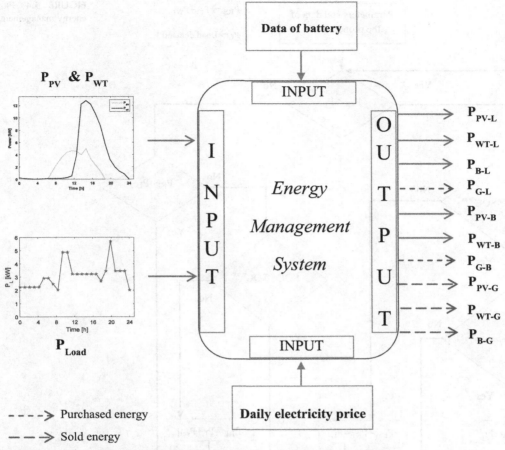

FIGURE 5.3   Configuration of the grid-connected photovoltaic—wind turbine—battery system power flows.

of the household, the energy needed is purchased from the grid (P_{buy}). On the other hand, if the PV and WT productions exceed the energy demand, the RES charge the battery ($P_{\text{PV-B}}$, $P_{\text{WT-B}}$). Thus if the battery is fully charged and the excess power is still generated, it is sold to the grid (P_{sell}). Besides, if the PV and WT systems are not available, the battery will be charged from the grid depending on the price of electricity. More details are given in Section 5.6.

5.4.2   Objective function

The objective of the control is to minimize the net electricity cost Fc of the system. Eq. (5.9) formulates the multiobjective function of the model operating over the given period. The first part of the function aims to minimize the purchased electricity from the grid for charging the battery and supplying the load. The second and the last part of the function maximizes the benefit from the excess of the PV, WT, and battery power sold to the grid, while maximizing the use of energy from renewable sources.

In this study, the daily operating cost of the RES and battery is not incorporated since it is not within the objective of the present study. It focuses principally on the optimal energy dispatching of any given system.

$$Fc = \sum_{t=1}^{N} p(t)(P_{G\text{-}B}(t) + P_{G\text{-}L}(t))\Delta t - \sum_{t=1}^{N} (C_{\text{PV}}P_{\text{PV-}G}(t) + C_{\text{WT}}P_{\text{WT-}G}(t))\Delta t - \sum_{t=1}^{N} (\rho_p P_{B\text{-}G}(t))\Delta t \qquad (5.9)$$

where $\rho(t)$ is the electricity tariff purchased from the utility grid depending on the applying program (TOU tariff, critical peak pricing, etc.), ρ_p is the peak electricity price (\$ kWh^{-1}), t is the tth sampling interval, N is the number of sampling intervals, Δt is the sampling time, and C_{PV} and C_{WT} are the cost of PV and wind FIT (\$ kWh^{-1}), respectively.

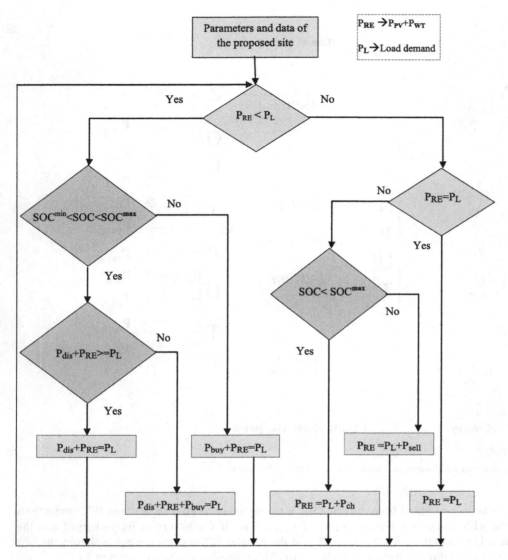

FIGURE 5.4 Flowchart of the energy management strategy.

5.4.3 Constraints and power flow limits

5.4.3.1 Power balance

Eq. (5.10) expresses the power balance where the load demand should be satisfied by all sources at any sampling time (t):

$$P_{PV-L(t)} + P_{WT-L(t)} + P_{B-L(t)} + P_{G-L(t)} = P_{L(t)} \qquad (5.10)$$

5.4.3.2 Constraints

The objective function is constrained by Eqs. (5.11) and (5.12), which the total of PV and WT power for charging the battery, supplying the demand, and feed into the main grid should be less than the generated PV and wind power.

$$P_{PV-B(t)} + P_{PV-L(t)} + P_{PV-G(t)} \leq P_{PV(t)} \qquad (5.11)$$

$$P_{WT-B(t)} + P_{WT-L(t)} + P_{WT-G(t)} \leq P_{WT(t)} \qquad (5.12)$$

TABLE 5.1 Lower and upper limits of the linear programming optimization method.

	P_{PV-B}	P_{PV-L}	P_{PV-G}	P_{WT-B}	P_{WT-L}	P_{WT-G}	P_{G-B}	P_{G-L}	P_{B-L}	P_{B-G}
lb	0	0	0	0	0	0	0	0	0	0
ub	P_{PV-B}^{max}	P_{PV-L}^{max}	P_{PV-G}^{max}	P_{WT-B}^{max}	P_{WT-L}^{max}	P_{WT-G}^{max}	P_{G-B}^{max}	P_{G-L}^{max}	P_{B-L}^{max}	P_{B-G}^{max}

5.4.3.3 Limitations of power flow

The SOC of the battery defined in Eq. (5.5) can be formulated for the studied system by the initial $SOC_{(t)}$ as follows:

$$SOC_{(t+1)} = SOC_{(t)} + \frac{\Delta t \times \eta_c}{E_{nom}} \times (P_{PV-B(t)} + P_{WT-B(t)} + P_{G-B(t)}) - \frac{\Delta t}{E_{nom}\eta_D} \times (P_{B-L(t)} + P_{B-G(t)}) \qquad (5.13)$$

Thus the battery SOC should be satisfying the following limits:

$$SOC^{min} \leq SOC_{(t)} \leq SOC^{max} \qquad (5.14)$$

The control variables are limited by upper boundary (*ub*) and lower boundary (*lb*) as defined in Table 5.1.

5.4.4 The applied algorithm

As the function is addressed as a linear function and the constraints are linear, the optimization problem can be solved with the LP algorithm. The objective function, presented by Eq. (5.15), is implemented in MATLAB environment. In this consideration, the main steps of the proposed algorithm for the open-loop control are summarized in the flowchart of Fig. 5.5.

$$\underset{x}{\text{Min}} / \underset{x}{\text{Max}} f(x), \quad \text{subject to} \begin{cases} Ax \leq b \\ A_{eq}x = b_{eq} \\ lb \leq x \leq ub \end{cases} \qquad (5.15)$$

where $f(x)$ represents the objective function; A and b denote the coefficients corresponding to inequality constraints; A_{eq} and b_{eq} denote the coefficient associated with equality constraints; and lb and ub are the lower and upper limits for each variable.

5.5 Model predictive control for the PV/wind turbine/battery system

The open-loop model designed in Fig. 5.3 involves, in this case, the MPC controller, which can optimally manage the energy flow of the system. This is due to the absence of feedback that makes the system sensitive to the disturbances in RE and load demand. In this work a multiinput–multioutput state-space model is applied for the discrete linear MPC control (Wang, 2009).

5.5.1 Multiinput–multioutput linear state-space model of the designed system

The predictive control problem can be expressed as follows:

$$\begin{cases} x(t+1) = Ax(t) + Bu(t) \\ y(t) = Cx(t) + Du(t) \end{cases} \qquad (5.16)$$

where $x(t)$, $y(t)$, and $u(t)$ are the state, the input, and the output respectively. A, B, C, and D are the system matrix, control matrix, output matrix, and the feed-forward matrix, which is null in the proposed system, respectively. Therefore $y(t)$ can be reformulated as follows:

$$y(t) = Cx(t) \qquad (5.17)$$

As the predictive horizon denoted N_p is equal to the control horizon N_c, the predictive state variables are calculated as follows:

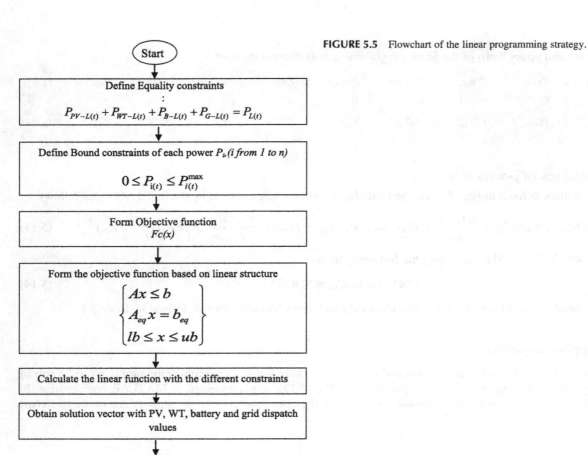

FIGURE 5.5 Flowchart of the linear programming strategy.

$$x(t_i + 1|t_i) = Ax(t_i) + Bu(t_i)$$
$$x(t_i + 2|t_i) = Ax(t_i + 1|t_i) + Bu(t_i + 1|t_i)$$
$$= A^2 x(t_i + 1|t_i) + Bu(t_i + 1|t_i)$$
$$\vdots$$

$$x(t + Np - 1|t_i) = A^{Np-1} x(t_i) + \sum_{j=1}^{Np-1} A^{Np-1-j} Bu(t_i + 1|t_i) \tag{5.18}$$

and the output vectors can be computed as:

$$Y(t) = \; = Fx(t) + \phi U \tag{5.19}$$

where

$$Y(t) = [C, C, \ldots C]X(t) \tag{5.20}$$

$$F = \begin{bmatrix} CA \\ CA^2 \\ . \\ . \\ CA^{Np} \end{bmatrix} \tag{5.21}$$

$$\Phi = \begin{bmatrix} CB & \cdots & 0 \\ \vdots & \ddots & \vdots \\ CA^{Np-1}B & \cdots & CA^{Np-Nc}B \end{bmatrix} \tag{5.22}$$

The generic form to express this kind of MPC problem is given by the quadratic equation as follows:

$$f_{\text{MPC}} = \sum_{j=1}^{Np} (Y - R_{ef})^T (Y - R_{ef}) \tag{5.23}$$

This function is subject to the following constraint:

$$NU(t) < \lambda \tag{5.24}$$

where R_{ef} is the predictive reference of the system for Y. N and λ are the matrices and vector, respectively. Then, Eq. (5.19) is substituted into Eq. (5.23) to get the optimal solution as follows:

$$f_{\text{MPC}}(t) = (Y(t) - R_{ef}(t))^T (Y(t) - R_{ef}(t)) \tag{5.25}$$

$$f_{\text{MPC}}(t) = (Fx(t) - R_{ef}(t))^T (Fx(t) - R_{ef}(t)) + 2(Fx(t) - R_{ef}(t))^T \phi U(t) + U(t)^T \phi^T \phi U(t) \tag{5.26}$$

Therefore the optimal $f_{\text{MPC}}(t)$ can be calculated as follows (Wang, 2009):

$$\min f_{\text{MPC}}(t) = \min(Y(t) - R_{ef}(t))^T (Y(t) - R_{ef}(t)) \tag{5.27}$$

$$\min f_{\text{MPC}}(t) = \min[2(Fx(t) - R_{ef}(t))^T \phi U(t) + U(t)^T \phi^T \phi U(t)] \tag{5.28}$$

$$\min f_{\text{MPC}}(t) = \min(U(t)EU(t) + 2HU(t)) \tag{5.29}$$

where

$$E(t) = \phi(t)^T \phi(t); H(t) = (Fx(t) - R_{ef}(t))^T \phi \tag{5.30}$$

5.5.2 Design of the model predictive control

Based on Eq. (5.16), the PV−WT−battery based system should be modeled as a linear state-space form to simplify the MPC model. Accordingly, the control input is defined as follows:

$$u(t) \stackrel{\Delta}{=} [P_{G\text{-}B}(t), P_{\text{PV-}B}(t), P_{\text{WT-}B}(t), P_{B\text{-}G}(t), P_{B\text{-}L}(t), P_{\text{PV-}G}(t), P_{\text{WT-}G}(t), P_{\text{PV-}L}(t), P_{\text{WT-}L}(t)]^T \tag{5.31}$$

The power flow $P_{G\text{-}L}$ can be formulated as the following equation:

$$P_{G\text{-}L}(t) = P_L(t) - P_{\text{PV-}L}(t) - P_{\text{WT-}L}(t) - P_{B\text{-}L}(t) \tag{5.32}$$

The control outputs are expressed as follows:

$$\begin{cases} y_a(t) = {}^{\Delta} \omega_1 (P_L(t) - P_{G\text{-}L}(t)) = \omega_1 \rho(t)(P_{\text{PV-}L}(t) + P_{\text{WT-}L}(t) + P_{G\text{-}L}(t)) \\ y_b(t) = {}^{\Delta} \omega_3 (\rho(t)(C_{\text{PV}} P_{\text{PV-}G}(t) + C_{\text{WT}} P_{\text{WT-}G}(t)) + \rho_p P_{B\text{-}G}(t)) \\ y_c(t) = {}^{\Delta} \omega_2 (P_{\text{PV-}B}(t) + P_{\text{WT-}B}(t) + P_{G\text{-}B}(t) + P_{B\text{-}G}(t) + P_{B\text{-}L}(t)) \end{cases} \tag{5.33}$$

where ω_1, ω_2, and ω_3 are the weight coefficients that should be selected by the consumer to obtain the desired performance of the system. Then, the output and the system states will be designed as follows:

$$y(t) \stackrel{\Delta}{=} [y_a(t-1), y_b(t-1), y_c(t-1)]^T \tag{5.34}$$

$$x(t) \stackrel{\Delta}{=} [\text{SOC}(t), y_a(t-1), y_b(t-1), y_c(t-1)]^T \tag{5.35}$$

Therefore to obtain the linear state-space system (5.34) representing the plant controlled by the proposed MPC managing system method, the system matrix is written as follows:

$$A = \begin{bmatrix} 1 & 0 & 0 & 0 \\ 0 & 0 & 0 & 0 \\ 0 & 0 & 0 & 0 \\ 0 & 0 & 0 & 0 \end{bmatrix}, \quad B = \begin{bmatrix} \eta_{\text{ch}} & \eta_{\text{ch}} & \eta_{\text{ch}} & -\eta_{\text{dis}} & -\eta_{\text{dis}} & 0 & 0 & 0 & 0 \\ 0 & 0 & 0 & 0 & \omega_1 & 0 & 0 & \omega_1 & \omega_1 \\ 0 & 0 & 0 & \omega_3 & 0 & \omega_3 & \omega_3 & 0 & 0 \\ \omega_2 & \omega_2 & \omega_2 & \omega_2 & \omega_2 & 0 & 0 & 0 & 0 \end{bmatrix}, \quad C = \begin{bmatrix} 0 & 1 & 0 & 0 \\ 0 & 0 & 1 & 0 \\ 0 & 0 & 0 & 1 \end{bmatrix} \tag{5.36}$$

5.5.2.1 The objective function of the MPC approach and constraints

The objective function of the MPC control system comprises three main parts. The first Eq. (5.37a) is given from $(y_a(t))$ and aims to minimize the energy purchased from the grid that can be expressed as follows:

$$f_1(t) = \min \sum_t^{t+Np} (P_L(t) - y_a(t))^2 \tag{5.37a}$$

The second Eq. (5.37b) maximizes the exported excess energy to the grid and it is formulated as follows:

$$f_2(t) = \min \sum_t^{t+Np} \omega_3 \left(\rho(t)(C_{PV}P_{PV\text{-}G}(t) + C_{WT}P_{WT\text{-}G}(t)) + \rho_p P_{B\text{-}G}(t)) - y_b(t) \right)^{20} \tag{5.37b}$$

The last Eq. (5.37c) allows minimizing the excessive usage of the battery; its expression is given as follows:

$$f_3(t) = \min \sum_t^{t+Np} \omega_3 (y_c(t))^2 \tag{5.37c}$$

Defining the reference value as:

$$R_{ef}(t) \triangleq \begin{bmatrix} \omega_1 P_L(t), \omega_3(C_{PV}P_{PV}(t) + C_{WT}P_{WT}(t) + \rho_p P_B(t)), 0, \omega_1 P_L(t+1), \\ \omega_3(\rho(t)(C_{PV}P_{PV}(t+1) + C_{WT}P_{WT}(t+1)) + \rho_p P_B(t+1)), 0, \dots, \omega_1 P_L(t), \\ \omega_3(C_{PV}P_{PV}(t+N_p-1) + C_{WT}P_{WT}(t+N_p-1) + \rho_p P_B(t+N_p-1)), 0 \end{bmatrix}^T \tag{5.38}$$

The global objective function is, then, rewritten as follows:

$$\min f_{MPC} = \min(f_1(t) + f_2(t) + f_3(t)) = \min \left(Y(t) - R_{ef}(t)\right)^T \left(Y(t) - R_{ef}(t)\right) \tag{5.39}$$

The multiobjective function of this closed-loop approach is constrained by the same constraints mentioned in open-loop method. They can be reformulated by the compact form of (5.24) as follows:

$$N_1 = \begin{bmatrix} N_{11} & \dots & N_{19} \\ \dots & \dots & \dots \\ N_{221} & \dots & N_{229} \end{bmatrix}, \lambda_1 = \begin{bmatrix} 0 \\ 0 \\ 0 \\ 0 \\ 0 \\ 0 \\ 0 \\ 0 \\ 0 \\ P_L(t) \\ P_{PV}(t) \\ P_{WT}(t) \\ P_2^{max} \\ . \\ . \\ P_{10}^{max} \\ P_{G\text{-}L}^{max}(t) - P_L(t) \end{bmatrix} \tag{5.40}$$

This constraint can be rewritten for the control vector $U(t)$ as follows:

$$\overline{N}_1 U(t) \leq \overline{\lambda}_1 \tag{5.41}$$

where $\overline{N}_1$ is the matrix constraint and $\overline{\lambda}_1$ is the vector constraint. The dimension relies on the dimension of the input of the studied system and the number of the constraints.

The model of the battery storage can be expressed as follows:

$$x(t) = x(t-1) + b_s u(t-1) \tag{5.42}$$

where

$$b_s = [\eta_{ch}, \eta_{ch}, \eta_{ch}, -\eta_{dis}, -\eta_{dis}, 0, 0, 0, 0] \qquad (5.43)$$

Therefore Eq. (5.42) can be reformulated as follows:

$$x(t+i|t) = x(t) + b_s \sum_{j=t}^{j \leq t+i} u(t) \qquad (5.44)$$

and

$$X(t) = x(t)[1, 1, 1, \ldots, 1]^T + B_s U(t) \qquad (5.45)$$

with

$$X(t) = [x(t), x(t+1|t), \ldots, x(t+N_c-1|t)]^T \qquad (5.46)$$

$x(t+i|t)$ presents the predicted value of x and B_s is a matrix that is expressed as follows:

$$B_s = \begin{bmatrix} bs & 0 & \ldots & 0 \\ bs & bs & \vdots & \vdots \\ \vdots & & \ddots & 0 \\ bs & bs & \ldots & bs \end{bmatrix} \qquad (5.47)$$

Subject to the constraint of the battery, it can be written as:

$$B_s^{min}[1, 1, \ldots, 1]^T \leq x(t)[1, 1, \ldots 1]^T + B_s U(t) \leq B_s^{max}[1, 1, \ldots, 1]^T \qquad (5.48)$$

Then, the constraint related to the SOC can be expressed as follows:

$$\overline{N}_2 U(t) \leq \overline{\lambda}_2 \qquad (5.49)$$

where

$$\overline{N}_2 = \begin{bmatrix} -B_s \\ B_s \end{bmatrix}, \overline{\lambda}_2 = \begin{bmatrix} (SOC(t) - SOC^{min})[1, 1, \ldots, 1]^T \\ (SOC^{max} - SOC(t))[1, 1, \ldots, 1]^T \end{bmatrix} \qquad (5.50)$$

Lastly, the final constraint after combining Eqs. (5.40) and (5.49) is formulated as:

$$\overline{N} = \left[\overline{N}_1^T, \overline{N}_2^T\right]^T, \overline{\lambda} = \left[\overline{\lambda}_1^T, \overline{\lambda}_1^T\right]^T \qquad (5.51)$$

5.5.3   Pseudo code of the model predictive control approach

As illustrated in Fig. 5.6, the MPC is considered as a discrete-time control of method to minimize the cost function related to the performance of the system during the defined future time steps. The linear MPC is a linearized model that

FIGURE 5.6 The diagram of the proposed model predictive control.

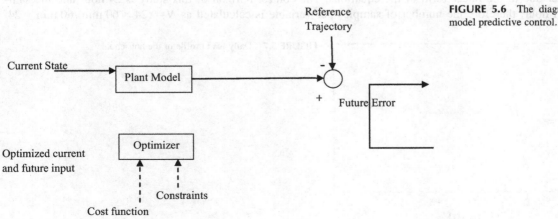

calculates the control sequence for N_p future time steps, with N_p denoting the prediction horizon, to reduce the weighted errors. Nevertheless, just the first input that is applied, then the others are updated in the next time step.

The execution of the linear MPC algorithm requires steps based on the objective function, the constraints analyzed previously and the instruction given in Wang (2009). Algorithm 1 describes the main steps of the proposed MPC approach.

Algorithm 1. Algorithm of the MPC method

begin

repeat from 1 to 4
  1: set $t = 0$;
  2: for a given time t, find the control horizon $N_c(t)$ and the predictive horizon $N_p(t)$;
  3: calculate the optimization approach by solving:
  $U(t)^T = EU(t) + 2HU(t)$ subject to the constraint in Eq. (5.51),
  calculate the MPC gains E and H by the objective function in Eq. (5.39);
  4: calculate the receding horizon control $u(t)$;
  5: set the next time $t = t + 1$ with updating the inputs, states, and outputs of the system;

until achieves $t=N_p$;

end

5.6 Results and discussion

5.6.1 Case study description

The hybrid system analyzed in this work is a grid-connected PV/WT/battery system supplying a household in Morocco (Tangier city). A typical daily profile of the load demand, PV, and WT output powers is illustrated in Figs. 5.7 and 5.8. The different parameters and data used in the simulation are listed in Table 5.2. The system is connected to the grid. For the studied system as a residential application and low voltage, the daily electricity price applied is shown in Fig. 5.2. As mentioned previously, the customers whose consumption exceeds 500 kWh month^{-1} should preferentially apply the TOU tariff than the fixed tariff by attributing the peak periods to the higher tariffs. The household in this work consumes 2250 kWh month^{-1}; therefore the TOU tariff is the appropriate one.

In this work, it is assumed that FITs are the PV and WT prices $C_{PV} = 0.043$ \$ kWh^{-1} and $C_{WT} = 0.032$ \$ kWh^{-1}, respectively (Feed-in tariff). For an attractive price, the FIT is increased to be equal approximately to the peak TOU electricity price.

5.6.2 Simulation results and discussion

This section represents the simulation results obtained from MATLAB code, which is developed for an optimal control and management system. The MPC is simulated using the "Quadprog" optimization solver in MATLAB optimization toolbox as the problem is formulated as a quadratic form. Besides, the open-loop control simulation uses the "Linprog" optimization solver due to the linear form of the equations. The control horizon of this study is 24 hour and the sampling time Δt is 1 hour; therefore the number of samples per variable is calculated as $N = (24 \times 60$ min$)/60$ min $= 24$.

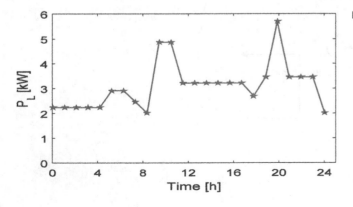

FIGURE 5.7 Daily load profile of the household.

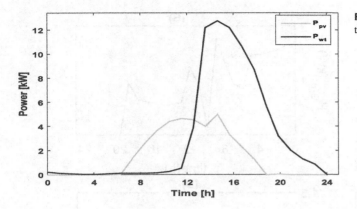

FIGURE 5.8 Daily output of the photovoltaic and wind turbine generation system.

TABLE 5.2 Simulation parameters of the photovoltaic (PV)—wind turbine (WT)—battery system.

Parameter	Meaning	Value
SOC_0	Battery initial SOC	90%
SOC^{max}	Battery maximum SOC	95%
SOC^{min}	Battery minimum SOC	40%
η_{Ch}	Battery charging efficiency	85%
η_{dis}	Battery discharging efficiency	95%
η_{PV}	PV coversion efficiency	85%
η_{WT}	WT coversion efficiency	95%
Δt	Sampling time	1 hour
ω_1	Weight coefficient	1
ω_2	Weight coefficient	0.8
ω_3	Weight coefficient	1
N_c	Control horizon	24
N_p	Predictive horizon	24
E_{nom}	Battery nominal capacity	5 kWh

SOC, State of charge.

The optimal energy flow dispatching of the PV, WT, battery, and the grid of the MPC approach and open-loop method are depicted in figures below to compare and verify the effectiveness of the proposed strategy.

Figs. 5.9–5.11 correspond to the MPC approach. Figs. 5.12–5.14 correspond to the open-loop control using LP algorithm.

In the MPC approach, Fig. 5.9C and D highlights that during off-peak periods from 00 to 09 hours morning; there is no power generated from RES to charge the battery. Therefore the battery bank is charged from the grid at a cheaper price (Fig. 5.9B). This permits storing the energy required for satisfying the peak of load demand. It can be seen in Fig. 5.10B that the load is mainly met by the battery storage system due to the low energy consumed from PV and WT (Fig. 5.11E and F). As a result, a small power is used from the grid to balance the deficit of the load power. However, even though the battery in the open-loop control stores an important energy from the utility grid to cover the required demand, a little power is delivered to the household (Fig. 5.12C). This can justify the purchasing power from the grid to supply the demand and battery since the PV and WT are not able to produce power during these periods. Hence, as shown in Fig. 5.14E, the priority is given to sell the energy from the battery to the grid than supplying the load.

From 10 to 17 hours, in the MPC case, there is no power purchased from the grid to meet the load (Fig. 5.9A) because of the availability of PV and WT power to satisfy the demand. As there is a peak demand approximately from

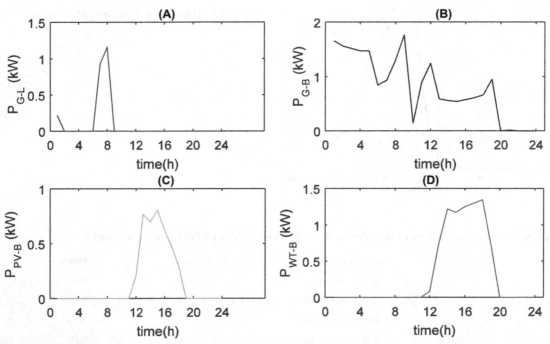

FIGURE 5.9 (A and B) Exported power from the grid to load and battery. (C and D) Charged battery power from photovoltaic and wind turbine.

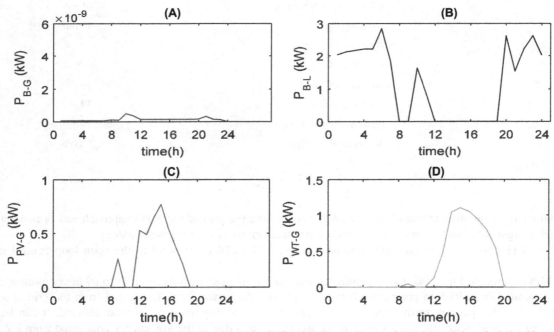

FIGURE 5.10 (A and B) Discharged battery power to the grid and load. (C and D) Exported power from photovoltaic and wind turbine to the grid.

10 to 11 hours, the battery is used to cover the remaining load demand. This explains the discharge rate of the battery through the SOC in Fig. 5.11G. It can be observed from Fig. 5.9B that the imported power from the grid to charge the battery is minimized gradually due to the important renewable output power for charging the battery bank. Meanwhile, there is an excess energy from PV and WT sold to the grid because of the large production during this period with a low contribution of the battery. As for the open-loop control, there is a small power purchased from the utility grid

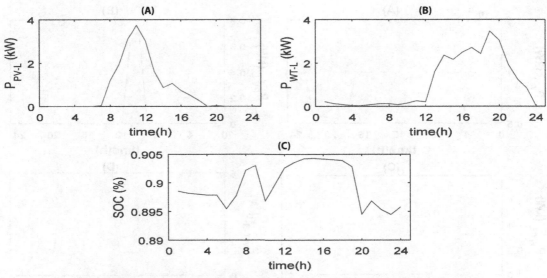

FIGURE 5.11   (A) and (B) Generated PV and wind turbine power to supply the load. (C) State of charge of the battery.

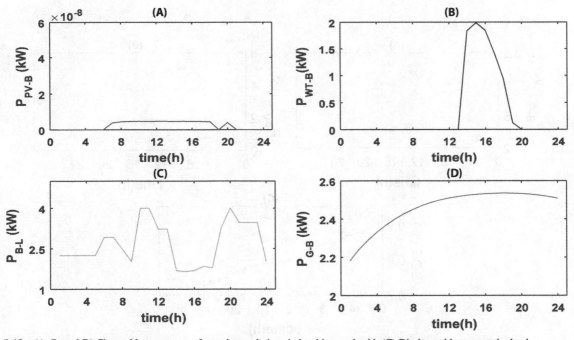

FIGURE 5.12   (A, B, and D) Charged battery power from photovoltaic, wind turbine, and grid. (C) Discharged battery to the load.

(Fig. 5.12A) to meet the load, which is successfully supplied by the battery and the RES as pointed out in Figs. 5.12C and 5.13B and C. The battery bank is discharged at its maximum power, which explains the high decrease of SOC in Fig. 5.12G as well as the excess power exported to the grid (Fig. 5.14E). This makes the battery operates excessively and the capacity degradation can occur. Furthermore, the battery is charged principally from the grid since a small output PV power is used. The reason is that the priority was given to sell from RES, which is not efficient.

During the peak price period, from 17 to 22 hours, in MPC strategy, there is no power imported from the main grid due to the high electricity price (Fig. 5.9A). This is because of the producing energy from PV and WT generators. As the load demand is high and reaches its peak, the battery bank has also enough energy to satisfy the required power.

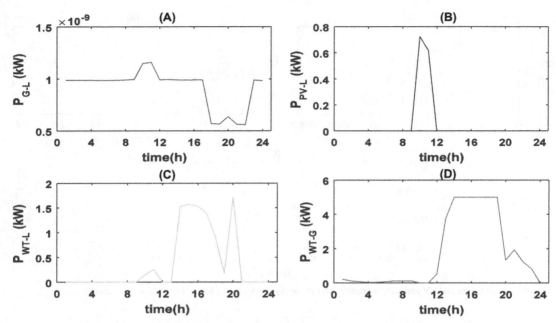

FIGURE 5.13   (A and B) The supply of the load from grid and photovoltaic. (C and D) Generated wind turbine power to the load and grid.

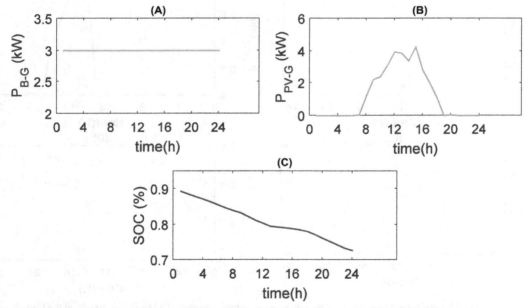

FIGURE 5.14   (E and F) Exported power from battery and wind turbine to the grid. (G) State of charge of the battery.

This can be seen from Fig. 5.11G by the decrease in the SOC. Likewise, a small amount is purchased from the grid to charge the battery. To reduce the net energy cost, the surplus energy from PV and WT is feed into the grid with poor excess of energy from the battery to prevent the deep discharge (Fig. 5.10A). For the open-loop control, it is worthy to note that the electricity price is high during this peak period; the power is purchased from the grid to supply the load and the battery. This can be explained by the following reasons: the battery is discharged at its maximum corresponding to the SOC decrease (Fig. 5.14G), the wind power only cannot satisfy enough the load, and the priority was given to sell the energy to the utility grid than satisfying the lack of energy demand.

5.6.3 Economic analysis

To further analyze the performance of the proposed closed-loop approach, a comparison with the open-loop control is performed in regard to a daily cost and energy savings. The obtained revenue from the energy sold to the grid is 1.00827 \$ day^{-1}. The baseline cost is the paid bill by the inhabitants before implementing the optimal control while the optimal cost is the electricity bill after applying the optimal control. Fig. 5.15 describes the basic and optimal cost obtained by each type of controller, and it can be observed that the implemented MPC system can reduce optimally the grid energy cost, which is calculated as 2.9689 \$ day^{-1} for the 24 hours. It can be also deduced from Fig. 5.16 that MPC achieves the maximum percentage of cost-savings (74.47%).

The calculation of the cost-saving (%) is formulated as follows:

$$\text{Cost-saving (\%)} = \left(\frac{\text{baseline} - \text{optimal cost}}{\text{baseline}} \right) \times 100 \tag{5.52}$$

In addition, another comparison is made on the cost function between the two approaches as illustrated in Fig. 5.17. From this figure, it is obvious that the MPC method reduces electricity cost, better than the open-loop control.

5.7 Conclusion

In this study, an MPC technique is developed for the energy management and control of a grid-tied residential PV, WT, and battery banks. The proposed optimal control is a closed-loop algorithm that is robust against the variation of renewable energy resources and prevents the deep discharge of the battery storage. Since the hybrid system assumed to import the power from the utility grid, a TOU tariff is applied for a residential household within the Moroccan demand side management program. The strategy can improve the overall performance index of the system by minimizing the energy purchased from the utility grid and profiting from the proposed PV and WT FIT, while taking into consideration the prevent of the deep discharge of the battery. A simulation of two methods of management is performed; the first one is considered as an open-loop control using LP algorithm and the second one is the MPC, to compare the effectiveness of the proposed approach. It has been observed from the results that MPC strategy has the best optimal cost and percentage

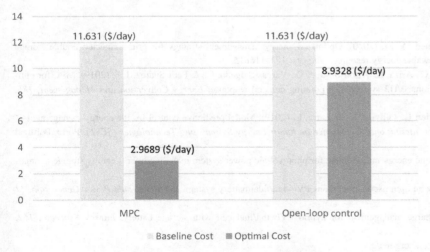

FIGURE 5.15 The baseline cost of model predictive control and open-loop strategy for the typical day.

FIGURE 5.16 Comparison of economic results between model predictive control and open-loop control.

Percentage Savings (%)

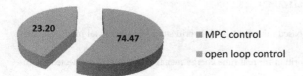

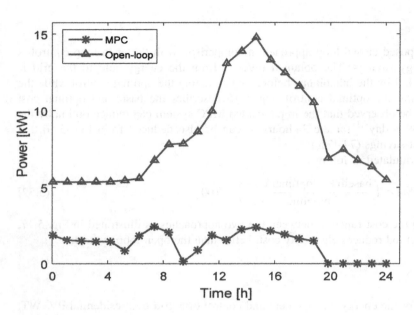

cost-saving of 2.9689 \$ day^{-1} and 74.47%, respectively. Indeed, the robustness and efficiency of the proposed control is due to the ability of MPC to predict the future behavior of the system while respecting the different constraints.

Future work of this research will be focusing on:

- The influence of hybrid system maintenance and battery storage degradation costs on the electricity bill saving.
- The impact of incorporating a subsidy price and more attractive FIT to encourage the green energies and increase the customer's benefits.

References

Abdelshafy, A. M., Jurasz, J., Hassan, H., & Mohamed, A. M. (2020). Optimized energy management strategy for grid connected double storage (pumped storage-battery) system powered by renewable energy resources. *Energy*, *192*, 116615.

Acevedo-Arenas, C. Y., Correcher, A., Sánchez-Díaz, C., Ariza, E., Alfonso-Solar, D., Vargas-Salgado, C., & Petit-Suárez, J. F. (2019). MPC for optimal dispatch of an AC-linked hybrid PV/wind/biomass/H2 system incorporating demand response. *Energy Conversion and Management*, *186*, 241–257.

Aguilera Gonzalez, A., Bottarini, M., Vechiu, I., Gautier, L., Ollivier, L., & Larre, L. (2019). Model predictive control for the energy management of a hybrid PV/battery/fuel cell power plant. In *2019 International Conference on Smart Energy Systems and Technologies (SEST)*, Porto, Portugal (pp. 1–6).

Aktas, A., Erhan, K., Özdemir, S., et al. (2018). Dynamic energy management for photovoltaic power system including hybrid energy storage in smart grid applications. *Energy*, *162*, 72–82.

Askarzadeh, A. (2016). Electrical power generation by an optimised autonomous PV/wind/tidal/battery system. *IET Renewable Power Generation*, *11*(1), 152–164.

Bendary, A. F., & Ismail, M. M. (2019). Battery charge management for hybrid PV/wind/fuel cell with storage battery. *Energy Procedia*, *162*, 107–116.

Bi-hourly electricity tariff, available from: http://www.one.org.ma/.

Bonthu, R. K., Aguilera, R. P., Pham, H., Phung, M. D., & Ha, Q. P. (2019, January). Energy cost optimization in microgrids using model predictive control and mixed integer linear programming. In *The 20th IEEE international conference on industrial technology (ICIT)* (pp. 1113–1118).

Chakir, A., Tabaa, M., Moutaouakkil, F., Medromi, H., Julien-Salame, M., Dandache, A., & Alami, K. (2020). Optimal energy management for a grid connected PV-battery system. *Energy Reports*, *6*, 218–231.

Dongol, D., Feldmann, T., Schmidt, M., et al. (2018). A model predictive control based peak shaving application of battery for a household with photovoltaic system in a rural distribution grid. *Sustainable Energy, Grids and Networks*, *16*, 1–13.

Feed-in tariff, available from: http://www.masen.ma/fr/projets/.

Garcia-Torres, F., & Bordons, C. (2015). Optimal economical schedule of hydrogen-based microgrids with hybrid storage using model predictive control. *IEEE Transactions on Industrial Electronics*, *62*(8), 5195–5207.

Hossain, M. A., Pota, H. R., Squartini, S., & Abdou, A. F. (2019). Modified PSO algorithm for real-time energy management in grid-connected microgrids. *Renewable Energy*, *136*, 746–757.

Hossain, M. A., Pota, H. R., Squartini, S., Zaman, F., & Muttaqi, K. M. (2019). Energy management of community microgrids considering degradation cost of battery. *Journal of Energy Storage*, *22*, 257−269.

Khan, M. R. B., Pasupuleti, J., Al-Fattah, J., & Tahmasebi, M. (2019). Energy management system for PV-battery microgrid based on model predictive control. *Indonesian Journal of Electrical Engineering and Computer Science*, *15*(1), 20−25.

Khawaja, Y., Giaouris, D., Patsios, H., & Dahidah, M. (2017, October). Optimal cost-based model for sizing grid-connected PV and battery energy system. In *2017 IEEE Jordan conference on applied electrical engineering and computing technologies (AEECT)* (pp. 1−6). IEEE.

Lagorse, J., Simões., Marcelo, G., & Miraoui, A. (2009). A multiagent fuzzy-logic-based energy management of hybrid systems. *IEEE Transactions on Industry Applications*, *45*(6), 2123−2129.

Luna, A. C., Diaz., Nelson, L., Graells, M., et al. (2016). Mixed-integer-linear-programming-based energy management system for hybrid PV-wind-battery microgrids: Modeling, design, and experimental verification. *IEEE Transactions on Power Electronics*, *32*(4), 2769−2783.

Lupangu, C., Justo, J. J., & Bansal, R. C. (2020). Model predictive for reactive power scheduling control strategy for PV−battery hybrid system in competitive energy market. *IEEE Systems Journal*.

Ma, T., Wu, J., Hao, L., et al. (2018). The optimal structure planning and energy management strategies of smart multi energy systems. *Energy*, *160*, 122−141.

Maleki, A., Rosen, M. A., & Pourfayaz, F. (2017). Optimal operation of a grid-connected hybrid renewable energy system for residential applications. *Sustainability*, *9*(8), 1314.

Marín, L. G., Sumner, M., Muñoz-Carpintero, D., Köbrich, D., Pholboon, S., Sáez, D., & Núñez, A. (2019). Hierarchical energy management system for microgrid operation based on robust model predictive control. *Energies*, *12*(23), 4453.

Mbungu, T., Naidoo, R., Bansal, R., et al. (2016). Smart SISO-MPC based energy management system for commercial buildings: Technology trends. In *2016 Future technologies conference (FTC)* (pp. 750−753). IEEE.

Momoh, J. A. (2012). *Smart grid: Fundamentals of design and analysis*. John Wiley & Sons.

Negi, S., & Mathew, L. (2014). Hybrid renewable energy system: A review. *International Journal of Electronic and Electrical Engineering*, *7*(5), 535−542.

Pachori, A., & Suhane, P. (2014). Modeling and simulation of photovoltaic/wind/diesel/battery hybrid power generation system. *International Journal of Electrical, Electronics and Computer Engineering*, *3*(1), 122.

Paul, T. G., Hossain, S. J., Ghosh, S., Mandal, P., & Kamalasadan, S. (2018). A quadratic programming based optimal power and battery dispatch for grid-connected microgrid. *IEEE Transactions on Industry Applications*, *54*(2), 1793−1805.

Seal, S., Boulet, B., & Dehkordi, V. R. (2019). Centralized model predictive control strategy for thermal comfort and residential energy management. *arXiv preprint arXiv:1912.06943*.

Tang, R., Li, X., & Lai, J. (2018). A novel optimal energy-management strategy for a maritime hybrid energy system based on large-scale global optimization. *Applied Energy*, *228*, 254−264.

Trifkovic, M., Sheikhzadeh, M., Nigim, K., et al. (2013). Modeling and control of a renewable hybrid energy system with hydrogen storage. *IEEE Transactions on Control Systems Technology*, *22*(1), 169−179.

Wang, L. (2009). *Model predictive control system design and implementation using MATLAB®*. Springer Science & Business Media.

Yan, H., Zhuo, F., Lv, N., Yi, H., Wang, Z., & Liu, C. (2019, June). Model predictive control based energy management of a household microgrid. In *2019 IEEE 10th international symposium on power electronics for distributed generation systems (PEDG)* (pp. 365−369). IEEE.

Yilmaz, U. C., Sezgin, M. E., & Gol, M. (2020). A model predictive control for microgrids considering battery aging. *Journal of Modern Power Systems and Clean Energy*, *8*(2), 296−304.

Zhang, Y., Fu, L., Zhu, W., et al. (2018). Robust model predictive control for optimal energy management of island microgrids with uncertainties. *Energy*, *164*, 1229−1241.

Chapter 6

Efficient maximum power point tracking in fuel cell using the fractional-order PID controller

K.P.S. Rana[1], Vineet Kumar[1], Nitish Sehgal[1], Sunitha George[1] and Ahmad Taher Azar[2,3]

[1]Department of Instrumentation and Control Engineering, Netaji Subhas University of Technology, Dwarka, India, [2]Faculty of Computers and Artificial Intelligence, Benha University, Benha, Egypt, [3]College of Computer and Information Sciences, Prince Sultan University, Riyadh, Kingdom of Saudi Arabia

6.1 Introduction

As the world population increases, so is the need and demand for energy. To cater to such requests, new energy production technologies have been emerging. So, the condition of the hour is the efficient energy production. The depletion of fossil fuels is the biggest challenge it is facing, and its only promising solution is shifting the energy dependence on renewable sources. These unconventional energy sources commonly include solar energy, wind energy, and fuel cell (FC). The major drawback of solar and wind energy is constant irradiation and wind, thereby making them dependent on the climate, which is very uncertain. Furthermore, for large energy generation, vast spaces of land are required to set up a farm. FC has an advantage over these two in that it is possible to provide constant input using hydrogen cylinders, which can be implemented without any geographical or climatic constraints. FC can also produce large amounts of energy in a compact space. Hence, FC proves itself to be a more viable alternative than its counterparts. Furthermore, some other advantages of FCs (Boudghene Stambouli & Traversa, 2002) are constant power production, energy security, choice of fuel, reliability, quiet operation, low operating cost, clean emissions, high efficiency, etc. There are numerous variants of FC available, namely, proton-exchange membrane fuel cell (PEMFC), direct methanol fuel cell, alkaline fuel cell, phosphoric acid fuel cell, molten carbonate fuel cell, solid oxide fuel cell, zinc air fuel cell, and photonic ceramic fuel cell (Gou, Na, & Diong, 2010). Out of all these types, PEMFC is the most promising one due to its compact size and comparatively low operating temperature ($60°C-80°C$).

The $P-I$ characteristics of PEMFC have only one maximum for a particular operating point, which is called maximum power point (MPP) (Amara et al., 2019; Ammar, Azar, Shalaby, & Mahmoud, 2019; Ben Smida, Sakly, Vaidyanathan, & Azar, 2018; Fekik et al., 2021a, 2021b; Ghoudelbourk, Azar, Dib, & Omeiri, 2020; Kamal & Ibrahim, 2018). Since the cell's output is generally used to drive a load or charge a battery therefore for maximum power generation, the cell is desired to operate at MPP only. To manage the cell at this point, the cell's internal impedance must match with the load impedance, which is not possible for everyday use due to continuous load variation. Therefore the impedance should not be kept fixed as it will allow the cell to operate only at a single point of operation. For varying the impedance, a DC/DC boost, buck, or buck−boost converter is used depending on the application, and it is inserted between load and cell to alter the net impedance seen by the cell. The converter disconnects and connects the cell's load at a particular duty cycle governed by a maximum power point tracking (MPPT) controller (2005) for load-matching purposes.

To drive PEMFC at its MPP, various control techniques have been proposed in the literature. These can be categorized into more giant buckets, namely, conventional approaches, adaptive strategies, and intelligent designs. Under the conventional techniques, perturb and observe (P&O) (Giustiniani et al., 2006, 2010) is one of the most straightforward methods based on the common hill-climbing approach. Though it is simple and has a low implementation cost, it is not useful in MPPT due to its fixed step size (FSS). Another technique that falls under this category is the incremental

Renewable Energy Systems. DOI: https://doi.org/10.1016/B978-0-12-820004-9.00017-6

conductance (IC) (Rezk, 2017) technique with similar limitations but uses the derivates to find maxima and can be implemented by only using two calculative steps three steps in P&O.

However, a variation in IC was created the using variable step size (VSS) rather than FSS that proved to perform better by increasing the MPPT speed (Harrag & Messalti, 2017). Still, both the techniques are prone to oscillations around MPP, and large undershoots at initial stages of operation. Further advancement in IC included a fraction order (FO) filter (Chen et al., 2017), which was added to reduce the ripple content around the MPP but somewhat increased the short response time due to introducing a filter. The second bucket of adaptive MPPT techniques, first in the category, is extremum seeking (ES) (Zhong et al., 2008), which tracks MPP effectively. Numerous advancements have been made in its case to increase the effectiveness of the controller, namely, higher-order ES (Bizon, 2010), modified ES (Bizon, 2013), global ES (Bizon, 2017), and ES with FO-HPF (Liu, Zhao, & Chen, 2016). Though it has higher tracking ability, this technique has a few drawbacks like output ripples observed in power and current, also known as chattering effect, which reduces the lifetime of FC due to high oscillations of current. Another popular technique is the sliding mode control (SMC), which is quite complex in operation but very effective (Abdous, 2009; Azar & Serrano, 2018a, 2018b, 2020; Azar, Serrano, Rossell, Vaidyanathan, & Zhu, 2020a, 2020b, 2020c, 2020d, 2020e; Djouima, Azar, Drid, & Mehdi, 2018; Kunusch, Puleston, & Mayosky, 2012; Mekki, Boukhetala, & Azar, 2015; Singh et al., 2017; Vaidyanathan et al. 2019; Vaidyanathan, Sampath, & Azar, 2015; Vaidyanathan & Azar, 2015a, 2015b). However, it is also prone to a similar chattering effect and its adverse effects. Various variations recorded so far in SMC are SMC with variable duty cycle, total SMC (Wai & Shin, 2011), and extension SMC. In general, adaptive techniques are more useful to attain MPP but are quite complex to understand and implement.

Third, in line are intelligent techniques that are further divided into three more categories of neural network (NN), fuzzy logic control (FLC), and evolutionary algorithm (EA)-based techniques. NN predictive control uses NN to calculate the duty cycle for load-matching purposes. FLC (Al-Dabbagh, Lu, & Mazza, 2010; Benchouia et al., 2015; Venkateshkumar, Sarathkumar, & Britto, 2014) is a three-step process of fuzzification, inference, and defuzzification, which uses error and rate of change of error as input to estimate the correct duty cycle for MPPT. Further advancement in this category is reported as 3-D FLC (Wang et al., 2016) and VSS FLC. Third and last division of intelligent techniques is EA-based techniques that use numerous EAs to calculate the reference MPP effectively or could be used to tune the PID controller gains, which is used for MPPT. Numerous methods listed in literature belonging to this category are PSO (Soltani, 2013), EaS (Sarvi et al., 2015), WCA–PID (Avanaki & Sarvi, 2016), PSO–PID (Ahmadi, Abdi, & Kakavand, 2017), EaS (coupled with PSO and cuckoo optimization)–PID (Sarvi et al., 2014), and gray wolf optimizer (GWO)–PID (Rana et al., 2019). These intelligent techniques are easier to understand than adaptive methods but require relatively higher computational abilities, which increase implementation cost. Since these techniques are based on software computation, they are reusable and rewritable and have higher accuracy than other approaches. Other techniques include a hybrid of techniques mentioned above, such as NN VSS IC, fuzzy PI, fuzzy PID, fuzzy SMC, and some unclassified techniques, such as resistance estimation (Bankupalli et al., 2019) and backstepping (Derbeli et al., 2017), which provide better efficiency than conventional techniques but have a lot of hidden potential for future research.

Nowadays, soft-computing techniques are getting lots of attention due to their high accuracy and low cost of implementation (Ammar, Azar, Tembi, Tony, & Sosa, 2018; Azar, El-Said, & Hassanien, 2013; Barakat, Azar, & Ammar, 2020; Djeddi, Dib, Azar, & Abdelmalek, 2019; Giove, Azar, & Nordio, 2013; Gorripotu, Samalla, Jagan Mohana Rao, Azar, & Pelusi, 2019; Kamal, Azar, Elbasuony, Almustafa, & Almakhles, 2020; Khettab, Bensafia, Bourouba, & Azar, 2018; Mjahed, Bouzaachane, Azar, Hadaj, & Raghay, 2020; Pilla, Azar, & Gorripotu, 2019; Pilla, Botcha, Gorripotu, & Azar, 2020; Sallam, Azar, Guaily, & Ammar, 2020; Soliman, Azar, Saleh, & Ammar, 2020). They can be either used in run time to calculate the effective V_{ref} or they can be used to tune the gain parameters of PID to get the best results. Since these plants are highly nonlinear, popular tuning methods, such as Ziegler–Nichols and Cohen–Coon, cannot be applied to this scenario directly. Therefore this problem is solved by treating this as an optimization problem. Now greater the number of tuning parameters, more is the freedom for the control engineer to increase the accuracy and robustness, but this comes at extra computational efforts. Furthermore, the number of tuning parameters can be increased by introducing new variables in the controller design; one such way of doing this is the inclusion of fractional-order integrator and differentiator. These fractional operators are modeled in the frequency domain as s^{μ} or $s^{-\mu}$ depending upon nature. By including these FO operators in the conventional PID, two new control variables are available for tuning, which increase the degree of freedom of the system. This inclusion of fractional-order calculus has allowed control engineers to make their designs more effective and robust (AbdelAty, Azar, Vaidyanathan, Ouannas, & Radwan, 2018; Alain et al., 2018; Alain, Azar, Kengne, & Bertrand, 2020; Azar, Radwan, & Vaidyanathan, 2018a, 2018b; Fekik et al., 2021; Ibraheem, Azar, Ibraheem, & Humaidi, 2020; Kammogne et al., 2020; Khan, Singh, & Azar, 2020a, 2020b;

Khennaoui, Ouannas, Boulaaras, Pham, & Azar, 2020; Mittal, Azar, & Kamal, 2021; Ouannas, Grassi, & Azar, 2020a, 2020b; Ouannas, Grassi, Azar, & Khennaoui, 2021; Pham, Gokul, Kapitaniak, Volos, & Azar, 2018; Pham et al., 2017; Shukla, Sharma, & Azar, 2018; Singh & Azar, 2020; Singh, Azar, Vaidyanathan, Ouannas, & Bhat, 2018). Fractional-order PID (FOPID) controllers have gained a lot of attention in the past two decades and proved themselves a promising solution for the industry. This concept was initially proposed by Podlubny (1999a, 1999b) in which it was confirmed that FOPID had a better response than the conventional PID controller. A frequency-domain approach for realizing various fractional-order systems was presented in (Vinagre et al., 2000). Also, numerous methods for tuning FOPID are available in the literature, which is based on different strategies. For example, (Feliu et al., 2004; Luo and Chen, 2009; Monje et al., 2008) used gain margin and phase margins for tuning. Ziegler−Nichols (Valério and da Costa, 2006) used some basic concepts, such as rise time and overshoot/undershoot, for deriving the standard rules of designing. Moreover, soft-computing methods (Bingul and Karahan, 2018; Tang et al., 2012) used various optimization techniques to find the best gains by minimizing the error function.

Fractional-order controllers have been applied to a wide variety of complex and nonlinear system, such as active suspension systems (Kumar et al., 2018), hybrid electric vehicles (Khaldi and Ammari, 2015; Kumar, Rana, & Mishra, 2016), binary distillation (Mishra, Kumar, & Rana, 2015), power systems (Nithilasaravanan et al., 2019), and robotic manipulators. Furthermore, real-time hardware implementation of these controllers has been carried out (Rana et al., 2016; Tolba et al., 2017a, 2017b) over FPGA, which in turn proves these controllers' effectiveness in high-speed applications. As technology progresses, the use of FOPID controllers in the field of energy generation is also increasing. Few of the examples include wind energy (Ghoudelbourk, Dib, Omeiri, & Azar, 2016; Meghni, Dib, & Azar, 2017a, 2017b) and solar energy in which FOPID was used to extract the maximum power at a particular operating point, which in turn served as the main inspiration for this work (Jeba and Immanuel Selvakumar, 2018; Kler, Rana, & Kumar, 2018; Sahin, Ayas, & Altas, 2014). Despite having high robustness and higher controlling action, these fractional-order controllers have not been implemented for MPPT in FCs, which is the most promising solution for future renewable energy sources.

Based on the above-presented literature survey, one can infer that fractional-order controllers effectively control various time-varying and nonlinear systems. Because of the nonlinearities, these systems become complex, further aggravating when exposed to external environmental disturbances. Also, fractional-order controllers are more effective on these types of nonlinear systems than their conventional counterparts. These facts served as the primary objective and motivation for this work to explore the option to enhance the efficiency of MPPT in PEMFC. Furthermore, to compare and investigate its relative efficacy with its other conventional counterparts over specific performance metrics that will assess these controllers' complete performance in both transient and steady-state duration had been the critical outcome of these efforts. Therefore this chapter is proposed to enhance the performance of dP/dI feedback-based PID controller by augmenting it with FOPID controller.

This chapter is organized as follows. Section 6.1 presents the background introduction and brief literature survey of various MPPT controllers applied to FC and general applications of FOPID controllers. Section 6.2 outlines the model description of PEMFC along with its mathematical formulations and different design constants. It also presents the sample characteristics plots of the PEMFC over varying membrane water content λ and temperature T values. Section 6.3 describes the general control methodology employed using DC/DC converter for MPPT and a short description of each component. The design and implementation of FOPID using the Grunwald−Letnikov (GL) technique are described in Section 6.4. Section 6.5 presents the tuning methodology used to tune PID and FOPID along with respective tuning parameters. Section 6.6 exhibits detailed performance investigations of all the controllers, P&O, and IC over different λ and T profiles. Finally, conclusions are drawn out in Section 6.7.

6.2 PEMFC system description

PEMFC plant has exponential, nonintegral powers and logarithmic nonlinearities in its mathematical model described in the following subsections.

6.2.1 Working principle

PEMFC works based on the chemical equation as shown in Eq. (6.1).

$$H_2(g) + \frac{1}{2}O_2(g) \rightleftharpoons H_2O(l) \tag{6.1}$$

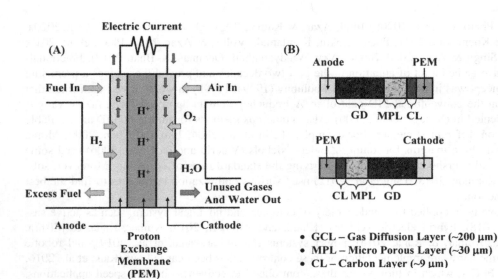

FIGURE 6.1 (A) Movement of electrons and reactions occurring in proton-exchange membrane fuel cell. (B) Expanded view of the membrane.

- GCL – Gas Diffusion Layer (~200 μm)
- MPL – Micro Porous Layer (~30 μm)
- CL – Carbon Layer (~9 μm)

This reaction is completed in two steps: first, $H_2(g)$ is oxidized to $H^+(l)$ at the anode and second, $O_2(g)$ is reduced to $H_2O(l)$ at the cathode. For the second step, the $H^+(l)$ has to reach the cathode from the anode for completing the chemical reaction; subsequently, a proton exchange membrane is used for the process. Fig. 6.1A represents the movement of electrons inside the FC, and Fig. 6.1B presents the expanded view of both anodic and cathodic membranes along with their respective widths.

Generally, for MPPT purposes, PEMFC is modeled via two methods, namely, polarization curve model (PCM) and dynamic gas transport model (DGTM). In this chapter, we will only focus on the PCM as researchers generally use it for MPPT using DC/DC converter, whereas fuel starvation control is used in the case of the DGTM. Since the proposed control scheme is based on the DC/DC converter, the PCM approach has been used for MPPT. Furthermore, this approach is based on two essential inputs, namely, temperature (T) and membrane water content (λ). The exact dependence of the working of PEMFC on the inputs is explained in further subsections along with a detailed mathematical model.

6.2.2 Mathematical model: PCM

The considered system is the Ballard Power System Mark-V PEMFC with Nafion 117 membrane. Standard temperature and pressure condition is 298.15K and 1atm. The output voltage $V_{cell}(V)$ of the PEMFC is described as (Rana et al., 2019):

$$V_{cell} = E_{nernst} - E_{act} - E_{ohmic} - E_{conc} \tag{6.2}$$

where $E_{nernst}(V)$ is the reversible voltage of the cell and $E_{act}(V)$, $E_{ohmic}(V)$, and $E_{conc}(V)$ are used for modeling losses of PEMFC. Here, E_{act} is the activation overpotential, $E_{ohmic}(V)$ represents the losses due to charge transport, and E_{conc} represents the losses due to mass transport. These voltages are given as:

$$E_{nernst} = 1.229 - 8.458 \times 10^{-4}(T - 298.15) + 4.308 \times 10^{-5}\left(\ln\left(p_{H_2}\right) + 0.5\ln\left(p_{O_2}\right)\right) \tag{6.3}$$

where p_{H_2} and p_{O_2} are the partial pressure (atm) of H_2 and O_2, respectively.

$$E_{act} = \xi_1 + \xi_2 T + \xi_3 T\ln\left(C_{O_2}^*\right) + \xi_4 T\ln(I_{FC}) \tag{6.4}$$

where $\xi_1(V)$, $\xi_2(V\,K^{-1})$, ξ_3 ($V\,K^{-1}$), and ξ_4 ($V\,K^{-1}$) are the parametric coefficients whose values are:

$$\xi_1 = 0.944V \tag{6.5}$$

$$\xi_2 = -0.00354VK^{-1} \tag{6.6}$$

$$\xi_3 = -7.80 * 10^{-5}VK^{-1} \tag{6.7}$$

$$\xi_4 = 0.000196 \text{VK}^{-1} \tag{6.8}$$

Therefore the net activation overpotential for the PEMFC is:

$$E_{act} = 0.944 - 0.00354T - 7.8 * 10^{-5}T\ln\left(C_{O_2}^*\right) + 0.000196\ln(I_{FC}) \tag{6.9}$$

where $C_{O_2}^*$ is the effective concentration of O_2 at cathode (mol cm^{-3}) is related to p_{O_2}(atm) by the following relation:

$$C_{O_2}^* = 1.97 \times 10^{-7} p_{O_2} e^{\frac{498}{T}} \tag{6.10}$$

$$R_m = \frac{181.6\left[1 + 0.03\left(\frac{I_{FC}}{A}\right) + 0.062\left(\frac{T}{303}\right)^2\left(\frac{I_{FC}}{A}\right)^{2.5}\right]}{\left[\lambda - 0.634 - 3\left(\frac{I_{FC}}{A}\right)\right]e^{1268\left(\frac{1}{303}-\frac{1}{T}\right)}} \tag{6.11}$$

Therefore total *ohmic* resistance R_{ohmic} and *ohmic* voltage loss E_{ohmic} are defined as:

$$R_{ohmic} = \frac{R_m t_m}{A} \tag{6.12}$$

where A is the area of cross-section of the FC.

$$E_{ohmic} = I_{FC}R_{ohmic} \tag{6.13}$$

$$E_{conc} = \frac{RT}{nF}\left(1 + \frac{1}{\alpha}\right)\ln\left(\frac{I_L}{I_L - I_{FC}}\right) \tag{6.14}$$

For an FC system, the output voltage $V_{FC}(V)$ and power $P_{FC}(W)$ are defined as follows:

$$V_{FC} = N_{FC}V_{cell} \tag{6.15}$$

$$P_{FC} = V_{FC}I_{FC} \tag{6.16}$$

6.2.3 Characteristic power versus current plots of the used PEMFC

PEMFC is more sensitive to current variations than voltage variations due to recent mathematical model feedback. Fig. 6.2A presents the P_{FC} versus current density curve for a fixed $T = 323$K with varying λ, whereas Fig. 6.2B presents the same for a fixed $\lambda = 15$ with varying temperature. These plots are based on the parameters listed in Table 6.1.

6.3 MPPT control configuration

The standard control configuration used for MPPT of PEMFC is as shown in Fig. 6.3. As the behavior and mathematical model of PEMFC have been defined in the above section, the used MPPT controller, PWM generator, DC/DC converter, and load are described in the following subsections.

6.3.1 MPPT controller

The MPPT controller takes in the measured variables, such as I_{FC} and V_{FC}. The control algorithm is used to calculate the appropriate duty cycle for extracting maximum power out of the cell.

The MPPT controller's output is fed into the PWM generator to convert these control signals into actual physical signals and thereby serve as a medium for load matching between the cell and load. In this chapter, the control scheme adopted is $\frac{dP}{dI}$ feedback-based GWO$-$FOPID controller, wherein the slope error is fed back into the control system as the error signal, and suitable action is performed afterward. The controller is known as GWO$-$FOPID because GWO is used to tune the various parameters of FOPID to get the best results and minimum power wastage. Fig. 6.4 represents the working of $\frac{dP}{dI}$ feedback-based control structure.

In this control scheme, the setpoint is taken to be the peak of the characteristic curve where the slope is zero, that is, $\frac{dP}{dI} = 0$. Furthermore, the error calculation block calculates the error or deviation from the calculated $\frac{dP}{dI}$, which is ultimately fed as an error to the MPPT controller block.

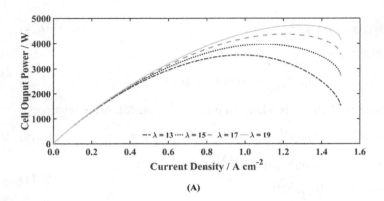

FIGURE 6.2 Characteristics of P_{FC} versus I_{FC} curves (A) for fixed T and varying λ and (B) for fixed λ and varying T.

TABLE 6.1 Fuel cell parameters' values for polarization curve model.

S. no.	Symbol	Description	Unit	Value
1.	n	Number of moles of electron transfer	mol e$^-$ (mol reactant)$^{-1}$	2
2.	F	Faraday's constant	C (mol e)$^{-1}$	96,485
3.	T_{ref}	Reference temperature	K	298.15
4.	R	Universal gas constant	J mol^{-1} K^{-1}	8.314
5.	α	Transfer coefficient	—	0.3–0.7
6.	A	Cross sectional area	cm^2	232
7.	ξ_1	Constant	V	0.944
8.	ξ_2	Constant	VK^{-1}	−0.00354
9.	ξ_3	Constant	VK^{-1}	-7.80×10^{-5}
10.	ξ_4	Constant	VK^{-1}	0.000196
11.	t_m	Nafion 117 membrane thickness	cm	1.78×10^{-2}
12.	I_L	Limiting current	A	2
13.	N_{FC}	Number of FCs	—	30

6.3.2   PWM generator

A PWM generator generates a square wave of the corresponding duty cycle as fed by the MPPT controller. In this work, a PWM generator with a switching frequency of 10kHz was used, and the sample time of the simulation was kept to be 10μs.

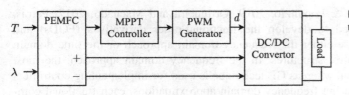

FIGURE 6.3 General configuration used for maximum power point tracking of proton-exchange membrane fuel cell.

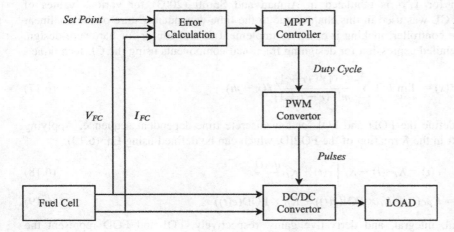

FIGURE 6.4 Implementation of $\frac{dP}{dI}$ feedback-based scheme.

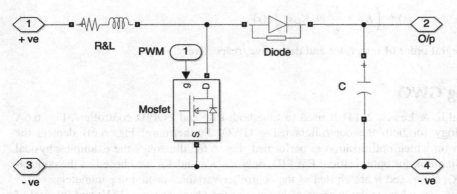

FIGURE 6.5 DC/DC convertor as employed for maximum power point tracking.

6.3.3   DC/DC converter

DC/DC converter performs the load matching by connecting and disconnecting the load from the cell according to the input duty cycle d. DC/DC convertor generally used can be of three types: buck, boost, and buck–boost whose selection depends on the load current requirements. The PWM generator's varying duty cycle is fed as input to DC/DC convertor, which satisfies the load's current condition and performs the load matching as directed by the MPPT controller. The DC/DC converter, as shown in Fig. 6.5, is used with the values of resistor, inductor, and capacitor as $R = 0.2\text{m}\Omega$, $L = 5\text{mH}$, and $C = 100\mu\text{F}$, respectively.

6.3.4   Load

A purely resistive load of value $R = 1\Omega$ is considered for simplicity.

6.4   Design and implementation of FOPID MPPT control technique

Fractional-order calculus is a domain of mathematics developed for studying noninteger-order derivatives and integrals' behavior. These mathematical operators have been defined in the literature using various methods, that is,

Riemann–Liouville, Caputo, and GL (Garrappa, Kaslik, & Popolizio, 2019; Ortigueira and Machado, 2017; Petráš, 2011; Sen, 2014). However, only GL definition is used to develop the fractional-order derivatives (FODs) and fractional-order integrals (FOIs) in this work. One can use either the frequency domain approach or the time domain approach to implement these fractional operators in control structures. In the frequency domain approach, the most renowned strategy is Oustaloup's successive approximation, whereas GL technique is used for implementing controllers in the time domain. To design a fractional-order system using frequency domain approximations, each fractional component is approximated using the integral-order counterpart, which is valid within a specific frequency range (Sabatier et al., 2002). A similar approximation for $1/s^q$ is tabulated in Ahmad and Sprott (2003) for various values of $q \in [0.1, 0.2, \ldots, 0.9]$. On the other hand, GL was used in this chapter due to the time-dependent nature of the nonlinear plant and discrete implementation of the controller, making it easier to implement in real life or for hardware design. Furthermore, Eq. (6.17) represents the detailed expression for designing fractional components using the GL technique.

$$D_t^r f(x) = \lim_{h \to 0;} h^{-r} \sum_{j=0}^{\frac{x-a}{h}} \frac{(-1)^j \Gamma(r+1)}{m! \Gamma(r-j+1)} f(x - jh) \tag{6.17}$$

Using the above operator, one can define the FOD and FOI for any discrete time-dependent sequence. Applying these operators over Eq. (6.18) will result in the formation of the FOPID, which can be defined using Eq. (6.19).

$$c(t) = K_P e(t) + K_I \int e(t) + K_D \frac{de(t)}{dt} \tag{6.18}$$

$$c^*(t) = K_P e(t) + K_I \times \text{FOI}(e(t)) + K_D \times \text{FOD}(e(t)) \tag{6.19}$$

where K_P, K_I, and K_D are proportional, integral, and derivative gains, respectively. FOI and FOD represent the fractional-order integral and derivative, respectively. In s-domain [Eq. (6.19)], it transforms into Eq. (6.20).

$$c^*(s) = \left(K_P + \frac{K_I}{s^\gamma} + K_D s^\mu \right) e(s) \tag{6.20}$$

Here, γ and μ represent the nonintegral order of integrator and derivative, respectively.

6.5 Controller tuning using GWO

In this chapter, GWO (Mirjalili, Mirjalili, & Lewis, 2014) is used to tune both PID and FOPID controllers. Fig. 6.6A clearly explains the tuning methodology for both the controllers using GWO. Furthermore, Fig. 6.6B depicts the expanded view of the physical system on which optimization is performed. Fig. 6.6C illustrates the example physical system of FOPID used in MATLAB/Simulink for optimization. For PID, only K_P, K_I, and K_D are chosen as the controller variables whereas for FOPID, K_P, K_I, K_D, γ, and μ are chosen as the controller variables or tuning parameters.

The fitness function to be minimized was taken as the average of integral of the absolute error (IAE) and integral of the time-weighted absolute error (ITAE).

$$\text{Fitness function} = 0.5 \underbrace{\int |e(t)| dt}_{\text{IAE}} + 0.5 \underbrace{\int |te(t)| dt}_{\text{ITAE}} \tag{6.21}$$

This fitness function was chosen to account for the errors coming at the later stages of operation, that is, steady state. Since the cell spends a major part of its lifetime at a steady state, therefore errors occurring at the steady-state operation must be heavily penalized. This tuning of parameters for both the controllers was performed at a fixed operating point or nominal condition for which T and λ were kept constant 333K and 14, respectively. Furthermore, both PID and FOPID were tuned for 20 iterations with population size kept as 30. Table 6.2 presents the upper and lower bounds for the gains and nonintegral factors for both the controllers. The upper and lower bounds for both the controllers were kept the same to provide a uniform comparative platform.

This tuning process was carried out 10 times, and the best gains were chosen as the final tuned value. The final fitness value obtained for PID and FOPID is 0.769905 and 0.76981, respectively. At first glance, there is not much difference between the above two values so both of them must give similar performance but this is not the case as the tuning is performed over a single nominal operating point. In contrast, the cell is expected to perform at numerous operating points dependent upon the climatic condition and external environment. The similar performance at the tuned

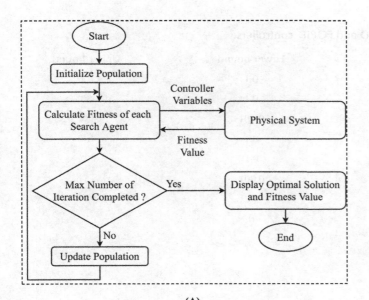

(A)

FIGURE 6.6 (A) Tuning methodology for the controllers. (B) Expanded view of physical system block. (C) Physical system model for the FOPID controller tuning.

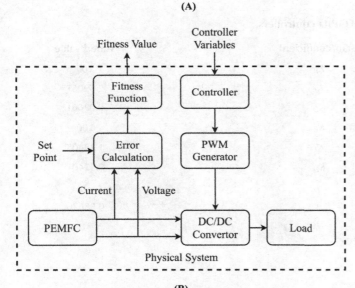

(B)

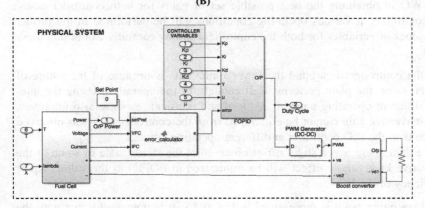

(C)

TABLE 6.2 Bounds for various gains of PID and FOPID controllers.

Controller	Coefficient	Lower bound	Upper bound
PID	K_P	0.1	100
	K_I	0.1	50
	K_D	0.001	5
FOPID	K_P	0.1	100
	K_I	0.1	50
	K_D	0.001	5
	γ	0	1
	μ	0	1

FOPID, fractional-order PID.

TABLE 6.3 GWO-tuned gains of PID and FOPID controllers.

Controller	Gain/coefficient	Tuned value
PID	K_P	100
	K_I	2.69061
	K_D	0.02661
FOPID	K_P	100
	K_I	1.23405
	K_D	0.51401
	γ	1
	μ	0.65826

FOPID, fractional-order PID; *GWO*, gray wolf optimizer.

operational end depicts the efficiency of GWO in obtaining the best possible set of gains for both controller cases. Thorough testing and analysis of both the controllers at various operating conditions will be performed in the further sections. Table 6.3 presents the final tuned values of variables for both the controllers and by carefully examining these values, few important points can be deduced.

1. The derivative value observed for both the controllers is around the lower limit. This is because of the nature of PEMFC. Using the proportional controller alone, the plant performs well enough for the operation. Using the integral and derivative control, the offset at different operating points is reduced and transient response also improves. Therefore despite having low value, this derivative gain cannot be eliminated from the controller as it was observed that PI controller had inferior performance than the PID controller at different operating points.
2. As explained earlier, due to the derivative gain, there was a slight offset observed at the steady state of some of the operating points. This is called the derivative kick, whose effect will be minimized in FOPID as the value of μ is less and will ultimately increase the efficiency of the FC.

The controllers' operation at the nominal operating point is represented in Fig. 6.7A. It is noteworthy that both the controllers perform almost similar at the tuned conditions, which is a desirable optimization trait. However, once variation in operating conditions is introduced, GWO—FOPID must perform better than the GWO—PID due to reduced

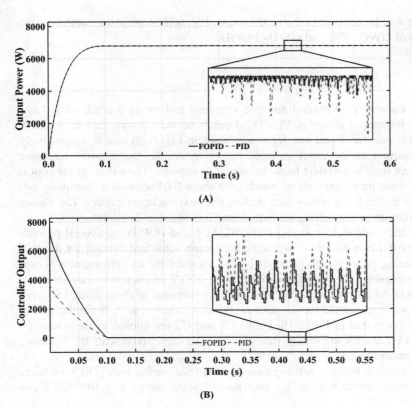

(A)

(B)

FIGURE 6.7 (A) Maximum power point tracking responses of gray wolf optimizer-tuned PID and fractional-order PID controllers at operating point. (B) Controller outputs for gray wolf optimizer-tuned PID and fractional-order PID.

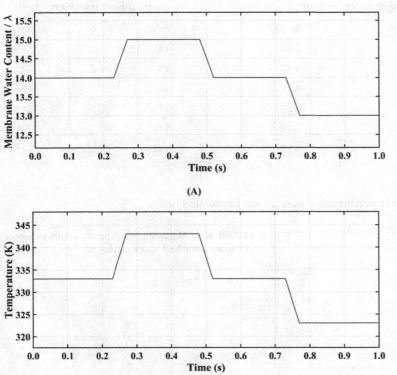

(A)

(B)

FIGURE 6.8 Profiles of (A) λ variation and (B) T variation.

derivative kick. A detailed analysis of the same will be covered in the next section. Fig. 6.7B depicts the controller output for $T = 333K$ and $\lambda = 14$ using tuned gains of GWO–PID and GWO–FOPID.

6.6 MPPT performance analysis

Two different cases are designed to assess the controller on several possible scenarios and create a practical and uniformly applicable comparative platform. Since the output power of PEMFC depends on two factors, that is, λ and T, therefore two different input profiles are chosen, one for λ and one for T as shown in Fig. 6.8A and B, respectively. The profiles of λ and T depict the slant transition in the input variable values to imitate the real-life scenarios. However, the magnitude of the variation was kept high to simulate harsh industrial conditions. The nature of the profile is the same for both λ and T, and one can see three transitions, all of which take about 0.04 seconds to complete, and enough time was provided to all the controllers to reach the steady state before the next variation occurs. The chosen transition period along with their respective variations in magnitude for both λ and T is listed in Table 6.4.

These transitions are so chosen to assess all the controllers, namely, P&O, IC, PID, and FOPID, on several possible scenarios. Each variation case is further analyzed for two different variations: increasing value and another for decreasing the variable's value from the nominal operating point. Each controller is also assessed for its performance at both transients and steady-state condition. Fig. 6.9 represents the considered controllers' MPPT responses at various points of inspections for the considered scenarios. Points A1 and A2 represent the points for transient analysis study for variation in λ whereas points B1 and B2 depict the positions for the steady-state analysis of variation in λ. A similar performance investigation is done for T variation, as illustrated in Fig. 6.10. Points C1 and C2 are chosen as the points for transient analysis study, whereas points D1 and D2 are for the steady-state analysis. Table 6.5 represents the summary of all the analysis, along with the nature of the transition.

For the transient analysis, two performance metrics, namely, settling time t_{set} and total energy loss (TEL) are taken into account. For the steady-state analysis, average power loss (APL), root mean square energy loss (RMSEL), and

TABLE 6.4 Chosen transition timings and magnitudes of λ and T.

Input profile (λ/T)	Transition duration (seconds)		Magnitude of transition	
	From	To	From	To
λ	0.23	0.27	14	15
	0.48	0.52	15	14
	0.73	0.77	14	13
$T(K)$	0.23	0.27	333	343
	0.48	0.52	343	333
	0.73	0.77	333	323

maximum power ripple (MPR) are used. Complete performance metrics are formulated below.

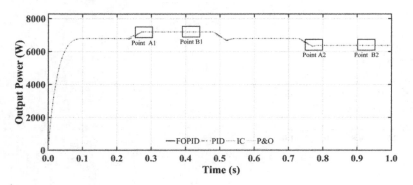

FIGURE 6.9 Maximum power point tracking responses of various controllers under varying λ.

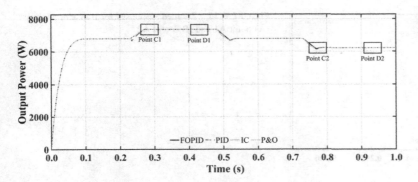

FIGURE 6.10 Maximum power point tracking responses of various controllers under varying T.

TABLE 6.5 Type of assessment performed for the points marked in Fig. 6.9 and Fig. 6.10.

Point index	Nature of profile	Type of analysis (transient/steady state)
Point A1	Increase in λ	Transient analysis
Point A2	Decrease in λ	Transient analysis
Point B1	Increase in λ	Steady-state analysis
Point B2	Decrease in λ	Steady-state analysis
Point C1	Increase in T	Transient analysis
Point C2	Decrease in T	Transient analysis
Point D1	Increase in T	Steady-state analysis
Point D2	Decrease in T	Steady-state analysis

$$\text{Settling time}(t_{\text{set}}) = \text{Time taken by FC output power to enter the 0.05\% band and then settles inside the margin} \tag{6.22}$$

$$\text{TEL} = \sum_{t_{\text{start}}}^{t_{\text{set}}} (P_{max} - P_{\text{controller}}) \times T_s \tag{6.23}$$

$$\text{APL} = \sum_{t_{\text{end}}-0.1}^{t_{\text{end}}} \frac{P_{max} - P_{\text{controller}}}{N_s} \tag{6.24}$$

$$\text{RMSEL} = \sqrt{\frac{\sum_{t_{\text{end}}-0.1}^{t_{\text{end}}} (P_{max} - P_{\text{controller}})^2 T_s^2}{N_s}} \tag{6.25}$$

$$\text{MPR} = max(P_{max} - P_{\text{controller}}) \text{ at steady state} \tag{6.26}$$

where t_{start} is the starting point of the analysis chosen by the user, P_{max} is the maximum power of FC at that particular setpoint, $P_{\text{controller}}$ is the instantaneous power output of FC, T_s is the sampling time chosen by the user and is equal to 10^{-5} seconds in this work, t_{end} is the time just before the next transition, which is chosen by the user according to the profile and N_s is the number of samples occurring in-between the chosen time frame having a value of $N_s = \frac{0.1}{T_s} = \frac{0.1}{10^{-5}} = 10,000$, in this work. Furthermore, to enhance user readability and improve the overall representation, this section is divided into two parts: the first for performance assessment on λ variations and the second for performance assessment over T variations.

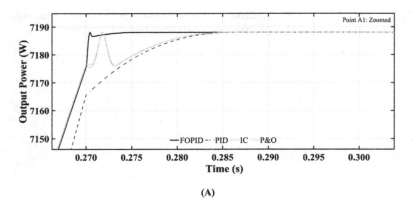

FIGURE 6.11 Zoomed responses for transient analysis: (A) point A1 and (B) point A2.

TABLE 6.6 Transient performance metrics at points A1 and A2: variation in λ.

Point	Performance metric	Measured value				Improvement offered by FOPID (%)		
		FOPID	PID	IC	P&O	PID	IC	P&O
A1	t_{set}(s)	*0.00025*	0.00936	0.00875	0.00870	97.33	97.14	97.13
	TEL (W)	*0.00211*	0.10500	0.06557	0.06420	97.98	96.77	96.70
A2	t_{set}(s)	*0.01386*	0.01386	0.01441	0.01439	0	3.82	3.68
	TEL (W)	*0.37916*	0.37916	0.41798	0.41671	0	9.29	9.01

FOPID, fractional-order PID; *IC*, incremental conductance; *P&O*, perturb and observe; *TEL*, total energy loss.

6.6.1 Case A: performance assessment: variation in λ

Fig. 6.9 shows that 4 different points are chosen, points A1 and A2 for transient and points B1 and B2 for steady-state performance assessment of all the controllers. The following subsections will provide a detailed investigation of performance improvement of GWO−FOPID over P&O, IC, and GWO−PID using various performance metrics defined in the previous section.

6.6.1.1 Transient analysis

Fig. 6.11A and B represents the zoomed version of points A1 and A2, where all the controllers' transient performance is analyzed. Settling points and TEL calculated using the Eqs. (6.22) and (6.23), respectively, are tabulated in Table 6.6.

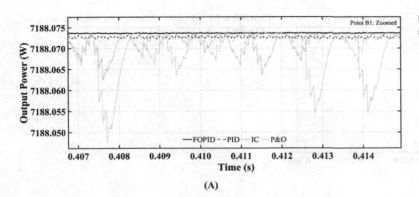

FIGURE 6.12   Zoomed responses for steady-state analysis: (A) point B1 and (B) point B2.

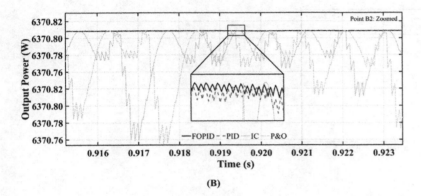

TABLE 6.7 Steady state performance metrics at points B1 and B2: variation in λ.

Point	Performance metric	Measured value				Improvement offered by FOPID (%)		
		FOPID	PID	IC	P&O	PID	IC	P&O
B1	APL (wM)	0.15988	1.15604	3.38779	3.53533	86.16	95.28	95.48
	RMSEL (J)	1.71×10^{-9}	1.18×10^{-8}	5.34×10^{-8}	5.70×10^{-8}	85.51	96.80	97.01
	MPR (mW)	0.31944	1.74193	24.82389	26.17514	81.66	98.71	98.78
B2	APL (wM)	0.15646	0.21339	15.13773	16.78990	26.67	98.96	99.06
	RMSEL (J)	1.77×10^{-9}	2.37×10^{-9}	2.09×10^{-7}	2.28×10^{-7}	25.17	99.15	99.22
	MPR (mW)	0.43252	0.53424	68.77506	67.31174	19.03	99.37	99.36

APL, average power loss; *FOPID*, fractional-order PID; *IC*, incremental conductance; *MPR*, maximum power ripple; *P&O*, perturb and observe; *RMSEL*, root mean square energy loss.

6.6.1.2   Steady-state analysis

The zoomed versions of points B1 and B2 are shown in Fig. 6.12A and B, respectively. Further corresponding APL, RMSEL, and MPR calculated using Eqs. (6.24)–(6.26), respectively, are listed in Table 6.7. The steady-state assessment is done in a window of 0.38–0.48 seconds for point B1 and from 0.9 to 1 seconds for point B2.

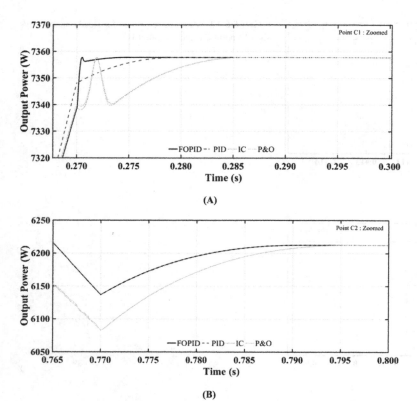

TABLE 6.8 Transient performance metrics at points C1 and C2: variations in T.

Point	Performance metric	Measured value				Improvement offered by FOPID (%)		
		FOPID	PID	IC	P&O	PID	IC	P&O
C1	t_{set}(s)	**0.00033**	0.00372	0.01019	0.01028	91.12	96.76	96.78
	TEL (W)	**0.40257**	4.48842	12.49020	12.59323	91.03	96.77	96.80
C2	t_{set}(s)	**0.01616**	0.01616	0.02142	0.02138	0	24.55	24.41
	TEL (W)	**0.47148**	0.47166	1.00782	1.00254	0.04	53.21	52.97

FOPID, fractional-order PID; *IC*, incremental conductance; *P&O*, perturb and observe; *TEL*, total energy loss.

6.6.2   Case B: performance assessment: variation in T

Fig. 6.10 depicts the four points of analysis chosen for the overall performance of all the controllers. Points C1 and C2 are for transient and points D1 and D2 are for steady-state investigations over which further subsections are created. All the performance metrics are calculated based on the previously mentioned equations and the performance improvement of GWO—FOPID over P&O, IC, and GWO—PID is tabulated.

6.6.2.1   Transient analysis

Zoomed versions of points C1 and C2 marked in Fig. 6.10 are depicted by Fig. 6.13. All the performances, that is, settling time and undershoot, are calculated using Eqs. (6.22) and (6.23), respectively, and tabulated in Table 6.8. For the calculation t_{set} was chosen as 0.27 and 0.77 seconds, respectively, for points A1 and A2, as it marked the transition's completion.

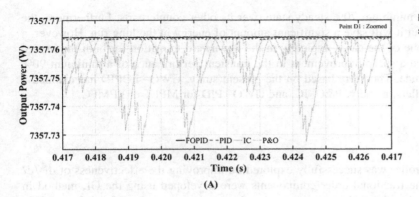

(A)

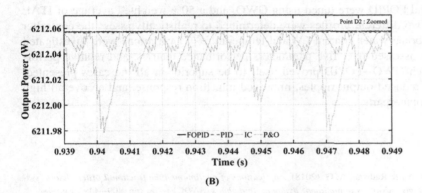

(B)

FIGURE 6.14   Zoomed responses for steady-state analysis: (A) point D1 and (B) point D2.

TABLE 6.9 Steady-state performance metrics at points D1 and D2: variations in *T*.

Point	Performance metric	Measured value				Improvement offered by FOPID (%)		
		FOPID	PID	IC	P&O	PID	IC	P&O
D1	APL (mW)	0.14161	0.63706	7.39280	7.10557	77.77	98.08	98.00
	RMSEL (J)	1.55×10^{-9}	6.61×10^{-9}	1.06×10^{-7}	1.01×10^{-7}	76.56	98.54	98.47
	MPR (mW)	0.29920	1.02644	37.13097	36.72162	70.81	99.19	99.18
D2	APL (mW)	0.32953	0.96724	13.43729	16.99106	65.93	97.54	98.06
	RMSEL (J)	3.44×10^{-9}	9.93×10^{-9}	1.99×10^{-7}	2.49×10^{-7}	65.38	98.27	98.61
	MPR (mW)	0.56585	1.51385	89.06499	78.66588	62.62	99.36	99.28

APL, average power loss; *FOPID*, fractional-order PID; *IC*, incremental conductance; *MPR*, maximum power ripple; *P&O*, perturb and observe; *RMSEL*, root mean square energy loss.

6.6.2.2   Steady-state analysis

Fig. 6.14A and B shows the zoomed version of point marked in Fig. 6.10 for points D1 and D2. All the steady-state performance metrics, namely, APL, RMSEL, and MPR, are calculated, and the performance improvement of GWO−FOPID over other controllers is tabulated in Table 6.9. The steady-state assessment is done in a window of 0.38−0.48 seconds for point D1 and from 0.9 to 1 seconds for point D2.

In all the investigations presented above, it inferred that the GWO-tuned FOPID controller's performance is superior to the rest of the considered controllers' overall given cases. Its transient operation or steady-state analysis FOPID showed a high-performance improvement of a minimum 70% steady state over its other counterparts. Furthermore, the magnitude of improvement seems to be less, but it will save a significant amount of energy in the long run. However, a similar performance to PID was observed in one of the cases, typically in the downward transition for both the inputs. Still, on the other hand, GWO−FOPID showed a sharp improvement in the transient performance of a minimum 90% for the cases of the upward rise in the magnitude. Therefore based on the present study, GWO−FOPID has claimed a superior option over other conventional controllers, namely, P&O, IC, and GWO−PID for MPPT of PEMFC.

6.7 Conclusion

In this chapter, the application of FOPID controller was successfully explored on improving the effectiveness of dP/dI feedback-based MPPT method in PEMFC. The fractional-order components were developed using the GL method in the discrete-time domain. FOPID was compared to three different conventional controllers, namely, P&O, IC, and PID, to create a comparative platform. Both PID and FOPID were tuned using GWO, and a 50% weighted average of ITAE and IAE was chosen as the fitness function. Two different profiles were determined to realistically assess the controller over all the possible real-life scenarios. Membrane water content (λ) and temperature (T) variations were investigated wherein these controllers' performances were assessed over five parameters performance metrics, and relative performance investigations were carried out in which GWO−FOPID proved itself to be superior in all the cases presented. GWO−FOPID offered the highest efficiency, reduced output ripples, improved transition response, and an overall high energy gain compared to other conventional counterparts.

References

AbdelAty, A. M., Azar, A. T., Vaidyanathan, S., Ouannas, A., & Radwan, A. G. (2018). *Applications of continuous-time fractional order chaotic systems. Mathematical techniques of fractional order systems, advances in nonlinear dynamics and chaos (ANDC) series* (pp. 409−449). Elsevier.

Abdous, F. (2009). Fuel Cell/DC-DC convertor control by sliding mode method. *World Academy of Science, Engineering and Technology, 37*(1), 1106−1110.

Ahmad, W. M., & Sprott, J. C. (2003). Chaos in fractional-order autonomous nonlinear systems. *Chaos, Solitons, and Fractals, 16*(2), 339−351. Available from https://doi.org/10.1016/S0960-0779(02)00438-1.

Ahmadi, S., Abdi, S., & Kakavand, M. (2017). Maximum power point tracking of a proton exchange membrane fuel cell system using PSO-PID controller. *International Journal of Hydrogen Energy, 42*(32), 20430−20443. Available from https://doi.org/10.1016/j.ijhydene.2017.06.208.

Alain, K. S. T., Azar, A. T., Kengne, R., & Bertrand, F. H. (2020). Stability analysis and robust synchronisation of fractional-order modified Colpitts oscillators. *Internationnal Journal of Automation and Control, 14*(1), 52−79.

Alain, K.S.T., Romanic, K., Azar, A.T., Vaidyanathan, S., Bertrand, F.H., & Adele, N.M. (2018). Dynamics analysis and synchronization in relay coupled fractional order colpitts oscillators. In *Advances in system dynamics and control* (pp. 317−356). USA: IGI-Global. doi:10.4018/978-1-5225-4077-9.ch011.

Al-Dabbagh, A. W., Lu, L., & Mazza, A. (2010). Modelling, simulation and control of a proton exchange membrane fuel cell (PEMFC) power system. *International Journal of Hydrogen Energy, 35*(10), 5061−5069. Available from https://doi.org/10.1016/j.ijhydene.2009.08.090.

Amara, K., Malek, A., Bakir, T., Fekik, A., Azar, A. T., Almustafa, K. M., ... Hocine, D. (2019). Adaptive neuro-fuzzy inference system based maximum power point tracking for stand-alone photovoltaic system. *International Journal of Modelling, Identification and Control, 33*(4), 311−321.

Ammar, H. H., Azar, A. T., Shalaby, R., & Mahmoud, M. I. (2019). Metaheuristic optimization of fractional order incremental conductance (FO-INC) maximum power point tracking (MPPT). *Complexity, 2019*, 1−13. Available from https://doi.org/10.1155/2019/7687891.

Ammar, H.H., Azar, A.T., Tembi, T.D., Tony, K., & Sosa, A. (2018). Design and implementation of fuzzy PID controller into multi agent smart library system prototype. In *The international conference on advanced machine learning technologies and applications (AMLTA2018). Advances in* intelligent systems and computing (Vol. 723, pp. 127−137). Cham: Springer.

Avanaki, N., & Sarvi, M. (2016). A new maximum power point tracking method for PEM fuel cell based on water cycle algorithm. *Journal of Renewable Energy and Environment, 3*(1), 35−42.

Azar, A. T., & Serrano, F. E. (2020). Stabilization of port hamiltonian chaotic systems with hidden attractors by adaptive terminal sliding mode control. *Entropy, 22*(1), 122. Available from https://doi.org/10.3390/e22010122.

Azar, A. T., & Serrano, F. E. (2018a). Adaptive decentralised sliding mode controller and observer for asynchronous nonlinear large-scale systems with backlash. *International Journal of Modelling, Identification and Control, 30*(1), 61−71.

Azar, A. T., & Serrano, F. E. (2018b). *Fractional order sliding mode PID controller/observer for continuous nonlinear switched systems with PSO parameter tuning, . The international conference on advanced machine learning technologies and applications (AMLTA2018). Advances in intelligent systems and computing* (Vol. 723, pp. 13−22). Cham: Springer.

Azar, A. T., Serrano, F. E., Rossell, J. M., Vaidyanathan, S., & Zhu, Q. (2020a). Adaptive self-recurrent wavelet neural network and sliding mode controller/observer for a slider crank mechanism. *International Journal of Computer Applications in Technology, 63*(4), 273−285.

Azar, A. T., Serrano, F. E., Vaidyanathan, S., & Albalawi, H. (2020b). *Adaptive higher order sliding mode control for robotic manipulators with matched and mismatched uncertainties, . The international conference on advanced machine learning technologies and applications (AMLTA2019). Advances in intelligent systems and computing* (Vol. 921, pp. 360–369). Cham: Springer.

Azar, A. T., Serrano, F. E., Koubaa, A., Kamal, N. A., Vaidyanathan, S., & Fekik, A. (2020c). *Adaptive terminal-integral sliding mode force control of elastic joint robot manipulators in the presence of hysteresis, . The international conference on advanced intelligent systems and informatics (AISI 2019). Advances in intelligent systems and computing* (Vol. 1058, pp. 266–276). Springer.

Azar, A.T., Serrano, F.E., & Koubaa, A. (2020d) Adaptive fuzzy type-2 fractional order proportional integral derivative sliding mode controller for trajectory tracking of robotic manipulators. In *2020 IEEE international conference on autonomous robot systems and competitions (ICARSC), April 15–17, Ponta Delgada, Portugal* (pp. 183–187). IEEE. doi: 10.1109/ICARSC49921.2020.9096163.

Azar, A.T., Serrano, F.E., & Koubaa, A. (2020e). Adaptive integral sliding mode force control of robotic manipulators with parametric uncertainties and time-varying loads. In *2020 IEEE international conference on autonomous robot systems and competitions (ICARSC), April 15–17, Ponta Delgada, Portugal* (pp. 188–193). IEEE. doi: 10.1109/ICARSC49921.2020.9096110.

Azar, A. T., Radwan, A. G., & Vaidyanathan, S. (2018a). *Fractional order systems: Optimization, control, circuit realizations and applications*. Elsevier, ISBN: 9780128161524.

Azar, A. T., Radwan, A. G., & Vaidyanathan, S. (2018b). *Mathematical techniques of fractional order systems*. Elsevier, ISBN 9780128135921.

Azar, A. T., El-Said, S. A., & Hassanien, A. E. (2013). Fuzzy and hard clustering analysis for thyroid disease. *Computer Methods and Programs in Biomedicine, 111*(1), 1–16.

Bankupalli, P. T., et al. (2019). A non-iterative approach for maximum power extraction from PEM fuel cell using resistance estimation. *Energy Conversion and Management, 187*(September), 565–577. Available from https://doi.org/10.1016/j.enconman.2019.02.091.

Barakat, M.H., Azar, A.T., & Ammar, H.H. (2020). Agricultural service mobile robot modeling and control using artificial fuzzy logic and machine vision. In *The international conference on advanced machine learning technologies and applications (AMLTA2019). Advances in intelligent systems and computing* (Vol. 921, pp. 453–465). Cham: Springer.

Benchouia, N. E., et al. (2015). An adaptive fuzzy logic controller (AFLC) for PEMFC fuel cell. *International Journal of Hydrogen Energy, 40*(39), 13806–13819. Available from https://doi.org/10.1016/j.ijhydene.2015.05.189.

Ben Smida, M., Sakly, A., Vaidyanathan, S., & Azar, A. T. (2018). *Control-based maximum power point tracking for a grid-connected hybrid renewable energy system optimized by particle swarm optimization. Advances in system dynamics and control* (pp. 58–89). USA: IGI-Global, 10.4018/978-1-5225-4077-9.ch003.

Bingul, Z., & Karahan, O. (2018). Comparison of PID and FOPID controllers tuned by PSO and ABC algorithms for unstable and integrating systems with time delay. *Optimal Control Applications and Methods, 39*(4), 1431–1450. Available from https://doi.org/10.1002/oc.2419.

Bizon, N. (2010). On tracking robustness in adaptive extremum seeking control of the fuel cell power plants. *Applied Energy, 87*(10), 3115–3130. Available from https://doi.org/10.1016/j.apenergy.2010.04.007.

Bizon, N. (2013). Energy harvesting from the FC stack that operates using the MPP tracking based on modified extremum seeking control. *Applied Energy, 104*, 326–336. Available from https://doi.org/10.1016/j.apenergy.2012.11.011.

Bizon, N. (2017). Energy optimization of fuel cell system by using global extremum seeking algorithm. *Applied Energy, 206*(March), 458–474. Available from https://doi.org/10.1016/j.apenergy.2017.08.097.

Boudghene Stambouli, A., & Traversa, E. (2002). Fuel cells, an alternative to standard sources of energy. *Renewable and Sustainable Energy Reviews, 6*(3), 297–306. Available from https://doi.org/10.1016/S1364-0321(01)00015-6.

Chen, P. Y., et al. (2017). A novel variable step size fractional order incremental conductance algorithm to maximize power tracking of fuel cells. *Applied Mathematical Modelling, 45*, 1067–1075. Available from https://doi.org/10.1016/j.apm.2017.01.026.

Derbeli, M., et al. (2017). A robust MPP tracker based on backstepping algorithm for proton exchange membrane fuel cell power system. In *2017 11th IEEE international conference on compatibility, power electronics and power engineering (CPE-POWERENG 2017)* (pp. 424–429). doi: 10.1109/CPE.2017.7915209.

Djouima, M., Azar, A. T., Drid, S., & Mehdi, D. (2018). Higher order sliding mode control for blood glucose regulation of type 1 diabetic patients. *International Journal of System Dynamics Applications, 7*(1), 65–84.

Djeddi, A., Dib, D., Azar, A. T., & Abdelmalek, S. (2019). Fractional order unknown inputs fuzzy observer for takagi-sugeno systems with unmeasurable premise variables. *Mathematics, 7*(10), 984. Available from https://doi.org/10.3390/math7100984.

Feliu, V., et al. (2004). Proposals for fractional PI λ D μ tuning 1, January.

Fekik, A., Azar, A.T., Kamal, N.A., Serrano, F.E., Hamida, M.L., Denoun, H., & Yassa, N. (2021a). Maximum power extraction from a photovoltaic panel connected to a multi-cell converter. In *Proceedings of the international conference on advanced intelligent systems and informatics 2020 (AISI 2020). Advances in intelligent systems and computing* (Vol. 1261). Cham: Springer.

Fekik, A., Azar, A. T., Kamal, N. A., Denoun, H., Almustafa, K. M., Hamida, M. L., & Zaouia, M. (2021b). *Fractional-order control of a fuel cell-boost converter system, . Advanced machine learning technologies and applications. AMLTA 2020. Advances in intelligent systems and computing* (Vol. 1141, pp. 713–724). Singapore: Springer.

Garrappa, R., Kaslik, E., & Popolizio, M. (2019). Evaluation of fractional integrals and derivatives of elementary functions: Overview and tutorial. *Mathematics, 7*(5), 1–21. Available from https://doi.org/10.3390/math7050407.

Ghoudelbourk, S., Azar, A. T., Dib, D., & Omeiri, A. (2020). Selective harmonic elimination strategy in the multilevel inverters for grid connected photovoltaic system. *International Journal of Advanced Intelligence Paradigms, 15*(3), 317–339.

Ghoudelbourk, S., Dib, D., Omeiri, A., & Azar, A. T. (2016). MPPT control in wind energy conversion systems and the application of fractional control (PIα) in pitch wind turbine. *International Journal of Modelling, Identification and Control (IJMIC), 26*(2), 140–151.

Giove, S., Azar, A. T., & Nordio, M. (2013). Fuzzy logic control for dialysis application. In A. T. Azar (Ed.), *Biofeedback systems and soft computing techniques of dialysis* (Vol. 405, pp. 1181–1222). GmbH Berlin/Heidelberg: Springer-Verlag, 10.1007/978-3-642-27558-6_9.

Giustiniani, A., et al. (2006). PEM fuel cells control by means of the perturb and observe technique. In *IECON proceedings (industrial electronics conference)* (pp. 4349–4354). doi: 10.1109/IECON.2006.347792.

Giustiniani, A., et al. (2010). Enhancing polymeric electrolyte membrane fuel cell control by means of the perturb and observe technique. *Journal of Fuel Cell Science and Technology*, 7(1), 011021. Available from https://doi.org/10.1115/1.3120275.

Gorripotu, T. S., Samalla, H., Jagan Mohana Rao, C., Azar, A. T., & Pelusi, D. (2019). TLBO algorithm optimized fractional-order PID controller for AGC of interconnected power system. In J. Nayak, A. Abraham, B. Krishna, G. Chandra Sekhar, & A. Das (Eds.), *Soft computing in data analytics. Advances in intelligent systems and computing* (Vol 758). Singapore: Springer.

Gou, B., Na, W. K., & Diong, B. (2010). *Fuel cells: Modeling, control and applications.* CRC Press Taylor & Francis Group.

Harrag, A., & Messalti, S. (2017). Variable step size IC MPPT controller for PEMFC power system improving static and dynamic performances'. *Fuel Cells*, 17(6), 816–824. Available from https://doi.org/10.1002/fuce.201700047.

Ibraheem, G. A. R., Azar, A. T., Ibraheem, I. K., & Humaidi, A. J. (2020). A novel design of a neural network based fractional PID controller for mobile robots using hybridized fruit fly and particle swarm optimization. *Complexity*, 2020, 1–18. Available from https://doi.org/10.1155/2020/3067024, Article ID 3067024.

Jeba, P., & Immanuel Selvakumar, A. (2018). FOPID based MPPT for photovoltaic system. *Energy Sources, Part A: Recovery, Utilization and Environmental Effects*, 40(13), 1591–1603, 10.1080/15567036.2018.1486480.

Kamal, N. A., Azar, A. T., Elbasuony, G. S., Almustafa, K. A., & Almakhles, D. (2020). *PSO-based adaptive perturb and observe MPPT technique for photovoltaic systems*, . The international conference on advanced intelligent systems and informatics AISI 2019. Advances in intelligent systems and computing (Vol. 1058, pp. 125–135). Springer.

Kamal, N. A., & Ibrahim, A. M. (2018). Conventional, intelligent, and fractional-order control method for maximum power point tracking of a photovoltaic system: A review. In A. T. Azar (Ed.), *Fractional order systems optimization, control, circuit realizations and applications, advances in nonlinear dynamics and chaos (ANDC)* (pp. 603–671). Elsevier.

Kammogne, A. S. T., Kountchou, M. N., Kengne, R., Azar, A. T., Fotsin, H. B., & Ouagni, S. T. M. (2020). Polynomial robust observer implementation based-passive synchronization of nonlinear fractional-order systems with structural disturbances. *Frontiers of Information Technology & Electronic Engineering*, 21(9), 1369–1386.

Khaldi, H.S. and Ammari, A.C. (2015) 'Fractional-order control of three level boost DC/DC converter used in hybrid energy storage system for electric vehicles'. In 2015 6th international renewable energy congress, IREC 2015. doi: 10.1109/IREC.2015.7110930.

Khan, A., Singh, S., & Azar, A.T. (2020a). Synchronization between a novel integer-order hyperchaotic system and a fractional-order hyperchaotic system using tracking control. In *The international conference on advanced machine learning technologies and applications (AMLTA2019). Advances in intelligent systems and computing* (Vol. 921, pp. 382–391). Cham: Springer.

Khan A., Singh S., Azar A.T. (2020b) Combination-combination anti-synchronization of four fractional order identical hyperchaotic systems. In *The international conference on advanced machine learning technologies and applications (AMLTA2019). Advances in intelligent systems and computing*, vol 921, pp. 406–414, Springer, Cham.

Khennaoui, A. A., Ouannas, A., Boulaaras, S., Pham, V. T., & Azar, A. T. (2020). A fractional map with hidden attractors: Chaos and control. *The European Physical Journal Special Topics*, 229, 1083–1093.

Khettab, K., Bensafia, Y., Bourouba, B., & Azar, A. T. (2018). *Enhanced fractional order indirect fuzzy adaptive synchronization of uncertain fractional chaotic systems based on the variable structure control: Robust H∞ design approach. Mathematical techniques of fractional order systems, advances in nonlinear dynamics and chaos (ANDC) series* (pp. 559–595). Elsevier.

Kler, D., Rana, K. P. S., & Kumar, V. (2018). A nonlinear PID controller based novel maximum power point tracker for PV systems. *Journal of the Franklin Institute*, 355(16), 7827–7864. Available from https://doi.org/10.1016/j.jfranklin.2018.06.003.

Kumar, J., Kumar, V., & Rana, K. P. S. (2018). A fractional order fuzzy PD + I controller for three-link electrically driven rigid robotic manipulator system. *Journal of Intelligent and Fuzzy Systems*, 35(5), 5287–5299. Available from https://doi.org/10.3233/JIFS-169812.

Kumar, V., Rana, K. P. S., & Mishra, P. (2016). Robust speed control of hybrid electric vehicle using fractional order fuzzy PD and PI controllers in cascade control loop. *Journal of the Franklin Institute*, 353(8), 1713–1741. Available from https://doi.org/10.1016/j.jfranklin.2016.02.018.

Kunusch, C., Puleston, P., & Mayosky, M. (2012). *Sliding-mode control of PEM fuel cells.* Available from https://doi.org/10.1007/978-1-4471-2431-3.

Liu, J., Zhao, T., & Chen, Y. (2016). Maximum power point tracking of proton exchange membrane fuel cell with fractional order filter and extremum seeking control. *CAA Journal of Automatica Sinica*, 4(1), 70–79. Available from https://doi.org/10.1115/detc2015-46633.

Luo, Y., & Chen, Y. Q. (2009). Fractional order [proportional derivative] controller for a class of fractional order systems. *Automatica*, 45(10), 2446–2450. Available from https://doi.org/10.1016/j.automatic.2009.06.022.

Meghni, B., Dib, D., & Azar, A. T. (2017a). A second-order sliding mode and fuzzy logic control to optimal energy management in wind turbine with battery storage. *Neural Computing and Applications*, 28(6), 1417–1434. Available from https://doi.org/10.1007/s00521-015-2161-z.

Meghni, B., Dib, D., Azar, A. T., Ghoudelbourk, S., & Saadoun, A. (2017b). *Robust adaptive supervisory fractional order controller for optimal energy management in wind turbine with battery storage*, . Studies in Computational Intelligence (Vol. 688, pp. 165–202). Germany: Springer-Verlag.

Mekki, H., Boukhetala, D., & Azar, A. T. (2015). *Sliding modes for fault tolerant control*, . Studies in computational intelligence (Vol. 576, pp. 407–433). Berlin/Heidelberg: Springer-Verlag GmbH, 10.1007/978-3-319-11173-5_15.

Mirjalili, S., Mirjalili, S. M., & Lewis, A. (2014). Grey wolf optimizer. *Advances in Engineering Software*, 69, 46–61. Available from https://doi.org/10.1016/j.advengsoft.2013.12.007.

Mishra, P., Kumar, V., & Rana, K. P. S. (2015). A fractional order fuzzy PID controller for binary distillation column control. *Expert Systems with Applications, 42*(22), 8533−8549. Available from https://doi.org/10.1016/j.eswa.2015.07.008.

Mjahed, S., Bouzaachane, K., Azar, A. T., Hadaj, S. E., & Raghay, S. (2020). Hybridization of fuzzy and hard semi-supervised clustering algorithms tuned with ant lion optimizer applied to Higgs Boson Search. *Computer Modeling in Engineering & Sciences, 125*(2), 459−494.

Mittal, S., Azar, A.T., & Kamal, N.A. (2021). Nonlinear fractional order system synchronization via combination-combination multi-switching. In A. E. Hassanien, A. Slowik, V. Snášel, H. El-Deeb, & F. M. Tolba (Eds.), *Proceedings of the international conference on advanced intelligent systems and informatics 2020. AISI 2020. Advances in intelligent systems and computing* (Vol. 1261). Cham: Springer, https://doi.org/10.1007/978-3-030-58669-0_75.

Monje, C. A., et al. (2008). Tuning and auto-tuning of fractional order controllers for industry applications. *Control Engineering Practice, 16*(7), 798−812. Available from https://doi.org/10.1016/j.conengprac.2007.08.006.

Nithilasaravanan, K., et al. (2019). Efficient control of integrated power system using self-tuned fractional-order fuzzy PID controller. *Neural Computing and Applications, 31*(8), 4137−4155. Available from https://doi.org/10.1007/s00521-017-3309-9.

Ortigueira, M., & Machado, J. (2017). Fractional definite integral. *Fractal and Fractional, 1*(1), 1−9. Available from https://doi.org/10.3390/fractalfract1010002.

Ouannas, A., Grassi, G., Azar, A.T., & Khennaoui, A.A. (2021). Synchronization control in fractional discrete-time systems with chaotic hidden attractors. In A. Hassanien, R. Bhatnagar, & A. Darwish (Eds.), *Advanced machine learning technologies and applications. AMLTA 2020. Advances in intelligent systems and computing* (Vol. 1141, pp. 661−669). Singapore: Springer,.

Ouannas, A., Grassi, G., & Azar, A.T. (2020a). A new generalized synchronization scheme to control fractional chaotic systems with non-identical dimensions and different orders. In *The international conference on advanced machine learning technologies and applications (AMLTA2019). Advances in intelligent systems and computing* (Vol. 921, pp. 415−424). Cham: Springer.

Ouannas, A., Grassi, G., & Azar, A.T. (2020b). Fractional-order control scheme for Q-S chaos synchronization. In *The international conference on advanced machine learning technologies and applications (AMLTA2019). Advances in intelligent systems and computing* (Vol. 921, pp. 434−441). Cham: Springer.

Petráš, I. (2011). An effective numerical method and its utilization to solution of fractional models used in bioengineering applications. *Advances in Difference Equations, 2011*. Available from https://doi.org/10.1155/2011/652789.

Pham, V. T., Gokul, P. M., Kapitaniak, T., Volos, C., & Azar, A. T. (2018). *Dynamics, synchronization and fractional order form of a chaotic system with infinite equilibria. Mathematical techniques of fractional order systems, advances in nonlinear dynamics and chaos (ANDC) series* (pp. 475−502). Elsevier.

Pham, V.T., Vaidyanathan, S., Volos, C.K., Azar, A.T., Hoang, T.M., & Yem, V.V. (2017). A three-dimensional no-equilibrium chaotic system: Analysis, synchronization and its fractional order form. In *Studies in computational intelligence* (Vol. 688, pp 449−470). Germany: Springer-Verlag.

Pilla, R., Azar, A. T., & Gorripotu, T. S. (2019). Impact of flexible AC transmission system devices on automatic generation control with a metaheuristic based fuzzy PID controller. *Energies, 12*(21), 4193. Available from https://doi.org/10.3390/en12214193.

Pilla, R., Botcha, N., Gorripotu, T.S., & Azar, A.T. (2020). Fuzzy PID controller for automatic generation control of interconnected power system tuned by glow-worm swarm optimization. In J. Nayak, V. Balas, M. Favorskaya, B. Choudhury, S. Rao, & B. Naik (Eds.), *Applications of robotics in industry using advanced mechanisms. ARIAM 2019. Learning and analytics in intelligent systems* (Vol. 5, pp 140−149). Cham: Springer.

Podlubny, I. (1999a). Fractional-order systems and PIλDμ-controllers. *IEEE Transactions on Automatic Control, 44*(1), 208−214.

Podlubny, I. (1999b). *Fractional differential equations: An introduction to fractional derivatives, fractional differential equations, to methods of their solution and some of their applications.* London: Academic Press.

Rana, K. P. S., et al. (2016). Implementation of fractional order integrator/differentiator on field programmable gate array. *Alexandria Engineering Journal, 55*(2), 1765−1773. Available from https://doi.org/10.1016/j.aej.2016.03.030.

Rana, K.P.S., et al. (2019). A novel dP/dI feedback based control scheme using GWO tuned PID controller for efficient MPPT of PEM fuel cell, *ISA Transactions* (93), 312−324. doi: 10.1016/j.isatra.2019.02.038.

Rezk, H. (2017). Performance of incremental resistance MPPT based proton exchange membrane fuel cell power system. In *2016 18th international middle-east power systems conference, MEPCON 2016—Proceedings* (pp. 199−205). doi: 10.1109/MEPCON.2016.7836891.

Sabatier, J., et al. (2002). CRONE control: Principles and extension to time-variant plants with asymptotically constant coefficients. *Nonlinear Dynamics, 29*(1−4), 363−385. Available from https://doi.org/10.1023/A:1016531915706.

Sahin, E., Ayas, M.S., & Altas, I.H. (2014). A PSO optimized fractional-order PID controller for a PV system with DC-DC boost converter. In *16th international power electronics and motion control conference and exposition, PEMC 2014* (pp. 477−481). doi: 10.1109/EPEPEMC.2014.6980539.

Sallam, O.K., Azar, A.T., Guaily, A., & Ammar, H.H. (2020). Tuning of PID controller using particle swarm optimization for cross flow heat Exchanger based on CFD System Identification. In *The international conference on advanced intelligent systems and informatics AISI 2019. Advances in intelligent systems and computing* (Vol. 1058, pp. 300−312). Springer.

Sarvi, M., et al. (2014). Optimal operation and output oscillations reduction of PEMFC by using an intelligent strategy. *International Journal of Electrochemical Science, 9*(8), 4172−4189.

Sarvi, M., et al. (2015). Eagle strategy based maximum power point tracker for fuel cell system. *International Journal of Engineering, 28*(4 A), 529−536. Available from https://doi.org/10.5829/idosi.ije.2015.28.04a.06.

Sen, M. (2014). *Introduction to fractional-order operators and their engineering applications*, 69.

Shukla, M. K., Sharma, B. B., & Azar, A. T. (2018). *Control and synchronization of a fractional order hyperchaotic system via backstepping and active backstepping approach. Mathematical techniques of fractional order systems, advances in nonlinear dynamics and chaos (ANDC) series* (pp. 597−624). Elsevier.

Singh, S., Azar, A.T., Ouannas, A., Zhu, Q., Zhang, W., & Na, J. (2017) Sliding mode control technique for multi-switching synchronization of chaotic systems. In *Proceedings of 9th international conference on modelling, identification and control (ICMIC 2017)* (pp. 880−885), July 10−12, 2017. Kunming, China: IEEE.

Singh, S., & Azar, A.T. (2020) Multi-switching combination synchronization of fractional order chaotic systems. In *Proceedings of the international conference on artificial intelligence and computer vision (AICV2020). Advances in intelligent systems and computing* (Vol. 1153, pp. 655−664). Cham: Springer.

Singh, S., Azar, A. T., Vaidyanathan, S., Ouannas, A., & Bhat, M. A. (2018). *Multiswitching synchronization of commensurate fractional order hyperchaotic systems via active control. Mathematical techniques of fractional order systems, advances in nonlinear dynamics and chaos (ANDC) series* (pp. 319−345). Elsevier.

Soliman, M., Azar, A.T., Saleh, M.A., & Ammar, H.H. (2020). Path planning control for 3-Omni fighting robot using PID and fuzzy logic controller. In *The international conference on advanced machine learning technologies and applications (AMLTA2019). Advances in intelligent systems and computing* (Vol. 921, pp. 442−452). Cham: Springer.

Soltani, I. (2013). An intelligent, fast and robust maximum power point tracking for proton exchange membrane fuel cell. *World Applied Programming*, *3*(July), 264−281.

Tang, Y., et al. (2012). Optimum design of fractional order PI λD μ controller for AVR system using chaotic ant swarm. *Expert Systems with Applications*, *39*(8), 6887−6896. Available from https://doi.org/10.1016/j.eswa.2012.01.007.

Tolba, M. F., AbdelAty, A. M., Soliman, N. S., Said, L. A., Madian, A. H., Azar, A. T., & Radwan, A. G. (2017a). FPGA implementation of two fractional order chaotic systems. *International Journal of Electronics and Communications*, *28*(2017), 162−172.

Tolba, M.F., AbdelAty, A.M., Said, L.A., Elwakil, A.S., Azar, A.T., Madian, A.H., ... Ouannas, A. (2017b). FPGA realization of Caputo and Grünwald-Letnikov operators. In *The 6th international conference on modern circuits and systems technologies (MOCAST)*, 4−6 May 2017, Thessaloniki, Greece.

Vaidyanathan, S., Azar, A. T., Akgul, A., Lien, C. H., Kacar, S., & Cavusoglu, U. (2019). A memristor-based system with hidden hyperchaotic attractors, its circuit design, synchronisation via integral sliding mode control and voice encryption. *International Journal of Automation and Control (IJAAC)*, *13*(6), 644−667.

Vaidyanathan, S., Sampath, S., & Azar, A. T. (2015). Global chaos synchronisation of identical chaotic systems via novel sliding mode control method and its application to Zhu system. *International Journal of Modelling, Identification and Control (IJMIC)*, *23*(1), 92−100.

Vaidyanathan, S., & Azar, A. T. (2015a). *Anti-synchronization of identical chaotic systems using sliding mode control and an application to Vaidyanathan-Madhavan chaotic systems, . Studies in computational intelligence* (Vol. 576, pp. 527−547). GmbH Berlin/Heidelberg: Springer-Verlag, 10.1007/978-3-319-11173-5_19.

Vaidyanathan, S., & Azar, A. T. (2015b). *Hybrid synchronization of identical chaotic systems using sliding mode control and an application to Vaidyanathan chaotic systems, . Studies in computational intelligence* (Vol. 576, pp. 549−569). Berlin/Heidelberg: Springer-Verlag GmbH, 10.1007/978-3-319-11173-5_20.

Valério, D., & da Costa, J. S. (2006). Tuning of fractional PID controllers with Ziegler-Nichols-type rules. *Signal Processing*, *86*(10), 2771−2784. Available from https://doi.org/10.1016/j.sigpro.2006.02.020.

Venkateshkumar, M., Sarathkumar, G., & Britto, S. (2014). Intelligent control based MPPT method for fuel cell power system. In *Proceedings—2013 International Conference on Renewable Energy and Sustainable Energy, ICRESE 2013* (pp. 253−257). doi: 10.1109/ICRESE.2013.6927825.

Vinagre, B., et al. (2000). Some approximations of fractional order operators used in control theory and applications. *Fractional Calculus and Applied Analysis*, *3*(3), 231−248.

Wai, R.J., & Shin, L.Z. (2011). Total sliding-mode voltage tracking control for DC-DC boost converter. In *Proceedings of the 2011 6th IEEE conference on industrial electronics and applications, ICIEA 2011* (Vol. 58, pp. 2676−2681). doi:10.1109/ICIEA.2011.5976049.

Wang, Y. X., et al. (2016). Temperature control for a polymer electrolyte membrane fuel cell by using fuzzy rule. *IEEE Transactions on Energy Conversion*, *31*(2), 667−675. Available from https://doi.org/10.1109/TEC.2015.2511155.

Zhong, Z., et al. (2008). Adaptive maximum power point tracking control of fuel cell power plants. *Journal of Power Sources*, *176*(1), 259−269. Available from https://doi.org/10.1016/j.jpowsour.2007.10.080.

Chapter 7

Robust adaptive nonlinear controller of wind energy conversion system based on permanent magnet synchronous generator

Rachid Lajouad[1], Abdelmounime El Magri[1], Abderrahim El Fadili[2], Aziz Watil[1], Lhoussaine Bahatti[1], Fouad Giri[3], Mohammed Kissaoui[1] and Ahmad Taher Azar[4,5]

[1]IESI Laboratory, ENSET Mohammedia, Hassan II University of Casablanca, Casablanca, Morocco, [2]FST Mohammedia, Hassan II University of Casablanca, Casablanca, Morocco, [3]LAC Laboratory, University of Caen Normandie, Caen, France, [4]Faculty of Computers and Artificial Intelligence, Benha University, Benha, Egypt, [5]College of Computer and Information Sciences, Prince Sultan University, Riyadh, Kingdom of Saudi Arabia

7.1 Introduction

Today, the use of energy is one of the clearest indicators of country development. The most industrialized and energy-consuming nations continue to rely on energy as a driver of growth and economic development. Its contribution to the creation of national wealth is not limited to its own added value, but affects all the other sectors of which it allows the activity.

The capital importance of investments in this sector and their life span mean that the strategic choices made today will shape the future energy landscape. In addition, our planet can no longer physically bear the greenhouse emissions that the world has known for the past years. Climate change and the disruption of many ecological balances would cause irreversible damage. We are therefore faced with the need to implement sustainable policies with three components: economic, social, and environmental (Global Wind Energy Council, 2019).

Wind power is the main pillar of tomorrow's energy supply (Ackermann & Söder, 2002; Anaya-Lara, Jenkins, Ekanayake, Cartwright, & Hughes, 2011; Hermann & Krener, 1977; Kesraoui, Korichi, & Belkadi, 2011). It generates clean and climate-friendly electricity, creates jobs, and reduces risks on several levels, such as exposure to particulate matter and susceptibility to the price volatility of imported fuel.

When comparing different technologies on the basis of all parameters, the cost of wind power declines considerably from the high price tag of energy production. This is primarily due to the very positive effect on local employment.

Historically, only fixed frequency and fixed speed systems have been used for the production of electricity in wind power plants (Anaya-Lara et al., 2011; Devaiah & Smith, 1975). However, with the advent of power electronics and the availability of semiconductor devices capable of withstanding high currents, variable speed systems are increasingly used. It is emphasized, however, that the optimal choice of the production system was not decided by considering the generator alone. The optimal choice is the one that minimizes the cost of energy produced by the wind power plant. The two types of electric machines that are mostly used in the wind industry are synchronous machines and asynchronous machines in their various variants.

The DC generator is rarely used. It is mainly encountered in micro-wind turbines (power less than 500 W) (Abouloifa, Giri, Lachkar, Chaoui, & El Magri, 2009). Although variable reluctance generators are mechanically robust machines offering good performance at all rotational speeds, they are not, until now, produced on a large scale, especially for larger powers, indeed these machines are very difficult to be controlled at a variable speed (Torrey, 1993; Torrey & Hassanin, 1995).

Renewable Energy Systems. DOI: https://doi.org/10.1016/B978-0-12-820004-9.00001-2
133

The asynchronous machine currently equips a large part of the wind farm. It has the advantage of being robust and less expensive. The variable speed control of this machine requires the use of two static converters (AC/DC) and (DC/AC) (El Fadili, Giri, El Magri, Lajouad, & Chaoui, 2012a, 2012b; El Fadili, Giri, Ouadi, et al., 2010; El Fadili, Giri, Ouadi, Dugard, & El Magri, 2014; Lajouad et al., 2013).

Synchronous generators, historically used in huge plant to produce electricity (thermal, hydraulic, or nuclear power plants), are now used coupled to wind energy conversion (WEC) system. In addition, when this type of machine is directly connected to the network, its rotation speed is fixed and proportional to the frequency of the network (El Magri, Giri, Abouloifa, & El Fadili, 2010; El Fadili, Giri, El Magri, & Ouadi, 2010). As a result of this high rigidity of generator-network connection, the fluctuations in the torque picked up by the aeroturbine propagate over the entire power train, up to the electrical power produced. This is why synchronous machines are not used in wind turbines directly connected to the network. They are, on the other hand, used when they are connected to the network by means of power converters (El Magri, Giri, Abouloifa, & Chaoui, 2010; El Magri, Giri, Abouloifa, & El Fadili, 2014; El Magri, Giri, Abouloifa, & El Fadili, 2009; El Magri, Giri, Abouloifa, & Haloua, 2007; El Magri, Giri, Abouloifa, Lachkar, & Chaoui, 2009a; El Magri, Giri, Abouloifa, Lachkar, & Chaoui, 2009b). In this configuration, the network frequency and the machine rotation speed are decoupled. This speed can therefore vary so as to optimize the aerodynamic efficiency of the wind turbine and dampen the torque fluctuations in the power train.

Presently, the focus is made on the WEC system of Fig. 7.1, which includes a permanent magnet synchronous generator (PMSG) that converts wind turbine power into electric power; the corresponding output voltage amplitude and frequency vary with wind speed. PMS generators offer several benefits in wind power applications due to their high power density, high efficiency (as the copper losses in the rotor disappear), absence of gearbox, and reduced active weight. These characteristics make it possible to achieve the high performances of variable speed control of PMSG and very loyal operating conditions (reduced need for maintenance). In variable speed operating mode, the PMSG is connected to the grid via a three-phase electronic power system (see the DC/AC part of Fig. 7.1). The three-phase voltage, generated by the PMS machine, of variable frequency and amplitude is adapted using a buck-to-buck insulated gate bipolar transistor (IGBT)-based rectifier—inverter [AC/DC/AC pulse wide modulation (PWM) converters] (Fig. 7.1). The AC side of the rectifier is connected to the PMSG stator; the output of the inverter (DC/AC) is directly connected to the power network.

Knowing that the power extracted from the turbine depends on its rotation speed, one major goal in the considered system is the control of PMSG rotation speed (Fig. 7.2A). It turns out that the achievement of maximum wind energy extraction in presence of varying wind speed conditions necessitates a varying turbine speed operation mode.

More specifically, the turbine rotating speed must be controlled so that its power-speed operating point is constantly maintained near the optimal position (Fig. 7.2A). This control objective is commonly referred to "maximum power point tracking (MPPT)" and its achievement guarantees optimal aerodynamic efficiency. Presently, we seek MPPT achievement with the WEC system of Fig. 7.1.

The global system (including wind turbine, PMSG, and AC/DC/AC power converter) has to be controlled in order to achieve a tight reference-speed tracking. Over and above that, the rotor speed reference should be overhauled online, following the variation of wind velocity (v_w), to fulfill the MPPT requirement. Several articles are published on this subject. There are mainly two methods to deal with this MPPT goal. The first one, introduced by Senjyu et al. (2009), based on the characteristics of wind turbine requires a measurement of the wind speed. However, the wind speed along a blade of the wind turbine is no longer constant; the measurement therefore is only an average, which makes this

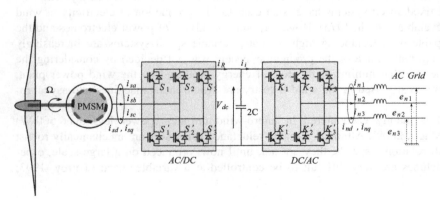

FIGURE 7.1 The AC/DC/AC converter power circuit wind energy conversion system.

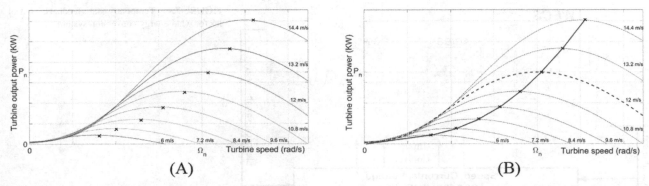

FIGURE 7.2 Turbine power versus rotor speed characteristics. (A) Turbine power characteristics ($\beta = 0$ degree) and (B) optimal power characteristics.

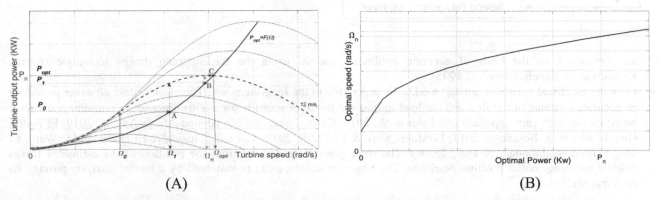

FIGURE 7.3 Speed-reference optimization. (A) Optimal speed seeking and (B) Function $F(.)$ defining the speed-reference optimizer.

method difficult to implement, especially for large wind turbines. This weakness has been overcome in Rocha (2011) by using a Kalman observer, which is able to estimate online load/turbine torque based on rotor speed measurements.

There, the output-feedback controller (including the Kalman predictor) is based on a linear approximation of the WEC systems, and no formal analysis is made there for the proposed control strategy (e.g., predictor estimator convergence is not proved). The second category of MPPT methods, using the so-called extremum-seeking or perturbation-observation technique, do not necessitate turbine characteristics (Heier, 2014; Jain, 2011; Khalil, 2009) [43]. These methods are most suitable for wind turbines with small inertia.

In this work, the aim is achieved (guarantee tight rotor speed-reference tracking and rotor speed-reference optimization for MPPT achievement, without necessitating mechanical sensors for wind and rotor speeds and load torque) based on the use of so-called power to optimal speed (POS) optimizer published in El Fadili et al. (2018), El Fadili, Giri, El Magri, Lajouad, and Chaoui (2014), El Magri, Giri, Besançon, et al. (2013), Lajouad, El Magri, El Fadili, Chaoui, and Giri (2015), Lajouad, Giri, Chaoui, Fadili, and Magri (2019), and Watil, El Magri, Raihani, Lajouad, and Giri (2020b). The sensorless feature is quite beneficial as it entails cost reduction due to no sensor implementation and maintenance. Interestingly, sensorless (output-feedback) controllers remain beneficial, even when sensors are available, for sensor fault detection and isolation and fault tolerant control (Fig. 7.3).

Presently, the sensorless aspect is tackled using the state observers. Hence, the proposed output-feedback control architecture involves three main types of components (Fig. 7.4):

- Speed-reference optimizer, regulators, and observers. The online speed-reference optimizer is presently designed, making full use of the nonlinear wind turbine aerofoil characteristic.
- Rotor speed control is performed using a nonlinear regulator that regulates the d-component of the stator current to zero, thus optimizes the delivered stator current. Besides, the control strategy regulates the DC-link voltage (between AC/DC rectifier and inverter) (Fig. 7.1) to a constant reference value, commonly equal to the nominal PMSG stator voltage. Indeed, this regulation loop controls the reactive power delivered to the grid. All previous regulation loops

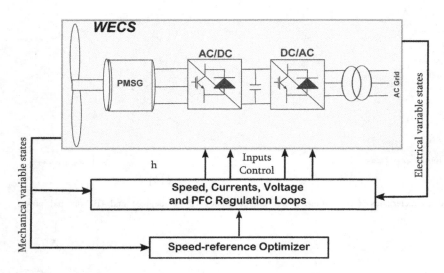

are developed on the basis of accurate nonlinear models, using the backstepping design technique (Krstic, Kokotovic, & Kanellakopoulos, 1995).

- Only the electrical variables are assumed to be accessible to the measurements. Therefore a state observer providing an accurate estimate of the flux and the load torque and the rotor speed is used. This observer is a nonlinear observer based on the high-gain approach (Al Tahir et al., 2016; Cuny, Lajouad, Giri, Ahmed-Ali, & Assche, 2019; El Fadili, Giri, El Magri, & Besancon, 2014; El Magri, Giri, & El Fadili, 2010; Lajouad, Chaoui, & El Fadili, 2015; Watil, El Magri, Raihani, Lajouad, & Giri, 2020a). The rotor position is derived from the position of the estimated fluxes without requiring initial position detection. The observer development is sustained by a formal analysis proving its convergence.

The implementation of the state observer, thus synthesized, is a challenge to be raised. In fact, apart from the linear systems for which there are established implementation procedures, this is not the case for nonlinear systems. Many works are published in this focus (Dabroom & Khalil, 2001; Khalil, 2017; Koutroulis & Kalaitzakis, 2006), all of them gives us a numeric approximation of the derivative of the estimated state variable vector. The first point that we will discuss is the choice of of the sampling period, which strongly depends on the dynamics of the system. Indeed, a wrong choice of this period can cause the instability of the observer and subsequently of the entire command. A general guideline is to choose this period for the dynamics of the fastest state of the observer. Then, the accurate estimation law is to switch from continuous time $\hat{x}(t)$ to discrete state $\hat{x}(kT)$. Finally, a technical framework to design this numeric observer in a digital signal processor (DSP) has to be discussed in the light of the distortion made on the estimated states.

The control strategy thus developed is illustrated in Fig. 7.4. It is optimal (in the MPPT sense), multiloop (speed, current, voltage regulation loops) and output-feedback (mechanical and electromagnetic variables observers). It is sustained by a theoretical analysis and a simulation study showing the achievement of quite satisfactory control performances, despite the varying wind velocity and the corresponding change in load torque.

This chapter is organized as follows: the speed-reference optimizer is designed in Section 7.2; the WEC system under study is modeled, and its representation in the state space is given in Section 7.3; the state-feedback controller is designed and analyzed in Section 7.4; the adaptive state observer is designed and analyzed, and its discretization is discussed in Section 7.5. Several simulations in MATLAB®/Simulink based on an accurate system model is made to show the controller performances.

7.2   Speed-reference optimization: power to optimal speed

In order to maximize the extracted power from the WEC system, an algorithm to generate an optimal speed reference will be designed. In this respect, the aim is a construction of a speed-reference optimizer that ensure the MPPT requirement. Specifically, to overcome the effect of wind speed variation, the optimizer is expected to compute online the optimal turbine speed value Ω_{opt}. In fact, if the current turbine rotor speed Ω is made equal to Ω_{opt}, then the maximal wind energy is captured and transmitted to the grid through the aerogenerator. In this work, the so-called approach POS (El

Magri, Giri, Besançon, et al., 2013) will be applied. In this adopted approach, the optimizer algorithm is based on the power characteristic of the turbine (Fig. 7.2A). One does not require any measurement of the wind speed.

7.2.1   Power characteristic of the turbine $P(\Omega, v_w)$

The turbine characteristic is illustrated in Fig. 7.2A. The stiffness of the drivetrain is infinite, and the friction factor and the inertia of the turbine are combined with those of the generator coupled to the turbine. The wind power acting on the swept area of blade A is a function of the air density ρ (kg/m^3) and the wind velocity $v_w(m/s)$. The transmitted power P (W) is generally deduced from the wind power, using the power coefficient C_p, as follows:

$$P = \frac{1}{2} C_p \rho A v_w^3 \tag{7.1}$$

where C_p is the performance coefficient of the turbine; it is a nonlinear function of the tip speed ratio of the rotor blade tip speed to wind speed $\lambda = R_t \Omega / v_w$ (with R_t the turbine radius), which depends on the wind velocity v_w and the rotation speed of the aerogenerator rotor $\Omega(rad/s)$.

Fig. 7.2A represents the transmitted power according to the aerogenerator rotor speed for various values of the wind speed v_w (with a blade pitch angle $\beta = 0$ degree). It is clear that for any wind speed, there is a rotor speed that maximizes the extracted power. Now, let us find the characteristic of the "extractable" power versus the speed of the turbine rotor.

7.2.2   Optimal power characteristic of the turbine $(P_{opt}, \ \Omega)$

The summits of the curves, of the Fig. 7.2A, give the maximum "extractable" power P_{opt} and so represent the optimal points. Each one of these points is characterized by the optimal speed Ω_{opt}. It is readily seen from Fig. 7.2B that for any wind velocity value, say v_w^i, there is a unique couple (Ω_i, P_i) that involves the largest extractable power. The set of all such optimal couples (Ω_i, P_i) is represented by the blue curve (parabolic curve) in Fig. 7.2B. A number of such couples have been collected from Fig. 7.2A and interpolated to get a polynomial function $\Omega_{opt} = F(P_{opt})$. Let the obtained polynomial be denoted as follows:

$$F(P) = h_n P^n + h_{n-1} P^{n-1} + \cdots + h_1 P + h_0 \tag{7.2}$$

The values of the n coefficients $h_i (i = 0 \cdots n)$ of the polynomial (7.2) can be determined based on the summit points in Fig. 7.2A. It is precisely this function $F(.)$ that defines the speed-reference optimizer (Fig. 7.3B). Indeed, suppose that, at some instant, the wind velocity is $v_w^0 = 14$ m/s and the rotor speed is Ω_0. One can see on Fig. 7.3A, which transmitted wind power P_0 that corresponds to this couple (v_w^0, Ω_0). The point is that P_0 is easily computed online (as it simply equals the product *rotor speed* $\times$ *torque*). Given only the value of P_0, the speed-reference optimizer gives a new rotor speed-reference value $\Omega_1 = F(P_0)$. This corresponds to point A on Fig. 7.2A. Assume that a speed regulator (to be determined) is available, which makes the machine rotor rotate at the new speed reference, that is, $\Omega = \Omega_1$. Then, according to Fig. 7.2A, the wind turbine provides a new extractable power value equal to P_1. Then, the speed-reference optimizer will suggest (to the speed regulator) a new speed reference Ω_2 (point B). This process will continue until the achievement of the optimal point (P_{opt}, Ω_{opt}) (point C on Fig. 7.2A).

Remark 7.1: *The polynomial interpolation yielding the function $F(.)$ has been obtained using the MATLAB functions* **max, polyval, spline,** *and* **polyfit.**

7.3   Modeling of the association "permanent magnet synchronous generator−AC/DC/AC converter"

The structure of a WEC system is presented in Fig. 7.1, which typically consists of two parts. The first one is the mechanical part that transforms the kinetic energy of the wind into mechanical energy by the blades of the turbine. The second one is the electrical part that converts the mechanical energy to electrical energy through the PMSG, and then it is injected into the grid via an AC/DC/AC converter. The controlled system (Fig. 7.1) is a series combination of an association of "synchronous aerogenerator−rectifier" and "inverter−grid." The rectifier is an AC/DC converter that is operating like the DC/AC inverter according to the known PWM principle. It consists of six IGBTs with antiparallel diodes for bidirectional power flow mode, displayed in three legs 1, 2, and 3. To avoid short-circuiting of the three-phase voltage source, only one switch on the same branch can be closed at a time.

7.3.1 Modeling of the combination "permanent magnet synchronous generator—AC/DC rectifier"

The fact that the three-phase symmetrical sinusoidal quantities are transformed into two $dc-$ components through the well-known Park's transformation makes the $dq-$ reference frame more suitable for developing control laws. According to El Magri, Giri, Besançon, et al. (2013), Fadili, Giri, and El Magri (2013), González, Figueres, Garcerá, and Carranza (2010), and Watil, Raihani, and Lajouad (2018), the electrical behavior of the association $PMSG-AC/DC$ converter can be described in the $dq-$ rotating reference frame by the following differential equations, where the d-axis is aligned with the direction of the permanent magnet flux vector.

$$\frac{d\Omega}{dt} = -\frac{F}{J}\Omega + \frac{K_\phi}{J}i_{sq} - \frac{T_g}{J}$$

$$\frac{di_{sq}}{dt} = -\frac{R_s}{L_s}i_{sq} - p\Omega i_{sd} - p\Omega\frac{\phi_r}{L_s} + \frac{v_{sq}}{L_s} \tag{7.3}$$

$$\frac{di_{sd}}{dt} = -\frac{R_s}{L_s}i_{sd} + p\Omega i_{sq} + \frac{v_{sd}}{L_s}$$

where R_s and L_s are the stator resistance and inductance; J, F, and p are the total rotor inertia, viscous coefficient, and number of pole pairs, respectively; Ω is the angular rotor speed; T_g is the aerodynamic torque; ϕ_r is the amplitude of the flux induced by the permanent magnets of the rotor in the stator phases; $K_\phi = (3/2)p\phi_r$ is a constant; (v_{sd}, v_{sq}) and (i_{sd}, i_{sq}) are the stator voltages and currents in the dq axis.

The three phases stator output voltage $v_{s_{abc}}$ and current $i_{s_{abc}}$ are also the input of AC/DC converter (Fig. 7.2A). The semiconductor switching devices (S_{123} and $\bar{S}_{123}$) of the rectifier are controlled by PWM signals. The rectifier is featured by the fact that the stator voltages can be controlled independently. To this end, the stator voltages v_{sabc} and the output current i_{rec} in AC/DC are expressed in function of the corresponding control action as follows (El Magri, Giri, El Fadili, & Chaoui, 2012; Franklin, Powell, & Workman, 1998; Lajouad et al., 2019; Magri, Assche, Fadili, Chaoui, & Giri, 2013):

$$v_{sa} = q_1 v_{dc}; \quad v_{sb} = q_2 v_{dc}; \quad v_{sc} = q_3 v_{dc}$$
$$i_{rec} = q_1 i_{sa} + q_2 i_{sb} + q_3 i_{sc} \tag{7.4}$$

where (q_1, q_2, q_3) are the switch position functions taking values in the discrete set $0, 1$. Specifically, one has:

$$q_i = \begin{cases} 1 & \text{if} \quad Q_i \quad \text{ON} \quad \text{and} \quad \bar{Q}_i \quad \text{OFF} \\ 0 & \text{if} \quad Q_i \quad \text{OFF} \quad \text{and} \quad \bar{Q}_i \quad \text{ON} \end{cases}, \quad i = 1, 2, 3 \tag{7.5}$$

Converting the system of Eq. (7.4) into rotating $dq-$ coordinates using Park's transformation, we obtain:

$$v_{sq} = u_1 v_{dc}; \quad v_{sd} = u_2 v_{dc}; \quad i_{rec} = u_1 i_{sq} + u_2 i_{sd} \tag{7.6}$$

where u_1 and u_2 represent the average control of the rectifier in $dq-$ frame.

7.3.2 Modeling of the combination "DC/AC inverter—grid"

The considered subsystem is a three-phase DC/AC inverter (Fig. 7.2A) associated with a three phases network supply represented by a series of a resistance R_g and an inductance L_g. The grid electromotive forces e_{gabc} are sinusoidal and three-phase net voltages. Applying Kirchhoff's laws and Park transformation ($abc \rightarrow dq$), the mathematical model of the association $DC/AC\ inverter-Grid$ can be described in the dq coordinates as follows (El Magri, Giri, Abouloifa, & Chaoui, 2010; El Magri, Giri, & El Fadili, 2010; Watil et al., 2018):

$$C\frac{dv_{dc}^2}{dt} = v_{dc}i_{rec} - (E_{gd}i_{gd} + E_{gq}i_{gq})$$

$$\frac{di_{gd}}{dt} = -\frac{R_g}{L_g}i_{gd} + \omega_g i_{gq} - \frac{E_{gd}}{L_g} + \frac{v_{gd}}{L_g} \tag{7.7}$$

$$\frac{di_{gq}}{dt} = -\frac{R_g}{L_g}i_{gq} - \omega_g i_{gd} - \frac{E_{gq}}{L_g} + \frac{v_{gq}}{L_g}$$

where (E_{gd}, E_{gq}), (i_{gd}, i_{gq}), and (v_{gd}, v_{gq}) are the grid voltages, the injected currents, and the DC/AC output voltages, respectively in $dq-$ frame. The output line voltages $v_{g_{dq}}$ and the input current of DC/AC inverter i_{inv} are given by the following equations (El Magri et al., 2014; Lajouad, El Magri, et al., 2015):

$$v_{gd} = u_3 v_{dc}; \quad v_{gq} = u_4 v_{dc}; \quad i_{inv} = u_3 i_{gd} + u_4 i_{gq} \tag{7.8}$$

where u_3 and u_4 represent the average dq-axis components of the three-phase switching functions k_1, k_2, and k_3, which is given by:

$$k_i = \begin{cases} 1 & \text{if} \quad K_i \quad \text{ON} \quad \text{and} \quad \overline{K}_i \quad \text{OFF} \\ 0 & \text{if} \quad K_i \quad \text{OFF} \quad \text{and} \quad \overline{K}_i \quad \text{ON} \end{cases}, \quad i = 1, 2, 3 \tag{7.9}$$

Now, let us introduce the state variables $x_1 = \Omega$, $x_2 = i_{sq}$, $x_3 = i_{sd}$, $x_4 = v_{dc}^2$, $x_5 = i_{gd}$, $x_6 = i_{gq}$. Then the state space representation of the association of the whole system including the synchronous generator combined with AC/DC/AC converters:

$$\dot{x}_1 = -\frac{F}{J} x_1 + \frac{K_\phi}{J} x_2 - \frac{T_g}{J} \tag{7.10a}$$

$$\dot{x}_2 = -\frac{R_s}{L_s} x_2 - p x_1 x_3 - \frac{p \phi_r}{L_s} x_1 + \frac{v_{dc}}{L_s} u_1 \tag{7.10b}$$

$$\dot{x}_3 = -\frac{R_s}{L_s} x_3 + p x_1 x_2 + \frac{v_{dc}}{L_s} u_2 \tag{7.10c}$$

$$\dot{x}_4 = -\frac{E_{gd}}{C} x_5 - \frac{E_{gq}}{C} x_6 + \frac{i_{rec} v_{dc}}{C} \tag{7.10d}$$

$$\dot{x}_5 = -\frac{R_g}{L_g} x_5 + \omega_g x_6 - \frac{E_{g_d}}{L_g} + \frac{v_{dc}}{L_g} u_3 \tag{7.10e}$$

$$\dot{x}_6 = -\frac{R_g}{L_g} x_6 - \omega_g x_5 - \frac{E_{gq}}{L_g} + \frac{v_{dc}}{L_g} u_4 \tag{7.10f}$$

7.4 State-feedback nonlinear controller design

This section aims to develop a nonlinear controller that is able to achieve four operational control objectives.

7.4.1 Control objectives

There are three operational control objectives:

CO1 Speed regulation: In order to ensure the MPPT requirement, the rotor aerogenerator speed Ω must track, as closely as possible, the reference signal Ω_{ref} generated by the optimizer (see Section 7.2).

CO2 The power factor correction (PFC) requirement: The injected currents (i_{g1}, i_{g2}, i_{g3}) into the grid must be sinusoidal with the same frequency as the grid voltages and the reactive power in the AC grid must be well regulated.

CO3 DC-link voltage control: Controlling the continuous voltage v_{dc} in order to track a given reference signal V_{dcref}. This reference is generally set to a constant value, equal to the nominal aerogenerator voltage.

Since there are four control inputs, there is a possibility to account an additional objective:

CO4. Knowing that only the q-axis reactance is involved in producing active power in the aerogenerator, that is, there is no direct magnetization or demagnetization of d-axis, only the permanent magnets contribute to producing the flux along this direction (see e.g., El Magri, Giri, Abouloifa, & Haloua, 2007; Rashid, 2001). Doing so, the current i_{sd} must be regulated to a reference I_{dref}, preferably equal to zero in order to optimize the aerogenerator stator currents.

To deal with these objectives, a nonlinear state-feedback controller will be designed in the next section. It includes speed and reactive power loops, which, together with the speed-reference generator designed in Section 7.2, lead to the state-feedback controller illustrated in Fig. 7.4.

7.4.2 Speed regulator design for synchronous generator

The regulator design is based on Eqs. (7.10a) and (7.10b) where the input signal stands as the actual input u_1, in order to guarantee speed-reference tracking. Following the backstepping technique (Krstic et al., 1995), let z_1 denote the speed tracking error:

$$z_1 = x_1 - \Omega_{\text{ref}} = x_1 - x_1^* \tag{7.11}$$

In view of Eq. (7.10a), the error in previous equation undergoes the following equation:

$$\dot{z}_1 = -\frac{F}{J}x_1 - \frac{K_\phi}{J}x_2 + \frac{T_g}{J} - \dot{x}_1^* \tag{7.12}$$

In Eq. (7.12), the quantity $\alpha = -(K_\phi/J)x_2$ stands up as a (virtual) control input for the z_1-dynamics. Let $\alpha*$ denotes the stabilizing function (yet to be determined) associated to α. It is easily seen from Eq. (7.12) that if $\alpha = \alpha*$ with:

$$\alpha* = \left(-c_1 z_1 + \frac{F}{J}x_1 - \frac{T_g}{J} + \dot{x}_1^* \right) \quad (c_1 \text{ is a design parameter}) \tag{7.13}$$

Indeed, if $\alpha = \alpha*$, one will have $\dot{z}_1 = -c_1 z_1$, which clearly is asymptotically stable with respect to the Lyapunov function:

$$V_1 = 0.5z_1^2 \tag{7.14}$$

Indeed, one then has:

$$\dot{V}_1 = z_1\dot{z}_1 = -c_1 z_1^2 < 0 \tag{7.15}$$

As $\alpha = -(K_\phi/J)x_2$ is just a virtual control input, one cannot set $\alpha = \alpha*$. Nevertheless, the expression of $\alpha*$ is retained as a first stabilization function, and a new error is introduced:

$$z_2 = \alpha - \alpha^* \tag{7.16}$$

Using Eqs. (7.13)–(7.16), it follows from Eq. (7.12) that the z_1-dynamics undergoes the following equation:

$$\dot{z}_1 = -c_1 z_1 + z_2 \tag{7.17}$$

The next step consists in determining the control input u_1 so that the (z_1, z_2) error system is asymptotically stable. First, let us obtain the trajectory of the error z_2. Deriving z_2 with respect to time and using Eq. (7.16) gives:

$$\dot{z}_2 = -(K_\phi/J)\dot{x}_2 - \dot{\alpha}* \tag{7.18}$$

Using Eq. (7.13), (7.10a), (7.10b) in Eq. (7.18), we get:

$$\dot{z}_2 = \chi(x,t) - c_1^2 z_1 + c_1 z_2 + \frac{K_\phi}{JL_s}u_1 v_{dc} \tag{7.19}$$

where

$$\chi(x,t) = \frac{K_\phi}{J}\left(\frac{R_s}{L_s} \cdot x_2 + px_1 x_3 - \frac{K_\phi}{L_s}x_1 \right) + \left(\frac{F^2}{J^2}x_1 + \frac{FK_\phi}{J^2} \cdot x_2 \right) - \frac{FT_g}{J^2} + \frac{\dot{T}_g}{J} - \ddot{x}_1^* \tag{7.20}$$

The error equations (7.16) and (7.19) are given the more compact form:

$$\dot{z}_1 = -c_1 z_1 + z_2 \tag{7.21}$$

$$\dot{z}_2 = \chi(x,t) - c_1^2 z_1 + c_1 z_2 + \frac{K_\phi}{JL_s}u_1 v_{dc} \tag{7.22}$$

To determine a stabilizing control law for Eq. (7.22), let us consider the quadratic Lyapunov function candidate:

$$V_2 = V_1 + 0.5z_2^2 = 0.5z_1^2 + 0.5z_2^2 \tag{7.23}$$

Using Eqs. (7.21) and (7.22), the time derivative of V_2 can be rewritten as:

$$\dot{V}_2 = -c_1 z_1^2 + z_1 z_2 + z_2 \dot{z}_2 \tag{7.24}$$

This shows that, for the (z_1, z_2)-system to be globally asymptotically stable, it is sufficient to choose the control u_1 so that:

$$\dot{V}_2 = -c_1 z_1^2 - c_2 z_2^2 \tag{7.25}$$

where $c_2 > 0$ is a new design parameter. In view of Eq. (7.24), Eq. (7.25) is ensured if:

$$\dot{z}_2 = -c_2 z_2 - z_1 \tag{7.26}$$

Comparing Eqs. (7.26) and (7.22) yields the following backstepping control law:

$$u_1 = -\frac{JL_s}{3K_\varphi} \frac{(c_1 + c_2)z_2 - (c_1^2 + 1)z_1 + \chi(x, t)}{v_{dc}} \tag{7.27}$$

7.4.3   d-Axis current regulation

The d-axis current x_3 undergoes Eq. (7.10c) in which the following quantity is introduced:

$$v = px_2 x_1 - u_2 v_{dc}/L_s \tag{7.28}$$

As the reference signal I_{dref} is null, it follows that the tracking error $z_3 = x_3 - I_{dref}$ undergoes the equation:

$$\dot{z}_3 = -\left(\frac{R_s}{L_s}\right)z_3 + v \tag{7.29}$$

To get a stabilizing control signal for this first-order system, consider the following quadratic Lyapunov function:

$$V_3 = 0.5z_3^2 \tag{7.30}$$

It is easily checked that, if the virtual control is let to be:

$$v = -(-R_s/L_s + c_3)z_3 \tag{7.31}$$

where $c_3 > 0$ is a new design parameter, then:

$$\dot{V}_3 = -c_3 z_3^2 \tag{7.32}$$

which is negative definite. Furthermore, substituting Eq. (7.31) in Eq. (7.29), we get the closed-loop equation:

$$\dot{z}_3 = -c_3 z_3 \tag{7.33}$$

Now, it is readily observed that the actual control input is obtained by substituting Eq. (7.31) into Eq. (7.28) and solving the resulting equation for u_2. Doing so, we get:

$$u_2 = \left(c_3 z_3 - \frac{R_s}{L_s}z_3 + px_2 x_1\right)\frac{L_s}{v_{dc}} \tag{7.34}$$

The control closed loops induced by the speed and d-axis current control laws thus defined by Eqs. (7.27) and (7.34) are analyzed in the following proposition.

Proposition 7.1: *Consider the control system consisting of the subsystem (7.10a) and(7.10b) and the control laws (7.27) and (7.34). The resulting closed-loop system undergoes, in the (z_1, z_2, z_3)-coordinates, the following equation:*

$$\begin{pmatrix} \dot{z}_1 \\ \dot{z}_2 \\ \dot{z}_3 \end{pmatrix} = B_1 \begin{pmatrix} z_1 \\ z_2 \\ z_3 \end{pmatrix} \quad \text{with } B_1 = \begin{pmatrix} -c_1 & 1 & 0 \\ -1 & -c_2 & 0 \\ 0 & 0 & -c_3 \end{pmatrix} \tag{7.35}$$

This equation defines a stable system and the error vector (z_1, z_2, z_3) converges exponentially fast to zero, whatever the initial conditions.

Proof 7.1: *Eq. (7.35) is directly obtained from Eqs. (7.21), (7.26), (7.33). It is clear that the matrix B_1 is Hurwitz, implying that the closed loop system (7.35) is globally exponentially stable (GES). This completes the proof of Proposition 7.1.*

7.4.4   Reactive power and DC voltage controller

The PFC requirement amounts to ensure a sinusoidal output current and the injection or extraction of a desired reactive power in the electric network. DC voltage regulation entails the control of the continuous voltage v_{dc} so that it takes a given reference value V_{dcref}. The achievement of these objectives necessitates two control loops. The first one ensures the regulation of the DC voltage x_4, and the second ensures the injection of the desired reactive power.

7.4.4.1 DC voltage loop

Based on Eqs. (7.10d) and (7.10e), a first equation involving the control input u_3 will now be designed, using the backstepping technique (Krstic et al., 1995), so that the squared DC-link voltage $x_4 = v_{dc}^2$ tracks well any reference signal $x_4^* = V_{dcref}^2 > 0$. As the subsystem (7.10d) and (7.10e) is of relative degree 2, the design toward that equation is performed in two steps.

Step 1: Let z_4 denote the speed tracking error:

$$z_4 = x_4 - x_4^* \tag{7.36}$$

In view of Eq. (7.10d), the error in previous equation undergoes the following equation:

$$\dot{z}_4 = -\frac{1}{C} E_{gd} x_5 + \beta(x_{i=1..6}, z_{i=1..3}) - \dot{x}_4^*$$

$$\beta(x_{i=1..6}, z_{i=1..3}) = -\frac{1}{C}\left(E_{gq} x_6 + \frac{JL_s}{3K_\varphi}((c_1 + c_2)z_2 - c_1^2 z_1 + \chi(x,t))x_2 \qquad - L_s(c_3 z_3 - R_s z_3/L_s + px_2 x_1)x_3 \right) \tag{7.37}$$

In Eq. (7.37), the quantity $\alpha_1 = -E_{gd}x_5/C$ stands up as a (virtual) control input for the z_4-dynamics because the actual control input u_3 acts on z_4 indirectly through α_1. Following the backstepping design technique, the Lyapunov function candidate is considered as: $V_4 = 0.5z_4^2$. Deriving V_4 along the trajectory of Eq. (7.37) yields:

$$\dot{V}_4 = z_4 \dot{z}_4 = -z_4 \left(\frac{1}{C} E_{gd} x_5 - \beta(x,z) + \dot{x}_4^* \right) \tag{7.38}$$

This suggests for the (virtual control) following control law:

$$\alpha_1^* = -c_4 z_4 - \beta(x,z) + \dot{x}_4^* \tag{7.39}$$

with $c_4 > 0$ a design parameter. Indeed, substituting α_1^* to $\alpha_1 = -E_{gd}x_5/C$ gives $\dot{V}_4 = -c_4 z_4^2$, which clearly is negative definite in z_4. As α_1 is just a virtual control input, one cannot set $\alpha_1 = \alpha_1^*$. Nevertheless, the previous expression of α_1^* is retained, and a new error is introduced:

$$z_5 = \alpha_1 - \alpha_1^* \tag{7.40}$$

Using Eq. (7.39), it follows from Eq. (7.37) that the z_4-dynamics undergoes the following equation:

$$\dot{z}_4 = -c_4 z_4 + z_5 \tag{7.41}$$

Step 2: Now, the aim is to make the couple of errors (z_4, z_5) vanish asymptotically. The trajectory of the error z_5 is obtained by the time-derivation of Eq. (7.40), that is,

$$\dot{z}_5 = -\frac{E_{gd}}{C} \dot{x}_5 + c_4 \dot{z}_4 + \dot{\beta}(x,z) - \ddot{x}_4^* \tag{7.42}$$

Using Eqs. (7.41), (7.10d), (7.10e) in Eq. (7.42) yields:

$$\dot{z}_5 = \beta_1(x_{i=1...6}, z_{i=1...5}) - \frac{E_{gd}}{CL_0} u_3 v_{dc} \tag{7.43}$$

with

$$\beta_1(x_{i=1...6}, z_{i=1...5}) = c_4 \dot{z}_4 + \dot{\beta}(x,z) - \ddot{x}_4^* + \frac{E_{gd}^2}{CL_0} - \frac{E_{gd}}{C} \omega_n x_6 \tag{7.44}$$

To determine a stabilizing control law for Eqs. (7.10d) and (7.10e), let us consider the quadratic Lyapunov function candidate:

$$V_5 = 0.5z_4^2 + 0.5z_5^2 \tag{7.45}$$

Using Eqs. (7.41)–(7.43), we get from Eq. (7.45) that:

$$\dot{V}_5 = z_4\dot{z}_4 + z_5\dot{z}_5 \tag{7.46}$$

$$= -c_4 z_4^2 + z_5\left(z_4 + \beta_1(x_{i=1...6}, z_{i=1...5}) - \frac{E_{gd}}{CL_0}u_3 v_{dc}\right) \tag{7.47}$$

This suggests for the control variable u_3 the following choice:

$$u_3 = (c_5 z_5 + z_4 + \beta_1(x_{i=1...6}, z_{i=1...5}))\frac{CL_0}{E_{gd}v_{dc}} \tag{7.48}$$

where $c_5 > 0$ is a new design parameter. Indeed, substituting Eq. (7.48) in Eq. (7.47) yields:

$$\dot{V}_5 = -c_4 z_4^2 - c_5 z_5^2 < 0 \tag{7.49}$$

Now, substituting Eq. (7.48) in Eq. (7.43), we obtain the DC voltage closed-loop control system:

$$\dot{z}_4 = -c_4 z_4 + z_5 \tag{7.50}$$

$$\dot{z}_5 = -c_5 z_5 - z_4 \tag{7.51}$$

7.4.4.2 Reactive power loop

Here, the focus is made on the control objective *CO3* that involves the reactive power Q_n, which is required to track its reference Q_n^*. The electrical reactive power injected in the grid is given by $Q_n = E_{gd}x_6 - E_{gq}x_5$. To harmonize notation throughout this section, the corresponding tracking error is denoted as $z_6 = Q_n - Q_n^*$. It follows from Eqs. (7.10e) and (7.10f) that z_6 undergoes the differential equation:

$$\dot{z}_6 = \beta_2(x_5, x_6) + \frac{v_{dc}}{L_0}\left(E_{gd}u_4 - E_{gq}u_3\right) \tag{7.52}$$

with

$$\beta_2(x_5, x_6) = -\omega_n(E_{gd}x_5 + E_{gq}x_6) - \dot{Q}_n^*$$

As Eq. (7.52) is a first-order one, it can be (globally asymptotically) stabilized using a simple proportional control law:

$$v_{dc}(E_{gd}u_4 - E_{gq}u_3)/L_0 = -c_6 z_6 - \beta_2(x_5, x_6) \quad \text{with} \quad c_6 > 0 \tag{7.53}$$

Then the control law u_4 is given as:

$$u_4 = (-L_0(c_6 z_6 + \beta(x_5, x_6))/v_{dc} + E_{gq}u_3)/E_{gd} \tag{7.54}$$

It can be easily checked that the dynamic of z_6 undergoes the following equation:

$$\dot{z}_6 = -c_6 z_6 \tag{7.55}$$

The control closed loops induced by the DC voltage and reactive power control laws thus defined by Eqs. (7.48) and (7.54) are analyzed in the following proposition.

Proposition 7.2: *Consider the control system consisting of the subsystem* (7.10d and (7.10f) *and the control laws* (7.48) *and* (7.54). *The resulting closed-loop system undergoes, in the* (z_4, z_5, z_6)*-coordinates, the following equation:*

$$\dot{Z}_2 = B_2 Z_2 \quad \text{with} \quad B_2 = \begin{pmatrix} -c_4 & 1 & 0 \\ -1 & -c5 & 0 \\ 0 & 0 & -c_6 \end{pmatrix} \tag{7.56}$$

This equation defines a stable system and the error vector (z_4, z_5, z_6) converges exponentially fast to zero, whatever the initial conditions.

Proof 7.2: Eq. (7.56) *is directly obtained from* Eqs. (7.50), (7.51), (7.55). *It is clear that the matrix* B_2 *is Hurwitz, implying that the closed loop system* (7.56) *is GES.*

This completes the proof of Proposition 7.2.

Remark 7.2: *(1) The generator speed and the d-component of its stator current both converge to their respective references because the errors* (z_1, z_3) *converge to zero, as a result of* Proposition 7.1. *(2)* Proposition 7.2 *also demonstrates that the tracking objectives are achieved for the DC-link squared voltage* $x_4 = \bar{v}_{dc}^2$ *and the reactive power* $Q_n = E_{gd}x_6 - E_{gq}x_5$.

7.5   Output-feedback nonlinear controller design

In Section 7.4, it has been demonstrated that the developed controller achieves all control objectives listed in Section 7.4.1. This controller is synthesized using the model in the frame dq, which requires several measurements such as the flux and the rotor position. Since there are no cheap and reliable sensors for these variables, the aforementioned controller will remain useless. Hence, in this section, an observer will be designed to provide an output-feedback controller.

7.5.1   Permanent magnet synchronous generator model in $\alpha\beta$-coordinates

Because a mechanical rotor position is practically unavailable for measurement devices, the PMSG model is considered in the $\alpha\beta$-frame, which is more suitable for observer design. According to El Magri, Giri, and El Fadili (2013), the PMSG model in the $\alpha\beta$-coordinates is given by:

$$\frac{di_s}{dt} = -\frac{p}{L_s}\Omega J_2\psi - \frac{R_s}{L_s}i_s + \frac{1}{L_s}v$$

$$\frac{d\psi}{dt} = p\Omega J_2\psi$$

$$\frac{d\Omega}{dt} = \frac{p}{J}i_s^T J_2\psi - \frac{F}{J}\Omega - \frac{1}{J}T_g \qquad (7.57)$$

$$\frac{dT_L}{dt} = \varepsilon(t)$$

where $i_s = [\begin{matrix} i_{s\alpha} & i_{s\beta} \end{matrix}]^T$, $\psi = [\begin{matrix} \psi_{s\alpha} & \psi_{s\beta} \end{matrix}]^T$, $v = [\begin{matrix} v_{s\alpha} & v_{s\beta} \end{matrix}]^T$ are the stator currents, the rotor fluxes and the voltages, respectively. Ω and T_g denote the rotor speed and the generator torque, respectively. $\varepsilon(t)$ is an unknown bounded function; J_2 is the (2 × 2) matrix defined as $J_2 = \begin{pmatrix} 0 & -1 \\ 1 & 0 \end{pmatrix}$; J is the total rotor inertia; p is the number of pole pairs. The electrical parameters R_s and L_s are the stator resistor and inductance, respectively. Notice that the time derivative of the turbine torque is described by an unknown bounded function $\varepsilon(t)$.

The first issue one must deal with is, under what conditions that all the state variables of this subsystem, i_s, ψ, Ω, and T_g can be determined using only measurements of the electrical variables, that is, the stator current and supply voltage measurements i_s and v, respectively. This is the observability issue. For clarifying our purposes, we introduce the following notations:

$$\xi = \begin{pmatrix} \xi_1 \\ \xi_2 \\ \xi_3 \end{pmatrix}$$

with,

$$\xi_1 = \begin{pmatrix} \xi_{11} \\ \xi_{12} \end{pmatrix} = \begin{pmatrix} i_{s\alpha} \\ i_{s\beta} \end{pmatrix},$$

$$\xi_2 = \begin{pmatrix} \xi_{21} \\ \xi_{22} \end{pmatrix} = \begin{pmatrix} \psi_{r\alpha} \\ \psi_{r\beta} \end{pmatrix}, \qquad (7.58)$$

$$\xi_3 = \begin{pmatrix} \xi_{31} \\ \xi_{32} \end{pmatrix} = \begin{pmatrix} \Omega \\ T_g \end{pmatrix}.$$

In the sequel, the notation I_k and 0_k will be used to denote the $(k \times k)$ identity matrix and the $(k \times k)$ null matrix, respectively. The rectangular $(k \times m)$ null matrix shall be denoted by $0_{k \times m}$. System (7.57) can then be rewritten under the following condensed form:

$$\begin{cases} \dot{\xi} &= f(\xi, v) + B\varepsilon(t) \\ y &= C\xi = \xi_1 \end{cases} \tag{7.59}$$

where

$$f(\xi, v) \triangleq \begin{pmatrix} f_1(\xi, v) \\ f_2(\xi, v) \\ f_3(\xi, v) \end{pmatrix} = \begin{pmatrix} -\dfrac{p}{L_s}\xi_{31}J_2\xi_2 - \dfrac{R_s}{L_s}\xi_1 + \dfrac{1}{L_s}v \\ p\xi_{31}J_2\xi_2 \\ \left(\dfrac{p}{J}\xi_1^T J_2\xi_2 - \dfrac{F}{J}\xi_{31} - \dfrac{1}{J}\xi_{32} \right) \\ 0 \end{pmatrix} \tag{7.60}$$

$$B = \begin{bmatrix} 0_{5 \times 1} & 1 \end{bmatrix}^T, \quad C = \begin{bmatrix} I_2 & 0_2 & 0_2 \end{bmatrix} \tag{7.61}$$

7.5.2 Model transformation and observability analysis

In this section, we will introduce a classical state transformation $\Phi \in \mathbb{R}^{6 \times 6}$ that puts system (7.57) under a known observable canonical form (Giri, 2013). The sufficient conditions under which the considered state transformation is a diffeomorphism. In particular, this analysis will emphasize the Jacobian matrix (of the considered state transformation) that is required to be full rank. Now, let us consider the following change of variables.

$$\Phi : \mathbb{R}^6 \to \mathbb{R}^6, \ \xi \to \eta = \begin{pmatrix} \eta_1 \\ \eta_2 \\ \eta_3 \end{pmatrix} = \Phi(\xi) = \begin{pmatrix} \Phi_1(\xi) \\ \Phi_2(\xi) \\ \Phi_3(\xi) \end{pmatrix} \tag{7.62}$$

One can illustrate that this state transformation puts the system (7.59) under the following canonical form:

$$\begin{cases} \dot{\eta}_1 &= \eta_2 + \varphi_1(\eta_1, v) \\ \dot{\eta}_2 &= \eta_3 + \varphi_2(\eta_1, \eta_2) \\ \dot{\eta}_3 &= \varphi_3(z) + b(\eta)\varepsilon(t) \\ y &= C\eta = \eta_1 \end{cases} \tag{7.63}$$

where $\eta = \begin{bmatrix} \eta_1 & \eta_2 & \eta_3 \end{bmatrix}^T$; $\eta_k = \begin{bmatrix} \eta_{k1} & \eta_{k2} \end{bmatrix}^T$ with $(k = 1, 2, 3)$; $\varepsilon(t)$ is given by Eq. (7.57) and the nonlinear functions η_k, $b(\eta)$ and $\varphi_k \in \mathbb{R}^2$, $k = 1, 2, 3$ are defined as follows:

$$\begin{cases} \eta_1 &= \Phi_1(\xi) = \xi_1 \\ \eta_2 &= \Phi_2(\xi) = -\dfrac{p}{L_s}\xi_{31}J_2\xi_2 \\ \eta_3 &= \Phi_3(\xi) = \dfrac{p^2}{L_s}\xi_{31}^2\xi_2 - \dfrac{p}{JL_s}(p\xi_1^T J_2\xi_2 - \xi_{32})J_2\xi_2 \end{cases} \tag{7.64}$$

and

$$\begin{cases} \varphi_1(\eta_1, v) &= -\dfrac{R_s}{L_s}\eta_1 + \dfrac{1}{L_s}v \\ \varphi_2(\eta_1, \eta_2) &= -\dfrac{F}{J}\eta_2 \\ \varphi_3(\eta) &\triangleq \dfrac{\partial \Phi_3}{\partial \xi_1}(\xi)\dot{\xi}_1 + \dfrac{\partial \Phi_3}{\partial \xi_2}(\xi)\dot{\xi}_2 + \dfrac{\partial \Phi_3}{\partial \xi_{31}}(\xi)\dot{\xi}_{31} \\ b(\eta) &\triangleq \dfrac{\partial \Phi_3}{\partial \xi_{32}}(\xi) \end{cases} \tag{7.65}$$

The transformed system (7.63) is well known to be observable in the rank sense regardless of the entry ν (Giri, 2013).

One can deduce that the system (7.63) is observable whatever input ν. As a result, the system (7.57) shall be observable in the rank sense on $\mathbb{R}^6$ as soon as the transformation $\Phi(\xi)$ exists and is regular almost everywhere (Hong, Lu, & Chiou, 2009). Accordingly, we will analyze the Jacobian matrix of that transformation and determine sufficient conditions for its full rank.

Let J_Φ be the Jacobean of $\Phi(\xi)$. According to Eq. (7.64), we have:

$$J_\Phi(\xi) = \begin{pmatrix} I_2 & 0_2 & 0_2 \\ 0_2 & \dfrac{\partial \Phi_2}{\partial \xi_2}(\xi) & \dfrac{\partial \Phi_2}{\partial \xi_3}(\xi) \\ \dfrac{\partial \Phi_3}{\partial \xi_1}(\xi) & \dfrac{\partial \Phi_3}{\partial \xi_2}(\xi) & \dfrac{\partial \Phi_3}{\partial \xi_3}(\xi) \end{pmatrix} \qquad (7.66)$$

It is clear that the matrix $J_\Phi(\xi)$ is of full rank if and only if the following square matrix is also of full rank:

$$G_\Phi(\xi) = \begin{pmatrix} \dfrac{\partial \Phi_2}{\partial \xi_2}(\xi) & \dfrac{\partial \Phi_2}{\partial \xi_3}(\xi) \\ \dfrac{\partial \Phi_3}{\partial \xi_2}(\xi) & \dfrac{\partial \Phi_3}{\partial \xi_3}(\xi) \end{pmatrix} \triangleq \begin{pmatrix} G_1(\xi) & G_2(\xi) \\ G_3(\xi) & G_4(\xi) \end{pmatrix} \qquad (7.67)$$

Then, we focus on the matrix G_Φ in order to exhibit a sufficient condition under which it is of full rank almost everywhere. Again, according to Eq. (7.64), one has:

$$\begin{cases} G_1(\xi) &= -\dfrac{p}{L_s}\xi_{31}J_2 \\[2mm] G_2(\xi) &= \left[-\dfrac{p}{L_s}J_2\xi_2 \quad 0_{2\times 1} \right] \\[2mm] G_3(\xi) &= \dfrac{p^2}{L_s}\xi_{31}^2 I_2 + \dfrac{p}{JL_s}(\xi_{32}I_2 - p\xi_1^T(J_2\xi_2 + \xi_2 J_2))J_2 \\[2mm] G_4(\xi) &= \left[2\dfrac{p^2}{L_s}\xi_{31}\xi_2 \quad -\dfrac{p}{JL_s}J_2\xi_2 \right] \end{cases} \qquad (7.68)$$

From Eq. (7.68), one can easily demonstrate that: $G_1 G_3 = G_3 G_1$. Then, the determinant of the matrix $G_\Phi(\xi)$ can be written as follows:

$$\det G_\Phi(\xi) = \det(G_1 G_4 - G_3 G_2) \qquad (7.69)$$

According to the expression of $G_i(\xi)$ ($i = 1, 2, 3, 4$) and the previous equation, after some computations:

$$\det G_\Phi(\xi) = \frac{p^5}{L_s J}\left(1 + \frac{2p}{J}\right)\xi_{31}^3 \xi_2^T \xi_2 \qquad (7.70)$$

Then, the considered state transformation is of full rank (and so the system 7.57) if:

$$\xi_{31}^3 \xi_2^T \xi_2 \neq 0 \qquad (7.71)$$

This condition is expressed, using the original motor variables, as follows:

$$\psi_r^2 \Omega^3 \neq 0 \qquad (7.72)$$

7.5.3 High-gain observer design and convergence analysis

The aim of this section is to propose a state observer for system (7.57). Such observer has to provide online estimation of i_s, ψ, Ω, and T_g using only the measurements of the stator currents $i_s = \begin{bmatrix} i_{s\alpha} & i_{s\beta} \end{bmatrix}^T$ and the voltages $v = \begin{bmatrix} v_{s\alpha} & v_{s\beta} \end{bmatrix}^T$. In this respect, note that the WEC system controllers are designed (Section 7.4) using the dq-model necessitating online measurements of several state variables including the mechanical rotor position. Interestingly, the mechanical and electrical rotor position can be estimated online, from the state estimates of the rotor flux $[\psi_{r\alpha} \quad \psi_{r\beta}]$, using the following well-known relation:

$$\theta_r = \frac{\theta_e}{p} = \frac{1}{p}\arctan\left(\frac{\psi_{r\beta}}{\psi_{r\alpha}}\right) \tag{7.73}$$

where θ_r and θ_e denote the rotor position and the electrical position, respectively.

7.5.4 Observer structure

For convenience, the system model (7.59) gives the following more compact form:

$$\begin{cases} \dot{\eta} &= A\eta + \varphi(\eta, v) + Bb(\eta)\varepsilon(t) \\ y &= C\eta = \eta_1 \end{cases} \tag{7.74}$$

where the state $\eta = [\eta_1 \quad \eta_2 \quad \eta_3]^T \in \mathbb{R}^6$, $b(\eta)$ is (2×2) bounded matrix, the matrix A is the following antishift block matrix:

$$A = \begin{bmatrix} 0_2 & I_2 & 0_2 \\ 0_2 & 0_2 & I_2 \\ 0_2 & 0_2 & 0_2 \end{bmatrix} \tag{7.75}$$

the matrix B and C are defined as follows:

$$B = \begin{bmatrix} 0_2 & 0_2 & I_2 \end{bmatrix}^T, \quad C = \begin{bmatrix} I_2 & 0_2 & 0_2 \end{bmatrix} \tag{7.76}$$

the function $\phi(\eta, v)$ has a triangular structure:

$$\phi(\eta, v) = \begin{pmatrix} \phi_1(\eta_1, v) \\ \phi_2(\eta_1, \eta_2, v) \\ \phi_3(\eta, v) \end{pmatrix} \in \mathbb{R}^6$$

The analysis of the proposed observer requires the following technical assumptions:

Assumption A1: *Function* $\varphi(\eta, v)$ *is globally Lipschitz with respect to* η *uniformly in* v, *that is,* $\exists \chi_1 > 0, \quad \forall(\eta, \hat{\eta}) \in \mathbb{R}^n: \quad |\phi(\hat{\eta}, v) - \phi(\eta, v)| \le \chi_1 |\hat{\eta} - \eta|.$

Assumption A2: *The function* $\varepsilon(t)$ *is bounded and the scalar* $\delta > 0$ *is its upper bound* ($|\varepsilon(t)| < \delta$).

The following nonlinear high-gain observer is considered:

$$\{ \dot{\hat{\eta}} = A\hat{\eta} + \varphi(\hat{\eta}, v) - \theta\Delta^{-1}K(C\hat{\eta} - y(t))y(t) = C\eta(t) = \eta_1(t) = \xi_1(t) \tag{7.77}$$

where $\theta > 0$ is a design parameter. The block diagonal matrix Δ is defined by:

$$\Delta = \mathrm{diag}\left(I_2, \frac{1}{\theta}I_2, \frac{1}{\theta^2}I_2\right) \tag{7.78}$$

The gain matrix $K \in \mathbb{R}^{6 \times 2}$ can be chosen as:

$$K^T = \begin{bmatrix} k_1 I_2 & k_2 I_2 & k_3 I_2 \end{bmatrix} \tag{7.79}$$

where $k_i(i = 1, 2, 3)$ is a positive scalar so that $(A - KC)$ is Hurwitz and satisfies the following algebraic Lyapunov function:

$$(A - KC)^T P + P(A - KC) = -Q \tag{7.80}$$

where $(P, Q) \in \mathbb{R}^{6 \times 6}$ is a pair of symmetric positive matrices.

The vector $\hat{\eta}$ is the continuous-time estimate of the system state η. The vector $y(t)$ represents the output.

7.5.5   Stability analysis of the proposed observer

In this section, we will study the convergence of the proposed observer (7.77). In the following theorem, we describe under which conditions (on the design parameters) the estimation errors converge.

Theorem 7.1 (Observer convergence): *Under* Assumptions A1 and A2, *consider the WEC system described by the nonlinear model (7.74), where the signal $\varepsilon(t)$ is assumed to be bounded. There exists a real positive constants $(\theta_0, \mu_1, \mu_2, \mu_3)$ such that for all $\theta > \theta_0$ and the real positive design parameters $k_{i=1,2,3}$ satisfy (7.80), the estimation error $e_\eta = \hat{\eta}(t) - \eta(t)$ is globally convergent whatever the initial condition $\hat{\eta}(0)$ and satisfies the following inequality*

$$||e_\eta(t)|| \leq \mu_1 \exp^{-\mu_2(t)} + \mu_3 \qquad t \in [0, \infty) \tag{7.81}$$

where $\hat{\eta}(t)$ is the unknown trajectory of the high-gain observer (7.77). Consequently, whatever its initial condition, the estimation error $\hat{\eta}(t)$ can be made arbitrarily small letting θ be sufficiently large.

Proof of Theorem 7.1: *Let us introduce the estimation errors $e_\eta(t) = \hat{\eta}(t) - \eta(t)$. According to , the dynamics of these observation errors undergo the following equation:*

$$\dot{e}_\eta = (A - \theta\Delta^{-1}KC)e_\eta + (\phi(\hat{\eta}, v) - \phi(\eta, v)) - Bb(\eta)\varepsilon(t) \tag{7.82}$$

Introduce the change of coordinates $\epsilon_\eta = \Delta e_\eta$. Then, it is readily checked that Eq. (7.82) can be rewritten in term of ϵ_η as follows:

$$\dot{\epsilon}_\eta = \Delta A \Delta^{-1}\epsilon_\eta - \theta KC\Delta^{-1}\epsilon_\eta + \Delta(\varphi(\hat{\eta}, v) - \varphi(\eta, v)) - \Delta Bb(\eta)\varepsilon(t) \tag{7.83}$$

Now, one can easily check the mathematical identities: $\Delta^{-1}A\Delta = \theta A$ and $C\Delta^{-1} = C\Delta = C$. Hence, the dynamic equation(7.82) becomes:

$$\dot{\epsilon}_\eta = \theta(A - KC)\epsilon_\eta + \Delta(\varphi(\hat{\eta}, v) - \varphi(\eta, v)) - \Delta Bb(\eta)\varepsilon(t) \tag{7.84}$$

Inspired by the works (Ahmed-Ali, 2012; Watil et al., 2020a), let us choose the following Lyapunov function:

$$V_\eta = \epsilon_\eta^T P \epsilon_\eta \tag{7.85}$$

Where the matrix P is the symmetric, positive definite matrix [solution of the algebraic Lyapunov function (7.80)].

In order to prove the convergence of the proposed observer, it is enough to find a condition for the following inequalities to be established:

$$\dot{V}_\eta < -cV_\eta + \mu\sqrt{V_\eta} \tag{7.86}$$

where c and μ are the real positives.

It is clear by time derivative of Eq. (7.85) that the dynamic of the Lyapunov function is $V_\eta = \epsilon_\eta^T P \epsilon_\eta$. Using the estimation dynamic errors (7.84), one can deduce that:

$$\begin{aligned}\dot{V}_\eta = {} & \theta\epsilon_\eta^T((A - KC)^T P + P(A - KC))\epsilon_\eta \\ & + 2\epsilon_\eta^T P\Delta(\varphi(\hat{\eta}, v) - \varphi(\eta, v)) \\ & - 2\epsilon_\eta^T P\Delta Bb(\eta)\varepsilon(t) \end{aligned} \tag{7.87}$$

We can remark, using the Lyapunov function defined in Eq. (7.80) that $-\epsilon_\eta^T Q\epsilon_\eta = \epsilon_\eta^T((A - KC)^T P + P(A - KC))\epsilon_\eta$. Then, from Eq. (7.87) we have:

$$\begin{aligned}\dot{V}_\eta = {} & -\theta\epsilon_\eta^T Q\epsilon_\eta + 2\epsilon_\eta^T P\Delta(\phi(\hat{\eta}, v) - \phi(\eta, v)) \\ & - 2\epsilon_\eta^T P\Delta Bb(\eta)\varepsilon(t) \end{aligned} \tag{7.88}$$

The largest and the smallest positive eigenvalues of the matrix X (X is Q, P or K), are denoted by λ_{Xm} and λ_{XM}, respectively. Using the Assumptions (A1 and A2) then: $\Delta||\varphi(\hat{\eta}, v) - \varphi(\eta, v)|| \leq \chi_1||\epsilon_\eta||$ and $\Delta B||b(\eta)\varepsilon(t)|| \leq \frac{\chi_2}{\theta^2}\delta$, with χ_1, χ_2 and δ are the known real positives. So, we have:

$$\begin{aligned}\dot{V}_\eta \leq {} & -(\theta\lambda_{Qm} - 2\chi_1\lambda_{PM})||\varepsilon_\eta||^2 \\ & + 2\lambda_{PM}\frac{\chi_2}{\theta^2}\delta||\varepsilon_\eta|| \end{aligned} \tag{7.89}$$

Then, the proposed observer is GES for whatever the initial conditions $\eta(0)$. The observation error vector $e_\eta(t) = \hat{\eta}(t) - \eta(t)$ converges exponentially to a compact neighborhood of the origin, and the size of this compact set can be made small enough by choosing the design parameter θ sufficiently large.

This ends the proof of Theorem 7.1.

7.5.6 Observer in ξ-coordinates

The previous section provided a state observer for system (7.59), in η-coordinates. For implementation purpose, the proposed observer needs to be expressed in terms of the original ξ-coordinates. Note that, from the transformation $\eta = \Phi(\xi)$, the derivative of estimate $\hat{\eta} = \Phi(\hat{\xi})$ can be written as:

$$\dot{\hat{\xi}} = J_\Phi^{-1}(\hat{\xi})\dot{\hat{\eta}} \tag{7.90}$$

On the other hand, using Eqs. (7.64) and (7.65) we can see that: $C(A\hat{\eta} + \phi(\hat{\eta}, v)) = f_1(\xi, v)$. As a result, from Eq. (7.77), the following dynamical system is an observer for the system represented by its model (7.59).

$$\left\{ \dot{\hat{\xi}} = f(\hat{\xi}, v) - \theta J_\Phi^{-1}\Delta^{-1}K(C\hat{\xi} - y(t))y(t) = C\xi(t) = \xi_1(t) \right. \tag{7.91}$$

where J_Φ^{-1} is given by Eq. (7.66); Δ, K, C are given by Eqs. (7.78), (7.79), (7.76), respectively.

7.5.7 Output-feedback controller

Following the output-feedback control architecture of Fig. 7.5, the mechanical states (not accessible to measurements), involved in the control laws (7.27), (7.34), (7.48), (7.54), are now replaced by their online estimates provided by the observer (7.91). Doing so, the output-feedback controller turns out to be defined by the following control laws:

$$u_1 = -\frac{JL_s}{3K_\varphi}\frac{(c_1 + c_2)z_2 - (c_1^2 + 1)z_1 + \chi(\hat{x}, t)}{v_{dc}} \tag{7.92}$$

$$u_2 = \left(c_3 z_3 - \frac{R_s}{L_s}z_3 + p\hat{x}_2\hat{x}_1 \right)\frac{L_s}{v_{dc}} \tag{7.93}$$

$$u_3 = (c_5 z_5 + z_4 + \beta_1(\hat{x}_{i=1...6}, z_{i=1...5}))\frac{CL_0}{E_{gd}v_{dc}} \tag{7.94}$$

$$u_4 = (-L_0(c_6 z_6 + \beta(\hat{x}_5, \hat{x}_6))/v_{dc} + E_{gq}u_3)/E_{gd} \tag{7.95}$$

7.5.8 Simulation results

7.5.8.1 Simulation protocols

The experimental setup (Fig. 7.5) has been simulated using the MATLAB/Simulink resources. The WEC system characteristics are summarized in Table 7.1. The output-feedback controller performances will be evaluated in the presence of (timevarying) wind speed described in Fig. 7.6A. The applied wind speed v_w is chosen so that the aerogenerator works in different wind velocity zones. The WEC system operates in low, medium, and high wind speed. According to the control design (Sections 7.4 and 7.5), the remaining closed loop inputs are kept constant, namely $i_{sdref} = 0$ A, the DC-link voltage reference $V_{dcref} = 500$ V, and the reactive power signal reference $Q_{gref} = 0$ VAR.

7.5.8.2 Construction of the speed-reference optimizer

The considered wind speed profile is shown in Fig. 7.6A. It is seen that wind velocity varies between low speed 2 m s^{-1} and high speed 14.2 m s^{-1}, over the time interval [0,120 seconds]. In response to this wind speed profile, the turbine generates the rotor torque shown in Fig. 7.6B. Then, using the generated active power of the WEC system, the POS optimizer designed in Section 7.2 generates the rotor speed reference as shown in Fig. 7.6C. The active power extracted by the synchronous aerogenerator is shown in Fig. 7.6D. Referring to Fig. 7.6D, it is clear that the power produced by the WEC system is optimal for each wind speed value concerned ($P_g \approx P_{max}$, $\forall v_w$).

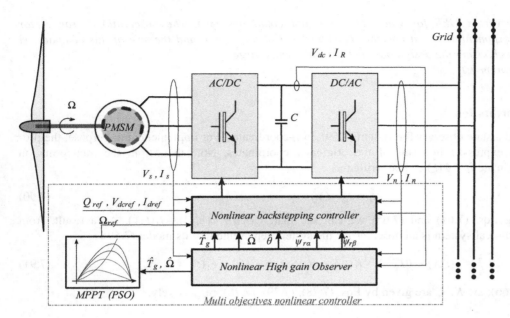

FIGURE 7.5 Practical scheme for the system output-feedback control.

TABLE 7.1 Wind energy conversion system characteristics.

Characteristics	Values	Characteristics	Values
Wind turbine		Total inertia	$J = 0.1 \text{ Nm}^{-1} \text{ rad s}^{-2}$
Nominal power	$P_t = 3$ kW	Total viscous friction	$f = 0.07 \text{ kg m}^{-2} \text{ s}^{-1}$
Turbine radius	$R = 1.3$ m	**AC/DC/AC converters**	
Blade pitch	$\beta = 2$ degrees	Capacitor	$C = 47 \text{ }\mu F$
Aerogenerator		Inductor	$L_g = 0.05$ H
Nominal power	$P_n = 3$ kW	Resistor	$R_g = 0.5 \text{ } \Omega$
Number of pole pairs	$p = 5$	DC-link voltage	$V_{dc} = 500$ V
Nominal speed	$\Omega_n = 60 \text{ rad s}^{-1}$	Modulation frequency	$F_m = 20$ kHz
Stator resistor	$R_s = 0.6 \text{ } \Omega$	**Three-phase network**	
Stator cyclic inductor	$L_s = 0.0094$ H	Voltage	$E_g = 230/400$ V
Rotor flux	$\varphi_r = 0.3$ Wb	Network frequency	$f_g = 50$ Hz

7.5.8.3 Illustration of the observer performances

The performances of proposed high-gain observer (7.91) are tested for a WEC system connected to three-phase grid (frequency $f_g = 50$ Hz and grid voltage $E_g = 280$ V). The test will be accomplished by considering quite tough operation conditions (wind velocity and generator torque) described in Figs. 7.6A and 7.7B. Accordingly, the applied generator torque and reference speed are profiled so that the WEC system operates successively in low, medium, and high speed modes while facing large generator torque changes. The following values of the observer design parameters proved to be suitable:

$$\theta = 220,$$
$$k_1 = 5, \quad k_2 = 10, \quad k_3 = 5.$$

In addition, the changes in the external torque and speed are not explicitly accounted in the observer design. That is, their effect assimilate to initial conditions reset is compensated by the observer due to its global convergence nature.

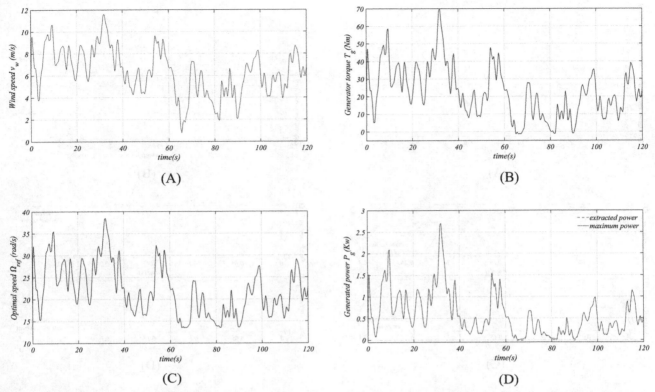

FIGURE 7.6 Performances of the power to optimal speed optimizer developed in Section 7.2. (A) Wind velocity profile v_w (m s^{-1}), (B) generator torque T_g (Nm), (C) optimized rotor speed reference Ω_{ref} (rad s^{-1}), and (D) extracted and maximum powers P_g (kW).

The initial values of the state variables are allowed to be arbitrary, thanks to the globality of the observer convergence. The resulting observer performances are illustrated in Fig. 7.7.

Fig. 7.7B shows the generator torque applied by the wind turbine to the PMSG rotor as well as its estimate is provided online by the observer (7.91). One can see that the estimation error vanishes after a short transient period, and the generator torque estimation error converges to zero. The rotor speed and its estimate are illustrated in Fig. 7.7A. It is seen that the observer speed matches well with the measured speed and both track well with the speed-reference trajectory. This is better emphasized in bottom Fig. 7.7A, which shows that the speed estimation error converges to zero after short periods.

The PMSG flux and current components ($\psi_{r\alpha}$, $\psi_{r\beta}$, and $i_{s\alpha}$, $i_{s\beta}$) and their observer values are illustrated in Fig. 7.7C and D. It is seen that the estimated values, provided by the observer, match well with the real values. Using Eq. (7.73), the rotor position can be estimated online from the state estimates of the rotor flux ($\psi_{r\alpha}$ and $\psi_{r\beta}$). It is seen from Fig. 7.7E that the observed rotor position tracks closely the real value after a short time.

7.5.8.4 Output-feedback controller performances

Recall that the nonlinear output-feedback controller [for the WEC system (Fig. 7.4)] to be illustrated is described by the control laws (7.27), (7.34), (7.48), (7.54). The following values of the controller design parameters are proved to be suitable:

$$c_1 = 15, \quad c_2 = 50, \quad c_3 = 100, \quad c_4 = 12, \quad c_5 = 100, \quad c_6 = 10$$

In the following simulations, the mechanical states (rotor speed Ω, the electrical position θ_e, and load torque T_g) are no longer available. They are estimated by the observer (7.91) under the assumptions (7.71).

The output-feedback controller performances are illustrated in Fig. 7.8. The curves (7.10a)–(7.10d) show that the tracking quality of the proposed controller/observer is quite satisfactory for all desired references. The disturbing effect, due to the wind velocity changes, is also well compensated by the control laws.

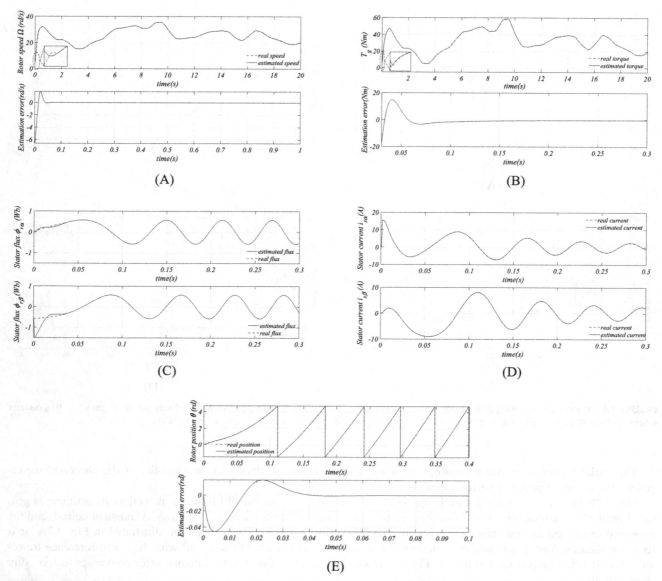

FIGURE 7.7 Tracking performances of the observer defined by Eq. (7.91) (aerogenerator side). (A) Rotor speed Ω (rad/s), (B) generator torque T_g (Nm), (C) $\alpha\beta$ stator flux components ϕ_α and ϕ_β (rad/s), (D) $\alpha\beta$ stator current components $I_{s\alpha}$ and $I_{s\beta}$ (A), and (E) rotor position θ (rad).

7.6 Digital implementation

7.6.1 Foreground general considerations

In addition to the great flexibility of implementing nonlinear functions, digital implementation of the observer offers distinct advantages over analog control that explain its popularity. Here are some of its many advantages:

- *Implementation errors*: Digital processing of control signals involves addition and multiplication by storing numerical values. The errors that result from digital representation and arithmetic are negligible.
- *Accuracy*: Digital signals are represented in terms of zeroes and ones. This involves a very small error as compared to analog signals, where noise and power supply drift are always present.
- *Flexibility*: An analog implementation is difficult to modify or redesign once fired in hardware. A digital observer is implemented in firmware or software, and its modification is possible without a complete replacement of the original controller.

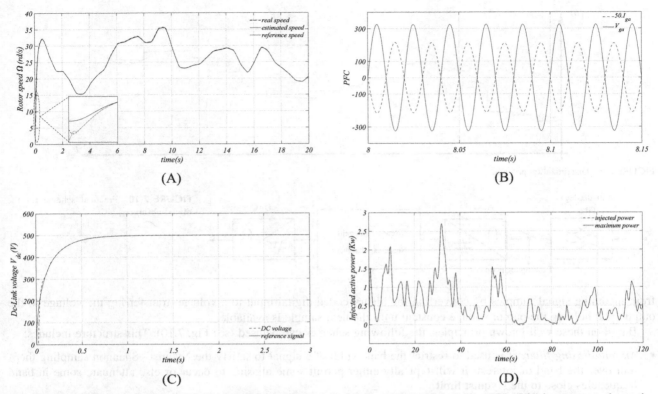

FIGURE 7.8 Tracking performances of the output-feedback controller defined by Eqs. (7.27), (7.34), (7.48), (7.54), (7.91) in response to the varying wind speed of Fig. 7.6A. (A) Rotor speed Ω (rad s^{-1}), (B) wave form line current and voltage, (C) DC-link voltage V_{dc} (V), and (D) injected power in the three-phase network.

- *Speed*: Since the 1980s, the speed of computer hardware has increased exponentially, which allowed a process control at very high speeds. Indeed, the sampling period can be made very small. Hence, digital controllers achieve performance that is essentially the same as that based on continuous monitoring of the controlled variable.
- *Cost*: The cost of digital circuitry continues to decrease. Advances in very large-scale integration dtechnology have made it possible to manufacture better, faster, and more reliable integrated circuits and to offer them to the consumer at a lower price.

This part describes the discretization of the high-gain observer (7.91) for the digital implementation. The nonlinear observer is discretized using the forward difference method. Properties similar to what we have seen for continuous-time controllers are established for discrete-time controller when the observer parameter and the sampling period are sufficiently small.

7.6.2 Practical scheme

Sampling is the process of deriving a discrete-time sequence from a continuous-time function. As shown in Fig. 7.9, an incoming continuous-time signal $f(t)$ is sampled by an analog-to-digital (A/D) converter to produce the discrete-time sequence $f(k)$. The A/D converter produces a binary representation, using a finite number of bits, of the applied input signal at each sample time. Using a finite number of bits to represent a signal sample generally results in quantization errors in the A/D process. For example, the maximum quantization error in 16-bit A/D conversion is $2^{-16} = 0.0015\%$, which is very low compared with typical errors in analog sensors. This error, if taken to be "noise," gives a signal-to-noise (SNR) of $20\log_{10}(2^{-16}) = -96.3$ db which is much better than that of most control systems.

Reconstruction is the formation of a continuous-time function from a sequence of samples. Many different continuous-time functions can have the same set of samples; so reconstruction is not unique. Reconstruction is performed using digital-to-analog (D/A) converters. Electronic D/A converters typically produce a step reconstruction

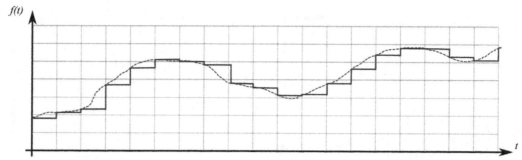

FIGURE 7.9 Discretization process.

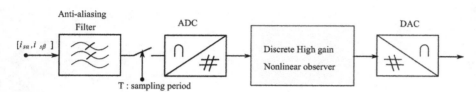

FIGURE 7.10 Practical scheme for the digital observer.

from incoming signal samples by converting the binary-coded digital input to a voltage, transferring the voltage to the output, and holding the output voltage constant until the next sample is available.

Based on these well-known principles, the following schemes can be used (see Fig. 7.10). This structure includes:

- *An antialiasing filter:* It is used to restrict the bandwidth of a signal to satisfy the Nyquist−Shannon sampling theorem over the band of interest. It will typically either permit some aliasing to occur or else attenuate some in-band frequencies close to the Nyquist limit.
- *An Analog-to-digital converter* (*ADC*): It converts a continuous-time and continuous-amplitude analog signal (in our case stator currents) to a discrete-time and discrete-amplitude digital signal. The sampling bloc over a sampling period T is generally integrated in the ADC bloc.
- *A digital-to-analog converter* (*DAC*): This block provides the reverse operation. It transforms the digital signal from the observer into an analog signal that can be used directly by the control.

7.6.3 Observer discretization

7.6.3.1 Technical discussion

Discrete-time equivalents of the continuous-time system can be found by standard methods, which are well documented in the literature. In Gauthier and Kupka (2001), five methods are presented to deal with this discretization: the bilinear transformation, the zero-order hold (ZOH), the first-order hold, the backward difference, and the forward difference (FD) methods. Formulas for discrete equivalents of state space models, for the first three methods, are given in Gauthier and Kupka (2001), and they are available in the control toolbox of MATLAB. It is important to remember that all these methods are approximations; there is no exact solution for all possible inputs because the continuous-time system responds to the complete time history of $y(t)$, whereas the discrete-time equivalent has access only to the samples $y(kT)$. In a sense, the various discretization methods make different assumptions about what happens to $y(t)$ between the sample points.

It is well known that the FD method has less steady-state error for the higher derivative, and also its transient behavior is more oscillatory than the other methods.

A finite difference method consists of a transient approximation of a state differentiation.

$$\frac{dx(t)}{dt} \approx \frac{x((k+1)T) - x(kT)}{T} \tag{7.96}$$

This allows rewriting the differential equation $\dot{x} = f(x)$ by:

$$\frac{x(k+1) - x(k)}{T} = f(x(k)) \tag{7.97}$$

then:

$$x(k+1) = x(k) + Tf(x(k)) \qquad (7.98)$$

Hence, to calculate the next value of state $x(k+1)$, we need the last value $x(k)$ and sampling period T. This value is maintained during a sampling period T.

The first challenge to be raised is the choice of this sampling period and subsequently the cut-off frequency of the antialiasing filter. A general rule consists in choosing this period less than the fastest dynamics of the system. Knowing that the observer settling time is all time less than the settling time of the system.

7.6.3.2 Digital synthesis of the observer

Proposition 7.3: *Consider the continuous form observer (7.77) and suppose a transformation $q = \Delta\hat{\eta}$, this observer can be rewritten as:*

$$\dot{q} = \theta A_d q + \Delta\varphi(\Delta^{-1}q, v) + \theta Ky(t)$$

If h is the sampling period of the deducted system, then the sampling period is $T = h\theta$.
Thus the digital observer represented by Eq. *(7.105) is an asymptotic observer of the system (7.57).*

Proof 7.3: *First, we recall the observer expression (7.77), and let us start by scaling the observation variables to avoid the poor conditioning inherent in the realization of* Eq. *(7.77) when θ is very large:*

$$q = \Delta\hat{\eta} \qquad (7.99)$$

then the observer (7.77) can be rewritten as:

$$\begin{aligned}\dot{q} &= \Delta A\Delta^{-1}q + \Delta\varphi(\Delta q, v) - \theta\Delta\Delta^{-1}KC\Delta^{-1}q + \theta\Delta\Delta^{-1}Ky(t)\\ &= (\Delta A\Delta^{-1} - \theta KC\Delta^{-1})q + \Delta\phi(\Delta^{-1}q, v) + \theta Ky(t)\\ &= \theta A_d q + \Delta\varphi(\Delta^{-1}q, v) + \theta Ky(t)\end{aligned} \qquad (7.100)$$

with $A_d = \begin{bmatrix} -k_1 I_2 & I_2 & O_2 \\ -k_2 I_2 & O_2 & I_2 \\ -k_3 I_2 & O_2 & O_2 \end{bmatrix}$ then, we can write:

$$\frac{dq}{\theta dt} = A_d q + \theta^{-1}\Delta\phi(\Delta^{-1}q, v) + Ky(t) \qquad (7.101)$$

The change of time variable from t to $\tau = \theta t$ puts Eq. (7.101) into the form:

$$\frac{dq}{d\tau} = A_d q + \theta^{-1}\Delta\phi(\Delta^{-1}q, v) + Ky(\tau) \qquad (7.102)$$

Hence, if h is an appropriate sampling period for Eq. (7.102), then the sampling period for Eq. (7.101) is taken as $T = h\theta$.

In addition to all that, if we recall the diffeomorphism allowing to switch from the variable η to ξ, we can write:

$$\frac{d\eta}{dt} = \frac{d\eta}{d\xi}\cdot\frac{d\xi}{dt} = J_\varphi\dot{\xi} \qquad (7.103)$$

which can be, using Euler approximation, written as:

$$\xi(k+1) = \xi(k) + J_\varphi^{-1}(k)(\eta(k+1) - \eta(k)) \qquad (7.104)$$

Then, using the FD method, we can deduce the discretized form of the observer (7.102):

$$\begin{cases} q(k+1) &= (I + hA_d)q + h\theta^{-1}\Delta\varphi(\Delta^{-1}q, v) + hKy(\tau)\\ \hat{\eta}(k+1) &= \Delta^{-1}q(k+1)\\ \xi(k+1) &= \xi(k) + J_\varphi^{-1}(k)(\hat{\eta}(k+1) - \hat{\eta}(k)) \end{cases} \qquad (7.105)$$

7.6.4 Digital output-feedback controller

Now, the state-feedback controller (7.27), (7.34), (7.48), (7.54), developed previously, uses the measurable signals together with the signals generated by the discrete observer (7.105). All involved components are simulated using MATLAB/Simulink that offers a quite accurate representation of the whole system. Actually, the ODE1 (Euler) solver type is selected with a fixed step time of 1 microsecond However, for practical implementation, we should further select a data sampling time and use a digital signal processing DSP card for doing interfacing with all parts.

7.6.5 Simulation results

The output-feedback controller, including the control laws (7.92)–(7.95) and digital observers (7.105), is implemented using MATLAB/Simulink. As a matter of fact, the control performances depend, among others, on:

- The configuration protocol shown in (Fig. 7.5) and described in Section 7.5.8.1.
- The system parameters given in Table 7.1.
- The observer parameters given in Section 7.5.8.3.
- The wind speed profile represented in Fig. 7.6.
- Using trial and error method, we can show that the sampling period can be taken equal to ($h = 500$ microsecond). This value allows an interesting dynamic of the whole output-feedback controller. Of course, we can use a value less than the proposed one.

Then, recall that the sensorless feature of the proposed output-feedback controller entails only the use of electrical (current/voltage) sensors, known to be quite reliable and cheap.

Fig. 7.11 shows that the tracking quality of the digital output-feedback controller is completely satisfactory for all the desired references.

It is seen from Fig. 7.11A that the rotation speed follows as closely as possible the desired speed Ω_{ref} generated by the MPPT (POS) block. This convergence allowed the extracted power to follow the maximum point of the inherent

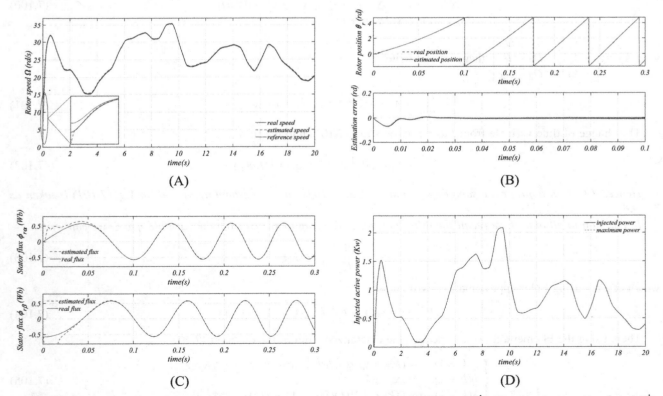

FIGURE 7.11 Tracking performances of the digital output-feedback controller. (A) Rotor speed Ω (rad s^{-1}), (B) turbine rotor position θ_r (rad s^{-1}), (C) permanent magnet synchronous generator rotor flux components $\psi_{r\alpha}$ $\psi_{r\beta}$, and (D) injected power in the three-phase network.

power in the wind (Fig. 7.11D). Fig. 7.11B and C show that the equivalent flows in the $\alpha\beta$-coordinates and the angular position θ_r converge asymptotically to their nominal values.

7.7 Conclusion

This chapter covers the synthesis of a nonlinear output-feedback backstepping controller based on the use of a nonlinear high-gain observer. The problem of online estimation of the rotor position, the rotor speed, and the load torque of the PMSG has been addressed. Maximum wind energy extraction is achieved by running the wind turbine generator in variable-speed mode without using a wind velocity sensor. The controlled system is an association including wind turbine, PMSG, and AC/DC/AC converter connected to a tri-phase network. The system dynamics have been described by the averaged sixth order nonlinear state space model (7.10a)—(7.10f). First, the multiloops nonlinear controller, defined by the control laws (7.27), (7.34), (7.48), (7.54), has been designed, assuming availability of all the states. Then, a nonlinear high-gain observer is proposed to get online estimates of all mechanical state variables in PMSG (rotor position, speed, and load torque). Only the electrical variables are supposed to be accessible to measurements. Based on this observer, the output-feedback controller defined by Eqs. (7.92)—(7.95) can then be built. The Lyapunov stability and backstepping design technique are used. The controller has been designed to: (1) satisfactory rotor speed-reference tracking for extracting maximum power; (2) tight regulation of the stator d-axis; (3) PFC; (4) well-regulated DC-link voltage (v_{dc}). These results have been confirmed by a simulation study.

A discretization approach has been introduced to facilitate the implementation of the considered controller. The observer discretization is based on Euler approximation, which considers also the ZOH discretization methods. Closed-loop simulation in the MATLAB/Simulink shows that the discrete observer/controller recovers the performances of the continuous-time controller as the sampling frequency and the observer gain become sufficiently large.

References

Abouloifa, A., Giri, F., Lachkar, I., Chaoui, F., & El Magri, A. (2009). DC motor velocity adaptive control through buck-boost ac/dc converter with power factor correction. *IFAC Proceedings Volumes (IFAC-PapersOnline)*, 42, 356—361. Available from https://doi.org/10.3182/20090705-4-SF-2005.00063, URL. Available from https://www.sciencedirect.com/science/article/pii/S1474667015302299?.

Ackermann, T., & Söder, L. (2002). An overview of wind energy-status 2002. *Renewable and Sustainable Energy Reviews*, 6, 67—127.

Ahmed-Ali, T. (2012). High gain observer design for some networked control systems. *IEEE Transactions on Automatic Control*, 57, 995—1000. Available from https://doi.org/10.1109/TAC.2011.2168049.

Al Tahir, A., Lajouad, R., Chaoui, F., Tami, R., Ahmed-Ali, T., & Giri, F. (2016). A novel observer design for sensorless sampled output measurement: Application of variable speed doubly fed induction generator. *IFAC-PapersOnLine*, 49, 235—240. Available from https://doi.org/10.1016/j.ifacol.2016.07.957, URL. Available from https://www.sciencedirect.com/science/article/pii/S2405896316312459.

Anaya-Lara, O., Jenkins, N., Ekanayake, J. B., Cartwright, P., & Hughes, M. (2011). *Wind energy generation: Modelling and control*. West Sussex, UK: John Wiley & Sons.

Cuny, F., Lajouad, R., Giri, F., Ahmed-Ali, T., & Assche, V. V. (2019). Sampled-data observer design for delayed output-injection state-affine systems. *International Journal of Control*, 0, 1—11. Available from https://doi.org/10.1080/00207179.2019.1569263, URL. Available from https://doi.org/10.1080/00207179.2019.1569263.

Dabroom, A. M., & Khalil, H. K. (2001). Output feedback sampled-data control of nonlinear systems using high-gain observers. *IEEE Transactions on Automatic Control*, 46, 1712—1725.

Devaiah, T. J., & Smith, R. T. (1975). Generation schemes for wind power plants. *IEEE Transactions on Aerospace and Electronic Systems*, AES-11, 543—550.

El Fadili, A., Boutahar, S., Dhorhi, I., Stitou, M., Lajouad, R., El Magri, A., Elm'kaddem, KH., (2018). Reference speed optimizer controller for maximum power tracking in wind energy conversion system involving DFIG. In *2018 Renewable energies, power systems & green inclusive economy (REPS-GIE)* (pp. 1—6). IEEE. <https://ieeexplore.ieee.org/abstract/document/8488782>. Available from https://doi.org/10.1109/REPSGIE.2018.8488782.

El Fadili, A., Giri, F., El Magri, A., & Besancon, G. (2014). Sensorless induction machine observation with nonlinear magnetic characteristic. *International Journal of Adaptive Control and Signal Processing*, 28, 149—168. Available from https://doi.org/10.3182/20130703-3-FR-4038.00113, URL. Available from https://www.sciencedirect.com/science/article/pii/S1474667016329640.

El Fadili, A., Giri, F., El Magri, A., Lajouad, R., & Chaoui, F. (2012a). Output feedback control for induction machine in presence of nonlinear magnetic characteristic. *IFAC Proceedings Volumes (IFAC-PapersOnline)*, 8, 600—605. Available from https://doi.org/10.3182/20120902-4-FR-2032.00105, URL. Available from https://www.sciencedirect.com/science/article/pii/S1474667016320377.

El Fadili, A., Giri, F., El Magri, A., Lajouad, R., & Chaoui, F. (2012b). Towards a global control strategy for induction motor: Speed regulation, flux optimization and power factor correction. *International Journal of Electrical Power and Energy Systems*, 43, 230—244. Available from https://doi.org/10.1016/j.ijepes.2012.05.022. Available from https://www.sciencedirect.com/science/article/abs/pii/S0142061512002128.

El Fadili, A., Giri, F., El Magri, A., Lajouad, R., & Chaoui, F. (2014). Adaptive control strategy with flux reference optimization for sensorless induction motors. *Control Engineering Practice*, 26, 91–106. Available from https://doi.org/10.1016/j.conengprac.2013.12.005, URL. Available from https://www.sciencedirect.com/science/article/abs/pii/S0967066113002384.

El Magri, A., Giri, F., Abouloifa, A., & Chaoui, F. (2010). Robust control of synchronous motor through ac/dc/ac converters. *Control Engineering Practice*, 18, 540–553. Available from https://doi.org/10.1016/j.conengprac.2010.02.005, URL. Available from https://www.sciencedirect.com/science/article/abs/pii/S0967066110000420.

El Magri, A., Giri, F., Abouloifa, A., & El Fadili, A. (2009). Nonlinear control of a variable speed wind generator. *IFAC Proceedings Volumes (IFAC-PapersOnline)*, 416–421. Available from https://doi.org/10.3182/20090705-4-SF-2005.00073, URL. Available from https://www.sciencedirect.com/science/article/pii/S1474667015302391.

El Magri, A., Giri, F., Abouloifa, A., & El Fadili, A. (2010). Wound rotor synchronous motor control through ac/dc/ac converters. *IFAC Proceedings Volumes (IFAC-PapersOnline)*, 1344–1349. Available from https://doi.org/10.3182/20100901-3-IT-2016.00082, URL. Available from https://www.sciencedirect.com/science/article/pii/S1474667015371524.

El Magri, A., Giri, F., Abouloifa, A., & El Fadili, A. (2014). Nonlinear control of associations including wind turbine, PMSG and AC/DC/AC converters. In *2009 European Control Conference, ECC 2009* (pp. 4338–4343). <http://www.scopus.com/inward/record.url?eid = 2-s2.0–84955190444partnerID = MN8TOARS>.

El Magri, A., Giri, F., Abouloifa, A., & Haloua, M. (2007). Nonlinear control of wound-rotor synchronous-motor. In *Proceedings of the IEEE international conference on control applications* (pp. 3110–3115). <https://ieeexplore.ieee.org/document/4067564>. Available from https://doi.org/10.1109/CCA.2006.286085.

El Magri, A., Giri, F., Abouloifa, A., Lachkar, I., & Chaoui, F. (2009a). Nonlinear control of associations including synchronous motors and AC/DC/AC converters: A formal analysis of speed regulation and power factor correction. In *Proceedings of the American Control Conference* (pp. 3470–3475). <https://ieeexplore.ieee.org/document/5160580>. Available from https://doi.org/10.1109/ACC.2009.5160580.

El Magri, A., Giri, F., Abouloifa, A., Lachkar, I., & Chaoui, F. (2009b). Speed regulation with power factor correction of synchronous motors including ac/dc/ac converters. *IFAC Proceedings Volumes (IFAC-PapersOnline)*, 344–349. Available from https://doi.org/10.3182/20090705-4-SF-2005.00061, URL. Available from https://www.sciencedirect.com/science/article/pii/S1474667015302275.

El Magri, A., Giri, F., Besançon, G., El Fadili, A., Dugard, L., & Chaoui, F. (2013). Sensorless adaptive output feedback control of wind energy systems with pms generators. *Control Engineering Practice*, 21, 530–543. Available from https://doi.org/10.1016/j.conengprac.2012.11.005, URL. Available from https://www.sciencedirect.com/science/article/abs/pii/S0967066112002365.

El Magri, A., Giri, F., & El Fadili, A. (2010). An interconnected observer for wind synchronous generator. *IFAC Proceedings Volumes (IFAC-PapersOnline)*, 765–770. Available from https://doi.org/10.3182/20100901-3-IT-2016.00255, URL. Available from https://www.sciencedirect.com/science/article/pii/S1474667015370567.

El Magri, A., Giri, F., & El Fadili, A. (2013). *Control models for synchronous machines*. AC electric motors control: Advanced design techniques and applications (pp. 41–56). West Sussex, UK: John Wiley & Sons, Ltd. Available from http://doi.org/10.1002/9781118574263.ch3.

El Magri, A., Giri, F., El Fadili, A., & Chaoui, F. (2012). An adaptive control strategy for wound rotor synchronous machines. *International Journal of Adaptive Control and Signal Processing*, 26, 821–847.

El Fadili, A., Giri, F., El Magri, A., & Ouadi, H. (2010). Sensorless induction machine observer in presence of nonlinear magnetic characteristic. *IFAC Proceedings Volumes (IFAC-PapersOnline)*, 1, 60–65. Available from https://doi.org/10.3182/20100826-3-TR-4015.00014, URL. Available from https://www.sciencedirect.com/science/article/pii/S1474667015323405.

El Fadili, A., Giri, F., Ouadi, H., Dugard, L., & El Magri, A. (2014). Induction machine control in presence of magnetic saturation speed regulation with optimized flux reference. In *2009 European Control Conference, ECC 2009* (pp. 2542–2547). <http://www.scopus.com/inward/record.url?eid = 2-s2.0–84955186658partnerID = MN8TOARS>.

El Fadili, A., Giri, F., Ouadi, H., El Magri, A., Dugard, L., & Abouloifa, A. (2010). Induction motor control through ac/dc/ac converters. In *Proceedings of the 2010 American Control Conference, ACC 2010* (pp. 1755–1760). <https://ieeexplore.ieee.org/abstract/document/5531480>. Available from https://doi.org/10.1109/ACC.2010.5531480.

Fadili, A., Giri, F., & El Magri, A. (2013). *Control models for induction motors*. AC electric motors controls: Advanced design techniques and applications. Oxford, UK: John Wiley & Sons. Available from http://doi.org/10.1002/9781118574263.ch2.

Franklin, G. F., Powell, J. D., & Workman, M. L. (1998). *Digital control of dynamic systems* (Vol. 3). Menlo Park, CA: Addison-Wesley.

Gauthier, J.-P., & Kupka, I. (2001). *Deterministic observation theory and applications*. Cambridge, UK: Cambridge University Press.

Giri, F. (2013). *AC electric motors control: Advanced design techniques and applications*. Oxford, UK: John Wiley & Sons.

Global Wind Energy Council. (2019). *Global wind energy report 2019*. Brussels, Belgium: Global Wind Energy Council.

González, L., Figueres, E., Garcerá, G., & Carranza, O. (2010). Maximum-power-point tracking with reduced mechanical stress applied to wind-energy-conversion-systems. *Applied Energy*, 87, 2304–2312.

Heier, S. (2014). *Grid integration of wind energy: Onshore and offshore conversion systems*. West Sussex, UK: John Wiley & Sons.

Hermann, R., & Krener, A. (1977). Nonlinear controllability and observabilityl. *IEEE Transactions on Automatic Control*, 22, 728–740. Available from https://doi.org/10.1109/TAC.1977.1101601.

Hong, Y.-Y., Lu, S.-D., & Chiou, C.-S. (2009). MPPT for PM wind generator using gradient approximation. *Energy Conversion and Management*, 50, 82–89.

Jain, P. (2011). *Wind energy engineering*. New York: McGraw-Hill.

Kesraoui, M., Korichi, N., & Belkadi, A. (2011). Maximum power point tracker of wind energy conversion system. *Renewable Energy*, *36*, 2655−2662.

Khalil, H. K. (2009). Analysis of sampled-data high-gain observers in the presence of measurement noise. *European Journal of Control*, *15*, 166−176.

Khalil, H. K. (2017). *High-gain observers in nonlinear feedback control*. Philadelphia: SIAM.

Koutroulis, E., & Kalaitzakis, K. (2006). Design of a maximum power tracking system for wind-energy-conversion applications. *IEEE Transactions on Industrial Electronics*, *53*, 486−494.

Krstic, M., Kokotovic, P. V., & Kanellakopoulos, I. (1995). *Nonlinear and adaptive control design*. New York: Wiley.

Lajouad, R., Chaoui, F.Z., & El Fadili, A. (2015). A simple high gain observer for the wind energy system based on the doubly fed induction generator. In Recent advances on systems, signals, control, communications and computers. <https://www.researchgate.net/publication/288835930_DNCOSE-47>.

Lajouad, R., El Magri, A., El Fadili, A., Chaoui, F., Giri, F., & Besancon, G. (2013). State feedback control of wind energy conversion system involving squirrel cage induction generator. *IFAC Proceedings Volumes (IFAC-PapersOnline)*, *11*, 299−304. Available from https://doi.org/10.3182/20130703-3-FR-4038.00116, URL. Available from https://www.sciencedirect.com/science/article/pii/S1474667016329615.

Lajouad, R., El Magri, A., El Fadili, A., Chaoui, F.-Z., & Giri, F. (2015). Adaptive nonlinear control of wind energy conversion system involving induction generator. *Asian Journal of Control*, *17*, 1365−1376. Available from https://doi.org/10.1002/asjc.1020, URL. Available from https://onlinelibrary.wiley.com/doi/abs/10.1002/asjc.1020.

Lajouad, R., Giri, F., Chaoui, F. Z., Fadili, A. E., & Magri, A. E. (2019). Output feedback control of wind energy conversion system involving a doubly fed induction generator. *Asian Journal of Control*, *21*, 2027−2037. URL: https://doi.org/10.1002.asjc.2116. Available from https://doi.org/10.1002/asjc.2116.

Magri, A., Assche, V., Fadili, A., Chaoui, F., & Giri, F. (2013). *Nonlinear state-feedback control of three-phase wound rotor synchronous motors. AC electric motors control: Advanced design techniques and applications* (pp. 429−451). Oxford: Wiley.

Rashid, M. (2001). *Power electronics handbook* (3rd ed.). Academic Press.

Rocha, R. (2011). A sensorless control for a variable speed wind turbine operating at partial load. *Renewable Energy*, *36*, 132−141.

Senjyu, T., Ochi, Y., Kikunaga, Y., Tokudome, M., Yona, A., Muhando, E. B., et al. (2009). Sensor-less maximum power point tracking control for wind generation system with squirrel cage induction generator. *Renewable Energy*, *34*, 994−999.

Torrey, D. A. (1993). Variable-reluctance generators in wind-energy systems. In *Proceedings of IEEE power electronics specialist conference - PESC '93* (pp. 561−567).

Torrey, D. A., & Hassanin, M. (1995). The design of low-speed variable-reluctance generators. In *IAS '95. Conference record of the 1995 IEEE industry applications conference thirtieth IAS annual meeting* (pp. 427−433, Vol. 1).

Watil, A., El Magri, A., Raihani, A., Lajouad, R., & Giri, F. (2020a). An adaptive nonlinear observer for sensorless wind energy conversion system with pmsg. *Control Engineering Practice*, *98*. Available from https://doi.org/10.1016/j.conengprac.2020.104356, URL. Available from https://doi.org/10.1016/j.conengprac.2020.104356.

Watil, A., El Magri, A., Raihani, A., Lajouad, R., & Giri, F. (2020b). Multi-objective output feedback control strategy for a variable speed wind energy conversion system. *International Journal of Electrical Power & Energy Systems*, *121*, 106081.

Watil, A., Raihani, A., & Lajouad, R. (2018). Nonlinear control of an aerogenerator including DFIG and AC/DC/AC converters. In *The Proceedings of the third international conference on smart city applications* (pp. 1122−1137). Cham: Springer. <https://link.springer.com/chapter/10.1007/978-3-030-11196-0_91>. Available from https://doi.org/10.1007/978-3-030-11196-0_91.

Chapter 8

Improvement of fuel cell MPPT performance with a fuzzy logic controller

Arezki Fekik[1,2], Ahmad Taher Azar[3,4], Hakim Denoun[2], Nashwa Ahmad Kamal[5], Naglaa K. Bahgaat[6], Tulasichandra Sekhar Gorripotu[7], Ramana Pilla[8], Fernando E. Serrano[9], Shikha Mittal[10], K.P.S. Rana[11], Vineet Kumar[11], Sundarapandian Vaidyanathan[12], Mohamed Lamine Hamida[2], Nacera Yassa[1] and Karima Amara[2]

[1]Akli Mohand Oulhadj University, Bouira, Algeria, [2]Electrical Engineering Advanced Technology Laboratory (LATAGE), Mouloud Mammeri University, Tizi Ouzou, Algeria, [3]Faculty of Computers and Artificial Intelligence, Benha University, Benha, Egypt, [4]College of Computer and Information Sciences, Prince Sultan University, Riyadh, Saudi Arabia, [5]Faculty of Engineering, Cairo University, Giza, Egypt, [6]Electrical Communication Department, Faculty of Engineering, Canadian International College (CIC), Giza, Egypt, [7]Department of Electrical & Electronics Engineering, Sri Sivani College of Engineering, Srikakulam, India, [8]Department of EEE, GMR Institute of Technology, Rajam, India, [9]Universidad Tecnolgica Centroamericana (UNITEC), Zona Jacaleapa, Tegucigalpa, Honduras, [10]Department of Mathematics, Jesus and Mary College, University of Delhi, New Delhi, India, [11]Division of Instrumentation and Control Engineering, Netaji Subhas University of Technology, Dwarka, India, [12]Research and Development Centre, Vel Tech University, Chennai, India

8.1 Introduction

Existence has provided us with an enormous amount of energy in different forms. The energy conversion rule stipulates that it already exists and that it is neither produced nor lost, converted and represented only in any other form. Energy has two main categories: nonrenewable and renewable energy. Nonfuels emit many ozonated substances, such as carbon dioxide, which is the main cause of the rise in global temperature and global warming, which significantly affects the atmosphere due to the limited storage of nonrenewable petroleum-based energy (Ahmed & Salam, 2016; Femia, Petrone, Spagnuolo, & Vitelli, 2005; Gow & Manning, 1999; Masoum, Dehbonei, & Fuchs, 2002; Sera, Teodorescu, & Rodriguez, 2007; Xiao, Huang, & Kang, 2015). The increasing use of sustainable or renewable energy sources, in particular photovoltaic cells, biomass, wind power, and geothermal energy, has increased considerably. Photovoltaic technology is one of the fastest-growing innovations among the various renewable resources due to its abundant source of solar lighting and its negative environmental impact. Due to the advancement of electronic energy devices and control rules, there is a substantial increase in the demand for energy generated from photovoltaic systems (Amara et al., 2018, 2019a; Arezki Fekik et al., 2017; Azar et al., 2019; Denoun, Hamida, Fekik, Dyhia, & Ghanes, 2018; Fekik, Denoun, Benamrouche, Benyahia, & Zaouia, 2015; Fekik et al., 2018a; Fekik, 2018; Fekik et al. 2018b, 2018c, 2019; Ghoudelbourk, Azar, Dib, & Omeiri, 2020; Hamida, Denoun, Fekik, & Vaidyanathan, 2019; Lamine, Hakim, Arezki, Nabil, & Nacerddine, 2018; Lamine et al., 2019). Nowadays, researchers around the world are studying and studying multiple sources of free energy to transform it into functional means.

As the supply of petrochemicals decreases, green and renewable technologies have become the first choice for their main electricity sources. Modern fuel cell (FC) technology is provided to overcome the limitations. FCs are used to convert energy from hydrogen to produce electricity. Several forms of FCs are based on their electrolytes. The following are widely used: (1) lower ambient temperature; (2) reduce operating pressure to increase safety; and (3) best convertibility ratio Bocci et al. (2014), Cipriani et al. (2014). FCs require a high construction cost if they are necessary for a high power application because they produce only a low output voltage.

A broad number of internal parameters can influence the FC's output voltage, but the IV curve represents a single point which represents the maximum output point (MPP) (Amara et al., 2019b; Ammar, Azar, Shalaby, & Mahmoud, 2019; Ben Smida, Sakly, Vaidyanathan, & Azar, 2018; Chavan & Talange, 2017; Fekik et al., 2021b; Ghoudelbourk,

Renewable Energy Systems. DOI: https://doi.org/10.1016/B978-0-12-820004-9.00023-1

Dib, Omeiri, & Azar, 2016; Hong, Xu, Li, & Ouyang, 2017; Kamal, Azar, Elbasuony, Almustafa, & Almakhles, 2020; Zhang, Fu, Sumner, & Wang, 2017). The FC reaches the optimum energy at this level. According to its capacity to produce energy from the fuel flow available, the FC must run at MPP. This will avoid leakage of fuel and slow service. Using the maximum power point tracking (MPPT) method, the MPP FC can be calculated. The MPPT technique is important in any environmental context to predict and monitor the MPP and then force the PV system to work at this stage. The MPPT algorithm can be used to find the converter's duty cycle by comparing the value of the FC's current, power, or voltage. There are several methods for MPPT which can be applied depending on the system requirements. The MPPT method varies in complexity, hardware implementation, and detected parameters.

There are many studies on different MPPT for FCs, such as escalation/disturb and observe (P&O), adaptive control of MPPT, adaptation of resistance, MPPT based on voltage and current, constant voltage, and adaptive extremum seeking control (Azri, Khanipah, Ibrahim, & Rahim, 2017; Makhloufi, Hatti, & Taleb, 2017; Nasiri Avanaki & Sarvi, 2016; Khanipah, Azri, & Ibrahim, 2018; Wang, Huang, Jiang, & Liou, 2016). The P&O is most commonly used due to its simple structure to use in the MPPT FC (Drissi, Khediri, Zaafrane, & Braiek, 2017; Roza, 2018; Sreejith & Singh, 2019).

Single control strategies cannot control the FCs; advanced algorithms are needed for robust control and fast response (Bansal et al., 2021; Drhorhi et al., 2021; Fekik et al., 2021c; Sambas et al., 2021a; Singh, Mathpal, Azar, Vaidyanathan, & Kamal, 2021; Vaidyanathan, Sambas, & Azar, 2021b; Vaidyanathan, Sambas, Azar, Rana, & Kumar, 2021c; Vaidyanathan, Sambas, Azar, Serrano, & Fekik, 2021e; Vaidyanathan, Sambas, Azar, & Singh, 2021f).

This chapter discusses the FC power source power management device's voltage attached to the DC voltage converter. The method suggested is based on the smooth logic for designing the full FC power extraction algorithm linked to the DC−DC converter. The effective use of simulation experiments is to test the closed-loop method. The power balance on the input and the DC−DC converter's output is faced with appropriate control of the output voltage and a good behavior about the variance in DC-output voltage.

This book chapter is organized into six sections. Section 8.1 is introductory. In Section 8.2, Mathematical model of proton-exchange membrane fuel cells (PEMFC) is given. In Section 8.3, mathematical model of DC−DC converter is described. In Section 8.4, the proposed algorithm is presented. Simulations results are presented in Section 8.5. Discussion is given in Section 8.6. Finally in Section 8.7, concluding remarks are given.

8.2 Modeling of proton-exchange membrane fuel cells

A simple PEMFC device can be seen as a two-electrode divided by a solid membrane serving as an electrolyte (anode and cathode) as shown in Fig. 8.1. The anode network draws hydrogen fuel through protons, which pass from the membrane to the cathode, and electrons captured by an internal circuit that binds the two electrodes. The oxidant (air in this research) moves through a similar network of channels to the cathode, which combines oxygen and electrons in the inner circuit to create gas.

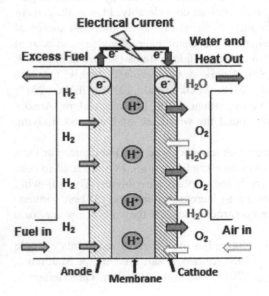

FIGURE 8.1 The operating principle of proton-exchange membrane fuel cell.

8.2.1 Static model of PEMFC

The electrons are free in the anode electrolyte via the hydrogen gas ionization, and the H + ions can be produced as (Derbeli, Farhat, Barambones, & Sbita, 2017; Fekik et al., 2021a; Khan & Mathew, 2019):

$$2H_2 \Leftrightarrow 4H^+ + 4e^- \tag{8.1}$$

Water is created by electrode cathode by the interaction of oxygen and electrode electrons and electrode H + ions from electrolytes, given by:

$$O_2 + 4H^+ + 4e^- \Leftrightarrow 2H_2O \tag{8.2}$$

General cell reaction

$$2H_2 + O_2 \Leftrightarrow 2H_2O \tag{8.3}$$

A variety of parameters are used in the mathematical model, the interpretation of which is important to obtain the best simulation performance. The output voltage of a single cell should be set at this point, which can be defined as (Derbeli et al., 2017; Fekik et al., 2021a; Khan & Mathew, 2019):

$$V_{cell} = E - \eta_{act} - \eta_{ohm} - \eta_{dif} \tag{8.4}$$

The first word in Eq. (8.4) applies to the open FC circuit voltage, while the other words reduce the effective cell voltage under a certain operational situation. Each one under Eq. (8.4) can be determined using the parameters described in Table 8.1 by the following equations.

- Reversible cell potential (E)

E is a cell's capacity for electrochemical thermodynamics, and it has the highest output voltage. The Nernst equation is determined by relation to

If the increase in Gibbs free reaction energetic (J mol^{-1}), ΔS means a shift in the reaction entropy (J mol^{-1}), ΔS is a rise in hydrogel and partial oxygen pressure in both (atm), and T and T_{ref} are the temperature and reference temperature of (K), respectively. Where ΔG is the increase in the reaction entropy (d mol^{-1}).

- Activation polarization loss η_{act}

The overvoltage activation is the voltage decrease induced by anode and cathode activation. In the following calculation, activation overvoltage is measured in an FC.

$$\eta_{act} = \epsilon + \epsilon_2 T + \epsilon_3 T.\ln(CO_2) + \epsilon_4 T.\ln(I) \tag{8.5}$$

where the cell load in (A) is I, CO_2 is the concentration of oxygen dissolved on the catalytic surface of the cathode on an interface in water (mol cm^{-2}) determined by the rule of Henry based on partial oxygen pressure and cell temperature.

$$CO_2 = \frac{P_{O_2}}{5.08 \times 10^6 \times e^{\frac{-498}{T}}} \tag{8.6}$$

TABLE 8.1 Fuel cell model parameters.

Symbol	Value	Unit
Stack power	5998.5	W
Fuel cell resistance	0.07833	Ω
Nernst voltage of one cell	1.1288	V
Number of cells	65	–
Operating temperature	65	°C

- Ohmic polarization loss η_{ohm}

    The lack of ohms polarization is a product of membrane proton transfer resistance and the electrical electrode communication resistance. The ohms losses can be described as follows:

$$\eta_{ohm} = I.(R_m + R_c) \tag{8.7}$$

where R_m equal electron flow resistance and R_c shall be considered a steady proton resistance

$$R_m = \frac{\rho_m.l}{A} \tag{8.8}$$

With ρ_m is the specific resistance of the membrane (Ω cm); A is the membrane active area (cm^2); and l is the thickness of the membrane (cm).

- Concentration polarization loss η_{con}

    The loss of the concentration results in a decrease in the reactants, oxygen, and hydrogen concentrations that change throughout the reaction. Therefore the failure of the chemical reaction may be interpreted.

$$\eta_{con} = B\ln\left(1 - \frac{J}{J_{max}}\right) \tag{8.9}$$

Although B is a parameter that is dependent on cell type, J represents the current cell density (A cm^{-2}) and is defined as:

$$J = \frac{I}{A} \tag{8.10}$$

- Fuel-cell stack V_{cell}

    FC stack is a single pile lumped together with the following equation to determine the performance voltage.

$$V_{cell} = E - \eta_{act} - \eta_{ohm} - \eta_{dif} \tag{8.11}$$

And, for N cells put in series, creating a stack, the voltage, V_{stack}, can be determined by:

$$V_{stack} = N_{cell} \times V_{cell} \tag{8.12}$$

The electrical power produced by the cell to the load can be provided by the equation:

$$P_{FC} = V_{FC} \times I \tag{8.13}$$

where V_{FC} is the tension of the cell for all the operational conditions using the PEMFC equations given in the document (Derbeli et al., 2017), the PEMFC system can be modeled as shown in Fig. 8.2.

8.2.2 Dynamic model of PEMFC

The two layers isolated from the membrane serve like a charged double in a PEMFC, which can store electric power, which can be treated as a condenser, thanks to this structure. The related tolerance to various forms of FC losses is R_{act},

FIGURE 8.2 Static model.

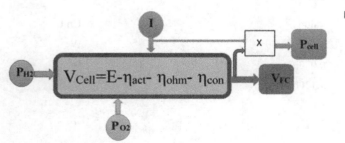

FIGURE 8.4 Proton-exchange membrane fuel cell scheme model.

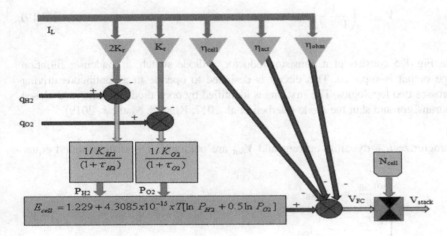

R_{con}, and R_U. The circuit FC model is shown in Fig. 8.3, taking into account all the above described impact. The following equation is given by Kirchhoff Law:

$$I = C\frac{dV_d}{dt} + \frac{V_d}{R_{con} + R_{act}}$$

(8.14)

The above equation can be described as follows:

$$\frac{V_d}{dt} = \frac{1}{C}I - \frac{1}{\tau}V_d$$

(8.15)

where V_d is the variable voltage of the corresponding capacitor (associated with η_{act} and η_{con}); C is the electric capacitance equivalent; and τ is the electric time of the combustible cell, depending on the cell's cell temperature:

$$\tau = C(R_{act} + R_{con}) = C\left(\frac{\eta_{act} + \eta_{con}}{I}\right)$$

(8.16)

The resulting FC voltage (*Reference Control PFMEC₄*) will then be modified, including this electric dynamic activity term, with the following equation:

$$V_{cell} = E - V_d - IR_{ohm}$$

(8.17)

While R_{ohm}, R_{act}, and R_{con} reflect the resistance of the ohmic, the activation, and the concentration and C is the ability of the membrane due to its double-layer impact. This influence is built into the PEMFC's output voltage. The transition function (8.18) was obtained using Eqs. (8.16) and (8.17) and Laplace transformations, in which S describes the Laplace operator:

$$V_{cell} = E - \left(\frac{R_{act} + R_{con}}{(R_{act} + R_{con})CS + 1} + R_{ohm} \right) I \tag{8.18}$$

where V_{FC} is the cell voltage for all operating conditions using the PEMFC equations given in the document (Derbeli et al., 2017) and the PEMFC system can be modeled as seen by Fig. 8.4.

8.3 Mathematical model of DC–DC converter

Booster converters are typically boost power converters that have a low voltage input and an output at a higher voltage. An ideal DC/DC overcurrent converter circuit is given by Fig. 8.5. The mathematical relationship between the input and output voltages is regulated by the duty cycle d, as given by the equation below Derbeli et al.(2017).

$$V_{out} = \left(\frac{1}{1 - d} \right) V_{in} \tag{8.19}$$

with $V_{in} = V_{stack}$

The converter boost circuit as shown in Fig. 8.5 consists of a transistor inductor, a diode switch, a condenser filtration device, and a load resistor. The DC voltage output is supplied. This circuit is designed to operate in a continuous driving mode. In this theory, the circuit cyclically crosses two topologies. The first one is identified by open diode switches and closed transistors. The second is a turn to open the transistor and shut the diode (Derbeli et al. 2017; Khan & Mathew, 2019).

- First topology (u is ON and d is OFF)

  The differential equations that characterize i_{cell} dynamic current and V_{out} are obtained. The differentiated equations are:

$$\begin{cases} \dfrac{di_{cell}}{dt} = \dfrac{1}{L} V_{Cell} \\ \dfrac{dV_{out}}{dt} = -\dfrac{1}{C} i_{out} \end{cases} \tag{8.20}$$

This topology will compose and notice the state-space equations as follows:

$$\dot{x} = A_1 x + B_1 v \tag{8.21}$$

where the matrices x, A_1, and B_1 are described as bellows:

$$\begin{cases} x = \begin{pmatrix} x_1 \\ x_2 \end{pmatrix} = \begin{pmatrix} I_{Cell} \\ V_{Out} \end{pmatrix} \\ A_1 = \begin{pmatrix} 0 \\ 0 \end{pmatrix} = \begin{pmatrix} 0 \\ -\dfrac{1}{RC} \end{pmatrix} \\ B_1 = \begin{pmatrix} \dfrac{1}{L} \\ 0 \end{pmatrix} \end{cases} \tag{8.22}$$

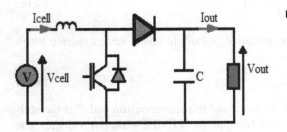

FIGURE 8.5 Proton-exchange membrane fuel cell scheme model.

- Second topology (u is OFF and d is ON)

Differential equations have been obtained to characterize the dynamic state of the inductance current iL and the output voltage V_{out}.

$$\begin{cases} \dfrac{di_{\text{cell}}}{dt} = \dfrac{1}{L}(V_{\text{Cell}} - V_{\text{out}}) \\ \dfrac{dV_{\text{out}}}{dt} = \dfrac{1}{C}(i_{\text{cell}} - i_{\text{out}}) \end{cases} \tag{8.23}$$

The spatial equations of state of this topology can be written and noted as follows:

$$\dot{x} = A_2.x + B_2.v \tag{8.24}$$

The matrices A_2 and B_2 are defined as follows:

$$\begin{cases} A_1 = \begin{pmatrix} 0 & -\dfrac{1}{L} \\ \dfrac{1}{C} & -\dfrac{1}{RC} \end{pmatrix} \\ B_1 = \begin{pmatrix} \dfrac{1}{L} \\ 0 \end{pmatrix} \end{cases} \tag{8.25}$$

The representation of the status space for the boost converter can be written and noted as follows:

$$\begin{cases} \dot{x} = Ax + Bv \\ y = Cx + Ev \end{cases} \tag{8.26}$$

Where the matrices A, B, C, and E are given by:

$$\begin{cases} A = uA_1 + (1 - u)A_2 \\ B = uB_1 + (1 - u)B_2 \\ C = [01] \\ E = 0 \end{cases} \tag{8.27}$$

The final representation of the state space for the boost converter can be expressed by the following expression:

$$\begin{cases} \dot{x} = \begin{pmatrix} 0 & \dfrac{u-1}{L} \\ \dfrac{1-u}{C} & -\dfrac{1}{RC} \end{pmatrix} x + \begin{pmatrix} \dfrac{1}{L} \\ 0 \end{pmatrix} v \\ y = [0 \quad 1]x \end{cases} \tag{8.28}$$

8.4 Proposed algorithm

Recently, different control approaches have been proposed for designing nonlinear systems for many practical applications, such as optimal control (Alimi, Rhif, Rebai, Vaidyanathan, & Azar, 2021; Azar, Ammar, Beb, Garces, & Boubakari, 2020; Meghni, Dib, Azar, Ghoudelbourk, & Saadoun, 2017b; Meghni, Dib, Azar, & Saadoun, 2018; Singh & Azar, 2020a) and nonlinear control (Abdul-Adheem, Azar, Ibraheem, & Humaidi, 2020a; Abdul-Adheem, Ibraheem, Azar, & Humaidi, 2020b; Abdul-Adheem, Ibraheem, Humaidi, & Azar, 2020c; Pham, Vaidyanathan, Azar, & Duy, 2021; Sambas et al., 2021b; Sambas et al., 2021c; Vaidyanathan & Azar, 2021; Vaidyanathan, Pham, & Azar, 2021a; Vaidyanathan, Sambas, Azar, Rana, & Kumar, 2021d; Humaidi, Ibraheem, Azar, & Sadiq, 2020).

In this section, a fuzzy logic controller (FLC) is applied to extract the maximum power and improve the performance of the PEMFC system. MPPT methods can be divided into traditional and smart methods based on the control algorithm. The conventional MPPT approach involves perturbation and observation (P&O), incremental conductance

(INC), voltage-feedback methods, and so on. The intelligent methods include fuzzy logic (FLC), neural network, and genetic algorithm and are used to achieve high performance and efficiency (Abdelmalek, Azar, & Dib, 2018a; Abdelmalek, Azar, & Rezazi, 2018b; Ajeil, Ibraheem, Azar, & Humaidi, 2020a; Ajeil, Ibraheem, Azar, & Humaidi, 2020b; Ibraheem, Azar, Ibraheem, & Humaidi, 2020; Ibrahim, Azar, Ibrahim, & Ammar, 2020; Liu, Ma, Azar, & Zhu, 2020; Mohamed, Azar, Abbas, Ezzeldin, & Ammar, 2020; Najm, Ibraheem, Azar, & Humaidi, 2020; Pilla, Azar, & Gorripotu, 2019; Sallam, Azar, Guaily, & Ammar, 2020; Sayed et al., 2020; Kammogne et al., 2020; Elkholy, Azar, Magd, Marzouk, & Ammar, 2020a,b). The closed-loop control is shown in Fig. 8.6.

FC's productivity is improved by the MPPT technique when they get to work with maximum power. There are several methods of MPPT. In this chapter, MPPT is used to extract the FC system's maximum power and transfer this power under continuous load via a DC−DC converter. The DC/DC converter is used to transfer the maximum power from the FC to the receiver. The DC/DC converter serves as an interface between the load and the battery. By modifying the duty cycle, the load impedance seen by the source is modified and adjusted to the peak power of the source to transfer the maximum power (Fig. 8.7).

The understanding of the problems and the knowledge associated with the linguistic variables must be well known (Ammar, Azar, Tembi, Tony, & Sosa, 2018; Barakat, Azar, & Ammar, 2020; Djeddi, Dib, Azar, & Abdelmalek, 2019; Giove, Azar, & Nordio, 2013; Khettab, Bensafia, Bourouba, & Azar, 2018; Kumar, Azar, Kumar, & Rana, 2018; Meghni, Dib, & Azar, 2017a; Pilla et al., 2019; Pilla, Botcha, Gorripotu, & Azar, 2020; Soliman, Azar, Saleh, & Ammar, 2020; Vaidyanathan & Azar, 2016e). In the intuition method, triangular membership functions are used to fuzzify net worth into membership value. In the proposed FLC-based MPPT, variables, such as error and error change, are grouped in membership value by the triangular membership function. The range of input variables (−5, 5) is defined. A triangular type membership function is proposed, which recognizes, for any specific entry, a single dominant fuzzy subset. Seven fuzzy sets are taken into account for the membership functions. These variables are expressed in terms of linguistic variables, such as negative large, negative medium, negative small, zero, positive small, medium positive, and large positive, being basic fuzzy sets as shown in Fig. 8.8.

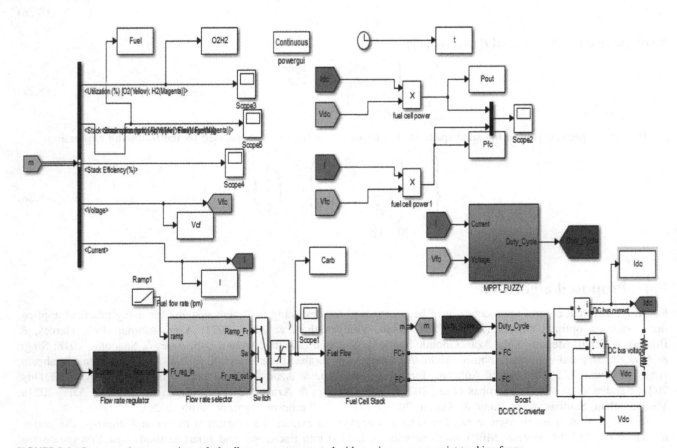

FIGURE 8.6 Proton-exchange membrane fuel cell power system control with maximum power point tracking-fuzzy.

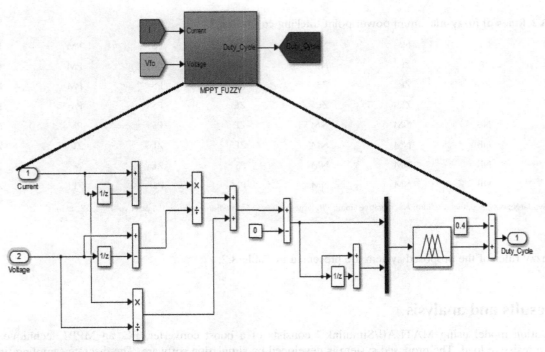

FIGURE 8.7 Proton-exchange membrane fuel cell power system control with maximum power point tracking-fuzzy.

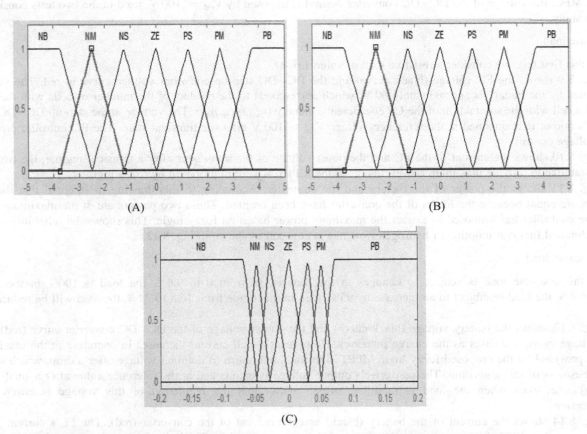

FIGURE 8.8 Membership functions for input variables error $E(k)$ (A), change in error $\Delta E(k)$ (B) and output variable $D(k)$ (C).

TABLE 8.2 Rules of fuzzy-maximum power point tracking controller.

$\Delta E(k)$	NB	NM	NS	ZE	PS	PM	PB
NB	ZE	ZE	NS	NM	PM	PM	PB
NM	ZE	ZE	ZE	NS	PS	PM	PB
NS	ZE	ZE	ZE	ZE	PS	PM	PB
ZE	NB	NM	NM	ZE	PS	PM	PB
PS	NB	NM	NM	ZE	ZE	ZE	ZE
PM	NB	NM	NM	PS	ZE	ZE	ZE
PB	NB	NM	NM	PM	PS	ZE	ZE

NB, negative large; *NM*, negative medium; *NS*, negative small; *PB*, large positive; *PM*, medium positive; *PS*, positive small; *ZE*, zero.

The fuzzy rules of the proposed system are presented in Table 8.2.

8.5 Results and analysis

The simulation model using MATLAB/Simulink™ consists of a boost converter, FC, an MPPT technique based on FLC, and a resistive load. The proposed system is developed by simulation software. The discrete sampling time for the simulation is selected as 50 ms. Two tests will be carried out to illustrate the validity and the reliability of the proposed algorithm, namely, a test with a static load and the other with a variation of the load. To obtain the maximum power of the PEMFC, the voltage of the DC−DC converter desired is imposed by $V_{ref} = 100$ V, used in the two tests considered in this study.

- Static load

In this first test, we consider a resistive load of value 1.67 Ω.

Fig. 8.9 shows the FC voltage (black curve) and the DC−DC step-up converter voltage curve in red. The voltage generated by the battery is approximately 60 V, which corresponds to the product of the number of cells with the voltage of a cell with the subtraction of the CF coefficients, namely ($\eta_{act}, \eta_{ohm}, \eta_{dif}$). The voltage at the step-up DC−DC converter's output is maintained at the reference voltage $V_{ref} = 100$ V after the transient state. The PI controller provides this voltage control.

Fig. 8.10 shows the current of the FC and the boost voltage of the transducer after a transient regime; the two currents stabilize to provide maximum power, thanks to the MPPT algorithm proposed by fuzzy logic.

Fig. 8.11 shows the power of the battery and the converter. According to this drawing, it is noted that the two powers are equal because the losses of the converter have been omitted. These two powers are at the maximum point that the controller has proposed to extract the maximum power based on fuzzy logic. This shows the reliability of this algorithm and fuel consumption of hydrogen flow rate is constant as shown in Fig. 8.12.

- Dynamic load

In this test, the load is subject to changes in the time range from 0 to 9.6 S, the load is 100% in the range 9.6−16.8 S, the load is subject to an increase by 95%, then in this range from 16.8 to 24 S, the load will be reduced by 95%.

Fig. 8.13 shows the battery voltage (black curve) and the output voltage of the DC−DC converter curve (red). The FC voltage curve decreases as the charge increases. However, the cell current increases in operating at the maximum power provided by the proposed fuzzy logic MPPT algorithm and return to nominal voltage after a drop, which shows the robustness of this algorithm. The converter's output voltage is maintained at the reference value after a small transient regime, even when the load is modified, which shows that the regulation of this voltage is ensured by compression.

Fig. 8.14 shows the current of the battery (black) and the current of the converter (red). The FC's current curve increases as the charge increases to guarantee its maximum output power. Similarly, for the current at the converter's

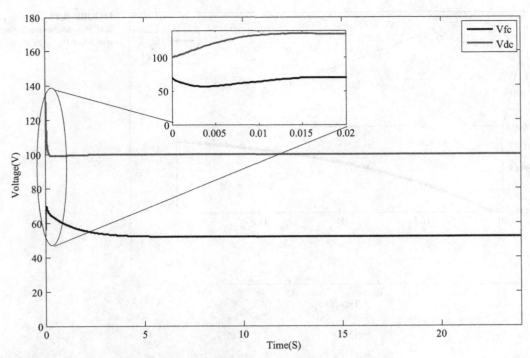

FIGURE 8.9 Ouput voltage of fuel cell V_{FC} and ouput voltage of DC−DC converter V_{dc} with static load.

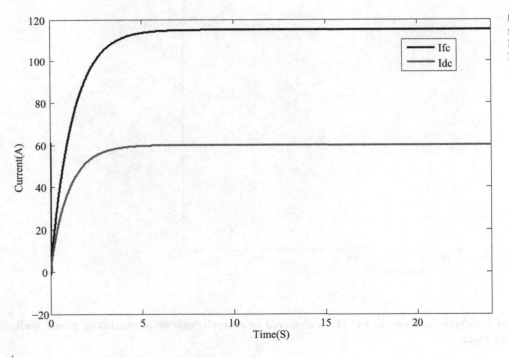

FIGURE 8.10 Ouput current of fuel cell I_{FC} and ouput current of DC−DC converter I_{dc} with static load.

output, the increase in the load causes an increase in the current to ensure the extraction of the maximum power transferred to the load, which proves that the algorithm $MPPT_{Fuzzy}$ is reliable and robust.

Fig. 8.15 shows that the maximum power is guaranteed when changing the load and confirms that both Figs. 8.13 and 8.14 because the power is equal to the product of voltage and current. In Fig. 8.16, when the load is variable, an

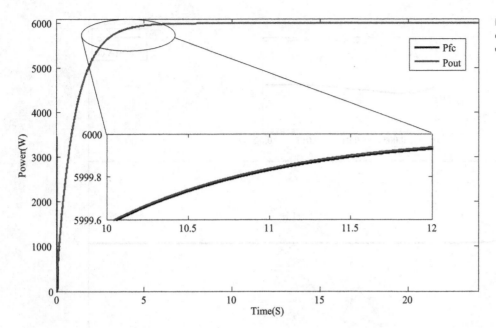

FIGURE 8.11 Ouput power of fuel cell P_{FC} and ouput power of DC−DC converter P_{dc} with static load.

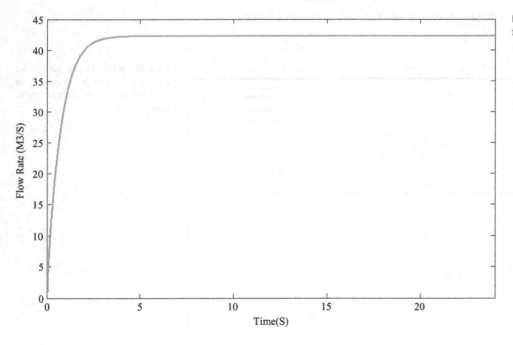

FIGURE 8.12 Fuel flow rate of fuel cell with static load.

increase in the consumption of hydrogen flow rate of the FC is observed to ensure operation at maximum power with the algorithm proposed in fuzzy logic.

8.6 Discussion

Control is a key technology allowing renewable energy systems to be used. The efficient use of advanced adaptive control technology is important for FCs (Vaidyanathan, Azar, & Ouannas, 2017a; Vaidyanathan, Azar, & Ouannas, 2017b; Vaidyanathan, Zhu, & Azar, 2017c; Vaidyanathan, Azar, & Boulkroune, 2018a; Vaidyanathan et al., 2018b;

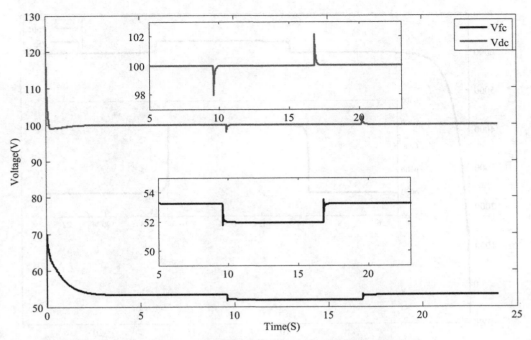

FIGURE 8.13 Ouput voltage of fuel cell V_{FC} and ouput voltage of DC−DC converter V_{dc} with dynamic load.

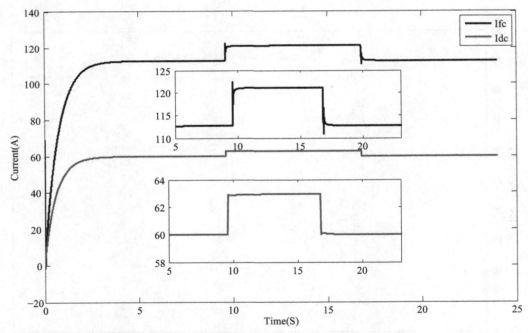

FIGURE 8.14 Ouput current of fuel cell I_{FC} and ouput current of DC−DC converter I_{dc} with dynamic load.

Vaidyanathan & Azar, 2016a, 2016b, 2016c, 2016d). Moreover, full functionality cannot be done until control systems are commonly used at all levels. This is why that robust control systems are recommended for these purposes (Azar & Serrano, 2020; Djouima, Azar, Drid, & Mehdi, 2018; Kumar et al., 2018; Mekki, Boukhetala, & Azar, 2015; Singh et al., 2017; Singh, Azar, & Zhu, 2018b; Vaidyanathan, Sampath, & Azar, 2015; Vaidyanathan et al., 2019; Vaidyanathan & Azar, 2015a, 2015b).

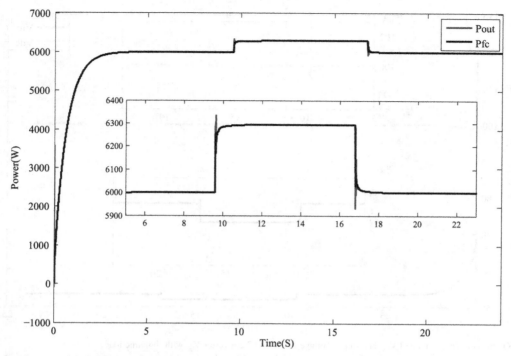

FIGURE 8.15 Ouput power of fuel cell P_{FC} and ouput power of DC–DC converter P_{dc} with dynamic load.

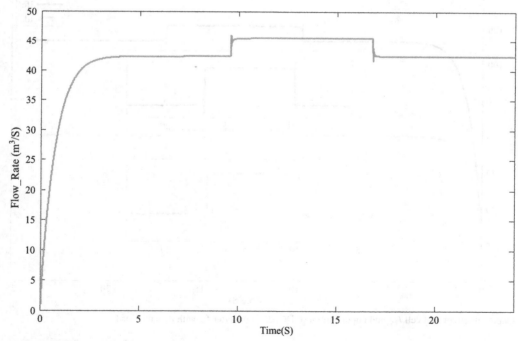

FIGURE 8.16 Fuel flow rate of fuel cell with dynamic load.

One of the main challenges for energy storage systems to be able, through a reduction of hydrogen consumption and a higher lifetime of PEMFCs, is to implement effective energy management control techniques. For better output and cost reduction per kilowatt hour advanced robust control strategies to address these issues are needed (Ammar & Azar,

2020; Gorripotu, Samalla, Jagan Mohana Rao, Azar, & Pelusi, 2019; Khan, Singh, & Azar, 2020a; Khan, Singh, & Azar, 2020b; Kumar et al., 2018; Mittal, Azar, & Kamal, 2021; Ouannas, Grassi, Azar, & Singh, 2019; Ouannas, Grassi, & Azar, 2020b; Ouannas, Grassi, & Azar, 2020c; Ouannas, Grassi, Azar, Khennaouia, & Pham, 2020d; Ouannas, Grassi, Azar, Khennaouia, & Pham, 2020e; Singh & Azar, 2020b; Singh, Azar, Vaidyanathan, Ouannas, & Bhat, 2018a; Ouannas, Grassi, Azar, & Khennaoui, 2021).

In particular, PEMFCs are electrochemicals that depend on slow electrochemical reactions. PEMFC should have relatively stable power during service in order to achieve its longer life cycle. The need for robust control strategies that can respond fast and can regulate the FC operating conditions is needed (Alain et al., 2018; Alain, Azar, Kengne, & Bertrand, 2020; Bouakrif, Azar, Volos, Muñoz-Pacheco, & Pham, 2019; Ibraheem et al., 2020; Khennaoui, Ouannas, Boulaaras, Pham, & Azar, 2020; Khettab et al., 2018; Ouannas, Azar, & Vaidyanathan, 2017a; Ouannas, Azar, & Vaidyanathan, 2017b; Ouannas, Azar, Ziar, & Radwan, 2017c; Ouannas, Azar, Ziar, & Radwan, 2017d; Ouannas, Azar, Ziar, & Vaidyanathan, 2017e; Ouannas, Azar, Ziar, & Vaidyanathan, 2017f; Ouannas, Azar, Ziar, & Vaidyanathan, 2017g; Ouannas, Azar, & Ziar, 2020a; Ounnas, Azar, & Radwan, 2016; Pham et al., 2017; Pham et al., 2018; Tolba et al., 2017). In this chapter, a fuzzy controller was designed to improve PEMFC system performance. For FCS, sensor locations are limited to the use of integral control and observation-based controller. In particular, it is not possible to directly measure the performance variable, that is, the excess oxygen ratio. Traditional feedback from the control is the compressor flow rate, which is located upstream from the stack. Our observability analysis shows that the measurement of stack voltage can be used to improve the performance and strength of the closed-loop system. Presently, voltage measurement is for safety surveillance only. We show however that the mean voltage of the FC stack can be used for active control of the starvation in the FC stack. This result shows how we define critical and economical sensor locations for the FCS.

Additional limitations arise if all the auxiliary equipment is directly fueled from an FC without secondary electricity sources under the FCS architecture. Due to its simplicity, compaction and low cost, this plant configuration is preferred. We have used optimum control design to identify frequencies at which the transient system net power and the stack starvation control are severely affected. In case of a fast transient response, the results can be used to determine the necessary size of the additional energy or oxygen storage systems. We have shown that the multivariable controller enhances the performance of the FCS and results in a new dynamic current−voltage relationship that is captured by the FCS impedance. The derived closed-loop impedance FCS is expected to be very useful for a systematic design of electronic FC components.

8.7 Conclusion and perspectives

In this chapter, a mathematical model of a proton exchange membrane FC is presented. The FC characteristics are obtained for different values of the input parameters. It should be noted that by operating the FC at higher values of the input variables, the voltage losses can be reduced. Besides, a fuzzy logic control is proposed to maximize the power generated by the PEMFC with the guarantee of the system's stability in the presence of noise and disturbances. The supertorsion algorithm is used to smooth the discontinuous control term in order to reduce the knock phenomenon. The simulation results indicate that the suggested approach has considerable advantages over control in conventional methods. In the context of future work, we wish to carry out an experimental study of the control approach on a real PEMFC system and to carry out a comparative study with the other algorithms, namely disturb and observe (P&O).

References

Abdelmalek, S., Azar, A. T., & Dib, D. (2018a). A novel actuator fault-tolerant control strategy of dfig-based wind turbines using takagi-sugeno multiple models. *International Journal of Control, Automation and Systems, 16*(3), 1415−1424.

Abdelmalek, S., Azar, A. T., & Rezazi, S. (2018b). An improved robust fault-tolerant trajectory tracking controller (ftttc). In: *2018 international conference on control, automation and diagnosis (ICCAD)* (pp. 1−6). Marrakech, Morocco: IEEE.

Abdul-Adheem, W., Azar, A. T., Ibraheem, I., & Humaidi, A. (2020a). Novel active disturbance rejection control based on nested linear extended state observers. *Applied Sciences, 10*(12), 4069.

Abdul-Adheem, W., Ibraheem, I., Azar, A. T., & Humaidi, A. (2020b). Improved active disturbance rejection-based decentralized control for mimo nonlinear systems: Comparison with the decoupled control scheme. *Applied Sciences, 10*(7), 2515.

Abdul-Adheem, W. R., Ibraheem, I. K., Humaidi, A. J., & Azar, A. T. (2020c). Model-free active input−output feedback linearization of a single-link flexible joint manipulator: An improved active disturbance rejection control approach. *Measurement and Control, 2020*, 1−16.

Ahmed, J., & Salam, Z. (2016). A modified p&o maximum power point tracking method with reduced steady-state oscillation and improved tracking efficiency. *IEEE Transactions on Sustainable Energy, 7*(4), 1506−1515.

Ajeil, F. H., Ibraheem, I. K., Azar, A. T., & Humaidi, A. J. (2020a). Autonomous navigation and obstacle avoidance of an omnidirectional mobile robot using swarm optimization and sensors deployment. *International Journal of Advanced Robotic Systems, 17*(3), 1−15.

Ajeil, F. H., Ibraheem, I. K., Azar, A. T., & Humaidi, A. J. (2020b). Grid-based mobile robot path planning using aging-based ant colony optimization algorithm in static and dynamic environments. *Sensors, 20*(7), 1880.

Alain, K. S. T., Azar, A. T., Kengne, R., & Bertrand, F. H. (2020). Stability analysis and robust synchronisation of fractional-order modified colpitts oscillators. *International Journal of Automation and Control, 14*(1), 52−79.

Alain, K. S. T., Romanic, K., Azar, A. T., Vaidyanathan, S., Bertrand, F. H., & Adele, N. M. (2018). Dynamics analysis and synchronization in relay coupled fractional order colpitts oscillators. In A. T. Azar, & S. Vaidyanathan (Eds.), *Advances in system dynamics and control. Advances in systems analysis, software engineering, and high performance computing (ASASEHPC)* (pp. 317−356). IGI Global.

Alimi, M., Rhif, A., Rebai, A., Vaidyanathan, S., & Azar, A. T. (2021). Optimal adaptive backstepping control for chaos synchronization of nonlinear dynamical systems. In S. Vaidyanathan, & A. T. Azar (Eds.), *Backstepping control of nonlinear dynamical systems. Advances in nonlinear dynamics and chaos (ANDC)* (pp. 291−345). Academic Press.

Amara, K., Bakir, T., Malek, A., Hocine, D., Bourennane, E.-B., Fekik, A., & Zaouia, M. (2019a). An optimized steepest gradient based maximum power point tracking for PV control systems. *International Journal on Electrical Engineering & Informatics, 11*(4).

Amara, K., Fekik, A., Hocine, D., Bakir, M. L., Bourennane, E.-B., Malek, T. A., & Malek, A. (2018). Improved performance of a PV solar panel with adaptive neuro fuzzy inference system anfis based MPPT. In: *2018 7th international conference on renewable energy research and applications (ICRERA)* (pp. 1098−1101). IEEE.

Amara, K., Malek, A., Bakir, T., Fekik, A., Azar, A. T., Almustafa, K. M., . . . Hocine, D. (2019b). Adaptive neuro-fuzzy inference system based maximum power point tracking for stand-alone photovoltaic system. *International Journal of Modelling, Identification and Control, 33*(4), 311−321.

Ammar, H. H., & Azar, A. T. (2020). Robust path tracking of mobile robot using fractional order pid controller. In A. E. Hassanien, A. T. Azar, T. Gaber, R. Bhatnagar, & M. F. Tolba (Eds.), *The international conference on advanced machine learning technologies and applications (AMLTA2019). Vol. 921 of advances in intelligent systems and computing* (pp. 370−381). Cham: Springer International Publishing.

Ammar, H. H., Azar, A. T., Shalaby, R., & Mahmoud, M. I. (2019). Metaheuristic optimization of fractional order incremental conductance (FO-INC) maximum power point tracking (MPPT). *Complexity, 2019*(7687891), 1−13.

Ammar, H. H., Azar, A. T., Tembi, T. D., Tony, K., & Sosa, A. (2018). Design and implementation of fuzzy pid controller into multi agent smart library system prototype. In A. E. Hassanien, M. F. Tolba, M. Elhoseny, & M. Mostafa (Eds.), *The international conference on advanced machine learning technologies and applications (AMLTA2018). Vol. 723 of advances in intelligent systems and computing* (pp. 127−137). Cham: Springer International Publishing.

Arezki Fekik, H., Zaouia, M., Benyahia, N., Benamrouche, N., Badji, A., & Vaidyanathan, S. (2017). Improvement of the performances of the direct power control using space vector modulation of three phases pwm-rectifier. *International Journal of Control Theory and Applications, 10*(34), 125−137.

Azar, A. T., Ammar, H. H., Beb, M. Y., Garces, S. R., & Boubakari, A. (2020). Optimal design of pid controller for 2-dof drawing robot using bat-inspired algorithm. In A. E. Hassanien, K. Shaalan, & M. F. Tolba (Eds.), *Proceedings of the international conference on advanced intelligent systems and informatics 2019. Vol. 1058 of advances in intelligent systems and computing* (pp. 175−186). Cham: Springer International Publishing.

Azar, A. T., & Serrano, F. E. (2020). Stabilization of port hamiltonian chaotic systems with hidden attractors by adaptive terminal sliding mode control. *Entropy, 22*(1), 122.

Azar, A. T., Serrano, F. E., Koubaa, A., Kamal, N. A., Vaidyanathan, S., & Fekik, A. (2019). *Adaptive terminal-integral sliding mode force control of elastic joint robot manipulators in the presence of hysteresis. International conference on advanced intelligent systems and informatics* (pp. 266−276). Springer.

Azri, M., Khanipah, N. H. A., Ibrahim, Z., & Rahim, N. A. (2017). Fuel cell emulator with MPPT technique and boost converter. *International Journal of Power Electronics and Drive Systems, 8*(4), 1852.

Bansal, N., Bisht, A., Paluri, S., Kumar, V., Rana, K., Azar, A. T., & Vaidyanathan, S. (2021). Single-link flexible joint manipulator control using backstepping technique. In S. Vaidyanathan, & A. T. Azar (Eds.), *Backstepping control of nonlinear dynamical systems. Advances in nonlinear dynamics and chaos (ANDC)* (pp. 375−406). Academic Press.

Barakat, M. H., Azar, A. T., & Ammar, H. H. (2020). Agricultural service mobile robot modeling and control using artificial fuzzy logic and machine vision. In A. E. Hassanien, A. T. Azar, T. Gaber, R. Bhatnagar, & M. F. Tolba (Eds.), *The international conference on advanced machine learning technologies and applications (AMLTA2019). Vol. 921 of advances in intelligent systems and computing* (pp. 453−465). Cham: Springer International Publishing.

Ben Smida, M., Sakly, A., Vaidyanathan, S., & Azar, A. T. (2018). Control-based maximum power point tracking for a grid-connected hybrid renewable energy system optimized by particle swarm optimization. In A. T. Azar, & S. Vaidyanathan (Eds.), *Advances in system dynamics and control. Advances in systems analysis, software engineering, and high performance computing (ASASEHPC)* (pp. 58−89). IGI Global.

Bocci, E., Carlo, Di, McPhail, A., Gallucci, S., Foscolo, K., Moneti, P., . . . M., Carlini, M. (2014). Biomass to fuel cells state of the art: A review of the most innovative technology solutions. *International Journal of Hydrogen Energy, 39*(36), 21876−21895.

Bouakrif, F., Azar, A. T., Volos, C. K., Muñoz-Pacheco, J. M., & Pham, V.-T. (2019). Iterative learning and fractional order control for complex systems. *Complexity, 2019*, 7958625.

Chavan, S. L., & Talange, D. B. (2017). Modeling and performance evaluation of pem fuel cell by controlling its input parameters. *Energy, 138*, 437−445.

Cipriani, G., Di Dio, V., Genduso, F., La Cascia, D., Liga, R., Miceli, R., & Galluzzo, G. R. (2014). Perspective on hydrogen energy carrier and its automotive applications. *International Journal of Hydrogen Energy, 39*(16), 8482−8494.

Denoun, H., Hamida, M. L., Fekik, A., Dyhia, K., & Ghanes, M., 2018. Petri nets modeling for two-cell chopper control using dspace 1104. *2018 6th international conference on control engineering & information technology (CEIT)* (pp. 1–6). IEEE.

Derbeli, M., Farhat, M., Barambones, O., & Sbita, L. (2017). Control of pem fuel cell power system using sliding mode and super-twisting algorithms. *International Journal of Hydrogen Energy*, 42(13), 8833–8844.

Djeddi, A., Dib, D., Azar, A. T., & Abdelmalek, S. (2019). Fractional order unknown inputs fuzzy observer for takagi–sugeno systems with unmeasurable premise variables. *Mathematics*, 7(10), 984.

Djouima, M., Azar, A. T., Drid, S., & Mehdi, D. (2018). Higher order sliding mode control for blood glucose regulation of type 1 diabetic patients. *International Journal of System Dynamics Applications*, 7(1), 65–84.

Drhorhi, I., El Fadili, A., Berrahal, C., Lajouad, R., El Magri, A., Giri, F., ... Vaidyanathan, S. (2021). Adaptive backstepping controller for dfig-based wind energy conversion system. In S. Vaidyanathan, & A. T. Azar (Eds.), *Backstepping control of nonlinear dynamical systems. Advances in nonlinear dynamics and chaos (ANDC)* (pp. 235–260). Academic Press.

Drissi, H., Khediri, J., Zaafrane, W., & Braiek, E. B. (2017). Critical factors affecting the photovoltaic characteristic and comparative study between two maximum power point tracking algorithms. *International Journal of Hydrogen Energy*, 42(13), 8689–8702.

Elkholy, H. A., Azar, A. T., Magd, A., Marzouk, H., & Ammar, H. H. (2020a). Classifying upper limb activities using deep neural networks. In A.-E. Hassanien, A. T. Azar, T. Gaber, D. Oliva, & F. M. Tolba (Eds.), *Proceedings of the international conference on artificial intelligence and computer vision (AICV2020)* (pp. 268–282). Cham: Springer International Publishing.

Elkholy, H. A., Azar, A. T., Shahin, A. S., Elsharkawy, O. I., & Ammar, H. H. (2020b). Path planning of a self driving vehicle using artificial intelligence techniques and machine vision. In A.-E. Hassanien, A. T. Azar, T. Gaber, D. Oliva, & F. M. Tolba (Eds.), *Proceedings of the international conference on artificial intelligence and computer vision (AICV2020)* (pp. 532–542). Cham: Springer International Publishing.

Fekik, A. (2018). *Commande directe de puissance d'un redresseur à mli par dsp* (Ph.D. thesis), Université Mouloud Mammeri.

Fekik, A., Azar, A. T., Kamal, N. A., Denoun, H., Almustafa, K. M., Hamida, L., & Zaouia, M. (2021a). Fractional-order control of a fuel cell–boost converter system. In A. E. Hassanien, R. Bhatnagar, & A. Darwish (Eds.), *Advanced machine learning technologies and applications* (pp. 713–724). Singapore: Springer Singapore.

Fekik, A., Azar, A. T., Kamal, N. A., Serrano, F. E., Hamida, M. L., Denoun, H., & Yassa, N. (2021b). Maximum power extraction from a photovoltaic panel connected to a multi-cell converter. In A. E. Hassanien, A. Slowik, V. Snášel, H. El-Deeb, & F. M. Tolba (Eds.), *Proceedings of the international conference on advanced intelligent systems and informatics 2020* (pp. 873–882). Cham: Springer International Publishing.

Fekik, A., Denoun, H., Azar, A. T., Kamal, N. A., Zaouia, M., Benyahia, N., ... Vaidyanathan, S. (2021c). Direct power control of three-phase pwm-rectifier with backstepping control. In S. Vaidyanathan, & A. T. Azar (Eds.), *Backstepping control of nonlinear dynamical systems. Advances in nonlinear dynamics and chaos (ANDC)* (pp. 215–234). Academic Press.

Fekik, A., Denoun, H., Azar, A. T., Hamida, M. L., Atig, M., Ghanes, M., ... Barbot, J.-P. (2018a). *Direct power control of a three-phase pwm-rectifier based on petri nets for the selection of switching states. 2018 7th international conference on renewable energy research and applications (ICRERA)* (pp. 1121–1125). IEEE.

Fekik, A., Denoun, H., Azar, A. T., Zaouia, M., Benyahia, N., Hamida, M. L., ... Vaidyanathan, S. (2018b). *Artificial neural network for pwm rectifier direct power control and dc voltage control. Advances in system dynamics and control* (pp. 286–316). IGI Global.

Fekik, A., Denoun, H., Hamida, M. L., Azar, A. T., Atig, M., & Zhu, Q. M. (2018c). *Neural network based switching state selection for direct power control of three phase PWM-rectifier. 2018 10th international conference on modelling, identification and control (ICMIC)* (pp. 1–6). IEEE.

Fekik, A., Denoun, H., Azar, A. T., Kamal, N. A., Zaouia, M., Yassa, N., & Hamida, M. L. (2019). *Direct torque control of three phase asynchronous motor with sensorless speed estimator. International conference on advanced intelligent systems and informatics* (pp. 243–253). Springer.

Fekik, A., Denoun, H., Benamrouche, N., Benyahia, N., & Zaouia, M. (2015). A fuzzy–logic based controller for three phase pwm rectifier with voltage oriented control strategy. *International Journal of Circuits, Systems and Signal Processing*, 9, 7.

Femia, N., Petrone, G., Spagnuolo, G., & Vitelli, M. (2005). Optimization of perturb and observe maximum power point tracking method. *IEEE Transactions on Power Electronics*, 20(4), 963–973.

Ghoudelbourk, S., Azar, A. T., Dib, D., & Omeiri, A. (2020). Selective harmonic elimination strategy in the multilevel inverters for grid connected photovoltaic system. *International Journal of Advanced Intelligence Paradigms*, 15(3), 317–339.

Ghoudelbourk, S., Dib, D., Omeiri, A., & Azar, A. T. (2016). MPPT control in wind energy conversion systems and the application of fractional control (piα) in pitch wind turbine. *International Journal of Modelling, Identification and Control*, 26(2), 140–151.

Giove, S., Azar, A. T., & Nordio, M. (2013). Fuzzy logic control for dialysis application. In A. T. Azar (Ed.), *Modeling and control of dialysis systems: Volume 2: Biofeedback systems and soft computing techniques of dialysis* (pp. 1181–1222). Berlin, Heidelberg: Springer.

Gorripotu, T. S., Samalla, H., Jagan Mohana Rao, C., Azar, A. T., & Pelusi, D. (2019). Tlbo algorithm optimized fractional-order pid controller for agc of interconnected power system. In J. Nayak, A. Abraham, B. M. Krishna, G. T. Chandra Sekhar, & A. K. Das (Eds.), *Soft computing in data analytics* (pp. 847–855). Singapore: Springer.

Gow, J., & Manning, C. (1999). Development of a photovoltaic array model for use in power-electronics simulation studies. *IEE Proceedings-Electric Power Applications*, 146(2), 193–200.

Hamida, M. L., Denoun, H., Fekik, A., & Vaidyanathan, S. (2019). Control of separately excited dc motor with series multi-cells chopper using PI-petri nets controller. *Nonlinear Engineering*, 8(1), 32–38.

Hong, P., Xu, L., Li, J., & Ouyang, M. (2017). Modeling and analysis of internal water transfer behavior of pem fuel cell of large surface area. *International Journal of Hydrogen Energy*, 42(29), 18540–18550.

Humaidi, A., Ibraheem, I., Azar, A., & Sadiq, M. (2020). A new adaptive synergetic control design for single link robot arm actuated by pneumatic muscles. *Entropy, 22*(7), 723.

Ibraheem, G. A. R., Azar, A. T., Ibraheem, I. K., & Humaidi, A. J. (2020). A novel design of a neural network-based fractional pid controller for mobile robots using hybridized fruit fly and particle swarm optimization. *Complexity, 2020*(2020), 1−18.

Ibrahim, H. A., Azar, A. T., Ibrahim, Z. F., & Ammar, H. H. (2020). A hybrid deep learning based autonomous vehicle navigation and obstacles avoidance. In A.-E. Hassanien, A. T. Azar, T. Gaber, D. Oliva, & F. M. Tolba (Eds.), *Proceedings of the international conference on artificial intelligence and computer vision (AICV2020)* (pp. 296−307). Cham: Springer International Publishing.

Kamal, N. A., Azar, A. T., Elbasuony, G. S., Almustafa, K. M., & Almakhles, D. (2020). Pso-based adaptive perturb and observe MPPT technique for photovoltaic systems. In A. E. Hassanien, K. Shaalan, & M. F. Tolba (Eds.), *Proceedings of the international conference on advanced intelligent systems and informatics 2019. Vol. 1058 of advances in intelligent systems and computing* (pp. 125−135). Cham: Springer International Publishing.

Kammogne, A. S. T., Kountchou, M. N., Kengne, R., Azar, A. T., Fotsin, H. B., & Ouagni, S. T. M. (2020). Polynomial robust observer implementation based passive synchronization of nonlinear fractional-order systems with structural disturbances. *Frontiers of Information Technology & Electronic Engineering, 21*(9), 1369−1386.

Khan, A., Singh, S., & Azar, A. T. (2020a). Combination-combination anti-synchronization of four fractional order identical hyperchaotic systems. In A. E. Hassanien, A. T. Azar, T. Gaber, R. Bhatnagar, & M. F. Tolba (Eds.), *The international conference on advanced machine learning technologies and applications (AMLTA2019). Vol. 921 of advances in intelligent systems and computing* (pp. 406−414). Cham: Springer International Publishing.

Khan, A., Singh, S., & Azar, A. T. (2020b). Synchronization between a novel integer-order hyperchaotic system and a fractional-order hyperchaotic system using tracking control. In A. E. Hassanien, A. T. Azar, T. Gaber, R. Bhatnagar, & M. F. Tolba (Eds.), *The international conference on advanced machine learning technologies and applications (AMLTA2019). Vol. 921 of advances in intelligent systems and computing* (pp. 382−391). Cham: Springer International Publishing.

Khan, M. J., & Mathew, L. (2019). Fuzzy logic controller-based MPPT for hybrid photo-voltaic/wind/fuel cell power system. *Neural Computing and Applications, 31*(10), 6331−6344.

Khanipah, N. H. A., Azri, M., & Ibrahim, Z. (2018). Simple MPPT technique for dc-dc converter in fuel cell system. *International Journal of Electrical Engineering and Applied Sciences, 1*(1), 59−70.

Khennaoui, A. A., Ouannas, A., Boulaaras, S., Pham, V.-T., & Azar, A. (2020). A fractional map with hidden attractors: Chaos and control. *The European Physical Journal Special Topics, 229*, 1083−1093.

Khettab, K., Bensafia, Y., Bourouba, B., & Azar, A. T. (2018). Enhanced fractional order indirect fuzzy adaptive synchronization of uncertain fractional chaotic systems based on the variable structure control: Robust h∞ design approach. In A. T. Azar, A. G. Radwan, & S. Vaidyanathan (Eds.), *Mathematical techniques of fractional order systems. Advances in nonlinear dynamics and chaos (ANDC)* (pp. 597−624). Elsevier.

Kumar, J., Azar, A. T., Kumar, V., & Rana, K. P. S. (2018). Design of fractional order fuzzy sliding mode controller for nonlinear complex systems. In A. T. Azar, A. G. Radwan, & S. Vaidyanathan (Eds.), *Mathematical techniques of fractional order systems. Advances in nonlinear dynamics and chaos (ANDC)* (pp. 249−282). Elsevier.

Lamine, H. M., Hakim, D., Arezki, F., Dyhia, K., Nacereddine, B., & Youssef, B. (2019). Control of three-cell inverter with a fuzzy logic-feedback linearization strategy to reduce the harmonic content of the output current. In: *2019 international conference on computer science and renewable energies (ICCSRE)* (pp. 1−5). IEEE.

Lamine, H. M., Hakim, D., Arezki, F., Nabil, B., & Nacerddine, B. (2018). Cyclic reports modulation control strategy for a five cells inverter. *2018 international conference on electrical sciences and technologies in maghreb (CISTEM)* (pp. 1−5). IEEE.

Liu, L., Ma, D., Azar, A., & Zhu, Q. (2020). Neural computing enhanced parameter estimation for multi-input and multi-output total non-linear dynamic models. *Entropy, 22*(5), 510.

Makhloufi, A. B., Hatti, M., & Taleb, R. (2017). Comparative study of photovoltaic system for hydrogen electrolyzer system. In: *2017 6th international conference on systems and control (ICSC)* (pp. 445−450). IEEE.

Masoum, M. A., Dehbonei, H., & Fuchs, E. F. (2002). Theoretical and experimental analyses of photovoltaic systems with voltageand current-based maximum power-point tracking. *IEEE Transactions on Energy Conversion, 17*(4), 514−522.

Meghni, B., Dib, D., & Azar, A. T. (2017a). A second-order sliding mode and fuzzy logic control to optimal energy management in wind turbine with battery storage. *Neural Computing and Applications, 28*(6), 1417−1434.

Meghni, B., Dib, D., Azar, A. T., Ghoudelbourk, S., & Saadoun, A. (2017b). Robust adaptive supervisory fractional order controller for optimal energy management in wind turbine with battery storage. In A. T. Azar, S. Vaidyanathan, & A. Ouannas (Eds.), *Fractional order control and synchronization of chaotic systems. Vol. 688 of studies in computational intelligence* (pp. 165−202). Cham: Springer International Publishing.

Meghni, B., Dib, D., Azar, A. T., & Saadoun, A. (2018). Effective supervisory controller to extend optimal energy management in hybrid wind turbine under energy and reliability constraints. *International Journal of Dynamics and Control, 6*(1), 369−383.

Mekki, H., Boukhetala, D., & Azar, A. T. (2015). Sliding modes for fault tolerant control. In A. T. Azar, & Q. Zhu (Eds.), *Advances and applications in sliding mode control systems* (pp. 407−433). Cham: Springer International Publishing.

Mittal, S., Azar, A. T., & Kamal, N. A. (2021). Nonlinear fractional order system synchronization via combination-combination multi-switching. In A. E. Hassanien, A. Slowik, V. Snášel, H. El-Deeb, & F. M. Tolba (Eds.), *Proceedings of the international conference on advanced intelligent systems and informatics 2020* (pp. 851−861). Cham: Springer International Publishing.

Mohamed, N. A., Azar, A. T., Abbas, N. E., Ezzeldin, M. A., & Ammar, H. H. (2020). Experimental kinematic modeling of 6-dof serial manipulator using hybrid deep learning. In A.-E. Hassanien, A. T. Azar, T. Gaber, D. Oliva, & F. M. Tolba (Eds.), *Proceedings of the international conference on artificial intelligence and computer vision (AICV2020)* (pp. 283–295). Cham: Springer International Publishing.

Najm, A. A., Ibraheem, I. K., Azar, A. T., & Humaidi, A. J. (2020). Genetic optimization-based consensus control of multi-agent 6-DOF UAV system. *Sensors, 20*(12), 3576.

Nasiri Avanaki, I., & Sarvi, M. (2016). A new maximum power point tracking method for pem fuel cells based on water cycle algorithm. *Journal of Renewable Energy and Environment, 3*(1), 35–42.

Ouannas, A., Azar, A. T., & Vaidyanathan, S. (2017a). A new fractional hybrid chaos synchronisation. *International Journal of Modelling, Identification and Control, 27*(4), 314–322.

Ouannas, A., Azar, A. T., & Vaidyanathan, S. (2017b). A robust method for new fractional hybrid chaos synchronization. *Mathematical Methods in the Applied Sciences, 40*(5), 1804–1812.

Ouannas, A., Azar, A. T., Ziar, T., & Radwan, A. G. (2017c). Generalized synchronization of different dimensional integer-order and fractional order chaotic systems. In A. T. Azar, S. Vaidyanathan, & A. Ouannas (Eds.), *Fractional order control and synchronization of chaotic systems. Vol. 688 of studies in computational intelligence* (pp. 671–697). Cham: Springer International Publishing.

Ouannas, A., Azar, A. T., Ziar, T., & Radwan, A. G. (2017d). A study on coexistence of different types of synchronization between different dimensional fractional chaotic systems. In A. T. Azar, S. Vaidyanathan, & A. Ouannas (Eds.), *Fractional order control and synchronization of chaotic systems. Vol. 688 of studies in computational intelligence* (pp. 637–669). Cham: Springer International Publishing.

Ouannas, A., Azar, A. T., Ziar, T., & Vaidyanathan, S. (2017e). Fractional inverse generalized chaos synchronization between different dimensional systems. In A. T. Azar, S. Vaidyanathan, & A. Ouannas (Eds.), *Fractional order control and synchronization of chaotic systems. Vol. 688 of studies in computational intelligence* (pp. 525–551). Cham: Springer International Publishing.

Ouannas, A., Azar, A. T., Ziar, T., & Vaidyanathan, S. (2017f). A new method to synchronize fractional chaotic systems with different dimensions. In A. T. Azar, S. Vaidyanathan, & A. Ouannas (Eds.), *Fractional order control and synchronization of chaotic systems. Vol. 688 of studies in computational intelligence* (pp. 581–611). Cham: Springer International Publishing.

Ouannas, A., Azar, A. T., Ziar, T., & Vaidyanathan, S. (2017g). On new fractional inverse matrix projective synchronization schemes. In A. T. Azar, S. Vaidyanathan, & A. Ouannas (Eds.), *Fractional order control and synchronization of chaotic systems. Vol. 688 of studies in computational intelligence* (pp. 497–524). Cham: Springer International Publishing.

Ouannas, A., Azar, A. T., & Ziar, T. (2020a). Fractional inverse full state hybrid projective synchronisation. *International Journal of Advanced Intelligence Paradigms, 17*(3-4), 279–298.

Ouannas, A., Grassi, G., & Azar, A. T. (2020b). Fractional-order control scheme for q-s chaos synchronization. In A. E. Hassanien, A. T. Azar, T. Gaber, R. Bhatnagar, & M. F. Tolba (Eds.), *The international conference on advanced machine learning technologies and applications (AMLTA2019)* (pp. 434–441). Cham: Springer International Publishing.

Ouannas, A., Grassi, G., & Azar, A. T. (2020c). A new generalized synchronization scheme to control fractional chaotic systems with non-identical dimensions and different orders. In A. E. Hassanien, A. T. Azar, T. Gaber, R. Bhatnagar, & M. F. Tolba (Eds.), *The international conference on advanced machine learning technologies and applications (AMLTA2019). Vol. 921 of advances in intelligent systems and computing* (pp. 415–424). Cham: Springer International Publishing.

Ouannas, A., Grassi, G., Azar, A. T., & Khennaoui, A. A. (2021). Synchronization control in fractional discrete-time systems with chaotic hidden attractors. In A. E. Hassanien, R. Bhatnagar, & A. Darwish (Eds.), *Advanced machine learning technologies and applications* (pp. 661–669). Singapore, Singapore: Springer.

Ouannas, A., Grassi, G., Azar, A. T., Khennaouia, A. A., & Pham, V.-T. (2020d). Chaotic control in fractional-order discrete-time systems. In A. E. Hassanien, K. Shaalan, & M. F. Tolba (Eds.), *Proceedings of the international conference on advanced intelligent systems and informatics 2019. Vol. 1058 of advances in intelligent systems and computing* (pp. 207–217). Cham: Springer International Publishing.

Ouannas, A., Grassi, G., Azar, A. T., Khennaouia, A.-A., & Pham, V.-T. (2020e). Synchronization of fractional-order discrete-time chaotic systems. In A. E. Hassanien, K. Shaalan, & M. F. Tolba (Eds.), *Proceedings of the international conference on advanced intelligent systems and informatics 2019* (pp. 218–228). Cham: Springer International Publishing.

Ouannas, A., Grassi, G., Azar, A. T., & Singh, S. (2019). New control schemes for fractional chaos synchronization. In A. E. Hassanien, M. F. Tolba, K. Shaalan, & A. T. Azar (Eds.), *Proceedings of the international conference on advanced intelligent systems and informatics 2018. Vol. 845 of advances in intelligent systems and computing* (pp. 52–63). Cham: Springer International Publishing.

Ounnas, A., Azar, A. T., & Radwan, A. G. (2016). On inverse problem of generalized synchronization between different dimensional integer-order and fractional-order chaotic systems. In: *2016 28th international conference on microelectronics (ICM)* (pp. 193–196).

Pham, V., P.M., G., Kapitaniak, T., Volos, C., & Azar, A. T. (2018). Dynamics, synchronization and fractional order form of a chaotic system with infinite equilibria. In A. T. Azar, A. G. Radwan, & S. Vaidyanathan (Eds.), *Mathematical techniques of fractional order systems. Advances in nonlinear dynamics and chaos (ANDC)* (pp. 475–502). Elsevier.

Pham, V.-T., Vaidyanathan, S., Azar, A. T., & Duy, V. H. (2021). A new chaotic system without linear term, its backstepping control, and circuit design. In S. Vaidyanathan, & A. T. Azar (Eds.), *Backstepping control of nonlinear dynamical systems. Advances in nonlinear dynamics and chaos (ANDC)* (pp. 33–52). Academic Press.

Pham, V.-T., Vaidyanathan, S., Volos, C. K., Azar, A. T., Hoang, T. M., & Van Yem, V. (2017). A three-dimensional no-equilibrium chaotic system: Analysis, synchronization and its fractional order form. In A. T. Azar, S. Vaidyanathan, & A. Ouannas (Eds.), *Fractional order control and synchronization of chaotic systems. Vol. 688 of studies in computational intelligence* (pp. 449–470). Cham: Springer International Publishing.

Pilla, R., Azar, A. T., & Gorripotu, T. S. (2019). Impact of flexible AC transmission system devices on automatic generation control with a metaheuristic based fuzzy PID controller. *Energies, 12*(21), 4193.

Pilla, R., Botcha, N., Gorripotu, T. S., & Azar, A. T. (2020). Fuzzy pid controller for automatic generation control of interconnected power system tuned by glow-worm swarm optimization. In J. Nayak, V. E. Balas, M. N. Favorskaya, B. B. Choudhury, S. K. M. Rao, & B. Naik (Eds.), *Applications of robotics in industry using advanced mechanisms* (pp. 140−149). Cham: Springer International Publishing.

Roza, F.S.D. (2018). A small wind generation system to the on-board control unit of an awe system.

Sallam, O. K., Azar, A. T., Guaily, A., & Ammar, H. H. (2020). Tuning of pid controller using particle swarm optimization for cross flow heat exchanger based on cfd system identification. In A. E. Hassanien, K. Shaalan, & M. F. Tolba (Eds.), *Proceedings of the international conference on advanced intelligent systems and informatics 2019. Vol. 1058 of advances in intelligent systems and computing* (pp. 300−312). Cham: Springer International Publishing.

Sambas, A., Vaidyanathan, S., Sukono., Azar, A. T., Hidayat, Y., Gundara, G., & Mohamed, M. A. (2021a). A novel chaotic system with a closed curve of four quarter-circles of equilibrium points: Dynamics, active backstepping control, and electronic circuit implementation. In S. Vaidyanathan, & A. T. Azar (Eds.), *Backstepping control of nonlinear dynamical systems. Advances in nonlinear dynamics and chaos (ANDC)* (pp. 485−507). Academic Press.

Sambas, A., Vaidyanathan, S., Zhang, S., Mohamed, M. A., Zeng, Y., & Azar, A. T. (2021b). A new 3-d chaotic jerk system with a saddle-focus rest point at the origin, its active backstepping control, and circuit realization. In S. Vaidyanathan, & A. T. Azar (Eds.), *Backstepping control of nonlinear dynamical systems. Advances in nonlinear dynamics and chaos (ANDC)* (pp. 95−114). Academic Press.

Sambas, A., Vaidyanathan, S., Zhang, S., Mohamed, M. A., Zeng, Y., & Azar, A. T. (2021c). A new 4-d chaotic hyperjerk system with coexisting attractors, its active backstepping control, and circuit realization. In S. Vaidyanathan, & A. T. Azar (Eds.), *Backstepping control of nonlinear dynamical systems. Advances in nonlinear dynamics and chaos (ANDC)* (pp. 73−94). Academic Press.

Sayed, A. S., Azar, A. T., Ibrahim, Z. F., Ibrahim, H. A., Mohamed, N. A., & Ammar, H. H. (2020). Deep learning based kinematic modeling of 3-rrr parallel manipulator. In A.-E. Hassanien, A. T. Azar, T. Gaber, D. Oliva, & F. M. Tolba (Eds.), *Proceedings of the international conference on artificial intelligence and computer vision (AICV2020)* (pp. 308−321). Cham: Springer International Publishing.

Sera, D., Teodorescu, R., & Rodriguez, P. (2007). PV panel model based on datasheet values. In: *2007 IEEE international symposium on industrial electronics* (pp. 2392−2396). IEEE.

Singh, S., & Azar, A. T. (2020a). Controlling chaotic system via optimal control. In A. E. Hassanien, K. Shaalan, & M. F. Tolba (Eds.), *Proceedings of the international conference on advanced intelligent systems and informatics 2019. Vol. 1058 of advances in intelligent systems and computing* (pp. 277−287). Cham: Springer International Publishing.

Singh, S., & Azar, A. T. (2020b). Multi-switching combination synchronization of fractional order chaotic systems. In A.-E. Hassanien, A. T. Azar, T. Gaber, D. Oliva, & F. M. Tolba (Eds.), *Proceedings of the international conference on artificial intelligence and computer vision (AICV2020)* (pp. 655−664). Cham: Springer International Publishing.

Singh, S., Azar, A. T., Ouannas, A., Zhu, Q., Zhang, W., & Na, J. (2017). Sliding mode control technique for multi-switching synchronization of chaotic systems. In: *9th international conference on modelling, identification and control (ICMIC 2017)* (pp. 1−6), July 10−12, 2017, Kunming, China: IEEE.

Singh, S., Azar, A. T., Vaidyanathan, S., Ouannas, A., & Bhat, M. A. (2018a). Multiswitching synchronization of commensurate fractional order hyperchaotic systems via active control. In A. T. Azar, A. G. Radwan, & S. Vaidyanathan (Eds.), *Mathematical techniques of fractional order systems. Advances in nonlinear dynamics and chaos (ANDC)* (pp. 319−345). Elsevier.

Singh, S., Azar, A. T., & Zhu, Q. (2018b). Multi-switching master−slave synchronization of non-identical chaotic systems. In Q. Zhu, J. Na, & X. Wu (Eds.), *Innovative techniques and applications of modelling, identification and control: Selected and expanded reports from ICMIC'17. Vol. 467 of lecture notes in electrical engineering* (pp. 321−330). Singapore: Springer.

Singh, S., Mathpal, S., Azar, A. T., Vaidyanathan, S., & Kamal, N. A. (2021). Multi-switching synchronization of nonlinear hyperchaotic systems via backstepping control. In S. Vaidyanathan, & A. T. Azar (Eds.), *Backstepping control of nonlinear dynamical systems. Advances in nonlinear dynamics and chaos (ANDC)* (pp. 425−447). Academic Press.

Soliman, M., Azar, A. T., Saleh, M. A., & Ammar, H. H. (2020). Path planning control for 3-omni fighting robot using pid and fuzzy logic controller. In A. E. Hassanien, A. T. Azar, T. Gaber, R. Bhatnagar, & M. F. Tolba (Eds.), *The international conference on advanced machine learning technologies and applications (AMLTA2019)* (pp. 442−452). Cham: Springer International Publishing.

Sreejith, R., & Singh, B. (2019). Intelligent nonlinear sensorless predictive field oriented control of pmsm drive for three wheeler hybrid solar pv-battery electric vehicle. In: *2019 IEEE transportation electrification conference and expo (ITEC)* (pp. 1−6). IEEE.

Tolba, M. F., AbdelAty, A. M., Soliman, N. S., Said, L. A., Madian, A. H., Azar, A. T., & Radwan, A. G. (2017). Fpga implementation of two fractional order chaotic systems. *AEU—International Journal of Electronics and Communications, 78*, 162−172.

Vaidyanathan, S., & Azar, A. T. (2015a). Anti-synchronization of identical chaotic systems using sliding mode control and an application to vaidyanathan-madhavan chaotic systems. In A. T. Azar, & Q. Zhu (Eds.), *Advances and applications in sliding mode control systems. Vol. 576 of studies in computational intelligence* (pp. 527−547). Berlin, Germany: Springer.

Vaidyanathan, S., & Azar, A. T. (2015b). Hybrid synchronization of identical chaotic systems using sliding mode control and an application to vaidyanathan chaotic systems. In A. T. Azar, & Q. Zhu (Eds.), *Advances and applications in sliding mode control systems. Vol. 576 of studies in computational intelligence* (pp. 549−569). Berlin, Germany: Springer.

Vaidyanathan, S., & Azar, A. T. (2016a). *A novel 4-D four-wing chaotic system with four quadratic nonlinearities and its synchronization via adaptive control method. Advances in chaos theory and intelligent control* (pp. 203−224). Berlin, Germany: Springer.

Vaidyanathan, S., & Azar, A. T. (2016b). *Adaptive control and synchronization of Halvorsen circulant chaotic systems. Advances in chaos theory and intelligent control* (pp. 225−247). Berlin, Germany: Springer.

Vaidyanathan, S., & Azar, A. T. (2016c). *Generalized projective synchronization of a novel hyperchaotic four-wing system via adaptive control method. Advances in chaos theory and intelligent control* (pp. 275−290). Berlin, Germany: Springer.

Vaidyanathan, S., & Azar, A. T. (2016d). Qualitative study and adaptive control of a novel 4-d hyperchaotic system withÂ three quadratic nonlinearities. In A. T. Azar, & S. Vaidyanathan (Eds.), *Advances in chaos theory and intelligent control* (pp. 179−202). Cham: Springer International Publishing.

Vaidyanathan, S., & Azar, A. T. (2016e). Takagi-Sugeno fuzzy logic controller for Liu-Chen four-scroll chaotic system. *International Journal of Intelligent Engineering Informatics, 4*(2), 135−150.

Vaidyanathan, S., & Azar, A. T. (2021). An introduction to backstepping control. In S. Vaidyanathan, & A. T. Azar (Eds.), *Backstepping control of nonlinear dynamical systems. Advances in nonlinear dynamics and chaos (ANDC)* (pp. 1−32). Academic Press.

Vaidyanathan, S., Azar, A. T., Akgul, A., Lien, C.-H., Kacar, S., & Cavusoglu, U. (2019). A memristor-based system with hidden hyperchaotic attractors, its circuit design, synchronisation via integral sliding mode control and an application to voice encryption. *International Journal of Automation and Control, 13*(6), 644−667.

Vaidyanathan, S., Azar, A. T., & Boulkroune, A. (2018a). A novel 4-d hyperchaotic system with two quadratic nonlinearities and its adaptive synchronisation. *International Journal of Automation and Control, 12*(1), 5−26.

Vaidyanathan, S., Azar, A. T., Sambas, A., Singh, S., Alain, K. S. T., & Serrano, F. E. (2018b). A novel hyperchaotic system with adaptive control, synchronization, and circuit simulation. In A. T. Azar, & S. Vaidyanathan (Eds.), *Advances in system dynamics and control. Advances in systems analysis, software engineering, and high performance computing (ASASEHPC)* (pp. 382−419). IGI Global.

Vaidyanathan, S., Azar, A. T., & Ouannas, A. (2017a). An eight-term 3-D novel chaotic system with three quadratic nonlinearities, its adaptive feedback control and synchronization. In A. T. Azar, S. Vaidyanathan, & A. Ouannas (Eds.), *Fractional order control and synchronization of chaotic systems. Vol. 688 of studies in computational intelligence* (pp. 719−746). Cham: Springer International Publishing.

Vaidyanathan, S., Azar, A. T., & Ouannas, A. (2017b). Hyperchaos and adaptive control of a novel hyperchaotic system with two quadratic nonlinearities. In A. T. Azar, S. Vaidyanathan, & A. Ouannas (Eds.), *Fractional order control and synchronization of chaotic systems. Vol. 688 of studies in computational intelligence* (pp. 773−803). Cham: Springer International Publishing.

Vaidyanathan, S., Zhu, Q., & Azar, A. T. (2017c). Adaptive control of a novel nonlinear double convection chaotic system. In A. T. Azar, S. Vaidyanathan, & A. Ouannas (Eds.), *Fractional order control and synchronization of chaotic systems. Vol. 688 of studies in computational intelligence* (pp. 357−385). Cham: Springer International Publishing.

Vaidyanathan, S., Pham, V.-T., & Azar, A. T. (2021a). A new chaotic jerk system with egg-shaped strange attractor, its dynamical analysis, backstepping control, and circuit simulation. In S. Vaidyanathan, & A. T. Azar (Eds.), *Backstepping control of nonlinear dynamical systems. Advances in nonlinear dynamics and chaos (ANDC)* (pp. 53−71). Academic Press.

Vaidyanathan, S., Sambas, A., & Azar, A. T. (2021b). A 5-d hyperchaotic dynamo system with multistability, its dynamical analysis, active backstepping control, and circuit simulation. In S. Vaidyanathan, & A. T. Azar (Eds.), *Backstepping control of nonlinear dynamical systems. Advances in nonlinear dynamics and chaos (ANDC)* (pp. 449−471). Academic Press.

Vaidyanathan, S., Sambas, A., Azar, A. T., Rana, K., & Kumar, V. (2021c). A new 4-d hyperchaotic temperature variations system with multistability and strange attractor, bifurcation analysis, its active backstepping control, and circuit realization. In S. Vaidyanathan, & A. T. Azar (Eds.), *Backstepping control of nonlinear dynamical systems. Advances in nonlinear dynamics and chaos (ANDC)* (pp. 139−164). Academic Press.

Vaidyanathan, S., Sambas, A., Azar, A. T., Rana, K., & Kumar, V. (2021d). A new 5-d hyperchaotic four-wing system with multistability and hidden attractor, its backstepping control, and circuit simulation. In S. Vaidyanathan, & A. T. Azar (Eds.), *Backstepping control of nonlinear dynamical systems. Advances in nonlinear dynamics and chaos (ANDC)* (pp. 115−138). Academic Press.

Vaidyanathan, S., Sambas, A., Azar, A. T., Serrano, F. E., & Fekik, A. (2021e). A new thermally excited chaotic jerk system, its dynamical analysis, adaptive backstepping control, and circuit simulation. In S. Vaidyanathan, & A. T. Azar (Eds.), *Backstepping control of nonlinear dynamical systems. Advances in nonlinear dynamics and chaos (ANDC)* (pp. 165−189). Academic Press.

Vaidyanathan, S., Sambas, A., Azar, A. T., & Singh, S. (2021f). A new multistable plasma torch chaotic jerk system, its dynamical analysis, active backstepping control, and circuit design. In S. Vaidyanathan, & A. T. Azar (Eds.), *Backstepping control of nonlinear dynamical systems. Advances in nonlinear dynamics and chaos (ANDC)* (pp. 191−214). Academic Press.

Vaidyanathan, S., Sampath, S., & Azar, A. T. (2015). Global chaos synchronisation of identical chaotic systems via novel sliding mode control method and its application to zhu system. *International Journal of Modelling, Identification and Control, 23*(1), 92−100.

Wang, M. H., Huang, M.-L., Jiang, W.-J., & Liou, K.-J. (2016). Maximum power point tracking control method for proton exchange membrane fuel cell. *IET Renewable Power Generation, 10*(7), 908−915.

Xiao, X., Huang, X., & Kang, Q. (2015). A hill-climbing-method-based maximum-power-point-tracking strategy for direct-drive wave energy converters. *IEEE Transactions on Industrial Electronics, 63*(1), 257−267.

Zhang, Y., Fu, C., Sumner, M., & Wang, P. (2017). A wide input-voltage range quasi-z-source boost dc−dc converter with high-voltage gain for fuel cell vehicles. *IEEE Transactions on Industrial Electronics, 65*(6), 5201−5212.

Chapter 9

Control strategies of wind energy conversion system-based doubly fed induction generator

Boaz Wadawa, Youssef Errami, Abdellatif Obbadi and Smail Sahnoun

Laboratory of Electronics, Instrumentation and Energy, Team of Exploitation and Processing of Renewable Energy, Department of Physics, Faculty of Science, University of Chouaib Doukkali, El Jadida, Morocco

9.1 Introduction

Today, the wind energy sector has grown rapidly and is the fastest growing renewable energy source (Berahab, 2019; Debouza, Al-Durra, Errouissi, & Muyeen, 2018). However, the integration of wind turbines into the electricity grid is not a sinecure. It highlights the major problems, namely the difficulty of forecasting production, the risk of untimely disconnection, and the degradation of the quality of electricity produced by wind turbines; on the other hand, the capacity of the electricity transmission system is often very limited (Robyns et al., 2006). Moreover, all these problems are intrinsically linked to the stochastic nature of disturbances from internal and/or external system elements. As a result, the latter are highly vulnerable to the risks of a possible failure and/or malfunction (Li, Wang, Tian, Aitouch, & Klein, 2016; Robyns et al., 2006). To effectively improve the quality of the energy produced and the energy efficiency of grid-connected wind energy systems, many researchers refer most to the field of automatic control of electrical machines, in particular the doubly fed induction generator (DFIG). The DFIG, unlike other so-called synchronous and asynchronous machines with their different variants, offers the following main advantages:

- Thanks to its dual power supply, several reconfiguration possibilities can be used and thus a wide range of DFIG applications (Shukla, Tripathi, & Thakur, 2017);
- The accessibility to its rotor and stator via static converters makes the measurement of currents possible, which allows great flexibility and precision in the control of electromagnetic flux and torque (Bounar, Labdai, & Boulkroune, 2019; Morshed & Fekih, 2017);
- The DFIG can develop a slightly higher power-to-weight ratio than other machines (Li et al., 2016); and
- The converter connected to the rotor can be dimensioned at least one-third of the rated power of the machine (Bedoud, Ali-Rachedi, Bahi, Lakel, & Grid, 2015).

In other words, this flexibility of the DFIG gives it four possible operating modes, stationary, synchronous, and hypo- and hypersynchronous. This allows the increase of energy efficiency and the reduction of the cost of energy conversion in a wind energy system (Bedoud et al., 2015; Morshed & Fekih, 2017). However, the DFIG has the characteristics of a highly complex machine. Its dynamic electrical model is nonlinear, multivariate, and strongly coupled. Similarly, its parameters, such as inductances, resistances, and friction coefficient, are not easily quantifiable with precision (Bounar et al., 2019). Moreover, as we can see in the generalized study scheme of Fig. 9.1, the DFIG is located as the main element of the wind turbine chain; the wind and grid disturbances directly or indirectly influence its rotor and/ or stator. Therefore the performance of a DFIG-based wind turbine connected to a grid depends entirely on the control strategy applied to the system (Bounar et al., 2019).

Depending on whether the system is operating in a steady-state and/or transient mode, the main objectives assigned to the control are: maximizing the power of the available wind energy and regulating the reactive power at a desirable power factor according to the grid requirements. To effectively achieve these objectives for operational reliability, the

Renewable Energy Systems. DOI: https://doi.org/10.1016/B978-0-12-820004-9.00020-6

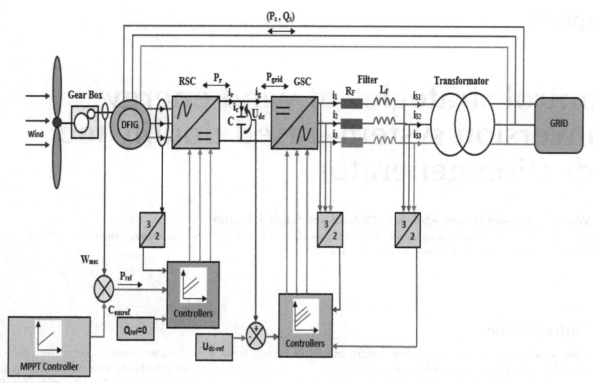

FIGURE 9.1 Structure for controlling a wind chain on a doubly fed induction generator-based electrical power grid.

controller is generally evaluated by four performance criteria: robustness, speed, stability, and simplicity or flexibility of implementation (Nazari, Seron, & De Doná, 2017; Van, 2018).

Currently, several control techniques are being developed to provide better solutions to problems related to wind energy conversion systems, especially those connected to power grids and using DFIG. Among recent work, the different strategies encountered can be grouped into three main families of controls as follows:

- Analog controls are usually described from representations by mathematical symbols with differential equations, for example. Examples include integral and derivative proportional or PI controllers, sliding mode controller (SMC), backstepping, infinite H (H∞), or H2/H∞. These can be found in works (Aroussi, Ziani, & Bossoufi, 2020; Bossoufi, Karim, Lagrioui, Taoussi, & Derouich, 2015; Islam, Seyedmahmoudian, & Stojcevski, 2019; Khan, Ansari, Chachar, Katyara, & Soomro, 2018; Lhachimi, Sayouti, & El Kouari, 2018; Xiong et al., 2019);
- Numerical commands, based on numerical computation or artificial intelligence methods. We have the controllers particle swarm optimization (PSO), artificial neuron network, Fuzzy logic, etc., as in studies (Ardjoun, Denai, & Abid, 2019; Bharti, Saket, & Nagar, 2017; Rocha-Osorio, Solís-Chaves, Rodrigues, & al, 2018; Tang, Ju, He, Qin, & Wu, 2013); and
- Mixed commands, on the other hand, are a combination of several types of commands. One can note the: SMC-PI, Fuzzy-PI, PSO-GA and SMC-Fuzzy, H∞-PI, Fuzzy-H∞, etc. (Bounar et al., 2019; Lazrak et al., 2017; Li et al., 2016; Yu & Li, 2012; Saihi, Berbaoui, & Glaoui, 2020).

Indeed, we have seen that all the commands developed in the above-mentioned works were based on the strategy of vector control (VC), to obtain a simplified dynamic model of the DFIG and then to be able to implement their different commands. In addition, to validate the performances of the new controls on systems, this work considers as a reference, the performances of the wind system based on this principle of the VC incorporating PI controllers. Consequently, they lead to conclusions that can contribute to the development of controllers with better performances. In particular, in terms of robustness against parameter uncertainties, good stability in the fast and precise monitoring of setpoint quantities. However, apart from VC using PI, these other high-performance controllers are relatively more complex in terms of the mathematical model used, the calculation time, and the practical implementation, which may give rise to additional costs. Thus due to the current trend to increasingly use the VC approach as an effective tool for wind turbine analysis and supervision, we believe it is appropriate to give more emphasis to this technique in this chapter.

VC, introduced by BLASCKE in the early 1970s, was to develop rapidly both in modeling and control of complex systems, such as electrical machines. The aim of the VC is to control the asynchronous machine as a direct current machine with independent excitation, where there is a natural decoupling between the variable controlling the excitation flux or current and that related to the torque or induced current (Allam, Dehiba, Abid, & al, 2014). Then, the application of the VC principle to a DFIG is essentially defined in the two-axis, direct (*d*) and quadratic reference frame of the Park transformation, and allows the following major advantages (Kai, Wenfeng, & Jinping, 2019; Li et al., 2016; Mehdipour, Hajizadeh, & Mehdipour, 2016):

- Facilitate the design or modeling, and simulation of the wind system. Based on the so-called simplified DFIG model obtained from the assumption of the flow orientation, either on the (*d*) axis or on the (*q*) axis;
- Offering a good condition for integration or connection to the grid. The purpose here is to ensure the control of the whole system. In other words, to allow the quantities to be controlled by the DFIG to follow their corresponding reference values.

However, the application of the VC strategy to a complex system, such as a wind turbine, can have significant limitations. These include strong linearization, sensitivity to disturbances, and reduced system efficiency. To this end, to assess the impact of using the VC approach in the study of grid-connected wind systems based on a DFIG, the majority of studies use integral proportional (PI) controllers for system control. Two main groups of studies can be distinguished among the few more recent ones.

In the first group, the study of the entire chain is focused solely on the simplified model of the DFIG. The different operating regimes of the system are materialized by the speeds that the reference quantities, such as the mechanical speed or electromagnetic torque and the active and reactive powers with their respective currents, can take (Allam et al., 2014; Benbouhenni, 2018; Ghoudelbourk, Bahi, & Mohammedi, 2012; Hamdi & Bouzid, 2013; Li et al., 2016; Mazouz, Belkacem, Harbouch, Abdessemed, & Ouchen, 2017; Senani, Rahab, & Benalla, 2018; Yasmine, Chakib, & Badre, 2017). While, for the second group, the studies are initially very few, given the great difficulty to simulate the whole wind turbine chain, when we want to consider constraints related to a real system and its environment. In this case, the work is limited to partial studies of the system (Aydin, Polat, & Ergene, 2016; Bourdoulis & Alexandridis, 2014; Kai et al., 2019; Li, Haskew, Williams, & Swatloski, 2012; Mehdipour et al., 2016; Mohammadi, Vaez-Zadeh, Afsharnia, & Daryabeigi, 2014).

In summary, we can say that in most of these works that the use of the VC approach, we can see the concealment of uncertainties and their influences. These are all the more remarkable when the application of this technique often lacks rigor. As a result, the strategy can be either poorly or weakly exploited.

To ensure the analysis of the proper functioning of wind energy systems using the VC approach, a more rigorous study is required. In this chapter, we propose a more detailed and efficient approach to guide future work by a good quantification of the system parameters and the quantities to be controlled. In addition, a methodology for modeling the whole system is considered, taking into account the constraints of an ideal and real environment. This is a valuable contribution by allowing, on the one hand, to restore an image a little closer to what can be observed experimentally and, on the other hand, to predict or anticipate on the behaviors of more varied phenomena of the system and also to better design, evaluate, and optimize the performance of controllers. In addition, we present and develop DFIG's direct and indirect VC (IVC) strategy, using PI controllers for energy conversion in a 2-MW wind system, connected to the power grid. Specifically, the main objective of the wind chain control strategy is to ensure the, maximum power point tracking (MPPT), decoupling and control of active and reactive powers, regulation of the DC bus voltage, control of filter currents, rejection of disturbances and/or overshooting of controlled quantities, and a unit power factor; under the influence of wind and grid disturbances. To do this, two main case studies are presented. The first case aims at a comparative evaluation between the performances of direct VC, and that of IVC in open loop (IVCOL) and IVC in closed loop (IVCCL), applied to an ideal wind energy system, and designed around the simplified model of the DFIG. This part also highlights the effects of model and parameter uncertainties, on the one hand, and the sudden and permanent variations of the network, such as wind, by reference quantity models, on the other hand.

The second part consists in reproducing the study of the first case in another, slightly more real operating environment of the system. Random wind models, the real DFIG model, and coupling effects between different blocks of the conversion system are taken into account. Validation studies of the modeling and simulation results are performed with MATLAB/Simulink. The document is structured as follows:

- The first part concerns the methodology. It details the mathematical simulation models of the different parts of the wind system, such as the turbine, the DFIG, the generator side (RSC) and grid side (GSC) pulse width modulation

(PWM) static converters, the DC bus, and the current filter. In addition, the synthesis model of VC with PI is determined for each block of the wind energy conversion system;

- The second part is the results and discussion phase of the simulation related to the evaluation of the PI performance. The second part is the results and discussion phase of the simulation related to the evaluation of the performance of the PI. Using on the one hand, the techniques of direct VC with PI (DVC-PI), the IVCOL-PI, and the IVCCL-PI, on the simplified model of the DFIG in the case of the ideal system. Then, on the other hand, on the real DFIG model for the whole so-called real system;
- The third part is devoted to conclusions and perspectives.

9.2 Modeling with syntheses of PI controllers of wind system elements

9.2.1 Mathematical model and identification of wind turbine parameters

In general, a wind turbine is characterized by its mechanical efficiency, also known as its power coefficient, which is rated C_p (λ, β). The latter depends on the wind conditions called relative or specific wind speed (λ); it is also a function of the geometric or aerodynamic shape of the rotor or turbine, in terms of blade length or radius (R), the orientation angle of the blades, also called the pitch angle (β), and the total number of blades on the rotor. There are several mathematical approximation models of C_p (λ, β) (Rechka et al., 2004). The one that seems the most appropriate for our case study is given by the following expression:

$$C_p(\lambda, \beta) = C_1\left(\frac{C_2}{\lambda_i} - C_3\beta - C_4\right)\exp\left(-\frac{C_5}{\lambda_i}\right) + C_6\lambda \tag{9.1}$$

where $C_1 = 0.73$, $C_2 = 151$, $C_3 = 0.002$, $C_4 = 13.2$, $C_5 = 18.4$, and $C_6 = 0$.

With,

$$\frac{1}{\lambda_i} = \frac{1}{\lambda + 0.08\beta} \cdot \frac{0.035}{\beta^3 + 1} \tag{9.2}$$

And,

$$\lambda = \frac{\Omega_{turbine}R}{V} \tag{9.3}$$

$\Omega_{turbine}$ is the turbine rotation speed and V is the speed of the air movement when crossing the blades.

The mechanical or aerodynamic power (P_{aero}) extracted from the wind and available on the wind turbine shaft is given by the relation:

$$P_{aero} = \frac{1}{2}C_p(\lambda, \beta)\rho SV^3 \tag{9.4}$$

where ρ is the density of the air $(\rho = 1.225$ kg m^{-3} at atmospheric pressure) and S is a surface swept by the propeller.

The model of the mechanical transmission can be summarized as follows:

$$J\frac{d\Omega_{mec}}{dt} = \sum C = C_g - C_{em} - C_{vis} \tag{9.5}$$

With C_{vis} is the viscous torque being proportional to the speed; C_{em} is the electromechanical torque; C_g is the total mechanical torque applied to the machine rotor; and J is the total inertia of the turbine distributed over the generator rotor (Belmokhtar & Doumbia, 2010).

The gearbox on the wind turbine in Fig. 9.1 houses the speed multiplier defined as the multiplier coefficient or gain (G). The respective expressions of torque and speed transmitted to the generator rotor as a function of the gain (G) are:

$$C_g = \frac{C_{aero}}{G} \tag{9.6}$$

$$\Omega_{mec} = \frac{\Omega_{turbine}}{G} \tag{9.7}$$

With C_{aero} is the mechanical or aerodynamic torque and Ω_{mec} is the rotational speed of the generator shaft.

In addition, the total moment of inertia (J), as a function of the gain (G), the turbine inertia (J_{turbine}), and the machine inertia (J_{machine}) (Belmokhtar & Doumbia, 2010) can be determined by the following relationship:

$$J = \frac{J_{turbine}}{G^2} + J_{machine} \tag{9.8}$$

9.2.2 Synthesis of wind turbine MPPT regulation

Based on the expressions of the turbine parameter equations identified in Section 9.2.1, the block diagram of the MPPT control model with wind turbine PI, illustrated in Fig. 9.2, is obtained.

The purpose of the MPPT control model in Fig. 9.2, is to allow the maximum amount of energy to be extracted from the wind and then transmitted to the machine in the form of energy available on the wind turbine shaft. This ratio between the maximum power extracted and the total power theoretically available reflects what is also known as the coefficient of maximum performance of the wind turbine denoted C_{pmax}. According to Betz's theoretical limit, the C_{pmax} must never reach or exceed the value of 0.593 (59.3%) (Abolvafaei & Ganjefar, 2020). However, to determine the value of C_{pmax}, we rely on considerations, such as the optimal orientation angle is set at $\beta_{opt} = 0$ degrees; the optimal specific wind speed (λ_{opt}); and the estimated wind speed (V_{est}), are given by the following relation:

$$V_{est} = \frac{R\Omega_{turbine}}{\lambda_{opt}} \tag{9.9}$$

9.2.2.1 Overview of the PI controller in the MPPT model

In the case of the functional model in Fig. 9.2, the PI speed control Ω_{mec} with wind speed control is used. The principle of synthesis of the law of rotation speed control is given in Fig. 9.3.

Applying the synthesis principle of Fig. 9.3, we can establish a relation between the transfer function $F(p)$ of the PI and the transfer function of the model $G(p)$, to determine the values of the coefficients K_p and K_i of the PI corrector. For this purpose, the closed-loop transfer function in Fig. 9.3 gives us the following expression:

$$\text{TFCL}_{\Omega_{mec}} = \frac{\left(\frac{K_p}{J} \times s + \frac{K_I}{J}\right)}{\left(s^2 + \left[(f + K_p)/J\right] \times s + K_I/J\right)} \tag{9.10}$$

By identifying the $\text{TFCL}_{\Omega_{mec}}$ to a transfer function of a second-order filter, the following expressions can be deduced:

$$\begin{cases} K_i = \omega_0^2 J \\ K_p = 2\xi\omega_0 J - f \end{cases} \tag{9.11}$$

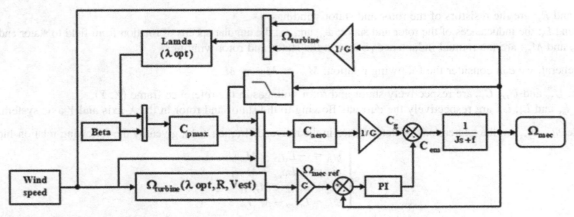

FIGURE 9.2 Maximum power point tracking control model with PI of the wind turbine.

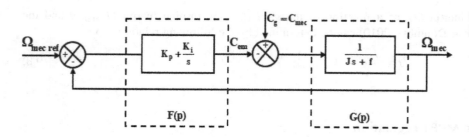

FIGURE 9.3 Synoptic of the law of PI regulation of the speed of rotation.

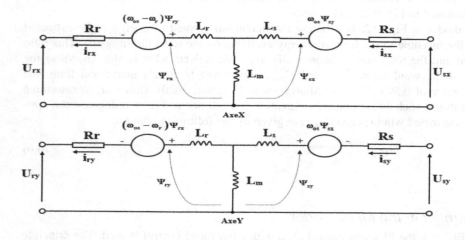

FIGURE 9.4 Wiring diagram of the equivalent model of the doubly fed induction generator in the Park (X, Y) datum.

The response time is given as follows:

$$\tau(n) = \frac{1}{\xi \omega}\left(\frac{100}{n}\right) \tag{9.12}$$

For our case study we took $n = 2$, that is, 2%, and $\xi = 1$, to correspond to a sufficiently fast response time value for the DFIG.

n is the actual value of depreciation rate; ξ is the amortization rate (without unit); and ω_0 is the filter cut-off pulse. The parameters K_p, K_i, τ_r, and ε of the PI controller are calculated and presented in Table A2.

9.2.3 Mathematical model and identification of DFIG parameters

Fig. 9.4 illustrates the electrical diagram of the equivalent DFIG model in the generalized Park (X, Y) reference frame (Bharti et al., 2017; Rocha-Osorio et al., 2018). The different constituent elements of Fig. 9.4 are defined as follows:

— R_r and R_s: are the resistors of the rotor and stator windings;
— L_r and L_s: the inductances of the rotor and stator; ω_{os} and ω_r: the angular speeds of rotation from field to stator and rotor;
— M_{sr} and M_{rs}: are the mutual inductances between the stator and rotor windings.

In general, we can consider the following relation: $M_{sr} = M_{rs} = M$.

— U_{sx}, U_{sy} and U_{rx}, U_{ry} are respectively stator and rotor voltages in the reference frame (X, Y);
— i_{sx}, i_{sy} and i_{rx}, i_{ry}: are respectively the currents flowing to the stator and rotor in the X-axis and Y-axis system.

while, Ψ_{sx}, Ψ_{sy} are the expressions of stator fluxes in the reference frame (X, Y), given by the following relationship:

$$\begin{cases} \psi_{sx} = -L_s i_{sx} + M_{sr} i_{rx} \\ \psi_{sy} = -L_s i_{sy} + M_{sr} i_{ry} \\ \psi_{rx} = L_r i_{rx} - M_{rs} i_{sx} \\ \psi_{ry} = L_r i_{ry} - M_{rs} i_{sy} \end{cases} \tag{9.13}$$

When the mesh law is applied to the diagram in Fig. 9.4, and using Eq. (9.13), we arrive at the expression of the general or real dynamic model of DFIG according to Park's frame of reference (X, Y), in the form:

$$
\begin{cases}
U_{sx} = -R_s i_{sx} - L_s \dfrac{di_{sx}}{dt} + M\dfrac{di_{rx}}{dt} + \omega_{os}L_s i_{sy} - \omega_{os}M i_{ry} \\[2mm]
U_{sy} = -R_s i_{sy} - L_s \dfrac{di_{sy}}{dt} + M\dfrac{di_{ry}}{dt} - \omega_{os}L_s i_{sx} - \omega_{os}M i_{rx} \\[2mm]
U_{rx} = R_r i_{rx} + L_r \dfrac{di_{rx}}{dt} - M\dfrac{di_{sx}}{dt} - (\omega_{os} - \omega_r)L_r i_{ry} + (\omega_{os} - \omega_r)M i_{sy} \\[2mm]
U_{ry} = R_r i_{ry} + L_r \dfrac{di_{ry}}{dt} - M\dfrac{di_{sy}}{dt} + (\omega_{os} - \omega_r)L_r i_{rx} - (\omega_{os} - \omega_r)M i_{sx}
\end{cases}
\tag{9.14}
$$

The electromagnetic torque of the machine is obtained by:

$$
C_{em} = \frac{3}{2}p\left(\psi_{sx} i_{sy} - \psi_{sy} i_{sx}\right)
\tag{9.15}
$$

where p is the number of pole pairs of the DFIG.

The active and reactive stator and rotor powers are written as follows:

$$
\begin{cases}
P_s = U_{sx} i_{sx} + U_{sy} i_{sy} \\
Q_s = U_{sy} i_{sx} - U_{sx} i_{sy} \\
P_r = U_{rx} i_{rx} + U_{ry} i_{ry} \\
Q_r = U_{ry} i_{rx} - U_{rx} i_{ry}
\end{cases}
\tag{9.16}
$$

9.2.4 Synthesis of direct and indirect vector commands with DFIG PI

In general, to facilitate on a DFIG, the application of a VC law with PI and/or other controls, the following assumptions of the VC principle are solicited:

For convenience, let us identify the $X-Y$ marker with the notation $d-q$ and then the orientation of the flow following q can be written:

$$
\begin{cases}
\psi_{qs} = \psi_s = V_s \omega_s = cst \\
V_{ds} = 0
\end{cases}
\tag{9.17}
$$

Assume the stable and powerful grid, the simple voltage V_s = cte and the frequency f_s = cte.

When the above assumptions are used on Eq. (9.14) of the real model, we find the following simplified mathematical model of the DFIG:

$$
\begin{cases}
V_{rd} = \left[R_r + \left(L_r - \dfrac{M^2}{L_s}\right)s\right]i_{rd} - \omega_m\left(L_r - \dfrac{M^2}{L_s}\right)i_{rq} \\[3mm]
V_{rq} = \left[R_r + \left(L_r - \dfrac{M^2}{L_s}\right)s\right]i_{rq} + \omega_m\left(L_r - \dfrac{M^2}{L_s}\right)i_{rd} + g\dfrac{MV_s}{L_s}
\end{cases}
\tag{9.18}
$$

In the same way, with the previous hypotheses, we can easily deduce from Eq. (9.16), the expressions of active and reactive stator powers as follows:

$$
\begin{cases}
P_s = \dfrac{MV_s}{L_s}i_{rq} \\[3mm]
Q_s = \left(\dfrac{MV_s}{L_s}i_{rd} - \dfrac{V_s^2}{L_s}\right)
\end{cases}
\tag{9.19}
$$

In addition, the expression of the electromagnetic torque of the machine can be written in the form:

$$C_{em} = P\left(\frac{MV_s}{L_s}\right) i_{rq} \tag{9.20}$$

9.2.4.1 Direct PI vector control synthesis with power loops

The principle is based on the simplified DFIG model of Eq. (9.18) with the following new assumptions:

- The coupling terms $(g\omega_s\sigma i_{rq})$ and $(g\omega_s\sigma i_{rd})$ (practical case, g low slip) should be neglected;
- Steady state operation (i_{rd} and i_{rq} derivatives are null); and
- Knowing that $\Psi_{qs} = \Psi_s = V_s\omega_s = $ cst and then $V_{ds} = 0$.

Thus the equations of the direct power VC model are obtained, in the forms:

$$\begin{cases} V_{rd}^* = R_r i_{rd}^* \\ V_{rq}^* = R_r i_{rq}^* + \dfrac{gMV_s}{L_s} \end{cases} \tag{9.21}$$

$$\begin{cases} P_s^* = \dfrac{MV_s}{L_s} i_{rq}^* \\ Q_s^* = \dfrac{V_s^2}{\omega_s} - \dfrac{MV_s}{L_s} i_{rd}^* \end{cases} \tag{9.22}$$

The effects of uncertainties due to the terms (V_s^2/ω_s) and (gMV_s/L_s), and to the transfer function of the DFIG model, will have to be compensated by the PI regulation. Accordingly, a representation of the functional structure of the DFIG powers VC is shown in Fig. 9.5. The dimensioning of the PI for the power control loops in Fig. 9.5 obeys the control law illustrated by the diagram in Fig. 9.6.

Thus in Fig. 9.6, the open-loop transfer function (TFOL$_{P,Q}$) and the closed-loop transfer function (TFCL$_{P,Q}$) of active and reactive powers are determined by the following expressions:

$$\text{TFOL}_{P,Q} = \frac{\left(s + \frac{K_I}{K_p}\right)}{\left(\frac{s}{K_p}\right)} \cdot \frac{\frac{MV_s}{L_s\left(L_r - \frac{M^2}{L_s}\right)}}{s + \frac{L_s.R_r}{L_s\left(L_r - \frac{M^2}{L_s}\right)}} \tag{9.23}$$

where

$$\sigma = L_r - M^2/L_s \tag{9.24}$$

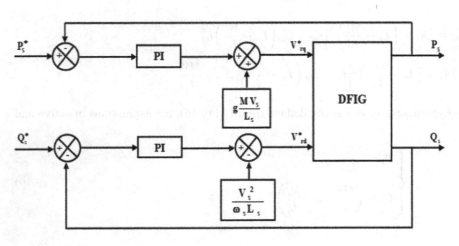

FIGURE 9.5 Doubly fed induction generator direct vector control model with power control loop.

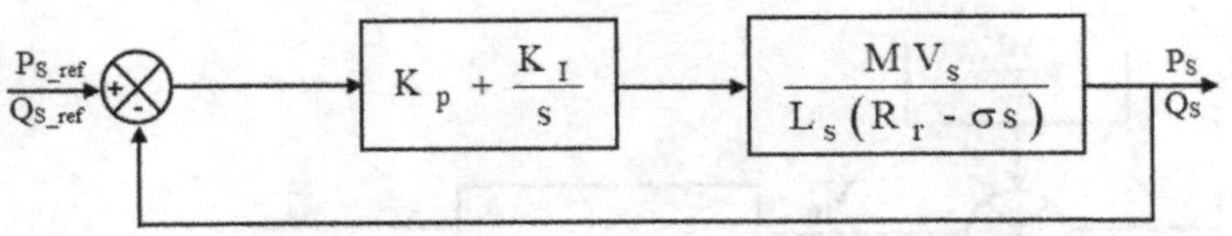

FIGURE 9.6 Active and reactive power loop control law $(P_s$ and $Q_s)$.

$$\text{TFCL}_{P,Q} = \frac{K_p M V_s}{K_p M V_s + L_s \sigma s} \tag{9.25}$$

By rearranging the equations of $\text{TFCL}_{P,Q}$ and $\text{TFOL}_{P,Q}$, the following expressions of K_i and K_p are derived:

$$\begin{cases} K_i = \dfrac{R_r}{\sigma} K_p \\[2mm] K_p = \dfrac{L_s \sigma}{\tau_r M V_s} \end{cases} \tag{9.26}$$

where τ_r is the response time of the system and can be conveniently selected.

The parameters K_p, K_i and τ_r of the PI controller are calculated and presented in Table A2.

9.2.4.2 Synthesis of indirect PI vector control with and without power loops

Also, using the same simplified DFIG model of Eq. (9.18), the following assumptions are considered:

- The terms $(g\omega_s\sigma i_{rq})$ and $(g\omega_s\sigma i_{rd})$ are no longer neglected (electromotive force as a function of ω_r) and
- Also (V^2_s/ω_s) and (gMV_s/L_s) will be compensated; in the order $\Psi_{qs} = \Psi_s = V_s\omega_s =$ cte and then $V_{ds} = 0$.

We thus obtain the equations of the IVC models with loops and without power loops, in the following forms:

$$\begin{cases} E_{rq} = g\omega_s\sigma i_{rq} \\[2mm] E_{rd} = g\omega_s\sigma i_{rd} + g\dfrac{MV_s}{L_s} \end{cases} \tag{9.27}$$

And,

$$\begin{cases} U_{rd1} = V_{rd} + E_{rq} \\ U_{rq1} = V_{rq} - E_{rd} \end{cases} \tag{9.28}$$

The control equation for the currents i_{rd} and i_{rq} is determined in the form:

$$\begin{cases} U_{rd1} = (R_r + \sigma p) i_{rd} \\ U_{rq1} = (R_r + \sigma p) i_{rq} \end{cases} \tag{9.29}$$

Eq. (9.29) can be written with the transfer function of the PI offset as follows:

$$\begin{cases} U_{rd1}\left(K_p + \dfrac{K_i}{s}.\right) = i_{rd} \\[3mm] U_{rq1}\left(K_p + \dfrac{K_i}{s}.\right) = i_{rq} \end{cases} \tag{9.30}$$

By rearranging Eq. (9.30) in Eqs. (9.28) and (9.27), with Eq. (9.19), the block diagram representations of the DFIG IVC model, without loops and with power loops, respectively, are obtained in Figs. 9.7 and 9.8. However, to improve the influences due to static disturbances in the schematic of Fig. 9.7, two additional PI correctors are added in the representation of Fig. 9.8.

The dimensioning of the additional PIs at the level of the active and reactive power loops of the DFIG outputs obeys the law of trial-and-error adjustment. Here, it consists of observing the system response as a function of the gain

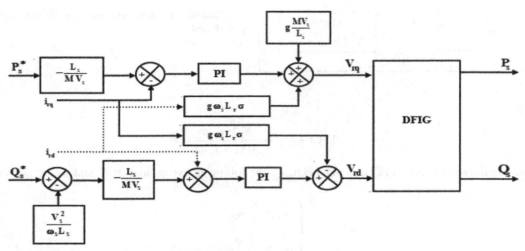

FIGURE 9.7 Indirect vector control block without doubly fed induction generator control loop.

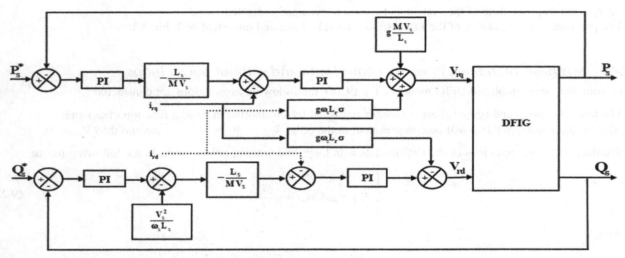

FIGURE 9.8 Indirect vector control block with doubly fed induction generator control loop.

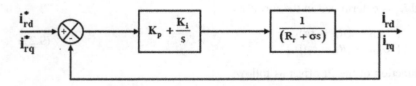

FIGURE 9.9 Active and reactive current control loop (i_{rd} and i_{rq}).

adjustment, with the time constant of the transfer function of the PI corrector, whereas the control PIs of the ird and irq currents in Figs. 9.7 and 9.8 are identical and obey the same control law illustrated in Fig. 9.9.

Thus from the diagram in Fig. 9.9, the open-loop and closed-loop transfer function of the currents is determined in the following ways:

$$\text{TFOL}_{i_{rd}, i_{rq}} = \frac{\frac{K_p}{s}\left(s + \frac{K_i}{K_p}\right)}{\sigma\left(\frac{R_r}{\sigma} + s\right)} \tag{9.31}$$

$$\text{TFCL}_{i_{rd}, i_{rq}} = \frac{1}{1 + \frac{\sigma}{K_p}p} \tag{9.32}$$

From the equations of $\text{TFCL}_{i_{rd},i_{rq}}$ and $\text{TFOL}_{i_{rd},i_{rq}}$, we deduce the expressions of K_i and K_p as follows:

$$\begin{cases} K_i = \dfrac{R_r}{\tau_r} \\[2mm] K_p = \dfrac{\sigma K_i}{R_r} \end{cases} \tag{9.33}$$

The parameters K_p, K_i, and τ_r of the PI controller are calculated and presented in Table A2.

9.2.5 Modeling and synthesis of the adjacent PWM control of the inverter

The inverter, thanks to the forced operation of its electronic switches (insulated gate bipolar transistor (IGBT), gate turn off (GTO) or metal oxide semiconductor field effect transistor (MOSFET) with adjacent controls (PWM, phase locked loop (PLL), or vector modulation), allows the transformation of the energy from the AC source to a DC source and vice versa, on one side on the rotor (RSC) and on the other side on the grid side (GSC) of the DFIG (Belmokhtar & Doumbia, 2010). The equivalent generalized electrical schematic representation of the inverter is given in Fig. 9.10 (Wadawa, Errami, Obbadi, Sahnoun, & Aoutoul, 2019). This backside corresponds to the following description: V_{ab}, V_{bc}, and V_{ac} are the expressions of the compound voltages between phases of the inverter; V_{an}, V_{bn}, and V_{an} are the phase-balanced single voltages of the converter; and E is the DC reference voltage at the input of the converter PWM.

Each string of the inverter in Fig. 9.10, is modeled by two switch states (K_i or $K_{a,b,c}$) and defined by the following logical connection function (f_i):

$$f_i = \begin{cases} 1, & K_i lead \ and \ \overline{K}_i blocked \\ 0, & \overline{K}_i blocked \ and \ K_i lead \end{cases} \tag{9.34}$$

Expressions of compound tensions are:

$$\begin{cases} V_{ab} = V_{an1} - V_{bn1} = E(f_1 - f_2) \\ V_{bc} = V_{bn1} - V_{cn1} = E(f_2 - f_3) \\ V_{ca} = V_{cn1} - V_{an1} = E(f_3 - f_1) \end{cases} \tag{9.35}$$

where f_1, f_2, and f_3 are the logic functions defining the inverter switch states.

The expression of the simple tensions of the balanced system gives:

$$V_{an1} + V_{bn1} + V_{cn1} = 0 \tag{9.36}$$

The resolution of the two previous Eqs. (9.35) and (9.36) results in the inverter model as the next matrix:

$$\begin{bmatrix} V_{an1} \\ V_{bn1} \\ V_{cn1} \end{bmatrix} = \frac{E}{3} \begin{bmatrix} 2 & -1 & -1 \\ -1 & 2 & -1 \\ -1 & -1 & 2 \end{bmatrix} \begin{bmatrix} f_1 \\ f_2 \\ f_3 \end{bmatrix} \tag{9.37}$$

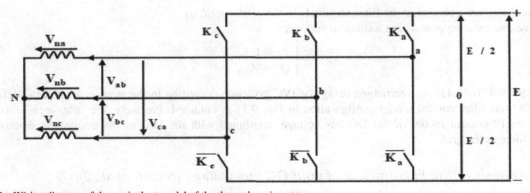

FIGURE 9.10 Wiring diagram of the equivalent model of the three-phase inverter.

9.2.5.1 Synthesis of control by sine-delta modulation

The synthesis of the adjacent command by selected PWM is defined by the following expression:

$$\begin{cases} V_{pm} = \dfrac{V_r}{r} \\[2mm] m = \dfrac{f_p}{f} \end{cases} \tag{9.38}$$

where r is the voltage setting coefficient; V_{pm} is the peak value of the ripple; m is the modulation index; f_p is the carrier frequency; and f is the frequency of the modulating signal.

According to some works in the literature, it is preferable to take $r = 0.8$ and f_p from the kilohertz order to acceptably reduce the harmonic rate (Wadawa et al., 2019). For our case study we obtain $f_p = 1350H$, by the law of trial and error adjustment.

9.2.6 Modeling and synthesis of the DC bus PI and the network filter

The control of the DC bus voltage around the capacitance (C) in Fig. 9.1 is based on the expression of the following energy balance (Wadawa et al., 2019):

$$P_R = P_c + P_g \tag{9.39}$$

where P_g is the power transmitted or received to the grid by the inverter; P_R is the active power of the bus on the rotor side of the DFIG; and P_c is the power stored by the capacitor C.

$$P_c = V_{dc}i_c \tag{9.40}$$

V_{DC} and i_c are the voltage across the capacitor and the current in the capacitor, respectively.

While, using the law of meshing, the matrix of the voltages at the terminals of the filter of RF resistors and LF inductances of the main Fig. 9.1, can be written:

$$\begin{pmatrix} V_{F1} \\ V_{F2} \\ V_{F3} \end{pmatrix} = R_F \begin{pmatrix} i_1 \\ i_2 \\ i_3 \end{pmatrix} + L_F \frac{d}{dt} \begin{pmatrix} i_1 \\ i_2 \\ i_3 \end{pmatrix} + \begin{pmatrix} V_{ps1} \\ V_{ps2} \\ V_{ps3} \end{pmatrix} \tag{9.41}$$

V_{F123} and V_{ps123} are the three-phase input and output voltages at the filter terminals and i_{123} is the three-phase currents through the filter.

Applying Park's transform to Eq. (9.13) and then rearranging the latter yield the equation for the control law of the active and reactive currents i_{fd} and i_{fq} of the filter.

$$\begin{cases} i_{fd} = \dfrac{1}{(R_F + L_F.s)}\left(V_{Fd} + L_F\omega_s i_{fq} - V_{sd}\right) \\[3mm] i_{fq} = \dfrac{1}{(R_F + L_F.s)}\left(V_{Fq} - L_F\omega_s i_{fd} - V_{sq}\right) \end{cases} \tag{9.42}$$

where V_{fdq} and V_{sdq} are voltages at the filter terminals in the marking (d, q).

The active and reactive powers are written as follows:

$$\begin{cases} P_f = V_s i_{fq} \\ Q_f = V_s i_{fd} \end{cases} \tag{9.43}$$

When Eqs. (9.39)–(9.43) are rearranged using the IVC principle according to the steps presented in Section 9.2.4.2 above, the DC bus filter control model configuration in Fig. 9.11 is obtained. Precisely, the latter represents the block diagram of the PI control model of the DC bus voltage, combined with the IVC of the active and reactive currents through the filter to the grid.

9.2.6.1 Synthesis of the PI controller of your DC bus voltage (Nazari et al., 2017)

The DC bus voltage regulation law is shown in Fig. 9.12.

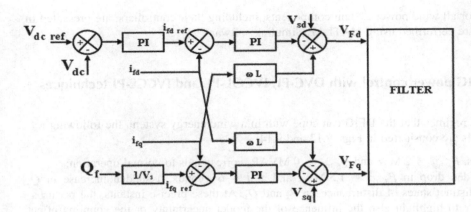

FIGURE 9.11 DC bus voltage control model with the indirect vector control of active and reactive filter currents.

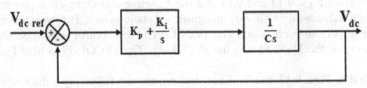

FIGURE 9.12 DC bus voltage regulation loop (V_{DC}).

The open- and closed-loop transfer functions in the diagram in Fig. 9.12 are defined by the expressions:

$$\text{TFOL}_{V_{dc}} = \frac{K_p + \frac{K_i}{s}}{Cs} \tag{9.44}$$

$$\text{TFCL}_{V_{dc}} = \frac{\frac{K_p s + K_i}{C}}{s^2 + \frac{K_p s}{C} + \frac{K_i}{C}} \tag{9.45}$$

By identifying $\text{TFCL}_{V_{dc}}$ to a second-order filter, we deduce:
The parameters K_p, K_i, τ_r, and ε of the PI controller are calculated and presented in Table A2.

9.2.6.2 Overview of the PI filter current controllers ifd and ifq

The synthesis of regulation by IVC of the filter currents follows the same steps as those set out in Section 9.2.4.2 of the DFIG above. Except that here, the coupling term between the i_{fd} and i_{fq} filter currents is of the form ($L\omega$).

$$\begin{cases} K_i = C\omega_0^2 \\ K_p = 2\xi\omega_0 C \end{cases} \tag{9.46}$$

The parameters K_p, K_i, and τ_r of the PI controller are calculated and presented in Table A2.

9.3 Results and discussions

In this part, the aim is to evaluate the efficiency of techniques, such as DVC-PI, IVCOL-PI, and IVCCL-PI, in the energy conversion of a 2-MW wind system connected to the electrical grid. To validate our study, two main simulation steps are considered. The first step is based on a so-called ideal system designed around the simplified DFIG model of Eq. (9.18); these are the cases of Figs. 9.5, 9.7, and 9.8. This is to establish good comparisons between the different types of VCs studied and to clearly deduce the superiority of a technique compared to other controllers. The second part consists of a system that is somewhat closer to reality than the one studied in the previous step. Thus under the influence of a random wind model, we consider the model in Fig. 9.2 for the turbine, the actual DFIG model of Eq. (9.14) associated with the GSC and GSC models of Eq. (9.37), and the block in Fig. 9.11 for the DC Bus and filter case. The whole is connected to the grid according to the provisions of the elements on the wind system diagram in

Fig. 9.1. The simulation parameters for all wind power system components, including their controllers, are presented in Tables A1 and A2. All these studies are performed using MATLAB/Simulink software.

9.3.1 Step 1: simulation of DFIG power control with DVC-PI, IVCOL-PI, and IVCCL-PI techniques in an ideal system

To materialize the different operating regimes that the DFIG can cope with in a wind energy system, the following reference active and reactive power levels are considered in Figs. 9.13 and 9.14:

- Between the times 0 and 3 seconds, $P_{s\ ref}$ = 2 MW and $Q_{s\ ref}$ = 0 MVAR, corresponds to normal operation;
- At t = 3 seconds, there is a sudden drop in $P_{s\ ref}$ = 1.5 MW, and at t = 6 seconds, a sudden increase in $Q_{s\ ref}$ = 1.5 MVAR, reflecting two distinct states of disturbance on P_s and Q_s. At these precise instants, the perturbations are taken in a very high way to highlight also the influence of the model uncertainty or the coupling effect between P_s and Q_s.

In general, it can be seen from Figs. 9.13 and 9.14 that the P_s active and Q_s reactive power curves for the three controllers follow their reference value well, with very insignificant static errors. However, each of them offers different levels of performance with respect to each other. The IVCCL-PI is the control technique that allows more the reach the power setpoints as quickly as the DVC-PI and the IVCOL-PI. The IVCOL-PI is also faster than the conventional DVC-PI.

The exceedances observed in Figs. 9.13 and 9.14 oscillate around the following values: 0% for IVCCL-PI, 4.8% for IVCOL-PI, and 17.6% for DVC-PI.

For perturbations related to model uncertainties with abrupt power changes present at times t = 3 seconds in Fig. 9.13 and at t = 0 seconds and t = 6 seconds in Fig. 9.14, it can be seen that, compared to the other two, the IVCCL-PI has the best perturbation rejection and a good decoupling between P_s and Q_s of the DFIG. Similarly, the IVCOL-PI is better than the classical DVC-PI.

For the results of the robustness test illustrated in Figs. 9.15 and 9.16, we take into account not only the previously studied setpoint conditions but also, to add the influence of the parametric uncertainty, in particular by choosing a 10% increase in the nominal value of the mutual inductance (m). On these figures, it can be seen that the power curves P_s and Q_s of the DVC-PI are strongly influenced by the perturbations and uncertainties compared to the other two

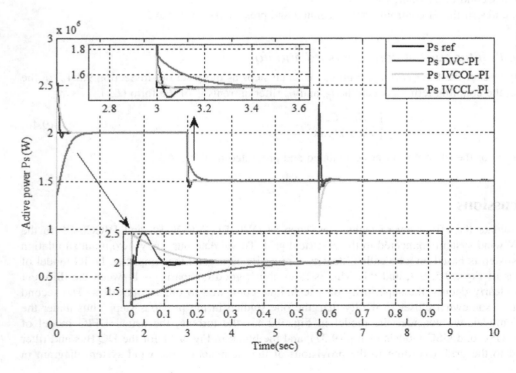

FIGURE 9.13 Active power regulation with the direct vector control with PI, indirect vector control in open loop-PI and indirect vector control in closed loop-PI techniques.

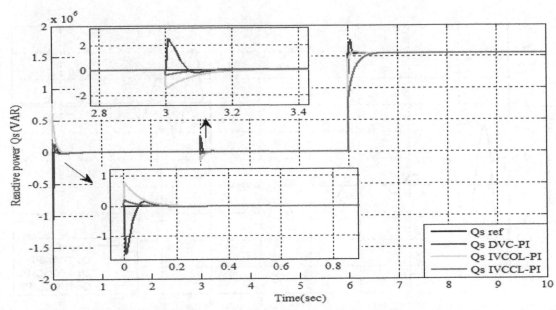

FIGURE 9.14   Regulation of reactive powers with the direct VC with PI, indirect vector control in open loop -PI and indirect vector control in closed loop-PI techniques.

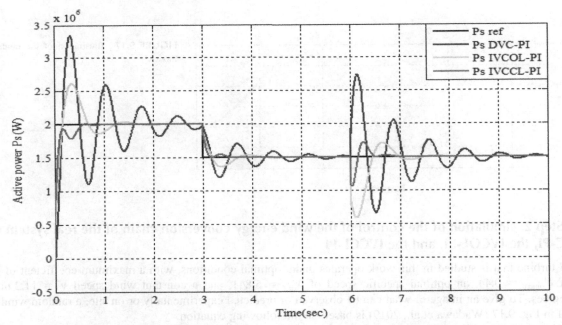

FIGURE 9.15   Robust regulation of active powers with direct VC with PI, indirect vector control in open loop -PI and indirect vector control in closed loop-PI techniques.

techniques. Also, the IVCOL-PI is more influenced than the IVCCL-PI. The observed static errors are negligible. While, the exceedances are around: 4% for IVCCL-PI, 30% with IVCOL-PI, and 60% for DVC-PI.

We see that the indirect control technique marks its superiority over the direct control technique. In other words, the IVCCL-PI offers a good rejection of disturbance, an assured decoupling of P_s and Q_s, good setpoint monitoring and good stability with respect to uncertainties. It is followed by IVCOL-PI that improves at least 20% the performance of conventional DVC-PI, which finds itself in a situation of a most uncomfortable robustness.

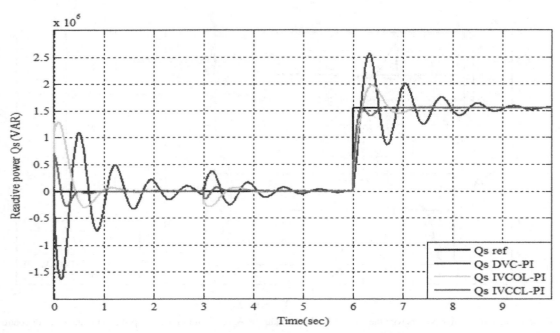

FIGURE 9.16 Robust regulation of reactive powers with direct VC with PI, indirect vector control in open loop -PI and indirect vector control in closed loop-PI techniques.

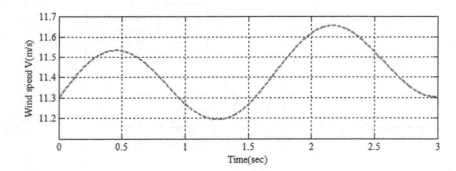

FIGURE 9.17 Illustration of the random wind model.

9.3.2 Step 2: simulation of the control of the wind energy conversion chain of the real system with the DVC-PI, the IVCOL-PI, and the IVCCL-PI

The wind turbine that is studied in this work operates under optimal conditions, with a maximum coefficient of performance of $C_{pmax} = 0.44$, an optimal specific speed of $\lambda_{opt} = 5.841$, and a constant wind speed $V = 11.2 \text{ m s}^{-1}$ at $\beta = 0$ degrees. To give an image of what can be observed or predicted experimentally or on site, a random wind model illustrated in Fig. 9.17 (Wadawa et al., 2019) is based on the following equation:

$$V = 11.3 + (0.2\sin{(0.1047t)} + 0.2\sin{(0.2665t)} + 0.2\sin{(3.6645t)}) \qquad (9.47)$$

In Fig. 9.18 with the fixed value $\beta = 0$ degrees and an overrun of 35%, it can be seen that the Cpmax curve stabilizes around its optimal value of 0.44. It can be said that, despite the disturbed wind conditions, the extraction of maximum power is ensured by the PI regulation. The latter, as shown in Fig. 9.19, also ensures the control of the mechanical speed transmitted to the DFIG machine with an overrun of 9%.

Based on the active and reactive powers in Figs. 9.20−9.25, it can be seen that, in terms of rapid follow-up of their setpoints, reduced displacement, and significant fluctuations, particularly at system start-up, the IVCCL-PI technique is superior to the other two. Thus although the IVCOL-PI is better in terms of speed, it, nevertheless, has greater amplitudes and fluctuations than the DVC-PI and IVCCL-PI. Moreover, the indirect controls of the active and reactive

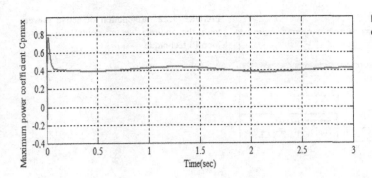

FIGURE 9.18 Appearance of the maximum performance coefficient with PI regulation.

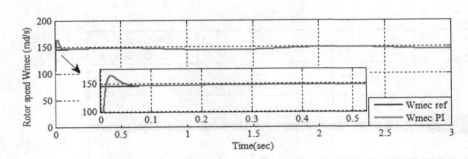

FIGURE 9.19 Regulation of the mechanical speed of rotation with PI regulation.

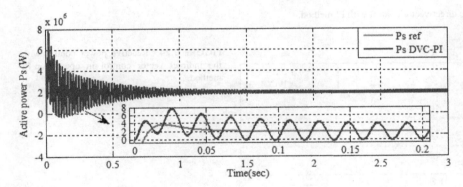

FIGURE 9.20 Regulation of active powers with the direct vector control with PI technique.

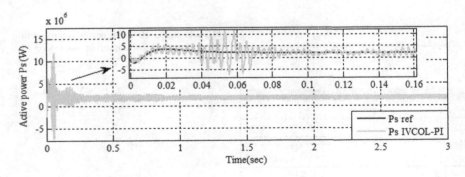

FIGURE 9.21 Regulation of active powers with the indirect vector control in open loop-PI technique.

currents of the DFIG illustrated in Figs. 9.26−9.29 confirm that these currents are the images of the active and reactive power controls of the IVCOL-PI and IVCCL-PI techniques.

The regulation of the DC bus voltage shown in Fig. 9.30 shows that the PI reassures the rapid tracking of the reference value with a displacement of 13%. When looking at the active currents of the filter in Figs. 9.31−9.33, one can see a good rapid follow-up of the reference value by the IVCCL-PI compared to the others. Furthermore, although fast,

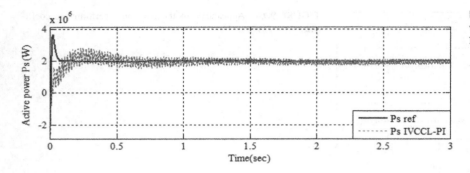

FIGURE 9.22 Regulation of active powers with the indirect vector control in closed loop-PI technique.

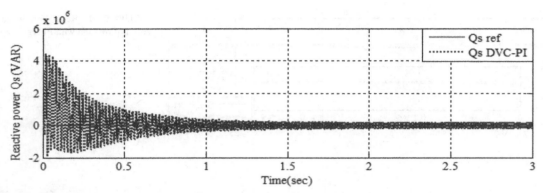

FIGURE 9.23 Reactive power control with the direct vector control with PI method.

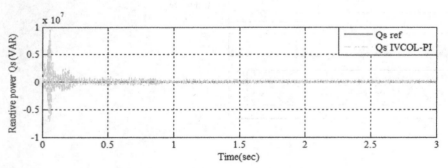

FIGURE 9.24 Reactive power control with the indirect vector control in open loop-PI method.

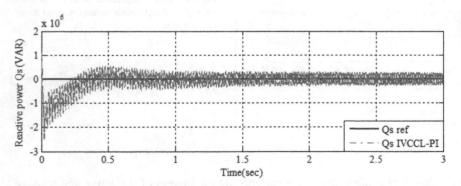

FIGURE 9.25 Reactive power control with the indirect vector control in open loop-PI method.

the IVCOL-PI has larger amplitudes and fluctuations than the DVC-PI and IVCCL-PI. Fig. 9.34 shows us that the Ifq reactive current curve of the IVCCL-PI oscillates closer to zero than the other two. This proximity to zero could improve cosφ on the power grid.

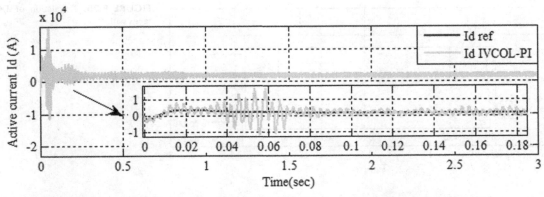

FIGURE 9.26   Doubly fed induction generator active current control with the indirect vector control in open loop-PI method.

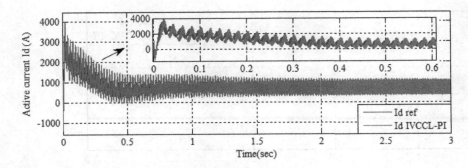

FIGURE 9.27   Control of doubly fed induction generator active currents with the indirect vector control in closed loop-PI method.

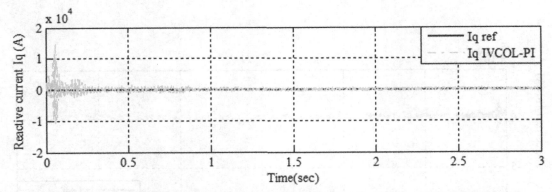

FIGURE 9.28   Doubly fed induction generator reactive current control with the indirect vector control in open loop-PI method.

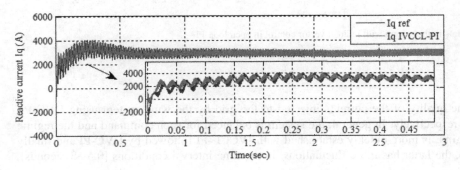

FIGURE 9.29   Control of reactive currents of doubly fed induction generator with the indirect vector control in closed loop-PI method.

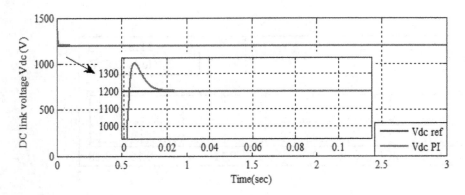

FIGURE 9.30 Regulation of DC bus voltages with PI.

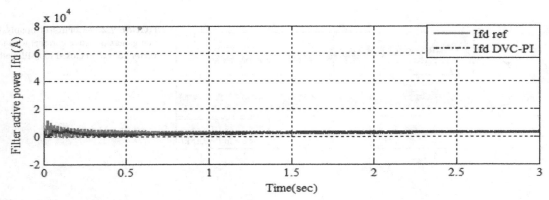

FIGURE 9.31 Regulation of active filter currents in the system with direct VC with PI.

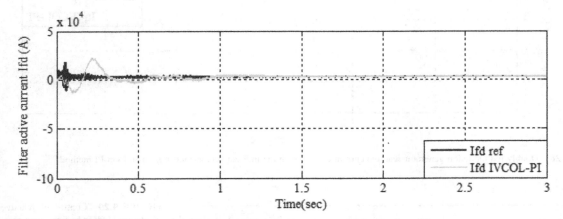

FIGURE 9.32 Regulation of active filter currents in the system with indirect vector control in open loop-PI.

Figs. 9.35, 9.37, and 9.39 show the quality of the currents injected to the grid by the wind system using DVC-PI, IVCOL-PI, and IVCCL-PI techniques respectively. In general, the system is balanced for each command and the regime stabilizes around 0.5 seconds. This regime is more quickly established with IVCCL-PI followed by DVC-PI and finally IVCOL-PI. This is because at start-up, the latter has more fluctuations. In the time interval conditions [0.5−3 seconds],

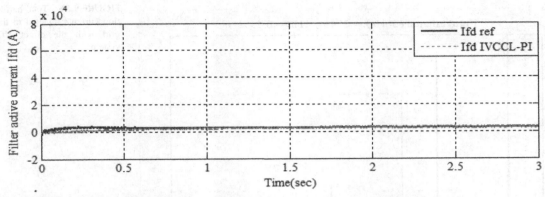

FIGURE 9.33 Regulation of active filter currents in the system with indirect vector control in closed loop-PI.

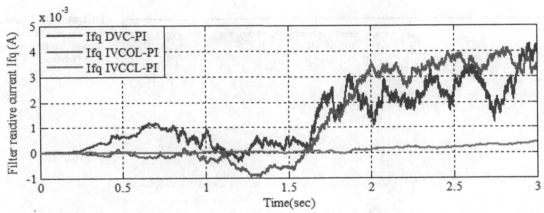

FIGURE 9.34 Regulation of reactive filter currents in direct VC with PI, indirect vector control in open loop-PI and indirect vector control in closed loop-PI systems.

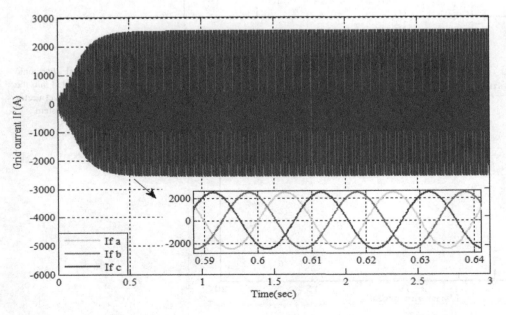

FIGURE 9.35 Currents injected into the power grid with the direct VC with PI system.

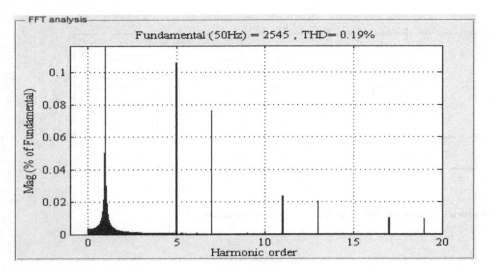

FIGURE 9.36 Total harmonic distortion currents injected to the electrical grid with the direct VC with PI system.

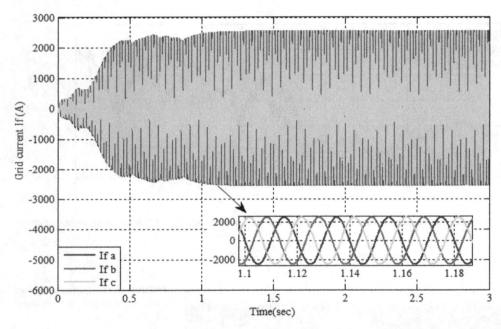

FIGURE 9.37 Currents injected into the electrical grid with the indirect vector control in open loop-PI system.

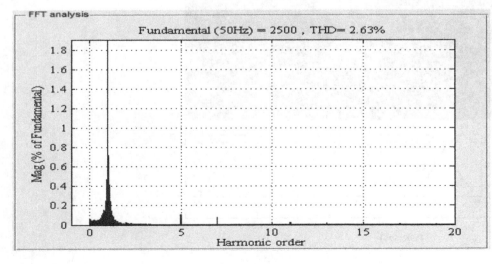

FIGURE 9.38 Total harmonic distortion currents injected into the power grid with the indirect vector control in open loop-PI system.

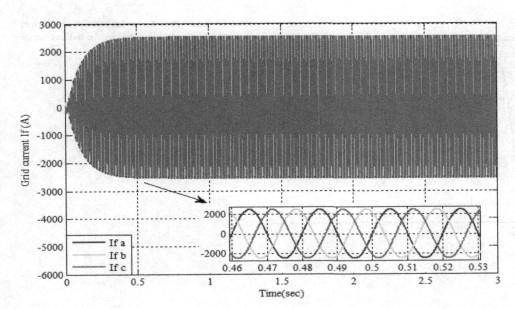

FIGURE 9.39 Currents injected into the power grid with the indirect vector control in closed loop-PI system.

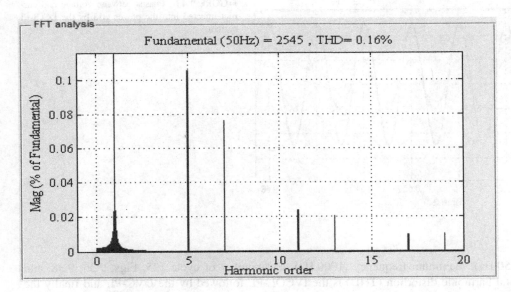

FIGURE 9.40 Total harmonic distortion of currents injected into the electrical grid with the indirect vector control in closed loop-PI system.

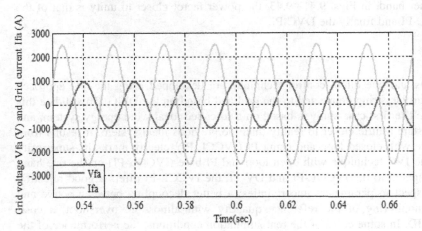

FIGURE 9.41 Phases between voltage and current injected to the power grid by the direct vector control with PI system.

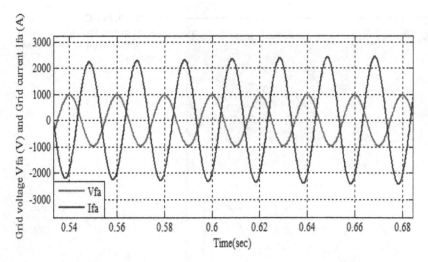

FIGURE 9.42 Phases between voltage and current injected into the electrical grid by the indirect vector control in open loop-PI system.

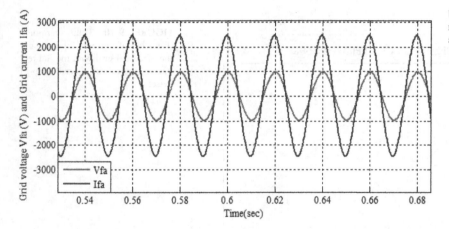

FIGURE 9.43 Phases between voltage and current injected into the power grid by the LCCI-PI system.

fundamental frequency ($f = 50$ Hz), maximum frequency (1000 Hz), and number of cycle 125, Figs. 9.36, 9.38, and 9.39 report that the highest total harmonic distortion (THD) is the IVCOL-PI, followed by the DVC-PI, and finally the IVCCL-PI is the best (Fig. 9.40). On the other hand, in Figs. 9.41–9.43, the power factor closer to unity is that of the IVCCL-PI, followed relatively by the IVCOL-PI and finally the DVC-PI.

9.4 Conclusion

In this study, two indirect control techniques and one direct control technique are developed using the VC approach with the use of PI to control a 2-MW wind turbine string connected to the grid and based on the DFIG. To validate this work, two main case studies were conducted. The first case study is based on a so-called ideal wind energy system, and the second study considers a wind energy system a little closer to reality than the previous ideal system. It results from the simulation results that, the IVC technique with closed loop and using PI (IVCCL-PI), clearly marks its superiority compared to the other two commands like the IVC technique with open loop and PI base (IVCOL-PI) on the one hand and DVC-PI on the other hand. Clearly, compared to the IVCOL-PI and DV-PI, the IVCCL-PI offers a good rejection of perturbation and good robustness with respect to parameters uncertainties; a better decoupling between active and reactive powers (P_s and Q_s), good rapid monitoring of the reference quantity with almost no overshoot, a $\cos\phi$ approaching unity and with a much lower THD. In some cases of the real simulation conditions, the performance of the

IVCOL-PI shows shortcomings in terms of fluctuations especially at system start-up. Thereby, it is clearly observed that the IVCCL-PI allows the improvement of the IVCOL-PI by reducing the effects of static disturbances or fluctuations. Moreover, it can be seen that the PI correctors cancel out the static error. However, the response time and overshoot are highly dependent on the setting of these correctors. It can be concluded that the control techniques we have developed in this document are very important for the control of wind energy systems because it is clear that the choice of a technique can not only significantly improve the control performance but also offer a wide range of possibilities for the implementation of other types of more powerful controls, whether analog, digital, and/or mixed.

Appendix

TABLE A1 Simulation parameters for DFIG, DC bus, filter, and turbine (Wadawa et al., 2019).

Parameters	Values
DFIG	
Stator peak phase voltage	$U_s = 690$ V
Stator resistance	$R_s = 1.69$ mΩ
Stator inductance	$L_s = 2.95$ mH
Rotor resistance	$R_r = 1.52$ mΩ
Rotor inductance	$L_r = 2.97$ mH
Magnetizing inductance	$L_m(M) = 2.91$ mH
Leakage coefficient	$f = 0.03$ (friction f)
Inertial of rotor	$J_{rotor} = 682.3$ kg m^2
Grid frequency	$F = 50$ Hz
DC-link and filter	
DC-link capacitor	$C = 38{,}000$ μF
DC-link voltage	$V_{DC} = 1200$ V
Filter resistance	$R_F = 6$ mΩ
Filter inductance	$L_F = 0.6$ mH
Turbine	
Pole pairs	$P = 2$
Gear box ratio	$G = 90$
Inertial of turbine	$J_{turbine} = 50.105$ kg m^2
Diameter of three blades	$D = 82$ m

DFIG, doubly fed induction generator.

TABLE A2 Summary of controller parameters DVC-PI, IVCOL-PI, IVCCL-PI, and PI.

Controlled quantities of models	Control parameters		
	K_p	K_i	τ_r (ms)
DCV-PI (P_s/Q_s of DFIG)	4.5829e − 4	7.7874e − 3	1.2
IVCOL-PI (P_s/Q_s of DFIG)	0.19891	3.04	0.5
IVCCL-PI (P_s/Q_s of DFIG)	0.995	15.2	0.1
	0.194891	3.04	0.5
PI (Ω_{mec} of turbine MPPT control)	−136,460	−6,823,000	40, $\varepsilon = 1$
PI (V_{DC} of DC-link control)	30.4	6080	10, $\varepsilon = 1$
PI (I_{fd}/I_{fq} of filter currents control)	24	240	2.5e − 2

DFIG, doubly fed induction generator; *DVC*, direct vector control; *IVCCL*, indirect vector control in closed loop; *IVCOL*, indirect vector control in open loop; *MPPT*, maximum power point tracking.

References

Abolvafaei, M., & Ganjefar, S. (2020). Maximum power extraction from fractional order doubly fed induction generator based wind turbines using homotopy singular perturbation method. *International Journal of Electrical Power & Energy Systems*, *119*, 105889. Available from https://doi.org/10.1016/j.ijepes.2020.105889.

Allam, M., Dehiba, B., Abid, M., et al. (2014). Etude comparative entre la commande vectorielle directe et indirecte de la Machine Asynchrone à Double Alimentation (MADA) dédiée à une application éolienne. *Journal of Advanced Research in Science and Technology*, *1*(2), 88−100.

Ardjoun, S., Denai, M., & Abid, M. (2019). Robustification du contrôle des éoliennes pour une meilleure intégration dans un réseau déséquilibré. In *CAGRE 2019: Proceedings of algerian large electrical network conference* (1−6). February 26−28. Algiers, Algeria: IEEE. https://doi.org/10.1109/cagre.2019.8713286.

Aroussi, H. A., Ziani, E., & Bossoufi B. (2020). Backstepping approach applied to the DFIG-WECS. In *QScience proceedings, international meeting on advanced technologies in energy and electrical engineering* (Vol. 1, p. 18). Hamad bin Khalifa University Press (HBKU Press). https://doi.org/10.5339/qproc.2019.imat3e2018.18.

Aydin, E., Polat, A., & Ergene, L. T. (2016). Vector control of DFIG in wind power applications. In *ICRERA 2016: Proceedings of the 5th international conference on renewable energy research and applications* (pp. 478−483). November 20−23. Birmingham, UK: IEEE. doi:10.1109/icrera.2016.7884383.

Bedoud, K., Ali-Rachedi, M., Bahi, T., Lakel, R., & Grid, A. (2015). Robust control of doubly fed induction generator for wind turbine under sub-synchronous operation mode. *Energy Procedia*, *74*, 886−899. Available from https://doi.org/10.1016/j.egypro.2015.07.82.

Belmokhtar, K., & Doumbia, M. L. (2010). Modeling and control of a wind turbine system based on an asynchronous machine with double power supply for the supply of power to the electrical network. *Journal of Scientific Research*, *2*.

Benbouhenni, H. (2018). Comparative study between different vector control methods applied to DFIG wind turbines. *Majlesi Journal of Mechatronic Systems*, *6*(4), 15−23. Available from http://journals.iaumajlesi.ac.ir/ms/index/index.php/ms/article/view/382.

Berahab, R. (2019). *Energies renouvelables en afrique: Enjeux, défis et opportunités/Renewable Energy in Africa: Issues, challenges and opportunities*.

Bharti, O. P., Saket, R. K., & Nagar, S. K. (2017). Controller design for doubly fed induction generator using particle swarm optimization technique. *Renewable Energy*, *114*, 1394−1406. Available from https://doi.org/10.1016/j.renene.2017.06.061.

Bossoufi, B., Karim, M., Lagrioui, A., Taoussi, M., & Derouich, A. (2015). Observer backstepping control of DFIG-Generators for wind turbines variable-speed: FPGA-based implementation. *Renewable Energy*, *81*, 903−917. Available from https://doi.org/10.1016/j.renene.2015.04.013.

Bounar, N., Labdai, S., & Boulkroune, A. (2019). PSO-GSA based fuzzy sliding mode controller for DFIG-based wind turbine. *ISA Transactions*, *85*, 177−188. Available from https://doi.org/10.1016/j.isatra.2018.10.020.

Bourdoulis, M. K., & Alexandridis, A. T. (2014). Direct power control of DFIG wind systems based on nonlinear modeling and analysis. *IEEE Journal of Emerging and Selected Topics in Power Electronics*, *2*(4), 764−775. Available from https://doi.org/10.1109/jestpe.2014.2345092.

Debouza, M., Al-Durra, A., Errouissi, R., & Muyeen, S. M. (2018). Direct power control for grid-connected doubly fed induction generator using disturbance observer based control. *Renewable Eenergy*, *125*, 365−372. Available from https://doi.org/10.1016/j.renene.2018.02.121.

Ghoudelbourk, S., Bahi, T., & Mohammedi, M. (2012). Improving the quality of energy supplied by a doubly-fed induction generator. In *EFEA 2012: Proceedings of 2nd international symposium on environment friendly energies and applications* (437−442). June 25−27. Newcastle Upnon Tyne, UK: IEEE. doi:10.1109/efea.2012.6294041.

Hamdi, N., & Bouzid, A. (2013). Active and reactive power control of a DFIG for variable speed wind energy conversion using a new controller. *International Journal of Computer Applications, 67*(16). Available from https://doi.org/10.5120/11476-6876.

Islam, K. S., Seyedmahmoudian, M., & Stojcevski, A. (2019). Decentralized robust mixed H2/H∞ reactive power control of DFIG cluster using SMES. *International Journal of Electrical Power & Energy Systems, 113*, 176−187. Available from https://doi.org/10.1016/j.ijepes.2019.05.017.

Kai, J. I., Wenfeng, L., & Jinping, H. E. (2019). Indirect vector control for stand-alone operation brushless doubly fed induction generator employing power winding stator flux orientated approach. *IOP Conference Series: Materials Science and Engineering, 612*(4), 042085. Available from https://doi.org/10.1088/1757-899X/612/4/042085.

Khan, B. D., Ansari, J. A., Chachar, F., Katyara, S., & Soomro, J. (2018). Selection of optimal controller for active and reactive power control of doubly fed induction generator (DFIG). In *iCoMET 2018: Proceedings of international conference on computing, mathematics and engineering technologies* (pp. 1−5). March 3−4. Sukkur, Pakistan: IEEE. doi:10.1109/icomet.2018.8346416.

Lazrak, A., & Abbou, A. (2017). Robust power control of DFIG based wind turbine without currents rotor sensor. In *IRSEC 2017: Proceedings of international renewable and sustainable energy conference* (1−6). December 4−9. Tangier, Morocco: IEEE. https://doi.org/10.1109/IRSEC.2017.8477340.

Lhachimi, H., Sayouti, Y., & El Kouari, Y. (2018). Optimal improvement of direct power control strategy based on sliding mode controllers. *Computers and Electrical Engineering, 71*, 637−656. Available from https://doi.org/10.1016/j.compeleceng.2018.08.013.

Li, S., Haskew, T. A., Williams, K. A., & Swatloski, R. P. (2012). Control of DFIG wind turbine with direct-current vector control configuration. *IEEE Transactions on Sustainable Energy, 3*(1), 1−11. Available from https://doi.org/10.1109/tste.2011.2167001.

Li, S., Wang, H., Tian, Y., Aitouch, A., & Klein, J. (2016). Direct power control of DFIG wind turbine systems based on an intelligent proportional-integral sliding mode control. *ISA Transactions, 64*, 431−439. Available from https://doi.org/10.1016/j.isatra.2016.06.003.

Mazouz, F., Belkacem, S., Harbouch, Y., Abdessemed, R., & Ouchen, S. (2017). Active and reactive power control of a DFIG for variable speed wind energy conversion. In *ICSC 2017: Proceedings of 6th international conference on systems and control* (pp. 27−32). May 7−9. Batna, Algeria: IEEE. doi:10.1109/icosc.2017.7958642.

Mehdipour, C., Hajizadeh, A., & Mehdipour, I. (2016). Dynamic modeling and control of DFIG-based wind turbines under balanced network conditions. *International Journal of Electrical Power & Energy Systems, 83*, 560−569. Available from https://doi.org/10.1016/j.ijepes.2016.04.046.

Mohammadi, J., Vaez-Zadeh, S., Afsharnia, S., & Daryabeigi, E. (2014). A combined vector and direct power control for DFIG-based wind turbines. *IEEE Transactions on Sustainable Energy, 5*(3), 767−775. Available from https://doi.org/10.1109/tste.2014.2301675.

Morshed, M. J., & Fekih, A. (2017). A new fault ride-through control for DFIG-based wind energy systems. *Electric Power Systems Research, 146*, 258−269. Available from https://doi.org/10.1016/j.epsr.2017.02.010.

Nazari, R., Seron, M. M., & De Doná, J. A. (2017). Actuator fault tolerant control of systems with polytopic uncertainties using set-based diagnosis and virtual-actuator-based reconfiguration. *Automatica, 75*, 182−190. Available from https://doi.org/10.1016/j.automatic.2016.09.012.

Rechka, S., Roy, G., Dennetiere, S., et al. (2004). Modeling of multi-mass electromechanical systems based on asynchronous machines, using MATLAB and EMTP tools with application to wind turbines. Technical report number: EPM-RT-2004-04. https://publications.polymtl.ca/2616/.

Robyns, B., Davigny, A., Saudemont, C., Ansel, A., Courtecuisse, V., François, B., ... Deuse, J. (2006). Impact de l'éolien sur le réseau de transport et la qualité de l'énergie. *J3eA, 5*, 003.

Rocha-Osorio, C. M., Solís-Chaves, J. S., Rodrigues, L. L., & al. (2018). Deadbeat−fuzzy controller for the power control of a doubly fed induction generator based wind power system. *ISA Transactions, 88*, 258−267. Available from https://doi.org/10.1016/j.isatra.2018.11.038.

Saihi, L., Berbaoui, B., & Glaoui, H. (2020). Robust control H∞ fuzzy of a doubly fed induction generator integrated to wind power system. *Majlesi Journal of Electrical Engineering, 14*(1), 59−69. Available from http://www.mjee.org/index/index.php/ee/article/view/2981.

Senani, F., Rahab, A., & Benalla, H. (2018). Vector control and direct power control of wind energy conversion system based on a DFIG. *Journal of Electrical and Electronics Engineering, 11*(1), 15−20.

Shukla, R. D., Tripathi, R. K., & Thakur, P. (2017). DC grid/bus tied DFIG based wind energy system. *Renewable Energy, 108*, 179−193. Available from https://doi.org/10.1016/j.renene.2017.02.064.

Tang, Y., Ju, P., He, H., Qin, C., & Wu, F. (2013). Optimized control of DFIG-based wind generation using sensitivity analysis and particle swarm optimization. *IEEE Transactions on Smart Grid, 4*(1), 509−520. Available from https://doi.org/10.1109/tsg.2013.2237795.

Van, M. (2018). An enhanced robust fault tolerant control based on an adaptive Fuzzy PID-nonsingular fast terminal sliding mode control for uncertain nonlinear systems. *IEEE/ASME Transactions on Mechatronics, 23*(3), 1362−1371. Available from https://doi.org/10.1109/tmech.2018.2812244.

Wadawa, B., Errami, Y., Obbadi, A., Sahnoun, S., & Aoutoul, M. (2019). Control of wind energy conversion system connected to the power grid and based on double fed induction generator. In *BDioT 2019: Proceedings of the 4th international conference on big data and internet of things* (1−8). October 23−24. Rabat, Morocco: BDioT. https://doi.org/10.1145/3372938.3373004.

Xiong, L., Li, P., Wu, F., Ma, M., Khan, M. W., & Wang, J. (2019). A coordinated high-order sliding mode control of DFIG wind turbine for power optimization and grid synchronization. *International Journal of Electrical Power & Energy Systems, 105*, 679−689. Available from https://doi.org/10.1016/j.ijepes.2018.09.008.

Yasmine, I., Chakib, B., & Badre, B. (2017). Power control of DFIG-generators for wind turbines variable-speed. *International Journal of Power Electronics and Drive System, 8*(1), 444−453. Available from https://doi.org/10.11591/ijpeds.v8i1.

Yu, C., & Li, D. (2012). Fuzzy-PI and feedforward control strategy of DFIG wind turbine. In *ISGT 2012: Proceedings of PES innovative smart grid technologies Asia* (pp. 1−5). May 21−24. Tianjin, China: IEEE. https://doi.org/10.1109/isgt-asia.2012.6303299.2012.

Chapter 10

Modeling of a high-performance three-phase voltage-source boost inverter with the implementation of closed-loop control

P. Arvind, Sourav Chakraborty, Deepak Kumar and P.R. Thakura[†]

Department of Electrical and Electronics Engineering, Birla Institute of Technology, Mesra, Ranchi, India

10.1 Introduction

With the sophistication of technology, electrical power usage has increased manifolds and has become more diversified, leading to a variety of unprecedented obstacles faced by the global electricity market and its customers since the onset of widespread electrification. With the precipitous depletion of conventional sources, the necessity for a cost-effective, clean, efficient, and uninterrupted supply of power comes into play. Also, in today's scenario, the need for energy-efficient and cost-effective power converters is significantly escalating (Bihari, Sadhu, Das, Arvind, & Gupta, 2019; Liserre, Teodorescu, & Blaabjerg, 2006; Pina, Silva, & Ferrão, 2012; Weisser, 2004). To counteract such problems, there has been tremendous research in the field of power electronics for decades. Power electronic converters can be speculated wherever the form of electrical energy (i.e., voltage, current, or frequency) needs to be modified, ranging between hundreds of megawatts and several mill watts. Its significance has annexed to various sectors, from aerospace to utility grids (Bose, 2007).

A sufficient amount of literature has been surveyed in topics related to power electronic switches, power modulators, their various topologies, applications, and their control. The findings were mostly in favor of the conventional Voltage Source Inverters (VSIs) to be by far among the most imperative converters when it comes to interfacing DC power to an AC load/grid. Among them, the three-phase VSI is more prominent in having applications in numerous commercial and industrial sectors (Holmes & Kotsopoulos, 1993; Twining & Holmes, 2003). The downside of this topology is that the voltage inverted to AC is of buck nature, which conveys that the output voltage is less than the input. It is a well-known fact that distributed energy resources like that of solar PV are nowadays broadly used due to its compatibility in various application fields. However, they are intermittent, which signifies that the power-delivering capacity depends on the climatic conditions. Moreover, the inherently low nature of voltage provided by these panels requires the role of advanced power converters for processing the input to make it available in the usable form (Cavallo, 2001). It is normally customary to see a VSI equipped with a preceding DC−DC converter operating in tandem when porting solar power or other inherently low DC voltage sources. Such multistage conversions lead to a substantial increase in the implementation cost and the complexity of the system based on the power delivered or the concerned voltage level (Chakraborty, Kumar, & Thakura, 2019; Christensen, Reddy, McKinstrie, Rottwitt, & Raymer, 2015; Singaravel & Daniel, 2015).

Current source inverters (CSIs) have appreciable boosting capabilities and their architecture has been discussed in (Dash & Kazerani, 2011; Rajeev & Agarwal, 2018). However, in such systems, the presence of a high value of inductance considerably increases the system size. The control of such inverters is highly complex.

[†] Dr. P.R. Thakura- deceased 13th September 2020.

Renewable Energy Systems. DOI: https://doi.org/10.1016/B978-0-12-820004-9.00012-7

Z-source inverters (Gajanayake, Vilathgamuwa, & Loh, 2007; Peng, Shen, & Qian, 2005) combine both the properties of VSIs and CSIs but likewise the increase in the number of storage elements again leads to difficulties in its control. Multilevel inverters, such as the diode clamp inverters and flying capacitor-type inverters, were also studied for their amplification capabilities (Barkati, Baghli, Berkouk, & Boucherit, 2008; Escalante, Vannier, & Arzandé, 2002). But the dependence on a large number of switches leads to excessive switching losses.

In 1995, Caceres and Barbi (1999) formulated the first Boost Inverter topology, which could amplify the output AC voltage to twice its DC input value. The topology was for single-phase. This chapter includes operation, analysis, modulation, control technique, and experimental findings. The working principle revolves around the fact that two separate boost converters where one is the complement of the other to supply power to the load. Based on this prototype, the control strategies of the boost inverters were devised in (Koushki, Khalilinia, Ghaisari, & Nejad, 2008; Sanchis, Ursæa, Gubía, & Marroyo, 2005). Here, the inductors are the boosting element and works similar to that of a Boost converter.

Koushki and Ghaisari (2009) have deduced a voltage reference design for the three-phase boost inverter where the reference set could be applied to any type of differential boost inverter. In addition to this, the reference set required a lower maximum voltage for reverse mode of switching as well as a lower maximum voltage for inverter capacitors. The problem with these topologies discussed in this part was the existence of shoot through voltage, which was necessary to be controlled.

Nguyen, Le, Park, and Lim (2014) and Ravindranath, Mishra, and Josh (2012), have presented a switched boost inverter that combined a boost converter with an inverter. Not only the presence of an additional switch with high frequency switching causes additional switching losses, but also the control for such topologies was complex. Moreover, there was an additional requirement of $L-C$ or $L-C-L$ filters that made the circuitry bulkier.

Raveendhra and Pathak (2018) presented a capacitor clamped inverter, which used small passive components. The state-space modeling was done taking into consideration of the one-leg operation. The prototype showed reduced voltage stress on the capacitors and was compared with Z-Source inverters and Switched Boost Inverters.

Among all the inverter topologies discussed above, the common points shared by these prototypes is the requirement of high frequency switching which makes it necessary for incorporating high-end gate driver and bootstrapping circuits due to which the issues that arise are: high switching losses that require a robust heat dissipating sinks and high probability of electromagnetic interferences due to which the requirement of proper enveloping persists. Moreover, the Total Harmonic Distortion (THD) in the output was seen to be high, thereby inviting the need of a proper filtration of the harmonic components.

IEEE Standard 519-2014 requires that the maximum THD should not exceed beyond 5% (Langella, Testa, & Alii, 2014). Therefore, a significant portion of discussion focusses on a new topology of high-performance boost inverter with output as three-phase AC greater in magnitude compared to the input DC voltage. The system works on fundamental frequency switching. The single-stage conversion reduces the number of conversion rate and also increases the efficiency, which would have been low compared to multistage conversion. The topology generates a negligible THD and is able to boost the voltage quadruple times at a duty cycle of 0.5.

The single-phase version of this topology has been discussed in Tripathi, Keshri, and Thakura (2019) where the authors successfully managed to achieve an ultra-low harmonic distortion and also the Peak Inverse Voltage (PIV) was seen to be equal to the magnitude of that of the DC voltage. Also, the Total Blocking Voltage (TBV) was quite low. An insignificant amount of semiconductor loss was perceived. The topology also claimed of negligible electromagnetic interference and hence requires simple packaging. Similarly, the three-phase version has been discussed in Arvind, Kumar, and Thakura (2019), Kumar Barwar, Tripathi, and Thakura (2018), and Ojha, Barwar, Tripathi, and Thakura (2018).

Batarseh and Harb (2018) and Bose and Bose (1997) have presented a comparative study of various power electronic switches. Out of which the use of IGBTs have been seen to be advantageous when it comes to designing high power converters with high voltage capabilities and frequency of switching below 20 kHz. The gate drive current is nearly zero. The requirement of commutation circuits is also not needed, which gives it an upper hand over SCRs in this case.

In this chapter, a detailed mathematical analysis of the three-phase high gain boost inverter has been provided along with its closed-loop control for a stand-alone system. The novelty of this topology lies in the fact that the boosting and filtering is done due to the inductor—capacitor combination being coupled to resonant frequency making the whole system behave like series resonant inverter. Furthermore, it is observed that the boost inverter is load dependent and the gain achieved by the system is similar to gain achieved by a two-pole filter at resonant frequency.

The following points sum up the objectives of this chapter:

1. Understanding the state-of-the-art high-power switches and power electronic converters, single-stage three-phase voltage source boost inverter.

2. Developing the state variable and transfer function model of the three-phase voltage source boost inverter.
3. Formulating a closed-loop control strategy for the three-phase boost inverter.
4. Applying the voltage source three-phase boost inverter in feeding a three-phase load and over the top.
5. Experimentation and validation of the performance of the converter in simulation.

The chapter has been organized in the following format:

Section 10.2 gives an in-depth mathematical analysis of the converter topology, considering both one-leg operation and six-state variable method.

Section 10.3 highlights the control strategy for the inverter topology feeding a three-phase resistive load

Section 10.4 presents the results and discussions following the obtained results

In particular, the contributions, conclusion, and future prospects addressed for the three-phase boost inverter are provided in Section 10.5.

10.2 Mathematical analysis of the three-phase boost inverter

The single-stage three-phase boost inverter resembles a three-phase VSI. The only difference is that each leg is coupled with an $L-C$ filter tuned at the resonant frequency. This $L-C$ pair not only boosts the voltage but also acts as a filter itself. Fig. 10.1 shows the circuit diagram of a three-phase boost inverter. The inverter works in 180-degree conduction mode. The three legs refer to the three respective phases A, B, and C. Each leg has two semiconductor switches (IGBTs). Switches 1, 3, and 5 will form the positive group, and switches 2, 4, and 6 will form the negative group. The values of all three inductors are the same, and the values of all the capacitors are equal. The junction of the $L-C$ tank is where the load is connected. The switching is done at 50 Hz, which is in proximation to that of the tuned resonant frequency.

Clearly, it can be seen that modes 4, 5, and 6 are compliments of 1, 2, and 3.

10.2.1 Mathematical analysis based on one-leg operation

For simplicity, the mathematical modeling of the boost inverter can be explained by studying the working of any one of the legs relevant to a single phase. Let us consider an arbitrary phase (say A). There are two operating modes per phase. As of now, L_1 and C_1 will be considered as L and C respectively (for ease of understanding).

10.2.1.1 Mode I operation

The operation of Mode I is depicted in Fig. 10.2. During this mode, S_1 is ON, while S_4 is OFF. The current flows through the inductor, capacitor, and load. The magnetizing current rises through the inductor, and the capacitor gets charged (Here V_{dc} and V_{in} mean the same).

From the laws of Kirchoff (current and voltage), we get the equation as:

$$V_{in} = V_L(t) + V_C(t) \tag{10.1}$$

We know that

$$V_L(t) = L\frac{di_L(t)}{dt} \tag{10.2}$$

And due to L and C are in series,
So,

$$i_L = i_C = i; \quad i_C(t) = C\frac{dVc(t)}{dt} = i_L(t)$$

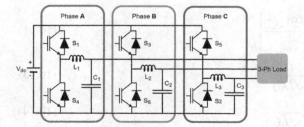

FIGURE 10.1 Power circuit diagram of the high-performance three-phase boost inverter.

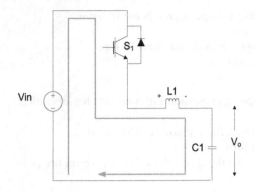

FIGURE 10.2   One-leg equivalent circuit-mode I.

Now Eq. (10.2) can be written as

$$V_L = LC \frac{d^2 V_C(t)}{dt^2}$$

Now replace $V_L(t)$ value in Eq. (10.1), then new equation form as

$$V_{in} = LC \frac{d^2 V_C(t)}{dt^2} + V_C(t) \tag{10.3}$$

$$V_C(t) = V_{in} - LC \frac{d^2 V_C(t)}{dt^2} \tag{10.4}$$

Taking Laplace to both side of Eq. (10.4):

$$V_C(S) = \frac{V_{in}}{S} - \left\{ LC \left(S^2 V_C(S) \right) - S V_C(0^-) - V_C'(0^-) \right\} \tag{10.5}$$

Due to the first mode taking initial condition

$$V_C(0^-) = 0 \text{ and } V_C'(0^-) = 0$$

Now Eq. (10.5) is written as

$$V_C(S) = \frac{V_{in}}{S} - \left\{ LC \left(S^2 V_C(S) \right) \right\} \tag{10.6}$$

$$V_C(S) \left[1 + LCS^2 \right] = \frac{V_{in}}{S} \tag{10.7}$$

$$V_C(S) = \frac{v_{in}}{S(1 + LCS^2)} = \frac{v_{in}/LC}{S\left(\frac{1}{LC} + S^2\right)} = \frac{1}{S} \left[\frac{v_{in}/LC}{\left(\frac{1}{LC} + S^2\right)} \right] \tag{10.8}$$

$$V_C(S) = \frac{1}{S} \left(V_{in}/\sqrt{LC} \right) \left[\frac{v_{in}/\sqrt{LC}}{S^2 + \left(\frac{1}{\sqrt{LC}}\right)^2} \right] \tag{10.9}$$

On taking the inverse Laplace, Eq. (10.9) will be:

$$V_C(t) = \left(V_{in}/\sqrt{LC} \right) \int_0^t \sin \frac{1}{\sqrt{LC}} t \, dt \tag{10.10}$$

$$V_C(t) = \left(V_{in}/\sqrt{LC} \right) \left[\frac{-\cos \frac{t}{\sqrt{LC}}}{\frac{1}{\sqrt{LC}}} \right]_0^t \tag{10.11}$$

$$V_C(t) = V_{in} \left[1 - \cos \frac{t}{\sqrt{LC}} \right] \tag{10.13}$$

$$V_C(t) = V_{in} - V_{in} \cos \frac{t}{\sqrt{LC}} \tag{10.14}$$

10.2.1.2 Mode II operation

In mode II, the upper switch S_1 is OFF, which disconnects the DC supply, and the lower switch S_4 is ON. The inductor gets demagnetised and its polarity is reversed. The capacitor gets discharged through the load. The equivalent circuit is displayed in Fig. 10.3.

Appling KVL in equivalent circuit as shown in Fig. 10.2, equation comes as:

$$V_L(t) - V_C(t) = 0 \tag{10.15}$$

$$V_L(t) = V_C(t) \tag{10.16}$$

We know that,

$$V_L(t) = L\frac{di_L(t)}{dt} \tag{10.17}$$

As the current flows in opposite direction,

$$V_L(t) = -C\frac{dV_C(t)}{dt} \tag{10.18}$$

Now Eq. (10.17) will be written as

$$V_L(t) = -LC\frac{d^2V_C(t)}{dt^2} \tag{10.19}$$

Now put the value of $V_L(t)$ in Eq. (10.16), then the equation

$$V_C(t) = -LC\frac{d^2V_C(t)}{dt^2} \tag{10.20}$$

Taking Laplace to both sides of the equation, Eq. (10.20) becomes:

$$V_C(S) = -\left\{LC\left(S^2V_C(S)\right) - SV_C(0^-) - V'_C(0^-)\right\} \tag{10.21}$$

Taking initial condition

$$V_C(0^-) = V_{in}; \quad V'_C(0^-) = 0; \quad i(0^-) = 0$$

Now Eq. (10.21) written as,

$$V_C(S) = -\left\{LC\left(S^2V_C(S) - SV_{in}\right)\right\} = -LCS2\ V_C(S) + SLCV_{in} \tag{10.22}$$

$$V_C(S)\left[1 + LCS^2\right] = SLCV_{in} \tag{10.23}$$

Now Eq. (10.23) written as,

$$V_C(S) = \frac{SLCV_{in}}{\left[1 + LCS^2\right]} \tag{10.24}$$

Dividing left side upper and lower part with LC, then the equation

$$V_C(S) = \frac{SV_{in}}{\left[\frac{1}{LC} + S^2\right]} \tag{10.25}$$

FIGURE 10.3 One-leg equivalent circuit-mode II.

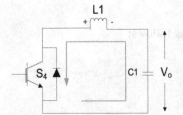

$$V_C(S) = V_{in}\sqrt{LC}S\frac{\frac{1}{\sqrt{LC}}}{\left[\left(\frac{1}{\sqrt{LC}}\right)^2 + S^2\right]} \tag{10.26}$$

Taking Inverse Laplace of Eq. (10.25), then the equation will be

$$V_C(t) = V_{in}\sqrt{LC}\frac{d}{dt}\left[\sin\frac{1}{\sqrt{LC}}t\right] \tag{10.27}$$

$$V_C(t) = V_{in}\sqrt{LC}\frac{1}{\sqrt{LC}}\cos\frac{t}{\sqrt{LC}} \tag{10.28}$$

$$V_C(t) = V_{in}\cos\frac{t}{\sqrt{LC}} \tag{10.29}$$

10.2.2   State space representation of the one-leg operation

Referring Figs. 10.2 and 10.3, respectively, and also the aforestated equations, we obtain the state equations and output equations for the respective modes by applying the Kirchoff's Laws.

For mode 1:

State equation:
$$\begin{bmatrix} \dfrac{dI_{L1}}{dt} \\ \dfrac{dV_{C1}}{dt} \end{bmatrix} = \begin{bmatrix} 0 & -\dfrac{1}{L_1} \\ \dfrac{1}{C_1} & 0 \end{bmatrix}\begin{bmatrix} I_{L1} \\ V_{c1} \end{bmatrix} + \begin{bmatrix} \dfrac{1}{L_1} \\ 0 \end{bmatrix}V_{in} \tag{10.30}$$

Output equation:
$$[V_o] = \begin{bmatrix} 0 & 1 \end{bmatrix}\begin{bmatrix} I_{L1} \\ V_{c1} \end{bmatrix} \tag{10.31}$$

where $A_1 = \begin{bmatrix} 0 & -\dfrac{1}{L_1} \\ \dfrac{1}{C_1} & 0 \end{bmatrix}$; $B_1 = \begin{bmatrix} \dfrac{1}{L_1} \\ 0 \end{bmatrix}$; $C_1 = \begin{bmatrix} 0 & 1 \end{bmatrix}$

For mode 2:

State equation:
$$\begin{bmatrix} \dfrac{dI_{L1}}{dt} \\ \dfrac{dV_{C1}}{dt} \end{bmatrix} = \begin{bmatrix} 0 & -\dfrac{1}{L_1} \\ \dfrac{1}{C_1} & 0 \end{bmatrix}\begin{bmatrix} I_{L1} \\ V_{c1} \end{bmatrix} + \begin{bmatrix} 0 \\ 0 \end{bmatrix}V_{in} \tag{10.32}$$

Output equation:
$$[V_o] = \begin{bmatrix} 0 & 1 \end{bmatrix}\begin{bmatrix} I_{L1} \\ V_{c1} \end{bmatrix} \tag{10.33}$$

where $A_2 = \begin{bmatrix} 0 & -\dfrac{1}{L_1} \\ \dfrac{1}{C_1} & 0 \end{bmatrix}$; $B_2 = \begin{bmatrix} 0 \\ 0 \end{bmatrix}$; $C_2 = \begin{bmatrix} 0 & 1 \end{bmatrix}$.

The averaged state space representation can be done by taking matrices A, B, and C as follows:

$$A = A_1 d + A_2(1 - \delta)$$
$$B = B_1 d + B_2(1 - \delta)$$
$$C = C_1 d + C_2(1 - \delta)$$

Substituting the values, the final averaged state-space model for the one-leg operation is obtained as follows:

State equation:
$$\begin{bmatrix} \dfrac{dI_{L1}}{dt} \\ \dfrac{dV_{C1}}{dt} \end{bmatrix} = \begin{bmatrix} 0 & -\dfrac{1}{L_1} \\ \dfrac{1}{C_1} & 0 \end{bmatrix} \begin{bmatrix} I_{L1} \\ V_{c1} \end{bmatrix} + \begin{bmatrix} \dfrac{\delta}{L_1} \\ 0 \end{bmatrix} V_{in} \tag{10.34}$$

Output equation:
$$[V_o] = [0 \quad 1]\begin{bmatrix} I_{L1} \\ V_{c1} \end{bmatrix} \tag{10.35}$$

where
$$A = \begin{bmatrix} 0 & -\dfrac{1}{L_1} \\ \dfrac{1}{C_1} & 0 \end{bmatrix}; \quad B = \begin{bmatrix} \dfrac{\delta}{L_1} \\ 0 \end{bmatrix}; \quad C_{out} = [0 \quad 1]$$

As the switching pulses are quarter-wave symmetric, hence, for an undistorted sine waveform, the maximum duty cycle δ can be 0.5 for a time period of $2\pi/\omega$.

10.2.3   State space analysis considering six state variables

The single-stage three-phase boost Inverter consists of three $L-C$ filters coupled with each leg (as discussed earlier). Therefore there are six state variables due to six storage elements. In this section, analysis on all the six operating modes and the final state space representation have been elucidated (Fig. 10.4).

From the above switching diagram, and referring to the following equivalent circuits from Table 10.1, the average state-space model for the proposed single-stage three-phase boost inverter topology can be deduced. Where:

$$\sum_{k=1}^{6} D_k T = \frac{1}{f} = \frac{2\pi}{\omega} \tag{10.36}$$

Here, the symbols have their usual meaning. For simplicity, the three-phase load in this analysis is taken to be resistive. There are six duty ratios from D_1 to D_6 for each corresponding operational mode.

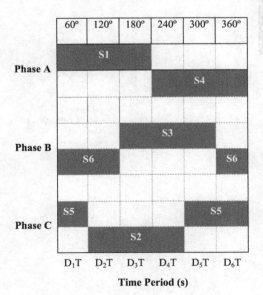

FIGURE 10.4   Switching diagram.

TABLE 10.1 Boost inverter-operating modes.

Mode	Equivalent circuit diagram	Switches ON	Switches OFF
1		1, 5, 6	2, 3, 4
2		1, 2, 6	3, 4, 5
3		1, 2, 3	4, 5, 6

(Continued)

TABLE 10.1 (Continued)

Mode	Equivalent circuit diagram	Switches ON	Switches OFF
4		2, 3, 4	1,5,6
5		3,4,5	1, 2, 6
6		4, 5, 6	1, 2, 3

For $t \in [0, D_1 T]$, the equivalent circuit is given as in Fig. 10.5:

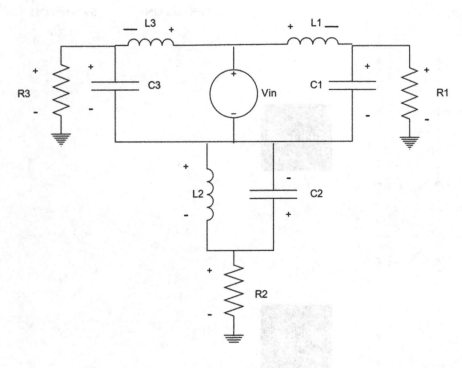

FIGURE 10.5 Equivalent circuit for mode 1.

For $t \in [D_1 T, D_2 T]$, the equivalent circuit is given as in Fig. 10.6:

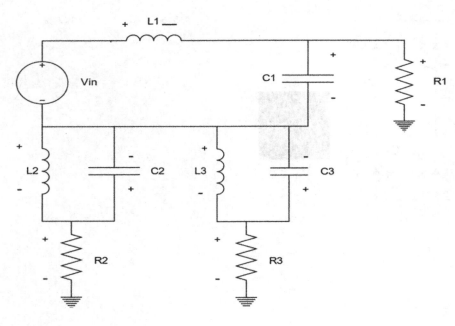

FIGURE 10.6 Equivalent circuit for mode 2.

For $t \in [D_2T, D_3T]$, the equivalent circuit is given as in Fig. 10.7:

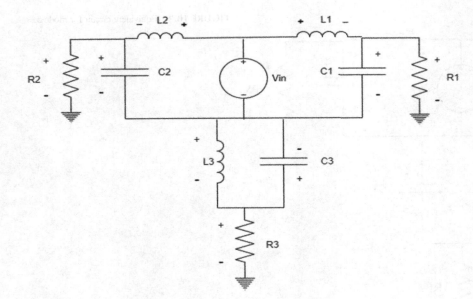

FIGURE 10.7 Equivalent circuit for mode 3.

For $t \in [D_3T, D_4T]$, the equivalent circuit is given as in Fig. 10.8:

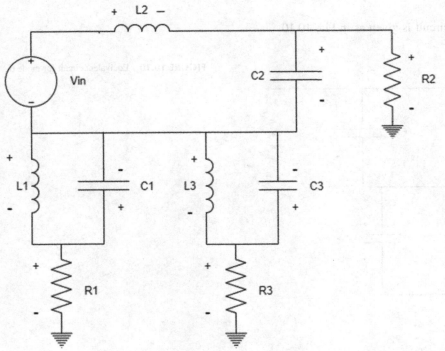

FIGURE 10.8 Equivalent circuit for mode 4.

For $t \in [D_4T, D_5T]$, the equivalent circuit is given as in Fig. 10.9:

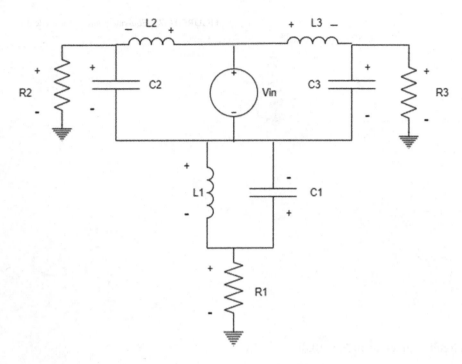

FIGURE 10.9 Equivalent circuit for mode 5.

For $t \in [D_5T, D_6T]$, the equivalent circuit is given as in Fig. 10.10:

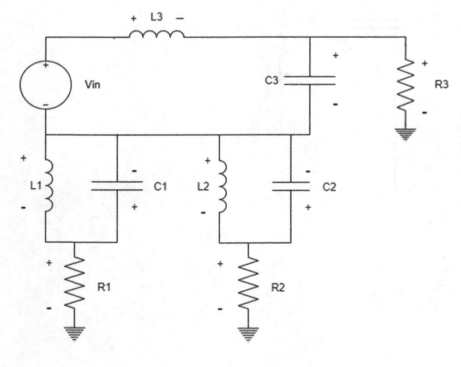

FIGURE 10.10 Equivalent circuit for mode 6.

From each of the circuits, it can be clearly verified that

$$i_{Ck} = i_{Lk} - i_{Rk}; \quad k = 1, 2, 3 \tag{10.37}$$

Also, from the ground node

$$i_{R1} + i_{R2} + i_{R3} = 0 \tag{10.38}$$

Moreover,

$$V_{C1} - V_{C2} = V_{R1} - V_{R2} \tag{10.39}$$

$$V_{C2} - V_{C3} = V_{R2} - V_{R3} \tag{10.40}$$

$$V_{C3} - V_{C1} = V_{R3} - V_{R1} \tag{10.41}$$

Based on the existence of V_{in} in the loop, for every inductor, the equation would be either

$$V_{Lk} = V_{in} - V_{Ck} \quad \text{or} \quad V_{Lk} = -V_{Ck}; \quad k = 1, 2, 3 \tag{10.42}$$

With these equations, it is possible to obtain the state space representation of every equivalent circuit using the capacitor voltages and the inductor currents as the states. The final part is to obtain a term i_{Rk} (which is the current through each phase) as a linear combination of the states. To do this, the system is solved as:

$$\begin{bmatrix} R_o & -R_p & 0 \\ 0 & R_p & -R_q \\ 1 & 1 & 1 \end{bmatrix} \begin{bmatrix} i_{Ro} \\ i_{Rp} \\ i_{Rq} \end{bmatrix} = \begin{bmatrix} V_{Co} - V_{Cp} \\ V_{Cp} - V_{Cq} \\ 0 \end{bmatrix} \tag{10.43}$$

where o, p, and q can be 1, 2, and 3 in any order. Which yields

$$\begin{bmatrix} i_{Ro} \\ i_{Rp} \\ i_{Rq} \end{bmatrix} = \frac{1}{\Delta} \left(V_{Co} \begin{bmatrix} R_p + R_q \\ -R_p \\ -R_q \end{bmatrix} + V_{Cp} \begin{bmatrix} -R_q \\ R_o + R_q \\ -R_o \end{bmatrix} + V_{Cq} \begin{bmatrix} -R_p \\ -R_o \\ R_o + R_p \end{bmatrix} \right) \tag{10.44}$$

$$\text{where} \quad \Delta = R_o R_p + R_p R_q + R_q R_o \tag{10.45}$$

Without the loss of generality, the indices are taken as $o = 1$, $p = 2$, and $q = 3$, and R^{-1} is defined as:

$$R^{-1} = \frac{1}{R_1 R_2 + R_2 R_3 + R_3 R_1} \begin{bmatrix} R_2 + R_3 & -R_3 & -R_2 \\ -R_3 & R_1 + R_3 & -R_1 \\ -R_2 & -R_1 & R_1 + R_2 \end{bmatrix} \tag{10.46}$$

I_R is the vector representing the load currents and s given as:

$$I_R = \begin{bmatrix} i_{R1} & i_{R2} & i_{R2} \end{bmatrix}^T \tag{10.47}$$

$$V_C = \begin{bmatrix} V_{C1} & V_{C2} & V_{C3} \end{bmatrix}^T \tag{10.48}$$

The matrix as per Ohm's law is given as:

$$I_R = R^{-1} V_C \tag{10.49}$$

Further it can be written as:

$$i_{Rk} = \sum_{j=1}^{3} R_{kj}^{-1} V_{Cj} \tag{10.50}$$

Therefore

$$i_{Ck} = i_{Lk} - \sum_{j=1}^{3} R_{kj}^{-1} V_{Cj}; \quad k = 1, 2, 3 \tag{10.51}$$

By definition,

$$i_{Ck} = C_k \frac{dV_{Ck}}{dt} \tag{10.52}$$

Thus in matrix form, it can be written as:

$$
\begin{bmatrix} C_3 & 0 & 0 \\ 0 & C_2 & 0 \\ 0 & 0 & C_3 \end{bmatrix} \begin{bmatrix} \dot{V}_{C1} \\ \dot{V}_{C2} \\ \dot{V}_{C3} \end{bmatrix}
$$

$$
= \begin{bmatrix} 1 & 0 & 0 & \dfrac{-(R_2+R_3)}{R_1R_2+R_2R_3+R_3R_1} & \dfrac{R_3}{R_1R_2+R_2R_3+R_3R_1} & \dfrac{R_2}{R_1R_2+R_2R_3+R_3R_1} \\ 0 & 1 & 0 & \dfrac{R_3}{R_1R_2+R_2R_3+R_3R_1} & \dfrac{-(R_1+R_3)}{R_1R_2+R_2R_3+R_3R_1} & \dfrac{R_1}{R_1R_2+R_2R_3+R_3R_1} \\ 0 & 0 & 1 & \dfrac{R_2}{R_1R_2+R_2R_3+R_3R_1} & \dfrac{R_1}{R_1R_2+R_2R_3+R_3R_1} & \dfrac{-(R_1+R_2)}{R_1R_2+R_2R_3+R_3R_1} \end{bmatrix} \begin{bmatrix} i_{L1} \\ i_{L2} \\ i_{L3} \\ V_{C1} \\ V_{C3} \\ V_{C3} \end{bmatrix} \quad (10.53)
$$

That is in the form of: $C\frac{dV_C}{dt} = \begin{bmatrix} I : -R^{-1} \end{bmatrix} \begin{bmatrix} I_L \\ \cdots \\ V_C \end{bmatrix}$

It is to be noted that this block will appear in every equivalent circuit. Here, I_L is the vector with all the average inductor currents and V_C is the vector with all the average capacitor voltages.

Now, for inductor voltages, a switching function $\sigma_k(t)$ is defined, which is 1 in the terms where V_{in} appears and 0 where it is absent. From Eq. (10.42), the equation can be re-written as:

$$
V_{Lk} = -V_{Ck} + \sigma_k(t); \quad k = 1, 2, 3 \tag{10.54}
$$

Again by definition,

$$
V_{Lk} = L_k \frac{di_{Lk}}{dt} \tag{10.55}
$$

By combining Eqs. (10.54) and (10.55) and writing it in the matrix form, Eq. (10.56) comes out to be as:

$$
\begin{bmatrix} L_1 & 0 & 0 \\ 0 & L_2 & 0 \\ 0 & 0 & L_3 \end{bmatrix} \begin{bmatrix} \dfrac{di_{L1}}{dt} \\ \dfrac{di_{L2}}{dt} \\ \dfrac{di_{L3}}{dt} \end{bmatrix} = \begin{bmatrix} 0 & 0 & 0 & -1 & 0 & 0 \\ 0 & 0 & 0 & 0 & -1 & 0 \\ 0 & 0 & 0 & 0 & 0 & -1 \end{bmatrix} \begin{bmatrix} i_{L1} \\ i_{L2} \\ i_{L3} \\ V_{C1} \\ V_{C3} \\ V_{C3} \end{bmatrix} + V_{in} \begin{bmatrix} \sigma_1(t) \\ \sigma_2(t) \\ \sigma_3(t) \end{bmatrix} \tag{10.56}
$$

Which is in the form: $L\frac{di_L}{dt} = \begin{bmatrix} O & \vdots & -I \end{bmatrix} \begin{bmatrix} I_L \\ \cdots \\ V_C \end{bmatrix} + V_{in} \cdot \sigma_k(t)$.

Here, O is defined as the null matrix. On appending Eqs. (10.53) and (10.56), the linearized differential equation is obtained as:

$$
\begin{bmatrix} L & \vdots & O \\ \cdots & \vdots & \cdots \\ O & \vdots & C \end{bmatrix} \frac{d}{dt} \begin{bmatrix} I_L \\ \cdots \\ V_C \end{bmatrix} = \begin{bmatrix} O & \vdots & -I \\ \cdots & \vdots & \cdots \\ I & \vdots & -R^{-1} \end{bmatrix} \begin{bmatrix} I_L \\ \cdots \\ V_C \end{bmatrix} + \begin{bmatrix} \sigma \\ \cdots \\ O \end{bmatrix} V_{in} \tag{10.57}
$$

Let $\delta_k = \sum_{k=1}^{k} D_k$, and then, σ is formally defined as:

$$
\sigma_1(t) = \begin{cases} 1 & t \in [nT, (n+\delta_3)T) \\ 0 & t \in [(n+\delta_3)T, (n+1)T) \end{cases}
$$

$$
\sigma_2(t) = \begin{cases} 1 & t \in [(n+\delta_2)T, (n+\delta_5)T) \\ 0 & t \in [nT, (n+\delta_2)T)U[(n+\delta_5)T, (n+1)T) \end{cases} \quad , n \in N \tag{10.58}
$$

$$
\sigma_3(t) = \begin{cases} 1 & t \in [nT, (n+\delta_1)T)U[(n+\delta_4)T, (n+1)T) \\ 0 & t \in [(n+\delta_2)T, (n+\delta_4)T) \end{cases}
$$

Due to the fact that the linear block differential equation only varies with time in the $\sigma_k(t)$ part; therefore only that part will be averaged. Hence, the averaged value of σ (1 for ON and 0 for OFF) is given as:

$$\langle \sigma \rangle = \frac{1}{T} \left[\int_0^{\delta_3 T} dt \begin{bmatrix} 1 \\ 0 \\ 0 \end{bmatrix} \right] + \left[\int_{\delta_2 T}^{\delta_5 T} dt \begin{bmatrix} 0 \\ 1 \\ 0 \end{bmatrix} + \left(\int_0^{\delta_1 T} dt + \int_{\delta_4 T}^{\delta_6 T} dt \right) \begin{bmatrix} 0 \\ 0 \\ 1 \end{bmatrix} \right] \tag{10.59}$$

Referring to the switching diagram displayed in Fig. 10.4, this yields as:

$$\sigma = (D_1 + D_2 + D_3) \begin{bmatrix} 1 \\ 0 \\ 0 \end{bmatrix} + (D_3 + D_4 + D_5) \begin{bmatrix} 0 \\ 1 \\ 0 \end{bmatrix} + (D_5 + D_6 + D_1) \begin{bmatrix} 0 \\ 0 \\ 1 \end{bmatrix} = \begin{bmatrix} U_1 \\ U_2 \\ U_3 \end{bmatrix} \tag{10.60}$$

where

$$\begin{bmatrix} U_1 \\ U_2 \\ U_3 \end{bmatrix} = \begin{bmatrix} (D_1 + D_2 + D_3) \\ (D_3 + D_4 + D_5) \\ (D_5 + D_6 + D_1) \end{bmatrix} = U \tag{10.61}$$

Therefore the differential equation assumes the form:

$$\frac{d}{dt} \begin{bmatrix} I_L \\ \cdots \\ V_C \end{bmatrix} = \begin{bmatrix} O & \vdots & -L^{-1} \\ \cdots\cdots & \vdots & \cdots\cdots \\ -C^{-1} & \vdots & -C^{-1}R^{-1} \end{bmatrix} \begin{bmatrix} I_L \\ \cdots \\ V_C \end{bmatrix} + \begin{bmatrix} \sigma \\ \cdots \\ 0 \end{bmatrix} V_{in} \tag{10.62}$$

The output voltages V_{O1}, V_{O2}, and V_{O3} can be obtained by multiplying the ground node equation (i.e., Eq. 10.38) by $R_1R_2R_3$. The linear system can be solved as:

$$\begin{bmatrix} 1 & -1 & 0 \\ 0 & 1 & -1 \\ R_2R_3 & R_1R_3 & R_1R_2 \end{bmatrix} \begin{bmatrix} V_{O1} \\ V_{O2} \\ V_{O3} \end{bmatrix} = \begin{bmatrix} V_{C1} - V_{C2} \\ V_{C2} - V_{C3} \\ V_{C1} - V_{C3} \end{bmatrix} \tag{10.63}$$

which yields

$$\begin{bmatrix} V_{O1} \\ V_{O2} \\ V_{O3} \end{bmatrix} = \frac{1}{R_1R_2 + R_2R_3 + R_3R_1} \left(V_{C1} \begin{bmatrix} R_1R_2 + R_3R_1 \\ -R_2R_3 \\ -R_2R_3 \end{bmatrix} + V_{C2} \begin{bmatrix} -R_1R_3 \\ R_1R_2 + R_2R_3 \\ -R_1R_3 \end{bmatrix} + V_{C3} \begin{bmatrix} -R_1R_2 \\ -R_1R_2 \\ R_1R_3 + R_2R_3 \end{bmatrix} \right) \tag{10.64}$$

This assumes the form:

$$V_O = R_{diag} R^{-1} V_C$$

$$\text{where} \qquad R_{diag} = \begin{bmatrix} R_1 & 0 & 0 \\ 0 & R_2 & 0 \\ 0 & 0 & R_3 \end{bmatrix} \quad \text{and} \quad V_O = \begin{bmatrix} V_{O1} \\ V_{O2} \\ V_{O3} \end{bmatrix} \tag{10.65}$$

The final output equation can be written as:

$$V_O = \begin{bmatrix} O & \vdots & R_{diag} R^{-1} \end{bmatrix} \begin{bmatrix} I_L \\ \cdots \\ V_C \end{bmatrix} \tag{10.66}$$

The value of R^{-1} can be found from Eq. (10.46).

On taking the values of L_1, L_2, $L_3 = L$ and C_1, C_2, $C_3 = C$ for a balanced load $(R_1 = R_2 = R_3 = R)$, from Eq. (10.62):

$$A = \begin{bmatrix} 0 & 0 & 0 & -\dfrac{1}{L} & 0 & 0 \\[2ex] 0 & 0 & 0 & 0 & -\dfrac{1}{L} & 0 \\[2ex] 0 & 0 & 0 & 0 & 0 & -\dfrac{1}{L} \\[2ex] \dfrac{1}{C} & 0 & 0 & -\dfrac{2}{3CR} & \dfrac{1}{3CR} & \dfrac{1}{3CR} \\[2ex] 0 & \dfrac{1}{C} & 0 & \dfrac{1}{3CR} & -\dfrac{2}{3CR} & \dfrac{1}{3CR} \\[2ex] 0 & 0 & \dfrac{1}{C} & \dfrac{1}{3CR} & \dfrac{1}{3CR} & \dfrac{2}{3CR} \end{bmatrix} \tag{10.67}$$

$$B = \begin{bmatrix} \dfrac{1}{L} & 0 & 0 \\[2ex] 0 & \dfrac{1}{L} & 0 \\[2ex] 0 & 0 & \dfrac{1}{L} \\[2ex] 0 & 0 & 0 \\[1ex] 0 & 0 & 0 \\[1ex] 0 & 0 & 0 \end{bmatrix} \tag{10.68}$$

The coefficient matrix C_{out} in the output equation can be obtained as:

$$C_{out} = \begin{bmatrix} 0 & 0 & 0 & \dfrac{2}{3} & -\dfrac{1}{3} & -\dfrac{1}{3} \\[2ex] 0 & 0 & 0 & -\dfrac{1}{3} & \dfrac{2}{3} & -\dfrac{1}{3} \\[2ex] 0 & 0 & 0 & -\dfrac{1}{3} & -\dfrac{1}{3} & \dfrac{2}{3} \end{bmatrix} \tag{10.69}$$

The transfer functions based on the above derivations are explained in the next subsection.

10.2.4   Transfer function modeling

From the above-obtained equations, it is to be noted that during averaging, the input in the model becomes the duty cycle, and V_{in} is just a parameter. The transfer function considering V_o with respect to U is explicated in Table 10.2.

10.2.5   Selection of inductor and capacitor values

The selection of L and C is tuned in such a way that the switching frequency is proximately cognate to the resonant frequency (Khoshkbar-Sadigh, Dargahi, Lakhera, & Corzine, 2019; Tali, Obbadi, Elfajri, & Errami, 2014).

All the inductor values per leg are identical and similarly all the capacitances per leg are the same.

$$f = \frac{1}{2\pi\left[\sqrt{(LC)}\right]} \tag{10.70}$$

The value of inductance and capacitance is given in Table 10.3.

TABLE 10.2 Transfer functions.

Transfer function	Relation	Expression
$G_{11}(S)$	$\frac{V_{O1}(S)}{U_1(S)}$	$\frac{2R(3C^2L^2R^2S^4 - CL^2RS^3 + 6CLR^2S^2 - 2L^2S^2 - LRS + 3R^2)V_{in}}{(9C^2L^2R^2S^4 - 3CL^2RS^3 + 18CLR^2S^2 - 4L^2S^2 - 3LRS + 9R^2)(CLRS^2 + LS + R)}$
$G_{12}(S)$	$\frac{V_{O1}(S)}{U_2(S)}$	$-\frac{R^2(3C^2L^2RS^4 - CL^2S^3 + 6CLRS^2 - LS + 3R)V_{in}}{(9C^2L^2R^2S^4 - 3CL^2RS^3 + 18CLR^2S^2 - 4L^2S^2 - 3LRS + 9R^2)(CLRS^2 + LS + R)}$
$G_{13}(S)$	$\frac{V_{O1}(S)}{U_3(S)}$	$-\frac{3R^2(CLS^2 + 1)V_{in}}{9C^2L^2R^2S^4 - 3CL^2RS^3 + 18CLR^2S^2 - 4L^2S^2 - 3LRS + 9R^2}$
$G_{21}(S)$	$\frac{V_{O2}(S)}{U_1(S)}$	$\frac{R^2(3C^2L^2RS^4 - CL^2S^3 + 6CLRS^2 - LS - LRS + 3R)V_{in}}{(9C^2L^2R^2S^4 - 3CL^2RS^3 + 18CLR^2S^2 - 4L^2S^2 - 3LRS + 9R^2)(CLRS^2 + LS + R)}$
$G_{22}(S)$	$\frac{V_{O2}(S)}{U_2(S)}$	$\frac{2R(3C^2L^2R^2S^4 - CL^2RS^3 + 6CLR^2S^2 - 2L^2S^2 - LRS + 3R^2)V_{in}}{(9C^2L^2R^2S^4 - 3CL^2RS^3 + 18CLR^2S^2 - 4L^2S^2 - 3LRS + 9R^2)(CLRS^2 + LS + R)}$
$G_{23}(S)$	$\frac{V_{O2}(S)}{U_3(S)}$	$-\frac{3R^2(CLS^2 + 1)V_{in}}{9C^2L^2R^2S^4 - 3CL^2RS^3 + 18CLR^2S^2 - 4L^2S^2 - 3LRS + 9R^2}$
$G_{31}(S)$	$\frac{V_{O3}(S)}{U_1(S)}$	$\frac{V_{in}(3CLRS^2 - 4LS + 3R)R}{9C^2L^2R^2S^4 - 3CL^2RS^3 + 18CLR^2S^2 - 4L^2S^2 - 3LRS + 9R^2}$
$G_{32}(S)$	$\frac{V_{O3}(S)}{U_2(S)}$	$\frac{V_{in}(3CLRS^2 - 4LS + 3R)R}{9C^2L^2R^2S^4 - 3CL^2RS^3 + 18CLR^2S^2 - 4L^2S^2 - 3LRS + 9R^2}$
$G_{33}(S)$	$\frac{V_{O3}(S)}{U_3(S)}$	$\frac{6R^2(CLS^2 + 1)V_{in}}{9C^2L^2R^2S^4 - 3CL^2RS^3 + 18CLR^2S^2 - 4L^2S^2 - 3LRS + 9R^2}$

TABLE 10.3 Inductor and capacitor values.

Component	Value
Inductor	25e − 3 H
Capacitance	300e − 6 F

10.3 System description

Fig. 10.11 shows the block diagram of a three-phase boost inverter interfaced stand-alone system. A DC source feeds the boost inverter connected to a three-phase load. A closed-loop control has been incorporated so as to regulate the system at a given setpoint.

In this case, a resistive load has been considered. As discussed earlier, the switching frequency is based on the fundamental frequency, which is 50 Hz. The load is of 5 kW. The closed-loop control has been discussed in Section 10.3.1.

10.3.1 Closed-loop control

Closed-loop control or feedback control alleviates the deficiencies of open-loop control, where the actual response of the system is compared continuously with that of the desired response. The control output to the process is adjusted and adapted to mitigate the deviations.

The proposed control strategy for the control of the boost inverter prototype is based on voltage control. The voltage on the output side is controlled by varying the duty cycle of the pulses fed to the inverter. Directly, a voltage control loop can provide a faster response. Fig. 10.12 shows the block diagram of the considered control strategy. The error signal is fed to the PI controller, which mitigates any steady-state error (Adhikari & Li, 2014). A constant, 1 is added such that if the DC source is a solar panel, the reference would be the replica of the input during low irradiance when compensated using a battery bank.

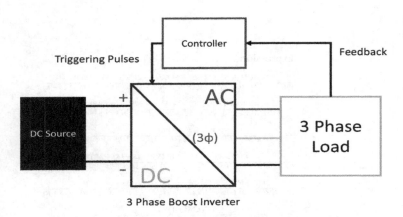

FIGURE 10.11 Schematic diagram of the system under study.

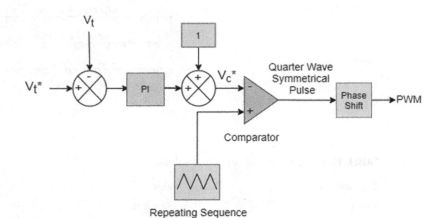

FIGURE 10.12 Block diagram of the closed-loop control strategy.

$$V_c^* = \left[1 + K_p\left(V_t^*(t) - V_t(t)\right) + K_I \int_0^t \left(\left(V_t^*(t) - V_t(t)\right)dt\right) \right] \qquad (10.71)$$

V_c^* is compared with a repeating sequence whose fundamental frequency is 50 Hz, and quarter-wave symmetric pulses are generated, which triggers the gates of the IGBTs.

10.4   Results and discussions

Fig. 10.13 shows the output waveform in the open-loop control mode of operation at a duty cycle of 0.5. The input voltage is taken to be 150 V. The peak AC voltage is seen to be around 600 V that is nearly four times that of the input DC voltage. It is also seen that the voltage waveform is purely sinusoidal.

The output current is shown in Fig. 10.13C, which assumes a peak value of 3.5 A. It is also seen that the output current is seen to be purely sinusoidal.

The THD plot for voltage and current is shown in Fig. 10.14. The voltage THD is 0.8% while the current THD is 2.9%, which is within the accepted range as per the guidelines of IEEE Standard 519-2014.

Fig. 10.15 shows the closed-loop output voltage waveform. The peak value of the output voltage waveform is seen to be four times the input voltage (for the same value of DC voltage as taken in the previous case).

Fig. 10.16 shows the closed-loop output current waveform. Slight variations could be seen in the amplitude. Ths is due to the fact that the voltage controller tries to maintain the output voltage constand by making a trade-off with current and power.

The THD plots obtained for the output voltage and current are shown in Fig. 10.17. Here the voltage THD is slightly reduced and attains the value of 0.79%, and the current THD is 2.28%. The dominant harmonics are mostly the 3rd, 5th, and 13th components. However, their effect on the sinusoidal waveform is quite minuscule.

Fig. 10.18 shows the bode plots obtained for each transfer functions (as discussed in Section 2.4).

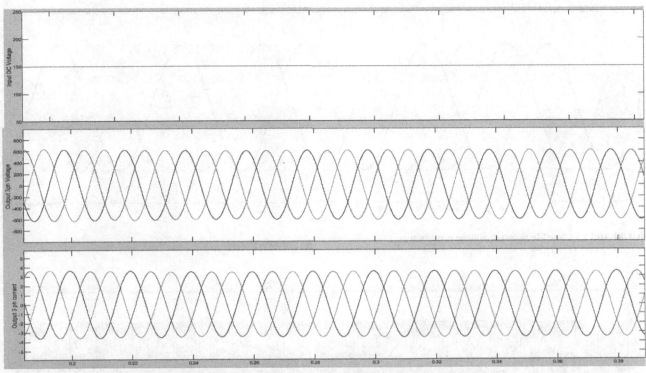

FIGURE 10.13   Open-loop waveforms: input voltage (*top*), open-loop output voltage waveform (*middle*), and output current (*bottom*).

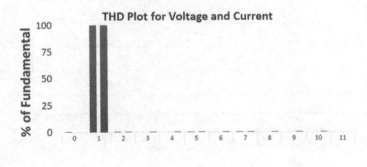

FIGURE 10.14   Open-loop voltage and current THD plots.

According to the first transfer function mentioned above, it can be inferred that the value of gain margin is −19.0 dB, which is considered as the positive gain margin when phase cross over frequency is at −180 degrees and the value of phase margin is −1 degree when gain cross over frequency is at 0-dB line. Hence, it can be concluded that the transfer function is stable condition.

In the second transfer function and the third transfer function, it can be deduced that the value of gain margin of both of the transfer function is infinity as phase cross over frequency is at −180 degrees and the value of phase margin is −180 degrees when gain cross over frequency is 0 dB. Therefore the transfer function derived is unstable condition.

From the fouth transfer function, it can be deducted that the value of gain margin is +40 dB when phase cross over frequency is at the position of −180 degrees and the value of phase margin is 0 degrees when gain cross over frequency is 0 dB. As a result, it can be concluded that the transfer function is conditionally stable condition.

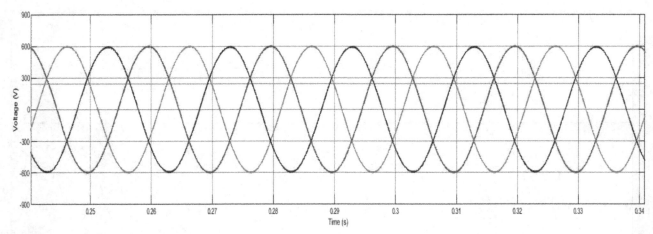

FIGURE 10.15 Closed-loop output voltage waveform.

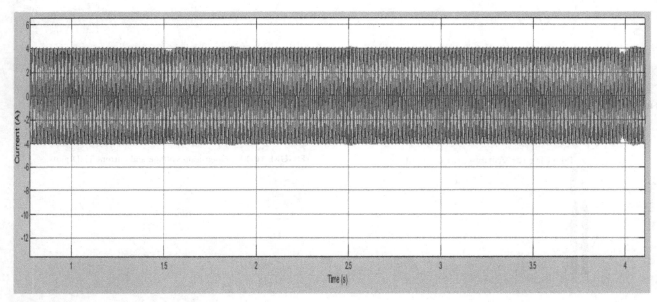

FIGURE 10.16 Closed-loop output current waveform.

FIGURE 10.17 Closed-loop voltage and current THD plot.

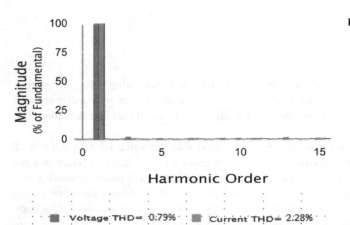

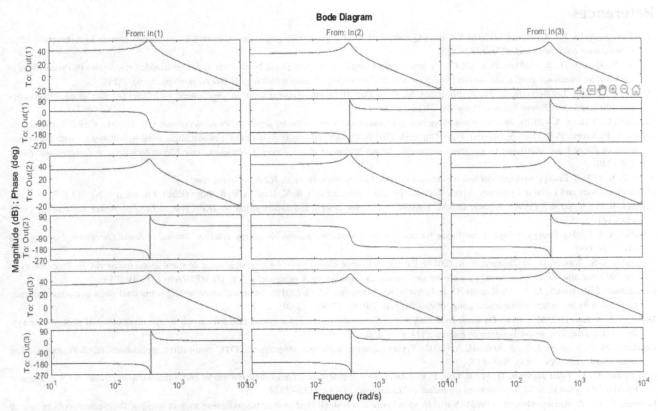

FIGURE 10.18   Bode plots for the obtained transfer function.

Considering the fifth transfer function, it can be deducted that the calculated value of gain margin having -19.1 dB when phase cross over frequency positioned at -180 degrees and the calculated value of phase margin which is equal to -180 degrees when gain cross over frequency is positioned at 0 dB. Therefore transfer function is stable condition.

From the sixth transfer function, the seventh transfer function, and the eighth transfer function, it can be inferred that the calculated value of these transfer functions of gain margin is $+30$ dB when phase cross over frequency just crossing at -180 degrees and the calculated-phase margin of these transfer functions equals to -2 degrees when gain cross over frequency positioned at 0 dB, which have resulted in stable condition.

While in the ninth transfer function, it can be inferred that that the calculated value of gain margin having -19 dB when phase cross over frequency is at -180 degrees and the calculated value calculated -2 degree phase margin where gain cross over frequency crosses at -180 dB. Therefore the obtained transfer function is stable at large.

10.5   Conclusion

In this chapter, a new topology for a three-phase boost inverter is implemented. The inverter produces a three-phase AC voltage with magnitude greater than the input DC voltage. The new converter also has a smaller number of elements in compare with other approaches. The working principle of the converter is based on the resonant frequency. The quadruple gain is obtained by suitable tuning of the inductor and capacitor. The dual stage amplification of the voltage can be replaced by the single stage as with the increase of power electronics stages, the THD increases, but in this work the high efficiency resonant converter is implemented in which an appreciably low harmonic distortion has been observed. Thus, this approach provides a ground breaking solution to reduce the harmonic content from the system. The MATLAB results show the validation of the converter. The future prospects of this work could extend to the use of genetic algorithms for frequency control, testing this prototype on dynamic loads and integration with the microgrid with the association of FACTS devices.

References

Adhikari, S., & Li, F. (2014). Coordinated V-f and PQ control of solar photovoltaic generators with MPPT and battery storage in microgrids. *IEEE Transactions on Smart Grid, 5*(3), 1270−1281.

Arvind, P., Kumar, D., & Thakura, P. R. (2019). Assessement of single stage three phase boost inverter for an islanded microgrid. In *Proceedings of 2019 3rd international conference on electronics, materials engineering & nano-technology (IEMENTech)* (pp. 1−6). IEEE.

Barkati, S., Baghli, L., Berkouk, E. M., & Boucherit, M. S. (2008). Harmonic elimination in diode-clamped multilevel inverter using evolutionary algorithms. *Electric Power Systems Research, 78*(10), 1736−1746.

Batarseh, I., & Harb, A. (2018). *Review of switching concepts and power semiconductor devices. Power electronics* (pp. 25−91). Cham: Springer.

Bihari, S. P., Sadhu, P. K., Das, S., Arvind, P., & Gupta, A. (2019). Design and implementation of a photovoltaic wind hybrid system with the assessment of fuzzy logic maximum power point technique. *Revue Roumaine des Sciences Techniques-Serie Electrotechnique et Energetique, 64*(3), 235−240.

Bose, B. K. (2007). Energy environment and importance of power electronics. In *Proc. IEEE powering conf.*

Power electronics and variable frequency drives: Technology and applications. In B. K. Bose, & B. K. Bose (Eds.), *Piscataway*. NJ: IEEE Press.

Caceres, R. O., & Barbi, I. (1999). A boost DC-AC converter: Analysis, design, and experimentation. *IEEE Transactions on Power Electronics, 14*(1), 134−141.

Cavallo, A. J. (2001). Energy storage technologies for utility scale intermittent renewable energy systems. *Journal of Solar Energy Engineering, 123*(4), 387−389.

Chakraborty, S., Kumar, D., & Thakura, P. R. (2019). PV-fed dual-stage closed loop V-f control scheme of stand-alone microgrid. In *Proceedings of the 2019 3rd international conference on electronics, materials engineering & nano-technology (IEMENTech)* (. 1−4). IEEE.

Christensen, J. B., Reddy, D. V., McKinstrie, C. J., Rottwitt, K., & Raymer, M. G. (2015). Temporal mode sorting using dual-stage quantum frequency conversion by asymmetric Bragg scattering. *Optics Express, 23*(18), 23287−23301.

Dash, P. P., & Kazerani, M. (2011). Dynamic modeling and performance analysis of a grid-connected current-source inverter-based photovoltaic system. *IEEE Transactions on Sustainable Energy, 2*(4), 443−450.

Escalante, M. F., Vannier, J. C., & Arzandé, A. (2002). Flying capacitor multilevel inverters and DTC motor drive applications. *IEEE Transactions on Industrial Electronics, 49*(4), 809−815.

Gajanayake, C. J., Vilathgamuwa, D. M., & Loh, P. C. (2007). Development of a comprehensive model and a multiloop controller for Z-source inverter DG systems. *IEEE Transactions on Industrial Electronics, 54*(4), 2352−2359.

Holmes, D. G., & Kotsopoulos, A. (1993). Variable speed control of single and two-phase induction motors using a three-phase voltage source inverter. In *Proceedings of the conference record of the 1993 IEEE industry applications conference twenty-eighth IAS annual meeting, October 2* (pp. 613−620). IEEE.

Khoshkbar-Sadigh, A., Dargahi, V., Lakhera, K., & Corzine, K. (2019). Analytical design of LC filter inductance for two-level inverters based on maximum ripple current. In *Proceedings of the IECON 2019—45th* annual conference of *the IEEE Industrial Electronics Society* (Vol. 1, pp. 1621−1626). IEEE.

Koushki, B., & Ghaisari, J. (2009). A voltage reference design for three-phase boost inverter. In *Proceedings of the IEEE EUROCON 2009* (pp. 650−654). IEEE.

Koushki, B., Khalilinia, H., Ghaisari, J., & Nejad, M. S. (2008). A new three-phase boost inverter: Topology and controller. In *Proceedings of the 2008 Canadian* conference on electrical and computer engineering (pp. 757−760). IEEE.

Kumar Barwar, M., Tripathi, P. R., & Thakura, P. R. (2018). Reduced THD model of single stage three-phase boost inverter. In *Proceedings of the 2018 4th international conference on electrical energy systems* (ICEES) (pp. 153−156). IEEE.

Langella, R., Testa, A., & Alii, E. (2014). IEEE recommended practice and requirements for harmonic control in electric power systems. *IEEE.*

Liserre, M., Teodorescu, R., & Blaabjerg, F. (2006). Stability of photovoltaic and wind turbine grid-connected inverters for a large set of grid impedance values. *IEEE Transactions on Power Electronics, 21*(1), 263−272.

Nguyen, M. K., Le, T. V., Park, S. J., & Lim, Y. C. (2014). A class of quasi-switched boost inverters. *IEEE Transactions on Industrial Electronics, 62*(3), 1526−1536.

Ojha, P. K., Barwar, M. K., Tripathi, P. R., & Thakura, P. R. (2018). Hardware implementation of three-phase single stage boost inverter for higher frequency. In *Proceedings of the 2018 second international conference on electronics, communication and aerospace technology (ICECA)* (pp. 1934−1938). IEEE.

Peng, F. Z., Shen, M., & Qian, Z. (2005). Maximum boost control of the Z-source inverter. *IEEE Transactions on Power Electronics, 20*(4), 833−838.

Pina, A., Silva, C., & Ferrão, P. (2012). The impact of demand side management strategies in the penetration of renewable electricity. *Energy., 41*(1), 128−137.

Rajeev, M., & Agarwal, V. (2018). Single phase current source inverter with multiloop control for transformer-less grid−PV interface. *IEEE Transactions on Industry Applications, 54*(3), 2416−2424.

Raveendhra, D., & Pathak, M. K. (2018). Three-phase capacitor clamped boost inverter. *IEEE Journal of Emerging and Selected Topics in Power Electronics, 7*(3), 1999−2011.

Ravindranath, A., Mishra, S. K., & Joshi, A. (2012). Analysis and PWM control of switched boost inverter. *IEEE Transactions on Industrial Electronics, 60*(12), 5593−5602.

Sanchis, P., Ursǽa, A., Gubía, E., & Marroyo, L. (2005). Boost DC-AC inverter: A new control strategy. *IEEE Transactions on Power Electronics*, *20* (2), 343−353.

Singaravel, M. R., & Daniel, S. A. (2015). MPPT with single DC−DC converter and inverter for grid-connected hybrid wind-driven PMSG−PV system. *IEEE Transactions on Industrial Electronics*, *62*(8), 4849−4857.

Tali, M., Obbadi, A., Elfajri, A., & Errami, Y. (2014). Passive filter for harmonics mitigation in standalone PV system for non-linear load. In *Proceedings of the 2014 international renewable and sustainable energy conference (IRSEC)* (pp. 499−504). IEEE.

Tripathi, P. R., Keshri, R. K., & Thakura, P. R. (2019). Low total harmonic distortion quadruple gain boost inverter. In *Proceedings of the IECON 2019—45th* annual conference *of the IEEE Industrial Electronics Society* (Vol. 1, pp. 1820−1825). IEEE.

Twining, E., & Holmes, D. G. (2003). Grid current regulation of a three-phase voltage source inverter with an LCL input filter. *IEEE Transactions on Power Electronics*, *18*(3), 888−895.

Weisser, D. (2004). Power sector reform in small island developing states: What role for renewable energy technologies? *Renewable and Sustainable Energy Reviews*, *8*(2), 101−127.

Chapter 11

Advanced control of PMSG-based wind energy conversion system applying linear matrix inequality approach

Chakib Chatri and Mohammed Ouassaid

Engineering for Smart and Sustainable Systems Research Center, Mohammadia School of Engineers, Mohammed V University in Rabat, Rabat, Morocco

11.1   Introduction

11.1.1   Context and problematic

Over the last decade, with the population and industrial growth, energy demands have increased significantly (Golnary & Moradi, 2019). Declining fossil fuel sources and concerns about pollution levels are the main motivations for electricity production from renewable energy sources, namely, biomass, solar, and wind (Chen, Xu, Gu, Schmidt, & Li, 2018; Naidu & Singh, 2017; Zheng, Feng, Han, & Yu, 2019). Among the renewable energy sources, wind energy has expanded rapidly and becomes the most vying form of renewable energies due to its inexhaustibility, cost-effectiveness, and eco-friendliness (Chan, Bai, & He, 2018; Nasiri & Mohammadi, 2017).

Three types of wind turbine (WT) are used: fixed speed WT, semivariable speed WT, and variable speed WT. Out of these three types of WT, the variable speed wind turbine (VS WT) is more advantageous thanks to its capability to improve power quality, reduce mechanical stress created by wind fluctuations, and its aptitude to attain maximum power point tracking (MPPT) of the controller to squeeze high power out at various wind rates (Li et al., 2019). There are many types of generators used for variable speed of wind energy conversion system (VS WECS), for instance, the doubly fed induction generator (El Karkri, Rey-Boué, El Moussaoui, Stöckl, & Strasser, 2019), the cage asynchronous generator (Elyaalaoui, Ouassaid, & Cherkaoui, 2019), and permanent magnet synchronous generator (PMSG) (Errami, Ouassaid, & Maaroufi, 2015a, 2015b). The PMSG is more favorable due to its self-excitation capability leading to increased operational efficiency, higher power factor, advanced precision, better grid compatibility, greater power density and lack of gearbox so it can operate at slow speed and reduce maintenance requirements (Errami et al., 2015a, 2015b; Jlassi & Cardoso, 2019; Jlassi, Estima, El Khil, Bellaaj, & Cardoso, 2014). Moreover, the performance of permanent magnet materials is improving and the cost is decreasing (Petrov & Pyrhonen, 2012). Hence, these advantages make the PMSG the most attractive machine in the application of small, medium, and high size in a wind energy conversion system (WECS). The VS WECS control remains remarkably complex due to many reasons, notably the nonlinearity, multivariability, and high coupling of the generators, disturbances of the external environment elements, and the impact of parametric variation. On the other hand, the quality and the amount of energy captured by a WECS is relying both on the characteristics of the wind regime at the site and regulatory approach used for the WECS. Besides, the conversion chain integrates power electronic devices, which increase the nonlinearity of the system. Diverse power conversion topologies are adopted in research papers for WECS based on PMSG (Tripathi, Tiwari, & Singh, 2015; Zhang, Zhao, Qiao, & Qu, 2015). Fig. 11.1 illustrates the general design of the WECS. It is composed of a PMSG connected by two variable frequency converters to the power grid. The created active and reactive powers of PMSG are adjusted with a pulse width modulation (PWM) rectifier in the generator-side, while the PWM inverter in grid side is used to control the exchanges of active and reactive powers, the regulation of the DC-link voltage, as well as that of the power factor.

Renewable Energy Systems. DOI: https://doi.org/10.1016/B978-0-12-820004-9.00002-4

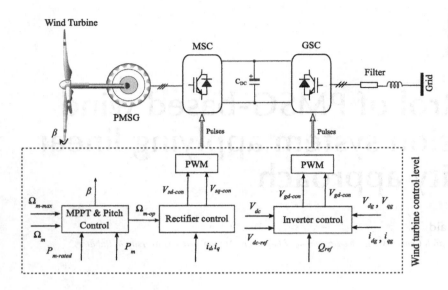

FIGURE 11.1 The general design of a wind energy conversion system based on permanent magnet synchronous generator.

To handle the different closed loop in the WECS, several control strategies have been developed recently. These strategies include conventional basic proportional–integral (PI) control, fuzzy logic method, and neural network approach. First, the conventional basic PI control method is easy to use and provides low cost, as it requires low maintenance. But its performance is limited by the lack of robustness due to the variations of the generator's internal parameters. Second, the fuzzy logic method has shown greater efficiency with robustness as long as it does not need knowledge about the system parameters. However, with huge real wind speed, this system shows difficulties in practice. Third, the neural network control acquires fast response. Nevertheless, all these strategies might encounter tracking and stability problems. The sliding mode controller has been introduced to solve the instability issues (Errami, Ouassaid, & Maaroufi, 2015a, 2015b; Zheng et al., 2019), but the chattering phenomenon is produced while the state trajectory achieves the sliding mode surface. Besides, this control fails by the difficulty of the implementation.

11.1.2   Contribution

To overcome all these disadvantages, the present chapter designs a new Takagi-Sugeno (T-S) fuzzy approach. The concept of this approach is to replace the nonlinear system model with linear subsystems. Indeed, the Takagi-Sugeno fuzzy control (T-SFC) strategy for the WECS has been the topic of previous research. These proposed studies highlight the benefit of the linear matrix inequality (LMI) gain design, but several permit values involve a complex coordinate transformation that requires stability with h infinity.The following points list the contributions of this chapter:

- The T-SFC conception by means of the modeling of PMSG-based WECS.
- Desired reference model (DRM) and nonlinear tracking regulator (N-LTR) are designed regarding an optimized speed.
- A new controller is constructed by using The parallel distributed compensation (PDC) strategy.
- The PDC controllers' gains are compiled by means of the Lyapunov s stability theory. The obtained LMIs guarantee the sufficient conditions.

11.1.3   Chapter organization

The remainder of this chapter is structured as follows. In Section 11.2, the literature review revolves around the recent research on the control developed for WECS. The model of the WECS is detailed in Section 11.3. Section 11.4 describes the proposed control strategy of the WECS. The simulation results showing the performance of the proposed T-S controller are reviewed in Section 11.5. Finally, Section 11.6 concludes the chapter.

11.2   Recent research on control in wind energy conversion systems

Linear conventional controllers have been largely used in the engineering for their simple and reliable design. During the last decades, numerous robust and nonlinear control techniques have been emerged. However, research is still limited and uncompleted in terms of results that prove the effectiveness of the new advanced controls to overcome the failures of conventional controls. In pursuit of their goal, most researchers are still investing their efforts to develop new complex control algorithms taking into account the difficulties related to the machine model nonlinearities. Hereafter, a brief review on the research work recently performed for the control of the WECs. These control methods are summarized as follows:

Conventional basic PI control: Chinchilla, Arnaltes, and Burgos (2006) represent the performance of a direct-driven PMSG-based WECS. With this system, it is possible to track the optimum power using generated power as input. To minimize power losses, the optimal current is calculated and imposed on the generator. The proposed controller is implemented using dSPACE.

In Matayoshi, Howlader, Datta, and Senjyu (2018), a comparative analysis is made for the conventional control method with the proposed control strategy method; both pitch angle and rotational speed control methods are conceived in the strategy method under extreme wind conditions. Consequently, the generator power is regulated with high exactness. Then, the algorithm reduces the mechanical torque as a way to typically decrease the centrifugal force and the power coefficient throughout the heavy wind loads.

Fuzzy logic in Mansour, Mansouri, Bendoukha, and Mimouni (2020): two controllers known as PI and fuzzy are proposed and examined for regulating the current from the PMSG. The fuzzy controller is employed to improve the tracking performance. The fuzzy logic method has shown greater efficiency with robustness as long as it does not need information about the system parameters. Then, a so-called flywheel energy storage system is connected to the DC bus to regulate the flow of power to the network and regulate the voltage of the DC bus. However, with huge real wind speed, this system shows difficulties in practice. This strategy handed the duty cycle of the control to attain the maximum power point. This method does not involve the actual measurement of the wind and information about the characteristics of the turbine, which minimize the cost of implementation of this controller. The proposed approach is implemented in real time using OPAL-RT 4510.

In Soliman, Hasanien, Azazi, El-Kholy, and Mahmoud (2018), a cascaded adaptive fuzzy logic control strategy was applied to regulate the generator side converter. The proposed adaptive controller determined by on-line updates the scaling factors of the FLC. This method is compared with a PT controller optimized by particle swarm optimization algorithm. The simulation results show that the proposed algorithm is extremely efficient for dealing with the nonlinear dynamic PMSG system.

Neural network approach: a neural network controller for MPPT in WECS is proposed in Wei, Zhang, Qiao, and Qu (2016). The main idea of this paper is to combine between the ANN and the Q-learning method. The suggested e-learning algorithm allows WECS to act as an intelligent agent with memory to learn from its own experience, thus improving learning efficiency. Also, the results of simulations and experiments have demonstrated the effectiveness of the NN-based RL MPPT algorithm for a PMSG-based WECS. Moreover, the neural network control ensures a fast response.

Nonlinear observer: Shotorbani, Mohammadi-Ivatloo, Wang, Marzband, and Sabahi (2019) propose an adaptive nonlinear observer for the estimation of the mechanical variables of PMSG-based WECS. The MPPT algorithm is developed by using the rotor speed estimation and wind speed estimation. The suggested controller stabilizes and tracks the optimum reference of the WECS based on PMSG in a short time with large setting time and robustizes against parameters' variations in the systems. The robustness and the stability in finite time are mathematically demonstrated. In addition, the performance of the proposed controller is shown by simulation results compared to a conventional controller.

An adaptive nonlinear observer is employed for estimating the mechanical state variables and grid parameters for the PMSG connected to the grid network using AC/DC/AC in Watil, El Magri, Raihani, Lajouad, and Giri (2020). First, the state space model of the WECS is obtained. Then, the analysis of the observability gives a sufficient condition, which normally satisfied in practice. Moreover, this study designs the high gain observer providing an on-line estimation of the rotor speed, position, and the generated torque. The high gain observer also gives online estimates of two more relevant grid parameters. Consequently, the observation of the frequency of the power network and the voltage are made, only, by measuring voltages and currents. Finally, simulation results show the effectiveness of the proposed high gain observer.

Backstepping control: Yin (2020) has implemented a nonlinear control algorithm for a VS WECS based on PMSG. On the other hand, an adaptive backstepping control based on the Lyapunov stability technique provides better results without using the dynamic model of the generator. In the same work, an adaptive update law is designed to compensate the system parameter uncertainties. The effectiveness of the adopted adaptive technique is confirmed by the simulation results showing that the PMSG produces more power in comparison with conventional controller.

Sliding mode control: Matraji, Al-Durra, and Errouissi (2018) applied an adaptive second-order sliding mode control for WECS based on the PMSG. The main idea of this adaptive strategy control is to produce the desired power by controlling the stator currents. Particularly, two algorithms are combined: the first is super twisting algorithm, which is built thanks to its robustness as opposed to the parameter uncertain. The second algorithm develops an adaptive super twisting to diminish the chattering problem. The effectiveness of the adaptive control is experimentally proved by using a WT and back-to-back converter. The experimental results show better performance, in terms of tracking and system rapidity.

A regulator for a VS WECS based on a five-phase PMSG connected to the large-scale grid proposed in Mousa, Youssef, and Mohamed (2020). The machine-side converter (MSC) is responsible for extracting the maximum power from the WT at various wind speeds by using the MPPT algorithm and the filed-oriented control applied to MSC, which composed of two control loops. An one-speed control loop uses a PI controller or integral sliding mode control, and another loop for current control. Then, voltage-oriented control is developed to the grid side converters. The objective is to control the DC link and the injection of the active power into the elecric network. Then, the ISMC stability is validated using the Lyapunov stability function. A comparison with the conventional PI controller regarding the closed speed control loop shows dynamic performance of the constructed controller.

In addition to the various control techniques reported from this brief literature survey, recent controllers are being applied to PMSG-based WECS. In this chapter, a T-SFC is constructed to control the states of the PMSG to keep them on their optimal trajectory and ensure maximum power production. The obtained results are compared with the conventional controllers using MATLAB/Simulink software.

11.3 Model of the PMSG-based WECS

In this section, the five main components of the WECS are detailed. First, the WT and the PMSG shown in Fig. 11.2 are modeled. Next,the modeling of two PWM converters and DC link intermediate DC circuit are studied. Finally, the model of the RL filter is presented.

11.3.1 Model of the wind turbine

The aerodynamic power output can be expressed as (Nguyen, Al-Sumaiti, Vu, Al-Durra, & Do, 2020):

$$P_{\text{aer}} = \frac{1}{2}\rho A V_w^3 C_p(\lambda, \beta) \tag{11.1}$$

where ρ is the air density (typically 1.225 kg m^{-3}), A is the swept area (m^2), V_w is the wind speed (m s^{-1}), C_p is the turbines power coefficient depending on the pitch angle of rotor blades β (in degrees), and λ is the tip−speed ratio (TSR) defined as

$$\lambda = \frac{\Omega_m R}{V_w} \tag{11.2}$$

FIGURE 11.2 The bolc diagramm the wind and the permanent magnet synchronous generator.

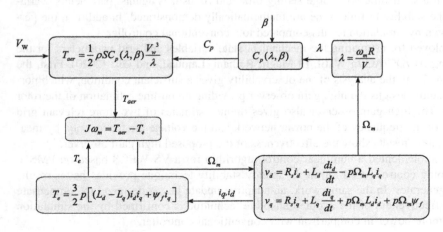

where Ω_m is the angular velocity of the rotor (in rad s^{-1}) and R defines the radium of the rotor (in m). The power coefficient is formulated as follows (Nguyen et al., 2020):

$$C_p(\lambda, \beta) = c_1 \left(\frac{c_2}{\lambda'} - c_3\beta - c_4 \right) \exp^{\frac{-c_5}{\lambda'}} + 0.0068\lambda \tag{11.3}$$

where

$$\frac{1}{\lambda'} = \frac{1}{\lambda + 0.08\beta} - \frac{0.035}{\beta^3 - 1} \tag{11.4}$$

The power coefficients c_1, c_2, c_3, c_4, and c_5 are given in Table 11.1.

The simulation of $C_p = (\lambda, \beta)$ shows the importance of this parameter when the blade pitch angle is small and reaches its maximum $C_{p\max} = 0.41$ with a zero angle and $\lambda_{op} = 8.1$ as Fig. 11.3 displays. On the WT generator power characteristic, for any wind speed value, there is a specific point, MPPT, at which the output power is boosted. Consequently, as seen in Fig. 11.4, as the wind speed varies while the peak power is retrieved from the wind continuously (MPPT control), the system may perform at the peak of the P_m curve.

TABLE 11.1 Power coefficient's values.

Symbol	c_1	c_2	c_3	c_4	c_5
Value	0.5176	116	0.4	5	21

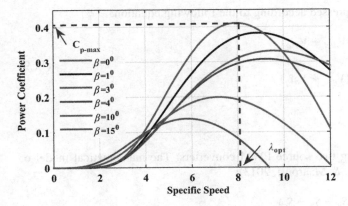

FIGURE 11.3 Evolution of power coefficient versus tip–speed ratio at different pitch angles.

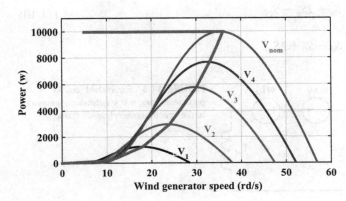

FIGURE 11.4 The power curves versus rotor speed with variable wind speeds.

11.3.2 Model of the PMSG

A simplified model is obtained by using the Park model of the PMSG in the rotor field synchronous reference frame as shown in Fig. 11.5. The resulting model in terms of its stator voltages can be expressed as (Chen, Yao, Zhang, Ren, & Jiang, 2019):

$$\begin{cases} V_d = R_s i_d + L_d \dfrac{di_d}{dt} - \omega_e L_q i_q \\[2mm] V_q = R_s i_q + L_q \dfrac{di_q}{dt} + \omega_e L_d i_d + \omega_e \psi_f \end{cases} \tag{11.5}$$

where V_d and V_q are respectively stator voltages of d-axis and q-axis. i_d and i_q stand for the respective stator currents of the d-axis and q-axis. R_s means the stator resistance. L_d and L_q represent together the generator inductances on the d-axis and q-axis. $\omega_e = p \times \Omega_m$ is the electrical angular velocity of the rotor; p is defined as the number of pole pairs, the mechanical angular velocity is stated by Ω_m, and ψ_f is the permanent magnetic flux.

The q-axis and d-axis inductances are equal ($L_d = L_q = L_s$) for salient pole PMSG. Then, the following equation is applied to derive the required electromagnetic torque:

$$T_{em} = \frac{3}{2} p \left[(L_d - L_q) i_d i_q + \psi_f i_q \right] = \frac{3}{2} p \psi_f i_q \tag{11.6}$$

The PMSG dynamic equation is as follows:

$$J \frac{d\Omega_m}{dt} = T_{em} - T_{aer} - F\Omega_m \tag{11.7}$$

Here, the viscous friction coefficient is symbolized by F. J defines the moment of inertia. T_{aer} is the mechanical torque outlined hereinafter:

$$T_{aer} = \frac{1}{2} C_p(\lambda, \beta) \rho \pi R^5 \frac{\omega_m^2}{\lambda^3} \tag{11.8}$$

The active and reactive powers of the PMSG can be expressed according to the following equation:

$$\begin{cases} P_m = \dfrac{3}{2}(V_d i_d + V_q i_q) \\[2mm] Q_m = \dfrac{3}{2}(V_q i_d + V_d i_q) \end{cases} \tag{11.9}$$

11.3.3 Model of the PWM converter

The variable frequency converters are constructed by using two source PWM converters. The mathematical model of PMW converters is expressed as follows (Errami, Ouassaid, & Maaroufi, 2013):

$$\begin{cases} V_a = \dfrac{V_{dc}}{3}(2S_a - S_b - S_c) \\[2mm] V_b = \dfrac{V_{dc}}{3}(-S_a + 2S_b - S_c) \\[2mm] V_c = \dfrac{V_{dc}}{3}(-S_a - S_b + 2S_c) \end{cases} \tag{11.10}$$

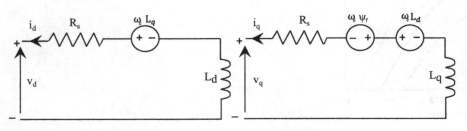

FIGURE 11.5 Equivalent circuit of the permanent magnet synchronous generator in the synchronous reference frame.

where v_{dc} is the DC-link voltage and S_a, S_b, and S_c are the states of the switches ($S_{1,2,3} = 0$ for the block state and $S_{1,2,3} = 1$ for the conduction state).

11.3.4 Model of the DC-link voltage

The DC-link consists of a capacitor that acts as a reservoir during the energy exchange and limits the undulation of the DC voltage. The model of the DC link is derived as follows:

$$C\frac{dV_{dc}}{dt}.V_{dc} = P_m - P_g \tag{11.11}$$

where P_m is the active power of the PMSG, P_g is the active power exchanged with the network, and i_{dc} is the capacitor current.

11.3.5 Model of the filter

The *RL* filter model is expressed as follows (Errami et al., 2013):

$$\begin{cases} V_{dg} = -R_f i_{d-f} - L_f \dfrac{di_{d-f}}{dt} + \varepsilon_d + e_d \\ V_{qg} = -R_f i_{q-f} - L_f \dfrac{di_{q-f}}{dt} - \varepsilon_q + e_q \end{cases} \tag{11.12}$$

Such that ε_d and ε_q are the coupling voltages given by:

$$\begin{aligned} \varepsilon_d &= \omega_g L_f i_{q-f} \\ \varepsilon_q &= \omega_g L_f i_{d-f} \end{aligned} \tag{11.13}$$

where V_{dg} and V_{qg} are grid voltages in the $d-q$ axis, respectively; i_{d-f} and i_{q-f} are filter currents in the $d-q$ axis, respectively; R_f and L_f are the filter resistance and inductance, respectively; $\omega_e = 2\pi f$ is the angular frequency of the grid voltage; f is the grid frequency; and e_d and e_q are components at the output of the inverter in the $d-q$ axis, respectively.

11.4 Controller design of the PMSG-based WECS

A variable speed WT works within four operation zones are indicated in Fig. 11.6, which represents the transmitted power of the WT versus the wind speed variation:

- Zone I: The wind is not powerful enough to start the WT; the rotation speed and the mechanical power are not available at this level.
- Zone II: The wind is at its minimum speed V_{W-min} required to start the WT. WT will operate in the MPPT mode and output power will increase with wind speed.
- Zone III: The wind speed reaches values higher than its nominal speed Vn. Rotation speed and mechanical power must be maintained at their nominal values to prevent any damage to the system.
- Zone IV: In this region, the wind speed becomes too high. An emergency device is needed to stop the WT (no electrical energy is produced) and bring it into standby mode to avoid any damage.

FIGURE 11.6 Operation zones of variable speed wind turbine.

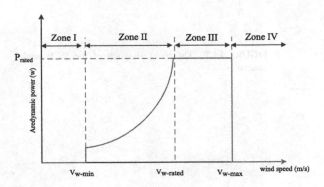

11.4.1 Maximum power point tracking and pitch angle control system

The MPPT systems objective is to adjust the turbine rotation speed Ω_m to the optimum rotational speed Ω_{m-op}, for extracting the most effective power from the existing wind energy.

The optimum rotation of the WT generator can be obviously assessed in this way:

$$\Omega_{m-op} = \frac{V_w \lambda_{op}}{R} \tag{11.14}$$

By substituting Eq. (11.14) into Eq. (11.1), the maximum aerodynamic power can be expressed by:

$$P_{aer-max} = \frac{1}{2}\rho A C_{p-max}\left(\frac{R\Omega_{m-op}}{\lambda_{opt}}\right)^3 \tag{11.15}$$

Thus the pitch control is applied in regions III and IV. The generator power is controlled using pitch angle control only as shown in Fig. 11.7. The angle β^* is taken from the conventional PI controller. The input of the regulator is the deviation between the maximum power P_{m-max} and the actual output power P_m of the PMSG. The philosophy of the pitch angle regulator is constructed by a hydraulic servo system. The servo system has a pitch angle control operating speed of ± 10 degrees second^{-1}. In addition, the angle of the blades β is zero when V_w is less than the rated velocity of the extraction of maximum power. Otherwise, if V_w is bigger than the rated velocity, then it is necessary to increase β, which degrades the C_p and consequently reduces the power.

11.4.2 Designing a T-S fuzzy control for the PMSG side rectifier

The first goal here is to conceive a fuzzy control that helps to monitor both stator current references and rotor speed. At first, the T-SFC is developed by applying the dynamic model of PMSG. Next, it is to develop the DRM and the N-LTR for the optimum rotational speed Ω_{m-op}. A new controller is then presented with T-SFC, DRM, and N-LTR in Fig. 11.8.

The T-S fuzzy model is a good approximation for nonlinear systems. It is designed to select a set of linear subsystems in accordance with a characterization of physical knowledge of the system properties (Takagi & Sugeno, 1985).

11.4.2.1 Model of the PMSG-WT fuzzy

The following nonlinear state space form describes the dynamic model of PMSG-WT (Chatri & Ouassaid, 2018):

$$\begin{cases} \dot{x} = A(\Omega_m(t))x(t) + Bv(t) + ET_{aer}(t) \\ y(t) = Cx(t) \end{cases} \tag{11.16}$$

where

$$A(\Omega_m(t))\begin{bmatrix} \dfrac{-F}{J} & \dfrac{3p\psi_F}{2J} & 0 \\[2ex] \dfrac{-p\psi_f}{L_s} & \dfrac{-R_s}{L_s} & -p\Omega_m(t) \\[2ex] 0 & p\Omega_m(t) & \dfrac{-R_s}{L_s} \end{bmatrix},$$

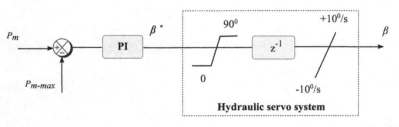

P_m PI β^* 90^0 $+10^0/s$ β

z^{-1}

P_{m-max} 0 $-10^0/s$

Hydraulic servo system

FIGURE 11.7 The block diagram of the Pitch angle controller.

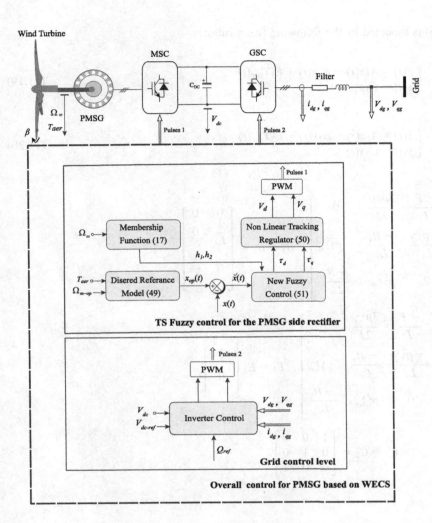

FIGURE 11.8 The proposed Takagi-Sugeno fuzzy control diagram for a permanent magnet synchronous generator-based wind energy conversion system.

$$x(t) = \begin{bmatrix} \Omega_m \\ i_q \\ i_d \end{bmatrix}, \quad B = \begin{bmatrix} 0 & 0 \\ \dfrac{1}{L_s} & 0 \\ 0 & \dfrac{1}{L_s} \end{bmatrix}, \quad v(t) = \begin{bmatrix} V_q \\ V_d \end{bmatrix},$$

$$E = \begin{bmatrix} \dfrac{-1}{J} \\ 0 \\ 0 \end{bmatrix}, \quad C = \begin{bmatrix} 1 & 0 & 0 \\ 0 & 1 & 0 \\ 0 & 0 & 1 \end{bmatrix}.$$

It is clearly seen that the three matrices B, C, and E are constant. On the other hand, the state matrix A is according to the speed rotor.

The fuzzy premise variable is bounded $\underline{\Omega}_m \leq \Omega_m(t) \leq \overline{\Omega}_m$. Thus it can be written as:

$$\Omega_m(t) = F_1 \overline{\Omega}_m + F_2 \underline{\Omega}_m \tag{11.17}$$

Denote by F_1 and F_2, the membership functions defined such as below:

$$F_1 = \frac{\Omega_m(t) - \underline{\Omega}_m}{\overline{\Omega}_m - \underline{\Omega}_m} \quad F_2 = 1 - F_1 = \frac{\overline{\Omega}_m - \Omega_m(t)}{\overline{\Omega}_m - \underline{\Omega}_m} \tag{11.18}$$

Therefore the nonlinear system (11.16) is modeled by the following fuzzy rules:
Rule 1: if $\Omega_m(t)$ is F_1 then

$$\begin{cases} \dot{x}(t) = A_1 x(t) + B_1 v(t) + E_1 T_{aer}(t) \\ y(t) = C_1 x(t) \end{cases} \tag{11.19}$$

Rule 2: if $\Omega_m(t)$ is F_2 then

$$\begin{cases} \dot{x}(t) = A_2 x(t) + B_2 v(t) + E_2 T_{aer}(t) \\ y(t) = C_2 x(t) \end{cases} \tag{11.20}$$

where

$$A_1 = \begin{bmatrix} \dfrac{-F}{J} & \dfrac{3p\psi_f}{2J} & 0 \\ \dfrac{-p\psi_f}{L_s} & \dfrac{-R_s}{L_s} & -p\overline{\Omega}_m \\ 0 & p\overline{\Omega}_m & \dfrac{-R_s}{L_s} \end{bmatrix}, \quad B_1 = B_2 = \begin{bmatrix} 0 & 0 \\ \dfrac{1}{L_s} & 0 \\ 0 & \dfrac{1}{L_s} \end{bmatrix}$$

$$A_2 = \begin{bmatrix} \dfrac{-F}{J} & \dfrac{3p\psi_f}{2J} & 0 \\ \dfrac{-p\psi_f}{L_s} & \dfrac{-R_s}{L_s} & -\underline{\Omega}_m \\ 0 & p\underline{\Omega}_m & \dfrac{-R_s}{L_s} \end{bmatrix}, \quad E_1 = E_2 \begin{bmatrix} \dfrac{-1}{J} \\ 0 \\ 0 \end{bmatrix},$$

$$C_1 = C_2 = \begin{bmatrix} 1 & 0 & 0 \\ 0 & 1 & 0 \\ 0 & 0 & 1 \end{bmatrix}.$$

11.4.2.2 T-S fuzzy controller

The trajectory tracking is a topical control issue. Indeed, most of the existing work deals with the T-S system stabilization. Nonetheless, this only guarantees the feedback of the nonlinear system to an equilibrium point of nonzero initial conditions (Khabou, Souissi, & Aitouche, 2020; Zhao, Xu, & Park, 2019).

The following T-S fuzzy model describes the PMSG-WT nonlinear system:

$$\dot{x}(t) = \sum_{i=1}^{2} h_i(\Omega_m(t))\{A_i x(t) + B_i v(t) + E_i T_{aer}(t)\} \tag{11.21}$$

where $h_i(\Omega_m(t)) = \frac{F_i}{\sum_{i=1}^{2} F_i}$ $\forall t > 0, 0 \le h_i(\Omega_m(t)) \le 1$ and $\sum_{i=1}^{2} h_i(\Omega_m(t)) = 1$.

The intent of a T-SFC is to fulfill the following condition:

$$x(t) - x_{op}(t) \to 0; t \to \infty \tag{11.22}$$

where $x_{op}(t)$ is the desired trajectory.

The tracking error should be set by $\tilde{x}(t) = x(t) - x_{op}(t)$.

The time derivative of $\tilde{x}(t)$ can be found:

$$\tilde{x}.(t) = \dot{x}(t) - x_{op}(t) \tag{11.23}$$

Applying Eq. (11.23), Eq. (11.21) turns into:

$$\tilde{x}.(t) = \sum_{i=1}^{2} h_i(\Omega_m(t))\{A_i \tilde{x}(t) + B_i v(t) + E_i T_{aer}(t)\} - \dot{x}_{op}(t) \tag{11.24}$$

By considering $x(t) = \tilde{x}(t) + x_{op}(t)$, then Eq. (11.24) becomes:

$$\dot{\tilde{x}}(t) = \sum_{i=1}^{2} h_i(\Omega_m(t))\left\{A_i\tilde{x}(t) + B_iv(t) + E_iT_{\mathrm{aer}}(t) + A_ix_{op}(t)\right\} - \dot{x}_{op}(t) \tag{11.25}$$

By calling for:

$$\sum_{i=1}^{2} h_i(\Omega_m(t))B_i\tau(t) = \sum_{i=1}^{2} h_i(\Omega_m(t))\{B_iv(t) + E_iT_{\mathrm{aer}}(t) + A_ix_{op}(t)\} - \dot{x}_{op}(t) \tag{11.26}$$

The derivative of $\tilde{x}(t)$ is rewritten as follows:

$$\dot{\tilde{x}}(t) = \sum_{i=1}^{2} h_i(\Omega_m(t))\{A_i\tilde{x}(t) + B_i\tau(t)\} \tag{11.27}$$

here $\tau(t)$ is the proposed controller's law.

The trajectory-tracking problem is considered as a generalization of the stabilization problem. The new objective is to converge asymptotically a regulator for the state $\tilde{x}(t)$ to zero.

The PDC approach is harnessed to develop this control system.

Controller rule j: if $\Omega_m(t)$ equals F_i, then

$$\tau(t) = - K_j\tilde{x}(t), j = 1, 2 \tag{11.28}$$

here K_j refers to the controller gains.

The after sum determines control system output:

$$\tau(t) = - \sum_{j=1}^{r} h_j(\Omega_m(t))K_j\tilde{x}(t) \tag{11.29}$$

By introducing the control system law (11.29) into Eq. (11.27), the new formula of the system's closed loop is

$$\dot{\tilde{x}}(t)) = \sum_{j=1}^{2} h_i(\Omega_m(t))h_j(\Omega_m(t))\{A_i - B_iK_j\}\tilde{x}(t) \tag{11.30}$$

with $G_{ij} = A_i - B_iK_j$, Eq. (11.30) becomes:

$$\dot{\tilde{x}}(t)) = \sum_{i=1}^{2}\sum_{j=1}^{2} h_i(\Omega_m(t))h_j(\Omega_m(t))G_{ij}\tilde{x}(t) \tag{11.31}$$

The Lyapunov approach stability survey is used to define the controller's gains.

Theorem 11.1: The balance of the system (11.31) is asymptotically stable via the PDC if there is a symmetric matrix $Q > 0$, with a scalar $\alpha > 0$ and matrices M_i with: $M_i = K_iQ^{-1}$ such that:

- The stability of each subsystem requires the state $i = j$:

$$QA_i^T + A_iQ - M_i^TB_i^T - B_iM_i + \alpha Q < 0 \tag{11.32}$$

- The performance of global system stability means that $i \neq j$

$$QA_i^T + A_iQ - M_j^TB_i^T - B_iM_j + \alpha Q < 0 \tag{11.33}$$

Proof: Function (11.34) is chosen to confirm the system stability:

$$V = \tilde{x}^TPx \tag{11.34}$$

where $V > 0$ and P is the Lyapunov matrix. For the system to be stable, in addition to improving convergence, it is necessary to add a convergence factor in finite time α.

$$\dot{V} = \dot{x}^T P x + x^T P \dot{x} + \alpha P \tag{11.35}$$

Introducing Eq. (11.31) in Eq. (11.35) yields to:

$$\dot{V} = \sum_{i=1}^{2} \sum_{j=1}^{2} h_i(\Omega_m(t)) h_j(\Omega_m(t)) \{x^T [G^T_{ij} P + PG_{ij} \tilde{x}] + \alpha P \tag{11.36}$$

Thus the Lyapunov function is given as follows

$$G^T_{ij} P + PG_{ij} + \alpha P < 0 \tag{11.37}$$

where $G_{ij} = A_i - B_i K_j$, then

$$A_i^T P + PA_i - K_j^T B_i^T P - PB_i K_j + \alpha P < 0 \tag{11.38}$$

Consequently, the way to move from a bilinear matrix inequality to an LMI is to multiply the previous equation by P^{-1} on the left and on the right.

$$P^{-1} A_i^T + A_i P^{-1} - P^{-1} K_j^T B_i^T - B_i K_j P^{-1} + \alpha P^{-1} < 0 \tag{11.39}$$

By asking: $Q = P^{-1}$

$$QA_i^T + A_i Q - QK_j^T B_i^T - B_i K_j Q + \alpha Q < 0 \tag{11.40}$$

Therefore the linear matrix inequality is expressed as follows:

$$QA_i^T + A_i Q - M_j^T B_i^T - B_i M_j + \alpha Q < 0 \tag{11.41}$$

- To enhance the stability of each subsystem, the $i = j$ is ensured:

$$QA_i^T + A_i Q - M_i^T B_i^T - B_i M_i + \alpha Q < 0 \tag{11.42}$$

- The global system stability performance calls for $i \neq j$:

$$QA_i^T + A_i Q - M_j^T B_i^T - B_i M_j + \alpha Q < 0 \tag{11.43}$$

11.4.2.3 DRM and N-LTR controller

Eq. (11.26) is used to establish two controllers; that of desired reference and the one of nonlinear tracking:

$$\sum_{i=1}^{2} h_i(\Omega_m(t)) B_i \tau(t) = \sum_{i=1}^{2} h_i(\Omega_m(t)) \{B_i v(t) + E_i T_{\text{aer}}(t) + A_i x_{op}(t)\} - \dot{x}_{op}(t) \tag{11.44}$$

Noting that:

$$\begin{cases} A(\Omega_m(t)) = \sum_{i=1}^{2} h_i A_i \\ B = \sum_{i=1}^{2} h_i B_i \\ E = \sum_{i=1}^{2} h_i E_i \end{cases} \tag{11.45}$$

Thus the new form of Eq. (11.44) is obtained:

$$B(v(t) - \tau(t)) = -A(\Omega_m) x_{op}(t) - E T_{\text{aer}}(t) + \dot{x}_{op}(t) \tag{11.46}$$

Enforcing Eq. (11.46) to the dynamic model of PMSG leads to the following matrix form:

$$
\begin{bmatrix} 0 & 0 \\ \dfrac{1}{L_a} & 0 \\ 0 & \dfrac{1}{L_a} \end{bmatrix}
\begin{bmatrix} V_q - \tau_q \\ V_q - \tau_d \end{bmatrix} = -
\begin{bmatrix} \dfrac{-F}{J} & \dfrac{3p\psi_F}{2J} & 0 \\ \dfrac{-p\psi_F}{L_s} & \dfrac{-R_s}{L_s} & -p\Omega_m \\ 0 & p\Omega_m & \dfrac{-R_s}{L_s} \end{bmatrix}
\tag{11.47}
$$

$$
\begin{bmatrix} \Omega_{m-op} \\ i_{q-op} \\ i_{d-op} \end{bmatrix} -
\begin{bmatrix} \dfrac{-1}{J} \\ 0 \\ 0 \end{bmatrix} T_{aer} + \dfrac{d}{dt}
\begin{bmatrix} \Omega_{m-op} \\ i_{q-op} \\ i_{d-op} \end{bmatrix}
$$

According to the vector control ($i_{d-op} = 0$), the first linear Eq. (11.47) is solved to obtain:

$$
i_{q-op} = \frac{2J}{3p\psi_F}\left(\dot{\Omega}_{m-op} + \frac{F}{J}\Omega_{m-op} + \frac{1}{J}T_{aer} \right)
\tag{11.48}
$$

As a result, the desired reference model is set by:

$$
x_{op}(t) = \begin{bmatrix} \Omega_{m-op} \\ \dfrac{2J}{3p\psi_F}\left(\dot{\Omega}_{m-op} + \dfrac{F}{J}\Omega_{m-op} + \dfrac{1}{J}T_{aer} \right) \\ 0 \end{bmatrix}
\tag{11.49}
$$

The N-LTR is obtained in (11.50) by using the second and third linear Eq. (11.47)

$$
\begin{cases} V_q = R_s i_{q-op} + L_s \dfrac{di_{q-op}}{dt} + p\psi_f \Omega_{m-op} + \tau_q \\ V_d = -L_s p\omega_m i_{q-op} + \tau_d \end{cases}
\tag{11.50}
$$

where τ_d and τ_q are the new control laws.

Therefore the law of the control system τ is indicated by the following sum:

$$
\tau(t) = -\sum_{i=1}^{2} h_i K_i (x(t) - x_{op}(t))
\tag{11.51}
$$

The T-SFC synoptic diagram for the PMSG-based WECS is displayed in Fig. 11.9. Finally, the control laws are obtained from Eqs. (11.50) and (11.51) via the new T-SFC and N-LTR. As a result, with these control laws, the WECS keeps its stability in the closed loop and a better tracking is achieved under different wind speed conditions.

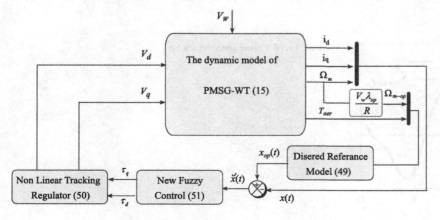

FIGURE 11.9 Takagi-Sugeno fuzzy controller synoptic diagram for the permanent magnet synchronous generator-based wind energy conversion system.

11.5   Simulation results and discussion

The performance evaluation of the enhanced control of WECS based on PMSG involves two simulation test using the MATLAB/Simulink environment. The first simulation aims to assess the T-S proposed strategy performance and the second one compares the conventional PI with the proposed control strategy. The WECS parameters are presented in Table 11.2 (Appendix hereunder.).

11.5.1   Simulation results of the proposed control

The T-S control of PMSG-WT presented in Fig. 11.8 is implemented. The wind speed profile and the rated wind speed versus time are shown in Fig. 11.10. Fig. 11.11 illustrates the rotor speed variation with time. As a first remark, it is easy to see that the rotor speed tracks very well its reference. Furthermore, its shape is the same as that of wind velocity profile.

TABLE 11.2 Paramerts of wind energy conversion system.

	Rated wind speed V_{w-op} 12 (m s^{-1})
	Blades radius R: 2.7 (m)
Wind turbine	Air density:ρ: 1.225 (kg m^{-3})
	Optimum tip$-$speed ratio λ: 8.1
	Optimum power coefficient C_{p-max}: 41
	Rated power P_m: 10 (kW)
	Rated rotor speed Ω_{m-op}: 36 (rad s^{-1})
	Stator resistance R_s: 0.05(ω)
PMSG	Stator inductance L_s: 0.000635(H)
	Moment of inertia J: 0.0089(kg m^{-2})
	Friction coefficient F: 0.005 (Nm s rad^{-1})
	Permanent magnet flux ψ_f: 0.192(wb)
	Pole pairs p: 4
	DC-link voltage V_{dc}: 850 (V)
	DC-link capacitor C: 0.0024 (F)
Grid	Filter resistance R_f: 0.1140 (ω)
	Filter inductance L_f: 0.0182 (H)
	Frequency f: 50 (Hz)

PMSG, permanent magnet synchronous generator.

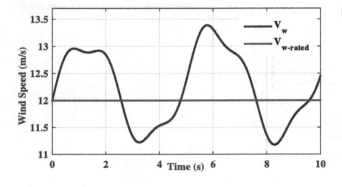

FIGURE 11.10   Wind speed trends (m s^{-1}).

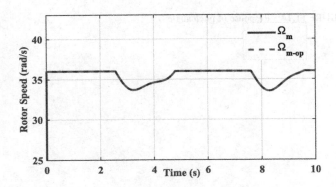

FIGURE 11.11 Evolution of rotor speed (rad s^{-1}).

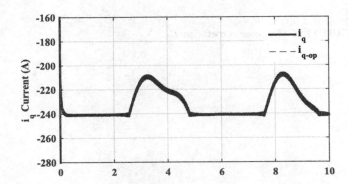

FIGURE 11.12 The q-axis current i_q and reference value i_{q-op}.

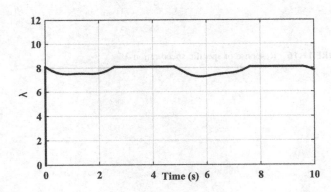

FIGURE 11.13 The d-axis current i_d and reference value i_{d-op}.

The stator currents of the q-axis and d-axis, shown in Figs. 11.12 and 11.13, respectively, seem to follow their references. Consequently, the proposed control can successfully bring the generator states to path their desired references.

The waveforms of pitch angle are shown in Fig. 11.14, and the power coefficient is presented in Fig. 11.15. The specific speed and generated power are shown in Figs. 11.16 and 11.17, respectively. All these figures depict that if the wind speed is lower than its rated speed ($V_w = 12$ m s^{-1}), the new value of the pitch control becomes null and, then, the power coefficient and the specific speed meet their maximum values ($C_{p-\max} = 0.41$ and $\lambda_{opt} = 8.1$, respectively). Otherwise, if the wind speed becomes higher than the nominal speed, the pitch control is activated. This reduces the power coefficient with an increased angle β, lowers the TSR and the extracted power is kept at the nominal value (10 kW).

11.5.2   Comparison of the proposed and PI controllers' performance

As mentioned before, the second simulation aims to compare the performance of the proposed controller with that of the PI linear controller. The waveforms of the wind speed illustrating the way the controllers react with unexpected shifts of the wind speed can be seen in Fig. 11.18. Figs. 11.19 and 11.20 show, respectively, the variation in the stator speed and in the power coefficient versus time for both controllers. On the basis of these figures and Table 11.3, the

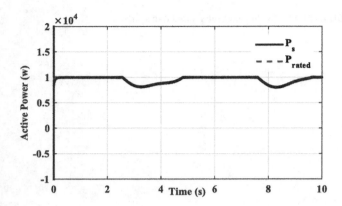

FIGURE 11.14 Response of pitch angle β.

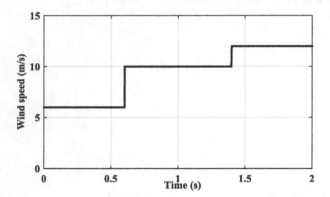

FIGURE 11.15 Response of coefficient C_p.

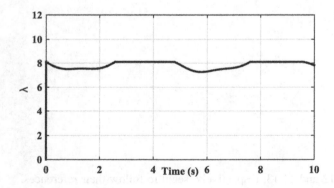

FIGURE 11.16 Response of specific speed ration λ.

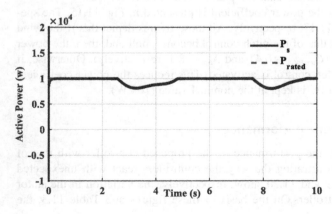

FIGURE 11.17 Power regulation (W).

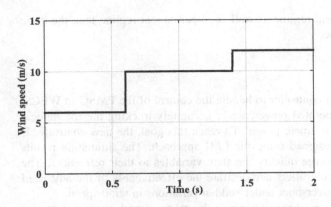

FIGURE 11.18 Wind speed (m s⁻¹).

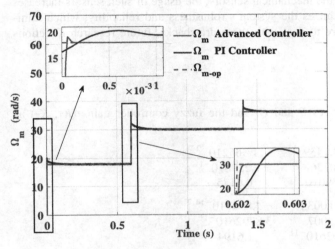

FIGURE 11.19 The Ω_m curve of the permanent magnet synchronous generator-wind turbine, by a proportional–integral controller (*blue*) and the proposed controller (*green*).

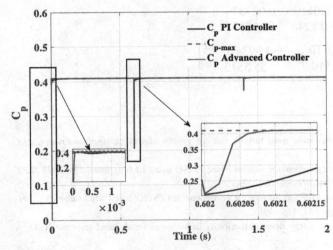

FIGURE 11.20 The C_p response by a proportional–integral controller (*blue*) and the proposed controller (*green*).

TABLE 11.3 Performance analysis of the PI controller and the advanced controller.

Parameters	Rotor speed		Power	
	PI controller	Advanced controller	PI controller	Advanced controller
Response time (ms)	4.2	0.47	74	6
Overshoot (%)	0.5	0.057	0	0

PI, proportional–integral.

proposed T-S controller can offer better performance in transient regime as well as in permanent regime than the basic PI controller. It ensures less overshoot and shorter reaction time.

11.6 Conclusion

This study has successfully developed and applied an advanced controller to handle the control of the PMSG in WECS. The proposed strategies' main objective is to enhance the generated power, while accurately tracking the mechanical speed of rotor and the generator stator currents relating to the ultimate power. To reach this goal, the new controller is constructed by using the PDC strategy, while the gains are designed using the LMI approach. The simulation results demonstrate that the proposed T-S controller succeeds to converge quickly the state variables to their references. The comparative simulation results of the advanced and linear PI controllers demonstrate the effectiveness of the advanced controller, which provides shorter reaction time and minimum overshoot under sudden variations in wind speed.

Furthermore, the control strategy for the WECS requires the mechanical sensors; the usage of such sensors increases costs and space and requires additional wiring, which minimizes the system's robustness and reliability, which minimizes the robustness and reliability of the system. Therefore to deal with these drawbacks, future research directions focus on the development of the sensorless control.

Appendix

The resolution of LMIs leads to the obtaining of the matrices P and Q, and the fuzzy controller gains (K_1, K_2) as follows:

$$P = \begin{bmatrix} 3.0366 10^5 & 440.1439 & -1.3652 10^{-9} \\ 440.1439 & 1.0726 & -1.2390 10^{-11} \\ -1.3652 10^{-9} & -1.2390 10^{-11} & 0.6175 \end{bmatrix}$$

$$Q = \begin{bmatrix} 8.1265 10^{-6} & -0.0033 & -4.8947 10^{-14} \\ -0.0033 & 2.3007 & 3.8792 6 10^{-11} \\ -4.8947 10^{-14} & 3.8792 6 10^{-11} & 1.6194 \end{bmatrix}$$

$$K_1 = \begin{bmatrix} 5.1418 10^4 & 1.2201 10^2 & 0.02337 \\ -98.1304 & -0.1248 & 55.5513 \end{bmatrix}$$

$$K_2 = \begin{bmatrix} 5.1418 10^4 & 1.2201 10^2 & -3.9934 10^{-10} \\ 1,2148 10^{-6} & 1,8545 10^{-9} & 55.5513 \end{bmatrix}$$

References
Chan, C. M., Bai, H., & He, D. (2018). Blade shape optimization of the savonius wind turbine using a genetic algorithm. *Applied Energy, 213*, 148–157.

Chatri, C., & Ouassaid, M. (2018). Design of fuzzy control TS for wind energy conversion system based PMSG using LMI approach. In: *2018 IEEE 5th international congress on information science and technology (CiSt)* (pp. 466–471). IEEE.

Chen, J., Yao, W., Zhang, C.-K., Ren, Y., & Jiang, L. (2019). Design of robust MPPT controller for grid-connected PMSG-based wind turbine via perturbation observation based nonlinear adaptive control. *Renewable Energy, 134*, 478–495.

Chen, Y., Xu, P., Gu, J., Schmidt, F., & Li, W. (2018). Measures to improve energy demand flexibility in buildings for demand response (DR): A review. *Energy and Buildings*.

Chinchilla, M., Arnaltes, S., & Burgos, J. C. (2006). Control of permanent-magnet generators applied to variable-speed wind-energy systems connected to the grid. *IEEE Transactions on Energy Conversion, 21*, 130–135.

El Karkri, Y., Rey-Boué, A. B., El Moussaoui, H., Stöckl, J., & Strasser, T. I. (2019). Improved control of grid-connected DFIG-based wind turbine using proportional-resonant regulators during unbalanced grid. *Energies, 12*, 4041.

Elyaalaoui, K., Ouassaid, M., & Cherkaoui, M. (2019). Dispatching and control of active and reactive power for a wind farm considering fault ride-through with a proposed PI reactive power control. *Renewable Energy, 28*, 56–65.

Errami, Y., Ouassaid, M., & Maaroufi, M. (2013). Modeling and variable structure power control of PMSG based variable speed wind energy conversion system. *Journal of Optoelectronics and Advanced Materials, 15*, 1248–1255.

Errami, Y., Ouassaid, M., & Maaroufi, M. (2015a). Optimal power control strategy of maximizing wind energy tracking and different operating conditions for permanent magnet synchronous generator wind farm. *Energy Procedia, 74*, 477–490.

Errami, Y., Ouassaid, M., & Maaroufi, M. (2015b). A performance comparison of a nonlinear and a linear control for grid connected pmsg wind energy conversion system. *International Journal of Electrical Power & Energy Systems, 68*, 180—194.

Golnary, F., & Moradi, H. (2019). Dynamic modelling and design of various robust sliding mode controls for the wind turbine with estimation of wind speed. *Applied Mathematical Modelling, 65*, 566—585.

Jlassi, I., & Cardoso, A. J. M. (2019). Fault-tolerant back-to-back converter for direct-drive pmsg wind turbines using direct torque and power control techniques. *IEEE Transactions on Power Electronics, 34*, 11215—11227.

Jlassi, I., Estima, J. O., El Khil, S. K., Bellaaj, N. M., & Cardoso, A. J. M. (2014). Multiple open-circuit faults diagnosis in back-to-back converters of pmsg drives for wind turbine systems. *IEEE Transactions on Power Electronics, 30*, 2689—2702.

Khabou, H., Souissi, M., & Aitouche, A. (2020). MPPT implementation on boost converter by using T-S fuzzy method. *Mathematics and Computers in Simulation, 167*, 119—134.

Li, P., Hu, W., Hu, R., Huang, Q., Yao, J., & Chen, Z. (2019). Strategy for wind power plant contribution to frequency control under variable wind speed. *Renewable Energy, 130*, 1226—1236.

Mansour, M., Mansouri, M., Bendoukha, S., & Mimouni, M. (2020). A grid-connected variable-speed wind generator driving a fuzzy-controlled pmsg and associated to a flywheel energy storage system. *Electric Power Systems Research, 180*, 106137.

Matayoshi, H., Howlader, A. M., Datta, M., & Senjyu, T. (2018). Control strategy of PMSG based wind energy conversion system under strong wind conditions. *Energy for Sustainable Development, 45*, 211—218.

Matraji, I., Al-Durra, A., & Errouissi, R. (2018). Design and experimental validation of enhanced adaptive second-order SMC for PMSG-based wind energy conversion system. *International Journal of Electrical Power & Energy Systems, 103*, 21—30.

Mousa, H. H., Youssef, A.-R., & Mohamed, E. E. (2020). Optimal power extraction control schemes for five-phase PMSG based wind generation systems. *Engineering Science and Technology, 23*, 144—155.

Naidu, N. S., & Singh, B. (2017). Grid-interfaced DFIG-based variable speed wind energy conversion system with power smoothening. *IEEE Transactions on Sustainable Energy, 8*, 51—58.

Nasiri, M., & Mohammadi, R. (2017). Peak current limitation for grid side inverter by limited active power in PMSG-based wind turbines during different grid faults. *IEEE Transactions on Sustainable Energy, 8*, 3—12.

Nguyen, H. T., Al-Sumaiti, A. S., Vu, V.-P., Al-Durra, A., & Do, T. D. (2020). Optimal power tracking of PMSG based wind energy conversion systems by constrained direct control with fast convergence rates. *International Journal of Electrical Power & Energy Systems, 118*, 105807.

Petrov, I., & Pyrhonen, J. (2012). Performance of low-cost permanent magnet material in PM synchronous machines. *IEEE Transactions on Industrial Electronics, 60*, 2131—2138.

Shotorbani, A. M., Mohammadi-Ivatloo, B., Wang, L., Marzband, M., & Sabahi, M. (2019). Application of finite-time control Lyapunov function in low-power PMSG wind energy conversion systems for sensorless MPPT. *International Journal of Electrical Power & Energy Systems, 106*, 169—182.

Soliman, M. A., Hasanien, H. M., Azazi, H. Z., El-Kholy, E. E., & Mahmoud, S. A. (2018). An adaptive fuzzy logic control strategy for performance enhancement of a grid-connected PMSG-based wind turbine. *IEEE Transactions on Industrial Informatics, 15*, 3163—3173.

Takagi, T., & Sugeno, M. (1985). Fuzzy identification of systems and its applications to modeling and control. *IEEE Transactions on Systems, Man, and Cybernetics*, 116—132.

Tripathi, S., Tiwari, A., & Singh, D. (2015). Grid-integrated permanent magnet synchronous generator based wind energy conversion systems: A technology review. *Renewable and Sustainable Energy Reviews, 51*, 1288—1305.

Watil, A., El Magri, A., Raihani, A., Lajouad, R., & Giri, F. (2020). An adaptive nonlinear observer for sensorless wind energy conversion system with PMSG. *Control Engineering Practice, 98*, 104356.

Wei, C., Zhang, Z., Qiao, W., & Qu, L. (2016). An adaptive network-based reinforcement learning method for MPPT control of PMSG wind energy conversion systems. *IEEE Transactions on Power Electronics, 31*, 7837—7848.

Yin, X. (2020). Adaptive backstepping control for maximizing marine current power generation based on uncertainty and disturbance estimation. *International Journal of Electrical Power & Energy Systems, 117*, 105329.

Zhang, Z., Zhao, Y., Qiao, W., & Qu, L. (2015). A discrete-time direct torque control for direct-drive PMSG-based wind energy conversion systems. *IEEE Transactions on Industry Applications, 51*, 3504—3514.

Zhao, J., Xu, S., & Park, J. H. (2019). Improved criteria for the stabilization of TS fuzzy systems with actuator failures via a sampled-data fuzzy controller. *Fuzzy Sets and Systems*.

Zheng, X., Feng, Y., Han, F., & Yu, X. (2019). Integral-type terminal sliding-mode control for grid-side converter in wind energy conversion systems. *IEEE Transactions on Industrial Electronics, 66*, 3702—3711.

Chapter 12

Fractional-order controller design and implementation for maximum power point tracking in photovoltaic panels

Ahmad Taher Azar[1,2], Fernando E. Serrano[3], Marco A. Flores[3], Nashwa Ahmad Kamal[4,5], Francisco Ruiz[3], Ibraheem Kasim Ibraheem[6], Amjad J. Humaidi[7], Arezki Fekik[8,9], Kammogne Soup Tewa Alain[10], Kengne Romanic[10], K.P.S. Rana[11], Vineet Kumar[11], Tulasichandra Sekhar Gorripotu[12], Ramana Pilla[13] and Shikha Mittal[14]

[1]*Faculty of Computers and Artificial Intelligence, Benha University, Benha, Egypt,* [2]*College of Computer and Information Sciences, Prince Sultan University, Riyadh, Saudi Arabia,* [3]*Instituto de Investigacion en Energia IIE, Universidad Nacional Autónoma de Honduras (UNAH), Tegucigalpa, Honduras,* [4]*Faculty of Engineering, Cairo University, Giza, Egypt,* [5]*International Group of Control Systems (IGCS), Riyadh, Saudi Arabia,* [6]*Department of Electrical Engineering, College of Engineering, University of Baghdad, Baghdad, Iraq,* [7]*Department of Control and Systems Engineering, University of Technology, Baghdad, Iraq,* [8]*Akli Mohand Oulhadj University, Bouira, Algeria,* [9]*Electrical Engineering Advanced Technology Laboratory (LATAGE), Mouloud Mammeri University, Tizi Ouzou Algeria,* [10]*Laboratory of Condensed Matter, Electronics and Signal Processing (LAMACETS), Department of Physic, Faculty of Sciences, University of Dschang, Dschang, Cameroon,* [11]*Division of Instrumentation and Control Engineering, Netaji Subhas University of Technology, Dwarka, India,* [12]*Department of Electrical & Electronics Engineering, Sri Sivani College of Engineering, Srikakulam, India,* [13]*Department of Electrical & Electronics Engineering, GMR Institute of Technology, Rajam, Srikakulam, India,* [14]*Department of Mathematics, Jesus and Mary College, University of Delhi, New Delhi, India*

12.1 Introduction

Due to the increase in applications and solutions related to renewable energies, specifically in solar energy, it is necessary to provide novel strategies for the maximum power point tracking (MPPT) of photovoltaic (PV) panels (Amara et al., 2019; Ammar, Azar, Shalaby, & Mahmoud, 2019; Ben Smida, Sakly, Vaidyanathan, & Azar, 2018; Ghoudelbourk, Dib, Omeiri, & Azar, 2016; Kamal, Azar, Elbasuony, Almustafa, & Almakhles, 2020; Kamal & Ibrahim, 2018). Due to an increase in the demand for solar energy generation and the fast-changing technologies related to solar energy and other renewable energies, new MPPT techniques are required to provide more robust strategies, accurate and with the capability to deal with uncertainties among other kinds of attributes, which are necessary for solar energy generation (Fekik et al., 2020).

First, it is important to mention some MPPT algorithms that are not necessarily based on automatic control because of the advantages to design and implement control techniques to maintain the maximum power extraction of PV panels when these kinds of algorithms do not provide the required robustness or disturbance rejection properties of control strategies. For example, in Kumar, Bhaskar, and Koti (2014), the comparison of the short-circuit method and the incremental conductance method is provided. Something important to remark is that in this study, the control circuit is simplified. Another interesting example can be found in Eltawil and Zhao (2013), in which some MPPT techniques are analyzed but considering other conditions, such as light, shade, and temperature and the grid connection in urban areas. In Visweswara (2014), the incremental conductance MPPT technique is evinced. This strategy consists of finding the optimum operating current for the maximum power extraction of PV panels and taking advantage of low-frequency switching. Besides, as shown in Logeswaran and SenthilKumar (2014), sometimes it is important to consider external conditions for MPPT, such as uniform and nonuniform irradiance. In Logeswaran and SenthilKumar (2014), some

MPPT techniques, such as particle swarm optimization, fuzzy logic, neural networks, and ant colony optimization, are designed, studied, and analyzed to solve this problem. Then, in Husain, Tariq, Hameed, Arif, and Jain (2017), several MPPT techniques are analyzed and discussed to provide solutions that can be selected to fit in the desired MPPT scheme.

Proportional integral derivative (PID) control is an important technique for the control and stabilization of linear and nonlinear systems (Ammar & Azar, 2020; Ammar, Azar, Tembi, Tony, & Sosa, 2018; Azar, Ali, Makarem, Diab, & Ammar, 2020; Azar, Ammar, Barakat, Saleh, & Abdelwahed, 2019; Azar, Ammar, Beb, Garces, & Boubakari, 2020; Azar et al., 2020; Azar, Hassan, Razali, de Brito Silva, & Ali, 2019; Azar, Sayed, Shahin, Elkholy, & Ammar, 2020; Azar & Serrano 2015; Gorripotu, Samalla, Jagan Mohana Rao, Azar, & Pelusi, 2019; Ibraheem, Azar, Ibraheem, & Humaidi, 2020; Pilla, Azar, & Gorripotu, 2019; Pilla, Botcha, Gorripotu, & Azar, 2020; Sallam, Azar, Guaily, & Ammar, 2020; Soliman, Azar, Saleh, & Ammar, 2020). In MPPT PID control, there are some interesting studies found in the literature in integer-order and fractional-order PID control to obtain the maximum power of PV panels. Example such as the reference Al-Dhaifallah, Nassef, Rezk, and Nisar (2018) provides an interesting technique taking advantage of fractional-order controller and the incremental conductance for MPPT to increase the robustness and accuracy. Another example can be found in Ahmadi, Abdi, and Kakavand (2017), in which a particle swarm optimization (PSO) and PID controller are used for MPPT purposes of a fuel cell system. Then, in Murtaza et al. (2017), an MPPT strategy is provided for uniformly irradiated PV arrays. In Yang et al. (2018a), a fractional-order PID controller is designed and implemented for a grid-connected PV inverter for MPPT.

Fractional-order sliding mode control for MPPT of PV panels has increased in its implementation. Still, despite this, there is a lot to investigate in this topic considering the recent solar energy technology. In papers like Dursun and Kulaksiz (2020), a second-order sliding mode voltage regulator is provided for the MPPT purpose of a permanent magnet synchronous generator for wind energy conversion system. Other examples can be found in Kchaou, Naamane, Koubaa, and Sirdi (2017) and Lamzouri, Boufounas, Brahmi, and Amrani (2020), where in the first case, a second-order sliding mode MPPT controller is presented. In the second case, a terminal sliding mode controller is designed for MPPT control, considering that the latest reference is essential for this study. In Yang et al. (2018b), a perturbation observer along with a fractional-order sliding mode controller for MPPT is presented for grid-connected PV inverters. Other examples of integer-order and fractional-order sliding mode controllers for other applications, such as aeronautics/astronautics and electrical and power systems, can be found in Ismail, Varatharajoo, and Chak (2020), Modiri and Mobayen (2020), and Zhu, Chen, Li, Zhang, and Wan (2020). Finally, another important technique for MPPT of PV panels is backstepping control. The backstepping methodology is a systemic and recursive design method for nonlinear feedback control (Azar, Serrano, Flores, Vaidyanathan, & Zhu, 2020; Vaidyanathan, Idowu, & Azar, 2015; Vaidyanathan, Jafari, Pham, Azar, & Alsaadi, 2018). The important interest brought to the procedure of backstepping is that all nonlinearities encountered can be treated in several ways. Useful nonlinearities that act for stabilization can be retained, and the rest of other nonlinearities can be treated with a linear control (Alimi, Rhif, Rebai, Vaidyanathan, & Azar, 2021; Azar, Serrano, Vaidyanathan, & Kamal, 2021; Bansal et al., 2021; Drhorhi et al., 2021; Fekik et al., 2021; Pham, Vaidyanathan, Azar, & Duy, 2021; Sambas et al., 2021a, 2021b; Vaidyanathan & Azar, 2021; Vaidyanathan, Pham, & Azar, 2021; Vaidyanathan, Sambas, Azar, Rana, & Kumar, 2021a, 2021b; Vaidyanathan, Sambas, Azar, Serrano, & Fekik, 2021; Vaidyanathan, Sambas, Azar, & Singh, 2021). Retaining nonlinearities instead of eliminating them requires less accurate models and also minimal control effort. We expect our control law simulation results to be optimal with respect to the performance index that guarantees certain robustness properties.

In references, such as Arsalan et al. (2018) and Naghmash et al. (2018), a nonlinear backstepping controller for MPPT for PV systems is presented considering an integral action. Then, in Yatimi, Ouberri, Chahid, and Aroudam (2020), the design of an off-grid backstepping controller is provided for MPPT purposes.

In this book chapter, two fractional-order controllers for the MPPT of PV panels are presented. The first one is a fractional-order PID controller with compensation in which an extra state is added to the system, which describes the PV panel and power electronic dynamics. Then, the compensation input is obtained by the Lyapunov theorem (Azar, Serrano, Koubaa, & Kamal, 2020; Fekik et al., 2021; Sambas et al., 2021; Shukla, Sharma, & Azar, 2018; Singh, Mathpal, Azar, Vaidyanathan, & Kamal, 2021; Vaidyanathan & Azar, 2016b, 2020; Vaidyanathan, Sambas, & Azar, 2021). The second strategy consists of a fractional-order terminal sliding mode controller in which the sliding variable is designed so that the state variables reach the equilibrium, in other words, that the power of the PV panel reaches the maximum power under a specific temperature condition, and by obtaining the appropriate control law.

The chapter is divided into the following sections. In Section 12.2, the related work of this study is presented. Then, in Sections 12.3 and 12.4, the problem formulation and the MPPT controller design are evinced. Finally, in Sections 12.5–12.7, the respective numerical experiments, discussion, and conclusion of this study are provided.

12.2 Related work

The energy received from the sun is unlimited; however, to take advantage of this resource, it is necessary to consider that it is intermittent energy and that it can vary due to meteorological, topographic, latitude, and orientation factors of the solar energy receivers, in addition to the hours of sunshine that depend on each region. To make the most of the panels' energy, we use MPPT regulators since this regulator allows us to find the PV panel's maximum operating powerpoint. During its operation, this point of maximum power changes (due to variation in radiation and temperature), this regulator seeks to optimize the power that the panel can provide us. An MPPT regulator comprises a DC (direct current) converter; besides, the MPPTs have an algorithm to determine the maximum power point during the panel's operation. To determine the maximum power point, the current and voltage are sampled, and the current sample and the previous sample are compared to determine the Pmax. The most common algorithms of MPPT regulators are:

- disturbance and observation (P&O);
- incremental conductance; and
- fractional open circuit voltage [voltage constant (CV) and current].

12.2.1 Perturb and observe (P&O)

This algorithm is the most widely used because it is easy to implement. This algorithm aims to find the point of maximum power according to the amount of radiation that the panel receives. To locate the point of maximum power, the voltage and power are constantly being sensed. At the time of sampling, the current measurement value is compared with the previous measurement. Depending on the value of voltage and power, it takes the criterion in which direction the sample moves. The measured value decides whether the reference value stays or changes to the new measured value. Fig. 12.1 shows the power−voltage curve of an arrangement of PV panels at different radiation levels, considering that the arrangement operates at point A, to reach the MPP, the sampling must be moved to the right (approaching MPP), it is observed that the maximum point in the different radiation graphs is at close voltage values. To locate the MPP, the algorithm considers certain criteria, including:

$$\text{If} \quad \Delta P = P(k) - P(k-1) > 0 \begin{cases} \Delta V = V(k) - V(k-1) > 0 & \text{Then} \quad V_r = V_r + \Delta V \\ \Delta V = V(k) - V(k-1) < 0 & \text{Then} \quad V_r = V_r - \Delta V \end{cases}$$

$$\text{If} \quad \Delta P = P(k) - P(k-1) < 0 \begin{cases} \Delta V = V(k) - V(k-1) > 0 & \text{Then} \quad V_r = V_r - \Delta V \\ \Delta V = V(k) - V(k-1) < 0 & \text{Then} \quad V_r = V_r + \Delta V \end{cases} \quad (12.1)$$

Depending on the curve's point where the array is operating, each of the above criteria locates the MPP. However, there is a disadvantage when locating this point. When it is located, the algorithm does not stop and continues sampling so it is unknown exactly when it locates it. It is constantly oscillating in it. Fig. 12.2 shows the algorithm's criteria. It senses the voltage and the current and, according to the difference in values in the parameters (if it is greater or less), it modifies the reference value of the voltage (Table 12.1).

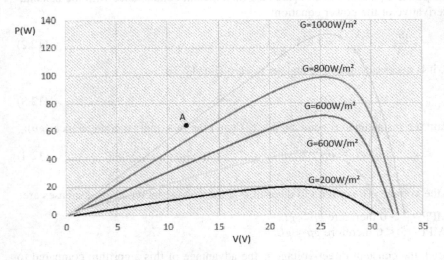

FIGURE 12.1 The power−voltage curve of an arrangement of photovoltaic panels.

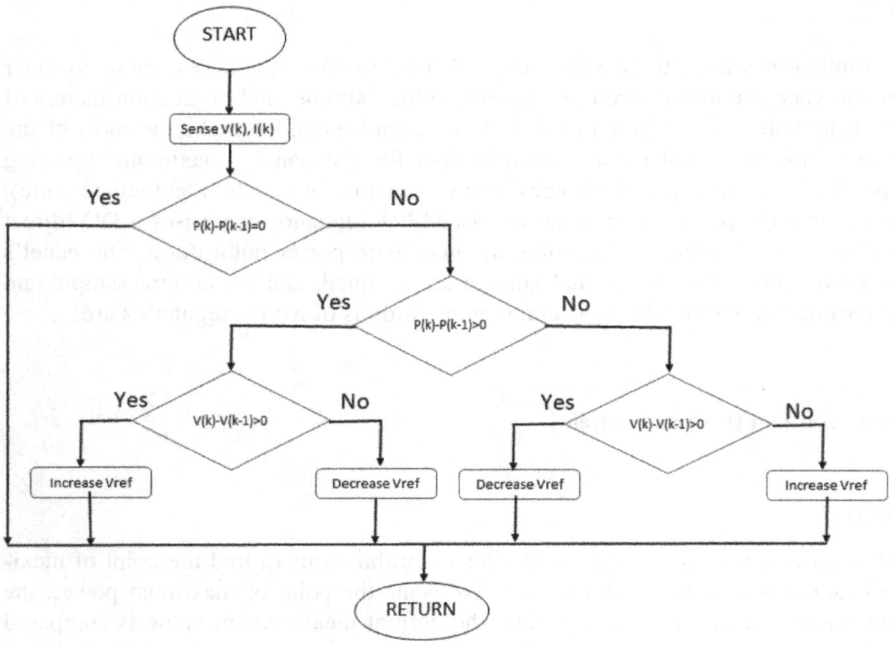

FIGURE 12.2 Algorithm of perturb and observe.

TABLE 12.1 Simulation parameters.

Parameter	Value
L	1.21 mH
C_1	1000 μF
C_2	1000 μF
R_c	39.6 Ω
R	25 Ω
E_g	1.1 eV

12.2.2 Incremental conductance

This algorithm is derived directly from the power equation and compares the incremental conductance with the instantaneous conductance in a PV system, the derivative of the power equation:

$$\frac{dP}{dV} = \frac{d(V.I)}{dV} = I + V.\frac{dI}{dV} \tag{12.2}$$

To obtain the maximum of a variable, it is necessary to derive and set to zero (Eq. 12.3):

$$\frac{dP}{dV} = 0 \tag{12.3}$$

Equating to zero and clearing, we obtain the instantaneous conductance (gL) and the incremental conductance (gp):

$$gL = -\frac{I}{V} \tag{12.4}$$

Therefore, $gL = gp$.

This algorithm increases or decreases the voltage. The criteria to determine if the voltage increases or decreases are:

If the working voltage is lower than MPP, $\frac{dP}{dV} > 0$ therefore $gp > gL$.
If the working voltage is higher than MPP, $\frac{dP}{dV} < 0$ therefore $gp < gL$.

Once the MPP ($gL = gp$) has been found, the constant direct voltage is the advantage of this algorithm compared to the P&O. Fig. 12.3 shows the areas where the incremental conductance algorithm operates.

12.2.3 Fractional open circuit voltage

This algorithm relates the maximum power point voltage and the open circuit voltage, this ratio being a constant K (Andujar, Bohorquez, & Enrique, 2006) (Fig. 12.4)

$$\frac{V_{mpp}}{V_{oc}} = K < 1 \tag{12.5}$$

To measure V_{oc}, the PV array is temporarily isolated. The MPPT regulator takes the K value as a reference and calculates the V_{mpp} and adjusts the array voltage until this value is reached. Fig. 12.5 shows a flow chart, which shows the logic that this algorithm follows. It is observed that to carry out the V_{oc} measurement, the PV panel arrangement must be isolated from the system.

Although this algorithm consists of adjusting the V_{mpp} taking as reference the V_{oc} and the constant k, it is complex to obtain the appropriate value of the constant k. The optimal value of k is not a fixed value, and it varies by 8%. By implementing this algorithm, it decreases the efficiency compared to the algorithms discussed previously. The selection of the value of k influences a lot. Another factor is when measuring V_{oc}, the arrangement must be disconnected from the system.

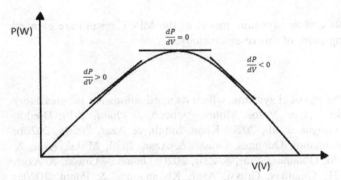

FIGURE 12.3 Areas where the incremental conductance algorithm operates.

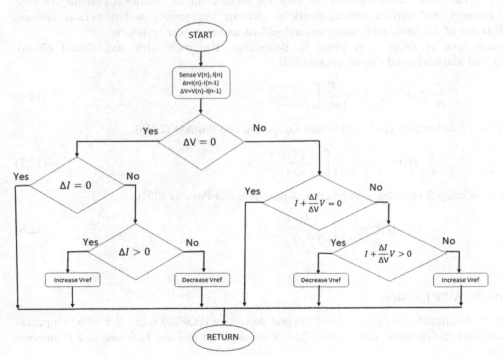

FIGURE 12.4 Algorithm of incremental conductance.

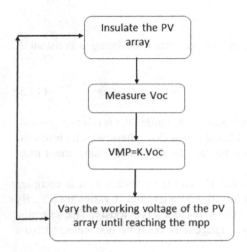

FIGURE 12.5 Algorithm fractional open circuit voltage.

12.3 Problem formulation

In this section, the fundamentals of fractional-order calculus and the dynamic model of the MPPT system are evinced to provide the theoretical background that serves as a starting point of this research study.

12.3.1 Fractional-order calculus

Fractional order attracted the interest of the community in the physical systems, which its applications are progressively extending (Alain, Azar, Kengne, & Bertrand, 2020; Bouakrif, Azar, Volos, Muñoz-Pacheco, & Pham, 2019; Djeddi, Dib, Azar, & Abdelmalek, 2019; Fekik et al., 2021; Kammogne et al., 2020; Khan, Singh, & Azar, 2020a, 2020b; Khennaoui, Ouannas, Boulaaras, Pham, & Azar, 2020; Khennaoui, Ouannas, Grassi, & Azar, 2020; Mittal, Azar, & Kamal, 2021; Ouannas, Azar, & Vaidyanathan, 2017a, 2017d; Ouannas, Azar, & Ziar, 2020; Ouannas, Grassi, & Azar, 2020a, 2020b; Ouannas, Grassi, Azar, & Khennaoui, 2021; Ouannas, Grassi, Azar, Khennaouia, & Pham, 2020a; Radwan, Azar, Vaidyanathan, Munoz-Pacheco, & Ouannas, 2017; Radwan, Emira, AbdelAty, & Azar, 2018; Singh & Azar, 2020b; Tolba et al., 2017). The main reasons behind the evolving enthusiasm for fractional systems are their extensive exposures, genetic features, and current developments in software engineering and numerical devices. Fractional-order control (FOC) is one of the fields with many research efforts and empowering results.

The Riemann−Liouville derivative of order α is given in Behinfaraz, Badamchizadeh, and Ghiasi (2016), Kubyshkin and Postnov (2016), and Matychyn and Onyshchenko (2018)

$$\frac{d^\alpha}{dt^\alpha}f(t) = \frac{1}{\Gamma(n-\alpha)}\frac{d^n}{dt^n}\int_0^t \frac{f^{(n)}(\tau)}{(t-\tau)^{n-\alpha+1}}d\tau \tag{12.6}$$

The Caputo derivative is given by Behinfaraz et al. (2016) and Kubyshkin and Postnov (2016):

$$\frac{d^\alpha}{dt^\alpha}f(t) = \frac{1}{\Gamma(n-\alpha)}\int_0^t \frac{f^{(n)}(\tau)}{(t-\tau)^{n-\alpha+1}}d\tau \tag{12.7}$$

and the left-side Riemann-Liouville integral of order α_i is given by Kubyshkin and Postnov (2016):

$$I_t^{\alpha_i}f_i(t) = \frac{1}{\Gamma(\alpha_i)}\int_0^t \frac{f_i(\tau)}{(t-\tau)^{1-\alpha_i}}d\tau \tag{12.8}$$

with $n-1 < \alpha \le n$.

12.3.2 Dynamic model of the MPPT system

The MPPT control systems for the fractional-order proportional integral derivative (FOPID) controller with compensation and the fractional-order terminal sliding mode control (FOTSMC) are depicted in Figs. 12.6 and 12.7 (Lamzouri et al., 2020; Yatimi et al., 2020).

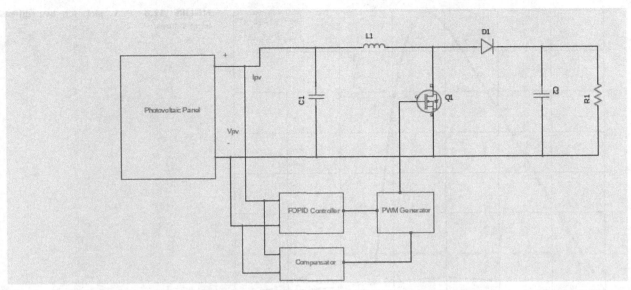

FIGURE 12.6 Block and circuit diagram of the fractional-order proportional integral derivative controller for maximum power point tracking.

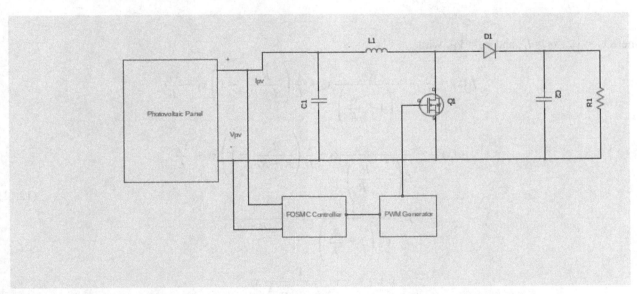

FIGURE 12.7 Block and circuit diagram of the fractional-order terminal sliding mode control for maximum power point tracking.

The current of the PV panel I_{pv} is given by (Lamzouri et al., 2020; Yatimi et al., 2020):

$$I_{pv} = N_p I_{ph} - N_p I_0 \left(\exp\left(\frac{q V_{pv}}{N_s AKT} \right) - 1 \right) \tag{12.9}$$

where V_{pv} is the voltage obtained from the PV panel, q is the electron charge, K is the Boltzmann constant, I_{ph} is the PV cell photocurrent, I_0 is the reverse saturation current, and N_p and N_s are the parallel and series cell number, respectively, T is the temperature, and the power is given by $P_{pv} = V_{pv} I_{pv}$. The open loop dynamics of the PV system is given by the following system (Lamzouri et al., 2020):

$$\dot{x}_1 = \frac{1}{C_1} (-x_2 + I_{pv}) = f_0(x)$$

$$\dot{x}_2 = f_1(x) + g_1(x)u \tag{12.10}$$

$$\dot{x}_3 = f_2(x) + g_2(x)u$$

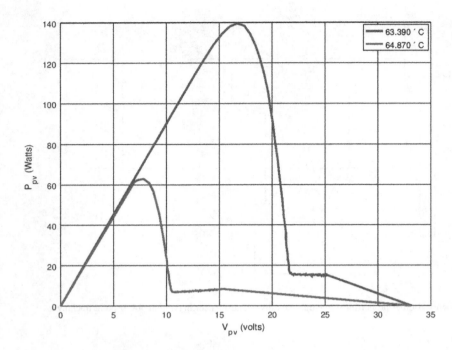

FIGURE 12.8 $P–V$ plot for two different temperatures.

where $x_1 = v_{pv}$, $x_2 = I_L$, and $x_3 = V_{C_2}$ with:

$$f_1(x) = \frac{x_1}{L} - \frac{R_C}{L\left(1 + \dfrac{R_C}{R}\right)}x_2 + \frac{1}{L}\left(\frac{R_C}{R + R_C} - 1\right)x_3 - \frac{V_D}{L}$$

$$g_1(x) = -\frac{R_C}{L\left(1 + \dfrac{R_C}{R}\right)}x_2 - \frac{1}{L}\left(\frac{R_C}{R + R_C} - 1\right)x_3 + \frac{V_D}{L}$$

$$f_2(x) = \frac{1}{C_2\left(1 + \dfrac{R_C}{R}\right)}x_2 - \frac{1}{C_2(R + R_C)}x_3$$

$$g_2(x) = -\frac{1}{C_2\left(1 + \dfrac{R_C}{R}\right)}x_2$$

(12.11)

The $P–V$ characteristics of the PV panel are shown in Fig. 12.8.

12.4 Fractional-order design techniques for MPPT of photovoltaic panels

As explained before, in this section, the derivation of an FOPID controller with compensation for MPPT of PV panels along with an FOTSMC for the same purpose is evinced. In the first case, considering that the FOPID controller is divided into two parts, it considers a compensator obtained by adding an extra variable to ensure the convergence to the desired maximum power extracted from the PV panel. For the second case, the FOTSMC for MPPT is designed by selecting an appropriate sliding variable and Lyapunov function to obtain the desired control law. Lyapunov functions have been used in various contexts (stability, convergence analysis, design of model reference adaptive systems, etc.). The main obstacle to the use of Lyapunov theory is in finding a suitable Lyapunov function (Abdul-Adheem, Azar, Ibraheem, & Humaidi, 2020; Abdul-Adheem, Ibraheem, Azar, & Humaidi, 2020; Ajeil, Ibraheem, Azar, & Humaidi,

2020; Alain, Azar, Bertrand, & Romanic, 2019; Azar, Adele, Alain, Kengne, & Bertrand, 2018; Humaidi, Ibraheem, Azar, & Sadiq, 2020; Najm, Ibraheem, Azar, & Humaidi, 2020; Ouannas, Azar, & Ziar, 2017, 2019; Vaidyanathan et al., 2018, 2019).

For the derivation of the two fractional-order controllers, the following voltage error variable is considered $e_1 = x_{1\text{ref}} - x_1$ where $x_{1\text{ref}}$ is the desired MPPT voltage.

12.4.1   FOPID MPPT controller design

For the FOPID MPPT controller design, consider the following controller u for system (12.10) where $u = u_{\text{FOPID}} + u_c$:

$$u_{\text{FOPID}} = K_P e_1 + K_I \frac{d^{-\lambda}}{dt^{-\lambda}} e_1 + K_D \frac{d^{\mu}}{dt^{\mu}} e_1 \tag{12.12}$$

for $K_P, K_I, K_D > 0$ and u_c is the compensator input shown later. To find the compensator input, consider the following theorem.

Theorem 12.1: *To find the compensator input u_c*

$$u_c = - K_P e_1 - \frac{1}{g_1(x)x_2 + g_2(x)x_3} x_2 f_1 - \frac{1}{g_1(x)x_2 + g_2(x)x_3} x_1 f_0(x)$$
$$- \frac{1}{g_1(x)x_2 + g_2(x)x_3} x_3 f_2 \tag{12.13}$$

The following augmented state x_4 and its derivative are needed in order to make the error e_1 to approximate zero in finite time:

$$\dot{x}_4 = - \frac{1}{x_4} \exp(2e_1) - \frac{1}{x_4}(g_1(x)x_2 + g_2(x)x_3)\left(K_I \frac{d^{-\lambda}}{dt^{-\lambda}} e_1 + K_D \frac{d^{\mu}}{dt^{\mu}} e_1 \right)$$
$$- \frac{1}{x_4} K_P |e_1| \tag{12.14}$$

and ensure the stability of the closed-loop system.

Proof: . Consider the following Lyapunov function:

$$V = \frac{1}{2}x_1^2 + \frac{1}{2}x_2^2 + \frac{1}{2}x_3^2 + \frac{1}{2}x_4^2 \tag{12.15}$$

Now taking the derivative of the Lyapunov function (12.15) along Eq. (12.10) with the FOPID controller (Eq. 12.12) is obtained (Vaidyanathan & Azar, 2015a, 2015b, 2015c, 2015d, 2016a, 2016c, 2016e):

$$\dot{V} = x_1 f_0(x) + x_2 (f_1(x) + g_1(x)u_{\text{FOPID}} + g_1(x)u_c)$$
$$+ x_3 (f_2(x) + g_2(x)u_{\text{FOPID}} + g_2(x)u_c) + x_4 \dot{x}_4 \tag{12.16}$$

Now substituting Eq. (12.14) and rearranging yields:

$$\dot{V} = x_1 f_0(x) + x_2 f_1(x) + (g_1(x)x_2 + g_2(x)x_3)K_P e_1 - \exp(2e_1)$$
$$- K_P |e_1| + (g_1(x)x_2 + g_2(x)x_3)u_c + x_3 f_2(x) \tag{12.17}$$

Now substituting Eq. (12.13) in Eq. (12.17) produces:

$$\dot{V} = - \exp(2e_1) - K_P |e_1| < 0 \tag{12.18}$$

So the stability of the system is ensured to make the variable e_1 to reach zero in finite time (Grassi et al., 2017; Ouannas et al., 2017; Singh et al., 2017; Singh, Azar, & Zhu, 2018; Vaidyanathan & Azar, 2016d, 2016f).

12.4.2 FOTSMC MPPT controller design

For the FOTSMC MPPT controller, consider system (12.10) and the error variable e_1 explained in this section. For the development of the control law, consider the following theorem:

Theorem 12.2: *Taking into consideration the following sliding variable*:

$$s_1 = D^{\alpha-1}e_1 + K_1 D^{-\alpha}(e_1 + \text{sign}(e_1)|e_1|^\rho) \tag{12.19}$$

with α is the fractional-order derivative constant, $K_1 > 0$ is the sliding gain, and $\rho > 0$ is the absolute value constant. By selecting an appropriate Lyapunov function, the following FOTSMC u is obtained (Alain et al., 2018; Pham et al., 2017; Vaidyanathan, Azar, & Ouannas, 2017a, 2017b; Vaidyanathan et al., 2018; Vaidyanathan, Zhu, & Azar, 2017; Volos, Pham, Azar, Stouboulos, & Kyprianidis, 2018):

$$u = \frac{1}{(x_2 g_1(x) + x_3 g_2(x))} \Phi$$
$$\Phi = -x_1 f_0 - x_2 f_1 - x_3 f_2 - s_1 D^\alpha e_1 - K_1 s_1 D^{-\alpha+1}(e_1 + \text{sign}(e_1)|e_1|^\rho)$$
$$- K|e_1| \tag{12.20}$$

where $K > 0$ is adjusting gain parameter.

Proof.: Consider the following Lyapunov function:

$$V = \frac{1}{2}x_1^2 + \frac{1}{2}x_2^2 + \frac{1}{2}x_3^2 + \frac{1}{2}s_1^2 \tag{12.21}$$

Taking the derivative of Eq. (12.21) along Eq. (12.10) yields (Ouannas, Grassi, Azar, & Gasri, 2019; Singh & Azar, 2020a; Vaidyanathan & Azar, 2016g; Vaidyanathan, Azar, Rajagopal, & Alexander, 2015; Vaidyanathan, Sampath, & Azar, 2015)

$$\dot{V} = x_1 f_0(x) + x_2 f_1(x) + (x_2 g_1(x) + x_3 g_2(x))u + x_3 f_2(x)$$
$$+ s_1 D^\alpha e_1 + K_1 s_1 D^{-\alpha+1}(e_1 + \text{sign}(e_1)|e_1|^\rho) \tag{12.22}$$

Now substituting Eq. (12.20) into Eq. (12.22) yields:

$$\dot{V} = -K|e_1| < 0 \tag{12.23}$$

and this completes the proof so the stability of the closed loop system is ensured (Ouannas, Azar, & Abu-Saris, 2017; Ouannas, Azar, & Vaidyanathan, 2017b, 2017c; Vaidyanathan, Azar, & Boulkroune, 2018; Wang, Volos, Kingni, Azar, & Pham, 2017).

12.5 Numerical experiments

In this section, two numerical experiments are performed, one for the FOPID MPPT controller and the second for the FOTSMC for MPPT with the following parameters:

12.5.1 Experiment 1: FOPID MPPT controller

Fig. 12.9 shows that the error variable V_{pv} reaches the final desired variable according to the desired power to obtain the optimum MPPT. Fig. 12.10 shows the error variable $e_1 = V_{pvref} - V_{pv}$. It can be noticed how the desired error reaches the zero value when times goes to infinity; this means that the desired voltage value according to the MPPT as shown in Fig. 12.8 is tracked accurately during the whole simulation time.

Meanwhile, in Fig. 12.11, the evolution in time of the desired tracked maximum power of the PV cell is obtained. As noted in Fig. 12.12 in which the maximum power extraction of the PV cell is extracted until this variable reaches the zero value in finite time $P_1 = P_{pvref} - P_{pv}$.

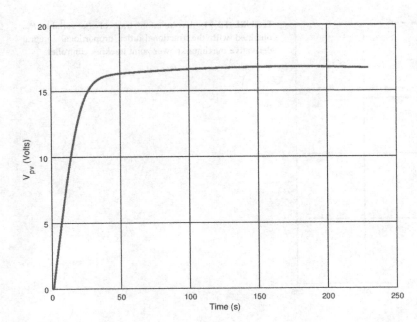

FIGURE 12.9 Evolution in time of the variable $x_1 = V_{pv}$ obtained with the fractional-order proportional integral derivative maximum power point tracking controller.

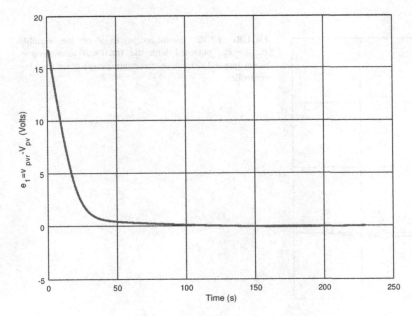

FIGURE 12.10 Evolution in time of the variable $e_1 = V_{pvref} - V_{pv}$ obtained with the fractional-order proportional integral derivative maximum power point tracking controller.

Finally, in Fig. 12.13, the input voltage generated by the MPPT controller is injected in the power electronics as a pulse width modulated signal. It can be noticed that the input voltage is not higher with a small overshoot, something that is desirable to avoid the saturation of the voltage input.

12.5.2 Experiment 2: FOTSMC for MPPT

In Fig. 12.14, the time evolution of the variable V_{pv} is shown. It is noted that this variable reaches the final value for MPPT at a finite time, taking into account the reference voltage as shown in Fig. As confirmed in Fig. 12.15, the error variable $e_1 = x_{pvref} - x_{pv}$ comes to zero at the final time in order to maintain the desired maximum power output of the PV panel.

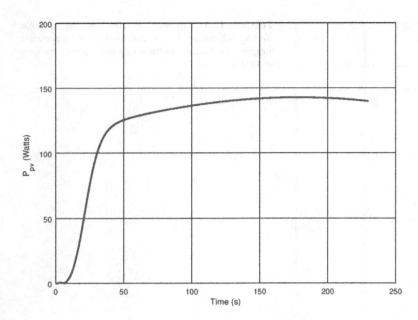

FIGURE 12.11 Evolution in time of the variable P_{pv} obtained with the fractional-order proportional integral derivative maximum power point tracking controller.

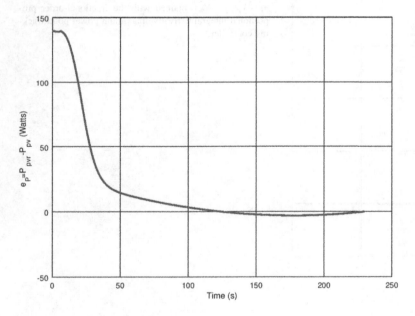

FIGURE 12.12 Evolution in time of the variable $P_{pvref} - P_{pv}$ obtained with the fractional-order proportional integral derivative maximum power point tracking controller.

In Fig. 12.16, the evolution of the MPPT reference power in time is shown, noting that the desired value is reached at a finite time, demonstrating the effectiveness of the MPPT FOTSMC. This result is confirmed in Fig. 12.17 in which the power error $P_{pvref} - P_{pv}$ is shown and, as noted, this error reaches the zero value in the finite time so that the desired power value is obtained by the FOTSMC for MPPT purposes.

In Fig. 12.18, the input voltage generated by the FOTSMC for MPPT is shown as the input voltage used for the pulse width modulated signal for the power circuit. At last, in Fig. 12.19, it is noted how the sliding variable s_1 reaches its origin at a finite time, ensuring the convergence of the desired maximum power of the PV panel.

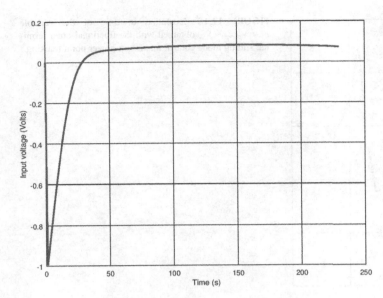

FIGURE 12.13 Evolution in time of the input variable u obtained with the fractional-order proportional integral derivative maximum power point tracking controller.

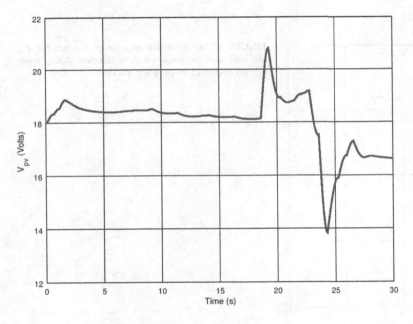

FIGURE 12.14 Evolution in time of the variable $x_1 = v_{pv}$ obtained with the fractional-order terminal sliding mode control maximum power point tracking.

12.6 Discussion

It has been established that fractional calculus gives a more realistic modeling of systems and the best performance is achieved when a fractional-order controller is employed for a fractional-order system (AbdelAty, Azar, Vaidyanathan, Ouannas, & Radwan, 2018; Alain et al., 2018; Azar, Kumar, Kumar, & Rana, 2018; Azar, Ouannas, & Singh, 2018; Azar & Serrano, 2018, 2019; Azar, Serrano, & Koubaa, 2020; Azar, Serrano, & Vaidyanathan, 2018; Fekik et al., 2020; Khettab, Bensafia, Bourouba, & Azar, 2018; Kumar, Azar, Kumar, & Rana, 2018; Meghni, Dib, Azar, Ghoudelbourk, & Saadoun, 2017; Ounnas, Azar, & Radwan, 2016; Ouannas, Azar, Ziar, & Radwan, 2017a, 2017b; Ouannas, Azar, Ziar, & Vaidyanathan, 2017a, 2017b, 2017c; Ouannas, Grassi, Azar, Khennaouia, & Pham, 2020b; Ouannas, Grassi, Azar, & Singh, 2019; Pham, G, Kapitaniak, Volos, & Azar, 2018; Pham et al., 2017; Singh, Azar, Vaidyanathan,

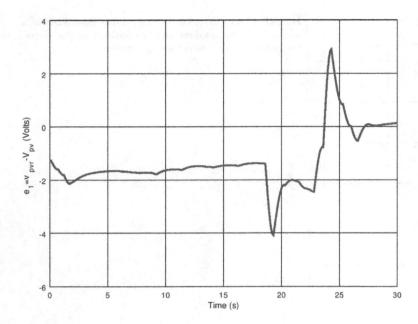

FIGURE 12.15 Evolution in time of the variable $e_1 = V_{pvref} - V_{pv}$ obtained with the fractional-order terminal sliding mode control maximum power point tracking.

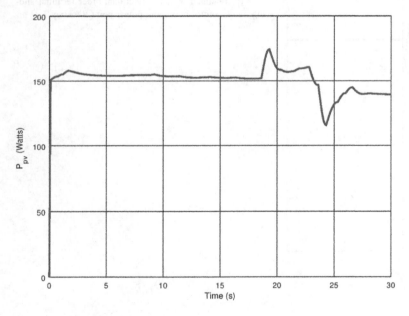

FIGURE 12.16 Evolution in time of the variable P_{pv} obtained with the fractional-order terminal sliding mode control maximum power point tracking.

Ouannas, & Bhat, 2018; Soliman et al., 2017). The theoretical and experimental results obtained in this research study show that the fractional-order techniques are a suitable and appropriate strategy considering the abundant nonlinearities found in a PV system's mathematical model, especially in its power electronics in which DC−DC converters are implemented for MPPT purposes. The FOPID MPPT controller provides an efficient and relatively simple control strategy for MPPT purposes. Considering that this study is developed an FOPID controller theoretically with compensation, it is proved that the stability conditions provide the additional compensation input considering that it will be helpful when there are changes in the climate or environmental conditions surrounding the PV systems. In the second case, the designed TMSC for MPPT purposes provides a strategy capable of dealing with uncertainties, which is an advantage compared to the FOPID MPPT controller. While observing the experimental results, it is concluded that both strategies

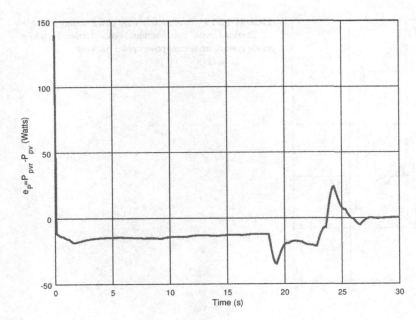

FIGURE 12.17 Evolution in time of the variable $P_{pvref} - P_{pv}$ obtained with the fractional-order terminal sliding mode control maximum power point tracking.

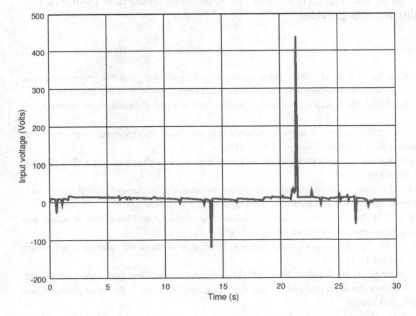

FIGURE 12.18 Evolution in time of the input variable u obtained with the fractional-order terminal sliding mode control maximum power point tracking.

are very accurate and with fast response. Still, it is important to remark that the FOTSMC for MPPT provides a smaller control action than the FOPID controller for MPPT purposes, which is an advantage considering the power electronics used in the DC–DC converter.

12.7   Conclusion

In this chapter, the design of two fractional-order control strategies for the MPPT of PV panels is presented. The first technique is an FOPID MPPT controller, in which the control law is divided into two parts, adding an extra input for compensation purposes. This controller is designed by adding an extra state variable to find the Lyapunov theorem's stability conditions and the respective compensation input. The FOTSMC for MPPT purpose consists of an appropriate

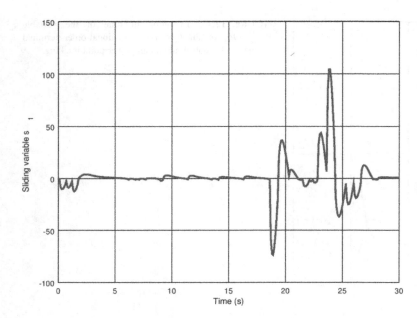

FIGURE 12.19 Evolution in time of the sliding variable s_1 obtained with the fractional-order terminal sliding mode control maximum power point tracking.

sliding variable, and by using the Lyapunov theorem, the respective control law is obtained. Numerical examples validate the theoretical results, and the respective discussion is provided.

References

AbdelAty, A. M., Azar, A. T., Vaidyanathan, S., Ouannas, A., & Radwan, A. G. (2018). *Applications of continuous-time fractional order chaotic systems. Mathematical techniques of fractional order systems. Advances in nonlinear dynamics and chaos (ANDC)* (pp. 409−449). Elsevier.

Abdul-Adheem, W., Azar, A. T., Ibraheem, I., & Humaidi, A. (2020). Novel active disturbance rejection control based on nested linear extended state observers. *Applied Sciences, 10*(12), 4069.

Abdul-Adheem, W., Ibraheem, I., Azar, A. T., & Humaidi, A. (2020). Improved active disturbance rejection-based decentralized control for mimo nonlinear systems: Comparison with the decoupled control scheme. *Applied Sciences, 10*(7), 2515.

Ahmadi, S., Abdi, S., & Kakavand, M. (2017). Maximum power point tracking of a proton exchange membrane fuel cell system using PSO-PID controller. *International Journal of Hydrogen Energy, 42*(32), 20430−20443.

Ajeil, F. H., Ibraheem, I. K., Azar, A. T., & Humaidi, A. J. (2020). Autonomous navigation and obstacle avoidance of an omnidirectional mobile robot using swarm optimization and sensors deployment. *International Journal of Advanced Robotic Systems, 17*(3), 1−15.

Alain, K. S. T., Azar, A. T., Bertrand, F. H., & Romanic, K. (2019). Robust observer-based synchronisation of chaotic oscillators with structural perturbations and input nonlinearity. *International Journal of Automation and Control, 13*(4), 387−412.

Alain, K. S. T., Azar, A. T., Kengne, R., & Bertrand, F. H. (2020). Stability analysis and robust synchronisation of fractional-order modified colpitts oscillators. *International Journal of Automation and Control, 14*(1), 52−79.

Alain, K. S. T., Romanic, K., Azar, A. T., Vaidyanathan, S., Bertrand, F. H., & Adele, N. M. (2018). *Dynamics analysis and synchronization in relay coupled fractional order colpitts oscillators. Advances in system dynamics and control, advances in systems analysis, software engineering, and high performance computing (ASASEHPC)* (pp. 317−356). IGI Global.

Al-Dhaifallah, M., Nassef, A. M., Rezk, H., & Nisar, K. S. (2018). Optimal parameter design of fractional order control based INC-MPPT for PV system. *Solar Energy, 159*, 650−664.

Alimi, M., Rhif, A., Rebai, A., Vaidyanathan, S., & Azar, A. T. (2021). *Optimal adaptive backstepping control for chaos synchronization of nonlinear dynamical systems. Backstepping control of nonlinear dynamical systems. Advances in nonlinear dynamics and chaos (ANDC)* (pp. 291−345). Academic Press.

Amara, K., Malek, A., Bakir, T., Fekik, A., Azar, A. T., Almustafa, K. M., ... Hocine, D. (2019). Adaptive neuro-fuzzy inference system based maximum power point tracking for stand-alone photovoltaic system. *International Journal of Modelling, Identification and Control, 33*(4), 311−321.

Ammar, H. H., & Azar, A. T. (2020). Robust path tracking of mobile robot using fractional order pid controller. In: *The international conference on advanced machine learning technologies and applications (AMLTA2019), Vol. 921 of advances in intelligent systems and computing* (pp. 370−381). Cham: Springer International Publishing.

Ammar, H. H., Azar, A. T., Shalaby, R., & Mahmoud, M. I. (2019). Metaheuristic optimization of fractional order incremental conductance (FO-INC) maximum power point tracking (MPPT). *Complexity, 2019*, 1−13.

Ammar, H. H., Azar, A.T., Tembi, T.D., Tony, K., & Sosa, A. (2018). Design and implementation of fuzzy pid controller into multi agent smart library system prototype. In: *The international conference on advanced machine learning technologies and applications (AMLTA2018), Vol. 723 of advances in intelligent systems and computing* (pp. 127−137). Cham: Springer International Publishing.

Andujar, J., Bohorquez, M. A., & Enrique, J. (2006). A new system for monitoring the maximum power point (mpp) in photovoltaic systems. *Comite Espanol de Automatica*, 633−640.

Arsalan, M., Iftikhar, R., Ahmad, I., Hasan, A., Sabahat, K., & Javeria, A. (2018). MPPT for photovoltaic system using nonlinear backstepping controller with integral action. *Solar Energy*, *170*, 192−200.

Azar, A. T., Adele, N. M., Alain, K. S. T., Kengne, R., & Bertrand, F. H. (2018). Multistability analysis and function projective synchronization in relay coupled oscillators. *Complexity*, *2018*, 1−12.

Azar, A. T., Ali, N., Makarem, S., Diab, M. K., & Ammar, H. H. (2020). Design and implementation of a ball and beam pid control system based on metaheuristic techniques. In: *Proceedings of the international conference on advanced intelligent systems and informatics 2019, Vol. 1058 of advances in intelligent systems and computing* (pp. 313−325). Cham: Springer International Publishing.

Azar, A. T., Ammar, H. H., Barakat, M. H., Saleh, M. A., & Abdelwahed, M. A. (2019). Self-balancing robot modeling and control using two degree of freedom pid controller. In: *Proceedings of the international conference on advanced intelligent systems and informatics 2018, Vol. 845 of advances in intelligent systems and computing* (pp. 64−76). Cham: Springer International Publishing.

Azar, A. T., Ammar, H. H., Beb, M. Y., Garces, S. R., & Boubakari, A. (2020). Optimal design of PID controller for 2-DOF drawing robot using bat-inspired algorithm. In: *Proceedings of the international conference on advanced intelligent systems and informatics 2019, Vol. 1058 of advances in intelligent systems and computing* (pp. 175−186). Cham: Springer International Publishing.

Azar, A. T., Ammar, H. H., Ibrahim, Z. F., Ibrahim, H. A., Mohamed, N. A., & Taha, M. A. (2020), Implementation of PID controller with pso tuning for autonomous vehicle. In: *Proceedings of the international conference on advanced intelligent systems and informatics 2019, Vol. 1058 of advances in intelligent systems and computing* (pp. 288−299). Cham: Springer International Publishing.

Azar, A. T., Hassan, H., Razali, M. S. A. B., de Brito Silva, G., & Ali, H. R. (2019). Two-degree of freedom proportional integral derivative (2-DOF PID) controller for robotic infusion stand. In: *Proceedings of the international conference on advanced intelligent systems and informatics 2018, Vol. 845 of advances in intelligent systems and computing* (pp. 13−25). Cham: Springer International Publishing.

Azar, A. T., Kumar, J., Kumar, V., & Rana, K. P. S. (2018). Control of a two link planar electrically-driven rigid robotic manipulator using fractional order sofc. In: *Proceedings of the international conference on advanced intelligent systems and informatics 2017, Vol. 639 of advances in intelligent systems and computing* (pp. 57−68). Cham: Springer International Publishing.

Azar, A. T., Ouannas, A., & Singh, S. (2018). Control of new type of fractional chaos synchronization. In: *Proceedings of the international conference on advanced intelligent systems and informatics 2017, Vol. 639 of advances in intelligent systems and computing* (pp. 47−56). Cham: Springer International Publishing.

Azar, A. T., Sayed, A. S., Shahin, A. S., Elkholy, H. A., & Ammar, H. H. (2020). PID controller for 2-DOFs twin rotor mimo system tuned with particle swarm optimization. In: *Proceedings of the international conference on advanced intelligent systems and informatics 2019, Vol. 1058 of advances in intelligent systems and computing* (pp. 229−242). Cham: Springer International Publishing.

Azar, A. T., & Serrano, F. E. (2015). *Design and modeling of anti wind up PID controllers* (pp. 1−44). Cham: Springer International Publishing.

Azar, A. T., & Serrano, F. E. (2018). Fractional order sliding mode PID controller/observer for continuous nonlinear switched systems with pso parameter tuning. In: *The international conference on advanced machine learning technologies and applications (AMLTA2018), Vol. 723 of advances in intelligent systems and computing* (pp. 13−22). Cham: Springer International Publishing.

Azar, A. T., & Serrano, F. E. (2019). Fractional order two degree of freedom pid controller for a robotic manipulator with a fuzzy type-2 compensator. In: *Proceedings of the international conference on advanced intelligent systems and informatics 2018, Vol. 845 of advances in intelligent systems and computing* (pp. 77−88). Cham: Springer International Publishing.

Azar, A. T., Serrano, F. E., Flores, M. A., Vaidyanathan, S., & Zhu, Q. (2020). Adaptive neural-fuzzy and backstepping controller for port-hamiltonian systems. *International Journal of Computer Applications in Technology*, *62*(1), 1−12.

Azar, A. T., Serrano, F. E., & Koubaa, A. (2020). Adaptive fuzzy type-2 fractional order proportional integral derivative sliding mode controller for trajectory tracking of robotic manipulators. In: *2020 IEEE international conference on autonomous robot systems and competitions (ICARSC)* (pp. 183−187).

Azar, A. T., Serrano, F. E., Koubaa, A., & Kamal, N. A. (2020). Backstepping h-infinity control of unmanned aerial vehicles with time varying disturbances. In: *2020 first international conference of smart systems and emerging technologies (SMARTTECH)* (pp. 243−248).

Azar, A. T., Serrano, F. E., & Vaidyanathan, S. (2018). *Sliding mode stabilization and synchronization of fractional order complex chaotic and hyperchaotic systems. Mathematical techniques of fractional order systems. Advances in nonlinear dynamics and chaos (ANDC)* (pp. 283−317). Elsevier.

Azar, A. T., Serrano, F. E., Vaidyanathan, S., & Kamal, N. A. (2021). *Backstepping control and synchronization of chaotic time delayed systems. Backstepping control of nonlinear dynamical systems. Advances in nonlinear dynamics and chaos (ANDC)* (pp. 407−424). Academic Press.

Bansal, N., Bisht, A., Paluri, S., Kumar, V., Rana, K., Azar, A. T., & Vaidyanathan, S. (2021). *Single-link flexible joint manipulator control using backstepping technique. Backstepping control of nonlinear dynamical systems. Advances in nonlinear dynamics and chaos (ANDC)* (pp. 375−406). Academic Press.

Behinfaraz, R., Badamchizadeh, M., & Ghiasi, A. (2016). An adaptive method to parameter identification and synchronization of fractional-order chaotic systems with parameter uncertainty. *Applied Mathematical Modelling*, *40*(7−8), 4468−4479.

Ben Smida, M., Sakly, A., Vaidyanathan, S., & Azar, A. T. (2018). *Control-based maximum power point tracking for a grid-connected hybrid renewable energy system optimized by particle swarm optimization. Advances in system dynamics and control, advances in systems analysis, software engineering, and high performance computing (ASASEHPC)* (pp. 58–89). IGI Global.

Bouakrif, F., Azar, A. T., Volos, C. K., Muñoz-Pacheco, J. M., & Pham, V.-T. (2019). Iterative learning and fractional order control for complex systems. *Complexity, 2019*, 7958625.

Djeddi, A., Dib, D., Azar, A. T., & Abdelmalek, S. (2019). Fractional order unknown inputs fuzzy observer for takagi–sugeno systems with unmeasurable premise variables. *Mathematics, 7*(10), 984.

Drhorhi, I., El Fadili, A., Berrahal, C., Lajouad, R., El Magri, A., Giri, F., ... Vaidyanathan, S. (2021). *Adaptive backstepping controller for DFIG-based wind energy conversion system. Backstepping control of nonlinear dynamical systems. Advances in nonlinear dynamics and chaos (ANDC)* (pp. 235–260). Academic Press.

Dursun, E. H., & Kulaksiz, A. A. (2020). Second-order sliding mode voltage-regulator for improving MPPT efficiency of PMSG-based WECS. *International Journal of Electrical Power and Energy Systems, 121*, 106149.

Eltawil, M. A., & Zhao, Z. (2013). MPPT techniques for photovoltaic applications. *Renewable and Sustainable Energy Reviews, 25*, 793–813.

Fekik, A., Azar, A. T., Denoun, H., Kamal, N. A., Hamida, M. L., Kais, D., & Amara, K. (2021). *A backstepping direct power control of three phase pulse width modulated rectifier. Soft computing applications* (pp. 445–456). Cham: Springer International Publishing.

Fekik, A., Azar, A. T., Kamal, N. A., Denoun, H., Almustafa, K. M., Hamida, L., & Zaouia, M. (2021). *Fractional-order control of a fuel cell–boost converter system. Advanced machine learning technologies and applications* (pp. 713–724). Singapore: Springer.

Fekik, A., Azar, A. T., Kamal, N. A., Serrano, F., Hamida, M. L., Denoun, H., & Yassa, N. (2020). Maximum power extraction from a photovoltaic panel connected to a multi-cell converter. In: *International conference on advanced intelligent systems and informatics* (pp. 873–882). Cham: Springer.

Fekik, A., Denoun, H., Azar, A. T., Kamal, N. A., Zaouia, M., Benyahia, N., ... Vaidyanathan, S. (2021). *Direct power control of three-phase PWM-rectifier with backstepping control. Backstepping control of nonlinear dynamical systems. Advances in nonlinear dynamics and chaos (ANDC)* (pp. 215–234). Academic Press.

Fekik, A., Denoun, H., Azar, A. T., Koubaa, A., Kamal, N. A., Zaouia, M., ... Yassa, N. (2020). Adapted fuzzy fractional order proportional-integral controller for dc motor. In: *2020 first international conference of smart systems and emerging technologies (SMARTTECH)* (pp. 1–6).

Ghoudelbourk, S., Dib, D., Omeiri, A., & Azar, A. T. (2016). Mppt control in wind energy conversion systems and the application of fractional control (piα) in pitch wind turbine. *International Journal of Modelling, Identification and Control, 26*(2), 140–151.

Gorripotu, T. S., Samalla, H., Jagan Mohana Rao, C., Azar, A. T., & Pelusi, D. (2019). *Tlbo algorithm optimized fractional-order PID controller for agc of interconnected power system. Soft computing in data analytics* (pp. 847–855). Singapore: Springer.

Grassi, G., Ouannas, A., Azar, A. T., Radwan, A. G., Volos, C., Pham, V.-T., ... Stouboulos, I. N. (2017). Chaos synchronisation of continuous systems via scalar signal. In: *6th International Conference on Modern Circuits and Systems Technologies (MOCAST)* (pp. 1–4), 4–6 May 2017. Thessaloniki, Greece: IEEE.

Humaidi, A. J., Ibraheem, I. K., Azar, A. T., & Sadiq, M. E. (2020). A new adaptive synergetic control design for single link robot arm actuated by pneumatic muscles. *Entropy, 22*(7), 723.

Husain, M. A., Tariq, A., Hameed, S., Arif, M. S. B., & Jain, A. (2017). Comparative assessment of maximum power point tracking procedures for photovoltaic systems. *Green Energy and Environment, 2*(1), 5–17.

Ibraheem, G. A. R., Azar, A. T., Ibraheem, I. K., & Humaidi, A. J. (2020). A novel design of a neural network-based fractional pid controller for mobile robots using hybridized fruit fly and particle swarm optimization. *Complexity, 2020*(2020), 1–18.

Ismail, Z., Varatharajoo, R., & Chak, Y.-C. (2020). A fractional-order sliding mode control for nominal and underactuated satellite attitude controls. *Advances in Space Research, 66*(2), 321–334.

Kamal, N. A., Azar, A. T., Elbasuony, G. S., Almustafa, K. M., & Almakhles, D. (2020). PSO-based adaptive perturb and observe mppt technique for photovoltaic systems. In: *Proceedings of the international conference on advanced intelligent systems and informatics 2019, Vol. 1058 of advances in intelligent systems and computing* (pp. 125–135). Cham: Springer International Publishing.

Kamal, N. A., & Ibrahim, A. M. (2018). *Conventional, intelligent, and fractional-order control method for maximum power point tracking of a photovoltaic system: A review. Fractional order systems. Advances in nonlinear dynamics and chaos (ANDC)* (pp. 603–671). Academic Press.

Kammogne, A. S. T., Kountchou, M. N., Kengne, R., Azar, A. T., Fotsin, H. B., & Ouagni, S. T. M. (2020). Polynomial robust observer implementation based passive synchronization of nonlinear fractional-order systems with structural disturbances. *Frontiers of Information Technology & Electronic Engineering, 21*(9), 1369–1386.

Kchaou, A., Naamane, A., Koubaa, Y., & Sirdi, N. M. (2017). Second order sliding mode-based MPPT control for photovoltaic applications. *Solar Energy, 155*, 758–769.

Khan, A., Singh, S., & Azar, A. T. (2020a). Combination-combination anti-synchronization of four fractional order identical hyperchaotic systems. In: *The international conference on advanced machine learning technologies and applications (AMLTA2019), Vol. 921 of advances in intelligent systems and computing* (pp. 406–414). Cham: Springer International Publishing.

Khan, A., Singh, S., & Azar, A. T. (2020b). Synchronization between a novel integer-order hyperchaotic system and a fractional-order hyperchaotic system using tracking control. In: *The international conference on advanced machine learning technologies and applications (AMLTA2019), Vol. 921 of advances in intelligent systems and computing* (pp. 382–391). Cham: Springer International Publishing.

Khennaoui, A. A., Ouannas, A., Boulaaras, S., Pham, V.-T., & Azar, A. (2020). A fractional map with hidden attractors: chaos and control. *The European Physical Journal Special Topics, 229*, 1083–1093.

Khennaoui, A. A., Ouannas, A., Grassi, G., & Azar, A. T. (2020). Dynamic analysis of a fractional map with hidden attractor. In: *Proceedings of the international conference on artificial intelligence and computer vision (AICV2020)* (pp. 731–739). Cham: Springer International Publishing.

Khettab, K., Bensafia, Y., Bourouba, B., & Azar, A. T. (2018). *Enhanced fractional order indirect fuzzy adaptive synchronization of uncertain fractional chaotic systems based on the variable structure control: Robust h∞ design approach. Mathematical techniques of fractional order systems. Advances in nonlinear dynamics and chaos (ANDC)* (pp. 597–624). Elsevier.

Kubyshkin, V. A., & Postnov, S. S. (2016). Optimal control problem investigation for linear time-invariant systems of fractional order with lumped parameters described by equations with Riemann-Liouville derivative. *Journal of Control Science and Engineering, 2016*, 1–12.

Kumar, J., Azar, A. T., Kumar, V., & Rana, K. P. S. (2018). *Design of fractional order fuzzy sliding mode controller for nonlinear complex systems. Mathematical techniques of fractional order systems. Advances in nonlinear dynamics and chaos (ANDC)* (pp. 249–282). Elsevier.

Kumar, K. K., Bhaskar, R., & Koti, H. (2014). Implementation of MPPT algorithm for solar photovoltaic cell by comparing short-circuit method and incremental conductance method. *Procedia Technology, 12*, 705–715.

Lamzouri, F.-E., Boufounas, E.-M., Brahmi, A., & Amrani, A. E. (2020). Optimized TSMC control based MPPT for PV system under variable atmospheric conditions using PSO algorithm. *Procedia Computer Science, 170*, 887–892.

Logeswaran, T., & SenthilKumar, A. (2014). A review of maximum power point tracking algorithms for photovoltaic systems under uniform and non-uniform irradiances. *Energy Procedia, 54*, 228–235.

Matychyn, I., & Onyshchenko, V. (2018). On time-optimal control of fractional-order systems. *Journal of Computational and Applied Mathematics, 339*, 245–257.

Meghni, B., Dib, D., Azar, A. T., Ghoudelbourk, S., & Saadoun, A. (2017). *Robust adaptive supervisory fractional order controller for optimal energy management in wind turbine with battery storage. Fractional order control and synchronization of chaotic systems', Vol. 688 of studies in computational intelligence* (pp. 165–202). Cham: Springer International Publishing.

Mittal, S., Azar, A. T., & Kamal, N. A. (2021). Nonlinear fractional order system synchronization via combination-combination multi-switching. In: *Proceedings of the international conference on advanced intelligent systems and informatics 2020* (pp. 851–861). Cham: Springer International Publishing.

Modiri, A., & Mobayen, S. (2020). Adaptive terminal sliding mode control scheme for synchronization of fractional-order uncertain chaotic systems. *ISA Transactions, 105*, 33–50.

Murtaza, A., Chiaberge, M., Spertino, F., Shami, U. T., Boero, D., & Giuseppe, M. D. (2017). MPPT technique based on improved evaluation of photovoltaic parameters for uniformly irradiated photovoltaic array. *Electric Power Systems Research, 145*, 248–263.

Naghmash., Armghan, H., Ahmad, I., Armghan, A., Khan, S., & Arsalan, M. (2018). Backstepping based non-linear control for maximum power point tracking in photovoltaic system. *Solar Energy, 159*, 134–141.

Najm, A. A., Ibraheem, I. K., Azar, A. T., & Humaidi, A. J. (2020). Genetic optimization-based consensus control of multi-agent 6-DOF UAV system. *Sensors, 20*(12), 3576.

Ouannas, A., Azar, A. T., & Abu-Saris, R. (2017). A new type of hybrid synchronization between arbitrary hyperchaotic maps. *International Journal of Machine Learning and Cybernetics, 8*(6), 1887–1894.

Ounnas, A., Azar, A. T., & Radwan, A. G. (2016). On inverse problem of generalized synchronization between different dimensional integer-order and fractional-order chaotic systems. In: *2016 28th international conference on microelectronics (ICM)* (pp. 193–196).

Ouannas, A., Azar, A. T., & Vaidyanathan, S. (2017a). A new fractional hybrid chaos synchronisation. *International Journal of Modelling, Identification and Control, 27*(4), 314–322.

Ouannas, A., Azar, A. T., & Vaidyanathan, S. (2017b). New hybrid synchronisation schemes based on coexistence of various types of synchronisation between master-slave hyperchaotic systems. *International Journal of Computer Applications in Technology, 55*(2), 112–120. Available from https://www.inderscienceonline.com/doi/abs/10.1504/IJCAT.2017.082868.

Ouannas, A., Azar, A. T., & Vaidyanathan, S. (2017c). On a simple approach for q-s synchronization of chaotic dynamical systems in continuous-time. *International Journal of Computing Science and Mathematics, 8*(1), 20–27.

Ouannas, A., Azar, A. T., & Vaidyanathan, S. (2017d). A robust method for new fractional hybrid chaos synchronization. *Mathematical Methods in the Applied Sciences, 40*(5), 1804–1812.

Ouannas, A., Azar, A. T., & Ziar, T. (2017). On inverse full state hybrid function projective synchronization for continuous-time chaotic dynamical systems with arbitrary dimensions. *Differential Equations and Dynamical Systems, 28*, 1045–1058. Available from https://doi.org/10.1007/s12591-017-0362-x.

Ouannas, A., Azar, A. T., & Ziar, T. (2019). Control of continuous-time chaotic (hyperchaotic) systems: F-m synchronisation. *International Journal of Automation and Control, 13*(2), 226–242.

Ouannas, A., Azar, A. T., & Ziar, T. (2020). Fractional inverse full state hybrid projective synchronisation. *International Journal of Advanced Intelligence Paradigms, 17*(3–4), 279–298.

Ouannas, A., Azar, A. T., Ziar, T., & Radwan, A. G. (2017a). *Generalized synchronization of different dimensional integer-order and fractional order chaotic systems. Fractional order control and synchronization of chaotic systems, Vol. 688 of studies in computational intelligence* (pp. 671–697). Cham: Springer International Publishing.

Ouannas, A., Azar, A. T., Ziar, T., & Radwan, A. G. (2017b). *A study on coexistence of different types of synchronization between different dimensional fractional chaotic systems. Fractional order control and synchronization of chaotic systems, Vol. 688 of studies in computational intelligence* (pp. 637–669). Cham: Springer International Publishing.

Ouannas, A., Azar, A. T., Ziar, T., & Vaidyanathan, S. (2017a). *Fractional inverse generalized chaos synchronization between different dimensional systems. Fractional order control and synchronization of chaotic systems', Vol. 688 of studies in computational intelligence* (pp. 525−551). Cham: Springer International Publishing.

Ouannas, A., Azar, A. T., Ziar, T., & Vaidyanathan, S. (2017b). *A new method to synchronize fractional chaotic systems with different dimensions. Fractional order control and synchronization of chaotic systems, Vol. 688 of studies in computational intelligence* (pp. 581−611). Cham: Springer International Publishing.

Ouannas, A., Azar, A. T., Ziar, T., & Vaidyanathan, S. (2017c). *On new fractional inverse matrix projective synchronization schemes. Fractional order control and synchronization of chaotic systems, Vol. 688 of studies in computational intelligence* (pp. 497−524). Cham: Springer International Publishing.

Ouannas, A., Grassi, G., & Azar, A. T. (2020a). Fractional-order control scheme for q-s chaos synchronization. In: *The international conference on advanced machine learning technologies and applications (AMLTA2019)* (pp. 434−441). Cham: Springer International Publishing.

Ouannas, A., Grassi, G., & Azar, A. T. (2020b). A new generalized synchronization scheme to control fractional chaotic systems with non-identical dimensions and different orders. In: *The international conference on advanced machine learning technologies and applications (AMLTA2019), Vol. 921 of advances in intelligent systems and computing* (pp. 415−424). Cham: Springer International Publishing.

Ouannas, A., Grassi, G., Azar, A. T., & Gasri, A. (2019). A new control scheme for hybrid chaos synchronization. In *Proceedings of the international conference on advanced intelligent systems and informatics 2018, Vol. 845 of advances in intelligent systems and computing* (pp. 108−116). Cham: Springer International Publishing.

Ouannas, A., Grassi, G., Azar, A. T., & Khennaoui, A. A. (2021). *Synchronization control in fractional discrete-time systems with chaotic hidden attractors. Advanced machine learning technologies and applications* (pp. 661−669). Singapore: Springer Singapore.

Ouannas, A., Grassi, G., Azar, A. T., Khennaouia, A. A., & Pham, V.-T. (2020a). Chaotic control in fractional-order discrete-time systems. In: *Proceedings of the international conference on advanced intelligent systems and informatics 2019, Vol. 1058 of advances in intelligent systems and computing* (pp. 207−217). Cham: Springer International Publishing.

Ouannas, A., Grassi, G., Azar, A. T., Khennaouia, A.-A., & Pham, V.-T. (2020b). Synchronization of fractional-order discrete-time chaotic systems. In: *Proceedings of the international conference on advanced intelligent systems and informatics 2019* (pp. 218−228). Cham: Springer International Publishing.

Ouannas, A., Grassi, G., Azar, A. T., Radwan, A. G., Volos, C., Pham, V.-T., Ziar, T., Kyprianidis, I. M., & Stouboulos, I. N. (2017), Dead-beat synchronization control in discrete-time chaotic systems. In: *6th international conference on modern circuits and systems technologies (MOCAST)* (pp. 1−4), 4−6 May 2017. Thessaloniki, Greece: IEEE.

Ouannas, A., Grassi, G., Azar, A. T., & Singh, S. (2019). New control schemes for fractional chaos synchronization. In: *Proceedings of the international conference on advanced intelligent systems and informatics 2018, Vol. 845 of advances in intelligent systems and computing* (pp. 52−63). Cham: Springer International Publishing.

Pham, V, G, P. M., Kapitaniak, T., Volos, C., & Azar, A. T. (2018). *Dynamics, synchronization and fractional order form of a chaotic system with infinite equilibria. Mathematical techniques of fractional order systems. Advances in nonlinear dynamics and chaos (ANDC)* (pp. 475−502). Elsevier.

Pham, V.-T., Vaidyanathan, S., Azar, A. T., & Duy, V. H. (2021). *A new chaotic system without linear term, its backstepping control, and circuit design. Backstepping control of nonlinear dynamical systems. Advances in nonlinear dynamics and chaos (ANDC)* (pp. 33−52). Academic Press.

Pham, V.-T., Vaidyanathan, S., Volos, C. K., Azar, A. T., Hoang, T. M., & Van Yem, V. (2017). *A three-dimensional no-equilibrium chaotic system: Analysis, synchronization and its fractional order form. Fractional order control and synchronization of chaotic systems, Vol. 688 of studies in computational intelligence* (pp. 449−470). Cham: Springer International Publishing.

Pilla, R., Azar, A. T., & Gorripotu, T. S. (2019). Impact of flexible AC transmission system devices on automatic generation control with a metaheuristic based fuzzy PID controller. *Energies, 12*(21), 4193.

Pilla, R., Botcha, N., Gorripotu, T. S., & Azar, A. T. (2020). *Fuzzy PID controller for automatic generation control of interconnected power system tuned by glow-worm swarm optimization. Applications of robotics in industry using advanced mechanisms* (pp. 140−149). Cham: Springer International Publishing.

Radwan, A. G., Azar, A. T., Vaidyanathan, S., Munoz-Pacheco, J. M., & Ouannas, A. (2017). Fractional-order and memristive nonlinear systems: Advances and applications. *Complexity, 2017*, 1−2.

Radwan, A. G., Emira, A. A., AbdelAty, A. M., & Azar, A. T. (2018). Modeling and analysis of fractional order dc-dc converter. *ISA Transactions, 82*, 184−199.

Sallam, O. K., Azar, A. T., Guaily, A., & Ammar, H. H. (2020). Tuning of pid controller using particle swarm optimization for cross flow heat exchanger based on cfd system identification. In: *Proceedings of the international conference on advanced intelligent systems and informatics 2019, Vol. 1058 of advances in intelligent systems and computing* (pp. 300−312). Cham: Springer International Publishing.

Sambas, A., Vaidyanathan, S., Sukono., Azar, A. T., Hidayat, Y., Gundara, G., & Mohamed, M. A. (2021). *A novel chaotic system with a closed curve of four quarter-circles of equilibrium points: dynamics, active backstepping control, and electronic circuit implementation. Backstepping control of nonlinear dynamical systems. Advances in nonlinear dynamics and chaos (ANDC)* (pp. 485−507). Academic Press.

Sambas, A., Vaidyanathan, S., Zhang, S., Mohamed, M. A., Zeng, Y., & Azar, A. T. (2021a). *A new 3-d chaotic jerk system with a saddle-focus rest point at the origin, its active backstepping control, and circuit realization. Backstepping control of nonlinear dynamical systems. Advances in nonlinear dynamics and chaos (ANDC)* (pp. 95−114). Academic Press.

Sambas, A., Vaidyanathan, S., Zhang, S., Mohamed, M. A., Zeng, Y., & Azar, A. T. (2021b). *A new 4-d chaotic hyperjerk system with coexisting attractors, its active backstepping control, and circuit realization. Backstepping control of nonlinear dynamical systems. Advances in nonlinear dynamics and chaos (ANDC)* (pp. 73–94). Academic Press.

Shukla, M. K., Sharma, B. B., & Azar, A. T. (2018). *Control and synchronization of a fractional order hyperchaotic system via backstepping and active backstepping approach. Mathematical techniques of fractional order systems. Advances in nonlinear dynamics and chaos (ANDC)* (pp. 559–595). Elsevier.

Singh, S., & Azar, A. T. (2020a). Controlling chaotic system via optimal control. In: *Proceedings of the international conference on advanced intelligent systems and informatics 2019, Vol. 1058 of advances in intelligent systems and computing* (pp. 277–287). Cham: Springer International Publishing.

Singh, S., & Azar, A. T. (2020b). Multi-switching combination synchronization of fractional order chaotic systems. In: *Proceedings of the international conference on artificial intelligence and computer vision (AICV2020)* (pp. 655–664). Cham: Springer International Publishing.

Singh, S., Azar, A. T., Ouannas, A., Zhu, Q., Zhang, W., & Na, J. (2017). Sliding mode control technique for multi-switching synchronization of chaotic systems. In: *9th international conference on modelling, identification and control (ICMIC 2017)* (pp. 1–6), July 10–12, Kunming, China: IEEE.

Singh, S., Azar, A. T., Vaidyanathan, S., Ouannas, A., & Bhat, M. A. (2018). *Multiswitching synchronization of commensurate fractional order hyperchaotic systems via active control. Mathematical techniques of fractional order systems. Advances in nonlinear dynamics and chaos (ANDC)* (pp. 319–345). Elsevier.

Singh, S., Azar, A. T., & Zhu, Q. (2018). Multi-switching master–slave synchronization of non-identical chaotic systems. In: *Innovative techniques and applications of modelling, identification and control: selected and expanded reports from ICMIC'17, Vol. 467 of lecture notes in electrical engineering* (pp. 321–330). Singapore: Springer.

Singh, S., Mathpal, S., Azar, A. T., Vaidyanathan, S., & Kamal, N. A. (2021). *Multi-switching synchronization of nonlinear hyperchaotic systems via backstepping control. Backstepping control of nonlinear dynamical systems. Advances in nonlinear dynamics and chaos (ANDC)* (pp. 425–447). Academic Press.

Soliman, M., Azar, A. T., Saleh, M. A., & Ammar, H. H. (2020). Path planning control for 3-omni fighting robot using pid and fuzzy logic controller. In: *The international conference on advanced machine learning technologies and applications (AMLTA2019)* (pp. 442–452). Cham: Springer International Publishing.

Soliman, N.S., Said, L.A., Azar, A.T., Madian, A.H., Radwan, A.G., & Ounnas, A. (2017). Fractional controllable multi-scroll v-shape attractor with parameters effect. In: *6th international conference on modern circuits and systems technologies (MOCAST)* (pp. 1–4), 4–6 May 2017, Thessaloniki, Greece: IEEE.

Tolba, M. F., AbdelAty, A. M., Soliman, N. S., Said, L. A., Madian, A. H., Azar, A. T., & Radwan, A. G. (2017). Fpga implementation of two fractional order chaotic systems. *AEU International Journal of Electronics and Communications, 78*, 162–172.

Vaidyanathan, S., & Azar, A. T. (2015a). *Analysis and control of a 4-D novel hyperchaotic system. Chaos modeling and control systems design, Vol. 581 of studies in computational intelligence* (pp. 19–38). Berlin, Germany: Springer.

Vaidyanathan, S., & Azar, A. T. (2015b). *Analysis, control and synchronization of a nine-term 3-D novel chaotic system. Chaos modeling and control systems design, Vol. 581 of studies in computational intelligence* (pp. 3–17). Berlin, Germany: Springer.

Vaidyanathan, S., & Azar, A. T. (2015c). *Anti-synchronization of identical chaotic systems using sliding mode control and an application to vaidyanathan-madhavan chaotic systems. Advances and applications in sliding mode control systems, Vol. 576 of studies in computational intelligence* (pp. 527–547). Berlin, Germany: Springer.

Vaidyanathan, S., & Azar, A. T. (2015d). *Hybrid synchronization of identical chaotic systems using sliding mode control and an application to vaidyanathan chaotic systems. Advances and applications in sliding mode control systems, Vol. 576 of studies in computational intelligence* (pp. 549–569). Berlin, Germany: Springer.

Vaidyanathan, S., & Azar, A. T. (2016a). *A novel 4-D four-wing chaotic system with four quadratic nonlinearities and its synchronization via adaptive control method. Advances in chaos theory and intelligent control* (pp. 203–224). Berlin, Germany: Springer.

Vaidyanathan, S., & Azar, A. T. (2016b). *Adaptive backstepping control and synchronization of a novel 3-D jerk system with an exponential nonlinearity. Advances in chaos theory and intelligent control* (pp. 249–274). Berlin, Germany: Springer.

Vaidyanathan, S., & Azar, A. T. (2016c). *Adaptive control and synchronization of Halvorsen circulant chaotic systems. Advances in chaos theory and intelligent control* (pp. 225–247). Berlin, Germany: Springer.

Vaidyanathan, S., & Azar, A. T. (2016d). *Dynamic analysis, adaptive feedback control and synchronization of an eight-term 3-D novel chaotic system with three quadratic nonlinearities. Advances in chaos theory and intelligent control* (pp. 155–178). Berlin, Germany: Springer.

Vaidyanathan, S., & Azar, A. T. (2016e). *Generalized projective synchronization of a novel hyperchaotic four-wing system via adaptive control method. Advances in chaos theory and intelligent control* (pp. 275–290). Berlin, Germany: Springer.

Vaidyanathan, S., & Azar, A. T. (2016f). *Qualitative study and adaptive control of a novel 4-D hyperchaotic system with three quadratic nonlinearities. Advances in chaos theory and intelligent control* (pp. 179–202). Cham: Springer International Publishing.

Vaidyanathan, S., & Azar, A. T. (2016g). Takagi-Sugeno fuzzy logic controller for Liu-Chen four-scroll chaotic system. *International Journal of Intelligent Engineering Informatics, 4*(2), 135–150.

Vaidyanathan, S., & Azar, A. T. (Eds.), (2020). *Backstepping control of nonlinear dynamical systems. Advances in nonlinear dynamics and chaos (ANDC).* Elsevier.

Vaidyanathan, S., & Azar, A. T. (2021). *An introduction to backstepping control. Backstepping control of nonlinear dynamical systems. Advances in nonlinear dynamics and chaos (ANDC)* (pp. 1−32). Academic Press.

Vaidyanathan, S., Azar, A. T., Akgul, A., Lien, C.-H., Kacar, S., & Cavusoglu, U. (2019). A memristor-based system with hidden hyperchaotic attractors, its circuit design, synchronisation via integral sliding mode control and an application to voice encryption. *International Journal of Automation and Control, 13*(6), 644−667.

Vaidyanathan, S., Azar, A. T., & Boulkroune, A. (2018). A novel 4-D hyperchaotic system with two quadratic nonlinearities and its adaptive synchronisation. *International Journal of Automation and Control, 12*(1), 5−26.

Vaidyanathan, S., Azar, A. T., & Ouannas, A. (2017a). *An eight-term 3-D novel chaotic system with three quadratic nonlinearities, its adaptive feedback control and synchronization. Fractional order control and synchronization of chaotic systems, Vol. 688 of studies in computational intelligence* (pp. 719−746). Cham: Springer International Publishing.

Vaidyanathan, S., Azar, A. T., & Ouannas, A. (2017b). *Hyperchaos and adaptive control of a novel hyperchaotic system with two quadratic nonlinearities. Fractional order control and synchronization of chaotic systems, Vol. 688 of studies in computational intelligence* (pp. 773−803). Cham: Springer International Publishing.

Vaidyanathan, S., Azar, A. T., Rajagopal, K., & Alexander, P. (2015). Design and spice implementation of a 12-term novel hyperchaotic system and its synchronisation via active control. *International Journal of Modelling, Identification and Control, 23*(3), 267−277.

Vaidyanathan, S., Azar, A. T., Rajagopal, K., Sambas, A., Kacar, S., & Cavusoglu, U. (2018). A new hyperchaotic temperature fluctuations model, its circuit simulation, FPGA implementation and an application to image encryption. *International Journal of Simulation and Process Modelling, 13*(3), 281−296.

Vaidyanathan, S., Azar, A. T., Sambas, A., Singh, S., Alain, K. S. T., & Serrano, F. E. (2018). *A novel hyperchaotic system with adaptive control, synchronization, and circuit simulation. Advances in system dynamics and control. Advances in systems analysis, software engineering, and high performance computing (ASASEHPC)* (pp. 382−419). IGI Global.

Vaidyanathan, S., Idowu, B. A., & Azar, A. T. (2015). *Backstepping controller design for the global chaos synchronization of sprott's jerk systems. Chaos modeling and control systems design, Vol. 581 of studies in computational intelligence* (pp. 39−58). Berlin, Germany: Springer.

Vaidyanathan, S., Jafari, S., Pham, V.-T., Azar, A. T., & Alsaadi, F. E. (2018). A 4-D chaotic hyperjerk system with a hidden attractor, adaptive backstepping control and circuit design. *Archives of Control Sciences, 28*(2), 239−254.

Vaidyanathan, S., Pham, V.-T., & Azar, A. T. (2021). *A new chaotic jerk system with egg-shaped strange attractor, its dynamical analysis, backstepping control, and circuit simulation. Backstepping control of nonlinear dynamical systems. Advances in nonlinear dynamics and chaos (ANDC)* (pp. 53−71). Academic Press.

Vaidyanathan, S., Sambas, A., & Azar, A. T. (2021). *A 5-D hyperchaotic dynamo system with multistability, its dynamical analysis, active backstepping control, and circuit simulation. Backstepping control of nonlinear dynamical systems. Advances in nonlinear dynamics and chaos (ANDC)* (pp. 449−471). Academic Press.

Vaidyanathan, S., Sambas, A., Azar, A. T., Rana, K., & Kumar, V. (2021a). *A new 4-D hyperchaotic temperature variations system with multistability and strange attractor, bifurcation analysis, its active backstepping control, and circuit realization. Backstepping control of nonlinear dynamical systems. Advances in nonlinear dynamics and chaos (ANDC)* (pp. 139−164). Academic Press.

Vaidyanathan, S., Sambas, A., Azar, A. T., Rana, K., & Kumar, V. (2021b). *A new 5-d hyperchaotic four-wing system with multistability and hidden attractor, its backstepping control, and circuit simulation. Backstepping control of nonlinear dynamical systems. Advances in nonlinear dynamics and chaos (ANDC)* (pp. 115−138). Academic Press.

Vaidyanathan, S., Sambas, A., Azar, A. T., Serrano, F. E., & Fekik, A. (2021). *A new thermally excited chaotic jerk system, its dynamical analysis, adaptive backstepping control, and circuit simulation. Backstepping control of nonlinear dynamical systems. Advances in nonlinear dynamics and chaos (ANDC)* (pp. 165−189). Academic Press.

Vaidyanathan, S., Sambas, A., Azar, A. T., & Singh, S. (2021). *A new multistable plasma torch chaotic jerk system, its dynamical analysis, active backstepping control, and circuit design. Backstepping control of nonlinear dynamical systems. Advances in nonlinear dynamics and chaos (ANDC)* (pp. 191−214). Academic Press.

Vaidyanathan, S., Sampath, S., & Azar, A. T. (2015). Global chaos synchronisation of identical chaotic systems via novel sliding mode control method and its application to zhu system. *International Journal of Modelling, Identification and Control, 23*(1), 92−100.

Vaidyanathan, S., Zhu, Q., & Azar, A. T. (2017). *Adaptive control of a novel nonlinear double convection chaotic system. Fractional order control and synchronization of chaotic systems, Vol. 688 of studies in computational intelligence* (pp. 357−385). Cham: Springer International Publishing.

Visweswara, K. (2014). An investigation of incremental conductance based maximum power point tracking for photovoltaic system. *Energy Procedia, 54*, 11−20.

Volos, C. K., Pham, V.-T., Azar, A. T., Stouboulos, I. N., & Kyprianidis, I. M. (2018). *Synchronization phenomena in coupled dynamical systems with hidden attractors. Nonlinear dynamical systems with self-excited and hidden attractors, Vol. 133 of studies in systems, decision and control* (pp. 375−401). Cham: Springer International Publishing.

Wang, Z., Volos, C., Kingni, S. T., Azar, A. T., & Pham, V.-T. (2017). Four-wing attractors in a novel chaotic system with hyperbolic sine nonlinearity. *Optik International Journal for Light and Electron Optics, 131*, 1071−1078.

Yang, B., Yu, T., Shu, H., Zhu, D., An, N., Sang, Y., & Jiang, L. (2018a). Energy reshaping based passive fractional-order pid control design and implementation of a grid-connected pv inverter for MPPT using grouped grey wolf optimizer. *Solar Energy, 170*, 31−46.

Yang, B., Yu, T., Shu, H., Zhu, D., An, N., Sang, Y., & Jiang, L. (2018b). Perturbation observer based fractional-order sliding-mode controller for MPPT of grid-connected PV inverters: Design and real-time implementation. *Control Engineering Practice*, *79*, 105−125.

Yatimi, H., Ouberri, Y., Chahid, S., & Aroudam, E. (2020). Control of an off-grid pv system based on the backstepping MPPT controller. *Procedia Manufacturing*, *46*, 715−723.

Zhu, P., Chen, Y., Li, M., Zhang, P., & Wan, Z. (2020). Fractional-order sliding mode position tracking control for servo system with disturbance. *ISA Transactions*, *105*, 269−277.

Chapter 13

Techno-economic modeling of stand-alone and hybrid renewable energy systems for thermal applications in isolated areas

M. Edwin[1], M. Saranya Nair[2] and S. Joseph Sekhar[3]

[1]Department of Mechanical Engineering, University College of Engineering, Nagercoil, Anna University Constituent College, Nagercoil, India,
[2]School of Electronics Engineering, Vellore Institute of Technology, Chennai Campus, Chennai, India, [3]Department of Engineering, Shinas College of Technology, University of Technology and Applied Sciences, Shinas, Sultanate of Oman

13.1   Introduction

In isolated areas, approximately 22% of agricultural produce, particularly fruit and vegetables, is spoiled, owing to a deficiency of suitable storage and transportation facilities. The provision of cooling units for short-term preservation could prevent this spoilage at the local village level. Many refrigeration systems need electrical power, and the diesel generator is used as the ideal source for electric power at most of the isolated locations. Connection to the power grid for such places is either complicated or expensive (Ramesh & Saini, 2020). The rising uncertainties in gasoline price and the concern in lowering greenhouse gas emissions lead researchers to believe otherwise: fulfilling minimally harmful energy requirements (Kartite & Cherkaoui, 2019). Besides, the postharvest losses and wastages can be minimized by keeping the cooling facility near to the source of the raw materials. The huge bio and solar energy resources available in countries like India may help to overcome such issues by developing a hybrid-powered, thermally controlled cold storage. A hybrid renewable energy (HRE) system can be highly effective by integrating multiple renewable energy sources and is seen as a potential solution to the above-mentioned issue (Guo, Liu, Sund, & Jin, 2018).

Several studies have been published in the literature related to power generation from HRE sources. Multiple energy sources through a suitable energy conversion system could supply power to local loads. It also increases the reliability of resources and the total performance and reduces the life cycle expense (Ashok, 2007). Renewable energy sources, such as solar and biomass (BM), can be combined appropriately to meet the energy requirements and enhance the environment and socioeconomic conditions of remote locations in countries like India (Singal, Varun, & Singh, 2007). Renewable resources may have intermittent characteristics which render many hurdles to reach the effect of complementarity on hybrid system efficiency. This efficiency limit may be set with an idealization of the mathematical functions that characterize the energy supply of the renewable resources being studied (Frederico, Filho, & Beluco, 2019). A sustainable hybrid system is an economically viable option for achieving electricity decarbonization and greenhouse gas mitigation (Mazzeo, Baglivo, Matera, Congedo, & Oliveti, 2020). Hybridization of power supplies is an important requirement for power production in the age of decarbonization of the energy grid by the usage of renewable energies (Babatunde, Munda, & Hamam, 2020). Despite the strong potential of integrated sun, wind and BM energy systems, their realistic deployment is constrained because of the difficulties involved with their efficient architecture, utilization, and planning (Bagheri, Shirzadi, Bazdar, & Kennedy, 2018).

An H_2O-NH_4 vapor absorption system, driven by solar and biogas (BG) energy sources, has been implemented to preserve milk in remote dairy farms of Brazil. The studies in such forms show that the solar-powered absorption refrigeration system integrated with BG would be an economically attractive option (Alvares & Trepp, 1987a, 1987b). Solar

Renewable Energy Systems. DOI: https://doi.org/10.1016/B978-0-12-820004-9.00013-9

and BM combined LiBr—water absorption chiller was also developed and studied (Prasartkaew & Kumar, 2010, 2013). A hybrid system with solar and BM can work with high reliability and performance when it is integrated with or without an auxiliary heater. The methods of optimizing hybrid energy sources, their size and devices selection are to be provided to the community with a cost-effective power solution (Suresh, Muralidhar, & Kiranmayi, 2020). The integrated technological optimization criteria based on the efficiency of four main device components (energy production, battery storage, building demand, and grid relief) and the enhanced LCOE (Levelized Cost of Energy) taking into account the comprehensive renewable energy benefits (FiT subsidy, transmission failure savings, network extension savings and carbon mitigation benefits) were built to optimize the design of renewable energy systems (Liu et al., 2020). The PV/diesel/small hydro/battery with an energy expenditure of 0.443\$ kWh^{-1} is the most feasible economic framework (Muh & Tabet, 2019). The combination of solar PV, wind turbine, bio-generator and battery bank backup are capable of fulfilling the expected demand stable at a cost of Rs 10.18 kWh^{-1} (Murugaperumal & Vimal Raj, 2019). To overcome the issues to satisfy the viable electrification in inaccessible areas, renewable energy sources, such as solar, wind, BM, and hydro, can be combined as hybrid systems. Besides, it could be a cost-effective and reliable strategy to overcome the power crisis (Nahid-ur-Rahman, Syed, Tofaeel, & Mahabub, 2012).

HRE solutions have been introduced to address the volatility and randomness of a single form of renewable energy. To enhance the overall efficiency of HRE systems, more optimization work is also needed (Krishan & Suhag, 2019). Technoeconomic analysis of hybrid energy system is important to reflect its supremacy and to determine what kind of framework and related factors are appropriate for a specific circumstance. To ensure that there is no excess energy consumption when incorporating RE sources into a network, it is necessary to establish a maximum limit for the share of feasible renewable energies (Ma, Xue, & Liu, 2018). Thus maximizing the size of the components and adopting an energy management strategy is important for decreasing the cost of the system and limiting its negative effects. Hence, energy management is a term that collects all the systematic procedures to control and minimize the quantity and the cost of energy used to provide its requirements (Olatomiwa, Mekhilef, Ismail, & Moghavvemi, 2016). Due to the fluctuating nature of renewable energy outputs, it is important to plan as much buffer as that portion of renewable energy in the energy system. It is important to note that the development of such versatility in the energy system also provides further research opportunities. This current piece of research would be a good guide to give them an insight (Yu, Ryu, & Lee, 2019).

Renewable energy resources have advantages, such as efficiency and ecofriendliness, besides their drawbacks, such as higher maintenance costs and inadequate device flexibility, often to cater for continuous energy demand. Hybrid systems have been developed to overcome these drawbacks including the use of more than one form of renewable energy resource and/or the use of conventional energy resources and/or the combination of storage systems (Tezer, Yaman, & Yaman, 2017). Hybrid system with optimum combination of hydro and solar energies, derived from MATLAB simulation could eliminate the drawbacks of individual energy sources and guaranty the continuous power production (Alexandre, Paulo, & Arno, 2012; Lian, Zhang, Ma, Yang, & Chaima, 2019). The approach suggested is a simplified variant of the crow search algorithm (CSA), in which the likelihood of recognition is changed adaptively. In the suggested algorithm (CSA adaptive-AP), there is an adaptive balancing of the likelihood of knowledge of the initial CSA. Compared with CSA, CSA adaptive-AP can have a greater compromise between diversification and intensification, and therefore CSA adaptive-AP can produce stronger outcomes than CSA on an average (Ghaffari & Askarzadeh, 2020). HOMER is a software tool, consists of several energy component models to simulate hybrid energy systems. It can evaluate technology options, cost and availability of resources (Ismail, Moghavvemi, & Mahlia, 2013; Kumaravel & Ashok, 2012). An artificial neural network, fuzzy logic, and neuro-fuzzy, evolutionary algorithms, and few more similar artificial intelligence techniques are seriously considered for the simulation in recent years (Recep, 2013). Some popular algorithms, such as simulated annealing and response surface methodology, are also used for optimizing the size of a PV-wind hybrid energy system (Orhan & Banu, 2010). A hybrid-system optimization program, HOGA, which works based on genetic algorithm, can be applied with monoobjective or multiobjective concepts. Much other software packaged, such as HOMER, HYBRID2, and HYBRIDS, are used in the design and analysis of complex hybrid energy systems (Edwin, Nair, & Sekhar, 2020; Nitin & Anoop Varun, 2012).

It is seen from the analysis of the literature that most of the works on the HRE systems focus on electrical energy generation. Further systems should also be modeled for the efficient use of thermal energy and improve financial benefits from different renewable energy systems. This could be a potential problem for research. For the successful implementation of a hybrid system, the cost should be reduced, and it can be achieved by improving either the conversion efficiency of the energy source or appropriate selection of low-cost renewable energy systems. Moreover, the appropriate selection of components in the application can lead to higher overall system efficiency followed by cost-effectiveness. In the case of the refrigeration system used in a remote location, this strategy could be very attractive for cost reduction as well as run the system with 100% renewable energy sources.

This chapter is organized into six sections. Section 13.1 is introductory. Section 13.2 presents the load/demand assessment of the selected study areas, energy and economic modeling and simulation of the proposed hybrid energy system. Section 13.3 presents the summary and findings of the technical and economical simulation of the stand-alone and hybrid energy-based chilling system in different study regions. Section 13.4 describes identifying the suitable renewable energy combination of the hybrid energy-based chilling system in different study regions based on the technoeconomic evaluation. Section 13.5 presents the sensitivity analysis of the proposed hybrid energy-based chilling system. Finally, Section 13.6 summarizes the conclusions and discusses the scope for future research.

13.1.1   Objectives of the work

Well-established technologies commonly used in India for the cooling systems in remote areas are electrical energy-based vapor compression cooling systems in which power is produced either from fossil fuel or stand-alone solar PV systems. By comparison, a fairly recent idea is the hybrid thermal energy-based cooling system. Therefore the renewable energy conversion technologies for bioenergy and solar thermal energy can be selected with suitable combinations to supply the required thermal energy for an absorption cooling system to preserve the perishable produce in remote locations of India, where the grid power is not available continuously and lack of transport facilities to send the goods to the market immediately after the harvest. To provide some solutions to the above-discussed issues, goals and objectives are defined as follows.

Collection of data relevant the operation and the load profile of the systems; Performance analyses on milk-cooling systems dependent on BM, BG, and solar energy in study areas; Economic study of the milk-cooling device operated with sustainable energy sources that are accessible locally; carryout a technoeconomic analysis and predict an appropriate blend of existing renewable energy sources for milk-cooling application; and define the system for maximum overall system performance and minimum cost.

13.2   Materials and methods

13.2.1   Selection of study region

Agro-and-dairy sector has tremendous significance in India's growth due to the crucial connections and synergies, it supports between the two economic pillars, namely, industry and agriculture. It is predicted that the agro-and-dairy industries will play an important role to ensure rural sustainability. Hence, some villages of southern India, where 75% of the people depend on agricultural production, were chosen for this investigation. Initially, the possibility of using potential renewable energy resources to meet the energy requirement for milk chilling was explored with a systematic survey.

The survey was focused on the domestic energy usage in various purposes, such as agriculture, dairy-related activities, and food preservation. Besides, the economic viability of technologies and the effect of technology in rural systems were studied based on the past works reported by numerous researchers. In the villages, field surveys focused on households and direct interview were performed to collect data about the supply of the alternative energy sources, the current energy use, etc. This research work is using the database as a method for collecting contextual knowledge from the field of study. Each family's detailed information was gathered through personal communication with the individuals. The consistency of the data from the survey was verified with some information available in various government and nongovernment repositories available in district/state administration, local Gram panchayat, pediatric dispensary, etc.

It was observed that the prominent types of BM energy resources known are coconut shell, coconut coir pith, wood chips, rubber seed kernel shell, etc., are available for the generation of producer gas. The major source of BG energy resources reported is municipal solid waste (MSW), which is classified as waste from domestic, commercial, and some (nonhazardous) collected by private and public authorities (biochemical process is used in this conversion) (Hilkiah, Ayotamuno, Eze, Ogaji, & Probert, 2008). On an average, the volume of solid scraps produced per family is $3-6$ kg day^{-1}, and the volume of waste needed to yield 1 m^3 of BG is predicted to be 23 kg day^{-1} (Khambalkar, Dhiraj, Karale, & Shilpa, 2008). The major gobar gas (GG) energy source is assessed from the population of livestock present in the study area. The livestock population details were obtained from personal interaction with the respondents. The biowaste produced per day from livestock was found as $12-15$ kg per buffalo, $3.0-7.5$ kg per cattle, 0.1 kg per sheep, or goat. The quantity of cow manure needed to generate 1 m^3 of gas is predicted as 12 kg (Singal et al., 2007). The system depends mainly on energy resources from BM, BG, and GG. However, owing to the scarce supplies of wood, besides BG and GG, solar energy conversion devices are used to mitigate the energy deficit. It is important to realize that a hybrid system is preferred to use all bioenergy sources to their maximum capacity. The average solar radiation of the study region is taken as 800 W m^{-2}.

The research area has been divided into two agro-climate study regions 1 (SR1) and 2 (SR2). Based on data collected from several villages, SR1 comprises of rubber cultivation land and SR2 comprises of coastline area. Figs. 13.1 and 13.2 show the images taken from the research areas. Under each area, data collection was conducted for a minimum of seven villages. Throughout these rural areas, the majority of people are farm-workers, and irrigation is the primary cause of their revenue. Table 13.1 displays the societal background of certain regions and their potential for the study.

In this network, the milk is collected at the community level by the farmers in the collection centers and then this milk is processed and transported to the milk preservation centers where milk is chilled to 4°C. This activity is performed twice per day at 6-hour interval.

13.2.2 Assessment of load and demand

Based on the potential of energy sources, milk production, and cattle population, the cooling load of the chilling plant and the heat required for the generator of the vapor absorption chilling system are calculated.

The amount of milk produced in the selected area is assessed according to the following equation:

$$V_m = (N_{\text{cows}} \times \text{milk produced/livestock}) \tag{13.1}$$

where V_m is the milk volume generated in L day^{-1}.

The milk yield from each cow is taken as $8-16$ L day^{-1} (Carrie, Samuel, & Jonathan, 2014; Zehetmeier, Baudracco, Hoffmann, & HeiBenhuber, 2012), and from these data, the volume of milk to be cooled per day is

$$m_m = V_m \times \rho_m \tag{13.2}$$

where m_m is the milk produced in kg day^{-1} and ρ_m is the density of milk in kg m^{-3} (1027 kg m^{-3}) (Edgar, 1998; McSweeney & Fox, 2009).

FIGURE 13.1 Photographic views of study region 1 (rubber cultivated area).

FIGURE 13.2 Photographic views of study region 2 (coastline area).

TABLE 13.1 Details of survey data obtained from the study area.

Study area	Overall population (n)	Households density (n)	Livestock growth rate (n)	Areas under agricultural use (%)	Biomass significance rate (kg day^{-1})	Biogas significance rate (kg day^{-1})	Gobar gas significance rate (kg day^{-1})	The volume of milk produced (L day^{-1})
SR1	2350	558	135	77	180–320	920–1210	680–780	1210
SR2	4230	1040	95	29	140–170	1850–2110	440–530	890

SR1, study region 1; *SR2*, study region 2.

The required refrigerating effect (Q_e) is derived from the following equation

$$Q_e = m_m \times C_{p,m} \times \Delta T \tag{13.3}$$

where Q_e is the evaporator load in kJ kg^{-1}, $C_{p,m}$ is the specific heat of milk in kJ kg^{-1} K^{-1}, and ΔT is difference between the raw milk and chilled milk in the evaporator.

Based on the demand for evaporator load and generator thermal load in the absorption chiller, the mass of BM, BG, and GG sources is determined. The heat demand of the generator is estimated from the standard COP (Coefficent of Performance) and the evaporator capacity of the vapor absorption chiller.

The power supplied to the generator is calculated from

$$Q_g = Q_e/\text{COP} \tag{13.4}$$

where Q_g is the heat load of the generator in kW. COP of the VARS is taken as 0.5 (Kim & Infante Ferreira, 2008).

The mass of the energy sources, such as BM, BG, and GG, is calculated form the following relations (Edwin & Sekhar, 2014b, 2015, 2016, 2018):

$$m_{\text{BM}} = (Q_g/CV_{\text{BM}}) \times \eta_{c,\text{BM}} \tag{13.5}$$

$$m_{\text{BG}} = (Q_g/CV_{\text{BG}}) \times \eta_{c,\text{BG}} \tag{13.6}$$

$$m_{\text{GG}} = (Q_g/CV_{\text{GG}}) \times \eta_{c,\text{GG}} \tag{13.7}$$

The quantity of available energy source and the quantity needed to supply the required heat in the generator to meet the expected cooling demand are shown in Fig. 13.3.

The figure shows that the maximum energy supplied from each energy resource is not sufficient to meet the energy needed to cool the milk produced per day. Therefore more than one energy source should be used with an appropriated combination to fulfill the minimum energy needs of the cooling device. Moreover, all energy sources must be used in suitable combinations. The proportions of the energy sources in the combinations of this study are listed in Table 13.2.

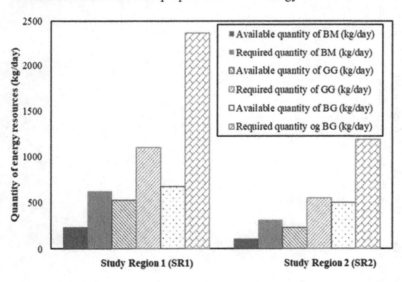

FIGURE 13.3 The available and the required amount of biomass, gobar gas, and BG in study regions 1 and 2.

TABLE 13.2 Proposed mixtures of renewable energy resources for the study.

Base energy sources	Energy resources in mixture components					
	BM–BG	BM–GG	BG–GG	BM–BG–SO	BM–GG–SO	BG–GG–SO
BM			X			X
BG		X			X	
GG	X			X		

BG, biogas; *BM*, biomass; *GG*, gobar gas.

These combinations are focused on the possibility of using multiple energy sources to operate the cooling facility in the two different regions, SR1 and SR2. The Base Component is any one of the energy sources and the mixture component shall be the mixture of the other sources of energy. The constant ratio of energy sources in the mixing element is also seen in Table 13.3. The selected mixture element ratios are 25%, 50%, and 75%. For a particular hybrid combination, the composition of the components in the mixture is kept constant for all combinations.

The hybrid energy cooling system was simulated using MATLAB software. To find the COP_{sys}, all of the elements are integrated according to the components in a specific system. Based on COP_{sys}, the best mixture of renewable energy sources was chosen.

13.2.3 Proposed system

The suggested hybrid energy-based chilling system involves the energy conversion system and the vapor absorption chilling system, and it is shown in Fig. 13.4. The conversion systems considered for renewable energy sources are BM gasifier, bio-digesters, and evacuated tube solar collector. Those systems convert the energy present in the feedstock to thermal energy for heating the generator in VARS. Moreover, they are chosen according to the potential for resource utilization and the cooling load, besides using one of the cooling pairs, LiBR-water or aqua-ammonia. The evaporator of VARS takes care of the overall cooling load for milk cooling.

13.2.4 Energy modeling

To study the chilling system, which is working with vapor absorption chiller and the combination of various renewable energy sources available in a region, a mathematical model was formulated in MATLAB software. The major parameter taken for this analysis is the overall system efficiency and the same has been calculated for various predefined combinations. The important assumptions used in the modeling of the proposed chilling system are constant heat capacity for energy resources, stable conversion efficiency for a single source of energy, unremitting process of the chilling facility, steady COP for the vapor absorption device, consistent year-round cooling conditions, marginal increases in labor costs, constant milk production, and marginal extra diesel costs attributable to variability in fuel prices.

TABLE 13.3 Details of the combinations for renewable energy sources used in this study.

Sl. no	The mixture of sources	Energy ratios	Sl. no	The mixture of sources	Mixture ratios
1	BM:BG + GG	1:0.50 + 0.50	10	BM:BG + GG + SO	1:0.50 + 0.25 + 0.25
2	BM:BG + GG	1:0.75 + 0.25	11	BM:BG + GG + SO	1:0.25 + 0.50 + 0.25
3	BM:BG + GG	1:0.25 + 0.75	12	BM:BG + GG + SO	1:0.25 + 0.25 + 0.50
4	BG:BM + GG	1:0.50 + 0.50	13	BG:BM + GG + SO	1:0.50 + 0.25 + 0.25
5	BG:BM + GG	1:0.75 + 0.25	14	BG:BM + GG + SO	1:0.25 + 0.50 + 0.25
6	BG:BM + GG	1:0.25 + 0.75	15	BG:BM + GG + SO	1:0.25 + 0.25 + 0.50
7	GG:BM + BG	1:0.50 + 0.50	16	GG:BM + BG + SO	1:0.50 + 0.25 + 0.25
8	GG:BM + BG	1:0.75 + 0.25	17	GG:BM + BG + SO	1:0.25 + 0.50 + 0.25
9	GG:BM + BG	1:0.25 + 0.75	18	GG:BM + BG + SO	1:0.25 + 0.25 + 0.50

BG, biogas; BM, biomass; GG, gobar gas.

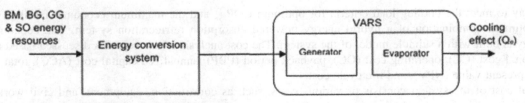

FIGURE 13.4 Proposed model of hybrid energy-based cooling system.

The conversion efficiencies of BM ($\eta_{c,\text{BM}}$), BG ($\eta_{c,\text{BG}}$), and GG ($\eta_{c,\text{GG}}$) are the important parameters to evaluate the heat energy supplied to the generator, and those standards are taken as 0.45, 0.26, and 0.35, respectively (Edwin, Nair, & Sekhar, 2019; Edwin & Sekhar, 2014b, 2016; Klaas, Michele, Rob, Steven, & Ouwens Jeroen, 2010). The data from the survey have been used to find the BM available per day. In this study, the values are taken as 180–320 and 140–170 kg day^{-1} in the regions SR1 and SR2, respectively. The usable energy extracted by the conventional burning of solid BM in the open atmosphere can be calculated by the following equation:

$$Q_{\text{BM}} = m_{\text{BM}} \times CV_{\text{BM}} \times \eta_{c,\text{BM}} \tag{13.8}$$

BG usually refers to a gas that is formed by the biochemical conversion of biowastes and organic matter in the absence of oxygen. Apart from kitchen waste and municipal solid waste, livestock manure is also used in anaerobic digesters to produce BG. In this generation process, the quantity and the quality of the organic matter play a vital role. The basic characteristics of the biowastes are taken to find the quality of the producer gas, which is specified by the methane content in the BG. The quantity and quality of available MSW were estimated based on the data from the investigation conducted in the study areas. The energy available from the BG energy source (Q_{BG}) is given as (Edwin & Sekhar, 2014b, 2015, 2016, 2018)

$$Q_{\text{BG}} = m_{\text{BG}} \times CV_{\text{BG}} \times \eta_{c,\text{BG}} \tag{13.9}$$

The GG is produced in an anaerobic digester which is loaded everyday with the dung of the livestock. The survey showed the quantity of livestock manure available per day and the same is used to determine the available energy produced from the GG as (Edwin & Sekhar, 2014b, 2015, 2016, 2018):

$$Q_{\text{GG}} = m_{\text{GG}} \times CV_{\text{GG}} \times \eta_{c,\text{GG}} \tag{13.10}$$

In the study area, the normal solar insolation (I) is measured as 800 W m^{-2}, and the normal thermal conversion efficiency of the solar conversion system can be taken as 45% (Edwin & Sekhar, 2015, 2018). Advanced evacuated tube collector is considered in the present study because it can be efficient for refrigeration applications. The effective collector energy output (Q_{SO}) in the steady-state is determined as follows (Edwin & Sekhar, 2015):

$$Q_{\text{so}} = A_{co}F_R[(\tau\alpha)I - I_{\text{SC1}}(T_i - T_a) - I_{\text{SC2}}(T_i - T_a)^2] \tag{13.11}$$

where τ is the transmissivity of the glass tube (0.84), α is the absorptivity of the surface (0.94), F_R is the heat removal factor (1). I_{SC1} and I_{SC2} are considered as solar constants taken as 0.65 and 0.06, respectively. T_i and T_a are the inlet and ambient temperatures, respectively.

The major objective of the proposed study is to eliminate the use of fossil energy systems to supply the energy for various applications relevant to rural communities, instead increase the readily available renewable energy resources to meet the energy needed for the operation of the chiller in food cooling. A trade-off between cost and overall system performance has been taken as the criteria for choosing the proper blends of renewable energy resources in a hybrid energy system has been identified. The uncertainties in the continuous supply of energy from renewable energy sources are a big challenge in the planning, design and construction of a hybrid energy-based system for a particular application. In the proposed system, meeting the cooling load of the milk chilling plant by HRE sources is complex, and specific simulations are needed to optimize the combination of energy sources. Moreover, the hybrid systems are interdependent and specific strategies are needed to optimize the overall performance (COP$_{\text{sys}}$ of the system). The overall performance (COP$_{\text{sys}}$) of the proposed hybrid energy powered chilling system is calculated from the following equation (Edwin & Sekhar, 2014b, 2015, 2016, 2018)

$$\text{COP}_{\text{sys}} = Q_e / (k \times IA_{co} + x \times m_{\text{BM}}CV_{\text{BM}} + y \times m_{\text{BG}}CV_{\text{BG}} + z \times m_{\text{GG}}CV_{\text{GG}}) \tag{13.12}$$

where x, y, z, and k are the composition of BM, BG, GG, and solar energy (SO) in the proposed cooling system.

13.2.5 Economic modeling

The best way to meet the cooling load criteria for optimum COP$_{\text{sys}}$ and the maximum economic feasibility is to consider an optimum combination of a hybrid energy powered absorption refrigeration system. The economic analysis methodology is built with a suitable model of the system. The cost analysis of this report is discussed based on the principle of capital cost (CC), operating cost (OC), payback period (PBP), annualized capital cost (ACC), total annual cost (TAC), net present value (NPV), and life cycle cost.

The total cost of the system consists of various costs, such as construction, equipment, and civil works. In some regions, subsidies are given by the government agencies to encourage the installation of the renewable energy-based application. The effect of subsidies is incorporated in the capital cost of the system. The major components considered for measuring the CC are the conversion devices, burners, heat exchangers, energy conversion plants, etc. Dependents on the current market rates as available from the suppliers, the cost factors are determined in the present context. Until the measurement is made, the total cost of each part is taken into account as the construction expense (Edwin et al., 2019; Edwin & Sekhar, 2014b).

$$\text{CC of HRES} = (\text{capital costs of BM, BG, and GG conversion system} + \text{solar collector}$$
$$+ \text{ chilling system} + \text{gas burner and heat exchanger}) - \text{govt.subsidies} \tag{13.13}$$

The running cost (RC) is calculated from the operational and maintenance cost and annual depreciation. The following assumptions are taken into the cost calculations. Some standard values taken in this study are maintenance cost—2% of the capital cost, lifetime of the system—18 years, annual interest rate (d)—12%, the cost of BM—Rs 0.95 kg^{-1}, the cost of cow dung—Rs 0.40 kg^{-1}, and the cost of BG sources—Rs 0.50 kg^{-1}. Transportation and labor costs are also suitably selected for this study (Edwin et al., 2019; Edwin & Sekhar, 2014b).

$$\text{RC of HES} = \text{costs of(energy source} + \text{operation and maintenance of energy conversion device}$$
$$+ \text{operation and maintenance of VARS} + \text{labor} + \text{depreciation}) \tag{13.14}$$

The performance of a hybrid energy-powered cooling system requires measuring the PBP to replace the traditional fossil fuel (diesel)-powered vapour compression refrigeration system (VCRS) (Edwin et al., 2019; Edwin & Sekhar, 2014b, 2016).

$$\text{Payback period} = \frac{\text{Incremental value of capital cost}}{\text{Annual savings(profit)}} \tag{13.15}$$

where incremental value is the difference between CCs of diesel generator-operated vapor compression cooling system and hybrid energy-operated vapor absorption cooling system (Edwin & Sekhar, 2014a).

$$\text{CC of fossil} - \text{fuel} - \text{operated VCRS} = \text{costs of diesel genset and VCRS} \tag{13.16}$$

Annual savings is the reduction in operating cost due to the implementation of hybrid renewable energy-based chilling system.

$$\text{RC of conventional VCRS} = \text{diesel cost} + \text{OMCof VCRS} + \text{labor cost} + \text{depreciation value} \tag{13.17}$$

TAC consists of ACC and the annual care and replacement costs. It can be stated as:

$$\text{TAC} = \text{ACC} + \text{annual cost of operation and maintenance cost} \tag{13.18}$$

Equivalent annual cost is the annual cost of possessing, operating and maintaining an asset over its entire life span, and it is calculated by:

$$\text{ACC} = \text{CC} \times \text{CRF}(i, n) \tag{13.19}$$

where n is the life span of the component in years, "I" is the annual interest rate, CRF is the capital recovery factor. The CRF can be calculated from

$$\text{CRF}(i, n) = \frac{i.(1+i)^n}{(1+i)^n - 1} \tag{13.20}$$

NPV is the actual value of the capital and RCs of a device over its lifespan. NPV is used as a primary economic measure for the evaluation of an energy system. The discrepancy between the actual value of the profits and the expenses resulting from an investment is the net present value of the system (Edwin et al., 2019; Edwin & Sekhar, 2014b, 2016). It is expressed as,

$$\text{NPV} = \left[S \times \left(\frac{(1+i)^n - 1}{i(1+i)^n} \right) \right] - \text{CC} \tag{13.21}$$

where "S" defines for benefits at the end of the period. Conditions for accepting an investment proposal, as evaluated from the NPV method are:

- NPV > 0, proposal can be accepted;
- NPV = 0, Proposal is indifferent; and
- NPV < 0, proposal cannot be accepted.

13.2.6 Simulation of proposed chilling system

A simulation model was developed to analyze the performance of the chilling system for various combinations of renewable energy sources in the hybrid system and the best model configuration was defined by analyzing the overall efficiency and financial constraints. To reduce the complications in the calculation some simple assumptions, such as the system is 100% reliable and no sudden changes in the performance parameters, are used. The designed simulation model has the provision to alter the sizing of the cooling system or heating system and process variables of the hybrid energy system.

The key renewable energy sources used in this study, based on the availability in the region of study are BM, BG, and GG. In this approach, to manage the energy needs to satisfy the cooling demand of the chilling system, renewable energy sources are combined in the form of hybrid systems. In some situations, where the main energy sources are not adequate for the necessary cooling demand, the solar heaters are used along with the other three energy sources discussed previously. The simulation program is defined to incorporate the above conditions. It checks the cooling demand and the availability of heat from the energy sources to operate the vapor absorption system and fix the need for solar heaters. Keeping these two scenarios, the regions of study is separated into two cases, such as case 1 and case 2.

Case 1: In some parts of the study area, the possible thermal energy from the renewable energy sources, such as BM, BG, and GG, is sufficient to manage the cooling load of the vapor absorption system to preserve the milk produced in that area. Therefore solar heating is not considered in the hybrid system and the hybrid energy sources are defined by combining the energy sources BM, BG, and GG at various propositions. For this scenario, the optimum combination is selected based on the overall efficiency of the total cooling system which is obtained from the simulation. The combinations for the three energy sources are predefined and the best one is finally selected based on efficiency and economic viability. The procedure used to model this case 1 is schematically represented in Fig. 13.5.

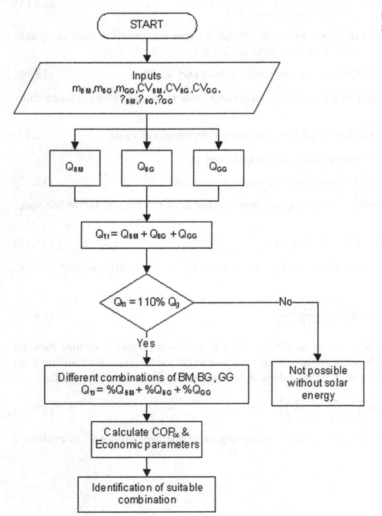

FIGURE 13.5 Combinations of the energy sources biogas, biomass, and gobar gas without SO (case 1).

Case 2: In some areas, the thermal energy supplied by the three energy sources discussed above are not sufficient to handle the milk-cooling load of the vapor absorption chiller as per the demand. Therefore solar heaters are implemented to meet the heat requirement of the generator in the vapor absorption cooling system. In case 2, BM, BG, GG, and solar thermal energy are the energy sources used in various combinations. The possible combinations are predefined and based on the simulation, the appropriate combinations are identified, and the final one is identified from the maximum overall efficiency and economic advantages. The procedure used to analyze the hybrid system with a combination of four renewable energy sources (case 2) is shown in Fig. 13.6.

In both cases, hybrid energy is used to run a LiBr-water vapor absorption refrigeration system. The parameter to define the vapor absorption cooling device is taken from the standard value available in the literature. All the required energy balance and efficiency terms are suitably incorporated in the simulation and COP_{sys} is calculated. Similarly, the parameter needed for the economic study, such as cost of various components, cost of energy source, and cost of conversion device, are taken from the manufacturer's catalogs and other published data. The data from the field survey are also used for this purpose.

13.3 Results and discussions

13.3.1 Thermal and economic performance

The results of the simulation outcomes of hybrid renewable energy-operated milk preservation systems in the study area are reported in this section. The typical milk production in most of the areas studied is between 2000 and

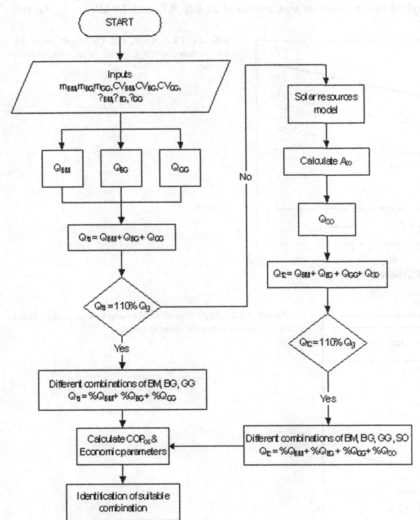

FIGURE 13.6 Combinations of the energy sources biogas, biomass, gobar gas, and SO (case 2).

5000 L day^{-1}, which should be cooled in 4 hours to the preservation temperature of 4°C. Hence, in this analysis, a 5TR refrigeration system was considered. Initially, the study was conducted in the stand-alone renewable energy-based chilling system to preserve the milk in the remote areas.

COP$_{sys}$ and CC of the stand-alone-energy-based chilling system for the different cooling loads shown in Figs. 13.7 and 13.8 indicate an improvement in COP$_{sys}$ due to the rise in cooling capacity and plant capacity (Fig. 13.7). A related pattern is found in previous studies also (Edwin et al., 2019; Edwin & Sekhar, 2014b, 2016). The supply of milk per day in the regions is surveyed between 2000 and 5000 L. Hence, the maximum milk quantity in the X-axis is fixed as 8000 L day^{-1} to take care of the extreme conditions. In Fig. 13.8, the capital cost of the stand-alone energy-based cooling device to replace the traditional VCRS is plotted. The solar energy-based cooling system has the highest capital cost of all energy sources and the diesel-powered plant has the lowest capital cost. Even though the diesel-powered plant has the lowest CC, it has large maintenance costs, environmental effects, and nonavailability in rural areas, besides it cannot be regarded as an acceptable source of electricity. Among the other three energy sources, GG-based refrigeration system has the lowest investment cost.

Fig. 13.9 shows the payback period for the stand-alone energy-based chilling system for an increase in milk volumes. This indicates that the payback time reduces as the volume of milk and the plant capacity increases. In previous research, a related pattern was also found (Edwin et al., 2014b; Singh & Sooch, 2004).

13.3.2 Thermal performance of cooling system working with hybrid energy

In SR1, the average available renewable energy sources are shown in Table 13.1, whereas the necessary cooling load identified is 10 kW. The total cooling power of hybrid energy systems was measured as 6.6, 8.7, and 7.9 kW using the full

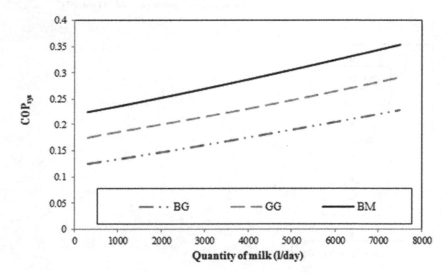

FIGURE 13.7 COP$_{sys}$ of the single renewable energy-based cooling system with milk quantity per day.

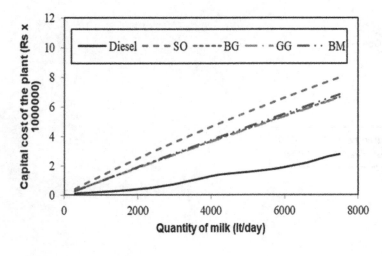

FIGURE 13.8 Capital cost of the stand-alone energy-based chilling system with milk quantity per day.

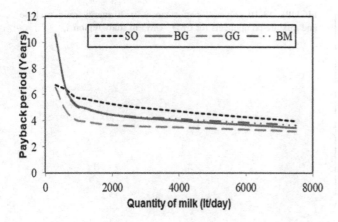

FIGURE 13.9 Payback period of the stand-alone energy-based chilling system with milk quantity per day.

TABLE 13.4 Identified combinations and the details of energy ratios and system performance for SR1.

Identified combinations (CM)	Base energy source	Energy sources in mixture	The proportion of energy sources in mixture component	Overall system COP (COP$_{sys}$)
CM1	BM	BG:GG	0.50:0.50	0.155−0.229
CM2	BM	BG:GG	0.75:0.25	0.144−0.229
CM3	BM	BG:GG	0.25:0.75	0.168−0.229
CM4	BG	BM:GG	0.50:0.50	0.127−0.195
CM5	BG	BM:GG	0.75:0.25	0.127−0.203
CM6	BG	BM:GG	0.25:0.75	0.127−0.181
CM7	GG	BM:BG	0.50:0.50	0.166−0.180
CM8	GG	BM:BG	0.75:0.25	0.180−0.190
CM9	GG	BM:BG	0.25:0.75	0.147−0.180

BG, biogas; *BM*, biomass; *GG*, gobar gas; *SR1*, study region.

utilization of usable energy resources in different configurations, such as BM−BG, BM−GG, and BG−GG. Therefore to fulfill the cooling load, a minimum of three renewable resources must be combined. Table 13.4 shows the COP$_{sys}$ of various configurations appropriate for the SR1. To estimate the outcome of different mixtures of existing energy resources in the hybrid system, one of the energy sources has been taken as the base-energy-source and the others is taken as mixture component. In the mixture component, a constant composition of the other components is used. However, the combination of the base component and mixture components is changed in this study. For this purpose the term named as energy ratio for base and mixture component (ERBMC) is introduced in this study, which is to state the variation of base and mixture components of energy sources in the hybrid system. The ratios of these two components are used to find the impacts of energy sources when they are used in different propositions in the chilling system as the source of energy supply.

In all the plots, ERBMC is given in the *X*-axis, in which the ratio of base component increase from left to right whereas the ratio of mixture component decreases. The ranges of COP$_{sys}$ for the selected blends of energy sources, CM1−CM9, are also given in Table 13.4. Moreover, the variations of COP$_{sys}$ with ERBMC for the selected combinations are depicted in Figs. 13.10−13.12. The configurations of the two elements in the mixture have been taken in three steps, such as 25%, 50%, and 75%. In all the selected hybrid combinations, the constant ratio is kept for the mixture component for a particular combination of hybrid energy.

Fig. 13.10 indicates that CM3 provides better COP$_{sys}$ while the highest value is held for the inputs from BM and GG. If the energy ratio in the element combination, BG and GG are modified to 0.5:0.5 (CM1) or 0.75:0.25 (CM2), the COP$_{sys}$ is found lower than that of CM3 if the base variable is over 75%; however, all the variations indicate that the COP$_{sys}$ values are similar to each other.

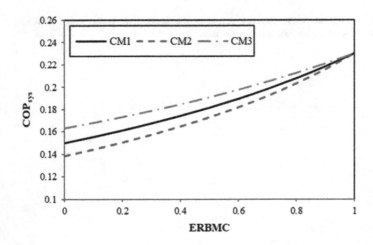

FIGURE 13.10 Overall performances with energy ratio of base and mixture components for CM1−CM3 in study region 1.

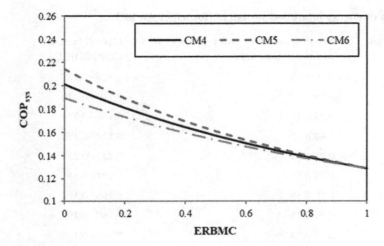

FIGURE 13.11 Overall performances with energy ratio of base and mixture components for CM4−CM6 in study region 1.

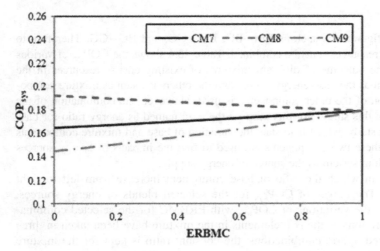

FIGURE 13.12 Overall performances with energy ratio of base and mixture components for CM7−CM9 in study region 1.

Fig. 13.11 reveals a rather small difference in COP_{sys} between CM4, CM5, and CM6 (less than 12%). So, any grouping can be used. However, the high COP_{sys} may be retained if the base portion is held below 10%−15%. Fig. 13.12 indicates that for an increase in COP_{sys} the base energy portion should be held to the limit. CM8 COP_{sys} is 25%−35% higher than other mixtures if the base portion is less than 45%. Figs. 13.10−13.12 indicate the COP_{sys}

declines as the source of BG energy rises. This is attributed to low calorific value and urban solid waste recovery performance compared with energy sources, such as BM and GG.

In SR2, the average of the obtainable renewable resources is quantified in Table 13.1. For the selected conditions, the potential of the renewable energy sources to meet the cooling loads are measured as 5.9, 6.2, 7.8, and 8.6 kW with the complete use of available resources in different configurations, such as BM–BG, BM–GG, BG–GG, and BM–BG–GG respectively. Therefore to reach the necessary cooling load of 9.2 kW, solar energy must be used along with the other three energy sources. Solar energy serves as a secondary source of energy in the SR2 where there is an inadequate supply of primary energy supplies, such as BM, GG, and BG.

Table 13.5 provides the various mixtures, their formulations, and COP_{sys} specific to SR2. The base component is used as any single source of energy, and the mixture component is a blend of the three other sources of energy. Table 13.5 also displays the constant ratio of the energy sources in the mixture component. The proportion of the base product in ERBMC decreases in the plots from left to right, while that of the part of the combination rises in the X-axis from right to left. Table 13.5 shows the minimum and maximum COP_{sys} obtained for variations CM10 to CM18. The three specific combinations of the three elements included in the mixture are estimated to be 25%, 50%, and 75%, as seen in Table 13.5. In a specific combination, the ratio of energy sources in a mixture is held constant in both proportions. Moreover, the variation of COP_{sys} along with the proportion of energy sources is shown in Figs. 13.13–13.15.

Fig. 13.13 indicates that COP_{sys} is high for CM12, while the highest amount of BM and SO contributions are preserved. If the energy ratios of BG, GG, and SO are changed to 0.50:0.25:0.25 (CM10) or 0.25:0.50:0.25 (CM11), the spectrum of COP_{sys} measured is smaller than that of CM12. However, if the base variable is more than 80%, the

TABLE 13.5 Identified combinations and the details of energy ratios and system performance for SR2.

Identified combinations	Base energy source	Energy sources in mixture	The proportion of energy sources in mixture component	Overall system COP (COP_{sys})
CM10	BM	BG:GG:SO	0.50:0.25:0.25	0.159–0.230
CM11	BM	BG:GG:SO	0.25:0.50:0.25	0.172–0.230
CM12	BM	BG:GG:SO	0.25:0.25:0.50	0.176–0.230
CM13	BG	BM:GG:SO	0.50:0.25:0.25	0.128–0.202
CM14	BG	BM:GG:SO	0.25:0.50:0.25	0.128–0.186
CM15	BG	BM:GG:SO	0.25:0.25:0.50	0.128–0.191
CM16	GG	BM:BG:SO	0.50:0.25:0.25	0.180–0.189
CM17	GG	BM:BG:SO	0.25:0.50:0.25	0.163–0.180
CM18	GG	BM:BG:SO	0.25:0.25:0.50	0.180–0.186

BG, biogas; *BM*, biomass; *GG*, gobar gas; *SR*, study region.

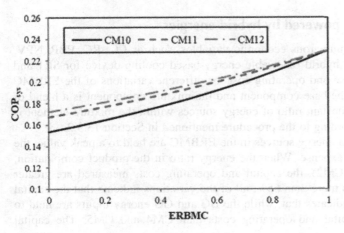

FIGURE 13.13 Overall performances with energy ratio of base and mixture components for CM10–CM12 in study region 2.

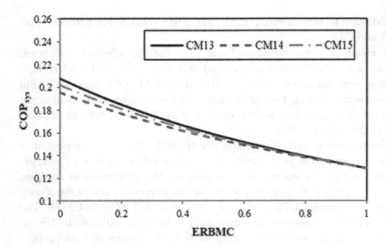

FIGURE 13.14 Overall performances with energy ratio of base and mixture components for CM13–CM15 in study region 2.

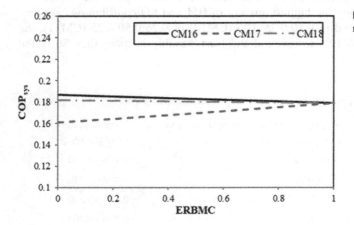

FIGURE 13.15 Overall performances with energy ratio of base and mixture components for CM16–CM18 in study region 2.

COP_{sys} in all variations is similar to each other. Fig. 13.14 displays the COP_{sys} of CM13, CM14, and CM15 and suggests that the variance for all the variations is very small (less than 10%). Therefore any mixture could be acceptable. If the base variable is kept below 18%, a high device COP_{sys} may be retained. Fig. 13.15 reveals that COP_{sys} of CM16 and CM17 are 10% and 40% higher than that of CM18, respectively. This is possible when the ratio is less than 40% for the base component. In the SR2, COP_{sys} is strongly affected when the input BM and GG are the inputs. The mixing of solar thermal energy with other forms of energy resources overcomes the inadequacy of energy demand and even impact the COP_{sys}.

13.3.3 Economic aspects of the chilling system—powered by hybrid energies

The findings of this chapter have been evaluated for the numerous economic variables, such as CC, RC, PBP, NPV, and TAC, and their effect on the selection of the suitable hybrid renewable energy-based cooling device for SR1 and SR2 is discussed. In Figs. 13.16–13.18, the capital expense and operating cost for different variations of the ERBMC in SR1 are shown. One of the energy sources is taken as the base component and the mixture component is a blend of two different sources of energy. Also, in Table 13.4, the constant ratio of energy sources within the mixture element is given. The X-axis of the all the figures has been taken according to the procedure mentioned in Section 13.3.2.

Fig. 13.16 indicates that while the inputs of BM and GG energy sources in the ERBMC are held to a peak value, the mixture CM3 provides the low capital cost and operating expense. When the energy ratio in the product combination, BG and GG, is modified to 0.5:0.5 (CM1) or 0.75:0.25 (CM2), the capital and operating costs measured are greater than that of CM3. But if the base component in ERBMC is more than 65%, all of the variations indicate that the capital and operating costs are similar to each other. Fig. 13.17 indicates that, while the BG and GG energy inputs are held to the maximum, the mixture CM6 provides the lower capital and operating costs than CM4 and CM5. The capital

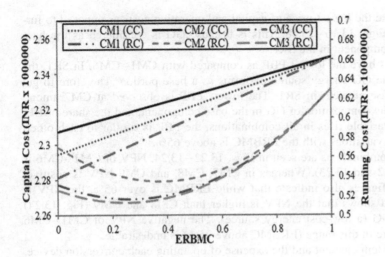

FIGURE 13.16 Variation of capital cost and running cost with energy ratio of base and mixture components for CM1–CM3 in study region 1.

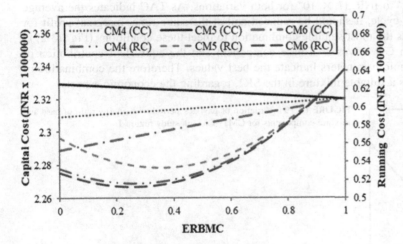

FIGURE 13.17 Variation of capital cost and running cost with energy ratio of base and mixture components for CM4–CM6 in study region 1.

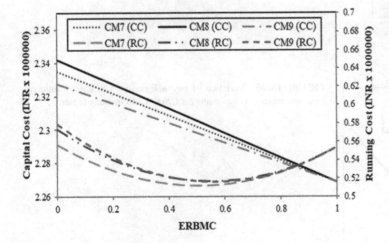

FIGURE 13.18 Variation of capital cost and running cost with energy ratio of base and mixture components for CM7–CM9 in study region 1.

expense of CM9 is 15% lower relative to CM7 (Fig. 13.18) as the GG serves as a base portion in BM and BG with 25% and 75%, respectively. The operating rate of CM9 is 17% higher than that of CM7 and the capital expense of all three variations is similar to each other when the base factor ratio is greater than 75%. The operating costs also show a similar pattern. The mixtures CM7−CM9 demonstrate the lowest price of capital and operating costs as opposed to further mixtures, such as CM1−CM6, as depicted in Figs. 13.16−13.18. This is because GG is considered as the base component. Therefore GG can be chosen as a base component in the SR1.

Figs. 13.19−13.21 indicate that CM7−CM9 pairs have the lowest PBP as compared with CM1−CM6. In SR1, the lowest payback duration has been detected where the GG energy source performs as a base portion. Therefore to get the low payback time, GG can be chosen as the base element in SR1. The lowest PBP is observed at CM7 among CM7, CM8, and CM9. This is due to the impact of the composition of GG in the base component, and the share of BM and BG in the mixture component. When the base variable rises in all combinations, the PBP is similar to each other. There is no significant gap in PBP between the three variations with the ERBMC is above 65%.

The net present values for the various ERBMC combinations are seen in Figs. 13.22−13.24. NPV of CM1−CM6 is positive until the ERBMC is below 80% (Figs. 13.22 and 13.23). Whereas in CM7, CM8, and CM9 NPV is positive, besides independent of ERBMC (Fig. 13.24). Both figures also indicate that while ERBMC is over 65%, the NPV is similar to each other. In the case of CM7, the trend shows that the NPV is higher than CM8 and CM9 (Fig. 13.24) which may be the impact of the base component, GG (a low-cost energy source). The negative NPV of CM1−CM6 when the ERBMC reaches 70%; therefore any mixture in this range (ERBMC above 75%) is undesirable.

TAC estimates the annualized expense of each system element and the expense of operating each conversion device, which will thus be held at a minimal value. The TAC of various combinations is plotted in Figs. 13.25−13.27, which indicate that the TAC ranges from INR 36×10^4 to INR 37×10^4 for both variations. As TAC indicates the average expense of operating a device over its entire life cycle, it should be reduced and at the same time, marginal profit for the mixture should be achievable. The CM7 seems the most suitable combination to meet these conditions (Fig. 13.27).

The simulation results demonstrate that when GG serves as a base component and the equivalent contribution of BM and BG in the mixture system, all the economic parameters indicate the best values. Therefore the combination of GG:(BM + BG) with 50:50 will be considered as a suitable mixture in the SR1, regarding the economic aspects.

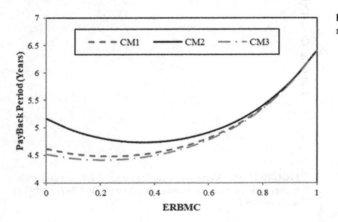

FIGURE 13.19 Variation of payback period with energy ratio of base and mixture components for CM1−CM3 in study region 1.

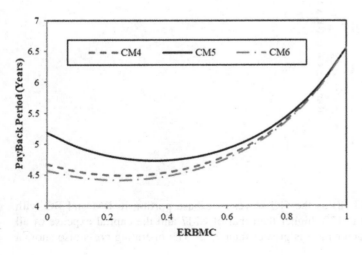

FIGURE 13.20 Variation of payback period with energy ratio of base and mixture components for CM4−CM6 in study region 1.

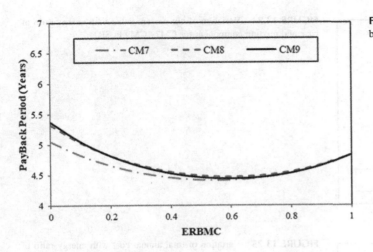

FIGURE 13.21 Variation of payback period with energy ratio of base and mixture component for CM7−CM9 in study region 1.

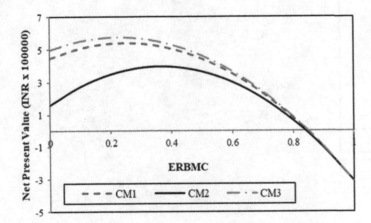

FIGURE 13.22 Variation of net present value with energy ratio of base and mixture components for CM1−CM3 in SR1.

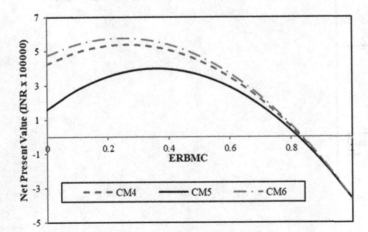

FIGURE 13.23 Variation of net present value with energy ratio of base and mixture components for CM4−CM6 in SR1.

BM and BG supplies are limited in the SR2. Therefore solar energy along with other renewable sources must be considered to fulfill the energy demand. In this regard, four energy sources in the hybrid configurations are considered. In the combinations, anyone energy resource has been taken as the base component and the blend of other three energy sources are taken as a mixture component. The details of the configurations of various combinations used in this study are given in Table 13.5.

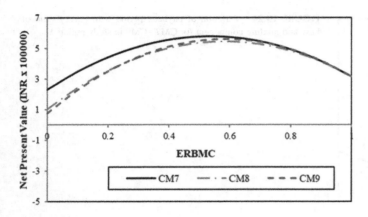

FIGURE 13.24 Variation of net present value with energy ratio of base and mixture component for CM7−CM9 in SR1.

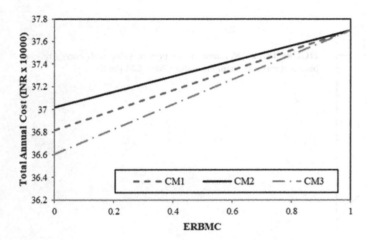

FIGURE 13.25 Variation of total annual cost with energy ratio of base and mixture components for CM1−CM3 in study region 1.

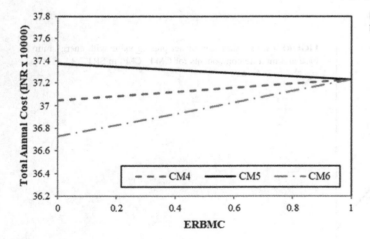

FIGURE 13.26 Variation of total annual cost with energy ratio of base and mixture components for CM4−CM6 in study region 1.

In Figs. 13.28 and 13.29, the capital expense and operating costs for different variations of the ERBMC in SR2 are shown. The X-axis has been taken in all the statistics according to the procedure mentioned in Section 13.3.2. Fig. 13.28 shows that CM11 mixture has low CC and OC due to the combined input of BM and GG energy sources in the mix of renewable sources. For both ERBMC, the total expense of CM10 is similar to CM11. This is due to the identical nature of elements in the blend. The RCs of CM10 and CM11 are found to rise in similar proportions while ERBMC is over 60% and the RCs of each one is identical to others as well. It is triggered in turn by the supremacy of BM. Fig. 13.29 reveals that CM14 and CM15 are having operating costs lower than CM13. This is because of the

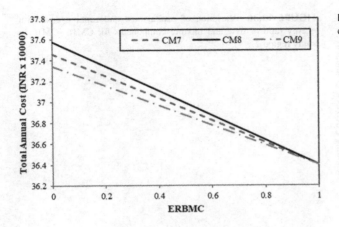

FIGURE 13.27 Variation of TAC with energy ratio of base and mixture component for CM7−CM9 in study region 1.

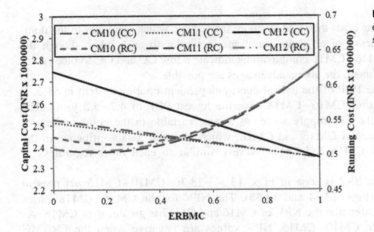

FIGURE 13.28 Variation of capital cost and running cost with energy ratio of base and mixture components for CM10−CM12 in study region 2.

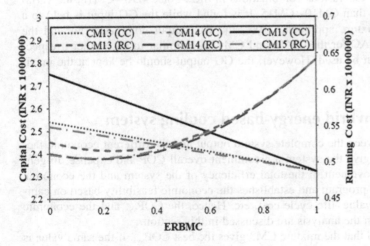

FIGURE 13.29 Variation of capital cost and running cost with energy ratio of base and mixture component for CM13−CM15 in study region 2.

composition of BG in the base portion. The variance in CC and OC of CM13−CM15 with ERBMC is close to the pattern demonstrated by CM10−CM12.

Fig. 13.30 reveals that the capital cost of CM16 and CM17 is similar to each other and closely matches the capital cost of CM10, CM11, CM13, and CM14. In comparison, it is less than the other combinations (CM12, CM15, and CM18) where the influence of solar energy in the mixture element dominates. CM18's operating expense displays the lowest benefit as the combination is kept at 35% of ERBMC. This is also attributed to the highest solar energy contribution in CM18.

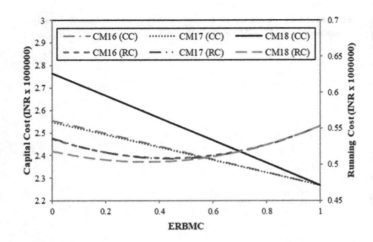

FIGURE 13.30 Variation of capital cost and running cost with energy ratio of base and mixture components for CM16–CM18 in study region 2.

The variations CM16 and CM17 indicate higher operating costs than CM18 (more than 15%), which attributes to the dominance of nonsolar energy sources. The operating expense of all three variations is similar to each other, though the ERBMC is over 60%. Figs. 13.28–13.30 reveal that CM16–CM18 combinations indicate a low CC and OC to other combinations of CM10–CM15. Since GG is the base component, the above advantages are possible.

Figs. 13.31–13.33 demonstrate the difference in the PBP of the hybrid energy-dependent cooling system in SR2. It is found that as compared to CM10–CM15, the mixtures CM16–CM18 have the lowest PBP of 4.7–5.5 years. The lowest PBP (4.6 years) was observed in SR2, where the GG supply serves as a base variable in the combinations of energy resources. The PBP of CM17 is shorter than that of CM16 and CM18, which is due to GG's influence in the base portion. In comparison, the variation in PBP of CM16 and CM17 is very similar to each other regardless of ERBMC (PBP is 4.6–5.3 years).

The NPV for various configurations of ERBMC in SR2 is seen in Figs. 13.34–13.36. CM10–CM15 net present values are positive when the ERBMC is below 80% (Figs. 13.34 and 13.35). The NPV for the CM16–CM18 varies positively independent of the ERBMC. Fig. 13.36 indicates that the NPV of CM16 and CM17 is greater than CM18. As the base component, GG contributes to this effect. In CM10–CM15, NPV values are negative when the ERBMC exceeds 75% since there is no suitable mixture in this range (ERBMC greater than 75%).

Figs. 13.37–13.39 indicate the TAC differences for the various combinations in SR2. For CM16–CM18, the variation in total annual expense indicates the lower benefit than CM10–CM15. It is found while the GG input is held to a limit. If the energy ratio for the product base and mixture approaches 70%, the GG output dominates this and the CM16–CM18 TAC is very small. The deviations of TAC for the CM16 and CM17 are similar to each other irrespective of ERBMC. Therefore all variations on this list can be used. However, the GG output should be kept at the maximum to get the optimum TAC.

13.4 Technoeconomic analysis of the hybrid energy-based cooling system

The technoeconomic analysis aims to find stability between the complete system output and the different economic factors within a given framework of a system, which will give the system a preeminent overall COP and expense. In general, the total system needs two competing goals: improvement in the total efficiency of the system and the economic feasibility. COP_{sys} determines the total efficiency of the program and establishes the economic feasibility based on capital expense, operating cost, payback time, net present value, life cycle cost, etc. Hence, the COP_{sys} and the economic constraints for the described combinations obtained from the analysis are discussed in this section.

From Figs. 13.10–13.12 and 13.16–13.27, it is found that the mixture CM3 gives the best COP_{sys}, if the same value is held for the BM and GG inputs. In comparison, the CM3 capital expenses, operating costs, payback time, and life cycle costs are also high. For the variations CM4–CM6, the poor COP_{sys} is noted with a higher value of capital expense, operating expense and payback time. Such combinations may not be acceptable in SR1. The CM7–CM9 variations suggest a reasonable set of COP_{sys}, and relatively decent quantities of economic factors. For CM9 mixture, COP_{sys} and other economic influences are inferior to CM8. Relative to CM7, the findings reveal that all the values of CM7 economic variables are 8%–10% higher than CM8. Nonetheless, it has been established as the appropriate combination in SR1 because of the 15% higher COP_{sys}. The NPV of CM8 over all of its ERBMC is also good, so this mixture is appropriate.

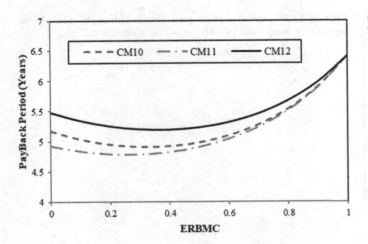

FIGURE 13.31 Variation of payback period with energy ratio of base and mixture components for CM10−CM12 in study region 2.

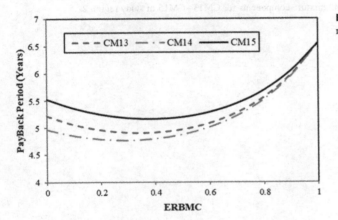

FIGURE 13.32 Variation of payback period with energy ratio of base and mixture components for CM10−CM12 in study region 2.

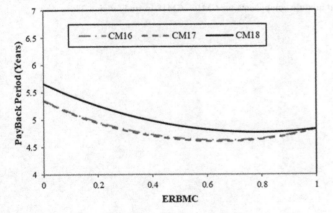

FIGURE 13.33 Variation of payback period with energy ratio of base and mixture components for CM10−CM12 in study region 2.

From Figs. 13.13−13.15 and 13.28−13.39, it is found that there are higher COP_{sys} and weaker economic influences in the combinations CM10−CM15. And such variations are not appropriate. CM13−CM15 demonstrated the lowest COP_{sys} and highest benefit of capital expense, operating cost, payback time, and duration of the life cycle. Thus these combinations are not appropriate for SR2. The variations in CM16−CM18 indicate low-cost of resources, operating expense, payback time and cost of the life cycle while COP_{sys} varies. In CM16 the variation COP_{sys} (0.17−0.19) is better than other blends, such as CM17 (0.15−0.17) and CM18 (0.180−0.184). The net present value of CM16 over its whole ERBMC is good, so this mixture may be an appropriate one. The discussion shows that the combination CM16 can be considered as the most suitable combination for SR2.

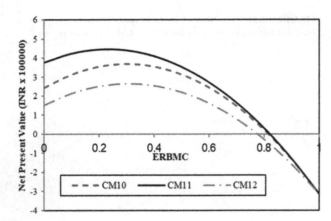

FIGURE 13.34 Variation of net present value with energy ratio of base and mixture components for CM10−CM12 in study region 2.

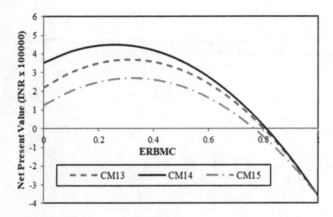

FIGURE 13.35 Variation of net present value with energy ratio of base and mixture components for CM13−CM15 in study region 2.

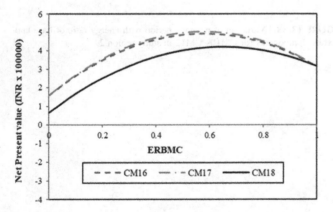

FIGURE 13.36 Variation of net present value with energy ratio of base and mixture components for CM16−CM18 in study region 2.

13.5 Sensitivity analysis

The uncertainties in the input parameter may influence the results of the analysis. Therefore the basic input parameters, such as cooling demand, availability of the renewable energy resources, fuel cost, conversion efficiency, and interest rate, and the output parameters, such as COP_{sys}, CC, RC, and PBP, are used in the sensitivity analyses. The standard procedure used in the literature has been used for this study.

In SR1, the combination of GG:(BM + BG) with 60:40(30 + 10) as the ratio for base and mixture parts has been recommended for better performance. For this proposed specification a sensitivity study has been performed to determine the effect of adjustments in input parameters. Fig. 13.40 depicts the sensitivity study of COP_{sys} for the proposed

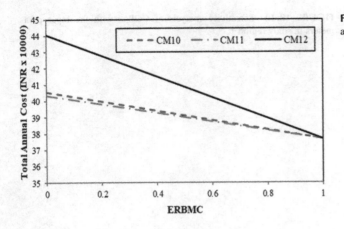

FIGURE 13.37 Variation of total annual cost with energy ratio of base and mixture components for CM10–CM12 in study region 2.

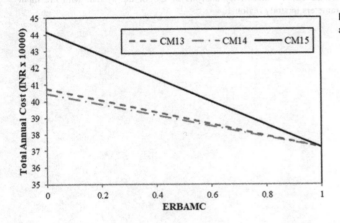

FIGURE 13.38 Variation of total annual cost with energy ratio of base and mixture components for CM13–CM15 in study region 2.

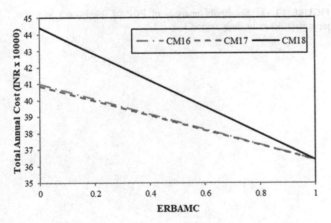

FIGURE 13.39 Variation of total annual cost with energy ratio of base and mixture components for CM16–CM18 in study region 2.

cooling system in SR1. The figure reveals that the uncertainty in cooling load and conversion efficiencies have a major effect on COP_{sys}. This indicates that the market price fluctuations of the energy for cooling is symmetrical and has a downside (0.185) relative to an upside (0.187). In this region, the dominant parameters are the conversion efficiencies of GG, BM, and BG, since their uncertainty is at the maximum. This trend reveals that a small variation in the conversion efficiencies of GG, BM, and BG may lead to a major fluctuation on the COP_{sys}.

The sensitivity analysis related to the capital cost of the proposed chilling system in SR1 region is depicted in Fig. 13.41. The figure shows that the uncertainty in the cost of vapor absorption chilling system has a major impact on the capital expense of the proposed hybrid energy-dependent cooling device. Variability in the costs of GG conversion

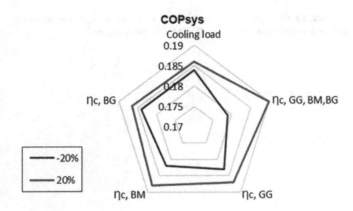

FIGURE 13.40 Sensitivity analysis of COP$_{sys}$ with the input parameters in study region 1.

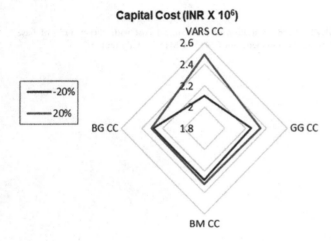

FIGURE 13.41 Sensitivity analysis of CC of the system with the input parameters in study region 1.

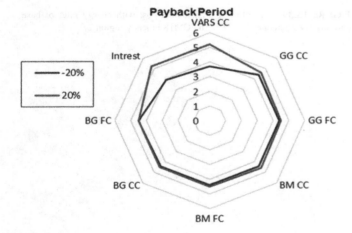

FIGURE 13.42 Sensitivity analysis of PBP of the system with the input parameters in study region 1.

device, BM gasifier, and BG plant are responsible for a small occurrence in the variance spectrum (less than 1%). Fig. 13.42 shows the sensitivity analysis of PBP for the proposed cooling system in SR1. It is found that the interest rate variance of ± 20% has a substantial effect on the operating expense and payback time of the hybrid energy-based cooling device. Changes in the costs of GG plant, gasifier, BG plant, and electricity are responsible for a small occurrence in the variability spectrum of PBP (less than 1.5%).

In SR2, GG:(BM + BG + SO) with 50:(25 + 12.5 + 12.5) as the ratio of base and mixture portion has been suggested, and for this mixture, a sensitivity study is performed to determine the effect of variations in input parameters.

The influence of input parameters on COP_{sys} of the cooling system in SR2 that works on hybrid energy has been depicted in Fig. 13.43. It shows that the uncertainties in the efficiencies due to cooling load have the maximum influence on the COP_{sys}. It also specifies that the change in COP is symmetrical and has a maximum deviation between 0.181 (upper side) and 0.183 (lower side). Moreover, the figure indicated that the conversion efficiencies of GG, BM, and BG sources have significant influence over COP_{sys}. An energy source's variability in conversion efficiency is responsible for a low incidence in the variance range (less than 2%).

Fig. 13.44 displays the sensitivity analysis of CC for the hybrid energy-based refrigeration device in SR2. The variability in the cost of VARS is found to have a major impact on the CC of the hybrid energy-dependent cooling device. The cost variability in the conversion system for GG, BG, BM, and solar energy is responsible for a small occurrence in the variance spectrum (less than 2%). Fig. 13.45 shows the sensitivity analysis of the PBP for the hybrid energy-based cooling system in SR2. The variability in the interest rate is found to have a major impact on operating expense and payback time. The uncertainty in the cost of the conversion systems used for the selected renewable energy sources and the cost of the energy sources are responsible for a low incidence in the operating cost and PBP variance spectrum.

13.6   Conclusion

The present research focuses on creating an environmentally sustainable, a 100% clean energy-efficient cooling device for the preservation and value addition of dairy products in rural regions. Alternate refrigeration systems are studied to integrate viable green energy options including BM, BG, gobar coal, and solar energy with a vapor absorption refrigeration device to resolve the growing demand for electricity and rising environmental issues from vapor compression cooling systems. Considering the types of activities, cooling requirements and accessibility energy sources, the study area has been divided as SR1 and SR2. The technoeconomic analysis is undertaken to determine the correct composition of

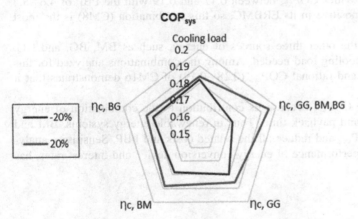

FIGURE 13.43   Sensitivity analysis of COP_{sys} with the input parameters in study region 2.

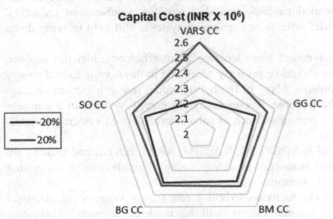

FIGURE 13.44   Sensitivity analysis of capital cost of the system with the input parameters in study region 2.

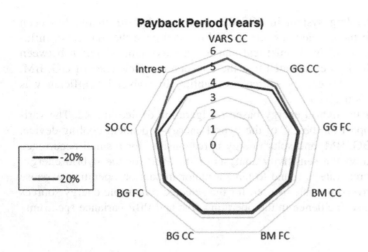

Payback Period (Years)

FIGURE 13.45 Sensitivity analysis of payback period of the system with the input parameters in study region 2.

available renewable energy sources to fulfill a given area's cooling demand. The following conclusions are taken based on the study.

In all the regions studied, a single source of energy could not fulfill the energy needed to develop the required cooling capacity. Therefore a mixture of renewable resources may lead to the effective execution of 100% renewable energy-powered refrigeration systems.

A minimum of three renewable energy resources shall be combined in SR1 to satisfy the required cooling demand. The mixture CM8 gives modest COP_{sys}, capital cost, operating expense, payback time, and duration of the life cycle. Therefore this mixture can be preferable to others. It provides COP_{sys} between 0.17 and 0.19 with the PBP of 4.6−5.4 years. Added to this, the NPV of CM8 combination is positive in its ERBMC, so this combination (CM8) is the most appropriate for meeting the cooling demand in SR1.

Solar energy must be used in the SR2 along with the other three sources of energy, such as BM, BG, and GG, because their availability is not adequate to satisfy the cooling load needed. Among the combinations analyzed for this region, appealing PBP (4.7−5.4 years), favorable NPV and rational COP_{sys} (0.18−0.19) of CM16 demonstrates that it is the most suitable one.

The study reveals that high BM usage will improve COP_{sys} because its contribution to the combination of energy sources decreases the production cost, operating costs, and payback time. Thus in renewable energy systems, BM and GG should be used to the maximum extent to boost COP_{sys} and reduce all the related costs and PBP. Sensitivity analysis indicates that uncertainty in variable inputs, such as performance of energy conversion device and interest rates, has a stronger effect on economic factors.

13.6.1 Scope for future work

Solar thermal installations are efficient, but due to the high capital costs, prospective consumers prefer to go for traditional systems. Since the high CC contributes to an extended payback period it requires a substantial reduction. Therefore more work on successful hybridization of local RES with solar conversion systems will help to bring down these costs.

Practical issues, including promising technologies for conversion, plant location, plant efficiency, logistics support, and a combination of green energy technologies and systems, should be properly addressed in developing hybrid energy systems based on BM energy to address the deployment problems. Many of the hybrid model research focuses on stagnant environment conditions. But in real-time it may not be possible. Therefore hybrid energy systems in a transient stage can be analyzed through phase variations in the vector, constraints, such as solar intensity, wind velocity, variability in the supply of bioenergy resources, and load requirement.

Some software packages, such as HYSIM, HYSYS, SOMES, SOLSTOR, RAPSIM, ARES, IPSYS, and INSEL, are developed for specific applications and they should be updated or merged to address today's complexities of integrating sustainable energy solutions for complex applications in remote locations.

The review of the current literature shows that most works focus on hybrid green energy systems for electrical energy generation. Further methods involving the use of thermal energy will be modeled as well. Alternative refrigeration and air-conditioning devices, multiple-effect reverse osmosis facilities, dryers, and dairy-processing absorption cooling systems may be proposed.

References

Alexandre, B., Paulo, K. S., & Arno, K. (2012). A method to evaluate the effect of complementarity in time between hydro and solar energy on the performance of hybrid hydro PV generating plants. *Renewable Energy, 45*, 24−30.

Alvares, S. G., & Trepp, Th (1987a). Simulation of a solar driven aqua ammonia absorption refrigeration system Part 1: Mathematical description and system optimization. *International Journal of Refrigeration, 10*, 40−48.

Alvares, S. G., & Trepp, Th (1987b). Simulation of a solar-driven aqua ammonia absorption refrigeration system Part 2: Viability for milk cooling at remote Brazilian dairy farms. *International Journal of Refrigeration, 10*, 70−76.

Ashok, S. (2007). Optimised model for community-based hybrid energy system. *Renewable Energy, 32*, 1155−1164.

Babatunde, O. M., Munda, J. L., & Hamam, Y. (2020). A comprehensive state-of-the-art survey on hybrid renewable energy system operations and planning. *IEEE Access, 8*, 75313−75346.

Bagheri, M., Shirzadi, N., Bazdar, E., & Kennedy, C. A. (2018). Optimal planning of hybrid renewable energy infrastructure for urban sustainability: Green Vancouver. *Renewable and Sustainable Energy Reviews, 95*, 254−264.

B. Prasartkaew, S., & Kumar. (2010). A low carbon cooling system using renewable energy resources and technologies. *Energy and Buildings, 42*, 1453−1462.

Prasartkaew, B., & Kumar, S. (2013). Experimental study on the performance of a solar-biomass hybrid air-conditioning system. *Renewable Energy, 57*, 86−93.

Carrie, H., Samuel, G., & Jonathan, W. (2014). Evaluation of energy efficiency and renewable energy generation opportunities for small scale dairy farms: A case study in Prince Edward Island, Canada. *Renewable Energy, 67*, 20−29.

Mazzeo, D., Baglivo, C., Matera, N., Congedo, P. M., & Oliveti, G. (2020). A novel energy-economic-environmental multi-criteria decision-making in the optimization of a hybrid renewable system. *Sustainable Cities and Society, 52*, 101780.

Edgar, S. (1998). *Milk and dairy product technology*. New York: Marcel Dekker Inc.

Edwin, M., Nair, M. S., & Sekhar, J. S. (2020). A comprehensive review for power production and economic feasibility on hybrid energy systems for remote communities. *International Journal of Ambient Energy*. Available from https://doi.org/10.1080/01430750.2020.1712252.

Edwin, M., Nair, M. S., & Sekhar, J. S. (2019). *Hybrid energy based chilling system for food preservation in remote areas, innovation in energy systems -new technologies for changing paradigms in operation, protection, business and development*. London: Intech Open Pub, 10.5772/intechopen.88264.

Edwin, M., & Sekhar, J. S. (2014a). Hybrid thermal energy based cooling system for a remote seashore villages. *Advanced Materials Research, 984−985*, 719−724.

Edwin, M., & Sekhar, J. S. (2018). Techno-economic evaluation of milk chilling unit retrofitted with hybrid renewable energy system in coastal province. *Energy, 151*, 66−78.

Edwin, M., & Sekhar, J. S. (2015). Thermal performance of milk chilling units in remote villages working with the combination of biomass, biogas and solar energies. *Energy, 91*, 842−851.

Edwin, M., & Sekhar, J. S. (2016). Thermo-economic assessment of hybrid renewable energy based cooling system for food preservation in hilly terrain. *Renewable Energy, 87*, 493−500.

Edwin, M., & Sekhar, J. S. (2014b). Techno-economic studies on hybrid energy based cooling system for milk preservation in isolated regions. *Energy Conversion & Management, 86*, 1023−1030.

Frederico, A., Filho, D., & Beluco, A. (2019). Simulating hybrid energy systems based on complementary renewable resources. *MethodsX, 6*, 2492−2498.

Ghaffari, A., & Askarzadeh, A. (2020). Design optimization of a hybrid system subject to reliability level and renewable energy penetration. *Energy, 193*, 116754.

Guo, S., Liu, Q., Sund, J., & Jin, H. (2018). A review of the utilization of hybrid renewable energy. *Renewable and Sustainable Energy Reviews, 91*, 1121−1147.

Hilkiah, I. A., Ayotamuno, M. J., Eze, C. L., Ogaji, S. O. T., & Probert, S. D. (2008). Designs of anaerobic digesters for producing biogas from municipal solid-waste. *Applied Energy, 85*, 430−438.

Ismail, M. S., Moghavvemi, M., & Mahlia, T. M. I. (2013). Design of a PV/diesel stand-alone hybrid system for a remote community in palestine. *Journal of Asian Scientific Research, 2*(11), 599−606.

Singh, K. J., & Sooch, S. S. (2004). Comparative study of economics of different models of family size biogas plants for state of Punjab, India. *Energy Conversion and Management, 45*, 1329−1341.

Kartite, J., & Cherkaoui, M. (2019). Study of the different structures of hybrid systems in renewable energies: A review. *Energy Procedia, 157*, 323−330.

Khambalkar, V. P., Dhiraj, S., Karale, S. R. G., & Shilpa, B. D. (2008). Assessment of bio resources potential of a rural village for self energy generation. *Bioresources, 3*(2), 566−575.

Kim, D. S., & Infante Ferreira, C. A. (2008). Solar refrigeration options—A state-of-the-art review. *International Journal of Refrigeration, 31*, 3−15.

Klaas, K., Michele, K., Rob, B., Steven, W., & Ouwens Jeroen, D. (2010). Evaluation of improvements in end-conversion efficiency for bioenergy production. *Ecofys*.

Krishan, O., & Suhag, S. (2019). Techno-economic analysis of a hybrid renewable energy system for an energy poor rural community. *Journal of Energy Storage, 23*, 305–319.

Kumaravel, S., & Ashok, S. (2012). An optimal stand-alone biomass/solar-PV/pico-hydel hybrid energy system for remote rural area electrification of isolated village in Western-Ghats Region of India. *International Journal of Green Energy, 9*, 398–408.

Lian, J., Zhang, Y., Ma, C., Yang, Y., & Chaima, E. (2019). A review on recent sizing methodologies of hybrid renewable energy systems. *Energy Conversion and Management, 199*, 112027.

Liu, J., Wang, M., Peng, J., Chen, X., Cao, S., & Yang, H. (2020). Techno-economic design optimization of hybrid renewable energy applications for high-rise residential buildings. *Energy Conversion and Management, 213*, 112868.

Ma, W., Xue, X., & Liu, G. (2018). Techno-economic evaluation for hybrid renewable energy system: Application and merits. *Energy, 159*, 385–409.

McSweeney, P. L. H., & Fox, P. F. (2009). *Advanced dairy chemistry, . (3).* New York: Springer.

Muh, E., & Tabet, F. (2019). Comparative analysis of hybrid renewable energy systems for off-grid applications in Southern Cameroons. *Renewable Energy, 135*, 41–54.

Murugaperumal, K., & Vimal Raj, D. A. (2019). Feasibility design and techno-economic analysis of hybrid renewable energy system for rural electrification. *Solar Energy, 188*, 1068–1083.

Nahid-ur-Rahman, C., Syed, E. R., Tofaeel, A. N., & Mahabub, A.-A.-F.-I. (2012). Present scenario of renewable energy in bangladesh and a proposed hybrid system to minimize power crisis in remote areas. *International Journal of Renewable Energy Research, 2*(2), 280–288.

Nitin, A., & Anoop Varun, K. (2012). Optimal design of hybrid PV–diesel–battery system for power generation in Moradabad district of Uttar Pradesh, India. *International Journal of Ambient Energy, 33*(1), 23–34.

Olatomiwa, L., Mekhilef, S., Ismail, M. S., & Moghavvemi, M. (2016). Energy management strategies in hybrid renewable energy systems: A review. *Renewable and Sustainable Energy Reviews, 62*, 821–835.

Orhan, E., & Banu, Y. E. (2010). Size optimization of a PV/wind hybrid energy conversion system with battery storage using simulated annealing. *Applied Energy, 87*, 592–598.

Ramesh, M., & Saini, R. P. (2020). Dispatch strategies based performance analysis of a hybrid renewable energy system for a remote rural area in India. *Journal of Cleaner Production, 259*, 120697.

Recep, Y. (2013). Role of energy management in hybrid renewable energy systems: case study-based analysis considering varying seasonal conditions. *Turkish Journal of Electrical Engineering & Computer Sciences, 21*, 1077–1091.

Suresh, V., Muralidhar, M., & Kiranmayi, R. (2020). Modelling and optimization of an off-grid hybrid renewable energy system for electrification in a rural areas. *Energy Reports, 6*, 594–604.

Singal, S. K., Varun., & Singh, R. P. (2007). Rural electrification of a remote island by renewable energy sources. *Renewable Energy, 32*, 2491–2501.

Tezer, T., Yaman, R., & Yaman, G. (2017). Evaluation of approaches used for optimization of stand-alone hybrid renewable energy systems. *Renewable and Sustainable Energy Reviews, 73*, 840–853.

Yu, J., Ryu, J.-H., & Lee, I.-B. (2019). A stochastic optimization approach to the design and operation planning of a hybrid renewable energy system. *Applied Energy, 247*, 212–220.

Zehetmeier, M., Baudracco, J., Hoffmann, H., & HeiBenhuber, A. (2012). Does increasing milk yield per cow reduce greenhouse gas emissions? A system approach. *Animal, 6*(1), 154–166.

Chapter 14

Solar thermal system—an insight into parabolic trough solar collector and its modeling

Anubhav Goel and Gaurav Manik

Department of Polymer and Process Engineering, Indian Institute of Technology Roorkee, Roorkee, India

14.1 Introduction

14.1.1 Motivation

Global energy consumption is increasing rapidly; growth of 2.9% in 2018 was the highest noted since 2010. Fossil fuels are still the primary source of energy in the world (BP, 2019). However, they are diminishing and will not be able to cater the energy requirements in the future. Based on the current rate of energy consumption, it is estimated that they will subside completely in 2100s (Shafiee & Topal, 2009). Moreover, their use poses a threat to the environment due to emission of greenhouse gases and aerosol particles. Thus it has become necessary to replace them by nature friendly energy resources, such as solar energy. PTSCs have the huge potential to reduce the dependency on fossil fuels based energy. They are capable of succeeding the fossil fuels in many fields, such as power generation, desalination, process heat in industries, and others (Abdulhamed, Adam, Ab-Kadir, & Hairuddin, 2018). So, the studies to enhance their efficiency and evaluate their performance become unavoidable. Modeling and simulation provide a suitable option to pursue the same. Therefore to conduct the research in this field, it is critical to gather knowledge about PTSC modeling and possible types of its models.

14.1.2 Background

The use of solar energy to generate heat is one of the fruitful and widely suitable ways to replace fossil fuels from industrial installations to power generating units. All the techniques to serve this conversion of solar energy to thermal energy comes under solar thermal systems and can be classified as: active and passive (Solar Thermal and Student Energy). A passive system does not require any external help to initiate the conversion or to propagate the heat produced, such as heat builds up inside a car parked in the sun. This kind of system is generally utilized for temperature maintenance in buildings, swimming pools, and others by using strategical designs for building them. An active system, however, comprises various components to collect, transform, transmit, and store the energy. The most widely used solar thermal systems that are capable of serving numerous applications in high-temperature ranges are known as concentrating solar power (CSP) systems/technologies.

CSP systems consist of a mirror (collector) to collect and concentrate sun rays over a point or line (receiver), from there, the acquired energy is converted to heat and utilized further in different applications (Goel, Manik, & Mahadeva, 2020). Depending on the focusing strategy adopted, they can be categorized as: line focusing—parabolic trough solar collectors (PTSCs) and linear fresnel reflectors (LFRs) and point focusing—parabolic dish (PD) and solar tower (ST). Each of these varies geometrically, depending upon the type and arrangement of collector and receiver. The systems are illustrated in Fig. 14.1, to reflect their components and focusing strategy.

Each of the CSP systems has different capabilities in terms of useful temperature range, cost, efficiency, and others. Depending upon the requirements of a system (industrial or power generation), a CSP technology can be picked,

Renewable Energy Systems. DOI: https://doi.org/10.1016/B978-0-12-820004-9.00021-8

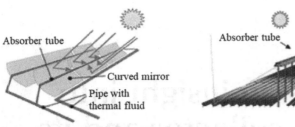

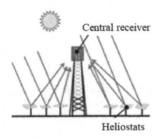

Parabolic dish   **Solar tower**

FIGURE 14.1 Schematic representation of different concentrating solar power systems. *Adapted with permission from Romero, M., & Steinfeld, A. (2012). Concentrating solar thermal power and thermochemical fuels.* Energy & Environmental Science, 5, 9234–9245. doi: 10.1039/c2ee21275g.

TABLE 14.1 Typical performance criterion for CSP technologies (Weinstein et al., 2015; Zhang, Baeyens, Degrève, & Cacères, 2013).

CSP system	Levelized energy cost (US$ kWh⁻¹)	Peak efficiency (%)	Operating temperature range (°C)	Concentration ratio
Parabolic trough solar collector	0.16 – 0.40	25	20–400	80
Linear fresnel reflector	0.14–0.45	18	50–300	30
Solar tower	0.13–0.30	22	400–600	1000
Parabolic dish	NA	32	550–750	1500

CSP, concentrating solar power.

although at the commercial level, majority of applications involve the usage of PTSCs. It holds around 90% of the share among the commercially used CSP systems because it is the most evolved one till date and is also supported by the presence of a small industry that manufactures and markets it. A comparison of performance indices for different CSP systems is seen in Table 14.1, wherein the values mentioned are characteristic of a specific type of system. There could be further variation for different systems considered. Here, levelized energy cost is the cost of electrical energy produced per kilowatt hour that includes the capital cost along with operation and maintenance cost (Dersch et al., 2007) while peak efficiency is the highest solar to electrical conversion efficiency achieved (Weinstein et al., 2015). Furthermore, concentration ratio (C) referred to as the ratio of collector aperture area (A_{cs}) to the receiver area (A_r) and can be formulated as:

$$C = \frac{A_{cs}}{A_r} \tag{14.1}$$

All these systems have the potential to serve in solar electric generating systems (SEGS) for power generation or in an industrial scenario to drive the heat-intensive processes. In process industries, heat is required at different temperature levels, generally between 50°C and 260°C, for sustaining various processes, such as drying, steam generation, cleaning or degreasing, and others. STs and PDs are meant for high operating temperatures and can serve well in SEGS while PDs are still under development. LFRs have discretely distributed mirrors, requiring tedious interconnections and have lower optical efficiency compared to PTSCs (Serrano, 2017). On the other hand, PTSCs have continuously interconnected mirrors, useful temperature range, and efficiency. Being most mature and proven, it makes them a convenient option to be used in industries for providing necessary process heat.

14.1.3 Problem statement

PTSC consists of a parabolic shape mirror ensuring that sun rays falling on it converge to a heat collector element (HCE) placed at its focal line. Heat transfer fluid (HTF), also referred to as working fluid (air, water, oil, or some organic solvent), flows through HCE and absorbs the energy concentrated on it. PTSCs, either used for power generation or industrial process heat (IPH), require a large setup, not only involving huge capital and land area but also are technically absorbing with a large number of parameters involved (Yılmaz & Mwesigye, 2018). It would be better to analyze such a system beforehand so that economic or operational losses can be minimized. This kind of analysis is possible using an appropriate modeling approach. Since the dynamics of the PTSC is well known and can be realized using simple laws of science, the modeling approach can be quite effective.

Modeling and simulation approach has been attempted to analyze the PTSCs for thermal as well as optical strength. Thermal and optical properties of PTSC are considered to be reasonably temperature independent, thus making it convenient to analyze them separately (Güven, Mistree, & Bannerot, 1984). Optical analysis includes ray tracing methods and analytical approaches to evaluate the optical performance of the PTSC. Thermal analysis may consist of many aspects, but in general, is related to the determination of thermal losses, photo-thermal efficiency, temperature profile of receiver surface, and fluid temperature. This indicates that there are various spheres of modeling a PTSC. So far, a noticeable number of models are available in the literature to analyze different aspects of PTSC performance. A problem occurs in choosing the kind of analysis or model that would be fruitful and sufficient to scrutinize the PTSC for a particular application.

14.1.4 Chapter outline

This chapter provides insights about PTSC by going through the following:

- A brief history of PTSCs.
- A detailed description of PTSC components and advances.
- Description of optical and thermal evaluation of PTSC. Both how and why are included.
- Introduction to different kinds of models for PTSC available in the literature. Analyses of their potential to predict the thermal performance of PTSC.
- Development of a heat transfer model for PTSC. Advantages and limitations of it over other prevailing models.
- Discussion and comments over the potential applications of PTSC with attention to its application in serving heat-intensive industrial processes.

14.2 Related work

Modeling of a system provides a suitable approach to analyze its possible performance over different parameters without troubling the practical setup or even before its installation. Modeling studies for PTSC have gained momentum over the last decade as the need for a sustainable energy option has become paramount. A statistical investigation for the same can be seen from Fig. 14.2A, which shows the number of research articles searched with the keyword

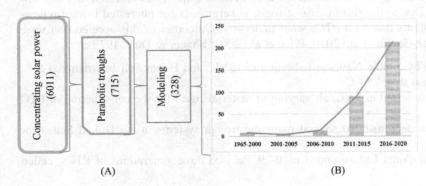

(A)   (B)

FIGURE 14.2 (A) Statistics of research articles by keywords concentrating solar power, parabolic troughs, and modeling. (B) Distribution of studies on the modeling of parabolic troughs from 1965 to 2020. *From Web of Science, Accessed 29.04.20.*

"concentrating solar power" on Web of Science and then refined using keywords "parabolic trough" and "modeling" subsequently. To further evaluate the progress of modeling studies, year-wise data with different spans were taken and plotted in Fig. 14.2B. As discussed in the history of PTSC, initially more emphasis was on developing different working prototypes so there were almost negligible studies published related to modeling. With the advancements in technology and commercial feasibility of PTSC, the interest has grown to make it more efficient, either by providing the methods to assess its thermal performance or to test modifications using the modeling approach.

Modeling-based studies on PTSC majorly concentrates on the evaluation of its thermal and optical performance under variation of different parameters or enhancements; few important ones are quoted hereafter. A model presented by Edenburn (1976) was found to be the quite aged one and includes energy balance equations solved by finite difference method. Yılmaz and Mwesigye (2018) have presented a comprehensive review on performance analysis of PTC while covering a rich portion of available literature on PTSCs. Forristall (2003) has presented an in-depth study using the 1-D approach and examined the effect of various geometrical and operating parameters on PTC performance. Gong, Huang, Wang, and Hao (2010) have used a 1-D model first to determine major heat loss from evacuated HCE and factors influencing the efficiency, then a 3-D model built using CFD software was utilized to further investigate the heat transfers involved. A comparative study considering various models differing in assumptions is conducted by Liang, You, and Zhang (2015). A receiver with half portion insulated and other half filled with air was proposed by Al-Ansary and Zeitoun (2011), using the numerical modeling. A detailed model covering the various heat transfer mechanisms in PTC is given by Kalogirou (2014). A model by García-Valladares and Velázquez (2009) on the study of two flow patterns using spatial discretization has delivered consistent results. A rigorous analysis to examine both optical and thermal performance of the PTC is given by Yılmaz and Söylemez (2014). Behar, Khellaf, and Mohammedi (2015) have suggested a heat transfer model based on energy balance along with a review on the design and manufacturing details of PTC. Effect of incidence angle on heat flux distribution and insertion of helical screws on the performance of HCE is studied using the 3-D models by Song et al. (2014). Three different collector kinds, evacuated, nonevacuated, and bare tube receiver were analyzed to examine the thermal improvement margin with the implementation of a nanofluid in PTSC (Bellos, Tzivanidis, & Said, 2020). To improve the thermal conductivity of HTFs, nanosized particles are dispersed in it, making them nanofluids. Many more studies on modeling and simulation of PTSC are mentioned and discussed throughout the chapter as and where found relevant.

14.3 Parabolic trough solar collector—history

The first known PTSC or parabolic trough collector (PTC) was constructed in 1870 by John Ericsson, had a collector area of 3.25 m^2 to concentrate sun rays over a tube, and was built to drive a small 373-W engine. Several variants of the same, varying in geometry and construction materials, were built and presented by him till 1886, and air was used as the heat transfer medium in these prototypes (Pytilinski, 1978). In 1907 two Germans—Wilhelm Meier and Adolf Remshardt, got the first patent for steam generation using parabolic trough technology (Fernández-García, Zarza, Valenzuela, & Pérez, 2010). Various PTC-based solar engines were built and tested by an American engineer, Frank Shuman, between 1906 and 1911; the knowledge gained was utilized by him in collaboration with Charles Vernon Boys, to construct a pumping plant for irrigation in Meadi, Egypt, in 1912 (Günther, Joemann, & Csambor, 2011). The pumps were driven by steam motors, for which the steam was generated using an array of PTCs; whole system was successfully running in 1913 but was shut down in 1915 due to World War I.

Development of PTC gone through a dull phase and no major contribution to the field was attained until in the mid-1970s, when the United States Department of Energy as well as the German Federal Ministry of research and technology started to fund projects based on PTCs. Due to the rise in cost of fuels, governments got interested toward renewable energy, and development of several systems based on PTCs went underway. Outcomes of this acceleration were positive and can be summarized as (Fernández-García et al., 2010; Price et al., 2002; Shaner & Duff, 1979):

- Two conceptually similar collectors, each by Sandia National Laboratories (SNL) and Honeywell International Inc., were developed to work at temperatures less than 250°C.
- In 1975 three more collectors for IPH were tested at SNL, all varying in materials used for the construction of PTC components.
- From 1977 to 1982, the company Acurex demonstrated several parabolic trough systems in the United States for process heat applications.
- The Israeli-American company Luz International Ltd., founded in 1979, devised three generations of PTCs, called LS-1, LS-2, and LS-3.

- Nine members of the International Energy Agency constituted a project with an objective of electricity generation using solar energy at Plataforma Solar de Almeria and it was operational in 1981. It was later taken over by the government of Spain.
- In the 1980s the PTCs entered the commercial market and some American companies, namely, Acurex Solar Corp., Solar Kinetics Corp., Jacobs Del. Corp., Honeywell Inc., General Electric Co., and Suntec Systems Corp.—Excel Corp., began to manufacture and market it.

The breakthrough for PTC technology came in 1983 when Southern California Edison (SCE) signed an agreement with Luz International Limited to obtain power from PTC-based power plants to be constructed in California. These plants were termed as SEGS I and II and were operational by 1985–86. This led the base for credibility for Luz, which later signed several contracts with SCE to develop SEGS III to IX plants. These plants are still in operation and vary in capacity from 14 to 80 MW and constitute a total installed capacity of 354 MW (Philibert, 2004). The operational and constructional experience gained through these plants propelled the PTC technology and laid the base for further technological advancements and project planning.

Another lean phase for the PTCs went on till 2007, where negligible expansion in installed capacity of PTC-based plants and no vital innovation in its technology were seen. But in 2007 Nevada Solar One with a capacity of 64 MW was started in Nevada. Further expansions took place in Europe, Andasol I in Granada generating electricity since 2008 with a capacity of 50 MW, and another power plant Andasol II is working since 2009; these were the first plants with thermal storage systems (Geyer et al., 2007).

With the advancements in technology and diminishing nonrenewable energy resources, there is a shift in the government's mindset globally to support PTSC-based systems. Although most of the PTSC-based power plants are in the United States and Spain, but several projects in India, Australia, Egypt, China, Algeria, Morocco, and other countries are in construction or planning phase. According to data provided by the National Renewable Energy Laboratory, United States, the population of such plants is increasing rapidly, with currently around 100 plants at different phases (Parabolic Trough Projects). In addition to power generation, parabolic troughs have the potential to serve heat-intensive industrial processes. PTSCs were initially tested for IPH applications in the mid-1970s and 1980s but could not meet the industrial criteria at that time due to technology constraints and high cost (Price et al., 2002), but with the lowering cost of PTSC components and improved technology they can fit in certainly.

14.4 Parabolic trough solar collector—an overview

As the name suggests, PTSC consists of a parabolic-shaped mirror as its core, and other supporting components make it collect solar energy to be converted into heat. An illustration of PTSC showing its various components has been done in Fig. 14.3. A sheet of reflective material is bent into a parabolic shape to collect and converge sun rays falling on to it toward its focal line; this is known as collector or parabolic trough (PT). It is made of explicit materials such as silvered low-iron glass and aluminum that provides low absorptance and high reflection (Solar Energy Research Institute, 1985). Other key factors that impact the production of PTSC mirrors are their cost and abrasive properties. Besides bending a mirror into the desired shape, it is required to go through processes, such as silvering, gluing, and protective coating application, for the proper fabrication and enhancement of optical properties (Behar et al., 2015).

Another critical component of PTSC, where the solar to thermal conversion takes place, is placed at the focal line of the collector and known as HCE (or receiver). An axial view of HCE and a radial view of its cross-section are shown in Fig. 14.4 to make a better understanding of components discussed hereafter. It is explicitly placed at the focal line of mirror so that it receives maximum solar concentration. HCE comprises a metal absorber tube generally made of stainless steel, encircled by a borosilicate glass cover. For an effective solar-thermal conversion, it is expected that an absorber surface possesses high solar absorbance and low thermal emittance at operating temperature. Typical values for effective performance of PTSC are as follows: absorbance >0.95 and emittance around 0.1 at 400ₒC (Selvakumar & Barshilia, 2012). So, absorber is layered with a selective coating for better absorptance (Barriga, Ruiz-De-Gopegui, Goikoetxea, Coto, & Cachafeiro, 2013), and to improve the transmission of solar heat to a working fluid flowing through it, which is also known as HTF.

For PTSCs to perform well in power generation, it is required that it reaches higher temperatures to drive higher power-cycle efficiencies and to favor lower costs of energy. Solar selective absorbing coatings (SSAC) play an important role in that by making absorber capable of absorbing maximum solar energy (Xu et al., 2020). As it is difficult to achieve maximum absorptance with a single layer, SSAC are generally a stack of layers that have the capability of

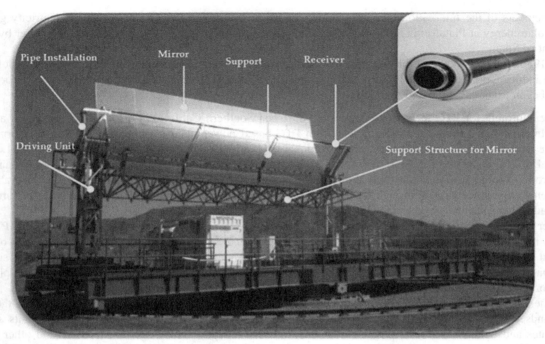

FIGURE 14.3 Illustration of parabolic trough solar collector. *Reprinted with permission from Abed, N., & Afgan, I. (2020). An extensive review of various technologies for enhancing the thermal and optical performances of parabolic trough collectors.* International Journal of Energy Research, *1−48. doi: 10.1002/er.5271.*

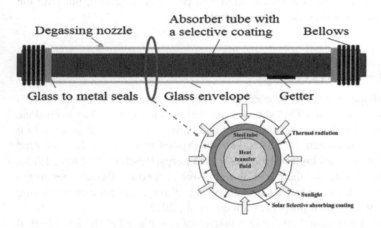

FIGURE 14.4 Axial view of heat collector element and a radial view of its cross-section. *Adapted with permission from Bijarniya, J. P., Sudhakar, K., & Baredar, P. (2016). Concentrated solar power technology in India: A review.* Renewable & Sustainable Energy Reviews, 63, *593−603. doi: 10.1016/j.rser.2016.05.064; Xu, K., Du, M., Hao, L., Mi, J., Yu, Q., & Li, S. (2020). A review of high-temperature selective absorbing coatings for solar thermal applications.* Journal of Materials, 6, *167−182. doi: 10.1016/j.jmat.2019.12.012.*

being spectral selective. It is viable to have an application specific and optimized SSAC by modifying the structure of coatings, deposition mechanism, and choosing different materials (Du et al., 2011).

Generally used HTFs for PTSCs are water, synthetic oil, and molten salts; at present, operating temperatures up to 400ₒC with the use of thermal oil as HTF are easily achievable (Kennedy & Price, 2005). Most of the conventional HTFs have inadequate heat transfer capacity and low photo-thermal conversion properties. That is why, there is an opportunity to enhance the thermophysical properties of HTF, especially if PTSCs are to be utilized for power generation. Advanced generation of HTFs include nanofluids and molten salts, which can attain higher temperatures (Krishna, Faizal, Saidur, Ng, & Aslfattahi, 2020). A detailed categorization of HTFs is shown in Fig. 14.5; depending upon the application, a wise choice of HTF can enhance the performance of the PTSC system.

To protect the absorber from oxidation and to minimize the convective heat losses due to wind, a glass envelope is placed around it. The envelope is laminated with antireflective coating with high transmittance and low absorptance so that it does not absorb much of solar radiation reaching HCE. The ends of HCE are provided with glass-to-metal seals

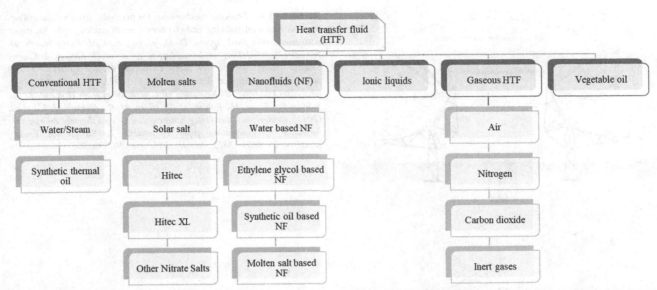

FIGURE 14.5 Classification of heat transfer fluids for parabolic trough solar collector system. *Adapted with permission from Krishna, Y., Faizal, M., Saidur, R., Ng, K. C., Aslfattahi, N. (2020). State-of-the-art heat transfer fluids for parabolic trough collector.* International Journal od Heat and Mass Transfer, 152, *119541. doi: 10.1016/j.ijheatmasstransfer.2020.119541.*

and metal bellows to account for differential expansion between metal tube absorber and a glass cover. Annulus region formed between absorber and the envelope is generally evacuated to diminish the radiative and convective losses in it (Price et al., 2006). After evacuation, the remaining gas pressure in annulus is suggested to be $<10^{-3}$ mbar or even $<10^{-4}$ mbar (Günther et al., 2011).

PTs can be arranged as modules connected via a common shaft; several PTs can be interconnected in rows and columns to attain a required output capacity. PTSCs can be classified as stationary and tracking type collectors depending upon the tracking mechanism adopted to receive maximum solar radiation. In stationery collectors, troughs are seasonally kept permanently in a particular alignment to maximize the solar concentration. On the other hand, the tracking collectors consist of an automated tracking mechanism that tracks the movement of the sun. There are different possible ways of solar tracking, but in PTSCs, single-axis tracking mechanism is sufficient, and it allows production of long collector modules (Kalogirou, 2014). The trough can be placed in east—west direction and track the sun from north to south, or it can be placed in north-south direction to track the sun from east to west. Fig. 14.6 gives the illustration of both the positions; each of them has pros and cons. Collector oriented in east—west direction requires minimum adjustment as full aperture always faces the sun during the afternoon. However, during morning and evening, a notable reduction in performance can be seen. For other orientation, optical loss is high at noon and lowest during morning or evening as the sun is due east or due west respectively. Placement of collector is not a single day affair and cannot be changed every day; it is observed that over a period of year, north-south oriented trough generally collects more solar radiation than an east—west one. Former receives most of the energy in summer, while the latter gets a significant share in winter (John & Duffie, 2013; Kalogirou, 2012). So, depending upon the location and need of an application to receive more energy in winter or summer, a tracking system can be opted.

14.5 Performance evaluation of PTSC

Before knowing how to examine PTSC, one crucial aspect, that is, the kind of solar concentration required for its operation should be understood. PTSC or any other CSP system requires direct normal irradiance (DNI) as input for its working (Serrano, 2017). As depicted in Fig. 14.7, sun rays coming toward earth can be categorized as diffused irradiance and DNI. Solar energy reaching the earth's surface after being scattered through clouds is known as diffused irradiance and the other that reaches the surface directly is DNI. Quality of DNI affects the performance of PTSC, and if a certain level of DNI is not available, then the output of the system will be null.

To comment upon the capability of PTSC, it is necessary to evaluate its performance over certain parameters. Both the optical and thermal parameters are required to be analyzed and a synopsis of such evaluations is presented in

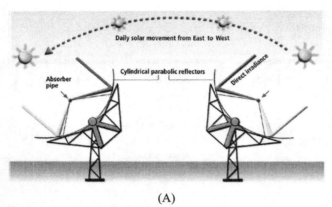

(A)

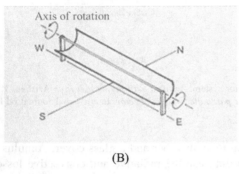

(B)

FIGURE 14.6 Tracking mechanisms for parabolic trough solar collector: (A) east-west tracking and (B) north—south tracking. *(A) Reprinted with permission from Nation, D. D., Heggs, P. J., & Dixon-Hardy, D. W. (2017). Modelling and simulation of a novel Electrical Energy Storage (EES) Receiver for Solar Parabolic Trough Collector (PTC) power plants.* Applied Energy, 195, 950–973. doi: 10.1016/j.apenergy.2017.03.084. *(B) Reprinted with permission from Hossain, M. S., Saidur, R., Fayaz, H., Rahim, N. A., Islam, M. R., Ahamed, J. U., et al. (2011). Review on solar water heater collector and thermal energy performance of circulating pipe.* Renewable & Sustainable Energy Reviews, 15, 3801–3812. doi: 10.1016/j.rser.2011.06.008.

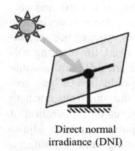

Direct normal
irradiance (DNI)

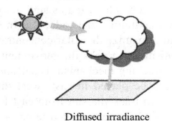

Diffused irradiance

FIGURE 14.7 Types of solar irradiation. *Reprinted with permission from Meyer, R. (2013). CSP & solar resource assessment CSP today South Africa. In* Proceedings of the 2nd concentrated solar thermal power conference and expo.

forthcoming section. Emphasis is on the modeling and simulation perspective of the same. Generally, the following evaluations are carried out to examine the performance of a PTSC.

14.5.1 Optical evaluation

Optical analysis of a PTSC includes determination of optical efficiency (η_o) and the factors affecting it. η_o can be defined as the ratio of solar energy absorbed by HCE to the energy incident over the collector surface, and it is mandatory to evaluate it first if thermal performance is to be tested. Various modeling studies have been carried out pertaining to the evaluation of η_o only or using it for thermal analysis. Mathematically, it can be represented as:

$$\eta_O = \gamma_{if}\rho_{rc}\tau_{ge}\alpha_{as} \tag{14.2}$$

Here, ρ_{rc} is the reflectivity of the clean collector surface, τ_{ge} represents the transmittance of the glass envelope, and α_{as} is the absorptance of the absorber surface. γ_{if} is known as intercept factor and is a product of many optical factors that can be summarized in Table 14.2.

Optical efficiency of PTSC is generally calculated at normal solar incidence, which means that the angle of incidence to the surface of collector is zero. The angle that is made between the sun rays hitting the collector surface and

TABLE 14.2 Optical factors constituting intercept factor or losses ($\in$) of different nature (Forristall, 2003; Price et al., 2006).

Optical term	Remark (typical value)
Geometric error ($\in_g$)	Due to misalignment of collector (0.98)
Tracking error ($\in_t$)	Due to imperfections in the solar tracking system (0.994)
Collector shadowing ($\in_{sh}$)	Shadow produced over collector by HCE and its supports causes losses (0.974)
Dirt on mirror ($\in_{dom}$)	Due to the accumulation of dust and contamination of collector surface (reflectivity/ρ_{rc})
Dirt on HCE ($\in_{dor}$)	Dust and contamination on receiver surface $(1 + \in_{dom})/2$
Miscellaneous aspects ($\in_{misc}$)	Scattering effects due to property of optical material used, slope errors of the collector surface and others (0.96)

HCE, heat collector element.

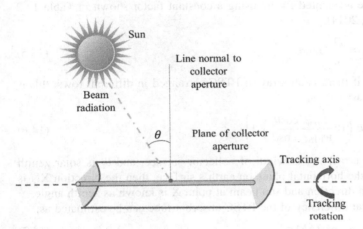

FIGURE 14.8 Angle of incidence.

TABLE 14.3 Correlations of IAM for some prominent collector types (Behar et al., 2015)

PTSC	IAM
LS-2	$IAM = \frac{1}{\cos\theta}\left(\cos\theta + 0.0008840 - 0.00005369\theta^2\right)$
LS-3	$IAM = 1 - 2.2307e^{-4}\theta - 1.1e^{-4}\theta^2 + 3.18596e^{-6}\theta^3 - 4.85509e^{-8}\theta^4$
ET	$IAM = \frac{1}{\cos\theta}\left(\cos\theta - 5.25097e^{-4}\theta - 2.859621e^{-5}\theta^2\right)$
IST	$IAM = \frac{1}{\cos\theta}\left(\cos\theta + 0.0003178\theta - 0.000039958\theta^2\right)$

IAM, incidence angle modifier.

the normal to that surface is called angle of incidence (θ), and the same is shown in Fig. 14.8. The factors mentioned in Table 14.2 and Eq. (14.2) are generally evaluated at $\theta = 0$; mostly, the evaluations in modeling studies and in experimental prototypes are made with this assumption.

Practically, the tracking system tries to capture the sun rays with normal incidence but cannot keep the same all day so θ varies. With this variation, the optical properties lose their credibility so to account for it a modifier term known as incidence angle modifier (IAM) is used. It is difficult to estimate IAM term theoretically, so it is expressed as a correlation in terms of θ using the experimental data (Behar et al., 2015). Thus it will depend upon the PTSC setup and cannot be generalized. Some of the correlations derived for well-known PTSCs are shown in Table 14.3.

With the uncontrolled movement of solar irradiation, few more factors affecting the η_O emerge, one such factor is end effect (or end loss). At $\theta = 0$, the sun rays hitting the collector surface are concentrated over the complete length of HCE, while at $\theta > 0$, a fraction of HCE is not illuminated by solar irradiance, as shown in Fig. 14.9. Studies have been carried out to account for these phenomena and correlations to calculate the same are given as:

$$\in_{end} = 1 - \left\{ \left(\frac{f_{cs}}{l_{cs}} \right) \left[1 + \left(\frac{A_{cs}^2}{48 f_{cs}^2} \right) \right] \tan\theta \right\} \tag{14.3}$$

by Gaul and Rabl (1980).

$$\in_{end} = 1 - \frac{f_{cs}}{l_{cs}} \tan\theta \tag{14.4}$$

by Lippke (1995).

Both the above equations give reasonable results but recent studies have shown that correlation by Gaul and Rabl is relatively better.

Another factor that affects the optical efficiency of PTSC is shading loss, which occurs due to its geometry. A shadow of HCE and its supports falls over the collector surface; thus a part of collector surface is underutilized as it is not subjected adequately to sun rays. This factor can be accounted for by using a constant factor shown in Table 14.2 or using the following correlation (Yılmaz & Söylemez, 2014):

$$\in_{sh} = \frac{D_{ao}}{l_{cs}} \tan\theta \tag{14.5}$$

Besides the mentioned case, shading loss can occur if there is an array of PTSCs arranged in different rows; this is illustrated in Fig. 14.10A. In this case, $\in_{sh}$ is evaluated as:

$$\in_{sh} = min \left[max \left(0; \frac{L_{row}}{w_{PTSC}} \frac{\cos\theta_z}{\cos\theta} \right); 1 \right] \tag{14.6}$$

where L_{row} is the spacing between two rows of PTSC array, w_{cs} is width of collector surface, and θ_z is solar zenith angle. As shown in Fig. 14.10B, consider a point X on the horizontal plane on earth's surface, then the direction XN is called zenith direction. The angle formed between zenith direction and sun beam at point X is known as zenith angle.

After considering all the mentioned factors, the optical efficiency of the PTSC absorber tube can be estimated as:

$$\eta_O = \gamma_{if} \rho_{rc} \tau_{ge} \alpha_{as} \text{IAM} \in_{end} \tag{14.7}$$

Above equation estimates η_O of absorber tube and can be used to determine solar energy absorbed by it. If the value of solar energy absorbed by glass envelope (Q_{o_ge}) and absorber tube (Q_{o_as}) is required to be calculated separately, then these values can be estimated as:

$$Q_{o\_ge} = \eta_{O-ge} Q_S \tag{14.8}$$

Here, η_{O-ge} is optical efficiency at glass envelope and can be calculated by replacing α_{as} by α_{ge} (absorptance of the glass envelope) in Eq. (14.7) ($\eta_{O-ge} = \gamma_{if} \rho_{rc} \alpha_{ge} \text{IAM} \in_{end}$), and Q_S is solar irradiance received per unit length of absorber, given by:

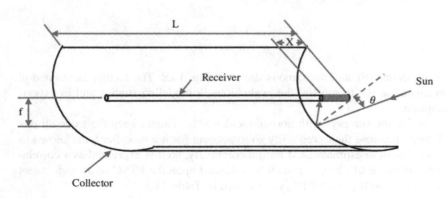

FIGURE 14.9 End loss. *Reprinted with permission from Abdulraheem-Alfellag, M. A. (2014). Modeling and experimental investigation of parabolic trough solar collector. Embry-Riddle Aeronautical University.*

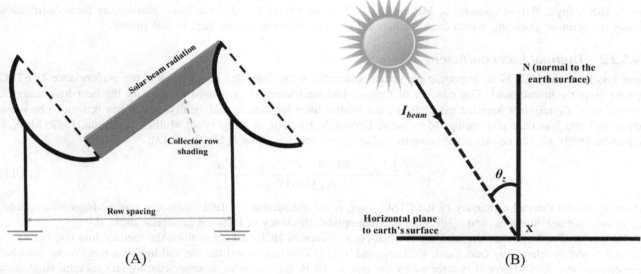

FIGURE 14.10 Illustration of (A) shading loss and (B) zenith angle.

$$Q_S = (\text{DNI})A_{cs}/l_r \tag{14.9}$$

A_{cs} and l_r represent the area of collector surface and length of receiver (HCE), respectively. Furthermore, the optical efficiency at absorber can be calculated by Eq. (14.7) or by:

$$\eta_{O-as} = \eta_{O-ge}\tau_{ge}\alpha_{as} \tag{14.10}$$

Finally, the solar energy absorbed by absorber tube can estimated as:

$$Q_{o\_as} = \eta_{O-as}Q_S \tag{14.11}$$

Thus with the input to PTSC known, and with the evaluation of the thermal efficiency of HCE completed, the overall proficiency of the system can be examined.

14.5.2 Thermal evaluation

It is aimed at the calculation of thermal attributes of PTSC, which are the fluid temperature, surface temperature profile of HCE, photo-thermal efficiency, and HCE thermal loss. Studies have focused on the calculation of any of these quantities or all in a single study, although these studies can be categorized into two broad categories confined to the evaluation of:

14.5.2.1 Heat flux and temperature profile

In thermal evaluation of PTSC, determination of heat flux distribution over an HCE plays an important role. Studies have shown that the flux distribution in a practical setup is nonuniform around the receiver and its realistic estimation is necessary to examine the thermal performance of PTSC. After obtaining the heat flux profile, the temperature profile around the HCE can be obtained. To make flux evaluations, a term, local concentration ratio (LCR), is used widely (Cheng, He, & Cui, 2013; Jeter, 1986; Yang, Zhao, Xu, & Zhu, 2010) and given by:

$$\text{LCR} = \frac{\text{heat flux}(Q')}{\text{DNI}} \tag{14.12}$$

The geometrical parameters, such as concentration ratio, optical errors, and angle of incidence, affect the LCR. It has been observed that an increase in incident angle, LCR falls gradually and is affected more significantly at higher θ (Jeter, 1986) (particularly >45 degrees). The effect of optical errors on the heat flux is evident, as the increase in optical errors or losses will degrade the quality of energy collected by HCE (Thomas & Guven, 1994). Different optical errors, tracking errors, shading, and others have different impacts on the LCR and studies confirming the same have been executed. The same trend of variation is followed by temperature profile around the HCE as by heat flux discussed above (Mwesigye, Bello-Ochende, & Meyer, 2014). It is important to calculate these profiles, as their distribution causes temperature gradients, which could be harmful to PTSC efficiency if not kept in safe limits.

14.5.2.2 Thermal loss coefficient

Heat loss from the HCE is an important parameter and needs to be evaluated wisely so that the performance of PTSC can be properly investigated. The concept of thermal loss coefficient (C_L) is used to simplify the heat loss analysis. Prediction of C_L requires iterative calculations, but studies have handled it analytically, which has proven to be more convenient and less time consuming (Mazumder, Bhowmik, Hussain, & Huq, 1986; Mullick & Nanda, 1982; Mullick & Nanda, 1989). C_L can be estimated using the following correlation (John & Duffie, 2013):

$$\eta_{TL} = \frac{\int A_{cs}\left[\eta_o(\theta)\mathrm{DNI} - C_L(T_{\mathrm{HCE}} - T_s)\right]dt}{A_{cs}\int \mathrm{DNI}dt} \tag{14.13}$$

where η_{TL} is the thermal efficiency of the PTSC, T_{HCE} is the temperature of HCE, T_s is the surroundings temperature, A_{cs} is collector aperture area, and $\eta_o(\theta)$ represents the optical efficiency of PTSC at incidence angle, θ.

Various studies have characterized the thermal performance of HCE and determined the thermal loss coefficient. A thermal model developed by Gee, Gaul, Kearney, and Rabl (1979) considered the thermal loss as a function of absorber temperature and showed how it is affected by the type of HCE. Emittance of absorber coating and vacuum conditions in the annulus region of HCE are shown to have a significant effect on heat loss (Wang, Huang, Gong, Hao, & Yin, 2011). HTFs get decomposed at higher temperatures and release hydrogen, which permeates through absorber to annulus; this increases the annulus pressure and reduces the HCE efficiency (Burkholder, 2011). Studies in the past explored the effect of replacing air by inert gases (argon or xenon) have demonstrated that it reduces the heat loss in the case of hydrogen permeation. Evaluation of heat loss to examine the thermal performance of HCE is a potent option, but the more significant and precise evaluation is possible by heat transfer analysis discussed next.

14.5.3 Heat transfer evaluation

It focuses mainly on the heat transfer analysis of HCE, being a core component of PTSC, since its performance may speak about the whole system. Modeling studies confined to performance testing of PTSC are majorly based on heat transfer evaluation. The heat transfer modeling of HCE involves energy balance of each component and requires some assumptions, which can vary from model to model, but some mutual ones can be listed as:

- Flow of HTF is fully developed;
- Dimensions of the system remain unchanged, such as constant diameters (unaffected by thermal expansion);
- Fluid is incompressible;
- The sky is considered as a black body; and
- Collector surface is specularly reflecting.

Based on the abovementioned assumptions, various models are present in literature. For a more straightforward understanding, their broad classification can be done depending upon the flow pattern studied (Yılmaz & Mwesigye, 2018), which is as follows:

14.5.3.1 Single-phase flow

Most of the PTSC systems use the HTFs, which do not change their phase during operation, thus said to have a single-phase flow. Despite being in single-phase, the flow could be considered as steady or transient. Handling the single-phase steady flow in heat transfer evaluations is relatively easier and mature analyses are available for it. Modeling studies for this type of flow can vary depending upon the assumptions other than aforementioned, direction of temperature gradients considered for HCE and others. Classification of heat transfer models based on different perspectives with single-phase steady flow is discussed later in the chapter; the same can be realized for other flow patterns discussed here.

Considering the flow pattern to be transient is relatively more realistic because the PTSC systems essentially operate in transient conditions. The variable nature of environmental conditions and solar irradiance does not allow PTSC to operate in steady conditions. Inclusion of transient flow in the model is difficult and should be considered while studying or developing PTSC model for power generation applications. Various studies have considered transient flow in

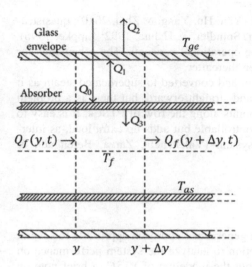

FIGURE 14.11 Lumped capacitance analysis of heat collector element. *Adapted with permission from Rongrong, Z., Yongping, Y., Qin, Y., & Yong, Z. (2013). Modeling and characteristic analysis of a solar parabolic trough system: Thermal oil as the heat transfer fluid.* Journal of Renewable Energy, 2013, 1–8. doi: 10.1155/2013/389514.

their model by using lumped capacitance analysis (Rohsenow & Hartnett, 1999; Rongrong, Yongping, Qin, & Yong, 2013; Xu et al., 2013). It involves transient analysis energy balance equations developed for various parts of the PTSC. This kind of analysis is done using control volumes for HCE as shown in Fig. 14.11. The general equations used for this analysis are:

$$\left(\rho_f c_f A_f\right)\Delta y \frac{\partial T_f(y,t)}{\partial t} = Q_3(y,t)\Delta y + Q_f(y,t) - Q_f(y+\Delta y,t) \text{(for HTF)} \tag{14.14}$$

$$\left(\rho_{as} c_{as} A_{as}\right)\Delta y \frac{\partial T_a(y,t)}{\partial t} = Q_0(y,t)\Delta y - Q_1(y,t) - Q_3(y,t) \text{(for absorber tube)} \tag{14.15}$$

$$\left(\rho_{ge} c_{ge} A_{ge}\right)\Delta y \frac{\partial T_{ge}(y,t)}{\partial t} = Q_1(y,t) - Q_2(y,t) \text{(for glass envelope)} \tag{14.16}$$

Different models have considered the transient nature of PTSC operation and shown the variation of different parameters on its performance. The variation in HTF type (Marif, Benmoussa, Bouguettaia, Belhadj, & Zerrouki, 2014; Mohamed, 2003), solar irradiance (Mohamed, 2003), mass flow rate of HTF (Eskin, 1999), energy collected (Liang, Zheng, Zheng, You, & Zhang, 2017; Sivaram, Nallusamy, & Suresh, 2016), and others has been studied.

14.5.3.2 Double phase flow

In PTSC systems meant for direct steam generation (DSG), the water gets boiled and converts to steam in the absorber tube itself, resulting in the two-phase flow. This flow regime is complicated to analyze as compared to a single-phase steady flow since both liquid and vapor flow together through the absorber tube. DSG systems are generally used where higher temperatures are required; this is due to the fact that most of the HTFs can attain temperatures up to 400°C only. Thermal oil, the most widely used HTF, has a limiting temperature (400°C), known as film temperature, beyond which it starts to degrade. So, in applications, such as SEGS, where higher operating temperatures are needed, the DSG systems and molten salts serve the purpose.

The operational concepts related to DSG systems using the numerical approach have been discussed by Lippke (1996). The two-phase flow can attain different flow patterns, namely, stratified, slug, wavy, annular, intermittent, and a possible dryout (Elsafi, 2015). It is important to determine the flow pattern, as a change in flow regime can cause a large temperature difference and, if not identified in time, can lead to physical damage, such as bending of the absorber (Almanza, Jiménez, Lentz, Valdés, & Soria, 2002). Various double phase flow maps have been developed for predicting the flow patterns, but it is complex to handle them and results provided are unreliable. Studies related to the effect of parameter variation on DSG systems and the application of performance enhancement techniques are also available in literature.

Abovementioned details are for double phase steady flow only; if transient flow is considered, here the flow characteristics become more complex. It is difficult to perceive transient flow characteristics in DSG systems due to flow instabilities caused by perturbations in solar irradiation. Several studies have tried to realize these characteristics in

PTSC systems; these studies are mainly based on lumped-capacitance model (Yan, Hu, Yang, & Zhai, 2010), quasisteady dynamic models (Xu & Wiesner, 2015), and energy balance (Heidemann, Spindler, & Hahne, 1992; Lippke, 1996). Both the steady and transient flows in DSG systems are studied under three operational concepts. The schematics for the same are shown in Fig. 14.12. A brief description of each of them is given thereafter.

In once-through concept, feed water goes through preheating, evaporation, and converted to superheated steam as it circulates in and out through the different PTSCs in a row, it is economic and straightforward, but there is a problem with its control. The injection concept involves water injections at various points along the row of PTSCs. It is easy to control but is complex and expensive. The recirculation concept is highly controllable but adds up extra load, as intermediate steam separator and the recirculation pump are added to the system (Valenzuela, Zarza, Berenguel, & Camacho, 2005).

14.6 Analytical thermal models

Modeling a system to analyze its performance over different parameters is a generalized approach applicable to several scientific operations. It offers many advantages, such as low cost and an option to analyze the system performance on number of parameters as compared to experimentation. Before knowing about the modeling of PTSC, a brief note on the general modeling approach is given here. There are three types of modeling, one that is wholly based on the laws of science and is known as white box modeling or first-principle modeling. The other is known as black box modeling and is completely based on input—output data of the system. In this approach, data are provided to a mathematical tool that derives a functional relation between the input and output. The last approach is grey box modeling, which involves some part to be developed using physical laws, and few parameters that cannot be presented using scientific equations are derived through mathematical tools. For example, the density of fluid varies with temperature and it may be required to use its values at different temperatures so an empirical equation that relates its values with temperature derived through software tool is used. Fig. 14.13 shows the same in a summarized form.

Various thermal models are available in the literature for evaluating the performance of PTSC; this section provides an introduction to such models and analyses their potential to examine the thermal performance of PTSC. Any specific method to categorize the models for PTSC is not available, so a categorization of various analytical models based on different technical aspects is shown in Fig. 14.14. It should be kept in mind that any of the model categories discussed here is not exclusive; one model can come under many categories. Categorization is made to simplify the understanding of various aspects related to PTSC modeling. For example, a 1-D model could be a uniform or nonuniform and also could be a dynamic thermal model.

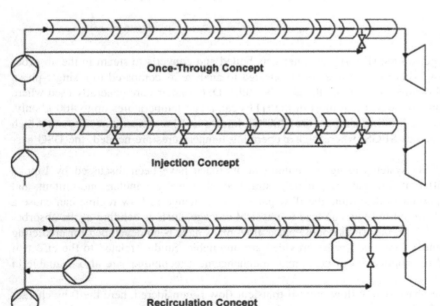

FIGURE 14.12 Basic operational concepts of parabolic trough solar collector-based direct steam generation systems. *Reprinted with permission from Valenzuela, L., Zarza, E., Berenguel, M., & Camacho, E. F. (2005). Control concepts for direct steam generation in parabolic troughs. Solar Energy, 78, 301–311. doi: 10.1016/j. solener.2004.05.008.*

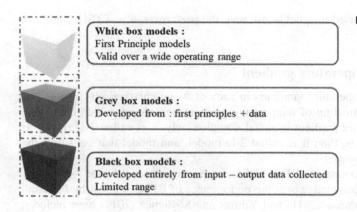

FIGURE 14.13 Different types of modeling approaches.

White box models :
First Principle models
Valid over a wide operating range

Grey box models :
Developed from : first principles + data

Black box models :
Developed entirely from input – output data collected
Limited range

FIGURE 14.13 Different types of modeling approaches.

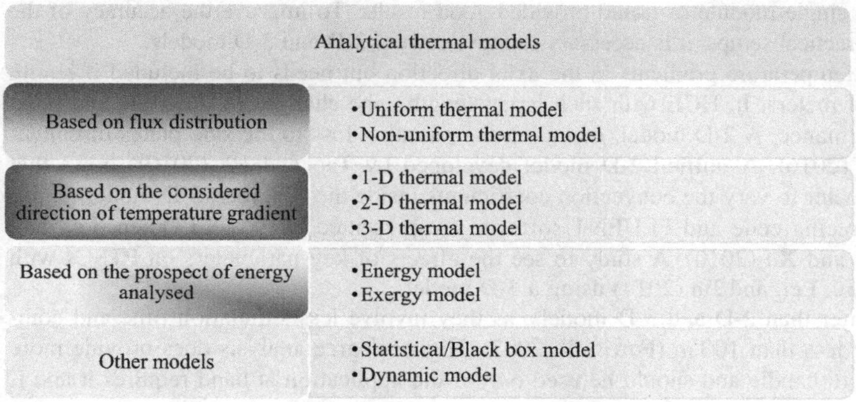

FIGURE 14.14 Classification of thermal models for parabolic trough solar collector.

Analytical thermal models	
Based on flux distribution	• Uniform thermal model • Non-uniform thermal model
Based on the considered direction of temperature gradient	• 1-D thermal model • 2-D thermal model • 3-D thermal model
Based on the prospect of energy analysed	• Energy model • Exergy model
Other models	• Statistical/Black box model • Dynamic model

All the models available in literature are more or less based on the energy balance equations. They can vary due to assumptions considered or any other prospect of PTSC. The categorization of PTSC models is done as below:

14.6.1 Based on flux distribution

In the context of PTSCs, the flux represents the heat flux (or thermal flux) on HCE, which is simply the amount of heat transferred per unit area of the surface. In modeling studies, even solar absorption is considered as heat flux to simplify the model, although it is physically false (Thomas & Guven, 1994). It is due to the fact that the HTF inside the absorber is unevenly heated, and thereby, the temperature profile of HCE becomes nonuniform (Cheng, He, & Qiu, 2015). This assumption gives acceptable results and is kind of necessary, as the correlations to calculate various heat transfers in PTSC are based on the same. Depending upon the flux distribution considered in the model, they can be classified as uniform and nonuniform models.

Uniform thermal models consider that flux distribution is uniform throughout the HCE, whether considered along the length or circumference. These models are simple to analyze and they provide reasonable results to estimate the performance of shorter length PTSC. As the length of HCE is increased, the error induced by uniform flux assumption also increases. So, this should be avoided in applications, such as power generation or high capacity desalination plants, where HCE lengths are quite high.

Heat flux is considered to be asymmetrically distributed throughout the HCE in nonuniform thermal models. They are more complex compared to uniform models and take more time for simulation. A study considering the nonuniformity of flux distribution and its effects on various governing factors of PTSC has been done earlier by Okafor, Dirker, and Meyer (2017). Model based on nonuniform flux by Wang, Liu, Xu, Zhou, and Xia (2016) shows that this approach is more reliable to estimate circumferential temperature distribution and local heat transfers. Examination of studies by Cheng et al. (2015) and Lu, Ding, Yang, and Yang (2013) suggest that the results obtained by uniform and nonuniform models match neck to neck. So, uniform models can be used to quickly estimate the performance of PTSCs and with less complexity.

14.6.2 Based on the considered direction of temperature gradient

HCE being subjected to solar thermal energy possess temperature gradients in each of its geometrical dimension. The models developed for it can be classified based on the direction of temperature gradient considered, which are radial, axial, and circumferential. If a model considers temperature gradient in radial direction only, it is called 1-D model. If both radial and axial directions are considered for analysis, then it is called 2-D model, and model that considers the temperature gradients in all three directions is called 3-D model.

1-D modeling is quite mature and often used approach to examine PTSC performance compared to 2-D and 3-D modeling. A detailed model, testing the effect of various design parameters over the performance of PTSC is given by Forristall (2003). Padilla, Demirkaya, Goswami, Stefanakos, and Rahman (2011) and Yılmaz and Söylemez (2014) have included the detailed radiative heat transfer analysis in their models. Different 1-D models under varying assumptions are compared by Liang et al. (2015) on the basis of different performance parameters. 1-D models developed to test the performance of PTSC have been majorly applied to test a single module of it and provided good results. To improve the accuracy of the PTSC models and to bring them close to practical setups, it is necessary to move toward 2-D and 3-D models.

It is difficult to take into account the temperature gradients in the axial direction but needs to be included if length of HCE reaches hundreds or thousands of meters. In HCE with such larger lengths, the changes in flow rate and pressure drop can appreciably affect its performance. A 2-D model, including the radiation loss to the side plates of collector is presented by Huang, Xu, and Hu (2016). A unified 2-D model developed by Tao and He (2010) shows that diameter ratio could be an effective parameter to vary the convection coefficients inside the absorber. 3-D models simulated by combining a Monte Carlo ray tracing code and FLUENT software are presented by Wu, Li, Yuan, Lei, and Wang (2014) and Cheng, He, Xiao, Tao, and Xu (2010). A study to see the effects of key parameters on PTSCs with molten salt as HTF was done by Wang, Liu, Lei, and Jin (2014) using a 3-D model.

Overall, 1-D models are relatively better than 2-D and 3-D models, as they involve less simulation time and complexity, especially when HCE lengths are less than 100 m (Forristall, 2003). Higher-degree analysis does provide more precise results but they are quite difficult to handle and should be used only if the application at hand requires it and if time is not a constraint.

14.6.3 Based on the prospect of energy analyzed

Energy analysis is a quantitative analysis that accounts for energies entering and leaving the system. It is based on the first law of thermodynamics. Exergy analysis is based on the second law of thermodynamics and is a qualitative analysis that identifies the magnitude of system inefficiencies along with its causes and locations (Park, Pandey, Tyagi, & Tyagi, 2014). Most of the models presented for PTSC involve energy analysis and there is an opportunity for its exergy analysis too. Some models proposed earlier (Al-Sulaiman, 2013; Yılmaz & Söylemez, 2014) have been simulated to evaluate the useful heat gained by HTF during its flow through the absorber. This represents energy analysis, as the quantity of energy gained by HTF is estimated. Kumaresan, Sridhar, and Velraj (2012) analyzed the energy collected by the PTSC for a fixed time period of 1 hr to comment upon its performance. A similar study was attempted by Ouagued, Khellaf, and Loukarfi (2013), but they calculated the heat gained by HTF for a day.

The exergy models are developed with a goal to identify the irreversibility sources, which can be used to improve the design of PTSC and its performance (Dincer & Rosen, 2007). These models are based on the control volumes, mass flow rate of HTF, variation in inlet temperature, solar irradiance, and wind speed (Conrado, Rodriguez-pulido, & Calderón, 2017). An exergy analysis for HCE, defining the equations for calculation of exergy losses and exergy output, is proposed by Padilla, Fontalvo, Demirkaya, Martinez, and Quiroga (2014). Khakrah, Shamloo, and Kazemzadeh Hannani (2018) have presented an exergy analysis of PTSC using nanofluid as HTF. An exergetic and energetic evaluation of PTSC with the inclusion of internal fins in absorber is presented by Bellos, Tzivanidis, and Daniil (2017). Not only for PTSC, the exergy analysis is gaining popularity to analyze various renewable energy systems.

14.6.4 Other models

Models that are implicit in nature and cannot be directly linked to the classifications discussed so far in the section come under this category. Such models are:

1. **Dynamic models**

Conventionally most of the models for PTSC work with an assumption that system is in steady state, which means constant flow rate and consistent temperatures along the HCE. Although, in practical setups, the system is dynamic, so if the model wants to perform in a more realistic situation, it should include this aspect as well. Dynamic models are also based upon the energy balance but they include transient behavior of the system. A dynamic model was developed previously by Xu et al. (2013) to provide the insightful analysis of actual large-scale experimental setup. Luo, Yu, Hou, and Yang (2015) have presented the lumped parameter dynamic model for a field of PTSCs. As discussed in Section 14.5.3, the dynamic or transient behavir of the system is difficult to include in model but should be included when analysis is required for an SEGS or very large setups. Any kind of model discussed previously in this section can also be extended to a dynamic model.

2. **Statistical models**

In a statistical model, the input and output data are fed to a black box (generally a machine learning tool) that returns a functional relationship between them (Zhang, 2010). This approach is relatively new in the field of PTSC and provides reasonable results. Liu et al. (2012) have used the least square support vector machine method to develop the relationship between PTSC efficiency and solar irradiance, the flow rate, and HTF temperature. This model has a major drawback that the input-output relationship is completely dependent upon the data fed for a particular system. The same relationship cannot be used for any other system and will be valid only for the range of data fed to the black box. Thus this model will have a lower range of validity as compared to other models that are based on heat transfer principles.

14.7   1-D heat transfer model

So far, different models and technical aspects related to them have been discussed. It is ascertaining that 1-D model is most likely the preferred choice if complexity and time are constraints. Also, it provides good results if HCE lengths are under 100 m, which covers many applications, such as IPH, heating and cooling systems, and small desalination systems. 1-D models meant for mentioned applications are normally developed with the assumption of uniform flux and examined using energy analysis. In this section, owing to their popularity the development and features of the same are discussed.

14.7.1   Development

As discussed in the previous section, in 1-D model, temperature gradients are considered and evaluated in a radial direction. An HCE can be discretized into "N" parts and kth part of it can be analyzed radially for the same. An illustration of this kth part is shown in Fig. 14.15.

Moving in the radial direction for the kth part of HCE and considering the energy balance at each surface, the set of equations representing the system mechanism can be built. At absorber surface, solar energy absorbed by absorber tube (Q_{o_as}) conducts (Q_{cond_ao-ai}) through it and is transferred to HTF via convection (Q_{cov} $1pt\_a-f$) while a part of it is lost toward glass cover by convection (Q_{cov} $1pt\_ao-ge$) and radiation (Q_{rad_ao-ge}). A portion of solar energy is absorbed by glass envelope (Q_{o_ge}); also, the energy lost by absorber reaches it and conducts (Q_{cond_ge}) through it to be lost to the environment through radiation to the sky (Q_{rad_ge-sky}) and through convection to ambient

FIGURE 14.15 Radial view of kth section of heat collector element.

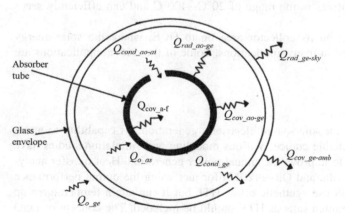

Absorber tube

Glass envelope

Q_{cond_ao-ai} Q_{rad_ao-ge} Q_{rad_ge-sky}

Q_{cov_a-f} Q_{cov_ao-ge}

Q_{o_as} Q_{cond_ge} Q_{cov_ge-amb}

Q_{o_ge}

$(Q_{Cov} \quad 1pt_{\_ge-amb})$. Developed equations are also known as heat transfer equations, as they are based on heat transfer mechanisms (conduction, convection, and radiation). A set of such equations is given below:

$$Q_{cov\_a-f} = Q_{cond\_ao-ai} \tag{14.17}$$

$$Q_{o\_as} = Q_{cond\_ao-ai} + Q_{cov\_ao-ge} + Q_{rad\_ao-ge} \tag{14.18}$$

$$Q_{cov\_ao-ge} + Q_{rad\_ao-ge} = Q_{cond\_ge} \tag{14.19}$$

$$Q_{cond\_ge} + Q_{o\_ge} = Q_{Cov\_ge-amb} + Q_{rad\_ge-sky} \tag{14.20}$$

Movement of heat represented via different heat transfer terms can be seen in Fig. 14.15. Each of these terms needs to be estimated to solve the developed set of equations. Heat transfer terms are dependent upon temperatures at different surfaces of the kth section of HCE and on other thermal properties. Generally, the values for HTF and ambient temperature are available, while values for other temperatures are estimated by making an initial guess and solving the equation set iteratively until the convergence criteria are met. Other parameters whose values are normally available for a system are HTF flow rate, wind speed around the PTSC, and DNI. The value of any other property required to estimate different heat transfer terms is calculated based on given and guessed temperature profile and standard conditions. The most commonly used correlations used for estimation of conduction, convection, and radiation heat transfers in PTSC are summarized in Table 14.4.

After evaluating the heat transfer terms and solving the energy balance equations, along with the optical analyses discussed in Section 14.5.1, the thermal efficiency of the PTSC can be estimated as follows:

$$Q_{hg} = \dot{m}_f c_p (T_o - T_i) \tag{14.21}$$

$$\eta_T = \frac{\text{Heat gain}}{\text{Heat input}} = \frac{Q_{hg}}{\text{DNI} \times A_{cs}} \tag{14.22}$$

Here, the most common and simple way of developing the 1-D model and to examine the performance of PTSC by using it is covered. However, more advanced studies are available, whose references are given in related work, Sections 14.5 and 14.6.

14.7.2 Advantages and limitations

Compared to other models discussed in this chapter, the 1-D model is relatively simple to understand and fabricate. For shorter HCEs, results provided by it are competitive with complex models, such as 2-D and nonuniform models. Higher-degree models are more precise but they are not only complex to develop, but it is time-consuming and difficult to solve them. 1-D models are suitable for many applications of PTSC and are simple to analyze and solve. HCE lengths limit the use of 1-D models, as they become larger, the model starts to give more deviations in performance estimation. 1-D models are generally based on uniform flux distribution and could underestimate or overestimate the PTSC performance in such cases.

14.8 Potential applications

As mentioned earlier, PTSCs can easily reach the temperatures in the range of 20°C−400°C and can efficiently serve many industrial applications.

As shown in Fig. 14.16, PTSC converges rays falling on its collector area on to HCE, where the solar energy absorbed is gained as heat by HTF, this heat can be utilized in many applications. Some of the critical applications are described below:

14.8.1 Power generation

In the past few decades PTSCs are quite extensively used for solar-based electricity generation. Its capability to attain high temperatures and concentration ratio has made it a reliable choice. Various modeling and simulation studies have been performed by researchers on PTSCs for power plants to improve the annual power generation. Heat transfer analysis of HCE has been done by Hachicha, Rodríguez, Capdevila, and Oliva (2013) for increasing the output performance of PTSC-based power plant. Most of the commercial SEGS use synthetic oil as HTF but it can attain temperatures up to 400°C. To attain higher temperatures DSG systems or molten salts as HTF should be included. The concept of DSG

TABLE 14.4 Commonly used equations for 1-D modeling of HCE.

Heat transfer term	Correlation along the heat transfer coefficients and parameters	Reference(s)
Convection from absorber tube to HTF	$Q_{cov}\ 1pt\_{a-f} = \pi D_{ai} h_f (T_{ai} - T_f)$ where, $h_f = Nu_{D_{ai}} \frac{k_f}{D_{ai}}$ and $Nu_{D_{ai}} = 4.36$ (for laminar flow) $Nu_{D_{ai}} = \frac{\frac{f}{8}(Re_{D_{ai}} - 1000)Pr_f}{1 + 12.7\sqrt{f/8}\left(Pr_f^{2/3} - 1\right)}\left(\frac{Pr_f}{Pr_{ai}}\right)^{0.11}$ (for turbulent flow) $f = \left[1.58 ln\left(Re_{D_{ai}}\right) - 3.28\right]^{-2}$ (friction factor)	Cengel and Ghajar (2015), Forristall (2003), and Gnielinski (2013)
Conduction across the absorber tube	$Q_{cond\_ao-ai} = \frac{2\pi k_{pipe}(T_{ai} - T_{ao})}{ln\left(\frac{D_{ao}}{D_{ai}}\right)}$	Incoropera and Dewitt (2006)
Convection in annulus region	Pressure <0.013 Pa $Q_{cov\_ao-ge} = \pi D_{ao} h_{ao-ge}(T_{ao} - T_{gei})$ $h_{ao-ge} = \frac{k_{ang}}{\frac{D_{ao}}{2ln\left(\frac{D_{gei}}{D_{ao}}\right)} + b\lambda\left(\frac{D_{ao}}{D_{gei}} + 1\right)}$ Where, $\lambda = \frac{2.331 \times 10^{-20}(T_{ao-gei} + 273)}{(P_{an}\delta^2)}$ and $b = \frac{(2-a)(9\gamma - 5)}{2a(\gamma + 1)}$ Pressure > 0.013 Pa $Q_{cov}\ 1pt\_{ao-ge} = \frac{2\pi k_{ef}}{ln\left(\frac{D_{gei}}{D_{ao}}\right)}\left(T_{gei} - T_{ao}\right)$ Where, $k_{ef} = 0.386 k_{ang}\left(\frac{Pr_{ao-gei}}{0.861 + Pr_{ao-gei}}\right)^{\frac{1}{4}}\left(Ra_y\right)^{\frac{1}{4}}$ $Ra_y = \frac{\left[ln\left(\frac{D_{gei}}{D_{ao}}\right)\right]^4}{\left((D_{gei} - D_{ao})/2\right)^3\left(D_{gei}^{-3/5} + D_{ao}^{-3/5}\right)^5} Ra_{D_{ao}}$	Cengel and Ghajar (2015), Raithby and Hollands (1976), and Ratzel, Hickox, and Gartling (1979)
Radiation in annulus region	$Q_{rad\_ao-ge} = \frac{\sigma\pi D_{ao}\left(T_{ao}^4 - T_{gei}^4\right)}{\left(\frac{1-\varepsilon_{ao}}{\varepsilon_{ao}} + \frac{1}{T_{ao-gei}} + \left(\frac{(1-\varepsilon_{gei})D_{ao}}{\varepsilon_{gei}D_{gei}}\right)\right)}$	Incoropera and Dewitt (2006)
Conduction across glass envelope	$Q_{cond\_ge} = \frac{2\pi k_{ge}(T_{gei} - T_{geo})}{ln\left(\frac{D_{geo}}{D_{gei}}\right)}$	Cengel and Ghajar (2015)
Convection from glass cover to ambient around PTSC	$Q_{Cov}\ 1pt\_{ge-amb} = h_{geo-amb}\pi D_{geo}\left(T_{geo} - T_{amb}\right)$ Where, $h_{geo-amb} = \frac{k_{air}}{D_{geo}} Nu_{D_{geo}}$ No wind $Nu_{D_{geo}} = \left[0.60 + \frac{0.387 Re_{D_{geo}}^{1/6}}{\left\{1 + (0.559/Pr_{geo-s})^{\frac{9}{16}}\right\}^{\frac{8}{27}}}\right]^2$ With wind $Nu_{D_{go}} = C Re_{D_{geo}}^m Pr_{amb}^n\left(\frac{Pr_{amb}}{Pr_{geo}}\right)^{\frac{1}{4}}$	Churchill and Chu (1975) and Incoropera and Dewitt (2006)
Radiation from glass envelope to sky	$Q_{rad\_ge-sky} = \sigma\varepsilon_{geo}\pi D_{geo}\left(T_{geo}^4 - T_{sky}^4\right)$	Cengel and Ghajar (2015)

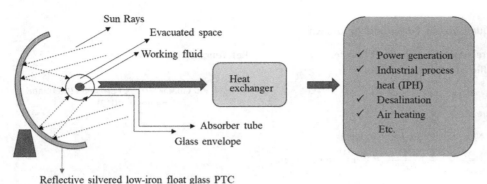

FIGURE 14.16 Solar to thermal conversion in parabolic trough solar collector for different applications. *Adapted with permission from Goel, A., Manik, G., Mahadeva, R. (2020). A review of parabolic trough collector and its modeling. In M. Pant, T. Sharma, O. Verma, R. Singla, & A. Sikander (Eds.), Soft comput. theor. appl. adv. intell. syst. comput (pp. 803−813), Singapore: Springer. doi: 10.1007/978-981-15-0751-9_73.*

was proposed and analyzed earlier by Odeh, Morrison, and Behnia (1998) to improve the efficiency of SEGS. Eck, Zarza, Eickhoff, Rheinländer, and Valenzuela (2003) investigated and demonstrated the DSG process in PTSCs under actual solar conditions. The concept was analyzed through six different packages. The main challenge for PTSC-based power plants is ineffective supply to the loads. This is because SEGS have peak efficiency hours in day time, which do not coincide with peak electricity demand, which normally occurs after sunset (Jebasingh & Herbert, 2016). So, the provision of thermal storage systems, which can provide output up to 6 hours after the sunset, becomes necessary for these systems. A review to examine the thermal energy storage prospects applied to PTSC-based power plants is presented by Herrmann and Kearney (2002). The thermal performance of 100-MW solar thermal power plant with thermal energy storage was evaluated using system advisor model by Bishoyi and Sudhakar (2017).

14.8.2 Industrial processes

Industries require a reasonable amount of heat to drive processes, such as pasteurization, drying, cooking, preheating of water, curing, and others; all these are normally executed using electricity. Generally, the electricity demands are fulfilled using fossil fuel based thermal power plants. If the heat requirement of these processes is catered by PTSCs, then solar energy can substitute a huge portion of electricity, thereby, providing a clean and sustainable option of power and reducing the dependency of industries over fossil fuels. Temperatures required by IPH applications are in the range of 50°C−260°C (Kalogirou, 2003). Some of these applications along with required temperatures are shown in Table 14.5 and can be easily served by PTSCs. Many modeling and simulation studies have been pursued to analyze the efficiency of PTSCs in IPH applications. A study on the use of PTSCs based plant to generate process heat for food processing industry by Silva, Javier, and Pérez-garcía (2014) evaluates the performance of the said system over various operational parameters. Pietruschka, Fedrizzi, Orioli, Söll, and Stauss (2012) have demonstrated the potential of different solar collector technologies for supplying IPH in different climatic conditions. Dynamic simulation of small sized PTSC is conducted by Ghazzani et al. with the objective of generating heated air for the food industry.

14.8.3 Air heating systems

Solar energy can be utilized to heat the air to utilize it for drying purposes in agriculture, marine foods, and textiles. It is a clean and sustainable option of energy, which can also be utilized in the heating of buildings and to renew dehumidifying agents. Solar energy-based heating systems can be classified into three types: active, hybrid, and passive systems. Since solar energy is time dependent, thermal storage systems are included with air heating systems, which involves two methods, namely, phase change materials (PCMs) and rock bed (Abdulhamed et al., 2018). A simulation-based study to study the effects of PCM-based energy storage on the performance of a solar air heating system was conducted by Morrison and Abdel-Khalik (1978). Researchers have shown that provision of thermal storage in solar air heating systems improves the overall performance of the system and helps to harness solar energy in a better way. PCMs, which store thermal energy by changing phase from solid to liquid, have proved worthy for thermal storage systems. Some of the commonly used PCMs for said purpose are hydrated salts (calcium chloride hexahydrate), paraffins, nonparaffins, and fatty acids (El Khadraoui, Bouadila, Kooli, Guizani, & Farhat, 2016; Lv, Liu, & Rao, 2016; Pielichowska & Pielichowski, 2014). An analytical study for analyzing PTSC-based air heating system for cut tobacco drying system was done by Jiang, Liu, Li, Yu, and Li (2014). The potential of PTSCs to provide heat for a district at

TABLE 14.5 Temperatures required by IPH applications (Kalogirou, 2003).

Industry	Process	Required temperature (°C)
Dairy	Pressurization	60–80
	Sterilization	100–120
	Drying	120–180
	Concentrates	60–80
	Boiler feed water	60–90
Tinned food	Sterilization	110–120
	Pasteurization	60–80
	Cooking	60–90
	Bleaching	60–90
Textile	Bleaching and dyeing	60–90
	Drying and degreasing	100–130
	Dyeing	70–90
	Fixing	160–180
	Pressing	80–100
Paper	Cooking, drying	60–80
	Boiler feed water	60–90
	Bleaching	130–150
Bricks and blocks	Curing	60–140
Chemical	Soaps	200–260
	Synthetic rubber	150–200
	Processing heat	120–180
	Preheating water	60–90
Beverages	Washing, sterilization	60–80
	Pasteurization	60–70
Chemical	Soaps	200–260
	Synthetic rubber	150–200
	Processing heat	120–180
	Preheating water	60–90
Meat	Washing, sterilization	60–90
	Cooking	90–100
Flours and by-products	Sterilization	60–80
Timber by-products	Thermodifussion beams	80–100
	Drying	60–100
	Preheating water	60–90
	Preparation pulp	120–70
Plastics	Preparation	120–140
	Distillation	140–150
	Separation	200–220
	Extension	140–160
	Drying	180–200
	Blending	120–140

high altitude is investigated by Krueger et al. (2000). Simulations were carried out using TRNSYS software and results suggested that PTSC can be useful to serve the mentioned purpose.

14.8.4   Desalination processes

Desalination is the process of converting brackish water into potable water. Normally, there are two approaches used for desalination, one is thermal energy based (multistage flash, multieffect distillation, etc.) and the other is membrane based (electrodialysis, reverse osmosis, etc.). Solar energy through PTSCs can be used directly in thermal-based systems by providing required heat through photo-thermal conversion and indirectly in membrane-based systems as electricity produced by SEGS (Goel et al., 2020). Several studies are conducted to prove that the assistance of PTSCs for desalination can make these systems sustainable, more effective, and economical. Bataineh (2016) investigated the performance of and driven by a solar steam generation plant. Palenzuela, Zaragoza, Alarcón-padilla, Guillén, and Ibarra (2011) have presented a thermodynamic evaluation of coupled system containing PT solar plant and desalination facility in the Middle East and North Africa region. Exergy and thermo-economic analysis for an MED system are done by Sharaf, Nafey, and García-Rodríguez (2011); two schemes were evaluated, direct usage of solar energy to provide heat to MED and solar to electricity conversion to drive the evaporator in MED.

14.9 Discussion

A generous number of studies have been covered over the modeling of PTSC in the preceding portions of this chapter. Techniques related to enhancing both the optical and thermal performance of PTSC are discussed, studies related to optical aspect are limited but a numerous number of investigations are available for improving thermal performance. There is an availability of a large number of effective and versatile tools available for thermal analysis. This could be potentially the reason for such a large number of studies related to it. Various schemes have been suggested to increase the efficiency of the PTSC, which are based on manipulation in optical design or through the variation in thermal properties. It should be kept in mind that small improvement in the performance can lead to significant benefits, especially for large setups.

Through the chapter, one can go through various models available for analysis of PTSC based on different technical aspects, namely, heat flux distribution, direction of temperature gradients, energy prospect analyses, and others. This gives a better idea about which model can be chosen to study the PTSC depending upon the application it intends to serve. It can be inferred that PTSCs with HCE lengths less than 100 m could be easily analyzed using 1-D models, uniform models, and exergy models. If the length of HCE is in hundreds of meters, which is in case of SEGS and large setups, then higher degree models, such as 2-D, 3-D, nonuniform, dynamic, and exergy models, should be employed. Statistical models can be applied for any application but suffer from a major drawback that it is mostly system specific. Model accurate for one system cannot be utilized for other system. The choice of a model not only depends upon the capacity of the system but also on the nature of analysis requested for it. For example, if intention is to just know about the performance of PTSC over the different parameters, then energy analysis is fine but if the location of insufficiencies or irreversibilities in the system is needed, then exergy analysis will be better.

PTSCs have the potential to deliver for many applications; some major applications, such as in power generation, desalination, IPH, and air heating systems, are discussed. The modeling studies attempted to analyze and improve them are also mentioned. Particular emphasis is given to the application of PTSCs in IPH applications. It is capable of serving these applications but commercially unsuccessful so far. Temperatures required by various processes in industries summarized to show the possible scope of PTSCs in the sectors. Modeling studies related to this aspect of PTSC are also limited. More detailed studies, supported with results from experimental prototypes, are required. However, in the field of power generation, PTSCs have proved their caliber and studies related to it are in good numbers. In the field of desalination, solar collectors do have the run, but still there is more to be achieved.

14.10 Conclusion

PTSCs have emerged as a viable option to harness solar energy efficiently. This is due to the availability of mature technical support and commercial production. Modeling and simulation provide a reliable option to analyze its performance over various operational parameters and to test its effectiveness with the inclusion of possible improvements. Modeling studies are also capable of identifying the magnitude and location of inefficiencies in the system. Various modeling and simulation-based investigations have been carried out for PTSC using different statistical tools to suggest the advancements and to stimulate its usage at large scale. The chapter has covered numerous studies contributing to the same and provides insights into PTSC modeling for a better understanding of prevailing methods and possible improvements. In a more generalized way, it is seen that most of the prevailing models for PTSC are grey box models and black box models have started to make their way in the field.

Modeling for both thermal and optical performance enhancement has also been attempted by researchers, considering them together as well as separately. Classification of modeling studies based on the fundamentals, such as flux distribution, direction of temperature gradients, energy, or exergy, is provided so that reader can easily relate to these concepts. A historical overview of PTSC technology is offered to sketch its journey to become a leading technology for CSP. Development of 1-D model is covered to equip the reader with base for modeling study of PTSC. It is evident from the chapter details that 1-D model can be extended to higher degree models also. Various potential applications of PTSC are discussed along with the modeling studies related to them. It can be concluded that for applications requiring the temperatures in the range of 50°C–260°C, 1-D model can provide reasonable results. Application of PTSC in power generation and large setups requires a higher degree of analysis for which more realistic models, such as 2-D and 3-D, with consideration of nonuniformity of heat flux will be the key. Also, exergy analysis of the system should be preferred while analyzing the large setups with HCE length in hundreds of meters, while energy analysis is sufficient for shorter length HCE.

Usage of PTSCs in power generation units is apparent but its commercial adaptation in the industrial scenario is still underway. Details about the temperature requirements of various processes in industries suggest that PTSCs can efficiently serve them. Studies exploring this dimension of PTSC are scarce and should gain pace over time. Expertise in other important applications of PTSC in desalination and air heating systems is also in a naive stage. However, advancements in technology and commercial viability should support the PTSC to gain a share in industrial usage.

A considerable amount of research is accomplished in this field but there is a scope of improvement. A lot of possible enhancements are suggested in the literature to improve the performance of PTSCs, but their practical realization is lacking due to existing challenges in technology. Some future directions for enhancing the PTSCs are suggested as follows:

- The reviewed articles reflect that the proficiency of PTSC can be enhanced further if process/design modifications suggested through optical and thermal modeling are incorporated in a practical setup.
- Advanced research approaches, such as the use of machine learning tools for analysis of PTSCs, are in infancy and should be given more chance to prove their worth.
- Modeling approaches are working under a load of assumptions and there is a scope to minimize them to be analytically prepared for futuristic high-performance collectors.
- Significant improvements are still required in selective absorber coatings and HTFs.
- Evacuated HCE is prone to leakage and physical damage at higher temperature gradients, also it is costly. Air-filled HCEs or air replaced by alternatives, such as inert gases, needs more emphasis.

Nomenclature

a	accommodation coefficient (−)
A	area (m^2)
b	interaction coefficient (−)
C	concentration ratio (−)
C_L	thermal loss coefficient ($W\,m^{-2}\,°C$)
c_p	specific heat capacity ($J\,kg^{-1}\,°C$)
D	diameter (m)
f	friction factor (−)
f_{cs}	focal length of PTSC (m)
F_{ao-gei}	view factor for radiation in annulus (−)
h	convection heat transfer coefficient ($W\,m^{-2}\,°C$)
IAM	incident angle modifier (−)
k_{pipe}	thermal conductivity of absorber tube ($W\,m^{-1}\,°C$)
k_{ang}	thermal conductivity of annulus gas at standard temperature and pressure ($W\,m^{-1}\,°C$)
k_{ge}	thermal conductivity of glass envelope ($W\,m^{-1}\,°C$)
k_{air}	thermal conductivity of surrounding air ($W\,m^{-1}\,°C$)
l_{cs}	length of the collector (m)
l_r	length of HCE (m)
L_{row}	spacing between two rows of PTSC array (m)
$\dot{m}_f$	mass flow rate of HTF ($kg\,s^{-1}$)
Nu	Nusselt number (−)
P_{an}	annulus gas pressure (mmHg)
Pr	Prandtl number (−)
Q	heat flux per unit length ($W\,m^{-1}$)
Q'	heat flux ($W\,m^{-1}$)

Q_{o_ge}	solar energy absorbed by glass envelope (W m^{-1})
Q_{o_as}	solar energy absorbed by absorber tube (W m^{-1})
Q_{hg}	useful heat gained (W)
Q_s	solar irradiance received per unit length of absorber (W m^{-1})
Ra	Rayleigh number (−)
Re	Reynolds number (−)
T	temperature (°C)
T_o	outlet temperature of fluid (°C)
T_i	inlet temperature of fluid (°C)
w_{cs}	width of collector surface (m)

Greek letters

α	absorptance (−)
γ	ratio of specific heats of annulus gas (−)
γ_{if}	intercept factor (−)
δ	molecular diameter of annulus gas (cm)
ε	emissivity (−)
$\in_{end}$	end loss (−)
$\in_{sh}$	shadowing loss (−)
η_O	optical efficiency of PTSC (−)
η_{O-as}	optical efficiency at absorber surface (−)
η_{O-ge}	optical efficiency at glass envelope (−)
η_T	thermal efficiency of PTSC (−)
θ	angle of incidence (degrees)
θ_z	solar zenith angle (degrees)
λ	mean-free-path between collisions of a molecule (m)
ρ	density (kg m^{-3})
ρ_{rc}	reflectivity of the clean collector surface (−)
σ	Stefan−Boltzmann constant $(= 5.67 \times 10^{-8} \text{Wm}^{-2}\text{K}^{-4})$
τ	transmittance (−)

Subscripts

$a\text{-}f$	between inner surface of absorber and heat transfer fluid
ai	inner surface of absorber
$an\text{-}cond$	heat transfer in annulus via conduction
an_conv	heat transfer in annulus via convection
ao	outer surface of absorber
$ao\text{-}ai$	between outer surface and inner surface of absorber
$ao\text{-}ge$	between outer surface of absorber and inner surface of glass envelope
as	absorber surface
$cond$	conduction
cov	convection
cs	collector surface
f	heat transfer fluid
ge	glass envelope
$ge\text{-}amb$	
gei	inner surface of glass cover
geo	outer surface of glass cover
$ge\text{-}sky$	between outer surface of glass envelope and sky
r	receiver
rad	radiation

Abbreviations

CFD	computational fluid dynamics
CSP	concentrating solar power
DNI	direct normal irradiation
DSG	direct steam generation
HCE	heat collector element
HTF	heat transfer fluid
IAM	incidence angle modifier
IPH	industrial process heat
LCR	local concentration ratio
LFR	linear fresnel reflector
MED	multi effect desalination
MED-TVC	multi effect desalination with a thermo-compressor
NREL	National Renewable Energy Laboratory
PCMs	phase change materials
PD	parabolic dish
PT	parabolic trough
PTC	parabolic trough collector
PTSC	parabolic trough solar collector
SEGS	solar electric generating systems
SNL	Sandia National Laboratory
SSAC	solar selective absorbing coatings
ST	solar tower

References

Abdulhamed, A. J., Adam, N. M., Ab-Kadir, M. Z. A., & Hairuddin, A. A. (2018). Review of solar parabolic-trough collector geometrical and thermal analyses, performance, and applications. *Renewable & Sustainable Energy Reviews*, 91, 822–831. Available from https://doi.org/10.1016/j.rser.2018.04.085.

Abdulraheem-Alfellag, M. A. (2014). *Modeling and experimental investigation of parabolic trough solar collector*. Embry-Riddle Aeronautical University.

Abed, N., & Afgan, I. (2020). An extensive review of various technologies for enhancing the thermal and optical performances of parabolic trough collectors. *International Journal of Energy Research*, 1–48. Available from https://doi.org/10.1002/er.5271.

Al-Ansary, H., & Zeitoun, O. (2011). Numerical study of conduction and convection heat losses from a half-insulated air-filled annulus of the receiver of a parabolic trough collector. *Solar Energy*, 85, 3036–3045. Available from https://doi.org/10.1016/j.solener.2011.09.002.

Almanza, R., Jiménez, G., Lentz, A., Valdés, A., & Soria, A. (2002). DSG under two-phase and stratified flow in a steel receiver of a parabolic trough collector. *Journal of Solar Energy Engineering, Transactions of the ASME*, 124, 140–144. Available from https://doi.org/10.1115/1.1463734.

Al-Sulaiman, F. A. (2013). Energy and sizing analyses of parabolic trough solar collector integrated with steam and binary vapor cycles. *Energy*, 58, 561–570. Available from https://doi.org/10.1016/j.energy.2013.05.020.

Barriga, J., Ruiz-De-Gopegui, U., Goikoetxea, J., Coto, B., & Cachafeiro, H. (2013). Selective coatings for new concepts of parabolic trough collectors. *Energy Procedia*, 49, 30–39. Available from https://doi.org/10.1016/j.egypro.2014.03.004.

Bataineh, K. M. (2016). Multi-effect desalination plant combined with thermal compressor driven by steam generated by solar energy. *Desalination*, 385, 39–52. Available from https://doi.org/10.1016/j.desal.2016.02.011.

Behar, O., Khellaf, A., & Mohammedi, K. (2015). A novel parabolic trough solar collector model—Validation with experimental data and comparison to Engineering Equation Solver (EES). *Energy Conversion and Management*, 106, 268–281. Available from https://doi.org/10.1016/j.enconman.2015.09.045.

Bellos, E., Tzivanidis, C., & Daniil, I. (2017). Energetic and exergetic investigation of a parabolic trough collector with internal fins operating with carbon dioxide. *International Journal of Energy and Environmental Engineering*, 8, 109–122. Available from https://doi.org/10.1007/s40095-017-0229-5.

Bellos, E., Tzivanidis, C., & Said, Z. (2020). A systematic parametric thermal analysis of nanofluid-based parabolic trough solar collectors. *Sustainable Energy Technologies and Assessments*, 39, 100714. Available from https://doi.org/10.1016/j.seta.2020.100714.

Bijarniya, J. P., Sudhakar, K., & Baredar, P. (2016). Concentrated solar power technology in India: A review. *Renewable & Sustainable Energy Reviews*, 63, 593–603. Available from https://doi.org/10.1016/j.rser.2016.05.064.

Bishoyi, D., & Sudhakar, K. (2017). Modeling and performance simulation of 100MW PTC based solar thermal power plant in Udaipur India. *Case Studies in Thermal Engineering*, 10, 216–226. Available from https://doi.org/10.1016/j.csite.2017.05.005.

BP. (2019). *BP statistical review of world energy*.

Burkholder, F. W. (2011). *Transition regime heat conduction of argon/hydrogen and xenon/hydrogen mixtures in a parabolic trough receiver.* University of Colorado.

Cengel, Y. A., & Ghajar, A. J. (2015). *Heat and mass transfer* (5th ed.). McGraw-Hill Education.

Cheng, Z. D., He, Y. L., & Cui, F. Q. (2013). A new modelling method and unified code with MCRT for concentrating solar collectors and its applications. *Applied Energy, 101*, 686−698. Available from https://doi.org/10.1016/j.apenergy.2012.07.048.

Cheng, Z. D., He, Y. L., & Qiu, Y. (2015). A detailed nonuniform thermal model of a parabolic trough solar receiver with two halves and two inactive ends. *Renewable Energy, 74*, 139−147. Available from https://doi.org/10.1016/j.renene.2014.07.060.

Cheng, Z. D., He, Y. L., Xiao, J., Tao, Y. B., & Xu, R. J. (2010). Three-dimensional numerical study of heat transfer characteristics in the receiver tube of parabolic trough solar collector. *International Communications in Heat and Mass Transfer, 37*, 782−787. Available from https://doi.org/10.1016/j.icheatmasstransfer.2010.05.002.

Churchill, S. W., & Chu, H. H. S. (1975). Correlating equations for laminar and turbulent free convection from a horizontal cylinder. *International Journal od Heat and Mass Transfer, 18*, 1049−1053. Available from https://doi.org/10.1016/0017-9310(75)90222-7.

Conrado, L. S., Rodriguez-pulido, A., & Calderón, G. (2017). Thermal performance of parabolic trough solar collectors. *Renewable & Sustainable Energy Reviews, 67*, 1345−1359. Available from https://doi.org/10.1016/j.rser.2016.09.071.

Dersch, J., Milow, B., Téllez, F., Ferriere, A., Langnickel, U., Steinfeld, A., et al. (2007). *Robert Pitz-Paal development steps for parabolic trough solar power technologies with maximum impact on cost reduction.* doi: 10.1115/1.2769697.

Dincer, I., & Rosen, M. A. (2007). *Exergy, energy and environment and sustainable development* (1st ed.). Elsevier.

Du, M., Hao, L., Mi, J., Lv, F., Liu, X., Jiang, L., et al. (2011). Optimization design of Ti0.5Al0.5N/Ti 0.25Al0.75N/AlN coating used for solar selective applications. *Solar Energy Materials & Solar Cells, 95*, 1193−1196. Available from https://doi.org/10.1016/j.solmat.2011.01.006.

Eck, M., Zarza, E., Eickhoff, M., Rheinländer, J., & Valenzuela, L. (2003). Applied research concerning the direct steam generation in parabolic troughs. *Solar Energy, 74*, 341−351. Available from https://doi.org/10.1016/S0038-092X(03)00111-7.

Edenburn, M. W. (1976). Performance analysis of a cylindrical parabolic focusing collector and comparison with experimental results. *Solar Energy, 18*, 437−444. Available from https://doi.org/10.1016/0038-092X(76)90010-4.

El Khadraoui, A., Bouadila, S., Kooli, S., Guizani, A., & Farhat, A. (2016). Solar air heater with phase change material: An energy analysis and a comparative study. *Applied Thermal Engineering, 107*, 1057−1064. Available from https://doi.org/10.1016/j.applthermaleng.2016.07.004.

Elsafi, A. M. (2015). On thermo-hydraulic modeling of direct steam generation. *Solar Energy, 120*, 636−650. Available from https://doi.org/10.1016/j.solener.2015.08.008.

Eskin, N. (1999). Transient performance analysis of cylindrical parabolic concentrating collectors and comparison with experimental results. *Energy Conversion and Management, 40*, 175−191. Available from https://doi.org/10.1016/S0196-8904(98)00035-1.

Fernández-García, A., Zarza, E., Valenzuela, L., & Pérez, M. (2010). Parabolic-trough solar collectors and their applications. *Renewable & Sustainable Energy Reviews, 14*, 1695−1721. Available from https://doi.org/10.1016/j.rser.2010.03.012.

Forristall, R. (2003). *Heat transfer analysis and modeling of a parabolic trough solar receiver implemented in engineering equation solver. NREL/TP-550−34169.*

García-Valladares, O., & Velázquez, N. (2009). Numerical simulation of parabolic trough solar collector: Improvement using counter flow concentric circular heat exchangers. *International Journal od Heat and Mass Transfer, 52*, 597−609. Available from https://doi.org/10.1016/j.ijheatmasstransfer.2008.08.004.

Gaul, H., & Rabl, A. (1980). Incidence-angle modifier and average optical efficiency of parabolic trough collectors. *Journal of Solar Energy Engineering, 102*, 16−21. Available from https://doi.org/10.1115/1.3266115.

Gee, R., Gaul, H., Kearney, D., & Rabl, A. (1979). Long-term average performance benefits of parabolic trough improvements. In *Sol. Ind. Process heat Conf.*, Oakland.

Geyer, M., Mehos, M., Meier, A., Meyer, R., Richter, C., & Weiss, W. (2007). *Solar power and chemical energy systems solarPACES annual report.*

Gnielinski, V. (2013). On heat transfer in tubes. *International Journal od Heat and Mass Transfer, 63*, 134−140. Available from https://doi.org/10.1016/j.ijheatmasstransfer.2013.04.015.

Goel, A., Manik, G., & Mahadeva, R. (2020). A review of parabolic trough collector and its modeling. In M. Pant, T. Sharma, O. Verma, R. Singla, & A. Sikander (Eds.), *Soft comput. theor. appl. adv. intell. syst. comput* (pp. 803−813). Singapore: Springer, 10.1007/978-981-15-0751-9_73.

Gong, G., Huang, X., Wang, J., & Hao, M. (2010). An optimized model and test of the China's first high temperature parabolic trough solar receiver. *Solar Energy, 84*, 2230−2245. Available from https://doi.org/10.1016/j.solener.2010.08.003.

Günther, M., Joemann, M., & Csambor, S. (2011). Parabolic trough technology. In *Adv. CSP teach. mater.* (p. 106).

Güven, H. M., Mistree, F., & Bannerot, R. B. (1984). Design synthesis of parabolic trough solar collectors for developing countries. *Engineering Optimization, 7*, 173−194. Available from https://doi.org/10.1016/j.chaos.2004.02.060.

Hachicha, A. A., Rodríguez, I., Capdevila, R., & Oliva, A. (2013). Heat transfer analysis and numerical simulation of a parabolic trough solar collector. *Applied Energy, 111*, 581−592. Available from https://doi.org/10.1016/j.apenergy.2013.04.067.

Heidemann, W., Spindler, K., & Hahne, E. (1992). Steady-state and transient temperature field in the absorber tube of a direct steam generating solar collector. *International Journal od Heat and Mass Transfer, 35*, 649−657. Available from https://doi.org/10.1016/0017-9310(92)90124-B.

Herrmann, U., & Kearney, D. W. (2002). Survey of thermal energy storage for parabolic trough power plants. *Journal of Solar Energy Engineering, Transactions of the ASME, 124*, 145−152. Available from https://doi.org/10.1115/1.1467601.

Hossain, M. S., Saidur, R., Fayaz, H., Rahim, N. A., Islam, M. R., Ahamed, J. U., et al. (2011). Review on solar water heater collector and thermal energy performance of circulating pipe. *Renewable & Sustainable Energy Reviews, 15*, 3801−3812. Available from https://doi.org/10.1016/j.rser.2011.06.008.

Huang, W., Xu, Q., & Hu, P. (2016). Coupling 2D thermal and 3D optical model for performance prediction of a parabolic trough solar collector. *Solar Energy, 139*, 365−380. Available from https://doi.org/10.1016/j.solener.2016.09.034.

Incoropera, F., & Dewitt, D. (2006). *Fundamentals of heat and mass transfer* (6th ed.). John wiley & sons.

Jebasingh, V. K., & Herbert, G. M. J. (2016). A review of solar parabolic trough collector. *Renewable & Sustainable Energy Reviews, 54*, 1085−1091. Available from https://doi.org/10.1016/j.rser.2015.10.043.

Jeter, S. M. (1986). Calculation of the concentrated flux density distribution in parabolic trough collectors by a semifinite formulation. *Solar Energy, 37*, 335−345. Available from https://doi.org/10.1016/0038-092X(86)90130-1.

Jiang, T., Liu, M., Li, F., Yu, D., & Li, L. (2014). A novel parabolic trough concentrating solar heating for cut tobacco drying system. *International Journal of Photoenergy*. Available from https://doi.org/10.1155/2014/209028.

John, A., & Duffie, W. A. B. (2013). *Solar engineering of thermal processes* (4th ed.). John Wiley & Sons, Inc.

Kalogirou, S. (2003). The potential of solar industrial process heat applications. *Applied Energy, 76*, 337−361. Available from https://doi.org/10.1016/S0306-2619(02)00176-9.

Kalogirou, S. A. (2012). A detailed thermal model of a parabolic trough collector receiver. *Energy, 48*, 298−306. Available from https://doi.org/10.1016/j.energy.2012.06.023.

Kalogirou, S. A. (2014). *Solar energy engineering: Processes and systems* (2nd ed.). Academic Press, 10.1016/b978-0-12-374501-9.00014-5.

Kennedy, C. E., & Price, H. (2005). Progress in development of high-temperature solar-selective coating. In *ASME 2005 Int Sol Energy Conf 2005* (pp. 749−755). doi: 10.1115/ISEC2005-76039.

Khakrah, H., Shamloo, A., & Kazemzadeh Hannani, S. (2018). Exergy analysis of parabolic trough solar collectors using Al2O3/synthetic oil nanofluid. *Solar Energy, 173*, 1236−1247. Available from https://doi.org/10.1016/j.solener.2018.08.064.

Krishna, Y., Faizal, M., Saidur, R., Ng, K. C., & Aslfattahi, N. (2020). State-of-the-art heat transfer fluids for parabolic trough collector. *International Journal od Heat and Mass Transfer, 152*, 119541. Available from https://doi.org/10.1016/j.ijheatmasstransfer.2020.119541.

Krueger, D., Heller, A., Hennecke, K., Duer, K., Energietechnik, S., Zentrum, D., et al. (2000). Parabolic trough collectors for district heating systems at high latitudes. In *Proc. Eurosun*.

Kumaresan, G., Sridhar, R., & Velraj, R. (2012). Performance studies of a solar parabolic trough collector with a thermal energy storage system. *Energy, 47*, 395−402. Available from https://doi.org/10.1016/j.energy.2012.09.036.

Liang, H., You, S., & Zhang, H. (2015). Comparison of different heat transfer models for parabolic trough solar collectors. *Applied Energy, 148*, 105−114. Available from https://doi.org/10.1016/j.apenergy.2015.03.059.

Liang, H., Zheng, C., Zheng, W., You, S., & Zhang, H. (2017). *Analysis of annual performance of a parabolic trough solar collector, . Energy Procedia* (Vol. 105, pp. 888−894). , 10.1016/j.egypro.2017.03.407.

Lippke, F. (1996). Direct steam generation in parabolic trough solar power plants: Numerical investigation of the transients and the control of a once-through system. *Journal of Solar Energy Engineering, Transactions of the ASME, 118*, 9−14. Available from https://doi.org/10.1115/1.2847958.

Lippke, F. (1995), *Simulation of the part-load behaviour of a 30MW SEGS plant*.

Liu, Q., Yang, M., Lei, J., Jin, H., Gao, Z., & Wang, Y. (2012). Modeling and optimizing parabolic trough solar collector systems using the least squares support vector machine method. *Solar Energy, 86*, 1973−1980. Available from https://doi.org/10.1016/j.solener.2012.01.026.

Lu, J., Ding, J., Yang, J., & Yang, X. (2013). Nonuniform heat transfer model and performance of parabolic trough solar receiver. *Energy, 59*, 666−675. Available from https://doi.org/10.1016/j.energy.2013.07.052.

Luo, N., Yu, G., Hou, H. J., & Yang, Y. P. (2015). Dynamic modeling and simulation of parabolic trough solar system. *Energy Procedia, 69*, 1344−1348. Available from https://doi.org/10.1016/j.egypro.2015.03.137.

Lv, P., Liu, C., & Rao, Z. (2016). Experiment study on the thermal properties of paraffin/kaolin thermal energy storage form-stable phase change materials. *Applied Energy, 182*, 475−487. Available from https://doi.org/10.1016/j.apenergy.2016.08.147.

Marif, Y., Benmoussa, H., Bouguettaia, H., Belhadj, M. M., & Zerrouki, M. (2014). Numerical simulation of solar parabolic trough collector performance in the Algeria Saharan region. *Energy Conversion and Management, 85*, 521−529. Available from https://doi.org/10.1016/j.enconman.2014.06.002.

Mazumder, R. K., Bhowmik, N. C., Hussain, M., & Huq, M. S. (1986). Heat loss factor of evacuated tubular receivers. *Energy Conversion and Management, 26*, 313−316. Available from https://doi.org/10.1016/0196-8904(86)90010-5.

Meyer, R. (2013). CSP & solar resource assessment CSP today South Africa. In *Proceedings of the 2nd concentrated solar thermal power conference and expo*.

Mohamed, A. M. (2003). Numerical investigation of the dynamic performance of parabolic trough solar collectors using different working fluids. *Port-Said Engineering Research Journal, 7*, 100−115.

Morrison, D. J., & Abdel-Khalik, S. I. (1978). Effects of phase-change energy storage on the performance of air-based and liquid-based solar heating systems. *Solar Energy, 20*, 57−67. Available from https://doi.org/10.1016/0038-092X(78)90141-X.

Mullick, S. C., & Nanda, S. K. (1989). An improved technique for computing the heat loss factor of a tubular absorber. *Solar Energy, 42*, 1−7. Available from https://doi.org/10.1016/0038-092X(89)90124-2.

Mullick, S. C., & Nanda, S. K. (1982). Heat loss factor for linear solar concentrators. *Applied Energy, 11*, 1−13. Available from https://doi.org/10.1016/0306-2619(82)90044-7.

Mwesigye, A., Bello-Ochende, T., & Meyer, J. P. (2014). Minimum entropy generation due to heat transfer and fluid friction in a parabolic trough receiver with non-uniform heat flux at different rim angles and concentration ratios. *Energy, 73*, 606−617. Available from https://doi.org/10.1016/j.energy.2014.06.063.

Nation, D. D., Heggs, P. J., & Dixon-Hardy, D. W. (2017). Modelling and simulation of a novel Electrical Energy Storage (EES) Receiver for Solar Parabolic Trough Collector (PTC) power plants. *Applied Energy, 195*, 950−973. Available from https://doi.org/10.1016/j.apenergy.2017.03.084.

Odeh, S. D., Morrison, G. L., & Behnia, M. (1998). Modelling of parabolic trough direct steam generation solar collectors. *Solar Energy, 62*, 395−406. Available from https://doi.org/10.1016/S0038-092X(98)00031-0.

Okafor, I. F., Dirker, J., & Meyer, J. P. (2017). Influence of non-uniform heat flux distributions on the secondary flow, convective heat transfer and friction factors for a parabolic trough solar collector type absorber tube. *Renewable Energy, 108*, 287−302. Available from https://doi.org/10.1016/j.renene.2017.02.053.

Ouagued, M., Khellaf, A., & Loukarfi, L. (2013). Estimation of the temperature, heat gain and heat loss by solar parabolic trough collector under Algerian climate using different thermal oils. *Energy Conversion and Management, 75*, 191−201. Available from https://doi.org/10.1016/j.enconman.2013.06.011.

Padilla, R. V., Demirkaya, G., Goswami, D. Y., Stefanakos, E., & Rahman, M. M. (2011). Heat transfer analysis of parabolic trough solar receiver. *Applied Energy, 88*, 5097−5110. Available from https://doi.org/10.1016/j.apenergy.2011.07.012.

Padilla, R. V., Fontalvo, A., Demirkaya, G., Martinez, A., & Quiroga, A. G. (2014). Exergy analysis of parabolic trough solar receiver. *Applied Thermal Engineering, 67*, 579−586. Available from https://doi.org/10.1016/j.applthermaleng.2014.03.053.

Palenzuela, P., Zaragoza, G., Alarcón-padilla, D. C., Guillén, E., & Ibarra, M. (2011). Assessment of different con fi gurations for combined parabolic-trough (PT) solar power and desalination plants in arid regions. *Energy, 36*, 4950−4958. Available from https://doi.org/10.1016/j.energy.2011.05.039.

Parabolic Trough Projects. By technology. https://solarpaces.nrel.gov/by-technology/parabolic-trough. Accessed 20.01.20.

Park, S. R., Pandey, A. K., Tyagi, V. V., & Tyagi, S. K. (2014). Energy and exergy analysis of typical renewable energy systems. *Renewable & Sustainable Energy Reviews, 30*, 105−123. Available from https://doi.org/10.1016/j.rser.2013.09.011.

Philibert, C. (2004). *International energy technology collaboration and climate change mitigation. Case study 1: Concentrating solar power technologies.*

Pielichowska, K., & Pielichowski, K. (2014). Phase change materials for thermal energy storage. *Progress in Materials Science, 65*, 67−123. Available from https://doi.org/10.1016/j.pmatsci.2014.03.005.

Pietruschka, D., Fedrizzi, R., Orioli, F., Söll, R., & Stauss, R. (2012). Demonstration of three large scale solar process heat applications with different solar thermal collector technologies. *Energy Procedia, 30*, 755−764. Available from https://doi.org/10.1016/j.egypro.2012.11.086.

Price, H., Forristall, R., Wendelin, T., Lewandowski, A., Moss, T., & Gummo, C. (2006). Field survey of parabolic trough receiver thermal performance. In *Proc. ISEC2006* (pp. 109−116). doi: 10.1115/ISEC2006-99167.

Price, H., Lü, E., Kearney, D., Zarza, E., Cohen, G., Gee, R., et al. (2002). Advances in parabolic trough solar power technology. *Journal of Solar Energy Engineering, 124*, 109−125. Available from https://doi.org/10.1115/1.1467922.

Pytilinski, J. T. (1978). Solar energy installations for pumping irrigation water. *Solar Energy, 21*, 255−262. Available from https://doi.org/10.1016/0038-092X(78)90001-4.

Raithby, G. D., & Hollands, K. G. T. (1976). Laminar and turbulent free convection from elliptic cylinders, with a vertical plate and horizontal circular cylinder as special cases. *Journal of Heat Transfer, 98*, 72−80. Available from https://doi.org/10.1115/1.3450473.

Ratzel, A. C., Hickox, C. E., & Gartling, D. K. (1979). Techniques for reducing thermal conduction and natural convection heat losses in annular receiver geometries. *Journal of Heat Transfer, Transactions of the ASME, 101*, 108−113. Available from https://doi.org/10.1115/1.3450899.

Rohsenow, W. M., & Hartnett, J. R. (1999). *Handbook of heat transfer* (3rd ed.). McGraw-Hill Education.

Romero, M., & Steinfeld, A. (2012). Concentrating solar thermal power and thermochemical fuels. *Energy & Environmental Science, 5*, 9234−9245. Available from https://doi.org/10.1039/c2ee21275g.

Rongrong, Z., Yongping, Y., Qin, Y., & Yong, Z. (2013). Modeling and characteristic analysis of a solar parabolic trough system: Thermal oil as the heat transfer fluid. *Journal of Renewable Energy, 2013*, 1−8. Available from https://doi.org/10.1155/2013/389514.

Selvakumar, N., & Barshilia, H. C. (2012). Review of physical vapor deposited (PVD) spectrally selective coatings for mid- and high-temperature solar thermal applications. *Solar Energy Materials & Solar Cells, 98*, 1−23. Available from https://doi.org/10.1016/j.solmat.2011.10.028.

Serrano, M. I. R. (2017). Concentrating solar thermal overview. *Green Energy and Technology*. Available from https://doi.org/10.1007/978-3-319-45883-0.

Shafiee, S., & Topal, E. (2009). When will fossil fuel reserves be diminished? *Energy Policy, 37*, 181−189. Available from https://doi.org/10.1016/j.enpol.2008.08.016.

Shaner, W. W., & Duff, W. S. (1979). Solar thermal electric power systems: Comparison of line-focus collectors. *Solar Energy, 22*, 49−61. Available from https://doi.org/10.1016/0038-092X(79)90059-8.

Sharaf, M. A., Nafey, A. S., & García-Rodríguez, L. (2011). Exergy and thermo-economic analyses of a combined solar organic cycle with multi effect distillation (MED) desalination process. *Desalination, 272*, 135−147. Available from https://doi.org/10.1016/j.desal.2011.01.006.

Silva, R., Javier, F., & Pérez-garcía, M. (2014). Process heat generation with parabolic trough collectors for a vegetables preservation industry in Southern Spain. *Energy Procedia, 48*, 1210−1216. Available from https://doi.org/10.1016/j.egypro.2014.02.137.

Sivaram, P. M., Nallusamy, N., & Suresh, M. (2016). Experimental and numerical investigation on solar parabolic trough collector integrated with thermal energy storage unit. *International Journal of Energy Research, 40*, 1564−1575. Available from https://doi.org/10.1002/er.3544.

Solar Energy Research Institute. (1985). *Silver/glass mirrors for solar thermal systems.*

Solar Thermal and Student Energy. https://www.studentenergy.org/topics/solar-thermal. Accessed 02.05.20.

Song, X., Dong, G., Gao, F., Diao, X., Zheng, L., & Zhou, F. (2014). A numerical study of parabolic trough receiver with nonuniform heat flux and helical screw-tape inserts. *Energy, 77*, 771−782. Available from https://doi.org/10.1016/j.energy.2014.09.049.

Tao, Y. B., & He, Y. L. (2010). Numerical study on coupled fluid flow and heat transfer process in parabolic trough solar collector tube. *Solar Energy*, *84*, 1863–1872. Available from https://doi.org/10.1016/j.solener.2010.07.012.

Thomas, A., & Guven, H. M. (1994). Effect of optical errors on flux distribution around the absorber tube of a parabolic trough concentrator. *Energy Conversion and Management*, *35*, 575–582. Available from https://doi.org/10.1016/0196-8904(94)90040-X.

Valenzuela, L., Zarza, E., Berenguel, M., & Camacho, E. F. (2005). *Control concepts for direct steam generation in parabolic troughs*, . *Solar Energy* (Vol. 78, pp. 301–311). , 10.1016/j.solener.2004.05.008.

Wang, J., Huang, X., Gong, G., Hao, M., & Yin, F. (2011). A systematic study of the residual gas effect on vacuum solar receiver. *Energy Conversion and Management*, *52*, 2367–2372. Available from https://doi.org/10.1016/j.enconman.2010.12.043.

Wang, P., Liu, D. Y., Xu, C., Zhou, L., & Xia, L. (2016). Conjugate heat transfer modeling and asymmetric characteristic analysis of the heat collecting element for a parabolic trough collector. *International Journal of Thermal Sciences*, *101*, 68–84. Available from https://doi.org/10.1016/j.ijthermalsci.2015.10.031.

Wang, Y., Liu, Q., Lei, J., & Jin, H. (2014). A three-dimensional simulation of a parabolic trough solar collector system using molten salt as heat transfer fluid. *Applied Thermal Engineering*, *70*, 462–476. Available from https://doi.org/10.1016/j.applthermaleng.2014.05.051.

Weinstein, L. A., Loomis, J., Bhatia, B., Bierman, D. M., Wang, E. N., & Chen, G. (2015). Concentrating solar power. *Chemical Reviews*, *115*. Available from https://doi.org/10.1021/acs.chemrev.5b00397.

Wu, Z., Li, S., Yuan, G., Lei, D., & Wang, Z. (2014). Three-dimensional numerical study of heat transfer characteristics of parabolic trough receiver. *Applied Energy*, *113*, 902–911. Available from https://doi.org/10.1016/j.apenergy.2013.07.050.

Xu, K., Du, M., Hao, L., Mi, J., Yu, Q., & Li, S. (2020). A review of high-temperature selective absorbing coatings for solar thermal applications. *Journal of Materials*, *6*, 167–182. Available from https://doi.org/10.1016/j.jmat.2019.12.012.

Xu, L., Wang, Z., Li, X., Yuan, G., Sun, F., & Lei, D. (2013). Dynamic test model for the transient thermal performance of parabolic trough solar collectors. *Solar Energy*, *95*, 65–78. Available from https://doi.org/10.1016/j.solener.2013.05.017.

Xu, R., & Wiesner, T. F. (2015). Closed-form modeling of direct steam generation in a parabolic trough solar receiver. *Energy*, *79*, 163–176. Available from https://doi.org/10.1016/j.energy.2014.11.004.

Yan, Q., Hu, E., Yang, Y., & Zhai, R. (2010). Dynamic modeling and simulation of a solar direct steam-generating system. *International Journal of Energy Research*, *34*, 1341–1355. Available from https://doi.org/10.1002/er.1678.

Yang, B., Zhao, J., Xu, T., & Zhu, Q. (2010). Calculation of the concentrated flux density distribution in parabolic trough solar concentrators by Monte Carlo Ray-Trace method. In *2010 symp. photonics optoelectron. SOPO 2010—Proc*. doi: 10.1109/SOPO.2010.5504452.

Yılmaz, H., & Söylemez, M. S. (2014). Thermo-mathematical modeling of parabolic trough collector. *Energy Conversion and Management*, *88*, 768–784. Available from https://doi.org/10.1016/j.enconman.2014.09.031.

Yılmaz, İ. H., & Mwesigye, A. (2018). Modeling, simulation and performance analysis of parabolic trough solar collectors: A comprehensive review. *Applied Energy*, *225*, 135–174. Available from https://doi.org/10.1016/j.apenergy.2018.05.014.

Zhang, H. L., Baeyens, J., Degrève, J., & Cacères, G. (2013). Concentrated solar power plants: Review and design methodology. *Renewable & Sustainable Energy Reviews*, *22*, 466–481. Available from https://doi.org/10.1016/j.rser.2013.01.032.

Zhang, P. (2010). *Industrial control system simulation routines. Advanced Industrial Control Technology* (pp. 781–810). , 10.1016/b978-1-4377-7807-6.10019-1.

Chapter 15

Energy hub: modeling, control, and optimization

Nouman Qamar[1], Tahir Nadeem Malik[1], Farhan Qamar[2], Mudassar Ali[2] and Muhammad Naeem[3]

[1]*Department of Electrical Engineering, University of Engineering and Technology, Taxila, Taxila, Pakistan,* [2]*Department of Telecommunication and Information Engineering, University of Engineering and Technology, Taxila, Taxila, Pakistan,* [3]*Department of Electrical Engineering, COMSATS University Islamabad, Wah Campus, Wah Cantt, Pakistan*

15.1 Introduction

Energy is the most important factor in the development of modern human societies. With the development of technology and automation in today's lifestyle, the use of energy is increasing very rapidly (Khan et al., 2019). Therefore countries need cheap, clean, safe, and sustainable energy for their development. Previously, fossil fuels were the main source of energy and generally converted into electrical energy by burning them at thermal power plants. This causes huge environmental concerns as they release hazardous gases into the atmosphere (Oruc & Dincer, 2019). Since, these thermal plants locate far away from the consumers, it results in high transmission and distribution losses which also increases its cost. To avoid these losses and increase energy efficiency, the concept of distributed generation (DG) has evolved (Kashyap, Mittal, & Kansal, 2019).

Instead of one far away centralized generation, DGs have gain popularity due to their numerous benefits over the conventional system. DGs or more generally distributed energy resources (DERs) have led to a more efficient power system and more competitive energy markets (Hou, Wang, & Luo, 2020). The idea of DER has led to the possibility of new power generators called prosumers (consumers who can generate their own electricity and can sell it back to the grid when needed) in the power grid. DGs have several advantages which include reduction in real power losses, emission reduction, grid stability enhancement, increased capacity of transmission lines and reduction in voltage deviation (Jalili & Taheri, 2020). So, with the active participation of the prosumers along with the demand response programs (DRPs), they are not only reducing their energy bills but also improving the grid reliability and stability.

The main source of DER is renewable energy resources (RERs), which can provide clean, economical, and sustainable energy. However, the intermittent nature of RERs limits their use, and to overcome this limitation, they are always accompanied with the energy storage systems (ESS). These storages can store excess amount of energy when generation is more than the load and provide stored energy back to the system in case of low RER generation or high demand. The maximum advantage of ESS can be obtained when used along with the demand side management (DSM) program. DSM program enables prosumers to interact with grid in a positive manner. By using DSM, prosumers reduce their demand when the grid in under stress by transferring their shift-able load to the off-peak hours. The DSM program also enables ESS to charge in off-peak hours and provide power back in peak hours or in times of need.

A significant amount of energy utilizes in heating and cooling of buildings either residential or commercial. Also, a lot of industrial processes requires heating and cooling for different purposes. Other than electrical energy, this heating and cooling can be obtained from different forms of energies. For example, gas can be directly used to heat up buildings in addition to the generation of electricity. Through the development of the devices that efficiently convert one form of energy to another form, it led the scientists to think of an intelligent hybrid system that can manage multiple energy sources. For this purpose, modeling, and optimization of a multienergy system (MES) is a required to increase the overall efficiency of the energy system. The inputs to the MES are electrical grid, gas grid, heating network, solar energy, wind energy, etc., and on the output side, it can provide electrical power, heating, cooling, gas, etc., simultaneously (Huang, Zhang, Yang, Wang, & Kang, 2017). The integration of MES can significantly improve the energy efficiency, increase renewable accommodation, and meet various types of energy demands.

Renewable Energy Systems. DOI: https://doi.org/10.1016/B978-0-12-820004-9.00018-8

Energy hub is a new concept in which integration of different forms of energies, their conversion from one form to another, their storage, and consumptions are carried out at one place. This concept can be applied at different levels with different sizes of energy systems. For smaller energy systems, we use the term micro energy hub, such as residential, industrial, commercial, and agriculture micro hubs. These micro energy hubs can be combined to form macro energy hub. Macro energy hub is a group of several small micro hubs, which are integrated and control together in a coordinated way. Different kinds of micro energy hubs can be combined and controlled centrally by a centralized macro energy hub. In this way, large industrial area, huge residential complexes or even a whole city can be modeled as a macro energy hub. For the optimal operation of macro energy hub, it requires a lot of data communication between different micro energy hubs, such as weather forecasting data, price market information, and adjacent systems. The schematic representation of macro energy hub is shown in Fig. 15.1.

The depletion of fossil fuels for the generation of electrical energy and their adverse effect on the environment initiates the need of some clean and sustainable energy that can meet the ever-increasing demand of the world. The solution to this is the RERs that can provide clean and cheap energy which is also sustainable. So, the integration of RERs to the energy hub is a good idea but the uncertain behavior of renewable energy makes it difficult to accurately model them. This makes it difficult to solely depend upon them rather they can be useful for providing a portion of energy demand, which can reduce the overall cost of energy production and dependence upon fossil fuels. The integration of RERs to the energy hub greatly reduces the operational cost as it is not only clean but also very cheap compared to conventional energy resources.

The main types of RERs are wind energy, solar energy, biomass energy, geothermal energy, and hydroelectricity. Solar energy is the most rapidly growing type of renewable energy among all the others (Moriarty & Honnery, 2019). It has the potential to be the major shareholder of the global energy production. Wind energy is also gaining popularity as it can produce electricity even at night when the solar is not available. But its disadvantage is its intermittency as it stops generating electricity when the speed of wind falls below the threshold level. Biomass is a unique type of RER that resembles biofuels in a sense that it can be stored and used at the time of desire unlike solar and wind energies, which are obtainable only when sun or wind are present respectively. Major portion of electricity from biomass is generated by burning them which is less efficient and limits their use. Geothermal energy comes from interior heat of earth

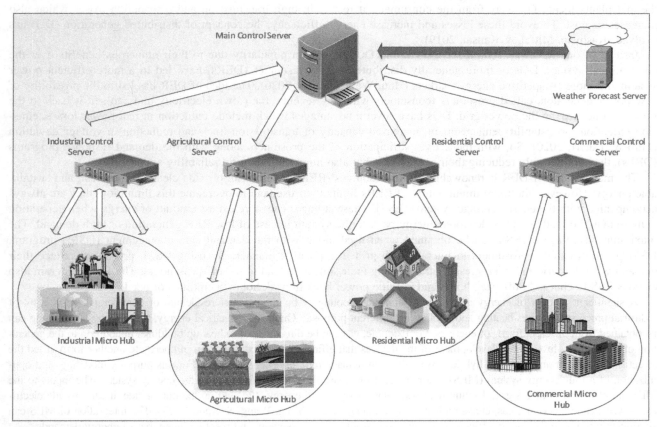

FIGURE 15.1 Schematic representation of macro energy hub.

or radioactive decay of different isotopes. Hydroelectricity is the production of electricity from running water and till now dominates the RERs' production despite advancements in solar and wind energy setups (Dudley, 2018).

RERs along with the storages can be used to meet demand in standalone or in a grid connected mode. This type of system is called hybrid renewable energy systems (HRES) (Rullo, Braccia, Luppi, Zumoffen, & Feroldi, 2019). The more the size of renewable energy and electrical storage (ES), the less will be the running operating cost of HRES, but for capital investment and system reliability, we need optimum sizing of RERs and ES. ES can store energy when renewable energy is in excess and can provide it back to the system when needed. The more the size of ES, more will be its capacity to store energy, but the capital cost of ES is high, and its life is relatively short. So, the proper sizing of ES is needed to optimize the system. HRES not only benefits consumers by reducing their energy bills but can also be helpful for the energy providers in a grid connected mode. For HRES to be connected and interacting with the grid, it needs to be smart enough for this purpose. For this, the term "Smart Grid" has repeatedly used in the literature.

Smart grid is a type of modern electrical grid that has the capability to accommodate and manage RERs, DGs, DRP, smart meters, smart appliances, storages, and communication modules and is considered as cyber secured. Smart grid is an umbrella term under which all features of futuristic modern grid come. For the grid to be smart it needs all components to communicate and respond with each other. In smart grid, consumers are well informed and responsive to the grid requirements. It can handle all types of DERs and storage options as well. It assesses condition of grid in real time, predicts behavior and also anticipates for any outages. With the inclusion of smart appliances, it can handle the stochastic demand by shifting less critical load in peak time. Thus we can get a lot of advantages by making our grid smart, but this would require great effort, new policies, a lot of money for research, and development.

This chapter focuses on the solution to the problem of optimization of energy hub. It includes the mathematical model of energy hub including different constraints associated with it. The mathematical models of electrical storage, heat storage (HS), and cold storage (CS) are also presented in this chapter. To make the energy hub more efficient, economical, and environment friendly, different types of renewable energies have been integrated to the energy hub. The problem of optimization of energy hub has been modeled as mixed integer linear programming (MILP) and solved in GAMS (General Algebraic Modeling System) software. Different case studies are analyzed to check the impact of storage capacities and RERs on the energy hub.

This chapter is divided into nine sections. Section 15.1 is an introductory part. Sections 15.2 and 15.3 describe the concepts of energy management systems and the energy hub, respectively. In Sections 15.4–15.6, mathematical modeling of energy hub, storage capacities, and RERs is presented. Section 15.7 deals with the system parameters whereas Section 15.8 is related to simulations, results, and discussion. Finally, the chapter is concluded in Section 15.9.

15.2 Energy management systems

Energy management has taken huge attention of researchers in the recent times due to increase in power demand by the growing population, which led to many problems, such as load shedding and power quality issues. A good energy management system can significantly reduce the energy requirements, thus leading to the cost savings and an additional advantage of environment protection. Energy management is equally important in distribution systems as well as in generation systems. In addition to these two systems, the applications of energy management are also required at different scales, that is, in industrial, commercial, and residential sectors to see the combined positive impact at national level. Fig. 15.2 shows the architecture of energy management system.

An efficient energy management system cannot be implemented without adopting energy management strategies. Following are the important points that should be considered while implementing energy management strategies.

1. Implementation of energy management strategies requires teamwork. An efficient team who knows the responsibilities can contribute to significant progress. This team should consist of energy manager, energy auditor, plant manager, maintenance experts, etc.
2. Energy management strategies should not be implemented in the same fashion across the plant. For example, a single plant may contain following systems to perform its operation (Kini, 2011):
   a. Heating systems,
   b. Boiler systems,
   c. Power system/electricity,
   d. Furnaces,
   e. Uninterruptable power supply systems,
   f. Temperature control systems,
   g. Refrigeration systems,

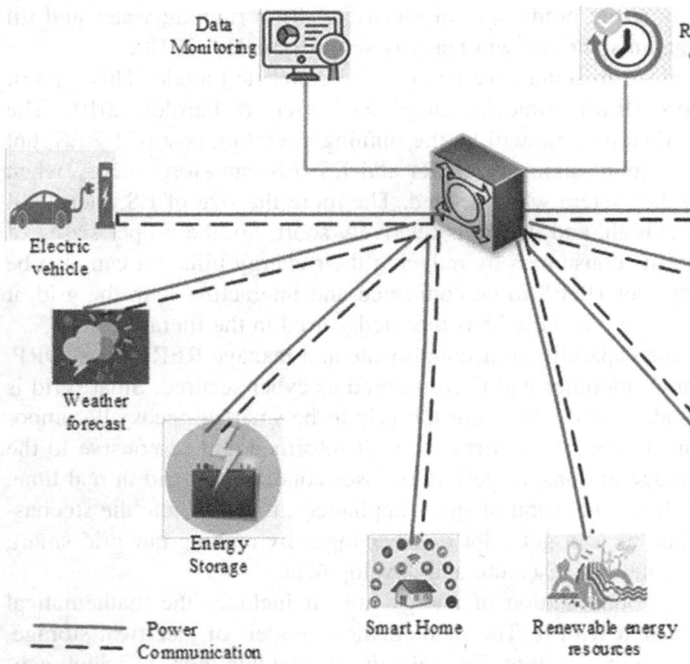

FIGURE 15.2 Energy management system architecture.

   h. Illumination systems,
   i. Ventilation systems,
   j. Steam systems,
   k. Motor systems,
   l. Instrumentation systems,
   m. Air-conditioning systems, and
   n. Auxiliary systems, etc.
     Different systems or sections of plant require different energy management strategies for effective savings.

3. All data of each system must be recorded for future planning. These data not only require defining the energy management strategies but also give the per unit cost of finished product.

15.2.1 Energy management information system

It is a performance management system that helps an individual or an organization in analyzing planning, and in decision-making regarding effective use of energy. It can also be considered as an IT-based unit that makes use of software applications to gather, record, and analyze energy data, which can also be used in future for planning purposes.

Data collection, processing, and analysis can be done on continuous basis by using the energy management information system. In modern industries, energy management systems are integrated along with the systems or subsystems of a plant. This helps in runtime monitoring of processes, which overall results in better control and management. As energy management information system processes all the critical and sensitive information, therefore it can be considered as the most important element of energy management systems. A good energy management information system should provide energy breakdown usage and product cost at different levels or processes. Main responsibilities of energy management information system include,

- Poor performance detection,
- Support in decision making,
- Performance reporting,
- Auditing,
- Justification of energy demand,
- Cost reduction, and
- Record management.

15.2.2 Energy management constraints

Energy management is related with planning, monitoring, and controlling energy-associated processes with the aim to preserve energy resources and energy cost savings and to protect environment. It is not so simple to manage energy as several constraints are associated with it. To apply energy management strategies on systems, it requires huge cost to bear at the start. Although it saves cost by managing energy, the capital cost of the system deployment cannot be ignored. The inclusion of many new devices and systems for energy management decreases the overall reliability of system because any device or system can malfunction at any time. The major part of energy management systems is renewables, but these renewables cannot be introduced beyond certain levels. Similarly, DRP for energy management cannot be used extensively because it will cause discomfort to the consumers. The balance between energy consumption and generation must be ensured all the time to avoid any disturbance. Reactive power in such systems must be within the range to ensure its stability. Energy management strategies requires huge amount of ESS for proper functioning, but their limited capacities make it difficult to apply properly for the energy management strategies. The summary of constraints associated with the energy management is shown in Fig. 15.3 (Alam & Arefifar, 2019).

15.3 Concept of energy hub

Energy hub can be considered as a multicarrier energy infrastructure that combines different types of DERs in which different forms of energies, such as electrical energy, gas energy, and renewables, are used as an input whereas the output of energy hub includes electrical demand, heating demand, and cooling demand. Integration of different types of energies at input side enhances the flexibility of continuous, cheap, and optimized energy provision. Energy hubs have the capability of receiving, sending, converting, and storing different kinds of energies. A combination of different types of energies can be integrated through the deployment of energy hub to optimally meet the demand (Fig. 15.4).

The main components of energy hub are,

1. Combined heat and power (CHP) unit,
2. Boiler,
3. Transformer,
4. Electric chiller,
5. Absorption chiller,
6. ES,
7. CS, and
8. HS.

CHP is the main part of energy hub that uses gas to produce electricity and heat. Its higher energy conversion efficiency makes it suitable to effectively couple the electrical and heating parts of energy hub (Wang, Zhang, Gu, & Li, 2017). Boiler on the other hand produces heat energy from gas to supply the heating load of energy hub. Transformer efficiently converts the high voltages from the grid to low voltages for the operation of electrical devices and appliances. Electric chiller converts electrical energy to cool energy for meeting the cooling demand. Absorption chiller uses

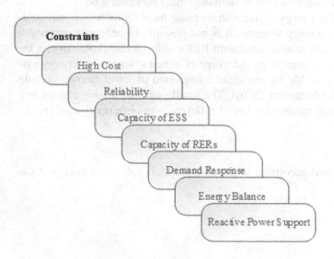

FIGURE 15.3 Constraints associated with energy management.

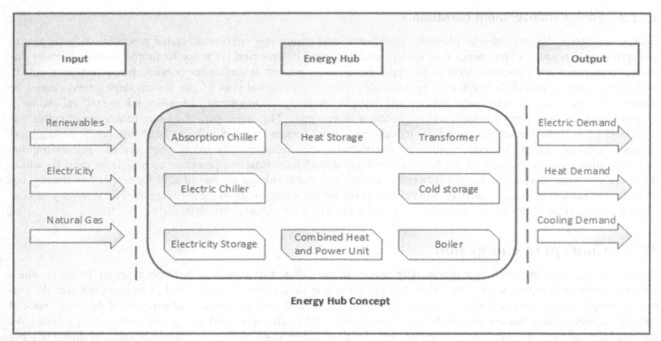

FIGURE 15.4 Energy hub concept.

the excess heat from the heating hub for the cooling process. Its combination with CHP makes it very cheap to run. The electrical storage, HS, and CS store excess amount of respective energies in them in case of less load demands. During the short supply, the same stored energy is given back from the storage to meet the demand. These storages can also be used for storing energies when the cost of purchasing energies is low and supply back when the cost is high. In this way a lot of cost can be saved.

15.3.1 Necessity of energy hub

The main disadvantage of conventional grid is that they mainly focus on electrical energy and consider only the electrical energy generation and demand. But the actual energy systems have multiple energy carriers, such as solar, wind, and natural gas, and multiple demands, such as electrical loads, heating loads, and cooling loads. Similarly, the concept of smart homes that focuses only on the controlling of electrical devices is incomplete without dealing with other types of devices. The sectors other than residential, such as industrial, commercial, and agriculture, also need various energy carriers to meet their demands. So, the conventional strategies that focus on single energy carrier need to be reviewed and a new model that can handle multiple energy carriers and different energy demands must be developed.

Recently, new developments on renewables have pushed the energy systems to increase their limits to accommodate the maximum amount of renewable energy. With our existing energy systems, it is not possible to penetrate renewable energy beyond certain limit. Therefore we need a system that can handle maximum RERs without compromising on the power quality and reliability. A lot of research has been done to maximize the usage of renewables in our energy systems. The dream of 100% renewable energy system is only possible by the smart integration of multi energy systems that combine in an intelligent framework (Connolly, Lund, & Mathiesen, 2016). The intelligent sustainable energy system to meet the future demand of world is only possible through the integration of different types of energy carriers.

15.3.2 Types of energy hub

The concept of energy hub can be applied at different levels and according to the consumption pattern of energy it can be classified into following four types.

1. Residential energy hub,
2. Commercial energy hub,
3. Industrial energy hub, and

4. Agricultural energy hub.

15.3.2.1 Residential energy hub

A significant amount of energy is consumed by the residential sector of any country. But the energy losses in this sector are significant due to poor distribution and transmission network, lack of responsive devices, and inefficient energy equipment. To make the consumption of energy efficient in residential sector, we need a smart and efficient multienergy management system. For this purpose, many countries have started using multiple energy resources, such as electricity and natural gas, to form residential energy hubs. A simple energy hub model for residential building has been proposed in Yuan et al. (2019), which includes RER and meets the electrical and heat demands of the building.

15.3.2.2 Commercial energy hub

Commercial buildings include offices, institutions, malls, hotels, etc., and are among the major consumers of energy through out the world. These buildings should be paid special attention for the efficient use of energy in their operation. Generally, commercial buildings purchase electricity from the grid for their electrical demand and depend upon gas for their heating load but with the increased prices of electricity and advancements in DER, the trend toward the onsite energy production has increased. The performance of combined cooling and heating power for commercial buildings and offices has been evaluated in Hanafizadeh, Eshraghi, Ahmadi, and Sattari (2016) and found very efficient in saving energy.

15.3.2.3 Industrial energy hubs

The development of any country can be measured by the strength of its industrial sector. By far the largest energy is consumed by this sector and is responsible for the major shareholder of greenhouse gas emission. So, the best way to save energy and reduce greenhouse gas emission is to make the consumption of our industrial sector more efficient. The prediction of industrial load is relatively easier and accurate as compared to commercial sector so it is easier to plan and integrate RERs in industry. Industrial energy hubs need to be optimally managed based upon several factors, such as cost minimization, profit maximization, and reduction in the consumption of energy (Paudyal, Cañizares, & Bhattacharya, 2014). For this purpose, multienergy systems with the integration of renewable energy along with the DSM program are required in industrial energy hub. The implementation of DSM programs in industrial energy hub is more difficult because it depends not only upon electricity and gas but also on other factors, such as water or raw material for their production (Alipour, Mohammadi-Ivatloo, & Zare, 2014). Also, some processes in industries are chain processes in which for the initiation of one process it needs completion of another process.

15.3.2.4 Agricultural energy hubs

Agriculture is the most important sector of any country, ensuring the food security and sustainability for it. Although the consumption of energy in agricultural sector is less than the other sectors, it is one of the fundamental factors for the profitability of agricultural products (Ball, Färe, Grosskopf, & Margaritis, 2015). Also, the consumption of energy in agricultural sector is increasing day by day due to the advancements in technology, use of automated machines, and need of more food across the world. Instead of expanding expensive transmission network, it is very beneficial to use RERs in agricultural sector as they are abundantly available on agricultural fields. Since the sun is readily available, solar energy can be used to generate not only electricity but also heating and drying agricultural crops. A large amount of agricultural waste can be used as a supply to biomass energy generation (Bilandzija et al., 2018). The wind energy in agricultural farms can be used for traditional grinding and water pumping and irrigation. Thus agricultural energy hub can be used to enhance the productivity of agriculture by efficiently utilizing the energy resources.

These micro energy hubs can be integrated together for their centralized control and operation, which makes the concept of macro energy hub (Walker, Labeodan, Maassen, & Zeiler, 2017). The integration of different micro energy hubs leads to increase in the overall efficiency of system and allows more penetration of RERs. The coordination between different micro energy hubs can efficiently handle the fluctuations in their individual demands and control the generation outages in a better way to increase the stability and reliability of macro energy hub.

15.4 Mathematical modeling of energy hub

The model presented here for energy hub includes two conventional energy sources, that is, electrical energy and natural gas. Two important forms of renewable energies, that is, solar and wind are also integrated in the proposed model.

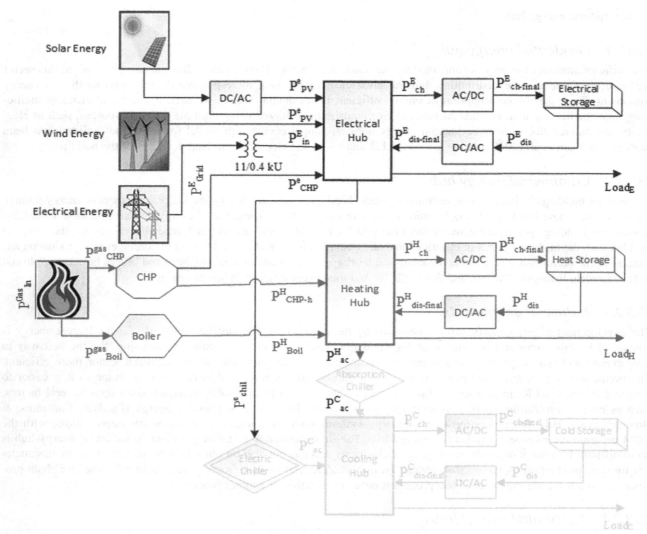

FIGURE 15.5 Proposed energy hub.

For simplicity energy hub model is divided into three subsections: electrical hub, heating hub, and cooling hub. The subenergy hubs are interconnected with each other for the conversion of one form of energy into another. The electrical hub includes solar energy, wind energy, electricity purchased from grid through transformer, electrical storage, and electrical load. The heating hub includes natural gas purchased from grid, CHP unit, boiler, HS, and heating load. The cooling hub contains electric chiller, absorption chiller, CS, and cooling loads. The electrical, heating, and cooling hubs are interconnected together to form a micro energy hub to meet the electrical, heating, and cooling demands efficiently. The proposed model of energy hub is shown in Fig. 15.5.

The proposed objective function to minimize the operational cost of energy hub during the 24 hours' time includes the cost of purchasing electricity from grid, cost of purchasing natural gas from grid, operational cost of wind turbine (WT), charging, discharging, and operational cost of electrical storage, HS, and CS and is given by Eq. (15.1).

$$\text{Cost} = \sum_{t=1}^{24} \left\{ \lambda_e(t) P_{Grid}^E(t) + \lambda_g(t) P_{in}^{GAS}(t) + \lambda_w P_{wind}^e(t) \right.$$

$$\left. + \left\{ \lambda_{OC}^E \left(\frac{P_{ch-final}^E}{\eta_{\frac{AC}{DC}}^E}(t) + \frac{P_{dis-final}^E}{\eta_{\frac{DC}{AC}}^E}(t) \right) + \lambda_e(t) \left(\frac{P_{ch-final}^E}{\eta_{\frac{AC}{DC}}^E}(t) - \frac{P_{dis-final}^E}{\eta_{\frac{DC}{AC}}^E}(t) \right) \right\} \right\}$$

$$+ \left\{ \lambda_{OC}^H \left(\frac{P_{ch-final}^H}{\eta_{Pr_{ch}^H}^H}(t) + \frac{P_{dis-final}^H}{\eta_{Pr_{dis}}^H}(t) \right) + \lambda_g(t) \left(\frac{P_{ch-final}^H}{\eta_{Pr_{ch}^H}^H}(t) - \frac{P_{dis-final}^H}{\eta_{Pr_{dis}}^H}(t) \right) \right\}$$

$$+ \left\{ \lambda_{OC}^C \left(\frac{P_{ch-final}^C}{\eta_{Pr_{ch}^C}^C}(t) + \frac{P_{dis-final}^C}{\eta_{Pr_{dis}}^C}(t) \right) + (0.5\lambda_e(t) + 0.5\lambda_g(t)) \left(\frac{P_{ch-final}^C}{\eta_{Pr_{ch}^C}^C}(t) - \frac{P_{dis-final}^C}{\eta_{Pr_{dis}}^C}(t) \right) \right\} \right\} \qquad (15.1)$$

The objective function calculates the total operational cost of energy hub during the 24 hours, the breakup of overall operational cost includes cost of purchasing electrical energy from electrical grid, cost of purchasing gas from the gas grid, operating cost of WT for production of electrical energy, operating cost of electrical storages, cost of purchasing electrical energy for final charging of electrical storages, operating cost of HS, cost of purchasing gas for final charging of HS, operating cost of CS, and cost of purchasing of electrical and gas energy for final charging of CS. The rate of electrical energy is denoted by $\lambda_e(t)$ and electrical energy purchased from the grid is denoted by $P_{Grid}^E(t)$. The rate of gas for which it is purchased from the gas grid is denoted by $\lambda_g(t)$ and gas purchased from the grid is denoted by $P_{in}^{GAS}(t)$. The cost of operation of WT is denoted by λ_w and the electrical energy produced from turbine is denoted by $P_{wind}^e(t)$. The operating cost of electrical storage is denoted by λ_{OC}^E, the final charge and discharge are denoted by $P_{ch-final}^E$ and $P_{dis-final}^E$, respectively, converting efficiency from AC to DC is denoted by $\eta_{\underline{AC}}^E$, and converting efficiency form DC to AC is denoted by $\eta_{\underline{DC}}^E$. The operating cost of HS is denoted by λ_{OC}^H, the final charge and discharge are denoted by $P_{ch-final}^H$ and $P_{dis-final}^H$, respectively, and charging and discharging efficiencies of pressure and temperature control unit of HS are denoted by $\eta_{Pr_{ch}}^H$ and $\eta_{Pr_{H}}^H$, respectively. The operating cost of CS is denoted by λ_{OC}^C, the final charge and discharge are denoted by $P_{ch-final}^{C_{ch}}$ and $P_{dis-final}^{C_{dis}}$, respectively, and charging and discharging efficiencies of pressure and temperature control unit of CS are denoted by $\eta_{Pr_{ch}}^C$ and $\eta_{Pr_{dis}}^C$, respectively. In our model, it is assumed that for charging of CS, 50% of electrical energy and 50% of gas energy are used.

For simplicity, the energy hub has been divided into three subparts, namely, electrical hub, heating hub, and cooling hub. The detailed mathematical modeling related to these subhubs is given below.

15.4.1 Modeling of electrical hub

The modeling of energy hub components, which deals mainly with electrical energy, is described in this section.

15.4.1.1 Electrical grid energy

The electrical energy purchased from the electrical grid is fed to the electrical hub by converting high voltage to low voltage through the transformer and is given by Eq. (15.2).

$$P_{in}^E(t) = \eta_{Trans}^E \cdot P_{Grid}^E(t) \qquad (15.2)$$

The efficiency of transformer is denoted by η_{Trans}^E.

15.4.1.2 Solar energy

The DC electrical energy generated from photovoltaic (PV) cells is converted into AC electrical energy and fed to the electrical hub and is given by Eq. (15.3).

$$P_{pv-final}^e(t) = \eta_{DC/AC}^{pv} \cdot P_{pv}^e(t) \qquad (15.3)$$

The DC electrical energy produced by PV cell is denoted by $P_{pv}^e(t)$, the efficiency of inverter that converts it into AC is denoted by $\eta_{DC/AC}^{pv}$, and the final AC electrical power that inputs to the energy hub by the solar is denoted by $P_{pv-final}^e(t)$.

15.4.1.3 Conversion of gas to electricity

The electrical energy produced from natural gas through CHP plant is given by Eq. (15.4).

$$P_{CHP}^e(t) = \eta_{CHP}^e \times P_{CHP}^{gas}(t) \qquad (15.4)$$

The gas input to the CHP from the grid is denoted by $P_{CHP}^{gas}(t)$, the efficiency of CHP in converting gas energy into electrical energy is denoted by η_{CHP}^e and electrical power produced from the CHP is denoted by $P_{CHP}^e(t)$.

15.4.1.4 Electrical load balance constraint

The electrical load and electrical generation through all means must be equal at each time interval, which is given by Eq. (15.5).

$$Load^E(t) + P^e_{chill}(t) + P^E_{ch}(t) = P^E_{in}(t) + P^e_{CHP}(t) + P^e_{wind}(t) + P^e_{pv-final}(t) + P^E_{dis-final}(t) \tag{15.5}$$

The electrical demand of energy hub is denoted by $Load^E(t)$, the electrical power consumed by chiller to produce cool energy is denoted by $P^e_{chill}(t)$, the electrical energy consumed in charging of electrical storage is denoted by $P^E_{ch}(t)$, and the input electrical energy from transformer is denoted by $P^E_{in}(t)$.

15.4.1.5 Electrical grid constraint

The electrical energy purchased from the grid at any time interval must be within allowable grid limits, which is given by Eq. (15.6).

$$0 \le P^E_{Grid}(t) \le P^E_{Gridmax} \tag{15.6}$$

$P^E_{Grid}(t)$ denotes the electrical power purchased from the grid and $P^E_{Gridmax}$ denotes the maximum power that can be imported from the grid at any time interval.

15.4.1.6 Electric chiller constraint

The electric chiller must consume electrical power from the hub in allowable limits, which is given by Eq. (15.7).

$$0 \le P^e_{chill}(t) \le P^e_{chillmax} \tag{15.7}$$

$P^e_{chill}(t)$ denotes the electrical power consumed by the chiller and $P^e_{chillmax}$ denotes the maximum amount of power that can be consumed by the electric chiller at any time interval.

15.4.1.7 CHP constraint

The electrical power consumes by CHP must be within the allowable limits, which is given by Eq. (15.8).

$$0 \le P^e_{CHP}(t) \le P^e_{CHPmax} \tag{15.8}$$

P^e_{CHP} denotes the electrical power consumed by the CHP and P^e_{CHPmax} denotes the maximum amount of power that can be consumed by the CHP at any time interval.

15.4.2 Modeling of heating hub

The modeling of heating hub components, which deals with the heat energy, is given in this section.

15.4.2.1 Gas balance constraints

The gas purchased from natural gas grid must be equal to the gas consumed by CHP and boiler at any instant of time and is given by Eq. (15.9).

$$P^{GAS}_{in}(t) = P^{gas}_{CHP}(t) + P^{gas}_{Boil}(t) \tag{15.9}$$

15.4.2.2 CHP

The CHP takes input from the gas grid and converts it into heat power that is required by the heating hub. It is given by the Eq. (15.10).

$$P^H_{CHP_h}(t) = \eta^H_{CHP_h} \times P^{gas}_{CHP}(t) \tag{15.10}$$

$\eta^H_{CHP_h}$ denotes the heating efficiency of CHP, which shows that how efficiently it converts the gas power into heating power. $P^H_{CHP_h}(t)$ denotes the heating power produced by the CHP unit at any time interval.

15.4.2.3 Boiler

The boiler converts gas energy into heat energy to meet a part of heating demand of energy hub and is given by Eq. (15.11).

$$P_{Boil}^H(t) = \eta_{Boil}^H \cdot P_{Boil}^{gas}(t) \tag{15.11}$$

η_{Boil}^H denotes the heating efficiency of boiler, which shows that how efficiently it converts the gas power into heating power. $P_{Boil}^H(t)$ denotes the heating power produced by the boiler.

15.4.2.4 Heating load balance constraint

The heating generation and heating demand must be equal at each time interval and is given by Eq. (15.12).

$$Load^H(t) + P_{ac}^H(t) + P_{ch}^H(t) = P_{CHP}^H(t) + P_{Boil}^H(t) + P_{dis-final}^H(t) \tag{15.12}$$

$Load^H(t)$ denotes the heating demand of the energy hub. $P_{ac}^H(t)$ denotes the heating power required by the absorption chiller. $P_{CHP}^H(t)$ and $P_{Boil}^H(t)$ denote the heating power produced by CHP and boiler respectively. $P_{ch}^H(t)$ and $P_{dis-final}^H(t)$ denote the charging and discharging of heating power at any time interval respectively.

15.4.2.5 Gas grid constraint

The gas energy purchased from the grid at any time interval must be within the allowable grid limits and is given by Eq. (15.13).

$$0 \le P_{in}^{GAS}(t) \le P_{in\ max}^{GAS} \tag{15.13}$$

$P_{in\ max}^{GAS}$ is the maximum amount of gas that can be purchased from the gas grid at any time interval.

15.4.2.6 Boiler constraint

The boiler must take the input from gas grid in allowable limits and is given by Eq. (15.14).

$$0 \le P_{Boil}^H(t) \le P_{Boilmax}^H \tag{15.14}$$

$P_{Boilmax}^H$ is the maximum amount of power that boiler can produce at any time interval.

15.4.3 Modeling of cooling hub

The modeling of cooling hub components, which deals with the cool energy, is given in this section.

15.4.3.1 Absorption chiller

The conversion of heat energy into cool energy by the absorption chiller is given by Eq. (15.15).

$$P_{ac}^C(t) = \eta_{ac}^C \times P_{ac}^H(t) \tag{15.15}$$

$P_{ac}^H(t)$ is the heat power consumed by the absorption chiller to produce the cooling power $P_{ac}^C(t)$ at any time interval. η_{ac}^C is the conversion efficiency of absorption chiller.

15.4.3.2 Electric chiller

The conversion of electrical energy into cool energy by the electric chiller is given by Eq. (15.16).

$$P_{ec}^C(t) = \eta_{chill}^C \cdot P_{chill}^e(t) \tag{15.16}$$

$P_{chill}^e(t)$ is the electrical power consumed by the electric chiller to produce the cooling power $P_{ec}^C(t)$ at any time interval. η_{chill}^C is the conversion efficiency of electric chiller.

15.4.3.3 Cooling load balance constraint

The cooling generation and cooling load must be equal at each time interval and is given by Eq. (15.17).

$$Load^C(t) + P_{ch}^C(t) = P_{ac}^C(t) + P_{ec}^C(t) + P_{dis-final}^C(t) \tag{15.17}$$

$Load^C(t)$ denotes the cooling demand of the energy hub. $P_{ch}^C(t)$ and $P_{dis-final}^C(t)$ denote the charging and discharging of cooling power at any time interval.

15.4.3.4 Absorption chiller constraint

The cooling power produced by absorption chiller must be in the allowable limits and is given by Eq. (15.18).

$$0 \le P_{ac}^C(t) \le P_{acmax}^C \tag{15.18}$$

P_{acmax}^C is the maximum amount of cooling power that can be produced by the absorption chiller.

15.5 Energy hub with storage capacities

The use of DER specifically RERs are increasing day by day, this shifts the paradigm from centralized power system to the distributed energy systems. The higher depletion rate of fossil fuel reserves and never-ending nature of RERs make it feasible to shift greatly toward renewables, but their intermittent and fluctuating nature limits their use. The production of energy from RERs mainly depends on the time and location which decreases the reliability of the power system (Mohammadi, Noorollahi, & Mohammadi-Ivatloo, 2018). The leading solution to the intermittency problem of RERs is the addition of ESS which store energy during the high production of renewables and provide the energy back to the system during low production. The additional advantage of ESS is that they can store energy in off-peak hours (low unit price) and provide energy back to the system in peak-hours (high unit price). The other advantages of ESS are less operational cost, less consumption of fossil fuels, less emissions and increased overall system efficiency (Barberis, Rivarolo, Traverso, & Massardo, 2016).

The three main types of ESS used in energy hub are (Rakipour & Barati, 2019),

1. ES,
2. HS, and
3. CS.

15.5.1 Mathematical modeling of ESS

The detailed modeling of ESS used in proposed energy hub is given in this section.

15.5.1.1 Electrical storages

The electrical power level in ES can be calculated by Eq. (15.19).

$$P_{es}(t) = P_{es}(t-1) + P_{ch-final}^E(t) - P_{dis}^E(t) - P_{es}^{loss}(t) \tag{15.19}$$

$P_{es}(t)$ and $P_{es}(t-1)$ are the current and previous power levels in ES, respectively. $P_{ch-final}^E(t)$ and $P_{dis}^E(t)$ are the charging and discharging powers of ES, respectively.

$P_{es}^{loss}(t)$ is the loss in electrical power stored in ES for any time interval and is given by Eq. (15.20).

$$P_{es}^{loss}(t) = \gamma_e^{loss} P_{es}(t) \tag{15.20}$$

γ_e^{loss} is the loss factor of ES that shows the loss of charge during storage.

The electrical charge must be stored within the specified limits and is given by Eq. (15.21).

$$\psi_{min}^e P_{es}^{cap^{max}} \le P_{es}(t) \le \psi_{max}^e P_{es}^{cap^{max}} \tag{15.21}$$

$P_{es}^{cap^{max}}$ denotes the maximum capacity of ES charge. ψ_{min}^e and ψ_{max}^e denote the allowable minimum and maximum limits of ES power level factors, respectively.

The allowable rate of charging of ES is given by Eq. (15.22).

$$\phi_{min}^e P_{es}^{cap^{max}}(1/\eta_{AC/DC}^E)I_{ch-final}^e(t) \le P_{ch-final}^E(t) \le \phi_{max}^e P_{es}^{cap^{max}}(1/\eta_{AC/DC}^E)I_{ch-final}^e(t) \tag{15.22}$$

ϕ_{min}^e and ϕ_{max}^e denote the minimum and maximum ES charge level factors, respectively. $\eta_{AC/DC}^E$ denotes the efficiency of rectifier.

The allowable rate of discharging of ES is given by Eq. (15.23).

$$\phi_{min}^e P_{es}^{cap^{max}} \eta_{DC/AC}^E I_{dis}^e(t) \leq P_{dis}^E(t) \leq \phi_{max}^e P_{es}^{cap^{max}} \eta_{DC/AC}^E I_{dis}^e(t) \tag{15.23}$$

$\eta_{DC/AC}^E$ denotes the efficiency of inverter.

To prevent the charge and discharge of ES simultaneously, Eq. (15.24) is used.

$$0 \leq I_{ch-final}^e(t) + I_{dis}^e(t) \leq 1 \tag{15.24}$$

$I_{ch-final}^e(t)$ and $I_{dis}^e(t)$ denote the binary variables for final charge and discharge of ES respectively.

15.5.1.2 Heat storages

The heat power level in HS can be calculated by using Eq. (15.25).

$$P_{hs}(t) = P_{hs}(t-1) + P_{ch-final}^H(t) - P_{dis}^H(t) - P_{hs}^{loss}(t) \tag{15.25}$$

$P_{hs}(t)$ and $P_{hs}(t-1)$ are the current and previous heat power levels in HS, respectively. $P_{ch-final}^H(t)$ and $P_{dis}^H(t)$ are the charging and discharging powers of HS, respectively.

$P_{hs}^{loss}(t)$ is the loss in heating power stored in HS for any time interval and is given by Eq. (15.26).

$$P_{hs}^{loss}(t) = \gamma_h^{loss} P_{hs}(t) \tag{15.26}$$

γ_h^{loss} is the loss factor of HS that indicates heat power dissipated during the storage.

The heat must be stored within the specified limits and is given by Eq. (15.27).

$$\psi_{min}^h P_{hs}^{cap^{max}} \leq P_{hs}(t) \leq \psi_{max}^h P_{hs}^{cap^{max}} \tag{15.27}$$

$P_{hs}^{cap^{max}}$ denotes the maximum capacity of HS. ψ_{min}^h and ψ_{max}^h denote the allowable minimum and maximum limits of HS power level factors, respectively.

The allowable rate of charging of HS is given by Eq. (15.28).

$$\phi_{min}^h P_{hs}^{cap^{max}} (1/\eta_{Pr_{ch}^H}^H) I_{ch-final}^h(t) \leq P_{ch-final}^H(t) \leq \phi_{max}^h P_{hs}^{cap^{max}} (1/\eta_{Pr_{ch}^H}^H) I_{ch-final}^h(t) \tag{15.28}$$

ϕ_{min}^h and ϕ_{max}^h denote the minimum and maximum HS charge level factors, respectively. $\eta_{Pr_{ch}^H}^H$ denotes the charging efficiency of pressure and temperature control unit of HS.

The allowable rate of discharging heat from HS is given by Eq. (15.29).

$$\phi_{min}^h P_{hs}^{cap^{max}} \eta_{Pr_{dis}^H}^H I_{dis}^h(t) \leq P_{dis}^H(t) \leq \phi_{max}^h P_{hs}^{cap^{max}} \eta_{Pr_{dis}^H}^H I_{dis}^h(t) \tag{15.29}$$

$\eta_{Pr_{dis}^H}^H$ denotes the discharge efficiency of pressure and temperature control unit of HS.

To prevent the charging and discharging of HS simultaneously, Eq. (15.30) is used.

$$0 \leq I_{ch-final}^h(t) + I_{dis}^h(t) \leq 1 \tag{15.30}$$

$I_{ch-final}^h(t)$ and $I_{dis}^h(t)$ denote the binary variables for final charge and discharge of HS, respectively.

15.5.1.3 Cold storages

The cooling power level in CS can be calculated by using Eq. (15.31).

$$P_{cs}(t) = P_{cs}(t-1) + P_{ch-final}^C(t) - P_{dis}^C(t) - P_{cs}^{loss}(t) \tag{15.31}$$

$P_{cs}(t)$ and $P_{cs}(t-1)$ are the current and previous cooling power levels in CS, respectively. $P_{ch-final}^C(t)$ and $P_{dis}^C(t)$ are the charging and discharging powers of CS, respectively.

$P_{cs}^{loss}(t)$ is the loss in cooling power stored in CS for any time interval and is given by Eq. (15.32).

$$P_{cs}^{loss}(t) = \gamma_c^{loss} P_{cs}(t) \tag{15.32}$$

γ_c^{loss} is the loss factor of CS that shows how much cooling power is dissipated during storage.

The cooling power must be stored within the specified limits and is given by Eq. (15.33).

$$\psi_{min}^c P_{cs}^{cap^{max}} \leq P_{cs}(t) \leq \psi_{max}^c P_{cs}^{cap^{max}} \tag{15.33}$$

$P_{cs}^{cap^{max}}$ denotes the maximum capacity of CS. ψ_{min}^c and ψ_{max}^c denote the allowable minimum and maximum limits of CS power level factors, respectively.

The allowable rate of charging of CS is given by Eq. (15.34).

$$\phi_{min}^c P_{cs}^{cap^{max}} (1/\eta_{Pr_{ch}^C}) I_{ch-final}^c(t) \leq P_{ch-final}^C(t) \leq \phi_{max}^c P_{cs}^{cap^{max}} (1/\eta_{Pr_{ch}^C}) I_{ch-final}^c(t) \tag{15.34}$$

ϕ_{min}^c and ϕ_{max}^c denote the minimum and maximum CS charge level factors, respectively. $\eta_{Pr_{ch}^C}$ denotes the charging efficiency of pressure and temperature control unit of CS.

The allowable rate of discharging cool energy from CS is given by Eq. (15.35).

$$\phi_{min}^c P_{cs}^{cap^{max}} \eta_{Pr_{dis}^C} I_{dis}^c(t) \leq P_{dis}^C(t) \leq \phi_{max}^c P_{cs}^{cap^{max}} \eta_{Pr_{dis}^C} I_{dis}^c(t) \tag{15.35}$$

$\eta_{Pr_{dis}^C}$ denotes the discharge efficiency of pressure and temperature control unit of CS.

To prevent the charging and discharging of CS simultaneously, Eq. (15.36) is used.

$$0 \leq I_{ch-final}^c(t) + I_{dis}^c(t) \leq 1 \tag{15.36}$$

$I_{ch-final}^c(t)$ and $I_{dis}^c(t)$ denote the binary variables for final charge and discharge of CS, respectively.

15.6 Integration of renewable resources to energy hub

Majority of energy suppliers produce electricity from the fossil fuels, which normally locates far away from the consumer side. Due to inefficient conversion of fossil fuels to electricity and transmission losses, a large amount of energy is being wasted; furthermore, long-distance transmission system decreases the overall reliability of the system. The easiest way to cater these issues is to use DER, more specifically RERs. Mathematical modeling of solar and wind energies is given in following sections.

15.6.1 Modeling of solar energy

PV cell produces electrical energy from the sunlight, which depends upon the solar irradiance. The solar irradiance further depends upon many factors, such as time of year, position of sun, cloud covering, intensity of sun, temperature, humidity, and de-rating factor of solar panel. The output from PV panels can be calculated by using Eq. (15.37) (Baneshi & Hadianfard, 2016; Gökçek, 2018).

$$P_{pv}^e(S_i(t)) = PV^{array} df^{PV} \left(\frac{S_i(t)}{S_{i,STC}} \right) [1 + \alpha p (T_c - T_{c,STC})] \tag{15.37}$$

P_{pv}^e is the electrical power produced by PV cell at solar irradiance $S_i(t)$. $S_{i,STC}$ denotes the solar irradiance at standard test conditions, df^{PV} is the de-rating factor of PV cell, αp is the temperature coefficient, PV^{array} is the rated capacity of solar panels, T_c is the temperature of PV cell, and $T_{c,STC}$ is the temperature of PV cell at standard test conditions.

15.6.2 Modeling of wind energy

Electrical power produced from the WT depends upon the wind speed. Eq. (15.38) is used for the production of electrical power from the wind (Shams, Shahabi, & Khodayar, 2018; Soroudi, Aien, & Ehsan, 2011).

$$P_{wind}^e(\omega(t)) = \begin{cases} P_{out}^w; & \omega_r^w \leq \omega(t) \leq \omega_o^{cut} \\ \dfrac{\omega(t) - \omega_i^{cut}}{\omega_r^w - \omega_i^{cut}} (P_{out}^w); & \omega_i^{cut} \leq \omega(t) \leq \omega_r^w \\ 0; & \omega(t) \leq \omega_i^{cut} or \ \omega(t) \geq \omega_o^{cut} \end{cases} \tag{15.38}$$

P_{wind}^e is the electrical power produced from the WT, P_{out}^w is the rated capacity of WT. ω_r^w, ω_i^{cut}, and ω_o^{cut} are the rated, cut-in, and cut-out speed of WT, respectively.

15.7 Simulations

GAMS is high level optimization tool for modeling and solving complex engineering problems. It is widely used in literature for the optimization of linear, nonlinear, and mixed integer problems (AlRafea, Fowler, Elkamel, & Hajimiragha, 2016; Andrei, 2017; Ha, Zhang, Huang, & Thang, 2016; Ha, Zhang, Thang, & Huang, 2017; Soroudi, 2017). It includes several solvers to solve different kinds of mathematical problems. In this section, the proposed energy hub has been modeled in GAMS and solved as an optimization problem for reducing the operational cost of energy hub during 24 hours' time horizon. The inputs to the energy hub are electricity from the electrical grid, gas from the gas grid, and RERs (solar and wind). The outputs from the energy hub are electrical, heating, and cooling demands, which must be fulfilled. The optimization problem is further modeled as an MILP to run simulations in GAMS software. The simulation parameters for the optimization of energy hub are given in Table 15.1.

The price of electricity and gas purchased from the grid is shown in Fig. 15.6. Graph shows prices of electricity and gas in different time hours.

The electrical, heating, and cooling demands of energy hub are shown in Fig. 15.7. These demands must be fulfilled at any time interval. Due to change in demands all the time, it is possible to optimally select the low peak hour energy from the different inputs to minimize the overall cost.

The wind speed and solar irradiance used in the simulations are shown in Fig. 15.8.

Four case studies have been designed and performed as shown in Table 15.2, to check the effectiveness of storage capacities and RERs in energy hub.

15.8 Optimization of energy hub in GAMS

In the first case, the energy hub has been optimized without any storage capacities and RERs. The electricity and gas are purchased from the grid to meet the electrical, heating, and cooling demands. There is no storage available, so the

TABLE 15.1 Simulation parameters.

Parameter	Value	Unit	Parameter	Value	Unit	Parameter	Value	Unit
λ_{OC}^{E}	2.5	Cent kWh^{-1}	η_{chill}^{C}	0.85	–	ψ_{max}^{c}	0.92	–
λ_{OC}^{H}	2.5	Cent kWh^{-1}	$P_{Gridmax}^{E}$	1200	kW	ϕ_{min}^{e}	0.05	–
λ_{OC}^{C}	2.5	Cent kWh^{-1}	$P_{in\ max}^{GAS}$	2000	kW	ϕ_{min}^{h}	0.05	–
λ_{w}	0.2	Cent kWh^{-1}	$P_{Boil\ max}^{H}$	800	kW	ϕ_{min}^{c}	0.05	–
$\eta_{\frac{AC}{DC}}^{E}$	0.92	–	$P_{CHP\ max}^{e}$	800	kW	ϕ_{max}^{e}	0.14	–
$\eta_{\frac{DC}{AC}}^{E}$	0.92	–	$P_{ac\ max}^{C}$	800	kW	ϕ_{max}^{h}	0.15	–
$\eta_{Pr_{ch}}^{H}$	0.92	–	$P_{CHP\ max}^{e}$	800	kW	ϕ_{max}^{c}	0.16	–
$\eta_{Pr_{dis}}^{H}$	0.92	–	γ_{e}^{loss}	0.02	–	P_{out}^{W}	300	kW
$\eta_{Pr_{ch}}^{H}$	0.92	–	$P_{es}^{cap^{max}}$	600	kW	ω_{r}^{w}	10	m s^{-1}
$\eta_{Pr_{dis}}^{H}$	0.92	–	$P_{hs}^{cap^{max}}$	600	kW	ω_{i}^{cut}	4	m s^{-1}
$\eta_{DC/AC}^{pv}$	0.95	–	$P_{cs}^{cap^{max}}$	600	kW	ω_{o}^{cut}	22	m s^{-1}
η_{Trans}^{E}	0.90	–	ψ_{min}^{e}	0.05	–	df^{PV}	80	%
η_{CHP}^{e}	0.40	–	ψ_{min}^{h}	0.05	–	$S_{i,STC}$	1	kW m^{-2}
$\eta_{CHP_{h}}^{H}$	0.35	–	ψ_{min}^{c}	0.05	–	αp	−0.5	% degree^{-1}
η_{Boil}^{H}	0.85	–	ψ_{max}^{e}	0.92	–	T_{c}	60	°C
η_{ac}^{C}	0.92	–	ψ_{max}^{h}	0.92	–	$T_{c,STC}$	25	°C

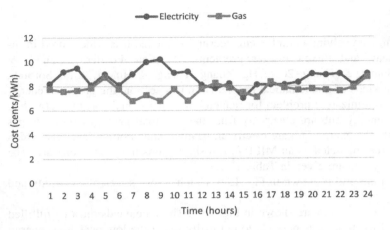

FIGURE 15.6 Cost of electricity and gas purchased from grid.

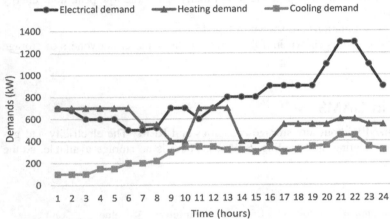

FIGURE 15.7 Electrical, heating, and cooling demands of energy hub.

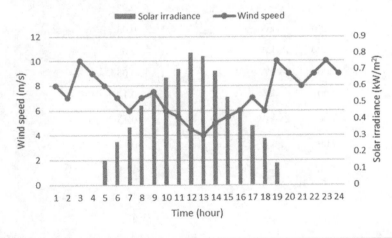

FIGURE 15.8 Wind speed and solar irradiance.

TABLE 15.2 Case studies.

Case studies	Demands	Storage capacities	Renewable energy resources
1	Electrical + heating + cooling	–	–
2	Electrical + heating + cooling	ES + HS + CS	–
3	Electrical + heating + cooling	–	Solar + wind
4	Electrical + heating + cooling	ES + HS + CS	Solar + wind

CS, cold storage; *HS*, heat storage.

electricity or gas must be purchased to meet the instantaneous demand, irrespective of the cost of electricity or gas. The total operational cost of energy hub for this case is 295,877.60$, which includes the cost of purchasing electricity and gas from the grid. In this case, no RERs are available, so no operational cost of renewables is included; similarly, the operational cost of storage capacities is also zero. The time in which the cost of electrical energy is high, dependency on gas increases due to its low price as shown in Fig. 15.9. Heat power generated by boiler is shown in Fig. 15.10. This gas is converted into electricity by CHP and more cooling power is produced by absorption chiller rather than electric chiller as shown in Fig. 15.11.

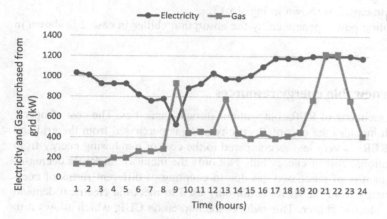

FIGURE 15.9   Electricity and gas purchased from grid in case 1.

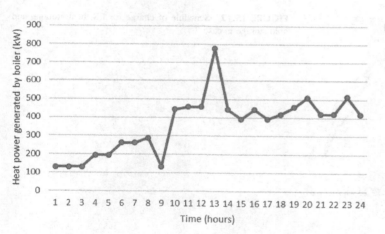

FIGURE 15.10   Heat power generated by the boiler in case 1.

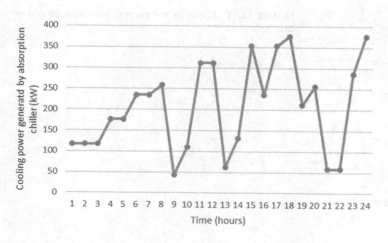

FIGURE 15.11   Electrical power generated by the absorption chiller in case 1.

15.8.1   Optimization of energy hub with storage capacities

In case 2, effect of storage capacities on the operational cost of energy hub has been analyzed whose function is to store energy when the energy cost is low and then provide the stored energy back to the hub when the energy cost is high. The total operational cost of energy hub in this case is 277,103.5$, which includes the cost of purchasing electricity and gas from the grid and operational cost of storage capacities. Although there is some operational cost of storage capacities, which is included in the overall operational cost of energy hub, this cost is far less than the savings done during the peak hours, in which either no or minimum energy is purchased from the commercial grid. The schedule of charge of the electrical storage, HS, and CS are shown in Fig. 15.12.

The electricity and gas purchased from the grid in case 2 is shown in Fig. 15.13.

The heat power generated by the boiler and cooling power generated by the absorption chiller in case 2 is shown in Figs. 15.14 and 15.15, respectively.

15.8.2   Optimization of energy hub with renewable energy resources

In case 3, energy hub has been optimized with the inclusion of RERs but without storage capacities. The cost of operation of energy hub in this case is 224,722.8$, which includes the cost of electricity and gas purchased from the grid and the operating cost of RERs. The operating cost of RERs is very less as compared to the cost of purchasing energy from the grid; this significantly reduces the overall operational cost of energy hub. Not only the inclusion of RERs decreases the quantity of purchased electricity but also the quantity of purchased gas due to coupling of different forms of energies in the energy hub. The additional electrical energy from the RERs is not only used to meet the electrical demand but also to convert it into the cooling power by the electric chiller. This reduces the burden on CHP, which allows it to

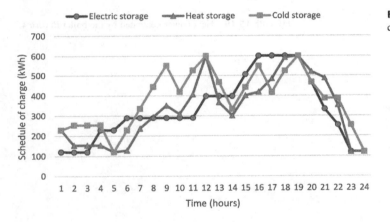

FIGURE 15.12   Schedule of charge for ES, heat storage, and cold storage in case 2.

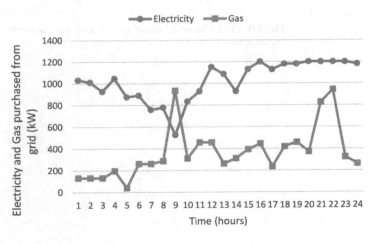

FIGURE 15.13   Electricity and gas purchased from the grid in case 2.

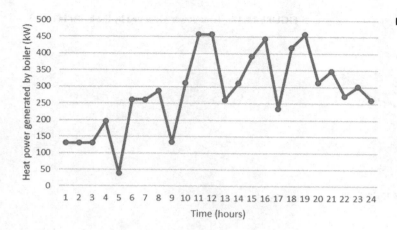

FIGURE 15.14   Heat power produced by the boiler in case 2.

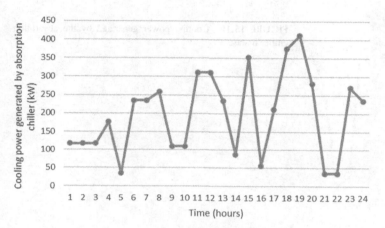

FIGURE 15.15   Cooling power generated by absorption chiller in case 2.

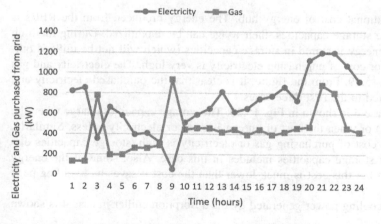

FIGURE 15.16   Electricity and gas purchased from the grid in case 3.

produce more heating power to meet the heating demand. The electricity and gas purchased from the grid in case 3 is shown in Fig. 15.16.

The heat power generated by the boiler and cooling power generated by the absorption chiller in case 3 is shown in Figs. 15.17 and 15.18, respectively.

15.8.3   Optimization of energy hub with storage capacities including renewable energy resources

In case 4, energy hub has been optimized with storage capacities including RERs. The total operational cost of energy hub in this case is 206,983.2$. The storage capacities enable the full utilization of RERs, which not only increases the

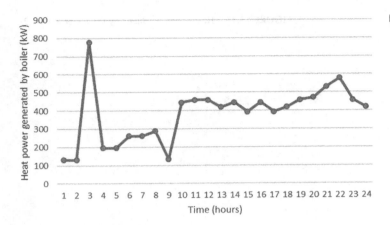

FIGURE 15.17 Heat power generated by the boiler in case 3.

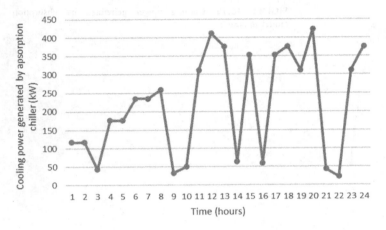

FIGURE 15.18 Cooling power generated by the absorption chiller in case 3.

reliability of the system but also reduces the operational cost of energy hub. The energy produced from the RERs is nondispatchable, which limits their use but with the storage capacities, their usage can be maximized. During the high energy production of RERs, the excess amount of energy is stored in storage capacities (which will not be utilized otherwise) to use at later time when generation is low or cost of purchasing electricity is very high. The electricity and gas purchased from the grid in case 4 is shown in Fig. 15.19. From the figure, it is clear that the purchased electricity and gas from the grid is significantly reduced as compared to the first three cases.

The schedule of charge for ES, HS, and CS in case 4 is shown in Fig. 15.20. The storage capacities charge when the excess energy is available (in case of low demands) or when the cost of purchasing gas or electricity is less. Similarly, when the generation is less, demand is high or the cost of purchasing gas or electricity is high, storage capacities discharge to meet the demand. The operating cost of storage capacities includes in this case. Also, some of the energy wastes in charging/discharging of storage capacities but this cost is much lower than the cost it saves by avoiding peak hour purchasing of gas and electricity.

The heating power generated by the boiler and cooling power generated by the absorption chiller in case 4 is shown in Figs. 15.21 and 15.22, respectively.

15.8.4 Discussion

In this chapter, energy hub model has been optimized under different scenarios to check the effectiveness of different components associated with energy hub. Four different cases are presented based on the combination of different inputs, which can be added to energy hub. In case 1, which is considered as the base case, the energy hub is connected with only conventional sources of energy, such as electricity and gas, as inputs. Some basic components, such as CHP, boiler, absorption chiller, and electric chiller, are used along with these energy sources to meet the electrical, heating, and cooling demands. The electrical demand is met by purchasing electrical energy from the grid or electrical energy

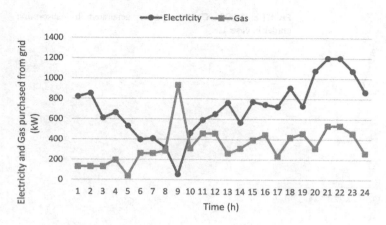

FIGURE 15.19 Electricity and gas purchased from the grid in case 4.

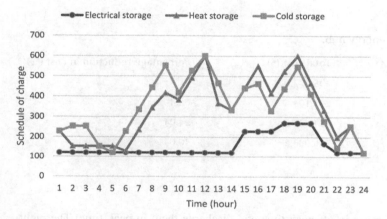

FIGURE 15.20 Schedule of charge for ES, heat storage, and cold storage in case 4.

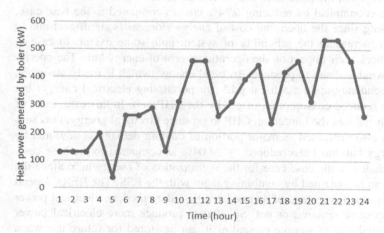

FIGURE 15.21 Heat power generated by the boiler in case 4.

produced by CHP. The gas consumed by the CHP is very high because it is cheaper to produce electrical energy from the CHP instead of purchasing electrical energy from the grid in peak hours as there are no alternatives available in case 1. To meet the cooling demand, the electric chiller and absorption chiller utilized the electrical energy and heat energy respectively. Similarly, heating demand is met by heat produced by the CHP unit and boiler; however, some part of heat produced is also consumed by absorption chiller to meet the cooling demand. The overall cost of operation of energy hub is higher as no cheap alternatives are available. In case 2, storages are added, which helped to store the energies during off-peak hours and this stored energy is supplied back to the energy hub in the peak hours for cost-saving purpose. However, this is not very simple to charge the batteries in off-peak hours and discharge in peak-hours as it includes the operating cost of charging/discharging and losses due to the conversion efficiencies. Despite this, in

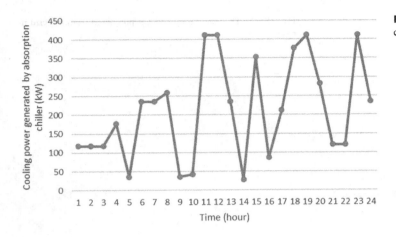

FIGURE 15.22 Cooling power generated by absorption chiller in case 4.

TABLE 15.3 Cost comparison of different cases of energy hub.

Cases	Energy hub	Total cost ($)	Percentage reduction in cost (%)
Case 1	Base case	295,877.6	–
Case 2	Base case + storages	277,103.5	6.34
Case 3	Base case + RERs	224,722.8	24.04
Case 4	Base case + storages + RERs	206,983.2	30.04

RERs, renewable energy resources.

some cases, it is more economical to charge the storages in off-peak time and discharge them in peak time. The inclusion of ESS capacities to energy hub makes it more economical by reducing 6.34% cost as compared to the base case. The percentage reduction in cost is not very promising since the operating cost of energy storages is significant due to charging/discharging. However, it can be useful in increasing the reliability of system upto some extent. In case 3, RERs are added to the base case of energy hub to check their impact on the operational cost of energy hub. The operating cost of RERs is very low as they depend upon wind and sun to produce electrical energy, which is freely available. The inclusion of RERs not only minimizes the dependency upon electrical grid (for purchasing electrical energy) but also decreases the dependency upon gas grid due to reduced consumption of gas by the CHP unit. In this case, most of the electrical energy demand is met by RERs, which reduces the burden on CHP to produce electrical energy, thus saving gas. The dependency upon absorption chiller is also decreased as major portion of cooling demand is now met by electric chiller. The overall operational cost of energy hub has been reduced by 24.04% as compared to the base case. In case 4, both RERs and storage capacities are included in the base case for the optimization of energy hub. Since the RERs are nondispatchable, their maximum benefit can be obtained by combining them with the ESS. The RERs depend upon the wind speed and solar irradiance for their production instead of demand profile—they produce electrical power when wind and solar irradiance are available irrespective required or not. So, if they produce more electrical power than the demand, it will be wasted but with the availability of storage capacities it can be stored for future use when required. The use of storage capacities with RERs increases the operational cost of storage capacities as they participate more in meeting the demands, but the overall operational cost of energy hub decreases significantly, that is, 30.04% compared to the base case. So, the best use of RERs is to use them with storage capacities to decrease the overall cost and increase the reliability of the system (Table 15.3).

15.9 Conclusion

In this chapter, the use and effect of RERs and storage capacities in energy hub are evaluated. The ever-increasing demand of energy with inadequate amount of fossil fuels and growing environmental concerns increases the need of optimal usage of energy sources and more penetration of renewables in existing system. The conventional inputs to the energy hub are electricity and gas from the grid supply, which are mainly from the fossil fuels so their utilization must be curtailed. The best way to do so is to maximize the penetration of renewables in existing system. But due to the unpredictable nature, their increased share can cause the imbalance between supply and demand, which effect the reliability of the system. This problem can be handled by using ESS, which stores the excess energy in case of low demand than the production. To check the effectiveness of RERs usage along with ESS, the energy hub is modeled in this chapter, which includes solar and wind energies as RERs and electrical storage, HS, and CS as ESS. The inclusion of RERs reduces the operational cost of energy hub, but its real advantage can be seen in combination with the ESS. So, the use of RERs in energy hub reduces the dependence on fossil fuels, which is further minimized by adding ESS with RERs. In future, some other kinds of renewables, that is, biogas, biomass, and hydrogen can be added to the energy hub model to make it even more optimal. RERs and load demands can be modeled stochastically to make the results more realistic. The evolving concept of electrical vehicles can be integrated to the energy hub model to charge vehicles when we have excess amount of energy. Similarly, these vehicles can be discharged in critical conditions to make the energy hub more economical, resilient, and reliable.

References

AlRafea, K., Fowler, M., Elkamel, A., & Hajimiragha, A. (2016). Integration of renewable energy sources into combined cycle power plants through electrolysis generated hydrogen in a new designed energy hub. *International Journal of Hydrogen Energy, 41*(38), 16718−16728.

Alam, M. S., & Arefifar, S. A. (2019). Energy management in power distribution systems: Review, classification, limitations and challenges. *IEEE Access, 7*, 92979−93001.

Alipour, M., Mohammadi-Ivatloo, B., & Zare, K. (2014). Stochastic risk-constrained short-term scheduling of industrial cogeneration systems in the presence of demand response programs. *Applied Energy, 136*, 393−404.

Andrei, N. (2017). *Continuous nonlinear optimization for engineering applications in GAMS technology*. Springer.

Ball, V. E., Färe, R., Grosskopf, S., & Margaritis, D. (2015). The role of energy productivity in United States agriculture. *Energy Economics, 49*, 460−471.

Baneshi, M., & Hadianfard, F. (2016). Techno-economic feasibility of hybrid diesel/PV/wind/battery electricity generation systems for non-residential large electricity consumers under southern Iran climate conditions. *Energy Conversion and Management, 127*, 233−244.

Barberis, S., Rivarolo, M., Traverso, A., & Massardo, A. (2016). Thermo-economic analysis of the energy storage role in a real polygenerative district. *Journal of Energy Storage, 5*, 187−202.

Bilandzija, N., et al. (2018). Evaluation of Croatian agricultural solid biomass energy potential. *Renewable and Sustainable Energy Reviews, 93*, 225−230.

Connolly, D., Lund, H., & Mathiesen, B. (2016). Smart Energy Europe: The technical and economic impact of one potential 100% renewable energy scenario for the European Union. *Renewable and Sustainable Energy Reviews, 60*, 1634−1653.

Dudley, B. (2018). BP statistical review of world energy. *BP Statistical Review, 6*, 2018, *London, UK, accessed Aug.*

Gökçek, M. (2018). Integration of hybrid power (wind-photovoltaic-diesel-battery) and seawater reverse osmosis systems for small-scale desalination applications. *Desalination, 435*, 210−220.

Ha, T., Zhang, Y., Thang, V., & Huang, J. (2017). Energy hub modeling to minimize residential energy costs considering solar energy and BESS. *Journal of Modern Power Systems and Clean Energy, 5*(3), 389−399.

Ha, T.-T., Zhang, Y.-J., Huang, J.-A., & Thang, V. (2016). Energy hub modeling for minimal energy usage cost in residential areas. In *2016 IEEE international conference on power and renewable energy (ICPRE)* (pp. 659−663). IEEE.

Hanafizadeh, P., Eshraghi, J., Ahmadi, P., & Sattari, A. (2016). Evaluation and sizing of a CCHP system for a commercial and office buildings. *Journal of Building Engineering, 5*, 67−78.

Hou, J., Wang, C., & Luo, S. (2020). How to improve the competiveness of distributed energy resources in China with blockchain technology. *Technological Forecasting and Social Change, 151*, 119744.

Huang, W., Zhang, N., Yang, J., Wang, Y., & Kang, C. (2017). Optimal configuration planning of multi-energy systems considering distributed renewable energy. *IEEE Transactions on Smart Grid, 10*(2), 1452−1464.

Jalili, A., & Taheri, B. (2020). Optimal sizing and sitting of distributed generations in power distribution networks using firefly algorithm. *Technology and Economics of Smart Grids and Sustainable Energy, 5*(1), 1−14.

Kashyap, M., Mittal, A., & Kansal, S. (2019). *Optimal placement of distributed generation using genetic algorithm approach. Proceeding of the Second International Conference on Microelectronics, Computing & Communication Systems (MCCS 2017)* (pp. 587−597). Springer.

Khan, Z. A., Zafar, A., Javaid, S., Aslam, S., Rahim, M. H., & Javaid, N. (2019). Hybrid *meta*-heuristic optimization based home energy management system in smart grid. *Journal of Ambient Intelligence and Humanized Computing, 10*(12), 4837−4853.

G. Kini, (2011). *Energy management systems*. BoD−Books on Demand,.

Mohammadi, M., Noorollahi, Y., & Mohammadi-Ivatloo, B. (2018). *Impacts of energy storage technologies and renewable energy sources on energy hub systems. Operation, Planning, and Analysis of Energy Storage Systems in Smart Energy Hubs* (pp. 23−52). Springer.

Moriarty, P., & Honnery, D. (2019). *Global renewable energy resources and use in 2050. Managing Global Warming* (pp. 221−235). Elsevier.

Oruc, O., & Dincer, I. (2019). Environmental impact assessment of using various fuels in a thermal power plant. *International Journal of Global Warming*, *18*(3−4), 191−205.

Paudyal, S., Cañizares, C. A., & Bhattacharya, K. (2014). Optimal operation of industrial energy hubs in smart grids. *IEEE Transactions on Smart Grid*, *6*(2), 684−694.

Rakipour, D., & Barati, H. (2019). Probabilistic optimization in operation of energy hub with participation of renewable energy resources and demand response. *Energy*, *173*, 384−399.

Rullo, P., Braccia, L., Luppi, P., Zumoffen, D., & Feroldi, D. (2019). Integration of sizing and energy management based on economic predictive control for standalone hybrid renewable energy systems. *Renewable Energy*, *140*, 436−451.

Shams, M. H., Shahabi, M., & Khodayar, M. E. (2018). Stochastic day-ahead scheduling of multiple energy carrier microgrids with demand response. *Energy*, *155*, 326−338.

Soroudi, A. (2017). *Power system optimization modeling in GAMS*. Springer.

Soroudi, A., Aien, M., & Ehsan, M. (2011). A probabilistic modeling of photo voltaic modules and wind power generation impact on distribution networks. *IEEE Systems Journal*, *6*(2), 254−259.

Walker, S., Labeodan, T., Maassen, W., & Zeiler, W. (2017). A review study of the current research on energy hub for energy positive neighborhoods. *Energy Procedia*, *122*, 727−732.

Wang, H., Zhang, H., Gu, C., & Li, F. (2017). Optimal design and operation of CHPs and energy hub with multi objectives for a local energy system. *Energy Procedia*, *142*, 1615−1621.

Yuan, Y., Bayod-Rújula, A. A., Chen, H., Martínez-Gracia, A., Wang, J., & Pinnarelli, A. (2019). An advanced multicarrier residential energy hub system based on mixed integer linear programming. *International Journal of Photoenergy*, *2019*.

Chapter 16

Simulation of solar-powered desiccant-assisted cooling in hot and humid climates

D.B. Jani

Gujarat Technological University—GTU, Government Engineering College, Dahod, India

16.1 Introduction

In developing country like India, present day facing various obstacts due to many challenges in the path of reaching rapid economic growth. One of them is due to producing electricity in fossil fuel-based thermal power plant leads to remarkable CO_2 emission. Since the last decade the energy consumption in India has grown remarkably high about 129%. In hot and humid enviornment in most part of India, air conditioning system that provides thermal comfort in various residential and industrial buildings mostly makes use of high grade electricity that creates many environmental pollution problems while generating the electricity in fossil fuel based thermal power stations. So, emphasis on energy requirement by the air conditioners for producing indoor thermal comfort is becoming great priority in the current energy scenario as exponential rise in cooling demand and associated environmental pollution. Air conditioners in building consume almost about more than 57% total building energy requirements. The main concern of an air conditioner in building is to handle the cooling load (sensible and latent heat) requirement and to maintain necessary thermal comfort. Accoridng to ASHRAE standard 55, the dry bulb temperature (DBT) should be in the range of 23°C−27°C and relative humidity in the range of 50%−55% is considered as the standard comfort condition.

In tropical climate, high ambient humidity coupled with increased ventilation flow requirement of modern building design is a major contributor to poor energy performance by the traditonally used heating, ventilation, and air conditioning (HVAC) systems.

Vapor compression refrigeration (VCR) systems are presently used to carry out the thermal comfort of various residential and industrial applications. The reason behind the same is its performance stability, compactness, and ease of availbility. However, the VCR systems operating on the use of vapor compressor, which consumes high-grade electrical power that creates environment pollution by releasing many green house gases during the production of electricity in fossil fuel-based thermal power stations. An option to the traditional air conditioners may be absorption cooling that makes use of low-grade solar thermal energy or industrial process waste heat to make energy efficient system for cooling. Sorption cooling mostly makes use of either ammonia−water or either lithium bromide−water system. In both the systems, the latent cooling load cannot meet efficiently as it requires cooling of air below its dew point temperature (DPT) and postreheating to meet desired room supply conditions. To overcome above problems desiccant-assisted evaporative cooling technology can be a good option. According to the types of desiccant materials used in the dehumidifier the desiccant cooling can be classified either in to the solid desiccant- or liquid desiccant cooling system. Even though the liquid desiccant cooling can be very popular, it has many operating problems, such as carry over, corrosion, and crystallization, while the solid desiccant cooling has many advantages, such as environmental friendly, durable, and compactness. Solid desiccant systems may be available in many types, such as fixed bed, rotary wheel, or desiccant-coated heat exchangers.

Especially for the hot climate, dadicated evaporative coolers can give better performance, but with an increase in humidity their performance degraded. This limitation of evaprotive cooler can be overcame by coupling it with desiccant wheel for the effective handling the air moisture seperately. According to the type of its configuration, it divided into two types either direct evaproative cooling or indirect evaproative cooling. In case of direct evaprorative cooling,

Renewable Energy Systems. DOI: https://doi.org/10.1016/B978-0-12-820004-9.00005-X

the process air comes in direct contact with cooling sprayed water increases its supply moisture in room on its continuous operation. While in case of indirect evaporative cooler, room moisture content remains unaffected for its entire duration of operation. Even though the evaproative coolers requires much lesser electrical power for its operation of components, such as air circulation fan or small water circulation pump, they cannot effectively handle the moisture in humid climates. So, the integration of desiccant dehumidification system with the evaproative coolers increases its climatic applicability. Furthermore, it provides amelioration in saving of cost, enegy, and power as compared with the traditionally used HVAC systems.

Air taken from room for air-conditioning is known as process air, enters initially into process air section of rotary desiccant dehumidifier. Moisture is driven off the room process air due to vapor pressure difference between hot desiccant and room process air. The water vapor is adsorbed by the desiccant surface due to its low vapor pressure. This makes the room process air warm and dry to supply for futher cooling in sensible cooling coil before delivering it back to the room supply. The saturated desiccant surface has been provided with hot reactivation air about temperature 70°C–80°C at the regeneration air side of the desiccant wheel to drive off moisture from the wheel to make it use for the next cycle. There are many sources of reactivation heat supply for regenerating the desiccant in dehumidifier, such as electrical heater, renewable solar energy, or industrial waste heat. For the enconomical and environmental point of view, thermal energy supplied for regeneration purpose from the freely availble renewable solar energy or industrial waste heat can be beneficial. In hybrid cooling, the desiccant dehumidifier gets coupled to the down sized traditional vapor compression unit. The desiccant wheel takes care of latent heat load (moisture removal) only while the evaproator cooling coil of vapor compression unit can sensibly cool the dehumidified air. In this way, the hybrid cooling can handle both latent and sensible heat load separately and effectively. This combined system substantially reduces energy required to produce the building cooling as compared to the conventional HVAC system.

In vapor compression-based traditional cooling system, the temperature of evaproator cooling coil must be maintained below the DPT DPT of incoming room process air for effective moisture removal. At the same time very low cooling coil temperature, sometime results in to over cooling of supply air. So that the supply air can be preheated to meet the design supply room conditions before reaching to the conditioned space. This provides the penalty for excess energy consumption for the conventional cooling system. By the use of desiccant dehumidifier, system gets rid of the maintenance of very low temperature cooling coil for effective air dehumidification. It occurs almost near ambient temperature and pressure conditions. The humidity of air (latent heat load) effectively handled separetaly by the desiccant dehumidifier, sensible cooling coil later to handle only sensible cooling load near temperature range 15°C–17°C, which is quite high as compared to the traditional vapor compression system evaporator temperature. As the rotary desiccant dehumidifier and the sensible cooling coil work individually, can be controlled separetly for their operation. In idle operation mode, the whole system shuts down automatically to save energy for its running. As the operation of the system carried out nearly at the ambient pressure, can prevent any leackages in the system. The temperature requirement of different desiccant groups for effective regeneration can be found different, such as temperature for silica gel 60°C–100°C, zeolite molecular sieves 175°C–250°C, and advced desiccant super polymer 45°C–75°C.

So, the application of freely available solar heat found to be a viable option for comfort space cooling of building as the cooling requirement almost matches the availablity of solar radiation. In solar power-based hybrid desiccant cooling cycle, freely available renewable solar heat is used for the purpose of regenerating the desiccant dehumidifier. For hot and humid country like India, solar energy is available freely in abundant amount during most period of the year. Use of freely available renewable solar energy in regeneration of desiccant dehumidification, can make system cost effective and eco-friendly.

Thus it can be seen as the desiccant-assisted cooling systems are increasingly developed as an alternative to the conventional vapor compression systems. The conventional vapor compression system consumes large amount of energy and causes environmental problems. Emphasis on desiccant systems is becoming a priority in the light of continuing rise in energy demand and increasing cost and various environmental problems, most notably the climate change. In hot and humid climates, HVAC device becomes inefficient due to increased humidity level of the supply air. Humidity of the outside air combined with ventilation requirement increases the latent load. Conventional VCR systems are not effective in handling the temperature and humidity separately. The use of desiccant cooling systems can improve the humidity control independent of temperature of supply air. It also improves overall energy efficiency and reduces energy costs. Besides, the desiccant cooling systems allow higher percentage of fresh air to achieve better air quality at lower energy cost. Energy consumption has increased in recent years with the development of worldwide economy. The energy required for the cooling and air conditioning is estimated between 30% and 40% of total energy use. Because of increased living standards and occupants demands, cooling energy demand will further increase. The peak load on the electricity grid increases in hot summer days because of high cooling requirement. This could cause blackouts and grid failure. Conventional vapor compression systems increase green house gases in the environment responsible for depleting the ozone layer. Solar-assisted desiccant cooling can help to alleviate this problem. The peak cooling demand in summer is associated with high solar radiation availability giving an excellent opportunity to exploit solar-assisted desiccant cooling technology. The optimal use of solar energy is based on the operational strategy that consumes the least electrical energy and uses thermal energy efficiently while maintaining the indoor comfort level at optimum operating cost. Desiccant cooling can be a perfect supplement to the traditional vapor compression system by controlling temperature and humidity independently.

This chapter presents a simulation comparison between ventilation and recirculation operating modes of solar-assisted solid desiccant-based dehumidification and cooling system. According to temperature and humidity ratio (HR) results of simulation models, the air properties, such as enthalpy for all important state points in both modes, were obtained to determine coefficient of performance (COP).

16.2 Literature survey

Use of freely availble renewable solar energy for the purpose of building cooling was under active investigation since the last many decades (Dunkle, 1965). Most of the earlier efforts were mainly concentrated on the modeling (Banks, 1972) of individual parts of the solid desiccant—vapor compression hybrid space cooling system, especially the solid desiccant integrated rotary dehumidifier (Maclaine-Cross & Banks, 1972) and the sensible heat exchanger wheel (Kays & London, 1984). Research was concentrated to determine viability of solid desiccant-integrated thermally cooled comfort space cooling (Nelson, Beckman, Mitchell, & Close, 1978) as a good option to a conventional space cooling techniques, such as vapor compression system and evaporative cooling, particularly in humid climate of tropical region (Jurinak, Mitchell, & Beckman, 1984). Similar other works treated configuration investigation (Kang & Maclaine-Cross, 1989) and augmentation in components' energy efficiency (Maclaine-Cross & Banks, 1981). In last decades recent works were mainly conducted (Alizadeh, 2008; Baniyounes, Liu, Rasul, & Khan, 2013; Bourdoukan, Wurtz, Joubert, & Sperandio, 2008a, 2008b; Halliday, Beggs, & Sleigh, 2002; Henning, Erpenbeck, Hindenburg, & Santamaria, 2001; Ma, Saha, Miller, & Guan, 2017; Tu, Liu, Hwang, & Ma, 2016) to investigate the potential of renewable solar heat in reactivation of the rotary dehumidifiers used in desiccant systems for economic vability of hybrid space cooling.

The mathematical model of solar-assisted solid desiccant-integrated hybrid space thermal cooling system was developed by Smith, Hwang, and Dougall (1993) while evaluating the potential for applicability of desiccant cooling at different locations in tropical climate. It is found that the suggested system can meet the requirement for producing necessary thermal comfort in building. Dai, Wang, and Xu (2002) developed numerical model of a solar regenerated hybrid solid-powered adsorption—desiccant-based thermally cooled air conditioning system for dehumidifying and cooling crops grain. The obtained results predict that the solar powered thermally cooling system can be used economically for grain storage. Performance evaluation of desiccant-integrated hybrid space comfort air-conditioning system using reactivation energy from solar heat was carried out by Kabeel (2007) at different outdoor climatic conditions of and variations in solar radiation intensity. A model of desiccant integrated thermally comfort room cooling system was suggested by Hürdoğan, Büyükalaca, Yilmaz, Hepbasli, and Uçkan (2012) to validate the application of regenerative solar power for reactivating the dehumidifier used in the system. The results available from experimental tests were compared with the anlytical model and it is seen that the application of solar heat in desiccant reactivation ameliorates the cooling performance of the system. Ando and Kodama (2005) carried out experimental investigation on double stage dehumidification by the use of 4-rotor desiccant cooling process. Desiccant reactivation temperature of 70°C was achieved by the use of solar heat, which is found sufficient for dehumidifying the efficiency of desiccant wheel when outdoor humidity is extensively high. Ahmed, Pasaran, and Wipke (1994) suggested a model of solid desiccant air-conditioning system to evaluate the unglazed type transpired based solar air collector used for the purpose of reactivating the rotary dehumidifier and compared its performance with the ordinary flat plate type solar collector.

A performance study on solar-integrated desiccant-assisted thermally cooling comfort space conditioning system was conducted by Khalid, Mahmood, Asif, and Muneer (2009) for carrying out air conditioning of industrial and residential buildings of tropical region. Simulation of the system was conducted using the TRaNsient SYStem (TRNSYS) software. It is concluded that a significant energy saving can be obtained by the use of desiccant-integrated hybrid thermally cooling as compared to the traditional HVAC systems. Ge, Ziegler, Wang, and Wang (2010) suggested solar-assisted two-stage rotary desiccant-integrated thermally cooling hybrid air-conditioning system. Thermodynamic and cost analysis of the desiccant-based thermally cooled hybrid comfort space air-conditioning system is conducted. It is found that the solid-integrated desiccant-powered dehumidification and thermally cooled hybrid space conditioning system can handle latent cooling load better than conventional system. Performance evaluation of multiple stage desiccant dehumidification-based thermally space cooling system powered by freely available renewable solar energy was carried out experimentally by Li, Dai, Li, La, and Wang (2011). The coefficient of performance of the system was found 0.95 for hot and humid ambient conditions in tropical region.

Comparisons of various desiccant cooling cycles for air conditioning in hot and humid climates have been done by Jain, Dhar, and Kaushik (1995). Psychrometric evaluation is carried out to simulate exact room conditions. Computer simulations for evaporative coolers are compared with the actual performance data of commercially available solid desiccant coolers. Influence of outdoor conditions on the effectiveness of coolers is investigated. It is found that for the wider range of outdoor conditions Dunkle cycle is better. Davanagere, Sherif, and Goswami (1999) simulated solar-assisted solid desiccant cooling system with a back up cooling coil to evaluate its performance and check its feasibility for four cities in the USA. Thermal performance has been optimized for varying design parameters, such as supply temperature, air flow rates, sensible, and latent cooling loads. Davanagere, Sherif, and Goswami (1999) further had done an economic analysis of solar-assisted solid desiccant cooling system with back up cooling coil to assess economic parameters, such as life cycle costs, life cycle savings, and payback periods. It was found that the systems with higher COP are required for the locations having higher humidity. Dhar and Singh (2001) analyzed the performance of four hybrid cycles for hot-dry and hot-humid weather based on the analogy method. They give substantial saving of energy (up to 30%) compared to VCR systems. Influence of room sensible heat factor, mixing ratio, and regeneration temperature on cycle performance has also been studied. Halliday et al. (2002) discussed the feasibility of desiccant cooling system using solar energy to evaluate the installations located at various places in the UK. Solar energy supplied is 72% of the thermal energy required to operate the desiccant system. Solar heating coils in summer to save energy up to 39%. Desiccant wheel-integrated vapor compression system is studied by Subramanyam, Maiya, and Murthy (2004) for cooling at low humidity conditions. Effect of various parameters, such as air flow rate, compressor capacity, and wheel speed, on system performance is evaluated. Camargo, Godoy, and Ebinuma (2005) analyzed the influence of ambient condition on the air-conditioning system performance for several cities of tropical climate. It analyzes the effect of some operational parameters, such as reactivation temperature and the thermodynamic conditions of the entering air flow. It shows the conditions for the best operation point with regard to the thermal comfort conditions and to the energy used in the process. Evaporative cooling technique is not much effective when ambient humidity is markedly high (Daou, Wang, & Xia, 2006). Desiccant cooling systems have feasibility in different climates in terms of energy and cost savings. Its energy saving potential can be increased by the regeneration using low grade thermal energy, such as solar or waste heat. Panaras, Mathioulakis, and Belessiotis (2007) suggested a model to investigate performance of a system by examining the effect of parameters, such as weather conditions, cooling load, air flow rate, and the regeneration temperature. The novel system is proposed having greater potential for covering the space requirements, design improvement, and flexible control strategies. Kim and Infante Ferreira (2008) made comparisons of different technologies, such as solar electric and solar thermal, based on the energy efficiency and economic feasibility. It is concluded that desiccant dehumidification is efficient than the other available technologies. Solar-driven desiccant dehumidification system proves to be economical when there is a large ventilation and dehumidification demand. Performance of solar-assisted solid desiccant cooling system assessed by Sabatelli, Cardinale, Copeta, and Marano (2009) for a residential building used energy plus simulation program and TRNSYS as input file. For a given building thermal COP and solar fractions calculated, the obtained results indicate significant reduction in the primary energy consumption. Khalid, Mahmood, Asif, and Muneer (2009) had done experiment on solar-assisted precooled hybrid desiccant cooling system. TRNSYS model is validated using measured data sets from test. Life cycle and economic assessments of solar air collector were performed and payback period was found about 14 years. TRNSYS simulation for desiccant cooling system is done to determine energy savings. The effect of ambient design conditions on the COP and exit air temperature of desiccant cooling cycle for ventilation and makeup modes are investigated by Heidarinejad and Pasdarshahri (2010). When the ambient air temperature is high, the COP of makeup mode is higher than the ventilation mode. This is due to lower energy needed in heater to obtain required regeneration air temperature for desorption. These analyses can be useful for a multiclimate country where a wide range of temperature and humidity is available as outdoor conditions. The potential of simple desiccant evaporative cooling cycle is examined by Parmar and Hindoliya (2010) in warm and humid climatic zones of India. The COP has been computed for different locations and compared. It is concluded that COP for different zones varies in the range of 0.14–0.21 and is highly influenced by ambient air HR. Higher is the ambient air HR, lower is the COP. Sphaier and Nobrega (2012) employed numerical procedure for designing desiccant cooling cycle for analyzing the impact of component effectiveness on the overall performance. Koronaki, Rogdakis, Tinia, and Kakatsiou (2012) developed neural network model to predict the dehumidification capacity and the outlet conditions in different climatic conditions for a silica gel desiccant system. Energy-saving potential of solar-assisted solid desiccant cooling system for an institutional building is studied by Baniyounes, Liu, Rasul, and Khan (2012) using TRNSYS. Technical and economical parameters, such as COP, solar fraction, life cycle, analysis and payback periods, were studied to check system's viability. Life cycle and economic assessments of the system are performed. Experimental investigation of solar-assisted hybrid desiccant cooling system was done by Dezfouli et al. (2012) for hot and humid weather of Malaysia. The COP of 0.6 has been achieved by using solar hybrid desiccant cooling system in hot and humid weather and considerable energy saving was obtained in comparison to the conventional VCR system. Dezfouli, Sohif, and Sopian (2013) further modeled a solar-assisted solid desiccant cooling system for hot and humid environment of Malaysia. The effects of reactivation temperature and HR on the system performance were studied.

Thus a number of investigations were conducted previously by the investigators to evaluate the study on solar-powered solid desiccant-based thermally cooling room air-conditioning system. But, most of the prior studies have been done for mild climate. So, an effort is necessary to assess the overall behavior of solid desiccant-based hybrid thermally cooling air-conditioning system regenerated mainly from freely available renewable solar energy particularly in North Himalayan hot and humid ambient for the tropical climate. Moreover, the impact of outside ambient climate on the system performance was also taken into consideration with reduced experimental work by the use of simulation study carried out in TRNSYS environment. Also, the availability and potential for using the freely available intensive renewable solar radiations to meet the required solid desiccant regeneration was also investigated.

16.3 System description

A lecture hall having sitting capacity of 100 persons has been selected (Kulkarni, Sahoo, & Mishra, 2011). The cooling load is estimated as 30 kW out of which the sensible cooling load is 24 kW and the latent cooling load is 6 kW (Jani, Mishra, & Sahoo, 2013b). The inside room conditions are assumed as 50% RH and 25°C DBT (ASHRAE, 2009). Outdoor conditions for the present case are 35°C DBT and 85% RH. The mass flow rates of process and regeneration air are 2.5 kg s^{-1}. Effectiveness of desiccant wheel and heat wheel are assumed as 70% and 80%, respectively. Saturation efficiency of evaporative cooling is 0.85 while effectiveness of process and regeneration air heat exchanger is assumed 1 (Hussain, Kalendar, Rafique, & Oosthuizen, 2020; Jani, Mishra, & Sahoo, 2015a, 2016a).

In the ventilation mode of solar-assisted solid desiccant cooling system, outdoor air enters into the desiccant wheel at point 1 (Fig. 16.1).

After dehumidification, temperature of process air increases due to adsorption. Process air is sensibly cooled between state points 2 and 3 in heat recovery wheel. It is further cooled between state points 3 and 4 by sensible cooling coil before entering to the room. After leaving the room, return air passes through evaporative cooler and heat recovery wheel before heating in water to air heating coil at 7–8. Between points 8–9, regeneration air removes moisture from dehumidifier by desorption process before leaving to atmosphere.

In the recirculation mode, the room air is recirculated to the process line 5–1 while the outside air is drawn into the regeneration line before sensible cooling coil as shown in Fig. 16.2. Room air is recirculated to remove moisture by passing through desiccant wheel. Heat wheel removes sensible heat and further it is cooled by sensible cooling coil

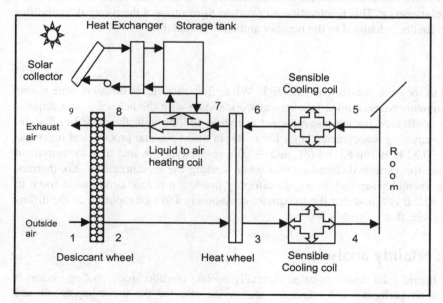

FIGURE 16.1 Solar-assisted solid desiccant cooling (ventilation mode).

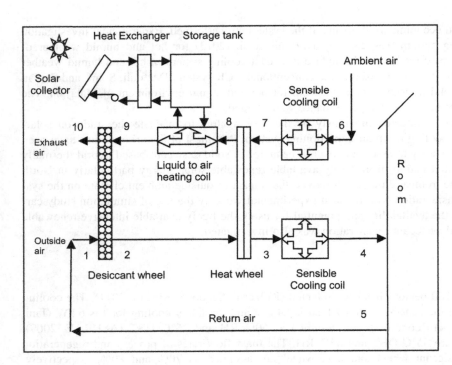

FIGURE 16.2 Solar-assisted solid desiccant cooling (recirculation mode).

before supplying to the room. Ambient air in regeneration side is cooled by evaporative cooler. Heat wheel preheats the exhaust air before it passes through the heating coil so as to achieve the required regeneration temperature.

A photographic view of solid desiccant coupled thermally cooling system has been illustrated as shown in Fig. 16.3. It mainly consists of a rotary solid desiccat dehumidifier, a heat recovery wheel, and a sensible cooling unit. In case of recirculation operating mode, humidity of indoor room supplied as process room moist air (at state 1) is effectively eliminated as it circulates via rotating desiccant dehumidifier wheel (at state 2). Thermally comfort cooling of hot and dehumidified air is possible by passing it through enthalpy wheel (at state 3). The required design room air temperature is obtained by making its flow over sensible cooling coil to cool it further (at state 4), and then it makes to enter to the test room.

Outside fresh air as regeneration air (at state 6) introduces in regeneration side through heat recovery wheel that conveys heat from hot and dry dehumidified air in exhaust process air outlet at the exit of rotary dehumidifier. The DBT of thermally heated regeneration air gets raised (at state 7), but it is insufficient for desorpting the dehumidifier. The required regeneration temperature of reactivation air is obtained after flowing it over liquid to air heating coil in which heating takes place by the use of solar energy. This reactivation air (at state 8) reactivates the rotary dehumidifier for its continual operation and it has been finally exhausted to the outdoor ambient (at state 9).

16.4 Measurements

Various measuring intruments were used to measure parameters like DBT, WBT, flow rate etc. at different state points in the system. All digital measuring instruments were coupled to a central work station with the help of a data acquisition digital scanner and so interfaces are configured for the measurement at hybrid cooling unit, the solar installation, and the outdoor ambient conditions. The measuring inaccuracies in air DBT, RH of moist air, and process and regeneration air steam flow rates are obtained as $+0.32$ K at 296 K, $+2.0\%$, and $+3.0\%$ respectively. A real-time measurement is recorded by the work station to evaluate the various deduced entities while starting the measurements. Six thermohygrometers are incorporated in the experimental setup and arranged uniformly inside the indoor conditioned room to monitor variations in the room DBT and RH. It is found that the maximum deviation in DBT of room air at the different locations inside the test room is acceptable, that is, within 3°C.

16.5 Data reduction and uncertainty analysis

The overall evaluation of solar integrated hybrid solid desiccant-based thermally cooling comfort space cooling system is determined by measuring its dehumidification performance by knowing the drying rate for the moist supply air and

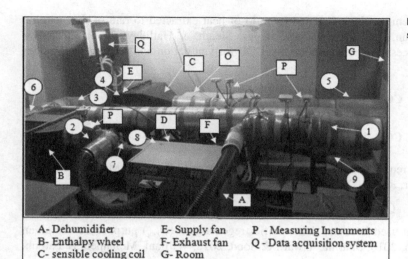

FIGURE 16.3 Photographic view of the experimental test set up.

A- Dehumidifier	E- Supply fan	P - Measuring Instruments
B- Enthalpy wheel	F- Exhaust fan	Q - Data acquisition system
C- sensible cooling coil	G- Room	
D- Heating coil	O- Outdoor unit	

dehumidification effectiveness in terms of moisture removal (ε_{dw}). The moisture transfer rate (Jani, Mishra, & Sahoo, 2016f; Das & Jain, 2017; Hwang & Radermacher, 2008) from moisture laden process air by the use of desiccant surface impregnated on matrix of solid desiccant wheel during the process air flowing through dehumidifier is represented as below

$$MRR = m_{pa}(\omega_1 - \omega_2) \tag{16.1}$$

where m_{pa} can be denoted as the process air flow rate at the entry to rotary dehumidifier and ω_1 and ω_2 are the specific HRs of indoor room supply process air to the entry and exit parts of rotary dehumidifier.

The dehumidification power of desiccant impregnated enthalpy wheel or rotary dehumidifier is calculated by deriving the ratio of the variation in actual specific HR of the process air stream at entry to the dehumidifier at process air side to the ideally possible elimination in ratio of specific humidity of the supply room process air at dehumidifier outlet. The dehumidification or drying effectiveness of the desiccant laden rotary dehumidifier (Fu, 2016; Jani, Mishra, & Sahoo, 2015b; Khan, Singh, Mathur, Bhandari, & Srivastava, 2017; Panchal & Mohan, 2017) can be calculated by

$$\varepsilon_{dw} = \frac{\omega_1 - \omega_2}{\omega_1 - \omega_{2,ideal}} \tag{16.2}$$

Here, $\omega_{2,ideal}$ is the theoretically/ideally eliminated moisture content of water vapor present in the totally dehumidified process air stream at the desiccant wheel process air outlet. Assuming that the proportion of water vapor or moisture present in the room process air is totally eliminated at state point 2 ideally, that is dehumidifier exit side of process air, the value of $\omega_{2,ideal}$ can be assumed to be nil.

In the same way, the effectiveness (Ge, Li, Dai, & Wang, 2010; Jani, Mishra, & Sahoo, 2016b, 2016d; Wang, Ge, Chen, Ma, & Xiong, 2009) of air-to-air heat recovery wheel (ε_{hrw}) is obtained as

$$\varepsilon_{hrw} = \frac{T_2 - T_3}{T_2 - T_6} \tag{16.3}$$

In the present study, two modes, ventilation and recirculation, are simulated with the same conditions and the same effectiveness of components. The effectiveness of the both evaporative coolers for the process and regeneration line are given by Henning et al. (2001), Hassan and Beliveau (2008), Henning (2007), Jani, Mishra, and Sahoo (2016e), and Hirunlabh, Charoenwat, Khedari, and Teekasap (2007).

$$\varepsilon_{DECpro} = \frac{T_3 - T_4}{T_3 - T_{3w}} \tag{16.4}$$

$$\varepsilon_{DECreg} = \frac{T_5 - T_6}{T_5 - T_{5w}} \tag{16.5}$$

The COP of the solar desiccant cooling system can be calculated by ratio of heat extracted to rate of regeneration (Jani, Mishra, & Sahoo, 2017; Joudi & Dhaidan, 2001; Mavroudaki, Beggs, Sleigh, & Halliday, 2002; Panchal & Sathyamurthy, 2017). Based on the mass of process and regeneration air, COP for ventilation and recirculation modes can be given as follows.

$$\text{COP}_V = \frac{m_p(h_5 - h_4)}{m_r(h_8 - h_7)} \tag{16.6}$$

$$\text{COP}_R = \frac{m_p(h_5 - h_4)}{m_r(h_9 - h_8)} \tag{16.7}$$

The Q_{cc} is the cooling capacity to produce required cooling of dehumidified process air (Jani, Mishra, & Sahoo, 2016c; Verma & Murugesan, 2017; Kumar, Dhar, & Jain, 2011; Zhang & Niu, 1999) and it is obtained as

$$Q_{cc} = m_{pa}(h_1 - h_4) \tag{16.8}$$

The thermal energy demand for the desiccant used in rotary dehumidifier regeneration (Q_{reg}) purpose can be given (Porumb, Ungureşan, Tutunaru, Şerban, & Bălan, 2016; Sharma, Tiwari, & Sood, 2012; William, Mohamed, & Fatouh, 2015; Yadav & Bajpai, 2013) by

$$Q_{reg} = m_{ra}(h_8 - h_7) \tag{16.9}$$

where m_{ra} denotes the flow rate of reactivation air stream on its mass basis, while h_7 and h_8 are the air enthalpies of the reactivation air supply at entrance and exit the regeneration thermal heater at process and regeneration air sides, respectively.

Inaccuracies calculated while measuring the DBT, RH, air steam flow volume, pressure drop along the process, and regeneration sides of desiccant wheel and running power requirement are ± 0.31 K at 296 K, $\pm 2.0\%$, $\pm 3.0\%$, $\pm 1.5\%$, and $\pm 2.0\%$, respectively.

16.6 Results and discussion

Fig. 16.4 depicts the influence of deviation in DBT of outdoor ambient air on outlet HR of dehumidified process air stream at rotary dehumidifier process air-leaving side. It seems that as the DBT of ambient air increases with HR of process air so the required regeneration temperature of the desiccant used in matrix surface of the rotary dehumidifier is also found increased. The reason behind this is that as temperature of room process air entering the dehumidifier increases, the vapor pressure of the same while passing through desiccant flute decreases, which ultimately resulted in diminishing the adsorption process of water vapor effectively from the room moist air. This in turn requires higher regeneration temperature to increase that pressure difference between circulated process air and desiccant laden matrix

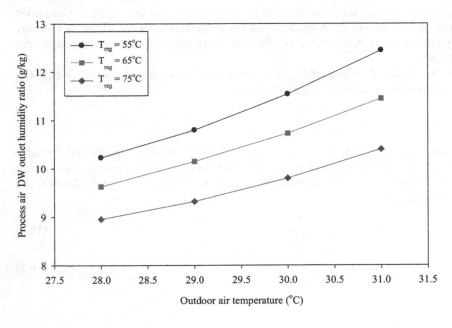

FIGURE 16.4 Influence of deviation in outdoor ambient air dry bulb temperature on outlet HR of outlet process air leaving the rotary dehumidifier.

surface. This is necessary to enlarge the difference of vapor pressure among the two for the effective separation of water vapor from the moist process air entering the rotary dehumidifier from the test room. So, minimized regeneration temperature decreases the rate of adsorption that ultimately lowers the effectiveness for moisture removal in the rotary desiccant dehumidifier. It is seen that the humidity ratio of process air at the exit of dehumidifier increases due to rise in regeneration temperature above 65°C for ambient temperature above 30°C.

Fig. 16.5 illustrates the effect of rise in the DBT of ambient air on ε_{dw} at successively increased regeneration temperature requirement by the dehumidifier. It is clear from the graph that the rotary dehumidifier effectiveness diminishes with successive amelioration in the DBT of the outdoor ambient air. The difference of vapor pressure necessary to attract the moisture from the process air by the desiccant surface gets lowered at higher ambient temperature. So, the rate of dehumidification also gets decreases. This in turn decreases the effectiveness of the dehumidifier as increment takes place in DBT of outdoor ambient air. Thus to increase the transfer of moisture from humid air to unsaturated desiccant layer ultimately demands higher regeneration temperature. This in turn ameliorates the dehumidifier effectiveness with rise in regeneration temperature requirements at higher outdoor DBT at dehumidifier entry side of the process air. During moisture adsorption, the latent heat for condensation is also added to the dehumidified air rises its temperature at dehumidifier exit.

Fig. 16.6 illustrates the effect of variations in DBT of outdoor air on moisture removal rate (MRR) of the rotary desiccant dehumidifier with successive greater demand in reactivation temperature for the desiccant regeneration. It is obvious from the graph that the rotary desiccant dehumidifier demands higher reactivation temperature for its desiccant regeneration for ameliorated outdoor ambient DBT to maintain its necessary MRR for the same effectiveness.

It is clear from the above discussion that when the DBT of ambient air gets increased, vapor pressure of water vapor present in the moist room process air, which is supplied to the desiccant wheel for dehumidification, gets lowered. This ultimately diminishes the difference between vapor pressure among the desiccant laden flute surface of the matrix and the incoming moist process air, which is supplied from the test room. So, the rate of dehumidification gets diminishes. This lowers the efficiency of rotary desiccant wheel in turn. It is also responsible for the reduction in MRR by the adsorption of dehumidifier. To maintain constant dehumidification rate, it demands higher deviation among the vapor pressure difference between the flow of cold process air and rotating desiccant matrix which is comparatively hot. It demands higher reactivation temperature to increase the temperature of desiccant material surface to lower its vapor pressure. This is the reason why higher regeneration temperature of dehumidifier increases its effectiveness and ultimately the MRR of dehumidifier.

16.7 Prediction of system performance by use of TRNSYS simulation

Since past few decades, investigators apply many efforts to solve a problem by use of costlier experimentations, but it is still challengeable to espouse this rigorous, extortionate, and time-consuming technique due to several techno-

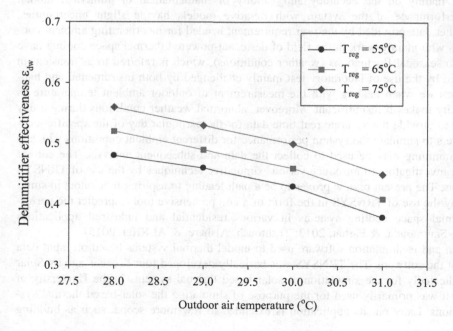

FIGURE 16.5 Influence of amelioration in dry bulb temperature of ambient air on ε_{dw} at successively increased regeneration temperature requirement by the dehumidifier.

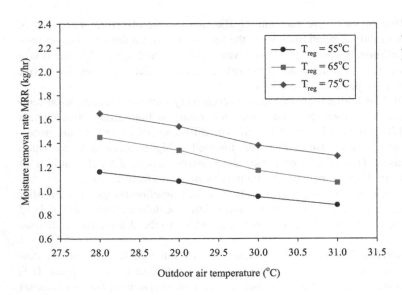

FIGURE 16.6 Effect of variations in dry bulb temperature of outdoor air on moisture removal rate of the rotary desiccant dehumidifier with successive greater demand in reactivation temperature for the desiccant regeneration.

commercial limitations. Real-world experiments in most cases face many limitations because of cost and time. Furthermore, it is impossible many a times to simulate the real word owing to extreme environment. Moreover, the same can also lead to many experimental uncertainties due to man and measuring instrument limitations. While working on simulation environment, the researchers and professionals have come across to many such problems for the design and optimization of desiccant powered thermal space cooling system in various air-conditioning applications. There is a fast growing market of building cooling requirement for both commercial and residential over worldwide. As alternative laborious experimentation technique, modeling and simulation is established as the predictive procedures to reduce the cost of experimentations substantially. So, the process of simulation is may be conducted in place of experimental evaluation, which is a far better process of interpolation through iteration conditions that is finally prove to save the cost and time of experimental study. Sometimes, physical experimentation may not be possible in the absence of laboratory, costlier equipment, and rigorous experimental procedure considering time constraint; instead it may get simulated by the use of numerical experimentation (Alahmer, Alsaqoor, & Borowski, 2019; Reddy, Priya, Carollo, & Kumar, 2017; TRNSYS, 2006).

Conducting simulations, can save cost and time as compared to rigorous lab experiments. Predictive results obtained by making use of simulation depend mainly on the accuracy and validity of mathematical or numerical model. Simulation study helps to predict performance of the system with iterative models having slight uncertainties. Simulation allows us synthesizing weather data provided by the user requirement needed for incorporating ambient conditions during experiments. Researchers who mainly work in the field of desiccant-powered thermal space cooling have come across similar problems related to seasonal fluctuations (weather conditions), which is referred to as transient in its nature. Accuracy of the data obtained by the use of laboratory test mainly challenged by both instrumental and handling uncertainties mostly found in the case when dealing with the measurement of outdoor ambient temperature as well as consideration of relative humidity makes it insignificant. Moreover, abnormal weather conditions during a day due to cloudy weather make it difficult to provide the accurate real time data for the particular day of the specific location on earth. It is difficult for researchers to simulate the system performance for different ambient conditions which is variable in nature. The computer programming may be used to collect the data and subsequent analysis. The current review article aims to acknowledge the investigators to introduce various simulative techniques by the use of TRNSYS in various thermally cooling applications. The present review proves to be a path leading to aspiring innovators in simulating the outdoor weather conditions by the use of TRNSYS in the form of a comprehensive tool to predict the performance of the desiccant-powered thermal space cooling systems in various residential and industrial applications (Chargui & Sammouda, 2014; Chargui, Sammouda, & Farhat, 2013; Tashtoush, Alshare, & Al-Rifai, 2015).

TRNSYS means TRaNsient SYStem and is simulation software used to model thermal systems based on input data for the component used in the library of the software. The TRNSYS was basically developed four decades ago by Solar Energy Lab members who worked dedicatedly for the simulation of solar based thermal systems in the University of Wisconsin-Madison. In its earlier days it was primarily used for the purpose of simulating the solar-based thermal systems in various water-heating applications. Later on, its application is extended to few more scope, such as building thermal analysis, electrical, and HVAC, with the inclusion of several component libraries related to each system (Khan, Badar, Talha, Khan, & Butt, 2018).

The TRNSYS consists of various libraries for particular components used in thermal system, such as solar collector, water heater, pump, and fan. These components are designated as types that consist of a set of governing equations or a mathematical modeling in the TRNSYS simulation program. These types are divided into groups; each one has number of types representing specific applications, such as solar, HVAC, and building load. The user can create TRNSYS simulation studio project as per the requirement of the component to be pick up it from the available library in the TRNSYS environment. The TRNSYS project in simulation studio environment designed basically by using the group of components, which are integrated or coupled in such a fashion that the input of one component can be the output of the other. The interconnection among them can be represented as the physical connection between the actual parts of the system. The forcing function is used to control the operation of the particular component during the simulation process. The weather data files are used to avail the ambient data, such as temperature, humidity, and wind speed, to the model while simulation process is running. Weather data file contains the data based on weather conditions for the particular day for the specific locations on the earth. Various configurations, such as ventilation or recirculation of the comfort cooling systems, were simulated in TRNSYS software. The modules typically related to solid desiccant-powered thermally cooling and air-conditioning systems are rotary regenerators, heat wheel, cooling coil, fans, and thermal heater to simulate the system in TRNSYS environment (Antoniadis & Martinopoulos, 2019; Persson, Perers, & Carlsson, 2011).

The TRNSYS is used as a visual interface to simulate the model of various solar thermal heating systems in terms of its energy efficiency. The components used in the assembly panel can take part in simulation by calling it from library. Various menus are provided in the simulation studio to select the particular component and run the simulation program in studio. The information may flow from one component to the other while simulation is running. The proforma mainly is a file for documenting information about the part of assembly in the project. It helps in simulation run of the studio. The TMF file or TRNSYS Model File of individual component is mainly used for the purpose of modeling the documentation standard for each type used in project assembly. This assembly is ultimately divided into proforma format for each type used in the model and is stored while accessing the same project. The components that are inserted and assembled in environment of the simulation studio program can make a proforma that is in the form of set of mathematical governing equations of the individual components. The proforma window visually displays the proforma for individual component used in the studio project. A very important feature in simulation studio is capability to provide governing equation for each particular type as input file to the project. This input file provides conditions at the inlet of that type, such as temperature, pressure, velocity, and humidity, to take part in simulation. Each time step and solver scheme can be selected for each simulation from information window provided in simulation studio. The output of each component can be saved as simulation result in a separate text file in the excel format in studio projects. The simulation can be run for the particular desired duration according to the given time frame that generates very large output files. A simulation result available from the output data can be postprocessed in spread sheet data file called deducing (Shrivastava, Kumar, & Untawale, 2017).

The particular governing equations can be used to define each component in project by the set of equations written as computer code. Each equation has been iteratively solved by the use of simulation studio over the selected time period. The component of the system assembly may call during the time of simulation as per the order they are saved in input or deck file of the project. The numerical solver scheme, such as successive substitution method or Powell's method, is used to solve each set of governing equations used in the program. The user can make choice of a particular solver scheme during running the simulation in the information provided in global information window available with the simulation studio. Thus each set of governing equations can be solved iteratively during each time step while running the simulation.

The development of desiccant cooling modeling with the help of TRNSYS was carried out initially by Sheridan and Mitchell (1985). The model was simulated for variable ambient in the case of evaluating the temperature and specific HR of outdoor air. The results show that when the outdoor ambient is humid, the desiccant cooling can capable to save around 40%−45% energy for cooling the building as compared to the traditional air conditioners.

White, Kohlenbach, and Bongs (2009) conducted tests on a solar-powered desiccant dehumidification-integrated thermally cooling system using TRNSYS simulations. It was shown that the ventilation-based desiccant cooling cycle is inferior to recirculation-based system in hot and humid transient climatic conditions of tropical region. It was also illustrated that at lower regeneration temperature, it requires large air flow rate to achieve the required comfort conditions.

An experimental measurement was carried out on desiccant-integrated thermally cooling system by Ouazia, Barhoun, Daddad, Armstrong, and Szadkwski (2009). TRNSYS simulations were also carried out along with it to examine the humidity control and the room comfort.

Khalid et al. (2009) carried out measuring tests on practical setup of a desiccant-coupled thermally cooling system provided regeneration heat by solar energy in humid climate of tropical region. TRNSYS model was developed and validated with experimental results. Life cycle assessment has also been done to predict the payback period of 14 years for solar-assisted desiccant-based hybrid cooling system designed by them.

La et al. (2011) carried out simulation tests on a hybrid cooling system made up of two-stage desiccant and VCR air-cooling hybrid cycle, which gets reactivated heat demanded for the effective desiccant regeneration by the use of freely available renewable solar energy. The substantial energy saving can be achieved by use of the hybrid cooling system and the performance of it can be predicted by the use of TRNSYS simulation. The thermal COP of overall system was found as 0.95 and electric power saving rate as 31% under especially hot and humid tropical climatic conditions.

La et al. (2011) also estimated the performance of solar-assisted desiccant cooling cycle in TRNSYS environment. From the experimental results, it was found that the required indoor comfort can be maintained by using the solar heat for desiccant regeneration during summer.

The performance of solid desiccant integrated thermally cooling system using TRNSYS simulations was also evaluated by Baniyounes, Liu, Rasul, and Khan (2012). The thermoeconomic analysis of the hybrid comfort space cooling cycle has been carried out for the total annual building cooling load of 6428 kWh. It showed a significant energy saving while using the solar energy for the regeneration of dehumidifier.

A solar-powered desiccant-integrated evaporative cooling system for both open and closed cycle was simulated using TRNSYS by Jani, Mishra, and Sahoo (2013a). The reactivation heat needed for the regeneration of desorbing the rotary regenerator was provided by use solar radiations. The COP in the ventilation and in recirculation mode were obtained as 0.492 and 0.692, respectively. The simulation results showed that for the same capacity, recirculation (closed) cycle is found more efficient than the ventilation (open) cycle.

Taweekun and Akvanich (2013) experimentally studied a solid desiccant-powered dehumidification and cooling system to handle excessive moisture load in humid climatic conditions to maintain necessary indoor comfort. TRNSYS was also used to predict the overall performance of the system. It was found that the use of solid desiccant-based cooling and dehumidification system reduces the HR of indoor space and the cooling load (moisture load) of the air conditioner by 14.3% and 19.23%, respectively.

A TRNSYS simulation studio model of the desiccant-assisted dehumidification followed with evaporative cooling system was first simulated and then compared with the experimental data by Khoukhi (2013). It was seen from the test results that the system reduces the relative humidity significantly. Furthermore, it was determined that the simulation results predicted that such system is useful for the air conditioning of buildings in hot-humid climates.

Dezfouli et al. (2014) conducted TRNSYS simulation of four different types of desiccant-integrated building cooling systems for excess humid weather of tropical Malaysian region. The influence of several parameters, such as room temperature, HR, and solar fraction, over cooling performance of different designs of desiccant-assisted air conditioning systems has been analyzed. It is found that the dual dehumidifier cycle operated in ventilation mode under humid ambient condition can improve system efficiency about 38% as compared to the other types of desiccant-assisted air-conditioning cycles.

The desiccant-coupled hybrid cooling cycle-supplied reactivated heat by freely available solar energy was modeled in TRNSYS to simulate it for the moist tropical condition (Heleyleh, Nejadian, Mohammadi, & Mashhodi, 2014). The influence of various operating variables, such as ambient conditions, process air outlet temperature and HR of dehumidifier, and energy consumption in vapor compression system, on the COP of overall system was investigated. It is observed from the simulation results that the removal of excess humidity in terms of water vapor from supply air by using desiccant dehumidifier leads to a 13.2% decrease in building moisture cooling load in summer cooling season. Moreover, the obtained result indicated that the application of freely available renewable solar heat for the effective regeneration of dehumidifier lowers the fissile fuel consumption up to 50%.

A solid desiccant and vapor compression hybrid cooling system was modeled and simulated by the application of TRNSYS simulation studio by Jani et al. (2013b) in a North Indian Himalayan transit hot and humid ambient. It is found that the results obtained by TRNSYS simulation show good agreement with that of the experimental measurements. This shows suitability of desiccant-assisted cooling in humid climate for maintaining necessary indoor thermal comfort.

The TRNSYS simulation is employed to predict the performance of solar-integrated solid desiccant-assisted vapor compression-based hybrid cooling system in terms of system performance. Studio project was simulated in TRNSYS environment for the hot and humid transient conditions of west Himalayan region by the use of TMY file of the location (Roorkee). Air stream flow rates, DBT, HR, etc., of the process and regeneration air are taken as the input parameters to the developed TRNSYS model. The output of the model is obtained in terms of the performance of the system. Furthermore, it provides effectiveness of various components, such as rotary dehumidifier, air-to-air heat recovery wheel, cooling coil, and regeneration heater, with regard to their interaction among the solar assisted hybrid cooling system. The major reason behind the simulating system in TRNSYS is owing to its capability for easy and quick change over in the system configuration and in obtaining the graphical output on time variant scale. The reason behind conducting the TRNSYS simulation was to estimate the overall system performance in terms of its COP as per the deviation in outdoor weather conditions to minimize the large amount of rigorous experimental tests, which are economically costlier and at the same time avoid unnecessary engineering effort.

Figs. 16.7 and 16.8 illustrate the layout of internal arrangement of components in studio project of TRNSYS© for the solar-integrated desiccant-based thermal-cooled hybrid air-conditioning system for ventilation- and recirculation-operating modes. The components selected for making system model in TRNSYS environment are as follows: type 683 used for modeling the dehumidifier, test room sensible and latent cooling load can be provided to the model as type 690, sensible cooling coil of traditional air conditioner can be modeled as type 665-9, air-to-air HRW can be modeled by selecting component 760b, solar collector used for component type 73, 3b used for pump for water transport, 33e used for psychrometrics, data file TMY used for calculation of the weather data, and type 4a stratified tank while component 65d is used for graphic plotter. Thus the TRNSYS library consists of more than 250 models that include basic HVAC equipment, various pumps and fans, multizone buildings, weather data processors, plotters, etc., which help dynamic simulation to define convergence and plot system variables. Simulation studio is a flexible, component-based software package accommodating the needs in various applications of the energy simulation. Thus TRNSYS helps to set modular, black-box component approach to develop and solve a transient simulation on graphical output. Furthermore, the application of user-defined modules can make it more convenient for providing transient simulation of the building models with the variable climatic ambient conditions of the surroundings. The components used mainly in the simulation environment of studio project are mainly three types of variables used in modeling the component type, namely, inputs, parameters, and outputs of system. The inputs in simulation studio defined as the variables that may change during a process of simulation and the component has assigned the input value from output of the preceding components. The temperature change may be taken as input makes change in the output of the preceding component in each simulation time step. Various operating paramters like temperature, humidity ratio, pressure etc. are entered as input to the simulation studio project model before starting the simulation.

Fig. 16.7 shows the simulation studio modeling of the solar-assisted solid desiccant cooling system using TRNSYS 16 in the ventilation mode. TRNSYS is a TRNSYS program developed at the University of Wisconsin to assess the performance of thermal energy systems (Khalid et al., 2009). In the present simulation studio project, rotary desiccant dehumidifier is modeled as type 683, type 760b is heat wheel, type 506c is supply air sensible cooler, type 506c-2 is regeneration air-sensible cooler (ambient air cooler in case of recirculation mode), type 690 is room, type 3b is pump, type 73 is flat plate collector, type 652 is heat exchanger, type 109-TMY2 is weather data file for Delhi, type 33e is psychrometrics, type 65d is plotter, and type 4a is stratified tank.

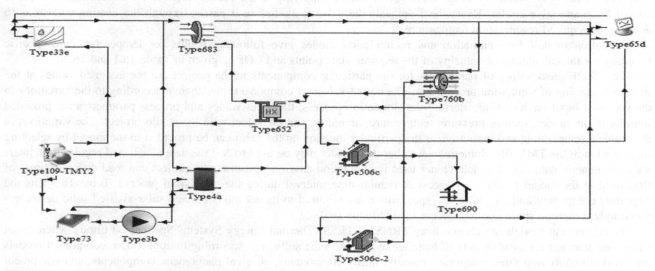

FIGURE 16.7   TRNSYS simulation studio project (ventilation mode).

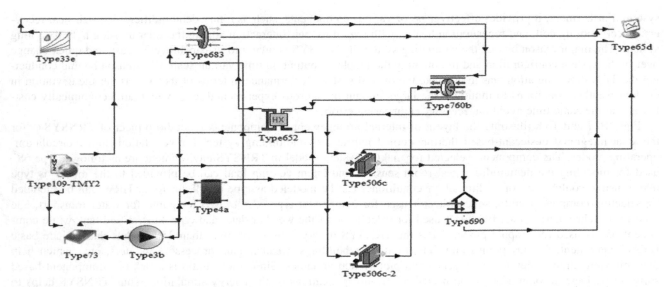

FIGURE 16.8 TRNSYS simulation studio project (recirculation mode).

TABLE 16.1 The conditions at main state point in ventilation mode.

Temperature (°C)	Sp. humidity (kg kg^{-1})	Enthalpy (kJ kg^{-1})	COP
$T_4 = 18$	0.01003	43.44	
$T_5 = 25$	0.01110	53.16	0.494
$T_7 = 63$	0.01473	102.62	
$T_8 = 82$	0.01473	122.26	

Fig. 16.8 shows the simulation studio modeling of the solar-assisted solid desiccant cooling system using TRNSYS 16 in the recirculation mode. Process air side in recirculation is closed loop while regeneration side is open to atmosphere. Room air is recirculated to remove moisture by passing through desiccant wheel. Heat wheel removes sensible heat and further it is cooled by evaporative cooler before supplying to the room. Ambient air in regeneration side is cooled by evaporative cooler. Heat wheel preheats the exhaust air before it passes through the heating coil so as to achieve the required regeneration temperature.

Simulation models for ventilation and recirculation modes give following results for temperature and specific humidity for the calculation of enthalpy of the required state points and COP as given in Table 16.1 and 16.2.

The effectiveness values of parameters for the particular components in the project are the assigned values at the time of beginning of simulation process run. The output values of component may change according to the variations in the provided input values. While starting the simulation process, the input values and process parameters are provided initially to the model, such as pressure, temperature, humidity, and air velocity, in the studio project. The variations in the ambient temperature and humidity for the particular location on the earth can be provided to the model by selecting that location in the TMY file. Sometimes weather data reader may be used to call the data at selected regular time interval from generic data file. The data reader used in simulation environment of studio project can read different types of files used in the model to allow its access at regular time interval during the simulation process. Table 16.3 lists the important component and its TRNSYS types, which are required to model the combined solar-assisted solid desiccant-based hybrid thermal space cooling system in simulation studio.

The component models are chosen from **TRNSYS TESS** (Thermal Energy Systems Specialists) library where input values are selected for weather data of Roorkee available in the software. According to application, component models are divided mainly into three categories, namely, utility components, physical phenomena components, and equipment components.

TABLE 16.2 The conditions at main state point in recirculation mode.

Temperature (°C)	Sp. humidity (kg kg^{-1})	Enthalpy (kJ kg^{-1})	COP
$T_4 = 17$	0.01150	44.95	
$T_5 = 25$	0.01248	55.71	0.693
$T_8 = 53$	0.01647	94.89	
$T_9 = 68$	0.01647	110.41	

TABLE 16.3 List of important component Type in studio project.

Component name	TRNSYS type
Cooling load	Type 690
Rotary regenerator	Type 683
Psychrometrics data file	Type 33e
Sensible cooler	Type 506c
Weather data file	Type 109-TMY2
Air-to-air sensible heat wheel	Type 760b
Tank	Type 4a
Heat exchanger	Type 652
Pump	Type 3b
Solar collector	Type 73
Plotter	Type 65d

16.7.1 Weather data reader—type 109 TMY2

Weather data reader is meant for calling weather data at spatial time duration from an earlier saved weather data for the particular location on earth. It is represented in the particular file format in simulation studio. Type 109 reads a weather data that are saved as file in a standard TMY format in TRNSYS library. The TMY files can be generated by the use of several programs, such as Meteonorm that generated generic data file according to geographic locations of different places on the earth.

16.7.2 Online graphical plotter—type 65d

The online graphics plotter is meant to display system output variable, such as temperature and HR, during the time of simulation. This component is very important as it displays the sets of output data and allows users to immediately monitor if the system output is performing as desired or not. The defined output variables, such as flow rate and pressure drop, have been shown on graphical window of simulation studio project. In case of online application of the type 65d, the output data file can be displayed during running the project in simulation studio program.

16.7.3 Psychrometrics—type 33e

Psychrometrics helps to input various psychrometric properties, such as DPT, ambient air temperature, and humidity, and property of the moist room air and calls the same during running of the simulation by the room cooling load.

16.7.4 Heat recovery wheel—type 760b

Type 760b is used to model a sensible heat wheel in which two air streams, namely, process air and reactivation air, pass so that only energy is interchanged between the two air streams. It may evaluate the sensible heat exchange by solving the model of heat recovery wheel for different configurations of air stream flow rates, such as parallel, cross, or counter flow, of process and regeneration air.

16.7.5 Sensible cooler—type 506c

Type 506c models as sensible cooler. The effectiveness of sensible cooler in simulation studio can be evaluated by the use of external weather data files for the specific geographic locations on earth.

16.7.6 Room load—type 690

This component makes access of sensible and latent loads as per the existing ambient conditions for the particular geographic locations on earth and converts them to temperatures and humidity for TRNSYS simulations by allocating the loads on a simple building cooling model. The user provides the calculated sensible or latent heating or cooling loads of the buildings as well as an estimate of thermal and moisture capacitance of the building and the model evaluates the resultant temperature, humidity, etc. In simulation studio, the user having capability to introduce the ventilation or the recirculation air configuration or mode to the model the desiccant-based comfort space cooling system, which may be externally controlled and conditioned, to offset the exposed sensible or latent heating or cooling loads.

16.7.7 Rotary desiccant dehumidifier—type 683

This component represents the proforma data file for rotary regenerator cum dehumidifier. Water vapor or moisture elimination performance (dehumidification) of the same is based on the set of mathematical governing equations for F_1-F_2 potentials. The model determines the reactivation temperature at outdoor HR, which will dry exactly up to the supply HR set point. The process stream outlet temperature of desiccant wheel is also determined by the use of this component in simulation studio.

Desiccant dehumidifier is the main component of the desiccant cooling system that is responsible for eliminating the water vapor from supply room process air above its DPT. It mainly depends upon adsorption of water on to the thin desiccant layer laden over matrix while the moist process air passing through individual flutes of desiccant matrix. Eventually, the desiccant gets saturated with the moisture level adsorbed over its surface and has to be regenerated through a desorption process to make it working for the next cycle. A working principle of the solid desiccant wheel is shown in Fig. 16.9. Type 683 represents desiccant wheel model in simulation studio.

The process and reactivation air flow oppositely over sections continuously in slowly rotating desiccant wheel at approximate speed around 20 rph. The total area of wheel at any time is divided between process and reactivation section in terms of 1/3 or 2/3. The regeneration temperature is the most influencing factor that controls the rate of dehumidification.

To evaluate the demand of exact reactivation temperature, which makes efficient desorption of wheel, can be modeled through simulation studio for various ambient conditions and seasonal changes. Moreover, the process-to-regeneration area ratio can also be an important evaluation in determining air stream flow rates for both process and regeneration air. Type 683 has been modeled as desiccant wheel in simulation studio and can be solved iteratively for given supply condition of moist process at (state point D) inlet as shown in Fig. 16.10.

The users have flexibility for availing any specific HR to the model while the humidifier is working in simulation according to its working capacity specified in model. The model evaluate the specific reactivation temperature

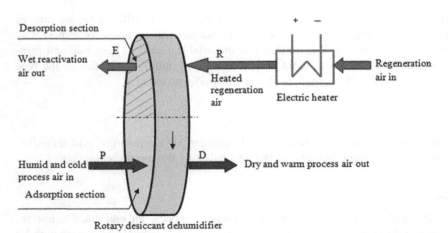

FIGURE 16.9 Working principle of rotary regenerating desiccant dehumidifier.

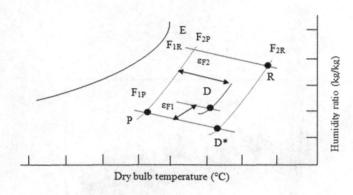

FIGURE 16.10 Ideal desiccant dehumidification process in rotary dehumidifier.

necessary to produce desorption of the desiccant used in the dehumidifier. Humidity at the process air exit of dehumidifier can be calculated by the model after providing input for air flow in process and reactivation air to the model in simulation studio.

F1 and F2 are defined as potentials because they are property functions of desiccant-fluid system for the specific value of temperature and humidity at the inlet of dehumidifier. Lines of constant F1 resemble lines of constant enthalpy while those of F2 resemble lines of constant relative humidity on psychrometric chart. The simulation studio model initially calculates the values of potential functions F1 and F2 of the system inlet. Owing to the similarity of potential function F1 at points P and D* (Fig. 16.9), the exit temperature of process air at desiccant wheel out let can be determined later.

The isotherm functions F_1 and F_2 for silica gel can be given by

$$F_1 = \frac{-2865}{T^{1.490}} + 4.344\omega^{0.86441}$$

(16.10)

$$F_2 = \frac{T^{1.490}}{6360} - 1.127\omega^{0.079691}$$

(16.11)

The functions F_1 and F_2 can be calculated for nonlinearities during the process of dehumidification within dehumidifier by two effectiveness values ε_{F1} and ε_{F2} given by

$$\varepsilon_{F1} = \frac{F_{1D} - F_{1P}}{F_{1R} - F_{1P}}$$

(16.12)

$$\varepsilon_{F2} = \frac{F_{2D} - F_{2P}}{F_{2R} - F_{2P}}$$

(16.13)

The rotary regenerator (type 683) evaluate temperature of process air at the end of adsorption process (state point D*) as per simulated value of potential function F_2. F_2 helps to calculate the value of required regeneration temperature (state point R) of desiccant used in dehumidifier during the process of simulation. The value of potential function at state point D can be simulated by the model using the value of F_1 and F_2. Furthermore, the value of F_1 and F_2 at state point D can iterates the exit humidity of the rotary regenerator for process air outlet. If the calculated value and assumed value of the humidity can be achieved within tolerance band, then the corresponding values of temperature of adsorbed air at dehumidifier exit can be predicted later by the model. The effectivenesses of ε_{F1} and ε_{F2} of desiccant wheel makes sure about approximation of adsorption as adiabatic and the degree to which dehumidification takes place within dehumidifier.

The final condition of process air leaving the desiccant wheel can be predicted by the simulation on the basis of iterative solution of model with regard to the heat and mass balance between the intake and exit of dehumidifier. Effectiveness values ε_{F1} and ε_{F2} in terms of higher and lower effectiveness of dehumidifier should be provided as fixed parameter while starting the simulation process. In the present case, the values of ε_{F1} and ε_{F2} were considered as 0.05 and 0.95, respectively, which correspond to the condition ε_{F1}, $\varepsilon_{F2} \le 1$ always valid for actual condition, and the hypothetical one refers to as $\varepsilon_{F1} = 0$ and $\varepsilon_{F2} = 1$.

Figs. 16.11 and 16.12 illustrate the variations in temperatures at important state points of solar-integrated hybrid solid desiccant-assisted thermal cooled hybrid air conditioning on time variant scale. The simulation studio model is simulated for variations in temperature at various state points in the hybrid cooling, such as the dehumidifier, test room, solar collector, reactivation heater, water thermal tank, and outdoor atmosphere. It is depicted that, among all the other

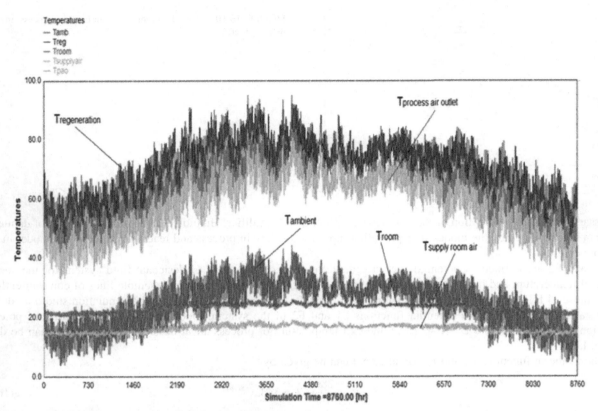

FIGURE 16.11 Temperature at various state point (ventilation mode).

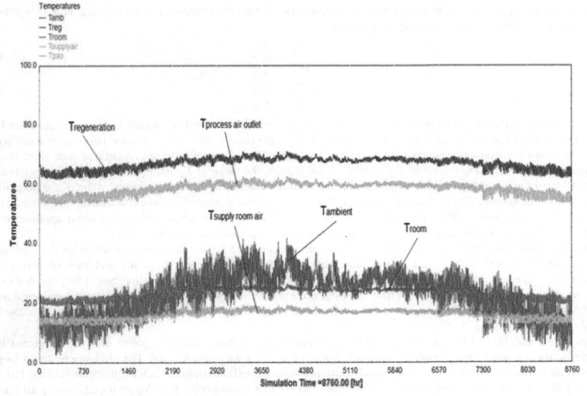

FIGURE 16.12 Temperature at various state point (recirculation mode).

parameters, the required regeneration temperature of desiccant is the dominant parameter, which is responsible for performance alteration in the solar-integrated desiccant-based hybrid cooling system. Figs. 16.11 and 16.12 show simulations results for the variation of temperatures in the ventilation mode and recirculation mode of solar-assisted solid desiccant cooling system. It is observed that the required regeneration temperature for the ventilation mode is 82°C and for the recirculation mode, it is 68°C. So, lower energy is needed in heater to obtain the required regeneration air temperature for the desorption of desiccant wheel in recirculation mode. Variations in ambient temperature, supply air temperature, process air outlet temperature, and room temperature also shown in the figure. Temperature of supply air is almost same in both the cases.

Variations in the HR of ambient air, process air at outlet of desiccant wheel, room air, and supply air are shown in Figs. 16.13 and 16.14 for the ventilation mode and the recirculation mode, respectively. HR is the main parameter indicating the removal of moisture from room in terms of latent heat to obtain desired comfort conditions inside the room. It is observed that the specific humidity of supply air and room air in the recirculation mode is little more than that in the ventilation mode. So, for the same capacity, the recirculation mode requires greater mass flow rate of supply air to maintain desired comfort conditions inside the room.

Figs. 16.13 and 16.14 illustrate variations in HR at important state points of solar-integrated hybrid solid desiccant-assisted thermal-cooled air conditioning on time variant scale. The simulation studio model is simulated for variations in HR at various state points in the hybrid cooling, such as the test room, dehumidifier, reactivation heater, and outdoor atmosphere. It is depicted that among all the other parameters, the supply room air HR is the dominant parameter, which is responsible for maintaining necessary thermal comfort inside the specified test room and performance alteration in the combined solar-assisted solid desiccant-based hybrid thermally cooling system.

It is observed from the graph that the drastic rise in HR for the month of May and June imparts the excess rise in the outdoor humidity for the cooling period of those months. So, the rise in HR of room air demands high reactivation heat for efficiently controlling the room humidity. Furthermore, it is responsible for increasing the electrical energy consumption in terms of raising the cooling demand in the form of higher sensible cooling load over traditional evaporator coil. It is seen that about more than 40% decrement is found in the humidity of the warm and dehumidified process air, which is exiting of desiccant laden dehumidifier as compared to moist supply air.

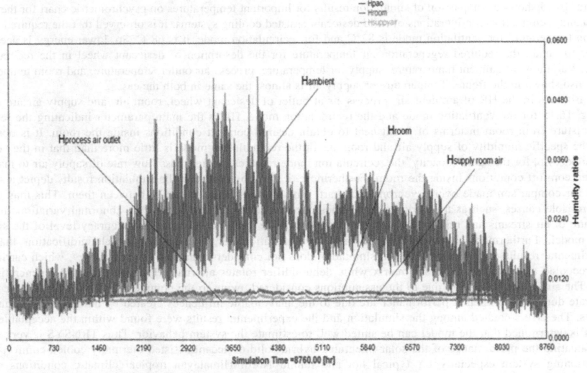

FIGURE 16.13 Specific humidity at various state points (ventilation mode).

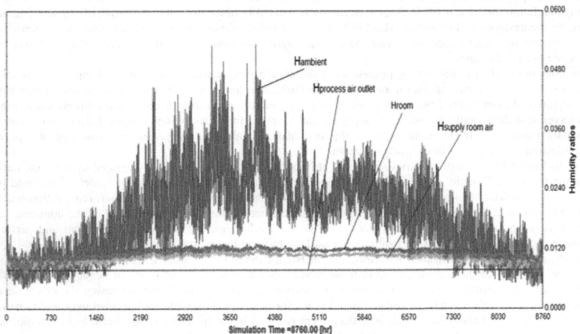

FIGURE 16.14 Specific humidity at various state points (recirculation mode).

Fig. 16.15 shows a comparison of simulation results for important temperatures on psychrometric chart for the ventilation and recirculation configurations of solid desiccant assisted cooling systems. It is observed that the required regeneration temperature for ventilation mode is 82°C and for recirculation mode, it is 68°C. So, lower energy is needed in heater to obtain the required regeneration air temperature for the desorption of desiccant wheel in the recirculation mode. Variations in ambient temperature, supply air temperature, process air outlet temperature, and room temperature have also shown in the figure. Temperature of supply air is almost the same in both the cases.

Variations in the HR of ambient air, process air at outlet of desiccant wheel, room air, and supply air are shown in Fig. 16.15 for the ventilation mode and the recirculation mode. HR is the main parameter indicating the removal of moisture from room in terms of latent heat to obtain desired comfort conditions inside the room. It is observed that the specific humidity of supply air and room air in the recirculation mode is little more than that in the ventilation mode. So, for the same capacity, the recirculation mode requires greater mass flow rate of supply air to maintain desired comfort conditions inside the room. Furthermore, it was observed that the simulation results depict a similar trend for comparison made over psychrometric chart, but small discrepancies exist between them. This may due to the probable causes, such as measuring uncertainties due to fluctuating test conditions, abnormal variations in temperature of air streams and relative humidity of outdoor air due to transit climate, and accuracy level of the simulation model. Furthermore, it is determined that the simulation results give better dehumidification than the experimental results owing to the fact that simulation does not considers a part of thermal energy, which carried out by precooled air from the desiccant matrix while dehumidifier rotates alternatively in process and regeneration air side. The simulation errors are due to the assumptions considered owing to the complexity of model while the errors generate during experimental performance are due to the unavoidable influences created by human and instrumental efforts. The errors created among the simulation and the experimental results were found within the acceptable limit and it is determined that the model can be suited well to estimate the system behavior. Thus TRNSYS is very useful in evaluating the performance of the solar-integrated hybrid solid desiccant-assisted thermally cooled comfort space conditioning system especially for typical hot and humid North Himalayan tropical climatic conditions of the Roorkee.

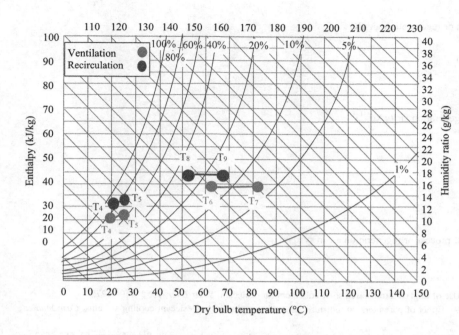

FIGURE 16.15 Comparison among the TRNSYS simulation for ventilation and recirculation modes.

Furthermore, as illustrated by the psychrometric chart, line 4—5 represents heating and humidification of the room air while line 8—9 represents desorption process in the desiccant dehumidifier. Both the processes follow the paths diverging slightly from the path of constant enthalpy in on psychrometric chart. During travel of desiccant wheel from regeneration to process air section, adds up little heat to slightly rise temperature of dehumidified air because of hot matrix surface. This is due to the fact that the desiccant matrix is found hotter as compared to the supplied process air from the conditioned room. In the same way during travel of desiccant wheel from process to regeneration air section, taken up little heat to slightly lower temperature of regeneration air because of cool matrix surface.

16.8 Conclusion

Temperature and specific humidity variations at important state point of a solar-assisted solid desiccant cooling system are studied for the ventilation and recirculation modes using TRNSYS simulation software. It is observed that COP for the recirculation mode is little higher than that in the recirculation mode. Also, regeneration temperature required for the desorption of dehumidifier is less in the recirculation mode. Thus recirculation cycle proves to be better and more efficient than the ventilation mode because of reduction in thermal energy required for reactivation and successively higher COP.

The future research on the investigations and use of novel desiccant materials (Bhabhor & Jani, 2019; Jani, 2019; Jani et al., 2019) that can be effectively regenerated near about lower temperature (near-ambient) so that the same can be used successfully by the application of renewable and economical solar (green) energy is the key of augmenting even greater the contribution that the desiccant cooling can bring to the amelioration of comfort, energy, and cost savings.

Nomenclature

COP	coefficient of performance
DBT	dry bulb temperature (K)
E	electricity consumption for overall system (kW)
h	enthalpy of air (kJ kg^{-1})
m	mass flow rate of process or regeneration air (kg s^{-1})
MRR	moisture removal or elimination rate (kg h^{-1})
Q_{cc}	cooling capacity (kW)
Q_{reg}	regeneration heat required for reactivating the desiccant (kW)
RH	relative humidity of the process or regeneration air (%)

T	dry bulb temperature of process or regeneration air (K)
TRNSYS	TRaNsient SYStem
VCR	vapor compression refrigeration

Greek letters

ε	desiccant wheel effectiveness
ω	specific humidity ratio of process or regeneration air (g kg^{-1})

Subscripts

a	dehumidified process air
dw	desiccant wheel
hrw	air-to-air heat recovery wheel
p	process air
r	reactivation air
1,2, etc.	state points for given condition of process or regeneration air in system

References

Ahmed, A., Pasaran, P., & Wipke, X. B. (1994). Use of unglazed transpired solar collectors for desiccant. *Solar Energy*, 52, 419–427.

Alahmer, A., Alsaqoor, S., & Borowski, G. (2019). Effect of paramters on moisture removal capactiy in the desiccant cooling systems. *Case Studies in Therm Engineering*, 13, 1–9.

Alizadeh, S. (2008). Performance of a solar liquid desiccant air conditioner—An experimental and theoretical approach. *Solar Energy*, 82, 563–572.

Ando, K., & Kodama, A. (2005). Experimental study on a process design for adsorption desiccant cooling driven with a low-temperature heat. *Adsorption*, 11, 631–636.

Antoniadis, C. N., & Martinopoulos, G. M. (2019). Optimization of a building integrated solar thermal system with seasonal storage using TRNSYS. *Renewable Energy*, 137, 56–66.

ASHRAE. (2009). Thermal comfort. In *ASHRAE handbook, fundamentals (SI)*. Atlanta, GA: The American Society of Heating, Refrigerating and Air-Conditioning Engineers, Inc. (Chapter 9).

Baniyounes, A., Liu, G., Rasul, M., & Khan, M. (2012). Analysis of solar desiccant cooling system for an institutional building in subtropical Queensland, Australia. *Renewable and Sustainable Energy Reviews*, 16, 6423–6431.

Baniyounes, A., Liu, G., Rasul, M., & Khan, M. (2012). Analysis of solar desiccant cooling system for an institutional building in subtropical Queensland, Australia. *Renewable and Sustainable Energy Reviews*, 16, 6423–6431.

Baniyounes, A. M., Liu, G., Rasul, M. G., & Khan, M. M. (2013). Comparison study of solar cooling technologies for an institutional building in sub-tropical Queensland, Australia. *Renewable and Sustainable Energy Reviews*, 23, 421–430.

Banks, P. J. (1972). Coupled quilibrium heat and single adsorbate transfer in fluid flow through a porous medium—I. Characteristic potential and specific capacity ratios. *Chemical Engineering Science*, 27(5), 1143–1155.

Bhabhor, K., & Jani, D. B. (2019). Progressive development in solid desiccant cooling: A review. *International Journal of Ambient Energy*. Available from https://doi.org/10.1080/01430750.2019.1681293.

Bourdoukan, P., Wurtz, E., Joubert, P., & Sperandio, M. (2008a). Potential of solar heat pipe vacuum collectors in the desiccant cooling process: Modelling and experimental results. *Solar Energy*, 82, 1209–1219.

Bourdoukan, P., Wurtz, E., Joubert, P., & Sperandio, M. (2008b). Overall cooling efficiency of a solar desiccant plant powered by direct flow vacuum tube collectors: Simulation and experimental results. *JBPS*, 1, 149–162.

Camargo, J. R., Godoy, E., & Ebinuma, C. D. (2005). An evaporative and desiccant cooling system for air conditioning in humid climates. *Journal of the Brazilian Soiety of Mechancal Sciences & Engineering.*, 3, 243–247.

Chargui, R., & Sammouda, H. (2014). Modeling of a residential house coupled with a dual source heat pump using TRNSYS software. *Energy Conversion and Management*, 81, 384–399.

Chargui, R., Sammouda, H., & Farhat, A. (2013). Numerical simulation of a cooling tower coupled with heat pump system associated with single house using TRNSYS. *Energy Conversion and Management*, 75, 105–117.

Dai, Y. J., Wang, R. Z., & Xu, Y. X. (2002). Study of a solar powered solid-adsorption-desiccant cooling system used for grain storage. *Renewable Energy*, 25, 417–430.

Daou, K., Wang, R. Z., & Xia, Z. Z. (2006). Desiccant cooling air conditioning: A review. *Renewable and Sustainable Energy Reviews*, 10, 55–77.

Das, R. S., & Jain, S. (2017). Experimental investigations on a solar assisted liquid desiccant cooling system with indirect contact dehumidifier. *Solar Energy*, 53, 289–300.

Davanagere, B. S., Sherif, S. A., & Goswami, D. Y. (1999). A feasibility study of a solar desiccant air-conditioning system-part-II: Transient simulation and economics. *International Journal of Energy Research*, 3, 103–116.

Davanagere, B. S., Sherif, S. A., & Goswami, D. Y. (1999). A feasibility study of a solar desiccant air-conditioning system-part-I: Psychrometrics and analysis of the conditioned zone. *International Journal of Energy Research*, 3, 7–21.

Dezfouli, M. M., Bakhtyar, M. H., Sopian, K., Zaharim, A. Mat, S., & Rachman, A. (2012). Experimental investigation of solar hybrid desiccant cooling system in hot and humid weather of Malaysia. In *Proceedings of the 3rd international conference on development, energy, environment, economics, DEEE'12, Paris, France, Eslamian, S., NAUU, Mathematics and Computers in Science and Engineering Series, 6* (pp. 172–176). WSEAS Press.

Dezfouli, M. M. S., Mat, S., Pirasteh, G., Sahari, K. S. M., Sopian, K., & Ruslan, M. H. (2014). Simulation analysis of the four configurations of solar desiccant system using evaporative cooling in tropical weather in Malaysia. *International Journal of Photoenergy, 14*, 1–14.

Dezfouli, M. M. S., Sohif, M., & Sopian, K. (2013). Comparison simulation between ventilation and recirculation of solar desiccant cooling system by TRNSYS in hot and humid area. In *Proceedings of the 7th International conference on Renewable Energy Sources, RES'13, Kaulalumpur, Malaysia, Zaharim, A., and Sopian, K., Energy Environmental and Structural Engineering Series, 8* (89–93). WSEAS Press.

Dhar, P. L., & Singh, S. K. (2001). Studies on solid desiccant based hybrid air-conditioning. *Applied Thermal Engineering, 21*, 119–134.

Dunkle, R. V. (1965). Method of solar air conditioning. *Inst. Engrs., Australia, Mech. and Chem. Eng. Trans., 1*, 73–78.

Fu, H. X. (2016). A dual-scale analysis of a desiccant wheel with a novel organic–inorganic hybrid adsorbent for energy recovery. *Applied Energy, 163*, 167–179.

Ge, T. S., Li, Y., Dai, Y. J., & Wang, R. Z. (2010). Performance investigation on a novel two-stage solar driven rotary desiccant cooling system using composite desiccant materials. *Solar Energy, 84*(2), 157–159.

Ge, T. S., Ziegler, F., Wang, R. Z., & Wang, H. (2010). Performance comparison between a solar driven rotary desiccant cooling system and conventional vapor compression system (performance study of desiccant cooling). *Applied Thermal Engineering, 30*(6–7), 724–731.

Halliday, S. P., Beggs, C. B., & Sleigh, P. A. (2002). The use of solar desiccant cooling in the UK: A feasibility study. *Applied Thermal Engineering, 22*, 1327–1338.

Hassan, M. M., & Beliveau, Y. (2008). Modeling of an integrated solar system. *Build Environment, 43*, 804–810.

Heidarinejad, G., & Pasdarshahri, H. (2010). The effects of operational conditions of the desiccant wheel on the performance of desiccant cooling cycles. *Energy and Buildings, 42*, 2416–2423.

Heleyleh, B. B., Nejadian, A. K., Mohammadi, A., & Mashhodi, A. (2014). Simulation of hybrid desiccant cooling system with utilization of solar energy. *Trends in Applied Sciences Research, 9*, 290–302.

Henning, H. M. (2007). Solar assisted air conditioning of buildings—An overview. *Applied Thermal Engineering, 27*, 1734–1749.

Henning, H. M., Erpenbeck, T., Hindenburg, C., & Santamaria, I. S. (2001). The potential of solar energy use in desiccant cooling cycles. *International Journal of Refrigeration, 24*(3), 220–229.

Hirunlabh, J., Charoenwat, R., Khedari, J., & Teekasap, S. (2007). Feasibility study of desiccant air-conditioning system in Thailand. *Building and Environment, 42*(2), 572–577.

Hürdoğan, E., Büyükalaca, O., Yılmaz, T., Hepbasli, A., & Uçkan, İ. (2012). Investigation of solar energy utilization in a novel desiccant based air conditioning system. *Energy and Buildings, 55*, 757–764.

Hussain, S., Kalendar, A., Rafique, M. Z., & Oosthuizen, P. (2020). Numerical investigations of solar-assisted hybrid desiccant evaporative cooling systems for hot and humid climate. *Advances in Mechanical Engineering, 12*(6), 1–16.

Hwang, Y., & Radermacher, R. (2008). Review of solar cooling technologies. *HVAC & Research, 14*, 507–528.

Jain, S., Dhar, P. L., & Kaushik, S. C. (1995). Evaluation of solid based evaporative cooling cycles for typical hot and humid climates. *International Journal of Refrigeration, 18*(5), 287, –29.

Jani, D. B., et al. (2019). A review on use of TRNSYS as simulation tool in performance prediction of desiccant cooling cycle. *Journal of Thermal Analysis and Calorimetry*. Available from https://doi.org/10.1007/s10973-019-08968-1.

Jani, D. B. (2019). An overview on desiccant assisted evaporative cooling in hot and humid climates. *Algerian Journal of Engineering and Technology, 1*, 1–7. Available from https://doi.org/10.5281/zenodo.3515477.

Jani, D. B., Mishra, M., & Sahoo, P. K. (2013a). Simulation of solar assisted solid desiccant cooling systems using TRNSYS. In *Proceedings of the 22nd national and 11th international ISHMT-ASME heat and mass transfer conference* (pp. 1–7), Decembrer 28–31. Kharagpur: IIT.

Jani, D. B., Mishra, M., & Sahoo, P. K. (2013b). Solid desiccant cooling—an overview. *International Conference on Advances in Chemical Engineering (ACE-2013), IIT, Roorkee, 61*, 1–5.

Jani, D. B., Mishra, M., & Sahoo, P. K. (2015a). Performance studies of hybrid solid desiccant–vapor compression air-conditioning system for hot and humid climates. *Energy and Buildings, 102*, 284–292.

Jani, D. B., Mishra, M., & Sahoo, P. K. (2015b). Experimental investigations on hybrid solid desiccant—Vapor compression air-conditioning system for Indian climate. In *The proceedings of the 24th IIR international congress of refrigeration, Yokohama, Japan* (pp. 1–9), August 16–22.

Jani, D. B., Mishra, M., & Sahoo, P. K. (2016a). Performance prediction of rotary solid desiccant dehumidifier in hybrid air-conditioning system using artificial neural network. *Applied Thermal Engineering, 98*, 1091–1103.

Jani, D. B., Mishra, M., & Sahoo, P. K. (2016b). Exergy analysis of solid desiccant—Vapor compression hybrid air conditioning system. *International Journal of Exergy, 20*, 517–535.

Jani, D. B., Mishra, M., & Sahoo, P. K. (2016c). Solid desiccant air conditioning—A state of the art review. *Renewable and Sustainable Energy Reviews, 60*, 1451–1469.

Jani, D. B., Mishra, M., & Sahoo, P. K. (2016d). Performance prediction of solid desiccant—Vapor compression hybrid air-conditioning system using artificial neural network. *Energy, 103*, 618–629.

Jani, D. B., Mishra, M., & Sahoo, P. K. (2016e). Experimental investigation on solid desiccant—Vapor compression hybrid air-conditioning system in hot and humid weather. *Applied Thermal Engineering, 104*, 556–564.

Jani, D. B., Mishra, M., & Sahoo, P. K. (2016f). Performance analysis of hybrid solid desiccant—Vapor compression air-conditioning system in hot and humid weather of India. *Building Services Engineering Research & Technology, 37*, 523−538.

Jani, D. B., Mishra, M., & Sahoo, P. K. (2017). Application of artificial neural network for predicting performance of solid desiccant cooling systems—A review. *Renewable and Sustainable Energy Reviews, 80*, 352−366.

Joudi, K. A., & Dhaidan, N. S. (2001). Application of solar assisted heating and desiccant cooling systems for a domestic building. *Energy Conversion and Management, 42*, 995−1022.

Jurinak, J. J., Mitchell, J. W., & Beckman, W. A. (1984). Open-cycle desiccant air conditioning as an alternative to vapor compression cooling in residential applications. *Journal of Solar Energy Engineering, 106*(3), 252−260.

Kabeel, A. E. (2007). Solar powered air conditioning system using rotary honeycomb desiccant wheel. *Renewable Energy, 32*, 1842−1857.

Kang, T. S., & Maclaine-Cross, I. L. (1989). High performance solid desiccant open cooling cycles. *Journal of Solar Energy Engineering, 111*, 176−183.

Kays, W. M., & London, A. L. (1984). *Compact heat exchangers*. New York: McGraw-Hill.

Khalid, A., Mahmood, M., Asif, M., & Muneer, T. (2009). Solar assisted, pre-cooled hybrid desiccant cooling system for Pakistan. *Renewable Energy, 34*(1), 151−157.

Khalid, A., Mahmood, M., Asif, T., & Muneer, T. (2009). Solar assisted, pre-cooled hybrid desiccant cooling system for Pakistan. *Renewable Energy, 34*, 151−157.

Khan, M. S. A., Badar, A. W., Talha, T., Khan, M. W., & Butt, F. S. (2018). Configuration based modeling and performance analysis of single effect solar absorption cooling system in TRNSYS. *Energy Conversion and Management, 157*, 351−363.

Khan, Y., Singh, G., Mathur, J., Bhandari, M., & Srivastava, P. (2017). Performance assessment of radiant cooling system integrated with desiccant assisted DOAS with solar regeneration. *Applied Thermal Engineering, 124*, 1075−1082.

Khoukhi, M. (2013). The Use of desiccant cooling system with IEC and DEC in hot-humid climates. *International Journal of Energy Engineering, 3*, 107−111.

Kim, D. S., & Infante Ferreira, C. A. (2008). Solar refrigeration options—A state-of-the-art review. *International Journal of Refrigeration, 31*, 3−15.

Koronaki, I. P., Rogdakis, E., & Tinia Kakatsiou, T. (2012). Experimental assessment and thermodynamic analysis of a solar desiccant cooling system. *International Journal of Sustainable Energy, 32*, 1−16.

Kulkarni, K., Sahoo, P. K., & Mishra, M. (2011). Optimization of cooling load for a lecture theater in a composite climate in India. *Energy and Building, 43*, 1573−1579.

Kumar, R., Dhar, P. L., & Jain, S. (2011). Development of new wire mesh packing for improving the performance of zero carry over spray tower. *Energy, 36*, 1362−1374.

La, D., Dai, Y., Li, H., Li, Y., Kiplagat, J. K., & Wang, R. (2011). Experimental investigation and theoretical analysis of solar heating and humidification system with desiccant rotor. *Energy and Buildings, 43*, 1113−1122.

La, D., Dai, Y., Li, H., Li, Y., Kiplagat, J. K., & Wang, R. (2011). Case study and theoretical analysis of a solar driven two-stage rotary desiccant cooling system assisted by vapour compression air-conditioning. *Solar Energy, 85*, 2997−3009.

Li, H., Dai, Y. J., Li, Y., La, D., & Wang, R. Z. (2011). Experimental investigation on a one-rotor two-stage desiccant cooling/heating system driven by solar air collectors. *Applied Thermal Engineering, 31*(17−18), 3677−3683.

Ma, Y., Saha, S. C., Miller, W., & Guan, L. (2017). Parametric analysis of design parameter effects on the performance of a solar desiccant evaporative cooling system in brisbane, Australia. *Energies, 10*(7), 8−49.

Maclaine-Cross, I. L., & Banks, P. J. (1972). Coupled heat and mass transfer in regenerators—prediction using an analogy with heat transfer. *International Journal of Heat and Mass Transfer, 15*(6), 1225−1242.

Maclaine-Cross, I. L., & Banks, P. J. (1981). A general theory of wet surface heat exchangers and its application to regenerative evaporative cooling. *ASME Journal of Heat Transfer, 103*, 579−585.

Mavroudaki, P., Beggs, C. B., Sleigh, P. A., & Halliday, S. P. (2002). The potential for solar powered single-stage desiccant cooling in southern Europe. *Applied Thermal Engineering, 22*(10), 1129−1140.

Nelson, J. S., Beckman, W. A., Mitchell, J. W., & Close, D. J. (1978). Simulations of the performance of open cycle desiccant systems using solar energy. *Solar Energy, 21*(4), 273−278.

Ouazia, B., Barhoun, H., Daddad, K., Armstrong, M., & Szadkwski, F. (2009). Desiccant evaporative cooling system for residential building. In *Proceedings of the 12th Canadian conference on building science and technology, Montreal, Quebec* (pp. 1−12), May 6−9.

Panaras, G., Mathioulakis, E., & Belessiotis, V. (2007). Achievable working range for solid all-desiccant air-conditioning systems under specific space comfort requirements. *Energy and Buildings, 39*, 1055−1060.

Panchal, H. N., & Mohan, I. (2017). Various methods applied to solar still for enhancement of distillate output. *Desalination, 415*, 76−89.

Panchal, H. N., & Sathyamurthy, R. (2017). Experimental analysis of single-basin solar still with porous fins. *International Journal of Ambient Energy, 1*, 1−7.

Parmar, H., & Hindoliya, D. A. (2010). Performance of solid desiccant based evaporative cooling system in warm and humid climatic zone of India. *International Journal of Engineering Science and Technology, 2*(10), 5504−5508.

Persson, H., Perers, B., & Carlsson, B. (2011). Type12 and Type56: A load structure comparison in TRNSYS. In *World Renewable Energy Congress-Sweden* (pp. 3789−3796). Linkoping University Electronic Press, Sweden No. 57.

Porumb, B., Unguresan, P., Tutunaru, L. F., Şerban, A., & Bălan, M. (2016). A review of indirect evaporative cooling technology. *Energy Procedia, 85*, 461−471.

Reddy, S., Priya, S. S., Carollo, A. J., & Kumar, S. H. (2017). TRNSYS simulation for solar-assisted liquid desiccant evaporative cooling. *International Journal of Ambient Energy*, *1*, 1−7.

Sabatelli, V., Cardinale, N., Copeta, C., & Marano D. (2009). Energy performance assessment of solar-assisted desiccant cooling systems. In *Proceedings of the 4th European solar thermal energy conference, ESTEC-2009* (pp. 57−63). Munich, Germany: ESTIF.

Sharma, N. K., Tiwari, P. K., & Sood, Y. R. (2012). Solar energy in India: Strategies, policies, perspectives and future potential. *Renewable and Sustainable Energy Reviews*, *16*(1), 933−941.

Sheridan, J. C., & Mitchell, J. W. (1985). A hybrid solar desiccant cooling system. *Solar Energy*, *34*, 187−193.

Shrivastava, R. L., Kumar, V., & Untawale, S. P. (2017). Modeling and simulation of solar water heater: A TRNSYS perspective. *Renewable and Sustainable Energy Reviews*, *67*, 126−143.

Smith, R. R., Hwang, C. C., & Dougall, R. S. (1993). Modeling of solar assisted desiccant air conditioner for a residential building. In *Proceedings of the SME winter annual meeting* (pp. 409−418), November 28−December 3. New Orleans, LA, USA.

Sphaier, L. A., & Nobrega, C. E. L. (2012). Parametric analysis of component effectiveness on desiccant cooling system performance. *Energy*, *38*, 157−166.

Subramanyam, N., Maiya, M. P., & Murthy, S. S. (2004). Application of desiccant wheel to control humidity in air-conditioning systems. *Applied Thermal Engineering*, *24*, 2777−2788.

Tashtoush, B., Alshare, A., & Al-Rifai, S. (2015). Hourly dynamic simulation of solar ejector cooling system using TRNSYS for Jordanian climate. *Energy Conversion and Management*, *100*, 288−299.

Taweekun, J., & Akvanich, V. (2013). The experiment and simulation of solid desiccant dehumidification for air-conditioning system in a tropical humid climate. *Engineering*, *5*, 146−153.

TRNSYS. (2006). *TRNSYS 16, a transient system simulation program*. Madison: The Solar Energy Laboratory, University of Wisconsin-Madison,.

Tu, R., Liu, X. H., Hwang, Y., & Ma, F. (2016). Performance analysis of ventilation systems with desiccant wheel cooling based on exergy destruction. *Energy Conversion and Management*, *123*, 265−279.

Verma, V., & Murugesan, K. (2017). Experimental study of solar energy storage and space heating using solar assisted ground source heat pump system for Indian climatic conditions. *Energy and Buildings*, *139*, 569−577.

Wang, R. Z., Ge, T. S., Chen, C. J., Ma, Q., & Xiong, Z. Q. (2009). Solar sorption cooling systems for residential applications: Options and guidelines. *International Journal of Refrigeration*, *32*(4), 638−660.

White, S. D., Kohlenbach, P., & Bongs, C. (2009). Indoor temperature variations resulting from solar desiccant cooling in a building without thermal backup. *International Journal of Refrigeration*, *32*, 504−695.

William, G. E., Mohamed, M. H., & Fatouh, M. (2015). Desiccant system for water production from humid air using solar energy. *Energy*, *90*, 1707−1720.

Yadav, A., & Bajpai, V. K. (2013). An experimental investigation of solar powered desiccant wheel with different rotational speeds. *International Journal of Ambient Energy*, *34*, 3−26.

Zhang, H., & Niu, J. L. (1999). Two-stage desiccant cooling system using lowtemperature heat. *Building Services Engineering Research & Technology*, *20*, 51−55.

Chapter 17

Recent optimal power flow algorithms

Mahrous A. Taher[1], Salah Kamel[1] and Francisco Jurado[2]

[1]Department of Electrical Engineering, Faculty of Engineering, Aswan University, Aswan, Egypt, [2]Department of Electrical Engineering,
University of Jaén, Jaén, Spain

17.1 Introduction

The contemporary life of human requires electricity supply and experience different demands. Optimum power flow (OPF) is defined as the tool that is capable of solving the intended objective functions (Obj Fun) through finding solutions for the control variables or dependant variables with obeying different constraints. Recently, OPF is considered as an essential tool even to plan and operate a power system. The considerable aim of OPF is finding the right amendment of control variables to optimize particular Obj Fun with obeying the operation constraints sufficiently at predefined loading and system regulations (Chowdhury & Rahman, 1990; Huneault & Galiana, 1991; Shaheen, El-Sehiemy, & Farrag, 2016). OPF has been employed to adjust both of the produced active power, reactive power, and the voltage at the terminals of generators. In addition, OPF has been employed to tune the taps of transformer, adjust shunt compensators and various control variables that arranged previously for the enhancement of power system prerequisites. Hence, various objective functions can be achieved such as; minimization of the produced fuel cost (FC), real power losses, and promote voltage stability and voltage profile. All these objectives are attained with satisfying the operation constraints, such as load bus voltage, load flow, and different state variables. It can be considered that the optimization methods give the computer orders to seek for solving an optimization problem. Therefore the essential merits of these methods are the least mistakes, lower personal participation, and less cost for designing (Mirjalili, Jangir, Mirjalili, Saremi, & Trivedi, 2017). The premier generation of optimization algorithms is characterized by the utilization of coming down of gradient (Ochs, Ranftl, Brox, & Pock, 2016), where the derivative of a problem starts from a primary point is obtained. Two demerits of the methods that depend upon gradient are reported: the first demerit is most of actual problems suffering from anonymous derivation of the equations and the second demerit is that the coming down of gradient may procure to local optima with the dependence upon primary solution, which is known as local optima stagnation (Du & Swamy, 2016). A number of traditional methods have been proposed, such as nonlinear programming (Dommel & Tinney, 1968), linear programming (Mota-Palomino & Quintana, 1986), quadratic programming (Burchett, Happ, & Vierath, 1984), Newton's method (Santos & Da Costa, 1995), and interior point (Yan & Quintana, 1999). Various heuristic or artificial intelligence or evolutionary techniques that are characterized by accurate and fast solutions have been proposed for solving OPF problems. Examples of these techniques are genetic algorithm (GA) (Chung & Li, 2001; Paranjothi & Anburaja, 2002), differential evolution (El Ela, Abido, & Spea, 2010; Varadarajan & Swarup, 2008), particle swarm optimization (PSO) (Abido, 2002), and modified grasshopper optimization algorithm (Taher, Kamel, Jurado, & Ebeed, 2019b). GA is considered as a famous evolution algorithm, which mimics the method of selecting the cream offspring that represent the auspicious individualist from the consecutive generations. Evolutionary techniques can perform particular power flow for each candidate during execution of the intended Obj Fun for checking that the constraints are feasible using penalty term that is added to the Obj Fun (Reddy & Bijwe, 2016). To solve the OPF problem, it is essential to use two contrasted definitions known as exploration and exploitation. Exploration refers to the capability to estimate nominee solutions that are neighbor to the existing solution; however, the exploitation refers to the capability to estimate nominee solutions that are close to the solution (Eiben & Schippers, 1998). This chapter aims to assess valuable knowledge about solving OPF problems using various recent evolutionary methods. In this chapter, the comparison depends upon

Renewable Energy Systems. DOI: https://doi.org/10.1016/B978-0-12-820004-9.00026-7

previously published comprehensive work have been done to solve OPF problems using both moth-flame optimization (MFO) and improved moth-flame optimization (IMFO) (Taher, Kamel, Jurado, & Ebeed, 2019a). Also, a comprehensive work has been done to solve OPF problems using moth swarm algorithm (MSA) (Mohamed, Mohamed, El-Gaafary, & Hemeida, 2017). However, the same algorithm is implemented in this work with different population sizes and maximum iteration number and new work for MSA. Both multiverse optimization (MVO) (Mirjalili, Mirjalili, & Hatamlou, 2016) and wale optimization algorithm (WOA) (Mirjalili & Lewis, 2016) algorithms are newly implemented as recent methods in this work to extend the comparison. The comparison is held between the aforementioned methods and the well-known evolutionary techniques, including GA, PSO, and teaching learning-based optimization (TLBO) to solve the same problems under the same conditions. In this chapter, single Obj Fun is implemented to solve OPF problems for the minimization of FC, FC with valve-point loadings (VPL), emission, and power loss. In addition, multi-Obj Fun are implemented to solve OPF problems for the minimization of FC with voltage stability (L_{max}) enhancement, FC with active power losses, FC with emission, FC with active power losses, and FC with voltage deviation (VD). This chapter contributes the following items:

- Implement MFO, IMFO and MSA to avoid local optima stagnation and enhance the convergence characteristics (CCs) of the traditional methods.
- Both WOA and MVO are new implemented as recent methods in this work to extend the comparison
- Various single and multi-Obj Fun are considered to emphasize the ability of the recent optimization methods for solving different kinds of OPF problems.
- Comparison among recent optimization methods and other renowned optimization methods is held to emphasize the effectiveness of the recent optimization methods.
- Standard IEEE 30-bus test system is used to validate and prove the efficiency and robustness recent optimization methods to solve OPF problems.

The rest of this chapter is arranged as follows: Section 17.2 describes MFO and IMFO algorithms. Section 17.3 describes MSA algorithm. Section 17.4 describes MVO algorithm. Section 17.5 describes WOA algorithm. Section 17.6 explains the Obj Fun formulation. Section 17.7 discusses the acquired results. Section 17.8 outlines the conclusion of this work.

17.2 Moth-flame optimization technique

MFO (Mirjalili, 2015) is considered as one of the most efficient and effective optimization techniques that is newly introduced. MFO is a population dependent method that based on the concept of simulatating the motion of moth's at night. The motion type of moths is known as transverse orientation, where it moves with stationary angle correlating to moon due to large space between them. The actual motion of moths correlating to artificial light source or flame follows the same concept with moving in spiral path.

17.2.1   Mathematical representation of moth-flame optimization

Updating of every moth related to flame can be represented by the following spiral functions:

$$Mi = S(Mi, Fj) \tag{17.1}$$

where S refers to spiral function. Mi and Fj refers to the moth of the order i and flame of the order j.

Moth improves its position related to flame through logarithmic spiral function that is expressed by this equation:

$$S(Mi, Fj) = Di \times e^{bt} \times \cos(2\pi t) + Fj \tag{17.2}$$

where Di represents space that separates Mi from Fj.

The following equation represents Di:

Di can be represented by the following equation:

$$Di = |Fj - Mi|) \tag{17.3}$$

17.2.2   Improved moth-flame optimization concept

IMFO is a modified version of MFO by Soliman, Khorshid, and Abou-El-Enien (2016). IMFO follows the same idea of MFO accompanied by amending of moth's path surrounding flame. The main equation of MFO (Eq. 17.2) is reformulated to represent the IMFO algorithm.

17.2.3   Improved moth-flame optimization mathematical formulation

The equation that represents the motion of moths is hyperbolical spiral function, which composed of trigonometric functions of sine and cosine. This equation is considered as surpassing function type, which is composed of more than one function; hyperbolic function is one of this family. Hyperbolic function is the inverse of Archimedean spiral. The base function of IMFO is formulated as follows:

$$S(Mi, Fj) = Di \times t \times \cos(2\pi t) + Fj \qquad (17.4)$$

17.3   Moth swarm algorithm

17.3.1   Inspiration

MSA follows the same concept of MFO; however, the movement of moths is subjected to transverse direction because of fruitless of spiral routes that are close to artificial lighting sources (Frank, 1988; Gaston, Bennie, Davies, & Hopkins, 2013). The expected solution of optimization problem using MSA is considered as the location of artificial lighting source and its strength represents the best solution. Also, the moth swarm is classified into three categories that can be outlined as follows:

Pathfinders: It is a little group of moths (n_p), which can explore new regions through the search space. The role of pathfinders is to distinguish the best locations that are lighting sources orientating the essential bevy.
Prospectors: This moth's class moves randomly in spiral route close to lighting sources.
Onlookers: It is a group of moths, which drifted in direct orientation to the moon lighting, hence this group is considered as the best obtained solution so far.

17.3.2   Mathematical modeling of moth swarm algorithm

To represent MSA mathematically in brief, the following steps are explained:

17.3.2.1   Reconnaissance phase

The pathfinder moths are executing important task for improving the exploration process, which negatively affected with the concentration of moths in areas causing stagnation into local optima. To avoid premature convergence, some swarms are forced to explore the areas which characterized by the least congestion.

17.3.2.1.1   Suggested diversity index

It is suggested a new scheme for the diversity index to find crossover points. All members of pathfinder moths are subjected to lower class of dispersal and will be considered with crossover points. The set of crossover points varies in a dynamic manner with the proceeding of the scheme.

17.3.2.1.2   Lévy flights

Lévy flights are considered as unplanned scheme founded upon α stable distribution that can fly for long span with various step sizes.

17.3.2.1.3   Difference vectors Lévy mutation

The group of alternative indicators (r^1, r^2, r^3, r^4, r^5, p) is chosen from pathfinders only.

17.3.2.1.4   Suggested acclimatized crossover process

To finalize the tentative solution, the pathfinders that considered as the steward vector updates their location via the crossover process by combining the mutated vectors with less dispersion to its identical variables of steward vector.

17.3.2.1.5 Selection strategy

Comparison is held between the congruous and steward solutions. Filtering process is made for exceeding to the following peer group and is represented according to the following equation:

$$P_p = \frac{fit_p}{\sum_{p=1}^{n_p} fit_p} \tag{17.5}$$

The luminescence strength can be computed from the Obj Fun f_b, which is represented as follows:

$$fit_p = \begin{cases} \dfrac{fit_p}{1+f_p} & \text{for } f_p \geq 0.5 \\ 1 + |f_p| & \text{for } f_p < 0.5 \end{cases} \tag{17.6}$$

17.3.2.2 Transverse orientation

The category of moths that is capable of reaching to the following best luminescence strength is selected as the prospectors. The count of prospectors n_f through the iterations T is decreasing according to the following equation:

$$n_f = \text{round}\left((n - n_p) \times \left(1 - \frac{t}{T}\right)\right) \tag{17.7}$$

17.3.2.3 Heavenly navigation

Through the progress of optimization, the count of prospectors decreases leading to increasing the count of onlookers, which is probable to make MSA faster. Onlookers' moths are characterized by their lower luminescent, so that they tend to move into the moon. In MSA, onlookers are obliged to seek more of the prospector. Onlookers are classified into the next two schemes.

17.3.2.3.1 Gaussian walks

Gaussian stochastic distribution is characterized by its capability to find regions that are favorable for best solutions

17.3.2.3.2 Assistive educating scheme with instant recollection

Evolution of moth's memory is poor so they are fall into light source or fire. The social relationship between moths affects the poor communication between them (Fan, Anderson, & Hansson, 1997; Skiri, Stranden, Sandoz, Menzel, & Mustaparta, 2005). The second category of onlookers are subjected to driftage in the direction of moon lighting, this process based on the concept of social learning operators that possess instant memory which capable of simulating the realistic attitude of moths. The equation that simulates this update can be formulated as follows:

$$X_i^{t+1} = X_i^t + 0.001 \times G\left[X_i^t - X_i^{min}, X_i^{max} - X_i^t\right] + \left(1 - \frac{g}{G}\right) \times r_1 \times \left(best_p^t - X_i^t\right)$$
$$+ \frac{2g}{G} \times r_2 \times \left(best_g^t - X_i^t\right) \tag{17.8}$$

17.4 Multiverse optimization

17.4.1 Inspiration

Multiverse (MV) is related to the presence of another universe along with our known universe (Tegmark, 2004). These universes react with themselves, where collision may be occurred among them. MVO is mainly inspired from the MV theory including both:

1. *White hole*: Nobody can see it where scientists believe that white hole is occurred when the big bang had been created our universe. Big bangs and white holes are formed when parallel universes collide to each other (Eardley, 1974; Steinhardt & Turok, 2002)

2. *Black holes*: Occurrence is repeatedly. They have the capability to attract everything even the light by their massive force (Davies, 1978).
3. *Wormholes* are similar to the channel for connection between universes. For MVO white and black holes are implemented for exploration for new universes, whereas wormholes are implemented for exploitation.

17.4.2   Mathematical modeling of multiverse optimization

Roulette wheel is implemented for the modeling of exchanging universe objects. Through course of iterations, universes are sorted with respect to the enlarge rates, and then a universe is chosen for processing white hole using roulette wheel.

The universe can be represented by the following matrix:

$$U = \begin{bmatrix} X_1{}^1 & \cdots & X_1{}^d \\ \vdots & \ddots & \vdots \\ X_n{}^1 & \cdots & X_n{}^d \end{bmatrix} \tag{17.9}$$

where d represents parameters' count and n represents of universe's count.

$$X_i{}^j = \begin{cases} X_k{}^j \text{ for } r1 < NI(Ui) \\ X_i{}^j \text{ for } r1 \geq NI(Ui) \end{cases} \tag{17.10}$$

where $X_k{}^j$ represents the parameter of the order j for the universe of the order Ui, $NI(Ui)$ refers to normalize inflation rate for Ui, $r1$ is a number that varies randomly within [0, 1], and $X_i{}^j$ represents the parameter of the order j for the universe of the order k.

When the inflation rate has lower values, the expected transmission of objects via white/black hole is increased.

To improve the inflation rate, wormhole is created between a universe and the best solution. This scheme can be formulated as follows:

$$X_i{}^j = \begin{cases} \begin{cases} X_j + \text{TDR} \times ((ub_j - lb_j) \times X_j + lb_j) & r3 < 0.5 \\ X_j + \text{TDR} \times ((ub_j - lb_j) \times X_j + lb_j) & r3 \geq 0.5 \end{cases} r2 < \text{WEP} \\ X_i{}^j \quad r2 \geq \text{WEP} \end{cases} \tag{17.11}$$

where X_j represents the parameter of the order j for the best solution universe, traveling distance rate (TDR), and wormhole existence probability (WEP), and $r2$, $r3$, and $r4$ are values that vary randomly within [0, 1].

The values of WEP and TDR are increased with the proceeding of optimization process to gain accurate exploitation. Acclimated formulation for WEP and TDR are represented as follows:

$$\text{WEP} = min + l \times \left(\frac{max - min}{L} \right) \tag{17.12}$$

$$\text{TDR} = 1 - \frac{l^{1/p}}{L^{1/p}} \tag{17.13}$$

where min and max refer to minimum and maximum (their values are chosen to be 0.2 and 1.0, respectively), l and L refer to current and maximum iterations, respectively, and p refers to the precision of exploitation (its value is chosen to be 6.0).

17.5   Wale optimization algorithm

17.5.1   Inspiration

Whales are deemed as the largest animal size on the earth planet due to their huge length and size. This animal is classified to seven types (Hof & Van der Gucht, 2007). The whalebone of humpback type is very big and its adult's size is analogous to autobus. Humpback follows a tactic known as bubble net for hunting the food (Watkins & Schevill, 1979). Hunting process is executed near the superficies of water through formation of large quantum of bubbles shapes as the nine numbers (Goldbogen et al., 2013). Mathematical representation of WOA is outlined in brief in the next subsection.

17.5.2   Mathematical modeling of wale optimization algorithm

To represent the hunting process of WOA mathematically in brief, the following steps are explained as follows:

17.5.2.1   Circling prey

First, the location of the prey is defined and the whales encircle it; this location represents the best solution that is near to the global best solution inside. The search agents can circle the victim in multidimensional search space.

17.5.2.2   Bubble-net attacking method

This stage can be executed according to the following schemes:

- Dwindling circling scheme

  This scheme is carried out by updating the search agents of their positions from the main to new location of the best agent.

- Update location in spiral path

  This path of movement is expressed as follows:

$$\overrightarrow{X}(t+1) = \vec{D'} \times e^{bl} \times \cos(2\pi l) + \vec{X}*(t) \tag{17.14}$$

$$\vec{D'} = \left| \vec{X}*(t) - \vec{X}(t) \right| \tag{17.15}$$

where D' represents spacing between the victim and best solution, b denotes the spiral motion form, and l is a random value lies in the interval $[-1, 1]$. The probability of option for spiral path and circling scheme is similar so that the spiral motion can be represented according to the following equation:

$$\vec{X}(t+1) = \begin{cases} \vec{X}*(t) - \vec{A} \times \vec{D} & \text{if } p < 0.5 \\ \vec{D'} \times e^{bl} \times \cos(2\pi l) + \vec{X}*(t) & \text{if } p \geq 0.5 \end{cases} \tag{17.16}$$

The subscript p is a random value lies within $[-1, 1]$.

17.5.2.3   Search for prey

Searching process of the humpback whales for the prey deems the exploration milestone that is executed by varying the values for vector $\vec{A}$. The random searching is based on the location of the search agents related to themselves. Exploitation milestone is executed by varying location of the search agents related to the best solution. Mathematical representation of this model is represented according to the following equations:

$$\vec{X}(t+1) = \left| \vec{C} \times \overrightarrow{X_{rand}}(t) - \vec{X} \right| \tag{17.17}$$

$$\vec{X}(t+1) = \overrightarrow{X_{rand}} - \vec{A} \times \vec{D} \tag{17.18}$$

where $\overrightarrow{X_{rand}}$ represents a randomly changed position vector related to whale that is selected in random regime.

17.6   Objective functions

OPF aims to solve power flow equations and find the independent variables. An Obj Fun is selected; the best solutions of this function are computed with satisfying the required constraints that can be expressed as follows (Sivasubramani & Swarup, 2011a, 2011b; Varadarajan & Swarup, 2008):

$$MinF(x, u) \tag{17.19}$$

Subjected to

$$g_j(x, u) = 0 \, j = 1, 2, \ldots, m \tag{17.20}$$

$$h_j(x, u) \leq 0 \, j = 1, 2, \ldots, p \tag{17.21}$$

where $F(x, u)$ is the Obj Fun, x and u are the state and control variables, $g(x, u)$ and $h(x, u)$ are the equality and inequality constraints, and m and p are the number of equality and inequality constraints, respectively. In this work, the proposed algorithms effectuate the chosen Obj Fun that required to implement the selected cases.

17.6.1 Single objective function

In this work, four cases of single Obj Fun are studied that can be outlined as follows.

17.6.1.1 Quadratic fuel cost

This case aims to minimize the quadratic FC, and its computation is assigned according to the next polynomial equation (Yuryevich & Wong, 1999):

$$f = \left(\sum_{i=3}^{NG} a_i P_{Gi}^2 + b_i P_{Gi} + C_i \right) + \text{penalty}(\$ \ h^{-1}) \tag{17.22}$$

where a_i, b_i, and c_i are the cost coefficients for the generator number i that can be obtained from Table 17.A3. NG represents the count of generators.

17.6.1.2 Optimum power flow for fuel cost with valve-point loadings

For both generator 1 and generator 2, the FC is deemed with the VPL due to multifuel operation as a separate term from the reminder generators. For thermal power plants that basically depend upon steam for energy production process, it is essential to consider multivalve that mainly obeys the rule of ripple-like effect of the generator heat rate curve (Hardiansyah, 2013). The Obj Fun for this case is represented according to the following equation (Alsumait, Sykulski, & Al-Othman, 2010):

$$f = \left(\sum_{i=1}^{2} a_i P_{Gi}^2 + b_i P_{Gi} + C_i + \left| d_i \sin\left(e_i\left(P_{Gi}^{min} - P_{Gi}\right)\right) \right| \right) + \left(\sum_{i=3}^{NG} a_i P_{Gi}^2 + b_i P_{Gi} + C_i \right) + \text{penalty} \tag{17.23}$$

where d_i and e_i represent the cost coefficients of VPL that are found in Table 17.A3.

17.6.1.3 Optimum power flow for emission

This case aims to solve OPF problem to reduce the emitted gases to the environment as a result of gases' emission from thermal power plants mainly dependent on fossil combustion. The essential pollutants are carbon dioxide (CO_2), nitrogen oxides (NO_x), and sulfur oxide (SO_x). The intended function is expressed according to the following equation (Abido, 2003):

$$f = \sum_{i=1}^{NG} \gamma_i P_{Gi}^2 + \beta_i P_{Gi} + \alpha_i + \zeta_i e^{(\lambda_i P_{Gi})} + \text{penalty} \ (\text{ton} \ h^{-1}) \tag{17.24}$$

where γ_i, β_i, α_i, ζ_i, and λ_i are the emission coefficients of the ith generator that can be obtained from Table 17.A3.

17.6.1.4 Optimum power flow for power loss minimization

The goal of this case is to solve OPF problem for minimizing the real power loss through the transmission lines (TLs). The intended function is expressed according to the following equation:

$$f = \sum_{i=1}^{nl} \sum_{j=1}^{nl} \left(G_{ij} V_i^2 + V_i^2 - 2V_i V_i \cos(\delta_i - \delta_j) \right) + \text{penalty} \tag{17.25}$$

17.6.2 Multiobjective function

The implemented multi-objective function aims to minimize both of FC, emission, total power system loss, and enhancing voltage profile. Formulation of this function is achieved in a mathematical form where it is transformed to single Obj Fun according to the following equation (Agrawal, Bharadwaj, & Kothari, 2016):

$$MinF = Min(F_1, F_2, F_3, \ldots F_n) \tag{17.26}$$

where F_1, $F_2, \ldots$, F_n represent the required Obj Fun. Equality and inequality constraints are taken into consideration.

17.6.2.1 Optimum power flow for fuel cost with voltage stability index

Voltage stability indicator L-index or L_{max} is calculated to indicate that all voltage buses are still within the allowable limits for operation. As the value of L_{max} is reduced much more, the system is far from voltage collapse. The goal of this case is to solve the OPF problem to optimize both FC and L_{max}, where both of them are minimized simultaneously. The intended function is expressed according to the following equation:

$$f = \left(\sum_{i=1}^{NG} a_i P_{Gi}^2 + b_i P_{Gi} + C_i) \right) + \lambda_{L_{max}} \max(L_J) + \text{penalty} \tag{17.27}$$

where λ_L is the weighting factor that is chosen to indicate the significance of L_{max} related to FC. Its value is set to 100.

17.6.2.2 Optimum power flow for fuel cost with emission

Emission gases' concept is aforementioned explained. The goal of this case is to solve the OPF problem to optimize both FC and emission where both of them are minimized simultaneously. The intended function is expressed according to the following equation:

$$f = \left(\sum_{i=1}^{NG} a_i P_{Gi}^2 + b_i P_{Gi} + C_i \right) + \lambda_i(\text{Emission}) + \text{penalty} \tag{17.28}$$

where λ_E is the weighting factor that is chosen to indicate the significance of emission gases related to FC. Its value is set to 1000.

17.6.2.3 Optimum power flow for fuel cost with active power losses

The active power transmission loss is a function of bus voltage magnitude and its phase angle. In this case, both FC and total power loss are minimized according to the following multi-Obj Fun:

$$f = \left(\sum_{i=1}^{NG} a_i P_{Gi}^2 + b_i P_{Gi} + C_i \right) + \lambda_P \sum_{i=1}^{nl} \sum_{j=1}^{nl} \left(G_{ij} V_i^2 + V_i^2 - 2V_i V_i \cos(\delta_i - \delta_j) \right) + \text{penalty} \tag{17.29}$$

where λ_P is the weighting factor that is chosen to indicate the significance of power loss related to FC. Its value is set to 40.

17.6.2.4 Optimum power flow for fuel cost with voltage deviation

VD can be minimized through adjusting load buses' voltages close to 1.0 p.u. This case aims for solving OPF problem to optimizing FC and VD, where both of them are minimized simultaneously (Yuryevich & Wong, 1999). The intended function is expressed according to the following equation (He et al., 2004):

$$f = \left(\sum_{i=1}^{NG} a_i P_{Gi}^2 + b_i P_{Gi} + C_i) \right) + \lambda_{VD} \sum_{i=1}^{NG} |V_{Li} - 1| + \text{penalty} \tag{17.30}$$

where λ_{VD} represents the weighting factor that is chosen to indicate the significance of VD related to FC. Its value is set to 100.

17.6.3 Constraints

The power system-operating constraints can be outlined in brief as follows.

17.6.3.1 State variables

The state variables x represent the inner rendition of the operating system that entirely depicts the state of the system and enough to calculate the control variables that the state variables can be formulated as follows:

$$x = [P_{G1}, V_{L1} \ldots V_{LNPQ}, Q_{G1} \ldots Q_{GNPV}, S_{TL1} \ldots S_{TLNTL}] \tag{17.31}$$

where P_{G1} is the real power of the slack bus, V_L is the voltage magnitude of PQ buses, Q_G is the reactive power of generators, and S_{TL} is the apparent power of the TLs, NPQ is the count of PQ buses, NPV is the count of PV buses, and NTL is the count of TLs.

17.6.3.2 Control variables

The control variables u can be outlined in brief as follows:

$$u = [P_{G2} \ldots P_{GNG}, V_{G1} \ldots V_{GNG}, Q_{C1} \ldots Q_{CNC}, T_1 \ldots T_{NT}] \tag{17.32}$$

where P_G is the real power of generators, V_G is the voltage magnitude of PV buses, Q_C is the reactive power of VAR (volt ampere reactive) compensators, T is the tap settings of regulating transformer, NG is the count of generators, NC is the count of VAR compensators, and NT is the count of transformers.

17.6.3.3 Operating constraints

To operate the power system safely and continuously, it is restricted with specified constraints. These constraints are mainly classified into two types as outlined in the following subsection.

17.6.3.4 Equality constraints

These are the continuous equations of load flow boundaries that can be expressed as follows:

$$P_{Gi} - P_{Di} = |V_i| \sum_{i=1}^{NB} |V_j| (G_{ij}\cos\delta_{ij} + B_{ij}\sin\delta_{ij}) \tag{17.33}$$

$$Q_{Gi} - Q_{Di} = |V_i| \sum_{j=1}^{NB} |V_j| (G_{ij}\sin\delta_{ij} - B_{ij}\cos\delta_{ij}) \tag{17.34}$$

where NB is the count of buses, P_{Di} is the real power demand load, Q_{Di} is the demanded reactive power load, G_{ij} is the conductance between bus i and bus j, and B_{ij} is the susceptance between bus i and bus j, respectively.

17.6.3.5 Inequality constraints

These boundaries can be outlined in brief as follows:

1. Generation constraints

    These are the generators limits.

$$P_{Gi}^{min} \leq P_{Gi} \leq P_{Gi}^{max} \quad i = 1, 2, \ldots, NPV \tag{17.35}$$

$$V_{Gi}^{min} \leq V_{Gi} \leq V_{Gi}^{max} \quad i = 1, 2, \ldots, NPV \tag{17.36}$$

$$Q_{Gi}^{min} \leq Q_{Gi} \leq Q_{Gi}^{max} \quad i = 1, 2, \ldots, NPV \tag{17.37}$$

2. Transformer constraints

    These are the transformers limits.

$$T_i^{min} \leq T_i \leq T_i^{max} \quad i = 1, 2, \ldots, N \tag{17.38}$$

3. Shunt VAR compensators constraints

    These are the limits of the injected shunt VAR compensators.

$$Q_{Ci}^{min} \leq Q_{Ci} \leq Q_{Ci}^{max} \quad i = 1, 2, \ldots, NC \tag{17.39}$$

4. Security constraints

    Security constraints include the equations of power flow through the TLs and the voltage magnitude limits of load buses.

$$S_{Li} \leq S_{Li}^{max} \quad i = 1, 2, \ldots, NTL \tag{17.40}$$

$$V_{Li}^{min} \leq V_{Li} \leq V_{Li}^{max} \quad i = 1, 2, \ldots, NPQ \tag{17.41}$$

$$\text{penalty} = K_P \left(P_{G1} - P_{G1}^{lim} \right)^2 + K_Q \sum_{i=1}^{NPQ} \left(Q_{Gi} - Q_{Gi}^{lim} \right)^2 + K_S \sum_{i=1}^{NTL} \left(S_{Li} - S_{Li}^{lim} \right)^2 + \text{penalty} \tag{17.42}$$

where K_P, K_Q, K_V, and K_S represent penalty factors. Control variables are limited by x^{lim}, which can be expressed as follows:

$$x^{lim} = \begin{cases} x^{max} \text{if} x > x^{max} \\ \qquad . \\ 0 \text{if} x < x^{max} \end{cases} \qquad (17.43)$$

17.7 Results and discussions

The IEEE 30-bus test system is chosen for proving the prevalence and performance of IMFO, MFO, MSA, MVO, WOA, GA, PSO, and TLBO algorithms. Also, the capability and performance of these algorithms is checked to accomplish promising OPF solutions with rapid convergence. Programming of the implemented methods in this work is carried out by MATLAB 2014a code on a PC that have the following properties: core i7 processor, 2.40 GHz, and 4 GB RAM. In this work, the convergence toleration is 10^{-6}, 100 MVA base. The system demanded real power is 283.4 MW, whereas the reactive power is 126.2 MVAR. System data are indicated in Alsac and Stott (1974). The search agent's number is chosen as 50. The maximum number of iterations is selected as 500 representing the stopping criterion for all the algorithms. The boundaries of control variables for the different case studies are included in (Yao, Liu, & Lin, 1999). The distinctive features' cost and emission coefficients for IEEE 30-bus system are stipulated in Tables 17.A1, 17.A2, and 17.A3, respectively. The results obtained by the applied methods are carried out under the same conditions. Simulation results are stipulated in Tables 17.1 and 17.2. The case studies in this work are represented and discussed as follows.

17.7.1 Case 5-1: Optimum power flow for fuel cost minimization

This is considered as the base case in this work, which aims to minimizing the FC of generation. The obtained OPF solutions are found in column 3 in Table 17.1. The results obtained by IMFO, MFO, MSA, MVO, WOA, GA, PSO, and TLBO algorithms are given in Table 17.1. The value of FC by IMFO is 800.3848 \$ h^{-1}, this is the smallest value with respect to values obtained by other methods. The values for MFO, MSA, MVO, WOA, GA, PSO, and TLBO are 800.6206, 800.7096, 800.7054, 800.7096, 800.4346, 800.4075, and 800.4104 \$ h^{-1}, respectively. Comparison among the CCs of the implemented methods is depicted in Fig. 17.1.

TABLE 17.1 OPF solution values obtained by different methods for Cases 5-1—5-4.

Case no.	Objective functions	Case 5-1	Case 5-2	Case 5-3	Case 5-4
IMFO	Fuel cost (\$ h^{-1})	800.3848	832.1023	944.2572	967.5900
	Emission (ton h^{-1})	0.3658	0.4382	0.20480	0.2073
	PLoss (MW)	8.9990	10.712	3.2601	3.0905
	QLoss (MVAR)	−10.638	−4.717	−30.983	−32.582
	VD (p.u.)	0.9035	0.8318	0.8881	0.9192
	L_{max} (p.u.)	0.1274	0.1281	0.1276	0.1274
MFO	Fuel cost (\$ h^{-1})	800.6206	832.3871	944.3942	967.6938
	Emission (ton h^{-1})	0.3663	0.4385	0.20483	0.2073
	PLoss (MW)	9.0790	10.800	3.254	3.139
	QLoss (MVAR)	−8.693	−2.923	−29.161	−33.041
	VD (p.u.)	0.6343	0.5417	0.9011	0.9080
	L_{max} (p.u.)	0.1310	0.1320	0.1287	0.1269

(Continued)

TABLE 17.1 (Continued)

Case no.	Objective functions	Case 5-1	Case 5-2	Case 5-3	Case 5-4
MSA	Fuel cost ($ h^{-1})	800.7054	832.4666	946.1300	967.7231
	Emission (ton h^{-1})	0.3659	0.4373	0.2049	0.2073
	PLoss (MW)	9.0877	10.7357	3.5826	3.1565
	QLoss (MVAR)	−11.937	−4.680	−29.886	−31.280
	VD (p.u.)	0.6842	0.8979	0.5748	0.6900
	L_{max} (p.u.)	0.1294	0.1263	0.1279	0.1293
	Fuel cost ($ h^{-1})	800.8067	831.8653	945.0071	959.3626
	Emission (ton h^{-1})	0.3589	0.4192	0.2049	0.2086
	PLoss (MW)	8.8505	10.4230	3.4333	3.3045
	QLoss (MVAR)	−11.975	−4.281	−27.497	−31.530
	VD (p.u.)	0.9105	0.3761	0.3568	0.8015
	L_{max} (p.u.)	0.1277	0.1348	0.1337	0.1307
WOA	Fuel cost ($ h^{-1})	800.7096	832.975	944.8076	967.7669
	Emission (ton h^{-1})	0.3676	0.4360	0.2049	0.2073
	PLoss (MW)	9.1348	10.8149	3.3713	3.1765
	QLoss (MVAR)	−11.167	1.441	27.074	−28.539
	VD (p.u.)	0.7692	0.3661	0.3766	0.8958
	L_{max} (p.u.)	0.1290	0.1405	0.1300	0.1311
GA	Fuel cost ($ h^{-1})	800.4346	832.4400	944.4295	967.6486
	Emission (ton h^{-1})	0.3683	0.4369	0.20483	0.2073
	PLoss (MW)	9.037	10.6479	3.266	3.118
	QLoss (MVAR)	−11.221	−6.131	−28.683	−31.490
	VD (p.u.)	0.9155	0.7282	0.8551	0.8325
	L_{max} (p.u.)	0.1275	0.1289	0.1280	0.1293
PSO	Fuel cost ($ h^{-1})	800.4075	832.1095	944.4481	967.6066
	Emission (ton h^{-1})	0.3660	0.4373	0.20483	0.2073
	PLoss (MW)	9.006	10.652	3.263	3.103
	QLoss (MVAR)	−11.924	−6.535	−30.898	−32.117
	VD (p.u.)	0.9129	0.8627	0.8815	0.9202
	L_{max} (p.u.)	0.1276	0.1290	0.1271	0.1277
TLBO	Fuel cost ($ h^{-1})	800.4104	832.0944	944.3278	967.5841
	Emission (ton h^{-1})	0.3663	0.4378	0.20482	0.2073
	PLoss (MW)	9.008	10.673	3.221	3.088
	QLoss (MVAR)	−10.326	−6.948	−30.000	−30.316
	VD (p.u.)	0.9068	0.8637	0.8988	0.9029
	L_{max} (p.u.)	0.1269	0.1280	0.1271	0.1274

GA, genetic algorithm; IMFO, improved moth-flame optimization; MFO, moth-flame optimization; MSA, moth swarm algorithm; OPF, optimum power flow; PSO, particle swarm optimization; TLBO, teaching learning-based optimization; VD, voltage deviation; WOA, wale optimization algorithm.

TABLE 17.2 OPF solution values obtained by different methods for Cases 5-5—5-8.

Case no.	Objective functions	Case 5-5	Case 5-6	Case 5-7	Case 5-8
IMFO	Fuel cost ($ h^{-1})	800.5392	836.4809	858.7455	803.6000
	Emission (ton h^{-1})	0.3655	0.2423	0.2291	0.3638
	PLoss (MW)	9.0190	5.374	4.582	9.8239
	QLoss (MVAR)	−12.403	−23.011	−25.144	−3.813
	VD (p.u.)	0.9305	0.9206	0.9559	0.0969
	L_{max} (p.u.)	0.1257	0.1280	0.1274	0.1376
MFO	Fuel cost ($ h^{-1})	800.9415	836.5734	858.9658	803.7983
	Emission (ton h^{-1})	0.3700	0.2422	0.2289	0.3628
	PLoss (MW)	9.222	5.400	4.556	9.8536
	QLoss (MVAR)	−4.768	−20.117	−26.216	−2.440
	VD (p.u.)	0.7789	0.9375	0.9549	0.1000
	L_{max} (p.u.)	0.1266	0.1279	0.1258	0.1390
MSA	Fuel cost ($ h^{-1})	800.7533	837.2442	855.8769	803.7048
	Emission (ton h^{-1})	0.3649	0.2421	0.2337	0.3579
	PLoss (MW)	9.0688	5.5191	4.8953	9.6512
	QLoss (MVAR)	−11.760	−20.665	−25.775	−6.184
	VD (p.u.)	0.7375	0.8447	0.3939	0.1126
	L_{max} (p.u.)	0.1256	0.1267	0.1335	0.1372
MVO	Fuel cost ($ h^{-1})	800.9656	835.7207	859.0555	803.9179
	Emission (ton h^{-1})	0.3668	0.2438	0.2357	0.3523
	PLoss (MW)	9.0560	5.5504	4.8403	9.5487
	QLoss (MVAR)	−11.455	−22.912	−25.224	−7.001
	VD (p.u.)	0.8775	0.4342	0.5917	0.1310
	L_{max} (p.u.)	0.1288	0.1341	0.1308	0.1378
WOA	Fuel cost ($ h^{-1})	800.8985	836.2389	856.2373	804.2670
	Emission (ton h^{-1})	0.3671	0.2428	0.2284	0.3521
	PLoss (MW)	9.1477	5.4763	4.7937	9.5013
	QLoss (MVAR)	−9.798	−23.267	−25.134	−2.562
	VD (p.u.)	0.8061	0.5718	0.4713	0.1845
	L_{max} (p.u.)	0.1280	0.1306	0.1341	0.1381
GA	Fuel cost ($ h^{-1})	800.4385	836.5425	859.0075	803.2347
	Emission (ton h^{-1})	0.3662	0.2422	0.2288	0.3644
	PLoss (MW)	9.013	5.396	4.5501	9.747
	QLoss (MVAR)	−10.543	−25.181	−27.497	−3.465
	VD (p.u.)	0.9272	0.8697	0.9339	0.1018
	L_{max} (p.u.)	0.1254	0.1273	0.1272	0.1374

(Continued)

TABLE 17.2 (Continued)

Case no.	Objective functions	Case 5-5	Case 5-6	Case 5-7	Case 5-8
PSO	Fuel cost ($ h^{-1})	800.5815	836.4357	858.8508	803.4736
	Emission (ton h^{-1})	0.3662	0.2423	0.2290	0.3639
	PLoss (MW)	9.056	5.365	4.537	9.821
	QLoss (MVAR)	−11.098	−25.595	−28.226	−2.899
	VD (p.u.)	0.8792	0.9296	0.9320	0.0978
	L_{max} (p.u.)	0.1280	0.1268	0.1272	0.1371
TLBO	Fuel cost ($ h^{-1})	800.4738	836.4141	859.0236	803.5675
	Emission (ton h^{-1})	0.3658	0.2423	0.2289	0.3636
	PLoss (MW)	9.012	5.363	4.5313	9.813
	QLoss (MVAR)	−13.075	−25.067	−27.161	−3.110
	VD (p.u.)	0.9432	0.9314	0.9317	0.0939
	L_{max} (p.u.)	0.1247	0.12741	0.1271	0.1369

GA, genetic algorithm; IMFO, improved moth-flame optimization; MFO, moth-flame optimization; MSA, moth swarm algorithm; MVO, multiverse optimization; OPF, optimum power flow; PSO, particle swarm optimization; TLBO, teaching learning-based optimization; VD, voltage deviation; WOA, wale optimization algorithm.

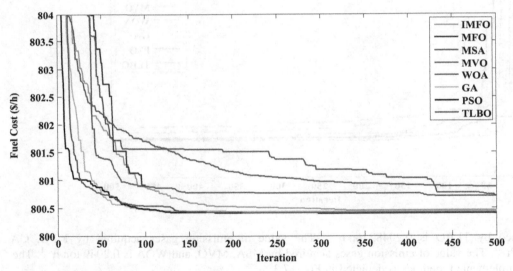

FIGURE 17.1 Convergence characteristics of various techniques for Case 5-1.

17.7.2 Case 5-2: Optimum power flow for minimization of quadratic fuel cost with valve-point loadings

The obtained solutions for this case are tabulated in column 4 in Table 17.1. For IMFO, the basic FC is increased from 800.3848 $ h^{-1} for Case 5-1 to 832.1023 $ h^{-1} for Case 5-2 due to the VPL effect. Also, the FC for all algorithms is increased where its value for MFO, MSA, MVO, WOA, GA, PSO and TLBO algorithms becomes 832.3871, 832.4666, 831.8653, 832.1095, 832.4400, and 832.1095 $ h^{-1}, respectively. The CCs' comparison of the implemented methods is depicted in Fig. 17.2.

17.7.3 Case 5-3: Optimum power flow for emission cost minimization

The obtained solutions for this case are tabulated in column 5 in Table 17.1. For IMFO, the value of emission gases is 0.20480 ton h^{-1}, which represents the smallest with respect to values obtained by other methods. The value of emission

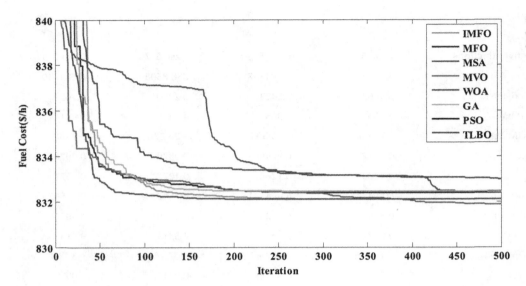

FIGURE 17.2 Convergence characteristics of various techniques for Case 5-2.

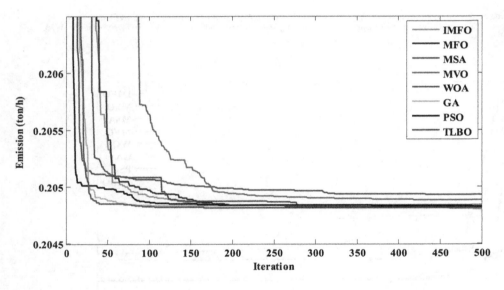

FIGURE 17.3 Convergence characteristics of various techniques for Case 5-3.

gases obtained by TLBO is 0.20482 ton h^{-1}. The value of emission gases acquired by MFO, GA, and PSO is 0.20483 ton h^{-1}. The value of emission gases acquired by MSA, MVO, and WOA is 0.2049 ton h^{-1}. The CCs' comparison of the implemented methods is depicted in Fig. 17.3.

17.7.4   Case 5-4: Optimum power flow for power loss minimization

The obtained solutions for this case are tabulated in column 6 in Table 17.1. The obtained value of power loss by TLBO is 3.088 MW, which is lower than the value of IMFO that is 3.0905 MW. The results obtained by MFO, MSA, MVO, WOA, GA, and PSO algorithms are 3.139, 3.1565, 3.3045, 3.1765, 3.118, and 3.103 MW, respectively. The CCs' comparison of the implemented methods is depicted in Fig. 17.4.

17.7.5   Case 5-5: Optimum power flow for minimization of fuel cost with voltage stability index

The obtained solutions for this case, which aims to minimize both FC and L_{max}, are tabulated in column 3 in Table 17.2. The obtained values of FC by IMFO, MFO, MSA, MVO, WOA, GA, PSO, and TLBO algorithms are 800.5392, 800.9415, 800.7533, 800.9656, 800.8985, 800.4385, 800.5815, and 800.4738 $ h^{-1}, respectively. The values

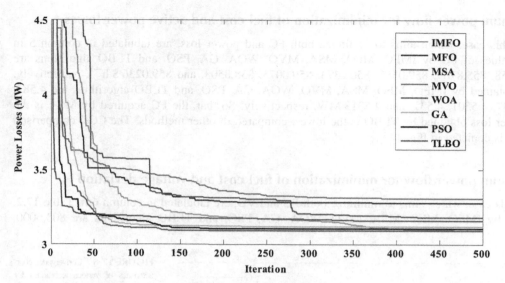

FIGURE 17.4 Convergence characteristics of various techniques for Case 5-4.

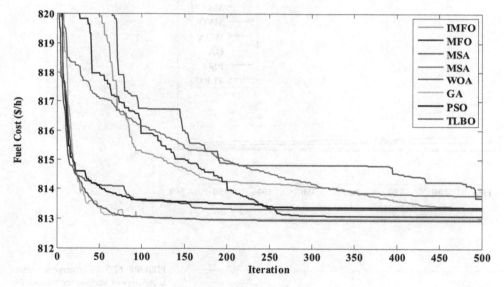

FIGURE 17.5 Convergence characteristics of various techniques for Case 5-5.

of L_{max} obtained by IMFO, MFO, MSA, MVO, WOA, GA, PSO, and TLBO algorithms are 0.1257, 0.1266, 0.1256, 0.1288, 0.1280, 0.1254, 0.1280, and 0.1247 p.u., respectively. It is noted that the value of FC acquired by GA is the lowest and the value of L_{max} obtained by TLBO is the lowest compared to other methods. The CCs' comparison of the implemented methods is depicted in Fig. 17.5.

17.7.6 Case 5-6: Optimum power flow for minimization of fuel cost with emission

The obtained solutions for this case, which aims to minimize both FC and emission gases, are tabulated in column 4 in Table 17.2. The obtained values of FC by IMFO, MFO, MSA, MVO, WOA, GA, PSO and TLBO algorithms are 836.4809, 836.5734, 837.2442, 835.7207, 836.2389, 836.5425, 836.4357, and 836.4141 $ h^{-1}, respectively. The value of emission gases acquired by both MFO and GA is 0.2422 ton h^{-1}. The value of emission gases obtained by IMFO, PSO and TLBO is 0.2423 ton h^{-1}. The value of emission gases acquired by both MVO and WOA is 0.2438 and 0.2428 p.u., respectively, whereas the lowest value acquired by MSA is 0.2421 ton h^{-1}, which is the lowest compared to other methods. The CCs' comparison of the implemented methods is depicted in Fig. 17.6.

17.7.7 Case 5-7: Optimum power flow for minimization of fuel cost and active power losses

The obtained solutions for this case, which aims to minimize both FC and power loss, are tabulated in column 5 in Table 17.2. The obtained values of FC by IMFO, MFO, MSA, MVO, WOA, GA, PSO, and TLBO algorithms are 858.7455, 800.9415, 858.9658, 855.8769, 859.0555, 856.2373, 859.0075, 858.8508, and 859.0236 $ h^{-1}, respectively. The values of power loss obtained by IMFO, MFO, MSA, MVO, WOA, GA, PSO, and TLBO algorithms are 4.582, 4.556, 4.8953, 4.8403, 4.7937, 4.5501, 4.537, and 4.5313 MW, respectively. So that, the FC acquired by MSA is the lowest and the value of power loss obtained by TLBO is the lowest compared to other methods. The CCs' comparison of the implemented methods is depicted in Fig. 17.7.

17.7.8 Case 5-8: Optimum power flow for minimization of fuel cost and voltage deviation

The obtained solutions for this case, which aims to minimize both FC and VD, are tabulated in column 6 in Table 17.2. The obtained values of FC by IMFO, MFO, MSA, MVO, WOA, GA, PSO, and TLBO algorithms are 803.6000,

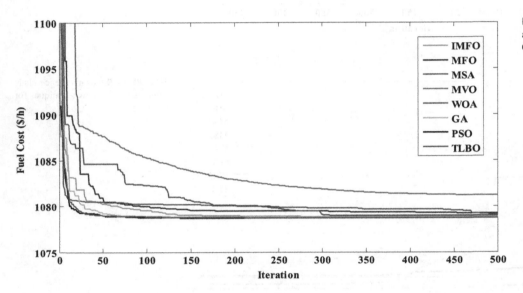

FIGURE 17.6 Convergence characteristics of various techniques for Case 5-6.

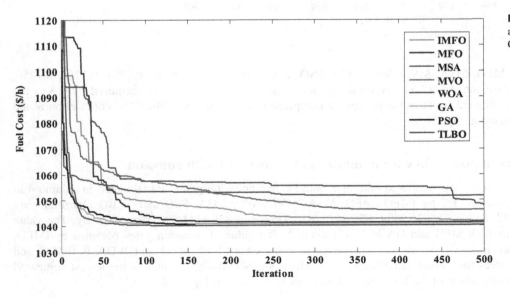

FIGURE 17.7 Convergence characteristics of various techniques for Case 5-7.

803.7983, 803.7048, 803.9179, 804.2670, 803.2347, 803.4736, and 803.5675 $ h^{-1}, respectively. The values of VD obtained by IMFO, MFO, MSA, MVO, WOA, GA, PSO, and TLBO algorithms are 0.0969, 0.1000, 0.1126, 0.1310, 0.1845, 0.1018, 0.0978, and 0.093 p.u., respectively. So that, the FC acquired by GA is the lowest and the value of power loss obtained by TLBO is the lowest compared to other methods. The CCs' comparison of the implemented methods is depicted in Fig. 17.8.

17.8 Conclusion

In this chapter, a comparative study has been applied between recent optimization techniques to solve OPF problems, which are latterly proposed. The applied recent techniques are IMFO, MFO, MSA, MVO, and WOA. All the applied recent methods have the capability for not to fall into local optima stagnation. For all the applied methods, single and multi-Obj Fun are taken into consideration to emphasize their capability for solving various OPF problems. IEEE 30-bus system is used to check the performance of the applied methods. The Comparison has been held among the applied recent optimization methods and famous optimization methods, including GA, PSO, and TLBO, to confirm the efficiency and quality of solutions. Comparison among the CCs has been held among all methods. It is possible in the future works to apply the latest proposed optimization methods to solve OPF, economic dispatch, and loadability problems with different scale power systems.

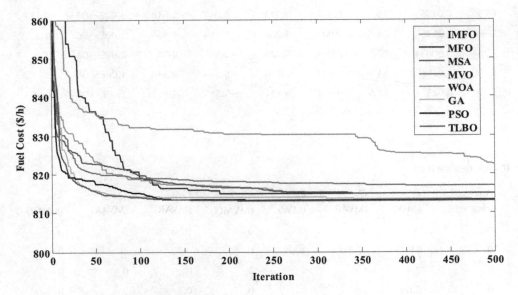

FIGURE 17.8 Convergence characteristics of various techniques for Case 5-8.

Appendix A (Tables 17.A1–17.A5)

TABLE 17.A1 Characteristics of IEEE 30-bus test system.

Characteristics	Value details	Details
Buses	30	Alsac and Stott (1974)
Branches	41	Alsac and Stott (1974)
Generators	6	Buses: 1, 2, 5, 8, 11, and 13
Load voltage limits	24	[0.95:1.05]
Shunts VAR compensation	9	Buses: 10, 12, 15, 17, 20, 21, 23, 24, and 29
Transformers	4	Branches: 11, 12, 15, and 36
Control variables	24	–

TABLE 17.A2 Generation limits for IEEE 30-bus test system.

Unit number	P_{min} (MW)	P_{max} (MW)	Q_{min} (MVAR)	Q_{max} (MVAR)
1	50	200	−20	43
2	20	80	−20	43
5	15	50	−20	50
8	10	35	−10	30
11	10	30	−10	45
13	12	40	−10	50

TABLE 17.A3 Cost and emission coefficients for the IEEE 30-bus test system.

Generator	Bus	a	b	c	d	e	α	β	γ	ω	μ
G1	1	0	2	0.00375	18	0.037	4.091	−5.554	6.49	2.00E−04	2.857
G2	2	0	1.75	0.0175	16	0.038	2.543	−6.047	5.638	5.00E−04	3.333
G3	5	0	1	0.0625	14	0.04	4.258	−5.094	4.586	1.00E−06	8
G4	8	0	3.25	0.00834	12	0.045	5.326	−3.55	3.38	2.00E−03	2
G5	11	0	3	0.025	13	0.042	4.258	−5.094	4.586	1.00E−06	8
G6	13	0	3	0.025	13.5	0.041	6.131	−5.555	5.151	1.00E−05	6.667

TABLE 17.A4 Bus data for 30-bus test system.

Bus no.	Bus type	V (p.u)	Angle (degrees)	P_L (MW)	Q_L (MVAR)	P_g (MW)	Q_g (MVAR)	Q_{min} (MVAR)	Q_{max} (MVAR)	Q_{inj} (MVAR)
1	Slack	1	0	0	0	0	0	0	0	0
2	PV	1	0	21.7	12.7	40.0	0	−40	50	0
3	PQ	1	0	2.4	1.2	0	0	0	0	0
4	PQ	1	0	7.6	1.6	0	0	0	0	0
5	PV	1	0	94.2	19	0	0	−40	40	0
6	PQ	1	0	0	0	0	0	0	0	0
7	PQ	1	0	22.8	10.9	0	0	0	0	0
8	PV	1	0	30	30	0	0	−10	40	0
9	PQ	1	0	0	0	0	0	0	0	0
10	PQ	1	0	5.8	2	0	0	0	0	19
11	PV	1	0	0	0	0	0	−6	24	0
12	PQ	1	0	11.2	7.5	0	0	0	0	0
13	PV	1	0	0	0	0	0	−6	24	0
14	PQ	1	0	6.2	1.6	0	0	0	0	0
15	PQ	1	0	8.2	2.5	0	0	0	0	0

(Continued)

TABLE 17.A4 (Continued)

Bus no.	Bus type	V (p.u)	Angle (degrees)	P_L (MW)	Q_L (MVAR)	P_g (MW)	Q_g (MVAR)	Q_{min} (MVAR)	Q_{max} (MVAR)	Q_{inj} (MVAR)
16	PQ	1	0	3.5	1.8	0	0	0	0	0
17	PQ	1	0	9	5.8	0	0	0	0	0
18	PQ	1	0	3.2	0.9	0	0	0	0	0
19	PQ	1	0	9.5	3.4	0	0	0	0	0
20	PQ	1	0	2.2	0.7	0	0	0	0	0
21	PQ	1	0	17.5	11.2	0	0	0	0	0
22	PQ	1	0	0	0	0	0	0	0	0
23	PQ	1	0	3.2	1.6	0	0	0	0	0
24	PQ	1	0	8.7	6.7	0	0	0	0	4.3
25	PQ	1	0	0	0	0	0	0	0	0
26	PQ	1	0	3.5	2.3	0	0	0	0	0
27	PQ	1	0	0	0	0	0	0	0	0
28	PQ	1	0	0	0	0	0	0	0	0
29	PQ	1	0	2.4	0.9	0	0	0	0	0
30	PQ	1	0	10.6	1.9	0	0	0	0	0

TABLE 17.A5 Transmission line data for 30-bus test system.

Line number	From bus	To bus	Impedance		Line charge	Max. limit (MVA)
			R (p.u.)	X (p.u.)	Y/2 (p.u.)	
1	1	2	0.0192	0.0575	0.0264	130
2	1	3	0.0452	0.1652	0.0204	130
3	2	4	0.057	0.1737	0.0184	65
4	3	4	0.0132	0.0379	0.0042	130
5	2	5	0.0472	0.1983	0.0209	130
6	2	6	0.0581	0.1763	0.0187	65
7	4	6	0.0119	0.0414	0.0045	90
8	5	7	0.046	0.116	0.0102	70
9	6	7	0.0267	0.082	0.0085	130
10	6	8	0.012	0.042	0.0045	32
11	6	9	0	0.208	0	65
12	6	10	0	0.556	0	32
13	9	11	0	0.208	0	65
14	9	10	0	0.11	0	65
15	4	12	0	0.256	0	65
16	12	13	0	0.14	0	65

(Continued)

TABLE 17.A5 (Continued)

Line number	From bus	To bus	Impedance		Line charge	Max. limit (MVA)
			R (p.u.)	X (p.u.)	Y/2 (p.u.)	
17	12	14	0.1231	0.2559	0	32
18	12	15	0.0662	0.1304	0	32
19	12	16	0.0945	0.1987	0	32
20	14	15	0.221	0.1997	0	16
21	16	17	0.0524	0.1923	0	16
22	15	18	0.1073	0.2185	0	16
23	18	19	0.0639	0.1292	0	16
24	19	20	0.034	0.068	0	32
25	10	20	0.0936	0.209	0	32
26	10	17	0.0324	0.0845	0	32
27	10	21	0.0348	0.0749	0	32
28	10	22	0.0727	0.1499	0	32
29	21	22	0.0116	0.0236	0	32
30	15	23	0.1	0.202	0	16
31	22	24	0.115	0.179	0	16
32	23	24	0.132	0.27	0	16
33	24	25	0.1885	0.3292	0	16
34	25	26	0.2544	0.38	0	16
35	25	27	0.1093	0.2087	0	16
36	28	27	0	0.396	0	65
37	27	29	0.2198	0.4153	0	16
38	27	30	0.3202	0.6027	0	16
39	29	30	0.2399	0.4533	0	16
40	8	28	0.0636	0.2	0.0214	32
41	6	28	0.0169	0.0599	0.0065	32

References

Abido, M. (2002). Optimal power flow using particle swarm optimization. *International Journal of Electrical Power & Energy Systems, 24*, 563–571.

Abido, M. A. (2003). Environmental/economic power dispatch using multiobjective evolutionary algorithms. *IEEE Transactions on Power Systems, 18*, 1529–1537.

Agrawal, R., Bharadwaj, S., & Kothari, D. (2016). Transmission loss and TCSC cost minimization in power system using particle swarm optimization. *International Journal of Innovative Research in Electrical, Electronics, Instrumentation and Control Engineering, 4*, 226–231.

Alsac, O., & Stott, B. (1974). Optimal load flow with steady-state security. *IEEE transactions on Power Apparatus and Systems*, 745–751.

Alsumait, J., Sykulski, J., & Al-Othman, A. (2010). A hybrid GA–PS–SQP method to solve power system valve-point economic dispatch problems. *Applied Energy, 87*, 1773–1781.

Burchett, R., Happ, H., & Vierath, D. (1984). Quadratically convergent optimal power flow. *IEEE Transactions on Power Apparatus and Systems*, 3267–3275.

Chowdhury, B. H., & Rahman, S. (1990). A review of recent advances in economic dispatch. *IEEE Transactions on Power Systems, 5*, 1248–1259.

Chung, T., & Li, Y. (2001). A hybrid GA approach for OPF with consideration of FACTS devices. *IEEE Power Engineering Review, 21*, 47–50.

Recent optimal power flow algorithms Chapter | 17 **409**

Davies, P. C. (1978). Thermodynamics of black holes. *Reports on Progress in Physics, 41*, 1313.

Dommel, H. W., & Tinney, W. F. (1968). Optimal power flow solutions. *IEEE Transactions on Power Apparatus and Systems*, 1866−1876.

Du, K.-L., & Swamy, M. (2016). *Particle swarm optimization. Search and optimization by metaheuristics* (pp. 153−173). Springer.

Eardley, D. M. (1974). Death of white holes in the early universe. *Physical Review Letters, 33*, 442.

Eiben, A. E., & Schippers, C. A. (1998). On evolutionary exploration and exploitation. *Fundamenta Informaticae, 35*, 35−50.

El Ela, A. A., Abido, M., & Spea, S. (2010). Optimal power flow using differential evolution algorithm. *Electric Power Systems Research, 80*, 878−885.

Fan, R.-J., Anderson, P., & Hansson, B. (1997). Behavioural analysis of olfactory conditioning in the moth Spodoptera littoralis (Boisd.) (Lepidoptera: Noctuidae). *Journal of Experimental Biology, 200*, 2969−2976.

Frank, K. D. (1988). Impact of outdoor lighting on moths: An assessment. *Journal of the Lepidopterists' Society (USA), 42*.

Gaston, K. J., Bennie, J., Davies, T. W., & Hopkins, J. (2013). The ecological impacts of nighttime light pollution: A mechanistic appraisal. *Biological Reviews, 88*, 912−927.

Goldbogen, J. A., Friedlaender, A. S., Calambokidis, J., Mckenna, M. F., Simon, M., & Nowacek, D. P. (2013). Integrative approaches to the study of baleen whale diving behavior, feeding performance, and foraging ecology. *Bioscience, 63*, 90−100.

Hardiansyah, H. (2013). A modified particle swarm optimization technique for economic load dispatch with valve-point effect. *International Journal of Intelligent Systems and Applications, 5*, 32−41.

He, S., Wen, J., Prempain, E., Wu, Q., Fitch, J., & Mann S. (2004). An improved particle swarm optimization for optimal power flow. In *2004 international conference on power system technology (PowerCon 2004)* (pp. 1633−1637).

Hof, P. R., & Van der Gucht, E. (2007). Structure of the cerebral cortex of the humpback whale, *Megaptera novaeangliae* (Cetacea, Mysticeti, Balaenopteridae). *The Anatomical Record: Advances in Integrative Anatomy and Evolutionary Biology, 290*, 1−31.

Huneault, M., & Galiana, F. (1991). A survey of the optimal power flow literature. *IEEE Transactions on Power Systems, 6*, 762−770.

Mirjalili, S. (2015). Moth-flame optimization algorithm: A novel nature-inspired heuristic paradigm. *Knowledge-Based Systems, 89*, 228−249.

Mirjalili, S., Jangir, P., Mirjalili, S. Z., Saremi, S., & Trivedi, I. N. (2017). Optimization of problems with multiple objectives using the multi-verse optimization algorithm. *Knowledge-Based Systems, 134*, 50−71.

Mirjalili, S., & Lewis, A. (2016). The whale optimization algorithm. *Advances in Engineering Software, 95*, 51−67.

Mirjalili, S., Mirjalili, S. M., & Hatamlou, A. (2016). Multi-verse optimizer: A nature-inspired algorithm for global optimization. *Neural Computing and Applications, 27*, 495−513.

Mohamed, A.-A. A., Mohamed, Y. S., El-Gaafary, A. A., & Hemeida, A. M. (2017). Optimal power flow using moth swarm algorithm. *Electric Power Systems Research, 142*, 190−206.

Mota-Palomino, R., & Quintana, V. (1986). Sparse reactive power scheduling by a penalty function-linear programming technique. *IEEE Transactions on Power Systems, 1*, 31−39.

Ochs, P., Ranftl, R., Brox, T., & Pock, T. (2016). Techniques for gradient-based bilevel optimization with non-smooth lower level problems. *Journal of Mathematical Imaging and Vision, 56*, 175−194.

Paranjothi, S., & Anburaja, K. (2002). Optimal power flow using refined genetic algorithm. *Electric Power Components and Systems, 30*, 1055−1063.

Reddy, S. S., & Bijwe, P. (2016). Efficiency improvements in meta-heuristic algorithms to solve the optimal power flow problem. *International Journal of Electrical Power & Energy Systems, 82*, 288−302.

Santos, A. J., & Da Costa, G. (1995). Optimal-power-flow solution by Newton's method applied to an augmented Lagrangian function. *IEE Proceedings − Generation, Transmission and Distribution, 142*, 33−36.

Shaheen, A. M., El-Sehiemy, R. A., & Farrag, S. M. (2016). Solving multi-objective optimal power flow problem via forced initialised differential evolution algorithm. *IET Generation, Transmission & Distribution, 10*, 1634−1647.

Sivasubramani, S., & Swarup, K. (2011a). Multi-objective harmony search algorithm for optimal power flow problem. *International Journal of Electrical Power & Energy Systems, 33*, 745−752.

Sivasubramani, S., & Swarup, K. (2011b). Sequential quadratic programming based differential evolution algorithm for optimal power flow problem. *IET Generation, Transmission & Distribution, 5*, 1149−1154.

Skiri, H., Stranden, M., Sandoz, J.-C., Menzel, R., & Mustaparta, H. (2005). Associative learning of plant odorants activating the same or different receptor neurones in the moth *Heliothis virescens*. *Journal of Experimental Biology, 208*, 787−796.

Soliman, G. M., Khorshid, M. M., & Abou-El-Enien, T. H. (2016). Modified moth-flame optimization algorithms for terrorism prediction. *International Journal of Application or Innovation in Engineering and Management, 5*, 47−58.

Steinhardt, P. J., & Turok, N. (2002). A cyclic model of the universe. *Science (New York, N.Y.), 296*, 1436−1439.

Taher, M. A., Kamel, S., Jurado, F., & Ebeed, M. (2019a). An improved moth-flame optimization algorithm for solving optimal power flow problem. *International Transactions on Electrical Energy Systems, 29*, e2743.

Taher, M. A., Kamel, S., Jurado, F., & Ebeed, M. (2019b). Modified grasshopper optimization framework for optimal power flow solution. *Electrical Engineering, 101*, 121−148.

Tegmark, M. (2004). In J. D. Barow, P. C. W. Davies, & C. L. Harper, Jr. (Eds.), *Science and ultimate reality*. Cambridge: Cambridge University Press.

Varadarajan, M., & Swarup, K. S. (2008). Solving multi-objective optimal power flow using differential evolution. *IET Generation, Transmission & Distribution, 2*, 720−730.

Watkins, W. A., & Schevill, W. E. (1979). Aerial observation of feeding behavior in four baleen whales: *Eubalaena glacialis*, *Balaenoptera borealis*, *Megaptera novaeangliae*, and *Balaenoptera physalus*. *Journal of Mammalogy*, *60*, 155–163.

Yan, X., & Quintana, V. H. (1999). Improving an interior-point-based OPF by dynamic adjustments of step sizes and tolerances. *IEEE Transactions on Power Systems*, *14*, 709–717.

Yao, X., Liu, Y., & Lin, G. (1999). Evolutionary programming made faster. *IEEE Transactions on Evolutionary Computation*, *3*, 82–102.

Yuryevich, J., & Wong, K. P. (1999). Evolutionary programming based optimal power flow algorithm. *IEEE Transactions on Power Systems*, *14*, 1245–1250.

Chapter 18

Challenges for the optimum penetration of photovoltaic systems

Antonio Colmenar-Santos[1], Ana-Rosa Linares-Mena[1], Enrique-Luis Molina-Ibáñez[1], David Borge-Diez[2] and Enrique Rosales-Asensio[3]

[1]*Department of Electric, Electronic and Control Engineering, UNED, Juan Del Rosal, 12 − Ciudad Universitaria, Madrid, Spain*, [2]*Department of Electrical and Systems Engineering and Automation, University of León, Campus de Vegazana, S / N, León, Spain*, [3]*Department of Electrical Engineering, University of Las Palmas de Gran Canaria, Campus de Tafira S / N, Las Palmas de Gran Canaria, Spain*

Nomenclature

CECOEL	Electricity Control Centre
CECRE	Control center of renewable energy
CCG	Generation control center
EN	European Norm
ES	Electrical system
EC	European Commission
EU	European Union
GSM	Global System for Mobile Communications
IEC	International Electrotechnical Commission
ICCP	Inter-Control Center Communications Protocol
LV	Low voltage
M2M	Machine to machine
MV	Medium voltage
P_{sc}	Short-circuit power
PV	Photovoltaic
P_{out}	Power output
PVVC	Process of verification, validation, and certification
RES	Renewable energy sources
rms	root-mean-square value
SO	System operator
TP	Testing point
UNE	Spanish Norm
U_n	Nominal voltage
U_{res}	Residual voltage

18.1 Introduction

In recent years, the worldwide photovoltaic (PV) market has experienced a significant increase because of technical improvements in component manufacturing and efficiency of the devices, which lead to cost reduction (Sener & Fthenakis, 2014). Furthermore, the development of policies supporting renewable energy sources (RES) has powered the integration of these technologies in electricity networks (Sahu, 2015).

Renewable Energy Systems. DOI: https://doi.org/10.1016/B978-0-12-820004-9.00015-2

The European Union (EU) has promoted the 20/20/20 targets inside the Directive 2009/28/EC of the European Parliament (European Commission, 2015c), expanding its commitment to 2030. Europe will cut its greenhouse gas emissions by 40% by 2030, compared with 1990 levels, the toughest climate change target of any region in the world, and will produce 27% of its energy from renewable sources by the same date. (De la Hoz, Martin, Miret, Castilla, & Guzman, 2016; De la Hoz, Martin, Montala, Matas, & Guzman, 2018; European Commission, 2015b). Consequently, EU Member States have enhanced the development of incentive policies (Hosenuzzamana et al., 2015) that allow the penetration of distributed generation at RES units in the energy mix (Dusonchet & Telaretti, 2015). Owing to its inherent characteristics, the integration of PV systems in power grids implies significant technical, legislative, and economic challenges (Abdmouleh, Alammari, & Gastli, 2015; Eltawila & Zhao, 2010; Etxegarai, Eguia, Torres, Iturregi, & Valverde, 2015). Furthermore, the geographical dispersion of power plants as well as the production variability owing to the weather conditions and location, and therefore the uncertainty in its prediction, makes it necessary to establish new strategies to ensure definite control of the system (Hammons, 2008), which will allow proper integration of RES in the electrical system (ES) without compromising the safety and the quality of supply (Moreno-Muñoz, De la Rosa, López, & Gil de Castro, 2010).

With all this, the expectation in the Spanish ES can be noticed (Colmenar-Santos, Molina-Ibañez, Rosales-Asensio, & Blanes-Peiró, 2018). One of the main objectives in Spain is the improvement of grid integration and the establishment of key parameters to obtain adequate performance of the PV plants as well as promotion of the industry-wide competitiveness of the necessary technology (Zubi, 2011). To tackle this challenge, Spain has developed a complex legal framework, which is constantly updated to allow the appropriate regulation and promotion of the satisfactory penetration of PV systems in the ES (Ministerio de, 2019).

From an economic point of view, the retribution mechanism during the last decades has been applied to the selling price of energy by a feed-in-tariff system (Ciarreta, Gutiérrez-Hita, & Nasirov, 2011; De la Hoz, Martín, Ballart, Córcoles, & Graells, 2013; Del Río & Mir-Artigues, 2012; Spanish Association of Renewable Energies, 2017). Because of this mechanism and other promotion politics, in 2017 the PV system installed power (Fig. 18.1; Red Eléctrica de España, 2017) reached 4675 MW, generating 8350 GWh, which accounted for 3.1% of the annual energy demand (Red Eléctrica de España, 2017); therefore PV technologies have a significant potential in the generation mix in Spain (Unión Española Fotovoltaica, 2017).

The main goals in the operation of the ES, managed by the system operator (SO), *Red Eléctrica de España*, are to ensure the security and the continuity of supply. In this sense, the PV systems, which are connected to the distribution network, must accomplish certain technical requirements to guarantee their correct operation in normal and special situations (Red Eléctrica de España). These requirements are defined within a compendium of rules that are included in a broad range of technical and legal documents, which represents a lack of standardization and update. Although there are numerous literature reports that refer to established technical requirements (Amundarain, Alberdi, Garrido, Garrido, & De la Sen, 2012; Carrasco et al., 2006; Gómez-Lázaro, Cañas, Fuentes, & Molina-García, 2007; Jiménez, Gómez-

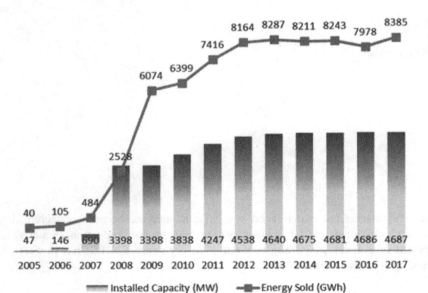

FIGURE 18.1 Evolution of installed capacity and energy sales of photovoltaic sector in Spain (Red Eléctrica de España, 2017).

Lázaro, Fuentes, Molina-García, & Vigueras-Rodríguez, 2013; Moreno-Muñoz et al., 2010; Veganzones et al., 2011; Zahedi, 2011), in the case of PV grid-connected systems, there is a lack of specific and updated documents. A different approach is required to gather singularities and specificities for the management and technical assessment of PV systems.

This chapter presents a comprehensive analysis of the main technical requirements and legal resources for the connection of PV systems to the Spanish electricity grid, but with the scope to be a reference model that could be extended worldwide. To promote the integration of PV technology in the generation mix, the different technical levels to be achieved are emphasized as well as the challenges must be addressed in the short and medium term.

In Figs. 18.2 and 18.3, the main PV facilities distributed in Spain can be noticed (Sistema de Información del Operador del Sistema (ESIOS)).

To perform this study, the information contained in laws, regulations, technical instructions, and other legislation were compiled and analyzed (Agencia Estatal Boletín Oficial del Estado (BOE); Asociación Española de Normalización y Certificación; Entidad Nacional de Acreditación (ENAC); Ministerio de, 2019). Court notes as well as reports published by industry associations and energy agencies were also taken into account, at both European (European Forum for Renewable Energy Sources (EUFORES); European Photovoltaic Industry Association (EPIA); European Photovoltaic Technology Platform (EU PVTP); International Energy Agency Photovoltaic Power System Programme (IEA-PVPS); The European Association for Renewable Energy) and national level (Asociación de Empresas de Energía Renovables (APPA); Asociación Nacional de Productores e Inversores de Energía Fotovoltaica (ANPIER); Instituto para la Diversificación y Ahorro de la Energía (IDEA); Unión Española Fotovoltaica (UNEF)). In addition, circulars, reports, queries, and recommendations published by advisory bodies in the field of electricity markets (Comisión Nacional de los Mercados y la Competencia (CNMC); Gómez-Lázaro et al., 2007) were also analyzed. This chapter gives an accurate and well-structured analysis that is focused on facilitating the optimal penetration of PV grid-connected systems in Spain.

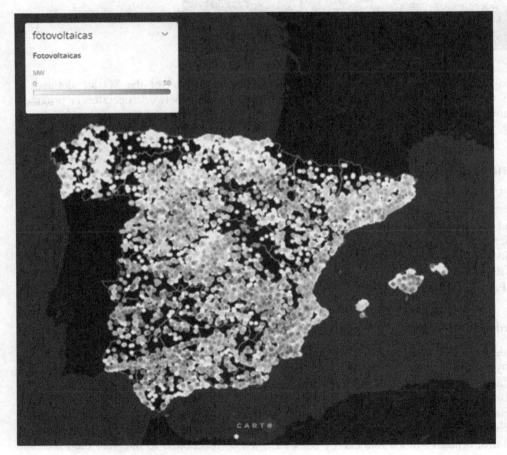

FIGURE 18.2 Distribution of photovoltaic installations in Spain, Iberian Peninsula (Sistema de Información del Operador del Sistema (ESIOS)).

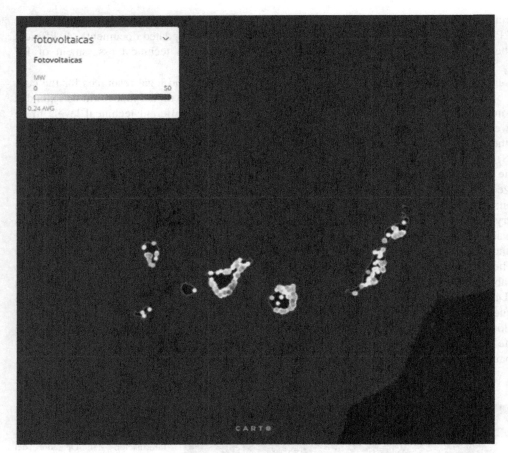

FIGURE 18.3 Distribution of photovoltaic installations in Spain, Canary Islands (Sistema de Información del Operador del Sistema (ESIOS)).

This chapter is organized as follows: in Section 18.2, the compulsory requirements of the SO are defined; in Section 18.3, a summary of the criteria and technical requirements defined by the network operator is provided; Section 18.4 describes the future recommendations at the legislative level. Finally, Section 18.5 concludes the chapter summarizing the main results.

18.2 PV system management

The fast development of PV systems has introduced new challenges in the management of the ES (European Photovoltaic Industry Association, 2012). The high complexity of this technology requires robust systems with real-time monitoring, analysis, and control (Zhang, Li, & Bhatt, 2010) to operate both the generation systems and transmission lines and match the generation units production scheduled with the consumer demand (Lund, Lindgren, Mikkola, & Salpakari, 2015). In addition, to ensure the proper technical management of the ES and to obtain the required data, the regulation and control of the measuring systems as well as the equipment that comprise and their characteristics are required (Boletín Oficial del Estado, number 224 of September 18th, 2007).

18.2.1 Control and monitoring

In Spain, coordinated operation and real-time monitoring of the national ES as well as the control of international trade are the functions performed by the SO in the electricity control center. The services are managed to adjust the requirements for quality, reliability, and safety of the system with production schedules, resulting from the daily and intraday electricity market (Liu, Li, Liu, Liu, & Mao, 2011). The solution of technical constraints, the allocation of additional services, and the deviation management are handled by setting operation points to the elements of the transport network to keep the control variables within the margins established by the operating procedures. To address these issues, in 2006 the SO launched the control center of renewable energy (CECRE), whose function is to integrate the maximum

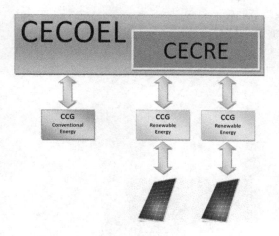

FIGURE 18.4 Interconnection with control center of renewable energy (Rodriguez et al., 2008).

energy production from renewable sources inside the ES, in both adequate safety and quality (Rodriguez, Alonso, Duvison, & Domingez, 2008). CECRE allows real-time monitoring and control of the transmission network to optimize its operation and ensure safety, reliability, flexibility, and efficiency (Liu et al., 2011). In particular, the interaction between PV generation units and CECRE is performed by connecting the units to the generation control centers (CCG) accredited by the SO (Fig. 18.4). With this powerful tool, Spain became the first country in the world to have all of its wind and solar farms over 10 MW in size connected to a control center (Gallardo-Calles, Colmenar-Santos, Ontañon-Ruiz, & Castro-Gil, 2013). In the first half of 2014, 37 CCG on the mainland and 6 CCG on islands, including 4 in remote regions, could communicate with CECRE. Of these, over 60% have been tested for production control during their operation.

18.2.2 Communications

The CECRE receives activity, reactive power, tension, connectivity, temperature, and wind speed data from each wind farm every 12 seconds. Based on this information, it calculates wind production that can be integrated into the ES at any time, depending on the characteristics of the generators and the state of the system itself. The CECRE needs at least the following information:

- the connection status
- produced active power (MW)
- produced/Absorbed Reactive Power (MVAr)
- status of connection with the distribution or transmission network (connectivity)
- voltage measurement (kV)

To do this, it is necessary to have a technical infrastructure with sufficient capacity to control, command, and monitor the generation of electricity connected to it, and to have appropriate training of human resources to ensure a secure dialogue and functionality 24/7 (Red Eléctrica de España, 2014a). As an example, Fig. 18.5 shows the structure of one CCG (Canary Island, Spain). There is a supervisory control and data acquisition in operation 24/7, covering single failures of equipment or functions, so that its annual availability sets the standard for this type of mission-critical system. Therefore a problem that affects a critical function can be solved at the maximum within 1 hour (Red Eléctrica de España, 2014b).

The communication protocol for the real-time data exchange is the Inter-Control Center Communications Protocol (ICCP or IEC 60870-6-503/TASE.2), which provides features for data transfer, monitoring, and control. ICCP functionality is specified as "conformance blocks," and implementation must support ICCP Block 1 and Block 2 (Toja-Silva, Colmenar-Santos, & Castro-Gil, 2013). Point-to-point redundant lines are used with independent paths. They do not share common infrastructure in terms of conduits and transmission equipment. The connection is permanent, bidirectional, and dedicated exclusively. The connections are TCP/IP type with n channels of 64 kbps ($4 < n < 32$) or 2 Mbps unstructured and must ensure full transparency in the transmitted information, without intermediate modification. The interface of the lines at the ends of the circuit must be of type V.35 or G703/E1 with BNC termination. Protocols and equipment must be Global System for Mobile Communications (GSM)-type machine to machine (M2M), excluding

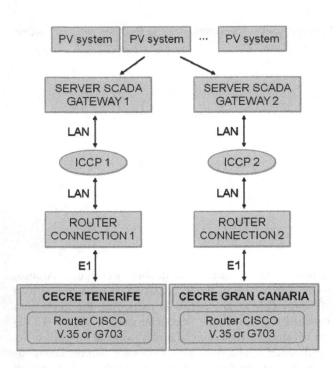

FIGURE 18.5 CCG-ITER in Tenerife, Canary Islands, Spain (Red Eléctrica de España, 2014a).

virtual private network, frame relay, integrated services digital network, asymmetric digital subscriber line connection type (Red Eléctrica de España, 2014b). This latter requirement may cause additional costs to the system and shortcomings in complying with new technologies (software and hardware), as these technologies are obsolete.

End and SO routers are CISCO1841, CISCO2800, or similar models. The bandwidth must be fixed and committed, ensuring correct exchange of information. Enabled ports for reception and transmission are 102TCP type, and IP addresses for ICCP remote servers must have, at minimum, two different and independent routable addresses for disjoint paths and network elements in the wide area network and local area network (Red Eléctrica de España, 2014b). CECRE (de la Torre, Juberías, Domínguez, & Rivas, 2012) remits to CCG the fundamental aspects for the attached generators, which ensures the compliance and maintenance of the operation points. The operation values of maximum power per node and the type of generator, with the indicator code of the cause of restriction, are received within a minimum period of 1 minute (Boletín Oficial del Estado, number 129 of May 28 th, 2009). To ensure the maintenance of the operation points for each CCG, deviations above 10% of the set point may be received in less than 5 minutes if it is permitted by the particular conditions of the operation of the system (FESCA ENDESA, 2009).

18.2.3   Metering

Remote management required for the PV generation units involves the fulfilment of meteorological controls to ensure the quality and accuracy of the measurements (Liu et al., 2011). In this context, PV generation units are classified in Table 18.1 according to the type of measuring point, establishing a growing number of technical requirements that affect the accuracy class of the measurement equipment, current—voltage transformers, redundant equipment, installation of recording, and the obligation to perform telemetry (Liu et al., 2011). In general, the measuring equipment consists, separately or integrally, of an active energy meter, a reactive energy meter, transformers, and other ancillary equipment such as recorders, elements of power control, modems, and schedule watches (Iberdrola Distribución Eléctrica, 2004).

Multifunctional static meters included in the same housing are used for recording the active energy in both directions of energy flow (buying and selling) and reactive energy in four quadrants programmed with current and time discrimination necessary for billing. Furthermore, they are enabled to close all contracts automatically at day 1. The meter has a verification LED indicator for both directions of energy flows. For installations with a capacity exceeding 15 kVA, it is mandatory that the meter registers the reactive power (Iberdrola Distribución Eléctrica, 2004). The accuracy of the electricity meters must be as indicated in Table 18.1 (Liu et al., 2011). The most stringent requirements are set for types 1, 2, and 3. Therefore these measuring devices enable remote reading and display the power, ensuring

TABLE 18.1 Accuracy class of the measurement equipment (Liu et al., 2011).

Point class	Rated apparent power (Sn)	Accuracy class					
		Transformers		Meters			
		Voltage	Current	Active	Reactive	Load curve	
1	12 MVA ≤ Sn	0.2	0.2 s	0.2 s	0.5	Required	
2	450 kVA ≤ Sn < 12 MVA	≤ 0.5	≤ 0.5 s	0.5 s	1	Required	
3	15 kVA < Sn < 450 kVA	≤ 1	≤ 1	0.5 s	1	Required	
5	15 kVA ≤ Sn			1	2	Optional	

reading even in the absence of voltage. The power control is accomplished by maximeters with an integration period of 15 minutes. Furthermore, available recorders are capable to parameterize integration periods up to 5 minutes as well as record and store the parameters required for the calculation of tariffs of access or supply. Likewise, it incorporates recording parameters related to the quality of service, storing at least the number and duration of each of the supply interruptions lasting less than 3 minutes and the time when the line voltage is outside the limits allowed by excess and by default (Liu et al., 2011).

18.3 PV system grid connection

In addition to the requirements of the PV systems monitoring, their integration in the network must be conducted to ensure that the connection settings are made in compliance with a series of technical and safety requirements (Colmenar-Santos et al., 2018; Del Amo, Martinez-Gracia, Bayod-Rujula, & Antoñanzas, 2017). To achieve this, apart from the requirements to withstand voltage sag that may occur in the network, the voltage at which the facility must be connected is defined, together with the requirements to be met by inverters, because of their role as connection interfaces between the PV generation unit and the network (Crăciun, Kerekes, Séra, & Teodorescu, 2012) and how protection systems should be provided.

18.3.1 General criteria

Generation units with power under 100 kVA must be connected to the low-voltage (LV) network. However, the generation unit can connect to the medium-voltage (MV) network if there are no LV facilities close by or if there is no enough capacity on the LV network to support the connection (FESCA ENDESA, 2009). The rated voltage for PV systems connected at LV is 230 V for single-phase and 400 V for three-phase electricity. If the connection is made to the MV network, the rated voltage is 25 kV. The facility should be designed for a short-circuit current of 10 and 20 kA in the LV and MV networks, respectively. Furthermore, the power factor reference range for the energy supplied to the network is set between 0.98 inductive and 0.98 capacitive (Boletín Oficial del Estado, number 140 of June 10, 2014).

18.3.2 Inverters

Three-phase inverters are used to avoid unbalanced energy generation. As an exception, generation units with a power rating below 5 kVA are allowed to connect with single-phase systems.

The inverter can inject into the network for the harmonic currents within the limits established by the following standards (FESCA ENDESA, 2009):

- IEC-EN 61000-3-2:2014: Limits for harmonic current emissions (equipment input current ≤ 16 A per phase)
- IEC-EN 61000-3-12:2011: Limits for harmonic currents produced by equipment connected to public low-voltage systems with input current > 16 A and ≤ 75 A per phase

In this sense, UNE 206007-1:2013 IN (Monedero et al., 2011) provides "minimum technical requirements for the connection of inverters to the power system" (International Energy Agency, 2013). Problems related to the quality of supply involve a wide range of electrical disturbances that are critical for system behavior (Awad, Svensson, & Bollen,

2003), such as waveshape faults, overvoltage, capacitor switching transients, harmonic distortion, and impulse transients (Monedero et al., 2011). The technical regulation (UNE 206007-1, 2013) includes the requirements for DC current injection (Salas, Olías, Alonso, Chenlo, & Barrado, 2006), behavior under isolation faults (Velasco, Trujillo, Garcera, & Figueres, 2010), detection of fault currents in the PV generator (Gautam & Kaushika, 2002), voltage and frequency shutdown (Amundarain et al., 2012; Awad et al., 2003), automatic reconnection (Ruiz-Romero, Colmenar-Santos, Mur-Pérez, & López-Rey, 2014), islanding (Ahmadn, Selvaraj, & Rahim, 2013; Velasco et al., 2010), overvoltage (Demirok et al., 2011), power quality, and reconnection out of synchronization (Hassaine, OLias, Quintero, & Salas, 2014; Zeng, Yang, Zhao, & Cheng, 2013)

18.3.3 Electrical protection systems

Islanding is a condition in which a portion of the utility system that contains both load and distributed resources remains energized while isolated from the remainder of the utility system. In this respect, the distributed resources supplying the loads within the island are not within the direct control of the power SO (IEEE Std 929-2000, 2000). Islanding represents a key security parameter, not only for the PV systems but also for the ES, compromising the security as well as the power restoration, degradation of power quality, and reliability of equipment (Raza, Mokhlis, Arof, Laghari, & Wang, 2015). Therefore it is necessary to provide appropriate security protection systems including a general cut off switch, permanently accessible by the distribution company, a residual current circuit breaker, and a circuit breaker for automatic shut connection of the facility in the event of voltage or grid frequency failure, together with a latching relay. After being disconnected, reconnection should be prevented before 3 minutes at the power recovery, even if disconnection occurred because of the action of a trigger with line reclosing. In addition, whenever possible, unwarranted disconnection must be avoided owing to normal variations in the operating parameters of the network and external faults of its connection line (FESCA ENDESA, 2009).

The system must have the following protections (Iberdrola Distribución Eléctrica, 2004), whose parameters are defined in Table 18.2:

- Maximum and minimum voltage protection, by controlling the voltage between phases.
- Maximum and minimum frequency protection, by controlling the frequency.
- Transient overvoltage protection: installing metal oxide lightning protection systems with a 25 kV voltage rating and 10 kA nominal discharge current, Provided it is advisable to install protections by the value of overloads and their frequencies.
- Fault current protection, both the phase currents and the earth fault current by overcurrent protections, being selective with the header line protections located at the substation level.
- Overload protection: regulation of delayed intensity protections, depending on the nominal power capacity of the PV system.
- Antiislanding protection (LV connections) through passive or active detection methods (phase jump detection, reactive power control, frequency shift) to avoid the operation of this equipment in terms of network loss, according to the UNE-EN 50438:2014 requirements for microgenerating plants to be connected in parallel with public low-voltage distribution networks. The islanding trigger signal will not disappear until their correct reference quantities remain uninterrupted for 3 minutes. During that time, the connections of the PV system to the network is prevented.

TABLE 18.2 Protections and settings for a photovoltaic system with an obligation to meet performance requirements against voltage sags (Iberdrola Distribución Eléctrica, 2004).

Type	Adjustment (islands)	Adjustment (mainland)
Minimum voltage protection	Trigger at 0.77 kV phase–phase, 1 s	Trigger at 0.85 kV phase–phase, 1.2 s
Maximum voltage protection	Trigger at 1.1 kV at MV (1.07 kV at LV) phase–phase, 0.5 s	Trigger at 1.1 kV at MV (1.07 kV at LV) phase–phase, 0.5 s
Minimum frequency protection	47 Hz, 3 s	48 Hz, 3 s
Overfrequency protection	51 Hz, 0.2 s	51 Hz, 0.2 s

The aforementioned described protections may act on the main switch or on the switch or switches on the equipment or generating equipment and may be integrated into the inverter. Similarly, they have galvanic isolation through transformers that are integrated or not integrated into the inverter (Iberdrola Distribución Eléctrica, 2004).

18.3.4 Voltage sags control

Voltage sags are one of the most severe failures of PV systems (Amundarain et al., 2012), as well as a major concern for SO because they have detrimental effects on the stability of the grid (Gómez-Lázaro et al., 2007). A voltage sag is defined as a sharp fall in the supply to a value between 90% and 1% of the voltage, followed by a value recovery after a short time (Asociación Empresarial Eólica, 2012; CEI 61000-4-30, 2008; UNE-EN 50160, 2011). Facilities connected to the distribution network must withstand voltage sags without disconnecting, avoiding cascade disconnections that could affect the continuity of electricity supply (Unión Española Fotovoltaica, 2017). In Spain, both facilities and groups of renewable energy installations exceeding 2 MW are required to comply with the operating procedure *PO 12.3 Requisitos de respuesta frente a huecos de tensión* (in English *Response requirements to voltage sags*) (Boletín Oficial del Estado, number 254 of October 24th, 2006). For this, facilities should be able to withstand voltage sags at the point of network connection, produced by three-phase, grounded two-phase, or single-phase short-circuits, with profiles of magnitude and duration as indicated in Fig. 18.6 (Boletín Oficial del Estado, number 254 of October 24th, 2006). That is, the installation disconnection will not occur for voltage sags in the main connection points included in the shaded area of Fig. 18.6. For simplicity and applicability, this study will focus on three-phase connections. In recent years, several simulation models have been developed (Amundarain et al., 2012; Camacho, Castilla, Miret, Guzman, & Borrel, 2014; Miret, Camacho, Castilla, Garcia de Vicuña, & Matas, 2013; Veganzones et al., 2011) that serve as a supporting tool for the modification and adaptation of the inverters.

If the inverter does not satisfy the requirements for voltage sags, it must be adapted by changing its hardware or software configuration (García-Gracia, El Halabi, Ajami, & Comech, 2012), along with the modification of the output relay parameters to ensure no power falls, or installing additional power electronics devices outside the inverters (Khazaie et al.) called flexible alternating current transmission systems to compensate for the effects of voltage sags on the facilities. The stakeholders have collaborated on the development of a particular process for the measurement and evaluation of PV conversion systems, given the complexity of requirements verification. The outcome was the process of verification, validation, and certification (PVVC) requirements of PO 12.3 regarding the response of wind and PV installations to voltage sags (PVVC10) (Asociación Empresarial Eólica, 2012). The document presents a verification system based on the compliance with the requirements for PV systems to have an adequate response to the voltage sag.

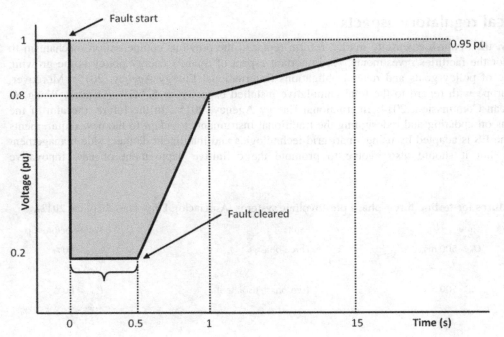

FIGURE 18.6 Voltage–time curve for a voltage sag in the connection point to the network in PO 12.3 (Boletín Oficial del Estado, 2006).

According to PVVC10 (Asociación Empresarial Eólica, 2012), the test is performed by applying a three-phase fault and an isolated two-phase fault, causing a dip in the affected phases. The voltage waveform should be obtained in three channels (phase-to-ground, phase-to-neutral, or phase-to-phase voltages). The one-cycle root-mean-square (rms) voltage is calculated every half cycle in every channel. The residual voltage (U_{res}) is the lowest rms voltage recorded in any of the channels during the event (UNE-EN IEC 61000-4-30).

The required tests that have to be conducted with three-phase converters are summarized in Table 18.3 (Asociación Empresarial Eólica, 2012), in which U_{res} is defined as a function of the nominal voltage (U_n), and P_{out} is the power output before an event. During the tests, the active and reactive power, currents, and voltages have to be recorded at the testing point (TP). Both in the test and in the process simulation, all registered data "(voltage and current) for each phase is performed with a sampling frequency of at least 5 kHz," according to Asociación Empresarial Eólica (2012). The moments before the beginning of the dip and 5 seconds after the recovery period are also registered. The instant of the voltage sag is randomly applied.

The criteria for test validation are the following:

1. Residual voltage and time during no-load test

Voltage sag profile that applies:

- If P_{sc} at TP $\geq$ 5 times the registered power, the voltage sag can be obtained by uncoupling the PV system at the dip generator (no-load test). Subsequent tests under load (PV system coupled) have to be performed with the same impedance adjustment of the dip generator equipment.
- If P_{sc} at TP $<$ 5 times registered power, it is compulsory to measure the dip profile under load.
2. Operating point
    According to Asociación Empresarial Eólica (2012), it is required "that the active power recorded prior to the implementation of the voltage sag is within the range that defines a partial load ($10\% < P_{out} < 30\%$) and full load ($P_{out} > 80\%$)."
3. Guarantee of continuity supply
    No PV system disconnection occurs during the application of voltage sag.
4. TP sharing power and energy conditions

According to Asociación Empresarial Eólica (2012), the value of the injected current by the PV System "during the failure must meet that specified in PO 12.3 in relation to the values of the reactive current as well as reactive and active power consumption." Measurements of the required voltage and current have to be registered at the TP.

18.4   Future technical regulatory aspects

Regarding renewable energy, the Spanish electricity market reform replaced the previous compensation mechanism to ensure a reasonable return for the facilities investments. An important aspect of Spain's energy policy is the growing role of the EU as the source of policy goals and related obligations (International Energy Agency, 2015). Moreover, Spain is still the third in Europe with regard to the total cumulative installed capacity, at 5.3 GW (Depuru, Wang, & Devabhaktuni, 2011; European Commission, 2014; International Energy Agency, 2015). In the future, the aim of the ES management should focus on updating and redesigning the traditional instruments to adapt to the new requirements for smart grids. Currently, the ES is adapted by using smart grid technologies and intelligent demand side management (Red Eléctrica de España), but it should also evolve to promote the definitive deployment of new innovative

TABLE 18.3 Voltage sag features for testing three-phase photovoltaic systems (Asociación Empresarial Eólica, 2012).

Voltage	Time	Faults	Power before dip
$U_{res} < 20\%$	$U_n > 500$ ms	Three-phase	$P_{out} > 80\%$
			$10\% < P_{out} < 30\%$
$U_{res} < 60\%$	$U_n > 500$ ms	Two-phase (isolated)	$P_{out} > 80\%$
			$10\% < P_{out} < 30\%$

mechanisms such as the integration of storage systems (Barbour & Gonzalez, 2018; Barelli, Desideri, & Ottaviano, 2015; Du & Lu, 2015; Kumar Selvaraju & Somaskandan, 2016), charging infrastructure (García-Villalobos, Zamora, San Martín, Asensio, & Aperribay, 2014), electrical mobility (Mwasilu, Justo, Kim, Do, & Jung, 2014), and the use of smart meters (Depuru et al., 2011), among others. To achieve these goals, legislation should advance by using new concepts and developments as well as generation and control systems that allow the shift from a centralized power generation model to a distributed electricity generation. It also should learn from the experience of previous research (Dusonchet & Telaretti, 2015; Martin & Grossmann, 2018; Mir-Artigues, Cerdá, & del Río, 2015; Reddy, Kumar, Mallick, Sharon, & Lokeswaran, 2014; Ruiz-Romero et al., 2014; Schiavo, Delfanti, Fumagalli, & Olivieri, 2013; Siano, 2014), identifying barriers and selecting the best mechanisms to ensure their applicability.

Both currently and toward the future, one of the possible avenues for the development of PV generation units lies in this electricity self-consumption model (Spanish Association of Renewable Energies, 2017) supported by instantaneous consumption into a net metering framework (Borlick & Wood, 2014; Eid, Reneses, Frias, & Hakvoort, 2014; Mir-Artigues, 2013). This model is an important technical and legislative challenge that countries such as the United States have already developed and widely applied (Aznar, 2014; Bogdanov & Breyer, 2016; Castillo-Cagigal et al., 2011; Chiaroni et al., 2014; Cucchiella, D'Adamo, & Gastaldi, 2016; European Commission, 2015a; Lund, 2012; Manrique Delgado et al., 2018; Roselli & Sasso, 2016). As recommended by the European Commission (2015a), Member States should promote the demand-side flexibility, including demand-side response (Alemany, Arendarski, Lombardi, & Komarnicki, 2018; Kinhekara, Padhy, & Gupta, 2014; López, de la Torre, Martín, & Aguado, 2015; Qazi & Flynn, 2016; Qiu, 2018; Rashidizadeh-Kermani, Vahedipour-Dahraie, Shafie-khah, & Catalão, 2019) and distributed energy storage (Rodrigues et al., 2014; Saboori, Hemmati, Ahmadi, & Jirdehi, 2015), by establishing simplified administrative and authorization procedures for guaranteeing the competiveness. Moreover, European Commission (EC) underlines the need to ensure the objective and nondiscriminatory criteria, while ensuring sufficient funding for grid and system costs (Eid et al., 2014).

In this sense, Spain must continue to develop a regulatory system aimed at facilitating a distributed energy system that allows the energy development of the local network. With the approval of *Real Decreto 244/2019*, of April 5, which gives continuity to *Real Decreto-Ley 15/2018*, it establishes three types of self-consumption, without surpluses, with surpluses that are subject to compensation, and with surplus not accepted as compensation. This law also indicates that the power installed in a PV installation will be considered the maximum power of the investor or, where appropriate, the sum of the maximum powers of the inverters. On the other hand, this law establishes the measurement equipment to be installed in different considerations:

- Generally speaking, only one bidirectional measuring device is needed at the boundary point.
- Collective self-consumption, with surpluses not covered by compensation with several supply contracts or nonrenewable technology, must have two teams. One for consumption and another that measures net generation.
- In certain cases, the measurement counter is allowed to be located outside the boundary point.

18.5   Conclusions

This chapter brings together and describes the technical requirements for the control and connection of PV systems to the electricity grid. It establishes a starting point to overcome the obstacles inherent to this technology and achieve greater penetration of PV technology in the energy mix. This is a significant technical and legislative challenge that must be faced, imperatively, by the institutional bodies owing to the future trend in smart grids. This integration involves significant technical considerations owing to the dispersion of the installations, the variability of their production, and uncertainty in their forecast, which makes it necessary to establish new strategies to ensure the control of these variables, and for the proper integration of PV systems in the ES without compromising safety and quality of supply. The conversion systems are under constant technological adaptation to ensure their operation without neglecting performance and reliability. The generation control procedures are well structured in which Spain is a pioneer in this field. However, the required communication systems are obsolete, causing unnecessary additional costs and deficiencies for adjustment to new technologies (software and hardware). The requirements in communication protocols force the use of outdated equipment type GSM that are type M2M voice; then the system is more expensive and does not take advantage of technological advances in this field. Regarding measurement systems, a powerful development within the framework of smart meters is needed to allow the integration of new technologies into the energy mix, such as charging systems, electric vehicles, and storage systems; they are essential elements to encourage the development of smart grids.

A self-consumption model based on distributed electricity generation into a net metering framework should be a reliable scenario for the integration of PV systems. The development of a consistent, uniform, and transparent regulatory framework is required to ensure proper access to the network at an optimum quality and safety, which evolved and adapts its characteristics to the consumers' needs including the optimal application of demand management owing to the dispersion of the generation units. After a significant electricity reform, the Spanish energy sector maintains its strengths, such as the quality and security of supply. However, the economic recession has resulted in new challenges to solve the tariff deficit issue. To achieve the targets set by the EC, substantial efforts are needed to continue the deployment and definitive penetration of the most cost-effective technologies, highlighting PV technologies, to boost policy measures (financial or technical), including support schemes, standards, procedures, and administrative rules. Introducing new mechanisms to encourage the successful integration of PV systems in the energy mix is a significant challenge, and a constant review and update of information are required in the future.

Acknowledgments

The authors gratefully acknowledge the assistance of Instituto Tecnológico y de Energías Renovables (ITER).

References

Abdmouleh, Z., Alammari, R. A. M., & Gastli, A. (2015). Review of policies encouraging renewable energy integration & best practices. *Renewable and Sustainable Energy Reviews, 45*, 249–262.

Agencia Estatal Boletín Oficial del Estado (BOE). http://www.boe.es/ Accessed April 2019.

Ahmadn, K. N. E. K., Selvaraj, J., & Rahim, N. A. (2013). A review of the islanding detection methods in grid-connected PV inverters. *Renewable and Sustainable Energy Reviews, 21*, 756–766.

Alemany, J. M., Arendarski, B., Lombardi, P., & Komarnicki, P. (2018). Accentuating the renewable energy exploitation: Evaluation of flexibility options. *International Journal of Electrical Power & Energy Systems, 102*, 131–151.

Amundarain, M., Alberdi, M., Garrido, A., Garrido, I., & De la Sen, M. (2012). Neural control for wave power plant during voltage sags. *Electric Power Systems Research, 92*, 96–105.

Asociación de Empresas de Energía Renovables (APPA). http://www.appa.es/ Accessed April 2019.

Asociación Empresarial Eólica. (2012). Procedimiento de verificación, validación y certificación de los requisitos del PO 12.3 sobre la respuesta de las instalaciones eólicas y fotovoltaicas ante huecos de tensión, version 10, January 26, 2012.

Asociación Española de Normalización y Certificación (AENOR). https://www.en.aenor.com/ Accessed April 2019.

Asociación Nacional de Productores e Inversores de Energía Fotovoltaica (ANPIER). http://anpier.org/ Accessed April 2019.

Awad, H., Svensson, J., & Bollen, M. H. J. (2003). Static series compensator for voltage sags mitigation. In *IEEE Bologna Power Tech conference proceedings* (Vol. 3), pp. 23–26, June 2003.

Aznar, A. (2014). *Weighing the costs and benefits of net metering and distributed solar.* National Renewable Energy Labortory (NREL). https://www.nrel.gov/tech_deployment/state_local_governments/blog/weighing-the-costs-and-benefits-of-net-metering-and-distributed-solar Accessed April 2019.

Barbour, E., & Gonzalez, M. C. (2018). Projecting battery adoption in the prosumer era. *Applied Energy, 215*, 356–370.

Barelli, L., Desideri, U., & Ottaviano, A. (2015). Challenges in load balance due to renewable energy sources penetration: The possible role of energy storage technologies relative to the Italian case. *Energy, 93*, 393–405.

Bogdanov, D., & Breyer, C. (2016). North-East Asian Super Grid for 100% renewable energy supply: Optimal mix of energy technologies for electricity, gas and heat supply options. *Energy Conversion and Management, 112*, 176–190.

Boletín Oficial del Estado, number 129 of May 28th. (2009). Resolution of the Energy General Secretary, by which the operation procedure PO 3.7 Programación de la generación de origen renovable no gestionable (Programming of intermittent RES generation), is approved. http://www.boe.es/diario_boe/txt.php?id = BOE-A-2009-8813 Accessed April 2019.

Boletín Oficial del Estado, number 140 of June 10. (2014). *Real Decreto 413/2014, of June 6, on electricity generation by means of renewable, cogeneration and waste facilities.* http://www.boe.es/diario_boe/txt.php?id = BOE-A-2014-6123 Accessed April 2019.

Boletín Oficial del Estado, number 224 of September 18th. (2007). *Real Decreto 1110/2007, of August 24, approving the unified Regulations of power grid metering points.* http://www.boe.es/diario_boe/txt.php?id = BOE-A-2007-16478 Accessed April 2019.

Boletín Oficial del Estado, number 254 of October 24th. (2006). Resolution of the Energy General Secretary, by which the operation procedure PO 12.3 Requisitos de respuesta frente a huecos de tensión de las instalaciones eólicas (Response requirements to voltage sags) is approved. http://www.boe.es/diario_boe/txt.php?id = BOE-A-2006-18485 Accessed April 2019.

Borlick, R., & Wood, L. (2014) *Net energy metering: subsidy issues and regulatory solutions.* Issue Brief. The Edison Foundation. http://www.edison-foundation.net/iei/documents/IEI_NEM_Subsidy_Issues_FINAL.pdf Accessed April 2019.

Camacho, A., Castilla, M., Miret, J., Guzman, R., & Borrel, A. (2014). Reactive power control for distributed generation power plants to comply with voltage limits during grid faults. *IEEE Transactions on Power Electronics, 29*, 6224–6234.

Carrasco, J. M., García, L., Bialasiewicz, J. T., Galván, E., Portillo, R. C., Martin, M. A., et al. (2006). Power-electronic systems for the grid integration of renewable energy sources: A survey. *IEEE Transactions on Industrial Electronics*, *53*(4).

Castillo-Cagigal, M., Gutiérrez, A., Monasterio-Huelin, F., Caamaño-Martín, F., Masa, D., & Jiménez-Leube, J. (2011). A semi-distributed electric demand-side management system with PV generation for self-consumption enhancement. *Energy Conversion and Management*, *52*(7), 2659–2666.

CEI 61000-4-30:2008. Electromagnetic compatibility (EMC)—Part 4-30: Testing and measurement techniques—Power quality measurement methods.

Chiaroni, D., Chiesa, V., Colasanti, L., Cucchiella, F., D'Adamo, I., & Frattini, F. (2014). Evaluating solar energy profitability: A focus on the role of self-consumption. *Energy Conversion and Management*, *88*, 317–331.

Ciarreta, A., Gutiérrez-Hita, C., & Nasirov, S. (2011). Renewable energy sources in the Spanish electricity market: Instruments and effects. *Renewable and Sustainable Energy Reviews*, *15*, 2510–2519.

Colmenar-Santos, A., Molina-Ibañez, E. L., Rosales-Asensio, E., & Blanes-Peiró, J. J. (2018). Legislative and economic aspects for the inclusion of energy reserve by a superconducting magnetic energy storage: Application to the case of the Spanish electrical system. *Renewable and Sustainable Energy Reviews*, *82*, 2455–2470.

Comisión Nacional de los Mercados y la Competencia (CNMC). https://www.cnmc.es/ Accessed April 2019.

Crăciun, B. I., Kerekes, T., Séra, D., & Teodorescu, R. (2012). Overview of recent grid codes for PV power integration. Optimization of electrical and electronic equipment (OPTIM). In *13th international conference* (pp. 959–965), Brasov, 24–26 May 2012.

Cucchiella, F., D'Adamo, I., & Gastaldi, M. (2016). A profitability assessment of small-scale photovoltaic systems in an electricity market without subsidies. *Energy Conversion and Management*, *129*, 62–74.

De la Hoz, J., Martín, H., Ballart, J., Córcoles, F., & Graells, M. (2013). Evaluating the new control structure for the promotion of grid connected photovoltaic systems in Spain: Performance analysis of the period 2008–2010. *Renewable and Sustainable Energy Reviews*, *19*, 541–554.

De la Hoz, J., Martin, H., Miret, J., Castilla, M., & Guzman, R. (2016). Evaluating the 2014 retroactive regulatory framework applied to the grid connected PV systems in Spain. *Applied Energy*, *170*, 329–344.

De la Hoz, J., Martin, H., Montala, M., Matas, J., & Guzman, R. (2018). Assessing the 2014 retroactive regulatory framework applied to the concentrating solar power systems in Spain. *Applied Energy*, *212*, 1377–1399.

de la Torre, M., Juberías, G., Domínguez, T., & Rivas R. (2012). The CECRE: supervision and control of wind and solar photovoltaic generation in Spain. In *Power and Energy Society general meeting*, San Diego, CA, July 22–26, 2012, pp. 1–6.

Del Amo, A., Martinez-Gracia, A., Bayod-Rujula, A. A., & Antoñanzas, J. (2017). An innovative urban energy system constituted by a photovoltaic/thermal hybrid solar installation: Design, simulation and monitoring. *Applied Energy*, *186*, 140–151.

Del Río, P., & Mir-Artigues, P. (2012). Support for solar PV deployment in Spain: Some policy lessons. *Renewable and Sustainable Energy Reviews*, *16*, 5557–5566.

Demirok, E., Casado, P., Frederiksen, K. H. B., Sera, D., Rodríguez, P., & Teodorescu, R. (2011). Local reactive power control methods for overvoltage prevention of distributed solar inverters in low-voltage grids. *IEEE Journal of Photovoltaics*, *1*(2), 174–182.

Depuru, S. S. S. R., Wang, L., & Devabhaktuni, V. (2011). Smart meters for power grid: Challenges, issues, advantages and status. *Renewable and Sustainable Energy Reviews*, *15*, 2736–2742.

Du, P., & Lu, N. (2015). *Energy storage for smart grids: Planning and operation for renewable and variable energy resources (VERs)* (1st ed.). Academic Press. Elsevier Inc.

Dusonchet, L., & Telaretti, E. (2015). Comparative economic analysis of support policies for solar PV in the most representative EU countries. *Renewable and Sustainable Energy Reviews*, *42*, 986–998.

Eid, C., Reneses, J., Frias, P., & Hakvoort, R. (2014). The economic effect of electricity net-metering with solar PV: Consequences for network cost recovery, cross subsidies and policy objectives. *Energy Policy*, *75*, 244–254.

Eltawila, M. A., & Zhao, Z. (2010). Grid-connected photovoltaic power systems: Technical and potential problems − A review. *Renewable and Sustainable Energy Reviews*, *14*, 112–129.

Entidad Nacional de Acreditación (ENAC). http://www.enac.es/web/enac/inicio Accessed April 2019.

Etxegarai, A., Eguia, P., Torres, E., Iturregi, A., & Valverde, V. (2015). Review of grid connection requirements for generation assets in weak power grids. *Renewable and Sustainable Energy Reviews*, *41*, 1501–1514.

European Commission. (2014). *JRC science and policy reports. PV status report.* http://iet.jrc.ec.europa.eu/remea/pv-status-report-2014 Accessed April 2019.

European Commission. (2015a). *Best practices on renewable energy self-consumption.* http://ec.europa.eu/energy/sites/ener/files/documents/1_EN_autre_document_travail_service_part1_v6.pdf Accessed April 2019.

European Commission. (2015b). *Climate action. 2030 framework for climate and energy policies.* http://ec.europa.eu/clima/policies/2030/index_en.htm Accessed April 2019.

European Commission. (2015c). Country report Spain 2015 including an in-depth review on the prevention and correction of macroeconomic imbalances {COM (2015) 85 final}. http://ec.europa.eu/europe2020/making-it-happen/country-specific-recommendations/index_en.htm Accessed April 2019.

European Forum for Renewable Energy Sources (EUFORES). http://www.eufores.org/ Accessed April 2019.

European Photovoltaic Industry Association. (2012). *Connecting the Sun: Solar photovoltaics on the road to large-scale grid integration.* http://pvtrin.eu/assets/media/PDF/Publications/other_publications/263.pdf Accessed April 2019.

European Photovoltaic Industry Association (EPIA). http://www.epia.org/home/ Accessed April 2019.

European Photovoltaic Technology Platform (EU PVTP). http://www.eupvplatform.org/ Accessed April 2019.

FESCA ENDESA. (2009). *Specific standards of connection for PV plants to the MV distribution network (NTP-FVMT).*

Gallardo-Calles, J. M., Colmenar-Santos, A., Ontañon-Ruiz, J., & Castro-Gil, M. (2013). Wind control centres: State of the art. *Renewable Energy, 51,* 93−100.

García-Gracia, M., El Halabi, N., Ajami, H., & Comech, M. P. (2012). Integrated control technique for compliance of solar photovoltaic installation grid codes. *IEEE Transactions on energy conversion, 27,* 792−798.

García-Villalobos, J., Zamora, I., San Martín, J. I., Asensio, F. J., & Aperribay, V. (2014). Plug-in electric vehicles in electric distribution networks: A review of smart charging approaches. *Renewable and Sustainable Energy Reviews, 38,* 717−731.

Gautam, N. K., & Kaushika, N. D. (2002). An efficient algorithm to simulate the electrical performance of solar photovoltaic arrays. *Energy, 27,* 347−361.

Gómez-Lázaro, E., Cañas, M., Fuentes, J. A., & Molina-García, A. (2007). *Characterization of measured voltage sags in wind farms in the light of the new grid codes. Power Tech* (pp. 2059−2064). IEEE Lausanne.

Hammons, T. J. (2008). Integrating renewable energy sources into European grids. *Electrical Power and Energy Systems, 30,* 462−475.

Hassaine, L., OLias, E., Quintero, J., & Salas, V. (2014). Overview of power inverter topologies and control structures for grid connected photovoltaic systems. *Renewable and Sustainable Energy Reviews, 30,* 796−807.

Hosenuzzamana, M., Rahima, N. A., Selvaraja, J., Hasanuzzamana, M., Maleka, A. B. M. A., & Nahar, A. (2015). Global prospects, progress, policies, and environmental impact of solar photovoltaic power generation. *Renewable and Sustainable Energy Reviews, 41,* 284−297.

Iberdrola Distribución Eléctrica. (2004). *Technical conditions of the installation of grid connected producers plants (MT3.53.01).*

IEEE Std 929-2000. (2000). *Recommended practice for utility interface of photovoltaic (PV) systems. IEEE Standards Coordinating Committee 21.*

Instituto para la Diversificación y Ahorro de la Energía (IDEA). *Ministerio para la transición ecológica.* http://www.idae.es/en Accessed April 2019.

International Energy Agency. (2013). *National survey report of PV power applications in Spain.* http://www.iea-pvps.org/index.php?id = 93&eID = dam_frontend_push&docID = 2101 Accessed April 2019.

International Energy Agency. (2015). *Energy policies of IEA countries − Spain review.* http://www.iea.org/Textbase/npsum/spain2015sum.pdf Accessed April 2019.

International Energy Agency Photovoltaic Power System Programme (IEA-PVPS). http://www.iea-pvps.org/ Accessed April 2019.

Jiménez, F., Gómez-Lázaro, E., Fuentes, J. A., Molina-García, A., & Vigueras-Rodríguez, A. (2013). Validation of a DFIG wind turbine model submitted to two-phase voltage sags following the Spanish grid code. *Renewable Energy, 57,* 27−34.

Khazaie, J., Nazarpour, D., Farsadi, M., Mokhtari, M., Khalilian, M., & Badkubi, S. Fault current limitation and contraction of voltage sags. Thanks to D-FACTS and FACTS Cooperation. In *7th international conference on electrical and electronics engineering* (ELECO) (pp. I-106−I-111), Bursa, Turkey, December 1−4.

Kinhekara, N., Padhy, N. P., & Gupta, H. O. (2014). Multiobjective demand side management solutions for utilities with peak demand deficit. *International Journal of Electrical Power & Energy Systems, 55,* 612−619.

Kumar Selvaraju, R., & Somaskandan, G. (2016). Impact of energy storage units on load frequency control of deregulated power systems. *Energy, 97,* 214−228.

Liu, J., Li, X., Liu, D., Liu, H., & Mao, P. (2011). Study on data management of fundamental model in control center for smart grid operation. *IEEE Transactions on Smart Grid, 2*(4).

López, M. A., de la Torre, S., Martín, S., & Aguado, J. A. (2015). Demand-side management in smart grid operation considering electric vehicles load shifting and vehicle-to-grid support. *International Journal of Electrical Power & Energy Systems, 64,* 689−698.

Lund, P. (2012). Large-scale urban renewable electricity schemes − Integration and interfacing aspects. *Energy Conversion and Management, 63,* 162−172.

Lund, P. D., Lindgren, J., Mikkola, J., & Salpakari, J. (2015). Review of energy system flexibility measures to enable high levels of variable renewable electricity. *Renewable and Sustainable Energy Reviews, 45,* 785−807.

Manrique Delgado, B., Kotireddy, R., Cao, S., Hasan, A., Hoes, P. J., Hensen, J. L. M., et al. (2018). Lifecycle cost and CO2 emissions of residential heat and electricity prosumers in Finland and the Netherlands. *Energy Conversion and Management, 160,* 495−508.

Martin, M., & Grossmann, I. E. (2018). Optimal integration of renewable based processes for fuels and power production: Spain case study. *Applied Energy, 213,* 595−610.

Mir-Artigues, P. (2013). The Spanish regulation of the photovoltaic demand-side generation. *Energy Policy, 63,* 664−673.

Mir-Artigues, P., Cerdá, E., & del Río, P. (2015). Analyzing the impact of cost-containment mechanisms on the profitability of solar PV plants in Spain. *Renewable and Sustainable Energy Reviews, 46,* 166−177.

Miret, J., Camacho, A., Castilla, M., Garcia de Vicuña, L., & Matas, J. (2013). Control scheme with voltage support capability for distributed generation inverters under voltage sags. *IEEE Transactions on Power Electronics, 28,* 5252−5262.

Monedero, I., León, C., Ropero, J., García, A., Elena, J. M., & Montaño, J. C. (2011). Classification of electrical disturbances in real time using neural networks. *IEEE Transactions on Power Delivery, 22*(3), 1288−1296.

Moreno-Muñoz, A., De la Rosa, J. J. G., López, M. A., & Gil de Castro, A. R. (2010). Grid interconnection of renewable energy sources: Spanish legislation. *Energy for Sustainable Development, 14,* 104−109.

Mwasilu, F., Justo, J. J., Kim, E. K., Do, T. D., & Jung, J. W. (2014). Electric vehicles and smart grid interaction: A review on vehicle to grid and renewable energy sources integration. *Renewable and Sustainable Energy Reviews, 34,* 501−516.

Qazi, H. W., & Flynn, D. (2016). Analysing the impact of large-scale decentralised demand side response on frequency stability. *International Journal of Electrical Power & Energy Systems, 80*, 1–9.

Qiu, J. (2018). How to build an electric power transmission network considering demand side management and a risk constraint? *International Journal of Electrical Power & Energy Systems, 94*, 311–320.

Rashidizadeh-Kermani, H., Vahedipour-Dahraie, M., Shafie-khah, M., & Catalão, J. P. S. (2019). A bi-level risk-constrained offering strategy of a wind power producer considering demand side resources. *International Journal of Electrical Power & Energy Systems, 104*, 562–574.

Raza, S., Mokhlis, H., Arof, H., Laghari, J. A., & Wang, L. (2015). Application of signal processing techniques for islanding detection of distributed generation in distribution network: A review. *Energy Conversion and Management, 96*, 613–624.

Ministerio de Industria. Energía y Turismo, Gobierno de España. http://www.minetur.gob.es/es-ES/Paginas/index.aspx Accessed April 2019.

Real Decreto 244/2019, of April 5. *By which the conditions are regulated administrative, technical and economic aspects of the self-consumption of electrical energy.* https://www.boe.es/boe/dias/2019/04/06/pdfs/BOE-A-2019-5089.pdf Accessed April 2019.

Real Decreto-Ley 15/2018, of October 5. *On urgent measures for the Energy transition and consumer protection.* https://www.boe.es/eli/es/rdl/2018/10/05/15 Accessed April 2019.

Red Eléctrica de España. http://www.ree.es/ Accessed April 2019.

Red Eléctrica de España. (2014a). *IEE.01 Especificaciones de conexión CCG-OS Canarias_Ed2.*

Red Eléctrica de España. (2014b). *IEE 05 Requerimientos enlaces REE Canarias_Ed2.*

Red Eléctrica de España. (2017). *The Spanish Electricity System. Preliminary report.* http://www.ree.es/en/statistical-data-of-spanish-electrical-system/annual-report/spanish-electricity-system-preliminary-report-2017 Accessed April 2019.

Reddy, K. S., Kumar, M., Mallick, T. K., Sharon, H., & Lokeswaran, S. (2014). A review of integration, control, communication and Metering (ICCM) of renewable energy based smart grid. *Renewable and Sustainable Energy Reviews, 38*, 180–192.

Rodrigues, E. M. G., Godina, R., Santos, S. F., Bizuayehu, A. W., Contreras, J., & Catalão, J. P. S. (2014). Energy storage systems supporting increased penetration of renewables in islanded systems. *Energy, 75*, 265–280.

Rodriguez, J. M., Alonso, O., Duvison, M., & Dominguez, T. (2008). The integration of renewable energy and the system operation: The Special Regime Control Centre (CECRE) in Spain. In *Power and Energy Society general meeting - conversion and delivery of electrical energy in the 21st century* (pp. 1–6), Pittsburgh, PA, July 20–24, 2008.

Roselli, C., & Sasso, M. (2016). Integration between electric vehicle charging and PV system to increase self-consumption of an office application. *Energy Conversion and Management, 130*, 130–140.

Ruiz-Romero, S., Colmenar-Santos, A., Mur-Pérez, F., & López-Rey, A. (2014). Integration of distributed generation in the power distribution network: The need for smart grid control systems, communication and equipment for a smart city — use cases. *Renewable and Sustainable Energy Reviews, 38*, 223–234.

Saboori, H., Hemmati, R., Ahmadi., & Jirdehi, M. (2015). Reliability improvement in radial electrical distribution network by optimal planning of energy storage systems. *Energy, 93*, 2299–2312.

Sahu, B. K. (2015). A study on global solar PV energy development and policies with special focus on the top ten solar PV power producing countries. *Renewable and Sustainable Energy Reviews, 43*, 621–634.

Salas, V., Olías, E., Alonso, M., Chenlo, F., & Barrado, A. (2006). DC current injection into the network at PV grid inverters. In *4th world conference on photovoltaic energy conversion, Conference record of the 2006 IEEE* (Vol. 2, pp. 2371–2374), Waikoloa, HI.

Schiavo, L. L., Delfanti, M., Fumagalli, E., & Olivieri, V. (2013). Changing the regulation for regulating the change: Innovation-driven regulatory developments for smart grids, smart metering and e-mobility in Italy. *Energy Policy, 57*, 506–517.

Sener, C., & Fthenakis, V. (2014). Energy Policy and financing options to achieve solar energy grid penetration targets: Accounting for external costs. *Renewable and Sustainable Energy Reviews, 32*, 854–868.

Siano, P. (2014). Demand response and smart grids — A survey. *Renewable and Sustainable Energy Reviews, 30*, 461–478.

Sistema de Información del Operador del Sistema (ESIOS). *Map of national photovoltaic installations.* https://www.esios.ree.es/es/mapas-de-interes/mapa-instalaciones-fotovoltaicas Accessed April 2019.

Spanish Association of Renewable Energies. (2017). *Study of the macroeconomic impact of renewable energies in Spain.* https://www.appa.es/es/wp-content/uploads/2018/10/Estudio_del_impacto_Macroeconomico_de_las_energias_renovables_en_Espa%C3%B1a_2017.pdf Accessed April 2019.

The European Association for Renewable Energy. http://www.eurosolar.de/en/ Accessed April 2019.

Toja-Silva, F., Colmenar-Santos, A., & Castro-Gil, M. (2013). Urban wind energy exploitation systems: Behaviour under multidirectional flow conditions — Opportunities and challenges. *Renewable and Sustainable Energy Reviews, 24*, 364–378.

UNE 206007-1:2013. IN. *Requirements for connecting to the power system. Part 1: Grid-connected inverters.*

UNE-EN 50160:2011. *Voltage characteristics of electricity supplied by public electricity networks.*

UNE-EN IEC 61000-4-30. *Electromagnetic compatibility (EMC) —Part 4-30: Testing and measurement techniques —Power quality measurement methods.*

Unión Española Fotovoltaica. (2017). *La energía fotovoltaica conquista el Mercado. Informe anual.* http://unef.es/ Accessed April 2019.

Unión Española Fotovoltaica (UNEF). http://unef.es/ Accessed April 2019.

Veganzones, C., Sánchez, J. A., Martínez, S., Platero, C. A., Blázquez, F., Ramírez, D., et al. (2011). Voltage dip generator for testing wind turbines connected to electrical networks. *Renewable Energy, 36*, 1588–1594.

Velasco, D., Trujillo, C. L., Garcera, G., & Figueres, E. (2010). Review of anti-islanding techniques in distributed generators. *Renewable and Sustainable Energy Reviews, 14*, 1608–1614.

Zahedi, A. (2011). A review of drivers, benefits and challenges in integrating renewable energy sources into electricity grid. *Renewable and Sustainable Energy Reviews, 15*, 4775−4779.

Zeng, Z., Yang, H., Zhao, R., & Cheng, C. (2013). Topologies and control strategies of multi-functional grid-connected inverters for power quality enhancement: A comprehensive review. *Renewable and Sustainable Energy Reviews, 24*, 223−270.

Zhang, P., Li, F., & Bhatt, N. (2010). Next-generation monitoring, analysis, and control for the future smart control center. *IEEE Transactions on Smart Grid, 1*(2).

Zubi, G. (2011). Technology mix alternatives with high shares of wind power and photovoltaics-case study for Spain. *Energy Policy, 39*, 8070−8077.

Chapter 19

Modeling and optimization of performance of a straight bladed H-Darrieus vertical-axis wind turbine in low wind speed condition: a hybrid multicriteria decision-making approach

Jagadish[1], Agnimitra Biswas[2] and Rajat Gupta[2]

[1]*Department of Mechanical Engineering, National Institute of Technology Raipur, Raipur, India,* [2]*Department of Mechanical Engineering, National Institute of Technology Silchar, Silchar, India*

19.1   Introduction

Wind turbines mean horizontal-axis wind turbine (HAWT) and vertical-axis wind turbine (VAWT). The axis of HAWT is horizontal and that of VAWT is vertical. Then plane of rotation of HAWT is vertical and that of VAWT is horizontal. HAWTs are installed in coastal regions where wind speed is high. HAWTs can be installed in onshore and offshore locations also due to very high wind speeds. However, VAWTs can be installed in urban locations, for example, on rooftop, and in building topologies, for example, in front of a building, in open locations between two buildings, and at the edge of a building (Francesco, Alessandro, Carnevale, Lorenzo, & Sandro, 2012). However, performance of VAWT is affected by the changing wind speed and its direction especially in urban areas. Moreover, in different parts of the world, wind has got its different regimes. Hence developing VAWT in low wind speed locations is a major task, which is being taken up by researchers, R&D sectors, project developers, and government agencies. Of late, straight-bladed VAWTs are finding increased applications for power generation in a befitting manner in urban locations. These turbines are simple in construction, devoid of yaw mechanism, less noisy, having less structural and vibrational problem, generator mounted at the bottom of the shaft, etc (Bhuyan & Biswas, 2014). However, urban locations have characteristically low wind speeds, which entail complicacies in harnessing wind power (Sengupta, Biswas, & Gupta, 2019). Tailoring of VAWT designs in sites having low wind speeds is a challenging task as the designers need to deal with the low Reynolds number effect and also as the VAWT needs to be small sized unlike the wind mills found in the coastal areas. Another problem confronting VAWT is its lack of efficiency in low wind speed conditions. Thus designing a straight-bladed VAWT is a challenge (Sengupta, Biswas, & Gupta, 2016; Singh, Biswas, & Misra, 2015). However, the opportunity for the research fraternity is that if a straight-bladed VAWT can be developed for low wind speed condition, then it can suffice the energy demands of small nuclear families, which can largely benefit community development and urbanization.

The aerodynamics of a small-sided VAWT is a complex feature that largely depends on the VAWT's blade design parameters and complicated operating parameters due to changing wind speeds. This chapter engages a small-sized airfoil bladed H-Darrieus VAWT having trialing edge blade twist, a type of straight-bladed VAWT, which although has good self-starting ability, its efficiency is less at a low wind speed (Gupta & Biswas, 2010). The performance of this straight-bladed VAWT can be termed as its efficiency. Efficiency of a specific VAWT is an important performance parameter. Efficiency of its torque generation depends on torque coefficient, and efficiency of power generation depends on its power coefficient. These coefficients depend on various VAWT design and operating parameters, for example, turbine-aspect ratio (i.e., height-to-diameter ratio), wind speed (WS), blade speed (BS), tip speed ratio (TSR), and blockage effects.

Renewable Energy Systems. DOI: https://doi.org/10.1016/B978-0-12-820004-9.00010-3

As these parameters have different scales of values and their effects can be counter-intuitive, hence enhancing one parameter may enhance or deteriorate the performance of the turbine due to the ill effect of the other parameter. Thus there is a certain level of uncertainty in the operation of a VAWT due to its influential and conflicting parameters. Further, the performance of VAWT is different under different environments due to fluctuations of wind in the low free stream because of the influence of the surrounding and also due to obstruction from the side walls of the wind tunnel test section where the VAWT's performance is tested. With these disturbances, the design and operating parameters of VAWT may alter, and designer needs to redesign the VAWT parameters, which may take more time and effort. And as such the performance of the VAWT is also affected. Therefore it can be considered to be a multiresponse or multiobjective optimization problem because of the varying design and operating parameters. Improper selection of operating and design parameters may lead to the error in the results, which ultimately affect the performance of VAWT. Hence, modeling and optimization of VAWT are essential. Besides, the application of multicriteria decision-making (MCDM) in the field of wind turbine specifically for material selection and design parameter selections are available, but the work related to modeling and optimization of the performance of wind turbines, especially straight-bladed H-Darrieus VAWT is hardly available. And there is an ardent requirement of the modeling of VAWT performance considering varied operational and design parameters that can affect its performance in low wind speed conditions of the urban environment. Therefore this work considers a straight-bladed NACA 0012 H-Darrieus VAWT to perform modeling and optimization under different input parameters for different low wind speed conditions by using an integrated method, that is, entropy with multicriteria ratio analysis (MCRA) method, which is hardly available in the existing literature

In this context for the considered straight-bladed NACA 0012 H-Darrieus VAWT with a trialing edge twist, its design and operating data are taken from that available in the literature. In this chapter, an integrated method that combines entropy with MCRA method is proposed. Here, entropy is employed for the determination of priority weights of the design and operating parameters of VAWT while MCRA method for optimal selection of operating parameters for VAWT. Next, a total of 55 trials are performed by varying five input parameters (WS, turbine rotational speed [TRS], BS, TSR, and wind blockage factor [WBF]), and the corresponding output or performance parameters, namely torque (*T*), power coefficient (PC), and torque coefficient (TC) are determined. From different combinations of input/output parameters, the most optimal parameters for improved performance of VAWT in low wind speed condition are determined via the proposed entropy−MCRA method. In addition, a parametric analysis is performed to determine the effect of each parameter on the performance of the considered VAWT. Then, the optimal performance value of the VAWT is verified by a confirmatory test.

The remaining parts this chapter are outlined in the following manner. In Section 19.2, the related works are discussed followed by the contribution of the present chapter. Next, the considered straight-bladed H-Darrieus VAWT design is elaborated along with discussing the experimental descriptions and formulations in Section 19.3. In Section 19.4, the proposed integrated entropy−MCRA method is introduced explaining the detailed methodology adopted in the method. It is then followed by the description of the modeling strategies of the VAWT in Section 19.5. In Section 19.6, the results of modeling and optimization of the considered VAWT are presented along with the results of a confirmatory test that is conducted after the optimization, then finally, in Section 19.7, the conclusions from this study are drawn and the scope for future work is enumerated.

19.2 Related work

The attempts to improve the performance of H-Darrieus VAWTs are many as can be understood by referring to the existing literature. The tangential thrust for the generation of torque is created on the blades of a VAWT when wind strikes the blades. The various researchers have given dedicated efforts to understand the VAWT aerodynamics and have worked on the blade design aspects to improve the PC and TC as well as the operating TSR of VAWT. The operating wind speed condition is dependent on the geographical location of a place, which is a basic input parameter that cannot be controlled by normal approach. On the contrary, by improving the VAWT's aerodynamic performance through blade design modifications the performance of a VAWT can be increased. In this section, a summary of some of the recent works done in this area is given. If looked back at the literature, it can be seen that the designs such as symmetrical/unsymmetrical blades, flexible blades, flexible leading and trialing edges, trialing edge blade twist, blade pitch angle, and solidity of VAWT configuration contribute to the augmentation of power and PC of VAWT.

Apart from that blade thickness, chord length and other similar design dimensions also affect the performance of VAWT. Sengupta, Biswas, and Gupta (2017a) investigated the effect of different blade parameters such as thickness-to-chord ratio and camber position on the static and dynamic performance of three different straight-bladed H-Darrieus turbines for low wind speed condition in the range 3.0−8.0 m/s. The maximum value of PC 0.19 was obtained for a straight unsymmetrical S815 blade H-Darrieus VAWT at a TSR 1.43. The PC was increased by 21.7% for their considered VAWT as compared to straight symmetric NACA blade VAWT. Bouzaher and Hadid (2017) did a numerical investigation on flexible blades. They reported that blades with flexible leading edge improve the turbine efficiency. The PC was found to be increased by 35% compared to rigid blade turbine. Bouzaher, Hadid, and Semch-Eddine (2017) also improved the performance of VAWT with deformable flexible blades. The PC in this case was increased by 38% compared to the original straight-bladed VAWT. Yang, Li, Zhang, Guo, and Yuan (2017) investigated the dynamic aerodynamic performance of VAWT with airfoil trialing edge flap and reported that the PC was increased by 10% with flap control compared to traditional VAWT. Lam, Liu, Peng, Lee, and Liu (2018) conducted a series of wind tunnel tests to investigate the effect of turbine design dimensions on the performance of a straight-bladed VAWT and reported that optimal design parameter selection leads to a broader range of TSR. Chen et al. (2017) investigated in detail a novel H-Darrieus VAWT with two numbers of blades. The performance of the VAWT was increased by increasing the distance between the two blades on the same radial arm. Sengupta, Biswas, and Gupta (2017b) investigated the aerodynamics of asymmetric and symmetric blade H-Darrieus turbine in low wind speed condition and obtained the maximum PC at wind speed 6.0 m/s. Ouro, Stoesser, and Ramirez (2018) investigated the effect of blade camber on asymmetric blade and found that the VAWT performance increased at low TSR. Qamar and Janajreh (2017) made a comprehensive investigation on cambered H-Darrieus VAWT and obtained higher PC at a lower TSR for high solidity of the VAWT design. Guo et al. (2019) investigated the aerodynamics of VAWT for different blade pitch angles and reported an increase of PC by 17% with blade pitch angle. Sagharichi, Zamani, and Ghasemi (2018) investigated the effect of solidity of VAWT design on the performance of variable pitch angle VAWT and reported positive effect of variable pitch on the PC.

Many studies in the literature were done on wind turbine optimization using MCDM method like Babu, Raju, Reddy, and Rao (2006) presented the application of MADM method, that is, TOPSIS with fuzzy linguistic variables for selection of optimal material for wind turbine blade. Their proposed method provided a best possible material, that is, composite material for wind turbine. Lee, Hung, Kang, and Pearn (2012) proposed a comprehensive evaluation model, which consists of interpretive structural modeling and fuzzy analytic network process for suitable selection of turbine for a wind farm. The work considered Taiwan as a place for their case study, which identified the most suitable turbine for installation there. The result showed that comprehensive model gave better and comparable turbine for Taiwan under the considered conditions. José et al. (2014) presented the optimization of the performance of a small horizontal-axis wind turbine (SHAWT) using a novel integrated method. The method was based on virtual genetic algorithm and the simulated annealing algorithm. The work performed experimentation on the SHAWT by varying three important parameters mainly, airfoil shape, chord length, twist angle, and thickness along the blade span position. It was concluded that aerodynamic efficiency of SHAWT was improved by changing the blade design. Lam and Meredith (2017) investigated the effect of design parameters on the performance of a VAWT subjected to realistic unsteady wind conditions. Thirteen VAWT design configurations were tested, and optimal design under urban/suburban environment was determined. The work considered four major input conditions such as height-to-diameter aspect ratio, blade airfoil shape, turbine solidity, and turbine moment of inertia, and the efficiency of the 13 turbine systems was calculated. The result indicated that VAWT could able to harvest the wind energy in the urban area as well as the suburban area and provided best results at a height of 9 m on the rooftops of commercial buildings. Mostafa (2017) employed three different MCDM methods for finding the best suitable location of wind turbine site. The work used data envelopment analysis method for prioritizing and ranking the cities, whereas analytical hierarchy process (AHP) and fuzzy technique for order of preference by similarity to ideal solution (FTOPSIS) methods were used to assess the validity of the results. According to the results obtained from these three methods, the city of Izadkhast was recommended as the best location for the construction of a wind farm. Harsh, Dipankar, Vlad, and Mihaela (2019) used MCDM methods to select the best possible alternatives for hybrid wind farm operation. Three methods mainly SAW (simple additive weighting), TOPSIS, and COPRAS (complex proportional assessment) were implemented by considering three set of criteria such as wake, wind curtailment, and forced outages. Comparative analysis was performed to judge the best results for hybrid wind farm and the same suggested some scope for future works. Okokpujie et al. (2020) presented the application of MCDM method for material selection for a HAWT blade. The work used AHP and TOPSIS methods for selection of optimal material among aluminum alloy, stainless steel, glass fiber, and mild steel. Results showed that the performance score for aluminum alloy was the best among all the materials.

19.2.1 Research gap and contribution of the present chapter

It can be observed from the aforementioned description of the related works that VAWT's performance had been investigated in the past. Parametric investigations of different design parameters, for example, turbine aspect ratio, chord length, airfoil shape, and turbine solidity, and operating parameters such as wind speed and TSR were performed to improve the performance of VAWT. But, the application of MCDM-based modeling approach to optimize VAWT performance is very few. At large, MCDM was applied for optimal selection of turbine blade material, selection of optimal site for sitting turbines, selection of a turbine type for a wind farm, etc. In addition, MCDM technique was applied to improve the aerodynamic efficiency of HAWT, considering different design parameters. However, as discussed earlier, for the complex aerodynamics of small-sized VAWT, the performance of VAWT is largely affected not only by the design parameters but also the changing operating parameters of VAWT in an urban location. Improper selection of operating and design parameters may lead to the error in the results, which ultimately affect the performance of VAWT. To handle the large complexities of operations under variable conditions, an MCDM-based modeling and optimization technique is deemed to be a promising approach to optimize small-sized VAWT performance. However, such a study is hardly available in the existing literature.

Thus this chapter contributes in the sense that it considers the modeling and optimization of a small-sized straight-bladed NACA 0012 H-Darrieus VAWT under different operating input parameters at different low wind speeds, which is performed by using an integrated method, that is, entropy with MCRA method, which is hardly available in the existing literature. In the operating conditions, blockage effects from the wind tunnel test section, as it comes into picture during experimentation on VAWT, are included, which have not been paid much attention in the optimization of VAWT performance done by using other approaches.

19.3 Turbine design and experimental description

The present VAWT is designed with two NACA 0012 blades having trialing edge twist and straight blade section along the vertical height as shown in Fig. 19.1. The VAWTs have different blade profiles of NACA series, which are classical blade forms developed by the National Advisory Committee for Aeronautics (NACA). Out of different NACA series blades for VAWT, NACA 0012 is a prominent form due to its high tangential thrust-to-weight ratio, useful for low wind speeds. An angular twist of 30 degrees for 10% of chord distance from the trialing edge is provided as shown in Fig. 19.1A. The blade twist is designed to make the turbine self-starting under low wind speed condition as twist reduces the negative blade wetted area when the wind sweeps past the blades. In a previous study on blade twist effect for NACA 0012 blade, it was reported that the trialing edge blade twist of 30 degrees results in a largest tangential thrust for a small-sized straight-bladed VAWT (Chen et al., 2017; Guo et al., 2019; Ouro et al., 2018; Qamar & Janajreh, 2017; Sagharichi et al., 2018; Sengupta et al., 2017b). The tip geometry can locally modify the inflow dynamic pressure and can control the performance of the turbine. The height of each blade is 20 cm, and chord length is 4.9 cm as shown in Fig. 19.1B. The blade thickness is considered as 5 mm. The blades made from lightweight aluminum are mounted on four numbers of mild steel bolts of 5 mm diameter and 12 cm length as shown in Fig. 19.1B. The central shaft of the turbine is 1.5 cm in diameter and 25 cm in length. By changing the overall diameter of the turbine but maintaining the height constant, 11 numbers of height-to-diameter (H/D) ratios, namely 0.85, 1.0, 1.10, 1.33, 1.54, 1.72, 1.80, 1.92, 2.0, 2.1, and 2.2 are created. Ball bearings are used to support the central shafts at the base of the turbines. The base is 7 cm wide and 2.4 cm thick as shown.

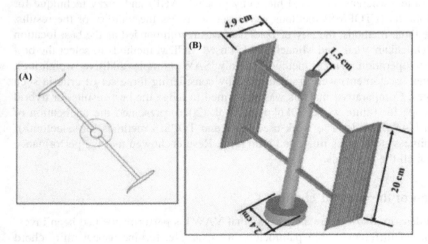

FIGURE 19.1 Straight-bladed NACA 0012 H-Darrieus VAWT with a trailing edge twist having Aluminium blades- (A) Top view, (B) Front view.

TABLE 19.1 Experimental results of vertical-axis wind turbine.

Trial no.	Input parameters					Output parameters		
	Turbine rotational speed (rpm)	Blade speed (m/s)	Wind speed (m/s)	Wind blockage factor	Tip speed ratio	Torque	Power coefficient	Torque coefficient
1	532	6.554	4.892	0.223	1.339	1.494	0.121	0.089
2	900	11.089	3.994	0.2233	2.776	2.411	0.167	0.058
3	820	10.103	2.824	0.2233	3.577	2.491	0.152	0.042
4	817	10.066	3.994	0.2233	2.521	2.568	0.147	0.061
5	777	9.573	4.892	0.2233	1.956	2.491	0.139	0.071
6	779	8.158	2.824	0.265	2.801	1.079	0.143	0.049
7	1224	12.819	3.994	0.265	3.209	1.287	0.202	0.063
8	1340	14.034	3.994	0.265	3.513	1.325	0.219	0.062
9	1377	14.421	4.892	0.265	2.947	1.571	0.202	0.069
10	1367	14.317	3.994	0.265	3.584	1.591	0.205	0.057
11	971	9.244	5.648	0.293	1.636	0.563	0.168	0.102
12	1394	13.271	5.648	0.293	2.349	0.751	0.217	0.092
13	1498	14.261	5.648	0.293	2.524	0.719	0.237	0.093
14	1434	13.652	4.892	0.293	2.791	0.735	0.232	0.083
15	1391	13.242	3.994	0.293	3.315	0.751	0.231	0.069
16	1375	10.829	6.315	0.361	1.714	0.567	0.144	0.084
17	1695	13.349	7.988	0.361	1.671	0.599	0.162	0.098
18	1849	14.562	8.931	0.361	1.631	0.771	0.157	0.097
19	1800	14.176	8.931	0.361	1.587	0.749	0.154	0.097
20	1774	13.971	9.367	0.361	1.491	0.846	0.144	0.096
21	1844	12.541	6.918	0.422	1.812	0.239	0.172	0.094
22	1858	12.636	6.918	0.422	1.833	0.271	0.167	0.091
23	1870	12.717	7.472	0.422	1.701	0.279	0.161	0.093
24	1857	12.629	6.918	0.422	1.825	0.271	0.167	0.091
25	1814	12.336	6.918	0.422	1.783	0.255	0.166	0.094
26	1338	8.146	5.648	0.476	1.442	0.102	0.129	0.089
27	1700	10.351	6.315	0.476	1.638	0.121	0.148	0.091
28	1831	11.148	6.315	0.476	1.765	0.147	0.153	0.087
29	1720	10.472	6.918	0.476	1.513	0.134	0.139	0.092
30	1828	11.131	6.315	0.476	1.762	0.121	0.159	0.091
31	1700	9.891	7.472	0.501	1.323	0.099	0.121	0.091
32	1733	10.082	7.472	0.501	1.349	0.099	0.123	0.091
33	1806	10.507	7.472	0.501	1.406	0.116	0.126	0.091
34	1831	10.652	6.918	0.501	1.532	0.116	0.136	0.088
35	1833	10.664	6.918	0.501	1.541	0.128	0.134	0.087
36	1226	6.687	6.315	0.537	1.058	0.077	0.088	0.083
37	1646	8.978	7.472	0.537	1.201	0.107	0.101	0.083
38	1745	9.518	7.472	0.537	1.273	0.066	0.109	0.086
39	1768	9.643	7.472	0.537	1.291	0.128	0.105	0.081

(Continued)

TABLE 19.1 (Continued)

Trial no.	Input parameters					Output parameters		
	Turbine rotational speed (rpm)	Blade speed (m/s)	Wind speed (m/s)	Wind blockage factor	Tip speed ratio	Torque	Power coefficient	Torque coefficient
40	1788	9.753	7.988	0.537	1.221	0.133	0.102	0.083
41	1335	6.991	8.931	0.562	0.782	0.094	0.064	0.075
42	1672	8.755	8.473	0.562	1.033	0.099	0.084	0.081
43	1603	8.394	8.473	0.562	0.991	0.118	0.081	0.082
44	1713	8.971	8.378	0.562	1.071	0.101	0.088	0.082
45	1664	8.713	7.988	0.562	1.091	0.104	0.087	0.081
46	1054	5.256	7.786	0.593	0.675	0.034	0.049	0.073
47	1333	6.648	8.931	0.593	0.744	0.048	0.055	0.074
48	1401	7.001	8.931	0.593	0.784	0.056	0.059	0.076
49	1447	7.216	9.195	0.593	0.831	0.131	0.064	0.077
50	1532	7.641	9.195	0.562	0.834	0.107	0.063	0.076
51	1164	5.541	9.108	0.625	0.608	0.081	0.044	0.073
52	1353	6.441	8.931	0.625	0.721	0.121	0.051	0.071
53	1300	6.188	9.108	0.625	0.679	0.107	0.049	0.072
54	1377	6.554	9.108	0.625	0.719	0.111	0.052	0.072
55	1385	6.592	9.367	0.625	0.703	0.105	0.051	0.072

The performance of this turbine model is investigated through experimental tests in an open circuit subsonic wind tunnel at different low wind speeds of less than 10 m/s. The details of the wind tunnel can be found in Gupta, Das, and Sharma (2006). The different operating parameters for the turbine as the input are considered such as WS, TRS, BS, TSR, and WBF. The output parameters studied are T, PC, and TC. By varying the input values of these parameters, a set of different output parameters are obtained, which are used in the modeling and optimization of the turbine performance. The details of the values obtained are highlighted in Table 19.1.

For each of the 55 trials as shown, the output parameters are determined by using the following expressions. The power coefficient of wind turbine can be expressed as:

$$C_p = \frac{P_{turbine}}{P_{max}} \tag{19.1}$$

where $P_{turbine}$ is the power output from the wind turbine, which is the product of aerodynamic torque and the blade tip speed and P_{max} is the useful power in the wind, which is given by:

$$P_{max} = \frac{1}{2} \rho A V_{free\_block}^3 \tag{19.2}$$

where V_{free_block} is the wind velocity under the effect of wind tunnel blockage, which is given by:

$$V_{free\_block} = V_{free}^*(1 + \varepsilon) \tag{19.3}$$

The torque coefficient of the turbine can be calculated from the following:

$$C_t = \frac{C_P}{\lambda} \tag{19.4}$$

The TSR is the ratio of blade tip velocity to the wind velocity, which is given by:

$$\lambda = \frac{R\omega}{V} \tag{19.5}$$

When a turbine is placed inside a wind tunnel, it increases the local free stream wind velocity in the test section by creating blockage to the flow. In wind tunnel testing, wind tunnel blockage effect should be taken into consideration to determine the actual power produced by the turbine. The total blockage correction factor is the sum of the correction factors for three major blockage effects namely, solid blockage, wake blockage, and tunnel sidewall interferences. The turbine creates blockage to the flow by its physical presence in the tunnel, called solid blockage. As the turbine rotates, there is a deficit of velocity downstream of turbine followed by a wake zone. The effect due to this is called wake blockage. In addition, sidewall interferences are also to be considered for the small-sized test section (30×30 cm). The solid blockage correction factor as given in Blackwell, Sheldahl, and Feltz (1978) can be expressed as:

$$\phi = \frac{A_F}{4A_{TS}} = \frac{A_F}{4H'W} \tag{19.6}$$

The wake blockage correction factor is given by:

$$\beta = \frac{q_c - q_u}{q_u} \tag{19.7}$$

Further,

$$\frac{q_c}{q_u} = \frac{C_{d,u}}{C_{d,c}} \tag{19.8}$$

where

$$\frac{C_{d,u}}{C_{d,c}} = \left[1 + \frac{1}{4}\left(\frac{A_F}{A_S}\right)\right]^2 \tag{19.9}$$

And the correction factor for sidewall boundary interference (γ) is derived by using the concept of Glauert correction methodology (Langer, Peterson, & Maier, 1996) for fixed pitch blades, which can be expressed for the present turbine as (Gupta & Biswas, 2011)-

$$\gamma = \frac{2n\delta_w A_S}{4A_{TS}} = \frac{\delta_w n^2 c/H}{2H'W} \tag{19.10}$$

where δ_w is the wall correction factor for open circuit rectangular wind tunnel. Its standard value for rectangular bladed turbine is approximately 0.1383 (Caliskan, Kursuncu, Kurbanoglu, & Guven, 2013).

Therefore the total blockage correction factor can be expressed as:

$$\varepsilon = \phi + \beta + \gamma \tag{19.11}$$

19.4   Integrated entropy—multicriteria ratio analysis method

An integrated entropy—MCRA method is proposed for modeling and optimization of the considered VAWT. In this method, entropy (Caliskan et al., 2013; Emma, Kieran, & Vida, 2013) is employed for extraction of precise priority weights of the criteria which influences the ranking of alternatives while MCRA (Brauers & Zavadskas, 2006; Karande & Chakraborty, 2012; Sharma, Muqeem, Sherwani, & Ahmad, 2018) for optimization of VAWT parameters and selecting the optimal experimental trials that gives better performance of VAWT. The priority weight for each of the criteria provides significance of the criteria among each on the performance of the system. If the criteria weights are higher, it means that criteria will be having more impact on the alternatives. The schematic flow chart of E-MCRA is shown in Fig. 19.2.

The detailed steps of E-MCRA method (Brauers & Zavadskas, 2006; Emma et al., 2013; Karande & Chakraborty, 2012; Sharma et al., 2018) are explained as follows:

Step 1: Identification of process attributes

In this step, first identification of process criteria, subcriteria, and alternatives required for evaluation of the performance of VAWT is done. This is done on the basis of previous studies and handbook of VAWT system. Sometime decision maker will decide the criteria, subcriteria, and alternatives for performance study based on the application

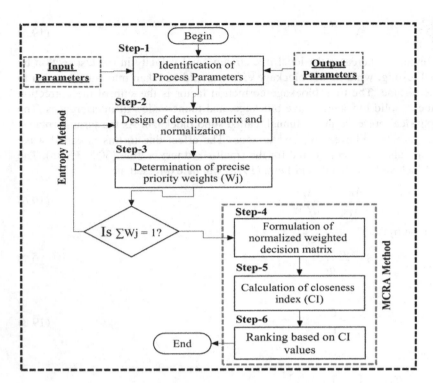

FIGURE 19.2 Schematic flow chart of E-MCRA method for VAWT.

conditions. The criteria and subcriteria of the VAWT consist of both beneficial and nonbeneficial criteria as well as qualitative and quantitative parameters that influence the VAWT performance. Beneficial criteria are those that have positive impact on system performance, that is, increment or decrement of the criteria will have always higher performance of the system, while nonbeneficial criteria are those that have negative impact on the performance.

Step 2: Design of decision matrix and normalization

After selecting the process parameters for the system, next is to design the decision matrix using Eq. (19.12). The decision matrix mainly consists of performance values of a number of criteria (each column element) and a corresponding number of alternatives (each row elements) (Emma et al., 2013; Karande & Chakraborty, 2012). The decision matrix is designed either by conducting the experimental trials or based on the opinion of the decision experts for the selected system. Then, normalization of the decision matrix is carried out using Eqs. (19.13)−(19.14). This process is carried out in order to convert the different dimensional units of performance values into comparable values in the range of 0−1. During the normalization process, higher the better performance values are considered as beneficial criteria while lower the better as nonbeneficial criteria. The decision matrix is formulated using the following expression:

$$\text{DM}_{ij} = \begin{bmatrix} & C_1 & C_2 & \ldots & C_n \\ A_1 & X_{11} & X_{12} & \ldots & X_{1n} \\ A_2 & X_{21} & X_{22} & \ldots & X_{2n} \\ \ldots & \ldots & \ldots & \ldots & \ldots \\ A_m & X_{m1} & X_{m2} & \ldots & X_{mn} \end{bmatrix} \tag{19.12}$$

where, $C_1, C_2, \ldots C_n$ denote the number of criteria, $A_1, A_2, A_3 \ldots A_m$ signify the number of alternatives, $X_{11}, X_{12}, \ldots X_{mn}$ denote performance values of each criteria corresponding to each alternative, n is the number of criteria, and m denotes the number of alternatives in the process.

$$[N_{ij}]_B = \frac{\text{DM}_{ij}}{\text{DM}_{bigger}} \tag{19.13}$$

$$[N_{ij}]_{NB} = \frac{\text{DM}_{small}}{\text{DM}_{ij}} \tag{19.14}$$

where $[N_{ij}]_B$ and $[N_{ij}]_{NB}$ represent normalized decision matrix for beneficial and nonbeneficial criteria, DM_{bigger} signifies the beneficial element of a column, and $DM_{smaller}$ is the nonbeneficial element of a column (Emma et al., 2013; Karande & Chakraborty, 2012).

Step 3: Determination of precise priority weights

Next, precise priority weights for each of the criteria are determined using Eqs. (19.15)–(19.17). First, the degree of divergence (d_j) for each of the criteria is calculated, which measures the distance of each criteria with respect to the comparable sequence. After that precise priority weights are determined using Eq. (19.17).

$$R_j = -k \sum_{i=1}^{m} \left([N_{ij}]_B \quad \text{and} \quad [N_{ij}]_{NB} \right) \ln \left([N_{ij}]_B \text{ or} [N_{ij}]_{NB} \right) \quad j = 1, 2, \ldots, n \tag{19.15}$$

$$d_j = |1 - R_j| \tag{19.16}$$

$$\theta_j = \frac{d_j}{\sum_{j=1}^{n} d_j} \tag{19.17}$$

where R_j is the entropy matrix, k is a constant term, and its value is determined by $[k = 1/\ln m, \quad 0 \le d_j \le 1]$, d_j denotes the degree of divergence for each of the criteria, θ_j is the precise priority weights for each of the criteria.

Step 4: Formulation of normalized weighted decision matrix

In this step, formulation of normalized weighted decision matrix is developed using Eq. (19.18). This matrix is developed by multiplying each precise priority weight obtained using Eq. (19.18) with normalized performance values of each criteria (Brauers & Zavadskas, 2006; Emma et al., 2013). This step is essential because a priority weight gives information about the most significant parameter that affects the alternatives thereby affecting the performance of the system.

$$W_{ij} = \theta_j \times \left([N_{ij}]_B \text{and} [N_{ij}^*]_{NB} \right) \tag{19.18}$$

where $[N_{ij}]_B$ and $[N_{ij}]_{NB}$ are the normalized decision matrix for beneficial and nonbeneficial criteria, respectively, W_{ij} represents the weighted normalized decision matrix, and θ_j denotes the precise priority weights for each of the criteria.

Step 5: Calculation of closeness index

In this step, closeness index (CI) values for each of the alternatives are determined using Eq. (19.21) followed by beneficial and nonbeneficial values for each of the criteria separately using Eqs. (19.19)–(19.20). This step converts the multiresponse problems into single response problems with considering the beneficial and nonbeneficial objects into account (Brauers & Zavadskas, 2006; Emma et al., 2013; Karande & Chakraborty, 2012; Sharma et al., 2018).

$$\zeta^+ = \sum_{i=1}^{b} W_{ij} \tag{19.19}$$

$$\zeta^- = \sum_{i=b+1}^{n} W_{ij} \tag{19.20}$$

$$CI = \frac{\zeta^-}{\zeta^+ + \zeta^-} \tag{19.21}$$

where ζ^+ and ζ^- are the sum of object values within the each of the alternatives, respectively; b denotes the number of beneficiary objects; n represents the total number of objects; CI is the closeness index.

After the determination of CI values, ranking of alternatives is done. Ranking of the alternatives will be done based on the CI values and based on the specified rank, the optimal alternatives will be established. The alternative with higher CI values yields optimal result compared to alternative with lower CI values.

19.5 Modeling of vertical-axis wind turbine using integrated entropy–multicriteria ratio analysis method

Modeling of VAWT using entropy–MCRA method (Brauers & Zavadskas, 2006; Emma et al., 2013; Karande & Chakraborty, 2012; Sharma et al., 2018) is discussed in this section. In modeling, first identification of process parameters, that is, input parameters and corresponding output parameters of the VAWT is done (Theodorsen, 1996). The present work considers mainly five input parameters, such as rotation rate (TRS) in rpm, BS in rpm, WS in m/s, WBF, TSR, and three output parameters, namely T, PC, and TC (Table 19.1). Second, design of decision matrix is done using Eq. (19.12). Here decision matrix is developed via experimentation done on the present VAWT at thermal lab NIT Silchar. A total 55 trials are performed for different input conditions of rotation rate (TRS) in rpm, BS in rpm, WS in m/s, WBF, TSR, and corresponding experimental results are tabulated in Table 19.1.

After that, normalization of the decision matrix is done using Eqs. (19.13)−(19.14), and the result of the normalization is given in Table 19.2. As the output parameters (T, PC, TC) of VAWT system have different units, normalization process is done to make them to a comparable sequence data. During the normalization process, all the output parameters of VAWT system are considered as higher the better performance values.

In the third step, evaluation of priority weights of output parameters, that is, T, PC, and TC are done using entropy method (Emma et al., 2013; Jagadish & Amitava, 2014a, 2014b, 2014c; Sandeep, Vidyapati, Jagadish, & Amitava, 2014; Sharma et al., 2018). During the calculation of priority weights, first reference range for each of the output parameters, that is, T, PC, and TC are calculated using Eq. (19.15), which measures the target value of the entire VAWT decision matrix data. After that, the distance of each VAWT parameter from the reference value, that is, degree of divergence (d) and corresponding precise priority weights (θ) are evaluated using Eqs. (19.16)−(19.17). The result of priority weights of VAWT is tabulated in Table 19.3. The results show that parameter T (0.5786) yields higher weights signifying higher impact on VAWT performance compared to the PC and TC. Hence, designer has to take more care of parameter, T, during the design and study of performance of VAWT system.

After the priority weights, formulation of a normalized weighted decision matrix is carried out using Eq. (19.18), in which precise priority weights obtained using entropy method for each of the VAWT parameters are multiplied with normalized performance values of the corresponding VAWT parameters (Jagadish & Amitava, 2014a, 2014b, 2014c; Sandeep et al., 2014). The result of weighted normalized decision matrix is tabulated in Table 19.4.

Similarly, evaluation of CI values for each of the VAWT parameters, that is, T, PC, and TC is done using Eq. (19.21). This converts the multiattribute problem into a single attribute problem with considering the effect of beneficial and nonbeneficial criteria of the process (Emma et al., 2013; Jagadish & Amitava, 2014a, 2014b, 2014c; Karande & Chakraborty, 2012; Sandeep et al., 2014; Sharma et al., 2018). At last, the ranking of CI values is done in ascending order; the ranks of each individual alternative are obtained.

19.6 Results and discussion

The performance of the present H-Darrieus VAWT can be studied by analyzing the parametric variations of the VAWT performance parameters, so that the performance can be modeled using the proposed MCDM method, and the same can ultimately be optimized for the given set of input parameters. In this section, the results of parametric analysis of the VAWT and the results of MCDM modeling are extracted for further analysis to understand the suitable performance of the present straight-bladed H-Darrieus VAWT.

19.6.1 Parametric analysis

In the parametric analysis, the variations of the performance parameters, for example, the PC, TC, and aerodynamic torque of the straight-bladed H-Darrieus VAWT are studied by varying the operating parameters such as TSR and TRS. Figs. 19.3 and 19.4 show the variations of PC and TC with respect to the changes in TSR. As can be seen, both PC and TC start increasing from a level less than 10% as the TSR is increased. Both the variations of output parameters have positive slope until they reach their corresponding maximum values at different operating TSR conditions. The maximum level of PC and TC corroborate to the rated performance, beyond which they are unable to maintain that performance as the values drop with the changes in operating TSR conditions from the level at the rated performance. This feature is also a common perception in the performance of VAWT that their performance is maintained at the highest level in a narrow range of operating TSR value. Next, in terms of the maximum value of PC and TC, it can be referred from Figs. 19.3 and 19.4 that maximum PC is greater than the maximum TC, and the TSR values at which these maxima occur are also different. More specifically, TSR is higher with regards to the rated PC than the rated TC. From this it can be inferred that attaining PC at its maximum or rated level signifies the highest attainable efficiency of the VAWT, as out of these two coefficients, PC is more significant in terms of its contribution in efficiency of the device.

Then, these performance coefficients are also dependent on the operating TRS. Figs. 19.5−19.7 show the variations of PC, T, and TC with respect to the TRS. The study of the effect of TRS is important from the perspective of required

TABLE 19.2 Normalized output parameters and priority weights of vertical-axis wind turbine.

Trial no.	Torque	Power coefficient	Torque coefficient	Trial no.	Torque	Power coefficient	Torque coefficient	Trial no.	Torque	Power coefficient	Torque coefficient
1	0.0353	0.1106	0.2415	21	0.0056	0.1572	0.2551	41	0.0022	0.0585	0.2035
2	0.0569	0.1526	0.1574	22	0.0064	0.1526	0.2470	42	0.0023	0.0768	0.2198
3	0.0588	0.1389	0.1140	23	0.0066	0.1471	0.2524	43	0.0028	0.0740	0.2225
4	0.0606	0.1343	0.1655	24	0.0064	0.1526	0.2470	44	0.0024	0.0804	0.2225
5	0.0588	0.1270	0.1927	25	0.0060	0.1517	0.2551	45	0.0025	0.0795	0.2198
6	0.0255	0.1307	0.1330	26	0.0024	0.1179	0.2415	46	0.0008	0.0448	0.1981
7	0.0304	0.1846	0.1710	27	0.0029	0.1352	0.2470	47	0.0011	0.0503	0.2008
8	0.0313	0.2001	0.1683	28	0.0035	0.1398	0.2361	48	0.0013	0.0539	0.2063
9	0.0371	0.1846	0.1873	29	0.0032	0.1270	0.2497	49	0.0031	0.0585	0.2090
10	0.0376	0.1873	0.1547	30	0.0029	0.1453	0.2470	50	0.0025	0.0576	0.2063
11	0.0133	0.1535	0.2768	31	0.0023	0.1106	0.2470	51	0.0019	0.0402	0.1981
12	0.0177	0.1983	0.2497	32	0.0023	0.1124	0.2470	52	0.0029	0.0466	0.1927
13	0.0170	0.2166	0.2524	33	0.0027	0.1151	0.2470	53	0.0025	0.0448	0.1954
14	0.0173	0.2120	0.2253	34	0.0027	0.1243	0.2388	54	0.0026	0.0475	0.1954
15	0.0177	0.2111	0.1873	35	0.0030	0.1224	0.2361	55	0.0025	0.0466	0.1954
16	0.0134	0.1316	0.2280	36	0.0018	0.0804	0.2253				
17	0.0141	0.1480	0.2660	37	0.0025	0.0923	0.2253				
18	0.0182	0.1435	0.2632	38	0.0016	0.0996	0.2334				
19	0.0177	0.1407	0.2632	39	0.0030	0.0959	0.2198				
20	0.0200	0.1316	0.2605	40	0.0031	0.0932	0.2253				

TABLE 19.3 Priority weights of vertical-axis wind turbine.

Weights (θ)	Torque	Power coefficient	Torque coefficient
	0.5786	0.2503	0.1711

shaft speed that is to be maintained for the turbo generator. These variations with respect to TRS follow the same trend as that seen with regards to the changing TSR. Here as well, the turbine will have its rated power and T if the turbine shaft revolves at the required TRS value. Now following the same pattern of variation like previous, the maximum PC and T occurs at different TRS; in fact the maximum PC occurs at a higher TRS than that of maximum T, as can be seen from Figs. 19.5 and 19.6. This happens when T is maximum, the rotational speed of the shaft drops to a lower value and when power is maximum, the same increases to a higher value. In the same manner, variation of TC with respect to changing TRS can be seen in Fig. 19.7. The peak value of TC is obtained at the same TRS as the peak value of T because TC is calculated from the results of T.

19.6.2　Optimization of vertical-axis wind turbine parameters

The optimal setting for VAWT system is determined based on the CI values for each of the trial number. The optimal setting or optimal trial number is selected based on the maximum CI value, and corresponding results are tabulated in Table 19.5.

Based on the results of ranking (Table 19.5), it can be observed that, experimental trial no. 13 has higher CI values compared to other trial numbers. The higher CI values (1.285) for the trial no. 13 shows optimal operating condition for the VAWT. The optimal setting is obtained as TRS = 1498 rpm, BS = 14.261 m/s, WS = 5.648 m/s, WBF = 0.293, and TSR = 2.524, which give optimal performance responses, that is, T = 0.719 Nm, PC = 0.237, and TC = 0.093 that directly or indirectly increase performance of the VAWT.

19.6.3　Utility of the optimization results

Initially in the experimental tests conducted on the present straight-bladed H-Darrieus VAWT, various operating parameters have been considered such as TRS, BS, and TSR. Based on the wide range of variations of such input parameters in Table 19.1, the effects of these input parameters on the performance parameters, that is, T, PC, and TC are obtained in the parametric investigations done in Section 6.1. The data in Table 19.1 have indicated that the variations of output parameters have occurred in different scales and levels due to wide variations of input parameters, which are the real conditions that a VAWT needs to adapt. In case of decisions to be taken by the VAWT developers for its effective performance, it is a tough decision to make when choosing the best combination of the input parameters. Thus the present optimization process has identified the required combinations of input parameters for improved performance of the VAWT under the dynamic situations of ever-changing operating input parameters. The present finding is again useful under another condition of operation of VAWT, which is the blockage condition. Here in this work, the turbine has been tested in the wind tunnel that has created different blockage values resulted from the varying operating condition of wind speed and design condition of turbine frontal area. There is hardly any study that had considered the WBF variations in the optimization of the turbine performance. In that respect, the present work has accounted for this parameter as well. In order to connect with real operation of the optimized VAWT, these results will be very useful as variations of all the design and operating parameters including blockage effects are considered in order to find VAWT's optimized performance. To connect with reality, the blockage conditions of the tunnel can also be correlated with the actual blockage situation that might be prevalent in different sites due to the design of the surrounding terrains and also the variable operating wind speeds. Further, it can be noticed from the trial no. 13, at which the optimized values are obtained, that the optimization process leads to the highest performance of the VAWT in terms of its PC but not in terms of TC or T. Thus this can again be corroborated with the result of the parametric analysis, and it can be said that the highest PC signifies the highest attainable efficiency of the VAWT, as out of the two performance coefficients, PC is more significant in terms of its contribution in efficiency of the device. Hence, it is recommended that optimal setting, that is, TRS = 1498 rpm, BS = 14.261 m/s, WS = 5.648 m/s, WBF = 0.293, and TSR = 2.524 should be used for this VAWT for getting the higher efficiency. Furthermore, the work suggested that, the entropy—MCRA method can be used as a systematic framework model for selection of optimal output parameters for other mechanical systems as well.

TABLE 19.4 Weighted normalized output parameters of vertical-axis wind turbine.

Trial no.	Torque	Power coefficient	Torque coefficient	Trial no.	Torque	Power coefficient	Torque coefficient	Trial no.	Torque	Power coefficient	Torque coefficient
1	0.2452	0.1504	0.0371	21	0.5318	0.0843	0.0228	41	0.5649	0.2244	0.0770
2	0.0358	0.0908	0.1255	22	0.5245	0.0908	0.0314	42	0.5637	0.1984	0.0599
3	0.0176	0.1102	0.1711	23	0.5226	0.0986	0.0257	43	0.5594	0.2023	0.0570
4	0.0000	0.1167	0.1169	24	0.5245	0.0908	0.0314	44	0.5633	0.1932	0.0570
5	0.0176	0.1271	0.0884	25	0.5281	0.0921	0.0228	45	0.5626	0.1945	0.0599
6	0.3400	0.1219	0.1512	26	0.5631	0.1401	0.0371	46	0.5786	0.2438	0.0827
7	0.2925	0.0454	0.1112	27	0.5587	0.1154	0.0314	47	0.5754	0.2360	0.0799
8	0.2838	0.0233	0.1141	28	0.5528	0.1089	0.0428	48	0.5736	0.2308	0.0742
9	0.2276	0.0454	0.0941	29	0.5557	0.1271	0.0285	49	0.5564	0.2244	0.0713
10	0.2231	0.0415	0.1283	30	0.5587	0.1012	0.0314	50	0.5619	0.2257	0.0742
11	0.4578	0.0895	0.0000	31	0.5637	0.1504	0.0314	51	0.5678	0.2503	0.0827
12	0.4149	0.0259	0.0285	32	0.5637	0.1478	0.0314	52	0.5587	0.2412	0.0884
13	0.4222	0.0000	0.0257	33	0.5599	0.1439	0.0314	53	0.5619	0.2438	0.0856
14	0.4185	0.0065	0.0542	34	0.5599	0.1310	0.0399	54	0.5610	0.2399	0.0856
15	0.4149	0.0078	0.0941	35	0.5571	0.1336	0.0428	55	0.5624	0.2412	0.0856
16	0.4569	0.1206	0.0513	36	0.5688	0.1932	0.0542				
17	0.4496	0.0973	0.0114	37	0.5619	0.1764	0.0542				
18	0.4103	0.1037	0.0143	38	0.5713	0.1660	0.0456				
19	0.4153	0.1076	0.0143	39	0.5571	0.1712	0.0599				
20	0.3932	0.1206	0.0171	40	0.5560	0.1751	0.0542				

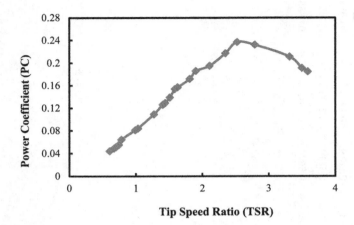

FIGURE 19.3 Variation of PC with regards to the changes in TSR.

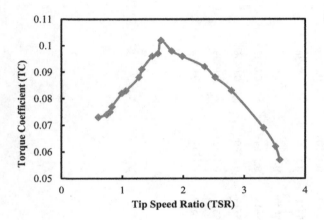

FIGURE 19.4 Variation of TC with regards to the changes in TSR.

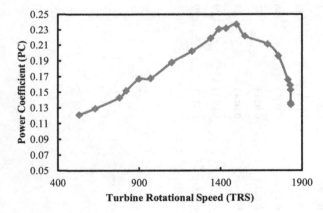

FIGURE 19.5 Variation of PC with regards to the changes in TRS (rpm).

19.6.4 Confirmatory test

In this section, confirmatory tests are conducted to verify the results of the optimal setting. The optimal setting, that is, TRS = 1498 rpm, BS = 14.261 m/s, WS = 5.648 m/s, WBF = 0.293, and TSR = 2.524 is used for confirmatory experiments, and the corresponding results are shown in Table 19.6. The confirmatory test results show that they are comparable and acceptable with regards to the experimental results for the optimal setting.

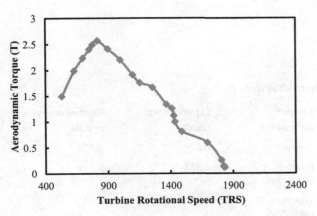

FIGURE 19.6 Variation of *T* with regards to the changes in TRS (rpm).

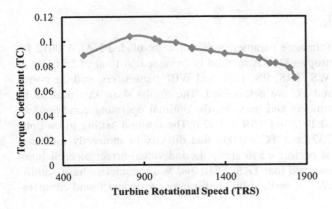

FIGURE 19.7 Variation of TC with regards to the changes in TRS (rpm).

TABLE 19.5 CI values of vertical-axis wind turbine.

Trial no.	CI values	Trial no.	CI values	Trial no.	CI values
1	0.067	21	1.225	41	1.246
2	0.037	22	1.232	42	1.260
3	0.734	23	1.239	43	1.239
4	0.624	24	1.251	44	1.246
5	0.604	25	1.251	45	0.915
6	0.475	26	1.242	46	0.896
7	0.464	27	1.177	47	0.908
8	0.071	28	1.160	48	0.857
9	0.056	29	1.156	49	1.241
10	0.924	30	1.246	50	1.249
11	1.006	31	1.268	51	1.248
12	0.907	32	1.235	52	1.245
13	1.285	33	1.233	53	1.277
14	1.247	34	1.160	54	1.273
15	1.253	35	1.168	55	1.234
16	1.251	36	1.249		
17	0.515	37	1.239		
18	0.907	38	1.242		
19	1.004	39	1.235		
20	0.987	40	1.262		

CI, closeness index.

TABLE 19.6 Confirmatory tests for vertical-axis wind turbine thermal system.

Input conditions	Output parameters	Experimental results	Confirmatory results
Test No-13 TRS = 1498 rpm, BS = 14.261 m/s, WS = 5.648 m/s, WBF = 0.293, and TSR = 2.524	Torque	0.729	0.726
	Power coefficient	0.237	0.232
	Torque coefficient	0.093	0.091

19.7 Conclusions and scope for future work

In this work, a detailed modeling and optimization of the performance parameters of a straight-bladed NACA 0012 H-Darrieus VAWT with a trialing edge twist using integrated entropy—MCRA method is discussed. A total of 55 trials of the experiments are performed on the VAWT by varying the WS, TRS, BS, TSR, and WBF parameters, and the corresponding output or performance parameters, namely T, PC, and TC are determined. The results show that trial no.13 yields higher CI value with rank 1 compared to other trial numbers and provides the optimal operating conditions as TRS = 1498 rpm, BS = 14.261 m/s, WS = 5.648 m/s, WBF = 0.293, and TSR = 2.524. The optimal setting gives optimal performance parameters, that is, $T = 0.719$ Nm, PC = 0.237, and TC = 0.093 that directly or indirectly increases performance of the VAWT. In addition, parametric analysis is performed to study the individual effect of each input parameters of VAWT on the output parameters. It has been observed that TRS, TSR, and WS parameters have significant effect on the PC. At last, optimal performance of the VAWT is verified by confirmatory tests and found comparable and acceptable with regards to its experimental results.

Thus the present study systematically underscores the effects of design and operating parameters to control the performance of the considered straight-bladed VAWT in order to give a direction for improvement of future VAWTs, which may be installed successfully in low wind speed of urban conditions. Future research in this direction further entails the optimization of other blade parameters, for example, blade twist as it affects the blade—fluid interactions near the trialing edge, which can also control the performance of the straight-bladed H-Darrieus VAWT. It is also observed that there is a huge scope of application of other MCDM methods for modeling and optimization of other thermal systems.

References

Babu, K. S., Raju, N. S., Reddy, M. S., & Rao, D. N. (2006). The material selection for typical wind turbine blades using a MADM approach & analysis of blades. In *Proceedings of 18th international conference on multiple criteria decision making* (MCDM) (pp. 19—23), Chania, Greece, June,.

Bhuyan, S., & Biswas, A. (2014). Investigations of self-starting and performances of simple H and hybrid H-Savonius vertical axis wind rotors. *Energy Conversion and Management, 87*, 859—867.

Blackwell, B. F., Sheldahl, R. E., & Feltz, L. V. (1978). Wind tunnel performance data for two & three bucket S-rotors. *International Journal of Energy, 2*, 160—164.

Bouzaher, M. T., & Hadid, M. (2017). Numerical investigation of a vertical axis tidal turbine with deforming blades. *Arabian Journal for Science and Engineering, 42*, 2167—2178.

Bouzaher, M. T., Hadid, M., & Semch-Eddine, D. (2017). Flow control for the vertical axis wind turbine by means of flapping flexible foils. *Journal of the Brazilian Society of Mechanical Sciences and Engineering, 39*, 457—470.

Brauers, W. K. M., & Zavadskas, E. K. (2006). The MOORA method and its application to privatization in a transition economy. *Control and Cybernetics, 35*(2), 445—469.

Caliskan, H., Kursuncu, B., Kurbanoglu, C., & Guven, S. Y. (2013). Material selection for the tool holder working under hard milling conditions using different multi criteria decision making methods. *Materials and Design, 45*, 473—479.

Chen, J., Liu, P., Xu, H., Chen, L., Yang, M., & Yang, L. (2017). A detailed investigation of a novel vertical axis Darrieus wind rotor with two sets of blades. *Journal of Renewable and Sustainable Energy, 9*, 013307. Available from https://doi.org/10.1063/1.4977004.

Emma, M., Kieran, S., & Vida, M. (2013). An assessment of sustainable housing affordability using a multiple criteria decision making method. *Omega, 41*(2), 270—279.

Francesco, B., Alessandro, B., Carnevale, E. A., Lorenzo, F., & Sandro, M. (2012). Feasibility analysis of a Darrieus vertical-axis wind turbine installation in the rooftop of a building. *Applied Energy, 97*, 921—929.

Guo, Y., Li, X., Sun, L., Gao, Y., Gao, Z., & Chen, L. (2019). Aerodynamic analysis of a step adjustment method for blade pitch of a VAWT. *Journal of Wind Engineering & Industrial Aerodynamics*, *188*, 90−101.

Gupta, R., & Biswas, A. (2010). Performance measurement of a twisted three-bladed airfoil-shaped H-rotor. *International Journal of Renewable Energy Technology*, *1*, 279−300.

Gupta, R., & Biswas, A. (2011). Comparative study of the performances of twisted two-bladed & three-bladed airfoil shaped H-Darrieus turbines by computational and experimental methods. *International Journal of Renewable Energy Technology*, *2*, 425−445.

Gupta, R., Das, R., & Sharma, K.K. (2006). Experimental study of a Savonius−Darrieus wind machine. In *Proceedings of the international conference on renewable energy for developing countries*, University of Columbia.

Harsh, S. D., Dipankar, D., Vlad, M., & Mihaela, L. U. (2019). Multi-criteria decision making approach for hybrid operation of wind farms. *Symmetry*, *11*, 675−680.

Jagadish., & Amitava, R. (2014a). Green cutting fluid selection using MOOSRA method. *International Journal of Research in Engineering and Technology*, *03*(03), 559−563.

Jagadish., & Amitava, R. (2014b). Green cutting fluid selection using multi-attribute decision making approach. *The Journal of Institute of Engineers India Series-C*, *96*(1), 35−39.

Jagadish., & Amitava, R. (2014c). Multi-objective optimization of green EDM: An integrated theory. *The Journal of Institute of Engineers India Series-C*, *96*(1), 41−47.

José, F., Herbert-Acero., Jaime, M. L., Oliver, P., Santos, M. D., Krystel, K., ... Réthoré, P.-E. (2014). A hybrid metaheuristic-based approach for the aerodynamic optimization of small hybrid wind turbine rotors. *Mathematical Problems in Engineering*, *746319*, 1−18.

Karande, P., & Chakraborty, S. (2012). Application of multi-objective optimization on the basis of ratio analysis (MOORA) method for materials selection. *Materials & Design*, *37*(1), 317−324.

Lam, H. F., Liu, Y. M., Peng, H. Y., Lee, C. F., & Liu, H. J. (2018). Assessment of solidity effect on the power performance of H-rotor vertical axis wind turbines in turbulent flows. *Journal of Renewable and Sustainable Energy*, *10*, 023304. Available from https://doi.org/10.1063/1.5023120.

Lam, N., & Meredith, M. (2017). Optimization of a vertical axis wind turbine for application in an urban/suburban area. *Journal of Renewable and Sustainable Energy*, *9*, 4−10.

Langer, H., Peterson, R.L., & Maier, T. H. (1996). An experimental evaluation of wind tunnel wall correction methods for helicopter performance. In *Proceedings of the 52nd annual forum of the American Helicopter Society*.

Lee, A. H., Hung, M. C., Kang, H. Y., & Pearn, W. (2012). A wind turbine evaluation model under a multi-criteria decision making environment. *Energy Conversion and Management*, *64*, 289−300.

Mostafa, R. S. (2017). The location optimization of wind turbine sites with using the MCDM approach: A case study. *Energy Equipment and system*, *5*, 165−187.

Okokpujie, I. P., Okonkwo, U. C., Bolu, C. A., Ohunakin, O. S., Agboola, M. G., & Atayero, A. A. (2020). Implementation of multi-criteria decision method for selection of suitable material for development of horizontal wind turbine blade for sustainable energy generation. *Heliyon*, *6*, 03142.

Ouro, P., Stoesser, T., & Ramirez, L. (2018). Effect of blade cambering on dynamic stall in view of designing vertical axis turbines. *Journal of Fluids Engineering*, *140*, 1−12.

Qamar, S. B., & Janajreh, I. (2017). A comprehensive analysis of solidity for cambered darrieus VAWTs. *International Journal of Hydrogen Energy*, *42*, 19420−19431.

Sagharichi, A., Zamani, M., & Ghasemi, A. (2018). Effect of solidity on the performance of variable-pitch vertical axis wind turbine. *Energy*, 1−4. Available from https://doi.org/10.1016/j.energy.2018.07.160.

Sandeep, S., Vidyapati, P., Jagadish., & Amitava, R. (2014). Multi attributes decision making for mobile phone selection. *International Journal of Research in Engineering and Technology*, *03*(03), 497−501.

Sengupta, A. R., Biswas, A., & Gupta, R. (2016). Studies of some symmetrical and unsymmetrical blade H-Darrieus rotors with respect to starting characteristics, dynamic performances and flow physics in low wind streams. *Renewable Energy*, *93*, 536−547.

Sengupta, A. R., Biswas, A., & Gupta, R. (2017a). Investigations of H-Darrieus rotors for different blade parameters at low wind speeds. *Wind and Structures, An International Journal*, *25*, 551−567.

Sengupta, A. R., Biswas, A., & Gupta, R. (2017b). The aerodynamics of high solidity unsymmetrical and symmetrical blade H-Darrieus rotors in low wind speed conditions. *Journal of Renewable and Sustainable Energy*, *9*. Available from https://doi.org/10.1063/1.4999965.

Sengupta, A. R., Biswas, A., & Gupta, R. (2019). Comparison of low wind speed aerodynamics of unsymmetrical blade H-Darrieus rotors- blade camber and curvature signatures for performance improvement. *Renewable Energy*, *139*, 1412−1427.

Sharma, A., Muqeem, M., Sherwani, A., & Ahmad, M. (2018). Optimization of diesel engine input parameters running on Polanga biodiesel to improve performance and exhaust emission using MOORA technique with standard deviation. *Energy Sources, Part A: Recovery, Utilization and Environmental Effects*, *40*(22), 1−18.

Singh, M. A., Biswas, A., & Misra, R. D. (2015). Investigation of self-starting and solidity on the performance of a three S1210 bladed H-type Darrieus rotor. *Renewable Energy*, *76*, 381−387.

Theodorsen, T. (1996). *Interference on an airfoil of finite span in an open rectangular wind tunnel*. Available via Langley Memorial Aeronautical Laboratory (Report No. 461). NASA. <http://naca.central.cranfield.ac.uk/reports/1934/naca-report-461.pdf>.

Yang, Y., Li, C., Zhang, W., Guo, X., & Yuan, Q. (2017). Investigation on aerodynamics and active flow control of a vertical axis wind turbine with flapped airfoil. *Journal of Mechanical Science and Technology*, *31*, 1645−1655.

Chapter 20

Maximum power point tracking design using particle swarm optimization algorithm for wind energy conversion system connected to the grid

Elmostafa Chetouani, Youssef Errami, Abdellatif Obbadi and Smail Sahnoun

Exploitation and Processing of Renewable Energy Team, Laboratory of Electronics, Instrumentation and Energy, Department of Physics, Faculty of Sciences, University of Chouaib Doukkali, El Jadida, Morocco

20.1 Introduction

In the two last decades, the electrical power produced by employing wind energy sources has been progressively growing, due to the important number of installed wind turbines by several countries. Besides, wind energy is renewable, omnipresent, and inexhaustible green energy resource, which justifies the significant importance according to this energy. Furthermore, the wind energy conversion system (WECS) has known significant technological developments that have improved the conversion efficiency and reduced the costs for wind energy production. The wind turbine, with high power in the order of megawatts, can operate with different generators technologies, such as permanent magnets synchronous generator, squirrel cage induction generator (SCIG), and doubly fed induction generator (DFIG). Several advantages make the DFIG the widely utilized generator in variable speed wind energy systems over any other configuration (Li & Haskew, 2009). Examples of such advantages are the ability to use a partial sized converter in the rotor to control the power, reducing power losses and cost, reducing efforts on mechanical parts, noise reduction, the control of active power and reactive, and a controllable power factor (Lamnadi, Trihi, Bossoufi, & Boulezhar, 2016; Ouassaid, Elyaalaoui, & Cherkaoui, 2016). The DFIG stator is directly connected to the grid and two bidirectional converters connect the rotor to the electrical network via a DC-link voltage, as shown in Fig. 20.1. The rotor side converter (RSC) is used to control active and reactive power exchanged with the grid and to ensure the regulation of the unit power factor (UPF). Yet, the grid side converter (GSC) is utilized to maintain the DC-link voltage constant (Rouabhi & Djerioui, 2014). The DFIG is characterized by a multivariable, nonlinear mathematical model. The magnetic field and the electromagnetic torque are strongly coupled. So, the DFIG control is more complex and difficult compared to the DC machine (Ihedrane, El Bekkali, Bossoufi, & Bouderbala, 2019). There are several techniques to control the DFIG that have been proposed. The field-oriented control (FOC) is commonly used in the WECS based on the DFIG and has shown satisfactory performance. In addition, this vector control permits to control of the DFIG as a separately excited DC machine, which makes it very popular in the industry. Lamnadi, Trihi, Bossoufi, and Boulezhar (2016) have proposed a direct vector control for regulating the active and reactive power. Likewise, Bouderbala et al. (2018) have proposed a comparative study between direct and indirect field-oriented control (IFOC) for WECS based on the doubly fed induction machine. Unlike direct vector control is simple and works only with two controllers to regulate the powers by controlling directly the rotor voltages, the indirect vector control is complex and needs four controllers regulating the stator powers and the rotor currents. But, the IFOC offers satisfactory performance in terms of efficiency and robustness (Ihedrane et al., 2019). Currently, other methods controlling the WECS are developed. Hassan, El-Sawy, and Kamel (2013) have applied the direct torque control (DTC) for DFIG driven by variable speed wind turbines. The authors have established the DTC strategy only for the RSC allowing very fast torque responses. The DTC technique controls the machine flux and the electromagnetic torque by the selection of the optimum inverter switching modes. Saravanan, Stalin, Sree, Renga, and Soundara Pandiyan

445

(2016) have modeled and simulated the DFIG integrated into WECS by using the direct power control strategy. The authors have utilized a Vienna rectifier as the RSC and a three-phase pulse width modulation inverter as GSC for DFIG. The results show that the proposed control gives a good performance in terms of reduction of the total harmonic distortion. Loucif, Boumediene, and Mechernene (2013) have proposed modeling of a DFIG driven by a wind turbine. They have established a FOC to regulate y the active and reactive power independently by synthesizing two types of controllers, a classical proportional–integral (PI) and a nonlinear controller backstepping. In addition, Ouassaid et al. (2016) have proposed a sliding mode strategy for variable speed wind turbine based on SCIG injecting the power into the grid. Their objective is to control the active and reactive power and maximization of power by the MPPT technique. The authors have demonstrated the robustness of the proposed control by variation of the machine parameters. More recently, many researchers have proposed optimization algorithms for controlling the WECS.

The extractable power from wind energy depends on the characteristics of each turbine and the wind variable speed eventually. Consequently, tracking the maximum power generated is required when the wind turbine operates in region II, as shown in Fig. 20.2 (Rouabhi & Djerioui, 2014). This strategy is known as maximum power point tracking (MPPT). To perform the MPPT algorithm, several control schemes are developed and can be classified into two categories. The first is that required knowledge of the characteristic aerodynamic curve of wind turbine speed, and the second is that no information about wind speed is necessary to generate the optimal speed reference. Examples of these schemes are the MPPT with optimal tip–speed ratio (TSR), the MPPT with optimal torque control, the MPPT with perturbation and observation, and the MPPT with power signal feedback (Ananth & Kumar, 2016; Chavero-Navarrete, Trejo-Perea, Jáuregui-Correa, Carrillo-Serrano, & Ríos-Moreno, 2019; Cortes-Vega, Ornelas-Tellez, & Anzurez-Marin, 2019). More details are given by Manonmani and Kausalyadevi (2014), which have proposed a comparative study for these algorithms. Besides, MPPT based on artificial intelligence has been proposed and used. Recently, particle swarm

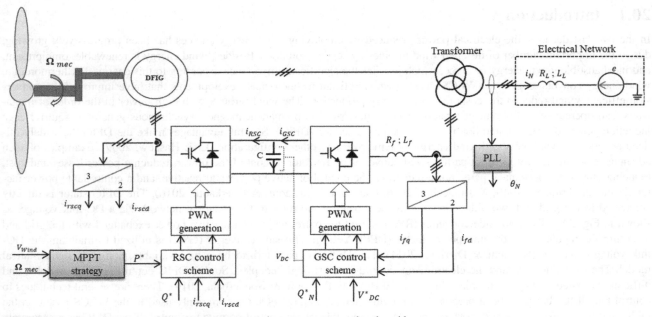

FIGURE 20.1 Synoptic scheme of the wind energy conversion system connected to the grid.

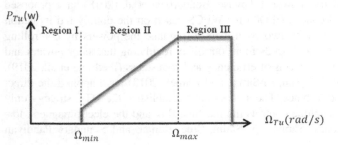

FIGURE 20.2 Ideal characteristic of a wind turbine at variable speed.

```
if  t < t1 then V_wind= 8;

else if  t >= t1 && t <= t2 then V_wind= 11;

else if  t >= t2 &&  t<= t3 then V_wind= 12.5;

else V_wind= 9;

end
```

FIGURE 20.3   Code of the chosen wind profile.

optimization (PSO) has been utilized to track the maximum point power. This control method minimizes the calculating time and keeps an excellent precision, and it can be implemented in a low-cost microcontroller (Ben Belghith, Sbita, & Bettaher, 2016). Ali Zeddini, Pusca, Sakly, and Mimouni (2016) have applied a new MPPT technique based on PSO for a standalone self-excited induction generator operating at variable wind speed and supplying an induction motor coupled to a centrifugal pump. In this chapter, a 5-MW DFIG driven by a variable speed wind turbine is modeled and simulated by the MATLAB/Simulink environment. The RSC and GSC are controlled by a classical pulse width modulation (PWM). The main contribution is the computing self-tuning controller for the MPPT based on a metaheuristic PSO. This optimization technique is compared with a classical PI controller. For this task, several criteria indices are employed. However, the IFOC is proposed to control the injected powers (active and reactive) into the electrical network, consequently regulating the power factor to value one. In addition, a combination of the PSO technique with vector control is proposed in order to maximize the advantages of both. Simulation results are presented and discussed to demonstrate the efficiency of the proposed optimization technique to extract the maximum power, especially for the variable wind speed. The chapter is organized as follows: Section 20.2 presents the modeling of WECS. Section 20.3 discusses the MPPT strategy established by the classical PI controller and smart PI controller based on PSO, and a comparative study between the two controllers type is presented. Section 20.4 proposes the control of the active and reactive powers by applying the vector control technique. Section 20.5 discusses and interprets the simulation results, and the conclusion is presented in Section 20.6.

20.2 Wind energy conversion system modeling

20.2.1 Wind profile modeling

The wind is the primary energy source for operating a wind turbine, so it is necessary to know its mathematical model. It can be very complex because it requires climatic and geographical data of the site concerned, as well as the period of the year concerned by the study. In this chapter, a wind profile is chosen and applied to the wind turbine to verify the proposed method. The script code shown in Fig. 20.3 defines it.

20.2.2 Wind turbine and gearbox modeling

The wind turbine is the essential element in the wind chain, which allows the conversion of wind energy into mechanical power. Albert Betz has demonstrated that wind turbines capture only 16/27 from wind energy. Eq. (20.1) shows the relation between the extracted power (P_{mech}) and the wind power (P_{wind}) (Errami, Ouassaid, & Maaroufi, 2015):

$$P_{mech} = C_P \times (\lambda, \ \beta) \times P_{wind} \tag{20.1}$$

The wind power can be determined by the following equation:

$$P_{wind} = \frac{1}{2} \times \rho \times \pi \times R^2 \times V_{wind}^3 \tag{20.2}$$

From Eqs. (20.1) and (20.2), the mechanical power as a function of wind speed is easily deduced as:

$$P_{mech} = \frac{1}{2} \times C_P(\lambda, \ \beta) \times \rho \times \pi \times R^2 \times V_{wind}^3 \tag{20.3}$$

where ρ is the air density (kg m^{-3}), R is the blade radius (m), C_P is the performance coefficient of the turbine, which is a function of the pitch angle of rotor blades β (degrees), and V_{wind} is the wind speed (m s^{-1}). The TSR λ equation is given by Errami, Maaroufi, and Ouassaid (2012) as:

$$\lambda = \frac{R \times \Omega_{mec}}{V_{wind}} \tag{20.4}$$

C_P (λ, β) is the performance coefficient of power representing the efficiency of the turbine. This coefficient can be estimated using Eq. (20.5) as follows (Dahbi, Nait-Said, & Nait-Said, 2016):

$$C_P(\beta, \lambda) = [0.5 - 0.0167 \times (\beta - 2)] \times \sin\left(\frac{\pi(\lambda + 0.1)}{18.5 - 0.3.(\beta - 2)}\right) - 0.00184 \tag{20.5}$$

The wind turbine mechanical torque output T_{Tu} is written as below:

$$T_{Tu} = \frac{P_{mech}}{\Omega_{mec}} \tag{20.6}$$

Considering the $C_P(\beta, \lambda)$ characteristic equation (Eq. 20.5), the C_P maximum value, that is $C_{Pmax} = 0.5$, corresponds to the maximum power that can be extracted and is obtained by maintaining $\beta = 2$degrees. As a result, the optimal value of the TSR (λ_{opt}), whcih is $\lambda_{opt} = 9.19$, has been obtained. Furthermore, as shown in Fig. 20.4, there is a maximum value of Cp for each value of angle pitch, which corresponds to an optimum speed ratio.

The turbine rotation speed is relatively slow and needs to be adapted to DFIG rotation speed. For this reason, a gearbox is used. Eq. (20.7) presents the mechanical equation of the system, taking into consideration that the overall mechanical dynamics of the system are brought back to the turbine shaft (Dahbi et al., 2016; Errami et al., 2015):

$$\Omega_{mec} \times (J \times s + f) = T_g - T_{em} \tag{20.7}$$

$$T_g = \frac{T_{Tu}}{G_B} \quad and \quad G_B = \frac{\Omega_{mec}}{\Omega_{Tu}} \tag{20.8}$$

where s is the Laplace operator, J is the overall inertia of WECS, T_{Tu} is the turbine torque, T_{Tem} is the electromagnetic torque of the DFIG, f is the overall viscous coefficient of friction, and Ω_{mec} is the mechanical speed at the rotor shaft of the wind turbine (rad s^{-1}). Fig. 20.5 presents a simple model of wind turbine established by using the previous equation and tested by MATLAB/Simulink.

20.2.3 Doubly fed induction generator modeling

The DFIG is characterized by its robustness and is applied in several industrial applications but has a complex equation system to study and control. Adopting some simplifying assumptions and using the Park transformation permit to get a simpler model of the DFIG and circumvent this complexity. Assuming the $d-q$-axis, which is rotating with the synchronous speed (ω_s), the generator voltage and flux equations are given by Rouabhi and Djerioui (2014).

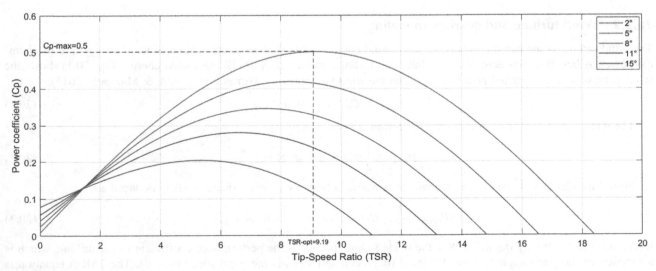

FIGURE 20.4 C_P characteristic as a function of λ.

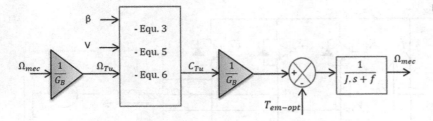

FIGURE 20.5 Wind turbine model.

The stator and rotor voltages equations can be formulated as follows:

$$
\begin{cases}
V_{sd} = R_s \times i_{sd} + \dfrac{d\varphi_{sd}}{dt} - \omega_s \varphi_{sq} \\[2mm]
V_{sq} = R_s \times i_{sq} + \dfrac{d\varphi_{sq}}{dt} - \omega_s \varphi_{sd} \\[2mm]
V_{rd} = R_r \times i_{rd} + \dfrac{d\varphi_{rd}}{dt} - \omega_r \varphi_{rq} \\[2mm]
V_{sq} = R_r \times i_{rq} + \dfrac{d\varphi_{rq}}{dt} - \omega_r \varphi_{rd}
\end{cases}
\tag{20.9}
$$

The stator and rotor field magnetic flux equations can be written as follows (Rouabhi & Djerioui, 2014):

$$
\begin{cases}
\varphi_{sd} = L_s \times i_{sd} + M \times i_{rd} \\
\varphi_{sq} = L_s \times i_{sq} + M \times i_{rq} \\
\varphi_{rd} = L_r \times i_{rd} + M \times i_{sd} \\
\varphi_{rq} = L_r \times i_{rq} + M \times i_{sq}
\end{cases}
\tag{20.10}
$$

The electromagnetic torque can be given as follows:

$$
T_{em} = -p \times \frac{M}{L_s} \times \left(i_{rq} \times \varphi_{sd} - i_{rd} \times \varphi_{sq} \right)
\tag{20.11}
$$

$$
S_j = 1, \; The \; K_i' \; is \; in \; the \; conducting \; state.
\tag{20.12}
$$

20.2.4 Modeling of the back-to-back converters

The bidirectional converters RSC and GSC are connected through a capacitor for coupling the rotor to the grid and for transferring the powers from the grid to the rotor, or back. Fig. 20.6 shows the configuration of the back-to-back converters, which are composed of the antiparallel diode and Insulated Gate Bipolar Transistor (IGBT) controlled by closing and opening by the classical PWM technique. The interrupt couple (K_i, K_i') is complementary controlled $(i = 1 \ldots 6)$. Their state is defined by the function given in Eqs. (20.12) and (20.13):

$$
S_j = -1, \quad The \; K_i' \; is \; in \; the \; conducting \; state.
\tag{20.13}
$$

Where S_j ($j = a, b,$ or c) presents the pulse generated by the PWM method.

The input voltages between phases of the converters can be described as a function of the voltage V_{DC} and the states of interrupt S_j (Bouderbala et al., 2018):

$$
\begin{aligned}
U_{ab} &= (S_a - S_b) \times V_{DC} \\
U_{bc} &= (S_b - S_c) \times V_{DC} \\
U_{ca} &= (S_c - S_a) \times V_{DC}
\end{aligned}
\tag{20.14}
$$

The simple input voltages (V_a, V_b, V_c) equations can be written as follows (Bouderbala et al., 2018):

$$
V_a = \frac{2 \times S_a - S_b - S_c}{3} \times V_{DC}
$$

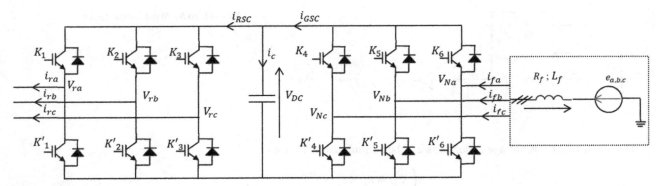

FIGURE 20.6 Back-to-back converters.

$$V_b = \frac{2 \times S_b - S_a - S_c}{3} \times V_{DC} \qquad (20.15)$$

$$V_c = \frac{2 \times S_c - S_a - S_b}{3} \times V_{DC}$$

From Eq. (20.15), the model of the GSC and the RSC converters can be given under the matrix presented below (Elazzaoui, 2015; Rouabhi & Djerioui, 2014):

$$\begin{bmatrix} V_a \\ V_b \\ V_c \end{bmatrix} = \frac{V_{DC}}{3} \begin{bmatrix} 2 & -1 & -1 \\ -1 & 2 & -1 \\ -1 & -1 & 2 \end{bmatrix} \times \begin{bmatrix} S_a \\ S_b \\ S_c \end{bmatrix} \qquad (20.16)$$

In this work, the RSC is modeled directly under MATLAB/Simulink by using Eq. (20.16). However, the GSC is associated with a serial Resistance-Inductance (RL) filter, as can you see in Fig. 20.6. Eq. (20.17) and (20.18)are modeled the converter GSC and the filter, which can be written in the rotating frame $d-q$ by applying the park transformation and the Laplace transformation as follows (Elazzaoui, 2015):

$$V_{Nd-Gsc} = -(R_f + L_f \times s) \times i_{fd} + w_N \times L_f \times i_{fq} + V_{Nd} \qquad (20.17)$$

$$V_{Nq-Gsc} = -(R_f + L_f \times s) \times i_{fq} - w_N \times L_f \times i_{fd} + V_{Nq} \qquad (20.18)$$

where $V_{Nd\text{-}Gsc}$ and $V_{Nq\text{-}Gsc}$ are the components of Park of the converter input voltages and V_{Nd} and V_{Nq} are the grid voltages in $d-q$ referee.

From Eqs. (20.16)−(20.18), the mathematical model can be deduced as follows (Qian, Gao, & Sheng, 2018):

$$(R_f + L_f \times s) \times i_{fd} - w_N \times L_f \times i_{fq} = V_{Nd} - V_{DC} \times S_d \qquad (20.19)$$

$$(R_f + L_f \times s) \times i_{fq} + w_N \times L_f \times i_{fd} = V_{Nq} - V_{DC} \times S_q \qquad (20.20)$$

where S_d and S_q are the switching functions of the GSC converter in the $d-q$ referee, which are expressed by applying the Park transformation as follows:

$$S_d = \frac{1}{\sqrt{6}}(2 \times s_a - s_b - s_c) \times \cos(\omega t) + \frac{1}{\sqrt{6}}(s_b - s_c) \times \sin(\omega t) \qquad (20.21)$$

$$S_q = \frac{1}{\sqrt{6}}(s_b - s_c) \times \cos(\omega t) - \frac{1}{\sqrt{6}}(2 \times s_a - s_b - s_c) \times \sin(\omega t) \qquad (20.22)$$

The output current of the GSC converter i_{GSC} can be expressed in the three-phase referee (a, b, c) as follows:

$$i_{GSC} = i_{fa} \times s_a + i_{fb} \times s_b + i_{fc} \times s_c \qquad (20.23)$$

Applying the Park transformation, the i_{GSC} can be written as follows:

$$i_{GSC} = i_{fd} \times S_d + i_{fq} \times S_q \qquad (20.24)$$

The equation of coupling alternative and continuous sides is expressed as follows:

$$C \times \frac{dV_{DC}}{dt} = i_{GSC} - i_{RSC} = (i_{fd} \times S_d + i_{fq} \times S_q) - i_{RSC} \tag{20.25}$$

The model of the GSC converter and the filter is designed as shown in Fig. 20.7 by using the previous equations.

20.2.5 Grid modeling

As illustrated in Fig. 20.1, the DFIG is connected to the grid through a transformer, which is characterized by a transformation ratio m. The grid three phases are considered a balanced system. R_L and L_L are the line transmission parameters, respectively. The electrical network equations are presented in the equation system (20.26), where e_1, e_2, and e_3 are the electromotive forces, V_{T1} is the transformer secondary voltage, and i_N is the network current (Camara, Camara, Dakyo, & Gualous, 2013).

$$\begin{cases} V_{T1} = e_1 + R_L \times i_{N1} + L_L \times \dfrac{di_{N1}}{dt} \\[2mm] V_{T2} = e_2 + R_L \times i_{N2} + L_L \times \dfrac{di_{N2}}{dt} \\[2mm] V_{T3} = e_3 + R_L \times i_{N3} + L_L \times \dfrac{di_{N3}}{dt} \end{cases} \tag{20.26}$$

Applying the Laplace transform to the differential equations in Eq. (20.26) allows the system currents to be determined as shown in the equation system (20.27). The latter is used to model the electrical grid.

$$i_{N1} = \frac{V_{T1} - e_1}{R_L + L_L \times s}$$

$$i_{N2} = \frac{V_{T2} - e_2}{R_L + L_L \times s} \tag{20.27}$$

$$i_{N3} = \frac{V_{T3} - e_3}{R_L + L_L \times s}$$

20.2.6 Phase-Locked Loop technique

Fig. 20.8 shows the phase-locked loop (PLL) technique, which is used to estimate the angle for transformation Park of the stator variables θ_N and determine the components of the voltage networks V_{Nd} and V_{Nq} (Elazzaoui, 2015). The rotor angle θ_r is computed by the following equation:

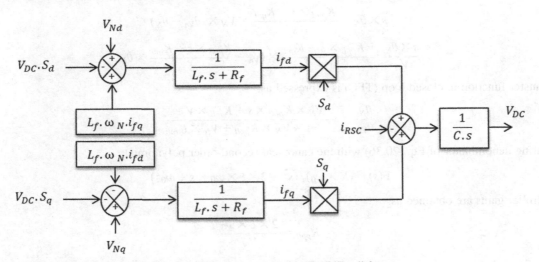

FIGURE 20.7 Model of the grid side converter and the filter implemented in MATLAB/Simulink.

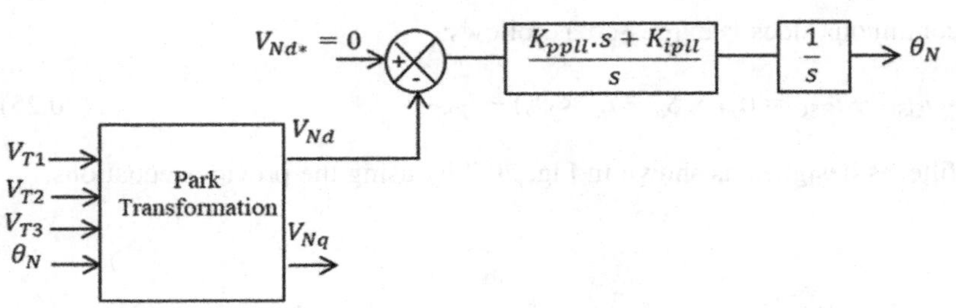

FIGURE 20.8 Phase-locked loop technique.

$$\theta_r = \theta_N - \theta_{mec} \tag{20.28}$$

The rotor angular position θ_{mec} is determined as below:

$$\theta_{mec} = \int p \times \omega_{mec} \tag{20.29}$$

where p is the number of pairs of poles of the DFIG and ω_{mec} is the rotor angular speed.

PLL aligns the stator voltage to the q-axis by the comparing d-axis load voltage with the zero reference voltage.

$$V_{Nd} = 0$$
$$V_{Nq} = V_N \tag{20.30}$$

20.2.6.1 Determination of the phase-locked loop controller parameters

According to the rotating coordinate system, the d-axis voltage can also be written as (Byunggyu, 2018):

$$V_{Nd} = - V_N \times \sin(\theta_{Vn} - \theta_N) \tag{20.31}$$

where the θ_N is the estimated phase angle by PLL technique and θ_{Vn} is the real phase angle.

By substituting $V_{Nd} = 0$ in Eq. (20.31), then we can deduce that: $\theta_{Vn} = \theta_N$.

The voltage error signal (ε) is defined and written as follows:

$$\varepsilon = V_{Nd-ref} - V_{Nd} = V_N \times \sin(\theta_{Vn} - \theta_N) \tag{20.32}$$

Considering $(\theta_{Vn} - \theta_n)$ is very small. So,

$$V_N \times \sin(\theta_{Vn} - \theta_N) \approx V_N \times (\theta_{Vn} - \theta_N) \tag{20.33}$$

$V_N (\theta_{Vn} - \theta_N)$ is the input of the PI controller and the output of the controller is $s \times \theta_N$. From the block diagram shown in Fig. 20.8, the corresponding equation can be given as:

$$s \times \theta_N = \frac{K_{ppll} \times s + K_{ipll}}{s} \times V_N \times (\theta_{Vn} - \theta_N) \tag{20.34}$$

$$\frac{s \times \theta_N}{V_N} = \frac{K_{ppll} \times s + K_{ipll}}{s} \times \theta_{Vn} - \frac{K_{ppll} \times s + K_{ipll}}{s} \times \theta_N \tag{20.35}$$

The transfer function in closed loop (TFC) is expressed as:

$$\frac{\theta_N}{\theta_{Vn}} = \frac{V_N \times K_{ppll} \times s + K_{ipll} \times V_N}{s^2 + s \times V_N \times K_{ppll} + V_N \times K_{ipll}} \tag{20.36}$$

Comparing denominator of Eq. (20.36) with the canonical second-order polynomial:

$$F(s) = (K \times \omega_0)/(s^2 + 2 \times \xi \times \omega_0 \times s + \omega_0{}^2) \tag{20.37}$$

PI controller gains are obtained as:

$$K_{ppll} = \frac{2 \times \xi \times \omega_0}{V_N} \tag{20.38}$$

$$K_{ipll} = \frac{\omega_0^2}{V_N} \tag{20.39}$$

Where ξ is the damping coefficient, which is chosen equal to 0.7, and ω_0 is the bandwidth frequency of the PLL controller. The latter can be calculated by knowing the switching frequency of the interrupts as: $\omega_0 = 2\pi \times f_0$

20.3   Control strategies of the maximum power point tracking

In this chapter, the MPPT algorithm is employed in region II to maximize the extracted power. This technique is based on the electromagnetic torque controlof the DFIG to regulate the rotation speed. There are two control modes of MPPT: Control without speed regulation and Control with speed regulation. The first depends on the rotation speed of the generator directly to calculate the optimal turbine speed for developing optimum generator torque. The wind speed is estimated. The second one consists of maintaining the DFIG rotation speed at a reference speed utilizing the PI controller. An anemometer is used to measure the wind speed in this mode. If the coefficient Cp is optimized, the reference speed is maximized. The relative speed λ is the most significant parameter to optimize in this regard.

In this work, the second strategy is studied. The electromagnetic torque T_{em} developed by the DFIG is equal to its optimal (reference) value $T_{em\text{-}opt}$ imposed by the command (Dahbi et al., 2016):

$$T_{em} = T_{em-opt} \tag{20.40}$$

As shown in Fig. 20.9, the optimal electromagnetic torque $T_{em\text{-}opt}$ for obtaining a rotation speed equal to the optimal speed is given as follows:

$$T_{em-opt} = \left[K_{pmppt} + K_{imppt} \times \frac{1}{s} \right] \times \left[\Omega_{mec-opt} - \Omega_{mec} \right] \tag{20.41}$$

where K_{pmppt} and K_{imppt} are the PI controller parameters, respectively. The optimal speed ($\Omega_{mec\text{-}opt}$) is given by the following equations:

$$\Omega_{mec-opt} = G_B \times \Omega_{Tu-opt} \tag{20.42}$$

$$\Omega_{Tu-opt} = \frac{V_{wind} \times \lambda_{opt}}{R} \tag{20.43}$$

where G_B is the coefficient of the gearbox used to adapt the slow speed of the turbine shaft to the speed one of the DFIG.

FIGURE 20.9 Maximum power point tracking scheme with speed regulation.

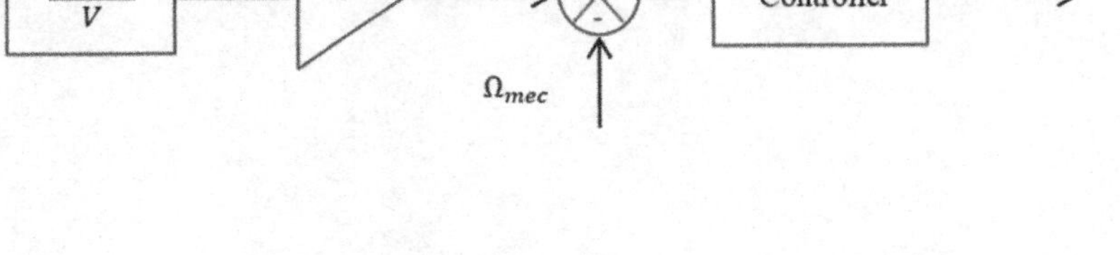

FIGURE 20.10 Rotation speed regulation loop.

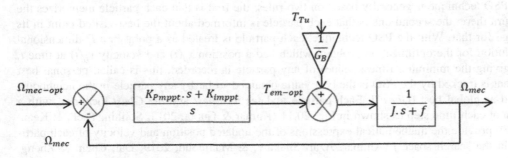

20.3.1 Classical proportional–integral for maximum power point tracking

Fig. 20.10 shows the regulation loop of the rotation speed. The turbine torque (T_{Tu}) is considered as a perturbation to compensate by the controller. The latter is employed to acheive a zero static error and reduce the response while keeping in mind the system stability, which must be conserved.

The PI controller parameters are determined by the pole compensation method. The transfer function in the open-loop (TFO) is expressed as follows:

$$\text{TFO} = K_{imppt} \times \frac{\frac{K_{pmppt}}{K_{imppt}} \times s + 1}{s} \times \frac{f}{\frac{J}{f} \times s + 1} = \frac{K_{imppt} \times f}{s}; \text{ With } \frac{K_{pmppt}}{K_{imppt}} = \frac{J}{f} \tag{20.44}$$

The TFC can be written as follows:

$$\text{TFC} = \frac{\text{TFO}}{\text{TFO} + 1} = \frac{1}{\frac{1}{K_{imppt} \times f} \times s + 1} \tag{20.45}$$

The time constant of the system is expressed as follows:

$$T_{sys} = \frac{J}{f} \tag{20.46}$$

The dynamic of the system is very slow due to J and f values. So, to have a rapid response, the system constant time is divided by 10^3.

$$K_{imppt} = \frac{1}{\tau \times f}; \text{ With } \tau = \frac{1}{10^3} \tag{20.47}$$

$$K_{pmppt} = \frac{K_{imppt} \times J}{f} \tag{20.48}$$

20.3.2 Particle swarm optimization for maximum power point tracking

20.3.2.1 Particle swarm optimization algorithm overview and concept

Recently, the collective intelligence metaheuristic PSO algorithm is one of the optimization algorithms used in MPPT control to tune the controller parameters optimally to have better performance. It was originally introduced and developed by Dr. Eberhat and Dr. Kennedy in 1995, basically inspired by bird or fish social behavior searching food or keeping safe away from enemies (Zahra, Salhi, & Mellit, 2017). The particles are the swarm individuals who are moving around and in a multidirectional search space to achieve their purpose and are evolving by collaboration and competition among themselves. The PSO technique is generally based on two rules: the first is that each particle memorizes the best point visited and can return there; the second one is that each particle is informed about the best-visited point in its neighbourhood and tends to go for that. With the PSO technique, each particle is treated as a point in a D-dimensional space and it is a potential solution for the optimization problem which had a position $x_{ij}(t)$ and velocity $v_{ij}(t)$ at time t. The best previous position (giving the minimum fitness value) of any particle is recorded, this is called personal best (*Pbest*). Another best value that is tracked by the PSO is the best value obtained so far by any particle in the neighbourhood of that particle [denoted as global best (*Gbest*)]. Each particle updates its *Pbest* and the *Gbest* locations, with a random weighted acceleration at each time step as shown in Fig. 20.11 (Kumar & Gupta, 2013; Solihin, Tack, & Kean, 2011). Eqs. (20.49) and (20.50) provide the mathematical expressions of the updated position and velocity of each particle i at each iteration t and in the search space j (Ammar, Azar, Shalaby, & Mahmoud, 2019; Dai, Chen, & Zheng, 2018).

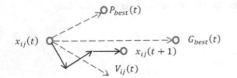

FIGURE 20.11 Update diagram of each particle in the swarm.

$$V_{ij}(t+1) = w \times V_{ij}(t) + C_1 \times r_1 \times \left(Pbest_{ij} - x_{ij}(t)\right) + C_2 \times r_2 \times \left(Gbest_{ij} - x_{ij}(t)\right) \qquad (20.49)$$

$$x_{ij}(t+1) = V_{ij}(t+1) + x_{ij}(t) \qquad (20.50)$$

where "w" is the inertia weight factor that is used as control parameter for the swarm velocity, "c_1" and "c_2" are the acceleration constants (positive constants) called the cognitive and social parameters and "r_1" and "r_2" are the random numbers between $[0-1]$. The particle swarm is also characterized by "d" (the dimension of the problem) and "n" (the size of the swarm).

20.3.2.2 Implementation of particle swarm optimization into proportional−integral controller for maximum power point tracking

To implement the MPPT control based on the PSO algorithm, the synoptic scheme shown in Fig. 20.12 is adopted. The rotation speed must be equal to the optimal rotation speed corresponding to the optimal ratio speed. A smart controller, parameters of which are self-tuning, ensures the regulation of the speed. For this task, the zero error e (t) is defined as an objective function (called fitness) to achieve by the PSO technique. Several performance indices are used to compute the error in the PSO algorithm, such as integral absolute error (IAE), integral square error (ISE), integral time square error (ITSE), and integral time absolute error (ITAE). In this work, the IAE criterion is used and compared with conventional PI. As mentioned by Solihin et al. (2011) and Kumar and Gupta (2013), these indices are defined by the following equations:

$$IAE = \int_0^\infty |e(t)| \times dt \qquad (20.51)$$

$$ISE = \int_0^\infty e^2(t) \times dt \qquad (20.52)$$

$$ITSE = \int_0^\infty t \times e^2(t) \times dt \qquad (20.53)$$

$$ITAE = \int_0^\infty t \times |e(t)| \times dt \qquad (20.54)$$

20.3.2.3 Algorithm steps and pseudo-code of basic particle swarm optimization

As mentioned by Vincent and Nersisson (2017), the steps of the searching procedure for the PSO technique can be listed as follows:

Step 1: Start
Step 2: Initialization:
    Particle parameters: C_1, C_2, W, Iter, dim;
    Particles with random position and velocity;
    Calculate the parameters of the PI controller randomly.
Step 3: Run the system WECS and evaluate the fitness function
Step 4: If the present value is better than Pbest
    Then Pbest is equal to present fitness value
    Else go to Step 3.

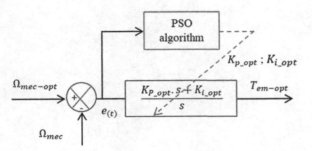

FIGURE 20.12 Synoptic scheme of particle swarm optimization−maximum power point tracking design.

Step 5: If the present fitness value is better than Gbest
   Then Gbest is equal to the present value of the fitness function
   Else update the position and velocity values of the particles and update the parameters of PI controller.
Step 6: Exit if a termination criterion is met
   Else go to step 3.
   End

As investigated by Dai, Chen, and Zheng (2018), the pseudo-code of basic PSO can be written as shown in Fig. 20.13.

The abovementioned steps can be organized by the flowchart given in Fig. 20.14.

20.4 Field-oriented control technique of the active and reactive power

20.4.1 Active and reactive power control

To solve the power decoupling problem and to control the powers of DFIG independently, the FOC method is adopted. Acting on the rotor voltages allows controlling the exchange powers active and reactive between the DFIG stator and the grid. As seen in Fig. 20.15, the stator flux is considered constant and is oriented according to the d-axis. The stator winding resistance is neglected, and the stator voltage equation can be simplified as follows (Elazzaoui, 2015; Rouabhi & Djerioui, 2014):

$$\begin{cases} \varphi_{sd} = \varphi_s; \ \varphi_{sq} = 0 \\ V_{sd} = 0; \ V_{sq} = V_s = \omega_s \times \varphi_s \end{cases} \tag{20.55}$$

The rotor voltages can be expressed as follows (Elazzaoui, 2015; Rouabhi & Djerioui, 2014):

$$V_{rd} = \left[R_r + \left(L_r - \frac{M^2}{L_s} \right) \times s \right] \times i_{rd} - g \times \omega_s \times \left(L_r - \frac{M^2}{L_s} \right) \times i_{rq} \tag{20.56}$$

$$V_{rq} = \left[R_r + \left(L_r - \frac{M^2}{L_s} \right) \times s \right] \times i_{rq} + g\omega_s \left(L_r - \frac{M^2}{L_s} \right) \times i_{rd} + g \times \frac{V_s \times M}{L_s} \tag{20.57}$$

FIGURE 20.13 Pseudo-code of particle swarm optimization.

```
Random generate an initial population

Repeat

        for i=1 to population size  do

                Calculate fitness value f(xij);

                if f(xij(t)) > f(pi(t)) then  pi(t) =  xij(t)

                f(pg(t))= min (f(pi(t)))

                end

                for d=1 to dimension do

                Velocity updating (Eq. (49))

                Position updating (Eq. (50))

                end

end

Until "Maximum iteration is met"
```

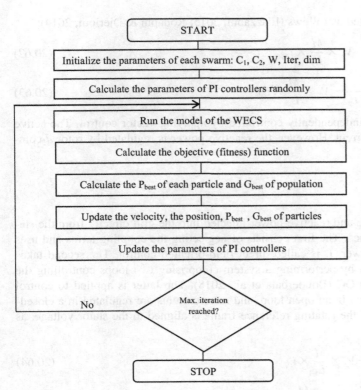

FIGURE 20.14 Flowchart of particle swarm optimization algorithm.

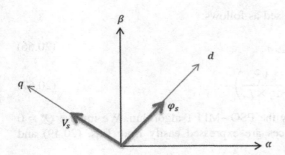

FIGURE 20.15 Stator flux orientation.

From Eqs. (20.56) and (20.57), the rotor currents expressions are deduced as follows:

$$i_{rd} = \left[V_{rd} + g \times \omega_s \left(L_r - \frac{M^2}{L_s}\right) \times i_{rq}\right] / \left[R_r + \left(L_r - \frac{M^2}{L_s}\right) \times s\right] \tag{20.58}$$

$$i_{rq} = \left[V_{rq} - g \times \omega_s \left(L_r - \frac{M^2}{L_s}\right) \times i_{rd} - g \times \frac{V_s \times M}{L_s}\right] / \left[R_r + \left(L_r - \frac{M^2}{L_s}\right) \times s\right] \tag{20.59}$$

By using the vector control simplifications cited in Eq. (20.55) and from Eq. (20.10), the stator currents expressions are deduced as follows:

$$i_{sd} = -\frac{M}{L_s} \times i_{rd} + \frac{\varphi_s}{L_s} \tag{20.60}$$

$$i_{sq} = -\frac{M}{L_s} \times i_{rq} \tag{20.61}$$

The stator active and reactive powers can be expressed as follows (Elazzaoui, 2015; Rouabhi & Djerioui, 2014):

$$P_s = -V_s \times \frac{M}{L_s} \times i_{rq} \tag{20.62}$$

$$Q_s = \frac{V_s^2}{\omega_s \times L_s} - V_s \times \frac{M}{L_s} \times i_{rd} \tag{20.63}$$

Eqs. (20.62) and (20.63) show that the powers are independently controlled by applying vector control. The active power is piloted by acting on the rotor q-component current. However, the reactive power is regulated by rotor d-component current.

20.4.2 Rotor side converter control

The RSC is designed to control independently the active and reactive power injected into the grid directly from the stator. The power control can be realized by two FOC types: The first consists of neglecting the coupling terms and uses the PI controller to regulate the active and reactive power. It is called direct field-oriented control. The second takes into account the coupling terms and compensates them by performing a system comprising two loops controlling the powers and the rotor currents. This method is called IFOC (Bouderbala et al., 2018). The latter is applied to control RSC. As exposed in Fig. 20.16, the powers are controlled in an open-loop, and rotor currents are regulated in a closed-loop. To establish the control of the RSC, the q-axis of the rotating reference frame is aligned to the stator voltage, as presented in Fig. 20.15.

$$P_s = -V_s \times \frac{M}{L_s} \times i_{rq} \tag{20.64}$$

$$Q_s = \frac{V_s^2}{\omega_s \times L_s} - V_s \frac{M}{L_s} \times i_{rd} \tag{20.65}$$

From Eqs. (20.64) and (20.65), the current references can be expressed as follows:

$$i_{rq\_ref} = -\frac{L_s}{M \times V_s} \times P_{s\_ref} \tag{20.66}$$

$$i_{rd\_ref} = -\frac{L_s}{M \times V_s} \times \left(Q_{sref} - \frac{V_s^2}{\omega_s \times L_s} \right) \tag{20.67}$$

P^* is the maximal power extracted by the turbine and obtained by the PSO−MPPT algorithm. We impose $Q^* = 0$ for having zero transfer of the reactive power. The voltage references are expressed easily from Eqs. (20.49) and (20.50) as follows:

$$V_{rq-ref} = \left[i_{rq-ref} - i_{rq} \right] \times \left[K_{p-rsc1} + K_{i-rsc1} \times \frac{1}{s} \right] + e_{rd} + g \frac{V_s M}{L_s} \tag{20.68}$$

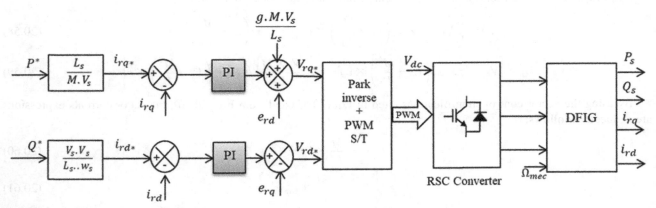

FIGURE 20.16 Indirect field-oriented control of powers scheme.

$$V_{rd-ref} = \left[i_{rd-ref} - i_{rd}\right] \times \left[K_{p-rsc2} + K_{i-rsc2} \times \frac{1}{s}\right] + e_{rq} \tag{20.69}$$

Where

$$e_{rd} = g \times \omega_s \left(L_r - \frac{M^2}{L_s}\right) \times i_{rd} \tag{20.70}$$

$$e_{rq} = g \times \omega_s \left(L_r - \frac{M^2}{L_s}\right) \times i_{rq} \tag{20.71}$$

$$T_s = (L_r - \frac{M^2}{L_s})/R_r \tag{20.72}$$

$$T_{rsc} = \frac{T_s}{100} \tag{20.73}$$

20.4.2.1 Determination of the proportional–integral controller parameters

Fig. 20.17 illustrates the regulation loop of the rotor currents i_{rd} and i_{rq}. To set the PI parameters (K_{prsc}, K_{irsc}), the pole compensation method is utilized to eliminate the transfer function zero. Eq. (20.72) shows the time constant of the controlled system. To have a satisfactory system dynamic, the time constant is divided by 10^2 as shown in Eq. (20.73):

Applying the proposed method, the equations of PI parameters (K_{prsc}, K_{irsc}) are given as follows:

$$K_{prsc} = \frac{1}{T_{rsc}} \times (L_r - \frac{M^2}{L_s}) \tag{20.74}$$

$$K_{irsc} = \frac{K_{prsc} \times R_r}{(L_r - \frac{M^2}{L_s})} \tag{20.75}$$

20.4.3 Grid side converter control

The purpose of the GSC control is to regulate the DC-link voltage and to control the reactive power exchanged with the grid to ensure that the power factor is equal to one. The GSC control scheme is given in Fig. 20.18. Using the Park transformation with an angle θ_N established using the PLL technique, as depicted in Fig. 20.8, and considering the converters' receptor convention, as shown in Fig. 20.1. The powers transferred between the grid and GSC are as follows (Elazzaoui, 2015):

$$P_N = V_{Nd} \times i_{fd} + V_{Nq} \times i_{fq} \tag{20.76}$$

$$Q_N = V_{Nq} \times i_{fd} - V_{Nd} \times i_{fq} \tag{20.77}$$

The equations used to model the bus continuous are given below:

$$i_C = i_{GSC} - i_{RSC} \tag{20.78}$$

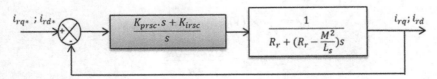

FIGURE 20.17 Inner closed loop of rotor currents regulation.

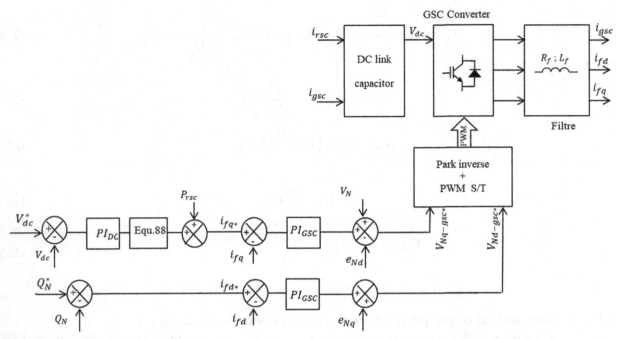

FIGURE 20.18 Grid side control scheme.

$$V_{DC} = \frac{1}{C \times s} \times i_C \qquad (20.79)$$

The q-axis is designated as an orientation axis, as shown in Fig. 20.15. The corresponding stator voltage can be written as follows:

$$V_{Nd} = 0; \quad V_{Nq} = V_N \qquad (20.80)$$

Therefore Eqs. (20.17), (20.18), (20.76), and (20.77) can be simplified as follows (Elazzaoui, 2015):

$$V_{Nd-Gsc} = -[R_f + L_f \times s] \times i_{fd} + w_N \times L_f \times i_{fq} \qquad (20.81)$$

$$V_{Nq-Gsc} = -[R_f + L_f \times s] \times i_{fq} - w_N \times L_f \times i_{fd} + V_N \qquad (20.82)$$

$$P_N = V_{Nq} \times i_{fq} \qquad (20.83)$$

$$Q_N = V_{Nq} \times i_{fd} \qquad (20.84)$$

Assuming that the back-to-back converter is lossless and neglecting the losses in the inductor resistor, the relation between powers can be expressed as follows:

$$V_{DC} \times i_C = P_{GSC} - P_{RSC} \qquad (20.85)$$

$$P_N = P_{GSC} = V_{DC} \times i_C + P_{RSC} \qquad (20.86)$$

$$P_{RSC} = V_{DC} \times i_{RSC} \qquad (20.87)$$

$$P_{c-ref} = V_{DC} \times i_{c-ref} \qquad (20.88)$$

From the above equations, the references of currents can be derived as follows:

$$i_{fq-ref} = \frac{1}{V_N} \times (V_{DC-ref} \times i_{c-ref} + P_{RSC}) \qquad (20.89)$$

$$i_{fd-ref} = \frac{Q_{N-ref} - Q_N}{V_N} \qquad (20.90)$$

The voltage references are expressed as follows:

$$V_{Nd-gsc*} = \left[i_{fd-ref} - i_{fd}\right] \times \left[K_{pgsc} + K_{igsc} \times \frac{1}{s}\right] + e_{Nq} \tag{20.91}$$

$$V_{Nq-gsc*} = \left[i_{fq-ref} - i_{fq}\right] \times \left[K_{pgsc} + K_{igsc} \times \frac{1}{s}\right] - e_{Nd} + V_N \tag{20.92}$$

where

$$e_{Nq} = \omega_N \times L_f \times i_{fq} \tag{20.93}$$

$$e_{Nd} = \omega_N \times L_f \times i_{fd} \tag{20.94}$$

Taking into consideration the model of the GSC in the referee $(d-q)$ and from Eqs. (20.19) and (20.20), the expressions of the grid currents can be deducted as follows:

$$i_{fq} = \frac{-1}{\left[R_f + L_f \times s\right]} \times \left(V_{Nq*} - \omega_N \times L_f \times i_{fd} - V_{DC} \times S_q\right) \tag{20.95}$$

$$i_{fd} = \frac{-1}{\left[R_f + L_f \times s\right]} \times \left(V_{Nd-ref} + \omega_N \times L_f \times i_{fq} + V_{DC} \times S_d\right) \tag{20.96}$$

20.4.3.1 Determination of the DC-link controller parameters

The PI controller parameters (K_{p-DC}, K_{i-DC}) are used to regulate the DC-link voltage, as shown in Fig. 20.19. From Eq. (20.79), the current flowing through the DC bus is deduced as below:

$$i_c = C \times s \times V_{DC} \tag{20.97}$$

From the block diagram shown in Fig. 20.19:

$$C \times s \times V_{DC} = \frac{K_{p-DC} \times s + K_{i-DC}}{s} \times \left(V_{DC-ref} - V_{DC}\right) \tag{20.98}$$

The TFC can be expressed as follows:

$$\frac{V_{DC}}{V_{DC-ref}} = \frac{\frac{1}{C} \times (K_{p-DC} \times s + K_{i-DC})}{s^2 + s \times \frac{K_{p-DC}}{C} + \frac{K_{i-DC}}{C}} \tag{20.99}$$

Comparing denominator of Eq. (20.99) with the canonical second-order polynomial given below:

$$F(s) = \frac{K \times \omega_0}{s^2 + 2 \times \xi \times \omega_0 \times s + \omega_0^2} \tag{20.100}$$

PI controller gains are obtained as:

$$K_{p-DC} = 2 \times \xi \times \omega \times c \tag{20.101}$$

where ξ is the damping coefficient

$$K_{i-DC} = \omega^2 \times c \tag{20.102}$$

20.4.3.2 Determination of the grid side converter controller parameters

The currents i_{Nq} and i_{Nd} are controlled by the same regulation loop, as shown in Fig. 20.20. The GSC is considered as a unit gain $(G_{GSC} = 1)$. The e_{Nq} and e_{Nd} are the coupling terms that are compensated by the PI controller.

FIGURE 20.19 Regulation loop of DC-link voltage.

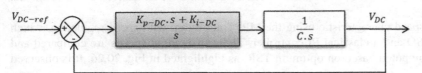

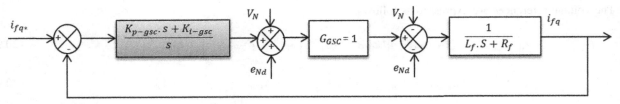

FIGURE 20.20 Control loop of currents flowing through the filter RL.

Eq. (20.103) shows the time constant of the controlled system. The time constant is divided by 10 to get an acceptable system dynamic, as shown in Eq. (20.104):

$$T_s = \frac{L_f}{R_f} \tag{20.103}$$

$$T = \frac{T_s}{10} \tag{20.104}$$

The TFO is expressed as follows:

$$\text{TFO} = K_{igsc} \times \frac{\frac{K_{pgsc}}{K_{igsc}} \times s + 1}{s} \times \frac{\frac{1}{R_f}}{\frac{L_f}{R_f} \times s + 1} = \frac{K_{igsc}}{R_f \times s}; \text{ With } \frac{K_{pgsc}}{K_{igsc}} = \frac{L_f}{R_f} \tag{20.105}$$

The TFC can be written as follows:

$$\text{TFC} = \frac{\text{TFO}}{\text{TFO} + 1} = \frac{1}{\frac{R_f}{K_{igsc}} \times s + 1} \tag{20.106}$$

By applying the pole compensator method, the PI controller gains (K_{pgsc}, K_{igsc}) are given as follows:

$$K_{pgsc} = \frac{L_f}{T} \tag{20.107}$$

$$K_{igsc} = \frac{R_f}{T} \tag{20.108}$$

20.5 Simulation results and discussion

The WECS connected to the grid is modeled by diverse equations established previously and is simulated employing the parameters presented in Tables A1–A3 under MATLAB/Simulink software. The wind profile illustrated in Fig. 20.21, that is, chosen variable, is considered to verify the superiority of the PSO method applied to the MPPT in the wind turbine system. The continuous bus voltage reference is taken constant and equal to 1200 V. The DFIG is driven by the output multiplier speed rotation and is fed by the produced mechanical, which is considered as a reference. To guarantee a UPF, the reactive power references Q^* and Q_N^* are set to 0 VAR (Volt-Ampere Reactive). The PI controller parameters for the MPPT algorithm are primarily set by the pole compensation method. Secondly, they are tuned by using the PSO algorithm. The following configurations are chosen to compute the PSO algorithm: population size = 8, number of parameters = 2, number of iterations = 5, and coefficients $W = 0.9$ and $C_1 = C_2 = 2$. The PI controller parameters used in the WECS for the vector control are calculated by the pole compensation technique.

Figs. 20.22–20.24 show the power coefficient (C_P) and the TSR. They are tested by the PI–MPPT and PSO–MPPT algorithms. Fig. 20.23 shows that the C_P is maintained at its maximal value after a perturbation when the wind speed suddenly changes. Remarkably, the PSO algorithm ensures a better pursuit and very good optimization of these parameters. That allows the mechanical power and the rotation speed of generator optimization regardless of the variation of the speed wind.

Fig. 20.25 presents the turbine rotation speed characteristic using the MPPT algorithm with speed regulation, which is regulated by an intelligent PI based on PSO and a classical PI controller. These two output speeds are compared and analyzed to the speed reference, which is computed based on optimum TSR, as highlighted in Fig. 20.26. It is observed

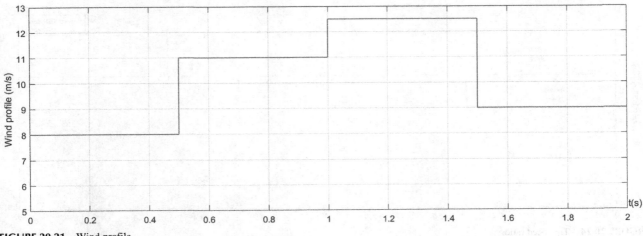

FIGURE 20.21 Wind profile.

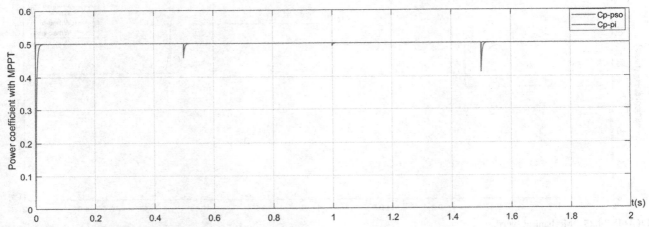

FIGURE 20.22 Power coefficient with maximum power point tracking.

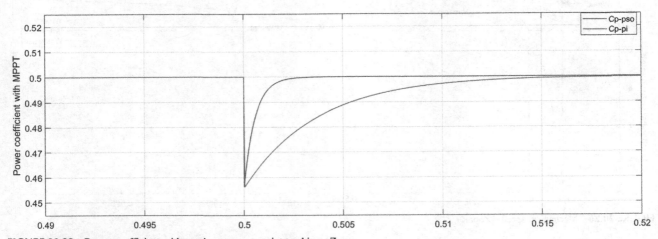

FIGURE 20.23 Power coefficient with maximum power point tracking - Zoom.

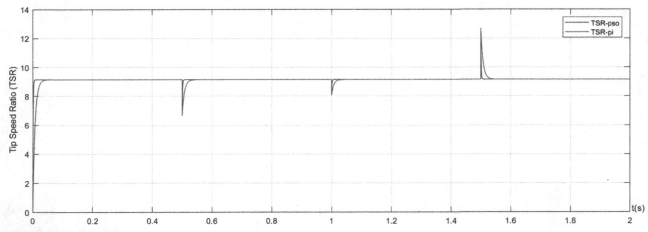

FIGURE 20.24 Tip-speed ratio.

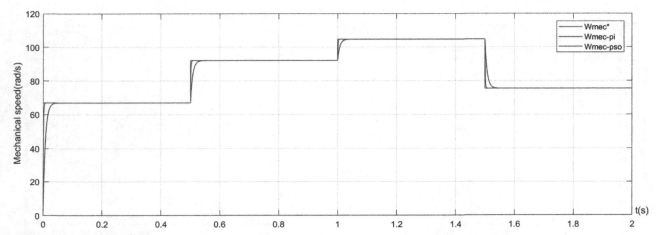

FIGURE 20.25 Mechanical speed.

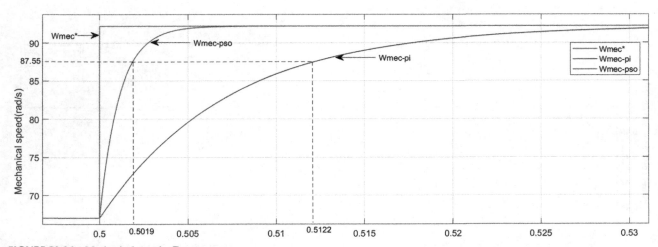

FIGURE 20.26 Mechanical speed—Zoom.

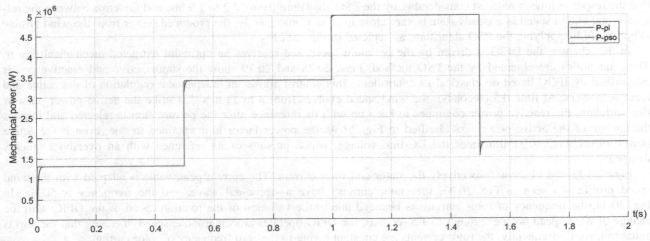

FIGURE 20.27 Mechanical power.

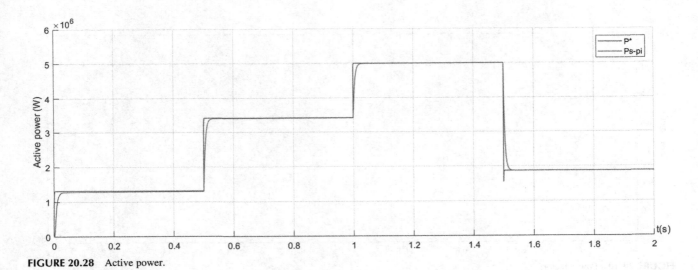

FIGURE 20.28 Active power.

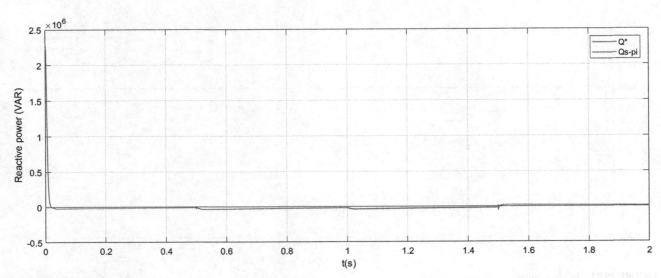

FIGURE 20.29 Reactive power.

that the response time is reduced considerably by the PSO algorithm from 12.2 to 1.9 ms, and the error between the reference and output speed in a steady state is very close to zero. Consequently, the produced power from the wind turbine is optimized by applying the PSO algorithm, as depicted in Fig. 20.27.

In this chapter, the DFIG is driven by the optimum speed and receives an optimum extracted mechanical power. These quantities are obtained by the PSO method. Figs. 20.28 and 20.29 show the stator active and reactive power established by IFOC based on classical PI controllers. This control allows an independent regulation of the active and reactive powers. At time 0.5 (seconds), the wind speed evolves from 8 to 11 m s^{-1}. Unlike the active power pursuits this variation, the reactive power continues to keep up with its reference after the perturbation is rejected and ignores the change of the active power. As clarified in Fig. 20.30, the power factor is maintained to one, even if the active power varies. Fig. 20.31 illustrates the DC-link voltage, which pursuits of its reference with an overshoot quickly damped.

Figs. 20.32−20.34 show respectively the stator and rotor current. The current peak value is adapted with the wind speed profile. As seen in Fig. 20.33, the stator currents have a sinusoidal wave, and the frequency is 50 Hz. In Fig. 20.34, the frequency of rotor currents is changed due to the variation of the rotation speed of the DFIG with the change of wind speed profile. From 1 to 1.5 seconds, the DFIG operates at synchronous speed, meaning that the slip is equal to zero. Consequently, the rotor currents are constant owing to the zero frequency of rotor current.

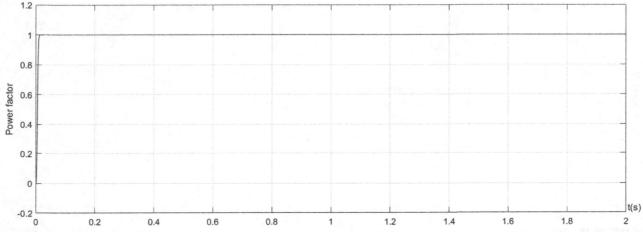

FIGURE 20.30 Power factor.

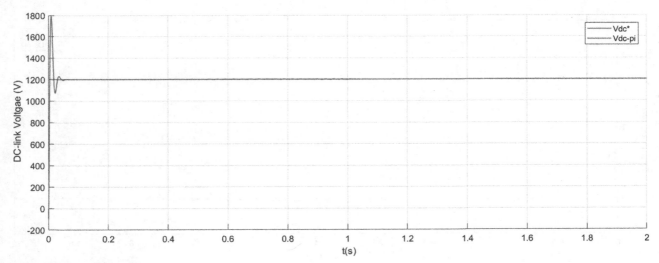

FIGURE 20.31 DC-link voltage.

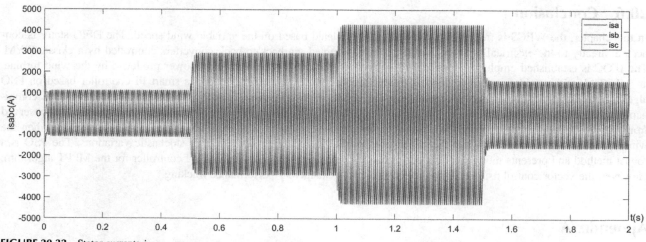

FIGURE 20.32 Stator currents $i_{s(a,b,c)}$.

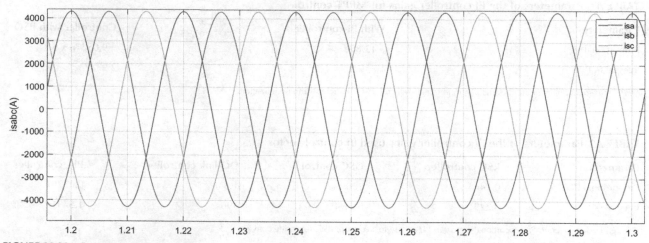

FIGURE 20.33 Stator currents $i_{s(a,b,c)}$—Zoom.

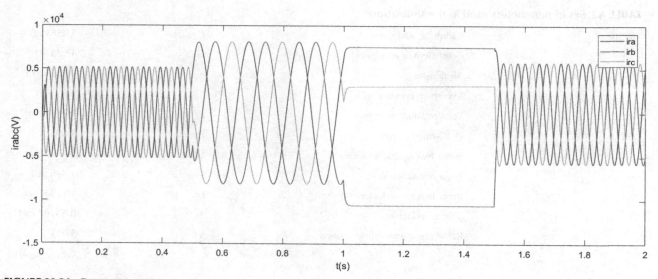

FIGURE 20.34 Rotor currents $i_{r(a,b,c)}$.

20.6 Conclusion

In this chapter, the WECS is modeled, controlled, and simulated based on the variable wind speed. The DFIG stator is connected directly to the electrical network, and the rotor is coupled via back-to-back converters controlled by a classic PWM. The IFOC is established employing classical PI controllers. To extract the maximum power produced by the wind turbine, an MPPT algorithm with speed regulation is established by dint of a classical PI and a smart PI controller based on PSO algorithm to tune the parameters of PI controllers. The results of the two types of MPPT techniques are compared. A combination of the PSO algorithm and the vector control applied to wind chain energy utilizing a DFIG is proposed. It is observed from the simulation results that the PSO tuning technique is very useful and gives interesting results, especially when the wind speed quickly changes or when the wind speed profile including a wide range of stochastic variations. The PSO is a robust method and presents more excellent performance compared to the conventional PI controller for the MPPT algorithm. However, the vector control using a PI controller presents a satisfactory performance tracking.

Appendix A

TABLE A1 Parameters of the PI controller gains for MPPT control.

Parameters	With PI controller	Controller with PSO
Proportional (K_p)	$-13.88e + 4$	$-9.0537e + 5$
Integral (K_i)	1	2.9337

MPPT, maximum power point tracking; PI, proportional–integral; PSO, particle swarm optimization.

TABLE A2 Parameters of the PI controller gains used in control vector.

Parameters	RSC controller	GSC control	DC-link controller	PLL controller
Proportional (K_p)	0.1444	200	1.848	0.02291
Integral (K_i)	0.2291	5e + 4	396	1.5791e + 5

GSC, grid side converter; PI, proportional–integral; PLL, phase-locked loop; RSC, rotor side converter.

TABLE A3 Set of parameters used in the simulation.

Turbine	Radius of blade	R	51.583 m
	Coefficient of multiplier	G_B	47.23
	Air density	ρ	1.225 kg m^{-3}
	Maximum power coefficient	C_{Pmax}	0.5
	Total moment of inertia	J	1000 kg m^2
DFIG	DFIG-rated power	Ps	5 MW
	Stator leakage inductance	Ls	1.2721 mH
	Rotor resistance	Rr	1.446 mΩ
	Rotor leakage inductance	Lr	1.1194 mH
	Mutual inductance	M	0.55187 mH
	Rated stator line to line voltage	Vs	950 V

(Continued)

Maximum power point tracking design **Chapter | 20** **469**

TABLE A3 (Continued)

Capacity	DC-link capacitance	C	4400 μF
	DC-link voltage reference	$V_{DC\text{-}ref}$	1200 V
Filter RL	Resistor of the filter	R_f	20 Ω
	Inductance of the filter	L_f	0.08 H
Transmission line	Resistance of the transmission line	R_L	20 mΩ
	Inductance of the transmission line	L_L	0.1 mH

DFIG, doubly fed induction generator.

References

Ali Zeddini, M., Pusca, R., Sakly, A., & Mimouni, M. F. (2016). PSO-based MPPT control of wind-driven Self-Excited Induction Generator for pumping system. *Renewable Energy*, *95*, 162–177. Available from https://doi.org/10.1016/j.renene.2016.04.008.

Ammar, H., Azar, A., Shalaby, R., & Mahmoud, M. (2019). Metaheuristic optimization of fractional order incremental conductance (FO-INC) maximum power point tracking (MPPT). *Complexity*, *2019*, 1–13. Available from https://doi.org/10.1155/2019/7687891.

Ananth, D., & Kumar, N. (2016). Tip speed ratio based MPPT algorithm and improved field oriented control for extracting optimal real power and independent reactive power control for grid connected doubly fed induction generator. *International Journal of Electrical and Computer Engineering (IJECE, 6*(3), 1319–1331. Available from https://doi.org/10.11591/ijece.v6i3.9306.

Ben Belghith, O., Sbita, L., & Bettaher F. (2016). MPPT design using PSO technique for photovoltaic system control comparing to fuzzy logic and P&O controllers. *Energy and Power Engineering*, *8*, 349–366. <https://doi.org/10.4236/epe.2016.811031>.

Bouderbala, M., Bossoufi, B., Lagrioui, A., Taoussi, M., Alami, H., & Ihedrane, Y. (2018). Direct and indirect vector control of a doubly fed induction generator based in a wind energy conversion system. *International Journal of Electrical and Computer Engineering (IJECE)*, *9*, 1531–1540. Available from https://doi.org/10.11591/ijece.v9i3.pp1531-1540.

Byunggyu, Y. (2018). An improved frequency measurement method from the digital PLL structure for single-phase grid-connected PV applications. *Electronics*, *7*(150), 1–11. Available from https://doi.org/10.3390/electronics7080150.

Camara, M. S., Camara, M. B., Dakyo, B., & Gualous, H. (2013). Modeling and control of the offshore wind energy system based on 5MW DFIG connected to grid. In *2013 AFRICON* (pp. 1–5). Pointe-Aux-Piments. <https://doi.org/10.1109/AFRCON.2013.6757850>.

Chavero-Navarrete, E., Trejo-Perea, M., Jáuregui-Correa, J. C., Carrillo-Serrano, R. V., & Ríos-Moreno, J. G. (2019). Expert control systems for maximum power point tracking in a wind turbine with PMSG: State of the art. *Applied Sciences*, *9*(12), 24–69. Available from https://doi.org/10.3390/app9122469.

Cortes-Vega, D., Ornelas-Tellez, F., & Anzurez-Marin, J. (2019). Comparative analysis of MPPT techniques and optimal control for a PMSG-based WECS. In *2019 IEEE 4th Colombian conference on automatic control (CCAC)* (pp. 1–6). Medellín, Colombia. <https://doi.org/10.1109/CCAC.2019.8921183>.

Dahbi, A., Nait-Said, N., & Nait-Said, M. (2016). A novel combined MPPT-pitch angle control for wide range variable speed wind turbine based on neural network. *International Journal of Hydrogen Energy*, *41*(22), 9427–9442. Available from https://doi.org/10.1016/j.ijhydene.2016.030.105.

Dai, H., Chen, D., & Zheng, Z. (2018). Effects of random values for particle swarm optimization algorithm. *Algorithms*, *11*, 23. Available from https://doi.org/10.3390/a11020023.

Elazzaoui, M. (2015). Modeling and control of a wind system based doubly fed induction generator: Optimization of the power produced. *Journal of Electrical & Electronic Systems*, *4*(1), 1–8. Available from https://doi.org/10.4172/2332-0796.1000141.

Errami, Y., Maaroufi, M., & Ouassaid, M. (2012). Variable structure direct torque control and grid connected for wind energy conversion system based on the PMSG. In *2012 international conference on complex systems (ICCS)* (pp. 1–6). <https://doi.org/10.1109/ICoCS.2012.6458524>.

Errami, Y., Ouassaid, M., & Maaroufi, M. (2015). Optimal power control strategy of maximizing wind energy tracking and different operating conditions for permanent magnet synchronous generator wind farm. In *International conference on technologies and materials for renewable energy, environment and sustainability, TMREES15* (Vol. 74, pp. 477–490). <https://doi.org/10.1016/j.egypro.2015.07.732>.

Hassan, A., El-Sawy, A., & Kamel, O. (2013). Direct torque control of a doubly-fed induction generator driven by a variable speed wind turbine. *Journal of Engineering Sciences, Assiut University*, *41*(1), 199–216.

Ihedrane, Y., El Bekkali, C., Bossoufi, B., & Bouderbala, M. (2019). Chapter 5: Control of power of a DFIG generator with MPPT technique for wind turbines variable speed. In N. Derbel & Q. Zhu (Eds.), *Modeling, identification and control methods in renewable energy systems. Green energy and technology* (pp. 105–129). Singapore: Springer. <https://doi.org/10.1007/978-981-13-1945-7_5>.

Kumar, A., & Gupta, R. (2013). Compare the results of tuning of PID controller by using PSO and GA Technique for AVR system. *International Journal of Advanced Research in Computer Engineering & Technology (IJARCET)*, *2*(6), 1–9.

Lamnadi, M., Trihi, M., Bossoufi, B., & Boulezhar, A. (2016). Modeling and control of a doubly-fed induction generator for wind turbine- generator systems. *International Journal of Power Electronics and Drive Systems (IJPEDS, 7*(3), 973–985. Available from https://doi.org/10.11591/ijpeds. v7.i3.pp982-994.

Li, S., & Haskew, T. (2009). Energy capture, conversion, and control study of DFIG wind turbine under Weibull wind distribution. In *2009 IEEE power & energy society general meeting* (pp. 1–9). Calgary, AB: IEEE. <https://doi.org/10.1109/PES.2009.5275870>.

Loucif, M., Boumediene, A., & Mechernene, A. (2013). Backstepping control of double fed induction generator driven by wind turbine. In *Third international conference on systems and control* (pp. 153–158). Algiers. <https://doi.org/10.1109/ICoSC.2013.6750851>.

Manonmani, N., & Kausalyadevi, P. (2014). A review of maximum power extraction techniques for wind energy conversion systems. *IJISET—International Journal of Innovative Science, Engineering & Technology, 1*(6), 1–8.

Ouassaid, M., Elyaalaoui, K., & Cherkaoui, M. (2016). Sliding mode control of induction generator wind turbine connected to the grid. In *Advances and applications in nonlinear control systems* (pp. 531–553). New York, NY: Springer International Publishing. <https://doi.org/10.1007/978-3-319-30169-3_23>.

Qian, K., Gao, G., & Sheng, Z. (2018). A maximum current control method for three-phase PWM rectifier for the ITER in-vessel vertical stability coil power supply. *IEEE Transactions on Plasma Science, 46*(5), 1689–1693. Available from https://doi.org/10.1109/TPS.2017.2787579.

Rouabhi, R., & Djerioui, A. (2014). Control of the power generated by variable speed wind turbine driving a doubly fed induction generator. *Journal of Electrical Engineering, 14*(3), 1–7.

Saravanan, K., Stalin, N., Sree, T., Renga R., & Soundara pandiyan B. (2016). Modelling and simulation of direct power control strategy for DFIG using Vienna rectifier. In *2016 international conference on emerging trends in engineering, technology and science (ICETETS)* (pp. 1–7). Pudukkottai. <https://doi.org/10.1109/ICETETS.2016.7603087>.

Solihin, M., Tack, L., & Kean, M. (2011). Tuning of PID controller using particle swarm optimization (PSO). *Proceeding of the International Conference on Advanced Science, Engineering and Information Technology*, 458–461. Available from https://doi.org/10.18517/ijaseit.1.4.93.

Vincent, A., & Nersisson, R. (2017). Particle swarm optimization based PID controller tuning for level control of two tank system. *IOP Conference Series: Materials Science and Engineering, 263*(5), 1–7. Available from https://doi.org/10.1088/1757-899X/263/5/052001.

Zahra, B., Salhi, H., & Mellit, A. (2017). Wind turbine performance enhancement by control of pitch angle using PID controller and particle swarm optimization. In *2017 5th international conference on electrical engineering—Boumerdes (ICEE-B)* (pp. 1–5). Boumerdes. <https://doi.org/10.1109/ICEE-B.2017.8192221>.

Chapter 21

Multiobjective optimization-based energy management system considering renewable energy, energy storage systems, and electric vehicles

Ismail El Kafazi[1], Rachid Bannari[2] and Ahmad Taher Azar[3,4]

[1]Laboratory SMARTILAB, Moroccan School of Engineering Sciences, EMSI Rabat, Rabat, Morocco, [2]Laboratory Systems Engineering, Ensa, Ibn Tofail University Kenitra, Kenitra, Morocco, [3]Faculty of Computers and Artificial Intelligence, Benha University, Benha, Egypt, [4]College of Computer and Information Sciences, Prince Sultan University, Riyadh, Kingdom of Saudi Arabia

21.1 Introduction

Energy management systems (EMSs) of microgrid (MG) are the most important systems' optimal operation (Khalid, Akram, & Shafiq, 2018; Lai, Lu, Yu, & Monti, 2019; Meghni, Dib, & Azar, 2017; Meghni, Dib, Azar, Ghoudelbourk, & Saadoun, 2017; Meghni, Dib, Azar, & Saadoun, 2018; Zia, Elbouchikhi, & Benbouzid, 2018). An efficient energy management strategy can improve MG to make full use of renewable energy (RE) and decrease the operation cost and CO_2 emission (Guo et al., 2016; Sedghi, Ahmadian, & Aliakbar, 2016). While ensuring the feasibility of the energy optimization scheme, the practical constraints should not be violated (Amara et al., 2019; Ammar, Azar, Shalaby, & Mahmoud, 2019; Ben Smida, Sakly, Vaidyanathan, & Azar, 2018; Fekik et al., 2021; Ghoudelbourk, Dib, Omeiri, & Azar, 2016; Li, Yang, Li, Zhao, & Tian, 2019; Pilla, Botcha, Gorripotu, & Azar, 2020; Schneider et al., 2017; Solanki, Raghurajan, Bhattacharya, & Cañizare, 2017; Zhang, Xu, Liu, Zang, & Yu, 2015). Normally, a MG may contain diverse sources, such as RE, distributed generations (DGs), and energy storage systems (ESSs), which leads to a large number of constraints. Moreover, the EMSs of MG is always a multiobjective optimization problem (Guo, Liu, Jiao, Hong, & Wang, 2014; Li et al., 2019; Liu et al., 2018; Mostafa, El-Shatshat, & Salama, 2013; Parhizi, Lotfi, Khodaei, & Bahramirad, 2015) in which more than one objective function requires to be optimized simultaneously, such as total operation cost, emission, and power supply reliability. This makes it difficult for the algorithms to reach the optimal solution as the energy optimization cost. However, to achieve this aim, the first step is to deal with various types of practical constraints in MG's multiobjective energy management (MEM-MG).

Several researchers have investigated the constraints handling strategies in solving MG energy management problems. In Tabar, Jirdehi, and Hemmati (2017), a MG-MEM problem is supposed to simultaneously reduce operational and emission costs, which is expressed as a constrained, linear, and mixed-integer programming, and an augmented Epsilon-constraint method is used to solve it. In Vergara, López, da Silva, and Rider (2017), a mixed-integer linear programming (MILP) model is suggested to reduce the operational cost of MGs, including all operational constraints to ensure the system's reliability as load balance, voltage and current magnitude limits, the system outages. The improved nonlinear model is converted into the MILP model by linearizing the objective function and the constraints. In Helal et al. (2017), an islanded MG's EMS problem is formulated as a mixed-integer nonlinear programming (MINLP), considering the converter constraints, DG output constraints, the operation limits of water desalination units, etc. Additionally, in Shena, Jianga, Liua, and Qianb (2016) and Olivares, Canizares, and Kazerani (2014), MILP and MINLP are also applied to solve the constrained MG energy management problems. The traditional mathematical programming may efficiently solve an optimization problem, whereas the optimization models' transformation method

Renewable Energy Systems. DOI: https://doi.org/10.1016/B978-0-12-820004-9.00016-4

may be difficult (Rezaei & Kalantar, 2015; Zia et al., 2018). In recent years, intelligent optimization algorithms are widely used in MG energy management problems (Liu et al., 2018; Lu, Lai, Yu, Wang, & Guerrero, 2018). In Chen, Duan, Cai, Liu, and Hu (2011), a smart EMS strategy is performed for a MG, which includes microturbine, fuel cell, photovoltaic (PV) array, and ESS. The economic load dispatch and operation optimization of DG are simplified into a single objective optimization problem. A matrix real-coded genetic algorithm (GA) is applied to achieve optimal solutions.

To avoid breaking the practical constraints such as spinning reserve constraint limits and system power balance, the violation value is introduced into the fitness function. Individuals with fewer violation values have the priority of being selected. This fitness function relaxation method is popular in dealing with optimization problems with all kinds of constraints. It is simple to convert the constrained optimization problem to an unconstrained one (Shahgholian & Movahedi, 2016). The constraint item in the fitness function is regularly transformed into a penalty term to estimate the constraints violations, which is described as the penalty function method (PFM). In Katiraei, Iravani, Hatziargyriou, and Dimeas (2008), a memory-based GA is suggested, optimally dealing with the power generation task among several microsources in MG. The method uses PFM to control balance constraints. The answer is how to perform the penalty function correctly based on the features of the optimization models (Lasnier & Ang, 1990). Since MG-MEM is a multiobjective particle, the Pareto-based multiobjective intelligent optimization algorithms, such as multiobjective optimization evolutionary algorithms, get more benefits to obtain the optimal solutions. In Li, Xi, and Li (2012), an original method based on multiobjective particle swarm optimization (MOPSO) is implemented to decrease the cost and maximize the MG's availability and reliability while satisfying the system limitations. The simulation results prove that the MOPSO algorithm has better performance and proficiency in finding optimal trade-off solutions. However, the constraints used are only boundary limits of DG outputs, which are easy for algorithms to deal with. In Lin, Yang, and Burnett (2002), the authors use nondominated sorting genetic algorithm II (NSGAII) to solve optimal 24 hours energy generation dispatch problems in MG, the constraints handling criteria proposed by Markvard (2000) is introduced taking account of both feasibility and domination. In Marra and Yang (2015), a hybrid constraint handling strategy is proposed for multiconstrained economic/environmental MEM problems of MG. The constraints violations can be removed in several steps during the evolutionary process by introducing the dimensionality reduction method, individual repair approach, and weights balancing process. The constraints handling strategy is utilized in NSGAII, and the results show that it can adapt to the change of MG-MEM problems. Otherwise, this hybrid strategy principally refers to several specific types of constraints. However, the constraints require to be specifically managed. For a MG-MEM problem with various types of constraints, the optimization models may change frequently, so the traditional programming methods cannot always adapt to the variations of the MG operation conditions. The smart optimization algorithms are independent of initial values and not sensitive to system models, which can adapt to the variations of MG-MEM constraints. Although, the commonly used constraints handling approaches, such as PFM and Deb's constraints handling criteria (Nejabatkhah & Li, 2015), may not always be qualified for the multiconstrained MG-MEM problems for the following challenges:

1. The MG-MEM problem has varied types of constraints. The constraints are usually very large, which make the search space extremely complicated, and the feasible solutions are difficult to obtain.
2. Different kinds of constraints violations in a MG-MEM problem cannot be compared directly, so it is hard to evaluate the overall violations by PFM or Deb's constraints handling criteria.
3. MG has various operating conditions, so the parameters, the microsources, and the combinations of constraints may always change. The current constraints handling methods may not adapt to all the MG-MEM problems. Generally, it is a multiconstraints approach strategy with strong flexibility required to deal with the constraints violations in different MG-MEM problems.

Driven by the concept of giving priority to energy saving, it is required to obtain the intelligence of devices, encouraging and helping the customers to use electric power carefully. A MG contains a mix of distributed energy resources (DERs), ESSs, loads, and tools (Katiraei et al., 2008; Nejabatkhah & Li, 2015). A MG operation requires an energy management strategy, which controls the MG's power by planning the power bought/sold to adjust their operation within the MG (Jiang, Xue, & Geng, 2013). The MG configuration integrated many types of converters used between different power sources and loads. Generally, DC−DC, DC−AC, and AC−DC converters are used in the power system due to the output voltages' different natures. Several techniques have been studied to solve EMS in MG, in the literature including dual decomposition (Zhang, Gatsis, & Giannakis, 2013) to develop a distributed EMS in MG, neural networks (Siano, Cecati, Yu, & Kolbusz, 2012), GAs (Fathima & Palanisamy, 2015), particle swarm optimization, and game theory (Zhao, 2012). The use of those methods does not prove the optimal solution. Besides, using linear and dynamic programming methods can ensure the optimal solution if it is feasible. However, they consider the RE as nondispatchable sources, and accordingly, ESSs are scheduled for balancing production and demand (Marra & Yang, 2015; Nejabatkhah & Li, 2015; Shi, Xie, Chu, & Gadh, 2015).

Several examples of ESSs could be used in MGs such as batteries, plug-in-hybrid vehicle batteries, supercapacitors, and flywheels (Tsikalakis & Hatziargyriou, 2008). Important research has been adopted in the integration of power sources and ESS into the MG. The output power of different REs and ESSs requires to be controlled and coordinated to improve the MG's reliability. While the ESSs and REs are used in the MG, the need for real-time power management starts and requires more attention. Optimizing energy management between different sources of MG, certain categories of research have been conducted out. A DC MG control strategy is developed (Xu & Chen, 2011), consisting of a hybrid system [wind turbine (WT), ESS, and DC load]. The obtained result presents different problems related to the integration of MG with the grid. The intermittent power requires ESSs in a decentralized generation system (Tani, Camara, & Dakyo, 2015). An EMS is suggested to keep a constant production between different sources such as solar energy (PV), power plants, and ESS. This strategy tries to control the state of charge (SOC) level to a reference value (Beltran et al., 2013). However, improving the batteries' performance, the charge and discharge cycle is recommended (ISCC21, 2008). In Hooshmand et al. (2014), an approach for limiting the deep of discharge to 30% has a disadvantage since the ESS charged from the grid is first utilized instead of using the surplus energy stored from the renewables. Considering ESS with REs will contribute to the cost minimization, energy-efficient, and secure solutions to the MG. Charging and discharging modes of ESS have a necessary function in power management. However, unplanned charging and discharging modes can affect MG's proper operation by significant problems such as power quality and voltage variations (Pahasa & Ngamroo, 2015). Proper scheduling for charging and discharging of ESS required a sound strategy for EMS. Newly, many methods have been published in the literature for the same. A mathematical model for electric vehicles (EVs) to grid integration is implied based on various charging and discharging efficiencies of ESS (Das et al., 2014). In this sense (Singh, Kumar, & Kar, 2013), to control the charging and discharging rates of ESS of EVs, a fuzzy-based controller is proposed. Controlling the SOC of ESS and limiting the energy generated by REs, a control strategy is suggested (De Matos et al., 2013). In Li et al. (2012), the authors used a standard droop control method to schedule the diesel generators and ESS in MG. By keeping the balance between SOC and ESS's output power, gain-scheduling techniques for power sharing and EMS based on voltage control are studied (Kakigano et al., 2013). A robust control strategy was proposed to manage the parameter uncertainties in heating a steel rod (Triki, Maidi, Belharet, & Corriou, 2017). Recently, in MGs, the trend is to increase the local consumption instead of exporting the surplus power generated by REs (Castillo-Cagigal et al., 2011). This trend can avoid the fluctuations in the voltage at the main grid (Sangwongwanich, Yang, & Blaabjerg, 2015).

The proposed study's ultimate purpose is to maximize the use of REs and to consider the ESS as a secondary source when the pick demand is reached. In addition, to minimize the grid's operation, and the excess of power produced from REs will be sold into the network. In this work, EMS's novel structure for ESS-based hybrid MGs is designed and tested to improve the optimal power references for DERs by taking account of their operation modes.

- The EMS includes the modeling and optimization problem that is proposed to minimize grid distribution costs, controlling the SOS for ESS and EV.
- ON/OFF control strategy of the grid, EVs, and ESS is modeled and tested.
- Adequate regulation of charging and discharging rates of batteries is secured for the system's proper running during events.

The proposed strategy can provide a proper and adjustable solution to avoid the grid operation's maximization and instability, based on the high penetrations of REs and ESSs of the grid. The scheduling strategy provides an order of power charge and discharge to the ESS and informs the SOC level requirements before the scheduled period. This study is directed as follows: Section 21.2 focuses on the MG operation with components specified as a case study, Section 21.3 comprises the modeling and the optimization, Section 21.4 introduces and interprets the results, and Section 21.5 concludes the work.

21.2 System description

Fig. 21.1 represents typical power systems with renewable energies, ESSs, and load demand. For the grid's principal operations, operating quantities (e.g., voltages, currents, and power demand) should be collected to the EMS through data communication infrastructure and monitoring devices. The demand is principally satisfied by the sum of the PV,

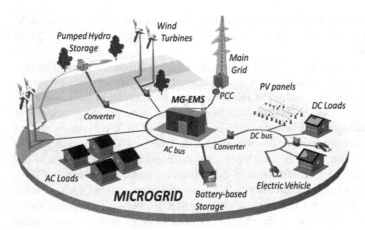

FIGURE 21.1 Structure of the microgrid based on renewable energy, energy storage system, and load demand (Adriana et al., 2017).

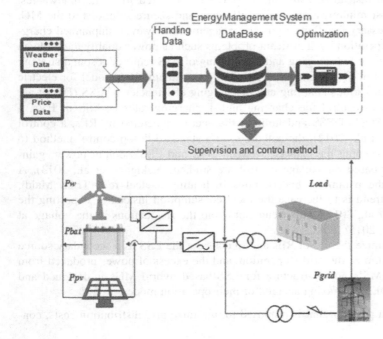

FIGURE 21.2 Illustration of the details of the microgrid in Fig. 21.1.

Wind Turbine, and the battery. If the sum of renewables (PV and WT generator) is over the demand, the surplus energy is stored in the storage system (Fig. 21.2).

The grid is used when the power from renewables and the battery cannot ensure the demand requirements. Regardless of the operation strategy selected, the grid can only supply the deficit of power needed by the load. The MG and operating quantities (e.g., voltages, currents, power generation, and weather data) will be collected, processed, and then saved in the EMS database as historical data. During the operation, data communication and devices perform this role with additional functions such as control actions. The EMS can secure these actions via data communication according to the device conditions (Choi et al., 2011).

21.2.1 Photovoltaic model

The PV generator is defined by the current I and the voltage V and by the equivalent circuit. Various mathematical models are formed to represent the behavior of PV (Borowy & Salameh, 1996; Lasnier & Ang, 1990; Markvard, 2000; Nikraz, Dehbonei, & Nayar, 2003; Zhou, Yang, & Fang, 2007). Different methods for PV models such as current and voltage at maximum power point tracking and open-circuit voltage are used (Lasnier & Ang, 1990).

21.2.2   Wind turbine system

Various factors that define the output generated from the wind energy are as follows: the output power can be determined as a purpose of aerodynamic power efficiency, mechanical transmission, power conversion, and wind speed distribution. There are many current models to measure the power generated from WTs such as linear (Bueno & Carta 2005), quadratic (Chou & Corotis, 1981; Lin et al., 2002), and cubic (Hocaoglu et al., 2009). Adaptive control theory is used for nonlinear dynamic systems for stability and tracking based on fuzzy logic systems (Chaoui & Gualous, 2017).

Proper forecasting requires knowledge of the RE generation and the load profile. Besides, it is important for the effectiveness of the operation of the EMS. However, RE generation and load profiles are generally forecasted the day before. EMS can decide on the reserve quantity and risk strategies for grid operation in advance with scheduled forecasting. An adequate method to correct such quantities is studied (Beyer et al., 2009) using real-time measurements. Besides, the ARIMA model, linear regression, and polynomial curve fitting are used (El Kafazi & Bannari, 2016; El Kafazi et al., 2017) to predict load demand and energy production. To improve the PV generator and WT, a scheduling strategy is proposed in El Kafazi et al. (2018). Besides, to reduce the intermittence impacts caused by the PV panel, an ESS is proposed to maintain the energy produced in Vieira, Pinheiro, Perez, and Bortoni (2018).

21.2.3   Electric vehicle system

Several studies have been applied to analyze the impact of the EVs charging/discharging on the grid (Amini, Moghaddam, & Karabasoglu, 2017; Dubey & Santoso, 2015; Rahman, Vasant, & Singh, 2016; Sachan & Kishor, 2016). For example, the authors (Rahman et al., 2016; Li, Davis, & Lukszo, 2016) summarized that the uncoordinated EVs charging could double the average load. Moreover, the studies in Gong, Midlam-Mohler, Marano, and Rizzoni (2012) and Maitra, Taylor, Brooks, Alexander, and Duvall (2009) noted that the EVs charging in peak hours could significantly decrease the high-voltage/low-voltage transformers' life duration. The uncoordinated EVs charging could lead to voltage drops, increased losses, transformer degradation, fuse blowouts, and feeders thermal limit violations reported in (Hoang, Wang, Niyato, & Hossain, 2017; Pieltain Fernandez, Gomez San Roman, Cossent, Mateo Domingo, & Frias, 2011). The authors in Turker, Bacha, Chatroux, and Hably (2012) proved that reducing transformer life duration is proportional to the EVs charging rate. Various studies confirm that smart charging strategies decrease the side effects of the massive EVs charging, affecting the distribution systems to deal with the previous results. The work (Papadopoulos, Skarvelis-Kazakos, Grau, Cipcigan, & Jenkins, 2010; Sortomme, Hindi, MacPherson, & Venkata, 2011) attested that smart charging could reduce losses. Besides, the authors (Rutherford & Yousef zadeh, 2011) state that a simple charging strategy could avoid the transformer loss of life duration.

21.3   Proposed scheduling and optimization model

The WT and PV generators are supposed to be connected, satisfying the entire load demand. However, the grid constraints and EVs are taken into consideration. The proposed model in this work solves the optimization problems of the minimization cost of the grid and economic scheduling simultaneously. The performance of the approach is dependent on the control parameters and constraints handling. Load demand is input; the estimated power of RE, the power of the ESS, and SOC of EVs is an obligation to verify if the ESS and EVs with renewables can meet the power load or not. On the one hand, if the renewables with storage systems are equal to power, demand is a regular operation. On the other hand, if the power load is superior or inferior to the renewables with ESS, we should use the grid's power. During the scheduling, SOC information from ESS, EVs, and forecasting from RE will inform the scheduler of the energy available. Based on this information, the optimization strategy-based controller can decide the import or export from the grid.

21.3.1   Optimization model

The optimization problem addressed in this work aims to minimize the grid cost by getting the optimal scheduling of power generation in the working day while responding to the load demand requirements. The selected study is a hybrid system composed of PV—wind—battery MG oriented to the self-consumption operation, and buying energy from the grid is allowed (Fig. 21.3).

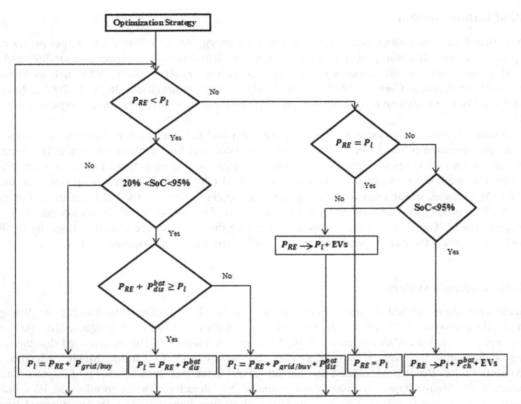

FIGURE 21.3 Operational process of the power system scheduling.

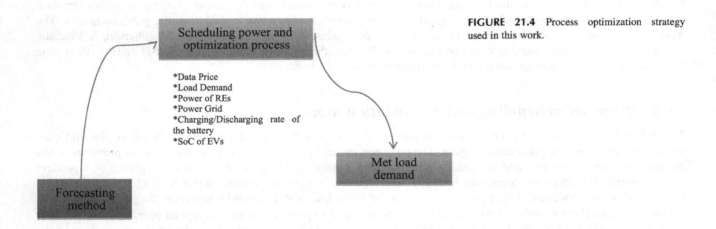

FIGURE 21.4 Process optimization strategy used in this work.

As mentioned, the control of the grid can be performed using "ON/OFF" control. In particular, the power variables to be scheduled are the power of the battery, the power used from the grid, defined as a dispatchable source, and the power for the REs, which are defined as nondispatchable sources, corresponding to the WT and PV, respectively (Fig. 21.4).

The scheduling power and optimization process are performed using real data of wind speed and solar irradiance of a summer day. The scheduling and optimization problem are fixed by using the software GAMS as an algebraic model language. The solver CPLEX is used to measure the power references obtained for the proposed model. The system is implemented to minimize the cost of consuming energy from the grid, generate the charging and discharging schedule of the battery, and charging/absorbing energy from EV (Table 21.1).

TABLE 21.1 Parameter description.

Components	Description
OF	Objective function
t	Time scheduling
i	Index of units of renewable energy
C_{sell}^{grid}	Cost of selling energy
C_{buy}^{grid}	Cost of buying energy
MG	Microgrid
ESS	Energy storage system
EMS	Energy management system
PV	Photovoltaic
SOC	State of charge
P_{pv}	Power from photovoltaic (kW)
P_{grid}	Power from the utility (kW)
P_{buy}^{grid}	Power bought from the utility (kW)
P_{sell}^{grid}	Power sold to the utility (kW)
P^{bat}	Power of the battery (kW)
P_{max}^{bat}	Maximum battery power level
P_{min}^{bat}	Minimum battery power level
$\eta^{bat,ch}$	Battery charging efficiency
$\eta^{bat,disch}$	Battery discharging efficiency
$\lambda_{grid}(t), B_t$	Binary variable
X_{load}	Connection load demand

21.3.2 Objective function

The objective is to find a proper solution of ON/OFF control of the grid that minimizes its operating cost. In addition, a method should be used to take the final optimal solution such that it can define the MG preferences. Different methods have been suggested in this area (Farzin et al., 2017). The REs work as the principal energy source, with the ESS, is considered an emergency supply source. If a deficit of power from the RE sources and the ESS battery is detected, the grid is turned on. Considering the grid is switched ON/OFF, and the REs and ESS battery are easily controlled to satisfy the demand, the objective function below is expressed as mixed-integer programming.

$$\text{OF} = \sum_{t=1}^{T} (P_{sell}^{grid}(t)\Delta t C_{sell}^{grid}(t) - P_{buy}^{grid}(t)\Delta t C_{buy}^{grid}(t)) \ \lambda_t \tag{21.1}$$

where λ_t is a discrete switch that enables to manage the energy by sending ON/OFF controls. $\lambda_t = 0$ signifies that the grid is turned off during the time t, while $\lambda_t = 1$ signifies that it is turned on.

Constraints: the power balance (Eq. 21.2) is the first condition that the model should include between generators, renewables, storage systems, and demand in the MG. The formulation can be shown as follows:

$$\sum_{t=1}^{T} P^{grid}(t)\Delta t\lambda_t + \sum_{i=1}^{2} P_{res,i}(t)\Delta t + P_{bat}(t)\Delta t + (EV_{Ch} - EV_{Dis})\Delta t = P_l(t)\Delta t X_{load} \tag{21.2}$$

The first term is related to the grid's energy and inserted into the grid, respectively. The second term is the energy provided by the renewables. In turn, the third expression corresponds to the cost related to disconnect the load, which is equal to 0 if the load is connected ($X_{load} = 1$).

Also, the boundaries related to the output of RE and utility grids are defined as follow:

$$0 \leq P_{res,i}(t) \leq P_{res,i}^{max}(t) \quad P_{res} = P_{pv} + P_w \tag{21.3}$$

$$0 \leq P_{sell}^{grid}(t) \leq \sum_{i=1}^{2} \lambda_t P_{max}^{grid}(t) \tag{21.4}$$

$$0 \leq P_{buy}^{grid}(t) \leq (1 - \lambda_t) P_{max}^{grid}(t) \tag{21.5}$$

$$0 \leq \lambda_t \leq 1 \tag{21.6}$$

21.3.2.1 Energy storage system

The generated power from the REs and the load demand at any interval, t, define whether the battery is charging or discharging. The constraints related to the ESS are defined as follows:

$$ES_t^{bat} = ES_{t-1}^{bat} + P_{ch,t}^{bat} \eta^{bat,ch} - P_{dis,t}^{bat}/\eta^{bat,dis} \tag{21.7}$$

where ES_t^{bat} is the energy storage in the battery, $P_{ch,t}^{bat}$ and $P_{dis,t}^{bat}$ are power charging and discharging of the battery, $\eta^{bat,ch}$ and $\eta^{bat,dis}$ are battery charging and discharging efficiency.

Moreover, the boundaries related to the output of the ESS are defined as follows:

$$(1 - B_t) + B_t = 1 \forall t \in T \tag{21.8}$$

$$0 \leq P_{ch,t}^{bat} \leq B_t P_{ch.max,t}^{bat} \quad B_t \in \{0, 1\} \quad \forall t \in T \tag{21.9}$$

$$0 \leq P_{dis,t}^{bat} \leq (1 - B_t) P_{dis.max,t}^{bat} \quad B_t \in \{0, 1\} \quad \forall t \in T \tag{21.10}$$

$$ES_{min}^{bat} \leq ES_t^{bat} \leq ES_{max}^{bat} \tag{21.11}$$

$$ES_{t=0}^{bat} = ES_{t=24}^{bat} \tag{21.12}$$

21.3.2.2 Electric vehicle system

In this part of the work, we have considered a residence where users own EVs and charge their vehicles in the parking. It is expected in this approach that the management system (MS) is apt to communicate with the EVs. The MS receives the EVs batteries' characteristics, mainly the battery's maximum capacity and the maximum power of each EV charger. Besides, The MS estimates the SOC using the statistical data of the EVs daily driving mileage. This charging policy aims to reduce EVs energy consumption cost.

$$SOC_{min} \leq SOC_t^{EV} \leq SOC_{max} \tag{21.13}$$

21.4 Results and discussion

The MG contains two renewables. First, a PV generator provides energy proportional to three factors: the size of the area covered by the PV, the efficiency of the solar panels, and the irradiance data. Second, a wind farm's simplified model generates electrical energy following a linear relationship with the wind. When the wind reaches a nominal value, the wind provides the nominal power. The MG is divided into four important parts: a grid, acting as the base power; a PV generator connected with a wind generator, to provide RE; a vehicle-to-grid (V2G) system situated next to the last part of the system, which is the load of the grid. The MG dimension represents approximatively a village of households during a low consumption day in spring or fall. There are 50 EVs in the base model. This is a possible scenario in the foreseeable future (Fig. 21.5).

The simulation lasts 24 hours. The solar intensity follows a normal distribution where the highest intensity is reached at midday. The wind varies greatly during the day and has multiple peaks and lows. The residential load

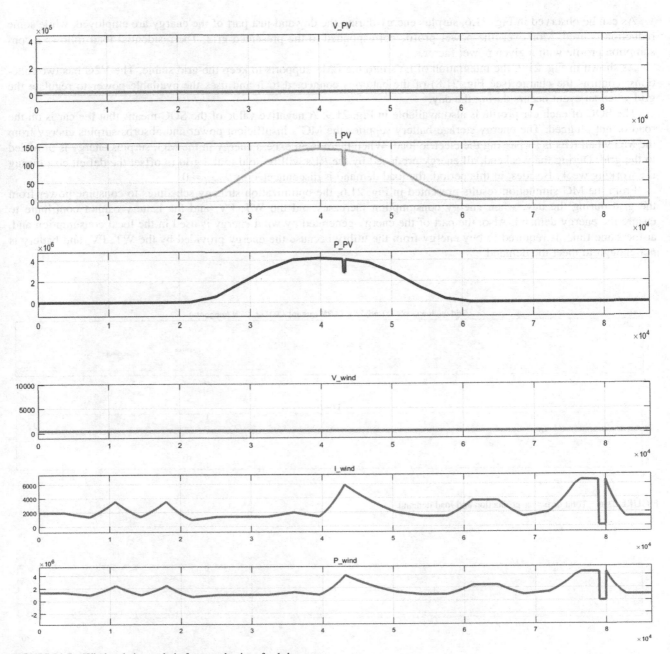

FIGURE 21.5   Wind and photovoltaic farm productions for 1 day.

follows a typical pattern similar to normal household consumption. The consumption is low during the day and increases to a peak during the evening and slowly decreases during the night. Three events will affect the grid frequency during the day:

- the kick-off of the asynchronous machine early at the third hour,
- a partial shading at noon affecting the production of solar power,
- a wind farm trip at 22 hours when the wind exceeds the maximum wind power allowed.

The findings illustrate that the PV generation has a huge part in supplying energy production, which has reached the highest excess energy percentage. Moreover, developing PV power performance and WT allocates a basic part of the energy generation, thereby increasing the battery and inverter sizes.

As can be observed in Fig. 21.6, surplus energy during the day and just part of the energy are employed, while some reduction is used. Moreover, the power profile corresponded to the presented grid. The residential load follows a consumption profile with a given power factor.

As shown in Fig. 21.7, the integration of EVs into the EMS supports to keep the grid stable. The V2G has two functions: controls the charge (see Fig. 21.8) of the batteries connected to it and uses the available power to regulate the grid when an event occurs during the day.

The SOC of each car profile is also available in Fig. 21.9. A negative value of the SOC means that the car is on the road or not plugged. The energy storage battery supplies the MG's insufficient power and absorbs surplus energy from the MG when REs is surpassing the electric load. When there is an excess energy in the MG, surplus energy is provided to the grid. During the weekend, all energy produced by the REs will be sold to the grid to offset the deficit cost during the working week. Besides, in this period, the load demand is disconnected by $X_{load} = 0$.

From the MG simulation results presented in Fig. 21.6, the optimization strategy schedules to consume power from the grid during the day, when energy consumption increases and the WT, PV, and the battery cannot contribute to ensure the energy demand. Also, the part of the energy generated by wind energy is used in the local consumption and, at the same time, is required to buy energy from the utility because the energy provided by the WT, PV, and battery is not enough to meet the demand.

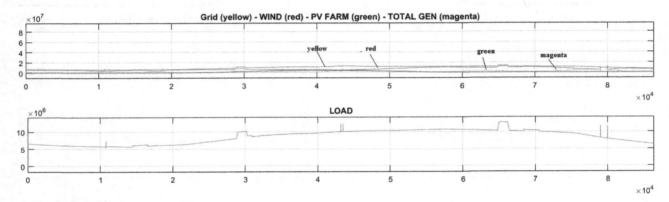

FIGURE 21.6 Total of power production and load demand.

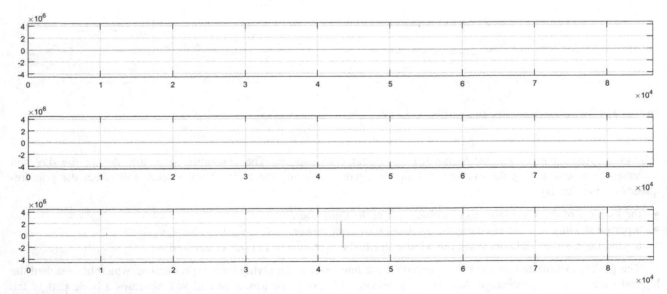

FIGURE 21.7 Grid regulations (current, voltage, and power).

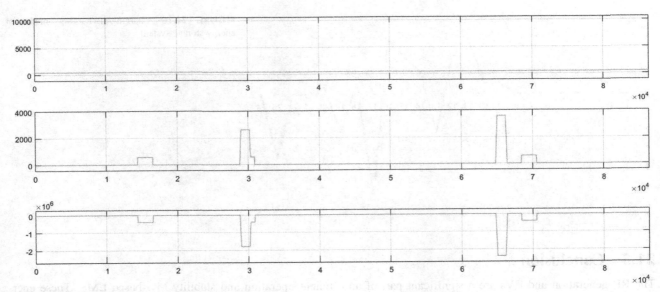

FIGURE 21.8 Grid regulations (current, voltage, and power).

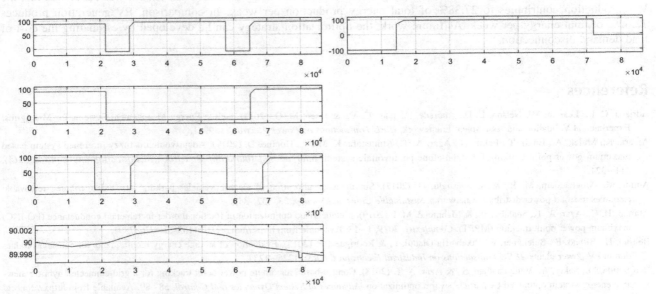

FIGURE 21.9 The state of charge of each car profile.

However, the battery is charged the day before, as shown in Fig. 21.9, for managing the imbalance between RE generation and load demand. The scheduled power profile, SOC, and the charging/discharging of the battery during the day are presented in Figs. 21.6, 21.9, and 21.10. As can be pointed out, the battery is discharged when the power stored from the REs is low, while it is charged when there is high power in the REs. As shown in Fig. 21.9, the scheduling keeps the battery level to manage the power at the end of the day when the WT and PV do not generate energy.

The technical methods were validated by performing the MG with the EMS. It is possible to conclude that the EMS reduces the utility grid's cost in the MG and includes a technical restriction for managing the storage system correctly.

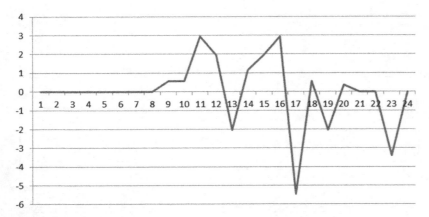

FIGURE 21.10 Charging/discharging mode of energy storage system.

21.5 Conclusion

The RE generation and EVs are a significant part of an efficient operation and stability MG-based EMS. These energies, connected with a storage system, would offer system reliability, making it proper for stand-alone applications. In this work, the MILP was used to achieve the optimal solution for the hybrid system's economic utilities. The proposed strategy has been used to set optimal power references for the distributed resources of the MG. Besides, it has been used to minimize the running cost of the utility grid. Related to the energy production results, it was concluded that the WT production contributes to 27.38% of total energy production per week. In comparison, PV generation produces 72.62% of total energy per week. As future work, the optimization strategy can be developed by estimating the cost of load demand disconnection.

References

Adriana, C. L., Lexuan, M., Nelson, L. D., Andrade, F., Juan, C. V., & Josep, M. G. (2017). Online Energy Management Systems for Microgrids: Experimental Validation and Assessment Framework. *IEEE Transactions on Power Electronics*, *99*, 1.

Amara, K., Malek, A., Bakir, T., Fekik, A., Azar, A. T., Almustafa, K. M., ... Hocine, D. (2019). Adaptive neuro-fuzzy inference system based maximum power point tracking for stand-alone photovoltaic system. *International Journal of Modelling, Identification and Control*, *33*(4), 311−321.

Amini, M., Moghaddam, M. P., & Karabasoglu, O. (2017). Simultaneous allocation of electric vehicles parking lots and distributed renewable resources in smart power distribution networks. *Sustainable Cities and Society*, *28*, 332−342.

Ammar, H. H., Azar, A. T., Shalaby, R., & Mahmoud, M. I. (2019). Metaheuristic optimization of fractional order incremental conductance (FO-INC) maximum power point tracking (MPPT). *Complexity*, *2019*, 1−13. Available from https://doi.org/10.1155/2019/7687891.

Beltran, H., Bilbao, E., Belenguer, E., Etxeberria-Otadui, I., & Rodriguez, P. (2013). Evaluation of storage energy requirements for constant production in PV power plants. *IEEE Transactions on Industrial Electronics*, *60*(3), 1225−1234.

Ben Smida, M., Sakly, A., Vaidyanathan, S., & Azar, A. T. (2018). Control-based maximum power point tracking for a grid-connected hybrid renewable energy system optimized by particle swarm optimization. *Advances in System Dynamics and Control*, 58−89. Available from https://doi.org/10.4018/978-1-5225-4077-9.ch003, IGI-Global.

Beyer, H., Martínez, J., Suri, M., Torres, J., Lorenz, E., Müller, S., Hoyer-Klick, C., & Ineichen, P. (2009). Report on benchmarking of radiation products. In Sixth framework programme MESOR, management and exploitation of solar resource knowledge. [Online]. Available at: http://www.mesor.org/docs/MESoRBench_marking_of_radiation_products.pdf.

Borowy, B. S., & Salameh, Z. M. (1996). Methodology for optimally sizing the combination of a battery bank and PV array in a wind/ PV hybrid system. *IEEE Transactions on Energy Conversions*, *11*(2), 367−375.

Bueno, C., & Carta, J. A. (2005). Technical-economic analysis of windpowered pumped hydrostorage systems. Part I: Model development. *Solar Energy*, *78*, 382−395.

Castillo-Cagigal, M., Caamao-Martin, E., Matallanas, E., Masa-Bote, D., Gutirrez, A., Monasterio-Huelin, F., et al. (2011). PV selfconsumption optimization with storage and active DSM for the residential sector. *Solar Energy*, *85*(9), 2338−2348.

Chaoui, H., & Gualous, H. (2017). Adaptive fuzzy logic control for a class of unknown nonlinear dynamic systems with guaranteed stability. *Journal of Control, Automation and Electrical Systems*. Available from https://doi.org/10.1007/s40313-017-0342-y.

Chen, C., Duan, S., Cai, T., Liu, B., & Hu, G. (2011). Smart energy management system for optimal microgrid economic operation. *IET Renewable Power Generation*, *5*(3), 258−267.

De Matos, J. G., Ribeiro, L. A. D. S., & Gomes, E. C. (2013). Power control in AC autonomous and isolated microgrids with renewable energy sources and energy storage systems. In Proceedings of the IEEE 39th annual conference on industrial electronics society, Vienna, Austria (pp. 1827–1832).

Dubey, A., & Santoso, S. (2015). Electric vehicle charging on residential distribution systems: Impacts and mitigations. *IEEE Access, 3*, 1871–1893.

El Kafazi, I., Bannari, R., & Abouabdellah, A. (2016). Modeling and forecasting energy demand. In 4th Edition of the international renewable and sustainable energy conference (IRSEC'16 November 14-17, 2016) (pp. 746–750).

El Kafazi, I., Bannari, R., Abouabdellah, A., Aboutafail, M. O., & Guerrero, J. M. (2017). Energy production a comparison of forecasting methods using the polynomial curve fitting and linear regression. In 5th Edition of the international renewable and sustainable energy conference (IRSEC'17December 04-07, 2017).

El Kafazi, I., Bannari, R., Lassioui, A., & Aboutafail, M. O. (2018). Power Scheduling for Renewable Energy Connected to the grid. In 2018 3rd international conference on power and renewable energy. https://doi.org/10.1051/e3sconf/20186408008.

Farzin, H., Firuzabad, M. F., & MoeiniAghtaie, M. (2017). A stochastic multi-objective framework for optimal scheduling of energy storage systems in microgrids. *IEEE Transactions on Smart Grid, 8*(1), 117–127.

Fathima, A. H., & Palanisamy, K. (2015). Optimization in microgrids with hybrid energy systems: A review. *Renewable and Sustainable Energy Reviews, 45*, 431–446.

Fekik, A., Azar, A. T., Kamal, N. A., Serrano, F. E., Hamida, M. L., Denoun, H., & Yassa, N. (2021). Maximum power extraction from a photovoltaic panel connected to a multi-cell converter. In A. E. Hassanien, A. Slowik, V. Snášel, H. El-Deeb, & F. M. Tolba (Eds.), *Proceedings of the international conference on advanced intelligent systems and informatics 2020. AISI 2020. Advances in Intelligent Systems and Computing* (Vol. *1261*). Cham: Springer. <https://doi.org/10.1007/978-3-030-58669-0_77>.

Ghoudelbourk, S., Dib, D., Omeiri, A., & Azar, A. T. (2016). MPPT Control in wind energy conversion systems and the application of fractional control (PI$^\alpha$) in pitch wind turbine. *International Journal of Modelling, Identification and Control (IJMIC), 26*(2), 140–151.

Gong, Q., Midlam-Mohler, S., Marano, V., & Rizzoni, G. (2012). Study of PEV charging on residential distribution transformer life. *IEEE Transactions on Smart Grid, 3*(1), 404–412.

Guo, L., Liu, W., Jiao, B., Hong, B., & Wang, C. (2014). Multiobjective stochastic optimal planning method for stand-alone microgrid system. *IET Generation, Transmission & Distribution, 8*(7), 1263–1273.

Guo, L., Liu, W., Li, X., Liu, Y., Jiao, B., Wang, W., ... Li, F. (2016). Energy management system for stand-alone wind powered desalination microgrid. *IEEE Transactions on Smart Grid, 7*(2), 1079–1087.

Helal, S. A., Najee, R. J., Hanna, M. O., Shaaban, M. F., Osman, A. H., & Hassan, M. S. (2017). An energy management system for hybrid microgrids in remote communities. In *Proceedings of the 30th IEEE Canadian conference on Electrical and Computer Engineering* (pp. 1–4).

Hoang, D. T., Wang, P., Niyato, D., & Hossain, E. (2017). Charging and discharging of plug-in electric vehicles (PEVs) in vehicle-to-grid (V2G) systems: A cyber insurance-based model. *IEEE Access, 5*, 732–754.

Hooshmand, A., Asghari, B., & Sharma, R. (2014). Experimental demonstration of a tiered power management system for economic operation of grid-tied microgrids. *IEEE Transactions on Sustainable Energy, 5*(4), 1319–1327.

ISCC21 (2008). Guide for optimizing the performance and life of leadacid batteries in remote hybrid power systems. IEEE Std 1561- 2007, pp. C1–25.

Jiang, Q., Xue, M., & Geng, G. (2013). Energy management of microgrid in grid-connected and stand-alone modes. *IEEE Transactions on Power Systems, 28*(3), 3380–3389.

Katiraei, F., Iravani, R., Hatziargyriou, N., & Dimeas, A. (2008). Microgrids management. *IEEE Power and Energy Magazine, 6*(3), 54–65.

Khalid, M., Akram, U., & Shafiq, S. (2018). Optimal planning of multiple distributed generating units and storage in active distribution networks. *IEEE Access, 6*, 55234–55244.

Lai, J., Lu, X., Yu, X., & Monti, A. (2019). Cluster-oriented distributed cooperative control for multiple AC microgrids. In *IEEE Transactions on Industrial Informatics*.

Lasnier, F., & Ang, T. G. (1990). *Photovoltaic engineering handbook.* Bristol: Routledge.

Li, G., Xi, F., Li, X., & Wang, C. (2012). Coordinated control of battery storage system and diesel generators in ac island microgrid. In *Proceedings of the 7th proceedings of the IEEE international power electronics and motion control conference*, China (pp. 112–117).

Li, Y., Davis, C., & Lukszo, Z. (2016). Electric vehicle charging in China power system: Energy, economic and environmental trade-offs and policy implications. *Applied Energy, 173*, 535–554.

Li, Y., Yang, Z., Li, G., Zhao, D., & Tian, W. (2019). Optimal scheduling of an isolated microgrid with battery storage considering load and renewable generation uncertainties. *IEEE Transactions on Industrial Electronics, 66*(2), 1565–1575.

Li, Y., Yang, Z., Zhao, D., Lei, H., Cui, B., & Li, S. (2019). Incorporating energy storage and user experience in isolated microgrid dispatch using a multiobjective model. *IET Renewable Power Generation, 13*(6), 973–981.

Lin, L., Yang, H., & Burnett, J. (2002). Investigation on wind power potential on Hong Kong islands—An analysis of wind power and wind turbine characteristics. *Renewable Energy, 27*, 1–12.

Liu, Z., Yang, J., Zhang, Y., Ji, T., Zhou, J., & Cai, Z. (2018). Multiobjective coordinated planning of active-reactive power resources for decentralized droop-controlled islanded microgrids based on probabilistic load flow. *IEEE Access, 6*, 40267–40280.

Lu, X., Lai, J., Yu, X., Wang, Y., & Guerrero, J. M. (2018). Distributed coordination of islanded microgrid clusters using a two-layer intermittent communication network. *IEEE Transactions on Industrial Informatics, 14*(9), 3956–3969.

Maitra, A., Taylor, J., Brooks, D., Alexander, M., & Duvall, M. (2009). Integrating plug-in electric vehicles with the distribution system. In *Proceedings of the International conference and exhibition on electricity distribution, CIRED*.

Markvard, T. (2000). *Solar electricity* (2nd ed.). London: Wiley.

Marra, F., & Yang, G. (2015). Decentralized energy storage in residential feeders with photovoltaics. In P. D. Lu (Ed.), *Energy storage for smart grids* (pp. 277–294). Boston: Academic Press.

Meghni, B., Dib, D., & Azar, A. T. (2017). A second-order sliding mode and fuzzy logic control to optimal energy management in PMSG wind turbine with battery storage. *Neural Computing and Applications, 28*(6), 1417–1434. Available from https://doi.org/10.1007/s00521-015-2161-z.

Meghni, B., Dib, D., Azar, A. T., Ghoudelbourk, S., & Saadoun, A. (2017). *Robust adaptive supervisory fractional order controller for optimal energy management in wind turbine with battery storage,* . Studies in computational intelligence (688, pp. 165–202). Germany: Springer-Verlag.

Meghni, B., Dib, D., Azar, A. T., & Saadoun, A. (2018). Effective supervisory controller to extend optimal energy management in hybrid wind turbine under energy and reliability constraints. *International Journal of Dynamics and Control, 6*(1), 369–383. Available from https://doi.org/10.1007/s40435-016-0296-0, Springer.

Mostafa, H. A., El-Shatshat, R., & Salama, M. M. A. (2013). Multiobjective optimization for the operation of an electric distribution system with a large number of single phase solar generators. *IEEE Transactions on Smart Grid, 4*(2), 1038–1047.

Nejabatkhah, F., & Li, Y. W. (2015). Overview of power management strategies of hybrid AC/DC microgrid. *IEEE Transactions on Power Electronics, 30*(12), 7072–7089.

Nikraz, M., Dehbonei, H., & Nayar, C. V. (2003). A DSP controlled PV system with MPPT. In *Australian power engineering conference,* Christchurch (pp. 1–6).

Olivares, D. E., Canizares, C. A., & Kazerani, M. (2014). A centralized energy management system for isolated microgrids. *IEEE Transactions on Smart Grid, 5*(4), 1864–1875.

Pahasa, J., & Ngamroo, I. (2015). PHEVs bidirectional charging/discharging and SoC control for microgrid frequency stabilization using multiple MPC. *IEEE Transactions on Smart Grid, 6*(2), 526–533.

Papadopoulos, P., Skarvelis-Kazakos, S., Grau, I., Cipcigan, L. M., & Jenkins, N. (2010). Predicting electric vehicle impacts on residential distribution networks with distributed generation. In *Proceedings of the IEEE vehicle power and propulsion conference,* VPPC.

Parhizi, S., Lotfi, H., Khodaei, A., & Bahramirad, S. (2015). State of the art in research on microgrids: A review. *IEEE Access, 3*, 890–925.

Pieltain Fernandez, L., Gomez San Roman, T., Cossent, R., Mateo Domingo, C., & Frias, P. (2011). Assessment of the impact of plug-in electric vehicles on distribution networks. *IEEE Transactions on Power Systems, 26*(1), 206–213.

Pilla, R., Botcha, N., Gorripotu, T. S., & Azar, A. T. (2020). Fuzzy PID controller for automatic generation control of interconnected power system tuned by glow-worm swarm optimization. In J. Nayak, V. Balas, M. Favorskaya, B. Choudhury, S. Rao, & B. Naik (Eds.), *Applications of robotics in industry using advanced mechanisms. ARIAM 2019. Learning and analytics in intelligent systems* (Vol. 5, pp. 140–149). Cham: Springer.

Rahman, I., Vasant, P. M., & Singh, B. S. M. (2016). Review of recent trends in optimization echniques for plug-in hybrid, and electric vehicle charging infrastructures. *Renewable and Sustainable Energy Reviews, 58*, 1039–1047.

Rezaei, N., & Kalantar, M. (2015). Stochastic frequency-security constrained energy and reserve management of an inverter interfaced islanded microgrid considering demand response programs. *International Journal of Electrical Power & Energy Systems, 69*, 273–286.

Rutherford, M. J., & Yousef zadeh, V. (2011). The impact of Electric Vehicle battery charging on distribution transformers. In *Proceedings of the IEEE applied power electronics conference and exposition,* APEC.

Sachan, S., & Kishor, N. (2016). Charging of electric vehicles under contingent conditions in smart distribution grids. In *Proceedings of the IEEE international conference on power electronics, drives and energy systems,* PEDES.

Sangwongwanich, A., Yang, Y., & Blaabjerg, F. (2015). High-performance constant power generation in grid-connected PV systems. *IEEE Transactions on Power Electronics, 99*, 1.

Schneider, K. P., Tuffner, F. K., Elizondo, M. A., Liu, C.-C., Xu, Y., & Ton, D. (2017). Evaluating the feasibility to use microgrids as a resiliency resource. *IEEE Transactions on Smart Grid, 8*(2), 687–696, Mar. 2017.

Sedghi, M., Ahmadian, A., & Aliakbar, M. (2016). Optimal storage planning in active distribution network considering uncertainty of wind power distributed generation. *IEEE Transactions on Power Systems, 31*(1), 304–316.

Shahgholian, G., & Movahedi, A. (2016). Power system stabiliser and flexible alternating current transmission systems controller coordinated design using adaptive velocity update relaxation particle swarm optimisation algorithm in multi-machine power system. *IET Generation, Transmission, & Distribution, 10*(8), 1860–1868.

Shena, J., Jianga, C., Liua, Y., & Qianb, J. (2016). A microgrid energy management system with demand response for providing grid peak shaving. *Electric Power Components and Systems, 44*(8), 843–852.

Shi, W., Xie, X., Chu, C.-C., & Gadh, R. (2015). Distributed optimal energy management in microgrids. *IEEE Transactions on Smart Grid, 6*(3), 1137–1146.

Siano, P., Cecati, C., Yu, H., & Kolbusz, J. (2012). Real time operation of smart grids via FCN networks and optimal power flow. *IEEE Transactions on Industrial Informatics, 8*(4), 944–952.

Singh, M., Kumar, P., & Kar, I. (2013). A multi charging station for electric vehicles and its utilization for load management and the grid support. *IEEE Transactions on Smart Grid, 4*, 1026–1037.

Solanki, B. V., Raghurajan, A., Bhattacharya, K., & Caæizare, C. A. (2017). Including smart loads for optimal demand response in integrated energy management systems for isolated microgrids. *IEEE Transactions on Smart Grid, 8*(4), 1739–1748.

Sortomme, E., Hindi, M. M., MacPherson, S. D. J., & Venkata, S. S. (2011). Coordinated charging of plug-in hybrid electric vehicles to minimize distribution system losses. *IEEE Transactions on Smart Grid, 2*(1), 198–205.

Tabar, V. S., Jirdehi, M. A., & Hemmati, R. (2017). Energy management in microgrid based on the multi objective stochastic programming incorporating portable renewable energy resource as demand response option. *Energy, 118*, 827–839.

Tani, A., Camara, M. B., & Dakyo, B. (2015). Energy management in the decentralized generation systems based on renewable energy ultra-capacitors and battery to compensate the wind/load power fluctuations. *IEEE Transactions on Industry Applications, 51,* 1817–1827.

Triki, A., Maidi, A., Belharet, K., & Corriou, J.-P. (2017). Robust control strategy for a conduction–convection system based on the scenario optimization. *Journal of Control, Automation and Electrical Systems.* Available from https://doi.org/10.1007/s40313-017-0317-z.

Tsikalakis, A. G., & Hatziargyriou, N. D. (2008). Centralized control for optimizing micro grids operation. *IEEE Transactions on Energy Conversion, 23,* 241–248.

Turker, H., Bacha, S., Chatroux, D., & Hably, A. (2012). Low-voltage transformer loss-of life assessments for a high penetration of plug-in hybrid electric vehicles (PHEVs). *IEEE Transactions on Power Delivery, 27*(3), 1323–1331.

Vergara, P. P., López, J. C., da Silva, L. C. P., & Rider, M. J. (2017). Security constrained optimal energy management system for three-phase residential microgrids. *Electric Power Systems Research, 146,* 371–382, May 2017.

Vieira, P. A. V., Pinheiro, B., Perez, F., & Bortoni, E. C. (2018). Sizing and evaluation of battery energy storage integrated with photovoltaic systems. *International Journal of Smart Grid and Sustainable Energy Technologies, 2,* 67–72.

Xu, L., & Chen, D. (2011). Control and operation of a DC micro grid with variable generation and energy storage. *IEEE Transactions on Industry Applications, 26,* 2513–2522.

Zhang, W., Xu, Y., Liu, W., Zang, C., & Yu, H. (2015). Distributed online optimal energy management for smart grids. *IEEE Transactions on Industrial Informatics, 11*(3), 717–727.

Zhang, Y., Gatsis, N., & Giannakis, G. (2013). Robust energy management for microgrids with high-penetration renewables. *IEEE Transactions on Sustainable Energy, 4*(4), 944–953.

Zhao, Z. (2012). *Optimal energy management for microgrids* (Ph.D. dissertation). Clemson, SC: Clemson University.

Zhou, W., Yang, H. X., & Fang, Z. H. (2007). A Novel model for photovoltaic array performance prediction. *Applied Energy, 84*(12), 1187–1198.

Zia, M. F., Elbouchikhi, E., & Benbouzid, M. (2018). Microgrids energy management systems: A critical review on methods, solutions, and prospects. *Applied Energy, 222,* 1033–1055.

Chapter 22

Fuel cell parameters estimation using optimization techniques

Ahmed S. Menesy[1,2], Hamdy M. Sultan[1], Salah Kamel[3], Najib M. Alfakih[2] and Francisco Jurado[4]

[1]Department of Electrical Engineering, Faculty of Engineering, Minia University, Minya, Egypt, [2]State Key Laboratory of Power Transmission Equipment & System Security and New Technology, School of Electrical Engineering, Chongqing University, Chongqing, P.R. China, [3]Department of Electrical Engineering, Faculty of Engineering, Aswan University, Aswan, Egypt, [4]Department of Electrical Engineering, University of Jaén, Jaén, Spain

22.1 Introduction

The sources of renewable energy such as photovoltaic (PV), wind turbine (WT), and fuel cell (FC) are becoming the most valuable sources of energy that are utilized in multiple scales (Ahme, Selim, Kamel, Yu, & Melguizo, 2018; Elkasem, Kamel, Rashad, & Melguizo, 2019; Ibrahim, Kamel, Rashad, Nasrat, & Jurado, 2019). FC is an electrochemical apparatus that changes energy from chemical form to electrical energy. FCs can be divided according to the type of electrolyte used (Corrêa, Farret, Canha, & Simoes, 2004; Larminie, Dicks, & Mcdonald, 2003; Mann et al., 2000; Motapon, Tremblay, & Dessaint, 2012; Sharaf & Orhan, 2014; Úbeda, Pinar, Canizares, Rodrigo, & Lobato, 2012). The most common FC types are alkaline FC, phosphoric acid FC, proton exchange membrane fuel cell (PEMFC), Solid FC, and molten carbonate FC (Sharaf & Orhan, 2014). Each type of them has different characteristics, restrictions, a group of advantages, and realistic applications (Sharaf & Orhan, 2014).

In recent years, the scientific researchers have become very interested about the technology of FC because it has several distinctive properties such as low temperature of operation, fast dynamic response, and high density of power (Wang, Chen, Mishler, Cho, & Adroher, 2011). In particular, the PEMFC attracted the majority of attention among the remaining FC kinds, due to its high efficiency, short start-up time, suitable weight, simple construction, easy to transport, and its good performance under low temperatures (Abdollahzadeh, Pascoa, Ranjbar, & Esmaili, 2014; Wang et al., 2011). The efficiency of the PEMFC changes from 30% to 60% depending on load condition, and the usual operating temperature occurs between 70°C and 85°C. There are three types of voltage drop in PEMFC, namely activation over potential, ohmic resistive voltage loss, and concentration over potential (Corrêa et al., 2004; Wang et al., 2011). The produced output voltage of the FC reduces sharply at first and then falls slowly and linearly. Finally, it decreases quickly with heavy loads (Corrêa et al., 2004).

Recently, many researchers have focused for optimal modeling of PEMFC and estimating their unknown parameters using different optimization techniques. An accurate mathematical PEMFC model leads to identify its performance, improving the cell design, reducing the cost, and saving time. The main target of this chapter is to build an accurate PEMFC mathematical model, which helps to simulate the actual characteristics of the FC at various operation conditions based on the mathematical model of Amphlett (Mann et al., 2000).

According to the modeling of PEMFC parameters, different optimization techniques are developed to find a precise solution to unknown parameters of PEMFC. For improving the estimated parameters' accuracy of the PEMFC, many researchers tried to model PEMFC's characteristics using various metaheuristic optimization techniques. In particular, a hybrid genetic algorithm was developed in Mo, Zhu, Wei, and Cao (2006). The particle swarm optimization has been proposed in Ye, Wang, and Xu (2009). The differential evaluation (DE) and hybrid adaptive DE (HADE) algorithm, have been presented to solve the studied optimization problem (Gong & Cai, 2013; Sun, Wang, Bi, & Srinivasan, 2015). Several novel techniques of optimization are suggested to solve the studied problem such as the harmony search algorithm (Askarzadeh & Rezazadeh, 2011), the seeker optimization algorithm (Dai et al., 2011), the multiverse optimizer (Fathy & Rezk, 2018), the adaptive RNA genetic algorithm (Zhang & Wang, 2013), Eagle strategy based on JAYA algorithm and Nelder-Mead simplex method (JAYA-NM) (Xu, Wang, & Wang, 2019), hybrid teaching learning based optimization-DE algorithm (TLBODE) (Turgut & Coban, 2016), shark smell optimizer (Rao, Shao, Ahangarnejad, Gholamalizadeh, & Sobhani, 2019), Cuckoo search algorithm with explosion operator (CS-EO) (Chen & Wang, 2019), chaotic Harris Hawks optimization (CHHO) (Menesy, Sultan, Selim, Ashmawy, & Kamel, 2019), selective hybrid stochastic strategy (Guarnieri, Negro, Di Noto, & Alotto, 2016), bird mating optimizer (Askarzadeh & Rezazadeh, 2013), modified artificial ecosystem optimization (MAEO) (Menesy et al., 2020), grasshopper optimizer (El-Fergany, 2017), equilibrium optimizer (EO) (Menesy, Sultan, & Kamel, 2020), and tree growth algorithm (Sultan, Menesy, Kamel, & Jurado, 2020).

Newly, efficient metaheuristic optimization techniques such as grey wolf optimizer (GWO) (Mirjalili, Mirjalili, & Lewis, 2014), salp swarm algorithm (SSA) (Mirjalili et al., 2017), and whale optimization algorithm (WOA) (Mirjalili & Lewis, 2016) have been utilized for solving different engineering problems. These algorithms have several advantages that encouraged the researchers to apply them in solving difficult optimization problems and encouraged us to apply them in this chapter. GWO is a very simple technique with a few number of parameters and needs less efforts in tuning procedure (Mirjalili et al., 2017). The two mainly used tuning parameters are the search agents' number and iterations number. Unlike other competing optimization techniques that have extra tuning parameters which cause some difficulties to the users during choosing a suitable values for such parameters (Mirjalili et al., 2014), SSA technique proved its ability in solving some engineering problems with fast convergence, lesser time, and gives good results compared with other optimization techniques (Mirjalili et al., 2017). WOA technique proved its superiority in solving several power system and engineering optimization problems with fast convergence; lesser number of parameters need to be adjusted to get the best solution (Mirjalili & Lewis, 2016).

This chapter seeks to demonstrate an accurate PEMFC model that simulates the operation of the commercial PEMFCs under various operating scenarios, that is, the cell temperature and reactants' pressure. The main contributions of this chapter could be drawn as follows:

- Different metaheuristic optimization techniques (GWO, SSA, and WOA) are applied to estimate the optimal values of the seven unknown parameters of the PEMFC model.
- The performance of developed PEMFC parameters' estimation algorithms are validated using three different commercial PEMFC stack models.
- Dynamic operation of PEMFC stack models is studied under the variation of cell temperature and reactants' pressures.
- Statistical tests are used to affirm the accuracy and viability of developed algorithms.
- Comprehensive comparison between the three developed algorithms is studied to determine the best one in solving the parameters' estimation of different PEMFC stack models.

The chapter is structured through some sections. Section 22.2 presents the mathematical modeling of PEMFCs and formulation of the optimization problem, Section 22.3 presents brief explanations of the proposed optimization techniques, Section 22.4 presents case studies included in the chapter, Section 22.5 provided simulation results and discussion upon these results, and finally Section 22.6 is dedicated for the main conclusions.

22.2 Mathematical model of proton exchange membrane fuel cell stacks

22.2.1 The concept of proton exchange membrane fuel cell

PEMFC consists of anode and cathode electrodes, and they are separated by a polymer electrolyte membrane to pass protons and prevent electrons. The schematic configuration of PEMFC is shown in Fig. 22.1.

The gas of hydrogen is applied continuously to anode side, while the gas of oxygen or air is applied continuously to cathode side. The electrochemical reactions occurring at sides of PEMFC electrodes is represented as (El-Tamaly, Sultan, & Azzam, 2014):

Anode side:

$$H_2 \rightarrow 2H^+ + 2e^-$$ (22.1)

Cathode side:

$$2H^+ + \frac{1}{2}O_2 \rightarrow H_2O$$ (22.2)

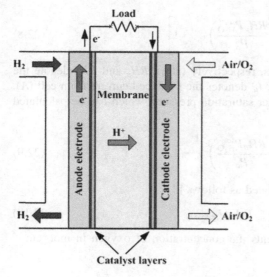

FIGURE 22.1 Schematic configuration of the proton exchange membrane fuel cell.

Overall reaction:

$$H_2 + \frac{1}{2}O_2 \rightarrow H_2O + \text{Energy} \tag{22.3}$$

The energy mentioned in Eq. (22.3) denotes the generated electrical energy, which results from the electron circulation from anode to cathode passing through an electrical load. The real output voltage and current of a single PEMFC are very small. Hence, if we need to get a high reasonable amount of voltage or current, many cells are combined together and connected in series and/or parallel.

Amphlett et al. (1995) proposed the mathematical model of PEMFC. When N_{cells} of identical FCs are combined in series, the output voltage will be calculated as depicted in Eq. (22.4):

$$V_{stack} = N_{cells} \cdot V_{cell} \tag{22.4}$$

where N_{cells} denotes the total number of cells, and V_{cell} represents the terminal voltage of each single FC.

The three kinds of voltages in PEMFC are activation overvoltage, ohmic resistive voltage drop, and concentration overvoltage. The produced voltage of each single FC is calculated using Eq. (22.5) (Amphlett et al., 1995; Aouali et al., 2017; Corrêa et al., 2004):

$$V_{cell} = E_{Nernst} - V_{act} - V_{ohmic} - V_{con} \tag{22.5}$$

where E_{Nernst} denotes the open-circuit voltage, V_{act} is the activation overvoltage per cell, V_{ohmic} represents the ohmic resistive voltage loss per cell, and V_{con} represents the concentration overvoltage per cell. E_{Nernst} is called reversible potential which can be expressed as:

$$E_{Nernst} = 1.229 - 0.85 \times 10^{-3} \left(T_{fc} - 298.15 \right)$$
$$+ 4.3085 \times 10^{-5} T_{fc} \times [\ln(P_{H_2}) - \frac{1}{2}\ln(P_{O_2})] \tag{22.6}$$

where T_{fc} defines the operating temperature in Kelvin; P_{O_2} and P_{H_2} represent the partial pressures of oxygen and hydrogen (atm), respectively.

When the inputs to PEMFC stack are pure oxygen/air and hydrogen, the partial pressures of oxygen and hydrogen can be expressed as (Panos, Kouramas, Georgiadis, & Pistikopoulos, 2012):

$$P_{O_2} = P_c - RH_c P_{H_2O}^{sat} - \frac{0.79}{0.21} P_{O_2} \times \exp\left(\frac{0.291 \left(I_{fc}/A \right)}{T_{fc}^{0.832}} \right) \tag{22.7}$$

$$P_{H_2} = 0.5RH_aP_{H_2O}^{sat}\left[\left(\exp\left(\frac{1.635(I_{fc}/A)}{T_{fc}^{1.334}}\right) \times \frac{RH_aP_{H_2O}^{sat}}{P_a}\right)^{-1} - 1\right] \tag{22.8}$$

where P_c and P_a represent the pressure at cathode and anode electrodes, respectively (atm); RH_c and RH_a denote the relative humidity of vapor around the two electrodes, respectively (atm); I_{fc} denotes the produced current from cell (A), A represents the area of membrane surface (cm^2), P_{H_2O} denotes the water saturation pressure, which can be calculated as (Panos et al., 2012):

$$P_{H_2} = 0.5RH_aP_{H_2O}^{sat}\left[\left(\exp\left(\frac{1.635(I_{fc}/A)}{T_{fc}^{1.334}}\right) \times \frac{RH_aP_{H_2O}^{sat}}{P_a}\right)^{-1} - 1\right] \tag{22.9}$$

The voltage loss resulting from the activation process (V_{act}) is formulated as follows [33]:

$$V_{act} = -\left[\xi_1 + \xi_2 T_{fc} + \xi_3 T_{fc}\ln(C_{O_2}) + \xi_4 T_{fc}\ln(I_{fc})\right] \tag{22.10}$$

where ξ_1, ξ_2, ξ_3, and ξ_4 denote semiempirical coefficients; C_{O_2} represents the concentration of oxygen in mol $\cdot$ cm^{-3} and can be represented as:

$$C_{O_2} = \frac{P_{O_2}}{5.08 \times 10^6 \times \exp^{-(498/T_{fc})}} \tag{22.11}$$

The ohmic resistive voltage loss (V_{ohmic}) is the second kind of voltage drops that occur in the FC, and it can be defined as:

$$V_{ohmic} = I_{fc}(R_M + R_C) \tag{22.12}$$

where R_M defines the membrane surface resistance in (Ω), R_C is the connection that the protons meet during flowing through membrane. The polymer resistance is calculated as:

$$R_M = \frac{\rho_M.l}{A} \tag{22.13}$$

where ρ_M represents the specific resistance of membrane material ($\Omega \cdot$ cm), l represents the thickness of membrane (cm), ρ_M can be expressed as follows:

$$\rho_M = \frac{181.6\left[1 + 0.03\left(\frac{I_{fc}}{A}\right) + 0.062\left(\frac{T_{fc}}{303}\right)^2\left(\frac{I_{fc}}{A}\right)^{2.5}\right]}{\left[\lambda - 0.634 - 3\left(\frac{I_{fc}}{A}\right)\right] \times \exp\left[4.18\left(\frac{T_{fc} - 303}{T_{fc}}\right)\right]} \tag{22.14}$$

where λ represents an adjustable empirical parameter. The last type of voltage losses occurred in the FC is the V_{con}, which is calculated as:

$$V_{con} = -b\ln\left(1 - \frac{J}{J_{max}}\right) \tag{22.15}$$

where b is a parametric coefficient that should to be identified; J and J_{max} denote the actual current and maximum current densities (A/cm^2), respectively.

From Eqs. (22.1) to (22.15), it can be noticed that the PEMFC mathematical model has seven unknown parameters that need to be estimated perfectly. In general, the unknown parameters are not mentioned in the datasheet of manufacturer. Therefore to guarantee a precise PEMFC modeling under simulation, effective and accurate identification of such parameters (ζ_1, ζ_2, ζ_3, ζ_4, λ, R_C, and b) is too significant. These parameters are optimized to obtain the optimal values in the range of their lower and upper limits using three different optimization techniques, namely GWO, WOA, and SSA.

22.2.2 Formulation of the objective function

The PEMFC mainly depends on seven variable parameters during its operation. The estimation of these parameters can be represented as an optimization problem. In this problem, the total squared deviations (TSD) between the measured terminal voltages and the estimated ones are considered as the main objective function (OF) (Ali, El-Hameed, & Farahat, 2017; Chen & Wang, 2019; El-Fergany, 2017; Menesy, Sultan, et al., 2020; Menesy, Sultan, Korashy, et al. 2020; Rao et al., 2019; Sultan et al., 2020; Turgut and Coban, 2016). However, this OF is represented as follows:

$$OF = minTSD(X) = \sum_{i=1}^{N} (V_{meas}(i) - V_{cal}(i))^2 \qquad (22.16)$$

where X is a vector of the seven parameters, N denotes the measured points number, i is an iteration counter, V_{meas} represents the measured voltage, and V_{cal} denotes the calculated PEMFC voltage.

22.3 Optimization techniques

22.3.1 Grey wolf optimizer

GWO simulates the hunting strategy and leadership hierarchy of grey wolves. Grey wolves can be categorized to four groups. The first group is the leader, which is called alphas (α) and considers the best one as well as responsible for taking decisions regarding organization of the group specially during attacking a prey. All other group's members must obey and follow the instructions of group leader; the second group is called beta (β), which monitors the remaining wolves of group and ensures that they obey and follow orders. Beta grey wolves contribute in attacking and hunting the prey. They will be the main candidates to replace the leader after his death. The third group is the delta (δ) wolves. They have several functions in the group such as attacking, monitoring, and paying attention to sick wolves. The other wolves are called omegas (ω), and they must obey and follow the α, β, and δ wolves (Mirjalili et al., 2014). The process of encircling is mathematically represented by Eq. (22.17).

$$\vec{D} = \left| \vec{C}.\vec{X}_p(t) - \vec{X}(t) \right|, \vec{X}(t+1) = \vec{X}_p(t) - \vec{A}.(\vec{D}) \qquad (22.17)$$

where t is the current iteration number $\vec{A}$ and $\vec{C}$ denote the coefficient vectors, $X_p(t)$ defines a vector that describes the prey position, and $\vec{X}$ is a vector that denotes the position of the grey wolf. $\vec{A}$ and $\vec{C}$ vectors can be represented by (22.18).

$$\vec{A} = 2\vec{a}.\vec{r}_1 - \vec{a}, \vec{C} = 2.\vec{r}_2 \qquad (22.18)$$

where $\vec{a}$ represents the vector components, which are reduced from 2 to 0, $\vec{r}_1$ and $\vec{r}_2$ are random vectors between 0 and 1. In GWO technique, the position of wolves is randomly changed around the prey following Eq. (22.18). The first three wolves' α, β, and δ possess some special data on the possible location of prey. Moreover, the first three optimal solutions are mathematically stored. Furthermore, according to the position of the optimal search agent, other agents update their position as presented in the following expressions:

$$\vec{D}_\alpha = \left| \vec{C}.\vec{X}_\alpha - \vec{X} \right|, \vec{D}_\beta = \left| \vec{C}.\vec{X}_\beta - \vec{X} \right|, \vec{D}_\delta = \left| \vec{C}.\vec{X}_\delta - \vec{X} \right| \qquad (22.19)$$

$$\vec{X}_1 = \vec{X}_\alpha - \vec{A}_1.(\vec{D}_\alpha), \vec{X}_2 = \vec{X}_\beta - \vec{A}_2.(\vec{D}_\beta), \vec{X}_3 = \vec{X}_\delta - \vec{A}_3.(\vec{D}_\delta) \qquad (22.20)$$

$$\vec{X}(t+1) = \frac{\vec{X}_1 + \vec{X}_2 + \vec{X}_3}{3} \qquad (22.21)$$

The search agent follows a special behavior during updating its position, and this has been described in Fig. 22.2 depending on Eqs. (22.19)–(22.21) (Mirjalili et al., 2014). According to the α, β, and δ positions, a proposed position randomly existed in the search space, where it is encircled by α, β, and δ locations. After α, β, and δ wolves identify the prey location, the remaining wolves start changing their positions around the prey

22.3.2 Salp swarm algorithm

Salps are from the family of Salpidae. Salps possess a transparent body, which is similar to the barrels style as well as their body tissues, and their motion is highly like jelly fish as shown in Fig. 22.3. SSA is a model that uses the salp strings to find a solution for complex problems. To mathematically model the salp chains, the salp population is classified into two groups: the leader and leader followers. The leader should be at the front of the chain and guides the

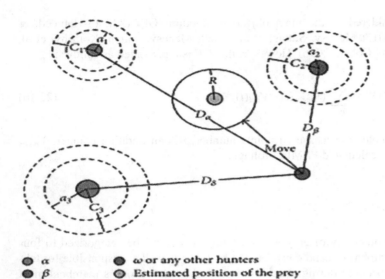

FIGURE 22.2 Position updating in grey wolf optimizer algorithm.

- ● α
- ● β
- ● δ
- ● ω or any other hunters
- ◉ Estimated position of the prey

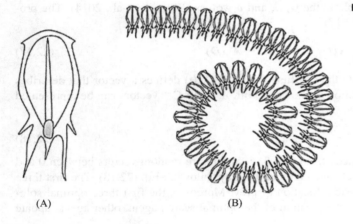

FIGURE 22.3 Aspect of (A) individual salp and (B) swarm of salps.

(A)   (B)

swarm to start searching for food, whereas the others of salps follow the leader (Mirjalili et al., 2017). The salp chain model can explore and exploit space about all fixed food sources (the position of local and global optimum) (Mirjalili et al., 2017).

Assume n is the variable number of a certain problem, x denotes the position of a certain salp, and y represents the food supply, which is the goal of salp swarm. The leader position in the search process has been updated according to the following formula:

$$x_i^1 = \begin{cases} y_i + C_1((ub_i - lb_i)C_2 + lb_i) & C_3 \geq 0 \\ y_i - C_1((ub_i - lb_i)C_2 + lb_i) & C_3 \leq 0 \end{cases} \tag{22.22}$$

where x_i^1 and y_i denote the position of leader and food source in the ith dimension, respectively. lb_i and ub_i are the lower and upper limits of the ith dimension, and C_1, C_2, C_3 denote random numbers.

The random number C_1 considers the main important parameter in SSA as it balances exploration and operation during the search process, and it can be represented as follows:

$$C_1 = 2e^{-\left(\frac{4l}{L}\right)^2} \tag{22.23}$$

where L represents the maximum number of iterations, l denotes the present iteration, C_2 and C_3 represent random numbers from 0 to 1. The positions of followers can be updated according to Newton's law of motion as follows:

$$x_i^j = \frac{1}{2} \beta t^2 + \sigma_0 t \tag{22.24}$$

where $j \geq 2$ and x_i^j represents the position of jth salp, t denotes the time, σ_0 represents an initial speed, and,

$$\beta = \frac{\sigma_{final}}{\sigma_0}, \quad \text{where} \quad \sigma = \frac{x - x_0}{t}. \tag{22.25}$$

Considering that the initial speed $\delta_0 = 0$ and the time interval t equal to the iteration, Eq. (22.24) can be described as follows:

$$x_i^j = \frac{1}{2} \left(x_i^j + x_i^{j-1} \right) \tag{22.26}$$

where $j \geq 2$. Eq. (22.26) represents the position of jth follower salp.

The following limitation formula is utilized to bring the salps, which go far away from the defined search area, back to the predefined limits.

$$x_i^j = \begin{cases} l^j & if \dots x_i^j \leq l^j \\ u^j & if \dots x_i^j \geq u^j \\ x_i^j & otherwise \end{cases} \tag{22.27}$$

22.3.3   Whale optimization algorithm

WOA technique depends on the whale hunting and encircling strategies. This technique is called bubble-net feeding mechanism. Humpback whales try to hunt little fish, which is near to the surface through making bubble net around the prey that rises along a circle path as shown in Fig. 22.4. The WOA technique assumes that the present optimal solution is the target prey. After the optimal search agent is defined, the rest search agents will hence seek to update their positions in the direction of the optimal search agent. The mathematical formulation of this phenomena can be described as follows (Mirjalili & Lewis, 2016):

$$\vec{D} = \left| \vec{C} . \vec{X}^*(t) - \vec{X}(t) \right| \tag{22.28}$$

$$\vec{X}(t + 1) = \vec{X}^*(t) - \vec{A} . \vec{D} \tag{22.29}$$

where t denotes the present iteration, $\vec{A}$ and $\vec{D}$ are coefficient vectors, $\vec{X}^*$ represents the position vector of optimal solution obtained so far, $\vec{X}$ denotes position vector. If there is a better solution, $\vec{X}^*$ must be updated in each individual iteration. The vectors $\vec{A}$ and $\vec{C}$ can be formulated as follows:

$$\vec{A} = 2 \vec{a} . \vec{r}_1 - \vec{a} \tag{22.30}$$

$$\vec{C} = 2 . \vec{r}_2 \tag{22.31}$$

where $\vec{a}$ is linearly decreased between [2, 0] through the number of iterations, and $\vec{r}$ is an arbitrary vector in [0, 1].

22.3.3.1   Bubble-net assaulting strategy (exploitation stage)

The air bubble net search mechanism of the humpback whales is performed in the following approaches:

1. Shrinking circling system

   This approach is described by the expression provided in Eq. (22.30) in which the fluctuation scope of A is controlled by the value of the factor a which occurs in the range between 2 and 0. In this case, the variable A will be randomly changed between $[-a, a]$. The updated search agent A position in the next iteration lies in between its current position and the position of best search agent. Fig. 22.4A shows the possible positions (X, Y) according (X^*, Y^*) which can be accomplished by $0 \leq A \leq 1$ in a two-dimension search space.

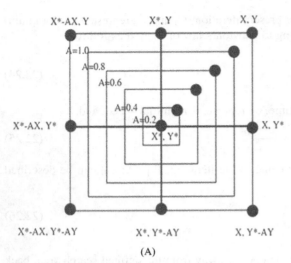

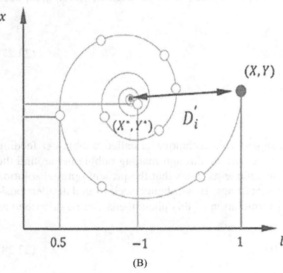

FIGURE 22.4 Bubble-net technique performed in whale optimization algorithm (X^* is the optimal solution): (A) shrinking encircling technique and (B) spiral shape for updating whale position. *Replotted from (Mirjalili & Lewis, 2016).*

2. Spiral updating position

This approach is described by graph shown in Fig. 22.4B, which is determined based on the spacing between the current position of whale (X, Y) and the position of its prey located at a certain point defined as (X^*, Y^*). The spiral path that controls the motion of the whale toward its prey can be expressed as:

$$\vec{X}(t+1) = \vec{D}'.e^{bl}.\cos(2\pi l) + \vec{X}^*(t) \tag{22.32}$$

where

$$\vec{D}' = \left| \vec{X}^*(t) - \vec{X}(t) \right|, \tag{22.33}$$

where $D^{\rightarrow'}$ gives the distance between the current position of ith whale and position of best search agent (prey position), b represents a constant number describing the condition of spiral logarithmic path, l is an arbitrary number that ranges from -1 to 1. Whales toward their way to the prey swim around inside shrinking circle forming a spiral shape. Two possible approaches are found that the whales may take, described in the following equation:

$$\vec{X}(t+1) = \begin{cases} \vec{X}^*(t) - \vec{A}.\vec{D} & \text{if } -p < 0.5 \\ \vec{D}'.e^{bl}.\cos(2\pi l) + \vec{X}^*(t) & \text{if } -p \geq 0.5 \end{cases} \tag{22.34}$$

where p represents random number that ranges from 0 to 1.

22.3.3.2 Scan for prey (investigation stage)

The whales take their position in a random manner with respect to the position of the intended prey. Therefore the vector $\vec{A}$ is formed randomly in the range $[-1, 1]$ to generate search agents that go outside the reference one. In the investigation stage, the position of whales is updated until the position of the best search agent is discovered. The mathematical model of this stage is described as follows:

$$\vec{D} = \left| \vec{C} . \vec{X}_{rand} - \vec{X}(t) \right| \tag{22.35}$$

$$\vec{X}(t+1) = \vec{X}_{rand} - \vec{A} . \vec{D} \tag{22.36}$$

22.4 Case study

In this chapter, GWO, WOA, and SSA are applied for estimating the seven parameters of various PEMFC stack models. The minimum (Min.) and maximum (Max.) values of these parameters are listed in Table 22.1 (El-Fergany, 2017; Menesy et al., 2019; Menesy, Sultan, et al., 2020; Menesy, Sultan, Korashy, et al., 2020; Sultan et al., 2020; Ye et al., 2009). The case study is performed on three various types of PEMFCs to show the efficiency and accuracy of GWO, WOA, and SSA in solving the problem of parameters' estimation of PEMFC depending on experimental data. The three PEMFC stacks are SR-12 500 W PEMFC, Temasek 1 kW PEMFC, and 250 W PEMFC stacks. The specifications and experimental data of SR-12 PEM 500 W (Corrêa et al., 2004; Jia, Li, Wang, Cham, & Han, 2009; Rajasekar et al., 2015), Temasek 1 kW (Mann et al., 2000), and 250 W PEMFC (Mo et al., 2006) are presented in Table 22.2.

The relative humidity of vapors at anode and cathode are 1.00 and 1.00, respectively. In this work, the same maximum numbers of 200 iterations and 20 search agents are used for all studied techniques. These techniques have a high level of haphazardness. Hence, the optimal results are executed over 30 individual runs.

TABLE 22.1 Practical limits of PEMFC unknown parameters.

Parameter	ξ_1	$\xi_2 \times 10^{-3}$	$\xi_3 \times 10^{-5}$	$\xi_4 \times 10^{-4}$	λ	$R_c \times 10^{-4}$	b
Min.	−1.1997	1	3.6	−2.6	10	1	0.0136
Max.	−0.8532	5	9.8	−0.954	23	8	0.5

TABLE 22.2 Datasheets of three different PEMFC stacks specifications under study.

PEMFC type	250 W stack	SR-12 PEM 500 W	Temasek 1 kW	Unit
N_{cells}	24	48	20	–
A	27	62.5	150	cm^2
l	127	25	51	μm
J_{max}	860	672	1500	mA/cm^2
P_{H_2}	1	1.47628	0.5	atm
P_{O_2}	1	0.2095	0.5	atm
T_{fc}	343.15	323	323	K

22.5 Results and discussion

22.5.1 Statistical measures

The developed PEMFC parameters' estimation algorithms based on GWO, WOA, and SSA are programmed using MATLAB® software simulation package on Laptop with processor Core i5-CPU@2.40 GHz and RAM of 6.00 MB. Statistical tests are demonstrated to assess the reliability of proposed algorithms. The accuracy of developed PEMFC parameters' estimation algorithms is assessed regarding OF values such as Min., Worst., relative error (RE), mean absolute error (MAE), mean, median, SD, root mean square error (RMSE), and efficiency. These metrics are calculated by Eqs. (22.37) to (22.41), respectively.

$$SD = \sqrt{\frac{\sum_{i=1}^{30} \left(TSD_i - \overline{TSD}\right)}{30 - 1}} \tag{22.37}$$

$$RE = \frac{\sum_{i=1}^{30} \left(TSD_i - TSD_{min}\right)}{TSD_{min}} \tag{22.38}$$

$$MAE = \frac{\sum_{i=1}^{30} \left(TSD_i - TSD_{min}\right)}{30} \tag{22.39}$$

$$RMSE = \sqrt{\frac{\sum_{i=1}^{30} \left(TSD_i - TSD_{min}\right)^2}{30}} \tag{22.40}$$

$$Eff. = \frac{TSD_{min}}{TSD_i} \times 100\% \tag{22.41}$$

where TSD_i represents the OF for every independent execution. TSD_{min} represents the Min. OF got over 30 runs. $\overline{TSD}$ represents the average of OF value.

To prove the performance of developed algorithms, the following parameters are selected; search agents = 20, the maximum number of iterations = 200, and the number of executions = 30 runs. The results of statistical tests for the three studied PEMFC models are listed in Tables 22.3–22.5. From these tables, it is evidently shown that the very minimal values of MAE, Min., and RMSE of GWO reveal the goodness of its performance compared with WOA and SSA. GWO shows the best matching between the estimated and measured data.

TABLE 22.3 Statistical results of GWO, SSA, and WOA techniques for 250 W PEMFC stack.

	Min.	Worst	Mean	Median	SD	RE	MAE	RMSE	Eff.
GWO	0.6444	0.8788	0.7069	0.6939	5.7510	2.9102	0.062	0.084	91.685
SSA	0.6500	1.7033	0.9540	0.8965	22.0054	14.029	0.304	0.373	71.157
WOA	0.7793	11.4491	4.4143	2.5445	372.817	139.94	3.635	5.1623	37.864

TABLE 22.4 Statistical results of GWO, SSA, and WOA techniques for SR-12 500 W PEMFC stack.

	Min.	Worst	Mean	Median	SD	RE	MAE	RMSE	Eff.
GWO	1.0597	1.2339	1.1086	1.0968	4.5369	1.386	0.0489	0.0662	95.731
SSA	1.0598	4.6964	1.6149	1.1432	100.267	15.71	0.5550	1.1313	79.698
WOA	1.0666	93.7643	14.021	5.34793	2076.81	364.3	12.954	24.181	39.509

TABLE 22.5 Statistical results of GWO, SSA, and WOA techniques for Temasek 1 kW PEMFC stack.

	Min.	Worst	Mean	Median	SD	RE	MAE	RMSE	Eff.
GWO	0.79510	1.0075	0.8433	0.8330	4.7615	1.8217	0.0482	0.0672	94.541
SSA	0.79681	1.2080	0.9061	0.8668	10.217	4.1175	0.1093	0.1484	88.908
WOA	0.82632	1.9270	1.0893	0.9875	29.691	9.5479	0.2629	0.3929	80.291

22.5.2 Parameters' estimation of proton exchange membrane fuel cell stacks

Three different optimization techniques namely, GWO, WOA, and SSA are used for estimating the best values of PEMFCs unknown parameters according to the Min. and Max. limits listed in Table 22.1. The results obtained by the three optimization techniques are compared with each other. The Min. values of the TSD prove the effectiveness of GWO in dealing with the studied problem to determine the best optimal unknown parameters values.

Fig. 22.5 presents the curves of TSD convergence for all the studied PEMFCs stacks obtained by the three developed algorithms. From this figure, it is clearly noticed that the GWO achieved the minimum value of the OF for all the studied PEMFCs compared with the other algorithms. In addition, the convergence curves obtained by the SSA are very good after GWO. As well as the WOA reaches its minimum value of TSD before GWO, but the best minimum TSD obtained by GWO is better than SSA as well as the WOA. Finally, it can be noticed that the convergence curves obtained by GWO indicate smooth, quick, and steady progress to the best last values for the three studied PEMFC stacks. From this, it is concluded that the GWO is the best technique compared to the remaining techniques that are suggested in this chapter; hence, we will complete the analysis using GWO technique. The developed three algorithms are executed over 30 independent runs, and the end values of TSD in every individual implementation are drawn. The convergence trends of the OF obtained by the best algorithm for the three studied PEMFC stacks over 30 runs are shown in Fig. 22.6.

To affirm the process of parameter identification, the optimal values of unknown parameters are estimated by GWO technique. The $I-V$ and $I-P$ curves of estimated values are plotted along with the experimental values for the three studied stacks as shown in Fig. 22.7. From this figure, it can be concluded that the obtained results are in good matching with the measured values.

Also, it is clear that when the load current or the current generated by the cell (I_{fc}) increases, the stack output voltage reduces and the stack output power increases as depicted in Fig. 22.7. First, the stack output voltage reduces with rapid voltage because of low activation voltage, and then slowly decreases due to ohmic voltage loss. Then, at heavy loads, the stack voltage rapidly decreases due to the overvoltage drop of diffusion.

22.5.3 Results of simulation under various operating conditions

In order to mimic the electrical characteristics of PEMFCs, different integrations of T_{fc}, P_{H_2}, and P_{O_2} are suggested. Therefore the polarization curves of PEMFCs are clarified, and the behavior of stack efficiency is observed. First, the FC stack efficiency can be calculated as follows (Srinivasulu, Subrahmanyam, & Rao, 2011):

$$\eta_{stack} = \mu_F \times \frac{V_{stack}}{N_{cells} \times v_{max}} \qquad (22.42)$$

where V_{max} denotes the maximum output voltage generated from the PEMFC and its value is 1.48 V per cell, and μ_F represents the utilization factor. It can be assumed that the hydrogen flow rate is adjusted according to the load state; therefore the utilization factor can be considered as a constant value, which equals to 95%. After estimating the seven PEMFC parameters, the $I-V$, $I-P$ and efficiency performance at various values of T_{fc}, while keeping the values of P_{H_2} and P_{O_2} constant at the values given in datasheets, are presented to affirm the robustness of three applied algorithms. In order to avoid article length and repetition of figures, only two PEMFCs stacks are studied. The 250 W PEMFC stack is operated under temperature variation, while the Temasek 1 kW PEMFC stack is examined under the variation of reactants' pressures.

The $I-V$, $I-P$ curves of 250 W PEMFC stack at 303, 313,323, and 333 K are shown in Fig. 22.8. The effect of varying the reactants pressures (P_{H_2}/P_{O_2}) with constant T_{fc}, mentioned in the manufacture's datasheet is shown.

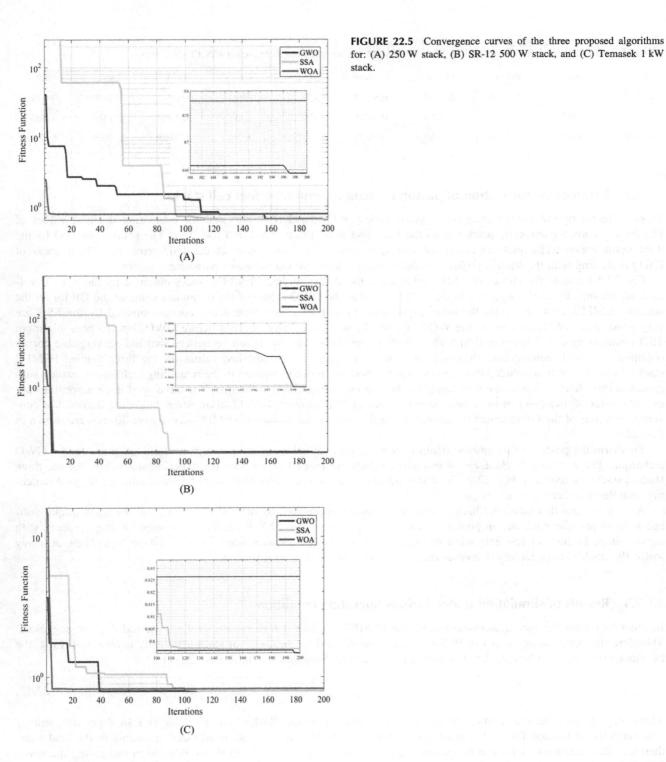

FIGURE 22.5 Convergence curves of the three proposed algorithms for: (A) 250 W stack, (B) SR-12 500 W stack, and (C) Temasek 1 kW stack.

Fig. 22.9 depicts the $I-V$, $I-P$, and efficiency of the Temasek 1 kW stack at pressures of 1/0.2075, 1.5/1, and 2.5/1.5 bar. It is observed that with increasing the inlet pressures and T_{fc}, the output voltage is improved as shown in Figs. 22.8 and 22.9.

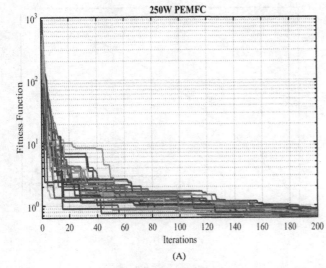

(A)

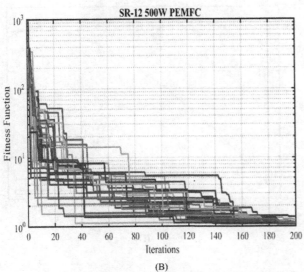

(B)

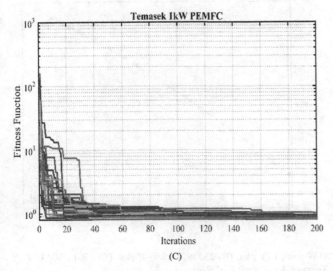

(C)

FIGURE 22.6 Convergence characteristics of the total squared deviations obtained by grey wolf optimizer algorithms over 30 runs for: (A) 250 W stack, (B) SR-12 500 W stack, and (C) Temasek 1 kW stack.

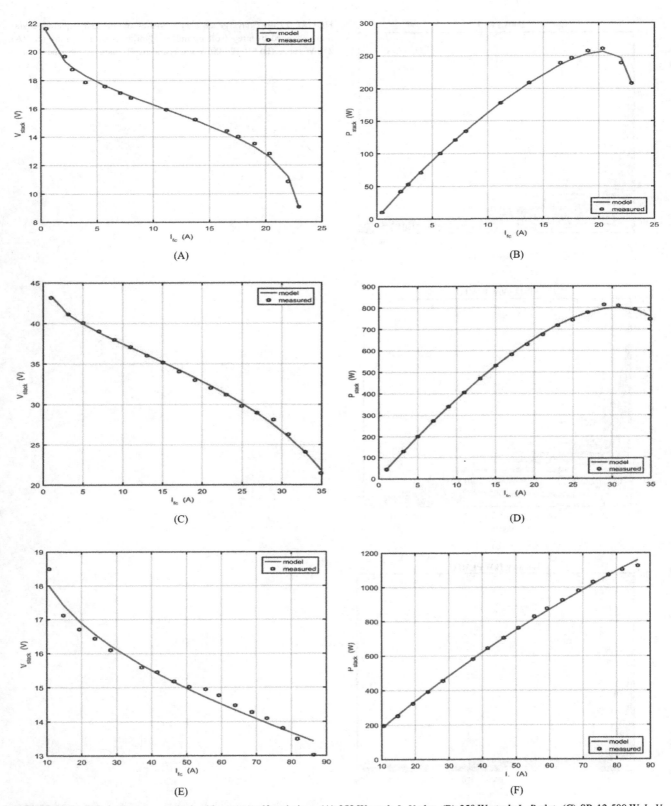

FIGURE 22.7   Polarization curves obtained by grey wolf optimizer: (A) 250 W stack $I-V$ plot, (B) 250 W stack $I-P$ plot, (C) SR-12 500 W $I-V$ plot, (D) SR-12 500 W $I-P$ plot, (E) Temasek 1 kW stack $I-V$ plot, and (F) Temasek 1 kW stack $I-P$ plot.

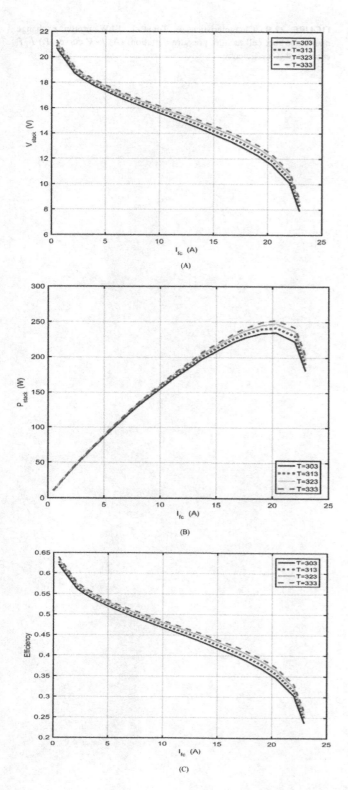

FIGURE 22.8 Characteristics of 250 W proton exchange membrane fuel cell at cell temperature change: (A) $I-V$ curve, (B) $I-P$ curve, and (C) efficiency.

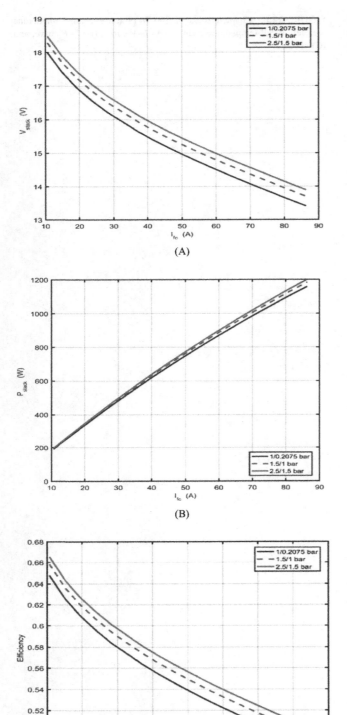

(A)

(B)

(C)

FIGURE 22.9 Characteristics of Temasek 1 kW proton exchange membrane fuel cell at inlet pressure variation: (A) $I-V$ curve, (B) $I-P$ curve, and (C) efficiency.

22.6 Conclusion

In this chapter, the unknown parameters of PEMFC models are estimated using three different optimization techniques, namely GWO, WOA, and SSA. A comprehensive comparison between these optimization techniques has been carried out to determine the best one that accurately solves the studied problem. Minimizing the absolute error between measured and estimated data of output voltages for three commercial PEMFC stacks is considered the main OF. The estimated parameters of PEMFC stacks that are obtained by developed algorithms have been compared with each other to determine the best algorithm. Furthermore, statistical measurements have been studied to prove the superiority, accurateness, and reliability of the proposed algorithms in extracting the parameters of PEMFC stacks. Based on the obtained results, it has been found that the GWO is the best optimization technique compared with WOA and SSA for estimating the PEMFC unknown parameters. It is characterized by fast convergence, minimum TSD, less complexity, easy implementation, and a stable solution.

References

Abdollahzadeh, M., Pascoa, J., Ranjbar, A., & Esmaili, Q. (2014). Analysis of PEM (polymer electrolyte membrane) fuel cell cathode two-dimensional modeling. *Energy*, *68*, 478–494.

Ahme, W., Selim, A., Kamel, S., Yu, J., & Melguizo, F. J. (2018). Probabilistic load flow solution considering optimal allocation of SVC in radial distribution system. *International Journal of Interactive Multimedia and Artificial Intelligence*, *5*, 152–161.

Ali, M., El-Hameed, M., & Farahat, M. (2017). Effective parameters' identification for polymer electrolyte membrane fuel cell models using grey wolf optimizer. *Renewable Energy*, *111*, 455–462.

Amphlett, J. C., Baumert, R. M., Mann, R. F., Peppley, B. A., Roberge, P. R., & Harris, T. J. (1995). Performance modeling of the Ballard Mark IV solid polymer electrolyte fuel cell I. Mechanistic model development. *Journal of the Electrochemical Society*, *142*, 1–8.

Aouali, F. Z., Becherif, M., Ramadan, H. S., Emziane, M., Khellaf, A., & Mohammedi, K. (2017). Analytical modelling and experimental validation of proton exchange membrane electrolyser for hydrogen production. *International Journal of Hydrogen Energy*, *42*, 1366–1374.

Askarzadeh, A., & Rezazadeh, A. (2011). A grouping-based global harmony search algorithm for modeling of proton exchange membrane fuel cell. *International Journal of Hydrogen Energy*, *36*, 5047–5053.

Askarzadeh, A., & Rezazadeh, A. (2013). A new heuristic optimization algorithm for modeling of proton exchange membrane fuel cell: bird mating optimizer. *International Journal of Energy Research*, *37*, 1196–1204.

Chen, Y., & Wang, N. (2019). Cuckoo search algorithm with explosion operator for modeling proton exchange membrane fuel cells. *International Journal of Hydrogen Energy*, *44*, 3075–3087.

Corrêa, J. M., Farret, F. A., Canha, L. N., & Simoes, M. G. (2004). An electrochemical-based fuel-cell model suitable for electrical engineering automation approach. *IEEE Transactions on Industrial Electronics*, *51*, 1103–1112.

Dai, C., Chen, W., Cheng, Z., Li, Q., Jiang, Z., & Jia, J. (2011). Seeker optimization algorithm for global optimization: a case study on optimal modelling of proton exchange membrane fuel cell (PEMFC). *International Journal of Electrical Power & Energy Systems*, *33*, 369–376.

El-Fergany, A. A. (2017). Electrical characterisation of proton exchange membrane fuel cells stack using grasshopper optimiser. *IET Renewable Power Generation*, *12*, 9–17.

Elkasem, A. H., Kamel, S., Rashad, A., & Melguizo, F. J. (2019). Optimal performance of doubly fed induction generator wind farm using multiobjective genetic algorithm. *International Journal of Interactive Multimedia and Artificial Intelligence*, *5*, 48–53.

El-Tamaly, H. H., Sultan, H. M., & Azzam, M. (2014). Control and operation of a solid oxide fuel-cell power plant in an isolated system. In *The international conference on electrical engineering* (pp. 1–13). Military Technical College.

Fathy, A., & Rezk, H. (2018). Multi-verse optimizer for identifying the optimal parameters of PEMFC model. *Energy*, *143*, 634–644.

Gong, W., & Cai, Z. (2013). Accelerating parameter identification of proton exchange membrane fuel cell model with ranking-based differential evolution. *Energy*, *59*, 356–364.

Guarnieri, M., Negro, E., Di Noto, V., & Alotto, P. (2016). A selective hybrid stochastic strategy for fuel-cell multi-parameter identification. *Journal of Power Sources*, *332*, 249–264.

Ibrahim, Y., Kamel, S., Rashad, A., Nasrat, L., & Jurado, F. (2019). Performance enhancement of wind farms using tuned SSSC based on artificial neural network. *International Journal of Interactive Multimedia and Artificial Intelligence*, *1*, 1–7.

Jia, J., Li, Q., Wang, Y., Cham, Y., & Han, M. (2009). Modeling and dynamic characteristic simulation of a proton exchange membrane fuel cell. *IEEE Transactions on Energy Conversion*, *24*, 283–291.

Larminie, J., Dicks, A., & Mcdonald, M. S. (2003). *Fuel cell systems explained*. Chichester: J. Wiley.

Mann, R. F., Amphlett, J. C., Hooper, M. A., Jensen, H. M., Peppley, B. A., & Roberge, P. R. (2000). Development and application of a generalised steady-state electrochemical model for a PEM fuel cell. *Journal of Power Sources*, *86*, 173–180.

Menesy, A. S., Sultan, H.M., & Kamel, S. (2020). Extracting model parameters of proton exchange membrane fuel cell using equilibrium optimizer algorithm. In *2020 international youth conference on radio electronics, electrical and power engineering* (REEPE), IEEE (pp. 1–7).

Menesy, A. S., Sultan, H. M., Korashy, A., Banakhr, F. A., Ashmawy, M. G., & Kamel, S. (2020). Effective parameter extraction of different polymer electrolyte membrane fuel cell stack models using a modified artificial ecosystem optimization algorithm. *IEEE Access*, *8*, 31892–31909.

Menesy, A. S., Sultan, H. M., Selim, A., Ashmawy, M. G., & Kamel, S. (2019). Developing and applying chaotic Harris Hawks Optimization technique for extracting parameters of several proton exchange membrane fuel cell stacks. *IEEE Access.*

Mirjalili, S., Gandomi, A. H., Mirjalili, S. Z., Saremi, S., Faris, H., & Mirjalili, S. M. (2017). Salp Swarm Algorithm: A bio-inspired optimizer for engineering design problems. *Advances in Engineering Software, 114,* 163−191.

Mirjalili, S., & Lewis, A. (2016). The whale optimization algorithm. *Advances in Engineering Software, 95,* 51−67.

Mirjalili, S., Mirjalili, S. M., & Lewis, A. (2014). Grey wolf optimizer. *Advances in Engineering Software, 69,* 46−61.

Mo, Z. J., Zhu, X. J., Wei, L. Y., & Cao, G. Y. (2006). Parameter optimization for a PEMFC model with a hybrid genetic algorithm. *International Journal of Energy Research, 30,* 585−597.

Motapon, S. N., Tremblay, O., & Dessaint, L.-A. (2012). Development of a generic fuel cell model: application to a fuel cell vehicle simulation. *International Journal of Power Electronics, 4,* 505−522.

Panos, C., Kouramas, K., Georgiadis, M., & Pistikopoulos, E. (2012). Modelling and explicit model predictive control for PEM fuel cell systems. *Chemical Engineering Science, 67,* 15−25.

Rajasekar, N., Jacob, B., Balasubramanian, K., Priya, K., Sangeetha, K., & Babu, T. S. (2015). Comparative study of PEM fuel cell parameter extraction using genetic algorithm. *Ain Shams Engineering Journal, 6,* 1187−1194.

Rao, Y., Shao, Z., Ahangarnejad, A. H., Gholamalizadeh, E., & Sobhani, B. (2019). Shark Smell Optimizer applied to identify the optimal parameters of the proton exchange membrane fuel cell model. *Energy Conversion and Management, 182,* 1−8.

Sharaf, O. Z., & Orhan, M. F. (2014). An overview of fuel cell technology: Fundamentals and applications. *Renewable and Sustainable Energy Reviews, 32,* 810−853.

Srinivasulu, G. N., Subrahmanyam, T., & Rao, V. D. (2011). RETRACTED: Parametric sensitivity analysis of PEM fuel cell electrochemical model. *International Journal of Hydrogen Energy, 36*(22), 14838−14844.

Sultan, H., Menesy, A., Kamel, S., & Jurado, F. (2020). Tree growth algorithm for parameter identification of proton exchange membrane fuel cell models. *International Journal of Interactive Multimedia and Artificial Intelligence.*

Sun, Z., Wang, N., Bi, Y., & Srinivasan, D. (2015). Parameter identification of PEMFC model based on hybrid adaptive differential evolution algorithm. *Energy, 90,* 1334−1341.

Turgut, O. E., & Coban, M. T. (2016). Optimal proton exchange membrane fuel cell modelling based on hybrid teaching learning based optimization−differential evolution algorithm. *Ain Shams Engineering Journal, 7,* 347−360.

Úbeda, D., Pinar, F. J., Canizares, P., Rodrigo, M. A., & Lobato, J. (2012). An easy parameter estimation procedure for modeling a HT-PEMFC. *International Journal of Hydrogen Energy, 37,* 11308−11320.

Wang, Y., Chen, K. S., Mishler, J., Cho, S. C., & Adroher, X. C. (2011). A review of polymer electrolyte membrane fuel cells: Technology, applications, and needs on fundamental research. *Applied Energy, 88,* 981−1007.

Xu, S., Wang, Y., & Wang, Z. (2019). Parameter estimation of proton exchange membrane fuel cells using eagle strategy based on JAYA algorithm and Nelder-Mead simplex method. *Energy, 173,* 457−467.

Ye, M., Wang, X., & Xu, Y. (2009). Parameter identification for proton exchange membrane fuel cell model using particle swarm optimization. *International Journal of Hydrogen Energy, 34,* 981−989.

Zhang, L., & Wang, N. (2013). An adaptive RNA genetic algorithm for modeling of proton exchange membrane fuel cells. *International Journal of Hydrogen Energy, 38,* 219−228.

Chapter 23

Optimal allocation of distributed generation/shunt capacitor using hybrid analytical/metaheuristic techniques

Amal Amin Mohamed[1], Salah Kamel[1], Ali Selim[1,2], Mohamed M. Aly[1] and Francisco Jurado[2]

[1]Electrical Engineering Department, Faculty of Engineering, Aswan University, Aswan, Egypt, [2]Department of Electrical Engineering, University of Jaén, Jaén, Spain

23.1 Introduction

Electric power distribution networks typology is considered the most appropriate form from control and protection sides. This part of electric power networks has high power losses due to the increase in the currents of branches and voltage drop, and this causes a decrease in the equality and efficiency of the power transmitted to the customers.

The continuous increase in electric distribution loads is ordinary phenomenon, but it has great challenge for electric power distribution network engineers to make the network adjustable for any increase in the load without exceeding the system constraints (Arulrag & Kumarappan, 2017).

The electric distribution network is mostly operated as radial due to its simplicity and cheap in operation. The load flows in the electric distribution line lead to high active power loss, increase voltage drop in the branches, and furthermore represent poor power factor at the loads. Reactive power compensation by shunt capacitor (SC) placement alone or combination between SC and distributed generation (DG) sources can avoid such problems (Biswal et al., 2018).

DGs and SCs' implementation can be purposefully integrated in electric power networks for increasing the efficiency and quality of the distribution network, grid strengthening, decreasing the system power losses, raising the energy efficiency, power factor correction, improving the voltage of system busses, and pollutant emissions' reduction in the environment (Georgilakis & Hatziargyriou, 2013).

Despite the great benefits of the DGs and SCs, it is difficult to choose the proper places and sizes. If the DGs or SCs are not placed at optimal locations with best sizes, it results in increased power losses and voltage fluctuations and increases the costs.

Researchers have used different techniques for solving the optimization problems in radial distribution networks, such as analytical techniques, numerical techniques (linear programming, quadratic programming, non-linear programming and mixed integer non-linear programming, Rueda-Medina et al., 2013; Zhang, Sharma, & Yanyi, 2012; Zhang, Karady, & Ariaratnam, 2013) and metaheuristic techniques (adaptive genetic algorithm, Ganguly & Samajpati, 2017; particle swarm optimization (PSO), Reddy, Dey, & Paul, 2012; teaching−learning-based optimization (TLBO), García & Mena, 2013; harmony search algorithm (HSA), Rao et al., 2012; artificial bee colony, Abu-Mouti & El-Hawary, 2011; and differential evolution, Arya, Koshti, & Choube, 2012).

There are several techniques have been applied in recent years to define the optimum placement of SCs, DGs, and SCs with DGs for voltage improvement and power loss reduction. In Khodabakhshian and Andishgar (2016), the researchers used intersect mutation differential evolution (IMDE), HSA (Muthukumar & Jayalalitha, 2016), direct search algorithm (Raju, Murthy, & Ravindra, 2012), binary particle swarm optimization (Baghipour & Hosseini, 2012), quasi-oppositional teaching learning-based optimization (Sultana & Roy, 2014), backtracking search optimization algorithm (BSOA) (El-Fergany, 2015), cuckoo search algorithm in Thangaraj (2018) and analytical technique in Amin et al. (2019).

Renewable Energy Systems. DOI: https://doi.org/10.1016/B978-0-12-820004-9.00019-X

In this chapter, the hybrid technique has been proposed to calculate the locations and sizes of the DGs and SCs, which have the advantage of efficient performance for finding the global optimum solution and handle complex problems; the analytical technique is proposed to calculate the sizes of the compensating units while the metaheuristic technique, which is known as whale optimization algorithm (WOA) and sine cosine algorithm (SCA), specifies the locations of the units.

The integration of analytical technique with *meta*heuristic techniques provides best performance with reduced number of iterations compared with analytical or *meta*heuristic technique alone.

This chapter is organized in six sections. Section 23.1 is an introductory section. Section 23.2 presents the objection function. Section 23.3 presents the mathematical formulation of the analytical technique. Section 23.4 presents the metaheuristic techniques. Section 23.5 presents the simulation results. Finally, Section 23.5 presents the conclusion.

23.2 Objective function

In this chapter, a backward—forward sweep algorithm is introduced for solving the problem of load flow in radial distribution system. Fig. 23.1 shows two simple networks in radial distribution system (RDS). Branch J connects between bus A and bus B.

The effective power P_A and the effective reactive power Q_A that flow from bus A to bus B can be determined backward as follows:

$$P_A = P_B + P_{LB} + R\left(\frac{(P_B + P_{LB})^2 + (Q_B + Q_{LB})^2}{|V_B|^2}\right) \tag{23.1}$$

$$Q_A = Q_B + Q_{LB} + X\left(\frac{(P_B + P_{LB})^2 + (Q_B + Q_{LB})^2}{|V_B|^2}\right) \tag{23.2}$$

The voltage of each bus can be calculated in the forward direction as:

$$V_B^2 = V_A^2 - 2(P_A R + Q_A X) + \frac{(R^2 + X^2)(P_A^2 + Q_A^2)}{V_A^2} \tag{23.3}$$

Therefore active power loss and reactive power loss in the branch J can be determined as:

$$P_{loss(J)} = R_J\left(\frac{P_A^2 + jQ_A^2}{|V_A|^2}\right) \tag{23.4}$$

$$Q_{loss(J)} = x_J\left(\frac{P_A^2 + jQ_A^2}{|V_A|^2}\right) \tag{23.5}$$

So, total active and reactive power losses in RDS can be estimated as:

$$\text{TPL} = \sum_{J=1}^{br} P_{loss(J)} \tag{23.6}$$

$$\text{TQL} = \sum_{J=1}^{br} Q_{loss(J)} \tag{23.7}$$

FIGURE 23.1 Representation of two buses in radial distribution system.

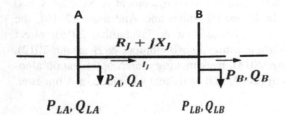

23.2.1   Equality and inequality constraints

1. Equality constraints:
  a. Power balance equation:

$$P_{slack} + \sum_{i}^{No} P_{DG(i)} = \sum_{i}^{No} P_{load(i)} + \sum_{i,i+1}^{br} P_{loss(i,i+1)} \tag{23.8}$$

$$Q_{slack} + \sum_{i}^{No} Q_{SC(i)} = \sum_{i}^{No} Q_{load(i)} + \sum_{i,i+1}^{br} Q_{loss(i,i+1)} \tag{23.9}$$

No is the number of buses, *br* is the number of branches, P_{Loss} and Q_{Loss} are the active and reactive power losses.
2. Inequality constraints
  b. Bus voltage limits:
      The voltage at each bus must be within 0.90 and 1.05

$$V_{min} \leq V_i \leq V_{max} \tag{23.10}$$

  c. SCs capacity limits:
      Total reactive power constraint:

$$\sum Q_{SC} \leq \sum_{i}^{No} Q_{load(i)} \tag{23.11}$$

  d. DGs capacity limits:
      Total active power constraint

$$\sum P_{DG} \leq \sum_{i}^{No} P_{load(i)} \tag{23.12}$$

23.3   Mathematical formulation of the analytical technique

The proposed analytical technique has been improvised for placement of multiple numbers of DGs and SCs. Active power loss is a function of power injection and buses voltage in RDS have *N*-number of buses; hence, the mathematical formulation can be calculated by the exact loss formula as presented in Elgerd and Happ (1972). The proposed analytical method was developed based on the BIBC and BCBV matrices, which can be illustrated in the following subsections (Fig. 23.2). P_A and P_B are the active injected power at nodes A and B, respectively. Q_A and Q_B are the reactive injected power at nodes A and B, respectively Where,

$$P_{loss} = \sum_{A=1}^{N} \sum_{B=1}^{N} \alpha_{AB}(P_A P_B + Q_A Q_B) + \beta_{AB}(Q_A P_B - P_A Q_B) \tag{23.13}$$

$$\alpha_{AB} = \frac{R_{AB}}{V_A V_B} \cos(\alpha_A - \alpha_B) \tag{23.14}$$

$$\beta_{AB} = \frac{R_{AB}}{V_A V_B} \sin(\alpha_A - \alpha_B) \tag{23.15}$$

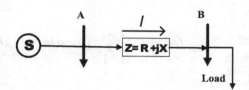

FIGURE 23.2   Simple two-bus network

$V_A V_B$ is the voltage at bus A, B, respectively. $Z_{AB} = R_{AB} + jX_{AB}$, where Z_{AB} is the bus impedance matrix.

Supposedly, n is the number of the DG, which injects only active power, and the locations of those DGs are $X_1, X_2, \ldots, X_n$, while $P_{(DG)_{X_1}}, P_{(DG)_{X_2}}, \ldots, P_{(DG)_{X_n}}$ are the capacity of the DGs, respectively

Therefore the generated power of the DG can be:

$$\left[P_{X_1} = P_{(SC)_{X_1}} - P_{(load)_{X_1}} \right], \ldots, \left[P_{X_n} = P_{(SC)_{X_n}} - P_{(load)_{X_n}} \right] \tag{23.16}$$

Likewise, let m is the numbers of the SC, which injects only reactive power, and the locations of those SCs are $Y_1, Y_2, \ldots, Y_m$, while $Q_{(SC)_{Y_1}}, Q_{(SC)_{Y_2}}, \ldots, Q_{(SC)_{Y_m}}$ are the capacity of the SCs.

And the generated reactive power of SC can be:

$$\left[Q_{Y_1} = Q_{(SC)_{Y_1}} - Q_{(load)_{Y_1}} \right], \ldots, \left[Q_{Y_m} = Q_{(SC)_{Y_m}} - Q_{(load)_{Y_m}} \right] \tag{23.17}$$

The active power losses become minimum if the first derivative of Eq. (23.18) with respect to the injection power from the DG become zero so it can written as

$$\alpha_{X_1 X_1} P_{X_1} + \alpha_{X_1 X_2} P_{X_2} + \cdots + \alpha_{X_1 X_n} P_{x_n} - \beta_{X_1 Y_1} Q_{Y_1} - \beta_{X_1 Y_2} Q_{Y_2} - \cdots - \beta_{X_1 Y_m} Q_{Y_m} \tag{23.18}$$

$$\frac{\partial P_{Loss}}{\partial P_{X_1}} = 2\alpha_{X_1 X_1} P_{X_1} + 2 \sum_{\substack{B=1 \\ B \neq X_1}}^{N} \left(\alpha_{X_1 B} P_B - \beta_{X_1 B} Q_B \right) = - \sum_{\substack{B=1 \\ B \neq X_1, X_2, \ldots, X_n \\ B \neq Y_1, Y_2, \ldots, Y_m}}^{N} \left(\alpha_{X_1 B} P_B - \beta_{X_1 B} Q_B \right) \tag{23.19}$$

Also, the derivative of the active power loss respect to P_{X_n} can be written as

$$\left[P_{X_1} = P_{(SC)_{X_1}} - P_{(load)_{X_1}} \right], \ldots, \left[P_{X_n} = P_{(SC)_{X_n}} - P_{(load)_{X_n}} \right] \tag{23.20}$$

$$\frac{\partial P_{loss}}{\partial P_{X_n}} = 2\alpha_{X_n X_n} P_{X_n} + 2 \sum_{\substack{B=1 \\ B \neq X_n}}^{N} \left(\alpha_{X_n B} P_B - \beta_{X_n B} Q_B \right) \tag{23.21}$$

$$\alpha_{X_n X_1} P_{X_1} + \alpha_{X_n X_2} P_{X_2} + \cdots + \alpha_{X_n X_n} P_{X_n} - \beta_{X_n Y_1} Q_{Y_1} - \beta_{X_n Y_2} Q_{Y_2} - \cdots - \beta_{X_n Y_m} Q_{Y_m} = - \sum_{\substack{B=1 \\ B \neq X_1, X_2, \ldots, X_n \\ B \neq Y_1, Y_2, \ldots, Y_m}}^{N} \left(\alpha_{X_n B} P_B - \beta_{X_n B} Q_B \right)$$

$$\tag{23.22}$$

While the derivative of the power loss respect to reactive power injection can be written as:

$$\frac{\partial P_{loss}}{\partial Q_{Y_1}} = 2\alpha_{Y_1 Y_1} Q_{Y_1} + 2 \sum_{\substack{B=1 \\ B \neq Y_1}}^{N} \left(\alpha_{Y_1 B} Q_B + \beta_{Y_1 B} P_B \right) \tag{23.23}$$

$$\beta_{Y_1 X_1} P_{X_1} + \beta_{Y_1 X_2} P_{X_2} + \cdots + \beta_{Y_1 X_n} P_{x_n} + \alpha_{Y_1 Y_1} Q_{Y_1} + \alpha_{Y_1 Y_2} Q_{Y_2} + \cdots + \alpha_{Y_1 Y_m} Q_{Y_m} = - \sum_{\substack{B=1 \\ B \neq X_1, X_2, \ldots, X_n \\ B \neq Y_1, Y_2, \ldots, Y_m}}^{N} \left(\alpha_{Y_1 B} Q_B + \beta_{Y_1 B} P_B \right) \tag{23.24}$$

Similarly, the derivative for Q_{Y_m}

$$\frac{\partial P_{loss}}{\partial Q_{Y_m}} = 2\alpha_{Y_m Y_m} Q_{Y_m} + 2 \sum_{\substack{B=1 \\ B \neq Y_m}}^{N} \left(\alpha_{Y_m B} Q_B + \beta_{Y_m B} P_B \right) \tag{23.25}$$

$$\beta_{Y_mX_1}P_{X_1} + \beta_{Y_mX_2}P_{X_2} + \cdots + \beta_{Y_mX_n}P_{X_n} + \alpha_{Y_mY_1}Q_{Y_1} + \alpha_{Y_mY_2}Q_{Y_2} + \cdots + \alpha_{Y_mY_m}Q_{Y_m} = - \sum_{\substack{B=1 \\ B \neq X_1, X_2, \ldots, X_n \\ B \neq Y_1, Y_2, \ldots, Y_m}}^{N} (\alpha_{Y_mB}Q_B + \beta_{Y_mB}P_B)$$

(23.26)

Let consider

$$
\begin{bmatrix}
\alpha_{X_1X_1} & \alpha_{X_1X_2} & \cdots & \alpha_{X_1X_n} & -\beta_{X_1Y_1} & -\beta_{X_1Y_2} & \cdots & \cdots & \cdots & -\beta_{X_1Y_m} \\
\alpha_{X_2X_1} & \alpha_{X_2X_2} & \cdots & \alpha_{X_2X_n} & -\beta_{X_2Y_1} & -\beta_{X_2Y_2} & \cdots & \cdots & \cdots & -\beta_{X_2Y_m} \\
\vdots & \vdots & & \vdots & \vdots & \vdots & & & & \vdots \\
\alpha_{X_nX_1} & \alpha_{X_nX_2} & \cdots & \alpha_{X_nX_n} & -\beta_{X_nY_1} & -\beta_{X_nY_2} & \cdots & \cdots & \cdots & -\beta_{X_nY_m} \\
\beta_{Y_1X_1} & \beta_{Y_1X_2} & \cdots & \beta_{Y_1X_n} & \alpha_{Y_1Y_1} & \alpha_{Y_1Y_2} & \cdots & \cdots & \cdots & \alpha_{Y_1Y_m} \\
\beta_{Y_2X_1} & \beta_{Y_2X_2} & \cdots & \beta_{Y_2X_n} & \alpha_{Y_2Y_1} & \alpha_{Y_2Y_2} & \cdots & \cdots & \cdots & \alpha_{Y_2Y_m} \\
\vdots & \vdots & & \vdots & \vdots & \vdots & & & & \vdots \\
\beta_{Y_mX_1} & \beta_{Y_mX_2} & \cdots & \beta_{Y_mX_n} & \alpha_{Y_mY_1} & \alpha_{Y_mY_2} & \cdots & \cdots & \cdots & \alpha_{Y_mY_m}
\end{bmatrix}
\begin{bmatrix}
P_{X_1} \\ P_{X_2} \\ \vdots \\ P_{X_n} \\ Q_{Y_1} \\ Q_{Y_2} \\ \vdots \\ Q_{Y_m}
\end{bmatrix}
= -
\begin{bmatrix}
D_{X_1} \\ D_{X_2} \\ \vdots \\ D_{X_n} \\ F_{Y_1} \\ F_{Y_2} \\ \vdots \\ F_{Y_m}
\end{bmatrix}
$$

(23.27)

Where

$$D_{X_i} = \sum_{\substack{B=1 \\ B \neq X_1, X_2, \ldots, X_n \\ B \neq Y_1, Y_2, \ldots, Y_m}}^{N} (\alpha_{X_iB}P_B - \beta_{X_iB}Q_B) \quad i = 1, 2, 3, \ldots, n$$

(23.28)

$$F_{Y_i} = \sum_{\substack{B=1 \\ B \neq X_1, X_2, \ldots, X_n \\ B \neq Y_1, Y_2, \ldots, Y_m}}^{N} (\alpha_{Y_iB}Q_B + \beta_{Y_iB}P_B) \quad i = 1, 2, 3, \ldots, m$$

(23.29)

And

$$\begin{bmatrix} [K_{11}]_{n \times n} & [K_{12}]_{n \times m} \\ [K_{21}]_{m \times n} & [K_{22}]_{m \times m} \end{bmatrix} \begin{bmatrix} [P_{X_i}]_{n \times 1} \\ [Q_{Y_i}]_{m \times 1} \end{bmatrix} = \begin{bmatrix} [D_{X_i}]_{n \times 1} \\ [F_{Y_i}]_{m \times 1} \end{bmatrix}$$

(23.30)

$$\begin{bmatrix} [P_{X_i}]_{n \times 1} \\ [Q_{Y_i}]_{m \times 1} \end{bmatrix} = \begin{bmatrix} [K'_{11}] & [K'_{12}] \\ [K'_{21}] & [K'_{22}] \end{bmatrix} \begin{bmatrix} [D_{X_i}]_{n \times 1} \\ [F_{Y_i}]_{m \times 1} \end{bmatrix}$$

(23.31)

$$[P_{X_i}]_{n \times 1} = [K'_{11}]_{n \times n}[D_{X_i}]_{n \times 1} + [K'_{12}]_{n \times m}[F_{Y_i}]_{m \times 1}$$

(23.32)

$$[Q_{Y_i}]_{m \times 1} = [K'_{21}]_{m \times n}[D_{X_i}]_{n \times 1} + [K'_{22}]_{m \times m}[F_{Y_i}]_{m \times 1}$$

(23.33)

Eqs. (23.34) and (23.35) are used to calculate the capacity of DG and SC, respectively.

Finally, the active power injection from the DG at bus X_i can be calculated from the following equations:

$$\left[P_{(DG)_{X_i}}\right]_{n \times 1} = [P_{X_i}]_{n \times 1} + \left[P_{(load)_i}\right]$$

(23.34)

$P_{(load)_i}$ is the active power load at bus X_i.

And the reactive power injection at bus Y_i can be written as:

$$\left[Q_{(SC)_{Y_i}}\right]_{m \times 1} = [Q_{Y_i}]_{m \times 1} + \left[Q_{(load)_i}\right]$$

(23.35)

$Q_{(load)_i}$ is the reactive power load at bus.

Note that:

In case of installing DG injecting only active power $X_A \neq Y_B \forall B$.

In case of installing SC injecting only reactive power $Y_A \neq X_B \forall B$.

In case of installing DG injecting active power and reactive power $X_A = Y_B$.

23.4 Metaheuristic technique

23.4.1 Sine cosine algorithm

Mirjalili (2016) proposed a new optimization technique called SCA which based on suggesting different random solutions and then vicissitude it around the optimal result using a mathematical form which based on sine and cosine functions. SCA grouped to the exploration and exploitation phases. Exploration phase can be used to create random solutions while exploitation phase requires them to fluctuate outwards or towards the best solution, and due to the utilization of the use of sine and cosine in this formulation it known as SCA and its effect on Eqs. (23.39) and (23.40) are explained in Fig. 23.3.

Procedure of SCA:

Step 1: Set the main parameters of the SCA, such as search agent number and maximum iterations numbers.

Step 2: Start randomly population of solutions considering upper and lower borders of variables as follows:

$$F(A,B) = \text{rand}\left(U_p(A,B) \times L_p(A,B)\right) + L_p(A,B) \tag{23.36}$$

where x is the search agent number and y is the variables number (dimension).

Step 3: The position for the population of solutions can be introduced as follows:

$$
\begin{bmatrix}
F_{1,1} & F_{1,2} & \cdots & F_{1,B} \\
F_{2,1} & F_{22} & \cdots & F_{2,1} \\
\vdots & \vdots & \ddots & \vdots \\
F_{A,1} & F_{A,2} & \cdots & F_{A,B}
\end{bmatrix}
\tag{23.37}
$$

Step 4: Calculate the values of objective function according to the search agent position as follows:

$$OS = [OS_1 OS_2 OS_3, \ldots, OS_N]^T \tag{23.38}$$

Step 5: Calculate the objective fitness that is known as a preferable fitness and the purpose situation that is known as a preferable position.

Step 6: Update the position of the search agent as introduced below:

$$F_i^{T+1} = F_i^T + w_1 \cos(w_2) \times |w_3 \times P_i^T - F_i^T| w_4 \geq 0.5 \tag{23.39}$$

$$F_i^{T+1} = F_i^T + w_1 \sin(w_2) \times |w_3 \times P_i^T - F_i^T| w_4 \geq 0.5 \tag{23.40}$$

where

$$w_1 = 2 - \frac{2T}{T_{\max}} \tag{23.41}$$

$$w_2 = 2\pi \times \text{rand}() \tag{23.42}$$

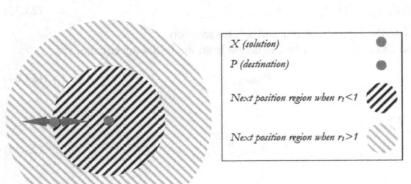

FIGURE 23.3 Sine cosine effect in Eqs. (23.45) and (23.46).

$$w_3 = 2 \times \text{rand}() \tag{23.43}$$

$$w_4 = \text{rand}() \tag{23.44}$$

T is the current iteration. $T_{\max}$ is the maximum iteration. P_i^T is the objective global best solution.

Step 7: Return the steps from 3 to 5 till the standing criteria is realized.

Step 8: Develop the optimal destination fitness and the best destination.

23.4.2 Whale optimization algorithm

Mirjalili and Lewis have recently introduced a new optimization technique named the WOA. Whales are the largest animal in the creation. Pursuant to Hof and Van (2007), the brain cells of whales are similar to human beings. The whales are considered to be moving animals as extremely intelligent. The WOA algorithm is influenced by humpback whales distinct hunting behavior. The humpbacks generally prefer to hunt krills or small fish near the ocean surface. Humpback whales use a special unique hunting method called bubble-net feeding method (Watkins & Schevill, 1979). In this technique, they swim around the prey and create a unique bubble along a circle or 9-shaped path. This behavior of searching is modeled mathematically to two phases (Mirjalili & Lewis, 2016)

- *Searching and encircling prey*

Searching prey represented using the following two equations:

$$D = |C \times X_{\text{rand}} - X| \tag{23.45}$$

$$X(t+1) = X_{\text{rand}} - A \times D \tag{23.46}$$

A and C are vectors and can be defined as:

$$A = 2 \times a \times r - a \tag{23.47}$$

$$C = 2 \times r \tag{23.48}$$

where "a" is linearly decreasing from 2 to 0 and "r" is the random number between [0, 1].

$$D = |C \times X * (t) - X(t)| \tag{23.49}$$

$$X(t+1) = X * (t) - A.D \tag{23.50}$$

In case $A \geq 1$, the process of searching prey is represented by Eqs. (23.45) and (23.46), otherwise encircling prey by shrinking mechanism is represented by Eqs. (23.49) and (23.50), where t is the current iteration, X is the position vector, and $X*$ is the best value of the position vector so far.

- *Spirally updating position*

Position updating is defined by Eq. (23.51):

$$X(t+1) = \begin{cases} X * (t) - A \times D & \text{if } P < 0.5 \\ D \times e^{bl} \times \cos(2\pi l) + X * (t) & \text{if } P \geq 0.5 \end{cases} \tag{23.51}$$

where P is the random number between [0, 1] and b is the constant for depicting the spiral shape.

Fig. 23.4 shows bubble-net searching mechanism in WOA

23.5 Simulation results

23.5.1 IEEE 33-bus RDS

This system consists of 33-buses and 32-branches as shown in Fig. 23.5 and the total active power load is 3715 kW and the total reactive power load is 2300 kVAR. The base voltage is 12.66 kV and the apparent power is 100 MVA.

- *Optimal allocation of SCs with DGs using hybrid (analytical—SCA) in IEEE 33-bus RDS*

A new efficient metaheuristic technique based on optimization algorithm known as SCA implemented with analytical technique to select optimal locations and sizes of DGs, SCs and SCs with DGs in RDS.

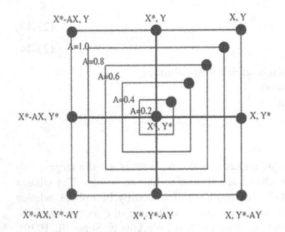

FIGURE 23.4 Bubble-net searching mechanism in WOA (Goldbogen et al., 2013).

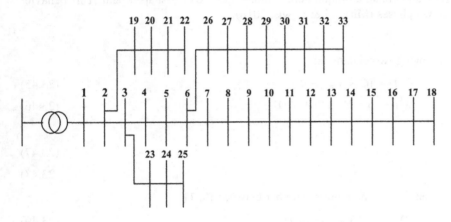

FIGURE 23.5 Single-line diagram of IEEE 33-bus RDS.

The hybrid (analytical–SCA) approach had introduced to solve the placement of DGs, SCs, and SCs with DGs, which presented in the following sections. In addition, to check the realization of the hybrid approach, it is compared with other techniques. The used parameters of algorithm and the operating constraints are as follows: the maximum iteration is 100 and the number of search agents is 30.

Case 1: base case

In this case, the power losses before any compensation is 210.9862 kW and the minimum voltage is 0.9038 p.u. at bus 18.

Case 2: one compensating unit

One SC: The power loss reduction of proposed method is 28.258% when the SC placed at bus 30 as described in Table 23.1. And the size of the SC by hybrid approach is 1258.001 kVAR.

One DG: The power loss obtained by the proposed method is 111.018 kW when the DG is placed at bus 6 and also the obtained minimum voltage value is 0.9424 at bus 18 as shown in Table 23.2, showing that the power loss by the hybrid approach is best than 112.18 in BSOA (El-Fergany, 2015), 111.03 kW in GA (Mittal & Kansal, 2017), and 111.029 kW in ALO (Ant Lion optimizer) (Ali et al., 2018).

One SC and one DG: Buses 6 and 30 are the best locations for DG and SC for the proposed method; the power loss obtained is 58.4428, which is best than 59.7 in PSO (Aman et al., 2013) as shown in Table 23.1.

Case 2: two compensating units

Two SCs: The power loss reduction is 32.78% and the minimum voltage profile is 0.9303 p.u. at bus 18 when the two SCs are placed at buses 12 and 30 in the proposed method, which is best than 31.83 in hybrid WIPSO–GSA (weight improved particle swarm optimization–gravitational search algorithm) (Rajendran & Narayanan, 2020) and 33.03% in FGA (fuzzy genatic algorithm) (Reddy, Prasad, & Laxmi, 2013) as described in Table 23.2.

Two DGs: The power loss obtained by the proposed method is 87.168 and the min voltage is 0.9683 p.u. for the proposed method when the DGs are placed at buses 13 and 30 Table 23.2. The power loss by the hybrid approach is best than 89.34 in hybrid WIPSO–GSA (Rajendran & Narayanan, 2020) and 119.7 kW in FGA (Reddy et al., 2013).

TABLE 23.1 Results of hybrid (analytical—SCA) for one DG, one SC, and one DG and one SC of 33-bus RDS.

Cases	Technique	Location/size	Total capacity	Power loss	V. min (bus)	Reduction
Case 1	Base case	–	–	210.986	0.9038 (18)	0
One SC	Proposed	30/1258.001	1258.001	151.365	0.9165 (18)	28.258
	Heuristic approach (Bayat & Bagheri, 2019)	30/1190	1190	151.55	NA	NA
	Hybrid (PSO—analytical) (Kansal, Kumar, & Tyagi, 2016)	30/1230	1230	151.4	NA	28.24
	Analytical (Naik, Khatod, & Sharma, 2013)	33/1000	1000	164.6	0.916	NA
One DG	Proposed	6/2590.213	2590.2	111.018	0.9424 (18)	47.381
	BSOA (El-Fergany, 2015)	8/1857.5	1857.5	118.12	0.9441	NA
	GA (Mittal & Kansal, 2017)	6/2600	2600	111.03	0.9424 (18)	NA
	ALO (Ali et al., 2018)	6/2590.2	2590	111.030	0.9425	47.38
	Hybrid TLBO—GWO (Nowdeh et al., 2019)	30/1000	2590.2	127.28	0.9285	NA
One SC and one DG	Proposed	30/1255.7885	1255.7885	58.4428	0.9537 (18)	72.3
		6/2531.7121	2531.7121			
	PSO (Aman et al., 2013)	30/1457	1457	59.7	0.9550(18)	71.7
		6/2511	2511			

ALO, Ant Lion optimizer; *BSOA*, backtracking search optimization algorithm; *DG*, distributed generation; *GA*, genetic algorithm; *GWO*, Grey Wolf optimizer; *NA*, numerical analysis; *PSO*, particle swarm optimization; *SC*, shunt capacitor; *SCA*, sine cosine algorithm; *TLBO*, teaching—learning-based optimization.

Two DGs and two SCs: Buses 10 and 30 are the best locations for two DGs and SCs obtained by the proposed method and the power loss reduction is 86.49%, while in case of FGA (Reddy et al., 2013), the power loss reduction is 71.8% and the minimum voltage obtained by the proposed method is 0.9804 at bus 25 as shown in Table 23.2.

Case 3: three compensating units

Three SCs: The power loss for the proposed method is 138.260 kW when buses 13, 24, and 30 with sizes 396.103, 524.809, and 1044.227 kVAR, respectively, are chosen as the best locations for the proposed method, which is the best than 144.04 in BFOA (bacterial foraging optimization algorithm) (Kowsalya et al., 2014) and 141.84 in hybrid (WIPSO—GSA) (Rajendran & Narayanan, 2020).

Three DGs: The power loss is reduced from 210.98627 to 72.789 kW for the proposed method, which is best than 89.90 kW in BFOA (Kowsalya et al., 2014) and 77.06 kW in WOA (Saleh, Mohamed, & Hemeida, 2019)

Three DGs and three SCs: The optimal power loss obtained in this case using the proposed method is 11.047 kW when three SCs and three DGs are installed at buses 13, 24, and 30 and the minimum voltage is 0.9921 at bus 8 as shown in Table 23.3.

The convergence characteristics of hybrid (analytical—SCA) for 1DG, 1SC, and 1DG&1SC are illustrated in Fig. 23.6, those for 2DG, 2SC, and 2DG&2SC are illustrated in Fig. 23.8, and those for 3DG, 3SC, and 3DG&3SC are illustrated in Fig. 23.10. The proposed approach obtains the best objective function in less computational time as observed in the figures. While Figs. 23.7, 23.9 and 23.11. Introduce the voltage profiles' improvement of DGs and SCs (Figs. 23.7—23.11).

- *Results of SCs and DGs considering load variation using hybrid (analytical—WOA) of 33-bus RDS*

In this section, a new *meta*heuristic technique, which is known as WOA, is implemented with analytical technique to calculate the optimal placement of three SCs and three DGs with considering the load demands varying from light load (50%), nominal load (100%), and peak load (150%), and the results are presented in Table 23.4; the results explain

TABLE 23.2 Results of hybrid (analytical–SCA) for two DGs, two SCs, and two DGs and two SCs of 33-bus RDS.

Cases	Technique	Location/size	Total capacity	Power loss	V. min (bus)	Reduction
Case 1	Base case	–	210.862		0.9038 (18)	0
Two SCs	Proposed	12/464.114	1530.45	141.83	0.9303 (18)	32.78
		30/1066.33				
	Hybrid WIPSO–GSA (Rajendran & Narayanan, 2020)	10/570	1380	143.84	NA	31.83
		30/810				
	Heuristic approach (Bayat & Bagheri, 2019)	13/405	1457		NA	NA
		30/1052		141.9		
	FGA (Reddy et al., 2013)	18/950	1650	141.3	0.929 (18)	33.03
		30/700				
Two DGs	Proposed	13/844.844	2011.1	87.168	0.9683 (18)	58.685
		30/1166.165				
	Hybrid WIPSO–GSA (Rajendran & Narayanan, 2020)	13/880	1804	89.34	0.9665	NA
		31/924				
	Heuristic approach (Bayat & Bagheri, 2019)	13/850	1990	87.18	NA	58.68
		30/1140				
	FGA (Reddy et al., 2013)	7/700	1800	119.7	0.935 (18)	43.27
		32/1100				
Two SCs and two DGs	Proposed	10/443	1499	28.496	0.9804 (25)	86.49
		30/1056				
		10/847	1989.52			
		30/1142.525				
	FGA (Reddy et al., 2013)	16/650	1450	59.5	0.9618 (18)	71.8
		33/800				
		7/600	1700			
		32/1100				

DGs, distributed generations; *FGA*, fuzzy genatic algorithm; *GSA*, gravitational search algorithm; *SCs*, shunt capacitors; *SCA*, sine cosine algorithm; *WIPSO*, weight improved particle swarm optimization.

as the loads decreasing the power losses and the compensating unity capacities and vice versa, while the locations are constant for three loading ratios. Fig. 23.15 shows the convergence characteristics for three loading ratios. Also, Figs. 23.12–23.14 show voltage profile improvement; the used parameters are as follows: max iteration is 100, search agent number is 30, and the simulation results are introduced next (Fig. 23.12–23.15).

23.5.2 IEEE 69-bus RDS

This system consists of bus 69 and branch 68 as shown in Fig. 23.16 and its total active load 3801 kW and reactive load 2694 kVAR. The base voltage is 12.66 kV and the apparent power is 100 MVA.

TABLE 23.3 Results of hybrid (analytical—SCA) and SCA for three DGs, three SCs, and three DGs and three SCs of 33-bus RDS.

Cases	Technique	Location/size	Total capacity	Power loss	V. min (bus)	Reduction
Three SCs	Proposed	13/396.103	1954.77	138.26	0.932(18)	
		24/524.809				34.47
		30/1044.227				
	BFOA	18/350	1447	144.04	0.9361	
		30/820				31.72
		33/277				
	Hybrid WIPSO—GSA (Rajendran & Narayanan, 2020)	12/470	1107	141.84	NA	32.78
		29/530				
		30/530				
	Heuristic approach (Bayat & Bagheri, 2019)	13/383	1769	138.65	NA	NA
		25/386				
		30/1000				
Three DGs	Proposed	13/805.714	2954.259	72.789	0.9685 (33)	65.5
		24/1104.318				
		30/1044.227				
	BFOA (Kowsalya et al., 2014)	14/652.1	1917.7	89.9	0.9705 (29)	57.38
		18/198.4				
		32/1067.2				
	WOA (Saleh et al., 2019)	31/748.15	2449.81	77.06	0.9686 (33)	NA
		6/1051.1				
		14/650.56				
Three SCs and three DGs	Proposed	24/517.6528	1901	11.7407	0.9921(8)	94.43
		13/370.3727				
		30/1012.989				
		24/1073.638	2897.09			
		13/792.5622				
		30/1030.932				
	BFOA (Kowsalya et al., 2014)	18/163	1509	41.41	0.9783	80.37
		30/541				
		33/338				
		17/542	1597			
		18/160				
		33/895				

BFOA, bacterial foraging optimization algorithm; *DGs*, distributed generations; *SCs*, shunt capacitors; *SCA*, sine cosine algorithm; *WOA*, whale optimization algorithm.

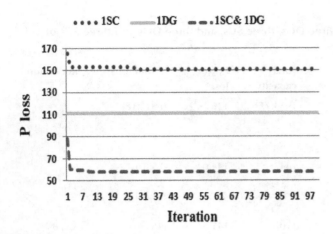

FIGURE 23.6 Convergence for one shunt capacitor, one distributed generation, and one distributed generation and one shunt capacitor of 33-bus RDS.

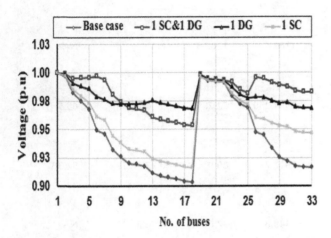

FIGURE 23.7 Voltages for one shunt capacitor, one distributed generation, and one distributed generation and one shunt capacitor of 33-bus RDS.

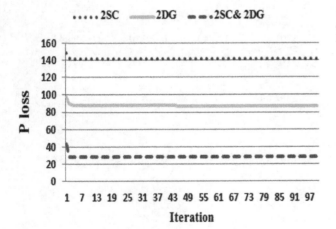

FIGURE 23.8 Convergence for two shunt capacitors, two distributed generations, and two distributed generations and two shunt capacitors of 33-bus RDS.

- *Optimal allocation of DGs and SCs using hybrid (analytical−SCA) of 69-bus RDS*

  Case 1: Base case
  In the base case, the active power loss before any compensation is 224.9599 kW and the minimum voltage is 0.9092 at bus 65.
  Case 2: One compensating unit
  In case of one SC: The power loss obtained by the proposed method is 152.0045 kW when the SC is placed at bus 61 with size 1329.93 kVAR and also it obtained the worst value of voltage as 0.9307 at bus 65 as presented in

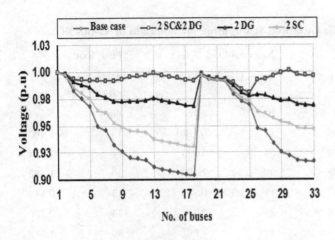

FIGURE 23.9 Voltages for two shunt capacitors, two distributed generations, and two distributed generations and two shunt capacitors of 33-bus RDS.

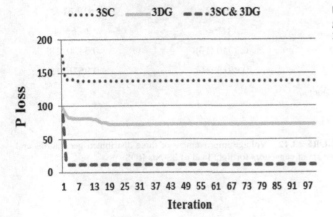

FIGURE 23.10 Convergence for three shunt capacitors, three distributed generations, and three distributed generations and three shunt capacitor of 33-bus RDS.

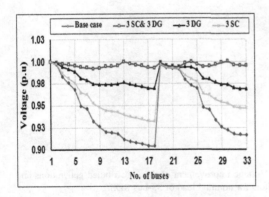

FIGURE 23.11 Voltages for three shunt capacitors, three distributed generations, and three distributed generations and three shunt capacitors of 33-bus RDS.

Table 23.5, also showing that the power loss by the proposed approach is best than 152.10 kW in heuristic approach (Bayat & Bagheri, 2019) and hybrid (PSO−analytical) (Kansal et al., 2016).

In case of one DG: The power loss of hybrid approach is 83.18 kW when the DG is placed at bus 61 with size 1872.64 kW. And the minimum voltage is 0.9683 at bus 27; this value of power losses is best than all compared techniques: 83.2231 kW in ALO (Ali et al., 2018), 111.56 kW in Hybrid TLBO−GWO (Nowdeh et al., 2019), 83.24 kW in GWO (Grey Wolf optimizer) (Sobieh et al., 2017), and 83.21 kW in hybrid (fuzzy−PSO) (Bala & Ghosh, 2019) as described in Table 23.5.

In case of one SC and one DG: Bus 61 is the best location for the DG and SC; hence, the power loss obtained by hybrid approach is 23.1463, which is best than 25.9 in case of PSO (Aman et al., 2013). The convergence characteristics of proposed method for one SC, one DG, and one DG and one SC are illustrated in Fig. 23.17, and the voltage profile improvement is presented in Fig. 23.18.

TABLE 23.4 Results of hybrid (analytical—WOA) for three DGs and three SCs in three loading ratio of 33-bus RDS.

Loading ratio	Light load 50% Loading ratio		Nominal load 100% Loading ratio		Peak load 150% Loading ratio	
Optimal DGs size (MW) and locations	Bus	Size	Bus	Size	Bus	Size
	13	0.394	13	0.7881	13	1.182
	24	0.534	24	1.071	24	1.611
	30	0.508	30	1.007	30	1.498
Optimal SCs size (MVAR) and locations	13	0.183	13	0.365	13	0.545
	24	0.258	24	0.516	24	0.776
	30	0.503	30	1.007	30	1.512
Base losses	48.7896		210.97		519.8142	
P-loss (kW) after installation	2.9217		11.766		26.6864	
Worst voltage (bus)	0.9540 (18)		0.9040 (18)		0.8480 (18)	
Worst voltage after installation	0.9962 (18)		0.9915 (18)		0.9876 (18)	

DGs, distributed generations; *SCs*, shunt capacitors; *WOA*, whale optimization algorithm,

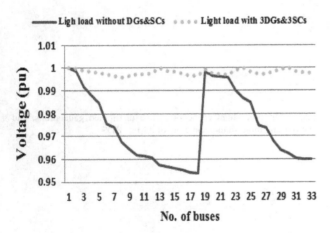

FIGURE 23.12 Voltage improvement of three distributed generations and three shunt capacitors for light load of 33-bus RDS.

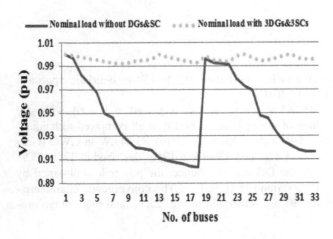

FIGURE 23.13 Voltage improvement of three distributed generations and three shunt capacitors for nominal load of 33-bus RDS.

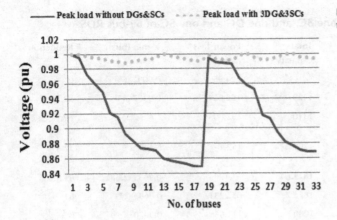

FIGURE 23.14 Voltage improvement of three distributed generations and three shunt capacitors for peak load of 33-bus RDS.

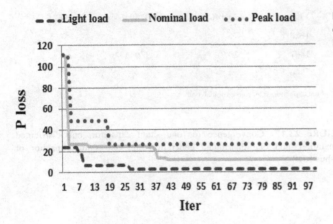

FIGURE 23.15 Active power loss with the number of iterations of three distributed generations and three shunt capacitors for three loading ratio of 33-bus RDS.

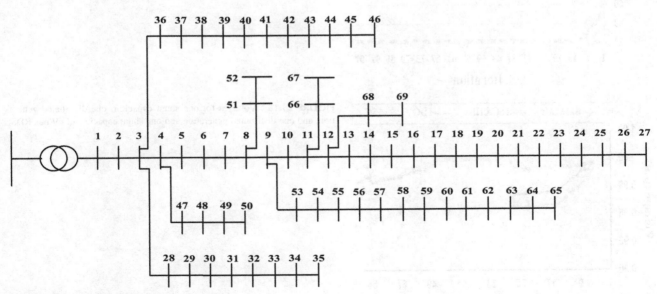

FIGURE 23.16 Single line diagram of 69-bus RDS.

TABLE 23.5 Results of hybrid (analytical–SCA) for one DG, one SC, and one DG and one SC of 69-bus RDS.

Case	Technique	Bus (size)	Total capacity	Power loss (kW)	V. min (bus)	Reduction
	Base case	–	–	224.959	0.9092 (65)	0
One SC	Proposed	61/1329.93	1329.927	152.0045	0.9307 (65)	32.45
	Heuristic Approach (Bayat & Bagheri, 2019)	61/1310	1310	152.005	NA	NA
	Hybrid (PSO–analytical) (Kansal et al., 2016)	61/1290	1290	152.1	NA	32.4
One DG	Proposed	61/1872.64	1872.64	83.18	0.9683 (27)	63.037
	ALO (Ali et al., 2018)	61/1872.7	1872.7	83.2231	0.9683 (27)	63
	Hybrid TLBO–GWO (Nowdeh et al., 2019)	61/1000	1000	111.56	0.9478	NA
	GWO (Sobieh et al., 2017)	61/1928.67	1928.67	83.24	0.9687 (27)	NA
	Hybrid (fuzzy–PSO) (Bala & Ghosh, 2019)	61/1870	1870	83.21	0.9683 (27)	NA
One SC and one DG	Proposed	61/1302	1302	23.1463	0.9788 (65)	89.71
		61/1828	1828			
	PSO (Aman et al., 2013)	61/1401	1401	25.9	0.9700 (27)	88.4
		61/1566	1566			

DG, distributed generation; *GWO*, grey wolf optimizer; *PSO*, particle swarm optimization; *SC*, shunt capacitor; *SCA*, sine cosine algorithm.

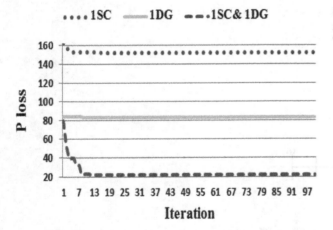

FIGURE 23.17 Convergence for one shunt capacitor, one distributed generation, and one distributed generation and one shunt capacitor of 69-bus RDS.

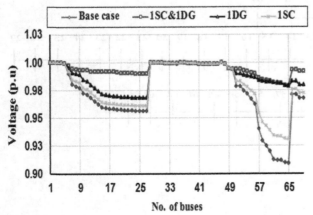

FIGURE 23.18 Voltages for one shunt capacitor, one distributed generation, and one distributed generation and one shunt capacitor of 69-bus RDS.

TABLE 23.6 Results of hybrid (analytical–SCA) for two DGs, two SCs, and two DGs and two SCs of 69-bus RDS.

Cases	Technique	Bus/size	Total capacity	Power loss	V. min (p.u.)	Reduction
	Base case	–	–	224.9599	0.9092 (65)	0
Two SCs	Proposed	17/361.53	163.6	146.41	0.9311 (65)	34.92
		61/1275.1				
	Heuristic approach (Bayat & Bagheri, 2019)	61/1224	1580	146.5	NA	NA
		17/356				
	Iterative-analytical (Forooghi Nematollahi et al., 2016)	17/357	1592	146.43	0.9305 (65)	34.88
		61/1235				
	Hybrid (PSO–analytical) (Kansal et al., 2016)	18/350	1590	146.52	NA	34.88
		61/1240				
Two DGs	Proposed	17/531.36	2312.84	71.656	0.9789 (65)	68.16
		61/1781.7				
	Hybrid TLBO–GWO (Nowdeh et al., 2019)	61/1000	1863	83.34	0.9682	NA
		62/863				
	GWO (Sobieh et al., 2017)	17/566.08	2382.5	71.74	0.9803 (65)	NA
		61/1816.2				
	Hybrid (fuzzy–PSO) (Bala & Ghosh, 2019)	18/510	2310	71.7	0.9795 (65)	NA
		61/1800				
Two SCs and two DGs	Proposed	17/352.51	1590.93	7.2013	0.9943 (69)	96.799
		61/1238.7				
		17/521.87	2256.6			
		61/1734.736				
	IMDE (Khodabakhshian & Andishgar, 2016)	63/109	1301	13.833	0.9915 (68)	93.84
		61/1192				
		62/1738	2262			
		24/479				

DG, distributed generation; IMDE, intersect mutation differential evolution; PSO, particle swarm optimization; SC, shunt capacitor; SCA, sine cosine algorithm.

Case 3: Two compensating units

In case of two SCs: The power loss reduction obtained by the hybrid method is 34.92% when the hybrid approach selects buses 17 and 61 as best locations as illustrated in Table 23.6, showing the power loss reduction of the hybrid method best than 34.88% for both iterative-analytical method (Forooghi Nematollahi et al., 2016) and hybrid (PSO–analytical) (Kansal et al., 2016).

In case of two DGs: The power loss obtained by the hybrid method is 71.656 kW when the DGs place buses 17 and 61 with sizes 531.36 and 1781.47, respectively, which is best than 71.74 kW in GWO (Sobieh et al., 2017) and 71.70 kW in hybrid (fuzzy–PSO) (Bala & Ghosh, 2019).

In case of two SCs and two DGs: Buses 17 and 61 are the best locations for hybrid approach; the power loss reduction obtained by hybrid approach is 96.799%, which is best than 93.84% in IMDE (Khodabakhshian & Andishgar, 2016) as described in Table 23.6.

The convergence characteristics of the proposed method for two SCs, two DGs, and two DGs and two SCs are illustrated in Fig. 23.19. And the voltage profile improvement is presented in Fig. 23.20.

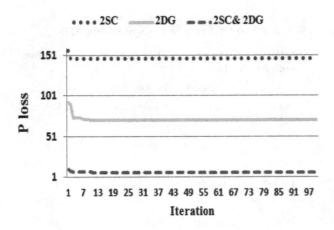

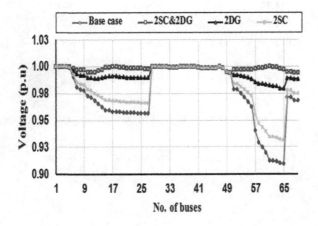

FIGURE 23.19 Convergence for two shunt capacitors, two distributed generations, and two distributed generations and two shunt capacitors of 69-bus RDS.

FIGURE 23.20 Voltages for two shunt capacitors, two distributed generations, and two distributed generations and two shunt capacitors of 69-bus RDS.

Case 3: Three compensating units

In case of three SCs: The power loss for the hybrid method is 145.09 when three SCs are installed at buses 11, 21, and 61, which is best than 145.29 kW in iterative-analytical method (Forooghi Nematollahi et al., 2016) and 145.26 kW in SSA (Salp Swarm algorithm) (Sambaiah & Jayabarathi, 2019) as shown in Table 23.7.

In case of three DGs: The power loss is reduced from 224.9599 to 69.417 kW for the proposed method when the three DGs are placed at buses 11, 17 and 61; also, this value is best than 70.19 kW in WOA (Saleh et al., 2019), and 72.37 kW in MFO (moth flame optimization) (Saleh et al., 2018) as shown in Table 23.7.

In case of three SCs and three DGs: The power loss obtained in this case using the hybrid method is 4.2692 kW when three DGs are installed at buses 11, 18, and 61. The convergence characteristics for three cases are illustrated in Fig. 23.21 and Fig. 23.22 shows the improvement of voltage profile of all busses after installing three SCs with three DGs.

- *Optimal allocation of DGs with SCs considering annual load growth using hybrid (analytical−WOA) of 69-bus RDS*

Also, the hybrid approach was applied to 69-bus RDS with considering the increasing in the loads up to 10 years, in the base year (at $y = 0$), the apparent power and voltage are 10 MVA and 12.66 kV. 224.9496 kW is the active power losses and 102.1456 kVAR is the reactive power loss before installing any unit in the network. This system is solved for installing one DG with one SC unit, two DGs with two SCs, and three DGs with three SCs using the proposed hybrid (method 3−WOA). Due to the annual load growth, the active power losses are increased gradually to 506.308 kW after 10 years as presented in Fig. 23.27. From Fig. 23.28, voltage buses decrease with increasing the years before installing the compensating units; also, the power losses become better after DGs and SCs installing as observed from Tables 23.9−23.11. Figs. 23.29−23.31 present the improvement of voltages after installing DGs and SCs in 69-bus RDS.

TABLE 23.7 Results of hybrid (analytical–SCA) for three DGs, three SCs, and three DGs and three SCs of 69-bus RDS.

Cases	Technique	Location/Size	Total capacity	Power loss	V. min (bus)	Reduction %
	Base case	–	224.9599		0.9092 (65)	0
Three SCs	Proposed	11/ 412.17	1871.998	145.09	0.9314 (65)	35.5
		21/229.24				
		61/1230.59				
	Heuristic approach (Bayat & Bagheri, 2019)	61/1210	1756	145.3	NA	NA
		21/226				
		12/320				
	Iterative-analytical (Forooghi Nematollahi et al., 2016)	11/232	1825	145.29	0.9314 (65)	35.34
		17/347				
		61/1246				
	SSA (Sambaiah & Jayabarathi, 2019)	17/300	1800	145.26	0.9308 (65)	NA
		60/1200				
		10/300				
Three DGs	Proposed	11/522.220	2629.861	69.417	0.9788 (65)	69.16
		17/396.241				
		61/1711.4				
	MFO (Saleh et al., 2018)	61/2000	2962.5	72.37	0.9939	64.29
		18/380.3				
		11/582.2				
	WOA (Saleh et al., 2019)	49/840.46	3182.38	70.19	NA	NA
		18/533.52				
		61/1808.4				
Three SCs and three DGs	Proposed	11/354.0436	1799.147	4.2692	0.9943 (50)	98.1
		17/251.0309				
		61/1194.147				
		11/505.0621	2554.24			
		17/376.0463				
		61/1673.209				

DGs, distributed generations; *SCs*, shunt capacitors; *SCA*, sine cosine algorithm; *SSA*, Salp Swarm algorithm; *MFO*, moth flame optimization; *WOA*, whale optimization algorithm.
● *Optimal allocation of SCs and DGs considering load variation using hybrid (analytical–WOA) (Table 23.8 and Figs. 23.23–23.26).*

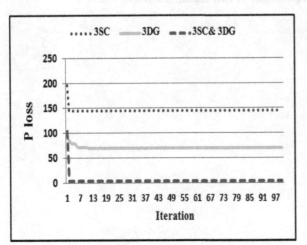

FIGURE 23.21 Convergence for three shunt capacitors, three distributed generations, and three distributed generations and three shunt capacitors of 69-bus RDS.

TABLE 23.8 Results of hybrid (Analytical— WOA) for 3DGs & 3SCs in three loading ratio of 69-bus RDS.

	Light loads	Nominal loads	Peak loads
	50% loading ratio	100% loading ratio	150% loading ratio
Power losses (kW)	51.6060	224.999	560.5258
Min voltage (p.u.)	0.9570	0.9091	0.8560
Min voltage bus	65	65	65
After compensation three DGs with three SCs			
DG size (bus)	249.34 (11)	504 (11)	767 (11)
DG size (bus)	188.74 (17)	376 (17)	562.6486 (17)
DG size (bus)	834.8716 (61)	1667 (61)	2494.8824 (61)
SC size (bus)	118.7658 (20)	237 (20)	356 (20)
SC size (bus)	183.5751 (11)	367 (11)	550.68 (11)
SC size (bus)	597.1383 (61)	1194 (61)	1789.919 (61)
Power losses (kW)	1.0613	4.2632	9.6366
Min voltage (p.u.)	0.9971	0.9943	0.9914
Min voltage (bus)	50	50	50

DGs, distributed generations; *SCs*, shunt capacitors; *WOA*, whale optimization algorithm.

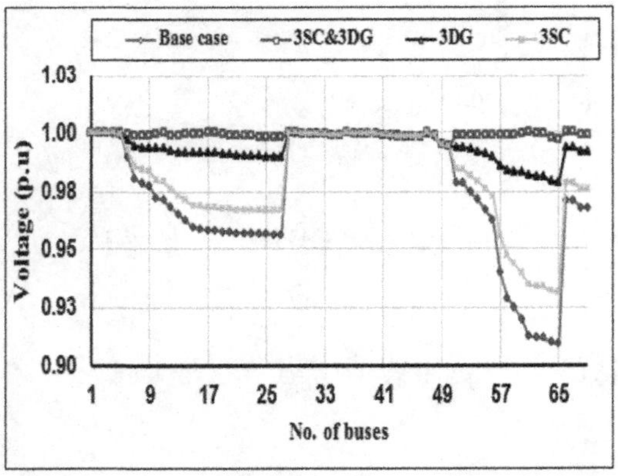

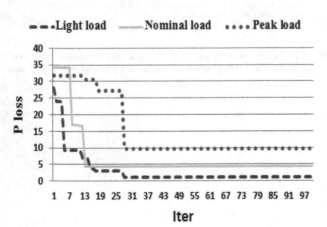

FIGURE 23.22 Voltages for three shunt capacitors, three distributed generations, and three distributed generations and three shunt capacitors of 69-bus RDS.

FIGURE 23.23 Loss with iteration of three shunt capacitors and three distributed generations for three loading ratio of 69-bus RDS.

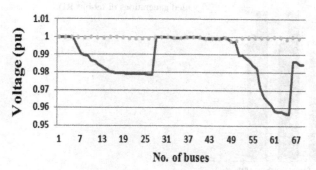

FIGURE 23.24 Voltage improvement of three shunt capacitors and three distributed generations for light load of 69-bus RDS.

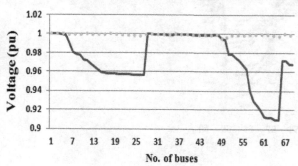

FIGURE 23.25 Voltage improvement of three shunt capacitors and three distributed generations for nominal load of 69-bus RDS.

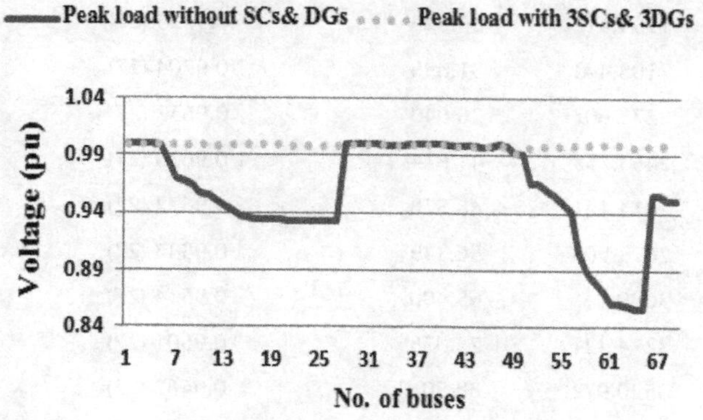

FIGURE 23.26 Voltage improvement of three shunt capacitors and three distributed generations for peak of 69-bus RDS.

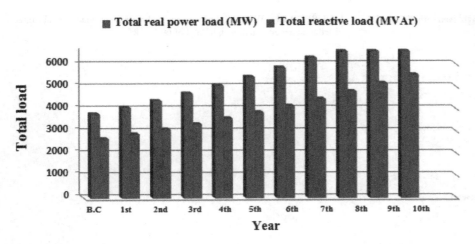

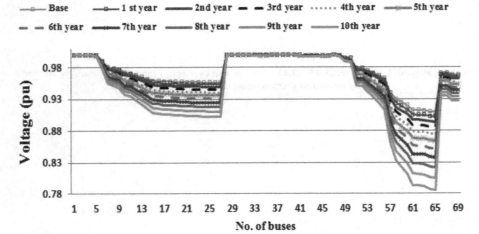

FIGURE 23.28 Voltage buses for 10 years before installing distributed generations of 69-bus RDS.

TABLE 23.9 Simulation results of hybrid (analytical–WOA) for one DG with one SC considering annual load growth of IEEE 69-bus RDS.

	Bus size (SC)	Bus size (DG)	Total (SC)	Total (DG)	Power loss (kW)	Min voltage (p.u.)
One SC and one DG						
Base	61 (1302)	61 (1828)	1302	1828	23.14	0.9788 (65)
1 year	61 (1400.897)	61 (1964.47)	1400.8968	1964.472	26.829	0.9725 (27)
2 years	61 (1512.188)	61 (2108.443)	1512.1882	2108.443	31.096	0.9704 (27)
3 years	61 (1620.5)	61 (2271.61)	1620.5336	2271.607	36.049	0.9657 (27)
4 years	61 (1727.74)	61 (2453.75)	1727.7364	2453.755	41.814	0.9630 (27)
5 years	61 (1892.51)	61 (2614.161)	1892.513	2614.161	48.528	0.9622 (27)
6 years	61 (2004.049)	61 (2873.90)	2004.0491	2873.903	56.349	0.9537 (27)
7 years	61 (2161.491)	61 (3000)	2161.4905	3000	65.400	0.9534 (27)
8 years	61 (2325.461)	61 (3274.31)	2325.4608	3274.321	75.875	0.9501 (27)
9 years	61 (2503.411)	61 (3520.972)	2503.4113	3520.972	88.50	0.9462 (27)
10 years	61 (2680.09)	61 (3790)	2680.09	3790	102.46	0.9419 (27)

DG, distributed generations; *SC*, shunt capacitors; *WOA*, whale optimization algorithm.

TABLE 23.10 Simulation results of hybrid (analytical–WOA) for two DGs with two SCs considering annual load growth of IEEE 69-bus RDS.

	Bus size (SC)	Bus size (DG)	Total (SC)	Total (DG)	Power loss (kW)	Min voltage (p.u.)
Two SCs and two DGs						
Base	17 (352.26) 61 (1238.6)	17 (521.86) 61 (1734.73)	1590.925	2256.6	7.2013	0.9943 (69)
1 year	17 (459.402) 61 (1303.59)	17 (551.831) 61 (1858.699)	1762.99	2410.53	8.5877	0.9943 (69)
2 years	17 (413.599) 61 (1449.766)	17 (565.4915) 61(2008.101)	1863.36	2573.6	9.666	0.9938 (50)
3 years	17 (433.279) 61 (1551.735)	17 (636.3654) 61 (2154.767)	1985	2791	11.151	0.9929 (69)
4 years	17 (545.184) 61 (1631.943)	17 (671.24) 61 (2319.523)	2177	2990	13.125	0.9929 (50)
5 years	17 (585.6913) 61 (1759.573)	17 (698.2038) 61 (2501.415)	2345.26	3199.6	15.25	0.9923 (50)
6 years	17 (554.0544) 61 (1910.805)	17 (802.2921) 61 (2680.395)	2464.85	3482.68	17.239	0.9915 (69)
7 years	17 (603.661) 61 (2042.67)	17 (877.6488) 61 (2884.482)	2646.33	3762.13	19.957	0.9911 (65)

DGs, distributed generations; SCs, shunt capacitors; WOA, whale optimization algorithm.

TABLE 23.11 Simulation results of hybrid (analytical–WOA) for three DGs with three SCs considering annual load growth of IEEE 69-bus RDS.

	Bus size (DG)	Bus size (SC)	Total (SC)	Total (DG)	Power loss (kW)	Min voltage (p.u.)
Three SCs and three DGs						
Base	11 (505) 17 (376) 61 (1673.206)	11 (354) 17 (251.03) 61 (1194.14)	1799	2554.24	4.26	0.9943 (50)
1 year	11 (392.4855) 17 (468.3924) 61 (1811.805)	11 (598.8325) 17 (110.7268) 61 (1277.874)	1987.43	2672.68	5.8	0.9938 (50)
2 years	11 (577.8023) 17 (423.0344) 61 (1943.799)	11 (598.7417) 17 (219.7366) 61 (1344.244)	2162.7	2944.63	6.02	0.9934 (50)
3 years	11 (500.75) 17 (524.559) 61(2092.56)	11 (524.32) 17 (292.67) 61 (1462.38)	2279.5	3117.64	6.7811	0.9929 (50)
4 years	11 (688.0303) 17 (511.5133) 61 (2225.86)	11 (614.7587) 17 (331.6917) 61 (1548.582)	2495	3425.40	7.8779	0.9923 (50)
5 years	11 (634.801) 17 (557.4456) 61 (2415.584)	11 (536.2432) 17 (341.8867) 61 (1726.196)	2604.32	3607.83	8.8891	0.9917 (50)

DGs, distributed generations; SCs, shunt capacitors; WOA, whale optimization algorithm.

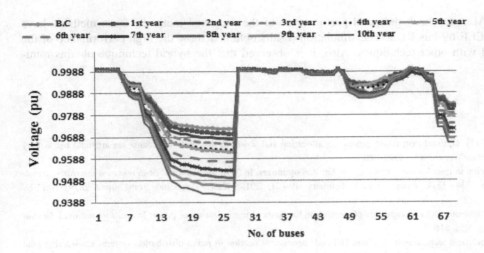

FIGURE 23.29 Voltage buses for 10 years after installing one distributed generation and one shunt capacitor of 69-bus RDS.

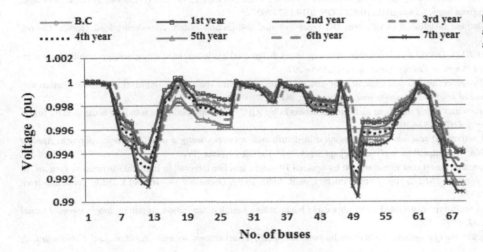

FIGURE 23.30 Voltage buses for 7 years after installing two distributed generations and two shunt capacitors of 69-bus RDS.

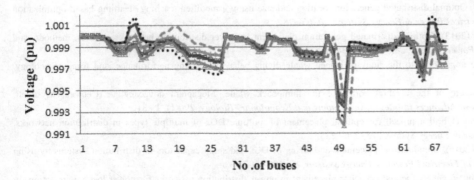

FIGURE 23.31 Voltage buses for 5 years after installing three distributed generations and three shunt capacitors of 69-bus RDS.

23.6 Conclusion

This chapter presents two hybrid techniques based on analytical and *meta*heuristic techniques for specifying the best locations and sizes of DG and SC to reduce power losses and improve the voltage in radial distribution system. The analytical is used to determine the size of compensating unit while the optimal locations are specified by the *meta*heuristic (SCA and WOA). Also, to check the good performance of the proposed methods, these methods have been tested on IEEE 33-bus and IEEE 69-bus RDS. The simulation results obtained show the high performance of the hybrid techniques when compared with other techniques. Also, it is observed that the hybrid technique obtains minimum losses in less time.

References

Abu-Mouti, F. S., & El-Hawary, M. E. (2011). Optimal distributed generation allocation and sizing in distribution systems via artificial bee colony algorithm. *IEEE Transactions on Power Delivery*, *26*(4), 2090−2101.

Ali, A. H., et al. (2018). Optimal DG allocation in distribution systems using Ant lion optimizer. In *2018 international conference on innovative trends in computer engineering (ITCE)* (pp. 324−331). Aswan, Egypt, February 19−21, 2018, IEEE. Available from https://doi.org/10.1109/ITCE.2018.8316645.

Aman, M. M., et al. (2013). Optimum simultaneous DG and capacitor placement on the basis of minimization of power losses. *International Journal of Computer and Electrical Engineering*, *5*(5), 516.

Amin, A., et al. (2019). Development of analytical technique for optimal DG and capacitor allocation in radial distribution systems considering load variation. In *IECON 2019—45th annual conference of the IEEE industrial electronics society* (pp. 2348−2353). Lisbon, Portugal, Portugal, October 14−17, 2019, IEEE. Available from https://doi.org/10.1109/IECON.2019.8927760.

Arulrag, R., & Kumarappan, N. (2017). Optimal installation of different DG types in radial distribution system considering load growth. *Electric Power Components and Systems*, 1−13.

Arya, L. D., Koshti, A., & Choube, S. C. (2012). Distributed generation planning using differential evolution accounting voltage stability consideration. *International Journal of Electrical Power & Energy Systems*, *42*(1), 196−207.

Baghipour, R., & Hosseini, S. M. (2012). Placement of DG and capacitor for loss reduction, reliability and voltage improvement in distribution networks using BPSO. *International Journal of Intelligent Systems and Applications*, *4*(12), 57.

Bala, R., & Ghosh, S. (2019). Optimal position and rating of DG in distribution networks by ABC−CS from load flow solutions illustrated by fuzzy-PSO. *Neural Computing and Applications*, *31*(2), 489−507.

Bayat, A., & Bagheri, A. (2019). Optimal active and reactive power allocation in distribution networks using a novel heuristic approach. *Applied Energy [Internet]*, *233*, 71−85. Available from https://doi.org/10.1016/j.apenergy.2018.10.030. Elseiver BV.

Biswal, S., et al. (2018). Cuckoo search algorithm based cost minimization by optimal DG and capacitor integration in radial distribution systems. In *2018 20th National Power Systems Conference (NPSC)* (pp. 1−6). Tiruchirappalli, India, India, December 14−16, 2018, IEEE. Available from https://doi.org/10.1109/NPSC.2018.8771773.

El-Fergany, A. (2015). Optimal allocation of multi-type distributed generators using backtracking search optimization algorithm. *International Journal of Electrical Power & Energy Systems*, *64*, 1197−1205.

Elgerd, O. I., & Happ, H. H. (1972). Electric energy systems theory: an introduction. *IEEE Transactions on Systems, Man, and Cybernetics*, *2*, 296−297.

Forooghi Nematollahi, A., et al. (2016). Optimal sizing and siting of DGs for loss reduction using an iterative-analytical method. *Journal of Renewable and Sustainable Energy*, *8*(5), 055301.

Ganguly, S., & Samajpati, D. (2017). Distributed generation allocation with on-load tap changer on radial distribution networks using adaptive genetic algorithm. *Applied Soft Computing*, *59*, 45−67.

García, J. A. M., & Mena, A. J. G. (2013). Optimal distributed generation location and size using a modified teaching−learning based optimization algorithm. *International Journal of Electrical Power & Energy Systems*, *50*, 65−75.

Georgilakis, P. S., & Hatziargyriou, N. D. (2013). Optimal distributed generation placement in power distribution networks: models, methods, and future research. *IEEE Transactions on Power Systems*, *28*(3), 3420−3428.

Goldbogen, J. A., et al. (2013). Integrative approaches to the study of baleen whale diving behavior, feeding performance, and foraging ecology. *Bioscience*, *63*(2), 90−100.

Hof, P. R., & Van, D. G. (2007). Structure of the cerebral cortex of the humpback whale, Megaptera novaeangliae (Cetacea, Mysticeti, Balaenopteridae). *The Anatomical Record: Advances in Integrative Anatomy and Evolutionary Biology*, *290*(1), 1−31.

Kansal, S., Kumar, V., & Tyagi, B. (2016). Hybrid approach for optimal placement of multiple DGs of multiple types in distribution networks. *International Journal of Electrical Power & Energy Systems*, *75*, 226−235.

Khodabakhshian, A., & Andishgar, M. H. (2016). Simultaneous placement and sizing of DGs and shunt capacitors in distribution systems by using IMDE algorithm. *International Journal of Electrical Power & Energy Systems*, *82*, 599−607.

Kowsalya, M., et al. (2014). Optimal distributed generation and capacitor placement in power distribution networks for power loss minimization. In *2014 international conference on advances in electrical engineering (ICAEE)* (pp. 1−6). Vellore, India, January 9−11, 2014. IEEE. Available from https://doi.org/10.1109/ICAEE.2014.6838519.

Mirjalili, S. (2016). SCA: A sine cosine algorithm for solving optimization problems. *Knowledge-Based Systems*, *96*, 120−133.

Mirjalili, S., & Lewis, A. (2016). The whale optimization algorithm. *Advances in Engineering Software*, *95*, 51−67.

Mittal, A. M. & Kansal, S. (2017). Optimal placement of distributed generation using genetic algorithm approach. In *Proceeding of the second international conference on microelectronics, computing & communication systems (MCCS)* (pp. 587−597).

Muthukumar, K., & Jayalalitha, S. (2016). Optimal placement and sizing of distributed generators and shunt capacitors for power loss minimization in radial distribution networks using hybrid heuristic search optimization technique. *International Journal of Electrical Power & Energy Systems*, *78*, 299−319.

Naik, S. G., Khatod, D. K., & Sharma, M. P. (2013). Optimal allocation of combined DG and capacitor for real power loss minimization in distribution networks. *International Journal of Electrical Power & Energy Systems*, *53*, 967−973.

Nowdeh, S. A., et al. (2019). Fuzzy multi-objective placement of renewable energy sources in distribution system with objective of loss reduction and reliability improvement using a novel hybrid method. *Applied Soft Computing*, *77*, 761−779.

Rajendran, A., & Narayanan, K. (2020). Optimal multiple installation of DG and capacitor for energy loss reduction and loadability enhancement in the radial distribution network using the hybrid WIPSO−GSA algorithm. *International Journal of Ambient Energy*, *41*(2), 129−141.

Raju, M. R., Murthy, K. R., & Ravindra, K. (2012). Direct search algorithm for capacitive compensation in radial distribution systems. *International Journal of Electrical Power & Energy Systems*, *42*(1), 24−30.

Rao, R. S., et al. (2012). Power loss minimization in distribution system using network reconfiguration in the presence of distributed generation. *IEEE Transactions on Power Systems*, *28*(1), 317−325.

Reddy, S. C., Prasad, P. V. N., & Laxmi, A. J. (2013). Placement of distributed generator, capacitor and DG and capacitor in distribution system for loss reduction and reliability improvement. *Journal of Electrical Engineering*, *13*(4), 329−337.

Reddy, S. S., Dey, S. H., & Paul, S. (2012). Optimal size and location of Distributed Generation and KVAR support in unbalanced 3-Φ distribution system using PSO. In *2012 international conference on emerging trends in electrical engineering and energy management (ICETEEEM)* (pp. 77−83). Chennai, India, December 13−15, 2012. IEEE. Available from https://doi.org/10.1109/ICETEEEM.2012.6494447.

Rueda-Medina, A. C., et al. (2013). A mixed-integer linear programming approach for optimal type, size and allocation of distributed generation in radial distribution systems. *Electric Power Systems Research*, *97*, 133−143.

Saleh, A. A., et al. (2018). Comparison of different optimization techniques for optimal allocation of multiple distribution generation. In *2018 international conference on innovative trends in computer engineering (ITCE)* (pp. 317−323). Aswan, Egypt, February 19−21 2018. IEEE. Available from https://doi.org/10.1109/ITCE.2018.8316644.

Saleh, A. A., Mohamed, A. A., & Hemeida, A. M. (2019). Optimal allocation of distributed generations and capacitor using multi-objective different optimization techniques. In *2019 international conference on innovative trends in computer engineering (ITCE)* (pp. 377−383). Aswan, Egypt, Egypt, February 2−4, 2019. IEEE. Available from https://doi.org/10.1109/ITCE.2019.8646426.

Sambaiah, K. S., & Jayabarathi, T. (2019). Optimal allocation of renewable distributed generation and capacitor banks in distribution systems using Salp Swarm algorithm. *International Journal of Renewable Energy Research*, *9*, 96−107.

Sobieh, A. A., et al. (2017). Optimal number size and location of distributed generation units in radial distribution systems using Grey Wolf optimizer. *International Electrical Engineering Journal (IEEJ)*, *7*(9), 2367−2376.

Sultana, S., & Roy, P. K. (2014). Multi-objective quasi-oppositional teaching learning based optimization for optimal location of distributed generator in radial distribution systems. *International Journal of Electrical Power & Energy Systems*, *63*, 534−545.

Thangaraj, Y. (2018). Simultaneous allocation renewable DGs and capacitor for typical indian rural distribution network using cuckoo search algorithm. *International Journal of Engineering and Technology*, *7*(3.12), 1011−1016.

Watkins, W. A., & Schevill, W. E. (1979). Aerial observation of feeding behavior in four baleen whales: Eubalaena glacialis, Balaenoptera borealis, Megaptera novaeangliae, and Balaenoptera physalus. *Journal of Mammalogy*, *60*(1), 155−163.

Zhang, X., Karady, G. G., & Ariaratnam, S. T. (2013). Optimal allocation of CHP-based distributed generation on urban energy distribution networks. *IEEE Transactions on Sustainable Energy*, *5*(1), 246−253.

Zhang, X., Sharma, R. & Yanyi, H. (2012). Optimal energy management of a rural microgrid system using multi-objective optimization. In *2012 IEEE PES innovative smart grid technologies (ISGT)* (pp. 1−8). Washington, DC, USA, January 16−20, 2012. IEEE. Available from https://doi.org/10.1109/ISGT.2012.6175655.

Chapter 24

Optimal appliance management system with renewable energy integration for smart homes

I. Hammou Ou Ali, M. Ouassaid and Mohamed Maaroufi

Engineering for Smart and Sustainable Systems Research Center, Mohammadia School of Engineers, Mohammed V University, Rabat, Morocco

24.1 Introduction

With the demographic increase, the industrialization and the technological developments, the existing power system has been facing many challenges in terms of environmental problems, sustainability, and service management. To cope with this mutation, changing the traditional grid into a smart one is a necessity. Smart grid (SG) represents a real challenge to take advantage of the new information and communication technologies to revolutionize the conventional electrical power system (Abujubbeh, Al-Turjman, & Fahrioglu, 2019; Dileep, 2020; Tuballa & Abundo, 2016). This revolution is illustrated in optimizing the energy loss, establishing two-way communication between the user and the utility, and increasing the user satisfaction by informing them about their energy consumption and electricity pricing. In SG, the energy management is mainly classified into two fields: supply-side management, which deals with the efficient generation, transmission, and distribution of the energy, and demand-side management (DSM), which aims at planning and monitoring the energy consumption in the user side (Gelazanskas & Gamage, 2014; Supply-side management). Taking advantage of SG, the user can actively operate in the electrical network. However, the user's interaction is assumed to be more important in dynamic pricing than in flat pricing. The electricity market provides different dynamic pricing such as time-of-use (TOU) pricing, critical peak pricing (CPP), real-time pricing (RTP), and inclining block rate (IBR) (Dutta & Mitra, 2017). Bearing in mind the electricity tariffs of the dynamic schemes, the user can schedule the load appliances into low peak hours to reduce the electricity bills. Indeed, the increasing use of distributed renewable energy sources (RESs) is one of the primary steps toward the global transition toward the SG paradigm, achieves energy conservation, and provides reliable and attractive solutions (Eltigani & Masri, 2015; Hossain et al., 2016). The main purpose behind SG is to promote an active smart consumer and even a smart consumer/producer participation and decision-making as well as to create the operating environment in which both utilities and consumers can interact with each other.

The importance of optimizing energy consumption in smart homes can be deduced from statistical information, which indicates that the building is one of the major energy consuming sectors. In Morocco, the electricity consumption in the residential sector represents over 33% of the global energy and records a strong growth of the annual energy consumption (Buildings, 2019). This is the reason why in recent years much attention is devoted to the application of DSM in the residential sector, as shown by the increasing number of publications. DSM offers various benefits to the future SG, such as cost savings by scheduling flexible power activities outside of price spikes, improving the load factor, managing demand—supply balance in systems with the local energy generation and storage system, and maintaining the system reliability of distribution networks through shifting or reducing demand. DSM may also achieve environmental goals by decreasing energy use, leading to reduced greenhouse gas emissions (Gudi, Wang, & Devabhaktuni, 2012; Trygg & Karlsson, 2005).

Managing and monitoring the residential loads are not the easy tasks; it needs optimal scheduling according to the user needs and comfort. In this context, the home energy management system (HEMS) is introduced. It is used not only for determining the optimal scheduling to minimize the electricity cost, increase the use of RESs, reduce the peak-to-average ratio (PAR), but also to guarantee the users' comfort. On the other hand, numerous constraints emerge. It includes the load categorization (fixed, shiftable, and elastic), the electricity market, the local energy generators, and the energy storage system (ESS) (Esther & Kumar, 2016). To tackle the appliance scheduling problem, various scheduling control approaches have been utilized: artificial intelligence (AI), and optimization techniques. Many optimization algorithms exist in the literature such as an integer linear programming (Amini, Frye, Ilić, & Karabasoglu, 2015; Jaramillo & Weidlich, 2016; Zhu, Tang, Lambotharan, Chin, & Fan, 2012), dynamic programming (Al-Jabery et al., 2016; Liu, Yuen, Yu, Zhang, & Xie, 2015; Muratori & Rizzoni, 2015), heuristic techniques (Awais et al., 2015; Logenthiran, Srinivasan, & Shun, 2012; Rahim et al., 2016; Rajarajeswari, Vijayakumar, & Modi, 2016), and hybrid approaches (Abushnaf, Rassau, & Górnisiewicz, 2016; Javaid et al., 2017). However, the controllers using AI techniques are based on artificial neural network (ANN) (Ahmed, Mohamed, Shareef, Homod, & Ali, 2016; Moon & Kim, 2010), fuzzy logic control (Hong, Lin, Wu, & Chuang, 2012; Wu, Zhang, Lu, & Du, 2011), and adaptive neural fuzzy inference system (Choi et al., 2015; Shahgoshtasbi & Jamshidi, 2014).

In the present work, a scheduling approach for managing the home appliances is developed. The objective of this approach is to minimize the electricity cost without affecting the user's comfort level in terms of waiting time. In this work, a single house with 12 loads is studied over 1 day. In addition to the main grid, solar panels and battery storage are added to the proposed system. The numerical simulation using mixed-integer linear programming (MILP) algorithm confirms the effectiveness of the proposed scheduling mechanism in terms of cost reduction and discomfort minimization as well as a desired trade-off between the electricity payment and the discomfort. The suggested approach is tested under three scenarios with or without solar panels and battery storage integration. The waiting time rate (WTR) is minimized without electricity overpayment. The main contributions of this work are as follows:

- A novel method to control residential load, taking into consideration the maximization of the users' comfort and minimization of the electricity cost, is proposed.
- Good allocation of the electrical power against the proposed objective functions and constraints, using MILP optimization under the three scenarios, is achieved.
- Efficient use of solar panels and battery storage systems to reduce further the electricity payment and the users' discomfort as well as to support the national grid, in times when solar energy is more than the demand, is established.

The remaining part of this chapter is structured as follows: Section 24.2 discusses related work. Section 24.3 describes the system architecture and its components. Section 24.4 elaborates the proposed approach for scheduling the residential appliances in a smart home. Section 24.5 deals with the results and discussion. Finally, Section 24.6 concludes the work and open new directions of future research.

24.2 Related work

Recently, much attention has been given to HEM strategies to tackle the appliance scheduling issue in smart houses. In this context, this section is organized into three axes consisting of different challenges associated with DSM. In the first axe, only the electricity cost is discussed. In Setlhaolo, Xia, and Zhang (2014), a mixed integer nonlinear optimization model under a TOU electricity tariff is used to schedule home appliances in order to minimize electricity cost. In this study, the authors propose 10 residential appliances in the home, and the obtained results show that the described scheduling model leads to more than 25% bill saving for the consumer. In the same context, Derakhshan, Shayanfar, and Kazemi (2016) proposes a teaching and learning-based optimization and shuffled frog leaping algorithms to optimize the electricity cost of four types of residential consumers under the TOU, RTP, CPP, and the last consumer under a flat tariff policy. The results demonstrate that the proposed optimization techniques reduce the total consumer's bills. The authors in Zhang, Evangelisti, Lettieri, and Papageorgiou (2016) integrate the microgrid system for scheduling the energy consumption in smart houses using MILP. The proposed model is implemented under three different price schemes for 30 houses with the same living habits and provides optimal results with a trade-off between economic cost and environmental emissions of CO_2. Similarly, the work in Samadi, Wong, and Schober (2015) integrates RESs for scheduling the residential loads using dynamic programming approach. Game theory is adopted to model the interaction between users with excess power generation in order to sell the electricity to local users and thus reduces the energy costs of the users. In the same context, the authors in Boynuegri, Yagcitekin, Baysal, Karakas, and Uzunoglu (2013) propose a HEM algorithm using loads shifting in smart homes with RESs and battery storage system. The simulation results show that the proposed algorithm can reduce the electricity bills in case the house is supplied with the grid only and with the RES integration.

The second axe is about the works that deal with the PAR minimization. The authors in Qayyum et al. (2015) propose branch-and-bound algorithm based on HEMS with RES integration for scheduling the home appliances in the smart home network. The simulation results confirm that the proposed method is proven to reduce both electricity bills and peak load. However, in Lee and Choi (2014) a linear programing method is used for shaving the peak load in smart home. The proposed method reduces both cost and PAR by charging the ESS from the utility in off-peak hours and is discharged in peak hours. In Ullah et al. (2015), the electricity cost and the PAR are minimized based on a combination of genetic algorithm (GA) and binary particle swarm optimization (BPSO). The proposed method has been proven to be efficient in shifting home appliances to operate in low peak hours. To achieve the balance load schedule in a smart home, the work in Bradac, Kaczmarczyk, and Fiedler (2015) proposes a MILP method with the aim of minimizing both the total energy paid and the power peaks. Similarly, the authors in Shakouri and Kazemi (2017) consider multiple scenarios using real data for minimizing both, the electrical peak and the cost. The problem is formulated as a multiobjective MILP considering the consumer preferences and the daily required energy as constraints. Besides, El Idrissi, Ouassaid, and Maaroufi (2018) proposes a predictive load shifting method using particle swarm optimization (PSO) and back tracking search algorithm in order to minimize both the cost and the peak load demand for the residential, commercial, and industrial users. On the other hand, Javaid et al. (2017) introduces an intelligent load management based on TOU pricing and RESs for scheduling the appliances of residential users. The authors propose algorithms such as BPSO, GA, and cuckoo search to reduce the electricity bill and peaks.

This third axe surveys the studies tied to the user's comfort. The authors in Ma, Yao, Yang, and Guan (2016) propose a power scheduling method under a day-ahead price for residential user to achieve a desired trade-off between the user discomfort and payments. Another work in Yang and Wang (2012) set up a multiagent technology to enable the building to interact with its occupants for achieving an effective energy and comfort management in a building. A MILP with solar resource is proposed in Ogunjuyigbe, Ayodele, and Oladimeji (2016) to effectively allocate electrical power in residential building based on a predefined priority of the user's loads. The authors in Muralitharan, Sakthivel, and Shi (2016) present a multiobjective evolutionary algorithm to minimize the electricity cost for energy usage and to minimize the delay time of the appliance execution. The work developed in Ogunjuyigbe, Ayodele, and Akinola (2017) puts forward a load-satisfaction management technique that controls residential loads to achieve a maximum satisfaction at a minimum cost. To perform the strength of the proposed DSM based on GA, the authors proposed a fixed daily budget. Moreover, the work presented in Garroussi, Ellaia, Talbi, and Lucas (2017) proposes a multiobjective model for managing the residential loads with solar panels and batteries integration. The proposed model proves its efficacy in reducing the user's discomfort, the total electricity cost, and the standard deviation of the consumed power. In Al Hasib, Nikitin, and Natvig (2015), the authors propose a multiobjective function for scheduling the residential appliances with the goal to maintain a balance between the user comfort and the electricity cost. The proposed model allows the user to sell the surplus energy, obtained from local RESs, so as to partially recover the electricity cost. Ahmed, Mohamed, Homod, and Shareef (2016) proposes a hybrid lightning search algorithm-based ANN to predict the use of home appliances according to the customer preferences and priority. The simulation results are compared to PSO-based ANN, and it has been demonstrated that the solution is better in reducing the energy consumption. In Rasheed et al. (2015), the authors propose an energy management model based on wind-driven optimization and PSO algorithms. which consider the electricity cost, the high peaks, and the user comfort requirements. Simulation results prove better performances of WDO than PSO in terms of appliance waiting time and electricity bill reduction. In Arshad et al. (2017), pigeon inspired optimization (PIO) and enhanced differential evolution are used under TOU pricing for scheduling the home appliances with maximum user comfort and minimum cost and PAR. Simulation results reveal that both of the proposed algorithms reduce the electricity payment and the PAR. However, with PIO the user can achieve more comfort. The work in Abbasi et al. (2017) emphasizes the home appliances management in order to minimize the cost, the PAR, and the waiting time under RTP tariffs. Flower pollination algorithm (FPA) and GA are used to solve the proposed problem. The results reveal better performance of FPA on reducing the cost and PAR while the waiting time of GA remains slightly better than FPA. In the same task, GA and crow search algorithm (CSA) are performed in Aslam et al. (2017) to manage different categories of appliances in a house under the RTP scheme with the aim to decline the electricity bills, the waiting time, and the PAR. The obtained results demonstrate that CSA performs better than GA in terms of cost reduction. Some recent works related to the user satisfaction are listed in Table 24.1.

24.3 System architecture

HEMS is a combination of hardware equipment and software algorithms with network connections to manage efficiently the energy consumption in a smart home. The energy management controller (EMC) is considered as the main element of the architecture. The householder enters the requirements through a smart interactive interface. The request is transferred to the EMC to schedule the loads according to the users' demand. In the other side, the data pricing sent

TABLE 24.1 Brief comparison of some works.

Technique used	Pricing scheme	Load classification	User comfort	Objective function	Limitation
Convex optimization (Ma et al., 2016)	Day-ahead price	Time flexible and power flexible	The discomfort is calculated using a delay function and Taguichi loss function.	Multiobjective function to minimize the electricity cost and the discomfort	Renewable energy integration is not considered.
Mixed-integer linear programming (Ogunjuyigbe et al., 2016)	Hybrid wind/ solar/ diesel system	Controllable and noncontrollable load	Priorities of loads are conserved as predefined by the user.	Uni-objective function to allocate loads based on the available solar resource	User comfort is not taken as objective parameter.
Multiobjective evolutionary algorithm (Muralitharan et al., 2016)	Time-of-use tariff	Permanent and schedulable devices	The user comfort is calculated using delay function.	Multiobjective function to minimize the cost and the delay	Renewable energy sources integration is not considered.
Genetic algorithm (Ogunjuyigbe et al., 2017)	Fixed tariff	Appliance load based on a single home	The user satisfaction is quantified based on certain rules.	Minimizing the user satisfaction at minimum cost	Load peak is not taken into consideration.
(NSGA-II) and an exact solver (CPLEX) (Garroussi et al., 2017)	Time-of-use tariff	Critical appliances, time shiftable appliances, and thermal appliances	The discomfort is calculated by the delay and the deviation of temperature function.	Reducing the total electricity cost and the discomfort	Renewable energy integration is not considered.
Integer linear programming (Al Hasib et al., 2015)	Real-time pricing	Fixed load, noninterruptible load, interruptible load, and comfort load	The user comfort is calculated using a comfort demand function	Minimizing the cost and maximizing the comfort level	Load's priority is not taken into consideration.
Hybrid lightning search algorithm-based artificial neural network (Ahmed et al., 2016)	Not mentioned	Air conditioner, water heater, refrigerator, and washing machine	Appliances are modeled according to customer preferences and priority of appliances.	Reducing the peak load while guaranteeing end-user comfort	The cost reduction is not taken as objective parameter.
Wind-driven optimization (Rasheed et al., 2015)	Time-of-use	Power adjustable, fixed power, and time scheduled	The user comfort function is calculated in terms of waiting time.	Comfort maximization along with minimum electricity cost	User activities and solar energy integration is not considered.

from the utility to the smart meter is transmitted to the EMC. The EMC computes the bill and makes the user aware of the energy consumption and cost as well. The user can make changes according to the preferences: cost minimization, comfort maximization, or a trade-off between them. RESs and battery storage can be integrated in the proposed architecture to reduce the energy obtained from the grid. The remaining energy generated locally by the solar panels can be used to supply the main grid in times when solar energy is more than the demand at home. In this case, the resident is not considered as a simple, smart user, but as a smart prosumer. The connection between the EMC and the appliances is established through different communication technology, which is detailed later.

In this work, a single home DSM system is proposed. Day-ahead electricity pricing as well as a predictive solar panels generation are used for scheduling the home appliances at low prices without affecting the customers' comfort level. Fig. 24.1. shows the HEMS used in this work.

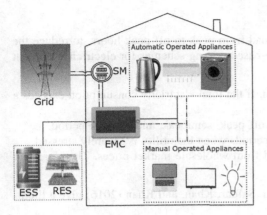

FIGURE 24.1 Architecture of the proposed home energy management system.

24.3.1 The home appliances

Managing the home appliances is not an easy task; it requires knowledge about the electrical characteristics of each appliance as well as behavior. The householder can allow scheduling the washing machine to turn on in a required time interval. However, the householder does not allow the remaining without light at night.

For optimal appliance scheduling in a smart home environment, in this work, the home appliances are divided into two categories:

- Automatic operated appliances (AOAs): AOAs are the loads that can be managed through the EMC without the interaction of the householder such as the air conditioner and the cloth dryer. However, the AOAs are also classified into two categories: flexible and nonflexible. The flexible loads are operating with flexible power usage while ensuring the total prescribed energy requirements such as the electric vehicle, water pump, and air conditioner. However, the nonflexible ones can shift their power consumption time within a preferred working period and cannot be interrupted when they are turned on. Washing machine, clothes dryer, dishwasher are the examples.
- Manual operated appliances (MOAs): MOAs are the loads that require the interaction of the householder to turn them on or off. Generally, the MOAs are strongly related to the unforeseeable behavior of the consumer. They cannot be delayed or not even forecasted and highly impact the user's comfort such as lighting and television.

24.3.2 Communication protocol technology

The HEMS components must be able to exchange information with each other through the same language, commonly called communication protocols. Information are sent from a device to another in order to trigger an action: turn the light on/off, increase the air conditioner temperature, start the washing machine, etc. The main purpose of communication in HEMS is monitoring, optimizing, and automating the appliance usage in a house and establishing a two-way communication between consumer and utility. Some of the major communication protocols for smart home devices are as given below:

- Bluetooth: Without physical connection, electronic devices can get connected through radio waves in a short range of 100 m maximum. Bluetooth technology is a relatively robust protocol that operates on low power and low operational cost.
- ZigBee: Similar to Bluetooth, over a limited area of 100 m. It requires relatively few frequencies low speed data exchange (2.4 GHz bandwidth). ZigBee Technology offers a large number of nodes in a wireless sensor network between machines and applications with high security and robustness at a low power consumption.
- Wi-Fi: Most commonly used technology in building and business houses. Wi-Fi is a fast data transfer protocol and have the capability to handle a large quantity of data at a high transmission speed (speed up to 1 Gbps) with a range of 50 m.
- Z-wave: Specially designed for residential load automation such as smart plugs, sensors, and controller. It connects up to 232 devices with a range of 30 m. The Z-wave technology operates on frequency of 900 MHz and allows faster and easier application development.

The authors in Baviskar, Baviskar, Wagh, Mulla, and Dave, 2015 and Ettalbi, Elabd, Ouassaid, and Maaroufi (2016) give a comparative study of communication technologies in smart homes.

24.3.3 Electricity tariffs

Since the main purpose behind DSM is to push the users to shift their load from peak to off-peak hours, to reduce the demand during peak hours, the electricity market provides different time-based pricing schemes for electricity payment during a day. Among these:

- IBR: The electricity price increases with the total energy consumption. If the total electricity consumption exceeds a certain value, then the electricity payment increases.
- TOU: The daily electricity pricing varies depending on three periods: off-peak, semipeak, and on-peak period.
- CPP: Is similar to the TOU. However, the prices can be very high in periods when the utility is severely stressed.
- RTP: In this scheme, electricity prices vary regularly and determined from wholesale market prices. There are two (RTP) schemes: day-ahead pricing and hourly pricing.

For the details of the pricing schemes, the authors of Khan, Mahmood, Safdar, Khan, and Khan (2016) give a literature review.

24.4 The proposed approach for scheduling the home appliances

24.4.1 Scheduling problem formulation

In this work, a smart home equipped with the HEMS and smart appliances is proposed to schedule the daily usage of the residential loads. Home appliances are categorized into two classes based on their characteristics as detailed earlier: MOA and AOA. Since the use of the MOA cannot be forecasted, the focus moves toward the AOA. The EMC can schedule the home appliances for the next day. Indeed, 1 day consists of 24 hours, which can be represented as follows:

$$h \in H = [1, 2, \ldots, 24] \tag{24.1}$$

In the proposed model, 1 hour is divided into 5 time slots, that is, each time slot is equivalent to 12 minutes. So, for a single day (24 hours) the scheduling horizon is composed of 120 times slots. The total slots can be represented as:

$$t \in T = [1, 2, \ldots, 120] \tag{24.2}$$

Any appliance of the previously cited category should run in time interval between the scheduling horizon {T}. For the AOA, one single load is represented by i, and the set of appliances by A. So, for each $i \in A$, the length operating time (LOT) is fixed, and it should be a multiple of 12. For example, if the operation time of the cloth dryer is 1 hour then the LOT should be set as 5 time slots. However, if the normal operation time is 46 minutes the LOT will be (46/12) with a markup of this number, the LOT should be set as 4 time slots.

Respecting the user comfort is an important issue for scheduling the residential appliances. In this context, the user should set up an operating time interval (OTI) for each appliance according to the daily activity. The OTI indicates the scheduling horizon in which the user wants to turn on the home appliances.

For each schedulable appliance, the user should set up its OTI. This interval of time represents the starting time and the ending time where the appliance should be scheduled; outside this interval of time, the appliance cannot operate. So, the OTI for each appliance can be represented as follows:

$$OTI = [ts_i \ te_i] \tag{24.3}$$

and

$$ts_i \leq te_i \tag{24.4}$$

Each schedulable appliance i has its length of operation time (LOT) denoted by l_i. The LOT refers to the number of time slots that each appliance needs to be run. The optimal starting time for each appliance scheduling interval is denoted by ta_i. ta_i should be greater or equal to ts_i, and less than or equal to $te_i - l_i$. In other words:

$$ts_i \leq ta_i \leq te_i - l_i \tag{24.5}$$

The overall operating time for n appliances can be represented as follows:

$$[ta_1, ta_2, \ldots, ta_n] \tag{24.6}$$

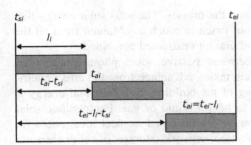

To illustrate these parameters, Fig. 24.2 represents a graphical representation, including the LOT, OTI, and the optimal staring time.

To address the problem of users' comfort, the concept of waiting time is introduced in this work. Usually residents hope that the appliances finish their work in the shortest interval of time, which can impact the increase of their electricity bills. The WTR has been defined in Zhao, Lee, Shin, and Song (2013) according to the following expression:

$$\text{WTR}_i = \frac{ta_i - ts_i}{te_i - l_i - ts_i} \tag{24.7}$$

The WTR indicates how quickly an appliance can perform within a desired scheduling horizon. For example, if the washing machine starts operating directly at ts_i, then the $\text{WTR}_i = 0$, and the user will be very satisfied with the service. However, if the appliance starts operating at time slot $te_i - l_i$, then the value of the WTR_i would be 1 but the user would be disappointed.

Therefore the waiting time function (WTF) can be expressed as follows:

$$\text{WTF} = \sum_{i=1}^{n} \gamma^{\text{WTR}_i} \tag{24.8}$$

where, γ is a delay parameter greater than 1. The WTF indicates that an increase in WTR_i will immediately increase the value of WTF. In order to ensure a maximum satisfaction of the user, the WTR_i should be as minimum as possible. Hence, the WTF should be minimized. For each single appliance, the power consumption value is fixed. z_i denotes the power consumption value per hour for the appliance i. In time slots, it is equivalent to:

$$P_i = \frac{z_i}{5} \tag{24.9}$$

For all the AOAs, the power consumtiom matrix can be formulated as follows:

$$P = \begin{cases} P_i^t = \dfrac{z_i}{5} \ \forall i \in A, \ t \in [ta_i, ta_i + l_i] \\ P_i^t = 0 \ \forall i \in A, \ t \in T / [ta_i, ta_i + l_i] \end{cases} \tag{24.10}$$

Each row of the matrix represents the power schedule of certain appliance, and each column refers to the interval of time from 1 to 120. So, for an appliance i, the power consumption in time interval between ta_i and $ta_i + l_i$ is equal to $(z_i/5)$; outside this interval, no power consumption is considered. The total power consumption at each time slot can be found by summing the power consumption value in each column. The daily total power consumption can be represented as:

$$P_{Total} = \sum_{i=1}^{A} P_i^t \tag{24.11}$$

24.4.2 Solar panels generation model

In the recent years, much attention has been given to the integration of RES in residential areas. RESs are considered as inexhaustible sources that can replace the traditional energy generation sources with zero carbon emission and low-price energy producers. Among the whole RESs that exist in nature, solar energy represents the most abundant source. According to a study in Solar Energy, the Earth receives 174 PW of incoming solar radiation. Thirty percent of the emitted radiation is reflected to space while the rest is absorbed by oceans and the masses. The total solar energy that reaches the Earth is so great that just in 1 year, the total solar energy is about twice as much as obtained from all the nonrenewable resources of the Earth, such as coal, petroleum, natural gas, and uranium extracted combined.

The techniques for directly capturing this energy can be classified between passive solar, photovoltaic solar, and thermal solar. Passive solar is the oldest use of solar energy that consists in taking advantage from the direct contribution of solar radiation to gain a significant part of the heating and lighting of the building. Solar thermal energy is a form of solar energy that use thermal from solar radiation with the aim to heat a liquid or gas. Photovoltaic solar energy is the electricity produced by transforming part of the solar radiation by a solar panel to meet local needs (in combination with storage means) or to be injected into an electrical distribution network (storage is being then not necessary).

Recent trends in renewable energy aim to produce locally the electricity needs. Photovoltaic system installation in the roofs of houses is one of the adopted solutions. In SG, the consumer participates actively in the local energy generation; the consumer becomes prosumer. Solar panels are integrated in residential areas to reduce the consumer electricity bills along with the grid stability. Photovoltaic system integration in a house requires only one-time investment; however, the energy produced is not regular; it varies according to the seasons, weather, and location.

In smart houses, solar panels are prioritized for providing electricity to loads in times of power generation to avoid the use of the storage system or the main grid. However, to cover the needs in electricity, when the power generated by the solar panels is insufficient or unavailable, the main grid takes over. In this regard, the proposed approach tries to maximize the benefits of the solar panels and minimize the electricity bills. Then, based on the hourly radiation and temperature, the total energy produced from the solar panels can be obtained as follows (Shirazi & Jadid, 2015):

$$E_t^{PV} = \eta^{PV} \times A^{PV} \times I_{r,t} \times \left(1 - 0.005\left(T_t^a - 25\right)\right) \tag{24.12}$$

where η^{PV} is the energy conversion efficiency of the PV system (%), A^{PV} is the area of the generator (m^2), $I_{r,t}$ is the solar irradiance (kW/m^2) at time t, and T_t^a is the outdoor temperature (°C) at time t.

24.4.3 Energy storage system model

Energy storage is at the heart of current challenges in SG, whether for optimizing energy resources or promoting access to them. Energy storage consists in preserving an amount of energy for later use. It adjusts the production and consumption of energy by limiting losses. The energy stored, when its availability exceeds needs, can be returned at a time when demand turns out to be greater. Faced with the intermittence or fluctuation in the production of certain energies, for example, renewable ones, this operation also makes it possible to meet a constant demand.

There are several technologies for energy storage, among others: storage in the form of chemical energy, thermal energy, mechanical energy, and storage in the form of electrochemical energy. Energy storage in electrochemical batteries is the most common technique for small amounts of electrical energy especially in case of electricity produced from renewable energies in smart homes. Depending on the type of battery (lead—acid, lithium-ion, sodium—sulfur, etc.), different chemical reactions are caused by electricity; this is the charge phase of the battery. Depending on demand, reverse chemical reactions then generate electricity and discharge the system.

Electrochemical batteries are often intended for portable applications. Of relatively low power, they nevertheless have a large storage capacity for long discharge times (up to several hours) with a yield rate of 70%—80%. With stored energy values of a few watt-hour up to 40 MWh, these devices can have backup functions to cover the user energy needs. Their main disadvantages are: high prices, possible environmental hazards, limited life cycle, and voltage and current limitations.

The proposed HEMS integrates battery storage system as a second source of energy to handle the residential loads, where solar energy is not available. The state of charge (SOC) of the battery storage can be modeled by the following equation (Adika & Wang, 2014).

$$\text{SOC}(t) = \text{SOC}(t-1) + \frac{E_{bc}^t}{E_{batt}} - \frac{E_{bd}^t}{E_{batt}} \tag{24.13}$$

This formulation indicates the level of charge of the battery related to its capacity. E_{bc}^t and E_{bd}^t are the energy charging from solar panels and discharged through the loads at time slot t. It can be calculated as follows:

$$E_{bc}^t = \eta_c \times P_{bc}^t \times \tau \tag{24.14}$$

TABLE 24.2 Battery specifications.

Battery storage parameters	Specifications
E_{batt}	3 kWh
$\eta_{c,d}$	85%, 95%
SOC^{min}, SOC^{max}	30%, 90%
$P_{bc}{}^{min}$, $P_{bc}{}^{max}$	0,20% of the rate capacity
$P_{bd}{}^{min}$, $P_{bd}{}^{max}$	0,20% of the rate capacity

and

$$E_{bd}^t = \frac{P_{bd}^t}{\eta_d} \times \tau \tag{24.15}$$

where η_c and η_d are the battery efficiency charging and discharging, respectively. P_{bc}^t and P_{bd}^t are the power battery charging and discharging. τ is the time slot.

In order to maintain a long life cycle of the batteries, the energy stored in the ESS should not exceed the limits defined by the manufacturer. In other words, the $SOC(t)$ is bounded within a maximum value denoted by SOC^{max} and a minimum state of charge denoted by SOC^{min} as in the following formulation:

$$SOC^{min} \leq SOC(t) \leq SOC^{max} \tag{24.16}$$

The battery parameters are listed in Table 24.2.

24.4.4 Objective function formulation

As it has been mentioned earlier, the purpose of the proposed work is to propose an optimal approach for scheduling the electricity usage pattern for residential appliances for 1 day. The householder's objectives change. There are users whose first and last goal is to minimize the electricity payment. Thus the objective function can be modeled as:

$$Min \sum_{t=1}^{120} prc_t \times P_{Total} \tag{24.17}$$

where prc_t presents the electricity price at time slot t, and P_{Total} is the total Power consumption. However, there are consumers whose main concern is to maximize their satisfaction without caring about the electricity payment. Accordingly, the objective function can be expressed as follows:

$$Min \sum_{i=1}^{A} \gamma^{WTR_i} \tag{24.18}$$

Combining Eqs. (24.17) and (24.18), the minimization of both the electricity payments and waiting time can be defined as follows:

$$Min \ \omega_1 \left(\sum_{t=1}^{120} prc_t \times P_{Total} \right) + \omega_2 \left(\sum_{i=1}^{n} \gamma^{WRT_i} \right) \tag{24.19}$$

where ω_1 and ω_2 are the weights that indicates respectively the importance of the electricity payments as shown in Eq. (24.17), and the WTR as shown in Eq. (24.18), respectively. $\omega_1 + \omega_2 = 1$ and $\omega_1, \omega_2 \in [0,1]$

For standardization needs, the final optimization formula can be expressed as:

$$Min \ \omega_1 \frac{\left(\sum_{t=1}^{120} prc_t \times P_{Total} \right)}{\left(\sum_{t=1}^{120} prc_t \times P_{Total} \right)_{max}} + \omega_2 \frac{\left(\sum_{i=1}^{n} \gamma^{WTR_i} \right)}{\left(\sum_{i=1}^{n} \gamma^{WTR_i} \right)_{max}} \tag{24.20}$$

Subject to:

$$P_{Total}(t) = E_{PV}(t) + E_{ESS}(t) + E_{grid}(t) \qquad (24.21)$$

$$ts_i \le ta_i \le te_i - l_i \qquad (24.22)$$

$$\forall t = 1, 2, \ldots, 120 \qquad (24.23)$$

$$\forall i = 1, 2, \ldots, n \qquad (24.24)$$

The objectives of the scheduling problem are to minimize the electricity cost as well as the waiting time to start operating the appliances, subject to the constraints (24.21)–(24.24). The hourly energy demand should be fulfilled by the grid, solar panels, and the battery storage (Eq. 24.21). Eq. (24.22) shows the interval of time to start turning any appliance. In Eqs. (24.23) and (24.24), the total time horizons of the simulation, as well as the total number of the appliances are given.

MILP has been chosen to deal with the optimization problem in this work using the MATLAB Optimization Toolbox, which includes intlinprog function. Based on the users' preferences, a day-ahead electricity pricing, and a predictive solar generation, the program is run to find an optimal pattern for power usage in a smart home during 1 day to optimally utilize the utility grid energy as well as solar panels and battery storage energy. The flowchart of the proposed approach is given in Fig. 24.3.

24.5   Results and discussion

In this section, the simulation results are discussed. The objective functions detailed previously such as cost and discomfort with or without solar panels and battery storage are treated. The proposed DSM technique performs optimal scheduling of the electricity usage pattern for the residential loads in a single smart home during 1 day. The proposed model is tested in MATLAB environment, which is suitable for optimization techniques, and the obtained results are presented in the form of graphs for better understanding.

In this work, seven types of AOAs are considered to simulate the proposed approach, since the residents may not use them all in 1 day. Therefore all of these appliances are considered in the simulation. The appliance specifications of these AOAs are listed in Table 24.3. Three other MOAs are added to the simulation. The taken MOAs are supposed to operate in a relative time slot during a probable OTI as detailed in Table 24.4. The fridge and the freezer that operate with a fixed power (2.5 and 2 kWh, respectively) throughout the day are also added to the simulation for more realistic results. A day-ahead electricity pricing is adopted in this work to anticipate the electricity usage pattern for residential appliances. Fig. 24.4 shows an example of day-ahead electricity pricing.

24.5.1   Basic scenario: the main grid provides the whole power need

In this scenario, only the main grid is considered to supply the residential electricity needs. Three cases are discussed in this scenario. In the first case, the householder is considered as a traditional user who does not own the EMC and the smart appliances, and thus the user does not aim at any optimization strategy. This case is considered as a reference one. Consequently, the user does not allow any restrictions for using the home appliances at any time the user wants, without waiting. In this case, the appliances are supposed to start operating at the beginning of the scheduling horizon without caring about the price of the electricity. In other words, if the householder wakes up in the morning and wants to have a shower immediately, the householder turns on the water heater without any attention to know in advance the suitable hourly time corresponding to a minimum electricity tariff. Fig. 24.5 depicts the loads scheduling according to the reference case.

In the second case, the user owns the EMC and the smart appliances. For economic reasons, the user switches to minimizing electricity bill mode. In this case, the cost weight attends its maximum value in Eq. (24.20) $\omega_1 = 1$. The program run to allocate the use of the appliances in time slots where the electricity cost is lower. Thing that influences a lot the behavior of the user, for example, instead of having the shower in the morning when the electricity price is high, the user can program it at night when the cost of electricity is low. And instead of turning on the rice cooker at the peak of electricity pricing, the user can delay its use until the price gets lower. The householder knows in advance that the user should wait; sometime from a few minutes to several hours, which negatively affects the user's comfort, but puts forward the user's electricity payment. Fig. 24.6 shows the energy consumption managed according to the cost minimization mode.

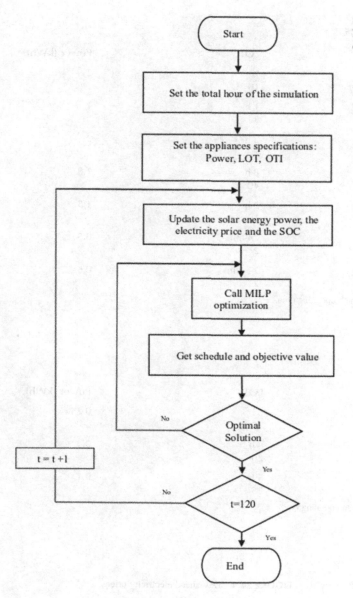

FIGURE 24.3 Flowchart of the proposed scheduling model. *LOT*, length operating time; *OTI*, operating time interval; *SOC*, state of charge; *MILP*, mixed integer linear programming.

Unlike the two previous cases, the third case aims to minimize both the electricity bills and the waiting time, therefore $\omega_1 = 0.5$ and $\omega_2 = 0.5$, and such case is provided to exhibit the performance of the proposed method as an optimal scheduling method for the householder that take into consideration both the waiting time and the electricity cost. The appliances do not work at the beginning of the proposed starting time and postponed several time slots to start operating but not as long as getting upset while waiting for an appliance to start operating. Fig. 24.7 depicts the optimal schedule for each appliance.

Figs. 24.8 and 24.9 show the simulation results for the daily energy consumption and daily electricity payment under the three cases, respectively. As a comparison between the three cases, it can be seen that, for the first case, the home appliances can operate at high pricing periods since the user has no attention of optimization. For the second case, the energy consumption is postponed from high pricing periods to low pricing periods. However, in the third case the user divides the energy consumption all over the day.

Table 24.5 summarizes the daily total cost and waiting time for the three cases. Yet, the traditional user maximizes the satisfaction in terms of avoiding the waiting time, but it is observed that the electricity payment is more expensive. On the other hand, the second case is characterized by the lowest electricity payment. In contrary, the highest waiting time is scored, which greatly influences the user's discomfort.

TABLE 24.3 Automatic operated appliances' specifications.

AOA	OTI	LOT	Power (kWh)
Washing machine	10–19 h 50–95	2 h 10	2
Cloth dryer	10–19 h 50–95	1 h 5	1
Dishwasher	12–23 h 60–115	2 h 10	2
Air conditioner	6–23 h 30–115	8 h 40	1.8
Water heater	6–23 h 30–115	2 h 10	1.5
Rice cooker 1	12–14 h 60–70	24 min 2	0.5
Rice cooker 2	17–20 h 85–100	24 min 2	0.5

AOA, automatic operated appliance; *OTI*, operating time interval; *LOT*, length operating time.

TABLE 24.4 Manual operated appliances' specifications.

MOA	OTI	LOT	Power (kWh)
Lighting	17–24 h 85–120	6 h 30	0.24
TV	18–24 h 90–120	4 h 20	0.4
Computer	18–24 h 90–120	5 h 25	0.1

MOA, manual operated appliances; *OTI*, operating time interval; *LOT*, length operating time.

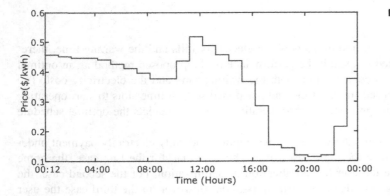

FIGURE 24.4 Day-ahead electricity prices.

It can be deduced from the two cases, that an increase in the daily electricity payment results in a reduction of the total daily waiting time. However, the third case achieves the desired trade-off between the electricity payment and the total waiting time. As a result, the user will not be affected neither by the high payment of electricity nor by the waiting time to start operating the appliances.

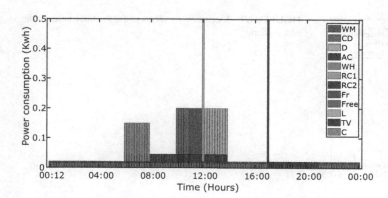

FIGURE 24.5 Power consumption of the appliances under the reference case in the first scenario.

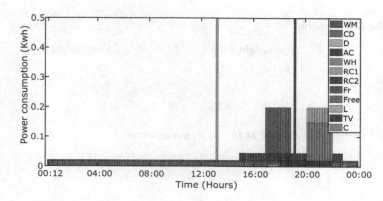

FIGURE 24.6 Power consumption of the appliances under the cost minimization case in the first scenario.

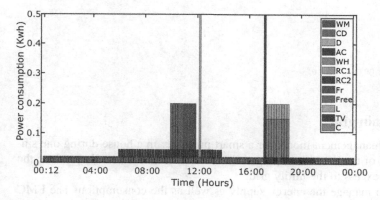

FIGURE 24.7 Power consumption of the appliances under the optimal case in the first scenario.

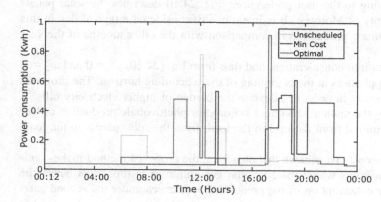

FIGURE 24.8 Power consumption under the three cases in the first scenario.

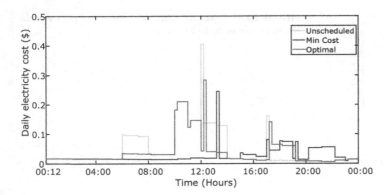

FIGURE 24.9   Electricity payments under the three cases in the first scenario.

TABLE 24.5 Comparison between the two cases in the first scenario.

	Cost weight = 0	Cost weight = 0.5	Cost weight = 1
Electricity cost ($)	5.79	4.65	3.2
Waiting time (h)	0	17	51.2

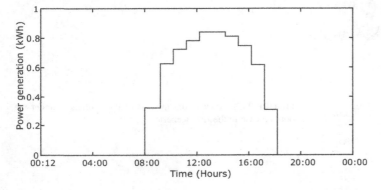

FIGURE 24.10   Solar panel generation.

24.5.2   Second scenario: solar panels and main grid

In this scenario, we proposed a demand-side energy management model for a smart prosumer in a house during one single day. In addition to the main grid supply, the roof of the house is equipped by the solar panels to locally produce the energy to cover the electrical needs and transmit the excess to the utility grid.

In addition, the householder disposes the EMC to manage the energy supply as well as the consumption. The EMC receives information about the electricity pricing for the next day as well as a prediction about the solar panel's generation. Then, it adjusts the energy consumption according to the user preferences. Fig. 24.10 describes the solar panels generation profile during a day of January in Rabat city of Morocco. It is intentionally considered a typical day in this month, which generally experiences minimal photovoltaic production in comparison with the other months of the year to test the effectiveness of the proposed approach.

In the first case, only the users' discomfort is taken into consideration, and thus from Eq. (24.20) $\omega_1 = 0$ and $\omega_2 = 1$. The program is run to schedule the operation of the appliances at the beginning of the scheduling horizon. The proposed algorithm supports the use of the energy produced locally in order to alleviate the burden of higher electricity bills. If the solar energy is not available, the EMC switches to the main grid without waiting for photovoltaic production or low electricity prices. Fig. 24.11 displays the power transmitted from the grid to the load, from the solar panels to the load, and from the solar panels to the main grid.

The second case aims at minimizing the electricity cost. To achieve this objective, the proposed method makes optimal schedules for the appliances to start operating in hours where there is solar generation or at time slots where the electricity price is low. Fig. 24.12 exhibits the power consumption of the predefined appliances under the second case.

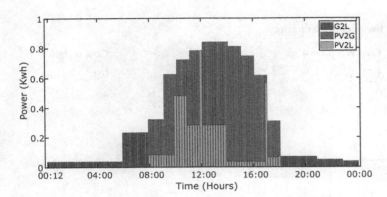

FIGURE 24.11 Power transmission under the first case in the second scenario.

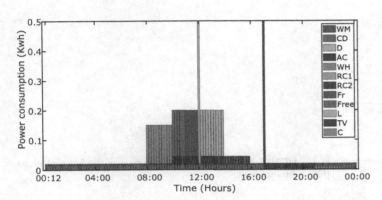

FIGURE 24.12 Power consumption of the appliances under the second case of the second scenario.

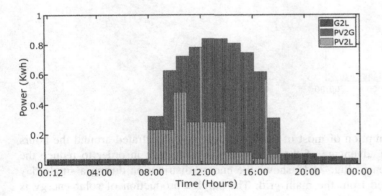

FIGURE 24.13 Power transmission under the second case in the second scenario.

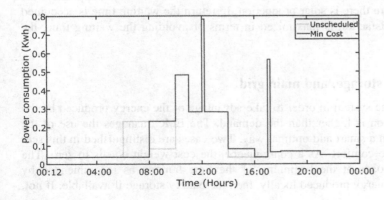

FIGURE 24.14 Power consumption under the two cases in the second scenario.

TABLE 24.6 Comparison between the two cases in the second scenario.

	Cost weight = 0	Cost weight = 1
Electricity cost ($)	1.71	0.9
Waiting time (h)	0	4

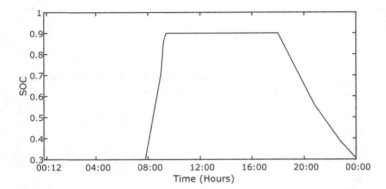

FIGURE 24.15 Battery SOC under first case in the third scenario.

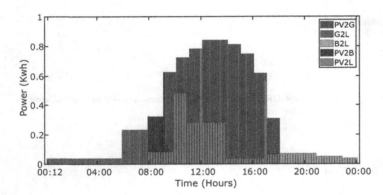

FIGURE 24.16 Power transmission under the first case in the third scenario.

It can be seen from the graph that the energy consumption of most of the appliances is concentrated around the hours when there is photovoltaic production so as to take advantage of the energy produced locally in order to reduce the electricity payment and achieves energy self-sufficiency. Fig. 24.13 shows the energy distribution during a single day for the second case. It displays the energy subtracted from the main grid. The surplus production of solar energy is injected into the utility grid.

As it can be seen on the energy consumption profiles exposed in Fig. 24.14, the use of the residential appliances for the second case is postponed into time intervals where there is solar production. In return the waiting time is accounted as detailed in Table 24.6. However, in case the satisfaction is maximized in terms of avoiding the waiting time, the electricity payment is still considered.

24.5.3   Third scenario: solar panels, battery storage, and main grid

In this scenario, battery storage system is added to the system in order to take advantage of the energy produced by the solar panels in time slots where the power generation is higher than the demand. The EMC manages the use of the sources in order to cover the user's electrical needs in a smart and optimal way. Two cases are distinguished in this scenario. In the first case, the user cares only about the comfort. As a consequence, the cost weight equals to zero. The scheduling program is run to start operating the loads at the beginning of the time horizon as recommended by the householder. The EMC promotes the use of the energy produced locally, then the battery storage if available. If not,

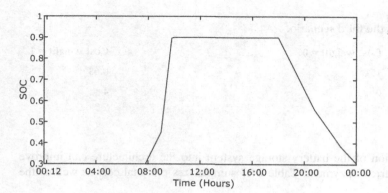

FIGURE 24.17 Battery SOC under second case in the third scenario.

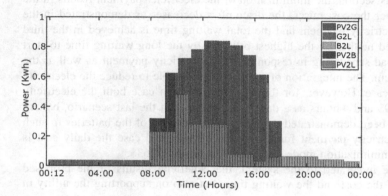

FIGURE 24.18 Power transmission under second case in the third scenario.

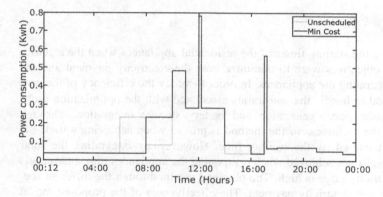

FIGURE 24.19 Power consumption under the two cases in the third scenario.

the EMC switches directly to the main grid without waiting. The batteries are supposed to be empty at the beginning of the day, then start charging from the solar panels and discharging to meet the appliance requirement where there is no solar generation. The battery SOC is shown in Fig. 24.15. To find out the energy distribution of the user all over the day, Fig. 24.16 illustrates the energy transmitted from the solar panels to the appliances, the battery storage, and grid as well as the energy transmitted from the batteries and the grid to the appliances.

In the second case, the user aims at maximizing the self-sufficiency and reducing the electricity bills. Unlike the previous case, the cost weight attends the maximum value. For that reason, the EMC runs to schedule the appliances at time horizon where there is solar panels generation or where the batteries can cover the appliance requirements. If not, the EMC can schedule the use of the appliances during the lowest price horizon. Fig. 24.17 shows the battery SOC; as it can be seen, the SOC is between maximum and minimum value. At the beginning of the day the batteries are empty and start charging during solar generation power to supply the loads during the highest price periods. Fig. 24.18 shows the energy used from the solar panels, the batteries, and the grid to satisfy the load requirement as well as the energy injected into the main to supply the utility in time slots between 10 and 18. Fig. 24.19 gives the energy consumption profiles under the two cases over the day.

TABLE 24.7 Comparison between the two cases in the third scenario.

	Cost weight = 0	Cost weight = 1
Electricity cost ($)	1.43	0.63
Waiting time (h)	0	4

In this scenario, it is observed that the integration of the battery storage system into the architecture can improve load scheduling in response to the minimizing electricity payment. Table 24.7 summarizes the total cost as well as the total waiting time for both cases.

As discussed through all the different parts of this section, the minimization of the electricity payment results in the users' discomfort in terms of waiting time. Whenever the cost raises, the discomfort decreases as demonstrated in the first scenario. However, a trade-off between the electricity payment and the total waiting time is achieved in the third case so that the householder will not be disappointed neither by the highest bills nor by the long waiting time to start operating the appliances. In order to improve the load scheduling in response to the electricity payment as well as the total waiting time, solar panels are added to the system. The integration of solar panels was able to reduce the electricity payment from 5.79$ to 1.71$ for the unscheduled case. However, for the cost minimization case both the electricity payment and the waiting time are reduced up to 0.9$ and 4 hours as a daily waiting time. In the last scenario, battery storage system is added into the architecture. It has been demonstrated that the implementation of the batteries in such a system ensures better results in reducing the electricity payment further. In the unscheduled case the daily cost is reduced up to 1.43$ per day. However, for the cost minimization mode, the daily cost reaches 0.63$.

In regard to scenarios where RES and ESS are integrated to the system, the beneficial results of the proposed method are not limited only on the minimization of the cost and the waiting time, but also on supporting the utility in time slots where solar energy is more than the demand.

24.6 Conclusion

In this research work, a novel approach for scheduling the starting times of the residential appliances when the electricity tariff is announced in advance is proposed. The objectives were to minimize both the electricity payment and the user discomfort caused by the waiting time to start operating the appliances. In order to verify the efficiency of the proposed model, a number of simulations were performed to handle the constraints associated with the optimization problem under different scenarios with and without solar panel generation and battery storage integration. Through simulation in case only the main grid is considered, the usefulness of the method is proved when achieving a trade-off between the electricity payment and the discomfort caused by the waiting time. However, by integrating the solar panels as well as the battery storage system, better results are achieved when postponing the appliances into solar panels production periods and using batteries when the electricity price is high. The obtained results through the different scenarios were compared in terms of daily waiting time and electricity payment. The effectiveness of the proposed model is not beneficial only for the user but also for the utility grid.

Future works are investigating in including other energy sources such as wind turbine, heat power combined by the solar panels, and battery storage system and extend the proposed model not only at a single house level but also at the district and city level.

References

Abbasi, B. Z., Javaid, S., Bibi, S., Khan, M., Malik, M. N., Butt, A. A., & Javaid, N. (2017). Demand side management in smart grid by using flower pollination algorithm and genetic algorithm. In *International conference on P2P, parallel, grid, cloud and Internet computing* (pp. 424–436).

Abujubbeh, M., Al-Turjman, F., & Fahrioglu, M. (2019). Software-defined wireless sensor networks in smart grids: An overview. *Sustainable Cities and Society*, *51*, 101754.

Abushnaf, J., Rassau, A., & Górnisiewicz, W. (2016). Impact on electricity use of introducing time-of-use pricing to a multi-user home energy management system. *International Transactions on Electrical Energy Systems*, *26*(5), 993–1005.

Adika, C. O., & Wang, L. (2014). Smart charging and appliance scheduling approaches to demand side management. *International Journal of Electrical Power & Energy Systems*, *57*, 232–240.

Ahmed, M. S., Mohamed, A., Homod, R. Z., & Shareef, H. (2016). Hybrid LSA-ANN based home energy management scheduling controller for residential demand response strategy. *Energies, 9*(9), 716.

Ahmed, M. S., Mohamed, A., Shareef, H., Homod, R. Z., & Ali, J.A. (2016). Artificial neural network based controller for home energy management considering demand response events. In *2016 international conference on advances in electrical, electronic and systems engineering (ICAEES)* (pp. 506—509).

Al Hasib, A., Nikitin, N., & Natvig, L. (2015). Cost-comfort balancing in a smart residential building with bidirectional energy trading. In *2015 sustainable Internet and ICT for sustainability (SustainIT)* (pp. 1—6).

Al-Jabery, K., Xu, Z., Yu, W., Wunsch, D. C., Xiong, J., & Shi, Y. (2016). Demand-side management of domestic electric water heaters using approximate dynamic programming. *IEEE Transactions on Computer-Aided Design of Integrated Circuits and Systems, 36*(5), 775—788.

Amini, M. H., Frye, J., Ilić, M. D., & Karabasoglu, O. (2015). Smart residential energy scheduling utilizing two stage mixed integer linear programming. In *2015 North American Power Symposium (NAPS)* (pp. 1—6).

Arshad, H., Batool, S., Amjad, Z., Ali, M., Aimal, S., & Javaid, N. (2017). Pigeon inspired optimization and enhanced differential evolution using time of use tariff in smart grid. In *International conference on intelligent networking and collaborative systems* (pp. 563—575).

Aslam, S., Bukhsh, R., Khalid, A., Javaid, N., Ullah, I., Fatima, I., & Hasan, Q. U. (2017). An efficient home energy management scheme using cuckoo search. In *International conference on P2P, parallel, grid, cloud and Internet computing* (pp. 167—178).

Awais, M., Javaid, N., Shaheen, N., Iqbal, Z., Rehman, G., Muhammad, K., & Ahmad, I. (2015). An efficient genetic algorithm based demand side management scheme for smart grid. In *2015 18th international conference on network-based information systems* (pp. 351—356).

Baviskar, A., Baviskar, J., Wagh, S., Mulla, A., & Dave, P. (2015). Comparative study of communication technologies for power optimized automation systems: A review and implementation. In *2015 fifth international conference on communication systems and network technologies* (pp. 375—380).

Boynuegri, A. R., Yagcitekin, B., Baysal, M., Karakas, A., & Uzunoglu, M. (2013). Energy management algorithm for smart home with renewable energy sources. In *4th international conference on power engineering, energy and electrical drives* (pp. 1753—1758).

Bradac, Z., Kaczmarczyk, V., & Fiedler, P. (2015). Optimal scheduling of domestic appliances via MILP. *Energies, 8*(1), 217—232.

Buildings and energy efficiency (2019). Available online: http://www.mhpv.gov.ma/?page_id = 3605 (accessed 01-05-2020).

Choi, I. H., Yoo, S. H., Jung, J. H., Lim, M. T., Oh, J. J., Song, M. K., & Ahn, C. K. (2015). Design of neuro-fuzzy based intelligent inference algorithm for energy-management system with legacy device. *Transactions of the Korean Institute of Electrical Engineers, 64*(5), 779—785.

Derakhshan, G., Shayanfar, H. A., & Kazemi, A. (2016). The optimization of demand response programs in smart grids. *Energy Policy, 94*, 295—306.

Dileep, G. (2020). A survey on smart grid technologies and applications. *Renewable Energy, 146*, 2589—2625.

Dutta, G., & Mitra, K. (2017). A literature review on dynamic pricing of electricity. *Journal of the Operational Research Society, 68*(10), 1131—1145.

El Idrissi, R. N., Ouassaid, M., & Maaroufi, M. (2018). Demand side management strategy by optimal day-ahead load shifting in smart grid. In *2018 6th international renewable and sustainable energy conference (IRSEC)* (pp. 1—6).

Eltigani, D., & Masri, S. (2015). Challenges of integrating renewable energy sources to smart grids: A review. *Renewable and Sustainable Energy Reviews, 52*, 770—780.

Esther, B. P., & Kumar, K. S. (2016). A survey on residential demand side management architecture, approaches, optimization models and methods. *Renewable and Sustainable Energy Reviews, 59*, 342—351.

Ettalbi, K., Elabd, H., Ouassaid, M., & Maaroufi, M. (2016). A comparative study of energy management systems for PV self-consumption. In *2016 international renewable and sustainable energy conference (IRSEC)* (pp. 1086—1010).

Garroussi, Z., Ellaia, R., Talbi, E. G., & Lucas, J. Y. (2017). Hybrid evolutionary algorithm for residential demand side management with a photovoltaic panel and a battery. In 2017 *international conference on control, artificial intelligence, robotics & optimization (ICCAIRO)* (pp. 4—10).

Gelazanskas, L., & Gamage, K. A. (2014). Demand side management in smart grid: A review and proposals for future direction. *Sustainable Cities and Society, 11*, 22—30.

Gudi, N., Wang, L., & Devabhaktuni, V. (2012). A demand side management based simulation platform incorporating heuristic optimization for management of household appliances. *International Journal of Electrical Power & Energy Systems, 43*(1), 185—193.

Hong, Y. Y., Lin, J. K., Wu, C. P., & Chuang, C. C. (2012). Multi-objective air-conditioning control considering fuzzy parameters using immune clonal selection programming. *IEEE Transactions on Smart Grid, 3*(4), 1603—1610.

Hossain, M. S., Madlool, N. A., Rahim, N. A., Selvaraj, J., Pandey, A. K., & Khan, A. F. (2016). Role of smart grid in renewable energy: An overview. *Renewable and Sustainable Energy Reviews, 60*, 1168—1184.

Jaramillo, L. B., & Weidlich, A. (2016). Optimal microgrid scheduling with peak load reduction involving an electrolyzer and flexible loads. *Applied Energy, 169*, 857—865.

Javaid, N., Javaid, S., Abdul, W., Ahmed, I., Almogren, A., Alamri, A., & Niaz, I. A. (2017). A hybrid genetic wind driven heuristic optimization algorithm for demand side management in smart grid. *Energies, 10*(3), 319.

Javaid, N., Ullah, I., Akbar, M., Iqbal, Z., Khan, F. A., Alrajeh, N., & Alabed, M. S. (2017). An intelligent load management system with renewable energy integration for smart homes. *IEEE Access, 5*, 13587—13600.

Khan, A. R., Mahmood, A., Safdar, A., Khan, Z. A., & Khan, N. A. (2016). Load forecasting, dynamic pricing and DSM in smart grid: A review. *Renewable and Sustainable Energy Reviews, 54*, 1311—1322.

Lee, J. Y., & Choi, S. G. (2014). Linear programming based hourly peak load shaving method at home area. In *16th international conference on advanced communication technology* (pp. 310—313).

Liu, Y., Yuen, C., Yu, R., Zhang, Y., & Xie, S. (2015). Queuing-based energy consumption management for heterogeneous residential demands in smart grid. *IEEE Transactions on Smart Grid, 7*(3), 1650—1659.

Logenthiran, T., Srinivasan, D., & Shun, T. Z. (2012). Demand side management in smart grid using heuristic optimization. *IEEE Transactions on Smart Grid, 3*(3), 1244–1252.

Ma, K., Yao, T., Yang, J., & Guan, X. (2016). Residential power scheduling for demand response in smart grid. *International Journal of Electrical Power & Energy Systems, 78*, 320–325.

Moon, J. W., & Kim, J. J. (2010). ANN-based thermal control models for residential buildings. *Building and Environment, 45*(7), 1612–1625.

Muralitharan, K., Sakthivel, R., & Shi, Y. (2016). Multiobjective optimization technique for demand side management with load balancing approach in smart grid. *Neurocomputing, 177*, 110–119.

Muratori, M., & Rizzoni, G. (2015). Residential demand response: Dynamic energy management and time-varying electricity pricing. *IEEE Transactions on Power systems, 31*(2), 1108–1117.

Ogunjuyigbe, A. S. O., Ayodele, T. R., & Akinola, O. A. (2017). User satisfaction-induced demand side load management in residential buildings with user budget constraint. *Applied Energy, 187*, 352–366.

Ogunjuyigbe, A. S. O., Ayodele, T. R., & Oladimeji, O. E. (2016). Management of loads in residential buildings installed with PV system under intermittent solar irradiation using mixed integer linear programming. *Energy and Buildings, 130*, 253–271.

Qayyum, F. A., Naeem, M., Khwaja, A. S., Anpalagan, A., Guan, L., & Venkatesh, B. (2015). Appliance scheduling optimization in smart home networks. *IEEE Access, 3*, 2176–2190.

Rahim, S., Javaid, N., Ahmad, A., Khan, S. A., Khan, Z. A., Alrajeh, N., & Qasim, U. (2016). Exploiting heuristic algorithms to efficiently utilize energy management controllers with renewable energy sources. *Energy and Buildings, 129*, 452–470.

Rajarajeswari, R., Vijayakumar, K., & Modi, A. (2016). Demand side management in smart grid using optimization technique for residential, commercial and industrial load. *Indian Journal in Science and Technology, 9*(43), 1–7.

Rasheed, M. B., Javaid, N., Ahmad, A., Khan, Z. A., Qasim, U., & Alrajeh, N. (2015). An efficient power scheduling scheme for residential load management in smart homes. *Applied Sciences, 5*(4), 1134–1163.

Samadi, P., Wong, V. W., & Schober, R. (2015). Load scheduling and power trading in systems with high penetration of renewable energy resources. *IEEE Transactions on Smart Grid, 7*(4), 1802–1812.

Setlhaolo, D., Xia, X., & Zhang, J. (2014). Optimal scheduling of household appliances for demand response. *Electric Power Systems Research, 116*, 24–28.

Shahgoshtasbi, D., & Jamshidi, M. M. (2014). A new intelligent neuro–fuzzy paradigm for energy-efficient homes. *IEEE Systems Journal, 8*(2), 664–673.

Shakouri, H., & Kazemi, A. (2017). Multi-objective cost-load optimization for demand side management of a residential area in smart grids. *Sustainable Cities and Society, 32*, 171–180.

Shirazi, E., & Jadid, S. (2015). Optimal residential appliance scheduling under dynamic pricing scheme via HEMDAS. *Energy and Buildings, 93*, 40–49.

Solar Energy. <https://en.wikipedia.org/wiki/Solar-energy> Accessed 05.05.20.

Supply-side management. <http://africa-toolkit.reeep.org/modules/Module13.pdf> Accessed 01.05.20.

Trygg, L., & Karlsson, B. G. (2005). Industrial DSM in a deregulated European electricity market—a case study of 11 plants in Sweden. *Energy Policy, 33*(11), 1445–1459.

Tuballa, M. L., & Abundo, M. L. (2016). A review of the development of smart grid technologies. *Renewable and Sustainable Energy Reviews, 59*, 710–725.

Ullah, I., Javaid, N., Khan, Z. A., Qasim, U., Khan, Z. A., & Mehmood, S. A. (2015). An incentive-based optimal energy consumption scheduling algorithm for residential users. *Procedia Computer Science, 52*, 851–857.

Wu, Y., Zhang, B., Lu, J., & Du, K. L. (2011). Fuzzy logic and neuro-fuzzy systems: A systematic introduction. *International Journal of Artificial Intelligence and Expert Systems, 2*(2), 47–80.

Yang, R., & Wang, L. (2012). Multi-agent based energy and comfort management in a building environment considering behaviors of occupants. In *2012 IEEE Power and Energy Society general meeting* (pp. 1–7).

Zhang, D., Evangelisti, S., Lettieri, P., & Papageorgiou, L. G. (2016). Economic and environmental scheduling of smart homes with microgrid: DER operation and electrical tasks. *Energy Conversion and Management, 110*, 113–124.

Zhao, Z., Lee, W. C., Shin, Y., & Song, K. B. (2013). An optimal power scheduling method for demand response in home energy management system. *IEEE Transactions on Smart Grid, 4*(3), 1391–1400.

Zhu, Z., Tang, J., Lambotharan, S., Chin, W. H., & Fan, Z. (2012). An integer linear programming based optimization for home demand-side management in smart grid. *In 2012 IEEE PES Innovative Smart Grid Technologies (ISGT)*, 1–5.

Chapter 25

Solar cell parameter extraction using the Yellow Saddle Goatfish Algorithm

K. Mohana Sundaram[1,*], P. Anandhraj[4], Ahmad Taher Azar[2,3] and P. Pandiyan[5]

[1]EEE Department, KPR Institute of Engineerimg and Technology, Coimbatore, [2]Faculty of Computers and Artificial Intelligence, Benha University, Benha, Egypt, [3]College of Computer and Information Sciences, Prince Sultan University, Riyadh, Kingdom of Saudi Arabia, [4]Research Scholar, Anna University, [5]Associate Professor, EEE Department, KPR Institute of Engineering and Technology, Coimbatore

*Corresponding author. Emails: kumohanasundaram@gmail.com, Chennaianandpannerselvam@gmail.com, ahmad.azar@fci.bu.edu.eg, aazar@psu.edu.sa, pandyyan@gmail.com

25.1 Introduction

The use of fossil fuels, such as coal, to generate power has a significant impact on global warming and greenhouse gas emissions; however, renewable energy sources produce electricity that is both safe and abundant in nature. Among the various sustainable energy sources, such as hydro, geothermal, tides, and wind energy, electrical energy obtained from solar is pollution free and less maintenance (Chan & Phang, 1987). In recent years the World's first fully green data center is solar-powered in India. The direct current voltage produced from the solar photovoltaic (PV) module can be connected either directly to the load or through various power converters. The perfect PV modeling is mandatory before the hardware installation part. The PV researchers faced major problems in solar PV cell modeling. The problems are mainly due to (1) identification of PV module parameter in complex and also they find difficult to understand the sudden spikes in irradiance that can affect solar power plants and (2) nonlinear $I-V$ (current−voltage) characteristic of PV module. The preliminary studies indicate that large-scale PV projects are not immune to such events, especially when the spikes last longer than a minute. The different PV models available in the literatures are (1) single-diode (SD) model, (2) double-diode (DD) model (Chan & Phang, 1987; Ishaque & Salam, 2011; Koad, Zobaa, & El-Shahat, 2016), (3) three-diode (TD) model (Liu et al., 2008), (4) SD model with parasitic capacitor (Sakar, Balci, Aleem, & Zobaa, 2018), (5) enhanced DD model (Jervase, Bourdoucen, & Al-Lawati, 2001; Villalva, Gazoli, & Ruppert Filho, 2009), (6) reverse DD model and diffusion-based model (Soon & Low, 2012), and (7) multidiode (MD) model (Askarzadeh & dos Santos Coelho, 2015). The accuracy of the abovementioned models is varied depending on the computed model parameters and also very difficult to figure out to set the global values for those model parameters because of the data changes and impossible to get the data from the manufacturer-provided datasheet. The scenario gets more worse when we have to estimate the model parameters using minimum data offered in the datasheet.

MATLAB/Simulink is identified as the best software for the PV modeling by worldwide researchers. Typically, a limited number of parameters are provided in the solar module datasheet. The parameters, such as series and shunt resistance (R_{se} and R_{sh}), diode saturation current (I_o), photo-generated current (I_{pv}), and ideality factor (a) would not available in the datasheets (Ishaque & Salam, 2011; Koad et al., 2016; Liu et al., 2008; Sakar et al., 2018). Initially, the researchers used numerical techniques, such as Gauss seidal and Newton−Raphson methods, to find these unknown parameters. However, these techniques are time consuming, tedious, and the number of iterations to reach the fitness function is quite high (Askarzadeh & dos Santos Coelho, 2015; Ismail, Moghavvemi, & Mahlia, 2013; Jervase et al., 2001; Soon & Low, 2012; Villalva et al., 2009). Along with the these two techniques, many other optimization techniques such as harmony search algorithm, pattern search algorithm, genetic algorithm (GA), and particle swarm optimization (PSO), were utilized by many of the researchers to find out the unknown parameters (AlHajri, El-Naggar, AlRashidi, & Al-Othman, 2012; Askarzadeh & Rezazadeh, 2013; El-Naggar, AlRashidi, AlHajri, & Al-Othman, 2012; Rajasekar, Kumar, & Venugopalan, 2013; Rao & More, 2017). In GA, the choice of chromosomes from the preliminary

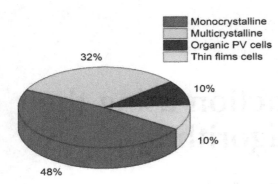

- Monocrystalline
- Multicrystalline
- Organic PV cells
- Thin flims cells

32%

10%

10%

48%

FIGURE 25.1 Usage of different types of solar panels.

population is identified as complicated task. Jaya algorithm has been employed to find the best results for some of the optimization problems (Anandhraj, Sundaram, Sanjeevikumar, & Holm-Nielsen, 2020; Rao, 2016; Rao, Rai, & Balic, 2017; Rao & Saroj, 2017a, 2017b). The variety of solar PV module types and their recent market potential is given in Fig. 25.1. The most commonly used solar panels are monocrystalline and multicrystalline cells. The efficiency and cost of these two panels will vary depending on the size. In fact, the monocrystalline panels are the most expensive and efficient whereas multicrystalline cells give moderate efficiency with reasonable cost.

As the market share and efficiency of both mono- and polycrystalline-type solar panels are high, the aforementioned type panels will be considered for parameter extraction in this chapter. Nevertheless, in the literature, Yellow Saddle Goatfish Algorithm (YSGA) has not been employed for the optimization of solar PV module to obtain the unknown parameters. The application of YSGA is to extract these parameters and it is explained in lucid manner in this chapter. Furthermore, the YSGA results are compared with Jaya algorithm and FPA and also validated with experimental data.

The rest of the section of this proposed chapter is structured as follows: mathematical modeling of the solar cell is presented in Section 25.2. In Section 25.3, YSGA-based solar cell parameter extraction is briefly explained. Results and discussions based on simulation work are discussed in Section 25.4. The experimental work using a SOLAR-4000 analyzer is reported in Section 25.5. Finally, a conclusion is drawn in the last section.

25.2 Solar cell mathematical modeling

The SD model has been used in this chapter, which is represented in Fig. 25.2. The mathematical modeling of solar module is done using MATLAB/Simulink. The practical solar PV model has R_{se} and R_{sh}, which are connected across the diode. The structural resistance (R_s) indicates the contact resistance between p semiconductor layer and metal base and resistance of p and n bodies. Furthermore, R_p depicts the loss in the device due to current leaked through resistive path in parallel (Villalva et al., 2009). The current produced from PV module is given by the following equation:

$$I = I_{PV} - I_D - \frac{V_D}{R_p} \tag{25.1}$$

The equation for diode current is represented as

$$I_D = I_0\left(\exp\left(\frac{V_D}{\alpha V_t}\right) - 1\right) \tag{25.2}$$

where I_0 is the saturation current of diode, and thermal voltage is denoted as V_t that can be written as

$$V_t = N_s kT/q \tag{25.3}$$

where k is the Boltzmann constant, T is the temperature (K), q is the electron charge, and N_s indicates the series-connected solar cells.

The expanded PV module current [Eqs. (25.6—25.9) is given as:

$$I = I_{PV} - I_0\left[\exp\left(\frac{V + R_s I}{V_t a}\right) - 1\right] - (V + R_s I)/R_p \tag{25.4}$$

Hence, these five unknown parameters, such as I_{pv}, R_p, R_s, a, and I_0 in the PV module, are mandatory for accurate solar PV cell modeling. In this proposed research, "I_{pv}" and "I_0" are calculated analytically to reduce the complication.

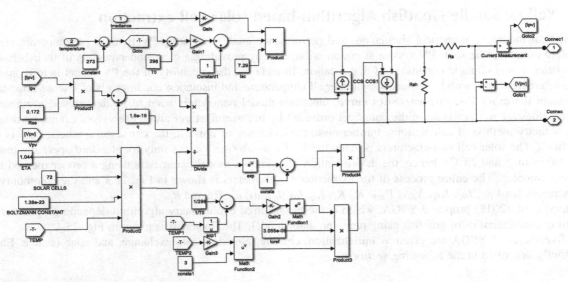

FIGURE 25.2 Mathematical modeling of photovoltaic cell.

The term "a" is selected randomly between 1 and 2 depending on the remaining parameters of the solar module (Askarzadeh & dos Santos Coelho, 2015). The remaining two parameters of "R_s" and "R_p" are computed using three optimization techniques. These two parameters are varying with respect to temperature and irradiance, which is then adjusted to the maximum point to minimize the error between actual and estimated power.

The photo-generated current (I_{pv}) of solar PV module (Villalva et al., 2009) is given by

$$I_{PV} = (I_{SC} + k_i dT) \times G/G_n \tag{25.5}$$

where G_n is the irradiance at standard operating condition and G is the actual solar irradiance. The magnitude of I_0 is based on the I_{PV} and V_{OC} (Ismail et al., 2013).

$$I_0 = I_{PV}/\exp\left(\frac{(V_{OC} + k_v dT) \times V_t}{a}\right) - 1 \tag{25.6}$$

The voltage produced from PV module is V_{MP}, and current through the PV module is I_{MP} when the solar PV curve attains its maximum power point (MPP). In addition to this point, $dp/dv = 0$, that is, the differentiation of power as a function of voltage becomes zero (Ismail et al., 2013). The unknown parameters are extracted using this condition:

$$\frac{dP}{dV} = 0 \tag{25.7}$$

$$\frac{d(V \times I)}{dV} = V\left(\frac{dI}{dV}\right) + I \tag{25.8}$$

$$\frac{dI}{dV} + \left(\frac{I}{V}\right) = 0 \tag{25.9}$$

Eq. (25.10) represents the derivative of current (I) as a function of voltage (V) at MPP where $Z_P = 1/R_P$, $\psi = 1/aV_t$

$$\left(\left(\frac{dI}{dV}\right)|_{(V_{mp}, I_{mp})} = (I_0 \psi \exp\{\psi(V_{mp} + I_{mp}R_s)\} - Z_P\right)/(1 + I_0 \psi R_s \exp\{\psi(V_{mp} + I_{mp}R_s)\} - Z_P R_s \tag{25.10}$$

$$J = |\left(\frac{dI}{dV}\right)|_{V_{mp}, I_{mp}} + \left(\frac{I_{mp}}{V_{mp}}\right) \tag{25.11}$$

The measured value is attained once the fitness function (J) approaches zero value approximately (Ismail et al., 2013).

25.3 Yellow Saddle Goatfish Algorithm-based solar cell extraction

The analytical modeling, numerical simulation, and performance analysis of solar PV module generator are significant tasks before mounting the solar PV system at any location, thereby providing a clear understanding of their behavior and characteristics in real climatic conditions of that location. In addition, the capability of the PV model is to reproduce the $V-I$ characteristics curve with higher accuracy under all temperature and insolation conditions is of intense importance.

To obtain the exact $V-I$ characteristics curve, unknown model parameters need to be determined accurately. The unknown values are not available in the datasheet provided by the manufacturer and these values cannot be determined through ordinary method of calculations. Furthermore, the existence of noise in the extracted synthetic data is made it very difficult. The solar cell manufacturers provide the $V-I$ characteristics curve only for standard operating test conditions (1000 W m^{-2} and $25°$C); hence, the determination of unknown model parameters using a proper method turn out to be more complex. The entire process of the extraction of parameters is shown in Fig. 25.3 and the commonly identified parameters are V_{OC}, I_{SC}, I_{MP}, V_{MP}, P_{MP}, K_i, K_v, I_{ph}, I_{01}, I_{02}, a_1, a_2, R_s, and R_p.

Zaldivar et al. (2018) proposed YSGA, which is the bio-inspired evolutionary algorithm depending upon the collective hunting characteristics of goatfish using prey—predator model. The flowchart is given in Fig. 25.4.

The five stages of YSGA are given as initialization, chasing, blocking, role exchange, and zone change. Each section is briefly described in the following sections:

25.3.1 Stage 1: initialization

In the initial phase, the population size of goatfish (m) in population (s) is distributed in n-dimensional search space as $\{S1, S2, \ldots, Sm\}$. The initialize stage can be made as

$$S_i^j = rand(b_j^{\max} - b_j^{\min}) + b_j^{\min} \tag{25.12}$$

where $i = 1, 2, 3, \ldots, m; j = 1, 2, 3, \ldots n$.

$b_j^{\max}$ and $b_j^{\min}$ represent the boundary ranges. The starting of supportive hunt through forming a cluster group within the neighborhood is the main motive of this stage. Fig. 25.5 gives the insight about the detection and separation between population and kth group, where γ_C and γ_G are the chaser and blocker fish.

25.3.2 Stage 2: chasing

Every cluster (C) chooses a chaser fish (γ_C) from the population (S), which in turn leads to the searching of the prey according to the best fitness value. To figure out the hidden prey, at times chaser fish (γ_C) will change its position from its old position, which can be obtained using the following equation:

$$\gamma_L^{t+1} = \gamma_L^t + \alpha \oplus Levy(\beta) \tag{25.13}$$

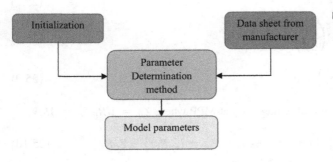

FIGURE 25.3 General process of parameter extraction.

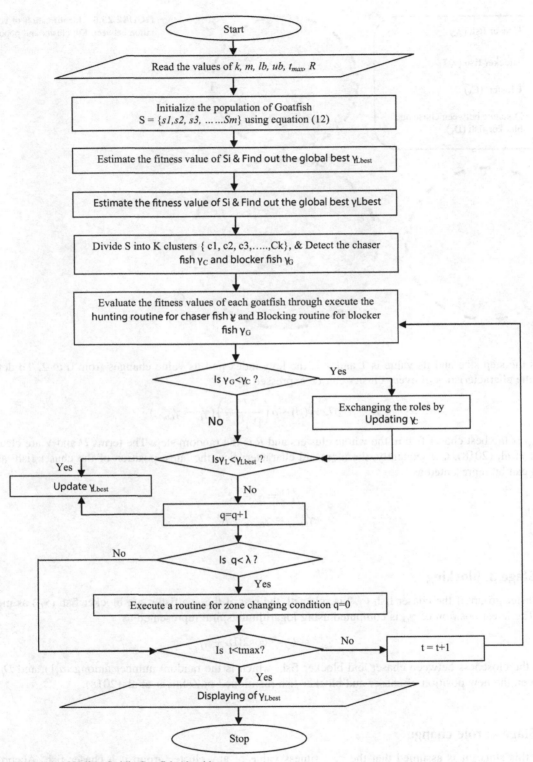

FIGURE 25.4 Flowchart of Yellow Saddle Goatfish Algorithm.

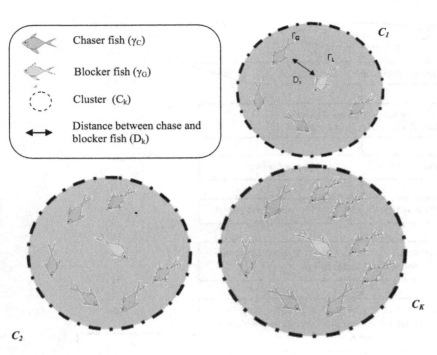

where α is the step size and its value is 1 and β is the levy index and its value changes from 0 to 2. To determine the best prey, the characteristics of every cluster can be expressed as

$$R = \alpha \oplus Levy(\beta) \sim a\left(\frac{U}{|V|^{1//\beta}}\right)(\gamma_L^t - \gamma_{Lbest}^t) \tag{25.14}$$

where γ_{Lbest}^t is the best chaser fish in the whole clusters and R is the random step. The terms U and V are clearly defined in Zaldivar et al. (2018). Consequently, the aforesaid characteristics, the latest position of the chaser fish, and the best chaser fish can be represented as

$$\gamma_L^{t+1} = \gamma_L^t + R \tag{25.15}$$

$$\gamma_{Lbest}^{t+1} = \gamma_{Lbest}^t + R^1 \tag{25.16}$$

25.3.3 Stage 3: blocking

In every cluster group, if the chaser fish (γ_C) is selected, the rest of the goatfishes are blocker fish (γ_G) as mentioned in Fig. 25.5. The latest position of γ_G is computed using logarithmic spiral represented as

$$\gamma_g^{t+1} = D_f e^{bp}\cos2\Pi\rho + \gamma_L \tag{25.17}$$

where ρ is the closeness between chaser and blocker fish, which is the random number among $(a,1)$, and D_f is the distance between the new position of chaser and blocker fish mentioned in Zaldivar et al. (2018).

25.3.4 Stage 4: role change

Initially in this stage, it is assumed that the best fitness value of any cluster group is a chaser fish. According to the behavior at the time of hunting process, the prey goes to the hunting area whereas blocker fish is nearby the prey and directs as a new chaser fish.

25.3.5 Stage 5: zone change

Zone change occurs in this stage when all the preys have been exploited in a cluster. The λ is the overexploitation parameters, which are taken into account during this stage. Therefore the zone change in this algorithm is written in the following equation:

$$S_g^{t+1} = \frac{\gamma_{Lbest} + S_g^t}{2} \tag{25.18}$$

where S_g^{t+1} is the latest position of goatfish.

25.4 Results and discussion

A 250 Wp SVL0250P solar module datasheet gives the data of open circuit voltage, short-circuit current, MPP current, and voltage, k_v and k_i. The unknown parameters are I_{pv}, I_0, R_s, a, and R_p. The parameters I_{pv} and I_0 are calculated analytically and R_s and R_p are accurately measured using YSGA, FPA, and Jaya algorithm.

The beginning value of R_s is selected at randomin between the span of $0-2$ and primary boundary range for R_p between 50 and 500. Table 25.1 represents the measured R_s and R_p values of SVL0250P PV Module using Jaya algorithm, FPO, and YSGA for different environmental conditions.

From Table 25.1, it can be noticed that the value of R_s is minute value and the value of R_p is more. Accordingly, lesser value of R_s and higher value of R_p move the $P-V$ curve toward MPP. At standard operating test condition (STC), the G and T are 1000 W m^{-2} and 25°C, respectively.

Fig. 25.6A–D depicts the simulated $I-V$ and $P-V$ characteristics curve of 250 Wp solar PV module (SVL0250P) for three atmospheric conditions, such as STC, $T = 47.4°C$ and $G = 525$ W m^{-2}, and $T = 45.9°C$ and $G = 368$ W m^{-2}.

From the above results, one can found that at $T = 47.4°C$ and $G = 525$ W m^{-2}, the MPP voltage and current are 27.63 V and 4.2 A, respectively. At $T = 45.9°C$ and $G = 368$ W m^{-2}, the MPP voltage and current are 27.18 V and 3 A, respectively. The graphs obviously indicate that the difference of significant solar PV module parameters, that is,

TABLE 25.1 Solar PV module unknown model parameters measured using YSGA, FPA, and Jaya algorithm for various irradiance (G) and temperature (T).

Parameters	SVL0250P		
	FPO	Jaya algorithm	YSGA
$T = 25°C$ and $G = 1000$ W m^{-2}			
R_s (Ω)	0.91778	0.916811	0.8589
R_p (Ω)	63.254	50.0028	321.821
$T = 47.4°C$ and $G = 525$ W m^{-2}			
R_s (Ω)	0.341673	0.772148	1.2689
R_p (Ω)	123.46722	50.00964	231.6698
$T = 45.9°C$ and $G = 368$ W m^{-2}			
R_s (Ω)	1.92697	1.485236	1.4789
R_p (Ω)	82.63984	50	402.8952

FPA, Flower Pollination Algorithm; *PV*, photovoltaic; *YSGA*, Yellow Saddle Goatfish Algorithm.

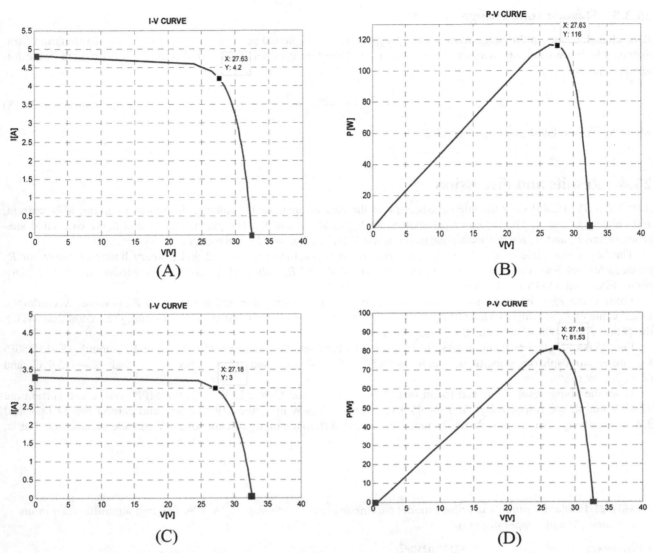

FIGURE 25.6 (A and B) $I-V$ and $P-V$ characteristics for irradiance (G) = 525 W m^{-2} and T = 47.4°C and (C and D) $I-V$ and $P-V$ characteristics for irradiance (G) = 368 W m^{-2} and T = 45.9°C.

V_{OC}, I_{sc}, MPP voltage, and current based on irradiance and temperature. Fig. 25.7 represents the fitness function as function of number iterations for various algorithms, such as YSGA, Jaya algorithm, and FPA.

Based on the convergence curve, it can be noticed that the convergence speed of proposed method (YSGA) is fast compared to Jaya algorithm and FPA. The YSGA has been converged at the 30th iteration, Jaya algorithm has been converged at the 160th iteration, and FPA has been converged at the 180th iteration. The qualitative performance comparison of YSGA, Jaya algorithm, FPA, GA, and PSO is depicted in Table 25.2.

From the preceding summarization of Table 25.2, it can be concluded that YSGA methodology requires fewer stages to provide accurate curve fit in a fair amount of time. The significant advantage of YSGA is that the uncertainty in the control variable is always available as a high merit in the FWA methodology. Moreover, the number of steps required is minimal, which is the secret to YSGA method and makes it an excellent optimization strategy for extracting unknown PV module characteristics.

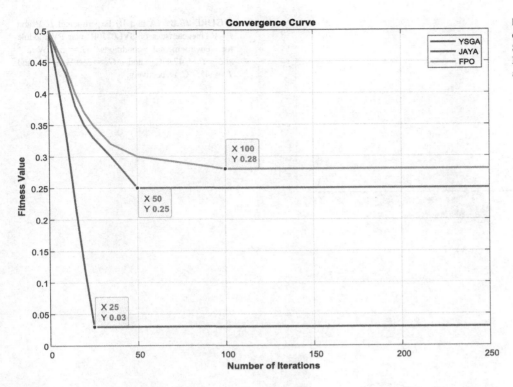

TABLE 25.2 Performance comparison of YSGA, Jaya algorithm, and FPA.

S. no	Evaluation criteria/method	GA	PSO	FPA	Jaya	YSGA
1	Convergence speed	Moderate	Moderate	Moderate	Moderate	High
2	Accuracy	Less	Less	High	High	High
3	Parameter dependency	High	High	Moderate	Moderate	Less
4	Attainment of local convergence	High	High	Moderate	Moderate	Less
5	Randomness in control variables	Less	Less	Moderate	Moderate	High
6	Steps involved	Moderate	Moderate	Moderate	Moderate	Less

GA, genetic algorithm; *PSO*, particle swarm optimization; *FPA*, Flower Pollination Algorithm; *YSGA*, Yellow Saddle Goatfish Algorithm.

25.5   Experimental data measurement of 250 Wp PV module (SVL0250P) using SOLAR-4000 analyzer

For analogy, real-time values are taken from Amprobe SOLAR-4000 analyzer for the same solar module. Fig. 25.8A and B depicts the measured $I-V$ and $P-V$ characteristics of $G = 525$ W m^{-2} and $T = 47.4°$C using 250 Wp PV module (SVL0250P) and $G = 368$ W m^{-2} and $T = 45.9°$C using Amprobe SOLAR-4000 analyzer.

From Fig. 25.8A and B, it can be concluded that the experimental $I-V$ and $P-V$ characteristics has been matched with the computed $I-V$ and $P-V$ characteristics for various environmental conditions. It clearly indicates that solar electrical characteristics obtained using YSGA is matching with the measured data using Amprobe SOLAR-4000 ana-

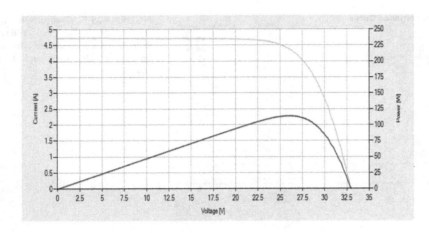

(A)

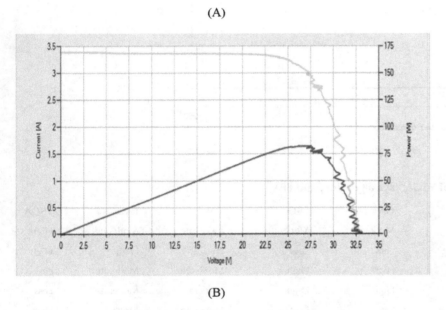

(B)

FIGURE 25.8 (A and B) Experimental $I-V$ and $P-V$ characteristics of SVL0250P solar PV module for environmental conditions $G = 525\ \text{W m}^{-2}$ and $T = 47.4°C$ and $G = 368\ \text{W m}^{-2}$ and $T = 45.9°C$, respectively.

lyzer. Error analysis effectively validates the accuracy of obtained results, and it is calculated for SVL0250P solar module $I-V$ curve at two different environmental conditions using the following equation:

$$\text{percent error} = (|(\text{measured current} - \text{calculated current})|/(\text{calculated current})) \times 100 \qquad (25.19)$$

The percent error has been taken between calculated and measured current of SVL0250P PV module for two environmental conditions, which is depicted in Tables 25.3 and 25.4.

From the results shown in Tables 25.3 and 25.4, it proves that solar module has the slightest percent error, which indicates the exactness of the proposed technique. The maximum percent error obtained between experimental and calculated current values is -0.245 and -0.246 for two environmental conditions ($T = 47.4°C$ and $G = 525\ \text{W m}^{-2}$ and $T = 45.9°C$ and $G = 368\ \text{W m}^{-2}$).

The verification of the relative error analysis has been made by curve fit between experimental and calculated values for solar PV module, which is presented in Fig. 25.9A and B. The curve fitting values for SVL0250P PV panel are estimated for every point in the $I-V$ and $P-V$ curve.

The above graphs clearly indicate very small marginal error between experimental and measured $I-V$ and $P-V$ characteristics.

TABLE 25.3 Error analysis between calculated and measured current of SVL0250P PV module at $T = 47.4°C$ and $G = 525 \text{ W m}^{-2}$.

S. no	Measured voltage (V)	Measured current (A)	Calculated current (A)	Percent error
1	3.6	4.73	4.792	− 0.062
2	3.71	4.73	4.79	− 0.06
3	4.28	4.73	4.784	− 0.054
4	5.47	4.73	4.774	− 0.044
5	8.85	4.717	4.74	− 0.023
6	11.62	4.711	4.712	− 0.001
7	14.12	4.714	4.68	0.034
8	19.94	4.694	4.64	0.054
9	23.09	4.641	4.62	0.021
10	27.63	3.996	4.2	− 0.204
11	31.04	2.027	2.272	− 0.245
12	31.89	1.22	1.193	0.027
13	32.88	0.063	0.06977	− 0.0067

PV, photovoltaic.

TABLE 25.4 Error analysis between calculated and measured current of SVL0250P PV module at $T = 45.90°C$ and $G = 368 \text{ W m}^{-2}$.

S. no	Measured voltage (V)	Measured current (A)	Calculated current (A)	Percent error
1	3.56	3.385	3.294	0.091
2	3.92	3.387	3.295	0.092
3	4.64	3.385	3.29	0.095
4	6.37	3.37	3.28	0.09
5	8.53	3.369	3.265	0.104
6	12.38	3.366	3.251	0.115
7	15.51	3.364	3.238	0.126
8	18.56	3.361	3.226	0.135
9	21.71	3.346	3.208	0.138
10	28.59	2.667	2.745	− 0.078
11	30.53	1.687	1.933	− 0.246
12	32.31	0.137	0.212	− 0.075
13	32.58	0.09	0.16	− 0.07

PV, photovoltaic.

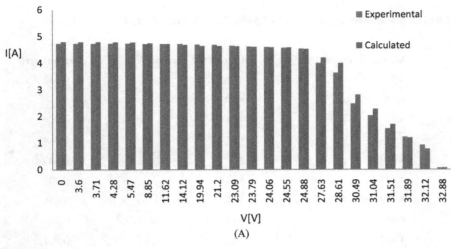

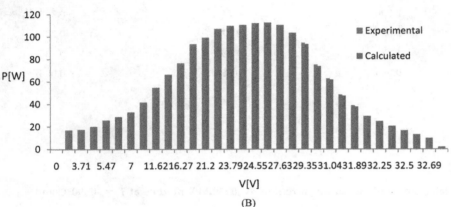

FIGURE 25.9 (A and B) Curve fit between experimental and calculated values of $I-V$ and $P-V$, respectively.

25.6 Conclusion

In this chapter, YSGA has been implemented to measure the electrical characteristics of SVL0250P solar PV module unknown parameters, such as R_s and R_p, accurately not including sub-optimal traps for various atmospheric conditions and its results are compared with FPA and Jaya algorithm. The electrical parameters of the proposed solar photovoltaic module are determined using a MATLAB/Simulink model and compared with experimentally obtained data using a SOLAR-4000 analyzer. The convergence graph has been plotted for YSGA in comparison to Jaya algorithm and FPA and the number of iteration to obtain fitness function in YSGA is lesser. The computational complexity in YSGA is lesser in comparison with the other two algorithms, such as FPA and Jaya algorithm. Thus it can be concluded that YSGA is a suitable algorithm that has less computational time for solar PV module unknown parameter extraction. The different types of algorithms with a change in temperature and irradiance profile will be implemented and compared their performance for each parameter identification.

References

AlHajri, M. F., El-Naggar, K. M., AlRashidi, M. R., & Al-Othman, A. K. (2012). Optimal extraction of solar cell parameters using pattern search. *Renewable Energy, 44*, 238–245.

Anandhraj, P., Sundaram, K. M., Sanjeevikumar, P., & Holm-Nielsen, J. B. (2020). *Extraction of solar module parameters using Jaya optimization algorithm. Advances in Greener Energy Technologies* (pp. 137–147). Singapore: Springer.

Askarzadeh, A., & dos Santos Coelho, L. (2015). Determination of photovoltaic modules parameters at different operating conditions using a novel bird mating optimizer approach. *Energy Conversion and Management, 89*, 608–614.

Askarzadeh, A., & Rezazadeh, A. (2013). Artificial bee swarm optimization algorithm for parameters identification of solar cell models. *Applied Energy, 102*, 943–949.

Chan, D. S., & Phang, J. C. (1987). Analytical methods for the extraction of solar-cell single- and double-diode model parameters from IV characteristics. *IEEE Transactions on Electron Devices, 34*(2), 286–293.

El-Naggar, K. M., AlRashidi, M. R., AlHajri, M. F., & Al-Othman, A. K. (2012). Simulated annealing algorithm for photovoltaic parameters identification. *Solar Energy, 86*(1), 266–274.

Ishaque, K., & Salam, Z. (2011). A comprehensive MATLAB Simulink PV system simulator with partial shading capability based on two-diode model. *Solar Energy, 85*(9), 2217–2227.

Ismail, M. S., Moghavvemi, M., & Mahlia, T. M. I. (2013). Characterization of PV panel and global optimization of its model parameters using genetic algorithm. *Energy Conversion and Management, 73*, 10–25.

Jervase, J. A., Bourdoucen, H., & Al-Lawati, A. (2001). Solar cell parameter extraction using genetic algorithms. *Measurement Science and Technology, 12*(11), 1922.

Koad, R. B., Zobaa, A. F., & El-Shahat, A. (2016). A novel MPPT algorithm based on particle swarm optimization for photovoltaic systems. *IEEE Transactions on Sustainable Energy, 8*(2), 468–476.

Liu, C. C., Chen, C. Y., Weng, C. Y., Wang, C. C., Jenq, F. L., Cheng, P. J., & Houng, M. P. (2008). Physical parameters extraction from current–voltage characteristic for diodes using multiple nonlinear regression analysis. *Solid-State Electronics, 52*(6), 839–843.

Rajasekar, N., Kumar, N. K., & Venugopalan, R. (2013). Bacterial foraging algorithm based solar PV parameter estimation. *Solar Energy, 97*, 255–265.

Rao, R. (2016). Jaya: A simple and new optimization algorithm for solving constrained and unconstrained optimization problems. *International Journal of Industrial Engineering Computations, 7*(1), 19–34.

Rao, R. V., & More, K. C. (2017). Design optimization and analysis of selected thermal devices using self-adaptive Jaya algorithm. *Energy Conversion and Management, 140*, 24–35.

Rao, R. V., & Saroj, A. (2017a). Constrained economic optimization of shell-and-tube heat exchangers using elitist-Jaya algorithm. *Energy, 128*, 785–800.

Rao, R. V., & Saroj, A. (2017b). Economic optimization of shell-and-tube heat exchanger using Jaya algorithm with maintenance consideration. *Applied Thermal Engineering, 116*, 473–487.

Rao, R. V., Rai, D. P., & Balic, J. (2017). A multi-objective algorithm for optimization of modern machining processes. *Engineering Applications of Artificial Intelligence, 61*, 103–125.

Sakar, S., Balci, M. E., Aleem, S. H. A., & Zobaa, A. F. (2018). Integration of large-scale PV plants in non-sinusoidal environments: Considerations on hosting capacity and harmonic distortion limits. *Renewable and Sustainable Energy Reviews, 82*, 176–186.

Soon, J. J., & Low, K. S. (2012). Photovoltaic model identification using particle swarm optimization with inverse barrier constraint. *IEEE Transactions on Power Electronics, 27*(9), 3975–3983.

Villalva, M. G., Gazoli, J. R., & Ruppert Filho, E. (2009). Comprehensive approach to modeling and simulation of photovoltaic arrays. *IEEE Transactions on Power Electronics, 24*(5), 1198–1208.

Zaldivar, D., Morales, B., Rodríguez, A., Valdivia-G, A., Cuevas, E., & Pérez-Cisneros, M. (2018). A novel bio-inspired optimization model based on yellow saddle goatfish behavior. *Bio Systems, 174*, 1–21.

Chapter 26

Reactive capability limits for wind turbine based on SCIG for optimal integration into the grid

Kamal Elyaalaoui[1], Mohammed Ouassaid[2] and Mohamed Cherkaoui[1]

[1]Mohammadia School of Engineers, Mohammed V University in Rabat, Rabat, Morocco, [2]Engineering for Smart and Sustainable Systems Research Center, Mohammadia School of Engineers, Mohammed V University in Rabat, Rabat, Morocco

26.1 Introduction

The production of energy through the burning of fossil fuels is continuing to increase, which increases CO_2 emissions. Therefore the development of other energy sources, such as the wind energy, is necessary (Abo-Khalil & Lee, 2008). The wind energy conversion systems (WECSs) are commonly based on three kinds of generators, such as doubly fed induction generator (DFIG), permanent magnet synchronous generator (PMSG), and squirrel cage induction generator (SCIG) (An, Ko, & Kim, 2012; Anon, 2015). The WECSs equipped with SCIG are mostly used, thanks to its advantages, such as low cost, simple construction, good reliability, and robustness (Anwar & Teimor, 2002; Bao, Huang, & Xu, 2003).

The reactive power can be generated to ensure the stability of the power system. It is required to maintain the grid voltage stability of the power network subjected to the voltage drop (Benlahbib & Bouchafaa, 2014). In Benlahbib, Ghennam, and Berkouk (2013), Bueno et al. (2006), Bueno, Rodriguez, and Espinosa (2008), Domínguez-García, Gomis-Bellmunt, Trilla-Romero, and Junyent-Ferré (2012), and Elyaalaoui, Ouassaid, and Cherkaoui (2019), the authors show that the reactive power reserve (RPR) is related to the margin of the voltage stability. In a heavily loaded system, a blackout and voltage collapse usually occur when the reactive power reserve of is insufficient. Furthermore, it is needed to support the voltage and transmit the active power demanded by the loads through the grid lines. Therefore a sufficient amount of reactive power should be available to satisfy the new grid code recommendations and reactive power demand (Folly & Sheetekela, 2009). The new grid code is imposed for the WF to produce or consume the reactive power (An et al., 2012).

The most wind turbines (WTs) installed in the world are the variable speed WT equipped with a back-to-back converter and a grid side system for injection of the required power into the power network. This full converter includes three-phase PWM rectifier, DC bus capacitor, and three-phase PWM inverter, as shown in Fig. 26.1.

Regarding the power production management, the WECS, based on SCIG, generates active power and consumes reactive power, which makes this generator more useful in wind power plant. This contributes to solve the problem of reactive power management and participates to service system (frequency and voltage control). Therefore the reactive power capability limit (RPCL) of WT based on SCIG must be developed. In addition, the capability limits of the grid side system (filter, transformer, and transmission line) are developed, for the system stability and security of, and also to guarantee secure transmission of the required reactive power.

The RPCL is a curve of reactive power versus active power under different constraints and operating conditions. This curve is given by manufacturers of generators to estimate the RPC and ensure the security and stability of the generators without exceeding the thermal limits (Ghennam, Francois, & Berkouk, 2009). The RPCL of DFIG and PMSG was developed, with stator resistance neglected and without power losses, in Hansen, Sorensen, Iov, and Blaabjerg (2006) and Hansen, Jauch, and Sørensen (2003), respectively. In Kieferndorf, Förster, and Lipo (2004), the operational limit of SCIG is given to not exceed the apparent power limit and without taking into consideration the thermal limits. The thermal limit is developed to avoid the heating of generator because of the reactive power increase. The reactive

Renewable Energy Systems. DOI: https://doi.org/10.1016/B978-0-12-820004-9.00029-2

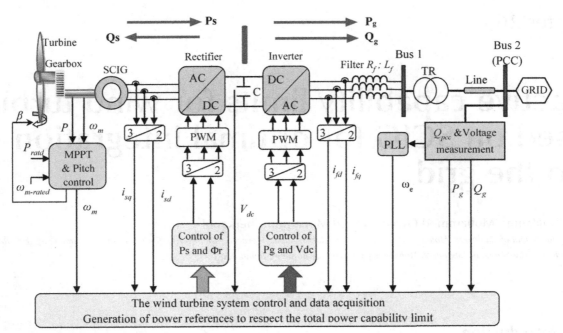

FIGURE 26.1 Scheme of wind energy conversion system and the converters control.

power can be investigated to develop the model of WPP as given in Kolar, Wolbank, and Schrodl (1999) and Kumar, Singh, and Aggarwal (2016).

In this context, this study establishes the total reactive power capability limit, taking into account the full converter limit and apparent power limit. These limits are developed taking into consideration the voltage and current constraints of the SCIG and the grid side system. The grid side system includes the filter, transformer, and transmission line, such as the limits of the stator current and rotor current, the stator voltage limit, the limits of filter current and filter voltage, and the steady-state stability limit of the generator. The capability limit of the full converter active and reactive powers control (PQ control) is also developed.

The reactive power generation depends on the DC bus capacitor parameters. In the variable WT system, the voltage ripples are reduced using an adequate parameter of the capacitor. It is also adopted to store energy, allowing the injection of the reactive power into the grid and magnetization of the generator windings. Hence, a DC bus capacitance, capable to satisfy the abovementioned requirements, should be designed.

The methodology adopted to reach the goal is tested and verified on a large-scale WT generator of 3 MW and the chapter objectives are summarized in the diagram presented in Fig. 26.2, where S_m is the slip, P_s, Q_s, S_s, and S_n indicate the active power, reactive power, apparent power, and nominal apparent power, at the stator side, T_m, ω_m, I_{r-mzax}, and Φ_{r-max} are the mechanical torque, mechanical speed, maximum rotor current, and maximum rotor flux.

This chapter is structured as follows. Section 26.2 presents the literature survey and grid code requirements. Section 26.3 presents the reactive capability limits of SCIG that are developed. The reactive capability limits for the grid side system is discussed in Section 26.4. In Section 26.5, the reactive capability limit for DC bus capacitance is given. Finally, validation result and conclusion are discussed in Sections 26.6 and 26.7, respectively.

26.2 Literature survey and grid code requirements

26.2.1 Reactive power capability curves in the grid code requirements

The safe operational of WT generators equipped with full converter is guaranteed when the terminal voltage is between 90% and 110% of rated value. This range allows the secure transmission of the power into the grid. The lagging power factor (PF) depends on the terminal voltage. An increase in lagging PF leads to the decrease in terminal voltage because of the voltage constraints and the operation conditions, but the leading PF capability increases with the voltage.

Fig. 26.3 presents the RPC of individual the full-converter wind generators. There are three types of reactive capability characteristic with a "triangular," "rectangular," or "D shape" (Leonardi & Ajjarapu, 2008).

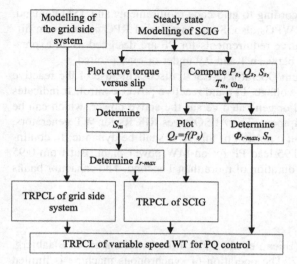

FIGURE 26.2 The diagram summarizing the content and objective of the chapter.

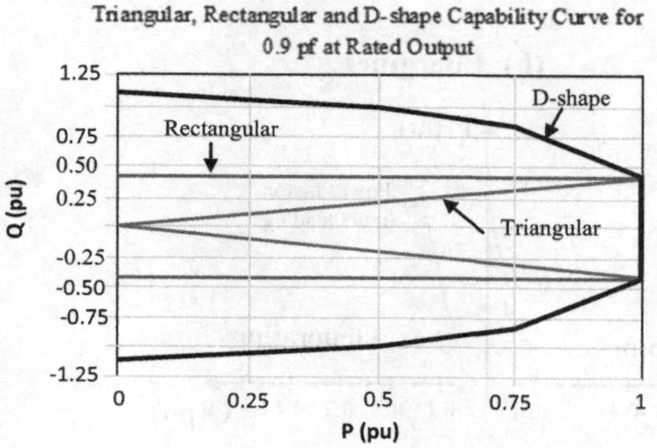

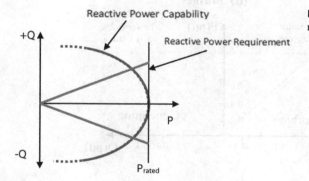

FIGURE 26.3 Reactive power capability limit curves for wind power production generator at nominal voltage (Li et al., 2012).

FIGURE 26.4 Typical comparison of reactive power capability and triangular reactive power requirement.

Fig. 26.4 presents the RPC curve for a renewable power plant for an optimal operation with unity PF (red line). The RPC curve is compared with a "triangular" curve (blue line), which indicates the reactive power capability requirement (RPCR) imposed by the transmission system operator (TSO). The triangular reactive power requirement shown in Fig. 26.4 should be needed for all transmission-connected inverter-based WPPs, which are capable to produce the required reactive power.

26.2.2   European grid codes for wind power production

The TSO imposes the recommendations for wind power production known as "grid codes." It is desirable that generators respect the grid codes for secured operation of grid without tripping of generators irrespective of amount of power generation and connection point to the grid.

The wind generators must generate or absorb reactive power according to grid code requirements and TSO demand. Occasionally, an additional reactive power should be generated by WPGs also. Some samples of RPC curves from different European TSOs are presented in Fig. 26.5. The capability curve requirements for PF are designed as an expression of Q versus P. The PF varies at the point of common coupling between 1 and 0.9 under or over excited.

Some grid code requirements, such as the Alberta Electric System Operator (AESO), discuss the use of the reactive power capability for the operation of the WPP and participation in voltage, PF, and reactive power control. It indicates that the maximum limits of the reactive power should be consumed or generated versus the active power, which can be investigated for the system design and optimal selection of the equipment. The AESO gives RPCRs for WT generators, as shown in Fig. 26.6. The reactive capability can be continuous, but a portion of this RPC shall be dynamic. In continuous RPC, the requirements shall meet the range from 0.9 lag to 0.95 lead PF for an MW power plant. But from 0.95 lag to 0.985 lead, RPC shall be dynamic and can be sustained for a duration of more than 1 second. The capacitor banks can be used to satisfy the total RPCR for dynamic RPC.

26.2.3 Reactive capability of synchronous generator

The RPC curve of synchronous generator is defined in literature (Bueno et al., 2008; Liserre, Klumper, & Blaabjerg, 2004) and the sample of capability curve is presented in Fig. 26.7. The operation of synchronous machine is limited because of the thermal constraints.

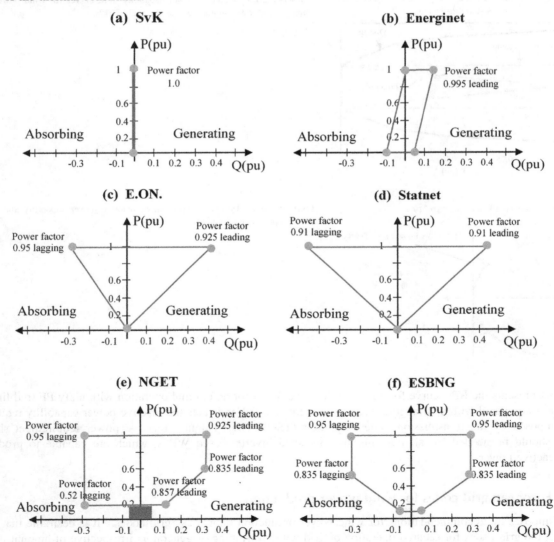

FIGURE 26.5 Some examples of reactive power capability and PQ diagrams from different European transmission system operators (Li et al., 2012).

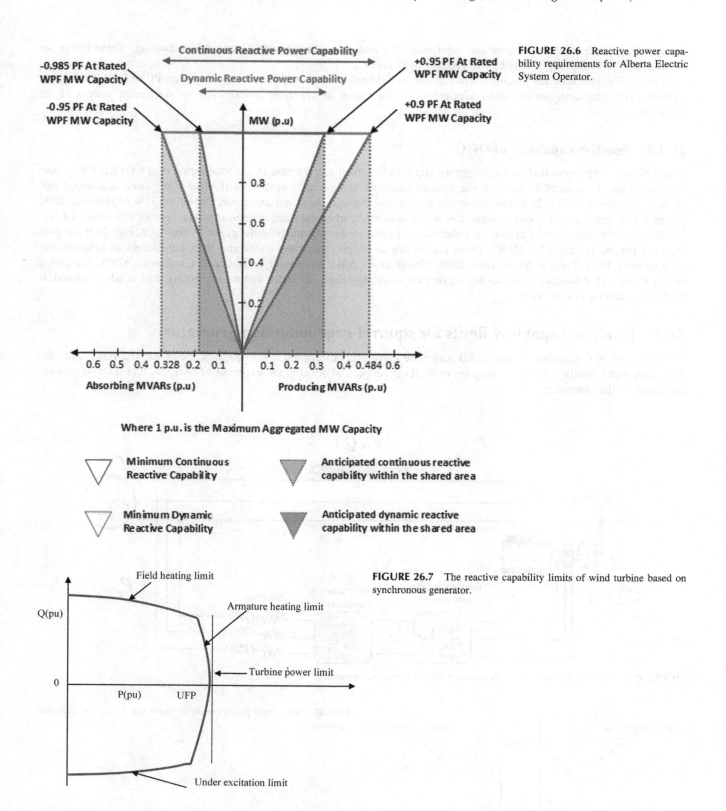

FIGURE 26.6 Reactive power capability requirements for Alberta Electric System Operator.

FIGURE 26.7 The reactive capability limits of wind turbine based on synchronous generator.

The operational constraints of the synchronous generator capability curves depend on stator current, excitation current, prime mover mechanical power, and end region heating, as given in Fig. 26.7. The typical curve has top side controlling limit given by a circle representing field current and on right hand side limiting circle is for stator current. These two characteristics intersect each other to give upper half of "D curve." Bottom side curve represents constant losses taking place in armature core and rotor magnetic material, which results in end region heating. These losses set heating limit, which is represented by lower part of D curve.

Most of the synchronous machines are capable of delivering rated output at reasonably high PF of 0.8−0.9 lagging without exceeding designed thermal considerations. The output of generator is restricted by mechanical output of the prime mover.

26.2.4 Reactive capability of DFIG

Power flow can be controlled by varying rotor slip power, which is a fraction of the total rating of the DFIG. The generator speed can be controlled to track the desired value by achieving bidirectional flow of slip power at reduced frequency. As shown in Fig. 26.8, the power can be injected through the stator and rotor. Therefore 33% of machine rated power is converted used the power converter which is considered as the main economic advantages of WT-based DFIG.

The stator reactive power in Fig. 26.8 should be limited to the capability limits given in Fig. 26.9. Fig. 26.9 presents the total capability limits for DFIG. These curves are developed using the stator and rotor constraints and the steady-state stability limit (Folly & Sheetekela, 2009; Liserre et al., 2004; Malamaki, Mushtaq, & Cvetkovic, 2020). The active power is converted to reactive power taking into the capability limits of stator and rotor currents, and steady-state stability limit as shown in Fig. 26.9.

26.3 Reactive capability limits for squirrel cage induction generator

The RPCL of WT generator is developed under the constraints of current and voltage of the stator and rotor, and also the steady-state stability. A vector diagram of SCIG is adopted to develop the expressions of active and reactive powers produced by the generator.

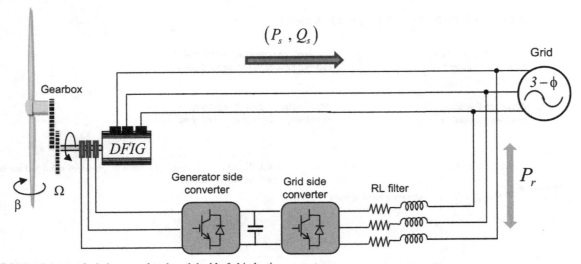

FIGURE 26.8 Scheme of wind power plant-based doubly fed induction generator.

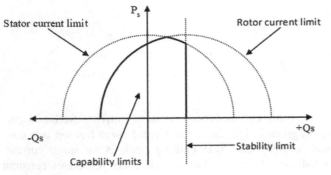

FIGURE 26.9 Total power capability curve for doubly fed induction generator.

26.3.1 Model of squirrel cage induction generator

The equivalent electrical scheme of induction generator is given in Fig. 26.10.

The mathematical model of SCIG is developed taking into account the following assumptions:

- The distribution of ampere turns in air gap is sinusoidal and the space harmonics are neglected.
- The stator and rotor three-phase windings are symmetrical and rotor windings are short-circuited (V_r 0).
- The iron losses (hysteresis and eddy current) will be ignored.
- The saturation will be neglected.

After simplification, the transient model of SCIG is expressed, in synchronous rotating reference frame according to per unit notation, by Eqs. (26.1) and (26.2) (Montilla-DJesus et al., 2017; Niazy et al., 2010):

$$\underline{V}_s = -R_s\underline{I}_s - jX_s\underline{I}_s - jX_m\underline{I}_r - \frac{1}{\omega_b}\frac{d}{dt}\left(L_s\underline{I}_s + L_m\underline{I}_r\right) \tag{26.1}$$

$$0 = -\frac{R_r}{s}\underline{I}_r - jX_r\underline{I}_r - jX_m\underline{I}_r - \frac{1}{\omega_b}\frac{d}{dt}\left(L_m\underline{I}_s + L_r\underline{I}_r\right) \tag{26.2}$$

where V_s and I_s are the stator voltage and current, respectively. I_r is the rotor current. X_s, X_r, and X_m are the stator, rotor, and mutual reactances of generator, respectively. s is the machine slip.

After simplification, Eqs. (26.1) and (26.2) are rewritten in vector notation for one phase of equivalent electrical scheme in the steady-state and synchronous rotating reference frame according to the unit notation. Hence, the new SCIG model shown in Fig. 26.11 is expressed as follows:

$$\underline{V}_s = -R_s\underline{I}_s - jX_s\underline{I}_s - jX_m\underline{I}_r \tag{26.3}$$

$$0 = -\frac{R_r}{s}\underline{I}_r - jX_r\underline{I}_r - jX_m\underline{I}_r \tag{26.4}$$

where $X_{\sigma s}$ and $X_{\sigma r}$ are leakage stator and rotor reactance, respectively.

26.3.2 Characteristics of SCIG and the maximum rotor flux

The machine internal electromagnetic force is given by:

$$E = -jX_m\underline{I}_r \tag{26.6}$$

Substituting Eq. (26.6) in Eq. (26.3) leads to Eq. (26.7).

$$E = \underline{V}_s + R_s\underline{I}_s + jX_s\underline{I}_s \tag{26.7}$$

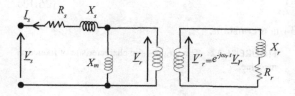

FIGURE 26.10 Equivalent electrical scheme of the induction generator (Montilla-DJesus, Arnaltes, Castronuovo, & Santos-Martin, 2017; Niazy et al., 2010).

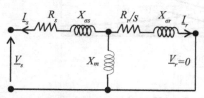

FIGURE 26.11 Generator equivalent scheme of one phase during steady state.

$$\begin{cases} X_s = X_{\sigma s} + X_m \\ X_r = X_{\sigma r} + X_m \end{cases} \tag{26.5}$$

The electromagnetic torque in steady state can be expressed, using rotor and stator current components, as follows:

$$T_e = X_m R_e \left[j \underline{I}_s^* \underline{I}_r \right] \tag{26.8}$$

where I_s^* is the conjugate of I_s.

The characteristic of the torque versus the slip, in steady state, is studied to determine the stable operation point. Substitution of rotor current expression derived from Eq. (26.4) in Eq. (26.8) leads to the following torque expression.

$$T_e = \frac{\frac{R_r}{s} X_m^2}{\left(\frac{R_r}{s}\right)^2 + X_r^2} |\underline{I}_s|^2 \tag{26.9}$$

From Fig. 26.11, the stator current is given by:

$$|\underline{I}_s| = \frac{|\underline{V}_s|}{|\underline{Z}|} \tag{26.10}$$

where Z is the equivalent impedance and expressed as follows:

$$\underline{Z} = \frac{\frac{R_s R_r}{s} + \left(X_m^2 - X_s X_r\right) + j\left(\frac{R_r}{s} X_s + R_s X_r\right)}{(R_r/s) + jX_r} \tag{26.11}$$

Hence, the torque of the machine is expressed, as follows:

$$T_e = \frac{X_m^2 (R_r \, s)}{\left[R_s R_r + s\left(X_m^2 - X_s X_r\right)\right]^2 + (R_r X_s + s R_s X_r)^2} V_s^2 \tag{26.12}$$

Eq. (26.4) shows that the rotor current (I_r) depends on the slip (s). To estimate the maximum slip value, during the stable operation, at the nominal values of the torque and stator voltage, the characteristic of the torque versus the slip is plotted in Fig. 26.12. Therefore the stable operation point is S_1 (-0.01, -1). Therefore the maximum value of rotor current, at the maximum slip $S_m = -0.01$ and the nominal voltage $V_s = 1$ pu, is equal to $I_{r\_max} = 0.9723$ pu.

For determining the characteristics of the SCIG during the steady-state operation, the generator is connected to power network without power converters. The aim is to determine the mechanical speed (ω_m), the mechanical torque (T_m), consumed reactive power (Q_s), produced active power (P_s), apparent power (S_s), and rotor flux (Φ_r). These quantities are given in Table 26.1, for different wind speed values (V_W). The apparent power varies considerably to reach 1 pu, which is the rated value.

The maximum reactive power consumed by generator windings is $Q_{s\_max} = -0.4243$ pu (equivalent to 1.4143 MW) whereas the maximum active power is $P_{s\_max} = 0.9$ pu (equivalent to 3 MW). The maximal rotor flux is $\Phi_{r\_max} = 0.9812$ pu. Note that the nominal apparent power is equal to 1 pu (equivalent to 3.33 MVA). These maximum limits can be investigated to improve the operation of variable speed WT. Using data given in Table 26.1, the relationship between reactive and active powers is described by the following expression.

$$Q_s = -\left(Q_{NL} + a \, P_s^2\right) \tag{26.13}$$

where $Q_{NL} = 0.1468$ (pu) is the no-load reactive power and $a = 0.34$ is a constant.

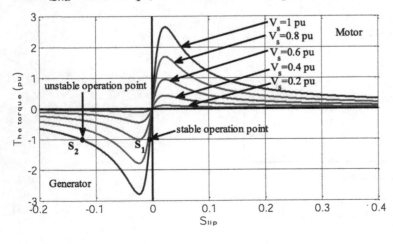

FIGURE 26.12 The torque–slip characteristic of induction machine.

TABLE 26.1 The characteristics of generator for various wind speed value.

VW (m s^{-1})	5	6	7	8	9
ω_m (pu)	0.9997	1	1.002	1.003	1.004
T_m (pu)	− 0.06654	− 0.1104	− 0.3677	− 0.6792	− 0.9148
P_s (pu)	0.0766	0.1003	0.3569	0.6666	0.9
Q_s (pu)	− 0.1468	− 0.1483	− 0.1859	− 0.2918	− 0.4243
S_s (pu)	0.1656	0.179	0.4024	0.7277	0.995
Φ_r (pu)	0.9812	0.9819	0.9773	0.9623	0.9424

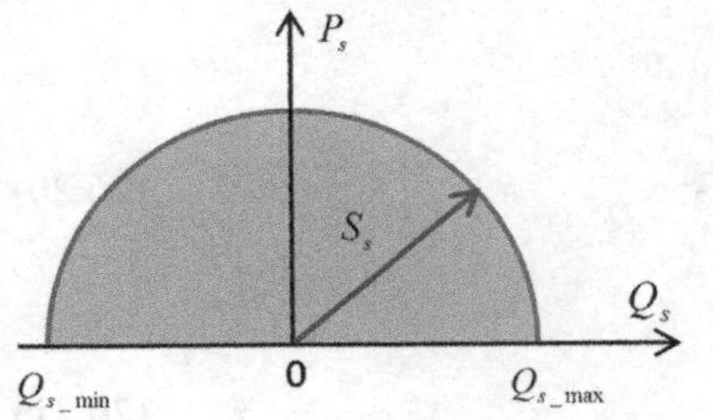

FIGURE 26.13 Reactive power capability limit under the stator current constraint.

To check that this operating point (Q_s, $P_{s\_}$) is within the allowable limits of the induction generator, a total reactive capability limit must be developed.

26.3.3 Reactive capability limits under constraints of stator voltage and current

The relationship between stator power (S_s), current (I_s), and voltage (V_s) is expressed as:

$$P_S^2 + Q_S^2 = S_S^2 = (V_s I_s)^2 \tag{26.14}$$

The active and reactive powers given in Eq. (26.14) describes a circle of center $C_{sc}(0,0)$ and radius $S_s = V_s I_s$, as shown in Fig. 26.13.

The RPCL under the stator current constraint is developed tacking into consideration the stator warming because of the winding Joule's losses. These limits are obtained under the conditions of the voltage and current to not exceed the rated values ($I_s \leq I_{s\text{-}nom}$ and $V_s \leq V_{s\text{-}nom}$). The minimal and maximal limits ($Q_{s\text{-}min}$ and $Q_{s\text{-}max}$) for reactive capacity are given by:

$$Q_{s-\max} = \sqrt{\left(V_{s\_nom} . I_{s\_nom}\right)^2 - P_s^2} \tag{26.15}$$

$$Q_{s-\min} = -\sqrt{\left(V_{s\_nom} . I_{s\_nom}\right)^2 - P_s^2} \tag{26.16}$$

26.3.4 Reactive capability limits under rotor current constraint

The reactive power limit, under rotor current constraints, takes into account the heating of the rotor windings due to Joule's effect. The magnitude of internal electromagnetic force Eq. (26.6) can be expressed as follows:

$$E = X_m I_r \tag{26.17}$$

The vector diagram of the generator voltages is presented in Fig. 26.14. Where δ is the phase shift between internal electromagnetic force (E) and stator voltage (V_s), and φ is the phase shift between stator voltage (V_s) and current (I_s).

The projection on the frame (x, y) in the vector diagram leads to write

$$\begin{cases} E \, \sin(\delta) = (-R_s \sin(\varphi) + X_s \, \cos(\varphi)) \, I_s \\ E \, \cos(\delta) = V_s + (R_s \cos(\varphi) + X_s \, \sin(\varphi)) I_s \end{cases} \tag{26.18}$$

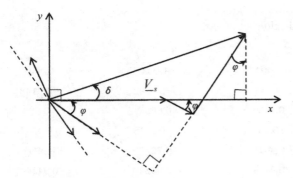

FIGURE 26.14 Vector diagram of machine voltages.

Then,

$$\begin{cases} (-R_s\sin(\varphi) + X_s\ \cos(\varphi))I_s = E\ \sin(\delta) \\ (R_s\cos(\varphi) + X_s\ \sin(\varphi))I_s = E\ \cos(\delta) - V_s \end{cases} \tag{26.19}$$

Then,

$$\begin{cases} -R_s V_s I_s\sin(\varphi) + X_s V_s I_s\cos(\varphi) = EV_s\sin(\delta) \\ R_s V_s I_s\cos(\varphi) + X_s V_s I_s\sin(\varphi) = EV_s\cos(\delta) - V_s^2 \end{cases} \tag{26.20}$$

The expressions of active and reactive powers are given by:

$$\begin{cases} P_s = V_s I_s\cos(\phi) \\ Q_s = V_s I_s\sin(\phi) \end{cases} \tag{26.21}$$

The substitution of Eq. (26.17) in Eq. (26.20) leads to the following expression:

$$\begin{cases} -R_s Q_s + X_f P_s = EV_s\sin(\delta) \\ R_s P_s + X_s Q_s = EV_s\cos(\delta)-V_s^2 \end{cases} \tag{26.22}$$

The new expressions of active and reactive power are;

$$\begin{cases} P_s = \dfrac{EV_s}{X_s^2 + R_s^2}(X_s\sin(\delta) + R_s\cos(\delta)) - \dfrac{R_s V_s^2}{X_s^2 + R_s^2} \\ Q_s = \dfrac{EV_s}{X_s^2 + R_s^2}(X_s\cos(\delta) - R_s\sin(\delta)) - \dfrac{X_s V_s^2}{X_s^2 + R_s^2} \end{cases} \tag{26.22}$$

Eq. (26.22) is used to determine the following relation:

$$\left(P_s + \frac{R_s V_s^2}{X_s^2 + R_s^2}\right)^2 + \left(Q_s + \frac{X_s V_s^2}{X_s^2 + R_s^2}\right)^2 = \left(\frac{X_m V_s}{\sqrt{X_s^2 + R_s^2}}I_r\right)^2 \tag{26.23}$$

Eq. (26.23) describes a circle of center $C_{Ir} = \left(\frac{-R_s V_s^2}{X_s^2 + R_s^2}, \frac{-X_s V_s^2}{X_s^2 + R_s^2}\right)^2$ and radius $S_{Ir} = \frac{X_m V_s}{\sqrt{X_s^2 + R_s^2}}I_r$ as shown in Fig. 26.15 where the radius depends on the rotor current constraint.

From Fig. 26.15, it can be seen that the reactive power capability depends on the maximum rotor current $(I_r \leq I_{r\_max})$. Therefore the diagram (P_s, Q_s) for RPC should be governed by the inequation (Eq. 26.24):

$$\left(P_s + \frac{R_s V_s^2}{X_s^2 + R_s^2}\right)^2 + \left(Q_s + \frac{X_s V_s^2}{X_s^2 + R_s^2}\right)^2 \leq \left(\frac{X_m V_s}{\sqrt{X_s^2 + R_s^2}}I_r\right)^2 \tag{26.24}$$

For an SCIG, if the generated active power increases the absorption of reactive power increases too (value of Q_s decreases). Therefore when P_s equals to its rated value (P_{s_nom}) the reactive power must be within the minimum limit (Q_{s_min}) and the maximum limit (Q_{s_max}). These RPCLs are given in Eqs. (26.25) and (26.26).

$$Q_{s\_\min} = -\sqrt{\left(\frac{X_m V_s I_{r\_\max}}{\sqrt{X_s^2 + R_s^2}}\right)^2 - \left(P_{s\_nom} + \frac{R_s V_s^2}{X_s^2 + R_s^2}\right)^2} - \frac{X_s V_s^2}{X_s^2 + R_s^2} \tag{26.25}$$

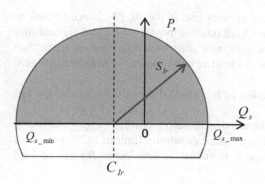

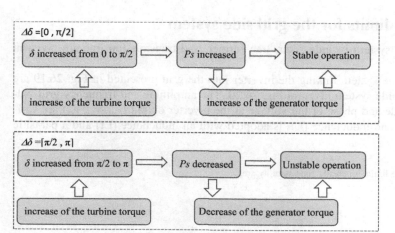

FIGURE 26.16 Diagram for steady-state operation.

$$Q_{s\_max} = \sqrt{\left(\frac{X_m V_s I_{r\_max}}{\sqrt{X_s^2 + R_s^2}}\right)^2 - \left(P_{s\_nom} + \frac{R_s V_s^2}{X_s^2 + R_s^2}\right)^2} - \frac{X_s V_s^2}{X_s^2 + R_s^2}$$
(26.26)

26.3.5 Steady-state stability limit

The steady-state stability limit is developed when the windings resistance is negligible. It is given by a vertical line in the PQ diagram. Therefore Eq. (26.22) is written as follows:

$$\begin{cases} P_s = \dfrac{E V_s}{X_s}(\sin\delta) \\[2mm] Q_s = \dfrac{E V_s}{X_s}(\cos\delta) - \dfrac{V_s^2}{X_s} \end{cases}$$
(26.27)

If consider that the magnitudes of rotor current and stator voltage are constant, Eq. (26.27) indicates that the power varies just with the load angle δ. Therefore two operation intervals of load angle ($\Delta\delta$) can be obtained as presented in Fig. 26.16. The active power, load angle and the torque of the turbine and generator are summarized in Fig. 26.16. When the angle δ varies from 0 to $\pi/2$, the active power increases, leading to the stable operation. But, the angle δ varies from $\pi/2$ to π, and the active power decrease, leading to the unstable operation. The increasing of generator torque produces the increase of the active power (Folly & Sheetekela, 2009).

From the diagram given in Fig. 26.16, the steady-state stability limit is obtained when $\delta = \pi/2$. Replacing δ by $\pi/2$ in Eq. (26.27), the steady-state stability limit is represented by vertical line. For the maximum value of the stator voltage (V_s), the reactive power limit (Q_{lim}) is expressed as:

$$Q_{lim} = -\frac{V_s^2}{X_s}$$
(26.28)

The generator stability area in steady state is defined by the left area of vertical line at the $[0, Q_{lim}]$ coordinates, as shown in Fig. 26.17. The Q_{lim} is considered as an approximation of the no-load reactive power (Q_{NL}) and the stability of generator depends on the absorption of reactive power. Therefore if the generator absorbs reactive power more than that of the no-load, it becomes unstable (Folly & Sheetekela, 2009). The no-load reactive power is obtained for zero active power.

The total RPCLs of the SCIG are defined in Fig. 26.18 by the intersection of the internal area of circles and the left area obtained by the constraint of the stability limit in steady state.

The maximum reactive power consumed by generator windings, for the maximum active power $P_{s\_max} = 0.9$pu, is equal to $Q_{s\_max} = 0.4243$ pu. This operating point (Q_{s_max}, P_{s_max}) is within the total capability limit of SCIG presented in Fig. 26.18. The reactive power consumed is less than that in the steady state (which is considered for the no-load case).

26.4 Estimation of reactive power limits for the grid side system

The single line electrical circuit of the grid side system include filter, transformer, and transmission line are given in Fig. 26.19.

The resistances and impedances of the grid side system linking the inverter into the grid presented in Fig. 26.19 are simplified using the single phase of the network side system given in Fig. 26.20. The amplitude and frequency grid voltages (v_{ag}, v_{bg}, v_{cg}) are constant. But the amplitude and phase of the voltage at the inverter output (v_{ai}, v_{bi}, v_{ci}) are variables. The RL filter (consisting of resistance (R) and inductance (L)) is adopted with an inductance (Lf) and a parasitic resistance (Rf).

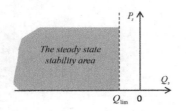

FIGURE 26.17 The limits for steady-state stability area.

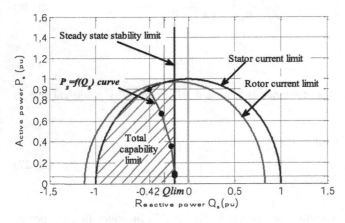

FIGURE 26.18 Total reactive power capability limit curves for squirrel cage induction generator.

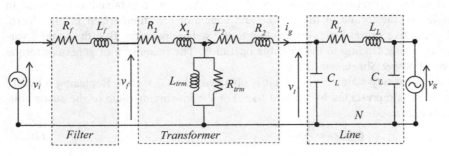

FIGURE 26.19 The electrical scheme of the filter, transformer, and transmission line.

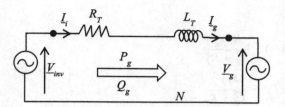

FIGURE 26.20 Simplified electrical scheme of single-phase grid side system.

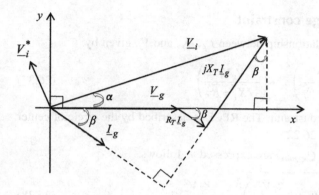

FIGURE 26.21 Vector diagram of grid side system voltages.

The maximum fundamental voltage at the inverter output is calculated using the DC bus (direct current bus) voltage as:

$$V_{i\_\max} = V_{dc}/\sqrt{3} \tag{26.29}$$

Then, the model of the grid side system used the developed the RPCL is given in Eq. (26.30), where R_T and X_T are the total resistance and reactance of the transmission connected WT, respectively.

$$\underline{V}_i = \underline{V}_g + (R_T + jL_T\omega_s)\underline{I}_g \tag{26.30}$$

Using Eq. (26.30), the RPCL can be developed, considering that the three-phase windings are identical and perfectly balanced.

The voltage vector diagram is resented in Fig. 26.21. Where α is the phase shift between V_i and V_g. The angle β is the phase shift between V_g and I_g.

Projecting Eq. (26.30) on the (x, y) base leads to write:

$$\begin{cases} V_i\sin(\delta) = (-R_T\sin(\varphi) + X_T\cos(\varphi))I_g \\ V_i\cos(\delta) = V_g + (R_T\cos(\varphi) + X_T\sin(\varphi))I_g \end{cases} \tag{26.31}$$

Therefore

$$\begin{cases} (-R_T\sin(\varphi) + X_T\cos(\varphi))I_g = V_i\sin(\delta) \\ (R_T\cos(\varphi) + X_T\sin(\varphi))I_g = V_i\cos(\delta) - V_g \end{cases} \tag{26.32}$$

Then,

$$\begin{cases} -R_TV_gI_g\sin(f) + X_TV_gI_g\cos(\varphi) = V_gV_i\sin(\delta) \\ R_TV_gI_g\cos(f) + X_TV_gI_g\sin(\varphi) = V_gV_i\cos(\delta) - V_g^2 \end{cases} \tag{26.33}$$

Multiply the equations by V_g and then introduce the equations of the active and reactive stator that are defined by:

$$\begin{cases} P_g = V_gI_g\cos(\varphi) \\ Q_g = V_gI_g\sin(\varphi) \end{cases} \tag{26.34}$$

Gives

$$\begin{cases} -R_TQ_g + X_TP_g = V_gV_i\sin(\delta) \\ R_TP_g + X_TQ_g = V_gV_i\cos(\delta) - V_g^2 \end{cases} \tag{26.35}$$

The new expressions of active and reactive power are:

$$\begin{cases} P_g = \dfrac{V_g V_i}{X_T^2 + R_T^2}(X_T \sin(\delta) + R_T \cos(\delta)) - \dfrac{R_T V_g^2}{X_T^2 + R_T^2} \\[4mm] Q_g = \dfrac{V_g V_i}{X_T^2 + R_T^2}(X_T \cos(\delta) - R_T \sin(\delta)) - \dfrac{X_f V_g^2}{X_T^2 + R_T^2} \end{cases} \tag{26.36}$$

26.4.1 Reactive capability limit under the filter voltage constraint

Projecting Eq. (26.25) on the (x, y) base leads to establish the relationship between P_g, Q_g, and V_i, given by:

$$\left(P_g + \frac{R_T V_g^2}{X_T^2 + R_T^2}\right)^2 + \left(Q_g + \frac{X_T V_g^2}{X_T^2 + R_T^2}\right)^2 = \left(\frac{V_g V_i}{\sqrt{X_T^2 + R_T^2}}\right)^2 \tag{26.37}$$

Eq. (26.37) is used to determine the RPCL with the voltage constraint. The RPCL is described by the circle of center $C_f = \left(\frac{-R_T V_g^2}{X_T^2 + R_T^2}, \frac{-X_T V_g^2}{X_T^2 + R_T^2}\right)$ and radius $S_g = \frac{V_g V_i}{\sqrt{X_T^2 + R_T^2}}$ as shown in Fig. 26.22.

The reactive power minimal and maximal limits (Q_{g_min} and Q_{g_max}) are expressed as follows:

$$Q_{g\_min} = -\sqrt{\left(\frac{V_g V_i}{\sqrt{X_T^2 + R_T^2}}\right)^2 - \left(P_g + \frac{R_T V_g^2}{X_T^2 + R_T^2}\right)^2} - \frac{X_T V_g^2}{X_T^2 + R_T^2} \tag{26.38}$$

$$Q_{g\_max} = \sqrt{\left(\frac{V_g V_i}{\sqrt{X_T^2 + R_T^2}}\right)^2 - \left(P_g + \frac{R_T V_g^2}{X_T^2 + R_T^2}\right)^2} - \frac{X_T V_g^2}{X_T^2 + R_T^2} \tag{26.39}$$

26.4.2 Reactive capability limit under the grid side current constraint

Using the apparent power expression, the relationship between P_g and Q_g, the grid side current (I_g), and the maximum inverter output voltage (V_{i_max}) is expressed by:

$$P_g^2 + Q_g^2 = S_g^2 = \left(V_{i\_max} I_g\right)^2 \tag{26.40}$$

Eq. (26.40) describes a circle of center C_{sc} (0, 0) and radius $S_g = V_{i\_max} I_g$ as illustrated in Fig. 26.23.

The total RPCLs of the grid side system are defined in Fig. 26.24 by the intersection of the internal area of the circles. The maximum value of reactive power injected through grid side system is less than or equal to 0.9 pu.

26.4.3 The constraints of AC/DC/AC full converter for PQ control

The stator side converter (SSC) of the WT is controlled to orient the rotor flux on d-axis and to convert the mechanical power to active power, consuming the reactive power. The reactive power needed to supply a rotating field (Q_s). Whereas the grid side converter (GSC) is controlled to maintain the DC bus voltage constant and inject the reactive

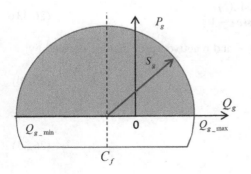

FIGURE 26.22 The diagram of reactive power capability limit under the inverter output voltage constraint.

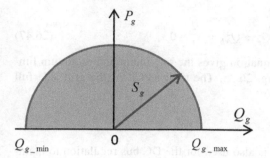

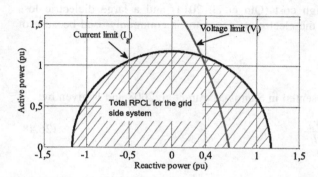

FIGURE 26.24 Diagram of total reactive power capability limit for the grid side system.

power into the grid. In this case, the stator active power is converted into the active and reactive powers (see Fig. 26.1). It can be concluded that the active power controlled by SSC and the reactive power by GSC. The control of the DC bus voltage allows active power flow to the grid.

When the $Q_{WG} = 0$, the grid side active power is equal to that of the generator side $P_{WG} = Ps$, but, during the PQ control, the reactive power (Q_{WG}) should be less or equal to that of produced by the generator (Ps) Eq. (26.41). Therefore despite that the GSC does not participate in the active power control, this power is injected into grid through the GSC.

$$Q_g \leq P_s \qquad (26.41)$$

Eq. (26.41) indicates that the maximum value of Q_g is obtained when it is equal to P_s $\left(Q_{g\_max} = P_s\right)$.

The WT shall satisfy the reactive power demand and must sacrifice an amount of active power to produce more reactive power, respecting Eq. (26.42) (Ouassaid, Elyaalaoui, & Cherkaoui, 2016; Qin, Wang, Han, & Du, 2011):

$$Q_{g\_max} = \sqrt{S_n^2 - P_g^2} - |Q_s| \qquad (26.42)$$

Substituting Eq. (26.13) in Eq. (26.42), the following expression is obtained

$$\left(Q_g + Q_{NL} + a\, P_s^2\right) + P_g^2 = S_n^2 \qquad (26.43)$$

To ensure the PQ control for the WT equipped with SCIG and DC bus voltage regulation, the grid power equals to that generated by the WT generator $P_g = P_s$. The equation becomes:

$$\left(Q_g + Q_{NL} + a\, Q_g{}^2\right)^2 + Q_g^2 = S_n^2 \qquad (26.44)$$

Therefore

$$\left(Q_{NL} + a\, Q_g^2\right)^2 + Q_g^2 + 2Q_g\left(Q_{NL} + a\, Q_g^2\right) + Q_g^2 = S_n^2 \qquad (26.45)$$

Therefore

$$Q_{NL}^2 + a^2 Q_g^4 + 2Q_{NL}a Q_g^2 + 2Q_g^2 + 2Q_g Q_{NL} + 2Q_g a\, Q_g^2 = S_n^2 \qquad (26.46)$$

Finally, the following equation is obtained:

$$f(Q_g) = a^2 Q_g^4 + 2a\,Q_g^3 + 2(Q_{NL}a + 1 + Q_{NL})Q_g^2 + Q_{NL}^2 - S_n^2 = 0 \qquad (26.47)$$

The curve of Eq. (26.47) is given in Fig. 26.25. The solution of the equation gives the maximum and minimum limits ($Q_{g\_min} = -0.6847$ pu, $Q_{g\_max} = 0.5806$ pu) and extracted from Fig. 26.25. The total RPCL of the grid side full converter that can be investigated in PQ control is presented in Fig. 26.26.

26.5   Reactive capability for DC bus capacitor

The DC bus capacitor behaves as a filter to reduce the voltage ripples. It is also used for the DC bus regulation to maintain the voltage constant in inverter input and produce the reactive power. The good mitigation of ripples is obtained for a large capacitance, but the large capacitance results a high cost (Qin et al. 2014) and a large dielectric loss (Salcone & Bond, 2009). In comparison with film capacitance, the electrolyte capacitor is commonly used because of its big capacitance per unit volume (Sandia, 2012).

26.5.1   DC capacitance power production

The reactive power can be produced by DC capacitance (Q_c) presented in Santos-Martin et al. 2008 and is given by:

$$Q_c = \frac{C_{dc}}{2}\frac{dv_{dc}^2}{dt} \qquad (26.48)$$

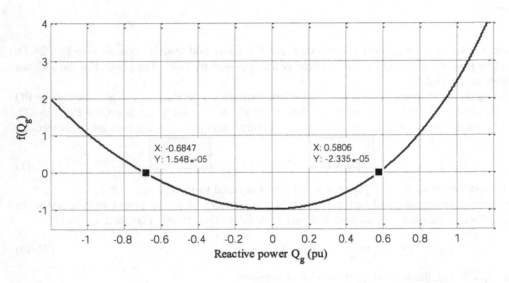

FIGURE 26.25 The curve of grid reactive power function for PQ control.

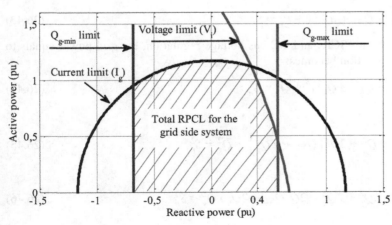

FIGURE 26.26 Total reactive power capability limit of the grid side full converter for PQ control.

where v_{dc} is the DC bus voltage taking into account the ripples. This voltage is expressed as:

$$v_{dc} = V_{dc} + V_{ac}\cos(\omega_{in}t) \tag{26.49}$$

$\omega_{in} = 2\pi F_{in}$, F_{in}, and V_{ac} are the frequency and the magnitude of voltage ripples, respectively. C_{dc} is the capacitance value capable to generate the quantity Q_c. The capacitance value (C_{dc}) depends on the quantity of power produced and DC-link voltage (V_{dc}). Expressions (26.48) and (26.49) offer a simple method for designing the full converter DC-bus capacitance for an optimal conversion of power generated by SCIG WT, but an efficient design must take into consideration the frequency of input voltage and converter switching.

26.5.2 Mitigation of the ripples and DC bus capacitance limit

The capacitance value C_{dc} should be sufficient enough to attenuate the current ripples at the DC bus, maintain a stiff voltage across the DC bus, and suppress the electromagnetic interference generated by pulsed inverter current Sarkar et al. (2019). The reduction of current ripple leads to a reduced capacitor power loss. This in turn can lead to an increase in lifetime of the capacitor and a decrease in warming losses in the capacitor due to the ripple current (Salcone & Bond, 2009; Tian et al., 2013; Vakula et al., 2015). For two-level converter connected to utility grid, the calculation of DC capacitance value depends on the application kind. In Sarkar et al. (2019) and Zhai and Liu, 2014, the authors illustrate that the converter can either work as active filter or as a rectifier.

The worst case is where the load changes from the negative sequence to the nominal power. Therefore the frequency of ripple is equal to rated frequency, which is the grid frequency (Zhao et al., 2011). In this case, the ripples are depending on the grid frequency and the DC capacitance value (C_{dc}) must be greater or equal than C_{dc1} (Eq. 26.50).

$$C_{dc} \geq C_{dc1} = \frac{S_b}{\omega_b V_{dc}\Delta V_{dc}} \tag{26.50}$$

where ΔV_{dc} is the allowable ripple of DC bus voltage. It is less than or equal to $10\% V_{dc}$. S_b and ω_b are the base apparent power and base frequency.

- Working as rectifier

In high frequency, due to the switching period, C_{dc} depends on DC bus current (i_{dc}) and the capacitor discharges from the nominal power during one-half of the commutation period. Generally, the discharge time is considered smaller (in practice, it depends on the load) (Ouassaid et al., 2016). C_{dc} should be expressed as follows:

$$C_{dc} \geq C_{dc2} = \frac{i_{dc}T_{sw}}{\Delta V_{dc}} = \frac{T_{sw}S_b}{V_{dc}\Delta V_{dc}} \tag{26.51}$$

where $T_{sw} = 0.1$ is the switching period IGBT (Insulated Gate Bipolar Transistor).

The DC bus capacitance should be large enough to maintain a good mitigation of current ripples. Taking into account the intersection of conditions (26.50) and (26.51), as shown in Fig. 26.27, its minimum value within the gray area is given in Eq. (26.52) where V_{dc_min} is the minimum value of V_{dc} (Eq. 26.53) (Niazy et al., 2010).

$$C_{dc} = \frac{S_b}{\omega_b V_{dc\_min}\Delta V_{dc\_min}} \approx 0.1F \tag{26.52}$$

$$V_{dc\_min} = 690\sqrt{2} \approx 1000\ V \tag{26.53}$$

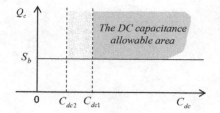

FIGURE 26.27 The allowable area for the DC bus capacitance.

26.6   Validation results

Using MATLAB/Simulink software, the simulation results are developed to verify the reactive capability and the secure injection of active and reactive power into the grid. The system parameters depicted in Fig. 26.1 are given in Tables A1 and A2. The maximum limits of powers given in above are investigated, and then, the maximum rotor flux (Φ_{r_max}) is considered as the reference value for the rotor flux orientation. The supervisory control generates the active and reactive power references for scenario including two kinds of control.

From 0 to 4 seconds: the maximum power point tracking control is applied, in which the reactive power equals to zero;

From 4 to 8 seconds: the PQ control is activated to produce and consume the reactive power according to the requested power plan production. The simulation results are given in Fig. 26.28. Fig. 26.28A shows that the active

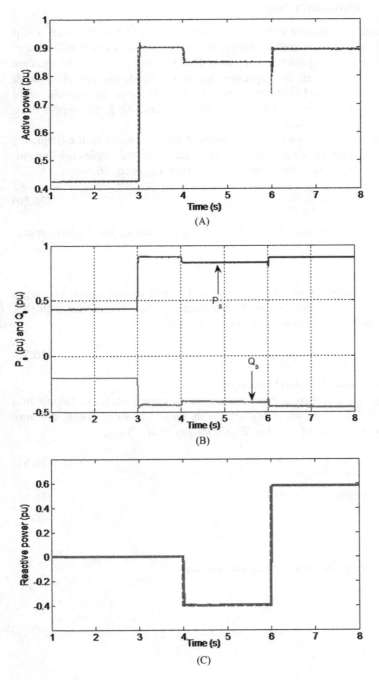

(A)

(B)

(C)

FIGURE 26.28   (A) Wind turbine active power injected into grid, (B) Active power generator output. (C) Reactive power injected into grid and that consumed by squirrel cage induction generator.

power is reduced, when the reactive power increases, to not exceed the apparent power. The generator output active power is proportional to the consumed reactive power (see Fig. 26.28B).

- From 4 to 6 seconds, the reference of reactive should be consumed (0.4 pu).
- From 6 to 8 seconds, the reference of reactive should be produced (0.7 pu), which is outside of total RPCL of the full converter. Then, the power reference is limited to $Q_{ref} = Q_{g\_max}$ Fig. 26.28B.

Fig. 26.28C shows that the produced reactive power required by the TSO is limited to a value inside the total RPCL, but, when the injected reactive power increased, the injected active power decreased to not exceed the apparent power. The reactive power consumed by the generator is limited to its maximum value when it could go as far as the full converter limit constraint during the PQ control.

26.7 Conclusion

The RPCLs for the WT must be developed to offer a plan of reactive power resources required to meet the new grid code recommendations, such as wind farm PF control and voltage control. In this chapter, the RPCLs are given for a large-scale WT to meet the power demands of the grid operator and ensure the security of SCIG and for optimal integration of WT in electrical power network with PQ control mode. These limits were obtained taking into account the maximum voltage and the maximum current of both the generator and the transmission line, and PQ control RPCL for a full converter. The DC bus capacitance is designed to produce the required reactive power and to attenuate the DC bus current ripples.

The results show that the generation and absorption of reactive power are always under capability limits during the PQ control mode.

As a perspective, it should be interesting to investigate the reactive power capability of a WF composed of large number of WTs to meet the grid code in a cooperative manner, using artificial intelligence techniques. These reactive power capability curves can also be investigated for an optimal distribution of the active and reactive power generation on WTs of the farm, using adequate optimization algorithms.

Nomenclature

Abbreviations

WPP	wind power plant
WT	wind turbine
WF	wind farm
SCIG	squirrel cage induction generator
DFIG	doubly fed induction generator
PMSG	permanent magnet synchronous generator
PCC	point of common coupling
RPCL	reactive power capability limit
SSC	stator side converter
GSC	grid side converter
WECSs	wind energy conversion systems
PWM	pulse width modulation
AESO	Alberta Electric System Operator
PF	power factor

Symbols

S_m	slip
P_s	stator active power
Q_s	stator reactive power
S_s	apparent power
S_n	nominal apparent power
T_m	mechanical torque
ω_m	mechanical speed
$I_{r\text{-}mzax}$	maximum rotor current and
$\Phi_{r\text{-}max}$	maximum rotor flux

TABLE A1 Parameters of squirrel cage induction generator.

Symbol	Quantity	Value
V_n	Nominal voltage (RMS Ph-Ph)	690 (V)
F	Nominal frequency	50 (Hz)
R_s	Stator resistance	0.004843 (pu)
L_s	Stator leakage inductance	0.1248 (pu)
R_r	Rotor resistance	0.004347 (pu)
L_r	Rotor leakage inductance	0.1791 (pu)
L_m	Mutual inductance	6.77 (pu)

TABLE A2 Parameters of grid side system.

Symbol	Quantity	Value (pu)
R_L	Cable resistance	0.019
L_L	Cable inductance	0.016
C_L	Cable capacitance	0.055
R_{tr}	Transformer resistance	0.0017
L_{tr}	Transformer inductance	0.05
R_f	Filter resistance	0.027
L_f	Filter inductance	0.186

Subscripts

s	the stator side of the induction generator
r	the rotor side of the induction generator
g	the grid side
m	the mechanical quantity

Appendix A

{{{Tables A1 and A2}}}

References

Abo-Khalil, A. G., & Lee, D.-C. (2008). DC-link capacitance estimation in AC/DC/AC PWM converters using voltage injection. *IEEE Transactions on Industry Applications, 44*(5).

An, H., Ko, H., Kim, H., et al. (2012). Modeling and voltage-control of variable-speed SCAG-based wind farm. *Renewable Energy, 42*, 28–35.

Anon (2015). Global wind report, annual market update 2014. Global Wind Energy Concil (GWEC).

Anwar, M. N., & Teimor, M. (2002). An analytical method for selecting DC-link-capacitor of a voltage stiff inverter. In 37th IAS annual meeting. Conference record of the industry applications conference, 2002 (pp. 803–810). IEEE.

Bao, L., Huang, Z., & Xu, W. (2003). On-line voltage stability monitoring using Var reserves. *IEEE Transactions on Power Systems, 18*(4), 1461–1469.

Benlahbib, B., & Bouchafaa, F. (2014). Centralized algorithm for wind farm supervision. In International conference on control, engineering and information technology (CEIT'14) proceedings (pp. 37–41).

Benlahbib, T., Ghennam, & Berkouk, E. (2013). Improvement algorithm for wind farm supervision based on proportional distribution. In *The international conference on electronics & oil: From theory to applications (ICEO'13)*, March 5–6.

Bueno, E. J. et al. (2006). Calculation of the dc-bus capacitors of the back-to-back NPC converters. Presented at 12th int. power electron. motion control conf. EPE-PEMC 2006, August 30/September 1.

Bueno, E. J., J., F., Rodriguez, A. H., & Espinosa, F. (2008). Design of a back-to-back NPC converter interface for wind turbines with squirrel-cage induction generator. *IEEE Transactions on Energy Conversion, 23*(3).

Domínguez-García, J. L., Gomis-Bellmunt, O., Trilla-Romero, L., & Junyent-Ferré, A. (2012). Indirect vector control of a squirrel cage induction generator wind turbine. *Computers and Mathematics with Applications, 64*, 102–114.

Elyaalaoui, K., Ouassaid, M., & Cherkaoui, M. (2019). Dispatching and control of active and reactive power for a wind farm considering fault ride-through with a proposed pi reactive power control. *Renewable Energy Focus, 28*, 56–65.

Folly, K.A. & Sheetekela, S.P.N., 2009. Impact of fixed and variable speed wind generators on the transient stability of a power system network (pp. 1–7).

Ghennam, T., Francois, B. & E.M. Berkouk, 2009. Local supervisory algorithm for reactive power dispatching of a wind farm. In 13th European conference on power electronics and applications EPE'09 (pp. 1–10), 8–10.

Hansen, A. D., Sorensen, P., Iov, F., & Blaabjerg, F. (2006). Centralised power control of wind farm with doubly fed induction generators. *Renewable Energy, 31*, 935–951.

Hansen, A., Jauch, C., & Sørensen, P. (2003). Dynamic wind turbine models in power system simulation tool DIgSILENT. RISO National Laboratory.

Kieferndorf, F. D., Förster, M., Lipo, T., et al. (2004). Reduction of DC-bus capacitor ripple current with PAM/PWM converter. *IEEE Transactions on Industry Applications, 40*(2), 607–614.

Kolar, J. W., Wolbank, T. M., & Schrodl, M. (1999). Analytical calculation of the RMS current stress on the dc link capacitor of voltage dc link PWM converter systems. In IEE conference on electrical machines and drives (pp. 81–89).

Kumar, S., Singh, O., & Aggarwal, S. K. (2016). A comparative study for reactive power capability of doubly fed induction generator and synchronous generator. In *International conference on control, computing, communication and materials (ICCCCM)*.

Leonardi, B., & Ajjarapu, V., (2008). Investigation of various generator reactive power reserve (GRPR) definitions for online voltage stability/security assessment. In Power and *energy society general meeting-conversion and delivery of electrical energy* in the 21st century (pp. 1–7). IEEE.

Li, H., et al. (2012). Analysis and estimation of transient stability for a grid-connected wind turbine with induction generator. *Renewable Energy, 36*, 1469–1476.

Liserre, M., Klumper, C., Blaabjerg, F. et al. (2004). Evaluation of the ride-through capability of an active-front-end adjustable speed drive under real grid conditions. In *30th annual conference of IEEE* Industrial Electronics Society, IECON 2004 (pp. 1688–1693).

Malamaki, K.-N., Mushtaq, U., & Cvetkovic, M. (2020). Enable ancillary services by RES. Report on the definition of the converter reactive power capability. EASY-RES.

Montilla-DJesus, M. E., Arnaltes, S., Castronuovo, E. D., & Santos-Martin, D. (2017). Optimal power transmission of offshore wind power using a VSC-HVdc interconnection. *Energies*.

Niazy, I. et al. (2010). Participation in reactive power market considering generator aging. In 2010 IEEE conference proceedings, IPEC (pp. 1067–1072).

Ouassaid, M., Elyaalaoui, K., & Cherkaoui, M. (2016). Reactive power capability of squirrel cage asynchronous generator connected to the grid. In Proceedings of 2015 IEEE international renewable and sustainable energy conference, IRSEC 2015.

Qin, W., Wang, P., Han, X., & Du, X. (2011). Reactive power aspects in reliability assessment of power systems. *IEEE Transactions on Power Systems, 26*(1).

Qin, Z., Wang, H., Blaabjerg, F., & Loh, P. C., 2014. Investigation into the control methods to reduce the DC-link capacitor ripple current in a back to-back converter. In *Proceedings of the 2014 IEEE energy conversion congress and exposition (ECCE)* (. 203–210). IEEE.

Salcone, M., & Bond, J. (2009). Selecting film bus link capacitors for high performance inverter applications. In International electric machines and drives conference (IEMDC'09) (pp. 1692–1699). IEEE.

Sandia, N. (2012). *Reactive power interconnection requirements for PV and wind plants (SAND2012–1098)*.

Santos-Martin, D., Arnaltes, S., & Amenedo, J. R. (2008). Reactive power capability of doubly fed asynchronous generators. *Electric Power Systems Research, 78*(111), 1837–1840.

Sarkar, M., Altin, M., Sørensen, P. E., & Hansen, A. D. (2019). Reactive power capability model of wind power plant using aggregated wind power collection system. *Energies, 12*(1607).

Tian, J., Su, C., & Chen, Z. (2013). Reactive power capability of the wind turbine with doubly fed induction generator. In IECON 2013—39th annual conference of the IEEE Industrial Electronics Society (pp. 5312–5317).

Vakula, V. S., Sandeep, G., & Krishna, S. V. (2015). Effective generator reactive power reserve management using fuzzy constrained optimization. *International Journal of Innovative Research in Advanced Engineering, 2*.

Zhai, J., & Liu, H. (2014). Reactive power control strategy of DFIG wind farms for regulating voltage of power grid. In *PES general meeting conference & exposition, 2014 IEEE* (pp. 1–5).

Zhao, J., Li, Z., & Li, D. (2011). Reactive power optimization algorithm of considering wind farm voltage control capability in distribution system. In *International conference on electrical machines and systems (ICEMS)* (pp. 1–4), Aug 20–23.

Chapter 27

Demand-side strategy management using PSO and BSA for optimal day-ahead load shifting in smart grid

Rajaa Naji E.L. Idrissi, Mohammed Ouassaid and Mohamed Maaroufi

Engineering for Smart and Sustainable Systems Research Center, Mohammadia School of Engineers, Mohammed V University in Rabat, Rabat, Morocco

27.1 Introduction

27.1.1 Context and problematic

During the 21st century, the renewable resources (RES) have received a big interest of many countries to establish the development of low greenhouse emissions. Integrating these clean resources into the power grid provides many advantages, such as improving the positive environmental impact, which leads to the grid decarbonization. Nevertheless, the random fluctuations and intermittent nature for both the load demand and production of renewable energy, even for a few minutes, produces various technical challenges in the power quality, power variation, storage, optimal placement of RES, and protection issues (Anees, 2012). Consequently, a smart additional control has to be integrated into the conventional energy units. In this context, the generation of smart grid presents an alternative solution to produce a permanent supply-demand balance in the real-time scale by incorporating more advancing technologies in the generation, transmission, and demand side of the network. Demand-side management (DSM) systems are one of the solutions focusing on the improvement of the efficiency of the renewable energy utility in the customer side. It includes the advanced mechanisms aiming to achieve an energy balance of the power network by reorganizing the energy consumption in the time pattern (Qureshi, Nair, & Farid, 2011; Zhou & Yang, 2015).

27.1.2 State of art

Commonly, DSM approaches are used to design a new load profile to profit from economic savings. This can motivate users to control their power flow and facilitate the energy management. As a result, the consumers have been more and more active by participating in the energy redistribution and ensuring their needs of power that eliminate the differentiation between producers and users to give a new entity called "prosumer" (Lee, Chuang, Chu, & Cheng, 2009).

DSM subject has been extensively studied in the scientific literature through three main approaches: (1) game theoretic (GT) approach, (2) multiagent systems (MAS), and (3) optimization techniques.

Many authors proposed the GT approach for solving DSM challenges. It is based on games between players to achieve the optimal output. Nekouei, Alpcan, and Chattopadhyay (2014) solve the issues of utility and consumer using a GT model. Similarly, the minimization of user's dissatisfaction and energy cost is solved in Yaagoubi and Mouftah (2014), while Saghezchi, Saghezchi, Nascimento, and Rodriguez (2014) deal with energy demand reduction based on user preferences and electricity pricing.

The energy management technique based on MAS was also studied in various papers (Asare-Bediako, Kling, & Ribeiro, 2013; Kyriakarakos, Piromalis, Dounis, Arvanitis, & Papadakis, 2013; Roche, Lauri, Blunier, Miraoui, & Koukam, 2013). Indeed, it focuses on interactions between various entities (controllers, power sources) to achieve the desired objectives. Nevertheless, most of the DSM problems were handled as an optimization problem; subject to minimize or maximize a certain objective functions under different constraints. These methods are divided into subcategories depending on: (1) user's interactions: individual or collective; (2) the time scale: either predictive or real-time control; or (3) the optimization scheme used to solve the problem: deterministic or stochastic. Siano (2014) reviews various DSM approaches in residential sector. In Huneke, Henkel, González, and Erdmann (2012), a linear programming is used to reduce the overall price of a hybrid off-grid system. Sarabi, Kefsi, Merdassi, and Robyns (2013) deal with dynamic programming to reduce the power supply from the distribution grid. A quadratic programming is proposed in Boonbumroong et al. (2009) to design wind-solar system for grid-connected and isolated application. In Logenthiran, Srinivasan, and Shun (2012), authors proposed a load reshapes in different sectors (commercial, industrial, and residential) by genetic algorithm. An identical model was solved using particle swarm optimization (PSO) in Logenthiran, Srinivasan, and Phyu (2015) and Gupta, Anandini, and Gupta (2016). Other studies considered a multiobjective optimization to maximize the renewable energy generation (Jiang & Low, 2011), to maintain high voltage stability (Wang et al., 2012), or to optimize simultaneously the peak load and energy cost (Pallotti, Mangiatordi, Fasano, & Del Vecchio, 2013).

Among the multiple tools deployed for the optimization of hybrid systems, HOMER Software was used in this work due to its advanced performances, including the ability of combining many sources of energy for both grid-connected and isolated designs, capability of performing optimization and sensitivity analysis, which evaluate various possible system configurations quickly and easily. Consequently, many researchers have used it to examine the technical and economic feasibility of hybrid system applications in the worldwide. An optimal sizing for rural electrification was presented in Kolhe, Iromi, Ranaweera, and Sisara Gunawardana (2015). The study performed in Rajbongshi, Borgohain, and Mahapatra (2017) is addressed to the design of hybrid system based on diesel, biomass, and photovoltaic (PV) connected to the grid to optimize the system configuration for several consumption profiles. The optimization, simulation, and computation of a fuel cell, solar, and biomass hybrid energy system were studied in Singh, Baredar, and Gupta (2015). The analysis of PV−diesel energy system performance, including different parameters, comprising ambient temperature, weather data, solar radiation intensity, and daily demand schemes, was applied for testing in Yousof Lau, Yousof, Arshad, Anwari, and Yatim (2010).

27.1.3 Contribution

In this chapter, the load shifting algorithm is proposed to handle an important number of loads from different types and multiple characteristics (time connection, power consumption...). For this objective, a novel algorithm named backtracking search algorithm (BSA) is used to solve two problems:

- Reduction of load peak;
- Minimization of the electricity bill.

The BSA was initiated by Civicioglu in 2013. It presents a simple and effective optimization algorithm by including new developed elements, for example, the employment of dual population, one control parameter to control the amplitude of search direction, though DE incorporates two parameters for fine-tuning, in addition of consulting the previous population at each iteration to obtain more efficient population (Zain, Kanesan, Kendall, & Chuah, 2018).

The main contribution of this work relies on the examination of the feasibility of the obtained load data after the required schedules of the day-ahead load shifting algorithm. Fig. 27.1 presents an overview of the main steps of this study. The study is carried out in two steps: (1) the reduction of the load demand that reduces the energy cost in less computational time and (2) the optimization of the distributed power based on a Hybrid Optimization of Multiple Electric Renewables (HOMER Pro Software) (Fig. 27.2).

27.1.4 Chapter organization

The rest of this chapter is organized as follows. Section 27.2 reflects the main driven of DSM approaches. Section 27.3 provides the mathematical formulation of the considered load shifting problem with constraints. In Section 27.4, the details of the proposed optimization methods for DSM programs are provided. The study of feasibility of the scheduled representative demand profiles is given in Section 27.5. Section 27.6 presents the obtained results from the proposed case study. Finally, Section 27.7 summarizes the conclusions and future venues of this study.

27.2 DSM driven approaches

DSM techniques are classified into six subcategories according to their purposes as given in Table 27.1 (Ouassaid et al., 2018). It affects the end users or utility activities to achieve three goals:

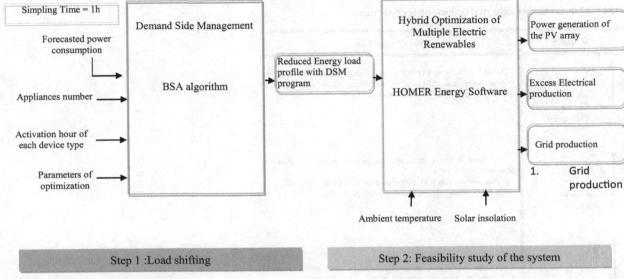

FIGURE 27.1   Overview of the proposed study.

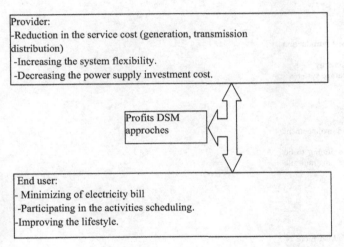

FIGURE 27.2   Benefits of demand-side management strategies.

27.2.1   Environmental goal

For example, in Lokeshgupta and Sivasubramani (2018), an optimization of gases emissions is proposed. The studied environmental dispatch was traited as a multiobjective DSM approach. A comparison of nondominated sorting genetic algorithm (NSGA-II) and multiobjective particle swarm optimization (MOPSO) demonstrates the effectiveness of the MOPSO in terms of optimum emissions.

The work presented in Sen, Sharma, Goyal, and Singh (2017) focused on pollution control. An MOPSO algorithm is formulated to minimize the fuel gases emissions (NO_x, CO_2, SO_2). The obtained results claim the total fuel dispatch minimization of 3.54%.

27.2.2   Economic dispatch

The tariff goal, such as avoiding high energy prices periods, was a subject of various studies. Neves and Silva (2015) optimize the total electricity bill of an isolated micro-grid application using genetic algorithm and linear programming to compare the representative day. The results prove the efficiency of the genetic optimization in the minimization of the operation cost scheduling.

TABLE 27.1 Demand side management techniques.

DSM Strategy	Main Characteristics
Peak Clipping	- Direct load control strategy to make minimization of the peak consumption, and valley filling constructs the off-peak power demand by exercising direct demand control.
Load Shifting	- The most effective demand management method in current distribution networks. -Advancing strategy with time independence of devices, and shifts loads from on-peak instants to off-peak time.
Strategic Conservation	- Plans to obtain load shape optimization over application of load reduction strategies directly at end-users premises.
Strategic Load Growth	- The daily response optimization of large request introduction beyond the valley filling process. -Increases the market share of loads supported by energy conversion and distributed energy resources or storage systems
Flexible Load Shape	- Principally linked to reliability of smart grid management systems. -Identify end-users with flexible loads which are aiming to be supervised during critical periods in exchange for multiple incentives.
Load Building	-The main purpose is to maximize the utility of load more or less equally during the day. -Presents the inverse of strategic conservation .

DSM problem for an economic dispatch is presented in Soares, Gomes, Antunes, and Cardoso (2013).The genrtic algorithm (GA) and time pricing are used to solve the consumer's bill issue. The shown results provide a 10% of minimization of unscheduled sheme compared to the scheduled one.

Bedi and Rajawat (2016) demonstrates the efficiency of a real time pricing in the load scheduling. An online scheduler shifts the loads according to the time pricing. About 20% of peak load were curtailed.

27.2.3 The network driven

It focused on minimizing problems of unstability in the electricity network through an optimal planning of the operation power systems to avoid an overload.

Khan, Mahmood, Safdar, Khan, and Khan (2016) review the mathematical models of demand forecasting applied in the DSM approaches. A modeling structure of renewable resources in smart grids is established in Et-Tolba, Maaroufi, and Ouassaid (2013) to ensure a balance supply—demand in real time.

In Cecati, Buccella, Siano, and Piccolo (2013), a DSM scheme is considered to optimize the operation of a smart grid. The integration of distributed renewable generation is taken into consideration as a motivating task.

Because of the RES data uncertainties, user's activities, and electricity market, stochastic optimization methods are widely used in solving DSM objectives.

27.3 Mathematical formulation of the problem

27.3.1 Problem formulation

In the following subsections, the optimal DSM is formulated with the main goal of minimizing the peak power demand. In addition; the optimization approach has to increase the energy efficiency of the system by reducing the energy bills.

First objective function:

As mentioned earlier, the developed DSM program considers a large number of controllable appliances. The main objective is to reschedule a forecasted load demand profile the nearly as possible to a desired load consumption curve. Eq. (27.1) provides the first objective function.

$$f_1(t): \sum_{t=1}^{24} [|P_{\text{forecast}}(t) - P_{\text{Target}}(t)| - |P_{\text{load}}(t)|] \tag{27.1}$$

Second objective function:

The scheduled demand profile is used to optimize the overall electricity cost. Accordingly, the second objective function to be optimized is expressed mathematically by Eq. (27.2).

$$f_2(t): \sum_{t=1}^{24} (f_1(t) \times C(t)) \tag{27.2}$$

$$\Delta P = P_{\text{forecast}} - P_{\text{Target}} \tag{27.3}$$

where P_{Target} is the target curve, that is, inversely proportional to the predicted electricity prices. In particular, at an instant, if the electricity price is high, the value of objective profile is low and vice versa. In this chapter, the P_{Target} is calculated using Eq. (27.4).

$$P_{\text{Target}}(t) = \left(\frac{C_{\text{avg}}}{C_{max}} \sum_{t=1}^{24} P_{\text{forecast}}(t) \right) \frac{1}{C(t)} \tag{27.4}$$

where C_{avg} refers to the average daily cost; C_{max} refers to the maximum cost throughout the day of 24 hours; and $C(t)$ refers to the cost of electricity at hth hour.

The $P_{\text{forecast}}(t)$ presents the predicted hourly demand. It has been taken from the data of the smart grid mentioned in Logenthiran et al. (2012) with three kinds of consumers, that is, commercial, residential, and industrial.

In addition, Eqs. (27.5) and (27.7) determine the total of the load P_{load} to be activated or disconnected at each hour t.

To approach an objective demand profile to a predicted one, the proportion ΔP determines the hourly load to be subtracted or added. Hence, two cases are presented:

Scenario 1: Time of connection ($\Delta P < 0$)

Note that, if for an hour the amount of expected load is lesser than the targeted demand value, the $P_{\text{load}}(t)$ is calculated by Eq. (27.5):

$$P_{\text{load}}(t) = \text{Connect}(t) \tag{27.5}$$

Hence, the task of appliances $P_{\text{load}}(t)$ need to be activated.

Scenario 2: Time of disconnection ($\Delta P > 0$)

In this case, the predicted demand is grater than the aimed consumption. Consequently, $P_{\text{load}}(t)$ is given as follows:

$$P_{\text{load}}(t) = \text{Disconnect}(t) \tag{27.6}$$

As result, the devices $P_{\text{load}}(t)$ should be disconnected.

In Eq. (27.7), Connect(t) presents the total of load required to be connected at hour h, so that the valley could be filled up. It is calculated for instants when the value of ΔP is negative. Otherwise the value of connected load equals to zero. Commonly, the connection times refers to the ones with low costs.

$$\text{Connect}(t) = \sum_{k=1}^{D} X_{kh}P_{1k} + \sum_{j=1}^{l-1}\sum_{k=1}^{D} X_{k(h+j)}P_{(1+j)k} \tag{27.7}$$

This equation is divided into two elements. The first element determines the power consumed by the X_{kh} number of appliances at the hth hour. In the second element, if the appliance operates in more than 1 hour, the consumed power is calculated at successive hours, that is, $h+1, h+2\ldots(h+l-1)$. X_{kh} and $X_{k(h+l)}$ are the quantity of appliances of type k assumed to be operating at hour h and $h+l$, P_{1k} is the power demand of appliance k in the first hour of operation, and $P_{(1+j)k}$ is the power demand of the same appliance at following hours.

Disconnect(t) indicates the total amount of appliances that needs to be shifted at the hth hour; thus load functioning time must be minimized. It is calculated at those times when ΔP has a positive value. Otherwise the disconnected load takes the value zero. Disconnect(t) is calculated by the following equation.

$$\text{Disconnect}(t) = \sum_{k=1}^{D} X_{kh}P_{1k} + \sum_{j=1}^{l-1}\sum_{k=1}^{D} X_{k(h+j)}P_{(1+j)k} \tag{27.8}$$

In this equation, the terms have a similar meaning to those of Eq. (27.7).

Constraints:

According to the operational constraints of the system, the following limitations are provided and must be assured over the operation of the studied system for any feasible solution.

- The number of appliances to be shifted at any time step must not be higher than the total of all the shiftable appliances controlled at the same time period.

$$\sum_{t=1}^{24} X_{kh} \leq \text{controllable}(k) \tag{27.9}$$

- The number of appliances displace away from an instant step cannot be a negative value. This constraint is expressed by Eq. (27.10):

$$X_{kh} \succ o \quad \forall_{k,h} \tag{27.10}$$

27.4 Proposed demand management optimization algorithm

In this section, the proposed day-ahead DSM algorithm is described. The objective is to redistribute and plan the user's demand activities based on the load shifting technique. This method presents the advantage of time independence of equipments and reshape the peak load from high energy price period to the lower one (Chuang & Gellings, 2008). The implementation of main steps of our proposed method is described in the flowchart given in Fig. 27.3 and can be summarized as follows:

Step 1: Gathering input data

In this first step, the algorithm reads the input data related to the connection time of devices, the number of each controllable devices the forecasted load data, and the optimization data.

Step 2: Target curve calculation

This second step is primordial. It gives the gap to be optimized between the forecasted values and objective curve.

Step 3: Evaluate the objective functions Eqs. (27.1) and (27.2)

The objective function is evaluated and updated at each iteration, until obtaining the best value.

$$bestfitness(i) = minobjective(i) \tag{27.11}$$

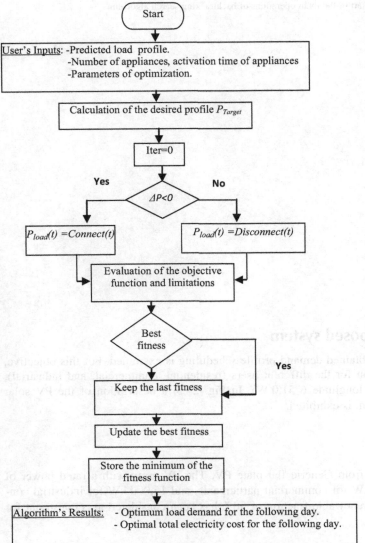

FIGURE 27.3 Flowchart of demand-side management method.

Step 4: Simulation outputs
The algorithm provides the results:

- -Optimum load demand of the following day;
- -Optimal total electricity cost of the following day.

%Load peak reduction

$$P = \frac{\text{reduced demand}}{\text{Max demand}} \qquad (27.12)$$

Several optimization techniques have been used to solve such a problematic. Among those techniques, deterministic programming approaches remain useful while the variable number is not important. However, the studied DSM algorithm should be effective and flexible enough to manage the important number of devices in different areas: residential, commercial, and industrial. Hence, in this study to minimize the load peak deficit for each profile, the BSA optimization algorithm is proposed. Basically, it established by generating three essential processes that are selection, mutation, and crossover as presented in Fig. 27.4.

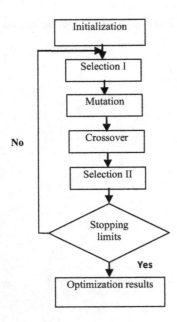

FIGURE 27.4 Flowchart of the main operations of backtracking search algorithm.

27.5 Energy management of the proposed system

In the second part of this work, a feasibility of the obtained demand profile scheduling is examined. For this objective, a hybrid system is designed for electricity generation for the different users (residential, commercial, and industrial), located in Rabat, Morocco (latitude 33°58.3′N and longitude 6°51.0′W). In Fig. 27.5, a description of the PV solar panels connected to the national electricity grid system is exhibited.

27.5.1 Solar PV modules

In this system, the PV-generated power is received from Generic flat plate PV. The PV array with a rated power of 3500 kW was used for residential consumers, 5000 kW for commercial participants, and 15,000 kW for industrial consumers. To ensure more usage of RES, a minimum renewable fraction of 50% has been considered as a constraint for this optimization problem.

27.5.2 Grid

In this designed model, grid is used to compensate electricity for the period in which the total consumer demand is higher than the available renewable energy power.

27.5.3 Battery

Though renewable resources provide an average amount of energy for a long period of time, the storage unit is considered to deliver the user request when the resources supply a low amount of energy. Thus generic lead acid energy storage units are incorporated, to store the energy supply at a nominal voltage of 12 V.

27.5.4 Converter

Converter can be used to convert AC to DC voltage and vice versa; for this purrpose, it can be used both as rectifier and inverter (Kashem & Arefin, 2017). In the designed hybrid system, the converter is used to convert the DC voltage supplied by batteries and PV module into AC voltage to feed the AC load. Data of characteristics and descriptions of each component are given in Table 27.2.

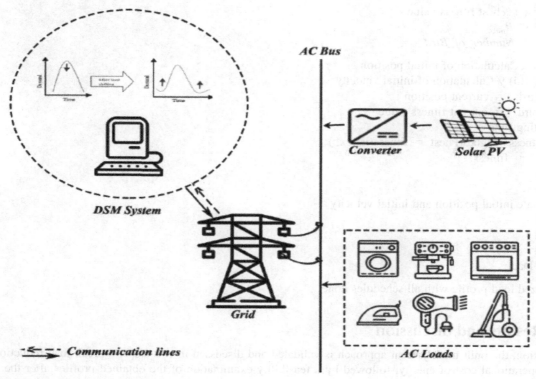

FIGURE 27.5 The architecture of the studied system.

TABLE 27.2 Components characteristics.

Components	Total		
	Residential zone	Commercial zone	Industrial zone
PV generic flat plate (kWh per year)	6,238,083	8,911,548	26,734,643
Battery generic lead acid (strings)	2	2	9
Converter system (kWh per year)	5,640,946	8,230,242	9,213,252
Grid (kWh per year)	2,599,361	3,836,064	24,574,080
PV, photovoltaic.			

Algorithm 27.1: Main PSO steps

1. Data inputs: (Number of devices, operation hours, forecasted load, starting time of each device, first hour of consumption…) **for** $t = 0:24$
2. Calculation of fitness function % Eq. (27.4) $t = t + 1$
3. Data initialization for PSO parameters:

Number_of_Bird $= 20\%$ Swarm size

$it_{max} = 100\%$ Maximun number of generation
$w_{min} = 0.9\%$ minimum inertia weight
$w_{max} = 0.4\%$ maximum inertia weight
$P_{best} = (\ )\%$ best position

G_{best} = ( ) %best fitness value
While $i < it_{max}$
for j = 1: *Number_of_Bird*

4. bird(:,j) %Calculation of initial position
5. Velocity(:,j) %Calculation of initial velocity
6. *Pbest*(bird) = current position
7. Gbest(bird) = current fitness value
8. Evaluating fitness
    if fitness < gbestPbest = fitness(index,:);
    G_{best} = fitness;
    end
    else
    Update initial position and initial velocity
    end
    end j
    end *t*
9. Output Data:

Optimized load profile with all schedules done.

27.6 Results and discussion

In this section, the built management approach is validated and discussed in two steps: a preak load reduction to minimize the operational cost of energy; followed by a feasibility examination of the obtained profiles after the load management using Homer energy software. An evaluation of the effectiveness of the results is carried out by comparing the PSO- and BSA-based solutions. The optimization process is repeated 100 times. Table 27.3 presents the main parameters of the optimization methods chosen appropriately. The scheduling program is tested for residential, commercial, and industrial users. As shown in the forecasted load profiles of each customer, each one has a unique model depending on the occupant consumption profile and weather conditions. On a typical day, the scheduling period is considered from the 8th hour of the present day to the 8th hour of the following day. For comparison performance, the load data subjected to shift are taken from Logenthiran et al. (2015).

27.6.1 Peak load reduction

Tables 27.4–27.6 illustrate the demand patterns of the loads presumed to be controlled for each areas. Three studied curves are compared in each case: the predictive one, desired one, and the obtained curve after implementing the optimal load shifting strategy.

For residential category, there are 14 different types of devices under time shifting forming 2600 equipments in the total. It is observed from the simulation results in Fig. 27.6 that the proposed DSM strategy has reduced peak load consumption by 33%, from 1336.6 to 907.34 kW. This leads to asignificant decrease of 16% in the utility bill.

Devices under control in the commercial zone have the upper power demand pattern than industrial and residential users. As mentioned in Table 27.5, it contains 808 appliances, belonging to 8 different types. As indicated in Fig. 27.7,

TABLE 27.3 Parameters of BSA and PSO.

PSO parameters		BSA parameters	
Swarm size	20	Population size	20
Maximum number of generation	100	Maximum number of generation	100
Maximum and minimum inertia weight	0.9, 0.4	Dimension rate	0.7
Cognitive and social constant	2		

BSA, backtracking search algorithm; *PSO*, particle swarm optimization.

TABLE 27.4 Details of shiftable loads for residential consumers.

Appliance type	Devices number	Start time	End time	Duration of load	Power (kW)
Kettle	406	9	15	2	2.00
Iron	340	9	19	1	1.00
Dish wacher	28	8	24	1	0.70
Fan	288	12	17	4	0.20
Oven	279	11	15	1	1.30
Washing machine	268	0	24	2	0.50
Dryer	189	0	24	1	1.20
Vacuum cleaner	158	10	17	1	0.40
Frying pan	101	11	18	1	1.10
Blender	66	9	20	1	0.30
Rice cooker	59	16	20	1	0.30
Hair dryer	58	20	22	1	1.50
Coffee maker	56	5	10	2	0.80
Toaster	48	6	9	1	0.90
Total	2604				

TABLE 27.5 Details of shiftable loads commercial consumers.

Appliance type	Devices number	Start time	End time	Duration of load	Power (kW)
Water dispenser	156	5	20	1	2.50
Dryer	117	0	24	1	3.50
Kettle	123	11	16	2	3.00
Oven	77	7	16	1	5.00
Coffee maker	99	6	14	2	2.00
Fan/AC	93	10	19	2	3.50
Aie Con	56	12	18	3	4.00
Lights	87	8	19	3	2.00
Total	808				

TABLE 27.6 Details of shiftable loads for industrial consumers.

Appliance type	Devices number	Start time	End time	Duration of load	Power (kW)
Water heater	39	6	20	4	12.50
Welding machine	35	8	23	5	25.00
Fan/AC	16	9	16	5	30.00
Arc furnace	8	0	24	6	50.00
Induction motor	5	8	18	6	100.00
DC motor	6	8	18	3	150.00
Total	109				

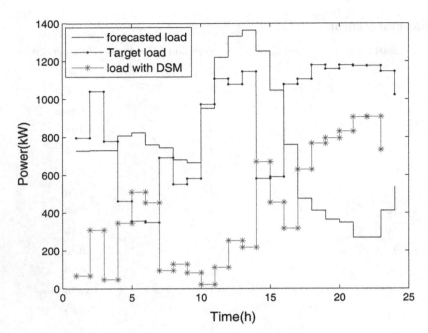

FIGURE 27.6 Demand-side management for residential users.

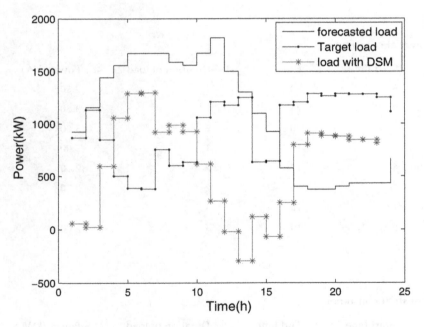

FIGURE 27.7 Demand-side management for commercial users.

the peak load in the predicted profile is minimized from 1818.2 to 1286.1 kW after implementing the DSM strategy. This reduction influenced the operational price, ensuring 21% saving.

Similarly, the obtained results of industrial sector demonstrates a decrease of the total energy request of 24% from 2727.7 kWh without DSM to 2278.6 kWh after applying the proposed DSM technique as shown in Fig. 27.8.

Fig. 27.9 illustrates a summary of the supply peak demand reductions for all user categories. It can be noticed that the saving rate is up to 21% in all areas. Fig. 27.9 illustrates a summary of the supply peak demand reductions for all user categories. It can be noticed that the saving rate is up to 21% in all areas.

Table 27.7 summarizes the performance analysis of DSM methodology. It is observed that both algorithms are identical in terms of best fitness. Nevertheless, the BSA provides solution in less computational time than the PSO

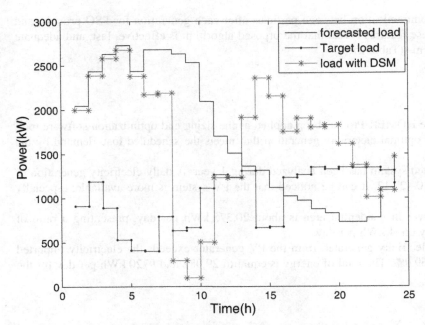

FIGURE 27.8 Demand-side management for industrial users.

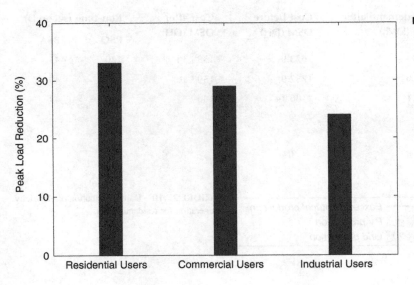

FIGURE 27.9 Peak demand minimization.

TABLE 27.7 Percentage minimization of peak demand (%) and cost (%).

Users	Peak demand minimization (%)			Cost minimization (%)		
	In Logenthiran et al. (2012)	In Gupta et al. (2016)	BSA	In Logenthiran et al. (2012)	In Gupta et al. (2016)	BSA
Residential	18.3	23.3	33	5	7.24	16
Commercial	18.3	18.5	29	5.8	9.65	21
Industrial	14.2	16.6	24	10	18.71	24

BSA, backtracking search algorithm.

simulations. This may be justified by the communication between particles after each generation for PSO process, and benefits of the simple structure of BSA. These features reveal that the prposed algorithm is effective, fast, and adequate to solve this category of optimization problems (Table 27.8).

27.6.2 Electricity generation

In the second stage of this work, the software HOMER Pro has been applied as the sizing and optimization software tool. The simulation was performed to obtain the optimal electricity generation that meets the scheduled load demand by the optimal load reshaping method studied first.

The power output of the PV-grid-connected system has been analyzed during 1 year. A daily electricity generation of each area category is presented in Figs. 27.10−27.12. It can be noticed that the PV system is more available, especially between 8 a.m. and 18 p.m.

The total amount of solar generated power in residential area is about 20,375 kWh per day, presenting a ratio of 70.6%; while the grid power ranks second by 6664 kWh per day.

From Fig. 27.11, it is obvious that the electricity generated from the PV generator exceeds the electricity imported from the grid with an important fraction of 69.9%. The total of energy is equal to 29,082 and 9720 kWh per day for the solar and grid power, respectively.

TABLE 27.8 Calculation of peak load and cost.

Users	Peak demand before DSM (kWh)	Peak demand after DSM (kWh)	Cost before DSM (DH)	Cost after DSM (DH)	Run-time (seconds)	
					PSO	BSA
Residential	1336.6	907.34	1673.9	1390.36	120	45
Commercial	2727.7	2278.6	1752.9	1390.36	109	49
Industrial	1818.2	1286.1	2386.14	1800	119	43

DSM, Demand-side management.

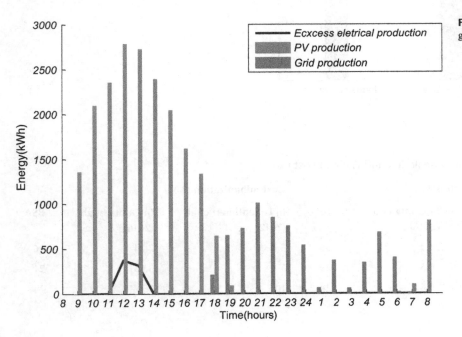

FIGURE 27.10 Daily distribution of electricity generation for residential users.

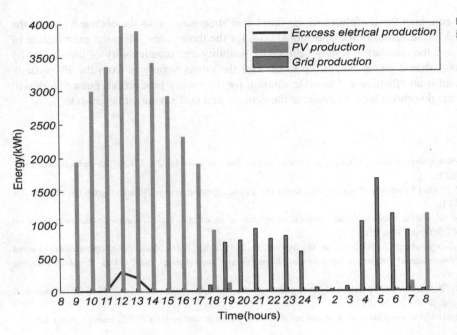

FIGURE 27.11 Daily distribution of electricity generation for commercial users.

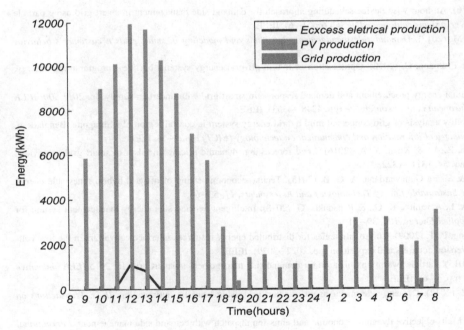

FIGURE 27.12 Daily distribution of electricity generation for industrial users.

As indicated from the hourly data for the industrial profile in Fig. 27.12, the renewable energy contribution surpasses the electricity grid generation.

Thus the obtained results prove the feasibility of the proposed power system to mitigate pollutant emissions from conventional power sources.

27.7   Conclusion

Among different developed solutions of energy management in smart grid, DSM is a primordial aspect permitting the customers to participate in making decision at consumption activities and utility bills. In this chapter, a day-ahead load shifting strategy was formulated for the optimal scheduling. The peak load reduction and energy cost problems are solved using the BSA- and PSO-based techniques. Then, the proposed algorithms are tested in three cases: residential, commercial, and industrial profiles, to enlighten the usefulness of the developed strategies. From the obtained results, the BSA guarantees about 16%, 21%, and 24% decreases in the operational cost, for the three cases. Also, the examination of the electricity generation contribution and the simulation results validate the viability and practicability of the system by ensuring the highest fraction of the total shared power output, up to 69% of the power supply is from the PV system. What is certify that the optimized system is an effective and feasible solution for the energy production. Future work will include the energy management in the neighborhood area, to evaluate the demand and cost saving, at large scale.

References

Anees, A. S. (2012). Grid integration of renewable energy sources: Challenges, issues and possible solutions. In: *2012 IEEE 5th India international conference on power electronics (IICPE)*. IEEE.

Asare-Bediako, B., Kling, W. L., & Ribeiro, P. F. (2013.) Integrated agent-based home energy management system for smart grids applications. In: *IEEE PES ISGT Europe 2013* (pp. 1–5). IEEE.

Bedi, A. S., & Rajawat, K. (2016). Online load scheduling under price and demand uncertainty in smart grid. In: *2016 international conference on signal processing and communications (SPCOM)* (pp. 1–5). IEEE.

Boonbumroong, U., Prtainthong, N., Thepa, S., Suwannakum, T., Pongchawee, D., Jivacate, C., ... Tia, S. (2009). Model-based optimization of stand alone hybrid power system. In: *Proc. 3rd international conference on "sustainable energy and environment" (SEE 2009), Bangkok, Thailand* (p. 514).

Cecati, C., Buccella, C., Siano, P., & Piccolo, A. (2013), Optimal operation of smart grids with demand side management. In: *2013 IEEE international conference on industrial technology (ICIT)* (pp. 2010–2015). IEEE.

Chuang, A. S., & Gellings, C. W. (2008). Demand-side integration in a restructured electric power industry. In: *CIGRE, number paper C6-105, session.*

Gupta, I., Anandini, G. N., & Gupta, M. (2016). An hour wise device scheduling approach for demand side management in smart grid using particle swarm optimization. *In 2016 national power systems conference (NPSC)* (pp. 1–6). IEEE.

Et-Tolba, E. H., Maaroufi, M., & Ouassaid, M. (2013). *Demand side management algorithms and modeling in smart grids a customer's behavior based study* (pp. 531–536). IEEE.

Huneke, F., Henkel, J., González, J. A. B., & Erdmann, G. (2012). Optimisation of hybrid off-grid energy systems by linear programming. *Energy, Sustainability and Society, 2*(1), 7.

Jiang, L., & Low, S. (2011). Multi-period optimal energy procurement and demand response in smart grid with uncertain supply. In: *2011 50th IEEE conference on decision and control and European control conference* (pp. 4348–4353). IEEE.

Kashem, F. B., & Arefin, M. S. (2017). Feasibility analysis of grid connected mini hybrid energy system in coastal region of Chittagong, Bangladesh. In: *2017 3rd international conference on electrical information and communication technology (EICT)* (pp. 1–5). IEEE.

Khan, A. R., Mahmood, A., Safdar, A., Khan, Z. A., & Khan, N. A. (2016). Load forecasting, dynamic pricing and dsm in smart grid: A review. *Renewable and Sustainable Energy Reviews, 54,* 1311–1322.

Kolhe, M. L., Iromi, K. M., Ranaweera, U., & Sisara Gunawardana, A. G. B. (2015). Techno-economic sizing of off-grid hybrid renewable energy system for rural electrification in sri lanka. *Sustainable Energy Technologies and Assessments, 11,* 53–64.

Kyriakarakos, G., Piromalis, D. D., Dounis, A. I., Arvanitis, K. G., & Papadakis, G. (2013). Intelligent demand side energy management system for autonomous polygeneration microgrids. *Applied Energy, 103,* 39–51.

Lee, C.-T., Chuang, C.-C., Chu, C.-C., & Cheng, P.-T. (2009). Control strategies for distributed energy resources interface converters in the low voltage microgrid. In: 2009 IEEE energy conversion congress and exposition (pp. 2022–2029). IEEE,.

Logenthiran, T., Srinivasan, D., & Phyu, E. (2015). Particle swarm optimization for demand side management in smart grid. In: *2015 IEEE innovative smart grid technologies-Asia (ISGT ASIA)* (pp. 1–6). IEEE.

Logenthiran, T., Srinivasan, D., & Shun, T. Z. (2012). Demand side management in smart grid using heuristic optimization. *IEEE Transactions on Smart Grid, 3*(3), 1244–1252.

Lokeshgupta, B., & Sivasubramani, S. (2018). Multi-objective dynamic economic and emission dispatch with demand side management. *International Journal of Electrical Power & Energy Systems, 97,* 334–343.

Nekouei, E., Alpcan, T., & Chattopadhyay, D. (2014). Game-theoretic frameworks for demand response in electricity markets. *IEEE Transactions on Smart Grid, 6*(2), 748–758.

Neves, D., & Silva, C. A. (2015). Optimal electricity dispatch on isolated mini-grids using a demand response strategy for thermal storage backup with genetic algorithms. *Energy, 82,* 436–445.

Ouassaid, M., Maaroufi, M., et al. (2018). Demand side management strategy by optimal day-ahead load shifting in smart grid. In: *2018 6th international renewable and sustainable energy conference (IRSEC)* (pp. 1–6). IEEE.

Pallotti, E., Mangiatordi, F., Fasano, M., & Del Vecchio, P. (2013). Ga strategies for optimal planning of daily energy consumptions and user satisfaction in buildings. In: *2013 12th international conference on environment and electrical engineering* (pp. 440–444). IEEE.

Qureshi, W. A., Nair, N.-K. C., & Farid, M. M. (2011). Impact of energy storage in buildings on electricity demand side management. *Energy Conversion and Management, 52*(5), 2110–2120.

Rajbongshi, R., Borgohain, D., & Mahapatra, S. (2017). Optimization of PV-biomass-diesel and grid base hybrid energy systems for rural electrification by using homer. *Energy, 126,* 461–474.

Roche, R., Lauri, F., Blunier, B., Miraoui, A., & Koukam, A. (2013). *Multi-agent technology for power system control. Power Electronics for Renewable and Distributed Energy Systems* (pp. 567−609). Springer.

Saghezchi, F. B., Saghezchi, F. B., Nascimento, A., & Rodriguez, J. (2014). Game theory and pricing strategies for demand-side management in the smart grid. In: *2014 9th international symposium on communication systems, networks & digital sign (CSNDSP)* (pp. 883−887). IEEE.

Sarabi, S., Kefsi, L., Merdassi, A., & Robyns, B. (2013). Supervision of plug-in electric vehicles connected to the electric distribution grids. *International Journal of Electrical Energy, 1*(4), 256−263.

Sen, G. D., Sharma, J., Goyal, G. R., & Singh, A. K. (2017). A multi-objective pso (mopso) algorithm for optimal active power dispatch with pollution control. *Mathematical Modelling of Engineering Problems, 4*(3), 113−119.

Siano, P. (2014). Demand response and smart grids—a survey. *Renewable and Sustainable Energy Reviews, 30*, 461−478.

Singh, A., Baredar, P., & Gupta, B. (2015). Computational simulation & optimization of a solar, fuel cell and biomass hybrid energy system using homer pro software. *Procedia Engineering, 127*, 743−750.

Soares, A., Gomes, Á., Antunes, C. H., & Cardoso, H. (2013). *Domestic load scheduling using genetic algorithms. European conference on the applications of evolutionary computation* (pp. 142−151). Springer.

Wang, D., Parkinson, S., Miao, W., Jia, H., Crawford, C., & Djilali, N. (2012). Online voltage security assessment considering comfort-constrained demand response control of distributed heat pump systems. *Applied Energy, 96*, 104−114.

Yaagoubi, N., & Mouftah, H. T. (2014). User-aware game theoretic approach for demand management. *IEEE Transactions on Smart Grid, 6*(2), 716−725.

Yousof Lau, K. Y. M. F. M., Yousof, M. F. M., Arshad, S. N. M., Anwari, M., & Yatim, A. H. M. (2010). Performance analysis of hybrid photovoltaic/diesel energy system under malaysian conditions. *Energy, 35*(8), 3245−3255.

Zain, M. Zb. M., Kanesan, J., Kendall, G., & Chuah, J. H. (2018). Optimization of fed-batch fermentation processes using the backtracking search algorithm. *Expert Systems with Applications, 91*, 286−297.

Zhou, K., & Yang, S. (2015). Demand side management in china: The context of china's power industry reform. *Renewable and Sustainable Energy Reviews, 47*, 954−965.

Chapter 28

Optimal power generation and power flow control using artificial intelligence techniques

Cheshta Jain Khare[1], H.K. Verma[1] and Vikas Khare[2]

[1]Shri G.S. Institute of Technology and Science, Indore, India, [2]School of Technology Management & Engineering (STME), NMIMS University, Indore, India

28.1 Introduction

An electric power system is a framework of electrical components that is used to supply and transmit electric power according to the consumer demand. Power system is one of the prominent part of electrical engineering that deals with the generation, transmission, distribution, and utilization of electric power. Power systems keep on increasing on the basis of changing environment, asset additions, and introduction of new technologies in generation, transmission, and distribution of electricity. In power system operation and control, the basic goal is to provide the users with quality electricity power in economic rational degree for power system and to ensure their stability and reliability. So it is necessary to carry out the planning of power system monitoring and control, but with the development of the electric power system, the demand is coming more and more, in order to ensure its safety, economic, and reliable operation of the workload increase; it has also increased the burden of staff operation. Now a day's artificial intelligence techniques have become popular for solving different problems in power systems such as control, planning, scheduling, and forecast. Artificial intelligence (AI) is a subfield of computer science that investigates how the thought and action of human beings can be mimicked by machines. Both the numeric and nonnumeric and symbolic computations are included in the area of AI (Kandel & Langholz, 1992). The mimicking of intelligence includes not only the ability to make rational decisions, but also to deal with the missing data, adapt to existing situations, and improve itself in the long-time horizon based on the accumulated experience. The computer software of the existing energy management system center is usually the numerical analysis software; it is difficult to have the target processing in operation of the power system, especially in the fault condition. Using AI techniques to assist operational personnel to check and judge greatly reduces the workload of operational staff and also improves the efficiency of dealing with incidents. This is one of the main causes in recent years that researches on application of electric power workers poured into AI (Madan & Bollinger, 1997).

AI techniques play a prominent role in power system management and control (Huajin & Hanping, 2002). The electric power industry is continuously searching for ways to improve the efficiency and reliability with which it supplies energy. Although the fundamental technologies of power generation, transmission, and distribution change quite slowly, the power industry has been quick to explore new technologies that might assist its search and to wholeheartedly adopt those that show benefits. This general tendency held true to form the various AI technologies: General planners, expert systems, artificial neural networks (ANNs), inductive learning, fuzzy logic (FL), genetic algorithms(GAs)—researchers have applied almost every form of AI tool in at least prototype form to one or more problem areas in the power industry, and new practical applications of AI appear with increasing frequency. In some cases, AI tools augment or replace existing techniques. In others, AI tools enable solutions to problems previously addressed only by natural intelligence, creating new applications for computers. The dual questions of which problems to attack with AI techniques and which AI techniques to use for a particular problem are abiding ones in the power industry, as they are in many other industries (Zhao & Zhang, 2016). Problem-centered approaches (problem-seeking solutions), tool-centered approaches (solution-seeking problems), and random-matching approaches (justified only by results) have all been employed in the power industry. In the last few years, many researchers have developed AI techniques in different areas of power system. In 1989, Lai developed and modeled an expert system for power system protection (Lai, 1989). He explained the methodology for identifying the problem, selecting the knowledge engineer, coding the rules, and for solving the problem of protection to coordinate with power industries. Further, Bretthauer, Handschin, and Hoffmannu (1992) illustrated the application of expert system in different areas of power system and power industry. They explained the methodology of software and hardware tool for designing expert system. Another method used is neural network. Nguyen (1997) demonstrated the performance of optimal power flow (OPF) using ANN. They developed parallel structure of ANN for interconnected power system to satisfy the load demand. Further, extension of using ANN is illustrated by Müller, Rider, and Castro (2010) to develop load flow studies with external equivalent. They compared the solution of basic load flow with all possible constraint as reactive power limit, voltage limit, and transformer tapping with the solution obtained ANN. They analyzed harsh conditions with computing external equivalent circuit. After development of ANN, FL method for power system protection was described by Liu (1997). Further, different AI techniques applicable to application of power system were illustrated by Saha, Rosolowski, and Izykowski (2001). Sumpavakup, Srikun, and Chusanapiputt (2010) described artificial bee colony (ABC) as one of the new computational intelligent method to solve the OPF problems. They compared the solution by ABC method with swarm intelligent−based methods to solve the problems. Further, Barbulescu, Kilyeni, Simo, and Oros (2015) developed new approaches for power system-related problem-solving (Müller et al., 2010). They elaborated an original mathematical model focused on GAs. They designed GA topology for OPF with considering some practical issues. Further they used this topology for transmission expansion planning. In this field, Imen, Djamel, Hassiba, Abdellah, and Selwa (2015) evaluated different configuration of transmission and generation of different load levels (Barbulescu et al., 2015). They compared conventional method such as Newton−Raphson technique with neural network method. Further, Pan and Zhao (2019) developed deep neural network (DNN) to illustrate DC OPF-based problems (Imen et al., 2015). This approach is inspired by observations to solving OPF of different power network, which is equivalent to high dimensional mapping between the load inputs and the dispatch. In this approach, a DNN model is constructed and trained to understand the mapping. After this mapping, optimized operating decision on the load point can be obtained. Recently, in power system, many of the researchers work on smart grid and microgrid. This becomes a new field for researchers to apply AI techniques. In this era, smart energy markets involve with AI to design efficient policy incentives and permit consumer to take the decision about their requirements with less CO_2 emission. Manage stand-alone microgrids to controllable loads; GA approach utilized by Neves, Pina, and Silca (2018). AI techniques are effectively resolving the limitations of traditional grid system (Pan & Zhao, 2019). The distributed computing methods for smart grid increase the number of security issues. Presently, cyber hacks are the threats that can increase the infrastructure failure (Neves et al., 2018; Pearson, 2011). This chapter analyzed advance challenges in the smart grid and distributed AI techniques for future generation policy. With the OPF and generation, AI techniques are also used in power system protection. Several new concepts including ANN and FL were used by Saha, Rosolowski, and Izykowski (2001) to design digital relays.

The main aim of OPF is to minimize the cost to meet the load demand with maintaining all security aspects of power system. The security of power system means maintaining all the devices within the desired range. This will comprise maximum and minimum limit of generation, maximum power flow on transmission lines with maintaining the bus voltages within the desired limits. Another aim of OPF is to determine the cost of MW transactions. There are various methodologies to solve OPF problems. These methods are grouped into two parts: conventional methods and intelligent methods. With the development of hybrid system, which may comprise conventional and nonconventional energy sources, optimized power generation and power flow become complex. In such system generation and demand, both are highly uncertain. Therefore precise optimization cannot be achieved using classical methods. This chapter explains how the AI techniques can be applied to find optimum power flow to meet the uncertain load demand. The main goal of optimum power flow is to generate optimum power in such a way that it will supply desired demand with minimum cost. Before applying AI techniques, conventional methods are discussed to understand the objective of optimum power flow problem in the next section.

This chapter is organized into six sections. In Section 28.1, introduction to artificial intelligence method for optimal power generation and power flow is given. In Section 28.2, basic OPF using conventional method is explained in brief. In this section classical methods such as gradient method, newton method, and linear programming are elaborated to find the optimum value of power flow. In Section 28.3 overview of ANN and FL is explained. In this section development algorithm of ANN is explained for optimal power generation. Concept of FL and membership function with VAr control is described in brief for optimum solution of power flow. Section 28.4 explains another intelligent method, that is, GA. This section designs an algorithm for optimum power flow with GA operators (selection, crossover, and mutation). Section 28.5 explains the future trend in artificial intelligence, which is an expert system. This section describes the application of expert system in different areas of power system. Development of expert system and their limitation is also covered in this section. Finally, Section 28.6 introduces the basic aspects of game playing in the field of OPF system through payoff function and mixed strategies.

28.2 Conventional methods

The classical methodologies include the well-known techniques such as gradient method, Newton method, quadratic method, and linear programming method. Within an OPF, to optimize (minimize or maximize) a predefined function, the values of any or all of the control variables need to be identified. Objective function comes in many forms, such as fuel costs, transmission losses, and allocation of reactive sources. The objective function of concern is usually to minimize the overall output of estimated generation units. This can be expressed for minimization of total cost as follows:

$$f(P_G) = \sum_{i=1}^{n} f_i(P_{Gi}) \tag{28.1}$$

where P_G is the power generation, and n is the number of generating bus.

28.2.1 Gradient method

Gradient method is based on the linear approximation to a gradient at any given point. To optimize the cost of generation using gradient method, the state vector x can be defined in terms of δ_i and $|v_i|$ of slack bus and δ_i of PV bus. Another vector of independent variable can be defined in terms of generator output and generator bus voltage with some fixed parameter such as real power (P) and reactive power (Q) at each bus. Therefore vector y can be written in terms of control variable say u and constant variable (Dommel & Tinney, 1968).

Now the objective function be defined as a function of control variable and state variable:

$$f(x, u) = \sum_{i=1}^{n} f_i(P_{Gi}) + f_{slack}(P_{slack}(v, \delta)) \tag{28.2}$$

To solve Eq. (28.2), define objective function with addition of Lagrangian function

$$L(x, u) = f(x, u) + \lambda^T g(x, u) \tag{28.3}$$

To minimize the objective function with constraints, the gradient of Lagrange function set to 0.
Therefore gradient vector can be written in terms of x, u, and λ:

$$\nabla L(x, u) = \frac{\partial L}{\partial x} = \frac{\partial f}{\partial x} + \left(\frac{\partial g}{\partial x}\right)^T \lambda = 0 \tag{28.4.1}$$

$$\nabla L(x, u) = \frac{\partial L}{\partial u} = \frac{\partial f}{\partial u} + \left(\frac{\partial g}{\partial u}\right)^T \lambda = 0 \tag{28.4.2}$$

$$\nabla L(x, u) = \frac{\partial L}{\partial \lambda} = g(x, u) = 0 \tag{28.4.3}$$

Eq. (28.4.1) is a deviation of Lagrange function with respect to state vector. Since in $f(x,u)$ the $f(P_{Gi})$ is not dependent on state vector, only $f_{slack}(P_{slack}(v, \delta))$ depend on state vector; therefore in Eq. (28.4.1), $\partial f/\partial x$ can be represented as:

$$\frac{\partial f}{\partial x} = \begin{bmatrix} \frac{\partial}{\partial P_{slack}} f_{slack}(P_{slack}) \left(\frac{\partial P_{slack}}{\partial \delta_k}\right) \\ \frac{\partial}{\partial P_{slack}} f_{slack}(P_{slack}) \left(\frac{\partial P_{slack}}{\partial |V_k|}\right) \end{bmatrix} \tag{28.5}$$

where k is the slack bus number. In Eq. (28.4.1) the term $\partial g/\partial x$ is actually the Jacobian matrix of Newton Raphson power flow which can be written as:

$$\frac{\partial g}{\partial x} = \begin{bmatrix} \dfrac{\partial P_1}{\partial \delta_1} & \dfrac{\partial P_1}{\partial V_1} & \cdots \cdots \\ \dfrac{\partial Q_1}{\partial \delta_1} & \dfrac{\partial Q_1}{\partial V_1} & \cdots \cdots \\ \cdots & \cdots & \cdots \end{bmatrix} \tag{28.6}$$

In Eq. (28.4.2), $\partial f/\partial u$ is a derivative of objective function with respect to control variable which is

$$\frac{\partial f}{\partial u} = \begin{bmatrix} \dfrac{\partial}{\partial P_1} f(P_1) \\ \dfrac{\partial}{\partial P_2} f(P_2) \\ \cdot \\ \cdot \end{bmatrix} \tag{28.7}$$

The later term in Eq. (28.4.2) $(\partial g/\partial u)$ is a matrix having some diagonal term as 1 and rest of the term as 0 and Eq. (28.4.3) having power flow equation.

Now using equation (28.7), the solution of OPF by gradient is expressed by the following steps:

Step 1: Set constant parameters and assume initial value of control variable u

Step 2: Solve power flow equations and obtain $\partial f/\partial x$ and $\partial g/\partial x$

Step 3: Solve for Lagrange parameter λ as

$$\lambda = - \left(\text{inverse of } \left(\frac{\partial g}{\partial x} \right)^T \right) \frac{\partial f}{\partial x}$$

Step 4: Substitute value of λ in Eq. (28.4.2) and obtain $\partial L/\partial u$

Step 5: Check if $\partial L/\partial u$ is equal to zero or not. If it is zero, mean minimum objective function is achieved, and then stop the procedure, otherwise

Step 6: Find new set of control variable u as:

$$u^{new} = u^{previous} + \Delta u \tag{28.8}$$

where $\Delta u = - \propto \nabla L(x, u)$ and repeat the procedure until minimum function is not achieved.

In this procedure, choosing of value $\propto$ is very important. If it is set at a very low value, then the possibility of convergence of the procedure is increased but time to obtain minimum function is also increased. If value of $\propto$ is too high, then there is a possibility to converge the procedure at the local minimum point rather than the global point.

The gradient method easily handles the nonlinear ties compared with other methods but due to higher dimension of the gradient the computational time increases. Further it suffers with slow convergence near the optimum point. Therefore to solve such problems, other alternative methods are required.

28.2.2 Newton method

Newton method is a well-known alternative to solve OPF problem in power system. This method is performing better with fast convergence characteristics near the optimum point. This method is easy designed with different OPF algorithms according to the requirements (Sun & Ashley, 1984).

The development of OPF using Newton method is explained in the following steps:

Step 1: Initialize all parameters such as bus voltages, angles, generator output power, and transformer tap changing ratio

Step 2: Evaluate all possible inequality constraints such as minimum and maximum value of reactive power and real power at all buses

Step 3: Choose one bus that has no limit on power and voltage

Step 4: Calculate Gradient and Hessian of the Lagrangian. For these, assume that the optimal point is set at $Z = [x_{optimal}, \lambda_{optimal}, u_{optimal}]$. Now according to Gradient method, gradient of Lagrange function is set to zero. Hence

$$\nabla L(Z) = \nabla L_x\left(x_{optimal}, \lambda_{optimal}, u_{optimal}\right) = 0 \tag{28.9}$$

similarly $\nabla L_\lambda\left(x_{optimal}, \lambda_{optimal}, u_{optimal}\right) = 0$ and

$$\nabla L_u\left(x_{optimal}, \lambda_{optimal}, u_{optimal}\right) = 0$$

With these limits we get

$$\lambda_{optimal} \geq 0; \quad \text{if inequality constraint is inactive mean } h\left(x_{optimal}\right) = 0 \tag{28.10}$$

$\lambda_{optimal} = 0$; if inequality constraint is active mean $h(x_{optimal}) \leq 0$

Step 5: Solve the Hessian Equation $[H]\ \Delta Z = \nabla L(Z)$

Step 6: Update optimum point Z

$$Z_{new} = Z_{old} - \Delta Z \tag{28.11}$$

Step 7: Check if $\Delta Z \leq \varepsilon$ then go to next step, otherwise go to step 4 and set Z to Z_{new}

Step 8: Check all constraints. If constraints are violated then go to step 2, otherwise stop the procedure

This method has several advantages such as fast convergence, easy handling of inequality constraints, and obtaining robust solution of practically large size system. On the other hand, penality factor near the optimum point may change the optimum value beyond the range.

28.2.3   Linear programming

This method effectively solves the problem associated with gradient and Newton method. In this method, objective function and constraints are assumed to be in linear form. This is based on the linear approximation of OPF problem. The objective function for linear programming is written as follows:

$$\begin{aligned} &min f(x_0 + \Delta x, u_0 + \Delta u)\\ &\text{with subject to } g(x_0 + \Delta x, u_0 + \Delta u) = 0\\ &\qquad\qquad h(x_0 + \Delta x, u_0 + \Delta u) \leq 0 \end{aligned} \tag{28.12}$$

where x_0 and u_0 are the initial values of x; u, ; Δx, and Δu are the change in x and u g; and h are the linear approximation to contraints.

The simple step to solve OPF by linear programming is as follows:

Step 1: Solve the power flow equation for given operating condition

Step 2: Set the change in control variable Δu as the change in the current control variable by shifting the cost curves

Step 3: Factorize the admittance matrix and set the change in limit Δu as obtained in step 2

Step 4: Solve OPF and compute the control variable u

Step 5: Update the control variable u as $u = u + \Delta u$

Step 6: If change in control variable is less than the tolerance, then go to step 4 otherwise stop the procedure.

Linear programming method tests the limits in the sparse form. To obtain exact solution of nonlinear problem, equality constraint g is necessary to provide accurate operating point x. This is set as starting point in optimization process to update the control variable u. However, solution of OPF can be obtained either by Newton–Raphson or Gauss method or fast decoupled method. Since in this method optimization problem is solved by linear approximation in every iteration, the constraint model for linear approximation can be derived either by Jacobian or by decoupled formulation.

This method easily solves the OPF problem with nonlinear constraints but accuracy of solution is less due to linear cost approximation. In conventional methods, optimum point is attained after a few iterations. Sometimes, these methods fail to converge. Conventional method is not able to get optimal solution as the complicity increases by considering contingency analysis. However, solution of OPF with contingency analysis in a small time is not an easy task, even with recent technologies.

To solve all these problems, several AI techniques have been developed and designed for many engineering applications. In the field of power system control, expert systems have been used for both real-time and offline applications. One of the AI methods is the ANN, which has been used extensively in pattern detection, classification, and tasks for prediction.

28.3 Artificial neural network and fuzzy logic to optimal power flow

28.3.1 Artificial neural network

The ANN is known as an AI technique that has been used in many fields of electrical power systems. The ANN will become a support device for Newton–Raphson's modern approach of being faster and requires specialized power flow problem solutions, owing to the difficulty of electrical systems nowadays.

ANN method is based on the function of human brain, and it has basic unit called the artificial neurons. When trained, it is able to get solution with less time and give output through arithmetic operations (Hagan & Menhaj, 1994). The goal of the training is to assign synaptic weights to appropriate values with the objective of producing the desired output array, or at least a compatible solution with an error range. The purpose of the learning process is to find a weight by applying a rule specifying that learning to learn. In ANN, each training process corresponds to an epoch. Through its form of supervised training with a very popular algorithm known as error backpropagation algorithm (BPA), multilayer perceptron (MLP) has been successfully applied to solve many difficult problems. Backpropagation is a technique that minimizes the error by using gradient descent.

The Hopfield network is using an effective linear programming strategy for classifying into ANN. The neuronal outputs are distinct (the form of MLP) due to the activation function being in line with saturation values and can only presume values −1 or +1, which is a violation in norms (Sharkawi & Niebur, 1996). The radial basis function (RBF) is originated as ANNs variety during the late 1980s. The RBF networks are composed of three connected layers: the input layer; hidden layer (or RBF), usually using the sigmoid base functions; and output sheet. It requires the first layer of radial base neural network works using input data set with a number of data set of nonlinear equations to connect input nodes to the neurons of the hidden layer, and the nth neuron output. The output layer classifies what has been obtained previous layer trends, hidden layer, by exit sum of the weights in each array.

28.3.1.1 Artificial neural network applied to optimal power flow

ANNs can be described as systems for processing information, with varying number of neurons or cells on layers. Neurons are the basic unit of system, and cells are receiving signal from input via dendrites. These neurons interact with each other through set of the details and are interconnected by weights. The ANN layout is abstract and general, with studying and memorizing and includes adaptation. A number of ANN methods have been projected for solution of OPF problem, such as multilayer perceptron, Hopfield model, and radial basis function (Paucar & Rider, 2002).

In ANN perceptron, basically there is a single layer neural network that gives single output as shown in Fig. 28.1. In this figure, $x_0, x_1 \ldots x_n$ are the various inputs to the network. Each input is multiplied with connection weight ($w_0, w_1, w_2 \ldots w_n$), which is called synapse. B is a bias value to move activation function. Further, these multiplied values are summed and forwarded to activation or transfer function.

The role of activation is necessary for an ANN to understand something very complicated and make sense of it. Its main purpose is to convert a node's input signal into an output signal in an ANN. This feedback is used as input to the next layer. There are various activation functions that are available such as threshold or binary step function, sigmoid or logistic function, hyperbolic tangent, and rectified linear unit. The expression of these activation functions is shown in Fig. 28.2 (Cox, 1992; Kim & Sohn, 2020).

FIGURE 28.1 Single-layer perceptron.

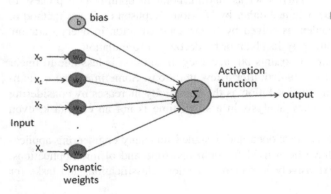

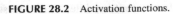
FIGURE 28.2 Activation functions.

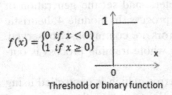

$$f(x) = \begin{cases} 0 \; if \; x < 0 \\ 1 \; if \; x \geq 0 \end{cases}$$

Threshold or binary function

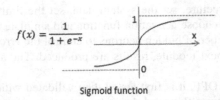

$$f(x) = \frac{1}{1+e^{-x}}$$

Sigmoid function

$f(x) = tanh x$

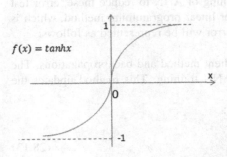

$$f(x) = \begin{cases} 0 \; if \; x < 0 \\ x \; if \; x \geq 0 \end{cases}$$

Hyperbolic tangent function

Rectified linear unit

FIGURE 28.3 Structure of multilayer perceptron.

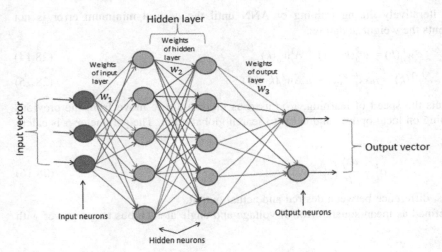

Single ANN to power system applications gives poor results. More than one neuron for magnitude of voltage, voltage phase angle for each bus are required for satisfactory result. One of the advantages of number of neurons is to train them individually for each bus.

In general, an ANN model has (1) input layer, (2) hidden layer with N number of neurons with hyperbolic tangent activation function, and (3) an output layer of single or multiple neurons with linear function. The number of neurons is selected according to the required errors for both training and simulation data (Kosko, 1992). The model of MLP that can be effectively used for OPF problem is shown in Fig. 28.3.

The neurons of MLP were trained through Levenberg–Marquardt method. This method prevents the convergence at local optima and searches the global optima. After selection, the number of neurons, number of hidden layer, and the learning process will proceed.

The learning of neurons has five modules:

1. define structure
2. fix training data set
3. formulate simulation and test the data set
4. run the simulation and control the error
5. generate the result

In the first module, to define structure, set the system data, set the limits of parameters, and set the generation of training data set. In the next module choose activation function and simulate the learning process. In module 4, heuristic method is used to determine the number of hidden neurons to control the errors. These errors are continuously observed, and the best value is stored. In the next module, results are produced. The algorithm for whole training process is outlined in the flowchart (Fig. 28.4).

To design the ANN structure for OPF, it is first tested then validated with minimum error, which is controlled using comparison between the obtained power flow and desired power flow. For training of ANN to reduce these, error test data can be obtained by any conventional methods such as Newton, gradient, or linear programming method, which is set as desired output. Once it is tested, it will validate to control the error. The error will be represented as follows:

Error (e) = desired output − actual output

There are various methods to control this error at each iteration such as gradient method and backpropagations. The back propagation learning algorithm is one of the most popular method for ANN training. This method updates the weights of the neurons by sending back the error signals to the network.

The output of jth neuron of lth hidden layer (hidden$_j^l$) is presented as follows:

$$\text{hidden}_j^l = \sigma \left(\sum_{i=1}^{N^{l-1}} w_{ij}^l x_i \right) \tag{28.13}$$

Where $\sigma(.)$ is the activation function that is choosen for a particular hidden layer, w_{ij}^l is the weight from ith input neuron to jth neuron of hidden layer (l), X_i is ith input neuron, and N^{l-1} is the number of neurons of previous layer $(l-1)$.

This algorithm updates the weights iteratively during training of ANN until the desired minimum error is not achieved. The following equation represents the weight updating:

$$w_{ij}^l(k) = w_{ij}^k(k-1) + \alpha_j^l w_{ij}^l(k) \tag{28.14}$$

$$w_{ij}^{l-1}(k) = \eta \alpha x_i^{l-1}(k) + \mu \Delta w_{ij}^l(k-1) \tag{28.15}$$

where η is the learning rate that affects the speed of learning, μ is used as momentum to move from the previous iteration. This parameter avoids the trapping on local optima and helps to search global point. The parameter α is calculated as follows:

$$\alpha_j^l = \frac{dx_j^l}{dt} \left(x_j^l - y_j \right) \tag{28.16}$$

where $(x_j^l - y_j)$ represents the error, that is, difference between desired and actual output.

Now for OPF, the error function is defined as mean square error of voltage and angle at each bus to get power with minimum transmission line loss.

$$\text{mean}(e_v) = \frac{1}{n_b . n_k} \sum_{i=1}^{n_b} \sum_{k=1}^{n_k} e_v^{ik} \tag{28.17}$$

Similarly expression of error of voltage angle can be written. In Eq. (28.17), i is the number of bus, k is the iteration count, n_b is the total number of bus, n_k is the total number of training, and e_v is the error between the bus voltage that is obtained through ANN and through any of the conventional methods such as Newton–Raphson method.

The basic neural network for power flow is shown in Fig. 28.5. The inputs to ANN are bus data such as conductance and susceptance. According to bus data, run the load flow and train the ANN for bus voltages and voltage angles. To train the ANN, run the load flow for the different loads of given minimum and maximum range of loads. This load range is applied on each bus to train the ANN. This load range may accord daily load curve or yearly load curve.

Once the voltage and angle are trained for different loads; then another ANN is trained for optimum power generation according to bus voltage and angle. Now to apply reactive power control, the input to ANN is increased. The inputs of ANN consist of limits on voltages at generator bus (v_{pv}), limits on transformer tap setting (tap), and limits of reactive power at generating bus (Q_{pv}).

Consideration of reactive power control at the generator bus is extension of basic ANN of power flow. In training the basic ANN of power flow the voltages and angle for each bus are computed then reactive power at generator bus is determined using load flow. If reactive power generated are within the limits then ANN of basic power flow as shown in Fig. 28.5 provides the optimum power generation with minimum losses. In case reactive power at generator bus is

FIGURE   28.4 Algorithm   for developing artificial neural network.

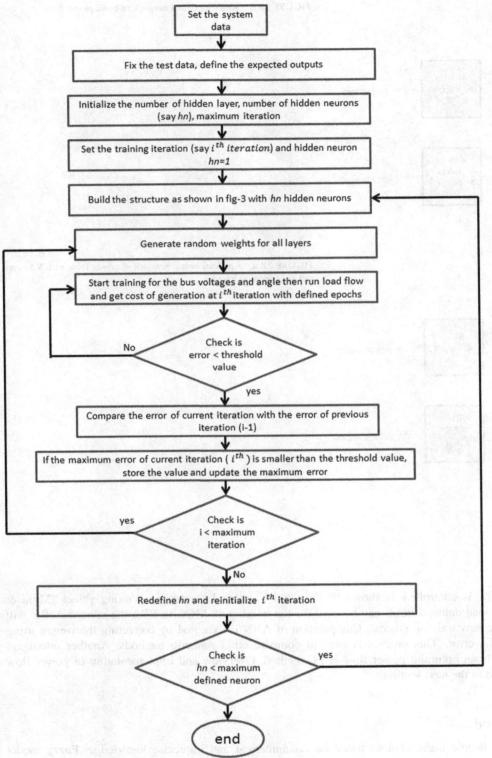

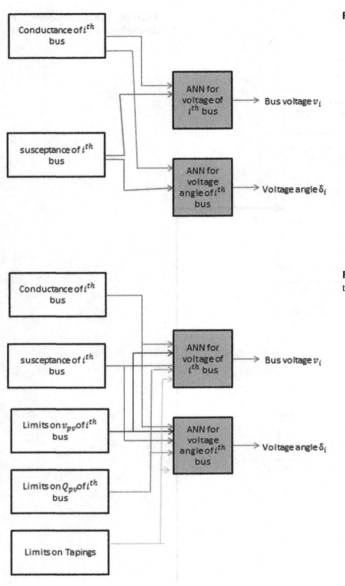

FIGURE 28.5 Artificial neural network of basic power flow.

FIGURE 28.6 Artificial neural network of power flow with VAr control at generation buses.

violated then the ANN of VAr is controlled as shown in Fig. 28.6 is utilized to obtain operating values (Mnih & Kavukcuoglu, 2015). Various load demand limits can be used for the training of ANN. In BPA the current ANNs will be used as initial point for the new training process. Computation of ANN is executed by correcting the weight using learning process to control the error. This process is easy to compare other numeric methods. Another intelligent method that can be used to obtain optimum power flow is FL method. Overview and implementation of power flow using FL method are explained in the next section.

28.3.2 Fuzzy logic method

FL is based on the fact that people make choices based on nonnumerical and imprecise knowledge. Fuzzy model reflects vagueness and imprecise knowledge as mathematical means of representation. Such models have the capacity to recognize, represent, and use ambiguous data and information. FL is a valued logic with true values of variables that has value between 0 and 1. The true value within the range may be completely true or false.

Basic of FL based on the concept of crisp set theory which is a collection of elements in the set and the degree to an element that belongs to a particular crisp set is known as the degree of membership of the set. The functions that

FIGURE 28.7 Membership functions.

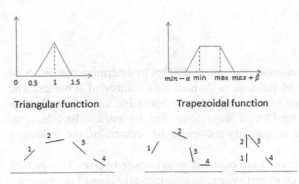

describe the degree of membership are called membership function. The method of determining the degree of member-ship function is called fuzzification. There are number of shapes available for membership function, which is normalized between 0 and 1 (Schwartz, Klir, Lewis, & Ezawa, 1994).

1. **Triangular membership function**: in this membership, fuzzy set is defined on real axis. Its maximum value will be 1 and decreases on both the side as value on real axis increases as shown in Fig. 28.7.
2. **Trapezoidal function**: a trapezoidal function of a fuzzy set S is defined in interval [min, max]; it has the following form:

$$S(t) = \begin{cases} 1 - \dfrac{min - t}{\alpha} & if\,min - \alpha \leq t \leq min \\ 1 & if\,min \leq t \leq max \\ 1 - \dfrac{t - max}{\beta} & if\,min \leq t \leq max + \beta \\ 0 & otherwise \end{cases} \qquad (28.18)$$

3. **Gaussian membership**: this function has the following form:

$$f(x) = \left(\frac{1}{2}\right)\exp\left(-\frac{(x-\mu)^2}{\sigma}\right) \qquad (28.19)$$

where μ is the mean value, and σ is the standard deviation.
4. **Generalized function**: in this method any shape of membership function can be used. It consists of at least four segments (1, 2, 3, and 4), which connect in any form. The length and angle of these segments are decided according to application in which it is used (Miloud, 2018).

In FL a base rule is used in such a way that no multiple rules are true at a time. If in any condition multiple rules have none of zero value then need a process to resolve this condition. Such output interprets after a process called defuzzification. In FL, optimum value can be obtained using the following procedures:

1. fuzzy input (fuzzification)
2. determine the rule
3. output composition
4. defuzzification

There are various methods available for fuzzy interference such as min−max method and max product method. Both has identical knowledge based-rule implementation. In min−max method truncated rule value is used to get output membership function. In place of truncating the output membership the scaling the function in the vertical direction is used in max product method. They combine the resultant shapes (Abaei & Ali, 2020) (Fig. 28.8).

Like fuzzy interference, defuzzification also have various methods. It is comprehended by a decision-making method to choose the optimum crisp value. The methods of defuzzyfication includes the following:

- basic defuzzyfication distribution
- center of gravity

- constraint decision defuzzyfication
- mean of maximum
- Center average method

The most useable method is the center of area, also known as centroid. This method helps to determine the center of area of selected set and return back to crisp value. The total area of function is divided into number of subareas. The centroid of each area is calculated then total of this centroid of each area is calculated to determine the defuzzyfication (Abusorrah, 2013; Yusuf Al-Turki, 2012). To develop this method for OPF, a fuzzy controller is needed to find the optimum generation. To minimize Eq. (28.1) both the constraints, that is, equality and inequality constraint, the following structure of FL can be used as shown in Fig. 28.9.

Fuzzy-based control system for OPF is rule based. Set of rules correct the outcome to get desired value. This control approach changes the strategy of basic control theory. This is based on expert system to automatically control the system.

The inequality constraint considered for power flow are

- limits on generator bus voltage ($V_{Gmin} \leq V_G \leq V_{Gmax}$),
- limits on reactive power ($Q_{min} \leq Q \leq Q_{max}$),
- limits on tapping ($T_{min} \leq T \leq T_{max}$),
- Equality constraint is simply expressed as $P_{Gi} - P_L - P_{Loss} = 0$ where P_L id load demand, P_{Loss} is loss of power transfer.

Implementation of FL controller with this limits is explained in the following steps:

Step 1: Set the input vector for fuzzy control. Each fuzzy has two variables, the first is error in voltage (difference in desired and actual voltages of all bus) and the second is deviation in voltage error

Step 2: Rules are generated based on real-power generation. Rules may be generated in five stages (negative large, negative small, zero, positive small, positive large) or may be in three stages (positive large, zero, negative large)

Step 3: Obtain reactive power and real power based on this rule

Step 4: Apply limits and check the error

Step 5: If error is greater than the threshold value change the rule based on this knowledge

Step 6: Apply defuzzyfication

Step 7: Obtain optimal power with minimum loss

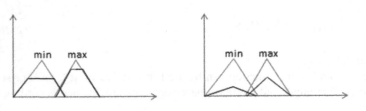

FIGURE 28.8 Fuzzy interference.

Fuzzy interference by Min-max method

Fuzzy interference by Max- product method

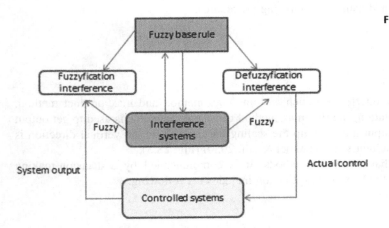

FIGURE 28.9 Basic structure of fuzzy logic controller.

FL approach is easy to design and implement various engineering applications. It can effectively solve the optimization of nonlinear system, expert system, which is difficult to design. It optimizes the efficiency of system with less time. On the other hand, it does not always give accurate solution because it is based on assumptions. Another intelligent method is more accurate and efficient, which is based on population, that is, GA and ant colony algorithm.

28.4 Genetic algorithm

GA is population-based heuristic method. This is based on nature genetics to search optimum point. GA work with multiple local solutions therefore required fitness function values to obtain global point. Each population of candidate is called the individual or chromosomes. These chromosomes are represented in binary strings of 1 and 0. GA is initialized with randomly generated population. In the next iteration, new population of same size is generated as the population of previous iteration using the three operators (Belew, 1989; Goldberg, 1989; Kitano, 1990), that is:

1. selection
2. crossover
3. mutation

New generated population is produced by selection operator. This operator selects the best chromosomes from the previous population. These new chromosomes have characteristics of best chromosomes of previous population. This procedure finds the best fitted chromosomes, which are forwarded for the next iteration, and less fitted chromosomes have higher possibility of dying out. It is possible that the best string may be presented in the new population or may be outside the pool of new population. Therefore this algorithm uses the concept of *Elitism* to hold the population corresponding to the best fitness function.

Before implementation the GA proper initialization, coding, and fitness function are required.

- *Initialization*: the ith population is initialized randomly within the search space of desired length of string (l) of 1 and 0 say S.
- *Coding*: in general optimization function has a number of parameters. After initialization of all parameters within the search space, binary string of all variables is joined to make one string of length l called chromosomes. Suppose d number of variable and n number of population are initialized randomly using Eq. (28.20). In coding binary string of the entire d variables is generating. Further combine binary string of all variables corresponding to each population and get n number of chromosomes, which is also called genes. For example, if the function to be minimized has two variables ($d = 2$), each with a binary string of 10 digits, the length of the chromosome's binary string becomes 20 digits.
- *Fitness function*: to optimize any problem, objective function is necessary. To get objective function of binary string, convert the string to decimal value within the desired search space using following mapping of ith individual:

$$x_i = x_{min,i} + \frac{x_{max,i} - x_{min,i}}{2^{l,i} - 1} \text{ (decoded decimal value } (S_i)) \tag{28.20}$$

where x_{min} is the minimum value, and x_{max} is the maximum value of ith variable in the desired search space. l is the length of binary string, and S_i is the binary string of ith individual. After decoding the individual in decimal objective function is calculated having variables, which are set to optimum value to get the global point of any function. This objective function basically measures the performance. This measure is transformed to distribution of chances by *fitness function*. The fitness function of any string is defined corresponding to current population. Fitness function of GA works for the maximization of the function of given variable. To optimize the function for minimization, the fitness function is mapped in such a way that the value that corresponds to the maximum value of fitness function gives the minimum value of the objective function. There is various mapping available to get the minimum objective function as follows (Davis, 1991; Hart & Belew, 1996):

$$\text{fitness function}(x) = \frac{1}{1 + \text{objective function}(x)}$$

or

$$\text{fitness function}(x) = \frac{1}{\text{objective function}(x)}$$

or

$$\text{fitness function}(x) = -\text{objective function}(x) \tag{28.21}$$

After evaluation of the fitness function, population for the next iteration is obtained. The population for next iteration is generated through tree operator, that is, selection, crossover and, mutation.

1. Selection:

   The aim of selection operator is to increase the possibility to get the best solution and to remove the worst value of a current population while maintaining the population size constant. This process is done after evaluating the fitness function. These fitness functions give the measures that which chromosome has worst or good value. The value of fitness function assigns a rank. The chromosome having high value of corresponding fitness function values have the highest chance to choose for the reproduction. There are various selection rules available to get the best chromosomes such as tournament selection, proportionate selection, rank selection, and roulette wheel selection.

   a. *Tournament selection*: In this selection method, many tournaments are organized with some number of individuals. These individuals are selected on random basis. The tournament winner represents the best fitness function value among the selected individuals and is chosen for the next generation. The individuals have the less fitness function value and have the less chance to choose for the next iteration if tournament size is small (Fig. 28.10).

   b. *Roulette wheel selection:* The chromosomes for the next iteration are selected according to the fitness function values (Walters & Sheble, 1993). The best fitness function has more chance to choose for the mating pool. For example the system has five chromosomes and the value of corresponding to each is as shown in Fig. 28.11. This figure clearly illustrates that the individual having greater value has large percent on pie of wheel and more possibility of quay in front of fixed point when it is rotated. To choose with this method follow following steps

   Step 1: Determine the sum of fitness function (say sum)
   Step 2: Generate random number between 0 and sum
   Step 3: Determine cumulative sum (C)
   Step 4: Determine the probability of choose ith chromosomes as

   $$P_i = \frac{fitness_i}{sum\ of\ fitness}$$

   Step 5: The chromosome for which $P_i >$ random number, choose for the next iteration

   c. Rank selection: In rank selection method, assign rank according to fitness function. In rank selection sum of rank is calculated then probability of selection is determined as

   $$sum\ of\ rank(rsum) = \sum_{i-1}^{total\ chromosomes} rank_i$$

   probability of rank $(P_{rank}) =$ rank of ith chromosome $(r_i)/rsum$

   Generate random number between 0 and *rsum*. If random number is greater than P_{rank}, then choose for next iteration.

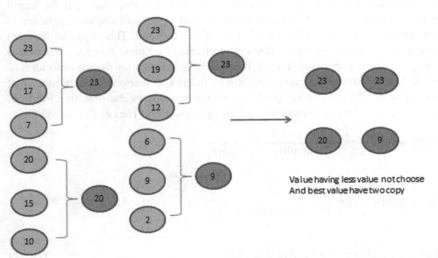

Value having less value not choose
And best value have two copy

FIGURE 28.10 Structure of tournament selection.

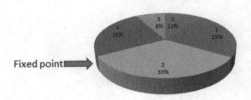

FIGURE 28.11   Roulette wheel selection.

FIGURE 28.11   Roulette wheel selection.

Roulette wheel

Fixed point ➡

2. Crossover: Another operator after the selection is crossover. After applying the selection criterion, mating pool is formed for next iteration then apply crossover that is used to create new chromosomes. This operator exchanges the information between the genes (Nanda & Narayanan, 2002; Sheble & Brittig, 1995). Every bit in a string actually represents the characteristics of output. This is based on the encoding of the string in binary form. String of chromosomes is in form of 1 and 0. There are various forms of crossover such as one-point crossover, two-point crossover, and multiple-point crossover that can be used. The point on which crossover between the string is selected is random selection. Some portion of the string, which is also selected randomly is exchanged between the strings.

Step 1: Combine strings of all variables, for example, fitness function has two variables of string length of 10 for each variable. Therefore string of chromosomes after combining the string of all variables has length 20.

Chromosome A	Variable 1:0011101011	Variable2:1000110001	String after combine: 00111010111000110001
Chromosome B	Variable 1:1100000001	Variable2: 0000110000	11000000010000110000

Step 2: Randomly select two chromosomes from the mating pool

Step 3: Generate random number between 0 and 1. If random number is greater the probability of crossover only then the cross over is applied and go to next step otherwise string is directly copied for the next operator.

Step 4: Generate random between 0 and l (l is the total length of chromosome after combining the string of all variables). For one point, generate one random number for each pair; similarly, generate two random number for two-point crossover.

Step 5: From crossing point, exchange the rest of the strings as shown in Fig. 28.12.

3. Mutation: After crossover, mutation is applied on new string. It is used to maintain diversity of chromosomes, which is forwarded to the next iteration. In mutation, bits of chromosomes alter from its initial state. Mutation is applied according to the probability of mutation. Basically, this probability is set at low value to get a new solution; otherwise it will trap to nascent random search. There are various ways to implement mutation operator. The most common method is to generate random number between 0 and 1 corresponding to each bit of selected string. If random number is greater than the probability of mutation, then flip the bit from initial state, that is, flip the bit from 1 to 0 or 0 to 1 as shown in Fig. 28.13. It is called point mutation. This operator reduces the possibility to trap on local optima and help to search global point.

Sometimes these operators loose the best solution of the mating pool. To prevent this, elitism is provided to prevent the local trapping (He & Hui, 2006) (Fig. 28.13).

To implement GA for OPF, the fitness function is obtained by modifying the basic equation of objective function as given in Eq. (28.1). In Eq.(28.1) apply constraint as a penalty factor as follows (Chen & Huang, 2020; Ma & Xia, 2017):

$$\text{objective function} f(P_{Gi}) = \sum_{i=1}^{N} f_i(P_{Gi}) + p1(P_{GN} - P_{GNmax})^2 + p2(P_{GNmin} - P_{GN})^2 \tag{28.22}$$

where $p1$ and $p2$ are penalty factors, P_{GNmax} and P_{GNmin} are the maximum and minimum power output of slack bus, and P_{GN} is the generated power of reference bus. Now the fitness function for minimization can be mapped using Eq. (28.22):

$$\text{fitness function}(f) = \frac{1}{f(P_{Gi})} \tag{28.23}$$

FIGURE 28.12 Crossover operators.

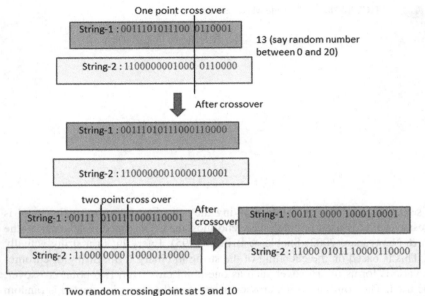

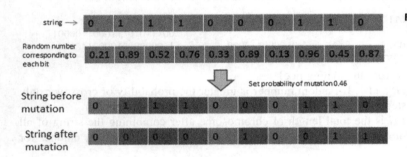

FIGURE 28.13 Mutation operator.

The random string is generated for $N-1$ bus of choosen length. The generated power of reference or slack bus is determined by

$$P_{GN} = P_D - \sum_{i=1}^{N-1} P_{Gi} \tag{28.24}$$

Each string represents the generating power of a unit. The length of each string depends on the limits of power generation and required accuracy. Large length string provides better solution; therfore the step size of a string can be computed as:

$$\text{stepsizeof}_{iunit} = (P_{Gmax,i} - P_{Gmin,i})/(2^n - 1) \tag{28.25}$$

where n is the length of string of ith unit.

Slection operator is applied corresponding to the fitness function and forms mating pool. First, exchange the information of chromosomes in the mating pool by crossover then apply mutation to maintain the diversity of process, and find the population for the next iteration. To implement these procedures, following initial data are required:

1. number of chromosomes (depending on the number of generator bus),
2. bit resolution (select the length of string for proper accuracy),
3. number of crossing point (choose one point or two point),
4. set probability of mutation and probability of crossover,
5. set minimum and maximum limit of power generation of each unit,

6. number of generation (7) coefficient of cost function,
7. set total demand.

The flowchart in Fig. 28.14 illustrates the procedure to implement GA for OPF.

GA is the probabilistic search algorithm in place of deterministic. It parallelly searches the global point of the population; therefore trapping on local point is reduced. On the other hand, it is time-consuming and slightly difficult to implement due to large heuristic nature. Another method that can be easily applied for different application of power system is expert system, which is explained in the next section.

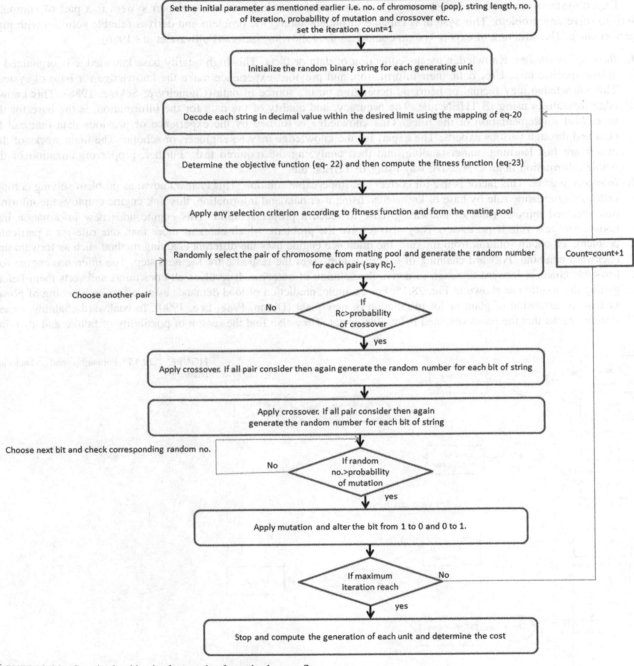

FIGURE 28.14 Genetic algorithm implementation for optimal power flow.

28.5 Application of expert system to power system

28.5.1 Overview of expert system

An expert system is another application area of artificial intelligence search algorithm. Expert system means performing special task in any area. It performs different tasks:

- define formula of problem from the available user data,
- identify the accurate solution of problem
- detail description of the procedure for solving the desired problem

Expert systems are used to design complex system using knowledge of rules. Expert system is a part of computer skill to solve any problem. This system is capable of understanding the problem and derives reliable solution with high performance. The structure of expert system comprises the following factors (Fujiwara et al., 1986):

1. *Base of knowledge*: Knowledge means collection of data or facts. This high quality base knowledge is organized to define specific task. User data, their information, and previous experience make the knowledge for base of system. This knowledge may factual or heuristic depending on the source of data (Choueiry & Sekine, 1988). This knowledge formalizes using IF THEN rule. The accuracy, and quality of the data for the information, is the base for the successful implementation of the rules. This knowledge is formed by the experience of previous data that can be obtained through various experts. The expert for the knowledge may be engineer, or scholar. The main works of this expert are fast learning, understanding, and then analyzing the required task. Further, proper organization of the whole information in an expressive way using IF THEN rule.

2. *Inference engine*: This factor is use for correct and impeccable solution. This is also known as problem solving components. After creating rule by base of knowledge using user data and information, this link engine employs the information obtained through rules of knowledge base to reach a proper solution. This engine adds new information into knowledge base rule if necessary. They also resolve the problem, which contains more than one rule for a particular problem. To obtain solution from the rule, the inference engine uses the different chaining method such as forward and backward chaining. Forward chaining of expert system gives the response for the next step. The inference engine follows this chain in forward direction to derive the solution of outcome. It resolves all constraints and sorts them before getting the solution as shown in Fig. 28.15. For example, prediction of load demands as an effect on planning of power system for extension of plant or for installation of new plant (Dillon, 1988; Lee, 1990). In backward chaining, expert system checks that the result obtained is feasible or not. They also find the reason of possibility of failure and also find

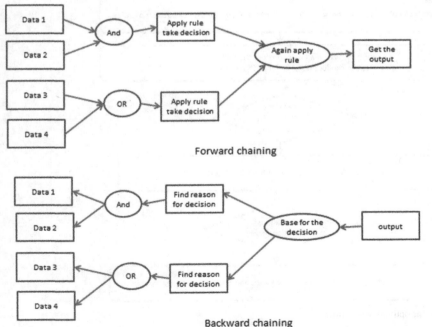

FIGURE 28.15 Forward and backward chaining.

the reason behind the failure. They analyze the conditions, which are occurred in past for this particular output, for example, occurrence of blackout in a particular area. Whenever problem of blackout occurs, experts analyze the reason behind this event. They analyze the system in backward direction to evaluate the reason.

3. *Knowledge acquiescing*: This factor updates the rules of base knowledge. It is used to understand the inputs of new knowledge. This easily represents the information of the rules and applies appropriate test to recognize the requirement of the system. This is mainly depending on the hardware and software.

4. *User interface*: It works as bridge between nonexpert user and expert user. It is simply a processing system for user who is knowledgeable to understand the task. Any nonexpert user in artificial intelligence expert system can access the task using this interfacing. This interfacing may be available in the following ways:

   a. display processing in simple language, which is understandable to user (Alpaslan & Tolun, 1994; Liu & Tomsovic, 1986).

   b. using simple narrations

   c. display the all-base rule on the screen.

This interface provides the shortest path to achieve desire goal. This interface makes in such a way that it is easily adaptable by the nonexpert user. This whole process of expert system is simple expressed in Fig. 28.16.

Nowadays expert system will perform important tasks such as *decision-making* to support the system for planning and implementation.

28.5.2   Application to power system

Recently, development on expert system for all the area of power system is illustrated by many researches. The main areas on which research on expert system is carried out are reactive power control, voltage control, unit commitment for conventional and nonconventional system, system scheduling, planning, protection, etc. The procedure for implementation of expert system for different application of power system is as follows (Leondes, 2002; Subba, 1999):

Step 1: To implement expert system, rules of knowledge need to be available. Not all the rules can be applied on a special system for a specific problem. Therefore main task in implementation is determining the rules for a problem. For a specific problem derive the rules using IF THEN rule. The input of the system is load and bus data

Step 2: Set the goal that need to be achieved with constraints

Step 3: Find the experts of specific domain of power system

Step 4: Update the rules because in power system loads are continuously changing therefore inputs of the system need to be updated according to the load demand. The rules need to be changed after a desired period of time. For these, mathematical model is required, which describes the relation between the inputs and rules. To achieve this, establish the iteration procedure (Giarratano & Riley, 2004).

Step 5: List the priority of the rules according to the load demand for a specific period of time.

Step 6: Further apply test and search strategies to check that the rules are appropriate for determining the solution or not. To verify the rules, any conventional method can be used.

FIGURE 28.16   Working structure of expert system.

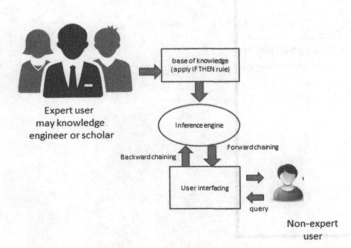

Step 7: According to the problem, select forward or backward chaining for inference between input and rules. For planning and extension use forward chaining and for protection use backward chaining.

Step 8: If optimum value reaches means, desire goal is achieved then display the output otherwise go to step 4

The expert system is the new field for researcher but on the other hand it has high maintenance cost and difficult to maintain (Abraham, 2005; Ball & Moody, 1993; Fernández-Isabel, Barriuso, & Cabezas, 2020). Due to limitations on the technology, knowledge acquisition is not an easy task. There are various technology options that are available to design expert system. A number of hardware tool are available such as workstation and minicomputer. Various tools such as high-performance editors, in-built model, knowledge representation, high-quality debugging tool are easily available. Due to development of recent technology and software, design of expert system for different domain become easy. Innovation of new tool reduces the cost of the system with higher speed of performance and minimum error.

28.6 Assessment of optimal power flow by game playing concept

In the process of optimal generation, transmission and distribution, it is necessary to adopt perfect decision-making in between the different parameters of power system, which effect directly or indirectly the process of AC OPF. Game playing is the prominent concept of the decision-making, which is based on the normal strategies of different games. A serious action including aptitude, possibility, or continuance with respect to at least two people who play as per a lot of rules, for the most part for their own beguilement or for that of onlookers. Game Playing is a learning of how the interactive choices of different expert systems produce results with respect to the preferences (or utilities) of those expert systems, where the consequences in demand may not have been intended by any of the expert systems.

GAMBIT is a software tool for game playing, and it is a set of tools for doing computation on finite, noncooperative games. This tool works in the extensive or strategic form and uses Nash equilibrium and other solution concepts in games. Interactive cross-platform graphical interface, command line tools for computing equilibrium, and extensibility and interoperability are the main features of GAMBIT software. The GAMBIT (Fig. 28.17) tool reads and writes file formats, which

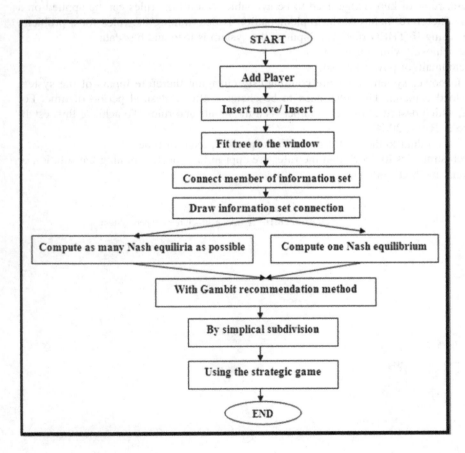

FIGURE 28.17 Flowchart of game playing-based GAMBIT technique.

are textual and documented, making them portable across systems and is able to interact with external instruments. The GAMBIT software is a framework for the definition and implementation of strategic interaction among player.

- Key features of GAMBIT

The GAMBIT has a number of features useful both for the researcher and the instructor:

- Interactive, cross-platform graphical interface:

All GAMBIT features are available through the use of a graphical interface, which runs under multiple operating systems: Windows, various flavors of UNIX (including Linux), and Mac OS X. The interface offers flexible methods for creating extensive and strategic games. It offers an interface for running algorithms to compute Nash equilibrium and for visualizing the resulting profiles on the game tree or table, as well as an interactive tool for analyzing the dominance structure of actions or strategies in the game. The interface is useful for the advanced researcher, but is intended to be accessible for students taking a first course in game playing as well. More advanced applications often require extensive computing time and/or the ability to script computations. All algorithms in GAMBIT are packaged as individual, command-line programs, whose operation and output are configurable.

- Extensibility and interoperability

The GAMBIT tools read and write file formats, which are textual and documented, making them portable across systems and are able to interact with external tools. It is therefore straightforward to extend the capabilities of GAMBIT by, for example, implementing a new method for computing equilibrium, implementing an existing one more efficient, or creating tools to programmatically create, manipulate, and transform games, or for econometric analysis of games. Fig. 28.17 presents flowchart of GAMBIT technique. This figure shows that the calculation of strategic decision-making of any player has to be done by Nash equilibrium.

How about we begin by characterizing a couple of terms ordinarily utilized in the investigation of game hypothesis:

1. Game: Game is a situational condition and circumstance in which consistently in any event, two conditions are fulfilled, for example, misfortune and win, aces and cens, and first and last.
2. Players: Player is the element of a particular game, for example in the game of Hockey, 22 players are played in a different circumstance. In the case of overall process of power system or conventional and nonconventional energy system have three key players, generation, transmission, and distribution.
3. Strategy: A general administration during the game, procedure, and marvel which is followed by the player or components.
4. Payoff: The consequence of game, procedure, or marvel regarding utility. In the game of Tennis, Serena win the match against Venus by 6−2, 6−3, 6−0 and in this case, Players are Serena and Venus and their payoff is 6−2, 6−3, 6−0 in the favor of Serena.
5. Information set: The data accessible at a given point in the distraction. The term enlightening record is most conventionally related when the redirection has a constant part. For the model in OPF framework, information of prefeasibility investigation, demonstrating, controlling, and unwavering quality is called as data set.
6. Equilibrium: The point in the division where the two players have chosen their decision and the outcome has been reached. In the AC OPF system if the system is 100% efficient and does not effect by the other technical and environmental factor then system is called equilibrium.
7. Nash equilibrium: In the game hypothesis, the Nash balance is the steady condition of the game strategy, which likewise distinguishes the immersed state of the various players. In the AC OPF analysis, the steady state characteristic of the different parameters is the Nash balance of the game playing concept.

Game playing has a wide extent of usages in the field of economic related angles, yet in the field of programming designing, especially in the improvement count is seldom used (Khare, Nema, & Baredar, 2012; Ogino, 1990; Sharp, 1981). Table 28.1 shows the comparison between AC OPF and game playing concept.

- Solution concepts:

A game technique is just a framework of coordinated effort; it does not decide in which manner players are ready to play, what moves should the players make, with a particular ultimate objective to achieve their objectives (boosting result for our circumstance).

There are two steps according to the functioning of OPF system:

- Prescriptive: Describe different circumstances, where players can work for OPF
- Descriptive: Description of strategies of each player

TABLE 28.1 Comparison between AC OPF and game playing.

Optimal energy system	Game playing
Value of voltage and voltage angle	Player
Unit generation for optimal power flow	Research space
Power flow constraints	Strategies
Economic assessment	Payoff function

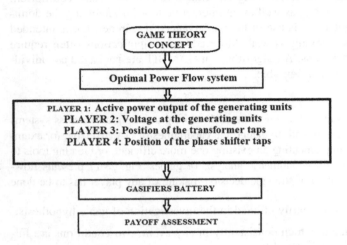

FIGURE 28.18 Key step of game playing in AC optimal power flow system.

These two philosophies are associated; we have to design beguilement using the prescriptive methodology and separate them using the depicted methodology. Finally, we need the reasonable examination to anticipate that players would play in the manner we proposed when we laid out the preoccupation. We are based on the prescriptive methodology. This is in light of the fact that our basic target is to setup amusements, which have a tolerable equilibrium(s), an incredible outcome for all players. An answer thought offers recommendations on what moves to make.

It is very necessary to make proper decision-making in AC OPF according to the different circumstances and in that case game playing plays very prominent role for proper functioning of AC OPF system. Game playing is the combination of different elements or parameters, which is also called the number of players. If we apply the concept of game playing in the field of OPF system then you can consider three factors GP = NO, SO, UO

where GP = Players of game playing

NO = Number of elements or parameters or players of OPF

SO = The number of strategies and decision for OPF

UO = Utility or payoff value of optimum value of OPF

When the outcome is random, the payoffs are usually weighed by their probabilities. In the optimal power gasification system, set of player N is equal to 4 (active power output of the generating units, voltage at the generating units, position of the transformer taps, position of the phase shifter taps). The active power output of the generating units, voltage at the generating units, position of the transformer taps, position of the phase shifter taps functioned as player 1, player 2, player 3, and player 4, respectively. Fig. 28.18 shows the significant step of game concept in OPF system.

Set the strategies: S

- linearize the problem around an operating point for OPF
- find the solution to this linearized optimization for OPF
- perform a full AC power flow at that solution to find the new operating point
- "Reactive power" aspects (VAr flows, voltages) are much harder to linearize than the "active power aspects" (MW flows)

If electricity consumer demand is greater than the electricity generation through the OPF system and in that case system goes toward the optimal performance rate.

In OPF system, game technique will be applied in the following way:

- Parity: Effective utilization of each input factor of OPF system

That is why an equal involvement of OPF and power grid system in the electrical energy production.

- Proportionality: There are different proportionality conditions between the different input and output parameters that are useful in OPF system.
- Normal form:

Normal forms of game playing show those strategies of each player or elements to find out the optimum value of different parameters of OPF system. In the normal form of game playing, only assess the parameter at the primary level and enter the initial value of each parameter. Normal form is also considered as the strategic form of game and is typically described by an array, which shows the different factors of energy system as a player, their strategic condition, and their optimum values. It is supposed that each element acts at the same time or without knowing the strategies of the other factors of OPF system (Lygeros, Godbole, & Sastry, 1996; Wu, Hamed, & Huang, 2012).

- Mixed strategy Nash equilibrium:

In the game definition, a player or an entity is said to use a mixed strategy if the player needs to randomize over the set of activity. On the other hand, a mixed approach is a probability allocation that assigns to every available event is a possibility of being selected. If only one event has a positive probability of being selected, the player is said to use a pure strategy. Table 28.2 presents blended technique of biomass sustainable power source framework.

- Let X is the probability of ample performance of generation expansion of proper functioning of OPF system. So that $1 - X$ is the probability of poor performance of generation expansion of proper functioning of OPF system.
- Let X is the probability of ample performance of distribution expansion of proper functioning of OPF system. So that $1 - X$ is the probability of poor performance of distribution expansion of proper functioning of OPF system.
- To find mixed strategies we add the P-mix and Q-mix strategies to the payoff matrix (Bayens, Bitar, Khargonekar, & Poolla, 2011; Mei, Wang, Liu, Zhang, & Sun, 2012; Saad & Han, 2011; Street, Lima, & Freire, 2011).
- Algebraically:

$50X + 10(1 - X) = 20X + 80(1 - X)$
$50X + 10 - 10X = 20X + 80 - 80X$
$40X + 10 = 80 - 60X$
$100X = 70$, SO
$X = 70/100 = 0.7$

If only the generation expansion is ideal, then the possibility of accomplishment of OPF system is 70%.

$50Y + 80(1 - Y) = 90Y + 20(1 - Y)$
$50Y + 80 - 80Y = 90Y + 20 - 20Y$
$80 - 30Y = 70Y + 20$
$60 = 100Y$, SO
$Y = 60/100 = 0.60$

If only the distributed expansion is ideal, then the possibility of accomplishment of OPF system is 60%.

TABLE 28.2 Mixed strategy of optimal power flow system.

Distribution expansion		Generation expansion		
		Zone 1	Zone 2	Y-mix
	Zone 1	50,50	80,20	$50Y + 80(1 - Y)$ $50Y + 20(1 - Y)$
	Zone 2	90,10	20,80	$90Y + 20(1 - Y)$ $10Y + 80(1 - Y)$
	X-Mix	$50X + 90(1 - X)$ $50X + 10(1 - X)$	$80X + 20(1 - X)$ $20X + 80(1 - X)$	

References

Abaei, G., & Ali, S. (2020). A fuzzy logic expert system to predict module fault proneness using un labelled data. *Journal of King Saud University - Computer and Information Sciences, 32*(6), 684−699.

Abraham, A. (2005). *Rule-based expert systems. Handbook of measuring system design.* John Wiley & Sons.

Abusorrah, A. M. (2013). Optimal power flow using adaptive fuzzy logic controllers. *Mathematical Problems in Engineering.*

Alpaslan, F. N., & Tolun, M. R. (1994). Connectionist expert systems: An approach to solve knowledge acquisition bottleneck in knowledge engineering. In *Proceedings of artificial neural networks and artificial life symposium*, METU, Ankara, Turkey, December 15.

Ball, J. T., & Moody, R. F. (1993). The future of expert system development tools. In The Fifteenth *annual ideas in science & electronic exposition and symposium'93*, ORION International Technologies, Inc..

Barbulescu, C., Kilyeni, S., Simo, A., & Oros, C. (2015). Artificial intelligence techniques for optimal power flow. *Soft Computing Applications, 357*, 1255−1269.

Bayens, E., Bitar, E. Y., Khargonekar, P. P., & Poolla, K. (2011). Coalitional aggregation of wind power. *IEEE Transaction on Power System, 28*, 1−11.

Belew, R. (1989). When both individuals and populations search: Adding simple learning to the genetic algorithm. In J. D. Schaffer (Ed.), *Proceedings of the third international conference on genetic algorithms.* Morgan-Kaufmann.

Bretthauer, G., Handschin, E., & Hoffmannu, W. (1992) Expert system application to power system state of the art and future trends. In *IFAC control of power plants and power systems*, Munich, Germany (pp. 463−468).

Chen, Q., & Huang, M. (2020). Reinforcement learning-based genetic algorithm in optimizing multidimensional data discretization scheme. *Mathematical Problems in Engineering.*

Choueiry, B., & Sekine Y. (1988). Knowledge based method for power generators maintenance scheduling, Preprints. In *Symposium on expert systems applications to power systems*, Stockholm/Helsinki (pp. 9−14).

Cox, E. (1992). Applications of fuzzy system models. *AI Expert, 6*(10), 34−39.

Davis, L. (1991). *Handbook of genetic algorithms.* Van Nostrand Reinhold.

Dillon, T. S. (1988). Expert systems potential and limitations in the application to power systems, Preprints. In *Symposium on expert systems application to power systems*, Stockholm/Helsinki (pp. 18−10).

Dommel, W., & Tinney, F. (1968). Optimal power flow solutions. *IEEE Transactions on Power Apparatus and Systems, PAS-87*(10), 1866−1876.

Fernández-Isabel, A., Barriuso, A., & Cabezas, J. (2020). Knowledge-based framework for estimating the relevance of scientific articles. *Expert Systems with Applications, 161.*

Fujiwara, R., et al. (1986). An intelligent load flow engine for power system planning. *IEEE Trans. on Power Systems, 1*(3), 302−307, 1986.

Giarratano, J., & Riley, G. (2004). *Expert systems - Principles and programming* (4th ed.). Course Technology.

Goldberg, D. (1989). *Genetic algorithms in search, optimization, and machine learning.* Addison-Wesley.

Hagan, M. T., & Menhaj, M. B. (1994). Training feed-forward networks with the Marquardt algorithm. *IEEE Transactions on Neural Networks, 5*(6), 989−993.

Hart, W., & Belew, R. (1996). *Optimization of genetic algorithm hybrids that use local search. Adaptive individuals in evolving populations: models and algorithms.* Boston: Addison-Wesley Longman Publishing Co..

He, Y., & Hui, C.-W. (2006). Dynamic rule-based genetic algorithm for large-size single-stage batch schedulingv. *Computer-Aided Chem Engineering, 21*, 1911−1916.

Huajin, T., & Hanping, C. (2002). Research on the application of artificial intelligence technology in electric power system. *Electric Power Construction, 23*(1), 42−44.

Imen, L., Djamel, L., Hassiba, S., Abdellah, D., & Selwa, F. (2015). Optimal power flow study using conventional and neural networks methods. In *IEEE international conference on renewable energy research and applications (ICRERA).*

Kandel, A., & Langholz, G. (Eds.), (1992). *Hybrid architectures for intelligent systems.* Boca Raton: CRC Press.

Khare, V., Nema, S., & Baredar, P. (2012). Application of game theory in solar wind hybrid energy system. *International Journal of Electrical and Electronics Engineering Research (IJEEER), 2*, 25−32.

Kim, K. H., & Sohn, S. Y. (2020). Hybrid neural network with cost-sensitive support vector machine for class-imbalanced multimodal data. *Neural Networks, 130*, 176−184, October.

Kitano, H. (1990). Designing neural networks using genetic algorithms with graph generation system. *Complex Systems, 4*, 461−476.

Kosko, B. (1992). *Neural networks and fuzzy systems: A dynamical approach to machine intelligence.* Englewood Cliffs, NJ: Prentice Hall.

Lai, L. L. (1989). An expert system used in power system protection. In *IFAC, power system and power plant control*, Korea (pp. 489−494).

Lee, S. J. (1990). An expert system for protective relay setting of transmission systems. *IEEE PWRD, 5*(2), 705−714.

Leondes, C. T. (2002). *Expert Systems. The technology of knowledge management and decision making for the 21st century.* Academic Press.

Liu, L. (1997). Fuzzy knowledge processing of power system relay protection expert system, High school, and Germany *Automation of Electric Power Systems, 21*(60), 34−37.

Liu, C., & Tomsovic, K. (1986). An expert system assisting decision-making of reactive power/voltage control. *IEEE Transactions on Power Systems, 1*(3), 195−201.

Lygeros, J., Godbole, D.N., & Sastry, S. (1996). Multiagent hybrid system design using game theory and optimal control. In *Proceedings of 35th IEEE conference on decision and control, Kobe* (Vol. 2, pp. 1190−1195).

Ma, B., & Xia, Y. (2017). A tribe competition-based genetic algorithm for feature selection in pattern classification. *Appl Soft Computing, 58*, 328–338.

Madan, S., & Bollinger, K. E. (1997). Applications of artificial intelligence in power systems. *Electric Power Systems Research, 41*(2), 117–131.

Mei, S., Wang, Y., Liu, F., Zhang, X., & Sun, Z. (2012). Game approaches for hybrid power system planning. *IEEE Transaction on Sustainable Energy, 3*, 1–12.

Miloud, S. (2018). Fuzzy logic expert system for classifying Solonchaks of Algeria. *Applied and Environmental Soil Science.*

Mnih, V., & Kavukcuoglu, K. (2015). Human-level control through deep reinforcement learning. *Nature, 518*(7540), 529–533.

Müller, H. H., Rider, M. J., & Castro, C. A. (2010). Artificial neural networks for load flow and external equivalents studies. *Electric Power Systems Research, 80*(9), 1033–1041.

Nanda, J., & Narayanan, R. B. (2002). Application of genetic algorithm to economic load dispatch with Line - flow constraints. *Electrical Power and Energy Systems, 24*, 723–729.

Neves, D., Pina, A., & Silca, C. (2018). Comparison of different demand response optimization goals on an isolated microgrid. *Sustainable Energy Technologies and Assessments, 30*, 209–215.

Nguyen, T. T. (1997). Neural network optimal-power-flow. In *Proceedings of the 4th international conference on advances in power system control, operation and management*, APSCOM-97, Hong Kong (pp.-266–271).

Ogino, K. (1990). Non-cooperative game approach for public participation and resource supply in electric generating system development. In *IEEE international conference on systems, man, and cybernetics conference proceedings*, Los Angeles, CA (pp.151–154).

Pan, X., & Zhao, T. (2019). *Deep neural network for DC optimal power flow, systems and control* (eess. SY).

Paucar, V. L., & Rider, M. J. (2002). Artificial neural networks for solving the power flow problem in electric power systems. *Journal of Electric Power Systems Research, 62*, 139–144.

Pearson, I. L. (2011). Smart grid cyber security for Europe. *Energy Policy, 39*(9), 5211–5218.

Saad, W., & Han, Z. (2011). Coalitional game theory for cooperative microgrid distribution network. In *ICC-IEEE*, Kyoto, 5–9 June.

Saha, M. M., Rosolowski, E., & Izykowski, J. (2001). Artificial intelligent application to power system protection. *Mathematical Modeling of Electric Power Systems.*

Schwartz, D. G., Klir, G. J., Lewis, H. W., III, & Ezawa, Y. (1994). Applications of fuzzy sets and approximate reasoning. *Proceedings of the IEEE, 82*(4), 482–498.

Sharkawi, M. E., & Niebur, D. (1996). Artificial neural networks with applications to power systems. In *IEEE PES Special Publication 96*, TP 112–0.

Sharp, J. K. (1981). Consumer incentives for solar energy. In *DCISAP-IEEE*, San Diego (pp. 849–851).

Sheble, G. B., & Brittig, K. (1995). Refined genetic algorithm - economic dispatch example. *IEEE Transactions on Power Systems, 10*(1).

Street, A., Lima, D. A., & Freire, L. (2011). Sharing quotas of a renewable energy hedge pool: A cooperative game theory approach. *Power Tech IEEE Trondheim, 1*, 19–23 June.

Subba, R. S. (1999). Artificial intelligence and expert systems applications in new product development-a survey. *Journal of Intelligent Manufacturing, 10*, 231–244.

Sumpavakup, C., Srikun, I. & Chusanapiputt, S. (2010). A solution to the optimal power flow using artificial bee colony algorithm. In *IEEE international conference on power system technology*, China.

Sun, D. I., & Ashley, B. (1984). Optimal power flow by Newton approach. *IEEE Transactions on Power Apparatus and Systems, PAS 103*(10), 2864–2880.

Walters, D. C., & Sheble, G. B. (1993). Genetic algorithm solution of economic dispatch with valve point loading. *IEEE Transactions on Power Systems, 8*(3).

Wu, C., Hamed, M. R., & Huang, J. (2012). Wind power integration via aggregator-consumer coordination: A game theoretic approach. In *PES-IEEE*, Washington, 16–20 January.

Yusuf Al-Turki, A. (2012). Optimization of fuzzy logic controller for supervisory power system stabilizers. *Acta Polytechnica, 52*(2), 7–16.

Zhao, X., & Zhang, X. (2016). Artificial intelligence applications in power system. *Advances in Intelligent Systems Research, 133*.

Chapter 29

Nature-inspired computational intelligence for optimal sizing of hybrid renewable energy system

Diriba Kajela Geleta[1,2] and Mukhdeep Singh Manshahia[1]

[1]Department of Mathematics, Punjabi University, Patiala, India, [2]Department of Mathematics, Madda Walabu University, Bale Robe, Ethiopia

29.1 Introduction

Optimization is commonly viewed as the technique of encountering hard real-world problem in all mathematics, statistics, and engineering disciplines. When proper decisions were under consideration, almost the concept of optimization was taking upper hand and almost it was everything in day-to-day activities. Everybody wants to maximize his/her efficiency, performance, profit, accuracy, etc., and minimize risk, wastage, distance travel, cost, and others. Subjects, such as management, accounting, finance, and economics, are frequently using optimization in their application area. Thus optimization literally means finding timely best solution among the given set of possible solutions. For real-world problems, no particular optimization algorithm exists that can be applied efficiently to all sorts of problems. Selecting and applying a particular algorithm in solving optimization problem depends on the nature of fitness function, its constraints and the quantity of different variables (Geleta & Manshahia, 2017a, 2017b).

Several difficulties, such as continuity, dimensionality, and differentiability, are the issues of optimization problems when the number of variables increases and search spaces are also vast. Conventional methods, such as steepest decent, Newton's, trust regions, linear programming, and dynamic programming, are usually not succeeded in resolving such large-scale problems particularly with not linear or convex optimization problems (Geleta & Manshahia, 2018).

Most of real-world optimization problems are not linear or convex. Thus the conventional methods that need gradient and/or hessian matrix information are not potential to solve these nondifferentiable functions with the help of such static methods. Furthermore, such methods frequently fail to resolve optimization problems that may have several feasible solutions (Geleta & Manshahia, 2017a, 2017b).

To avoid these problems, there is a need to develop more powerful optimization methods, which cannot give exact value, but timely good optimal solutions. Hence, a new era, *meta*-heuristic nature-inspired optimization techniques have been started for resolving these problems. *Meta*heuristic is a set of guided searches stochastic and population-based algorithms that proposed to resolve approximately an extensive range of hard optimization problems without having the knowledge about nature of the problem (Geleta & Manshahia, 2020a). The majority of these *meta*-heuristic's algorithms are nature inspired. Of course, since the variables in hybrid system entertain integral variables, classical methods, such as iteration method, discrete harmony search, and branch and bound methods, are used to solve it (Alireza, 2013; Kellog, Nehrir, Venkataramanan, & Gerez, 1998; Zong, 2012).

The problems of renewable energy systems are stochastic in nature. It needs vast searching space to escape the local optimal and get timely good solution with relatively less convergence rate. Applying nature-inspired algorithms leads to such solutions when those conventional methods fail to handle the problems. Nature-inspired algorithms were taken their inspiration from the process in nature (Alireza, 2013). They mimic either biological evolution, the activities swarm organisms perform during their natural movement or the effects observed from nature. A number of scholars employ different nature-inspired algorithms to solve the real-world problems. Effectiveness of these algorithms in solving a particular problem is varying based on nature of the problem and an experience of the researcher (Zong, 2012).

Renewable Energy Systems. DOI: https://doi.org/10.1016/B978-0-12-820004-9.00011-5

Here, in this research, we are going to make comparative analysis of nature-inspired algorithms, which we have applied to optimize hybrids of wind and solar renewable energy system in our previous works. Different scholars are applied such algorithms to optimize the hybrid of renewable energy problems in their own different ways. Based on this, we have reviewed the techniques; they have applied to optimize the hybrid wind and PV renewable energy system to get the optimum results as follows (Geleta & Manshahia, 2018).

Motaz (2013) has utilized particle swarm optimization (PSO) for the optimization of the power generation from a hybrid renewable energy system. They have formulated the objective function and constraints by taking certain technical aspects of the system into consideration. They have built the MATLAB code of the algorithm and run for the stated problem. Their results have shown that a typical load demand was satisfied and mentioned that PSO is preferred due to its advantages over the other techniques for reducing the levelized cost of energy and optimizing the system. Benatiallah, Kadi, and Dakyo (2010) have employed genetic algorithm (GA) for sizing standalone wind power system that is considered to produce a power to domestic load in Algeria. The iteration method applied by Geleta and Manshahia (2018) was discussed the optimal value of the hybrid of wind and solar renewable energy system by taking the load balance into consideration. The main objective they set up was optimizing the total annual cost of hybrid system, which can limit the numbers of wind turbine, solar panels, and batteries, which can satisfy the desired load through high reliability. For different iteration, they consider numbers of the components at which reliability of the system becomes zero. Their result has shown that iteration method can optimize hybrid energy system, which can satisfy the desired load.

Hipa et al. (2013) have made a comparative performance analysis on some nature-inspired optimization algorithms (DE, PSO, GA, and IWO) with a different basis for the complex high-dimensional curve/surface-fitting problems. Their results have shown that the methodologies entertain a number of advantages/disadvantages in solving the problem. Karaboga and Basturk (2007) have used artificial bee colony (ABC) algorithm for optimizing a large set of numerical test functions and they have shown that all the employed algorithms have their own advantages and disadvantages. These nature-inspired algorithms are effectively and efficiently applied to their problems, to solve the problems. The comparative studies performed by Hirpa et al. (2013) and Karaboga and Akay (2009) has also identified the performance of the algorithms when employed to solve different problems.

Even though the researchers have shown their results based on their objectives of study, we have identified some of the gaps related to these works. Most of them did not focus on the particular case. The issue of justifying the method by specific data was not given much attention. In some literature, the primery data for particular rural area were not mentioned and recommend whether the cost of the system was affordable for that area (Borowy & Salameh, 1996; Diaf, Belhamel, Haddadi, & Louche, 2007; Singh & Kaushik, 2016). As we mentioned in our previous works, the big challenging issue for the renewable energy projects, especially in developing counries, was budget. In the mentioned literature, the total cost of the project was not identfied with possible suggestion of its improvement. The possible type of renewable energy that is fruitfull and can affordable by rural settlers was not focused. The comparitive studies showing specific results of the employed algorithms need much effort.

By motivated to the short coming of the previous papers, the researchers wish to study the comparison of the results of grey wolf optimization (GWO), ABC, gravitational search algorithm (GSA), PSO, GA, and hybrids of grey wolf optimization and genetic algorithm (HGWOGA), which we have employed for minimizing the total annual cost of hybrid renewable energy system. In the previous works, we have applied the algorithms to solve the problem of renewable energy system. We have shown that all the mentioned nature-inspired algorithms can effectively solve the problem with their own optimal values and convergence rate. Here, we have aimed to show the differences observed among these algorithms with respect to optimal value and computational time. Hopefully this work can be used as bench mark for young researchers who want to study on comparisons of nature-inspired algorithms.

The organization of the remaining part of this chapter is as follows: Section 29.2 presents model of the system with brief reviews of the components. Section 29.3 discusses the developed fitness function and its possible constraints. Section 29.4 reviews nature-inspired algorithms in general and GWO, ABC, GSA, PSO, GA, and HGWOGA, in particular. Section 29.5 discusses some basic advantages and limitation of the algorithms. Section 29.6 presents numerical data in a tabular form and their graphic illustrations. Section 29.7 provides results. Section 29.8 elaborates main findings of the study. Finally, the conclusions and future directions are organized in Section 29.9.

29.2 Mathematical hybrid system model

Modeling is the physical representation of certain phenomena, which is under process. Here, modeling is the method of creating the physical representation of the of hybrid renewable energy system with all its components (Eltamaly & Mohamed, 2014; Etamaly, Mohamed, & Alo-lah, 2015). It is an important step in optimizing hybrid energy system. Effective modeling of the system with its components play a role in order to achieve stable, quality and maximum output. In this paper, we consider the model of off grid system comprising wind turbine, photovoltaic, and batteries. Convertors, investors, and certain load dump are also involved to control the generated power. Different components are outlined as follows.

29.2.1 Models of wind generator and PV panel

The power output obtained from each wind turbine is categorized by using the following equation in terms of maximum, minimum, and nominal wind speed (Javadi, Mazlumi, & Jalilvand, 2011; Saber-Arabi, Hadidian-Moghaddam, & Bigdeli, 2016):

$$P_{WT} = \begin{cases} 0 V_W \leq V_C, V_W \geq V_F \\ P_R \times \left(\dfrac{V_W - V_C}{V_R - V_C} \right), V_C \leq V_W \leq V_R \\ P_R, V_R \leq V_W \leq V_F \end{cases} \tag{29.1}$$

where P_{WT} the power output from wind turbine (W), P_R is the rated power from the wind turbine (W), V_W is the wind speed (m s^{-1}). V_C, V_F, and V_R are the minimum speed, the maximum speed, and rated or nominal speed of the wind turbine, respectively.

The wind power output can also be calculated based on turbine blade radius and wind speed of specific time as follows:

$$P_{WT} = \frac{\pi}{2} \eta \rho r^2 v^3 \tag{29.2}$$

where ρ is air density whose good average value is 1.2, η is a constant whose value is $\frac{16}{27}$, and r and v are radius of the turbine and wind speed, respectively.

The solar power output of certain solar radiation, which reached on specific area at time in to consideration, may calculate by the following equation.

$$P_{pv} = R_t A_t \mu_c(t) \tag{29.3}$$

where R_t is the solar radiation measured on the tilted panel, A_t is the total surface area of cell, and $\mu_c(t)$ is the constant value showing the efficiency of solar cell at time t.

Generally, the total power generated by wind turbine and solar panel at certain time, t was calculated by the following formulas:

$$P_{PV,T} = N_{PV} \times P_{PV}(t)$$
$$P_{WT,T} = N_{WT} \times P_{WT}(t) \tag{29.4}$$

The total generated power (PG) by the system is given by

$$P_{PG}(t) = N_{PV} P_{PV}(t) + N_{WT} P_{WT}(t) \tag{29.5}$$

where $P_{PV}(t)$ and $P_{WT}(t)$ are the power generated from solar panel and wind turbine, respectively, and $P_{PV,T}$ and $P_{WT,T}$ are total powers from their component.

29.2.2 Battery model

Since the power output from the PV cells and the wind turbines are vary in nature, proper sizing of battery is important in balancing the load. This leads us to the required storage capacity S_{Req} of the battery, which defined as the number of solar panels and wind turbines in the hybrid system can be obtained by using energy curve ΔW defined as (Borowy & Salameh, 1996; Diaf et al., 2007):

$$\Delta W = W_{Gen} - W_{Dem} = \int \Delta P dt = \int (P_{Gen} - P_{Dem}) dt \tag{29.6}$$

$$S_{Req}(N_{PV}, N_{WT}) = \text{Max}(\Delta W(N_{PV}, N_{WT})) - \text{Min}(\Delta W(N_{PV}, N_{WT})) \tag{29.7}$$

The storage capacity of the power curve is the difference between the generated power and demanded power. The charge and discharge states of the battery are depending on the positive and negative peak values of this curve. Thus the batteries' cycle varies between this value and depends on numbers of power-generating components.

The number of batteries N_{Batt} can be calculated by the following equation:

$$N_{Batt}(N_{pv}, N_{WT}) = Roundup\left(\frac{S_{Req}}{\eta S_{Batt}}\right) \quad (29.8)$$

where $Roundup(\cdot)$ function returns an integral value which indicate counting number; S_{Req} is the dependent storage capacity needed to maintain balance between demand and supply; η is the amount of rated capacity usage in %; and S_{Batt} is the rated capacity of each battery.

29.3 Optimization formulation

The main goal of sizing is to minimize the total annual cost (f_{TAC}) of the hybrid system. The total annual cost is taken as the sum of initial capital cost (C_{ICC}) and annual maintenance cost (Geleta & Manshahia, 2018; Yang, Zhou, & Lou, 2008; Zong, 2012). For size minimization, we are considering a fitness function, sign constraints, power-generated constraint, battery constraint, and system reliability constrain discussed in our previous paper (Geleta and Manshahia, 2021, 2018, 2020b).

To compare the results of optimal sizing the hybrid system by different nature-inspired algorithms here, the following fitness function has been considered (Geleta and Manshahia, 2021, 2018, 2020b).

$$\min f_{TAC} = C_{ICC} + C_{mnt} \quad (29.9)$$

29.4 Nature-inspired algorithms

Nature-inspired computing is the computing that has its source inspiration in nature, that is, humans and animals or activities observed from nature. As a result, these stochastic, population-based algorithms are termed as nature-inspired algorithms. The main sources of these algorithms are evolutionary, population-based, and ecology-based algorithms.

The most common families of evolutionary algorithms (EAs) are GA, genetic programming, differential evolution, and evolutionary strategy. The members of the EA family share a great number of characteristics in common (Back, 1996; Binitha & Sathya, 2012).

The second taxonomy of nature-inspired computational intelligence was swarm intelligence and was proposed by Kennedy and Eberhardt in 2001 (Manshahia, 2015, 2018, 2019). The third newly emerging category of nature inspired was ecology-inspired optimization (Manshahia, 2017). It is one of the latest groups of algorithms invented within the bioinspired optimization (Borkar & Patil, 2013; Eid & Grosan, 2018). It involves algorithms inspired by mimicking the activities observed in abstract nature (Holland, 1973; Kapoor, Dey, & Khurana, 2011).

The most commonly known families of the third group are water wave optimization, GSA, teaching—learning-based optimization, and flower pollination algorithm (Geleta & Manshahia, 2020c). These categories of nature-inspired algorithms are getting much popularity and hot research area when compared to the other two categories (Sharmistha, Bhattacharje, & Bhattacharya, 2016; Wang & Singh, 2009; Zelinka, Tomaszek, Vasant, Dao, & Hoang, 2018). The main nature-inspired algorithms with its source of inspiration were summarized as in the next diagram. General taxonomy of nature-inspired computational intelligence is given in Fig. 29.1.

In our previous works, we have applied ABC (Geleta & Manshahia, 2020a), GSA (Geleta & Manshahia, 2020b), GWO (Geleta and Manshahia, 2021), PSO (Geleta & Manshahia, 2020b), GA (Geleta and Manshahia, n.d.a) and HGWOGA (Geleta and Manshahia, n.d.a) to optimize hybrid renewable energy system. Here, we have aimed to compare the results obtained by these algorithms in case of total annual cost, reliability, loss of power supply probability, and convergence rates. To compare the results, we prefer to develop a common algorithm and flow chart for all above-mentioned algorithms. For each of the algorithms GWO, ABC, GSA, PSO, GA, and HGWOGA, we formulate the similarity, differences, and common algorithm as shown in Fig. 29.2.

As shown in Fig. 29.2, all the algorithms GWO, ABC, GSA, PSO, GA, and HGWOGA entertain similar features in common. In our previous works, we have applied all the algorithms to optimize the hybrid of wind and solar renewable energy system. Some variables and parameters of the algorithms were adjusted to fit them with the stated fitness function of our problem. Here, we develop the following common algorithm for these methods, which can be applied based on their unique variables, parameters, and vectors.

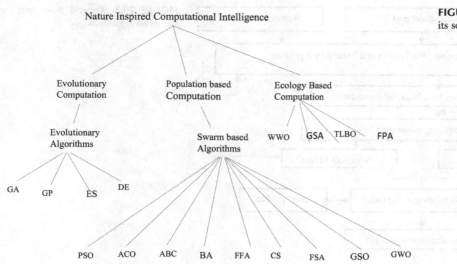

FIGURE 29.1 Computational algorithms with its source of inspiration.

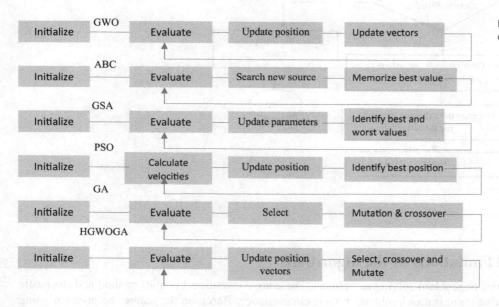

FIGURE 29.2 Similarities and differences among the algorithms.

The common algorithm for GWO, ABC, GSA, PSO, GA, and HGWOGA can be stated as follows:

1. Initialize the population,
2. Initialize parameters,
3. Repeat,
4. Calculate the fitness of each agent,
5. Update either vectors, parameters, or operators involved in the algorithm,
6. Compute the value of the updated, and
7. Until requirements are met.

When each of the algorithms are applied to find the total annual cost of hybrid wind and solar renewable energy system, all of the algorithms (GWO, ABC, GSA, PSO, GA, and HGWOGA) may employ the following flow chart in common. The update values for vectors, positions, constants, operators, and so on are based on the individual entire algorithms character. As far as comparison is concerned, the algorithms should carefully run for hybrid wind and solar renewable energy system and give the results. All the input data, value of parameters in the code should be given attention. After all, the following flow chart diagram shown in Fig. 29.3 can be considered as common flow chart.

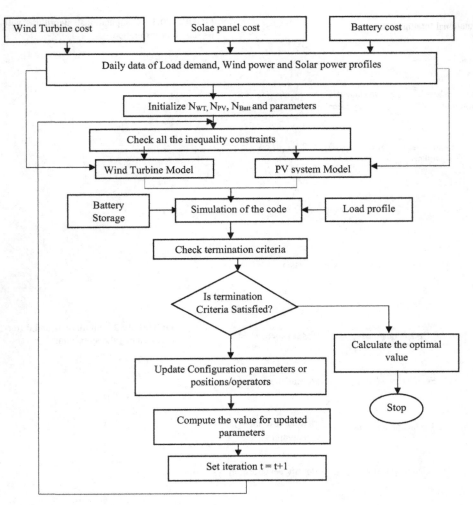

FIGURE 29.3 General flow chart for the stated algorithms.

29.5   Advantages and limitations of the algorithms

Basically, there are two categories of problem-solving techniques. These are conventional (static) method and stochastic method. Both categories can solve optimization problem in their circumstance. Based on the nature the problem going to be solved, the algorithms entertain different advantages and limitations. Conventional method is more of predesigned formula and has specific application area. Its advantage is less computational time and accuracy of optimal value. On the other hand, stochastic computational techniques are commonly called nature-inspired algorithms and have the ability to access multidimensional area of application. Due to their stochastic nature, nature-inspired algorithms have a lot of advantages when employed to solve a particular problem.

Here, under some comparison of computational and stochastic algorithms are given based on the advantages and disadvantages.

Effectiveness of the techniques (conventional or stochastic) depends on the nature of the problem it is employed to solve and experience of the user. In other words, there is no most effective algorithm chosen for all kinds of problems. As shown in Table 29.1, conventional methods are better and give accurate optimal value when the problem is continuously differentiable, whereas nature-inspired algorithms are more power full when the complexity of the problem increases.

Since the concern of this paper is studying the performance of the above-stated six nature-inspired algorithms, next we discuss some advantages and limitations of these algorithms. As discussed above, there is no clear demarcation of their performance between these algorithms. The algorithm that becomes good for one particular problem may fail to

TABLE 29.1 Comparison between conventional and nature-inspired algorithms.

Methodologies of conventional	Nature-inspired algorithms
Effective for easy problems	Timely good for hard problems
Computational time is less	Relatively it takes more time
It gives accurate result	It mostly gives local optimal value
It uses predesigned formulas	The formulas are designed based on the problem
Mostly applied for continuously differentiable functions	It was applied for noncontinuous/differentiable problems
Involve few numbers of variables	Number of variables are based on nature of the problem
Constraints frequently not available	Mostly stated with constraints
Employed by beginners	Employed by experienced scholars
Relatively less cost	More cost for simple problems

solve the other problems. There are several factors affecting the computational performance of nature-inspired algorithms. Some of them are a follows:

1. Nature of the problem to be solved,
2. Number of objective functions and constrains,
3. Number of variables and parameters involved to the problem,
4. Time to get the optimal solution, and
5. Experience of the designer of the problem.

Based on these factors, all nature-inspired algorithms have different performance in optimal value and convergence rate. The best way to employ one nature-inspired algorithm to get timely good optimal result is getting clearly understand about nature of the problem, number of variables, and parameters it has. All nature-inspired algorithms have their own advantages and limitations. Some basic advantages and limitation of GWO, ABC, GSA, PSO, GA, and HGWOGA are given in Table 29.2.

29.6 Numerical data

The comparison we have considered in this chapter was based on the algorithm's economic aspect and their computational time. The minimal total annual cost to satisfy the desired load was the main concern. The numerical data used in this paper are similar to those data proposed by Kellog et al. (1998). Zong (2012), and Geleta and Manshahia (2018, 2020a, 2020b) have improved these data by including some necessary data to simulate with MATLAB.

Table 29.3 shows the updated input numerical values of the decision variables and Fig. 29.4 shows the power difference each hour of a day. Power-gererating schemes generate less power between 5−10 hours and 20−24 hours. The peak value of generated power was given during 13−17 hours.

The values given in Table 29.4 are also taken from Geleta and Manshahia (2018, 2020a, 2020b) and were the updated values of annual average hourly demand P^t_{Dem}.

Since the graphical representation of the given data is more illustrated, the graph of power demand, wind energy profile, solar energy profile, and the power difference is given in Fig. 29.4, Fig. 29.5, Fig. 29.6, Fig. 29.7.

29.7 Results and discussion

29.7.1 Values used for the parameters

The GWO, ABC, GSA, PSO, GA, and HGWOGA have been used to optimize the hybrid system in their original versions. However, based on the developed fitness function, the parameters of each algorithm have to be adjusted to find the best optimal value from the given feasible solutions with in a reasonable amount of time. To do this, we have done adjustment of MATLAB codes for all the stated algorithms and observed the changes came in results. The results were

TABLE 29.2 Some advantages and limitation of the algorithms under study.

Methodology	Source of inspiration	Advantage	Limitations
GWO	Social hierarchy and hunting mechanism of grey Wolves	It has relatively fewer parameters to be adjusted	Less effective with large number of variables
		It was easy to apply for different problems	Relatively less convergence rate
		Can be improved by combining with other metaheuristic algorithms	Low capability to handle the difficulties of a multimodal search landscape
ABC	Intelligent behavior of honey bees	Few control parameters	Search space is limited by initial solution
		Fast convergence rate	Normal distribution sample was at the initial
		Involve both exploration and exploitations	High ignorance of the abundant source
GSA	Theory of gravitation	Has great ability to solve nonlinear optimization problems	Difficulty for appropriate selection of gravitational constant, G
		Stable convergence characteristics	Slow searching speed in the last iteration
		Memory less algorithm	It has relatively complex operator
PSO	Social behavior of birds flocking and food searching	Easy to implement for the problem at hand	It is easy to fall into local optimum in high-dimensional space
		Time of computation is relatively less	It has a low convergence rate in the iterative process.
		It has good capability in maintain the diversity of the population	Can converge prematurely and can be trapped in to local minima especially with complex problems
GA	Biological evolution	It does not need any information about the problem	The determination of the convenient parameters means, population size, mutation rate may be time consuming
		Can accommodate large searching area with multiobjective functions	Its fitness value is computed repeatedly
		Can be applied for complex and noisy problems	It only focusses on fittest value
HGWOGA	Composition	Improve the encircling and updating process of GWO by genetic operators	Entertain relatively large number of parameters
		Improve the convergence curve	

ABC, artificial bee colony; *GA*, genetic algorithm; *GSA*, gravitational search algorithm; *GWO*, grey wolf optimization; *HGWOGA*, hybrids of grey wolf optimization and genetic algorithm; *PSAO*, particle swarm optimization.

compared with our previously work and results of the same data in literature. The values of each parameter that chosen in running the code are shown in Table 29.5.

29.7.2 Experimental results and discussions

In this section, several experimental results from GWO, ABC, GSA, PSO, GA, and HGWOGA applied to solve hybrid renewable energy system were organized for different alternatives. The optimal value was recorded for different possible combinations of the system (Belfkira, Zhang, & Barakat, 2011; Chedid, Akiki, & Rahman, 1998; Eftichios, Dionissia, Antonis, & Kostas, 2006). The wind battery system, solar battery system, and hybrid of wind solar and

TABLE 29.3 Design variables used for solar and wind hybrid system.

Variables/parameters	Symbols	Values
Annual interest	i	6%
Life span of the system	n	20 years
Solar panel price	$nC_{PV,unit}$	$350/panel
Solar panel installation fee	$C_{inst,unit}$	50% of the price
Wind turbine price	$C_{WT,unit}$	$20,000/turbine
Wind turbine installation fee	$C_{inst,unit}$	25% of the price
Unit cost of the battery	$C_{Batt,unit}$	$170
Cost of backup generator	$C_{Backup,unit}$	$2000
Usage % of battery rated capacity	η	80%
Batteries rated capacity	S_{Batt}	2.1 kWh
Batteries life span	LS_{Batt}	4 years
Unit time	Δt	1 hour
Maintenance cost of PV array	$C_{PV,maint}$	0.5 cents kWh^{-1}
Maintenance cost of wind turbine	$C_{WT,maint}$	2 cents kWh^{-1}

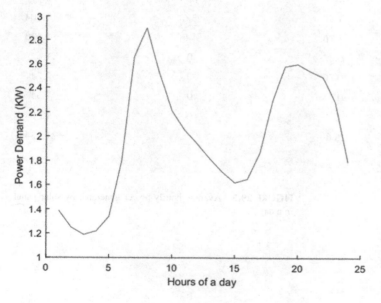

FIGURE 29.4   Daily average of power demand.

battery systems were taken to be tested their optimal values. The possible results of these algorithms for the three combinations were organized in Table 29.6, Table 29.7, Table 29.8.

From the results given in Table 29.6, we have seen that the value of LPSP is zero for all algorithms applied to solve the problem of wind battery renewable energy system. Here, the desired load shown in Table 29.4 can be satisfied when $N_{WT} = 2$ and $N_{Batt} = 2$ are used. The column LPSP is zero showing that all the algorithms can solve the problem effectively. For this alternative combination, GSA gives the optimal annual cost ($4671.30) of the system. The results in Table 29.7 show values of the algorithms for solar battery component only. The system was reliable with

TABLE 29.4 Updated daily power data for wind and solar system.

Time (t)	P_{Dem} (kW)	P_{WT} (kW)	P_{PV} (W)	ΔP (kW)
1	1.39	0.58	0	−0.81
2	1.25	0.49	0	−0.76
3	1.19	0.48	0	−0.71
4	1.22	0.53	0	−0.69
5	1.34	0.47	0	−0.87
6	1.8	0.51	0	−1.29
7	2.66	0.46	1.6	−2.198
8	2.9	0.46	3.4	−2.437
9	2.52	0.61	10.3	−1.899
10	2.21	0.76	24.6	−1.425
11	2.05	1.1	31.7	−0.918
12	1.94	1.53	35.3	−0.375
13	1.82	1.67	36.6	−0.113
14	1.71	1.89	37.4	0.217
15	1.62	2.43	36.8	0.847
16	1.65	2.45	33.5	0.833
17	1.87	1.91	24.2	0.064
18	2.29	1.76	13.4	−0.517
19	2.58	1.57	5.6	−1.004
20	2.6	1.16	1.5	−1.438
21	2.54	0.87	0	−1.67
22	2.49	0.76	0	−1.73
23	2.28	0.74	0	−1.54
24	1.79	0.7	0	−1.09
	47.72	25.89		−21.534

Here, P_{Dem} and P_{WT} are given in kilowatts.

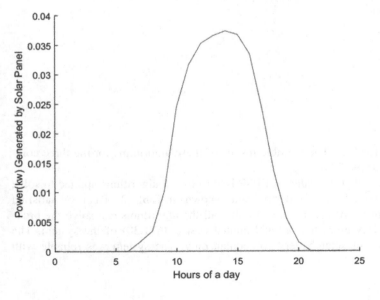

FIGURE 29.5 Average hourly power generated by solar panel in a day.

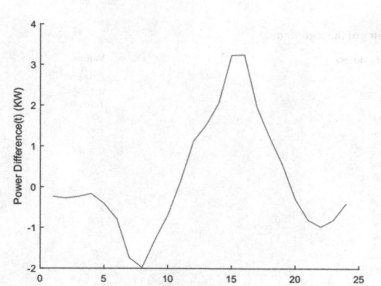

FIGURE 29.6 Hourly average wind profile.

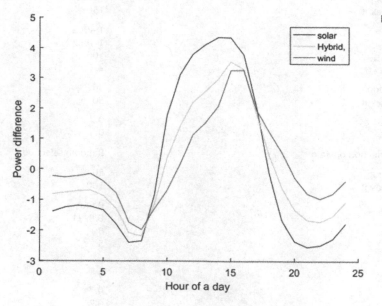

FIGURE 29.7 Average hourly power difference in a day.

$N_{PV} = 161$ and $N_{Batt} = 16$ and only 14 for HGWOGA with optimal total annual cost was given by GWO, which is $5572.00.

The Table 29.8 show results when the system was composed of wind, solar, and battery to satisfy the load. For this particular combination, the number of components is $N_{PV} = 74$, $N_{WT} = 1$, and $N_{Batt} = 11$ and less when HGWOGA was employed to $N_{PV} = 73$, $N_{PV} = 1$, and $N_{Batt} = 10$, which are needed to satisfy the load with the total optimal value given by both GWO and GSA, which is $5209.80.

Effectiveness of one algorithm when employed to solve a particular problem can be evaluated in a number of ways (Ashok, 2007; Elhadidy & Shaahid, 1999; Hadidian, Arabi, & Bigdeli, 2016). Even if we have solved the problem of hybrid renewable energy system to find its least total annual cost, computational time at which the algorithms tend to converge to the optimal value was also important. As we have discussed so far, the power obtained from renewable energy was less quality unless properly modeled and simulated.

Here, maximum iteration was taken as 50 for all algorithms to get the optimal value. After the developed MATLAB cod was run, the value at which each algorithm was converged to its optimal, which is given in Figs. 29.8 and 29.9 and summarized in Table 29.7.

TABLE 29.5 Tuning values of the optimized parameter of the algorithms.

Algorithms	Parameters	Values
GWO	c_1	0.914
	c_2	0.1386
	c_3	1.5694
	dim	10
	V_{max}	100
	V_{min}	0
	npop	20
	Max it	50
ABC	i_1	50
	i_2	2
	V_{max}	100
	V_{min}	2
	npop	20
	Max it	50
GSA	i	15
	V_{max}	100
	V_{min}	2
	npop	100
	Max it	150
PSO	c_1	1.4962
	c_2	1.4962
	V_{max}	100
	V_{min}	2
	npop	30
	Max it	50
GA	p_c	0.7
	p_m	0.3
	Selection operator	Random selection
	npop	50
	Max it	100
HGWOGA	r_1	0.667
	r_2	0.8721
	i_1	5
	i_2	5
	n_c	8
	n_m	2
	p_c	0.7
	p_m	0.2
	npop	50
	Max it	50

ABC, artificial bee colony; *GA*, genetic algorithm; *GSA*, gravitational search algorithm; *GWO*, grey wolf optimization; *HGWOGA*, hybrids of grey wolf optimization and genetic algorithm; *PSAO*, particle swarm optimization.

As shown in Table 29.9, the employed nature-inspired algorithms converge to its optimal solution at different iteration value. The convergence rate of all these algorithms was good except for GA, which runs 31 times to give its optimal value. The best iteration value was observed when hybrid algorithm HGWOGA was applied to solve the hybrid renewable energy optimization problem.

TABLE 29.6 Optimization results of wind-battery system.

Methodology	Number of panels	Number of Turbines	Number of Batteries	LPSP	Total annual cost ($)
GWO	0	2	9	0	5572.00
ABC	0	2	9	0	5572.01
GSA	0	2	9	0	4671.30
PSO	0	2	9	0	5572.04
GA	0	2	9	0	5572.00
HGWOGA	0	2	9	0	5572.00

ABC, artificial bee colony; *GA*, genetic algorithm; *GSA*, gravitational search algorithm; *GWO*, grey wolf optimization; *HGWOGA*, hybrids of grey wolf optimization and genetic algorithm; *PSAO*, particle swarm optimization.

TABLE 29.7 Optimization results of solar-battery system.

Methodology	Number of panels	Number of turbines	Number of batteries	LPSP	Total annual cost ($)
GWO	161	0	16	0	6395.40
ABC	161	0	16	0	8936.44
GSA	161	0	16	0	6395.80
PSO	161	0	16	0	8936.64
GA	161	0	16	0	8934.40
HGWOGA	162	0	14	0	8690.00

ABC, artificial bee colony; *GA*, genetic algorithm; *GSA*, gravitational search algorithm; *GWO*, grey wolf optimization; *HGWOGA*, hybrids of grey wolf optimization and genetic algorithm; *PSAO*, particle swarm optimization.

TABLE 29.8 Solar-wind hybrid-battery system optimization results.

Methodology	Number of panels	Number of turbines	Number of batteries	LPSP	Total annual cost ($)
GWO	74	1	11	0	5209.80
ABC	74	1	11	0	6827.91
GSA	74	1	11	0	5209.80
PSO	74	1	11	0	6827.91
GA	74	1	11	0	6827.91
HGWOGA	73	1	10	0	6517.00

ABC, artificial bee colony; *GA*, genetic algorithm; *GSA*, gravitational search algorithm; *GWO*, grey wolf optimization; *HGWOGA*, hybrids of grey wolf optimization and genetic algorithm; *PSAO*, particle swarm optimization.

As discussed above, different scholars solve this hybrid renewable energy by applying their own methodology (Kosmadakis, Sotirios, & Emmanuel, 2013; Koutroulis, Dionissia, Potirakis, & Kostas, 2006; Luna, Trejo, Vargas, & Os-Moreno, 2012). Geleta and Manshahia (2018), Zong (2012), Kellog et al. (1998), and Alireza (2013) have used different conventional optimization methods to solve the same problem with the same data. The results they have got are

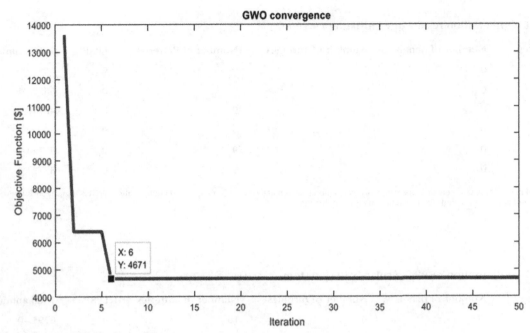

FIGURE 29.8 Convergence curve of grey wolf algorithm.

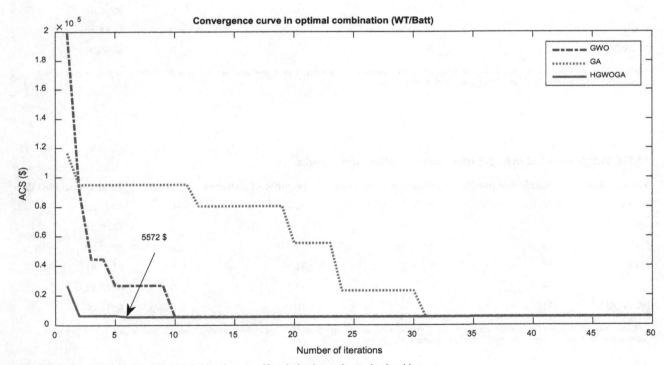

FIGURE 29.9 Convergence curve of hybrids of grey wolf optimization and genetic algorithm.

shown in Table 29.10, Table 29.11, Table 29.12. These scholars applied iteratiom method (Geleta & Manshahia, 2018; Kellog et al., 1998), branch and bound method (Zong, 2012), and discrite harmony search (Alireza, 2013) to find the least total annual cost. The detail results are given in Table 29.10, Table 29.11, Table 29.12.

Figs. 29.10 and 29.11 show the itration values and optimal values of the algorithms under study when wind battery system was used. From Fig. 29.10 one can easily understand the limitation of PSO as it needs more time of computation

TABLE 29.9 Convergency values of the algorithms for wind batter system.

Methodology	N_{PV}	N_{WG}	N_{BAT}	TAC ($)	Convergence iteration
GWO	0	2	9	4671.3	6
ABC	0	2	9	5572.01	3
GSA	0	2	9	4671.3	3
PSO	0	2	9	5572.04	4
GA	0	2	9	5572	31
HGWOGA	0	2	9	5572	2

ABC, artificial bee colony; *GA*, genetic algorithm; *GSA*, gravitational search algorithm; *GWO*, grey wolf optimization; *HGWOGA*, hybrids of grey wolf optimization and genetic algorithm; *PSAO*, particle swarm optimization.

TABLE 29.10 Optimization results of wind-battery system.

Methodology	Number of panels	Number of turbines	Number of batteries	LPSP	Total annual cost ($)
Geleta and Manshahia (2018)	0	2	9	0	5753.09
Alireza (2013)	0	2	11	0	5652.66
Zong (2012)	0	2	10	0	5652.3
Kellog et al. (1998)	0	2	9	0	5574

TABLE 29.11 Optimization results of solar-battery system.

Methodology	Number of panels	Number of turbines	Number of batteries	LPSP	Total annual cost ($)
Geleta and Manshahia (2018)	162	0	17	0	9242.79
Alireza (2013)	160	0	17	0	8844.09
Zong (2012)	160	0	17	0	8843.46
Kellog et al. (1998)	158	0	16	0	8677

TABLE 29.12 Solar-wind hybrid-battery system optimization results.

Methodology	Number of panels	Number of turbines	Number of batteries	LPSP	Total annual cost ($)
Geleta and Manshahia (2018)	74	1	12	0	7085.97
Alireza (2013)	72	1	11	0	6692.61
Zong (2012)	72	1	11	0	6692
Kellog et al. (1998)	72	1	11	0	6691

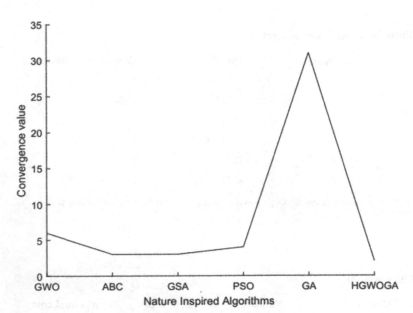

FIGURE 29.10 Convergence values of the algorithms.

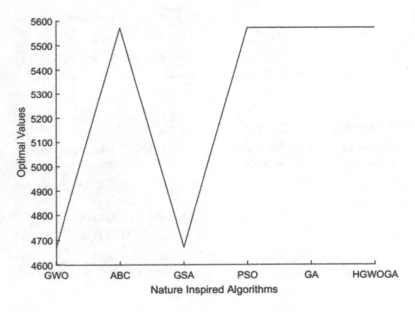

FIGURE 29.11 Optimal values of the algorithms under wind battery system.

to converge to optimal value. From the optimal results of Fig. 29.11, GWO and GSA are more prefereble for optimizing the total annual cost.

The optimal result of the system when wind battery was used in Table 29.10 was $5574.00 obtained by Kellog et al. (1998) by applying the iteration method. As shown in Table 29.4, this result was improved to $4671.30 when GSA was applied to optimize the problem. From Table 29.11, the optimal value when solar-battery component is considered was $8677.00, it was improved to $6395.40 while compared to Table 29.5, it was improved to $6395.40 which obtained when was employed. With the same fashion, the optimal value for the hybrids of wind solar battery system given in Table 29.12 was 6691.00 obtained by Kellog et al. (1998) with the iteration method. This value is also improved to $5209.80 as given in Table 29.8 by GWO algorithm. From these results, it is shown that iteration method approach was superior on the other conventional algorithms. Comparisons show that the results of traditional methods were failed to become optimal values along with those nature-inspired algorithms.

When optimal results of the conventional method were compared with optimal results of nature-inspired algorithms, as shown in Table 29.8, Table 29.9, Table 29.10, the iteration method employed by Geleta and Manshahia (2018) gives optima value from convetional method. From the numercal values of Table 29.4, Table 29.5, Table 29.6, GSA and GWO give the optimal value for all possible combinations of the systems. Here, we show results of optimal methods from conventional and stochastic algorithms.

Table 29.13 shows that the value ($5753.09) of iteration method, which is the optimal in convetional methods, is not mainatined its optimality when compared with the optimal value of the selected nature-inspired algorithms. The only advantage observed on convetional method is its computational time and efectivness when the problem was linear. Similar comparision results are observed in Tables 29.14 and 29.15.

29.8   Findings of the study

Since the result of our problem was integral value in the case of number of components and real number. As shown in Table 29.8, Table 29.9, Table 29.10, we have employed an iteration method to optimize the hybrid renewable energy problem stated in Section 29.3. The methods effectively solve the problem.

TABLE 29.13 Optimization results of wind-battery system.

Methodology	Number of panels	Number of turbines	Number of batteries	LPSP	Total annual cost ($)
Geleta and Manshahia (2018)	0	2	9	0	5753.09
GWO	0	2	11	0	5572.00
GSA	0	2	10	0	4671.30

GSA, gravitational search algorithm; *GWO*, grey wolf optimization.

TABLE 29.14 Optimization results of solar-battery system.

Methodology	Number of panels	Number of turbines	Number of batteries	LPSP	Total annual cost ($)
Geleta and Manshahia (2018)	162	0	17	0	9242.79
GWO	160	0	17	0	6395.40
GSA	160	0	17	0	6395.80

GSA, gravitational search algorithm; *GWO*, grey wolf optimization.

TABLE 29.15 Solar-wind hybrid-battery system optimization results.

Methodology	Number of panels	Number of turbines	Number of batteries	LPSP	Total annual cost ($)
Geleta and Manshahia (2018)	74	1	12	0	7085.97
GWO	72	1	11	0	5209.8
GSA	72	1	11	0	5209.8

GSA, gravitational search algorithm; *GWO*, grey wolf optimization.

In this chapter, the researchers want to compare the performance of GWO, ABC, GSA, PSO, GA, and HGWOGA; they have applied for optimal sizing of hybrid wind and solar renewable energy. All of the mentioned algorithms satisfy the stated objective of the project: minimizing the total annual cost of the system to satisfy the load. The data used to run all the algorithms were originally used by Kellog et al. (1998). As mentioned in Geleta and Manshahia (2018), we have improved the data to make it compatable with the present situation. We are focusing on these data is to study the comparative analysis of the performance of the algorithms among each other and also along with the performance of different optimization techniques stated in the literature.

Generally, from the study of comparative analysis of the results of convetional method with nature-inspired algorithm and among those nature-inspired algorithms under study, one can outline the following main findings:

- Computational algorithms are more effective and provide accurate optimal value for easy and continously differentialbe functions.
- As shown in Table 29.10, Table 29.11, Table 29.12, conventional methods can be employed for the optimization of hybrid renewable energy system, mosly when the number of variable is limited.
- Regarding the comparative study of nature-inspired algorithms, there is no mostly prefered algorithm in optimizing hybrid renewable energy system.
- Based on the performance of the algorithms to minimize the total annual cost, GWO and GSA have got an upper hand over the rest algorithms.
- As shown in Table 29.9, the hybrid nature-inspired algorithm (HGWOGA) has less convergence value. This means that the time needed to get the optimal result was inclaimed to hybrid nature-inspired algrithms.
- All the stated algorithms have their own advantages and limitations. Using these advantages by minimizing the limitation algirithms has the concer of beneficiaries.
- Hybrid nature-inspired algorithms are promising under improvments for the future.

29.9 Conclusion and future directions

This chapter presents a comparative study of results of GWO, ABC, GSA, PSO, GA, and HGWOGA employed to find the total annual cost of a hybrid. The key objective of this research compares the performance of these nature-inspired algorithms among themselves and along with some results in literature solved by conventional methods.

All nature-inspired algorithms employed to optimize the hybrid energy system have solved the problem efficiently. The loss of power supply probability (LPSP) is zero in all possible combination for all employed algorithms. This means that the desired load was effectively satisfied by the mentioned number of components and possible least total annual cost. The numerical values of each algorithms with their computation time called iteration values are organized in Table 29.4, Table 29.5, Table 29.6, Table 29.7. The results obtained by iteration method, branch and bound, and harmonic search utilized by different scholars are also organized in Table 29.8, Table 29.9, Table 29.10.

When the results of conventional methods (iteration method, branch and bound, and harmonic search) were compared with those nature-inspired algorithms, the latter methods have superior advantages in a number of ways including optimal value. From the numerical results of Table 29.8 and convergence rate values of Table 29.9, GSA has the least optimal solution and less iteration value ($4671.00 and 3), respectively. Second, employing hybrid algorithm (HGWOGA) has given relatively good optimal total annual cost value ($5572.00) and best computational iteration value (2). The way of organizing hybrid renewable energy system, employing different nature-inspired algorithms, organizing numerical values and iteration values, and also comparing their performance taken for optimization would be very beneficial and other researchers of the field can apply their methods and compare their results by taking different aspects into consideration will be for the future research.

Acknowledgments

The authors wish to thank all the research scholars cited in this paper and Punjabi University, Patiala, for providing library and Internet facility.

References

Alireza, A. (2013). Developing a discrete harmony search algorithm for size opti- mization of wind—photovoltaic hybrid energy system. *Solar Energy, 98*, 190—195.
Ashok, S. (2007). Optimised model for community-based hybrid energy system. *Renewable Energy, 32*, 1155—1164.

Back, T. (1996). *Evolutionary algorithms in theory and practice: Evolution strategies, evolutionary programming, genetic algorithms.* Oxford University Press.

Belfkira, R., Zhang, L., & Barakat, G. (2011). Optimal sizing study of hybrid wind/PV/diesel power generation unit. *Solar Energy, 85*(1), 100–110.

Benatiallah, A., Kadi, L., & Dakyo, B. (2010). Modeling and optimization of wind energy systems. *Jordan Journal of Mechanical and Industrial Engineering, 4*(1), 143–150.

Binitha, S., & Sathya, S. S. (2012). A survey of bio inspired optimization algorithms. *International Journal of Soft Computing and Engineering, 2*(2), 137–151.

Borkar, P. I., & Patil, L. H. (2013). Web information retrieval using genetic algorithm-particle swarm optimization. *International Journal of Future Computer and Communication, 2*(6).

Borowy, B. S., & Salameh, Z. M. (1996). Methodology for optimally sizing the combination of a battery bank and PV array in a wind/PV hybrid system. *IEEE Transactions on Energy Conversion, 11*(2), 367–375.

Chedid, R., Akiki, H., & Rahman, S. (1998). A decision support technique for the design of hybrid solar-wind power systems. *IEEE Transactions on Energy Conversion, 13*(1), 76–83.

Diaf, S., Belhamel, M., Haddadi, M., & Louche, A. (2007). A methodology for op-timal sizing of autonomous hybrid PV/wind system. *Energy Policy, 35*(11), 5708–5718.

Eftichios, K., Dionissia, K., Antonis, P., & Kostas, K. (2006). Methodology for optimal sizing of stand alone photovoltaic/wind generator systems using genetic algorithms. *Solar Energy, 80*(9), 1072–1088.

Eid, E. M. H., & Grosan, C. (2018). Experienced gray wolf optimization through reinforcement learning and neural networks. *IEEE Transactions on Neural Networks and Learning Systems, 29*(3), 681–694.

Elhadidy, M. A., & Shaahid, S. M. (1999). Optimal sizing of battery storage for hybrid (wind + diesel) power systems. *Renewable Energy, 18*(1), 77–86.

Eltamaly, A. M., & Mohamed, M. A. (2014). A novel design and optimization software for autonomous PV/wind/battery hybrid power systems. *Mathematical Problems in Engineering.*

Etamaly, A. M., Mohamed, M. A., & Alo-lah, A.I. (2015). A smart technique for optimization and simulation of hybrid photo-voltaic/wind/diesel/battery energy systems. In *Proceedings of 2015 IEEE international conference on smart energy grid engineering (SEGE)* (pp. 1–6).

Geleta, D. K., & Manshahia, M. S. (2021). A hybrid of grey wolf optimization and genetic algorithm for optimization of hybrid wind and solar renewable energy system. *Journal of the Operations Research Society of China.* Available from https://doi.org/10.1007/s40305-021-00341-0.

Geleta, D. K., & Manshahia, M. S. (2017a). Nature inspired computational intelligence: A survey. *International Journal of Engineering, Science and Mathematics, 6*(7), 769–795.

Geleta, D. K., & Manshahia, M. S. (2017b). Optimization of renewable energy systems: A review. *International Journal of Scientific Research in Science and Technology, 8*(3), 769–795.

Geleta, D. K., & Manshahia, M. S. (2018). Optimization of hybrid wind and solar renewable energy system by iteration method. In *Proceedings of international conference on intelligent computing and optimization* (pp. 98–107). Springer.

Geleta, D. K., & Manshahia, M. S. (2020a). *Artificial bee colony-based optimization of hybrid wind and solar renewable energy system. Handbook of research on energy-saving technologies for environmentally-friendly agricultural development* (pp. 429–453). IGI Global.

Geleta, D. K., & Manshahia, M. S. (2020b). Gravitational search algorithm based optimization of hybrid wind and solar renewable energy system. *Computational Intelligence.* Available from https://doi.org/10.1111/coin.12336.

Geleta, D. K., & Manshahia, M. S. (2020c). Teacher learning based optimization algorithm for optimal sizing of hybrid wind and solar renewable energy system. *International Journal of Innovative Technology and Exploring Engineering, 9*(6), 2278–3075.

Hadidian, M. J., Arabi, N. S., & Bigdeli, M. (2016). Optimal sizing of a stand-alone hybrid photovoltaic/wind system using new grey wolf optimizer considering reliability. *Journal of Renewable and Sustainable Energy, 8*(3).

Hipa, G., et al. (2013). A comparative analysis of nature-inspired optimization approaches to 2D geometric modelling for turbomachinery applications. *Mathematical Problems in Engineering.*

Holland, J. H. (1973). Genetic algorithms and the optimal allocation of trials. *SIAM Journal on Computing, 2*(2), 88–105.

Javadi, M. R., Mazlumi, K., & Jalilvand, A. (2011). Application of GA, PSO and ABC in optimal design of a stand-alone hybrid system for northwest of Iran. In *Electrical and electronics engineering (ELECO)* (pp. 1–203).

Kapoor, V., Dey, S., & Khurana, A. (2011). Genetic algorithm: An application to technical trading system design. *International Journal of Computer Applications, 36*(5).

Karaboga, D., & Basturk, B. (2007). A powerful and efficient algorithm for numerical function optimization: Artificial bee colony (ABC) algorithm. *Journal of Global Optimization, 39*(3), 459–47.

Karaboga, D., & Akay, B. (2009). A comparative study of artificial bee colony algorithm. *Applied Mathematics and Computation, 214*, 108–132.

Kellog, W., Nehrir, M., Venkataramanan, G., & Gerez, V. (1998). Generation unit sizing and cost analysis for stand-alone wind, phovoltaic, and hybrid wind/PV systems. *IEEE Transactions on Energy Conversion, 13*(1), 70–75.

Kosmadakis, G., Sotirios, K., & Emmanuel, K. (2013). *Renewable and conventional electricity generation systems, technologies and diversity of energy systems. Renewable Energy Governance* (pp. 9–13). London: Springer.

Koutroulis, E., Dionissia, K., Potirakis, A., & Kostas, K. (2006). Methodology for optimal sizing of stand-alone photovoltaic/wind-generator systems using genetic algorithms. *Solar Energy, 80*(9), 1072–1088.

Luna, R., Trejo, P. M., Vargas, D., & Os-Moreno, G. (2012). Optimal sizing of renewable hybrids energy systems: A review of methodologies. *Solar Energy, 88*(4), 1077–1088.

Manshahia, M. S. (2015). A firefly based energy efficient routing in wireless sensor networks. *IEEE African Journal of Computing and ICT, 8*(4), 27−32.

Manshahia, M. S. (2017). Water wave optimization algorithm based congestion control and quality of service improvement in wireless sensor networks. *Transactions on Networks and Communications, 5*(4), 31−39. Available from https://doi.org/10.14738/tnc.54.3567.

Manshahia, M. S. (2018). Swarm intelligence-based energy-efficient data delivery in WSAN to virtualise IoT in smart cities. *IET Wireless Sensor Systems, 8*(6), 256−259. Available from https://doi.org/10.1049/iet-wss.2018.5143.

Manshahia, M. S. (2019). Grey wolf algorithm based energy-efficient data transmission in internet of things. *Procedia Computer Science, 160*, 604−609. Available from https://doi.org/10.1016/j.procs.2019.11.040.

Motaz, A. N. (2013). *Optimization of hybrid renewable energy systems (HRES) using PSO for cost reduction* (42, pp. 318−327). Elsevier.

Saber-Arabi, N., Hadidian-Moghaddam, M. J., & Bigdeli, M. (2016). Optimal sizing of a stand-alone hybrid photovoltaic/wind system using new grey wolf optimizer considering reliability. *Journal of Renewable and Sustainable Energy, 8*(3), 035903.

Sharmistha, S., Bhattacharje, S., & Bhattacharya, A. (2016). Grey wolf optimisation for optimal sizing of battery energy storage device to minimise operation cost of microgrid. *IET Generation, Transmission and Distribution, 10*(3), 625−637.

Singh, S., & Kaushik, S. C. (2016). Optimal sizing of grid integrated hybrid PV-biomass energy system using artificial bee colony algorithm. *IET Renewable Power Generation, 10*(5), 642−650.

Wang, L., & Singh, C. (2009). Multicriteria design of hybrid power generation systems based on a modified particle swarm optimization algorithm. *IEEE Transactions on Energy Convesion, 24*(1), 163−172.

Yang, H., Zhou, W., & Lou, C. (2008). Optimal siz-ing method for stand-alone hybrid solar wind system with LPSP technology by using genetic algorithm. *Solar Energy, 82*(4), 354−367.

Zelinka, I., Tomaszek, L., Vasant, P., Dao, T. T., & Hoang, D. V. (2018). A novel approach on evolutionary dynamics analysis—A progress report. *Journal of Computational Science, 25*, 437−445.

Zong, W. G. (2012). Size optimization for a hybrid photovoltaic{wind energy system. *Electrical Power and Energy Systems, 42*(1), 448−451.

Chapter 30

Optimal design and techno-socio-economic analysis of hybrid renewable system for gird-connected system

Aashish Kumar Bohre[1], Yashwant Sawle[2] and Parimal Acharjee[1]

[1]Department of Electrical Engineering, National Institute of Technology (NIT), Durgapur, India, [2]School of Electrical Engineering, Vellore Institute of Technology (VIT), Vellore, India

30.1 Introduction

In recent days the usage of renewable based energy sources is more auspicious in power generation to plan the more effective and reliable hybrid renewable energy system (HRES) due to exhaustible nature of fossil fuel and global warming. The solar and wind sources have complementary nature to each other; therefore this combination of hybrid renewable systems have excessive potential to offer better power quality with upgraded reliability to consumers as compared to conventional single utility system. Nowadays, HRES have more attention worldwide. The HRES may be off-grid/isolated or grid-connected system. The off-grid or isolated HRES requires the sufficient storage and backup power capacity to maintain system reliability, which is consider as microgrid system. The grid-connected HRES are capable to deliver the electrical energy to local loads as well as to the grid by employing the net metering approach. In grid-connected HRES grid utility may be used as a backup power supply source; therefore the storage system capacity may be smaller. The standalone HRES based on the photovoltaic (PV), wind, and battery storage with hydrogen storage to accomplish load demand based on an economic predictive control method is implemented by genetic algorithms (GA) (Rullo, Braccia, Luppi, Zumoffen, & Feroldi, 2019). The technical and economic scrutiny of wind, PV, and fuel cell (FC) combined through electrolyzer is scrutinized using the firefly algorithm and the comparison of results is shown with respect to particle swarm optimization (PSO) and shuffled frog leaping algorithm (Samy, Barakat, & Ramadan, 2019). The object-oriented programming for optimum sizing of isolated PV−wind HRES and its techno-economic analysis is proposed in Belmili, Haddadi, Bacha, Almi, and Bendib (2014). The PV−wind-based HRES planning and techno-economic analysis for rural area electrification of distinct part of Indian states (Chatterjee & Rayudu, 2017; Suresh, Muralidhar, & Kiranmayi, 2020; Mazzola, Astolfi, & Macchi, 2016), Cambodia (Lao & Chungpaibulpatana, 2017), and China (Yang, Wei, & Chengzhi, 2009) are presented by using different strategy. The HRES design on the basis of various economic, social, and technical criteria, such as loss of power supply probability (LPSP), cost economy, particular matter (PM), emissions and human development using PSO technique for off-grid system, is reported (Sawle, Gupta, & Bohre, 2018; Sawle, Gupta, & Bohre, 2018). The multiobjective function-based approach, including minimization of distribution company (DisCom) energy cost and maximization of distributed generation owner's internal rate of return (IRR), is assessed for standalone HRES using the nondominated sorting genetic algorithm-II (Guo, Wang, Lu, Li, & Wang, 2016). The assessment of numerous potential benefits and motivations offered by the hybrid renewable energy sources (RES) are discussed in Elavarasan (2019) and Environmental Protection Agency (2010), which attract toward the more and more use of HRES globally. The recent development in socio-economic era, motivations, issues, system configuration, and planning methodologies using different objectives to evaluate the performance of HRES with isolated and grid-connected mode is discussed in Luna-Rubio, Trejo-Perea, Vargas-Vázquez, and Ríos-Moreno (2012), Babatunde, Munda, and Hamam (2020), and Bourennani, Rahnamayan, and Naterer (2015). The planning and design of low-cost energy solution for HRES including numerous technology options and availability of energy resources, reliability, and sensitivity analysis is very difficult task; therefore the use of simulation software platform, such as HOMER, HYBRID 2, RETScreen, SOME, and RAPSYS, is making it easier to decide the possible configuration of system and its analysis (HOMER Energy,

2020; Go, Kahrl, & Kolster, 2020). The HRES designs and evaluation consisting of PV/wind energy systems and its renewable penetration issues for the present distribution system are accessed based on LPSP and cost of energy (COE) approach (Deshmukh & Deshmukh, 2008). The multiobjective method for optimum design and analyses of HRES with different system components, such as FC-based micro-CHP (combined heat and power), electrolyzer PV and wind system using sunflower optimization (Fan et al., 2020), and PSO (Sharafi & EL Mekkawy, 2014), is employed. The GA- and PSO-based multiobjective approaches with system performance indices are investigated for the optimal planning of different types of DGs including renewables for the distribution network (Bohre, Agnihotri, & Dubey, 2016). The motivations toward installing the hybrid power supply backup as generator and battery storage for the bulk power system are described based on different cases in Gorman et al. (2020). The experimental study to manage energy within hybrid wind-solar together with storage unit for standalone and grid integration operation is presented by using controlled interconnection concept with LV grid (Dali, Belhadj, & Roboam, 2010). The study of grid-connected HRES comprising of PV, wind, and storage system is modeled in MATLAB environment to access system performances, such as voltage profile and power injected to grid (Saib & Gherbi, 2015). The independent microgrid performance evaluation, including PV, pumped-storage power generation, and FC, is reported in Obara, Morel, Okada, and Kobayashi (2016). Correspondingly, in Maleki and Askarzadeh (2014), the PV-, wind-, and FC-based HRES design with LPSP concept using artificial bee swarm optimization is explored. The optimum planning and analysis of hybrid solar−wind−battery-based power generation system to meet load demand on the basis of generic methodology is examined in Khalid, AlMuhaini, Aguilera, and Savkin (2018). The HRES optimum design and analysis for Nigerian coastline community with the multiple criteria assessments is implemented using hybrid optimization, including HOMER and TOPSIS algorithms (Diemuodeke, Hamilton, & Addo, 2016). The optimal sizing of HRES consisting of wind/solar/battery using artificial immune system technique considering frequency constraints due to the uncertainty nature of RES, which is maintained by battery energy storage system (Hafez, Hatata, & Aldl, 2019). The optimal design and expansion of HRES by using the PV, wind, and battery storage for experimental platform of a nanogrid system is developed in the HOMER design tool environment (Tudu, Mandal, & Chakraborty, 2019).

It is motivating that among the various renewable sources, wind and solar energy are growing very fast globally, which are available freely in large amount everywhere in the world. The intensity and strength of these resources may vary depending on the location of case study. The primary objective of presenting this case study is to increase the electrical energy production utilizing renewable sources, such as PV and wind, to minimize total system economy, cost of per unit energy, and harmful effects by burning fossil fuels and also to strengthen the overall efficiency of NIT Durgapur (NITD) system. Due to the aforesaid reasons, this chapter emphasizes on the development of HRES design and performance assessment based on PV and wind power for grid-connected power network of NITD.

This chapter is organized in eight different sections as follows: Section 30.1 is an introductory section. Section 30.2 summarizes the various motivation and potential benefits of hybrid RES. The detailed literature analysis in the area of HRES design and optimization is presented in Section 30.3. Section 30.4 gives the information about the availability of renewable sources and utilization for the proposed case study. Section 30.5 describes the modeling of different components of the HRES. Section 30.6 describes the problem and methodology employed for the case study. The final results and discussions for the proposed HRES are discussed in Section 30.7 and the major conclusions for the proposed work are summarized in Section 30.8.

30.2 Motivation and potential benefits of hybrid renewable sources

The energy is very important factor to assess the economic and social development of any country. At the present time, approximately 80% energy demand is supplied via fossil fuels worldwide. Solar, wind, geothermal, hydro, and other renewable sources are broadly used in worldwide, which are the abundant available renewable sources in the nature (Elavarasan, 2019; Environmental Protection Agency, 2010). Currently, whole world is facing many challenges, such as global warming, pollutant emission and availability of fossil fuels and conventional energy sources. Also, extraction of energy from renewable energy resources are very promising option nowadays by reason of its deep social and economic impact to modern society. The RES have various advantages over traditional fuel sources except intermittent nature and variable nature with geographic area. The RES have more positive impacts than negative impacts. The integration of hybrid renewable sources in the conventional grid/system offers many economic, social, and technical advantages. The key benefits of hybrid RES are as follows (Babatunde et al., 2020; Elavarasan, 2019; Environmental Protection Agency, 2010; Luna-Rubio et al., 2012):

- RES, for example, wind and sun, are nondepleting sources; thus utilization of these sources is more reliable than fossil fuel-based energy source.

- Reduces the environmental impact, that is, emissions of greenhouse gases and air pollution. Therefore they are environment and human friendly in nature.
- Expand the energy supply options and reduce dependency on conventional fuels.
- The peak hours energy demand can be managed efficiently.
- The hybrid renewable energy technology needs lower maintenance than conventional fuel sources.
- They provide improvement in the human health.
- These sources are available geographical everywhere, so decentralized power generation is possible to meet energy demand.
- The availability of wide range of renewable energy generation as per the requirement of user.
- RES can be easily designed and installed to meet the load demand of users/consumers, at both small scale and large scale.
- RES create new jobs significantly and support for better human development.
- The in-exhaustible energy supply from HRES.
- They offer low operating and maintenance cost.
- The global warming impacts are very less.
- The stable and lower per unit energy charges.
- HRES have more flexible, reliable and resilience operation because RES are less prone to large-scale failure.

30.3 Hybrid renewable energy system design and optimization

The development of efficient hybrid renewable system (HRES) needs an optimal design to achieve social, technical, and economic benefits and the feasibility analysis by using specified optimization approach of each component. The optimal size and feasibility analysis of HRES components should be economic, efficient, and reliable to encounter the defined objectives. The globally recent development in the area of HRES optimal design, optimization, and its components is demonstrated in Table 30.1.

In view of above-presented summary and discussion of various HRES design and optimization, the HRES can be classified into different systems based on the operating mode, such as off-grid/standalone/isolated and grid-integrated system. The essential requirement of isolated system is huge power supply backup by different mean, such as battery storage, generator, and other energy storage systems, to manage the load demand. Although the grid-connected system may need small storage and the deficit of load demand can be supplied through the grid. Numerous technical, economic, and other performances of system are focused and analyzed in different literatures as tabulated in above summary. Many researchers are trying to find the optimum sizing and the design of either proposed system or existing system by employing different remarkable methodologies, such as butterfly PSO, GA, PSO, and techno-economic approach using HOMER. However, there are many HRES design software tools and optimization algorithm available to design and analyze the proposed system but the HOMER software tool and MATLAB are widely used and globally accepted by the researchers. Hence, the HOMER-grid and MATLAB are utilized for the design and analysis of the proposed case study of NITD. The detailed technical, economic, and social benefits of the proposed grid-connected HRES are presented using HOMER optimizer and the comparison is given with respect to the existing base case power system.

30.4 Availability of renewable sources and utilization for case study

The availability of renewable resources for the proposed case study location NITD, is assessed by NASA prediction of Worldwide Energy Resource (Power) database with the HOMER-grid. The details of the proposed case study location as shown in Fig. 30.1 for NITD, are as follows (HOMER Energy, 2020):

- Location—Academic Building, NIT, A-Zone, Durgapur, West Bengal 713209, India
- Latitude—23°333′N
- Longitude—87°17.4′E
- Time zone—Asia/Kolkata

The availability of annual solar and wind data for the case study NITD are demonstrated in the Fig. 30.2 and Fig. 30.3, respectively. Also, the solar and wind intensity map of India is given in Fig. 30.4. The annual average solar global horizontal irradiance and wind speed are $4.824 \, \mathrm{kW \, m^{-2} \, day^{-1}}$ and $4.277 \, \mathrm{m \, s^{-1}}$, respectively, which are specified in Table 30.2. The available solar and wind resources for the NITD case study are updated in the HOMER-grid with the help of MATLAB file interface. And the final output power of solar and wind is estimated using HOMER with the available resources for case study.

TABLE 30.1 The summary for various HRES design and optimization.

Author/reference	HRES components	Objectives	Optimization method	System/case study
Rullo et al. (2019)	PV, wind, battery storage, and hydrogen storage	Minimize COE including investment and operation costs	Bi-level optimization framework by using mixed integer linear problem with GA	Off-grid system
Samy et al. (2019)	PV, wind, fuel cell, and battery	Minimum levelized COE	Firefly algorithm, shuffled frog leaping algorithm, and PSO	Isolated system and case study of remote area of Egypt
Belmili et al. (2014)	PV, wind, and battery	Minimize total cost and maximize reliability	Techno-economic algorithm programming	Off-grid system
Suresh et al. (2020)	PV, wind, fuel cell with battery	Minimize total net preset cost, COE, unmet load, CO_2 emissions	HOMER Pro Software and GA	Off-grid system with case study of three villages in Kollegal block of Chamarajanagar district, Karnataka, India
Chatterjee and Rayudu (2017)	PV, wind, and battery	Minimize COE	Techno-economic approach using HOMER Pro Software	Off-grid system with case study of Sapra, Jharkhand, India
Mazzola et al. (2016)	PV, biomass, battery, and diesel generator	Minimum levelized COE	Mixed integer linear programming	Standalone microgrid with case study of rural area of district Patna, Bihar, India
Lao and Chungpaibulpatana (2017)	PV, diesel generator with battery system	Minimize COE	Techno-economic approach using HOMER Pro Software	Isolated system with case study of Prasat Sambour in Kampong Thom districts, Cambodia
Yang et al. (2009)	PV, wind, and battery	Minimize total annualized cost of system	Techno-economic algorithm programming	Isolated system with the case study of telecommunication relay station along southeast coast of China
Sawle et al. (2018a; 2018b)	PV, wind, biomass, battery, and diesel generator	Minimize COE, LPSP, and particular matter and maximize renewable factor, Job creation, and HDI (Sawle et al., 2018a) Minimize COE and LPSP and maximize reliability (Sawle et al., 2018b)	Butterfly particle swarm optimization, GA, PSO, and teaching–learning-based optimization (Sawle et al., 2018a)HOMER Pro Software and PSO (Sawle et al., 2018b)	Standalone microgrid with case study of rural area of Barwani district, MP, India
Guo et al. (2016)	PV arrays, wind, ESS, and diesel generator	Minimize generating cost of DisCo and maximize internal rate of return of DGO	Multiobjective nondominated sorting genetic algorithm-II	Off-grid and grid-connected microgrid system with the case study of the aboriginal community in Canada
Bourennani et al. (2015)	Wind, PV, battery, and diesel generators	Minimize total net preset cost, COE, and pollutant emissions	Multiobjective optimization metaheuristics methods	Isolated microgrid system

(Continued)

TABLE 30.1 (Continued)

Author/reference	HRES components	Objectives	Optimization method	System/case study
Fan et al. (2020)	Heat pump, fuel cell, battery, and micro-CHP system	Minimizing the annual maintenance and capital costs	Multiobjective optimization with sunflower optimization algorithm	Isolated system
Sharafi et al. (2014)	PV, wind, diesel generator, battery, fuel cell, electrolyzer, and hydrogen tank	Minimize total cost, unmet load, and fuel emission.	PSO	Standalone system with case study of remote area at Zaragoza, Spain
Dali et al. (2010)	PV and wind experimental test bench	Maximize system reliability and economy	Techno-economic algorithm	Grid-connected hybrid system with experimental test bench
Maleki and Askarzadeh (2014)	Wind, PV, and fuel cell	Minimize total annual cost	Artificial bee swarm optimization algorithm	Isolated system with case study of Rafsanjan, South Iran
Khalid et al. (2018)	PV, wind, and battery	Minimize the demand-supply error and overall system costs	Generic optimization methodology	Isolated system with case study of solar and wind station of Australia
Diemuodeke et al. (2016)	PV, wind, diesel generator, and battery	Minimize COE	Multicriteria techno-economic approach using HOMER Pro Software	Isolated system with case study for coastline communities in Nigeria
Hafez et al. (2019)	PV, wind, and battery	Minimize total cost of system and COE	Artificial immune system technique	Isolated system with case study for Qena AL-Gadida, Qena, Egypt

COE, cost of energy; *DGO*, distributed generation owner's; *GA*, genetic algorithm; *HRES*, hybrid renewable energy system; *LPSP*, loss of power supply probability; *PSO*, particle swarm optimization; *PV*, photovoltaic; *CHP*, combined heat and power; *HDI*, human development index; *ESS*, energy storage system.

30.5   Modeling of hybrid renewable system components

Modeling of HRES involves the wind turbine (WT), PV array, battery storage converter, and other essential components. To plan and analyze the HRES performances, corresponding component modeling is required and the whole energy system component power output is utilized to fulfill energy demand of the system.

30.5.1   Solar−photovoltaic

The power output of PV solar-photovoltaic depends on the solar irradiation and temperature of the study, area which is determined by following equation (Belmili et al., 2014; Lao & Chungpaibulpatana, 2017; Samy et al., 2019; Suresh et al., 2020):

$$P_{PV}^t = P_{PV}^0 \times d_f \times \left(\frac{I_r^t}{I_r^{STC}} \right) \times \left\{ 1 + \tau_c \left(T^t - T^{STC} \right) \right\} \tag{30.1}$$

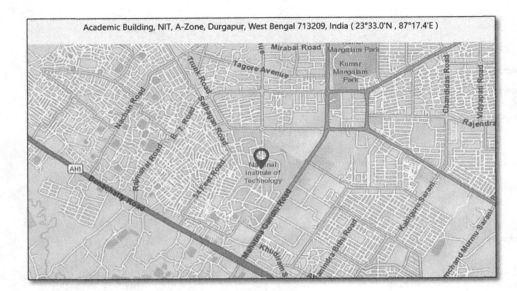

FIGURE 30.1 NIT Durgapur topographical location.

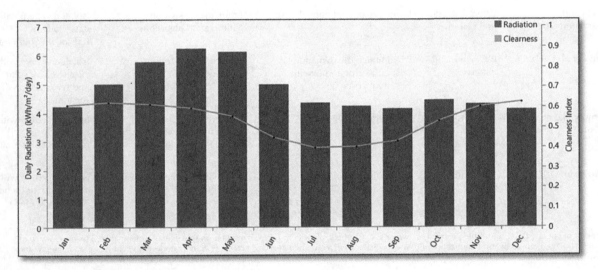

FIGURE 30.2 Availability of annual solar data for NIT Durgapur.

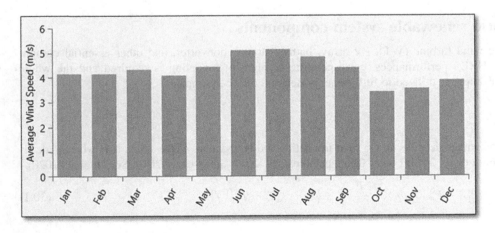

FIGURE 30.3 Availability of annual wind data for NIT Durgapur.

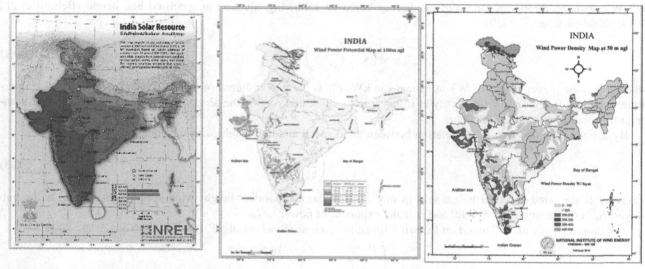

FIGURE 30.4 The solar and wind intensity map of India.

TABLE 30.2 Monthly average solar clearance index global horizontal irradiance, temperature, and wind speed.

Months	Clearance index	Radiation (kW m^{-2} day^{-1})	Temperature (°C)	Average wind speed
January	0.609	4.230	16.960	4.130
February	0.622	5.010	21.130	4.140
March	0.614	5.770	26.880	4.330
April	0.595	6.230	31.660	4.080
May	0.555	6.120	32.830	4.430
June	0.446	4.980	31.010	5.080
July	0.394	4.350	28.750	5.120
August	0.400	4.230	28.120	4.820
September	0.429	4.140	27.200	4.390
October	0.530	4.430	25.040	3.390
November	0.601	4.290	21.070	3.530
December	0.624	4.110	17.300	3.880
Annual average	0.535	4.824	25.663	4.277

where P_{PV}^t is the PV array output power at time t (kW); P_{PV}^0 is the PV array rated capacity for standard test conditions (STC) in kW; d_f is the derating factor of PV array in %; I_r^t is the solar radiation at time t in kW m^{-2}; I_r^{STC} is the solar radiation at STC, that is, 1 kW m^{-2}; τ_c is the temperature coefficient; T^t is the temperature at time t in °C; and T^{STC} is the temperature at STC, that is, 25°C.

30.5.2 Wind turbine

The WT is modeled considering the power curves at the standard pressure and temperature condition. The output of WT is varied with wind speed and air density variation at particular hub height. Therefore the actual output power of

WT is calculated through the ratio of air density based on following equation at specified hub height (Belmili et al., 2014; HOMER Energy, 2020):

$$P_w^t = \left(\frac{\alpha_d^t}{\alpha_d^{STP}} \right) \times P_W^0 \tag{30.2}$$

where P_w^t is the power output of WT at instant t in kW; P_W^0 is the power output of WT at standard pressure and temperature condition in kW; α_d^t is air density at instant t in kg m^{-3}; and α_d^{STP} is the air density at standard pressure and temperature condition, that is, 1.225 kg m^{-3}.

By applying the power law, the relation between wind speed and hub height with the equation given as:

$$\frac{v_h}{v_a} = \left(\frac{h_h}{h_a} \right)^{\gamma} \tag{30.3}$$

where v_h is the wind speed at hub height in m s^{-1}; v_a is the anemometer height wind speed in m s^{-1}; h_h is the hub height; h_a is the anemometer height; and γ is the exponent of power law.

The output power of WT based on the wind speed is given as (Sawle et al., 2018a, 2018b; Yang et al., 2009):

$$P_{WD}^t = \begin{cases} P_r & v_0 \leq v < v_{co} \\ P_{rw} \left(\dfrac{v - v_{ci}}{v_0 - v_{ci}} \right) & v_{ci} < v < v_0 \\ 0 & \text{else} \end{cases} \tag{30.4}$$

where P_{WD}^t is the power produced by turbine at instance t (kW); P_{rw} is the rated power of wind generator (kW); v_0 is the nominal wind speed of WT; v is the actual wind speed; v_{co} is the cut out speed; and v_{ci} is the cut in speed.

30.5.3 Battery storage system

The excess energy generated by means of hybrid renewable resources during light load condition is stored in battery storage system (BSS) and it dispatches suitable energy whenever required in the system. The mathematical equations for battery state of charge (SOC) during charging and discharging mode is given as (Sharafi et al., 2014):

$$B_{SOC}^t = B_{SOC}^{t-1} \pm \left(\frac{V_{bat}^t \times \eta_{bat}}{C_{bat}} \right) \tag{30.5}$$

where the $+$ve and $-$ve signs are considered for charging and discharging mode operation of battery energy storage system. Also, B_{SOC}^t and B_{SOC}^{t-1} are battery SOC for t and $t-1$ instance, respectively, η_{bat} is the round trip efficiency of battery, V_{bat}^t is charged or discharged power by battery at instance t, and C_{bat} is the battery nominal capacity in kWh. The BES can be charged upto B_{SOC}^{max} and discharged up to B_{SOC}^{min} where the minimum and maximum SOC limits are specified as:

$$B_{SOC}^{min} \leq B_{SOC} \leq B_{SOC}^{max}$$

30.5.4 System converter

The electronic system converter is required for maintaining the energy flow among the different AC and DC components. The electric energy transferred from one element to other element by different operating modes of system converter, that is, converters (AC to DC)/inverters (AC to DC), as per the needed load frequency. The efficiency of converter can be evaluated as (Sawle et al., 2018a, 2018b):

$$\eta_{imv} = \frac{P}{P + P_0 + kP^2} \tag{30.6}$$

where variables P, P_0, and k, respectively, obtained as:

$$P_0 = 1 - 99 \left(\frac{10}{\eta_{10}} - \frac{1}{\eta_{100}} - 9 \right)^2 , k = \frac{1}{\eta_{100}} - P_0 - 1, \text{ and } P = P_{out}/P_n \tag{30.7}$$

where η_{10} and η_{100} are the efficiency of the inverter at 10% and 100% of its insignificant energy, respectively, which are given by the makers.

30.5.5   Diesel generator

The diesel generator (DG) set is used for alternate backup supply power in the system. The DG fuel consumption is dependent on the size of generator and on connected load to which generator supplies the power. The DG fuel consumption can be calculated as (Sharafi et al., 2014):

$$F_{cons} = k_0 W_0 + k_1 W_{dg} \qquad (30.8)$$

where k_0 and k_1 are fuel consumption coefficients, W_0 and W_{dg} are DG-rated capacity and output power, respectively. Also, $k_0 = 0.081451$ L kWh^{-1} and $k_1 = 0.2461$ L kWh^{-1}. The efficiency of DG is defined by the ratio of the power output to the fuel consumption heating value (Sharafi et al., 2014).

$$\eta_{dg} = \frac{W_{dg}}{F_{cons} \times V_{lh}} \qquad (30.9)$$

where V_{lh} is the low heat value of gas/oil and it ranges from 10 to 11.6 kWh L^{-1} (Sharafi et al., 2014).

30.5.6   Load profile of system

The commercial load is considered in the study on the basis of daily energy consumption in kWh day^{-1} of NITD. The estimation of average load of microgrid system can be distinguished as area under load curve for specified duration. The per day and annual energy demand of NITD are given in Fig. 30.5 and Fig. 30.6, respectively. The peak load demand is 1488.52 kW. The average daily, monthly, and annual electric energy consumption/demand/load are 13,830.6 kWh day^{-1}, 420.7 MWh month^{-1}, and 5048.2 MWh, respectively. The total load of NITD case study is divided into critical or primary load and noncritical load. The critical load is primary load requiring power supply for primary official work and power supply in emergency for laboratory, health center, and other main priority works, which are supplied even in outages from backup supply. The load model of case study is implemented in HOMER-grid with the help of import MATLAB file interface. The HRES parameter values for case study are given in Table 30.3.

30.6   Explanation of problem and methodology for case study

The case study of the presented work is the current existing system of NITD, WB, India. The case study of NITD is considered with AC load, DG, and power supply by main grid, which is treated as base case system. The NITD system is connected with main grid, which is connected with Damodar Vally Corporation (DVC) power plant including the DVC tariff order for energy consumption (Tariff Order of DVC for the years 2015–16 and, 2017–18). The HRES containing PV, wind, and battery storage has been proposed for NITD system in this study, which offers lower per unit price of energy. The pollution in Durgapur is also increasing day by day due to many reasons, such as presence of large industries, use of conventional and fossil fuel-based power plants, IC vehicles, and many more. Thus the global warming, human health, and other social impacts are adversely affected. In recent days need is to develop a green campus globally; therefore the HRES together with PV panel, WT, and BSSs is proposed with existing NITD grid-connected power system. The proposed HRES will improve the social, economic, and technical stability of the grid-connected

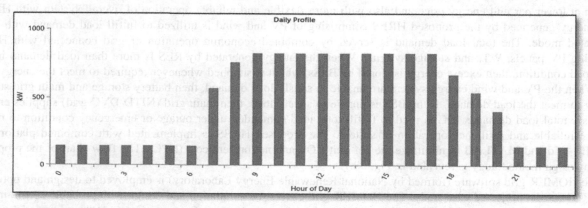

FIGURE 30.5   Load profile for a day of NIT Durgapur.

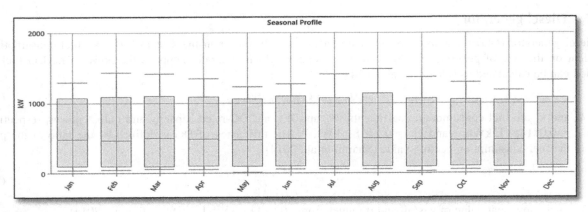

FIGURE 30.6   The seasonal load profile of NIT Durgapur.

TABLE 30.3 Hybrid renewable energy system parameter values for case study.

Solar—PV (generic flat PV)	Diesel generator
Capital cost—₹ 3000 kW^{-1}	Sizes considered—1500 kW
Replacement cost—₹ 3000 kW^{-1}	Maximum load ratio—25%
O&M cost—₹ 18 year^{-1}	Initial cost—₹ 52,500 kW^{-1}
Lifetime—25 years	Replacement cost—₹ 35,000 kW^{-1}
Sizes considered—300, 400, and 500 kW	O&M cost—₹ 0.7 hour^{-1}
Derating factor—80%	Lifetime—15,000 hours
	Diesel fuel price—₹ 66.0 L^{-1}
Wind turbine (generic)	Lower heating value—43.2 MJ kg^{-1}
Capital cost—₹ 7000 kW^{-1}	Density—820.00 kg m^{-3}
Replacement cost—₹ 7000 kW^{-1}	Carbon content—88.0%
O&M cost—₹ 70 year^{-1}	Sulfur content—0.4%
Lifetime—20 years	
Sizes considered—100, 150, 200, and 250 kW	System converter
Hub height—17 m	Capital cost—₹ 300 kW^{-1}
	Replacement cost—₹ 300 kW^{-1}
Battery storage (generic LA)	O&M cost—₹ 10 year^{-1}
Nominal voltage—6 V	Efficiency—90%
Min state of charge—20%	Lifetime—15 years
Capital cost—₹ 350 kWh^{-1}	Project lifetime—25 years
Replacement cost—₹ 350 kWh^{-1}	Inflation rate—2%
O&M cost—₹ 10 year^{-1}	Discount rate—8%
Lifetime throughput—3000 kWh	Interest rate—6%

PV, photovoltaic.

system at lower per unit energy price and also with more flexible and reliable operation of overall system with HRES. The energy generated by the proposed HRES comprising of PV and wind is utilized to fulfill load demand with grid-connected mode. The total load demand is served by combined economic operation of grid connected with HRES including PV panels, WT, and storage systems. When the energy generated by RES is more than load demand in the light load condition, then excess energy is stored by BSSs, which is supplied whenever required to meet the energy deficit. When the PV and wind energy system are unable to satisfy load demand, then battery storage and main grid supply energy to meet the load demand. If the BSS is unable to satisfy load, then main grid (NITD-DVC-grid) supplies energy to satisfy total load demands. DG is used to fulfill total load demanded under outage or emergency condition to maintain the reliable and resilience operation of system. The proposed HRES is implemented with combined platform of HOMER-grid and MATLAB to minimize the per unit COE and net present cost (NPC). The flow chart of the proposed system design methodology is revealed in Fig. 30.7.

The HOMER-grid software (formed by National Renewable Energy Laboratory) is employed to design and optimize the HRES for the projected case study. The HOMER-grid covers the simulation and design with probable constraints of system for the optimization. HOMER-grid is an innovative program design software to construct the new model and its

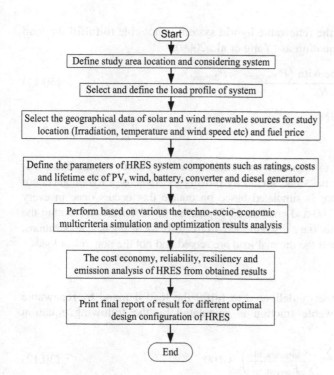

FIGURE 30.7 Flow chart for the proposed hybrid renewable energy system design methodology.

The flow chart contains the following boxes:

Start → Define study area location and considering system → Select and define the load profile of system → Select the geographical data of solar and wind renewable sources for study location (Irradiation, temperature and wind speed etc) and fuel price → Define the parameters of HRES system components such as ratings, costs and lifetime etc of PV, wind, battery, converter and diesel generator → Perform based on various the techno-socio-economic multicriteria simulation and optimization results analysis → The cost economy, reliability, resiliency and emission analysis of HRES from obtained results → Print final report of result for different optimal design configuration of HRES → End

operation for the HRES design including grid connection. The system simulation and analysis are fundamentally depended on the choice of system components selected in any study. The HOMER-grid allows the development of system components in huge amount and sizing of the successful operations of entire system and also evaluated the total quantity of WT, solar PV, battery storage, and system converter for study to be required. The optimum dynamic system design is selected during the HRES planning through the HOMER-grid optimization and simulation analysis. The reduction of demand charge and the other cost-saving benefits can be achieved by different options for reducing the electricity bills. The comparison of different costs parameters and cost savings for installing different combinations of batteries, solar panels, WT, and generators are presented. The HOMER-grid uses a powerful optimization to minimize system COE with minimizing NPC that will maximize the savings of the system. The bill from an electric utility can be comprised of a few different types of charges. The charges of energy are paid to utility for the quantity of total energy consumed for the month. The demand charges are paid to utility for the highest peak power draw in kilowatts (kW) or megawatts (MW) for the month. Finally, the fixed charge is a charge that is the same for every month and not varies with the peak load demand or consumption. Here, the optimization tool considers generators as a system for peak shaving and demand charge reduction. The total cost, replacement, capital, O&M, fuel costs, and other parameters are also obtained with HOMER-grid for the hybrid system. Finally, the HOMER-grid optimizes all the configuration of hybrid system basis of the multicriteria optimization approach based on various economic, social, and technical parameters, such as minimization of levelized COE (LCOE), total NPC, total annualized cost, reliability, and emissions and also maximization of reliability, resiliency, and renewable fraction. Therefore the economic, social, and technical parameters for the proposed HRES are defined as shown in Fig. 30.7.

30.6.1 Technical parameters

30.6.1.1 Reliability of system

The reliability is the ability of system to maintain operations during the outages of utility for short time, generally lasting a few hours or less. These outages may occur multiple times in a year. Maintaining operations during the longer outages comes under resiliency. The systems having energy storage or backup generator can continue the operations under the grid outage condition. In the proposed HRES, the reliability is simulated based on outages, which happen 10 times per year on average for 2-hour average repair time. The no electrical energy is purchased or sold by the grid during the outage time steps. Also, the reliability of system is evaluated by the concept of LPSP. The LPSP can be defined as the probability of deficit power supply consequences when the renewable hybrid system is not able to fulfill the load demand. The LPSP can be evaluated based on the following equation as (Yang et al., 2009):

$$\text{LPSP} = \frac{\sum_{t=0}^{T} P \times \text{Failure} \times \text{Time with} \left(P^t_{\text{Supplied}} < P^t_{\text{demand}}\right)}{N_t}$$

(30.11)

where T is the entire period and N_t is intervals for deficit energy.

30.6.1.2 Resilience of system

The resilience can be defined as the ability of system for a facility to respond to an extended, multiday utility outage like a natural disaster, such as hurricanes or wildfires. It provides an additional value stream by proving the value of a hybrid/generator project in extended outages. In the proposed HRES, the resilience is simulated based on outage that occurs once in every 1.0 years. The simulated grid shutdown starts on 21 September 03:00 and ends on 23 September 03:00 for 2 days. During the resilience outage duration, the HRES work as standalone systems (i.e., disconnected from grid) for resiliency applications. During a resilience event, only the critical/primary electrical load and the thermal load are served and not the noncritical loads.

30.6.1.3 Renewable factor

The renewable fraction can be defined on the basis of total energy delivered to fulfill the load demand by renewable sources with respect to conventional energy sources. Renewable fraction is calculated by the following equation (HOMER Energy, 2020; Sawle et al., 2018a):

$$\text{Renewable fraction} = \left(1 - \frac{\sum E_{\text{Non-renewable}}}{\sum E_{\text{served}}}\right) \times 100$$

(30.12)

30.6.2 Economic parameter

30.6.2.1 Total net present cost

The total net present cost (TNPC) of the systems can be defined as the total value of the present system cost components, including initial cost, replacement cost, operation and maintenance cost, emissions penalty, fuel costs, and costs of power purchasing from the grid over the project lifetime. The TNPC of a system is main economic parameters based on that the different system configurations are ranked, which is used to estimate LCOE and total annualized cost.

30.6.2.2 Levelized cost of energy

The LCOE is an eminent noteworthy factor to determine the cost effective analysis of distribution grid or network in the presence of renewable sources. The average COE or LCOE can be defined as (Sawle et al., 2018a; 2018b):

$$\text{LCOE} = \frac{C_{\text{TNPC}}}{\sum_{H=1}^{H=8760} P_{\text{demand}}} \times \text{CRF}$$

(30.13)

where P_{demand} is the energy consumed (kWh) hourly, C_{TNPC} is the TNPC, and CRF is the capital recovery factor, which is determines by the following equation:

$$\text{CRF} = \frac{r(1+r)^n}{(1+r)^n - 1}$$

(30.14)

where r is the actual rate of interest and n is the system/project lifetime

30.6.2.3 Total annualized cost

Total annualized cost (C_{annual}) of systems is depending on TNPC, and it is defined as the multiplication of CRF and the TNPC (C_{TNPC}) of the systems as illustrated in the following equation (Sawle et al., 2018b):

$$C_{\text{annual}} = \text{CRF} \times C_{\text{TNPC}}$$

(30.15)

where C_{TNPC} is the TNPC.

30.6.2.4   Annualized savings

The annualized saving is the difference in the annualized cost between the base case system and the proposed system. Generally, the annualized cost is the reparative cost for every year during project lifetime, which gives equivalent NPC based on the real cash flow sequences.

30.6.2.5   Capital investment

The capital investment is the additional installed cost of the proposed system related to base/current system case at initially for the proposed system.

30.6.2.6   Internal rate of return

The IRR is basically a discount rate as a result of that the NPC of the base case system and the proposed system are same. Here, the IRR by determining the discount rates makes zero difference in the present values of the two cash flows.

30.6.2.7   Return on investment

The return on investment (ROI) is yearly cost saving compared with initial investments. The ROI is defined as the ratio of annual average difference of nominal cash flow during project lifetime to the capital cost difference.

30.6.2.8   Simple payback

The simple payback is the number of years in which cumulative cash flows change moves toward positive values from the negative values for the proposed system relative to the system base case. The payback indicates the time it will take to recover the investment difference costs between the base case system and the proposed system.

30.6.3   Social parameters

The pollutant emission pollutant emission (CO_2, NO_x, and SO_2) and PM are taken in to consideration as social performances. The grid/utility majorly supplied with thermal power plants; therefore generally the pollutant emission component consists of CO (carbon monoxide), CO_2 (carbon dioxide), SO_2 (sulfur dioxide), NO_x (nitrogen oxides) and PM. These pollutant emission elements are badly affected the environment as well as the human health. Mainly PM effect to human health adversely which affect the respiratory system and other health issues. The equivalent PM consists of $M_{2.5}$ (air pollutant particles less than or equal to 2.5 μm diameter) and PM_{10} (air pollutant particles less than or equal to 10.0 μm diameter). The pollutant emission factors for CO_2 are 632.0 g kWh^{-1}, SO_2 are 2.74 g kWh^{-1} and NO_x are 1.34 g kWh^{-1} (HOMER Energy, 2020). Also, the PM is given on the basis of linear regression equation with correlation $r = 0.884$ of PM and carbon weight as (Sawle et al., 2018a, 2018b):

$$PM = (0.47 \times W_{carbon}) + 0.12 \text{ with correlation } r = 0.884 \qquad (30.10)$$

where PM = $PM_{2.5}$ + PM_{10} and W_{carbon} is the weight of total carbon emissions.

30.7   Results and discussion

The result analysis of the proposed HRES configuration based on HOMER-grid multicriteria optimization to supply the required load demand in the study area is presented in this section. Based on the available data of system parameters, the hourly simulation for different configurations is performed to decide the sizing and estimation of other technical, social, and economic parameters of the presented case study. From various configurations, the best configurations are analyzed and discussed based on its economic, social, and technical performances by using HOMER-grid multicriteria optimization. The optimal design solution of the proposed HRES is selected based on more feasible solution with respect to base case system. The analysis of two system cases is presented here, first is the base case system which includes only DG connected with DVC-NITD-grid and second is the proposed HRES comprising of solar PV, WT, battery storage, and DG connected with DVC-NITD-grid.

30.7.1 Analysis of base system (current system): diesel generator + DVC-NITD-grid

The base system or current system is the existing grid-connected (DVC-NITD-grid) system of NITD with DG. The average electric energy that needs current/existing system of NITD, WB, India, is 11,687.2 kWh day^{-1}, 355.5 MWh month^{-1}, and 4265.8 MWh year^{-1} with the peak hour demand of 1488.52 kW. The load profile of base case system for the day on which the largest demand occurs and the energy consumption pattern are shown in Fig. 30.8 and Fig. 30.9, respectively. Currently, it is connected with DVC-NITD-grid and DG units with the total capacity of 1500 kW. The NITD currently spends ₹ 23.97 M on the utility bill per year in which 24% of the utility bill is demand charges as given in Table 30.5. The total energy consumption in the presented case study is 11,687.2 kWh day − 1 with a peak of 1488.52 kW electrical load in the base case system, which is served by the only two sources, including outage conditions as tabulated in Table 30.4.

The total energy supply is 4,283,279 kWh year^{-1} to fulfill the load demand of 4,264,777 kWh year^{-1}. The monthly utility charge details of base case system are given in Table 30.5, which include total energy purchased from grid, energy sold to grid, net energy purchased, energy charges, and demand charges of utility grid. The total bill charge is ₹ 23,966,395.20 for base system in which ₹ 18,242,657.97 is total energy consumption charges and ₹ 5,723,737.24 is total energy demand charges. The net present and annual costs for base system are ₹ 318,722,339.23 and ₹ 24,654,568.26, respectively, as specified in Table 30.6. The total net present and annual costs of DG set are ₹ 8,896,368.62 and ₹ 688,173.06, which include the capital cost, replacement cost, O&M cost, fuel cost, and salvage cost. The tariff of utility grid (DVC) of base system for net present and annual costs are ₹ 309,825,970.61 and ₹ 23,966,395.20.

The annual and outage duration operational components of DG for base case system are tabulated in Table 30.7 and Table 30.8, respectively, which operates to maintain the reliability and resilience operation of the system under short and long outage conditions. The power output pattern of DG under outage, the grid outage, reliability (short outage) and resilience (long outage) pattern, and also the system operation under resilience (long outage) for base system under outage are illustrated in Fig. 30.10. The long outages correspond to whole day or multiple day utility outages for natural disaster, such as hurricanes or wildfires. It provides an additional value stream by proving the value of a hybrid power generation during outages, which are considered for the resilience analysis. The shorter outages are more frequent outages that are defined for the reliability analysis. In this case study, the large outage is defined and simulated that occurs once in every 1.0 years. It is assumed that the grid shutdown starts on 21 September 03:00 and ends on 23 September 03:00. Also, the short outages that occur 10 times per year on average with a 2.0-hour average repair time. The distributed generation system performance during these outages is shown in Fig. 30.10B and C.

The monthly electric consumption served by different sources, the electricity bill breakdown, and monthly peak day profile of base system are revealed in Figs. 30.11−30.13, respectively. The pollutant emissions of these environmental components are considering from total electricity production by generators and grid power purchasing. The generator pollutants depend on fuel consumption and its emission factors. The grid-related emissions of each pollutant are calculated on the basis of total grid purchasing, which is the net energy purchased from grid minus net energy sold to the grid. The pollutant emissions for the total grid power purchase can be given by multiplying its emission factor for each pollutant. The pollutant emission for the base system, such as carbon dioxide, carbon monoxide, and particulate matter, is 2,711,629, 153, and 0.927 kg year^{-1}, respectively, as demonstrated in Table 30.9.

30.7.2 Analysis of the proposed HRES: solar PV−wind−battery storage−diesel generator connected with DVC-NITD-grid

The proposed system includes the installation of solar PV, WT, battery storage, system converter, and DG connected with the DVC-NITD-grid system, which is obtained as feasible solution based on the minimum values of the COE and NPC. This system proposes adding 500 kW of solar PV, 3677 kWh of battery capacity, 250 kW of wind generation capacity, and 912 kW of system converter with the base case system. The proposed HRES reduce the system annual utility bill to ₹ 17.4 M. Also, the system investment has a payback of 0.60 years and an IRR of 167.76%. The proposed HRES has more economic, social, and technical benefits and savings. The installing component details of the proposed HRES are given in Table 30.10.

Solar PV: The detailed operational parameter values of solar PV system and operational profile for the proposed HRES are given in Table 30.11 and Fig. 30.14, respectively, specifying that the solar PV has a nominal capacity of

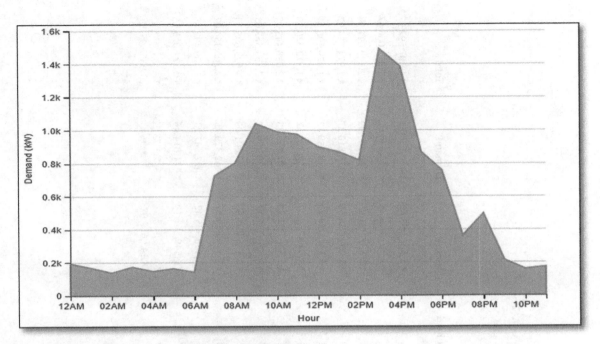

FIGURE 30.8 Load profile for the day on which the peak demand occurs.

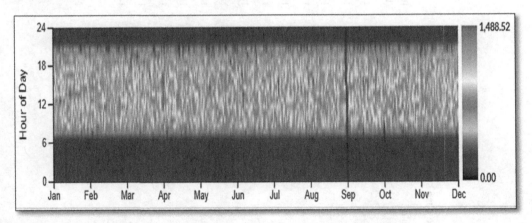

FIGURE 30.9 The energy consumption pattern.

TABLE 30.4 Base case energy supply and utilization summary (annual).

Total energy supply			Electrical energy consumptions	
Energy source	**kWh year^{-1}**	**%**	Total annual load (kWh year^{-1})	4,264,777
DG set	31,125	0.727	Excess energy (kWh year^{-1})	18,502
Grid purchases	4,252,154	99.3	Total	4,283,279
Total	4,283,279	100		

DG, diesel generator.

500 kW with the annual production of 762,863 kWh year^{-1}. The hours of operation are 4390 hours year^{-1} with the mean output of 2090 kWh day^{-1}.

WT: The complete operational component values and operational profile of WT system for the proposed HRES are given in Table 30.12 and Fig. 30.15, respectively. The WT-rated capacity is 250 kW with the mean output of 24.7 kW.

TABLE 30.5 Monthly utility charge details for base case system.

Month	Energy purchased (kWh)	Energy sold (kWh)	Net energy purchased (kWh)	Peak load (kW)	Energy charge	Demand charge	Total
January	357,626	0	357,626	1295	₹ 1,527,061.03	₹ 460,536.53	₹ 1,987,597.57
February	318,927	0	318,927	1437	₹ 1,361,818.45	₹ 510,848.02	₹ 1,872,666.47
March	376,548	0	376,548	1423	₹ 1,622,921.95	₹ 506,099.99	₹ 2,129,021.94
April	346,854	0	346,854	1349	₹ 1,494,938.60	₹ 479,816.37	₹ 1,974,754.97
May	357,550	0	357,550	1232	₹ 1,541,038.99	₹ 438,159.12	₹ 1,979,198.11
June	352,611	0	352,611	1273	₹ 1,519,755.07	₹ 452,765.52	₹ 1,972,520.59
July	358,211	0	358,211	1411	₹ 1,536,727.30	₹ 501,718.25	₹ 2,038,445.55
August	379,098	0	379,098	1489	₹ 1,626,331.12	₹ 529,250.92	₹ 2,155,582.03
September	334,707	0	334,707	1372	₹ 1,435,894.65	₹ 487,801.09	₹ 1,923,695.74
October	358,968	0	358,968	1300	₹ 1,539,974.27	₹ 462,379.75	₹ 2,002,354.02
November	348,228	0	348,228	1188	₹ 1,486,935.42	₹ 422,547.23	₹ 1,909,482.65
December	362,825	0	362,825	1327	₹ 1,549,261.10	₹ 471,814.46	₹ 2,021,075.56
Total annual	4,252,154	0	4,252,154	1489	₹ 18,242,657.97	₹ 5,723,737.24	₹ 23,966,395.20

TABLE 30.6 Net present and annual cost summary for base system.

Cost components	Total system	DG set	DVC tariff (grid utility)
Net present cost of system			
Capital	₹ 1,125,000.00	₹ 1,125,000.00	–
Replacement	–	₹ 0.00	–
O&M	₹ 309,842,065.36	₹ 16,094.76	₹ 309,825,970.61
Fuel	₹ 7,910,088.13	₹ 7,910,088.13	–
Salvage	₹ −154,814.26	₹ −154,814.26	–
Total	₹ 318,722,339.23	₹ 8,896,368.62	₹ 309,825,970.61
Annual cost of system			
Capital	₹ 87,023.68	₹ 87,023.68	–
Replacement	–	₹ 0.00	–
O&M	₹ 23,967,640.20	₹ 1245.00	₹ 23,966,395.20
Fuel	₹ 611,879.95	₹ 611,879.95	–
Salvage	₹ −11,975.56	₹ −11,975.56	–
Total	₹ 24,654,568.26	₹ 688,173.06	₹ 23,966,395.20

DG, diesel generator; *DVC*, Damodar Vally Corporation.

TABLE 30.7 Diesel generator operational components for base case.

Components	Value
Hours of operation	83.0 hours year^{-1}
Number of starts	11.0 starts year^{-1}
Operational life	181 years
Electrical production	31,125 kWh year^{-1}
Mean electrical output	375 kW
Fuel consumption	9271 L
Generator fuel price	66.0 ₹ L^{-1}
Maintenance cost	1245 ₹ year^{-1}

TABLE 30.8 The DG operation under outage for reliability (short outage) and resilience (long outage) of base system.

Components	During outage
DG runtime (hours day^{-1})	24.0
DG O&M cost (₹ day^{-1})	360
DG fuel consumption (L day^{-1})	2681
DG fuel cost (₹ day^{-1})	176,929

DG, diesel generator.

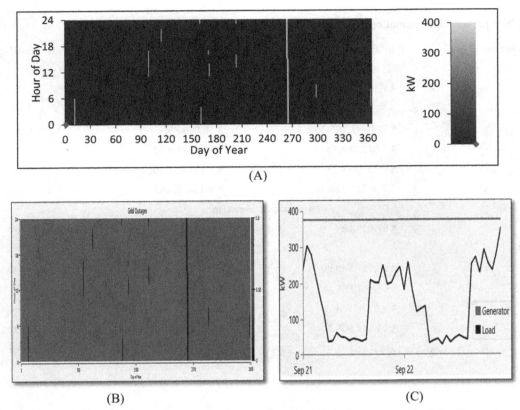

FIGURE 30.10 (A) Power output pattern of DG under outage; (B) grid outage and reliability (short outage), and resilience (long outage); and (C) system operation under resilience (long outage) for base system.

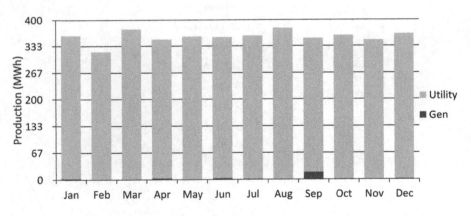

FIGURE 30.11 Monthly electric consumption served by sources for base case.

The total annual electricity produced by WT is 216,668 kWh year^{-1} and the annual operation hours for WT are 6321 hours year^{-1}.

BSS: The BSS has a nominal capacity of 3677 kWh and the annual throughput is 383,625 kWh year^{-1}. The autonomy for the battery is considered as 6 hours for this case study. The detailed operational parameter values and operational profile of BSS for the proposed HRES are shown in Table 30.13 and Fig. 30.16, respectively.

DG set: The total power output from the DG system, operating at 915 kW using diesel as fuel, is 5054 kWh year^{-1}. The DG set requires 10 operation hours per year with the total fuel consumption of 1425 L year^{-1}. The summary of

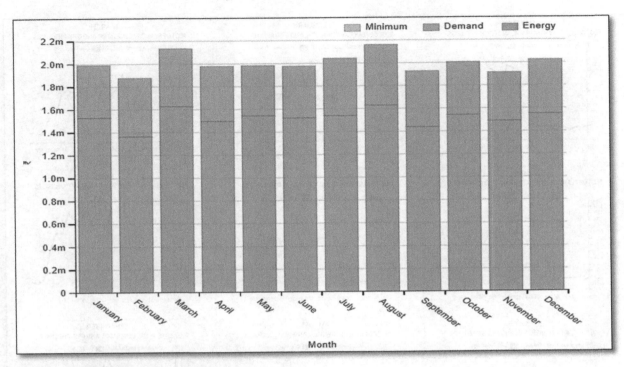

FIGURE 30.12 Monthly electricity bill breakdown for base case.

operational parameter values and operational profile of DG set for the proposed HRES are shown in Table 30.14 and Fig. 30.17, respectively.

System converter: The system converter has two operating modes, namely, inverter mode and rectifier mode, with the nominal capacity of 912 kW. The total operating hours per year for inverter mode and rectifier mode are 4740 and 3882 hours year^{-1} and also the operational losses for both modes of system converter are 51,914 and 19,038 kWh year^{-1}, respectively. The detailed operational parameters of system converter for the proposed HRES are given in Table 30.15.

The total electric energy supplied by distinct sources, such as solar PV, WT, DG set, and grid, is 4,424,768 kWh year^{-1}, which is utilized by load demand and different components as shown in Table 30.16. The total monthly utility charges for the proposed HRES are ₹ 17,391,702.05 year^{-1} as illustrated in Table 30.17. The total utility charges contain total energy purchased from grid and sold to grid, net energy purchased, energy charges, and demand charges of utility grid. The net energy purchased charge and the demand charge for the proposed HRES are ₹ 14,751,955.48 and ₹ 2,639,746.57, respectively. The net present and annual costs for the proposed HRES are ₹ 233,556,405.89 and ₹ 18,066,610.47, respectively, as listed in Table 30.18. The total net present and annual costs of system comprise the capital cost, replacement cost, O&M cost, fuel cost, and salvage cost of the system.

The NPC of solar PV, WT, battery storage, converter, DG set, and grid utility systems for the proposed HRES are ₹ 1,080,166.87, ₹ 2,219,724.72, ₹ 2,891,376.95, ₹ 367,732.81, ₹ 2,165,888.41, and ₹ 224,831,516.12, respectively. The annual costs of these components are ₹ 83,555.64, ₹ 171,705.43, ₹ 223,660.67, ₹ 28,445.74, ₹ 167,540.95, and ₹ 17,391,702.05, respectively. Resilience is considered for long duration outage while reliability for short outage. These are the ability of system for a facility to respond to an extended, multiday utility outage and short duration outage, for example, 2 hours. The considered outage assumptions in the presented case study are already discussed in the previous section. The annual operational parameters of DG set for outage duration are presented in Tables 30.19 and 30.20 to maintain the resilience (for long duration outage) and reliability (for short duration outage) of the proposed HRES. The operational performance of the proposed HRES during resilience (large outage) and reliability (short outage) is demonstrated in Figs. 30.18 and 30.19, showing that the renewable power, battery power, and generator power only fulfill the load demand under long and short duration outages to maintain the system resilience and reliability of the system. The

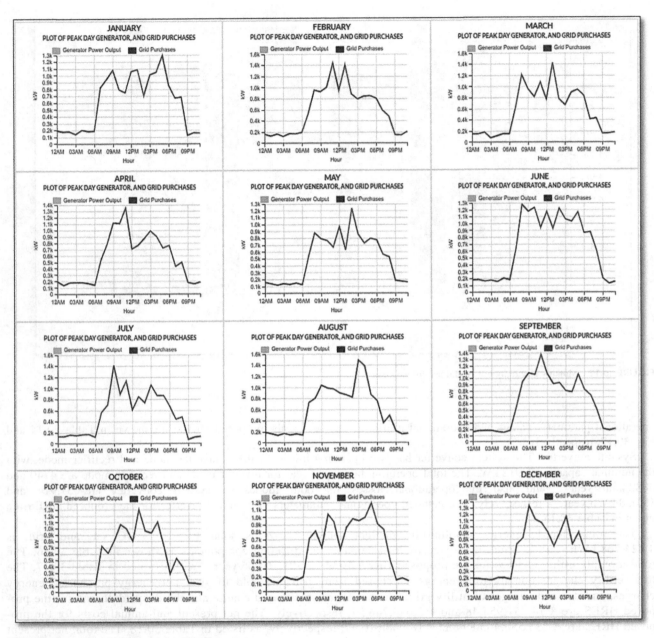

FIGURE 30.13 Monthly peak day profile of base system.

TABLE 30.9 The pollutant emissions for base case system.

Pollutant emission	Emission amount
Carbon dioxide	2,711,629 kg year^{-1}
Carbon monoxide	153 kg year^{-1}
Unburned hydrocarbons	6.68 kg year^{-1}
Particulate matter	0.927 kg year^{-1}
Sulfur dioxide	11,710 kg year^{-1}
Nitrogen oxides	5842 kg year^{-1}

TABLE 30.10 The Installing components for the proposed HRES.

Component	Installation size	Total installation/capital cost	Annual expenses/O&M
Solar PV	500 kW	₹ 1,069,394	₹ 833 year^{-1}
Wind turbine	250 kW	₹ 1,750,000	₹ 17,500 year^{-1}
Battery storage	3674 kWh (two parallel string of 1837 kWh)	₹ 1,102,200	₹ 36,740 year^{-1}
System converter	912 kW	₹ 273,525	---
DG set	1500 kW	₹ 1,125,000.00	₹ 94,183.78 year^{-1}

DG, diesel generator; *HRES*, hybrid renewable energy system; *PV*, photovoltaic.

TABLE 30.11 Solar PV details.

Parameters	Values
Rated capacity	500 kW
Mean output	87.1 kW
Mean output per day	2090 kWh day^{-1}
Capacity factor	17.4%
Total production	762,863 kWh year^{-1}
Minimum output	0 kW
Maximum output	498 kW
PV penetration	17.8%
Hours of operation	4390 hours year^{-1}
PV levelized cost	0.11 ₹ kWh^{-1}

PV, photovoltaic.

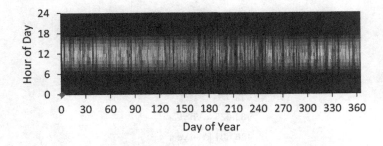

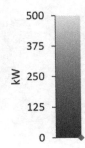

FIGURE 30.14 Solar photovoltaic operational pattern for the proposed hybrid renewable energy system.

pollutant emission for the proposed system, for example, carbon dioxide, carbon monoxide, and particulate matter, is 2,177,925, 23.5, and 0.142 kg year^{-1}, respectively, which are specified in Table 30.21.

The monthly electric consumption supplied by utilizing different energy sources and the monthly electricity bill breakdown for the proposed HRES are shown in Fig. 30.20 and Fig. 30.21, respectively. The annual saving cost summary and the

TABLE 30.12 Wind turbine details.

Parameters	Value
Total rated capacity	250 kW
Mean output	24.7 kW
Capacity factor	9.89%
Total production	216,668 kWh year^{-1}
Minimum output	0 kW
Maximum output	250 kW
Wind penetration	5.05%
Hours of operation	6321 hours year^{-1}
WT levelized cost	0.792 ₹ kWh^{-1}

WT, wind turbine.

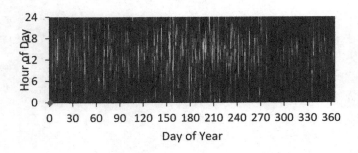
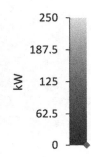

FIGURE 30.15 Wind turbine operational profile for the proposed hybrid renewable energy system.

TABLE 30.13 Battery storage details.

Parameters	Value
Rated capacity	3677 kWh
Annual throughput	383,625 kWh year^{-1}
Maintenance cost	36,740 ₹ year^{-1}
Expected life	7.50 years
Capital costs	₹ 1.10 M
Autonomy	6 hours
Annual throughput	383,625 kWh year^{-1}
Energy in	428,881 kWh year^{-1}
Energy out	343,125 kWh year^{-1}
Battery wear cost	0.419 ₹ kWh^{-1}
Losses	82,102 kWh year^{-1}
Storage depletion	−3655 kWh year^{-1}

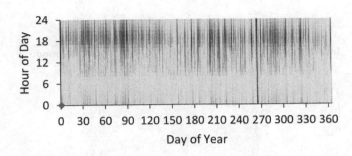
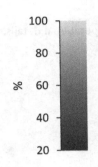

FIGURE 30.16 Battery storage operational profile for the proposed hybrid renewable energy system.

TABLE 30.14 Diesel generator details.

Parameters	Value
Hours of operation	10 hours year^{-1}
Number of starts	4 starts year^{-1}
Operational life	1500 years
Capacity factor	0.0385%
Fixed generation cost	1597 ₹ hour^{-1}
Marginal generation cost	15.6 ₹ kWh^{-1}
Electrical production	5054 kWh year^{-1}
Mean electrical output	505 kW
Minimum electrical output	375 kW
Maximum electrical output	915 kW
Fuel consumption	1425 L year^{-1}
Specific fuel consumption	0.282 L kWh^{-1}
Fuel energy input	14,020 kWh year^{-1}
Mean electrical efficiency	36.1%

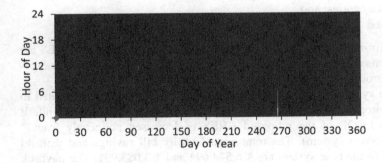
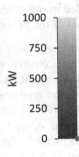

FIGURE 30.17 Diesel generator operational profile for the proposed hybrid renewable energy system.

different cast flows during the lifetime of the proposed HRES are illustrated in Fig. 30.22 and Fig. 30.23, respectively. Whereas Fig. 30.22 shows the summarize chart and the estimated annual savings in the following categories:

- Demand: savings from demand charge reduction;
- Energy: consumption reduction and self-consumption;

TABLE 30.15 The system converter operational details.

Parameters	Inverter mode	Rectifier mode
Capacity (kW)	912	912
Mean output (kW)	113	41.3
Minimum output (kW)	0	0
Maximum output (kW)	912	822
Capacity factor (%)	12.3	4.53
Hours of operation (hours year^{-1})	4740	3882
Energy out (kWh year^{-1})	986,361	361,718
Energy in (kWh year^{-1})	1,038,275	380,756
Losses (kWh year^{-1})	51,914	19,038

TABLE 30.16 The proposed HRES energy supply and utilization summary (annual).

Total energy supply			Electrical energy consumption	
Energy source	**kWh year^{-1}**	**%**	Total annual load (kWh year^{-1})	4,264,777
Solar PV	762,863	17.2	Excess energy (kWh year^{-1})	1706
DG set	5054	0.114	Grid sales (kWh year^{-1})	1577
Wind turbine	216,668	4.90	Average battery charge energy (kWh year^{-1})	156,708
Grid purchases	3,440,182	77.7	Total	4,424,768
Total	4,424,768	100		

DG, diesel generator; *HRES*, hybrid renewable energy system; *PV*, photovoltaic.

- Outages: cost savings related to utility service interruptions;
- O&M: operating and maintenance costs of the proposed components;
- Replacements: cost to replace the proposed system components over the project lifetime;
- Others: cost differences not included in the other categories; and
- Total: the total savings (annualized) of the proposed system.

The different system components output power profile and the monthly peak day profiles of the proposed HRES are shown in Fig. 30.24 and Fig. 30.25, respectively, demonstrating the detailed performance of different system components' output power profile for the proposed HRES throughout the year.

The comparative summary between the base case system and the proposed HRES for NITD system is presented in Table 30.22, clearly indicating that the NPC of the proposed HRES is significantly reduced to ₹ 233,556,400 while it was ₹ 318,722,300 for the base system. Also, the LCOE of the proposed HRES is decreased noticeably from ₹ 5.781 kWh^{-1} (base system) to ₹ 4.235 kWh^{-1} (proposed system). The total annual utility bill savings and demand charge savings for the proposed system as compared to the base system are ₹ 6,574,694 and ₹ 3,083,991. The payback time and the renewable fraction for the proposed system are 0.6 year and 19.25%. The pollutant emission components, such as CO_2 and particulate matter, are considerably reduced for the proposed HRES to 2177.9 t year^{-1} and 0.142 kg year^{-1} as compared to the base system. The comparison of utility bills and the comparative cash flow of NPC during the project lifetime for the current system (base system) and the proposed HRES are revealed in Fig. 30.26 and Fig. 30.27, respectively, confirming that the proposed HRES has the significant impacts on the base system performances and the overall performances are improved with the proposed HRES.

TABLE 30.17 Monthly utility charge details for the proposed HRES.

Month	Energy purchased (kWh)	Energy sold (kWh)	Net energy purchased (kWh)	Peak load (kW)	Energy charge	Demand charge	Total
January	286,466	200	286,266	1295	₹1,222,357.12	₹216,356.23	₹1,438,713.35
February	250,788	0	250,788	1437	₹1,070,865.98	₹185,664.65	₹1,256,530.63
March	296,388	299	296,089	1423	₹1,276,142.63	₹197,168.12	₹1,473,310.75
April	276,014	0	276,014	1349	₹1,189,620.41	₹207,760.33	₹1,397,380.74
May	280,252	22.2	280,230	1232	₹1,207,790.35	₹213,374.37	₹1,421,164.72
June	279,745	506	279,239	1273	₹1,203,521.91	₹242,617.17	₹1,446,139.07
July	290,912	243	290,669	1411	₹1,246,970.44	₹217,356.12	₹1,464,326.56
August	316,247	94.6	316,152	1489	₹1,356,293.30	₹234,637.59	₹1,590,930.88
September	279,964	19.5	279,945	1372	₹1,200,963.19	₹242,505.19	₹1,443,468.37
October	305,551	0	305,551	1300	₹1,310,814.76	₹205,260.03	₹1,516,074.78
November	287,341	0	287,341	1188	₹1,226,944.68	₹229,374.91	₹1,456,319.59
December	290,513	192	290,321	1327	₹1,239,670.71	₹247,671.88	₹1,487,342.59
Total annual	3,440,182	1577	3,438,606	1489	₹ 14,751,955.48	₹2,639,746.57	₹ 17,391,702.05

HRES, hybrid renewable energy system.

TABLE 30.18 Net present and annual cost summary for the proposed HRES.

Cost components	Total system	Solar PV	Wind turbine	DG set	DVC tariff (grid utility)	Battery storage	System converter
Net present cost of system							
Capital	₹ 5,320,118.94	₹ 1,069,393.94	₹ 1,750,000.00	₹ 1,125,000.00	₹ 0.00	₹ 1,102,200.00	₹ 273,525.00
Replacement	₹ 2,164,164.77	₹ 0.00	₹ 557,912.86	₹ 0.00	₹ 0.00	₹ 1,490,202.41	₹ 116,049.50
O&M	₹ 225,545,416.67	₹ 10,772.93	₹ 226,231.54	₹ 1939.13	₹ 22,4,831,516.12	₹ 474,956.96	₹ 0.00
Fuel	₹ 1,215,623.20	₹ 0.00	₹ 0.00	₹ 1,215,623.20	₹ 0.00	₹ 0.00	₹ 0.00
Salvage	₹ -688,917.70	₹ 0.00	₹ -314,419.68	₹ -176,673.91	₹ 0.00	₹ -175,982.42	₹ -21,841.69
Total	₹ 233,556,405.89	₹ 1,080,166.87	₹ 2,219,724.72	₹ 2,165,888.41	₹ 224,831,516.12	₹ 2,891,376.95	₹ 367,732.81
Annual cost of system							
Capital	₹ 411,534.49	₹ 82,722.30	₹ 135,370.16	₹ 87,023.68	₹ 0.00	₹ 85,260.00	₹ 21,158.36
Replacement	₹ 167,407.62	₹ 0.00	₹ 43,157.00	₹ 0.00	₹ 0.00	₹ 115,273.68	₹ 8976.94
O&M	₹ 17,446,925.39	₹ 833.33	₹ 17,500.00	₹ 150.00	₹ 17,391,702.05	₹ 36,740.00	₹ 0.00
Fuel	₹ 94,033.78	₹ 0.00	₹ 0.00	₹ 94,033.78	₹ 0.00	₹ 0.00	₹ 0.00
Salvage	₹ -53,290.80	₹ 0.00	₹ -24,321.74	₹ -13,666.50	₹ 0.00	₹ -13,613.01	₹ -1689.55
Total	₹ 18,066,610.47	₹ 83,555.64	₹ 171,705.43	₹ 167,540.95	₹ 17,391,702.05	₹ 223,660.67	₹ 28,445.74

DG, diesel generator; DVC, Damodar Vally Corporation; HRES, hybrid renewable energy system; PV, photovoltaic.

TABLE 30.19 The DG operation under outage for resilience (long outage) of the proposed HRES.

Parameters	During outage	No outage
DG set runtime (hours day^{-1})	3.00	—
DG set O&M cost (₹ day^{-1})	45.0	—
DG set fuel consumption (L day^{-1})	489	—
DG set fuel cost (₹ day^{-1})	32,273	—
Battery throughput (kWh day^{-1})	1730	777
Battery wear cost (₹ day^{-1})	725	326

DG, diesel generator; *HRES*, hybrid renewable energy system.

TABLE 30.20 The DG operation under outage for reliability (short outage) of the proposed HRES.

Parameters	During outage	No outage
DG set runtime (hours day^{-1})	2.89	—
DG set O&M cost (₹ day^{-1})	43.4	—
DG set fuel consumption (L day^{-1})	412	—
DG set fuel cost (₹ day^{-1})	27,190	—
Battery throughput (kWh day^{-1})	1357	1055

DG, diesel generator; *HRES*, hybrid renewable energy system.

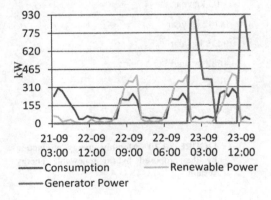

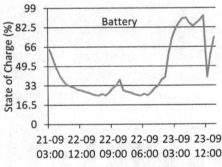

FIGURE 30.18 The proposed hybrid renewable energy system performance during resilience (large outage).

30.8 Conclusion

The proposed HRES design comprising of solar PV, WT, battery storage, and DG connected with DVC-NITD-grid is analyzed based on various system performances. The result analysis of two system cases, namely, base case system and the proposed HRES, is examined using HOMER-grid. The base system or the current system of NITD presently spends approximately ₹ 23,966,395.20 on the utility bill annually in which ₹ 5,723,737.24 is the demand charge that is approximately 24% of the utility bill. The proposed HRES reduce the system's annual utility bill to

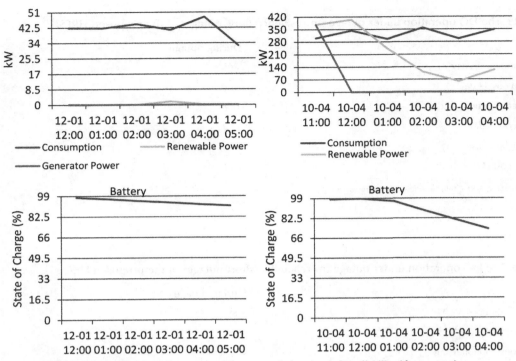

FIGURE 30.19 The proposed hybrid renewable energy system performances during outage for reliability (short outage).

TABLE 30.21 The pollutant emissions for the proposed HRES.

Pollutant emission	Emission amount
Carbon dioxide	2,177,925 kg year^{-1}
Carbon monoxide	23.5 kg year^{-1}
Unburned hydrocarbons	1.03 kg year^{-1}
Particulate matter	0.142 kg year^{-1}
Sulfur dioxide	9435 kg year^{-1}
Nitrogen oxides	4632 kg year^{-1}

HRES, hybrid renewable energy system.

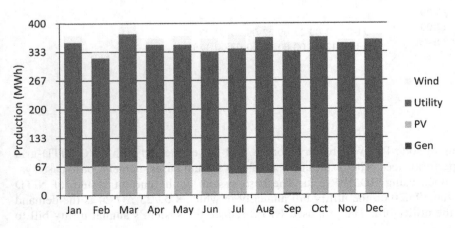

FIGURE 30.20 Monthly energy production by the proposed hybrid renewable energy system.

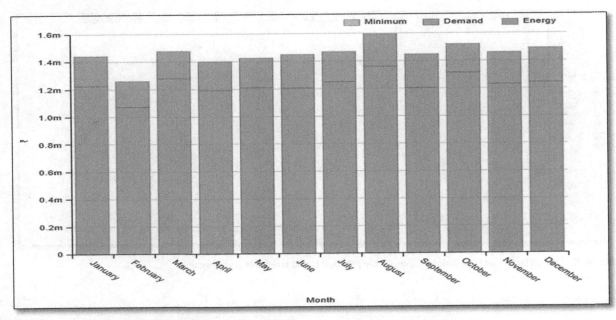

FIGURE 30.21 Monthly electricity bill breakdown for the proposed hybrid renewable energy system.

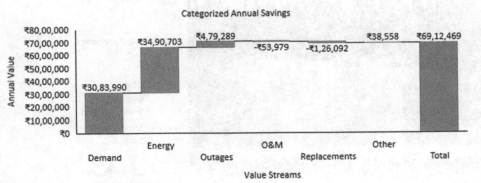

FIGURE 30.22 The annual savings cost summary for the proposed hybrid renewable energy system.

₹ 17,391,702.05 in which ₹ 2,639,746.57 are demand charges that are 15.18% of the utility bill with HRES. The demand charges by the utility are reduced from 24% to 15.18%. Similarly, the total energy purchased by grid utility is reduced from ₹ 18,242,657.97 (base system) to ₹ 14,751,955.48 (proposed HRES), which are reduced by 19.13%. As a result of that, the COE of the proposed system is reduced to ₹ 4.235 kWh^{-1} as compared to base case, which is ₹ 5.781 kWh^{-1}. In the same way, the NPC is also reduced for the proposed HRES. The comparative analysis of the proposed system and the base case system for NITD system clearly indicates that the total annual utility bill savings and demand charge savings for the proposed system as compared to the base system are ₹ 6,574,694 and ₹ 3,083,991, respectively. The payback time and the renewable fraction for the proposed system are 0.6 year and 19.25%, showing the effectiveness of the proposed system by utilizing the renewable sources. The relative assessment of utility bills and the NPC cash flow during the project lifetime for the current system and the proposed HRES are confirming that the proposed HRES has the significant impacts on the base system performances and the overall performances are improved with the proposed HRES. The reliability and resiliency of the HRES are enhanced than the base system by considering short outages for the reliability and long outages for the resilience. During the outage, the proposed system of the DG operating hours is less compared to the base

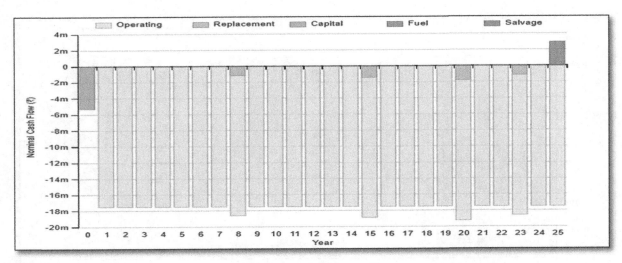

FIGURE 30.23 The different cast flow during lifetime for the proposed hybrid renewable energy system.

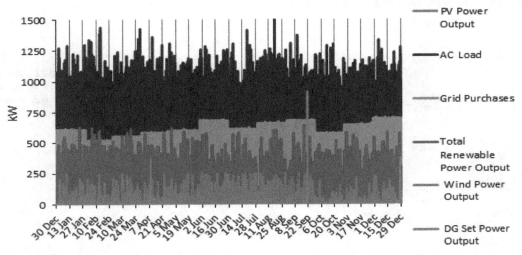

FIGURE 30.24 The different system components output power profile for the proposed hybrid renewable energy system.

case system. Also, in the proposed system, HRES has more option to supply the load demand efficiently; hence, the system reliability and resilience are improved. The required DG operation hours and the net grid energy purchases are reduced; therefore the pollutant emission components, such as the CO_2 and particulate matter, are reduced for the proposed HRES to 2177.9 t year^{-1} and 0.142 kg year^{-1}, respectively. The less pollutant emission decreases the environmental and social impacts, providing better human life with respect to bad impacts of pollutant on health. Finally, it is concluding that the proposed design of HRES is more feasible than the current or base case system of case study and the technical, social/environmental, and economic parameters of the proposed system are superior.

Further study related to this case study can be performed by considering the other energy sources, such as biomass, geothermal, and domestic waste-based sources, and also the impacts of different energy storage system can be analyzed. Also, the proposed HRES can be actually implanted at study location and the multiple tariff options and the demand response program for the case study can be modeled and its impacts on the technical, environmental, and economic parameters of the system can be investigated. In addition, this case study utilized

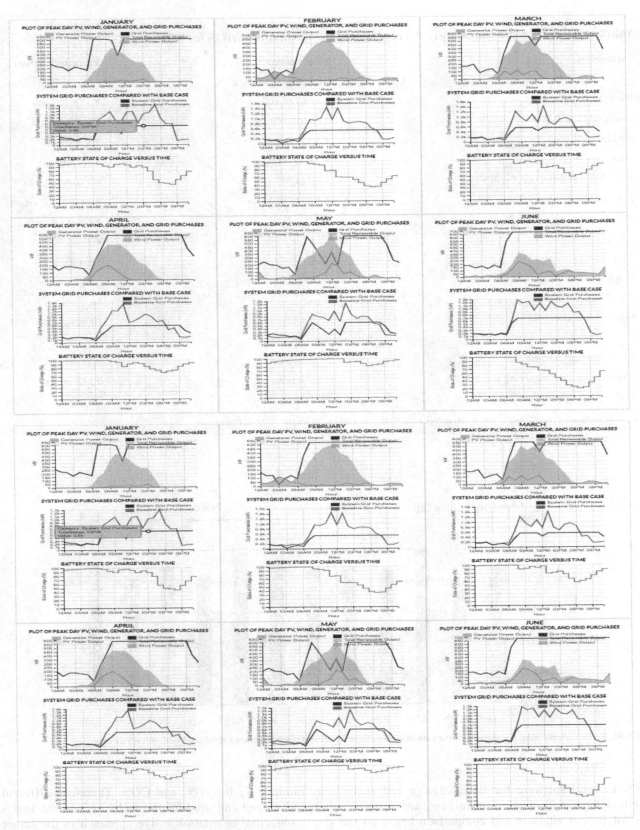

FIGURE 30.25 Monthly peak day profiles of the proposed hybrid renewable energy system components.

TABLE 30.22 Comparative summary of the base system and the proposed HRES for NITD.

Parameters	Base system	Proposed system
Costs, savings, and economic metrics		
Net present cost (₹)	₹ 318,722,300	₹ 233,556,400
Levelized COE (₹ kWh^{-1})	₹ 5.781 kWh^{-1}	₹ 4.235 kWh^{-1}
Initial capital cost (₹)	₹ 1,125,000	₹ 5,320,119
O&M cost (₹)	₹ 24,600,000	₹ 17,700,000
Annual demand charges (₹ year^{-1})	₹ 5,723,737 year^{-1}	₹ 2,639,747 year^{-1}
Annual energy charges (₹ year^{-1})	₹ 18,242,660 year^{-1}	₹ 14,751,960 year^{-1}
Annual utility bill savings (₹)	–	₹ 6,574,694
Annual demand charge savings (₹)	–	₹ 3,083,991
Annual energy charge savings (₹)	–	₹ 3,490,703
Simple payback time (years)	–	0.6
IRR (%)	–	167.76%
Renewable fraction (%)	–	19.25%
Environmental impact		
CO_2 emissions* (mt year^{-1})	2711.6 t year^{-1}	2177.9 t year^{-1}
Annual particulate matter (kg year^{-1})	0.927 kg year^{-1}	0.142 kg year^{-1}
Annual fuel consumption (L year^{-1})	9271	1425 L

COE, cost of energy; *HRES*, hybrid renewable energy system; IRR, internal rate of return; *NITD*, NIT Durgapur.

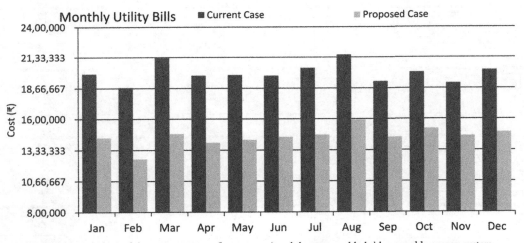

FIGURE 30.26 Utility bills' comparison of the current system (base system) and the proposed hybrid renewable energy system.

HOMER-grid software for HRES design and analysis by minimizing the NPC and COE. Therefore different methodology, optimization techniques, and software can be developed and used for the techno-economic analysis for further analysis and comparison based on the single objective or multiobjective approach in the future research.

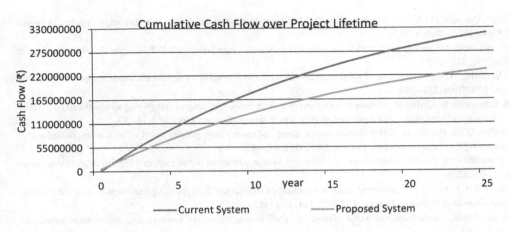

FIGURE 30.27 Comparative cash flow (net present cost) over project lifetime for the current system (base system) and the proposed hybrid renewable energy system.

Acknowledgment

First, the authors would like to thank NIT Durgapur for helping during the system study and to provide the facility to conducting the research. Moreover, the authors also would like to acknowledge HOMER Energy and also to acknowledge NASA Langley Research Center Atmospheric Science Data Center Surface meteorological and Solar Energy (SSE) web portal supported by the NASA to obtain the solar and wind data of case study.

References

Babatunde, O. M., Munda, J. L., & Hamam, Y. (2020). A comprehensive state-of-the-art survey on hybrid renewable energy system operations and planning. *IEEE Access*, 8(2020), 75313–75346.

Belmili, H., Haddadi, M., Bacha, S., Almi, M. F., & Bendib, B. (2014). Sizing stand-alone photovoltaic–wind hybrid system: Techno-economic analysis and optimization. *Renewable and Sustainable Energy Reviews*, 30(2014), 821–832.

Bohre, A. K., Agnihotri, G., & Dubey, M. (2016). Optimal sizing and sitting of DG with load models using soft computing techniques in practical distribution system. *IET Generation, Transmission & Distribution*, 10(6), 1–16, 2016.

Bourennani, F., Rahnamayan, S., & Naterer, G. F. (2015). Optimal design methods for hybrid renewable energy systems. *International Journal of Green Energy*, 12(2), 148–159.

Chatterjee, A., & Rayudu, R. (2017, December). Techno-economic analysis of hybrid renewable energy system for rural electrification in India. In *2017 IEEE innovative smart grid technologies—Asia (ISGT-Asia)* (pp. 1–5). IEEE.

Dali, M., Belhadj, J., & Roboam, X. (2010). Hybrid solar–wind system with battery storage operating in grid-connected and standalone mode: Control and energy management—experimental investigation. *Energy*, 35(6), 2587–2595. (2010).

Deshmukh, M. K., & Deshmukh, S. S. (2008). Modeling of hybrid renewable energy systems. *Renewable and Sustainable Energy Reviews*, 12(1), 235–249. (2008).

Diemuodeke, E. O., Hamilton, S., & Addo, A. (2016). Multi-criteria assessment of hybrid renewable energy systems for Nigeria's coastline communities. *Energy, Sustainability and Society*, 6(26), 1–12. (2016).

Elavarasan, R. M. (2019). The motivation for renewable energy and its comparison with other energy sources: A review. *European Journal of Sustainable Development Research*, 3(1), pem0076(1–19).

Environmental Protection Agency. (2010). Assessing the multiple benefits of clean energy: A resource for states (Chapter 5).

Fan, X., Sun, H., Yuan, Z., Li, Z., Shi, R., & Razmjooy, N. (2020). Multi-objective optimization for the proper selection of the best heat pump technology in a fuel cell-heat pump micro-CHP system. *Energy Reports*, 6(2020), 325–335.

Go, R., Kahrl, F., & Kolster, C. (2020). Planning for low-cost renewable energy. *The Electricity Journal*, 33(2). (2020), p. 106698(1–5).

Gorman, W., Mills, A., Bolinger, M., Wiser, R., Singhal, N. G., Ela, E., & O'Shaughnessy, E. (2020). Motivations and options for deploying hybrid generator-plus-battery projects within the bulk power system. *The Electricity Journal*, 33(5). (2020), p. 106739 (1–15).

Guo, L., Wang, N., Lu, H., Li, X., & Wang, C. (2016). Multi-objective optimal planning of the stand-alone microgrid system based on different benefit subjects. *Energy*, 116(2016), 353–363.

Hafez, A. A., Hatata, A. Y., & Aldl, M. M. (2019). Optimal sizing of hybrid renewable energy system via artificial immune system under frequency stability constraints. *Journal of Renewable and Sustainable Energy*, 11(1). (2019), pp. 015905 (1–17).

HOMER® Energy. (2020). HOMER grid v1.8 user manual. https://www.homerenergy.com/products/grid/docs/1.8/design.html.

Khalid, M., AlMuhaini, M., Aguilera, R. P., & Savkin, A. V. (2018). Method for planning a wind–solar–battery hybrid power plant with optimal generation-demand matching. *IET Renewable Power Generation*, 12(15), 1800–1806.

Lao, C., & Chungpaibulpatana, S. (2017). Techno-economic analysis of hybrid system for rural electrification in Cambodia. *Energy Procedia*, 138 (2017), 524–529.

Luna-Rubio, R., Trejo-Perea, M., Vargas-Vázquez, D., & Ríos-Moreno, G. J. (2012). Optimal sizing of renewable hybrids energy systems: A review of methodologies. *Solar Energy, 86*(4), 1077–1088. (2012).

Maleki, A., & Askarzadeh, A. (2014). Artificial bee swarm optimization for optimum sizing of a stand-alone PV/WT/FC hybrid system considering LPSP concept. *Solar Energy, 107*(2014), 227–235.

Mazzola, S., Astolfi, M., & Macchi, E. (2016). The potential role of solid biomass for rural electrification: A techno economic analysis for a hybrid microgrid in India. *Applied Energy, 169*(2016), 370–383.

Obara, S. Y., Morel, J., Okada, M., & Kobayashi, K. (2016). Performance evaluation of an independent microgrid comprising an integrated coal gasification fuel cell combined cycle, large-scale photovoltaics, and a pumped-storage power station. *Energy, 116*(2016), 78–93.

Rullo, P., Braccia, L., Luppi, P., Zumoffen, D., & Feroldi, D. (2019). Integration of sizing and energy management based on economic predictive control for standalone hybrid renewable energy systems. *Renewable Energy, 140*(2019), 436–451.

Saib, S., & Gherbi, A. (2015, May). Simulation and control of hybrid renewable energy system connected to the grid. In *2015 5th international youth conference on energy (IYCE)* (pp. 1–6). IEEE.

Samy, M. M., Barakat, S., & Ramadan, H. S. (2019). Techno-economic analysis for rustic electrification in Egypt using multi-source renewable energy based on PV/wind/FC. *International Journal of Hydrogen Energy, 45*(2020), 11471–11483.

Sawle, Y., Gupta, S. C., & Bohre, A. K. (2018a). Socio-techno-economic design of hybrid renewable energy system using optimization techniques. *Renewable Energy, 119*(4), 459–472, 2018.

Sawle, Y., Gupta, S. C., & Bohre, A. K. (2018b). Review of hybrid renewable energy systems with comparative analysis of off-grid hybrid system. *Renewable and sustainable energy reviews, 81*(2018), 2217–2235.

Sharafi, M., & EL Mekkawy, T. Y. (2014). Multi-objective optimal design of hybrid renewable energy systems using PSO-simulation based approach. *Renewable Energy, 68*(2014), 67–79.

Suresh, V., Muralidhar, M., & Kiranmayi, R. (2020). Modelling and optimization of an off-grid hybrid renewable energy system for electrification in a rural areas. *Energy Reports, 6*(2020), 594–604.

Tariff Order of DVC for the years 2015–16 and 2017–18. http://wberc.gov.in/sites/default/files/Tariff%20Order_DVC_2014-2017_pdf%20%281%29.pdf.

Tudu, B., Mandal, K. K., & Chakraborty, N. (2019). Optimal design and development of PV-wind-battery based nano-grid system: A field-on-laboratory demonstration. *Frontiers in Energy, 13*(2), 269–283. (2019).

Yang, H., Wei, Z., & Chengzhi, L. (2009). Optimal design and techno-economic analysis of a hybrid solar–wind power generation system. *Applied Energy, 86*(2), 163–169. (2009).

Chapter 31

Stand-alone hybrid system of solar photovoltaics/wind energy resources: an eco-friendly sustainable approach

Faizan Arif Khan[1], Nitai Pal[1] and Syed Hasan Saeed[2]

[1]*Department of Electrical Engineering, Indian Institute of Technology (Indian School of Mines), Dhanbad, India,* [2]*Department of Electronics & Communication Engineering, Integral University, Lucknow, India*

31.1 Introduction

Petroleum products, such as coal, oil, and gas, are right now the world's essential vitality sources. The overwhelming reliance on the petroleum product diminished its regular reserve. Researchers are progressively attempting to discover alternative vitality sources because of limited availability and unfriendly environmental impact of fossil fuel resources. Petroleum products are fairly available everywhere; it is expected that in future provincial or regional clashes may emerge from energy emergency if worldwide economy intensely relies upon it. The environment condition has been unfavorably affected worldwide, and the environment of certain areas has been harmed seriously by the usage of conventional power sources. Electric energy is indispensable for both financial and technical prosperities of a country. Growth of a region is exceptionally dependable and can be improved by maintaining the pace of clean energy production (Enslin, 1991).

It is seen that use of sustainable resources is incremented with small percentage; petroleum products, such as coal, gas, and petrol, still command everything. This situation insisted us to discover new strategies for increasing the energy demand (Wang, 2006). In case of remote locations, availability of national grid is minimum and grid extension is financially and technically limited. To overcome these constraints, sustainable power source is considered as an appealing option for remote locations. Generally, most utilized sustainable vitality sources are solar, wind biomass, and hydro energy system. These resources' framework can be used as individually or in the combination of more than one source. Such framework can have backup systems, such as battery storage and diesel generator (DG), to fulfill the load demand. This combination of more than one sustainable energy resource for electrical energy production is called as hybrid renewable energy system (HRES). The features of wind and solar energy have supplementing one another, except that both are unpredictable and having immediate changing nature of sun light and wind speed. Researchers had investigated the various aspects of solar/wind hybrid system in stand-alone and grid-connected operations for remote locations and users in small town. They also presented various suggestions and recommendations for students and consumers. To replace conventional sources, solar photovoltaics (PV)/wind hybrid system in association with battery storage and DG is highly recommended for remote locations (Elhadidy & Shaahid, 2000). The present work is based on the detailed study of solar PV/wind hybrid system. In this work, various aspects of hybrid system are present under the respective sections. The next section is the description of various renewable energy resources; after that HRES are presented. The importance of HRES and energy management of HRES are also discussed in consecutive sections. This chapter is organized in seven different sections. Section 31.1 gives the introductory details of the chapter, Section 31.2 discusses about the renewable energy generation from different resources, Section 31.3 describes the HRES with brief discussion on importance and energy management of HRES, Section 31.4 is about the modeling of SPV/wind HRES, Section 31.5 emphasizes on the optimization and sizing of SPV/wind HRES. The future aspect of hybrid systems is given in Section 31.6. The conclusion of this chapter is presented in Section 31.7.

Renewable Energy Systems. DOI: https://doi.org/10.1016/B978-0-12-820004-9.00030-9

31.2   Renewable energy sources

31.2.1   Solar energy

The sun accomplishes more than providing the light in daytime. The energy packets of sunlight known as photon arrive at earth surface in the form of electromagnetic radiations. These packets of energy contain vitality that energized our world. Sun-oriented vitality is an absolute source answerable for the living structures, climate frameworks, and vitality sources on the Globe (Chandel et al., 2016). The sun-oriented radiation hits the surface of planet in enough amounts at every hour to fulfill the worldwide vitality requirements. This vitality originates from the inner side of Sun that possesses a similar atomic reactor resembling system. Atomic amalgamation produces huge amount of vitality in the core of sun that emanates outward and scattered into the universe solar energy as light and heat. Latest technologies have capability to convert the solar vitality into the usable form by either PV or sun-powered thermal system. At present solar energy is exploited on very large scale but the contribution of solar vitality is very less as compared to conventional resources worldwide. The decreasing price with an increase in the techniques of SPV system implies that large amount of vitality can be achieved at lesser expenses. The energy-producing system of solar technology is a leading, sustainable resource and has a significant impact on the global vitality sector. There are numerous approaches to utilize vitality from the sun. The two fundamental approaches to utilize vitality from the sun are PV and sun-oriented thermal system. The PV approach is considerably more feasible for small power ventures, and thermal solar system is commonly just utilized for power creation for enormous scale. To deliver power, lower temperature varieties of sun-oriented thermal system can be utilized for the warming and cooling of domestic and commercial space. Solar-based systems are one of the fast emerging and lesser expensive sources of intensity on the planet. The improvement in technologies and financial advantages makes the PV technologies a perfect, sustainable power source with ecological advantages . Sunlight-based vitality is the most bountiful, endless, and clean sustainable power source till date. The solar energy received by the earth is greater than the current rate of solar vitality utilization. The solar energy is utilized as either in the form of PV conversion or in the form of thermal energy storage.

31.2.1.1   Sunlight-based PV

PV innovation is probably the best approach to utilize the sun power. PV transformation is the immediate change of solar light into power without any moving/rotating machine. PV panels are simple and rugged and straightforward in configuration requiring almost no support and their greatest favorable position being their development as independent frameworks to give outputs. Subsequently, they are utilized for power source, water siphoning, remote structures, sun-powered home frameworks, satellites and space vehicles, and even megawatt scale power plants. With such a huge swath of uses, the interest for PV is expanding each year. The simplified way for domestic consumers to exploit sun-based vitality is PV system. The PV panel framework converts the daylight directly into power that can be utilized quickly. The remaining power can be stored in a battery storage and used on the requirement in stand-alone system; in case of grid-tied system, the remaining energy can be transferred to the electric matrix, which can be for credited to the consumer's electric bill. The sun-powered panels convert sunlight-based vitality into usable power through a procedure known as the PV effects (Dennis Barley et al., 1996). The falling daylight strikes a semiconductor type material and knocks the electrons free, getting them movement and producing an electric flow that can be transferred to the wiring system of user. This current flow is known as direct current (DC) power and should be changed over to alternating current (AC) power utilizing a power electronics device inverter. This transformation is essential on the practical grounds that the Indian electric system works utilizing AC power, as do most domestic electric apparatuses. Sun-based vitality can be trapped at numerous scales utilizing PV and introducing SPV panels in a perceptive method to optimize electricity charges while diminishing consumer reliance on conventional energy resources. Many organizations and electrical utilities can be profited by PV vitality production by introducing huge SPV clusters that can energize the organization activities or consistent vitality supply to the conventional grid system.

31.2.1.2   Solar thermal energy

Sun thermal innovations, especially solar water-warming framework, sun-powered cookers and sunlight thermal frameworks are one of the most advanced sustainable power sources in India. Approaches are set to give further force to spread sun-powered advancements. A second way to use solar energy is to capture the heat from solar radiation directly and use that heat in a variety of ways. Thermal solar energy has an extensive scope for consumers than a PV framework. The utilization of solar warm vitality for power production at smaller scale is not as handy as utilizing SPV system. There are three basic categories of thermal vitality utilization, such as lower temperature, which is utilized for warming and cooling; middle level of temperature, which is utilized for warming water; and high temperature, which is utilized for electric power production. Sun-powered thermal vitality frameworks at lower temperature include the process of warming and cooling the airspace of a domestic or official place for atmosphere control. The design of passive buildings is the best example of sun-powered vitality utilization for any structure plan. The living space of a passive buildings can be make warm by solar energy. Middle-temperature solar warm vitality frameworks incorporate fluid or boiling water-based warming frameworks. In this system, the solar heat is tapped by collectors in terms of water or fluid at rooftop or open surfaces. This tapped heat is then moved to the water going through the piping system installed in entire residence; in this manner, consumers do not need to depend on conventional water-warming techniques. High temperature sunlight-based warm vitality frameworks are utilized for producing power for a bigger structure. In a sun-oriented warm power plant, mirrors are used to locate the sun beams around tubes containing a fluid that can hold heat vitality. This warmed liquid is utilized to transform the water into steam further that is utilized to rotate a turbine and produce the electrical power. This kind of innovation produces the electrical energy from solar radiation using different solar concentrator. The solar concentrators are classified as shown in Fig. 31.1.

31.2.2 Wind energy

The wind is produced as by-effect of sun-powered vitality; it is expected that around 2% of the sun's vitality arriving at the earth is changed over into wind vitality. In this process, the surface of the earth warms and cools unevenly, making pressure gradient zones that make wind flow from higher- to low-pressure regions. The wind has assumed a significant job throughout the entire existence of human development. People have been encumbering the vitality of the wind since early written history; previous exploitation of wind power was to sail the vessel, siphoning water, and granulating grains. An enormous rise in vitality demand is observed as fast exhaustion of the petroleum derivatives has constrained the whole world. It leads to consider the use of sustainable power source assets. The serious effects of environment degradation have additionally put the worldwide enthusiasm towards the sustainable power resource. However, the establishment cost of such sources is high yet low running expenses (Bhandari, Kyung, Lee, Cho, & Ahn, 2015). Wind power is one of the biggest sustainable power source assets because of its operational and availability. The first windmill used for the electricity production was started in the Scotland. Later around the same time, a wind-based machine to create power was presented in the USA (Bhandari et al., 2015). It has longer working duration that can deliver powers during shady days and night unlike sun-oriented energy. Wind power is controlled by small and huge wind turbines of diverse categories and arrangements. It is an emerging alternative vitality resource. So as to utilize stand-alone wind vitality framework viably and monetarily, the selected site ought to have great capability of wind vitality consistently (Sørensen, 2011). The restriction is that these turbines do not produce electricity when wind does not blow. During such condition to fulfill the consumer's need, another source is also required. Thus wind and sun-oriented energy-producing system necessities a storage system to stock excess vitality and use it in the absence of sufficient power created to fulfill the need. The capacity factor is the concluding parameter for the selection of a specific sort of wind turbine at the chosen site. The windmill with the most noteworthy average capacity factor has been suggested. Weibull wind speed dispersion parameter and Rayleigh wind speed dissemination are the technique utilized for the selection of wind vitality framework (Feijoo, Cidras, & Dornelas, 1999). Wind turbine power curve estimation utilized regression and artificial neural network (Li, Wunsch, O'Hair, & Giesselmann, 2001). The dominant and stirring highlights of wind resource over the other vitality sources are low venture cost as contrasted with traditional force plants, lesser arrangement time, no fuel cost, less upkeeping costs, no adverse effect on weather, zero reliance on petroleum products, decrease of greenhouse gas discharges, legitimate use of wind farm lands; in this way, the land underneath the pinnacles can be used particularly as horticultural regions (Enslin, 1991). In the beginning, the wind power generation was costlier than the present. Wind energy is the fastest developing innovation among all sustainable power sources. It is estimated that majority of world's new wind vitality establishments have been done by the few wind power commanding nations, for example, China, USA, Germany, India, Spain, UK, Canada, Italy, France, and Denmark (United States Energy Information Administration, 2014). India is at the fourth spot on the planet with absolute wind power creation up to June 2016. The quantities of wind power ventures have expanded altogether because of

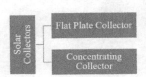

FIGURE 31.1 Classification of solar collectors (Researchgate, n.d.).

a sharp decrease in the expense of restraining wind vitality from inland wind (Wang, 2006). The most commonly used wind mill contains three blades around a horizontal axis introduced on a tall tower. This is known as horizontal axis wind turbine. In this turbine associated with a generator is fixed about the axis horizontally. This is the most well-known turbine scheme for wind. In expansion to being corresponding to the ground, the rotation axis of blade should be parallel to the stream of wind. Some machines are intended to work in an upwind mode, with the sharp edges upwind of the pinnacle. Usually, the horizontal hub wind factories are called as aero-turbine mills with 35% proficiency and homestead plants with 15% effectiveness. Another type of wind turbine is a known vertical axis wind turbines. Although vertical hub wind turbines have existed for centuries, these are not so much commonly utilized. They do not exploit the higher wind speeds at higher heights over the ground as horizontal turbines. The Darrieus turbine is basic vertical hub designs having curved blades with the productivity of 35%; other vertical axis turbines have straight blades with the effectiveness of 30%. A vertical hub machine needs not be arranged in the direction of wind propelling. The wind turbines are primarily classified as shown in Fig. 31.2.

31.2.3 Biomass

It is a substantial vitality resource for the any developing agricultural country due to the benefits offered. It is viable, usually reachable, pollution less and also has capability to offer employment in the country zones. These resources are utilized for electricity production incorporating bagasse, rice husk, straw, cotton tail, coconut shells, other agricultural wastes, and saw dust (Celik, 2003). Conventional sources, such as wood and animal excreta, have significant impact on vitality generation in India. They contribute in every domain of energy utilization like flexibly cooking vitality to practically the entirety of India's country populace (Varun, Prakash, & Bhat, 2009). Biogas is also an alternative source of vitality, obtained from the natural wastes in India, for more than three decades. Biogas is a perfect fuel created through anaerobic processing of an assortment of natural wastes: creature, farming, local, and mechanical. Solid biomass is utilized in India either in direct burning or gasification to create power or for the cogeneration of intensity and warmth. Biofuels in the liquid form (biodiesel and ethanol) are utilized to substitute the transportation fuel. Ethanol is produced at a great extent as a by product of sugar factories, by fermenting the molasses. Nonedible oil seeds are leading sources of biodiesel production in India. The development of the Jatropha and other locally biofuel-producing crops may become the regular source of income for rural populations (Renewable Energy Resources, 2014). According to MNRE that accessible animal excrement could strengthen roughly 12 million domestic biogas plants. The additional biomass power generation could encourage 25-MW power-generating limit (Shaji, 2014).

31.2.4 Small hydropower

Hydropower ventures are primarily organized as large and small hydropower. The hydropower plants up to 25 MW are considered as small hydropower ventures in India (Bifano, Ratajczak, Bahr, & Garrett, 1979). It has been perceived that small hydropower ventures may have an important role in improving the general vitality situation of the nation and specifically for remote zones. Small hydropower meets the criteria of inexhaustible resources; therefore, it is considered as a sustainable vitality resource. These ventures are built up on waterways, dam, and run-of stream locations. The greenhouse outflows and energy payback time are less in this framework as compared to the other power-generating systems (Varun, Bhat, & Prakash, n.d.). Small hydro advancement is based on the simple civil development work and economical equipment is required to set up and run the small hydropower ventures. In small hydro, no huge reservoir is made and population rehabilitation is not necessary. The extra advantages of small hydropower alongside power production are flood control, better irrigation, fisheries, and creation of tourism hotspot. An assessment of small hydropower against the parameters of sustainable vitality innovations viz. cost of power age, greenhouse discharges and energy payback time acknowledged that small hydro is a well-utilizable hotspot for the manageable improvement (Varun et al., 2009). The majority of the small hydro potential is in Himalayan states as waterway-based activities and in different states on water system trenches (Lim, 2012).

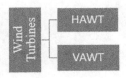

FIGURE 31.2 Classification of wind turbine rotors (Sørensen, 2011).

31.2.5 Other RES

31.2.5.1 Geothermal energy

Geothermal energy resource is based on heat energy stored in the earth crust. The heat energy of earth heats up the water located at the next upper layer. This heated water came out from the surface of the earth in the form of geysers at high temperature and pressure. The energy of geysers further utilizes in the generation of electricity. Geothermal vitality is at present contributing around 10 GW over the world and India's little assets can expand the above rate (Bifano et al., 1979). In India, this renewable resource is adopted in seven geothermal areas. The geothermal resource territories are found in India along the western coast of Gujarat and some places in Rajasthan and along the coastal border of Bengal; these are distributed in the 1500-km stretch of the Himalayas. The asset is minimally utilized right now; however, the administration has a desire to intensely increase the introduced producing limit (Lim, 2012).

31.2.5.2 Nuclear energy

Nuclear energy is obtained by the nuclear reactions of radioactive materials. This process is followed by two different phenomena called as nuclear fission and nuclear fusion. In the first event, a heavy atom is split into two tiny particles with a lot of energy. In contrast, two tiny atoms are merged into a single larger fraction in the second process and release a considerable amount of energy. The heat produced in the nuclear reaction is utilized in the power generation process at nuclear power plants. Atomic power stations utilize a fuel called uranium material. The vitality is discharged from uranium atom when a neutron split it into two parts. This atomic response is called the fission procedure. In an atomic station, the uranium is first framed into pellets and afterward into long bars. The uranium bars are kept cool by using the heavy water. The measure of power created in an atomic force station is comparable to that delivered by a fossil-fueled power station. The atomic stations do not consume petroleum products to deliver power and thus they do not create harming greenhouse gases. Numerous supporters of atomic force creation state that this kind of intensity is atmosphere supportive and clean.

31.2.5.3 Hydrogen energy resource

Hydrogen energy resource is a new renewable energy resource. At present, it is in the beginning phase of research and advancement. MNRE additionally financed research ventures on various parts of hydrogen vitality innovation improvement. Bringing down the expense of hydrogen energy in India incorporates a difficult task. The improvement in production rates from various strategies, advancement of stockpiling capacity and productivity improvement of various kinds of power device for the practical framework are still required. The specific guidelines are required to examine, improve, and exhibit exercises in different hydrogen energy-based energy advancements. (Bifano et al., 1979).

31.3 Hybrid renewable energy systems

31.3.1 Importance of HRES

This hybrid system combines various sustainable power sources in the combination of single sustainable power generation (United Nations, 2014). The sustainable power sources combined in the hybrid framework have different characteristics as per the seasons, the sun-oriented radiation is more prominent in summer than in winter, and wind speed is more noteworthy in other seasons. HRES frameworks serve to lessen the utilization of nonsustainable fuel. The first hybrid power frameworks were introduced in USA comprising of PV and DG (Bhandari et al., 2015; Bifano et al., 1979). In recent years SPV and wind source-based HRES are broadly utilized with or without vitality stockpiling support for giving electricity capacity the purchasers in remote territories. The renewable power sources are picked suitably for each area in the arrangement of HRES. This step diminishes the reliance on nonrenewable energy source, prompting an expansion in power supply consistently (Meyar-Naimi & Vaez-Zadeh, 2012). The most significant attribute of an HRES framework is the utilization of two sustainable resources minimum with or without interfacing with DGs/battery storage, which improves the framework productivity and overcomes the financial confinements emerging from single sustainable power source (Chauhan & Saini, 2015). Intermittency and irregularity are the fundamental attractions that represent renewable power sources. Intermittency incorporates both unsurprising and irregular varieties. The numerous disadvantages of discontinuity of sustainable sources can be overcome by some exceptional strategy. It has been observed that HRES has become a cost effective practical power generation system in off-grid applications. The improvement in the power converter and sustainable power technology also improves the framework proficiency (Ellabban, Abu-Rub, & Blaabjerg, 2014). Coordination of conventional sources with inexhaustible sources can increase the reliability of the electrical system particularly in natural conditions (Ismail, Moghavvemi, & Mahlia, 2013). Combination of SPV and wind energy has the advantage of less battery bank limit and diesel prerequisites as compared to individual renewable source. Stand-alone PV/wind hybrid framework produces power consistently to meet the load demand. Hybrid framework is most appropriate for off-grid system, the intake used for HRES is free and unlimited afterward electric vitality delivered by these frameworks is free of cost (Ahn et al., 2012). The fast worldwide effort in the improvement of sustainable power sources and the related advances that serve them as vitality is currently perceived as a strategic division.

In a developing nation, larger part of the population lives in the rural remote zones and various little segregated networks in such territories live without access to power from the grid. Expansion of the grid to remote areas is not practical or feasible. Along these lines, it is important to discover optional power arrangements that could replace the power grid. Potential choices for off-grid rural electricity incorporate stand-alone and hybrid power generation advancements (Siddaiah & Saini, 2016). It is widely accepted that sustainable hybrid power framework-based generation is an economical alternative for costly power grid extension. In an appropriated hybrid framework, power is delivered at or near the purpose of utilization. Stand-alone hybrid frameworks maintain a strategic alternative for the expenses and losses of transmission and appropriation.

Governments throughout the world have established new guidelines and arrangements empowering the work of sustainable power advances (Perera, Attalage, Perera, & Dassanayake, 2013). These endeavors incorporate advancing sustainable power source innovations, improving the effectiveness of energy uses, and building up energy preservation plans alongside their authoritative activities The improvement has secured different sustainable power sources, storage devices, and capacity frameworks (Mohammed, Mustafa, & Bashir, 2014). Consolidating them with traditional sources to empower isolated loads or smaller networks situated a long way from the primary grid is the solution for the continuous supply from the sustainable sources (Bansal, Saini, & Khatod, 2013). Hybrid renewable frameworks have demonstrated powerful in providing the necessary electrical supply in stand-alone application. It is evident that the choice of utilizing alone HRES frameworks or in association with diesel/petroleum fuel generator depends on the accessibility of the sustainable power source in that area. Subsequently, it is required to distinguish the potential areas for introducing PV and wind energy frameworks stand-alone and grid-connected operation so as to limit the cost of power without distressing the current system. Incorporating more than one sustainable power source and including storage arrangements and backup frameworks are among the couple of measures to overcome these drawbacks. These extra strategy parameters increment the general expense of the inexhaustible framework. This situation leads toward the optimization and sizing procedures (Abido, 2007). The existence of more than one energy sources, backup devices, and storage framework requires the energy management control system of energy stream among the different sources (Lanre, Saad, Ismail, & Moghavvemi, 2016).

31.3.2 Energy management of HRES

The energy management strategy (EMS) of HRES is required to acquire higher efficiencies than that could be gotten from a single source. Energy management is a term that gathers all the orderly techniques to control and limit the amount and the expenses of energy used to give a specific application. This procedure controls the progression of vitality through the flexibly framework. The HRES combines the SPV and wind vitality sources and works in two modes as sequential and simultaneous modes. In the sequential mode, both resources produce power alternatively and, in the simultaneous mode, the solar and wind vitality framework produces vitality simultaneously while in the sequential mode (United States Energy Information Administration, 2014; Wang, 2006).

The EMS technique arranges the two different resources to guarantying that the framework should work at high efficiency with unique execution. The energy management technique mostly relies upon the different features of framework and its segments. The disadvantages of discontinuity of inexhaustible sources can be overwhelmed by considering some exceptional plan contemplations. The EM system guarantees the energy balance in the generation by different components under the characterized requirements. One of the renewable resources is given priority as the fundamental source of generation, and the another framework assumes the optional job in HRES. In case of solar and wind resource inaccessibility due to shifting atmosphere conditions, the battery storage supports the wind and sun-based system. The additional objectives of the EM technique are reducing the energy cost for the consumers and improvement in the reliability. An increased level of framework reliability and diminished level of power interference with higher operating proficiency expands the lifetime of the system parts. Two most significant worries of hybrid framework are the reliability and the entire expenses of the framework. An EM system was constructed by various authors considering load shedding and diminishing the cost of energy (COE) with carbon discharge reduction simultaneously for a renewable hybrid framework (Zhang, Davigny, Colas, Poste, & Robyns, 2012). To guarantee the continuous electricity supply and to diminish the expense on energy creation, the EM system is normally coordinated with optimization and sizing techniques (Lanre et al., 2016). The EM system uses the monthly charging and discharging of storage portions of the capacity, start up/shut down restriction of the generator, and energy stockpiling SOCs, for getting the optimum arrangement (Atwa et al., 2010). To develop EM system for the grid-connected system, the control of energy flow to and from the grid and metering system is always considered as fundamental parameters. The flow of energy tracks the peak of load curve to use the minimal tariff cost in this grid-connected system (Lanre et al., 2016). The tilt angle of the solar PV panel and the optimal height of the wind turbine are providing the base for optimization and sizing with EM system. The EM system utilizes the state machine approach in its decision-making modes for various variable events of the system.

Different methodologies and procedures have been utilized to build up an effective EM system. A slightly different energy management strategy is needed for the intelligent grid-associated renewable hybrid system for better operation and control. This is not only required for an independent framework but also required for hybrid sustainable power frameworks associated with the smart grid network. The smart meters are fundamental equipment since it can give adequate data on immediate ongoing utilization and combined electrical utilization. Smart meters are used in this case that can be modified and programmed to manage any feed-in-levy or valuation of framework. Moreover, any beginning or stop signals for any hardware in smart grid system are started from the SCADA framework as per the energy management methodology. SCADA framework has prerecorded power and voltage settings of the vitality sources and storage capacity in the framework. Numerous examinations have received the conclusion that fuzzy logic-based EM system performs better in case of independent or grid-associated hybrid sustainable power source frameworks. Ipsakis et al. (2008) analyzed the performance of two different EM systems, utilizing the hysteresis band in the operation of renewable hybrid power framework for time period of four months. Abedi, Alimardani, Gharehpetian, Riahy, and Hosseinian (2012) introduced a novel technique for deciding the optimized EM strategy for hybrid power frameworks comprising different vitality source and storage frameworks. Pascual, Barricarte, Sanchis, and Marroyo (2015) presented an EM procedure for a microgrid comprising PV, wind, and battery associated with the grid network. HOMER is software that presents a computer-based model to assess different structure alternatives for both independent and grid-associated vitality frameworks (Lambert, Gilman, & Lilientha, 2006).

31.3.3 Operation modes of HRES

31.3.3.1 Grid-tied HRES

The hybrid sustainable power sources with grid coordination support the conventional electrical system, reduce the greenhouse gas emissions and overcome the inconsistent nature of renewable energy system. Grid-tied HRES is a mix of inexhaustible and conventional vitality source; it might likewise consolidate at least two sustainable power sources that work in independent or grid-connected approach. A grid-associated framework may associate with a small/bigger autonomous network and introduces the electrical energy reasonably into the framework directly. The addition of power into the network requires the change of DC into AC by a grid-associated intelligent inverter. The reason for the grid-associated approach is to expand the revenue that can be accomplished by boosting the energy taken from inexhaustible sources associated with the grid. The motivation behind the independent operation is to enhance the reliability of the energy framework in contrast to the depending just on the earnings.

31.3.3.2 Stand-alone HRES

This system is also called as off-grid frameworks. Practically, stand-alone HRES frameworks are planned and upgraded to fulfill the electricity need of remote places or off-grid power frameworks. The off-grid frameworks do not have an association with the principle electricity grid network. Independent frameworks fluctuate generally in size according to the utilization; it may be used from watches to remote structure or rocket/space ships.

31.3.3.3 Smart grid-based HRES

The requirement of a consistent and safe energy source that does not rely upon petroleum products is a crucial objective of most nations. Smart grids with integrated sustainable power sources fall under this objective (Alagoz, Kaygusuz, & Karabiber, 2012). In case of smart grid, brilliant meter, wise sensors, and a rapid dual energy data communication

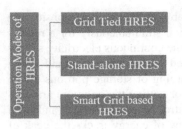

FIGURE 31.3 Operation modes of hybrid renewable energy system.

between the energy source and the purchasers ought to be accessible. This accessibility empowers the framework to effectively supervise energy conveyance to purchasers while permitting customers to have their own energy utilization decisions. Albeit numerous targets can be accomplished by smart grid networks, various difficulties may emerge in the incorporation of sustainable power sources in these smart grid frameworks (Fig. 31.3).

31.4 Modeling of SPV/wind HRES

31.4.1 System components of SPV/wind HRES

The immediate change in solar illumination and wind speed have exceptionally impacts on the energy creation; in this way, a cautious plan is required for hybrid framework for steady energy supply to the customers under fluctuating environmental condition. Similarly, a cautious structure ought to be made to keep the framework cost low. A hybrid power framework may comprise sustainable power systems, such as wind turbines, SPV panels, and storage component. The various segments and subsystems of an HRES are interconnected to enhance the entire framework as shown in the Fig. 31.4.

31.4.1.1 Solar photovoltaic array

It is one of the fundamental components of HRES. The PV system converts the solar energy into electrical energy (DC). Availability of solar energy is irregular, and it varies according to season and day duration. Solar panels are used according to the energy demand.

31.4.1.2 Wind turbine

Wind turbines are also used as renewable energy resource in HRES with solar PV. The wind turbine can operate 24 hours depending upon the availability of wind in the region. Wind turbines are available in various power ratings from small to higher ratings.

31.4.1.3 Battery storage

Depending upon the load demand, the renewable power is utilized to charge the battery storage arrangements. This battery storage is utilized to store the excess energy and provide the energy facility to the load required if there is lacking of electricity generation from the hybrid framework.

31.4.1.4 Inverter

The renewable resources are generating DC in most of cases while the conventional system has adopted AC system. This flaw between both the systems creates the requirement of inverter, for the transformation of DC supply into AC supply to fulfill the load demand.

31.4.1.5 Diesel generator

DG is used as backup resources. It is a small conventional power generation system. Small DG is using fossil fuels as input; consumers should avoid the use of DG.

31.4.2 Control strategies of SPV/wind HRES

Control strategies are utilized to maintain the coordination of different components of SPV/wind HRES, such as sources, load, and vitality stockpiling framework to guarantee a steady and safe activity. Control technique could be utilized for controlling the progression of vitality among the different sources, consumers, and storage frameworks.

Specific control of HRES with numerous renewable source/DG system and vitality stockpiling is critical to accomplish the framework reliability and activity productivity. Adoption of a specific control technique is required to control the progression of vitality among the different parts of the hybrid vitality framework (Dimeas & Hatziargyriou, 2005). Controller assumes fundamental characteristics of checking and directing the necessary energy to relieve the electricity requirement. A control framework is required to decide and relegate dynamic and responsive energy delivery by every vitality resource and maintain its voltage output at the anticipated level. The hybrid framework comprised of SPV and wind energy as its primary source of energy. The achievement of any created control procedure relies upon the need that all sources in the framework must be assimilated with the most recent innovation that encourages the interconnection among the different sources. The computerized innovation should incorporate consistent observation and automatic control and supervisions to improve the distribution of the generated power among different sources to save vitality through the most extreme usage of the different sources.

The other objective of the control procedure is to control the unidirectional and bi-directional DC—DC converter to work in suitable modes dependent on the battery state of charge and climate conditions. To execute any adopted vitality EM methodology, a centalized controller ought to be chosen, introduced, and modified to control the framework as indicated by an upgraded technique. This controller might be coordinated with a supervisory and monitory system, such as SCADA. The challenges in creating and utilizing an exact model for every one of the vitality sources or storing device, in anticipating the sun radiation or the breeze speed varieties, and in foreseeing the heap utilization or the status of the electrical matrix make fluffy rationale a very much adjusted instrument to perform vitality the executives and related control errands. Fluffy frameworks are control techniques that can be viably and effectively utilized for nonlinear frameworks.

The concept of control in the smart grid consistently required administrative and monitoring framework. The SCADA framework and ZigBee gadgets are the best examples of the supervisory framework. The complexity of a framework, the requirement of computational productivity, and the absence of information storage required improved intelligent controllers, such as artificial neural network and fuzzy logic controller (Wang, 1999). The control frameworks can be characterized into three stages, such as central controller, distributive controller, and hybrid controller. In all the conditions, each vitality source is required to have separate dedicated controller that can decide and instruct ideal action. For multiobjective targets, all vitality sources cannot work smoothly without a specific strategy. Therefore a comprising working strategy is required. Choice of assessment rules is one of the significant works essential for planning of HERS for a specified area given in (Kahraman, KayaI, & Cebi, 2009). Liao and Ruan (2009) proposed an energy control procedure for a PV/battery stand-alone hybrid system. In this technique, PV framework gives consistent energy, while the battery provides the backup energy. Hosseinzadeh and Salmasi (2015) built up an administrative control framework to oversee and work as microgrid. Similarly, (Berrazouane & Mohammedi, 2014) built up a cuckoo algorithm dependent on the fuzzy controller to work for independent hybrid framework comprising of SPV panels, DGs, and battery storage. Chedid, Akiki, and Rahman (1998) proposed choice-based procedure for strategy planers about the affecting components in the structure of grid-connected HRES framework. They utilized analytical hybridized process to evaluate different parameters of hybrid framework. Zhang et al. (2012) introduced an EM system to control the progression of energy of an inexhaustible framework of SPV and storage framework associated with the network. The framework expects to accomplish load shedding and decrease in both power bill and carbon emissions. Meng et al. (2016) summed up the controling objectives of monitoring controllers for microgrids. Basir Khan, Jidin, and Pasupuleti (2016) proposed a dispersed EM system engineering dependent on various parameters rather than centralize controller. Alsayed, Cacciato, Scarcella, and Scelba (2013) proposed a twofold layer-facilitated control approach, in which the main layer is liable for getting a monetary activity dependent on determining information and the subsequent layer is liable for giving vitality dependent on continuous information. Beccali, Cellura, and Mistretta (2003) utilized ELECTRE for evaluating an activity intended to diffuse sustainable power source innovations at provincial level. Goletsis, Psarras, and Samouilidis (2003) investigated the vitality arranging approach for positioning the activities. Topcu and Ulengin (2004) examined the conceivable vitality options relying upon their physical, ecological, economical, political, and other angles.

31.4.3 Mathematical modeling of SPV/wind HRES

The primary goal of the HRES scheme is the maximum usage of sustainable power sources upto extreme extents, economically, reliability, and durability. The initial phase in the enhancement of hybrid framework implementation is ensuring the representation of individual segments. Modeling is an essential primary step for optimum sizing (Celik, 2003) Three basic separate subsystem modeling is incorporated for an HERS, such as PV modeling, the wind turbine modeling, and the battery modeling. Bhandari, Poudel, Lee, and Ahn (2014) gave extensive numerical model of HRES frameworks. Modeling process empowers the understanding of circumstance, recognizing the issue and choice making. An adequately suitable model ought to be trade-off among multifaceted nature and precision. Accomplishment of different parts are demonstrated by either deterministive or probabilistic procedures (Bhandari et al., 2014). General strategy for displaying HRES, such as PV, wind, DG, and battery, is depicted underneath.

31.4.3.1 Modeling of PV array

The utmost renowned model used for observing the vitality creation in PV cell is the single diode circuit model appeared in Fig. 31.4 (Bhandari et al., 2015). The properties of PV array are depended on the SPV cell materials as it is the basic element of PV array. The intensity of solar radiation and the temperature on the PV panel are the basic parameters to decide the characteristics of solar cell. This single diode circuit model dictated the production intensity of the PV System with the area at which the sun-oriented radiation St (W m^{-2}) is falling (Markvart, 2000; Salam, Ishaque, & Taheri, 2010) (Fig. 31.5).

The energy produced by PV cell can be determined by given expression, where η_s is the PV generator effectiveness, A_s is area, and S_t is solar radiation (W m^{-2}) given by (Habib, Said, El-Hadidy, & Al-Zaharna, 1999).

$$P_g = \eta_s A_s S_t$$

The efficiency of PV generator (η_s) can be expressed in terms of reference module proficiency (η_{ref}) and the power generation productivity (η_p), which is equivalent to 1 for ideal MPPT utilization and the β_g is the generator proficiency.

$$\eta_s = \eta_{ref}\eta_p\left[1 - \beta_g(\theta_c - \theta_{cr})\right]$$

The θ_{cr} ref is the reference cell temperature and the cell temperature is θ_c that is function of θ_a the encompassing temperature, and NCT is the normal working temperature.

$$\theta_c = \theta_a + \left(\frac{NCT - 20}{800}\right)S_t$$

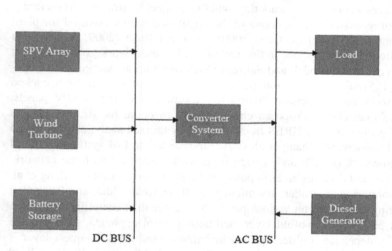

FIGURE 31.4   Schematic arrangement of SPV/wind hybrid renewable energy system.

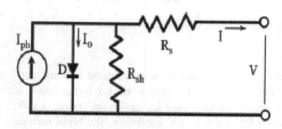

FIGURE 31.5   Equivalent model of solar cell.

31.4.3.2 Wind turbine modeling

Hongxing, Lu, and Wei (2007) presented the latest wind model to mimic the output power of a wind turbine. The power yield of wind generators for a particular location relies upon wind speed at tower tallness and speed qualities of the turbine. The output power can be given as.

$$P_w = \begin{cases} P_r \left(\dfrac{v - v_{in}}{v_r - v_{in}} \right) & v_{in} \le v \le v_r \\ P_r v_r & \le v \le v_o \\ 0 & v \le v_{in} \text{ and } v \ge v_o \end{cases}$$

where the rated power is P_r, v_{in} is the cut-in wind speed, v_r is the rated speed, and v_o is the cut-off wind speed (Ilinka, McCarthy, Chaumel, & Rétiveau, 2003). Wind speed at center point tallness can be determined by utilizing power-law condition.

$$v = v_r \left(\frac{h}{h_r} \right)^{\gamma}$$

where v is the breeze speed (m s^{-1}) estimated at the center point tallness H (m), v_r is the breeze speed (m s^{-1}) estimated at the reference stature $Href$ (m), and γ is the ground surface erosion coefficient (Ilinka et al., 2003).

31.4.3.3 Battery storage modeling

Battery bank is a backup vitality stockpiling framework. Battery measuring relies upon components, for example, most extreme profundity of release and temperature (Bhandari et al., 2015).

$$C_b(\tau) = C_b(\tau - 1) + (P_s(\tau) + P_w(\tau) - P_L(\tau))\eta_{con})(\Delta\tau \times \eta_{bc})$$

where $C_b(\tau)$ and $C_b(\tau - 1)$ are the available battery bank capacity (Wh) at times t and $t - 1$, respectively, $P_L(\tau)$ is the power consumed by the load at time t (Bin, Hongxing, Hui, & Xianbo, 2003; Bogdan & Salameh, 1996),

$$C_{bmin}(\tau) \le C_b(\tau) \le C_{bmax}$$

where $C_{bmin}(\tau)$ and Cbmax are the maximum and minimum allowable storage capacity. Using for C_{bmax} the storage nominal capacity $C_{bmin}(\tau)$, the minimum allowable storage capacity in terms of depth of battery discharge (DoD) can be determined by

$$C_{bmin} = \text{DoD} \times C_b$$

31.5 Optimization and sizing of SPV/wind HRES

31.5.1 Optimal design criteria for HRES

The structure of a hybrid framework has depended on the few requirements, such as area, grid-tied, or independent load. Ribeiro, Ferreira, and Araújo (2013) clarified different boundaries that can be supposed of while estimating a hybrid vitality framework. A simulation program permits to decide the ideal size of battery bank, PV panels, wind turbine, and other framework for an independent or grid-incorporated HRES for a given load on different standards. Many models are designed according to the least expense of the framework, least limit of framework and storage elements, and greatest energy production. Different enhancement strategies, such as graphical construction (Borowy & Salameh, 1996), probabilistic approach, iterative strategy, artificial intelligence (AI), dynamic programming, linear programming (Kellogg, Nehrir, Venkataramanan, & Gerez, 1996), and multiobjective, are utilized by analysts for optimal solution of hybrid PV/wind vitality framework. Numerous studies across the world have utilized the hybrid optimization model for electric renewable (HOMER) to enhance their recommended frameworks. HOMER was created by the National Renewable Energy Laboratory (NREL.HOMER, 2009). The HOMER examination results indicated that this independent hybrid sustainable power framework is a practical as an option in contrast to grid association. Nandi and Ghosh (2010), Bekele and Palm (2010), and Al-Karaghouli & Kazmerski, (2010) utilized HOMER to streamline their proposed independent hybrid frameworks. Along with these, some other parameters are also used for the optimization given as follows.

Technical parameters: These criteria cover the predefined functions, such as loss of power supply probability, equivalent loss factor, loss of load expected, loss of load probability (LOLP), total energy lost, level of autonomy, and battery state of charge (Bhandari et al., 2014; Upadhyay & Sharma, 2016).

Economic parameters: These criteria cover the economic predefined parameters, such as levelized COE, capital recovery factor, long-term capital cost, average generation of COE, net present value, annualized cost of system, fuel consumption cost, and cost of the energy life cycle (Chauhan & Saini, 2015; Renewable Energy Resources, 2014; Shaji, 2014).

Social and political parameters: Social acceptance, social performance evaluation criteria, social resistance against the HRES, visual impact, land utilization, electromagnetics interference, acoustical noises, shadow flickering, and ecosystem disturbance are some social parameters required to be considered in the modeling of HRES (Kellogg et al., 1996; Lanre et al., 2016).

Environment-related parameters: These are some environment-related parameters required to be considered in the modeling of HRES, such as emission function; the amount of CO_2 emission should be minimized by generating units (Mohammed et al., 2014).

31.5.2 Optimization problem

The process of finding economic solution or determining the optimal solution always required an objective function (Fc), which is to be minimized. This optimazion equation is formulated under the limitations all constraint parameters. The optimization equation can be solved by various techniques and methodologies as per convenience of the situation (Fulzele and Dutta, 2012). Here, a sample objective function $f_c(x)$ is proposed, which is required to minimize under the constraints given below. Therefore total cost and total energy-based expressions are established in the following equations, where capital cost is C_{cap} and the annual maintenance cost is C_{maint}. The proposed model of the SPV/wind HRES consists of SPV modules, wind turbine, battery storage, and inverters. To minimalize the total optimized cost C_{total} using the various methodologies, the generalized form of the objective function is expressed as follows (Tégani et al., 2014):

$$\text{Minimize objective function } f_c(x)$$

Following the equality constraint $g_c(x) = 0$ and inequality constraint $h_c \leq 0$, the optimization problem is formulated as

$$C_{total} = C_{cap} + C_{maint}$$

$$\sum_{\tau=1}^{\tau=24} E_{pv} + \sum_{\tau=1}^{\tau=24} E_w \geq \sum_{\tau=1}^{\tau=24} E_{dem}$$

$$E_{pv} = P_{pv} \times \Delta\tau = n_{pv} \times P_s \times \Delta\tau$$

$$E_w = P_w \times \Delta\tau = n_w \times P_w \times \Delta\tau$$

$$\text{Total installed wattage} \leq \sum_{dem} P_{in} \times n_{in}$$

$$\text{Photovoltaic power at STC} \leq \sum_d P_{cntr} n_{cntr}$$

$E_{pv}, E_w,$ and E_{dem} are energy from solar PV, wind system, and energy demand in the duration of 24 hours, and n_{pv} and n_w are the respective quantity variable for the PV panel and wind turbine (Mays & Tung, 1992). The P_{pv} and P_w are the power created by solar PV and wind turbine (Geem, 2012). The total power consumption can be determined by following the equipment's power those are connected as load.

$$\text{Total wattage installed} = \sum P_{load}$$

P_{in} is the maximum supplied power by the inverter and n_{in} is the number of inverter required. The number of the batteries can be determined by the following function where BS is the rated capability of every battery and BS_{req} is the required storage capability that can be calculated by the following equations (Geem, 2012):

$$N_b = \left[\frac{S_r}{\delta \times S_b} \right]$$

$$BS_{req} = \sum_{t-1}^{max} \left(P_{pv} + P_{wind} - P_{dem} \right) \times \tau - \sum_{t-1}^{min} \left(P_{pv} + P_{wind} - P_{dem} \right) \times \tau$$

where τ is the time for the energy generation (Mays & Tung, 1992). The capital-recovery factor is given by using initial capital cost I and is the annual interest rate (R).

$$C_{rf} = \frac{I\left((1+R)^l\right)}{(1+R)^l - 1}$$

The total capital cost function is expressed as follows:

$$C_t = \sum_s C_{pv} \times n_{pv} + \sum_s C_w \times n_w + \sum_s C_b \times n_b + \sum_s C_{in} \times n_{in} + \sum_s (C_{con} \times n_{con}) + C_{ex}$$

where, C_{pv}, C_w, C_b, C_{in}, C_{con}, and C_{ex}, are capital cost of PV, wind, battery, inverter, converter, and extra costs. The quantity of PV, wind, battery, inverter, and converter devices are represented by n_{pv}, n_w, n_b, n_{in}, n_{con}.

31.5.3   Optimization algorithm

The sizing of the components and adoption of a suitable EM methodology are basic requirement for optimization of the system cost and reducing the negative impacts. The vitality managing system is usually coordinated with optimization techniques to guarantee the continuation of supply to the load and to diminish the cost of vitality creation. The outcomes demonstrated the viability of the created algorithm regarding fitness execution as compared to other optimization techniques. Numerous optimization algorithms have been utilized for the energy management of hybrid frameworks utilizing intelligent strategies, for example, differential advancement (DE), genetic algorithm (GA), fuzzy logics, neural network, and neuro-fuzzy. The proper utilization of smart advancements prompts valuable frameworks with improved execution or different characteristics that could not be accomplished through conventional strategies (Li, Wei, & Xiang, 2012).

Designing the constraint problem for wind and PV hybrid system, number of factors, and expansive examination of the general framework are taking into consideration (Kaldellis, Zafirakis, & Kavadias, 2012). With the increase in optimizing factors, the simulation process increases exponentially, increasing the time and effort required. It is significant for designers to locate an attainable optimizing strategy to choose the ideal framework rapidly and precisely.

Dennis Barley, Byron, and Winn (1996) built up a specific EM system dependent on assumed information of future load demand in an independent wind/battery/DG hybrid power framework. Dahmane, Bosche, El-Hajjaji, and Dafarivar (2013) built up an algorithm for the optimization of an independent hybrid framework containing wind turbine, SPV panels, battery storage, and DG. Koutroulis, Kolokotsa, Potirakis, and Kalaitzakis (2006) utilized genetic algo-calculation to decide the expense of vitality of the overall framework and checked the utilization of hybridized PV/wind frameworks, which brings framework cost lower. Ekren and Ekren (2010) utilized annealing simulation with the software ARENA 12.0. The execution of the optimization process of the hybrid framework was gotten by considering the loss of load probability (LOLP) and self-governance investigation on an hourly premise. Gupta, Saini, and Sharma (2010) presented a calculation that is prepared to do effectively structuring a least-cost rural electric framework with high proficiency inspite of the fluctuating PV power. Analytical techniques require less time than Monte Carlo recreation in acquiring the necessary size for an unmistakable interest (Khatod, Pant, & Sharma, 2010). Markvart (2000) proposed a measuring strategy rewarding capacity vitality variety as an arbitrary manner. The likelihood thickness for every day addition or decrement of capacity level was approximated by a two-occasion probability distributive approach (Habib et al., 1999). Li et al. (2012) utilized this way for the optimization of wind/PV/battery HRES dependent on the minimization of life cycle cost. Mohammed et al. (2014) presented a multiobjective approach for generation development arranging and also used systematic hierarchy of command process-based model, helpful in taking care of a multiobjective issue comprising costs, ecological effect, imported fuel, and fuel cost dangers. Nasiraghdam and Jadid (2012) presented an analysis gotten by multiobjective bee-colony (ABC) calculation that has a decent quality and better assorted variety contrasted with NSGA-II and MOPSO techniques. Katsigiannis, Georgilakis, and Karapaidakis (2012) presented a hybrid SA (simulation annealing)−TS (Tabu search) methodology that improves the got solutions, in terms of value and convergence. Alsayed et al. (2013) described an optimal measuring of WT−PV by embracing distinctive multimodel choice investigation (MCDA) advancement methods. Pourmousavi, Nehrir, Colson, and Wang (2010) proposed another iterative strategy dependent on versatile input realizing, which was embraced to guarantee the quick convergence of the simulation algorithm. Xu, Ruan, Mao, Zhang, and Luo (2013) built up an algo-calculation appropriate for both independent and grid-associated mode where vitality filters are additionally applied to ensure the variance of intensity infused into the grid. An iterative advancement strategy was introduced by Tégani et al. (2014) to choose the PV module number and wind turbine size utilizing the effect between the power produced and power demand as near zero as conceivable over some undefined time frame. Ekren and Ekren (2010) proposed an improvement model of heavenly bodies by utilizing Artificial neural systems and hereditary calculations. The framework is displayed utilizing a TRNSYS program and the climatic states of Cyprus, remembered for a Common Meteorological Year document.

Different optimization procedures for hybrid wind–solar framework have been accounted here, for example, graphical development strategies, iterative strategy, probabilistic methodology, artificial intelligent techniques, and multicriteria structure. Utilizing feasible advancement strategy and ideal designs, which meet the energy necessity, can be acquired (Bhandari et al., 2015; Enslin, 1991). Iterative methodology is a scientific technique that created a grouping of improving appropriate answer for the advancement issue until an end model is reached (Bhandari et al., 2015). As the quantity of enhancement factors rises, the calculation time increments exponentially when utilizing this methodology. Enslin (1991) proposed an HSWSO model, following the LPSP model and levelized cost of vitality model for power unwavering quality and framework cost individually. Graphical development technique for calculating the optimal arrangement of PV panels and battery for an independent hybrid framework dependent on utilizing long time information of wind speed and solar radiation recorded for each hour of the day for a long time has been introduced by Upadhyay and Sharma (2016). Probabilistic methodology is a sizing method used for hybrid sun-powered wind framework accounting the impact of the sun-based radiation and wind speed fluctuations in the framework structure. AI is the part of software engineering that reviews and creates shrewd machines and programming. This reasoning in its broadest sense would mean the capacity of a machine or artefact to perform comparative sorts of capacities that describe human idea (Geem, 2012). AI comprises branches, such as GA fuzzy logics, artificial neural systems, and hybrid approaches, consolidating at least two of the above methods.

AI is a term that in its broadest sense would mean the capacity of a machine to perform comparative sorts of capacities that portray human idea. The advancement is completed by methods for GA and can be multiobjective or monoobjective. The control systems were optimized utilizing GAs. GAs are arranged as heuristics global approach. GA are a specific class of transformative algorithm that utilizes strategies enlivened by developmental science, for example, legacy, mutation, choice, and recombination.

Artificial neural system, frequently just called "Neural System," is a numerical model or computational model dependent on natural neural systems. It comprises an interconnected gathering of artificial neurons and procedures data utilizing a connectionist way to deal with calculation. In the ideal estimation of hybrid frameworks, it is wished to complete the design considering multiobjectives in any event (expenses and pollutant discharges). These two goals are in strife since a decrease in configuration costs infers an ascent in contamination discharges and the other way around (Kaldellis et al., 2012). Particle Swarm Optimization based system has rapid response time contrasted with consecutive quadratic programming. (Pourmousavi et al., 2010). Hybrid strategies are a viable mix of at least two unique procedures, which use the positive impact of these strategies in getting ideal outcome for a particular structure problem. A hybrid optimization approach can ideally solve multiobjective optimization problems.

31.5.4 Sizing techniques

Ideal sizing of PV/wind hybrid vitality framework is settled on hourly premise or average daily energy (Bhandari et al., 2015). The benefits of these techniques are the base expense of the framework with natural advantages. In sizing techniques, the normal wind, solar insolation, and energy demands are utilized for improvement of the framework estimating. This computation depends on the normal hourly, day by day, month to month, or yearly information of sun based and the wind vitality is utilized (El-Khadimi, Bchir, & Zeroual, 2004; Kellogg et al., 1996). Improvement programming and recreation programs are the most well-known apparatuses utilized for measuring the sunlight-based PV/wind hybrid frameworks. The ideal sizing of SPV/wind hybrid vitality sytem can be accomplished by different philosophies, such as realistic development techniques, probabilistic strategies, scientific strategies, iterative techniques, computerized reasoning strategies, and programming-based methodology. The ideal setup can be found by looking at the presentation and vitality creation cost of various framework designs by utilizing any of above procedures and software recreation. Graphical development techniques with two plan factors can be settled by observing graphically. In deterministic methodology, each arrangement of variable states is particularly controlled by boundaries in the model and sets of past conditions of these factors; in this manner, there is consistently special answer for given boundaries (Bhandari et al., 2015).

An estimating technique for a sun-based PV/wind hybrid independent framework is additionally used and tried for three distinct systems, such as cycle charging methodology, and load following technique by Upadhyay and Sharma (2016). Notton, Muselli, and Louche (1996) introduced a numerical model for estimating hybrid solar photovoltaic framework based on LOLP. Chauhan and Saini (2014) introduced a concentrateive review on estimation and sizing philosophies, stockpiling alternatives, and control techniques identified with independent sustainable power source frameworks. Markvart, Fragaki, and Ross (2006) explained the ideal measuring that is dictated by a superposition of commitments from climatic patterns of sunlight-based radiation. Researchgate (n.d.) presents an iterative procedure for measuring an HERS comprising wind and diesel as sources based on the complete cost. Bansal, Kumar, and Gupta (2013) and Kumar, Gupta, and Bansal (2013) presented another biogeography-based optimizing technique in sizing of HRES and compared the outcome and result got by HOMER. Bin et al. (2003) completed a techno-monetary investigation utilizing HOMER programming to plan an ideal combination of SPV/DG/battery framework for remote area in Jordan. Conversation of the previously mentioned strategies tends to be reasoned that estimating philosophies for hybrid framework give the most extreme adaptability. HRES framework sizing includes finding the least expensive combination of all generators' size and capacity limit that will fulfill the needed load.

A few programming softwares are accessible for the planning of hybrid frameworks, such as HOMER, HYBRID2, HOGA, and HYBRIDS. One of the well-known sizing programs for structuring and examining hybrid power framework is HOMER. Sunlight-based insolation, electrical demand, components costs, essentials ratings, controls parameters, and dispatch methodology are utilized as the contribution to the HOMER programming. It encourages the reorganization of sustainable power source frameworks dependent on the net present cost for a given arrangement. The ideal measuring technique can assist with ensuring the least venture with full utilization of the framework segment so the hybrid framework can work at the ideal conditions as far as speculation and framework power dependability necessity.

31.6   Future of SPV/wind HRES

Numerous researchers portrayed solar PV/wind hybrid renewable energy source in their work. HRES design and their assessment indicate that the hybrid PV/wind vitality frameworks are getting progressively famous (Notton et al., 1996). Renewable energy resources are increasing concern of the present world with the growing issue of energy security. In a developing nation, such as India, these are also major issue. It is preferable at this time to shift toward environment friendly, renewable, and sustainable energy source to counter the greenhouse gas emission. The utilization of renewable energy resources would be helpful in the enhancement of the economical growth and social progress. Solar PV and wind energy are the commonly used resources of clean renewable resources. These sustainable energy sources in combination together can produce large amount of energy that can resolve the issue of energy. The effectiveness of this combined hybrid system can be increased by providing storage system and DG, to the hybrid energy system. Renewable hybrid energy system is more economical than the individual resources those are running as a single energy-producing source. Projects of hybrid energy resources are at an initial stage across the world, which is same as every new innovation or technology. It may be a revolutionary scheme for human being. Except that several initial issues developer has not stopped adopting the hybrid renewable system for energy productions. In case of India, it has very bright future. The ministry responsible for renewable energy released a draft policy broadly that outlines the guidelines for establishing new hybrid solar/wind power schemes or updation of existing individual wind or solar system into hybrid power plants. The hybrid system would become a better alternative in grid-connected system. The rules, bidding process, and tariff process for hybrid renewable power mechanism are also required.

31.7   Conclusion

This chapter gives a concise coverage about the components and optimization techniques for SPV/Wind HRES that have been utilized for upgrading the HRES in the stand-alone mode. A comprehensive assessment of advancement of hybrid power framework comprising inexhaustible sources is explained in detail. A few advancement strategies and improvement plan are clarified with their eminence. A brief numerical model of a renewable resources and battery bank has been discussed.

The hybrid SPV/wind system offers a lot of advantages in comparison to the existing stand-alone DG system. The hybrid system would be more beneficial in comparison to the grid extention in some cases. The improvement in technology and enhancement in the efficiency of the system provide the hybrid system as good alternative for 24 × 7 supply to the consumer. The hybrid system also offers cheaper energy cost than the existing conventional system. The hybrid SPV−wind system is the environmental friendly system with 100% renewable power sources and zero harmful gas emissions; however, it has capacity shortage constraints. With a projection period of 25 years and 3% annual interest rate, it is found that the use of hybrid SPV−wind system could serve electricity with significantly lower COE as compared to the stand-alone diesel system. In sum up, the hybrid SPV—wind system is the best alternative in replacing or upgrading existing stand-alone diesel system in the studied village.

HRES is progressively being popularized for remote areas due to the lower cost of SPV and wind generator. It is felt that it will be instrumental for giving power to larger part of billion populace lacking power facility. HRES can accumulate vitality, rather than single renewable source, such as SPV or wind turbine, which improves power reliability of system. The energy optimizing situation, specifically the commitment of sustainable power sources, has been efficiently studied here. To achieve the targeted consumer demand, the power sector authorities play a prominent role in characterizing, planning, and actualizing the exploration of development projects. Research studies have identified two major thrust regions for sustainable improvement as solar and wind energy. Hence, the hybrid system of both the resources would be more effective for future demand. For the advancement and utilization of better sustainable hybrid system, the government is required to focus on the various projects and schemes. This is performed by offering different motivator plans and schemes' incentives and subsidies. Revision to the power demonstration and the relaxed tax approach is required for the flexible implementation and growth of sustainable hybrid system. According to the current status, the development and improvement of SPV/wind hybrid system is moving with a slow pace, which is required to be improved to achieve future targets.

References

Abedi, S., Alimardani, A., Gharehpetian, G., Riahy, G., & Hosseinian, S. (2012). A comprehensive method for optimal power management and design of hybrid RES-based autonomous energy systems. *Renewable & Sustainable Energy Reviews, 16*(3), 1577—1587.

Abido, A. M. (2007). Multi objective particle swarm optimization for environmental/ economic dispatch problem. In *Proceedings of the 8th international power engineering conference (IPEC-2007)* (pp. 138—590).

Ahn, S., Lee, K., Bhandari, B., Lee, G., Lee, C., et al. (2012). Formation strategy of renewable energy sources for high mountain off-grid system considering sustainability. *Journal of the Korean Society for Precision Engineering, 29*(9), 958—963.

Alagoz, B. B., Kaygusuz, A., & Karabiber, A. (2012). A user-mode distributed energy management architecture for smart grid applications. *Energy, 44*(1), 167—177.

Al-Karaghouli, A., & Kazmerski, L.L. (n.d.). Optimization and life-cycle cost of health clinic PV system for a rural area in Southern Iraq using HOMER Software. United States: N. p., 2010. Web. Available from: https://doi.org/10.1016/j.solener.2010.01.024.

Alsayed, M., Cacciato, M., Scarcella, G., & Scelba, G. (2013). Multicriteria optimal sizing of photovoltaic-wind turbine grid connected systems. *IEEE Transactions on Energy Conversion, 28*(2), 370—379.

Atwa, Y. M., El-Saadany, E. F., Salama, M. M. A., & Seethapathy, R. (2010). Optimal renewable resources mix for distribution system energy loss minimization. *IEEE Transactions on Power Systems, 25*(1), 360—370.

Bansal, A. K., Kumar, R., & Gupta, R. A. (2013). Economic analysis and power management of a small autonomous hybrid power system (SAHPS) using biogeography based optimization (BBO) algorithm. *IEEE Transactions on Smart Grid, 4*(1), 638—648.

Bansal, M., Saini, R. P., & Khatod, D. K. (2013). Development of cooking sector in rural areas in India—A review. *Renewable & Sustainable Energy Reviews, 17*(0), 44—53.

Basir Khan, M. R., Jidin, R., & Pasupuleti, J. (2016). Multi-agent based distributed control architecture for Microgrid energy management and optimization. *Energy Conversion and Management, 112*, 288—307.

Beccali, M., Cellura, M., & Mistretta, M. (2003). Decision-making in energy planning: Application of the ELECTRE method at region al level for the diffusion of renewable energy technology. *Renewable Energy, 28*, 2063—2087.

Bekele, G., & Palm, B. (2010). Feasibility study for a standalone solar—wind-based hybrid energy system for application in Ethiopia. *Applied Energy, 87*(2), 487—495.

Berrazouane, S., & Mohammedi, K. (2014). Parameter optimization via cuckoo optimization algorithm of fuzzy controller for energy management of a hybrid powersystem. *Energy Conversion and Management, 78*, 652—660.

Bhandari, B., Kyung, T. L., Lee, G. Y., Cho, Y. M., & Ahn, S. H. (2015). Optimization of hybrid renewable energy power systems: A review. *International Journal of Precision Engineering and Manufacturing-Green Technology, 2*(1), 99—112.

Bhandari, B., Poudel, S. R., Lee, K.-T., & Ahn, S.-H. (2014). Mathematical modeling of hybrid renewable energy system: A review on small hydro-solar-wind power generation. *International Journal of Precision Engineering and Manufacturing-Green Technology, 1*(2), 157—173.

Bifano, W.J., Ratajczak, A.F., Bahr, D.M., & Garrett, B.G. (1979). Social and economic impact of solar electricity at Schuchuli village. In *Presented at the seminar on solar technol. in rural settings: Assessments of field experiences*, Atlanta, GA, June 1—2. United Nations Univ.

Bin, A., Hongxing, Y., Hui, S., & Xianbo, L. (2003). Computer aided design for PV/Wind hybrid system. *Renewable Energy, 28*, 1491—1512.

Bogdan, S. B., & Salameh, Z. M. (1996). Methodology for optimally sizing the combination of a battery bank and PV array in a wind/PV hybrid system. *IEEE Transactions on Energy Conversion, 11*(2), 367—375.

Borowy, B. S., & Salameh, Z. M. (1996). Methodology for optimally sizing the combination of a battery bank and PV array in a wind/PV hybrid system. *IEEE Transactions on Energy Conversion, 11*(2), 367—375.

Celik, A. N. (2003). A simplified model for estimating the monthly performance of autonomous wind energy systems with battery storage. *Renewable Energy, 28*(4), 561—572.

Celik, A. K. (2003). Techno-economic analysis of autonomous PV–wind hybrid systems using different sizing methods. *Energy Conversion and Management, 44*(12), 1951–1968.

Chandel, S. S., Shrivastva, R., Sharma, V., & Ramasamy, P. (2016). Overview of the initiatives in renewable energy sector under the national action plan on climate change in India. *Renewable & Sustainable Energy Reviews, 54*, 866–873.

Chauhan, A., & Saini, R. P. (2014). A review on integrated renewable energy system based power generation for stand-alone applications: Configurations, storage options, sizing methodologies and control. *Renewable & Sustainable Energy Reviews, 38*, 99–120.

Chauhan, A., & Saini, R. P. (2015). Renewable energy based off-grid rural electrification in Uttarakhand state of India: Technology options, modelling method, barriers and recommendations. *Renewable & Sustainable Energy Reviews, 51*, 662–681.

Chedid, R., Akiki, H., & Rahman, S. (1998). A decision support technique for the design of hybrid solar-wind power systems. *IEEE Transactions on Energy Conversion, 13*(1), 76–83.

Dahmane, M., Bosche, J., El-Hajjaji, A., & Dafarivar, M. (2013). Renewable energy management algorithm for stand-alone system. In *Proceedings of the international conference on renewable energy research and applications, ICRERA* (pp. 621–6).

Dennis Barley, C., Byron., & Winn, C. (1996). Optimal dispatch strategy in remote hybrid power systems. *Solar Energy, 58*(4), 165–179.

Dimeas, A. L., & Hatziargyriou, N. D. (2005). Operation of a multi agent system for micro grid control. *IEEE Transactions on Power Systems, 20*(3), 1447–1455.

Ekren, O., & Ekren, B. Y. (2010). Size optimization of a PV/wind hybrid energy conversion system with battery storage using simulated annealing. *Applied Energy, 87*(1), 592–598.

Elhadidy, M., & Shaahid, S. (2000). Parametric study of hybrid (wind + solar + diesel) power generating systems. *Renewable Energy, 21*(2), 129–139.

El-Khadimi, A., Bchir, L., & Zeroual, A. (2004). Dimensionnementet Optimisation Technico-Economique D'un Système D'Energie Hybride Photovoltaïque-Eolien Avec Système de Stockage. *Revue des Énergies Renouvelables, 7*, 73–83.

Ellabban, O., Abu-Rub, H., & Blaabjerg, F. (2014). Renewable energy resources: Current status, future prospects and their enabling technology. *Renewable & Sustainable Energy Reviews, 39*, 748–764.

Enslin, J. (1991). Renewable energy as an economic energy source for remote areas. *Renewable Energy, 1*(2), 243–248.

Feijoo, A. E., Cidras, J., & Dornelas, J. G. (1999). Wind speed simulation in wind farms for steady-state security assessment of electrical power systems. *IEEE Transactions on Energy Conversion, 14*(4), 1582–1588.

Fulzele, J. B., & Dutta, S. (2012). Optimum planning of hybrid renewable energy system using HOMER. *Journal of Electrical & Computer Engineering, 1*, 68–74.

Geem, Z. W. (2012). Size optimization for a hybrid photovoltaic–wind energy system. *Electrical Power and Energy Systems, 42*, 448–451.

Goletsis, Y., Psarras, J., & Samouilidis, J. E. (2003). Project ranking in the Armenian energy sector using a multi criteria method for groups. *Annals of Operations Research, 120*, 135–157.

Gupta, A., Saini, R. P., & Sharma, M. P. (2010). Steady-state modelling of hybrid energy system for offgrid electrification of cluster of villages. *Renewable Energy, 35*(1), 520–535.

Habib, M. A., Said, S., El-Hadidy, M. A., & Al-Zaharna, I. (1999). Optimization procedure of a hybrid photovoltaic wind energy system. *Energy, 24*, 919–929.

Hongxing, Y., Lu, L., & Wei, Z. (2007). A novel optimization sizing model for hybrid solar wind power generation system. *Solar Energy, 81*, 76–84. Available from http://www.solarbuzz.com, S.

Hosseinzadeh, M., & Salmasi, F. R. (2015). Power management of an isolated hybrid AC/ DC micro-grid with fuzzy control of battery banks. *IET Renewable Power Generation, 9*(5), 484–493.

Ilinka, A., McCarthy, E., Chaumel, J. L., & Rétiveau, J. L. (2003). Wind potential assessment of Quebec Province. *Renewable Energy, 28*(12), 1881–1897.

Ipsakis, D., Voutetakis, S., Seferlis, P., Stergiopoulos, F., Papadopoulou, S., & Elmasides, C. (2008). The effect of the hysteresis band on power management strategies in a stand-alone power system. *Energy, 33*(10), 1537–1550.

Ismail, M. S., Moghavvemi, M., & Mahlia, T. M. I. (2013). Energy trends in Palestinian territories of West Bank and Gaza Strip: Possibilities for reducing the reliance on external energy sources. *Renewable & Sustainable Energy Reviews, 28*(0), 117–129.

Kahraman, C., KayaI., & Cebi, S. (2009). A comparative analysis for multi attribute selection among renewable energy alternatives using fuzzy axiomatic design and fuzzy analytic hierarchy process. *Energy, 34*, 1603–1616.

Kaldellis, J. K., Zafirakis, D., & Kavadias, K. (2012). Minimum cost solution of wind-photovoltaic based stand-alone power systems for remote consumers. *Energy Policy, 42*(1), 105–117.

Katsigiannis, Y. A., Georgilakis, P. S., & Karapaidakis, E. S. (2012). Hybrid simulated annealing-tabu search method for optimal sizing of autonomous power systems with renewables. *IEEE Transactions on Sustainable Energy, 3*(3), 330–338.

Kellogg, W., Nehrir, M., Venkataramanan, G., & Gerez, V. (1996). Optimal unit sizing for a hybrid wind/photovoltaic generating system. *Electric Power Systems Research, 39*(1), 35–38.

Khatod, D. K., Pant, V., & Sharma, J. (2010). Analytical approach for well-being assessment of small autonomous power systems with solar and wind energy sources. *IEEE Transactions on Energy Conversion, 25*(2), 535–545.

Koutroulis, E., Kolokotsa, D., Potirakis, A., & Kalaitzakis, K. (2006). Methodology for optimal sizing of stand-alone photovoltaic/wind-generator systems using genetic algorithms. *Solar Energy, 80*(1), 1072–1088.

Kumar, R., Gupta, R. A., & Bansal, A. K. (2013). Economic analysis and power management of a stand-alone wind/photovoltaic hybrid energy system using biogeography based optimization algorithm. *Swarm and Evolutionary Computation, 8*(1), 33–43.

Lambert, T., Gilman, P., & Lilientha, P. (2006). Micropower system modelling with HOMER. *Integration Alternative Sources of Energy*, *1*(1), 379–385.

Lanre, O., Saad, M., Ismail, M. S., & Moghavvemi, M. (2016). Energy management strategies in hybrid renewable energy systems: A review. *Renewable and Sustainable Energy Reviews*, *62*, 821–835.

Li, J., Wei, W., & Xiang, J. (2012). A simple sizing algorithm for stand-alone PV/wind/battery hybrid microgrids. *Energies*, *5*(12), 5307–5323.

Li, S., Wunsch, D. C., O'Hair, E., & Giesselmann, M. G. (2001). Comparative analysis of regression and artificial neural network models for wind turbine power curve estimation. *Journal of Solar Energy Engineering*, *123*(4), 327–332.

Liao, Z., & Ruan, X. (2009). A novel power management control strategy for stand-alone photovoltaic power system. In *Proceedings of the IEEE 6th international power electronics and motion control conference, IPEMC'09* (pp. 445–9).

Lim, J. H. (2012). Optimal combination and sizing of a new and renewable hybrid generation system. *International Journal of Future Generation Communication and Networking*, *5*(2), 43–59.

Markvart, T. (2000). *Solar electricity* (second (ed.)). USA: Wiley.

Markvart, T., Fragaki, A., & Ross, J. (2006). PV system sizing using observed time series of solar radiation. *Solar Energy*, *80*(1), 46–50.

Mays, L. W., & Tung, Y. K. (1992). *Hydro systems engineering and management*. New York: McGraw Hill.

Meng, L., Sanseverino, E. R., Luna, A., Dragicevic, T., Vasquez, J. C., & Guerrero, J. M. (2016). Microgrid supervisory controllers and energy management systems: A literature review. *Renewable & Sustainable Energy Reviews*, *60*, 1263–1273.

Meyar-Naimi, H., & Vaez-Zadeh, S. (2012). Sustainable development based energy policy making frameworks, a critical review. *Energy Policy*, *43*(0), 351–361.

Mohammed, Y. S., Mustafa, M. W., & Bashir, N. (2014). Hybrid renewable energy systems for off-grid electric power: Review of substantial issues. *Renewable & Sustainable Energy Reviews*, *35*, 527–539.

Nandi, S. K., & Ghosh, H. R. (2010). Prospect of wind–PV–battery hybrid power system as an alternative to grid extension in Bangladesh. *Energy*, *35*(7), 3040–3047.

Nasiraghdam, H., & Jadid, S. (2012). Optimal hybrid PV/WT/FC sizing and distribution system reconfiguration using multi-objective artificial bee colony (MOABC) algorithm. *Solar Energy*, *86*(1), 3057–3071.

Notton, G., Muselli, M., & Louche, A. (1996). Autonomous hybrid photovoltaic power plant using a back-up generator: A case study in a Mediterranean Island. *Renewable Energy*, *7*(4), 371–391.

NREL. HOMER. 2009. https://www.homerenergy.com.

Pascual, J., Barricarte, J., Sanchis, P., & Marroyo, L. (2015). Energy management strategy for a renewable-based residential microgrid with generation and demand forecasting. *Applied Energy*, *158*, 12–25.

Perera, A. T. D., Attalage, R. A., Perera, K. K. C. K., & Dassanayake, V. P. C. (2013). Designing standalone hybrid energy systems minimizing initial investment, life cycle cost and pollutant emission. *Energy*, *54*(0), 220–230.

Pourmousavi, S. A., Nehrir, M. H., Colson, C. M., & Wang, C. (2010). Real-time energy management of a stand-alone hybrid wind-micro turbine energy system using particle swarm optimization. *IEEE Transactions on Sustainable Energy*, *1*(3), 193–201.

Renewable Energy Resources. 2014. http://www.geni.org/globalenergy/library/renewableenergy-resources/world/Asia/bio-Asia/bio-india.shtml.

Researchgate. (n.d.). https://www.researchgate.net/figure/Types-of-Solar-Collectors-4_fig1_329337053.

Ribeiro, F., Ferreira, P., & Araújo, M. (2013). Evaluating future scenarios for the power generation sector using a Multi-Criteria Decision Analysis (MCDA) tool: The Portuguese case. *Energy*, *52*(1), 126–136.

Salam, Z., Ishaque, K., & Taheri, H. (2010). An improved two-diode photovoltaic (PV) model for PV system. In *Proceedings of the power electronics, drives and energy systems (PEDES) & 2010 Power India* (pp. 1–5).

Shaji, R. (2014). A review on India's renewable energy potential. *International Journal of Scientific & Engineering Research*, *5*(4), 985. Available from http://www.ijser.org.

Siddaiah, R., & Saini, R. P. (2016). A review on planning, configurations, modelling and optimization techniques of hybrid renewable energy systems for off grid applications. *Renewable and Sustainable Energy Reviews*, *58*, 376–396.

Sørensen, J., D. (2011). *Optimising Design and Construction for Safe and Reliable Operation.* **Wind Energy Systems**, (1st, pp. 167–207). (01, pp. 167–207). Woodhead Publishing.

Tégani, I., Aboubou, A., Ayad, M. Y., Becherif, M., Saadi, R., & Kraa, O. (2014). Optimal sizing design and energy management of stand-alone photovoltaic/wind generator systems. *The international conference on technologies and materials for renewable energy, environment and sustainability, TMREES14 Energy Procedia*, *50*, 163–170.

Topcu, Y. I., & Ulengin, F. (2004). Energy for the future: An integrated decision aid for the case of Turkey. *Energy*, *29*, 137–154.

United States Energy Information Administration. (2014). *International energy outlook 2011*. http://www.eia.gov/forecasts/archive/ieo11/. Accessed December 12, 2014.

United Nations. (2014). *Decade of sustainable energy for all*. http://www.un.org/News/Press/docs/2012/ga11333.doc.htm. Accessed 12.12.14.

Upadhyay, S., & Sharma, M. P. (2016). Selection of a suitable energy management strategy for a hybrid energy system in a remote rural area of India. *Energy*, *94*, 352–366.

Varun, Bhat, I. K., & Prakash, R. (n.d.). Life cycle analysis of run-of river small hydro.

Varun., Prakash, R., & Bhat, I. K. (2009). Energy economics and environmental impacts of renewable energy system. *Renewable and Sustainable Energy Reviews*, *13*, 2716–2721.

Wang, L.-X. (1999). *A course in fuzzy systems*. USA: Prentice-Hall Press.

Wang, C. (2006). *Modeling and control of hybrid wind/photovoltaic/fuel cell distributed generation systems* (Ph.D. thesis). Department of Electrical and Computer Engineering, Montant State University,.

Xu, L., Ruan, X., Mao, C., Zhang, B., & Luo, Y. (2013). An improved optimal sizing method for wind-solar-batteryhybridpowersystem. *IEEE Transactions on Sustainable Energy, 4*(3), 774—784.

Zhang, H., Davigny, A., Colas, F., Poste, Y., & Robyns, B. (2012). Fuzzy logic based energy management strategy for commercial buildings integrating photovoltaic and storage systems. *Energy and Buildings, 54*, 196—206.

Index